A Member of the International Code Family®

INTERNATIONAL

W9-BEV-402

RESIDENTIAL

CODE®

FOR ONE- AND TWO-FAMILY DWELLINGS

2006

Register your product and receive
FREE update services and much more.
Go to www.iccsafe.org/CodesPlus today.

2006 International Residential Code® for One- and Two-family Dwellings

First Printing: February 2006
Second Printing: November 2006
Third Printing: March 2007
Fourth Printing: March 2008
Fifth Printing: October 2008
Sixth Printing: July 2010
Seventh Printing: April 2012
Eighth Printing: May 2012
Ninth Printing: September 2013

ISBN-13: 978-1-58001-253-9 (soft-cover)
ISBN-10: 1-58001-253-1 (soft-cover)
ISBN-13: 978-1-58001-252-2 (loose-leaf)
ISBN-10: 1-58001-252-3 (loose-leaf)
ISBN-13: 978-1-58001-303-1 (e-document)
ISBN-10: 1-58001-303-1 (e-document)

PRINTED IN THE U.S.A.

PREFACE

Introduction

Internationally, code officials recognize the need for a modern, up-to-date residential code addressing the design and construction of one- and two-family dwellings and townhouses. The *International Residential Code®*, in this 2006 edition, is designed to meet these needs through model code regulations that safeguard the public health and safety in all communities, large and small.

This comprehensive, stand-alone residential code establishes minimum regulations for one- and two-family dwellings and townhouses using prescriptive provisions. It is founded on broad-based principles that make possible the use of new materials and new building designs. This 2006 edition is fully compatible with all the *International Codes®™* (I-Codes®) published by the International Code Council® (ICC)®, including the *International Building Code®*, ICC *Electrical Code®*, *International Energy Conservation Code®*, *International Existing Building Code®*, *International Fire Code®*, *International Fuel Gas Code®*, *International Mechanical Code®*, ICC *Performance Code®*, *International Plumbing Code®*, *International Private Sewage Disposal Code®*, *International Property Maintenance Code®*, *International Wildland-Urban Interface Code™* and *International Zoning Code®*.

The *International Residential Code* provisions provide many benefits, among which is the model code development process that offers an international forum for residential construction professionals to discuss prescriptive code requirements. This forum provides an excellent arena to debate proposed revisions. This model code also encourages international consistency in the application of provisions.

Development

The first edition of the *International Residential Code* (2000) was the culmination of an effort initiated in 1996 by the ICC and consisting of representatives from the three statutory members of the International Code Council at the time, including: Building Officials and Code Administrators International, Inc. (BOCA), International Conference of Building Officials (ICBO) and Southern Building Code Congress International (SBCCI) and representatives from the National Association of Home Builders (NAHB). The intent was to draft a stand-alone residential code consistent with and inclusive of the scope of the existing model codes. Technical content of the 1998 *International One- and Two-Family Dwelling Code* and the latest model codes promulgated by BOCA, ICBO, SBCCI and ICC was used as the basis for the development, followed by public hearings in 1998 and 1999 to consider proposed changes. This 2006 edition represents the code as originally issued, with changes reflected in the 2003 edition, and further changes developed through the ICC Code Development Process through 2005. Residential electrical provisions are based on the 2005 *National Electrical Code®* (NFPA-70). A new edition such as this is promulgated every three years.

Fuel gas provisions have been included through an agreement with the American Gas Association (AGA). Electrical provisions have been included through an agreement with the National Fire Protection Association (NFPA).

This code is founded on principles intended to establish provisions consistent with the scope of a residential code that adequately protects public health, safety and welfare; provisions that do not unnecessarily increase construction costs; provisions that do not restrict the use of new materials, products or methods of construction; and provisions that do not give preferential treatment to particular types or classes of materials, products or methods of construction.

Adoption

The *International Residential Code* is available for adoption and use by jurisdictions internationally. Its use within a governmental jurisdiction is intended to be accomplished through adoption by reference in accordance with proceedings establishing the jurisdiction's laws. At the time of adoption, jurisdictions should insert the appropriate information in provisions requiring specific local information, such as the name of the adopting jurisdiction. These locations are shown in bracketed words in small capital letters in the code and in the sample ordinance. The sample adoption ordinance on page v addresses several key elements of a code adoption ordinance, including the information required for insertion into the code text.

Maintenance

The *International Residential Code* is kept up to date through the review of proposed changes submitted by code enforcing officials, industry representatives, design professionals and other interested parties. Proposed changes are carefully considered through an open code development process in which all interested and affected parties may participate.

The contents of this work are subject to change both through the Code Development Cycles and the governmental body that enacts the code into law. For more information regarding the code development process, contact the Code and Standard Development Department of the International Code Council.

The maintenance process for the fuel gas provisions is based upon the process used to maintain the *International Fuel Gas Code*, in conjunction with the American Gas Association. The maintenance process for the electrical provisions is undertaken by the National Fire Protection Association.

While the development procedure of the *International Residential Code* assures the highest degree of care, ICC and the founding members of ICC and its members and those participating in the development of this code do not accept any liability resulting from compliance or noncompliance with the provisions because ICC and its founding members do not have the power or authority to police or enforce compliance with the contents of this code. Only the governmental body that enacts the code into law has such authority.

Marginal Markings

Solid vertical lines in the margins within the body of the code indicate a technical change from the requirements of the 2003 edition. Deletion indicators in the form of an arrow (➡) are provided in the margin where an entire section, paragraph, exception or table has been deleted or an item in a list of items or a table has been deleted.

ORDINANCE

The *International Codes* are designed and promulgated to be adopted by reference by ordinance. Jurisdictions wishing to adopt the 2006 *International Residential Code* as an enforceable regulation governing one- and two-family dwellings and townhouses should ensure that certain factual information is included in the adopting ordinance at the time adoption is being considered by the appropriate governmental body. The following sample adoption ordinance addresses several key elements of a code adoption ordinance, including the information required for insertion into the code text.

SAMPLE ORDINANCE FOR ADOPTION OF THE *INTERNATIONAL RESIDENTIAL CODE*

ORDINANCE NO._____

An ordinance of the **[JURISDICTION]** adopting the 2006 edition of the *International Residential Code*, regulating and governing the construction, alteration, movement, enlargement, replacement, repair, equipment, location, removal and demolition of detached one and two family dwellings and multiple single family dwellings (townhouses) not more than threes stories in height with separate means of egress in the **[JURISDICTION]**; providing for the issuance of permits and collection of fees therefor; repealing Ordinance No. _____ of the **[JURISDICTION]** and all other ordinances and parts of the ordinances in conflict therewith.

The **[GOVERNING BODY]** of the **[JURISDICTION]** does ordain as follows:

Section 1. That a certain document, three (3) copies of which are on file in the office of the **[TITLE OF JURISDICTION'S KEEPER OF RECORDS]** of **[NAME OF JURISDICTION]**, being marked and designated as the *International Residential Code*, 2006 edition, including Appendix Chapters **[FILL IN THE APPENDIX CHAPTERS BEING ADOPTED]** (see *International Residential Code* Section R102.5, 2006 edition), as published by the International Code Council, be and is hereby adopted as the Residential Code of the **[JURISDICTION]**, in the State of **[STATE NAME]** for regulating and governing the construction, alteration, movement, enlargement, replacement, repair, equipment, location, removal and demolition of detached one and two family dwellings and multiple single family dwellings (townhouses) not more than threes stories in height with separate means of egress as herein provided; providing for the issuance of permits and collection of fees therefor; and each and all of the regulations, provisions, penalties, conditions and terms of said Residential Code on file in the office of the **[JURISDICTION]** are hereby referred to, adopted, and made a part hereof, as if fully set out in this ordinance, with the additions, insertions, deletions and changes, if any, prescribed in Section 2 of this ordinance.

Section 2. The following sections are hereby revised:

Section R101.1. Insert: **[NAME OF JURISDICTION]**

Table R301.2 (1) Insert: **[APPROPRIATE DESIGN CRITERIA]**

Section P2603.6.1 Insert: **[NUMBER OF INCHES IN TWO LOCATIONS]**

Section P3103.1 Insert: **[NUMBER OF INCHES IN TWO LOCATIONS]**

Section 3. That Ordinance No. _____ of **[JURISDICTION]** entitled **[FILL IN HERE THE COMPLETE TITLE OF THE ORDINANCE OR ORDINANCES IN EFFECT AT THE PRESENT TIME SO THAT THEY WILL BE REPEALED BY DEFINITE MENTION]** and all other ordinances or parts of ordinances in conflict herewith are hereby repealed.

Section 4. That if any section, subsection, sentence, clause or phrase of this ordinance is, for any reason, held to be unconstitutional, such decision shall not affect the validity of the remaining portions of this ordinance. The **[GOVERNING BODY]** hereby declares that it would have passed this ordinance, and each section, subsection, clause or phrase thereof, irrespective of the fact that any one or more sections, subsections, sentences, clauses and phrases be declared unconstitutional.

Section 5. That nothing in this ordinance or in the Residential Code hereby adopted shall be construed to affect any suit or proceeding impending in any court, or any rights acquired, or liability incurred, or any cause or causes of action acquired or existing, under any act or ordinance hereby repealed as cited in Section 3 of this ordinance; nor shall any just or legal right or remedy of any character be lost, impaired or affected by this ordinance.

Section 6. That the **[JURISDICTION'S KEEPER OF RECORDS]** is hereby ordered and directed to cause this ordinance to be published. (An additional provision may be required to direct the number of times the ordinance is to be published and to specify that it is to be in a newspaper in general circulation. Posting may also be required.)

Section 7. That this ordinance and the rules, regulations, provisions, requirements, orders and matters established and adopted hereby shall take effect and be in full force and effect **[TIME PERIOD]** from and after the date of its final passage and adoption.

TABLE OF CONTENTS

Part I — Administrative

CHAPTER 1

ADMINISTRATION

SECTION R101
TITLE, SCOPE AND PURPOSE

R101.1 Title. These provisions shall be known as the *Residential Code for One- and Two-family Dwellings* of [NAME OF JURISDICTION], and shall be cited as such and will be referred to herein as "this code."

R101.2 Scope. The provisions of the *International Residential Code for One- and Two-family Dwellings* shall apply to the construction, alteration, movement, enlargement, replacement, repair, equipment, use and occupancy, location, removal and demolition of detached one- and two-family dwellings and townhouses not more than three stories above-grade in height with a separate means of egress and their accessory structures.

R101.3 Purpose. The purpose of this code is to provide minimum requirements to safeguard the public safety, health and general welfare through affordability, structural strength, means of egress facilities, stability, sanitation, light and ventilation, energy conservation and safety to life and property from fire and other hazards attributed to the built environment.

SECTION R102
APPLICABILITY

R102.1 General. Where, in any specific case, different sections of this code specify different materials, methods of construction or other requirements, the most restrictive shall govern. Where there is a conflict between a general requirement and a specific requirement, the specific requirement shall be applicable.

R102.2 Other laws. The provisions of this code shall not be deemed to nullify any provisions of local, state or federal law.

R102.3 Application of references. References to chapter or section numbers, or to provisions not specifically identified by number, shall be construed to refer to such chapter, section or provision of this code.

R102.4 Referenced codes and standards. The codes and standards referenced in this code shall be considered part of the requirements of this code to the prescribed extent of each such reference. Where differences occur between provisions of this code and referenced codes and standards, the provisions of this code shall apply.

> **Exception:** Where enforcement of a code provision would violate the conditions of the listing of the equipment or appliance, the conditions of the listing and manufacturer's instructions shall apply.

R102.5 Appendices. Provisions in the appendices shall not apply unless specifically referenced in the adopting ordinance.

R102.6 Partial invalidity. In the event any part or provision of this code is held to be illegal or void, this shall not have the effect of making void or illegal any of the other parts or provisions.

R102.7 Existing structures. The legal occupancy of any structure existing on the date of adoption of this code shall be permitted to continue without change, except as is specifically covered in this code, the *International Property Maintenance Code* or the *International Fire Code*, or as is deemed necessary by the building official for the general safety and welfare of the occupants and the public.

> **R102.7.1 Additions, alterations or repairs.** Additions, alterations or repairs to any structure shall conform to the requirements for a new structure without requiring the existing structure to comply with all of the requirements of this code, unless otherwise stated. Additions, alterations or repairs shall not cause an existing structure to become unsafe or adversely affect the performance of the building.

SECTION R103
DEPARTMENT OF BUILDING SAFETY

R103.1 Creation of enforcement agency. The department of building safety is hereby created and the official in charge thereof shall be known as the building official.

R103.2 Appointment. The building official shall be appointed by the chief appointing authority of the jurisdiction.

R103.3 Deputies. In accordance with the prescribed procedures of this jurisdiction and with the concurrence of the appointing authority, the building official shall have the authority to appoint a deputy building official, the related technical officers, inspectors, plan examiners and other employees. Such employees shall have powers as delegated by the building official.

SECTION R104
DUTIES AND POWERS OF THE
BUILDING OFFICIAL

R104.1 General. The building official is hereby authorized and directed to enforce the provisions of this code. The building official shall have the authority to render interpretations of this code and to adopt policies and procedures in order to clarify the application of its provisions. Such interpretations, policies and procedures shall be in conformance with the intent and purpose of this code. Such policies and procedures shall not have the effect of waiving requirements specifically provided for in this code.

R104.2 Applications and permits. The building official shall receive applications, review construction documents and issue permits for the erection and alteration of buildings and structures, inspect the premises for which such permits have been issued and enforce compliance with the provisions of this code.

R104.3 Notices and orders. The building official shall issue all necessary notices or orders to ensure compliance with this code.

R104.4 Inspections. The building official is authorized to make all of the required inspections, or the building official shall have the authority to accept reports of inspection by approved agencies or individuals. Reports of such inspections shall be in writing and be certified by a responsible officer of such approved agency or by the responsible individual. The building official is authorized to engage such expert opinion as deemed necessary to report upon unusual technical issues that arise, subject to the approval of the appointing authority.

R104.5 Identification. The building official shall carry proper identification when inspecting structures or premises in the performance of duties under this code.

R104.6 Right of entry. Where it is necessary to make an inspection to enforce the provisions of this code, or where the building official has reasonable cause to believe that there exists in a structure or upon a premises a condition which is contrary to or in violation of this code which makes the structure or premises unsafe, dangerous or hazardous, the building official or designee is authorized to enter the structure or premises at reasonable times to inspect or to perform the duties imposed by this code, provided that if such structure or premises be occupied that credentials be presented to the occupant and entry requested. If such structure or premises be unoccupied, the building official shall first make a reasonable effort to locate the owner or other person having charge or control of the structure or premises and request entry. If entry is refused, the building official shall have recourse to the remedies provided by law to secure entry.

R104.7 Department records. The building official shall keep official records of applications received, permits and certificates issued, fees collected, reports of inspections, and notices and orders issued. Such records shall be retained in the official records for the period required for the retention of public records.

R104.8 Liability. The building official, member of the board of appeals or employee charged with the enforcement of this code, while acting for the jurisdiction in good faith and without malice in the discharge of the duties required by this code or other pertinent law or ordinance, shall not thereby be rendered liable personally and is hereby relieved from personal liability for any damage accruing to persons or property as a result of any act or by reason of an act or omission in the discharge of official duties. Any suit instituted against an officer or employee because of an act performed by that officer or employee in the lawful discharge of duties and under the provisions of this code shall be defended by legal representative of the jurisdiction until the final termination of the proceedings. The building official or any subordinate shall not be liable for cost in any action, suit or proceeding that is instituted in pursuance of the provisions of this code.

R104.9 Approved materials and equipment. Materials, equipment and devices approved by the building official shall be constructed and installed in accordance with such approval.

R104.9.1 Used materials and equipment. Used materials, equipment and devices shall not be reused unless approved by the building official.

R104.10 Modifications. Wherever there are practical difficulties involved in carrying out the provisions of this code, the building official shall have the authority to grant modifications for individual cases, provided the building official shall first find that special individual reason makes the strict letter of this code impractical and the modification is in compliance with the intent and purpose of this code and that such modification does not lessen health, life and fire safety or structural requirements. The details of action granting modifications shall be recorded and entered in the files of the department of building safety.

R104.10.1 Areas prone to flooding. The building official shall not grant modifications to any provision related to areas prone to flooding as established by Table R301.2(1) without the granting of a variance to such provisions by the board of appeals.

R104.11 Alternative materials, design and methods of construction and equipment. The provisions of this code are not intended to prevent the installation of any material or to prohibit any design or method of construction not specifically prescribed by this code, provided that any such alternative has been approved. An alternative material, design or method of construction shall be approved where the building official finds that the proposed design is satisfactory and complies with the intent of the provisions of this code, and that the material, method or work offered is, for the purpose intended, at least the equivalent of that prescribed in this code. Compliance with the specific performance-based provisions of the International Codes in lieu of specific requirements of this code shall also be permitted as an alternate.

R104.11.1 Tests. Whenever there is insufficient evidence of compliance with the provisions of this code, or evidence that a material or method does not conform to the requirements of this code, or in order to substantiate claims for alternative materials or methods, the building official shall have the authority to require tests as evidence of compliance to be made at no expense to the jurisdiction. Test methods shall be as specified in this code or by other recognized test standards. In the absence of recognized and accepted test methods, the building official shall approve the testing procedures. Tests shall be performed by an approved agency. Reports of such tests shall be retained by the building official for the period required for retention of public records.

SECTION R105
PERMITS

R105.1 Required. Any owner or authorized agent who intends to construct, enlarge, alter, repair, move, demolish or change the occupancy of a building or structure, or to erect, install, enlarge, alter, repair, remove, convert or replace any electrical, gas, mechanical or plumbing system, the installation of which is regulated by this code, or to cause any such work to be done, shall first make application to the building official and obtain the required permit.

R105.2 Work exempt from permit. Permits shall not be required for the following. Exemption from permit requirements of this code shall not be deemed to grant authorization for any work to be done in any manner in violation of the provisions of this code or any other laws or ordinances of this jurisdiction.

Building:

1. One-story detached accessory structures used as tool and storage sheds, playhouses and similar uses, provided the floor area does not exceed 120 square feet (11.15 m²).

2. Fences not over 6 feet (1829 mm) high.

3. Retaining walls that are not over 4 feet (1219 mm) in height measured from the bottom of the footing to the top of the wall, unless supporting a surcharge.

4. Water tanks supported directly upon grade if the capacity does not exceed 5,000 gallons (18 927 L) and the ratio of height to diameter or width does not exceed 2 to 1.

5. Sidewalks and driveways.

6. Painting, papering, tiling, carpeting, cabinets, counter tops and similar finish work.

7. Prefabricated swimming pools that are less than 24 inches (610 mm) deep.

8. Swings and other playground equipment.

9. Window awnings supported by an exterior wall which do not project more than 54 inches (1372 mm) from the exterior wall and do not require additional support.

Electrical:

Repairs and maintenance: A permit shall not be required for minor repair work, including the replacement of lamps or the connection of approved portable electrical equipment to approved permanently installed receptacles.

Gas:

1. Portable heating, cooking or clothes drying appliances.

2. Replacement of any minor part that does not alter approval of equipment or make such equipment unsafe.

3. Portable-fuel-cell appliances that are not connected to a fixed piping system and are not interconnected to a power grid.

Mechanical:

1. Portable heating appliances.

2. Portable ventilation appliances.

3. Portable cooling units.

4. Steam, hot or chilled water piping within any heating or cooling equipment regulated by this code.

5. Replacement of any minor part that does not alter approval of equipment or make such equipment unsafe.

6. Portable evaporative coolers.

7. Self-contained refrigeration systems containing 10 pounds (4.54 kg) or less of refrigerant or that are actuated by motors of 1 horsepower (746 W) or less.

8. Portable-fuel-cell appliances that are not connected to a fixed piping system and are not interconnected to a power grid.

The stopping of leaks in drains, water, soil, waste or vent pipe; provided, however, that if any concealed trap, drainpipe, water, soil, waste or vent pipe becomes defective and it becomes necessary to remove and replace the same with new material, such work shall be considered as new work and a permit shall be obtained and inspection made as provided in this code.

The clearing of stoppages or the repairing of leaks in pipes, valves or fixtures, and the removal and reinstallation of water closets, provided such repairs do not involve or require the replacement or rearrangement of valves, pipes or fixtures.

R105.2.1 Emergency repairs. Where equipment replacements and repairs must be performed in an emergency situation, the permit application shall be submitted within the next working business day to the building official.

R105.2.2 Repairs. Application or notice to the building official is not required for ordinary repairs to structures, replacement of lamps or the connection of approved portable electrical equipment to approved permanently installed receptacles. Such repairs shall not include the cutting away of any wall, partition or portion thereof, the removal or cutting of any structural beam or load-bearing support, or the removal or change of any required means of egress, or rearrangement of parts of a structure affecting the egress requirements; nor shall ordinary repairs include addition to, alteration of, replacement or relocation of any water supply, sewer, drainage, drain leader, gas, soil, waste, vent or similar piping, electric wiring or mechanical or other work affecting public health or general safety.

R105.2.3 Public service agencies. A permit shall not be required for the installation, alteration or repair of generation, transmission, distribution, metering or other related equipment that is under the ownership and control of public service agencies by established right.

R105.3 Application for permit. To obtain a permit, the applicant shall first file an application therefor in writing on a form furnished by the department of building safety for that purpose. Such application shall:

1. Identify and describe the work to be covered by the permit for which application is made.

2. Describe the land on which the proposed work is to be done by legal description, street address or similar description that will readily identify and definitely locate the proposed building or work.

3. Indicate the use and occupancy for which the proposed work is intended.

4. Be accompanied by construction documents and other information as required in Section R106.1.

5. State the valuation of the proposed work.

6. Be signed by the applicant or the applicant's authorized agent.

7. Give such other data and information as required by the building official.

R105.3.1 Action on application. The building official shall examine or cause to be examined applications for permits and amendments thereto within a reasonable time after filing. If the application or the construction documents do not conform to the requirements of pertinent laws, the building official shall reject such application in writing, stating the reasons therefor. If the building official is satisfied that the proposed work conforms to the requirements of this code and laws and ordinances applicable thereto, the building official shall issue a permit therefor as soon as practicable.

R105.3.1.1 Determination of substantially improved or substantially damaged existing buildings in flood hazard areas. For applications for reconstruction, rehabilitation, addition or other improvement of existing buildings or structures located in an area prone to flooding as established by Table R301.2(1), the building official shall examine or cause to be examined the construction documents and shall prepare a finding with regard to the value of the proposed work. For buildings that have sustained damage of any origin, the value of the proposed work shall include the cost to repair the building or structure to its predamaged condition. If the building official finds that the value of proposed work equals or exceeds 50 percent of the market value of the building or structure before the damage has occurred or the improvement is started, the finding shall be provided to the board of appeals for a determination of substantial improvement or substantial damage. Applications determined by the board of appeals to constitute substantial improvement or substantial damage shall meet the requirements of Section R324.

R105.3.2 Time limitation of application. An application for a permit for any proposed work shall be deemed to have been abandoned 180 days after the date of filing unless such application has been pursued in good faith or a permit has been issued; except that the building official is authorized to grant one or more extensions of time for additional periods not exceeding 180 days each. The extension shall be requested in writing and justifiable cause demonstrated.

R105.4 Validity of permit. The issuance or granting of a permit shall not be construed to be a permit for, or an approval of, any violation of any of the provisions of this code or of any other ordinance of the jurisdiction. Permits presuming to give authority to violate or cancel the provisions of this code or other ordinances of the jurisdiction shall not be valid. The issuance of a permit based on construction documents and other data shall not prevent the building official from requiring the correction of errors in the construction documents and other data. The building official is also authorized to prevent occupancy or use of a structure where in violation of this code or of any other ordinances of this jurisdiction.

R105.5 Expiration. Every permit issued shall become invalid unless the work authorized by such permit is commenced within 180 days after its issuance, or if the work authorized by such permit is suspended or abandoned for a period of 180 days after the time the work is commenced. The building official is authorized to grant, in writing, one or more extensions of time, for periods not more than 180 days each. The extension shall be requested in writing and justifiable cause demonstrated.

R105.6 Suspension or revocation. The building official is authorized to suspend or revoke a permit issued under the provisions of this code wherever the permit is issued in error or on the basis of incorrect, inaccurate or incomplete information, or in violation of any ordinance or regulation or any of the provisions of this code.

R105.7 Placement of permit. The building permit or copy thereof shall be kept on the site of the work until the completion of the project.

R105.8 Responsibility. It shall be the duty of every person who performs work for the installation or repair of building, structure, electrical, gas, mechanical or plumbing systems, for which this code is applicable, to comply with this code.

SECTION R106
CONSTRUCTION DOCUMENTS

R106.1 Submittal documents. Construction documents, special inspection and structural observation programs and other data shall be submitted in one or more sets with each application for a permit. The construction documents shall be prepared by a registered design professional where required by the statutes of the jurisdiction in which the project is to be constructed. Where special conditions exist, the building official is authorized to require additional construction documents to be prepared by a registered design professional.

Exception: The building official is authorized to waive the submission of construction documents and other data not required to be prepared by a registered design professional if it is found that the nature of the work applied for is such that reviewing of construction documents is not necessary to obtain compliance with this code.

R106.1.1 Information on construction documents. Construction documents shall be drawn upon suitable material. Electronic media documents are permitted to be submitted when approved by the building official. Construction documents shall be of sufficient clarity to indicate the location, nature and extent of the work proposed and show in detail that it will conform to the provisions of this code and relevant laws, ordinances, rules and regulations, as determined by the building official.

R106.1.2 Manufacturer's installation instructions. Manufacturer's installation instructions, as required by this code, shall be available on the job site at the time of inspection.

R106.1.3 Information for construction in flood hazard areas. For buildings and structures located in whole or in part in flood hazard areas as established by Table R301.2(1), construction documents shall include:

1. Delineation of flood hazard areas, floodway boundaries and flood zones and the design flood elevation, as appropriate;

2. The elevation of the proposed lowest floor, including basement; in areas of shallow flooding (AO zones), the height of the proposed lowest floor, including basement, above the highest adjacent grade; and

3. The elevation of the bottom of the lowest horizontal structural member in coastal high hazard areas (V Zone); and

4. If design flood elevations are not included on the community's Flood Insurance Rate Map (FIRM), the building official and the applicant shall obtain and reasonably utilize any design flood elevation and floodway data available from other sources.

R106.2 Site plan. The construction documents submitted with the application for permit shall be accompanied by a site plan showing the size and location of new construction and existing structures on the site and distances from lot lines. In the case of demolition, the site plan shall show construction to be demolished and the location and size of existing structures and construction that are to remain on the site or plot.

R106.3 Examination of documents. The building official shall examine or cause to be examined construction documents for code compliance.

R106.3.1 Approval of construction documents. When the building official issues a permit, the construction documents shall be approved, in writing or by a stamp which states "APPROVED PLANS PER IRC SECTION R106.3.1." One set of construction documents so reviewed shall be retained by the building official. The other set shall be returned to the applicant, shall be kept at the site of work and shall be open to inspection by the building official or his or her authorized representative.

R106.3.2 Previous approvals. This code shall not require changes in the construction documents, construction or designated occupancy of a structure for which a lawful permit has been heretofore issued or otherwise lawfully authorized, and the construction of which has been pursued in good faith within 180 days after the effective date of this code and has not been abandoned.

R106.3.3 Phased approval. The building official is authorized to issue a permit for the construction of foundations or any other part of a building or structure before the construction documents for the whole building or structure have been submitted, provided that adequate information and detailed statements have been filed complying with pertinent requirements of this code. The holder of such permit for the foundation or other parts of a building or structure shall proceed at the holder's own risk with the building operation and without assurance that a permit for the entire structure will be granted.

R106.4 Amended construction documents. Work shall be installed in accordance with the approved construction documents, and any changes made during construction that are not in compliance with the approved construction documents shall be resubmitted for approval as an amended set of construction documents.

R106.5 Retention of construction documents. One set of approved construction documents shall be retained by the building official for a period of not less than 180 days from date of completion of the permitted work, or as required by state or local laws.

SECTION R107
TEMPORARY STRUCTURES AND USES

R107.1 General. The building official is authorized to issue a permit for temporary structures and temporary uses. Such permits shall be limited as to time of service, but shall not be permitted for more than 180 days. The building official is authorized to grant extensions for demonstrated cause.

R107.2 Conformance. Temporary structures and uses shall conform to the structural strength, fire safety, means of egress, light, ventilation and sanitary requirements of this code as necessary to ensure the public health, safety and general welfare.

R107.3 Temporary power. The building official is authorized to give permission to temporarily supply and use power in part of an electric installation before such installation has been fully completed and the final certificate of completion has been issued. The part covered by the temporary certificate shall comply with the requirements specified for temporary lighting, heat or power in the ICC *Electrical Code.*

R107.4 Termination of approval. The building official is authorized to terminate such permit for a temporary structure or use and to order the temporary structure or use to be discontinued.

SECTION R108
FEES

R108.1 Payment of fees. A permit shall not be valid until the fees prescribed by law have been paid. Nor shall an amendment to a permit be released until the additional fee, if any, has been paid.

R108.2 Schedule of permit fees. On buildings, structures, electrical, gas, mechanical and plumbing systems or alterations requiring a permit, a fee for each permit shall be paid as required, in accordance with the schedule as established by the applicable governing authority.

R108.3 Building permit valuations. Building permit valuation shall include total value of the work for which a permit is being issued, such as electrical, gas, mechanical, plumbing equipment and other permanent systems, including materials and labor.

R108.4 Related fees. The payment of the fee for the construction, alteration, removal or demolition for work done in connection with or concurrently with the work authorized by a building permit shall not relieve the applicant or holder of the permit from the payment of other fees that are prescribed by law.

R108.5 Refunds. The building official is authorized to establish a refund policy.

SECTION R109
INSPECTIONS

R109.1 Types of inspections. For onsite construction, from time to time the building official, upon notification from the permit holder or his agent, shall make or cause to be made any necessary inspections and shall either approve that portion of the construction as completed or shall notify the permit holder or his or her agent wherein the same fails to comply with this code.

R109.1.1 Foundation inspection. Inspection of the foundation shall be made after poles or piers are set or trenches or basement areas are excavated and any required forms erected and any required reinforcing steel is in place and supported prior to the placing of concrete. The foundation inspection shall include excavations for thickened slabs intended for the support of bearing walls, partitions, structural supports, or equipment and special requirements for wood foundations.

R109.1.2 Plumbing, mechanical, gas and electrical systems inspection. Rough inspection of plumbing, mechanical, gas and electrical systems shall be made prior to covering or concealment, before fixtures or appliances are set or installed, and prior to framing inspection.

Exception: Back-filling of ground-source heat pump loop systems tested in accordance with Section M2105.1 prior to inspection shall be permitted.

R109.1.3 Floodplain inspections. For construction in areas prone to flooding as established by Table R301.2(1), upon placement of the lowest floor, including basement, and prior to further vertical construction, the building official shall require submission of documentation, prepared and sealed by a registered design professional, of the elevation of the lowest floor, including basement, required in Section R324.

R109.1.4 Frame and masonry inspection. Inspection of framing and masonry construction shall be made after the roof, masonry, all framing, firestopping, draftstopping and bracing are in place and after the plumbing, mechanical and electrical rough inspections are approved.

R109.1.5 Other inspections. In addition to the called inspections above, the building official may make or require any other inspections to ascertain compliance with this code and other laws enforced by the building official.

R109.1.5.1 Fire-resistance-rated construction inspection. Where fire-resistance-rated construction is required between dwelling units or due to location on property, the building official shall require an inspection of such construction after all lathing and/or wallboard is in place, but before any plaster is applied, or before wallboard joints and fasteners are taped and finished.

R109.1.6 Final inspection. Final inspection shall be made after the permitted work is complete and prior to occupancy.

R109.2 Inspection agencies. The building official is authorized to accept reports of approved agencies, provided such agencies satisfy the requirements as to qualifications and reliability.

R109.3 Inspection requests. It shall be the duty of the permit holder or their agent to notify the building official that such work is ready for inspection. It shall be the duty of the person requesting any inspections required by this code to provide access to and means for inspection of such work.

R109.4 Approval required. Work shall not be done beyond the point indicated in each successive inspection without first obtaining the approval of the building official. The building official upon notification, shall make the requested inspections and shall either indicate the portion of the construction that is satisfactory as completed, or shall notify the permit holder or an agent of the permit holder wherein the same fails to comply with this code. Any portions that do not comply shall be corrected and such portion shall not be covered or concealed until authorized by the building official.

SECTION R110
CERTIFICATE OF OCCUPANCY

R110.1 Use and occupancy. No building or structure shall be used or occupied, and no change in the existing occupancy classification of a building or structure or portion thereof shall be made until the building official has issued a certificate of occupancy therefor as provided herein. Issuance of a certificate of occupancy shall not be construed as an approval of a violation of the provisions of this code or of other ordinances of the jurisdiction. Certificates presuming to give authority to violate or cancel the provisions of this code or other ordinances of the jurisdiction shall not be valid.

Exceptions:

1. Certificates of occupancy are not required for work exempt from permits under Section R105.2.

2. Accessory buildings or structures.

R110.2 Change in use. Changes in the character or use of an existing structure shall not be made except as specified in Sections 3406 and 3407 of the *International Building Code.*

R110.3 Certificate issued. After the building official inspects the building or structure and finds no violations of the provisions of this code or other laws that are enforced by the department of building safety, the building official shall issue a certificate of occupancy which shall contain the following:

1. The building permit number.

2. The address of the structure.

3. The name and address of the owner.

4. A description of that portion of the structure for which the certificate is issued.

5. A statement that the described portion of the structure has been inspected for compliance with the requirements of this code.

6. The name of the building official.

7. The edition of the code under which the permit was issued.

8. If an automatic sprinkler system is provided and whether the sprinkler system is required.

9. Any special stipulations and conditions of the building permit.

R110.4 Temporary occupancy. The building official is authorized to issue a temporary certificate of occupancy before the completion of the entire work covered by the permit, provided that such portion or portions shall be occupied safely. The building official shall set a time period during which the temporary certificate of occupancy is valid.

R110.5 Revocation. The building official shall, in writing, suspend or revoke a certificate of occupancy issued under the provisions of this code wherever the certificate is issued in error, or on the basis of incorrect information supplied, or where it is determined that the building or structure or portion thereof is in violation of any ordinance or regulation or any of the provisions of this code.

SECTION R111
SERVICE UTILITIES

R111.1 Connection of service utilities. No person shall make connections from a utility, source of energy, fuel or power to any building or system that is regulated by this code for which a permit is required, until approved by the building official.

R111.2 Temporary connection. The building official shall have the authority to authorize and approve the temporary connection of the building or system to the utility, source of energy, fuel or power.

R111.3 Authority to disconnect service utilities. The building official shall have the authority to authorize disconnection of utility service to the building, structure or system regulated by this code and the referenced codes and standards set forth in Section R102.4 in case of emergency where necessary to eliminate an immediate hazard to life or property or when such utility connection has been made without the approval required by Section R111.1 or R111.2. The building official shall notify the serving utility and whenever possible the owner and occupant of the building, structure or service system of the decision to disconnect prior to taking such action if not notified prior to disconnection. The owner or occupant of the building, structure or service system shall be notified in writing as soon as practical thereafter.

SECTION R112
BOARD OF APPEALS

R112.1 General. In order to hear and decide appeals of orders, decisions or determinations made by the building official relative to the application and interpretation of this code, there shall be and is hereby created a board of appeals. The building offi-

cial shall be an ex officio member of said board but shall have no vote on any matter before the board. The board of appeals shall be appointed by the governing body and shall hold office at its pleasure. The board shall adopt rules of procedure for conducting its business, and shall render all decisions and findings in writing to the appellant with a duplicate copy to the building official.

R112.2 Limitations on authority. An application for appeal shall be based on a claim that the true intent of this code or the rules legally adopted thereunder have been incorrectly interpreted, the provisions of this code do not fully apply, or an equally good or better form of construction is proposed. The board shall have no authority to waive requirements of this code.

R112.2.1 Determination of substantial improvement in areas prone to flooding. When the building official provides a finding required in Section R105.3.1.1, the board of appeals shall determine whether the value of the proposed work constitutes a substantial improvement. A substantial improvement means any repair, reconstruction, rehabilitation, addition or improvement of a building or structure, the cost of which equals or exceeds 50 percent of the market value of the building or structure before the improvement or repair is started. If the building or structure has sustained substantial damage, all repairs are considered substantial improvement regardless of the actual repair work performed. The term does not include:

1. Improvements of a building or structure required to correct existing health, sanitary or safety code violations identified by the building official and which are the minimum necessary to assure safe living conditions; or

2. Any alteration of an historic building or structure, provided that the alteration will not preclude the continued designation as an historic building or structure. For the purpose of this exclusion, an historic building is:

    2.1. Listed or preliminarily determined to be eligible for listing in the National Register of Historic Places; or

    2.2. Determined by the Secretary of the U.S. Department of Interior as contributing to the historical significance of a registered historic district or a district preliminarily determined to qualify as an historic district; or

    2.3. Designated as historic under a state or local historic preservation program that is approved by the Department of Interior.

R112.2.2 Criteria for issuance of a variance for areas prone to flooding. A variance shall only be issued upon:

1. A showing of good and sufficient cause that the unique characteristics of the size, configuration or topography of the site render the elevation standards in Section R324 inappropriate.

2. A determination that failure to grant the variance would result in exceptional hardship by rendering the lot undevelopable.

3. A determination that the granting of a variance will not result in increased flood heights, additional threats to public safety, extraordinary public expense, nor create nuisances, cause fraud on or victimization of the public, or conflict with existing local laws or ordinances.

4. A determination that the variance is the minimum necessary to afford relief, considering the flood hazard.

5. Submission to the applicant of written notice specifying the difference between the design flood elevation and the elevation to which the building is to be built, stating that the cost of flood insurance will be commensurate with the increased risk resulting from the reduced floor elevation, and stating that construction below the design flood elevation increases risks to life and property.

R112.3 Qualifications. The board of appeals shall consist of members who are qualified by experience and training to pass on matters pertaining to building construction and are not employees of the jurisdiction.

R112.4 Administration. The building official shall take immediate action in accordance with the decision of the board.

SECTION R113
VIOLATIONS

R113.1 Unlawful acts. It shall be unlawful for any person, firm or corporation to erect, construct, alter, extend, repair, move, remove, demolish or occupy any building, structure or equipment regulated by this code, or cause same to be done, in conflict with or in violation of any of the provisions of this code.

R113.2 Notice of violation. The building official is authorized to serve a notice of violation or order on the person responsible for the erection, construction, alteration, extension, repair, moving, removal, demolition or occupancy of a building or structure in violation of the provisions of this code, or in violation of a detail statement or a plan approved thereunder, or in violation of a permit or certificate issued under the provisions of this code. Such order shall direct the discontinuance of the illegal action or condition and the abatement of the violation.

R113.3 Prosecution of violation. If the notice of violation is not complied with in the time prescribed by such notice, the building official is authorized to request the legal counsel of the jurisdiction to institute the appropriate proceeding at law or in equity to restrain, correct or abate such violation, or to require the removal or termination of the unlawful occupancy of the building or structure in violation of the provisions of this code or of the order or direction made pursuant thereto.

R113.4 Violation penalties. Any person who violates a provision of this code or fails to comply with any of the requirements thereof or who erects, constructs, alters or repairs a building or structure in violation of the approved construction documents or directive of the building official, or of a permit or certificate issued under the provisions of this code, shall be subject to penalties as prescribed by law.

SECTION R114
STOP WORK ORDER

R114.1 Notice to owner. Upon notice from the building official that work on any building or structure is being prosecuted contrary to the provisions of this code or in an unsafe and dangerous manner, such work shall be immediately stopped. The stop work order shall be in writing and shall be given to the owner of the property involved, or to the owner's agent or to the person doing the work and shall state the conditions under which work will be permitted to resume.

R114.2 Unlawful continuance. Any person who shall continue any work in or about the structure after having been served with a stop work order, except such work as that person is directed to perform to remove a violation or unsafe condition, shall be subject to penalties as prescribed by law.

Part II — Definitions

CHAPTER 2
DEFINITIONS

SECTION R201
GENERAL

R201.1 Scope. Unless otherwise expressly stated, the following words and terms shall, for the purposes of this code, have the meanings indicated in this chapter.

R201.2 Interchangeability. Words used in the present tense include the future; words in the masculine gender include the feminine and neuter; the singular number includes the plural and the plural, the singular.

R201.3 Terms defined in other codes. Where terms are not defined in this code such terms shall have meanings ascribed to them as in other code publications of the International Code Council.

R201.4 Terms not defined. Where terms are not defined through the methods authorized by this section, such terms shall have ordinarily accepted meanings such as the context implies.

SECTION R202
DEFINITIONS

ACCESSIBLE. Signifies access that requires the removal of an access panel or similar removable obstruction.

ACCESSIBLE, READILY. Signifies access without the necessity for removing a panel or similar obstruction.

ACCESSORY STRUCTURE. A structure not greater than 3,000 square feet (279 m²) in floor area, and not over two stories in height, the use of which is customarily accessory to and incidental to that of the dwelling(s) and which is located on the same lot.

ADDITION. An extension or increase in floor area or height of a building or structure.

AIR ADMITTANCE VALVE. A one-way valve designed to allow air into the plumbing drainage system when a negative pressure develops in the piping. This device shall close by gravity and seal the terminal under conditions of zero differential pressure (no flow conditions) and under positive internal pressure.

AIR BREAK (DRAINAGE SYSTEM). An arrangement in which a discharge pipe from a fixture, appliance or device drains indirectly into a receptor below the flood-level rim of the receptor, and above the trap seal.

AIR CIRCULATION, FORCED. A means of providing space conditioning utilizing movement of air through ducts or plenums by mechanical means.

AIR-CONDITIONING SYSTEM. A system that consists of heat exchangers, blowers, filters, supply, exhaust and return-air systems, and shall include any apparatus installed in connection therewith.

AIR GAP, DRAINAGE SYSTEM. The unobstructed vertical distance through free atmosphere between the outlet of a waste pipe and the flood-level rim of the fixture or receptor into which it is discharging.

AIR GAP, WATER-DISTRIBUTION SYSTEM. The unobstructed vertical distance through free atmosphere between the lowest opening from a water supply discharge to the flood-level rim of a plumbing fixture.

ALTERATION. Any construction or renovation to an existing structure other than repair or addition that requires a permit. Also, a change in a mechanical system that involves an extension, addition or change to the arrangement, type or purpose of the original installation that requires a permit.

ANCHORS. See "Supports."

ANTISIPHON. A term applied to valves or mechanical devices that eliminate siphonage.

APPLIANCE. A device or apparatus that is manufactured and designed to utilize energy and for which this code provides specific requirements.

APPROVED. Acceptable to the building official.

APPROVED AGENCY. An established and recognized agency regularly engaged in conducting tests or furnishing inspection services, when such agency has been approved by the building official.

ASPECT RATIO. The ratio of the height to width (h/w) of a shear wall. The shear wall height is the maximum clear height from top of foundation or diaphragm to bottom of diaphragm framing above and the shear wall width is the sheathed dimension in the direction of applied force on the shear wall.

ATTIC. The unfinished space between the ceiling joists of the top story and the roof rafters.

BACKFLOW, DRAINAGE. A reversal of flow in the drainage system.

BACKFLOW PREVENTER. A device or means to prevent backflow.

BACKFLOW PREVENTER, REDUCED–PRESSURE-ZONE TYPE. A backflow-prevention device consisting of two independently acting check valves, internally force loaded to a normally closed position and separated by an intermediate chamber (or zone) in which there is an automatic relief means of venting to atmosphere internally loaded to a normally open position between two tightly closing shutoff valves and with means for testing for tightness of the checks and opening of relief means.

BACKFLOW, WATER DISTRIBUTION. The flow of water or other liquids into the potable water-supply piping from any sources other than its intended source. Backsiphonage is one type of backflow.

BACKPRESSURE. Pressure created by any means in the water distribution system, which by being in excess of the pressure in the water supply mains causes a potential backflow condition.

BACKPRESSURE, LOW HEAD. A pressure less than or equal to 4.33 psi (29.88 kPa) or the pressure exerted by a 10-foot (3048 mm) column of water.

BACKSIPHONAGE. The flowing back of used or contaminated water from piping into a potable water-supply pipe due to a negative pressure in such pipe.

BACKWATER VALVE. A device installed in a drain or pipe to prevent backflow of sewage.

BALCONY, EXTERIOR. An exterior floor projecting from and supported by a structure without additional independent supports.

BALL COCK. A valve that is used inside a gravity-type water closet flush tank to control the supply of water into the tank. It may also be called a flush-tank fill valve or water control.

BASEMENT. That portion of a building that is partly or completely below grade (see "Story above grade").

BASEMENT WALL. The opaque portion of a wall that encloses one side of a basement and has an average below grade wall area that is 50 percent or more of the total opaque and non-opaque area of that enclosing side.

BASIC WIND SPEED. Three-second gust speed at 33 feet (10 058 mm) above the ground in Exposure C (see Section R301.2.1) as given in Figure R301.2(4).

BATHROOM GROUP. A group of fixtures, including or excluding a bidet, consisting of a water closet, lavatory, and bathtub or shower. Such fixtures are located together on the same floor level.

BEND. A drainage fitting, designed to provide a change in direction of a drain pipe of less than the angle specified by the amount necessary to establish the desired slope of the line (see "Elbow" and "Sweep").

BOILER. A self-contained appliance from which hot water is circulated for heating purposes and then returned to the boiler, and which operates at water pressures not exceeding 160 pounds per square inch gage (psig) (1102 kPa gauge) and at water temperatures not exceeding 250°F (121°C).

BOND BEAM. A horizontal grouted element within masonry in which reinforcement is embedded.

BRACED WALL LINE. A series of braced wall panels in a single story constructed in accordance with Section R602.10 for wood framing or Section R603.7 or R301.1.1 for cold-formed steel framing to resist racking from seismic and wind forces.

BRACED WALL PANEL. A section of a braced wall line constructed in accordance with Section R602.10 for wood framing or Section R603.7 or R301.1.1 for cold-formed steel framing, which extend the full height of the wall.

BRANCH. Any part of the piping system other than a riser, main or stack.

BRANCH, FIXTURE. See "Fixture branch, drainage."

BRANCH, HORIZONTAL. See "Horizontal branch, drainage."

BRANCH INTERVAL. A vertical measurement of distance, 8 feet (2438 mm) or more in developed length, between the connections of horizontal branches to a drainage stack. Measurements are taken down the stack from the highest horizontal branch connection.

BRANCH, MAIN. A water-distribution pipe that extends horizontally off a main or riser to convey water to branches or fixture groups.

BRANCH, VENT. A vent connecting two or more individual vents with a vent stack or stack vent.

BTU/H. The listed maximum capacity of an appliance, absorption unit or burner expressed in British thermal units input per hour.

BUILDING. Building shall mean any one- and two-family dwelling or portion thereof, including townhouses, that is used, or designed or intended to be used for human habitation, for living, sleeping, cooking or eating purposes, or any combination thereof, and shall include accessory structures thereto.

BUILDING DRAIN. The lowest piping that collects the discharge from all other drainage piping inside the house and extends 30 inches (762 mm) in developed length of pipe, beyond the exterior walls and conveys the drainage to the building sewer.

BUILDING, EXISTING. Existing building is a building erected prior to the adoption of this code, or one for which a legal building permit has been issued.

BUILDING LINE. The line established by law, beyond which a building shall not extend, except as specifically provided by law.

BUILDING OFFICIAL. The officer or other designated authority charged with the administration and enforcement of this code.

BUILDING SEWER. That part of the drainage system that extends from the end of the building drain and conveys its discharge to a public sewer, private sewer, individual sewage-disposal system or other point of disposal.

BUILDING THERMAL ENVELOPE. The basement walls, exterior walls, floor, roof and any other building element that enclose conditioned spaces.

BUILT-UP ROOF COVERING. Two or more layers of felt cemented together and surfaced with a cap sheet, mineral aggregate, smooth coating or similar surfacing material.

CEILING HEIGHT. The clear vertical distance from the finished floor to the finished ceiling.

CHIMNEY. A primary vertical structure containing one or more flues, for the purpose of carrying gaseous products of combustion and air from a fuel-burning appliance to the outside atmosphere.

CHIMNEY CONNECTOR. A pipe that connects a fuel-burning appliance to a chimney.

CHIMNEY TYPES

Residential-type appliance. An approved chimney for removing the products of combustion from fuel-burning, residential-type appliances producing combustion gases not in excess of 1,000°F (538°C) under normal operating conditions, but capable of producing combustion gases of 1,400°F (760°C) during intermittent forces firing for periods up to 1 hour. All temperatures shall be measured at the appliance flue outlet. Residential-type appliance chimneys include masonry and factory-built types.

CIRCUIT VENT. A vent that connects to a horizontal drainage branch and vents two traps to a maximum of eight traps or trapped fixtures connected into a battery.

CLADDING. The exterior materials that cover the surface of the building envelope that is directly loaded by the wind.

CLEANOUT. An accessible opening in the drainage system used for the removal of possible obstruction.

CLOSET. A small room or chamber used for storage.

COMBINATION WASTE AND VENT SYSTEM. A specially designed system of waste piping embodying the horizontal wet venting of one or more sinks or floor drains by means of a common waste and vent pipe adequately sized to provide free movement of air above the flow line of the drain.

COMBUSTIBLE MATERIAL. Any material not defined as noncombustible.

COMBUSTION AIR. The air provided to fuel-burning equipment including air for fuel combustion, draft hood dilution and ventilation of the equipment enclosure.

COMMON VENT. A single pipe venting two trap arms within the same branch interval, either back-to-back or one above the other.

CONDENSATE. The liquid that separates from a gas due to a reduction in temperature, e.g., water that condenses from flue gases and water that condenses from air circulating through the cooling coil in air conditioning equipment.

CONDENSING APPLIANCE. An appliance that condenses water generated by the burning of fuels.

CONDITIONED AIR. Air treated to control its temperature, relative humidity or quality.

CONDITIONED AREA. That area within a building provided with heating and/or cooling systems or appliances capable of maintaining, through design or heat loss/gain, 68°F (20°C) during the heating season and/or 80°F (27°C) during the cooling season, or has a fixed opening directly adjacent to a conditioned area.

CONDITIONED FLOOR AREA. The horizontal projection of the floors associated with the conditioned space.

CONDITIONED SPACE. For energy purposes, space within a building that is provided with heating and/or cooling equipment or systems capable of maintaining, through design or heat loss/gain, 50°F (10°C) during the heating season and 85°F (29°C) during the cooling season, or communicates directly with a conditioned space. For mechanical purposes, an area, room or space being heated or cooled by any equipment or appliance.

CONFINED SPACE. A room or space having a volume less than 50 cubic feet per 1,000 Btu/h (4.83 L/W) of the aggregate input rating of all fuel-burning appliances installed in that space.

CONSTRUCTION DOCUMENTS. Written, graphic and pictorial documents prepared or assembled for describing the design, location and physical characteristics of the elements of a project necessary for obtaining a building permit. Construction drawings shall be drawn to an appropriate scale.

CONTAMINATION. An impairment of the quality of the potable water that creates an actual hazard to the public health through poisoning or through the spread of disease by sewage, industrial fluids or waste.

CONTINUOUS WASTE. A drain from two or more similar adjacent fixtures connected to a single trap.

CONTROL, LIMIT. An automatic control responsive to changes in liquid flow or level, pressure, or temperature for limiting the operation of an appliance.

CONTROL, PRIMARY SAFETY. A safety control responsive directly to flame properties that senses the presence or absence of flame and, in event of ignition failure or unintentional flame extinguishment, automatically causes shutdown of mechanical equipment.

CONVECTOR. A system-incorporating heating element in an enclosure in which air enters an opening below the heating element, is heated and leaves the enclosure through an opening located above the heating element.

CORROSION RESISTANCE. The ability of a material to withstand deterioration of its surface or its properties when exposed to its environment.

COURT. A space, open and unobstructed to the sky, located at or above grade level on a lot and bounded on three or more sides by walls or a building.

CRIPPLE WALL. A framed wall extending from the top of the foundation to the underside of the floor framing of the first story above grade plane.

CROSS CONNECTION. Any connection between two otherwise separate piping systems whereby there may be a flow from one system to the other.

DALLE GLASS. A decorative composite glazing material made of individual pieces of glass that are embedded in a cast matrix of concrete or epoxy.

DAMPER, VOLUME. A device that will restrict, retard or direct the flow of air in any duct, or the products of combustion of heat-producing equipment, vent connector, vent or chimney.

DEAD END. A branch leading from a DWV system terminating at a developed length of 2 feet (610 mm) or more. Dead ends shall be prohibited except as an approved part of a rough-in for future connection.

DEAD LOADS. The weight of all materials of construction incorporated into the building, including but not limited to walls, floors, roofs, ceilings, stairways, built-in partitions, finishes, cladding, and other similarly incorporated architectural and structural items, and fixed service equipment.

DECK. An exterior floor system supported on at least two opposing sides by an adjoining structure and/or posts, piers, or other independent supports.

DECORATIVE GLASS. A carved, leaded or Dalle glass or glazing material whose purpose is decorative or artistic, not functional; whose coloring, texture or other design qualities or components cannot be removed without destroying the glazing material; and whose surface, or assembly into which it is incor-porated, is divided into segments.

DESIGN PROFESSIONAL. See definition of "Registered design professional."

DEVELOPED LENGTH. The length of a pipeline measured along the center line of the pipe and fittings.

DIAMETER. Unless specifically stated, the term "diameter" is the nominal diameter as designated by the approved material standard.

DIAPHRAGM. A horizontal or nearly horizontal system acting to transmit lateral forces to the vertical resisting elements. When the term "diaphragm" is used, it includes horizontal bracing systems.

DILUTION AIR. Air that enters a draft hood or draft regulator and mixes with flue gases.

DIRECT-VENT APPLIANCE. A fuel-burning appliance with a sealed combustion system that draws all air for combustion from the outside atmosphere and discharges all flue gases to the outside atmosphere.

DRAFT. The pressure difference existing between the appliance or any component part and the atmosphere, that causes a continuous flow of air and products of combustion through the gas passages of the appliance to the atmosphere.

    Induced draft. The pressure difference created by the action of a fan, blower or ejector, that is located between the appliance and the chimney or vent termination.

    Natural draft. The pressure difference created by a vent or chimney because of its height, and the temperature difference between the flue gases and the atmosphere.

DRAFT HOOD. A device built into an appliance, or a part of the vent connector from an appliance, which is designed to provide for the ready escape of the flue gases from the appliance in the event of no draft, backdraft or stoppage beyond the draft hood; prevent a backdraft from entering the appliance; and neutralize the effect of stack action of the chimney or gas vent on the operation of the appliance.

DRAFT REGULATOR. A device that functions to maintain a desired draft in the appliance by automatically reducing the draft to the desired value.

DRAFT STOP. A material, device or construction installed to restrict the movement of air within open spaces of concealed areas of building components such as crawl spaces, floor-ceiling assemblies, roof-ceiling assemblies and attics.

DRAIN. Any pipe that carries soil and water-borne wastes in a building drainage system.

DRAINAGE FITTING. A pipe fitting designed to provide connections in the drainage system that have provisions for establishing the desired slope in the system. These fittings are made from a variety of both metals and plastics. The methods of coupling provide for required slope in the system (see "Durham fitting").

DUCT SYSTEM. A continuous passageway for the transmission of air which, in addition to ducts, includes duct fittings, dampers, plenums, fans and accessory air-handling equipment and appliances.

DURHAM FITTING. A special type of drainage fitting for use in the durham systems installations in which the joints are made with recessed and tapered threaded fittings, as opposed to bell and spigot lead/oakum or solvent/cemented or soldered joints. The tapping is at an angle (not 90 degrees) to provide for proper slope in otherwise rigid connections.

DURHAM SYSTEM. A term used to describe soil or waste systems where all piping is of threaded pipe, tube or other such rigid construction using recessed drainage fittings to correspond to the types of piping.

DWELLING. Any building that contains one or two dwelling units used, intended, or designed to be built, used, rented, leased, let or hired out to be occupied, or that are occupied for living purposes.

DWELLING UNIT. A single unit providing complete independent living facilities for one or more persons, including permanent provisions for living, sleeping, eating, cooking and sanitation.

DWV. Abbreviated term for drain, waste and vent piping as used in common plumbing practice.

EFFECTIVE OPENING. The minimum cross-sectional area at the point of water-supply discharge, measured or expressed in terms of diameter of a circle and if the opening is not circular, the diameter of a circle of equivalent cross-sectional area. (This is applicable to air gap.)

ELBOW. A pressure pipe fitting designed to provide an exact change in direction of a pipe run. An elbow provides a sharp turn in the flow path (see "Bend" and "Sweep").

EMERGENCY ESCAPE AND RESCUE OPENING. An operable exterior window, door or similar device that provides for a means of escape and access for rescue in the event of an emergency.

EQUIPMENT. All piping, ducts, vents, control devices and other components of systems other than appliances that are permanently installed and integrated to provide control of environmental conditions for buildings. This definition shall also include other systems specifically regulated in this code.

EQUIVALENT LENGTH. For determining friction losses in a piping system, the effect of a particular fitting equal to the friction loss through a straight piping length of the same nominal diameter.

ESSENTIALLY NONTOXIC TRANSFER FLUIDS. Fluids having a Gosselin rating of 1, including propylene glycol; mineral oil; polydimenthyoil oxane; hydrochlorofluorocarbon, chlorofluorocarbon and hydrofluorocarbon refrigerants; and FDA-approved boiler water additives for steam boilers.

ESSENTIALLY TOXIC TRANSFER FLUIDS. Soil, water or gray water and fluids having a Gosselin rating of 2 or more including ethylene glycol, hydrocarbon oils, ammonia refrigerants and hydrazine.

EVAPORATIVE COOLER. A device used for reducing air temperature by the process of evaporating water into an airstream.

EXCESS AIR. Air that passes through the combustion chamber and the appliance flue in excess of that which is theoretically required for complete combustion.

EXHAUST HOOD, FULL OPENING. An exhaust hood with an opening at least equal to the diameter of the connecting vent.

EXISTING INSTALLATIONS. Any plumbing system regulated by this code that was legally installed prior to the effective date of this code, or for which a permit to install has been issued.

EXTERIOR INSULATION FINISH SYSTEMS (EIFS). Synthetic stucco cladding systems typically consisting of five layers: adhesive, insulation board, base coat into which fiberglass reinforcing mesh is embedded, and a finish coat in the desired color.

EXTERIOR WALL. An above-grade wall that defines the exterior boundaries of a building. Includes between-floor spandrels, peripheral edges of floors, roof and basement knee walls, dormer walls, gable end walls, walls enclosing a mansard roof and basement walls with an average below-grade wall area that is less than 50 percent of the total opaque and nonopaque area of that enclosing side.

FACTORY-BUILT CHIMNEY. A listed and labeled chimney composed of factory-made components assembled in the field in accordance with the manufacturer's instructions and the conditions of the listing.

FENESTRATION. Skylights, roof windows, vertical windows (whether fixed or moveable); opaque doors; glazed doors; glass block; and combination opaque/glazed doors.

FIBER CEMENT SIDING. A manufactured, fiber-reinforcing product made with an inorganic hydraulic or calcium silicate binder formed by chemical reaction and reinforced with organic or inorganic non-asbestos fibers, or both. Additives which enhance manufacturing or product performance are permitted. Fiber cement siding products have either smooth or textured faces and are intended for exterior wall and related applications.

FIREBLOCKING. Building materials installed to resist the free passage of flame to other areas of the building through concealed spaces.

FIREPLACE. An assembly consisting of a hearth and fire chamber of noncombustible material and provided with a chimney, for use with solid fuels.

> **Factory-built fireplace.** A listed and labeled fireplace and chimney system composed of factory-made components, and assembled in the field in accordance with manufacturer's instructions and the conditions of the listing.

> **Masonry chimney.** A field-constructed chimney composed of solid masonry units, bricks, stones or concrete.

> **Masonry fireplace.** A field-constructed fireplace composed of solid masonry units, bricks, stones or concrete.

FIREPLACE STOVE. A free-standing, chimney-connected solid-fuel-burning heater designed to be operated with the fire chamber doors in either the open or closed position.

FIREPLACE THROAT. The opening between the top of the firebox and the smoke chamber.

FIRE SEPARATION DISTANCE. The distance measured from the building face to one of the following:

1. To the closest interior lot line; or

2. To the centerline of a street, an alley or public way; or

3. To an imaginary line between two buildings on the lot.

The distance shall be measured at a right angle from the face of the wall.

FIXTURE. See "Plumbing fixture."

FIXTURE BRANCH, DRAINAGE. A drain serving two or more fixtures that discharges into another portion of the drainage system.

FIXTURE BRANCH, WATER-SUPPLY. A water-supply pipe between the fixture supply and a main water-distribution pipe or fixture group main.

FIXTURE DRAIN. The drain from the trap of a fixture to the junction of that drain with any other drain pipe.

FIXTURE FITTING

> **Supply fitting.** A fitting that controls the volume and/or directional flow of water and is either attached to or accessible from a fixture or is used with an open or atmospheric discharge.

> **Waste fitting.** A combination of components that conveys the sanitary waste from the outlet of a fixture to the connection of the sanitary drainage system.

FIXTURE GROUP, MAIN. The main water-distribution pipe (or secondary branch) serving a plumbing fixture grouping such as a bath, kitchen or laundry area to which two or more individual fixture branch pipes are connected.

FIXTURE SUPPLY. The water-supply pipe connecting a fixture or fixture fitting to a fixture branch.

FIXTURE UNIT, DRAINAGE (d.f.u.). A measure of probable discharge into the drainage system by various types of plumbing fixtures, used to size DWV piping systems. The drainage fixture-unit value for a particular fixture depends on its volume rate of drainage discharge, on the time duration of a single drainage operation and on the average time between successive operations.

FIXTURE UNIT, WATER-SUPPLY (w.s.f.u.). A measure of the probable hydraulic demand on the water supply by various types of plumbing fixtures used to size water-piping systems. The water-supply fixture-unit value for a particular fixture depends on its volume rate of supply, on the time duration of a single supply operation and on the average time between successive operations.

FLAME SPREAD. The propagation of flame over a surface.

FLAME SPREAD INDEX. The numeric value assigned to a material tested in accordance with ASTM E 84.

FLOOD-LEVEL RIM. The edge of the receptor or fixture from which water overflows.

FLOOR DRAIN. A plumbing fixture for recess in the floor having a floor-level strainer intended for the purpose of the collection and disposal of waste water used in cleaning the floor and for the collection and disposal of accidental spillage to the floor.

FLOOR FURNACE. A self-contained furnace suspended from the floor of the space being heated, taking air for combustion from outside such space, and with means for lighting the appliance from such space.

FLOW PRESSURE. The static pressure reading in the water-supply pipe near the faucet or water outlet while the faucet or water outlet is open and flowing at capacity.

FLUE. See "Vent."

FLUE, APPLIANCE. The passages within an appliance through which combustion products pass from the combustion chamber to the flue collar.

FLUE COLLAR. The portion of a fuel-burning appliance designed for the attachment of a draft hood, vent connector or venting system.

FLUE GASES. Products of combustion plus excess air in appliance flues or heat exchangers.

FLUSH VALVE. A device located at the bottom of a flush tank that is operated to flush water closets.

FLUSHOMETER TANK. A device integrated within an air accumulator vessel that is designed to discharge a predetermined quantity of water to fixtures for flushing purposes.

FLUSHOMETER VALVE. A flushometer valve is a device that discharges a predetermined quantity of water to fixtures for flushing purposes and is actuated by direct water pressure.

FOAM BACKER BOARD. Foam plastic used in siding applications where the foam plastic is a component of the siding.

FOAM PLASTIC INSULATION. A plastic that is intentionally expanded by the use of a foaming agent to produce a reduced-density plastic containing voids consisting of open or closed cells distributed throughout the plastic for thermal insulating or acoustic purposes and that has a density less than 20 pounds per cubic foot (320 kg/m^3) unless it is used as interior trim.

FOAM PLASTIC INTERIOR TRIM. Exposed foam plastic used as picture molds, chair rails, crown moldings, baseboards, handrails, ceiling beams, door trim and window trim and similar decorative or protective materials used in fixed applications.

FUEL-PIPING SYSTEM. All piping, tubing, valves and fittings used to connect fuel utilization equipment to the point of fuel delivery.

FULLWAY VALVE. A valve that in the full open position has an opening cross-sectional area equal to a minimum of 85 percent of the cross-sectional area of the connecting pipe.

FURNACE. A vented heating appliance designed or arranged to discharge heated air into a conditioned space or through a duct or ducts.

GLAZING AREA. The interior surface area of all glazed fenestration, including the area of sash, curbing or other framing elements, that enclose conditioned space. Includes the area of glazed fenestration assemblies in walls bounding conditioned basements.

GRADE. The finished ground level adjoining the building at all exterior walls.

GRADE FLOOR OPENING. A window or other opening located such that the sill height of the opening is not more than 44 inches (1118 mm) above or below the finished ground level adjacent to the opening.

GRADE, PIPING. See "Slope."

GRADE PLANE. A reference plane representing the average of the finished ground level adjoining the building at all exterior walls. Where the finished ground level slopes away from the exterior walls, the reference plane shall be established by the lowest points within the area between the building and the lot line or, where the lot line is more than 6 ft (1829 mm) from the building between the structure and a point 6 ft (1829 mm) from the building.

GRIDDED WATER DISTRIBUTION SYSTEM. A water distribution system where every water distribution pipe is interconnected so as to provide two or more paths to each fixture supply pipe.

GROSS AREA OF EXTERIOR WALLS. The normal projection of all exterior walls, including the area of all windows and doors installed therein.

GROUND-SOURCE HEAT PUMP LOOP SYSTEM. Piping buried in horizontal or vertical excavations or placed in a body of water for the purpose of transporting heat transfer liquid to and from a heat pump. Included in this definition are closed loop systems in which the liquid is recirculated and open loop systems in which the liquid is drawn from a well or other source.

GUARD. A building component or a system of building components located near the open sides of elevated walking surfaces that minimizes the possibility of a fall from the walking surface to the lower level.

HABITABLE SPACE. A space in a building for living, sleeping, eating or cooking. Bathrooms, toilet rooms, closets, halls, storage or utility spaces and similar areas are not considered habitable spaces.

HANDRAIL. A horizontal or sloping rail intended for grasping by the hand for guidance or support.

HANGERS. See "Supports."

HAZARDOUS LOCATION. Any location considered to be a fire hazard for flammable vapors, dust, combustible fibers or other highly combustible substances.

HEATING DEGREE DAYS (HDD). The sum, on an annual basis, of the difference between 65°F (18°C) and the mean temperature for each day as determined from "NOAA Annual Degree Days to Selected Bases Derived from the 1960-1990 Normals" or other weather data sources acceptable to the code official.

HEAT PUMP. An appliance having heating or heating/cooling capability and that uses refrigerants to extract heat from air, liquid or other sources.

HEIGHT, BUILDING. The vertical distance from grade plane to the average height of the highest roof surface.

HEIGHT, STORY. The vertical distance from top to top of two successive tiers of beams or finished floor surfaces; and, for the topmost story, from the top of the floor finish to the top of the ceiling joists or, where there is not a ceiling, to the top of the roof rafters.

HIGH-TEMPERATURE (H.T.) CHIMNEY. A high temperature chimney complying with the requirements of UL 103. A Type H.T. chimney is identifiable by the markings "Type H.T." on each chimney pipe section.

HORIZONTAL BRANCH, DRAINAGE. A drain pipe extending laterally from a soil or waste stack or building drain, that receives the discharge from one or more fixture drains.

HORIZONTAL PIPE. Any pipe or fitting that makes an angle of less than 45 degrees (0.79 rad) with the horizontal.

HOT WATER. Water at a temperature greater than or equal to 110°F (43°C).

HURRICANE-PRONE REGIONS. Areas vulnerable to hurricanes, defined as the U.S. Atlantic Ocean and Gulf of Mexico coasts where the basic wind speed is greater than 90 miles per hour (40 m/s), and Hawaii, Puerto Rico, Guam, Virgin Islands, and America Samoa.

HYDROGEN GENERATING APPLIANCE. A self-contained package or factory-matched packages of integrated systems for generating gaseous hydrogen. Hydrogen generating appliances utilize electrolysis, reformation, chemical, or other processes to generate hydrogen.

IGNITION SOURCE. A flame, spark or hot surface capable of igniting flammable vapors or fumes. Such sources include appliance burners, burner ignitions and electrical switching devices.

INDIRECT WASTE PIPE. A waste pipe that discharges into the drainage system through an air gap into a trap, fixture or receptor.

INDIVIDUAL SEWAGE DISPOSAL SYSTEM. A system for disposal of sewage by means of a septic tank or mechanical treatment, designed for use apart from a public sewer to serve a single establishment or building.

INDIVIDUAL VENT. A pipe installed to vent a single-fixture drain that connects with the vent system above or terminates independently outside the building.

INDIVIDUAL WATER SUPPLY. A supply other than an approved public water supply that serves one or more families.

INSULATING CONCRETE FORM (ICF). A concrete forming system using stay-in-place forms of rigid foam plastic insulation, a hybrid of cement and foam insulation, a hybrid of cement and wood chips, or other insulating material for constructing cast-in-place concrete walls.

INSULATING SHEATHING. An insulating board having a minimum thermal resistance of R-2 of the core material.

JURISDICTION. The governmental unit that has adopted this code under due legislative authority.

KITCHEN. Kitchen shall mean an area used, or designated to be used, for the preparation of food.

LABEL. An identification applied on a product by the manufacturer which contains the name of the manufacturer, the function and performance characteristics of the product or material, and the name and identification of an approved agency and that indicates that the representative sample of the product or material has been tested and evaluated by an approved agency. (See also "Manufacturer's designation" and "Mark.")

LABELED. Devices, equipment or materials to which have been affixed a label, seal, symbol or other identifying mark of a testing laboratory, inspection agency or other organization concerned with product evaluation that maintains periodic inspection of the production of the above labeled items that attests to compliance with a specific standard.

LIGHT-FRAMED CONSTRUCTION. A type of construction whose vertical and horizontal structural elements are primarily formed by a system of repetitive wood or light gage steel framing members.

LISTED AND LISTING. Terms referring to equipment that is shown in a list published by an approved testing agency qualified and equipped for experimental testing and maintaining an adequate periodic inspection of current productions and whose listing states that the equipment complies with nationally recognized standards when installed in accordance with the manufacturer's installation instructions.

LIVE LOADS. Those loads produced by the use and occupancy of the building or other structure and do not include construction or environmental loads such as wind load, snow load, rain load, earthquake load, flood load or dead load.

LIVING SPACE. Space within a dwelling unit utilized for living, sleeping, eating, cooking, bathing, washing and sanitation purposes.

LOT. A portion or parcel of land considered as a unit.

LOT LINE. A line dividing one lot from another, or from a street or any public place.

MACERATING TOILET SYSTEMS. A system comprised of a sump with macerating pump and with connections for a water closet and other plumbing fixtures, that is designed to accept, grind and pump wastes to an approved point of discharge.

MAIN. The principal pipe artery to which branches may be connected.

MAIN SEWER. See "Public sewer."

MANIFOLD WATER DISTRIBUTION SYSTEMS. A fabricated piping arrangement in which a large supply main is fitted with multiple branches in close proximity in which water is distributed separately to fixtures from each branch.

MANUFACTURED HOME. Manufactured home means a structure, transportable in one or more sections, which in the traveling mode is 8 body feet (2438 body mm) or more in width or 40 body feet (12 192 body mm) or more in length, or, when erected on site, is 320 square feet (30 m²) or more, and which is built on a permanent chassis and designed to be used as a dwelling with or without a permanent foundation when connected to the required utilities, and includes the plumbing, heating, air-conditioning and electrical systems contained therein; except that such term shall include any structure that meets all the requirements of this paragraph except the size requirements and with respect to which the manufacturer voluntarily files a certification required by the secretary (HUD) and complies with the standards established under this title. For mobile homes built prior to June 15, 1976, a label certifying compliance to the Standard for Mobile Homes, NFPA 501, in effect at the time of manufacture is required. For the purpose of these provisions, a mobile home shall be considered a manufactured home.

MANUFACTURER'S DESIGNATION. An identification applied on a product by the manufacturer indicating that a product or material complies with a specified standard or set of rules. (See also "Mark" and "Label.")

MANUFACTURER'S INSTALLATION INSTRUCTIONS. Printed instructions included with equipment as part of the conditions of listing and labeling.

MARK. An identification applied on a product by the manufacturer indicating the name of the manufacturer and the function of a product or material. (See also "Manufacturer's designation" and "Label.")

MASONRY CHIMNEY. A field-constructed chimney composed of solid masonry units, bricks, stones or concrete.

MASONRY HEATER. A masonry heater is a solid fuel burning heating appliance constructed predominantly of concrete or solid masonry having a mass of at least 1,100 pounds (500 kg), excluding the chimney and foundation. It is designed to absorb and store a substantial portion of heat from a fire built in the firebox by routing exhaust gases through internal heat exchange channels in which the flow path downstream of the firebox includes at least one 180-degree (3.14-rad) change in flow direction before entering the chimney and which deliver heat by radiation through the masonry surface of the heater.

MASONRY, SOLID. Masonry consisting of solid masonry units laid contiguously with the joints between the units filled with mortar.

MASONRY UNIT. Brick, tile, stone, glass block or concrete block conforming to the requirements specified in Section 2103 of the *International Building Code*.

Clay. A building unit larger in size than a brick, composed of burned clay, shale, fire clay or mixtures thereof.

Concrete. A building unit or block larger in size than 12 inches by 4 inches by 4 inches (305 mm by 102 mm by 102 mm) made of cement and suitable aggregates.

Glass. Nonload-bearing masonry composed of glass units bonded by mortar.

Hollow. A masonry unit whose net cross-sectional area in any plane parallel to the loadbearing surface is less than 75 percent of its gross cross-sectional area measured in the same plane.

Solid. A masonry unit whose net cross-sectional area in every plane parallel to the loadbearing surface is 75 percent or more of its cross-sectional area measured in the same plane.

MASS WALL. Masonry or concrete walls having a mass greater than or equal to 30 pounds per square foot (146 kg/m²), solid wood walls having a mass greater than or equal to 20 pounds per square foot (98 kg/m²), and any other walls having a heat capacity greater than or equal to 6 Btu/ft² · °F [266 J/(m² · K)].

MEAN ROOF HEIGHT. The average of the roof eave height and the height to the highest point on the roof surface, except that eave height shall be used for roof angle of less than or equal to 10 degrees (0.18 rad).

MECHANICAL DRAFT SYSTEM. A venting system designed to remove flue or vent gases by mechanical means, that consists of an induced draft portion under nonpositive static pressure or a forced draft portion under positive static pressure.

Forced-draft venting system. A portion of a venting system using a fan or other mechanical means to cause the removal of flue or vent gases under positive static pressure.

Induced draft venting system. A portion of a venting system using a fan or other mechanical means to cause the removal of flue or vent gases under nonpositive static vent pressure.

Power venting system. A portion of a venting system using a fan or other mechanical means to cause the removal of flue or vent gases under positive static vent pressure.

MECHANICAL EXHAUST SYSTEM. A system for removing air from a room or space by mechanical means.

MECHANICAL SYSTEM. A system specifically addressed and regulated in this code and composed of components, devices, appliances and equipment.

METAL ROOF PANEL. An interlocking metal sheet having a minimum installed weather exposure of at least 3 square feet (0.28 m²) per sheet.

METAL ROOF SHINGLE. An interlocking metal sheet having an installed weather exposure less than 3 square feet (0.28 m²) per sheet.

MEZZANINE, LOFT. An intermediate level or levels between the floor and ceiling of any story with an aggregate floor area of not more than one-third of the area of the room or space in which the level or levels are located.

MODIFIED BITUMEN ROOF COVERING. One or more layers of polymer modified asphalt sheets. The sheet materials shall be fully adhered or mechanically attached to the substrate or held in place with an approved ballast layer.

MULTIPLE STATION SMOKE ALARM. Two or more single station alarm devices that are capable of interconnection such that actuation of one causes all integral or separate audible alarms to operate.

NATURAL DRAFT SYSTEM. A venting system designed to remove flue or vent gases under nonpositive static vent pressure entirely by natural draft.

NATURALLY DURABLE WOOD. The heartwood of the following species: Decay-resistant redwood, cedars, black locust and black walnut.

Note: Corner sapwood is permitted if 90 percent or more of the width of each side on which it occurs is heartwood.

NONCOMBUSTIBLE MATERIAL. Materials that pass the test procedure for defining noncombustibility of elementary materials set forth in ASTM E 136.

NONCONDITIONED SPACE. A space that is not a conditioned space by insulated walls, floors or ceilings.

OCCUPIED SPACE. The total area of all buildings or structures on any lot or parcel of ground projected on a horizontal plane, excluding permitted projections as allowed by this code.

OFFSET. A combination of fittings that makes two changes in direction bringing one section of the pipe out of line but into a line parallel with the other section.

OWNER. Any person, agent, firm or corporation having a legal or equitable interest in the property.

PELLET FUEL-BURNING APPLIANCE. A closed combustion, vented appliance equipped with a fuel feed mechanism for burning processed pellets of solid fuel of a specified size and composition.

PELLET VENT. A vent listed and labeled for use with a listed pellet fuel-burning appliance.

PERMIT. An official document or certificate issued by the authority having jurisdiction that authorizes performance of a specified activity.

PERSON. An individual, heirs, executors, administrators or assigns, and also includes a firm, partnership or corporation, its or their successors or assigns, or the agent of any of the aforesaid.

PITCH. See "Slope."

PLATFORM CONSTRUCTION. A method of construction by which floor framing bears on load bearing walls that are not continuous through the story levels or floor framing.

PLENUM. A chamber that forms part of an air-circulation system other than the occupied space being conditioned.

PLUMBING. For the purpose of this code, plumbing refers to those installations, repairs, maintenance and alterations regulated by Chapters 25 through 32.

PLUMBING APPLIANCE. An energized household appliance with plumbing connections, such as a dishwasher, food-waste grinder, clothes washer or water heater.

PLUMBING APPURTENANCE. A device or assembly that is an adjunct to the basic plumbing system and demands no additional water supply nor adds any discharge load to the system. It is presumed that it performs some useful function in the operation, maintenance, servicing, economy or safety of the plumbing system. Examples include filters, relief valves and aerators.

PLUMBING FIXTURE. A receptor or device that requires both a water-supply connection and a discharge to the drainage system, such as water closets, lavatories, bathtubs and sinks. Plumbing appliances as a special class of fixture are further defined.

PLUMBING SYSTEM. Includes the water supply and distribution pipes, plumbing fixtures, supports and appurtenances; soil, waste and vent pipes; sanitary drains and building sewers to an approved point of disposal.

POLLUTION. An impairment of the quality of the potable water to a degree that does not create a hazard to the public health but that does adversely and unreasonably affect the aesthetic qualities of such potable water for domestic use.

PORTABLE FUEL CELL APPLIANCE. A fuel cell generator of electricity, which is not fixed in place. A portable fuel cell appliance utilizes a cord and plug connection to a grid-isolated load and has an integral fuel supply.

POSITIVE ROOF DRAINAGE. The drainage condition in which consideration has been made for all loading deflections of the roof deck, and additional slope has been provided to ensure drainage of the roof within 48 hours of precipitation.

POTABLE WATER. Water free from impurities present in amounts sufficient to cause disease or harmful physiological effects and conforming in bacteriological and chemical quality to the requirements of the public health authority having jurisdiction.

PRECAST CONCRETE. A structural concrete element cast elsewhere than its final position in the structure.

PRESSURE-RELIEF VALVE. A pressure-actuated valve held closed by a spring or other means and designed to automatically relieve pressure at the pressure at which it is set.

PUBLIC SEWER. A common sewer directly controlled by public authority.

PUBLIC WATER MAIN. A water-supply pipe for public use controlled by public authority.

PUBLIC WAY. Any street, alley or other parcel of land open to the outside air leading to a public street, which has been deeded, dedicated or otherwise permanently appropriated to the public for public use and that has a clear width and height of not less than 10 feet (3048 mm).

PURGE. To clear of air, gas or other foreign substances.

QUICK-CLOSING VALVE. A valve or faucet that closes automatically when released manually or controlled by mechanical means for fast-action closing.

R-VALUE, THERMAL RESISTANCE. The inverse of the time rate of heat flow through a building thermal envelope element from one of its bounding surfaces to the other for a unit temperature difference between the two surfaces, under steady state conditions, per unit area (h · ft^2 · °F/Btu).

RAMP. A walking surface that has a running slope steeper than 1 unit vertical in 20 units horizontal (5-percent slope).

RECEPTOR. A fixture or device that receives the discharge from indirect waste pipes.

REFRIGERANT. A substance used to produce refrigeration by its expansion or evaporation.

REFRIGERANT COMPRESSOR. A specific machine, with or without accessories, for compressing a given refrigerant vapor.

REFRIGERATING SYSTEM. A combination of interconnected parts forming a closed circuit in which refrigerant is circulated for the purpose of extracting, then rejecting, heat. A direct refrigerating system is one in which the evaporator or condenser of the refrigerating system is in direct contact with the air or other substances to be cooled or heated. An indirect refrigerating system is one in which a secondary coolant cooled or heated by the refrigerating system is circulated to the air or other substance to be cooled or heated.

REGISTERED DESIGN PROFESSIONAL. An individual who is registered or licensed to practice their respective design profession as defined by the statutory requirements of the professional registration laws of the state or jurisdiction in which the project is to be constructed.

RELIEF VALVE, VACUUM. A device to prevent excessive buildup of vacuum in a pressure vessel.

REPAIR. The reconstruction or renewal of any part of an existing building for the purpose of its maintenance.

REROOFING. The process of recovering or replacing an existing roof covering. See "Roof recover."

RETURN AIR. Air removed from an approved conditioned space or location and recirculated or exhausted.

RISER. A water pipe that extends vertically one full story or more to convey water to branches or to a group of fixtures.

ROOF ASSEMBLY. A system designed to provide weather protection and resistance to design loads. The system consists of a roof covering and roof deck or a single component serving as both the roof covering and the roof deck. A roof assembly includes the roof deck, vapor retarder, substrate or thermal barrier, insulation, vapor retarder, and roof covering.

ROOF COVERING. The covering applied to the roof deck for weather resistance, fire classification or appearance.

ROOF COVERING SYSTEM. See "Roof assembly."

ROOF DECK. The flat or sloped surface not including its supporting members or vertical supports.

ROOF RECOVER. The process of installing an additional roof covering over a prepared existing roof covering without removing the existing roof covering.

ROOF REPAIR. Reconstruction or renewal of any part of an existing roof for the purposes of its maintenance.

ROOFTOP STRUCTURE. An enclosed structure on or above the roof of any part of a building.

ROOM HEATER. A freestanding heating appliance installed in the space being heated and not connected to ducts.

ROUGH-IN. The installation of all parts of the plumbing system that must be completed prior to the installation of fixtures. This includes DWV, water supply and built-in fixture supports.

RUNNING BOND. The placement of masonry units such that head joints in successive courses are horizontally offset at least one-quarter the unit length.

SANITARY SEWER. A sewer that carries sewage and excludes storm, surface and groundwater.

SCUPPER. An opening in a wall or parapet that allows water to drain from a roof.

SEISMIC DESIGN CATEGORY. A classification assigned to a structure based on its Seismic Group and the severity of the design earthquake ground motion at the site.

SEPTIC TANK. A water-tight receptor that receives the discharge of a building sanitary drainage system and is constructed so as to separate solids from the liquid, digest organic matter through a period of detention, and allow the liquids to discharge into the soil outside of the tank through a system of open joint or perforated piping or a seepage pit.

SEWAGE. Any liquid waste containing animal matter, vegetable matter or other impurity in suspension or solution.

SEWAGE PUMP. A permanently installed mechanical device for removing sewage or liquid waste from a sump.

SHALL. The term, when used in the code, is construed as mandatory.

SHEAR WALL. A general term for walls that are designed and constructed to resist racking from seismic and wind by use of masonry, concrete, cold-formed steel or wood framing in accordance with Chapter 6 of this code and the associated limitations in Section R301.2 of this code.

SIDE VENT. A vent connecting to the drain pipe through a fitting at an angle less than 45 degrees (0.79 rad) to the horizontal.

SINGLE PLY MEMBRANE. A roofing membrane that is field applied using one layer of membrane material (either homogeneous or composite) rather than multiple layers.

SINGLE STATION SMOKE ALARM. An assembly incorporating the detector, control equipment and alarm sounding device in one unit that is operated from a power supply either in the unit or obtained at the point of installation.

SKYLIGHT AND SLOPED GLAZING. See Section R308.6.1.

SKYLIGHT, UNIT. See Section R308.6.1.

SLIP JOINT. A mechanical-type joint used primarily on fixture traps. The joint tightness is obtained by compressing a friction-type washer such as rubber, nylon, neoprene, lead or special packing material against the pipe by the tightening of a (slip) nut.

SLOPE. The fall (pitch) of a line of pipe in reference to a horizonal plane. In drainage, the slope is expressed as the fall in units vertical per units horizontal (percent) for a length of pipe.

SMOKE-DEVELOPED RATING. A numerical index indicating the relative density of smoke produced by burning assigned to a material tested in accordance with ASTM E 84.

SOIL STACK OR PIPE. A pipe that conveys sewage containing fecal material.

SOLAR HEAT GAIN COEFFICIENT (SHGC). The solar heat gain through a fenestration or glazing assembly relative to the incident solar radiation (Btu/h · ft^2 · °F).

SOLID MASONRY. Load-bearing or nonload-bearing construction using masonry units where the net cross-sectional area of each unit in any plane parallel to the bearing surface is not less than 75 percent of its gross cross-sectional area. Solid masonry units shall conform to ASTM C 55, C 62, C 73, C 145 or C 216.

STACK. Any main vertical DWV line, including offsets, that extends one or more stories as directly as possible to its vent terminal.

STACK BOND. The placement of masonry units in a bond pattern is such that head joints in successive courses are vertically aligned. For the purpose of this code, requirements for stack bond shall apply to all masonry laid in other than running bond.

STACK VENT. The extension of soil or waste stack above the highest horizontal drain connected.

STACK VENTING. A method of venting a fixture or fixtures through the soil or waste stack without individual fixture vents.

STANDARD TRUSS. Any construction that does not permit the roof/ceiling insulation to achieve the required *R*-value over the exterior walls.

STATIONARY FUEL CELL POWER PLANT. A self-contained package or factory-matched packages which constitute an automatically-operated assembly of integrated systems for generating useful electrical energy and recoverable thermal energy that is permanently connected and fixed in place.

STORM SEWER, DRAIN. A pipe used for conveying rainwater, surface water, subsurface water and similar liquid waste.

STORY. That portion of a building included between the upper surface of a floor and the upper surface of the floor or roof next above.

STORY ABOVE GRADE. Any story having its finished floor surface entirely above grade, except that a basement shall be considered as a story above grade where the finished surface of the floor above the basement is:

1. More than 6 feet (1829 mm) above grade plane.

2. More than 6 feet (1829 mm) above the finished ground level for more than 50 percent of the total building perimeter.

3. More than 12 feet (3658 mm) above the finished ground level at any point.

STRUCTURAL INSULATED PANELS (SIPS). Factory fabricated panels of solid core insulation with structural skins of oriented strand board (OSB) or plywood.

STRUCTURE. That which is built or constructed.

SUMP. A tank or pit that receives sewage or waste, located below the normal grade of the gravity system and that must be emptied by mechanical means.

SUMP PUMP. A pump installed to empty a sump. These pumps are used for removing storm water only. The pump is selected for the specific head and volume of the load and is usually operated by level controllers.

SUNROOM. A one-story structure attached to a dwelling with a glazing area in excess of 40 percent of the gross area of the structure's exterior walls and roof.

SUPPLY AIR. Air delivered to a conditioned space through ducts or plenums from the heat exchanger of a heating, cooling or ventilating system.

SUPPORTS. Devices for supporting, hanging and securing pipes, fixtures and equipment.

SWEEP. A drainage fitting designed to provide a change in direction of a drain pipe of less than the angle specified by the amount necessary to establish the desired slope of the line. Sweeps provide a longer turning radius than bends and a less turbulent flow pattern (see "Bend" and "Elbow").

TEMPERATURE- AND PRESSURE-RELIEF (T AND P) VALVE. A combination relief valve designed to function as both a temperature-relief and pressure-relief valve.

TEMPERATURE-RELIEF VALVE. A temperature-actuated valve designed to discharge automatically at the temperature at which it is set.

THERMAL ISOLATION. Physical and space conditioning separation from conditioned space(s). The conditioned space(s) shall be controlled as separate zones for heating and cooling or conditioned by separate equipment.

THERMAL RESISTANCE, *R*-VALUE. The inverse of the time rate of heat flow through a body from one of its bounding surfaces to the other for a unit temperature difference between the two surfaces, under steady state conditions, per unit area (h · ft^2 · °F/Btu).

THERMAL TRANSMITTANCE, *U*-FACTOR. The coefficient of heat transmission (air to air) through a building envelope component or assembly, equal to the time rate of heat flow per unit area and unit temperature difference between the warm side and cold side air films (Btu/h · ft^2 · °F).

TOWNHOUSE. A single-family dwelling unit constructed in a group of three or more attached units in which each unit extends from foundation to roof and with open space on at least two sides.

TRAP. A fitting, either separate or built into a fixture, that provides a liquid seal to prevent the emission of sewer gases without materially affecting the flow of sewage or waste water through it.

TRAP ARM. That portion of a fixture drain between a trap weir and the vent fitting.

TRAP PRIMER. A device or system of piping to maintain a water seal in a trap, typically installed where infrequent use of the trap would result in evaporation of the trap seal, such as floor drains.

TRAP SEAL. The trap seal is the maximum vertical depth of liquid that a trap will retain, measured between the crown weir and the top of the dip of the trap.

TRIM. Picture molds, chair rails, baseboards, handrails, door and window frames, and similar decorative or protective materials used in fixed applications.

TRUSS DESIGN DRAWING. The graphic depiction of an individual truss, which describes the design and physical characteristics of the truss.

TYPE L VENT. A listed and labeled vent conforming to UL 641 for venting oil-burning appliances listed for use with Type L vents or with gas appliances listed for use with Type B vents.

_U_-FACTOR, THERMAL TRANSMITTANCE. The coefficient of heat transmission (air to air) through a building envelope component or assembly, equal to the time rate of heat flow per unit area and unit temperature difference between the warm side and cold side air films (Btu/h · ft^2 · °F).

UNCONFINED SPACE. A space having a volume not less than 50 cubic feet per 1,000 Btu/h (4.8 m^3/kW) of the aggregate input rating of all appliances installed in that space. Rooms communicating directly with the space in which the appliances are installed, through openings not furnished with doors, are considered a part of the unconfined space.

UNDERLAYMENT. One or more layers of felt, sheathing paper, nonbituminous saturated felt, or other approved material over which a roof covering, with a slope of 2 to 12 (17-percent slope) or greater, is applied.

UNUSUALLY TIGHT CONSTRUCTION. Construction in which:

1. Walls and ceilings comprising the building thermal envelope have a continuous water vapor retarder with a rating of 1 perm (5.7 · 10^{-11} kg/Pa · s · m^2) or less with openings therein gasketed or sealed.

2. Storm windows or weatherstripping is applied around the threshold and jambs of opaque doors and openable windows.

3. Caulking or sealants are applied to areas such as joints around window and door frames between sole plates and floors, between wall-ceiling joints, between wall panels, at penetrations for plumbing, electrical and gas lines, and at other openings.

VACUUM BREAKERS. A device which prevents backsiphonage of water by admitting atmospheric pressure through ports to the discharge side of the device.

VAPOR PERMEABLE MEMBRANE. A material or covering having a permeance rating of 5 perms (2.9 · 10^{-10} kg/Pa · s · m^2) or greater, when tested in accordance with the desiccant method using Procedure A of ASTM E 96. A vapor permeable material permits the passage of moisture vapor.

VAPOR RETARDER. A vapor resistant material, membrane or covering such as foil, plastic sheeting, or insulation facing having a permeance rating of 1 perm (5.7 · 10^{-11} kg/Pa · s · m^2) or less, when tested in accordance with the dessicant method using Procedure A of ASTM E 96. Vapor retarders limit the amount of moisture vapor that passes through a material or wall assembly.

VEHICULAR ACCESS DOOR. A door that is used primarily for vehicular traffic at entrances of buildings such as garages and parking lots, and that is not generally used for pedestrian traffic.

VENT. A passageway for conveying flue gases from fuel-fired appliances, or their vent connectors, to the outside atmosphere.

VENT COLLAR. See "Flue collar."

VENT CONNECTOR. That portion of a venting system which connects the flue collar or draft hood of an appliance to a vent.

VENT DAMPER DEVICE, AUTOMATIC. A device intended for installation in the venting system, in the outlet of an individual, automatically operated fuel burning appliance and that is designed to open the venting system automatically when the appliance is in operation and to close off the venting system automatically when the appliance is in a standby or shutdown condition.

VENT GASES. Products of combustion from fuel-burning appliances, plus excess air and dilution air, in the venting system above the draft hood or draft regulator.

VENT STACK. A vertical vent pipe installed to provide circulation of air to and from the drainage system and which extends through one or more stories.

VENT SYSTEM. Piping installed to equalize pneumatic pressure in a drainage system to prevent trap seal loss or blow-back due to siphonage or back pressure.

VENTILATION. The natural or mechanical process of supplying conditioned or unconditioned air to, or removing such air from, any space.

VENTING. Removal of combustion products to the outdoors.

VENTING SYSTEM. A continuous open passageway from the flue collar of an appliance to the outside atmosphere for the purpose of removing flue or vent gases. A venting system is usually composed of a vent or a chimney and vent connector, if used, assembled to form the open passageway.

VERTICAL PIPE. Any pipe or fitting that makes an angle of 45 degrees (0.79 rad) or more with the horizontal.

VINYL SIDING. A shaped material, made principally from rigid polyvinyl chloride (PVC), that is used to cover exterior walls of buildings.

WALL, RETAINING. A wall not laterally supported at the top, that resists lateral soil load and other imposed loads.

WALLS. Walls shall be defined as follows:

Load-bearing wall is a wall supporting any vertical load in addition to its own weight.

Nonbearing wall is a wall which does not support vertical loads other than its own weight.

WASTE. Liquid-borne waste that is free of fecal matter.

WASTE PIPE OR STACK. Piping that conveys only liquid sewage not containing fecal material.

WATER-DISTRIBUTION SYSTEM. Piping which conveys water from the service to the plumbing fixtures, appliances, appurtenances, equipment, devices or other systems served, including fittings and control valves.

WATER HEATER. Any heating appliance or equipment that heats potable water and supplies such water to the potable hot water distribution system.

WATER MAIN. A water-supply pipe for public use.

WATER OUTLET. A valved discharge opening, including a hose bibb, through which water is removed from the potable water system supplying water to a plumbing fixture or plumbing appliance that requires either an air gap or backflow pre-vention device for protection of the supply system.

WATER-RESISTIVE BARRIER. A material behind an exterior wall covering that is intended to resist liquid water that has penetrated behind the exterior covering from further intruding into the exterior wall assembly.

WATER-SERVICE PIPE. The outside pipe from the water main or other source of potable water supply to the water-distribution system inside the building, terminating at the service valve.

WATER-SUPPLY SYSTEM. The water-service pipe, the water-distributing pipes and the necessary connecting pipes, fittings, control valves and all appurtenances in or adjacent to the building or premises.

WET VENT. A vent that also receives the discharge of wastes from other fixtures.

WIND BORNE DEBRIS REGION. Areas within hurricane-prone regions within one mile of the coastal mean high water line where the basic wind speed is 110 miles per hour (49 m/s) or greater; or where the basic wind speed is equal to or greater than 120 miles per hour (54 m/s); or Hawaii.

WINDER. A tread with non-parallel edges.

WOOD STRUCTURAL PANEL. A panel manufactured from veneers; or wood strands or wafers; bonded together with waterproof synthetic resins or other suitable bonding systems. Examples of wood structural panels are plywood, OSB or composite panels.

YARD. An open space, other than a court, unobstructed from the ground to the sky, except where specifically provided by this code, on the lot on which a building is situated.

Part III — Building Planning and Construction

CHAPTER 3

BUILDING PLANNING

SECTION R301
DESIGN CRITERIA

R301.1 Application. Buildings and structures, and all parts thereof, shall be constructed to safely support all loads, including dead loads, live loads, roof loads, flood loads, snow loads, wind loads and seismic loads as prescribed by this code. The construction of buildings and structures in accordance with the provisions of this code shall result in a system that provides a complete load path that meets all requirements for the transfer of all loads from their point of origin through the load-resisting elements to the foundation. Buildings and structures constructed as prescribed by this code are deemed to comply with the requirements of this section.

R301.1.1 Alternative provisions. As an alternative to the requirements in Section R301.1 the following standards are permitted subject to the limitations of this code and the limitations therein. Where engineered design is used in conjunction with these standards the design shall comply with the *International Building Code.*

1. American Forest and Paper Association (AF&PA) *Wood Frame Construction Manual* (WFCM).

2. American Iron and Steel Institute (AISI) *Standard for Cold-Formed Steel Framing—Prescriptive Method for One- and Two-Family Dwellings (COFS/PM) with Supplement to Standard for Cold-Formed Steel Framing-Prescriptive Method for One- and Two-Family Dwellings.*

R301.1.2 Construction systems. The requirements of this code are based on platform and balloon-frame construction for light-frame buildings. The requirements for concrete and masonry buildings are based on a balloon framing system. Other framing systems must have equivalent detailing to ensure force transfer, continuity and compatible deformations.

R301.1.3 Engineered design. When a building of otherwise conventional construction contains structural elements exceeding the limits of Section R301 or otherwise not conforming to this code, these elements shall be designed in accordance with accepted engineering practice. The extent of such design need only demonstrate compliance of nonconventional elements with other applicable provisions and shall be compatible with the performance of the conventional framed system. Engineered design in accordance with the *International Building Code* is permitted for all buildings and structures, and parts thereof, included in the scope of this code.

R301.2 Climatic and geographic design criteria. Buildings shall be constructed in accordance with the provisions of this code as limited by the provisions of this section. Additional criteria shall be established by the local jurisdiction and set forth in Table R301.2(1).

R301.2.1 Wind limitations. Buildings and portions thereof shall be limited by wind speed, as defined in Table R301.2(1) and construction methods in accordance with this code. Basic wind speeds shall be determined from Figure R301.2(4). Where different construction methods and structural materials are used for various portions of a building, the applicable requirements of this section for each portion shall apply. Where loads for wall coverings, curtain walls, roof coverings, exterior windows, skylights, garage doors and exterior doors are not otherwise specified, the loads listed in Table R301.2(2) adjusted for height and exposure using Table R301.2(3) shall be used to determine design load performance requirements for wall coverings, curtain walls, roof coverings, exterior windows, skylights, garage doors and exterior doors. Asphalt shingles shall be designed for wind speeds in accordance with Section R905.2.6.

R301.2.1.1 Design criteria. Construction in regions where the basic wind speeds from Figure R301.2(4) equal or exceed 100 miles per hour (45 m/s) in hurricane-prone regions, or 110 miles per hour (49 m/s) elsewhere, shall be designed in accordance with one of the following:

1. American Forest and Paper Association (AF&PA) *Wood Frame Construction Manual for One- and Two-Family Dwellings* (WFCM); or

2. *Southern Building Code Congress International Standard for Hurricane Resistant Residential Construction* (SSTD 10); or

3. *Minimum Design Loads for Buildings and Other Structures* (ASCE-7); or

4. American Iron and Steel Institute (AISI), *Standard for Cold-Formed Steel Framing—Prescriptive Method For One- and Two-Family Dwellings (COFS/PM) with Supplement to Standard for Cold-Formed Steel Framing—Prescriptive Method For One- and Two-Family Dwellings.*

5. Concrete construction shall be designed in accordance with the provisions of this code.

TABLE R301.2(1)
CLIMATIC AND GEOGRAPHIC DESIGN CRITERIA

GROUND SNOW LOAD	WIND SPEED[d] (mph)	SEISMIC DESIGN CATEGORY[f]	SUBJECT TO DAMAGE FROM			WINTER DESIGN TEMP[e]	ICE BARRIER UNDERLAYMENT REQUIRED[h]	FLOOD HAZARDS[g]	AIR FREEZING INDEX[i]	MEAN ANNUAL TEMP[j]
			Weathering[a]	Frost line depth[b]	Termite[c]					

For SI: 1 pound per square foot = 0.0479 kPa, 1 mile per hour = 0.447 m/s.

a. Weathering may require a higher strength concrete or grade of masonry than necessary to satisfy the structural requirements of this code. The weathering column shall be filled in with the weathering index (i.e., "negligible," "moderate" or "severe") for concrete as determined from the Weathering Probability Map [Figure R301.2(3)]. The grade of masonry units shall be determined from ASTM C 34, C 55, C 62, C 73, C 90, C 129, C 145, C 216 or C 652.

b. The frost line depth may require deeper footings than indicated in Figure R403.1(1). The jurisdiction shall fill in the frost line depth column with the minimum depth of footing below finish grade.

c. The jurisdiction shall fill in this part of the table to indicate the need for protection depending on whether there has been a history of local subterranean termite damage.

d. The jurisdiction shall fill in this part of the table with the wind speed from the basic wind speed map [FigureR301.2(4)].Wind exposure category shall be determined on a site-specific basis in accordance with Section R301.2.1.4.

e. The outdoor design dry-bulb temperature shall be selected from the columns of $97^1/_2$-percent values for winter from Appendix D of the *International Plumbing Code*. Deviations from the Appendix D temperatures shall be permitted to reflect local climates or local weather experience as determined by the building official.

f. The jurisdiction shall fill in this part of the table with the seismic design category determined from Section R301.2.2.1.

g. The jurisdiction shall fill in this part of the table with (a) the date of the jurisdiction's entry into the National Flood Insurance Program (date of adoption of the first code or ordinance for management of flood hazard areas), (b) the date(s) of the currently effective FIRM and FBFM, or other flood hazard map adopted by the community, as may be amended.

h. In accordance with Sections R905.2.7.1, R905.4.3.1, R905.5.3.1, R905.6.3.1, R905.7.3.1 and R905.8.3.1, where there has been a history of local damage from the effects of ice damming, the jurisdiction shall fill in this part of the table with "YES". Otherwise, the jurisdiction shall fill in this part of the table with "NO".

i. The jurisdiction shall fill in this part of the table with the 100-year return period air freezing index (BF-days) from Figure R403.3(2) or from the 100-year (99%) value on the National Climatic Data Center data table "Air Freezing Index- USA Method (Base 32°Fahrenheit)" at www.ncdc.noaa.gov/fpsf.html.

j. The jurisdiction shall fill in this part of the table with the mean annual temperature from the National Climatic Data Center data table "Air Freezing Index-USA Method (Base 32°Fahrenheit)" at www.ncdc.noaa.gov/fpsf.html.

TABLE R301.2(2)
COMPONENT AND CLADDING LOADS FOR A BUILDING WITH A MEAN
ROOF HEIGHT OF 30 FEET LOCATED IN EXPOSURE B (psf)

	ZONE	EFFECTIVE WIND AREA (feet²)	BASIC WIND SPEED (mph—3-second gust)																					
			85		90		100		105		110		120		125		130		140		145		150		170	
Roof > 0 to 10 degrees	1	10	10.0	-13.0	10.0	-14.6	10.0	-18.0	10.0	-19.8	10.0	-21.8	10.5	-25.9	11.4	-28.1	12.4	-30.4	14.3	-35.3	15.4	-37.8	16.5	-40.5	21.1	-52.0
	1	20	10.0	-12.7	10.0	-14.2	10.0	-17.5	10.0	-19.3	10.0	-21.2	10.0	-25.2	10.7	-27.4	11.6	-29.6	13.4	-34.4	14.4	-36.9	15.4	-39.4	19.8	-50.7
	1	50	10.0	-12.2	10.0	-13.7	10.0	-16.9	10.0	-18.7	10.0	-20.5	10.0	-24.4	10.0	-26.4	10.6	-28.6	12.3	-33.2	13.1	-35.6	14.1	-38.1	18.1	-48.9
	1	100	10.0	-11.9	10.0	-13.3	10.0	-18.5	10.0	-18.2	10.0	-19.9	10.0	-23.7	10.0	-25.7	10.0	-27.8	11.4	-32.3	12.2	-34.6	13.0	-37.0	16.7	-47.6
	2	10	10.0	-21.8	10.0	-24.4	10.0	-30.2	10.0	-33.3	10.0	-36.5	10.5	-43.5	11.4	-47.2	12.4	-51.0	14.3	-59.2	15.4	-63.5	16.5	-67.9	21.1	-87.2
	2	20	10.0	-19.5	10.0	-21.8	10.0	-27.0	10.0	-29.7	10.0	-32.6	10.0	-38.8	10.7	-42.1	11.6	-45.6	13.4	-52.9	14.4	-56.7	15.4	-60.7	19.8	-78.0
	2	50	10.0	-16.4	10.0	-18.4	10.0	-22.7	10.0	-25.1	10.0	-27.5	10.0	-32.7	10.0	-35.5	10.6	-38.4	12.3	-44.5	13.1	-47.8	14.1	-51.1	18.1	-65.7
	2	100	10.0	-14.1	10.0	-15.8	10.0	-19.5	10.0	-21.5	10.0	-23.6	10.0	-28.1	10.0	-30.5	10.0	-33.0	11.4	-38.2	12.2	-41.0	13.0	-43.9	16.7	-56.4
	3	10	10.0	-32.8	10.0	-36.8	10.0	-45.4	10.0	-50.1	10.0	-55.0	10.5	-65.4	11.4	-71.0	12.4	-76.8	14.3	-89.0	15.4	-95.5	16.5	-102.2	21.1	-131.3
	3	20	10.0	-27.2	10.0	-30.5	10.0	-37.6	10.0	-41.5	10.0	-45.5	10.0	-54.2	10.7	-58.8	11.6	-63.6	13.4	-73.8	14.4	-79.1	15.4	-84.7	19.8	-108.7
	3	50	10.0	-19.7	10.0	-22.1	10.0	-27.3	10.0	-30.1	10.0	-33.1	10.0	-39.3	10.0	-42.7	10.6	-46.2	12.3	-53.5	13.1	-57.4	14.1	-61.5	18.1	-78.9
	3	100	10.0	-14.1	10.0	-15.8	10.0	-19.5	10.0	-21.5	10.0	-23.6	10.0	-28.1	10.0	-30.5	10.0	-33.0	11.4	-38.2	12.2	-41.0	13.0	-43.9	16.7	-56.4
Roof > 10 to 30 degrees	1	10	10.0	-11.9	10.0	-13.3	10.4	-16.5	11.4	-18.2	12.5	-19.9	14.9	-23.7	16.2	-25.7	17.5	-27.8	20.3	-32.3	21.8	-34.6	23.3	-37.0	30.0	-47.6
	1	20	10.0	-11.6	10.0	-13.0	10.0	-16.0	10.4	-17.6	11.4	-19.4	13.6	-23.0	14.8	-25.0	16.0	-27.0	18.5	-31.4	19.9	-33.7	21.3	-36.0	27.3	-46.3
	1	50	10.0	-11.1	10.0	-12.5	10.0	-15.4	10.0	-17.0	10.0	-18.6	11.9	-22.2	12.9	-24.1	13.9	-26.0	16.1	-30.2	17.3	-32.4	18.5	-34.6	23.8	-44.5
	1	100	10.0	-10.8	10.0	-12.1	10.0	-14.9	10.0	-16.5	10.0	-18.1	10.5	-21.5	11.4	-23.3	12.4	-25.2	14.3	-29.3	15.4	-31.4	16.5	-33.6	21.1	-43.2
	2	10	10.0	-25.1	10.0	-28.2	10.4	-34.8	11.4	-38.3	12.5	-42.1	14.9	-50.1	16.2	-54.3	17.5	-58.7	20.3	-68.1	21.8	-73.1	23.3	-78.2	30.0	-100.5
	2	20	10.0	-22.8	10.0	-25.6	10.0	-31.5	10.4	-34.8	11.4	-38.2	13.6	-45.4	14.8	-49.3	16.0	-53.3	18.5	-61.8	19.9	-66.3	21.3	-71.0	27.3	-91.2
	2	50	10.0	-19.7	10.0	-22.1	10.0	-27.3	10.0	-30.1	10.0	-33.0	11.9	-39.3	12.9	-42.7	13.9	-46.1	16.1	-53.5	17.3	-57.4	18.5	-61.4	23.8	-78.9
	2	100	10.0	-17.4	10.0	-19.5	10.0	-24.1	10.0	-26.6	10.0	-29.1	10.5	-34.7	11.4	-37.6	12.4	-40.7	14.3	-47.2	15.4	-50.6	16.5	-54.2	21.1	-69.6
	3	10	10.0	-25.1	10.0	-28.2	10.4	-34.8	11.4	-38.3	12.5	-42.1	14.9	-50.1	16.2	-54.3	17.5	-58.7	20.3	-68.1	21.8	-73.1	23.3	-78.2	30.0	-100.5
	3	20	10.0	-22.8	10.0	-25.6	10.0	-31.5	10.4	-34.8	11.4	-38.2	13.6	-45.4	14.8	-49.3	16.0	-53.3	18.5	-61.8	19.9	-66.3	21.3	-71.0	27.3	-91.2
	3	50	10.0	-19.7	10.0	-22.1	10.0	-27.3	10.0	-30.1	10.0	-33.0	11.9	-39.3	12.9	-42.7	13.9	-46.1	16.1	-53.5	17.3	-57.4	18.5	-61.4	23.8	-78.9
	3	100	10.0	-17.4	10.0	-19.5	10.0	-24.1	10.0	-26.6	10.0	-29.1	10.5	-34.7	11.4	-37.6	12.4	-40.7	14.3	-47.2	15.4	-50.6	16.5	-54.2	21.1	-69.6
Roof > 30 to 45 degrees	1	10	11.9	-13.0	13.3	-14.6	16.5	-18.0	18.2	-19.8	19.9	-21.8	23.7	-25.9	25.7	-28.1	27.8	-30.4	32.3	-35.3	34.6	-37.8	37.0	-40.5	47.6	-52.0
	1	20	11.6	-12.3	13.0	-13.8	16.0	-17.1	17.6	-18.8	19.4	-20.7	23.0	-24.6	25.0	-26.7	27.0	-28.9	31.4	-33.5	33.7	-35.9	36.0	-38.4	46.3	-49.3
	1	50	11.1	-11.5	12.5	-12.8	15.4	-15.9	17.0	-17.5	18.6	-19.2	22.2	-22.8	24.1	-24.8	26.0	-25.8	30.2	-31.1	32.4	-33.3	34.6	-35.7	44.5	-45.8
	1	100	10.8	-10.8	12.1	-12.1	14.9	-14.9	16.5	-16.5	18.1	-18.1	21.5	-21.5	23.3	-23.3	25.2	-25.2	29.3	-29.3	31.4	-31.4	33.6	-33.6	43.2	-43.2
	2	10	11.9	-15.2	13.3	-17.0	16.5	-21.0	18.2	-23.2	19.9	-25.5	23.7	-30.3	25.7	-32.9	27.8	-35.6	32.3	-41.2	34.6	-44.2	37.0	-47.3	47.6	-60.8
	2	20	11.6	-14.5	13.0	-16.3	16.0	-20.1	17.6	-22.2	19.4	-24.3	23.0	-29.0	25.0	-31.4	27.0	-34.0	31.4	-39.4	33.7	-42.3	36.0	-45.3	46.3	-58.1
	2	50	11.1	-13.7	12.5	-15.3	15.4	-18.9	17.0	-20.8	18.6	-22.9	22.2	-27.2	24.1	-29.5	26.0	-32.0	30.2	-37.1	32.4	-39.8	34.6	-42.5	44.5	-54.6
	2	100	10.8	-13.0	12.1	-14.6	14.9	-18.0	16.5	-19.8	18.1	-21.8	21.5	-25.9	23.3	-28.1	25.2	-30.4	29.3	-35.3	31.4	-37.8	33.6	-40.5	43.2	-52.0
	3	10	11.9	-15.2	13.3	-17.0	16.5	-21.0	18.2	-23.2	19.9	-25.5	23.7	-30.3	25.7	-32.9	27.8	-35.6	32.3	-41.2	34.6	-44.2	37.0	-47.3	47.6	-60.8
	3	20	11.6	-14.5	13.0	-16.3	16.0	-20.1	17.6	-22.2	19.4	-24.3	23.0	-29.0	25.0	-31.4	27.0	-34.0	31.4	-39.4	33.7	-42.3	36.0	-45.3	46.3	-58.1
	3	50	11.1	-13.7	12.5	-15.3	15.4	-18.9	17.0	-20.8	18.6	-22.9	22.2	-27.2	24.1	-29.5	26.0	-32.0	30.2	-37.1	32.4	-39.8	34.6	-42.5	44.5	-54.5
	3	100	10.8	-13.0	12.1	-14.6	14.9	-18.0	16.5	-19.8	18.1	-21.8	21.5	-25.9	23.3	-28.1	25.2	-30.4	29.3	-35.3	31.4	-37.8	33.6	-40.5	43.2	-52.0
Wall	4	10	13.0	-14.1	14.6	-15.8	18.0	-19.5	19.8	-21.5	21.8	-23.6	25.9	-28.1	28.1	-30.5	30.4	-33.0	35.3	-38.2	37.8	-41.0	40.5	-43.9	52.0	-56.4
	4	20	12.4	-13.5	13.9	-15.1	17.2	-18.7	18.9	-20.6	20.8	-22.6	24.7	-26.9	26.8	-29.2	29.0	-31.6	33.7	-36.7	36.1	-39.3	38.7	-42.1	49.6	-54.1
	4	50	11.6	-12.7	13.0	-14.3	16.1	-17.6	17.8	-19.4	19.5	-21.3	23.2	-25.4	25.2	-27.5	27.2	-29.8	31.6	-34.6	33.9	-37.1	36.2	-39.7	46.6	-51.0
	4	100	11.1	-12.2	12.4	-13.6	15.3	-16.8	16.9	-18.5	18.5	-20.4	22.0	-24.2	23.9	-26.3	25.9	-28.4	30.0	-33.0	32.2	-35.4	34.4	-37.8	44.2	-48.6
	5	10	13.0	-17.4	14.6	-19.5	18.0	-24.1	19.8	-26.6	21.8	-29.1	25.9	-34.7	28.1	-37.6	30.4	-40.7	35.3	-47.2	37.8	-50.6	40.5	-54.2	52.0	-69.6
	5	20	12.4	-16.2	13.9	-18.2	17.2	-22.5	18.9	-24.8	20.8	-27.2	24.7	-32.4	26.8	-35.1	29.0	-38.0	33.7	-44.0	36.1	-47.2	38.7	-50.5	49.6	-64.9
	5	50	11.6	-14.7	13.0	-16.5	16.1	-20.3	17.8	-22.4	19.5	-24.6	23.2	-29.3	25.2	-31.8	27.2	-34.3	31.6	-39.8	33.9	-42.7	36.2	-45.7	46.6	-58.7
	5	100	11.1	-13.5	12.4	-15.1	15.3	-18.7	16.9	-20.6	18.5	-22.6	22.0	-26.9	23.9	-29.2	25.9	-31.6	30.0	-36.7	32.2	-39.3	34.4	-42.1	44.2	-54.1

For SI: 1 foot = 304.8 mm, 1 square foot = 0.0929 m², 1 mile per hour = 0.447 m/s, 1 pound per foot = 0.0479 kPa.

NOTES: For effective areas between those given above the load may be interpolated, otherwise use the load associated with the lower effective area.

Table values shall be adjusted for height and exposure by multiplying by the adjustment coefficient in Table R301.2(3).

See Figure R301.2(7) for location of zones.

Plus and minus signs signify pressures acting toward and away from the building surfaces.

TABLE R301.2(3)
HEIGHT AND EXPOSURE ADJUSTMENT COEFFICIENTS FOR TABLE R301.2(2)

MEAN ROOF HEIGHT	EXPOSURE		
	B	C	D
15	1.00	1.21	1.47
20	1.00	1.29	1.55
25	1.00	1.35	1.61
30	1.00	1.40	1.66
35	1.05	1.45	1.70
40	1.09	1.49	1.74
45	1.12	1.53	1.78
50	1.16	1.56	1.81
55	1.19	1.59	1.84
60	1.22	1.62	1.87

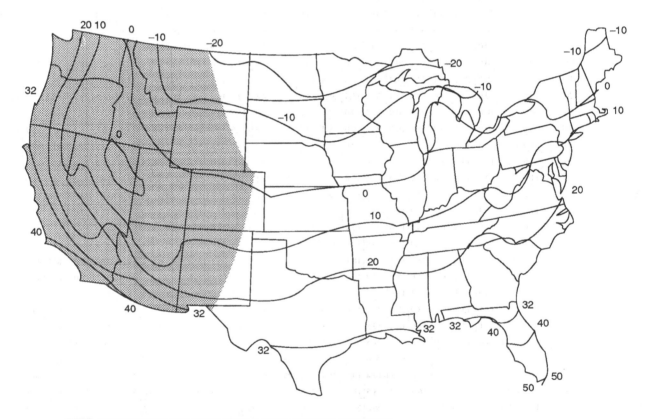

DESIGN TEMPERATURES IN THIS AREA MUST BE BASED ON ANALYSIS OF LOCAL CLIMATE AND TOPOGRAPHY

For SI: °C = [(°F)-32] /1.8.

FIGURE R301.2(1)
ISOLINES OF THE 97^1/$_2$ PERCENT WINTER (DECEMBER, JANUARY AND FEBRUARY) DESIGN TEMPERATURES (°F)

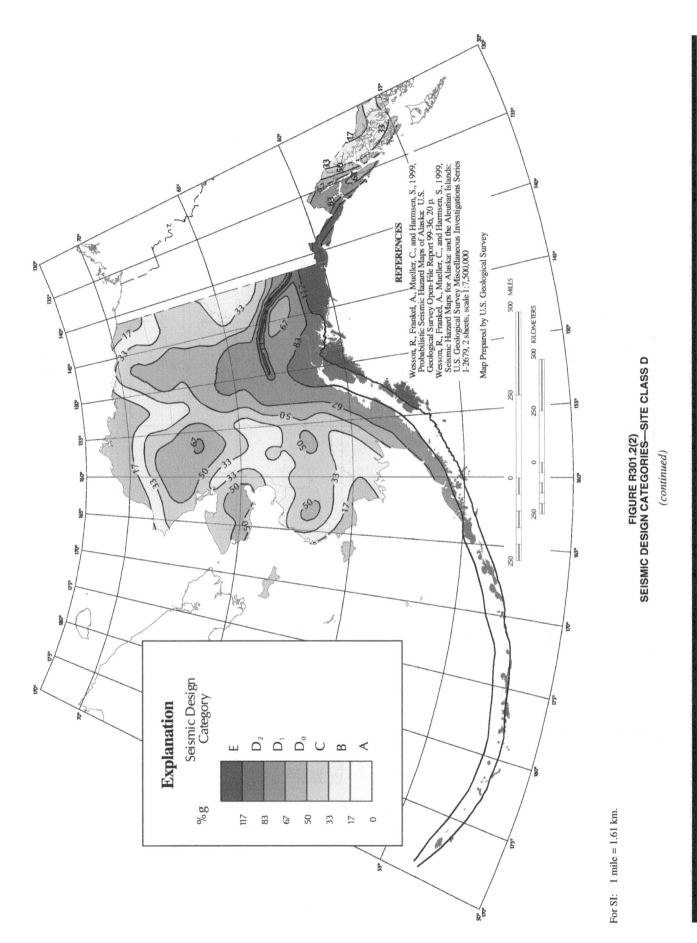

Explanation

Seismic Design Category

%g		Seismic Design Category
117		E
83		D₂
67		D₁
50		D₀
33		C
17		B
0		A

REFERENCES

Wesson, R., Frankel, A., Mueller, C., and Harmsen, S., 1999, Probabilistic Seismic Hazard Maps of Alaska: U.S. Geological Survey Open-File Report 99-36, 20 p.
Wesson, R., Frankel, A., Mueller, C., and Harmsen, S., 1999, Seismic Hazard Maps for Alaska and the Aleutian Islands: U.S. Geological Survey Miscellaneous Investigations Series I-2679, 2 sheets, scale 1:7,500,000

Map Prepared by U.S. Geological Survey

FIGURE R301.2(2)
SEISMIC DESIGN CATEGORIES—SITE CLASS D
(continued)

For SI: 1 mile = 1.61 km.

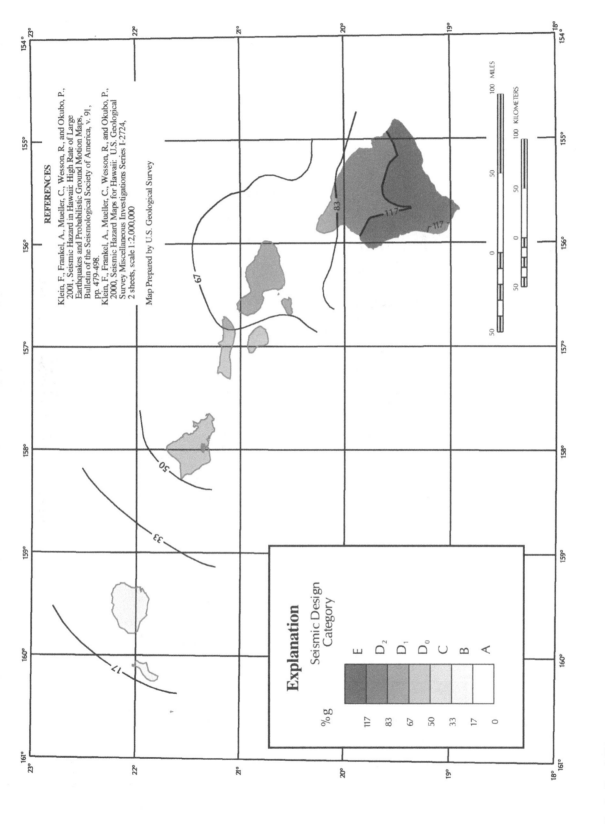

FIGURE R301.2(2)—continued
SEISMIC DESIGN CATEGORIES—SITE CLASS D

(continued)

For SI: 1 mile = 1.61 km.

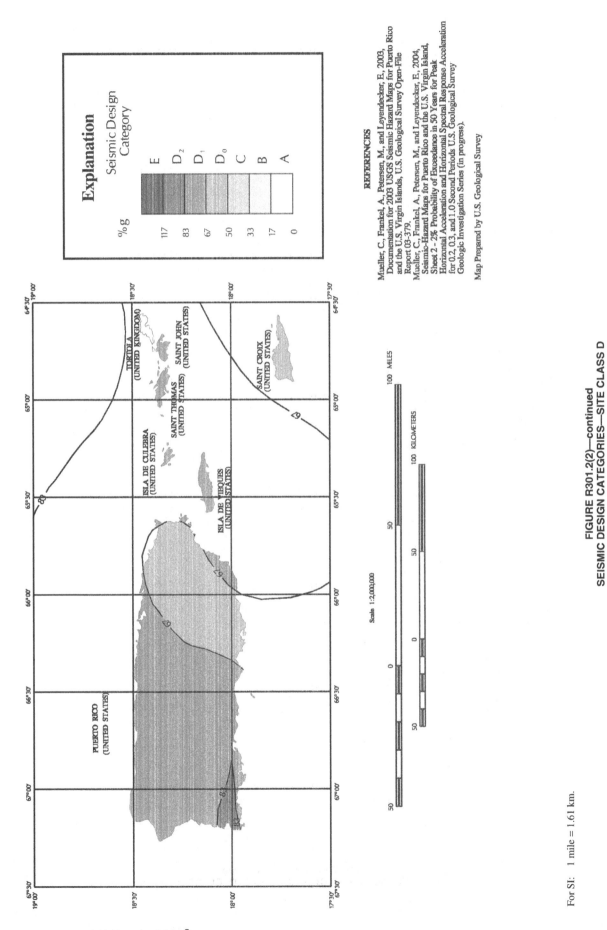

Explanation

Seismic Design Category

% g	
117	E
83	D_2
67	D_1
50	D_0
33	C
17	B
0	A

REFERENCES

Mueller, C., Frankel, A., Petersen, M., and Leyendecker, E., 2003, Documentation for 2003 USGS Seismic Hazard Maps for Puerto Rico and the U.S. Virgin Islands, U.S. Geological Survey Open-File Report 03-379.

Mueller, C., Frankel, A., Petersen, M., and Leyendecker, E., 2004, Seismic-Hazard Maps for Puerto Rico and the U.S. Virgin Island, Sheet 2 - 2% Probability of Exceedance in 50 Years for Peak Horizontal Acceleration and Horizontal Spectral Response Acceleration for 0.2, 0.3, and 1.0 Second Periods U.S. Geological Survey Geologic Investigation Series (in progress).

Map Prepared by U.S. Geological Survey

Scale 1:2,000,000

FIGURE R301.2(2)—continued
SEISMIC DESIGN CATEGORIES—SITE CLASS D

(continued)

For SI: 1 mile = 1.61 km.

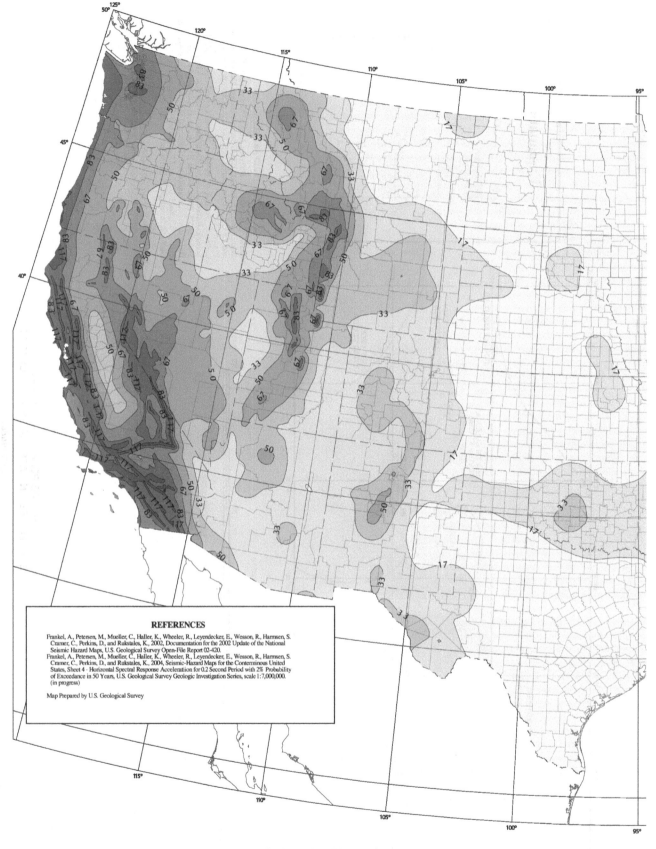

FIGURE R301.2(2)—continued
SEISMIC DESIGN CATEGORIES—SITE CLASS D

(continued)

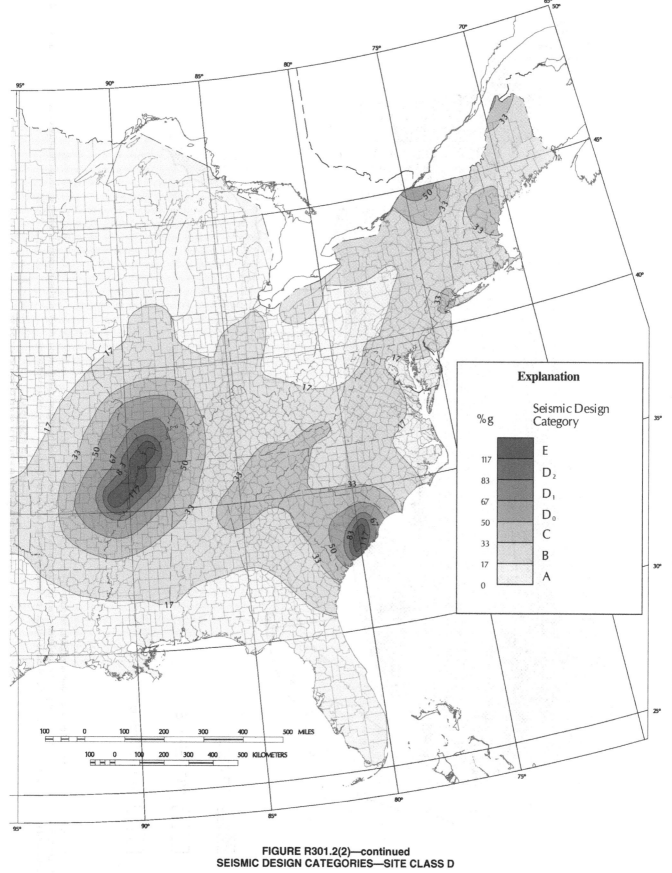

FIGURE R301.2(2)—continued
SEISMIC DESIGN CATEGORIES—SITE CLASS D

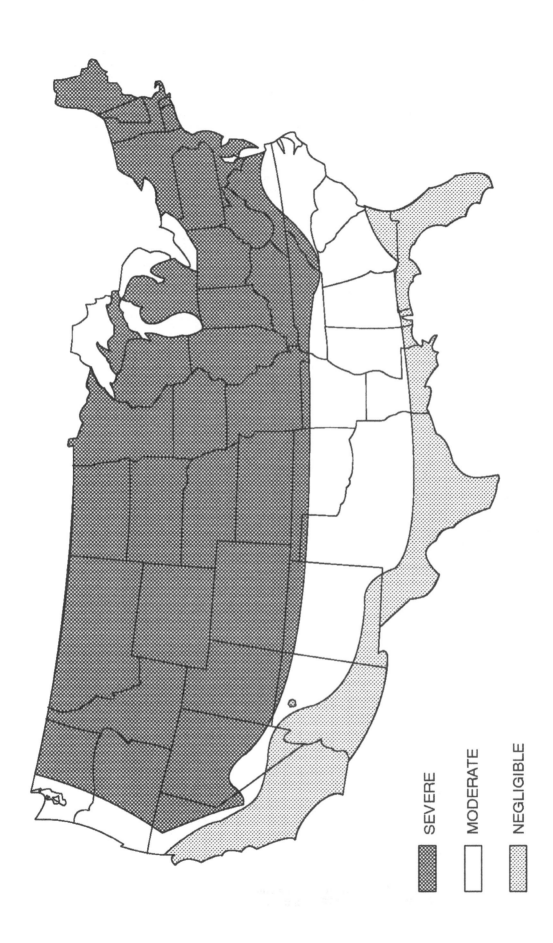

a. Alaska and Hawaii are classified as severe and negligible, respectively.
b. Lines defining areas are approximate only. Local conditions may be more or less severe than indicated by region classification. A severe classification is where weather conditions result in significant snowfall combined with extended periods during which there is little or no natural thawing causing deicing salts to be used extensively.

FIGURE R301.2(3)
WEATHERING PROBABILITY MAP FOR CONCRETE

SEVERE

MODERATE

NEGLIGIBLE

Location	V mph	(m/s)
Hawaii	105	(47)
Puerto Rico	145	(65)
Guam	170	(76)
Virgin Islands	145	(65)
American Samoa	125	(56)

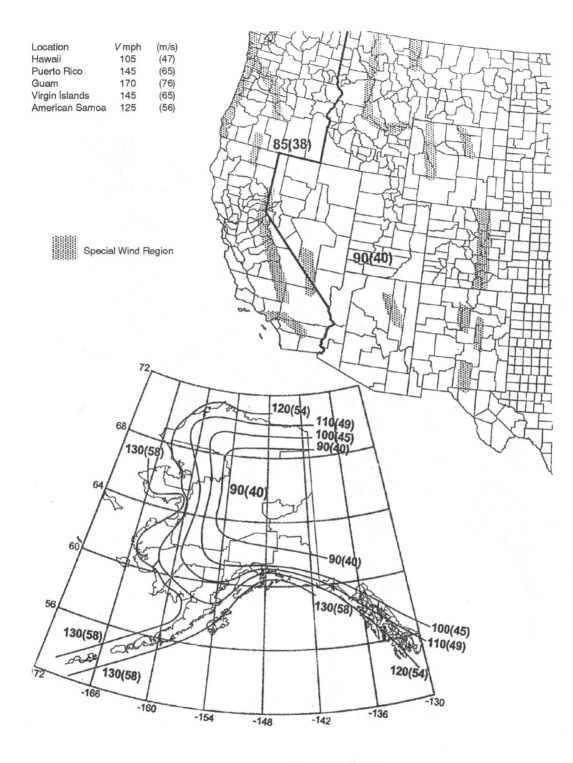

Special Wind Region

FIGURE R301.2(4)
BASIC WIND SPEEDS FOR 50-YEAR MEAN RECURRENCE INTERVAL

(continued)

For SI: 1 foot = 304.8 mm, 1 mile per hour = 0.447 m/s.

a. Values are nominal design 3-second gust wind speeds in miles per hour at 33 feet above ground for Exposure C category.

b. Linear interpolation between wind contours is permitted.

c. Islands and coastal areas outside the last contour shall use the last wind speed contour of the coastal area.

d. Mountainous terrain, gorges, ocean promontories and special wind regions shall be examined for unusual wind conditions.

e. Enlarged view of Eastern and Southern seaboards are on the following pages.

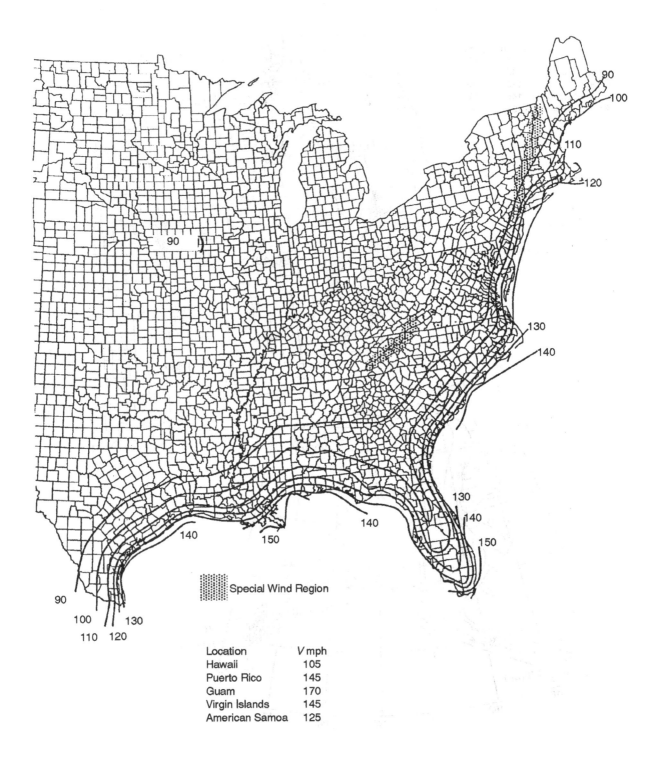

Location	V mph
Hawaii	105
Puerto Rico	145
Guam	170
Virgin Islands	145
American Samoa	125

FIGURE R301.2(4)—continued
BASIC WIND SPEEDS FOR 50-YEAR MEAN RECURRENCE INTERVAL

(continued)

For SI: 1 foot = 304.8 mm, 1 mile per hour = 0.447 m/s.

a. Values are nominal design 3-second gust wind speeds in miles per hour at 33 feet above ground for Exposure C category.

b. Linear interpolation between wind contours is permitted.

c. Islands and coastal areas outside the last contour shall use the last wind speed contour of the coastal area.

d. Mountainous terrain, gorges, ocean promontories and special wind regions shall be examined for unusual wind conditions.

e. Enlarged view of Eastern and Southern seaboards are on the following pages.

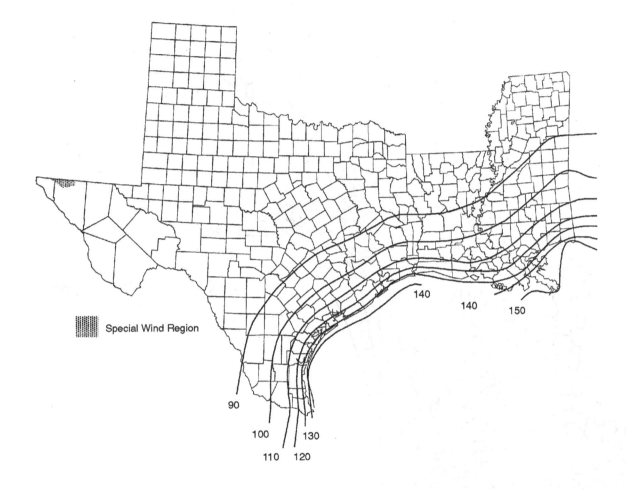

Special Wind Region

140

140

150

90

100 130

110 120

FIGURE R301.2(4)—continued
BASIC WIND SPEEDS FOR 50-YEAR MEAN RECURRENCE INTERVAL

(continued)

For SI: 1 foot = 304.8 mm, 1 mile per hour = 0.447 m/s.

a. Values are nominal design 3-second gust wind speeds in miles per hour at 33 feet above ground for Exposure C category.

b. Linear interpolation between wind contours is permitted.

c. Islands and coastal areas outside the last contour shall use the last wind speed contour of the coastal area.

d. Mountainous terrain, gorges, ocean promontories and special wind regions shall be examined for unusual wind conditions.

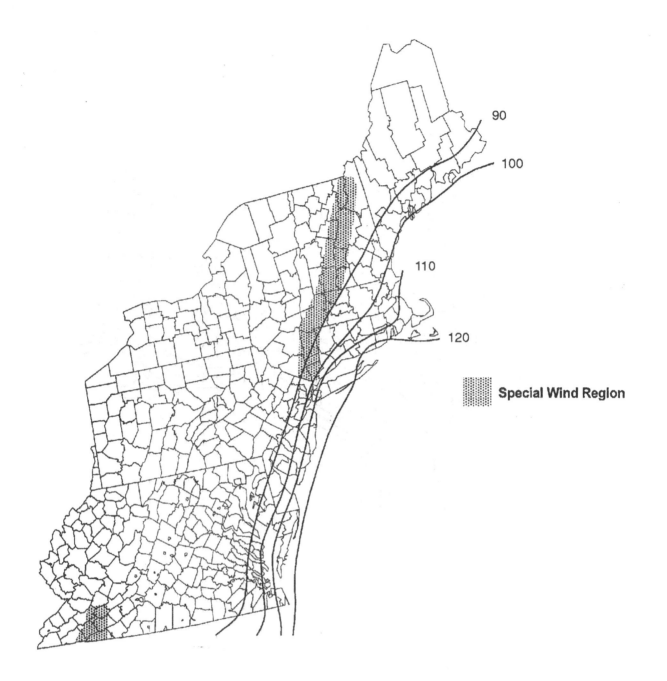

Special Wind Region

FIGURE R301.2(4)—continued
BASIC WIND SPEEDS FOR 50-YEAR MEAN RECURRENCE INTERVAL
(continued)

For SI: 1 foot = 304.8 mm, 1 mile per hour = 0.447 m/s.

a. Values are nominal design 3-second gust wind speeds in miles per hour at 33 feet above ground for Exposure C category.

b. Linear interpolation between wind contours is permitted.

c. Islands and coastal areas outside the last contour shall use the last wind speed contour of the coastal area.

d. Mountainous terrain, gorges, ocean promontories and special wind regions shall be examined for unusual wind conditions.

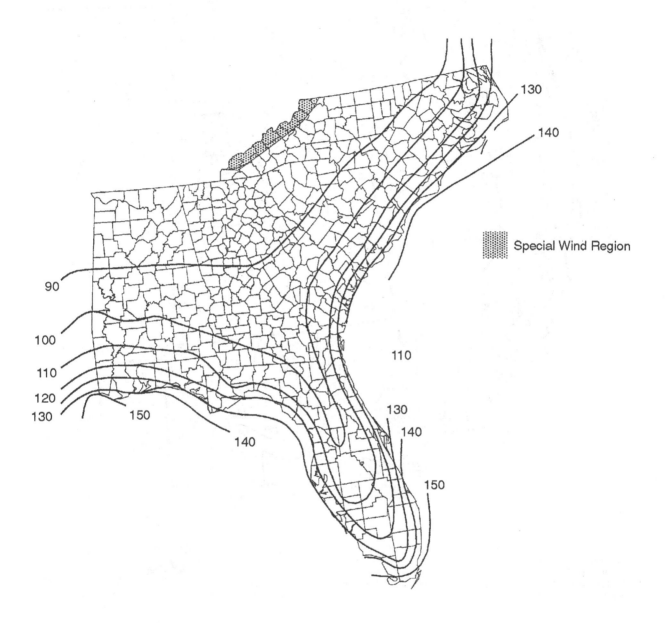

FIGURE R301.2(4)—continued
BASIC WIND SPEEDS FOR 50-YEAR MEAN RECURRENCE INTERVAL

For SI: 1 foot = 304.8 mm, 1 mile per hour = 0.447 m/s.

a. Values are nominal design 3-second gust wind speeds in miles per hour at 33 feet above ground for Exposure C category.

b. Linear interpolation between wind contours is permitted.

c. Islands and coastal areas outside the last contour shall use the last wind speed contour of the coastal area.

d. Mountainous terrain, gorges, ocean promontories and special wind regions shall be examined for unusual wind conditions.

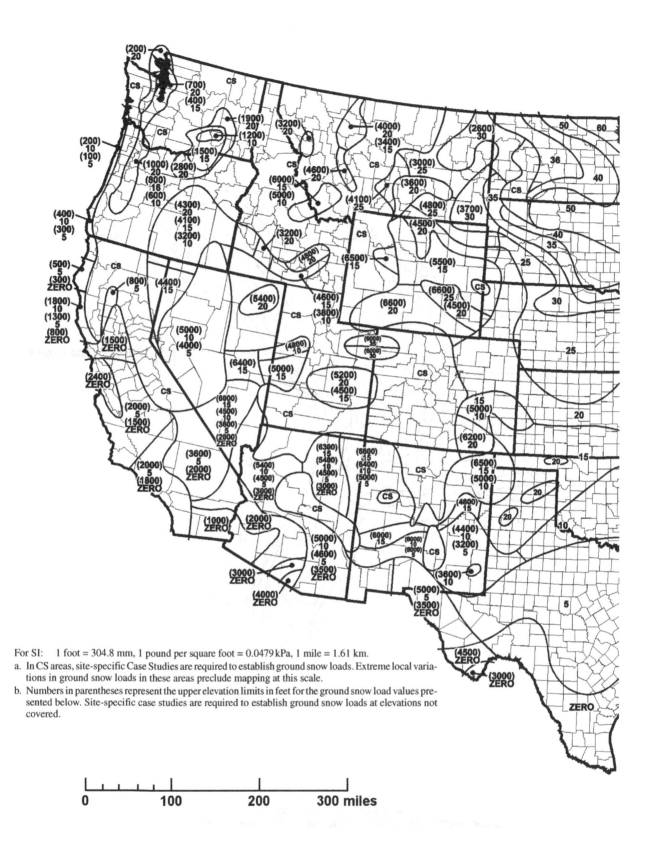

For SI: 1 foot = 304.8 mm, 1 pound per square foot = 0.0479 kPa, 1 mile = 1.61 km.

a. In CS areas, site-specific Case Studies are required to establish ground snow loads. Extreme local variations in ground snow loads in these areas preclude mapping at this scale.

b. Numbers in parentheses represent the upper elevation limits in feet for the ground snow load values presented below. Site-specific case studies are required to establish ground snow loads at elevations not covered.

FIGURE R301.2(5)
GROUND SNOW LOADS, P_g, FOR THE UNITED STATES (lb/ft^2)

(continued)

For SI: 1 foot = 304.8 mm, 1 pound per square foot = 0.0479 kPa.

FIGURE R301.2(5)—continued
GROUND SNOW LOADS, P_g, FOR THE UNITED STATES (lb/ft^2)

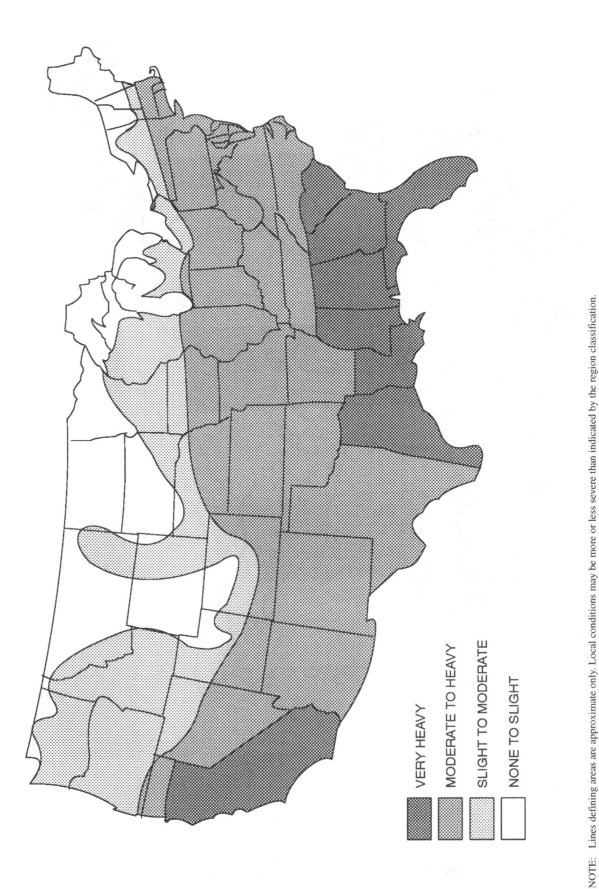

VERY HEAVY

MODERATE TO HEAVY

SLIGHT TO MODERATE

NONE TO SLIGHT

NOTE: Lines defining areas are approximate only. Local conditions may be more or less severe than indicated by the region classification.

FIGURE R301.2(6)
TERMITE INFESTATION PROBABILITY MAP

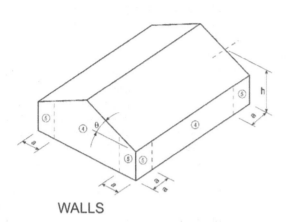

WALLS

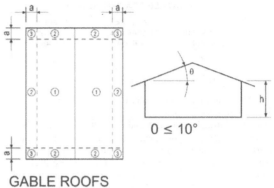

GABLE ROOFS
0 ≤ 10°

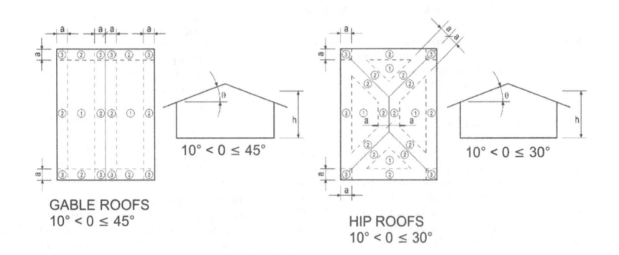

GABLE ROOFS
10° < 0 ≤ 45°

HIP ROOFS
10° < 0 ≤ 30°

For SI: 1 foot = 304.8 mm, 1 degree = 0.0175 rad.
NOTE: a = 4 feet in all cases.

FIGURE R301.2(7)
COMPONENT AND CLADDING PRESSURE ZONES

R301.2.1.2 Protection of openings. Windows in buildings located in windborne debris regions shall have glazed openings protected from windborne debris. Glazed opening protection for windborne debris shall meet the requirements of the Large Missile Test of ASTM E 1996 and ASTM E 1886 referenced therein.

> **Exception:** Wood structural panels with a minimum thickness of $^7/_{16}$ inch (11 mm) and a maximum span of 8 feet (2438 mm) shall be permitted for opening protection in one- and two-story buildings. Panels shall be precut so that they shall be attached to the framing surrounding the opening containing the product with the glazed opening. Panels shall be secured with the attachment hardware provided. Attachments shall be designed to resist the component and cladding loads determined in accordance with either Table R301.2(2) or ASCE 7. Attachment in accordance with Table R301.2.1.2 is permitted for buildings with a mean roof height of 33 feet (10 058 mm) or less where wind speeds do not exceed 130 miles per hour (58 m/s).

TABLE R301.2.1.2
WINDBORNE DEBRIS PROTECTION FASTENING SCHEDULE FOR WOOD STRUCTURAL PANELS[a, b, c, d]

FASTENER TYPE	FASTENER SPACING (inches)		
	Panel span ≤ 4 feet	4 feet < panel span ≤ 6 feet	6 feet < panel span ≤ 8 feet
No. 6 Screws	16″	12″	9″
No. 8 Screws	16″	16″	12″

For SI: 1 inch = 25.4 mm, 1 foot = 304.8 mm, 1 pound = 4.448N, 1 mile per hour = 0.447 m/s.

a. This table is based on 130 mph wind speeds and a 33-foot mean roof height.

b. Fasteners shall be installed at opposing ends of the wood structural panel. Fasteners shall be located a minimum of 1 inch from the edge of the panel.

c. Fasteners shall be long enough to penetrate through the exterior wall covering and a minimum of $1^1/_4$ inches into wood wall framing and a minimum of $1^1/_4$ inches into concrete block or concrete, and into steel framing a minimum of 3 exposed threads. Fasteners shall be located a minimum of $2^1/_2$ inches from the edge of concrete block or concrete.

d. Where screws are attached to masonry or masonry/stucco, they shall be attached using vibration-resistant anchors having a minimum ultimate withdrawal capacity of 490 pounds.

R301.2.1.3 Wind speed conversion. When referenced documents are based on fastest mile wind speeds, the three-second gust basic wind speeds, V_{3s}, of Figure R301.2(4) shall be converted to fastest mile wind speeds, V_{fm}, using Table R301.2.1.3.

R301.2.1.4 Exposure category. For each wind direction considered, an exposure category that adequately reflects the characteristics of ground surface irregularities shall be determined for the site at which the building or structure is to be constructed. For a site located in the transition zone between categories, the category resulting in the largest wind forces shall apply. Account shall be taken of variations in ground surface roughness that arise from natural topography and vegetation as well as from constructed features. For any given wind direction, the exposure in which a specific building or other structure is sited shall be assessed as being one of the following categories:

1. Exposure A. Large city centers with at least 50 percent of the buildings having a height in excess of 70 feet (21 336 mm). Use of this exposure category shall be limited to those areas for which terrain representative of Exposure A prevails in the upwind direction for a distance of at least 0.5 mile (0.8 km) or 10 times the height of the building or other structure, whichever is greater. Possible channeling effects or increased velocity pressures due to the building or structure being located in the wake of adjacent buildings shall be taken into account.

2. Exposure B. Urban and suburban areas, wooded areas, or other terrain with numerous closely spaced obstructions having the size of single-family dwellings or larger. Exposure B shall be assumed unless the site meets the definition of another type exposure.

3. Exposure C. Open terrain with scattered obstructions, including surface undulations or other irregularities, having heights generally less than 30 feet (9144 mm) extending more than 1,500 feet (457 m) from the building site in any quadrant. This exposure shall also apply to any building located within Exposure B type terrain where the building is directly adjacent to open areas of Exposure C type terrain in any quadrant for a distance of more than 600 feet (183 m). This category includes flat open country, grasslands and shorelines in hurricane prone regions.

4. Exposure D. Flat, unobstructed areas exposed to wind flowing over open water (excluding shorelines in hurricane prone regions) for a distance of at least 1 mile (1.61 km). Shorelines in Exposure D include inland waterways, the Great Lakes and coastal areas of California, Oregon, Washington and Alaska. This exposure shall apply only to those buildings and other structures exposed to the wind coming from over the water. Exposure D extends inland from the shoreline a distance of 1,500 feet (457 m) or 10 times the height of the building or structure, whichever is greater.

R301.2.2 Seismic provisions. The seismic provisions of this code shall apply to buildings constructed in Seismic Design Categories C, D_0, D_1 and D_2, as determined in accordance with this section. Buildings in Seismic Design Category E shall be designed in accordance with the *International Building Code*, except when the seismic design category is reclassified to a lower seismic design category in accordance with Section R301.2.2.1.

Exception: Detached one- and two-family dwellings located in Seismic Design Category C are exempt from the seismic requirements of this code.

The weight and irregularity limitations of Section R301.2.2.2 shall apply to buildings in all seismic design categories regulated by the seismic provisions of this code. Buildings in Seismic Design Category C shall be constructed in accordance with the additional requirements of Section R301.2.2.3. Buildings in Seismic Design Categories D_0, D_1 and D_2 shall be constructed in accordance with the additional requirements of Section R301.2.2.4

TABLE R301.2.1.3
EQUIVALENT BASIC WIND SPEEDS[a]

3-second gust, V_{3s}	85	90	100	105	110	120	125	130	140	145	150	160	170
Fastest mile, V_{fm}	71	76	85	90	95	104	109	114	123	128	133	142	152

For SI: 1 mile per hour = 0.447 m/s.

a. Linear interpolation is permitted.

R301.2.2.1 Determination of seismic design category. Buildings shall be assigned a seismic design category in accordance with Figure 301.2(2).

R301.2.2.1.1 Alternate determination of seismic design category. The Seismic Design Categories and corresponding Short Period Design Spectral Response Accelerations, S_{DS} shown in Figure R301.2(2) are based on soil Site Class D, as defined in Section 1613.5.2 of the *International Building Code.* If soil conditions are other than Site Class D, the Short Period Design Spectral Response Acceleration, S_{DS}, for a site can be determined according to Section 1613.5 of the *International Building Code.* The value of S_{DS} determined according to Section 1613.5 of the *International Building Code* is permitted to be used to set the seismic design category according to Table R301.2.2.1.1, and to interpolate between values in Tables R602.10.1, R603.7, and other seismic design requirements of this code.

TABLE R301.2.2.1.1
SEISMIC DESIGN CATEGORY DETERMINATION

CALCULATED S_{DS}	SEISMIC DESIGN CATEGORY
$S_{DS} \leq 0.17g$	A
$0.17g < S_{DS} \leq 0.33g$	B
$0.33g < S_{DS} \leq 0.50g$	C
$0.50g < S_{DS} \leq 0.67g$	D_0
$0.67g < S_{DS} \leq 0.83g$	D_1
$0.83g < S_{DS} \leq 1.17g$	D_2
$1.17g < S_{DS}$	E

R301.2.2.1.2 Alternative determination of Seismic Design Category E. Buildings located in Seismic Design Category E in accordance with Figure R301.2(2) are permitted to be reclassified as being in Seismic Design Category D_2 provided one of the following is done:

1. A more detailed evaluation of the seismic design category is made in accordance with the provisions and maps of the *International Building Code.* Buildings located in Seismic Design Category E per Table R301.2.2.1.1, but located in Seismic Design Category D per the *International Building Code,* may be designed using the Seismic Design Category D_2 requirements of this code.

2. Buildings located in Seismic Design Category E that conform to the following additional restrictions are permitted to be constructed in accordance with the provisions for Seismic Design Category D_2 of this code:

   2.1. All exterior shear wall lines or braced wall panels are in one plane vertically from the foundation to the uppermost story.

   2.2. Floors shall not cantilever past the exterior walls.

   2.3. The building is within all of the requirements of Section R301.2.2.2.2 for being considered as regular.

R301.2.2.2 Seismic limitations. The following limitations apply to buildings in all Seismic Design Categories regulated by the seismic provisions of this code.

R301.2.2.2.1 Weights of materials. Average dead loads shall not exceed 15 pounds per square foot (720 Pa) for the combined roof and ceiling assemblies (on a horizontal projection) or 10 pounds per square foot (480 Pa) for floor assemblies, except as further limited by Section R301.2.2. Dead loads for walls above grade shall not exceed:

1. Fifteen pounds per square foot (720 Pa) for exterior light-frame wood walls.

2. Fourteen pounds per square foot (670 Pa) for exterior light-frame cold-formed steel walls.

3. Ten pounds per square foot (480 Pa) for interior light-frame wood walls.

4. Five pounds per square foot (240 Pa) for interior light-frame cold-formed steel walls.

5. Eighty pounds per square foot (3830 Pa) for 8-inch-thick (203 mm) masonry walls.

6. Eighty-five pounds per square foot (4070 Pa) for 6-inch-thick (152 mm) concrete walls.

Exceptions:

1. Roof and ceiling dead loads not exceeding 25 pounds per square foot (1190 Pa) shall be permitted provided the wall bracing amounts in Chapter 6 are increased in accordance with Table R301.2.2.2.1.

2. Light-frame walls with stone or masonry veneer shall be permitted in accordance with the provisions of Sections R702.1 and R703.

3. Fireplaces and chimneys shall be permitted in accordance with Chapter 10.

TABLE R301.2.2.2.1
WALL BRACING ADJUSTMENT FACTORS BY
ROOF COVERING DEAD LOAD[a]

WALL SUPPORTING	ROOF/CEILING DEAD LOAD 15 psf or less	ROOF/CEILING DEAD LOAD 25 psf
Roof only	1.0	1.2
Roof plus one story	1.0	1.1

For SI: 1 pound per square foot = 0.049 kPa.
a. Linear interpolation shall be permitted.

R301.2.2.2.2 Irregular buildings. Prescriptive construction as regulated by this code shall not be used for irregular structures located in Seismic Design Categories C, D_0, D_1 and D_2. Irregular portions of structures shall be designed in accordance with accepted engineering practice to the extent the irregular features affect the performance of the remaining structural system. When the forces associated with the irregularity are resisted by a structural system designed in accordance with accepted engineering practice, design of the remainder of the building shall be permitted using the provisions of this code. A building or portion of a building shall be considered to be irregular when one or more of the following conditions occur:

1. When exterior shear wall lines or braced wall panels are not in one plane vertically from the foundation to the uppermost story in which they are required.

    Exception: For wood light-frame construction, floors with cantilevers or setbacks not exceeding four times the nominal depth of the wood floor joists are permitted to support braced wall panels that are out of plane with braced wall panels below provided that:

    1. Floor joists are nominal 2 inches by 10 inches (51 mm by 254 mm) or larger and spaced not more than 16 inches (406 mm) on center.

    2. The ratio of the back span to the cantilever is at least 2 to 1.

    3. Floor joists at ends of braced wall panels are doubled.

    4. For wood-frame construction, a continuous rim joist is connected to ends of all cantilever joists. When spliced, the rim joists shall be spliced using a galvanized metal tie not less than 0.058 inch (1.5 mm) (16 gage) and $1^1/_2$ inches (38 mm) wide fastened with six 16d nails on each side of the splice or a block of the same size as the rim joist of sufficient length to fit securely between the joist space at which the splice occurs fastened with eight 16d nails on each side of the splice; and

    5. Gravity loads carried at the end of cantilevered joists are limited to uniform wall and roof loads and the reactions from headers having a span of 8 feet (2438 mm) or less.

2. When a section of floor or roof is not laterally supported by shear walls or braced wall lines on all edges.

    Exception: Portions of floors that do not support shear walls or braced wall panels above, or roofs, shall be permitted to extend no more than 6 feet (1829 mm) beyond a shear wall or braced wall line.

3. When the end of a braced wall panel occurs over an opening in the wall below and ends at a horizontal distance greater than 1 foot (305 mm) from the edge of the opening. This provision is applicable to shear walls and braced wall panels offset in plane and to braced wall panels offset out of plane as permitted by the exception to Item 1 above.

    Exception: For wood light-frame wall construction, one end of a braced wall panel shall be permitted to extend more than 1 foot (305 mm) over an opening not more than 8 feet (2438 mm) wide in the wall below provided that the opening includes a header in accordance with the following:

    1. The building width, loading condition and framing member species limitations of Table R502.5(1) shall apply and

    2. Not less than one 2×12 or two 2×10 for an opening not more than 4 feet (1219 mm) wide or

    3. Not less than two 2×12 or three 2×10 for an opening not more than 6 feet (1829 mm) wide or

    4. Not less than three 2×12 or four 2×10 for an opening not more than 8 feet (2438 mm) wide and

    5. The entire length of the braced wall panel does not occur over an opening in the wall below.

4. When an opening in a floor or roof exceeds the lesser of 12 feet (3657 mm) or 50 percent of the least floor or roof dimension.

5. When portions of a floor level are vertically offset.

    Exceptions:

    1. Framing supported directly by continuous foundations at the perimeter of the building.

    2. For wood light-frame construction, floors shall be permitted to be vertically offset when the floor framing is lapped or tied together as required by Section R502.6.1.

6. When shear walls and braced wall lines do not occur in two perpendicular directions.

7. When stories above-grade partially or completely braced by wood wall framing in accordance with Section R602 or steel wall framing in accordance with Section R603 include masonry or concrete construction.

    Exception: Fireplaces, chimneys and masonry veneer as permitted by this code.

When this irregularity applies, the entire story shall be designed in accordance with accepted engineering practice.

R301.2.2.3 Seismic Design Category C. Structures assigned to Seismic Design Category C shall conform to the requirements of this section.

R301.2.2.3.1 Stone and masonry veneer. Stone and masonry veneer shall comply with the requirements of Sections R702.1 and R703.

R301.2.2.3.2 Masonry construction. Masonry construction shall comply with the requirements of Section R606.12.2.

R301.2.2.3.3 Concrete construction. Concrete construction shall comply with the requirements of Section R611 or R612.

R301.2.2.4 Seismic Design Categories D_0, D_1 and D_2. Structures assigned to Seismic Design Categories D_0, D_1 and D_2 shall conform to the requirements for Seismic Design Category C and the additional requirements of this section.

R301.2.2.4.1 Height limitations. Wood framed buildings shall be limited to three stories above grade or the limits given in Table R602.10.1. Cold-formed steel framed buildings shall be limited to two stories above grade in accordance with COFS/PM. Mezzanines as defined in Section 202 shall not be considered as stories.

R301.2.2.4.2 Stone and masonry veneer. Stone and masonry veneer shall comply with the requirements of Sections R702.1 and R703.

R301.2.2.4.3 Masonry construction. Masonry construction in Seismic Design Categories D_0 and D_1 shall comply with the requirements of Section R606.11.3. Masonry construction in Seismic Design Category D_2 shall comply with the requirements of Section R606.11.4.

R301.2.2.4.4 Concrete construction. Buildings with above-grade concrete walls shall be in accordance with Section R611, R612, or designed in accordance with accepted engineering practice.

R301.2.2.4.5 Cold-formed steel framing in Seismic Design Categories D_0, D_1 and D_2. In Seismic Design Categories D_0, D_1 and D_2 in addition to the requirements of this code, cold-formed steel framing shall comply with the requirements of COFS/PM.

R301.2.3 Snow loads. Wood framed construction, cold-formed steel framed construction and masonry and concrete construction in regions with ground snow loads 70 pounds per square foot (3.35 kPa) or less, shall be in accordance with Chapters 5, 6 and 8. Buildings in regions with ground snow loads greater than 70 pounds per square foot (3.35 kPa) shall be designed in accordance with accepted engineering practice.

R301.2.4 Floodplain construction. Buildings and structures constructed in whole or in part in flood hazard areas (including A or V Zones) as established in Table R301.2(1) shall be designed and constructed in accordance with Section R324.

> **Exception:** Buildings and structures located in whole or in part in identified floodways as established in Table R301.2(1) shall be designed and constructed as stipulated in the *International Building Code*.

R301.3 Story height. Buildings constructed in accordance with these provisions shall be limited to story heights of not more than the following:

1. For wood wall framing, the laterally unsupported bearing wall stud height permitted by Table R602.3(5) plus a height of floor framing not to exceed 16 inches.

   > **Exception:** For wood framed wall buildings with bracing in accordance with Table R602.10.1, the wall stud clear height used to determine the maximum permitted story height may be increased to 12 feet without requiring an engineered design for the building wind and seismic force resisting systems provided that the length of bracing required by Table R602.10.1 is increased by multiplying by a factor of 1.20. Wall studs are still subject to the requirements of this section.

2. For steel wall framing, a stud height of 10 feet, plus a height of floor framing not to exceed 16 inches.

3. For masonry walls, a maximum bearing wall clear height of 12 feet plus a height of floor framing not to exceed 16 inches.

   > **Exception:** An additional 8 feet is permitted for gable end walls.

4. For insulating concrete form walls, the maximum bearing wall height per story as permitted by Section 611 tables plus a height of floor framing not to exceed 16 inches.

Individual walls or walls studs shall be permitted to exceed these limits as permitted by Chapter 6 provisions, provided story heights are not exceeded. An engineered design shall be provided for the wall or wall framing members when they exceed the limits of Chapter 6. Where the story height limits are exceeded, an engineered design shall be provided in accordance with the *International Building Code* for the overall wind and seismic force resisting systems.

R301.4 Dead load. The actual weights of materials and construction shall be used for determining dead load with consideration for the dead load of fixed service equipment.

R301.5 Live load. The minimum uniformly distributed live load shall be as provided in Table R301.5.

R301.6 Roof load. The roof shall be designed for the live load indicated in Table R301.6 or the snow load indicated in Table R301.2(1), whichever is greater.

R301.7 Deflection. The allowable deflection of any structural member under the live load listed in Sections R301.5 and R301.6 shall not exceed the values in Table R301.7.

TABLE R301.5
MINIMUM UNIFORMLY DISTRIBUTED LIVE LOADS
(in pounds per square foot)

USE	LIVE LOAD
Attics with limited storage[b, g, h]	20
Attics without storage[b]	10
Decks[e]	40
Exterior balconies	60
Fire escapes	40
Guardrails and handrails[d]	200[i]
Guardrails in–fill components[f]	50[i]
Passenger vehicle garages[a]	50[a]
Rooms other than sleeping rooms	40
Sleeping rooms	30
Stairs	40[c]

For SI: 1 pound per square foot = 0.0479 kPa, 1 square inch = 645 mm², 1 pound = 4.45 N.

a. Elevated garage floors shall be capable of supporting a 2,000-pound load applied over a 20-square-inch area.

b. Attics without storage are those where the maximum clear height between joist and rafter is less than 42 inches, or where there are not two or more adjacent trusses with the same web configuration capable of containing a rectangle 42 inches high by 2 feet wide, or greater, located within the plane of the truss. For attics without storage, this live load need not be assumed to act concurrently with any other live load requirements.

c. Individual stair treads shall be designed for the uniformly distributed live load or a 300-pound concentrated load acting over an area of 4 square inches, whichever produces the greater stresses.

d. A single concentrated load applied in any direction at any point along the top.

e. See Section R502.2.2 for decks attached to exterior walls.

f. Guard in-fill components (all those except the handrail), balusters and panel fillers shall be designed to withstand a horizontally applied normal load of 50 pounds on an area equal to 1 square foot. This load need not be assumed to act concurrently with any other live load requirement.

g. For attics with limited storage and constructed with trusses, this live load need be applied only to those portions of the bottom chord where there are two or more adjacent trusses with the same web configuration capable of containing a rectangle 42 inches high or greater by 2 feet wide or greater, located within the plane of the truss. The rectangle shall fit between the top of the bottom chord and the bottom of any other truss member, provided that each of the following criteria is met:

　1. The attic area is accessible by a pull-down stairway or framed opening in accordance with Section R807.1; and

　2. The truss has a bottom chord pitch less than 2:12.

h. Attic spaces served by a fixed stair shall be designed to support the minimum live load specified for sleeping rooms.

i. Glazing used in handrail assemblies and guards shall be designed with a safety factor of 4. The safety factor shall be applied to each of the concentrated loads applied to the top of the rail, and to the load on the in-fill components. These loads shall be determined independent of one another, and loads are assumed not to occur with any other live load.

TABLE R301.6
MINIMUM ROOF LIVE LOADS IN POUNDS-FORCE
PER SQUARE FOOT OF HORIZONTAL PROJECTION

ROOF SLOPE	TRIBUTARY LOADED AREA IN SQUARE FEET FOR ANY STRUCTURAL MEMBER		
	0 to 200	201 to 600	Over 600
Flat or rise less than 4 inches per foot (1:3)	20	16	12
Rise 4 inches per foot (1:3) to less than 12 inches per foot (1:1)	16	14	12
Rise 12 inches per foot (1:1) and greater	12	12	12

For SI: 1 square foot = 0.0929 m², 1 pound per square foot = 0.0479 kPa, 1 inch per foot = 83.3 mm/m.

TABLE R301.7
ALLOWABLE DEFLECTION OF STRUCTURAL MEMBERS[a,b,c]

STRUCTURAL MEMBER	ALLOWABLE DEFLECTION
Rafters having slopes greater than 3/12 with no finished ceiling attached to rafters	L/180
Interior walls and partitions	H/180
Floors and plastered ceilings	L/360
All other structural members	L/240
Exterior walls with plaster or stucco finish	H/360
Exterior walls—wind loads[a] with brittle finishes	L/240
Exterior walls—wind loads[a] with flexible finishes	L/120

Note: L = span length, H = span height.

a. The wind load shall be permitted to be taken as 0.7 times the Component and Cladding loads for the purpose of the determining deflection limits herein.

b. For cantilever members, L shall be taken as twice the length of the cantilever.

c. For aluminum structural members or panels used in roofs or walls of sunroom additions or patio covers, not supporting edge of glass or sandwich panels, the total load deflection shall not exceed L/60. For sandwich panels used in roofs or walls of sunroom additions or patio covers, the total load deflection shall not exceed L/120.

R301.8 Nominal sizes. For the purposes of this code, where dimensions of lumber are specified, they shall be deemed to be nominal dimensions unless specifically designated as actual dimensions.

SECTION R302
EXTERIOR WALL LOCATION

R302.1 Exterior walls. Construction, projections, openings and penetrations of exterior walls of dwellings and accessory buildings shall comply with Table R302.1. These provisions shall not apply to walls, projections, openings or penetrations in walls that are perpendicular to the line used to determine the fire separation distance. Projections beyond the exterior wall shall not extend more than 12 inches (305 mm) into the areas where openings are prohibited.

Exceptions:

1. Detached tool sheds and storage sheds, playhouses and similar structures exempted from permits are not required to provide wall protection based on location on the lot. Projections beyond the exterior wall shall not extend over the lot line.

2. Detached garages accessory to a dwelling located within 2 feet (610 mm) of a lot line are permitted to have roof eave projections not exceeding 4 inches (102 mm).

3. Foundation vents installed in compliance with this code are permitted.

SECTION R303
LIGHT, VENTILATION AND HEATING

R303.1 Habitable rooms. All habitable rooms shall have an aggregate glazing area of not less than 8 percent of the floor area of such rooms. Natural ventilation shall be through windows, doors, louvers or other approved openings to the outdoor air. Such openings shall be provided with ready access or shall otherwise be readily controllable by the building occupants. The minimum openable area to the outdoors shall be 4 percent of the floor area being ventilated.

Exceptions:

1. The glazed areas need not be openable where the opening is not required by Section R310 and an approved mechanical ventilation system capable of producing 0.35 air change per hour in the room is installed or a whole-house mechanical ventilation system is installed capable of supplying outdoor ventilation air of 15 cubic feet per minute (cfm) (78 L/s) per occupant computed on the basis of two occupants for the first bedroom and one occupant for each additional bedroom.

2. The glazed areas need not be installed in rooms where Exception 1 above is satisfied and artificial light is provided capable of producing an average illumination of 6 footcandles (65 lux) over the area of the room at a height of 30 inches (762 mm) above the floor level.

3. Use of sunroom additions and patio covers, as defined in Section R202, shall be permitted for natural ventilation if in excess of 40 percent of the exterior sunroom walls are open, or are enclosed only by insect screening.

R303.2 Adjoining rooms. For the purpose of determining light and ventilation requirements, any room shall be considered as a portion of an adjoining room when at least one-half of the area of the common wall is open and unobstructed and provides an opening of not less than one-tenth of the floor area of the interior room but not less than 25 square feet (2.3 m²).

Exception: Openings required for light and/or ventilation shall be permitted to open into a thermally isolated sunroom addition or patio cover, provided that there is an openable area between the adjoining room and the sunroom addition or patio cover of not less than one-tenth of the floor area of the interior room but not less than 20 square feet (2 m²). The minimum openable area to the outdoors shall be based upon the total floor area being ventilated.

R303.3 Bathrooms. Bathrooms, water closet compartments and other similar rooms shall be provided with aggregate glazing area in windows of not less than 3 square feet (0.3 m²), one-half of which must be openable.

Exception: The glazed areas shall not be required where artificial light and a mechanical ventilation system are provided. The minimum ventilation rates shall be 50 cubic feet per minute (24 L/s) for intermittent ventilation or 20 cubic feet per minute (10 L/s) for continuous ventilation. Ventilation air from the space shall be exhausted directly to the outside.

TABLE R302.1
EXTERIOR WALLS

EXTERIOR WALL ELEMENT		MINIMUM FIRE-RESISTANCE RATING	MINIMUM FIRE SEPARATION DISTANCE
Walls	(Fire-resistance rated)	1 hour with exposure from both sides	0 feet
	(Not fire-resistance rated)	0 hours	5 feet
Projections	(Fire-resistance rated)	1 hour on the underside	2 feet
	(Not fire-resistance rated)	0 hours	5 feet
Openings	Not allowed	N/A	< 3 feet
	25% Maximum of Wall Area	0 hours	3 feet
	Unlimited	0 hours	5 feet
Penetrations	All	Comply with Section R317.3	< 5 feet
		None required	5 feet

N/A = Not Applicable.

R303.4 Opening location. Outdoor intake and exhaust openings shall be located in accordance with Sections R303.4.1 and R303.4.2.

R303.4.1 Intake openings. Mechanical and gravity outdoor air intake openings shall be located a minimum of 10 feet (3048 mm) from any hazardous or noxious contaminant, such as vents, chimneys, plumbing vents, streets, alleys, parking lots and loading docks, except as otherwise specified in this code. Where a source of contaminant is located within 10 feet (3048 mm) of an intake opening, such opening shall be located a minimum of 2 feet (610 mm) below the contaminant source.

For the purpose of this section, the exhaust from dwelling unit toilet rooms, bathrooms and kitchens shall not be considered as hazardous or noxious.

R303.4.2 Exhaust openings. Outside exhaust openings shall be located so as not to create a nuisance. Exhaust air shall not be directed onto walkways.

R303.5 Outside opening protection. Air exhaust and intake openings that terminate outdoors shall be protected with corrosion-resistant screens, louvers or grilles having a minimum opening size of $^1/_4$ inch (6 mm) and a maximum opening size of $^1/_2$ inch (13 mm), in any dimension. Openings shall be protected against local weather conditions. Outdoor air exhaust and intake openings shall meet the provisions for exterior wall opening protectives in accordance with this code.

R303.6 Stairway illumination. All interior and exterior stairways shall be provided with a means to illuminate the stairs, including the landings and treads. Interior stairways shall be provided with an artificial light source located in the immediate vicinity of each landing of the stairway. For interior stairs the artificial light sources shall be capable of illuminating treads and landings to levels not less than 1 foot-candle (11 lux) measured at the center of treads and landings. Exterior stairways shall be provided with an artificial light source located in the immediate vicinity of the top landing of the stairway. Exterior stairways providing access to a basement from the outside grade level shall be provided with an artificial light source located in the immediate vicinity of the bottom landing of the stairway.

> **Exception:** An artificial light source is not required at the top and bottom landing, provided an artificial light source is located directly over each stairway section.

R303.6.1 Light activation. Where lighting outlets are installed in interior stairways, there shall be a wall switch at each floor level to control the lighting outlet where the stairway has six or more risers. The illumination of exterior stairways shall be controlled from inside the dwelling unit.

> **Exception:** Lights that are continuously illuminated or automatically controlled.

R303.7 Required glazed openings. Required glazed openings shall open directly onto a street or public alley, or a yard or court located on the same lot as the building.

R303.7.1 Roofed porches. Required glazed openings may face into a roofed porch where the porch abuts a street, yard or court and the longer side of the porch is at least 65 percent open and unobstructed and the ceiling height is not less than 7 feet (2134 mm).

R303.7.2 Sunroom additions. Required glazed openings shall be permitted to open into sunroom additions or patio covers that abut a street, yard or court if in excess of 40 percent of the exterior sunroom walls are open, or are enclosed only by insect screening, and the ceiling height of the sunroom is not less than 7 feet (2134 mm).

R303.8 Required heating. When the winter design temperature in Table R301.2(1) is below 60°F (16°C), every dwelling unit shall be provided with heating facilities capable of maintaining a minimum room temperature of 68°F (20°C) at a point 3 feet (914 mm) above the floor and 2 feet (610 mm) from exterior walls in all habitable rooms at the design temperature. The installation of one or more portable space heaters shall not be used to achieve compliance with this section.

SECTION R304
MINIMUM ROOM AREAS

R304.1 Minimum area. Every dwelling unit shall have at least one habitable room that shall have not less than 120 square feet (11 m²) of gross floor area.

R304.2 Other rooms. Other habitable rooms shall have a floor area of not less than 70 square feet (6.5 m²).

> **Exception:** Kitchens.

R304.3 Minimum dimensions. Habitable rooms shall not be less than 7 feet (2134 mm) in any horizontal dimension.

> **Exception:** Kitchens.

R304.4 Height effect on room area. Portions of a room with a sloping ceiling measuring less than 5 feet (1524 mm) or a furred ceiling measuring less than 7 feet (2134 mm) from the finished floor to the finished ceiling shall not be considered as contributing to the minimum required habitable area for that room.

SECTION R305
CEILING HEIGHT

R305.1 Minimum height. Habitable rooms, hallways, corridors, bathrooms, toilet rooms, laundry rooms and basements shall have a ceiling height of not less than 7 feet (2134 mm). The required height shall be measured from the finish floor to the lowest projection from the ceiling.

Exceptions:

1. Beams and girders spaced not less than 4 feet (1219 mm) on center may project not more than 6 inches (152 mm) below the required ceiling height.

2. Ceilings in basements without habitable spaces may project to within 6 feet, 8 inches (2032 mm) of the finished floor; and beams, girders, ducts or other obstructions may project to within 6 feet 4 inches (1931 mm) of the finished floor.

3. For rooms with sloped ceilings, at least 50 percent of the required floor area of the room must have a ceiling

height of at least 7 feet (2134 mm) and no portion of the required floor area may have a ceiling height of less than 5 feet (1524 mm).

4. Bathrooms shall have a minimum ceiling height of 6 feet 8 inches (2036 mm) over the fixture and at the front clearance area for fixtures as shown in Figure R307.1. A shower or tub equipped with a showerhead shall have a minimum ceiling height of 6 feet 8 inches (2036 mm) above a minimum area 30 inches (762 mm) by 30 inches (762 mm) at the showerhead.

SECTION R306
SANITATION

R306.1 Toilet facilities. Every dwelling unit shall be provided with a water closet, lavatory, and a bathtub or shower.

R306.2 Kitchen. Each dwelling unit shall be provided with a kitchen area and every kitchen area shall be provided with a sink.

R306.3 Sewage disposal. All plumbing fixtures shall be connected to a sanitary sewer or to an approved private sewage disposal system.

R306.4 Water supply to fixtures. All plumbing fixtures shall be connected to an approved water supply. Kitchen sinks, lavatories, bathtubs, showers, bidets, laundry tubs and washing machine outlets shall be provided with hot and cold water.

SECTION R307
TOILET, BATH AND SHOWER SPACES

R307.1 Space required. Fixtures shall be spaced as per Figure R307.1.

R307.2 Bathtub and shower spaces. Bathtub and shower floors and walls above bathtubs with installed shower heads

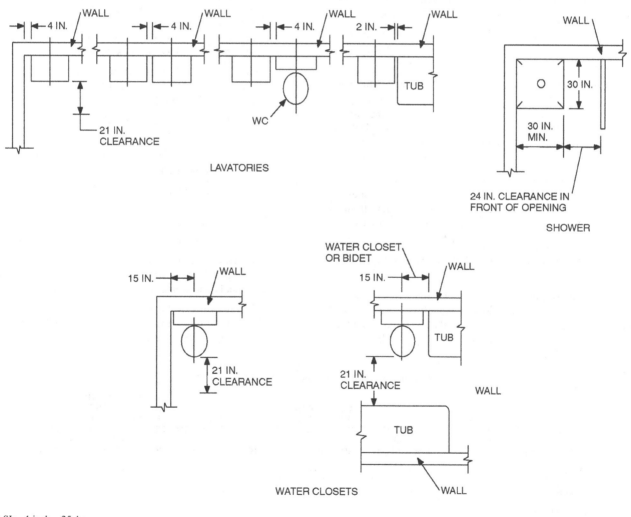

For SI: 1 inch = 25.4 mm.

FIGURE R307.1
MINIMUM FIXTURE CLEARANCES

and in shower compartments shall be finished with a nonabsorbent surface. Such wall surfaces shall extend to a height of not less than 6 feet (1829 mm) above the floor.

SECTION R308
GLAZING

R308.1 Identification. Except as indicated in Section R308.1.1 each pane of glazing installed in hazardous locations as defined in Section R308.4 shall be provided with a manufacturer's designation specifying who applied the designation, designating the type of glass and the safety glazing standard with which it complies, which is visible in the final installation. The designation shall be acid etched, sandblasted, ceramic-fired, laser etched, embossed, or be of a type which once applied cannot be removed without being destroyed. A label shall be permitted in lieu of the manufacturer's designation.

Exceptions:

1. For other than tempered glass, manufacturer's designations are not required provided the building official approves the use of a certificate, affidavit or other evidence confirming compliance with this code.

2. Tempered spandrel glass is permitted to be identified by the manufacturer with a removable paper designation.

R308.1.1 Identification of multipane assemblies. Multipane assemblies having individual panes not exceeding 1 square foot (0.09 m^2) in exposed area shall have at least one pane in the assembly identified in accordance with Section R308.1. All other panes in the assembly shall be labeled "16 CFR 1201."

R308.2 Louvered windows or jalousies. Regular, float, wired or patterned glass in jalousies and louvered windows shall be no thinner than nominal $^3/_{16}$ inch (5 mm) and no longer than 48 inches (1219 mm). Exposed glass edges shall be smooth.

R308.2.1 Wired glass prohibited. Wired glass with wire exposed on longitudinal edges shall not be used in jalousies or louvered windows.

R308.3 Human impact loads. Individual glazed areas, including glass mirrors in hazardous locations such as those indicated as defined in Section R308.4, shall pass the test requirements of CPSC 16 CFR, Part 1201. Glazing shall comply with CPSC 16

CFR, Part 1201 criteria for Category I or Category II as indicated in Table R308.3.

Exception: Louvered windows and jalousies shall comply with Section R308.2.

R308.4 Hazardous locations. The following shall be considered specific hazardous locations for the purposes of glazing:

1. Glazing in swinging doors except jalousies.

2. Glazing in fixed and sliding panels of sliding door assemblies and panels in sliding and bifold closet door assemblies.

3. Glazing in storm doors.

4. Glazing in all unframed swinging doors.

5. Glazing in doors and enclosures for hot tubs, whirlpools, saunas, steam rooms, bathtubs and showers. Glazing in any part of a building wall enclosing these compartments where the bottom exposed edge of the glazing is less than 60 inches (1524 mm) measured vertically above any standing or walking surface.

6. Glazing, in an individual fixed or operable panel adjacent to a door where the nearest vertical edge is within a 24-inch (610 mm) arc of the door in a closed position and whose bottom edge is less than 60 inches (1524 mm) above the floor or walking surface.

7. Glazing in an individual fixed or operable panel, other than those locations described in Items 5 and 6 above, that meets all of the following conditions:

    7.1. Exposed area of an individual pane larger than 9 square feet (0.836 m^2).

    7.2. Bottom edge less than 18 inches (457 mm) above the floor.

    7.3. Top edge more than 36 inches (914 mm) above the floor.

    7.4. One or more walking surfaces within 36 inches (914 mm) horizontally of the glazing.

8. All glazing in railings regardless of an area or height above a walking surface. Included are structural baluster panels and nonstructural infill panels.

9. Glazing in walls and fences enclosing indoor and outdoor swimming pools, hot tubs and spas where the bottom edge of the glazing is less than 60 inches (1524

TABLE R308.3
MINIMUM CATEGORY CLASSIFICATION OF GLAZING

EXPOSED SURFACE AREA OF ONE SIDE OF ONE LITE	GLAZING IN STORM OR COMBINATION DOORS (Category Class)	GLAZING IN DOORS (Category Class)	GLAZED PANELS REGULATED BY ITEM 7 OF SECTION R308.4 (Category Class)	GLAZED PANELS REGULATED BY ITEM 6 OF SECTION R308.4 (Category Class)	GLAZING IN DOORS AND ENCLOSURES REGULATED BY ITEM 5 OF SECTION R308.4 (Category Class)	SLIDING GLASS DOORS PATIO TYPE (Category Class)
9 sq ft or less	I	I	NR	I	II	II
More than 9 sq ft	II	II	II	II	II	II

For SI: 1 square foot = 0.0929 m^2.
NR means "No Requirement."

mm) above a walking surface and within 60 inches (1524 mm) horizontally of the water's edge. This shall apply to single glazing and all panes in multiple glazing.

10. Glazing adjacent to stairways, landings and ramps within 36 inches (914 mm) horizontally of a walking surface when the exposed surface of the glass is less than 60 inches (1524 mm) above the plane of the adjacent walking surface.

11. Glazing adjacent to stairways within 60 inches (1524 mm) horizontally of the bottom tread of a stairway in any direction when the exposed surface of the glass is less than 60 inches (1524 mm) above the nose of the tread.

Exception: The following products, materials and uses are exempt from the above hazardous locations:

1. Openings in doors through which a 3-inch (76 mm) sphere is unable to pass.

2. Decorative glass in Items 1, 6 or 7.

3. Glazing in Section R308.4, Item 6, when there is an intervening wall or other permanent barrier between the door and the glazing.

4. Glazing in Section R308.4, Item 6, in walls perpendicular to the plane of the door in a closed position, other than the wall toward which the door swings when opened, or where access through the door is to a closet or storage area 3 feet (914 mm) or less in depth. Glazing in these applications shall comply with Section R308.4, Item 7.

5. Glazing in Section R308.4, Items 7 and 10, when a protective bar is installed on the accessible side(s) of the glazing 36 inches ± 2 inches (914 mm ± 51 mm) above the floor. The bar shall be capable of withstanding a horizontal load of 50 pounds per linear foot (730 N/m) without contacting the glass and be a minimum of $1^1/_2$ inches (38 mm) in height.

6. Outboard panes in insulating glass units and other multiple glazed panels in Section R308.4, Item 7, when the bottom edge of the glass is 25 feet (7620 mm) or more above grade, a roof, walking surfaces, or other horizontal [within 45 degrees (0.79 rad) of horizontal] surface adjacent to the glass exterior.

7. Louvered windows and jalousies complying with the requirements of Section R308.2.

8. Mirrors and other glass panels mounted or hung on a surface that provides a continuous backing support.

9. Safety glazing in Section R308.4, Items 10 and 11, is not required where:

9.1. The side of a stairway, landing or ramp has a guardrail or handrail, including balusters or in-fill panels, complying with the provisions of Sections 1013 and 1607.7 of the *International Building Code*; and

9.2. The plane of the glass is more than 18 inches (457 mm) from the railing; or

9.3. When a solid wall or panel extends from the plane of the adjacent walking surface to 34 inches (863 mm) to 36 inches (914 mm) above the floor and the construction at the top of that wall or panel is capable of withstanding the same horizontal load as the protective bar.

10. Glass block panels complying with Section R610.

R308.5 Site built windows. Site built windows shall comply with Section 2404 of the *International Building Code*.

R308.6 Skylights and sloped glazing. Skylights and sloped glazing shall comply with the following sections.

R308.6.1 Definitions.

SKYLIGHTS AND SLOPED GLAZING. Glass or other transparent or translucent glazing material installed at a slope of 15 degrees (0.26 rad) or more from vertical. Glazing materials in skylights, including unit skylights, solariums, sunrooms, roofs and sloped walls are included in this definition.

UNIT SKYLIGHT. A factory assembled, glazed fenestration unit, containing one panel of glazing material, that allows for natural daylighting through an opening in the roof assembly while preserving the weather-resistant barrier of the roof.

R308.6.2 Permitted materials. The following types of glazing may be used:

1. Laminated glass with a minimum 0.015-inch (0.38 mm) polyvinyl butyral interlayer for glass panes 16 square feet (1.5 m²) or less in area located such that the highest point of the glass is not more than 12 feet (3658 mm) above a walking surface or other accessible area; for higher or larger sizes, the minimum interlayer thickness shall be 0.030 inch (0.76 mm).

2. Fully tempered glass.

3. Heat-strengthened glass.

4. Wired glass.

5. Approved rigid plastics.

R308.6.3 Screens, general. For fully tempered or heat-strengthened glass, a retaining screen meeting the requirements of Section R308.6.7 shall be installed below the glass, except for fully tempered glass that meets either condition listed in Section R308.6.5.

R308.6.4 Screens with multiple glazing. When the inboard pane is fully tempered, heat-strengthened or wired glass, a retaining screen meeting the requirements of Section R308.6.7 shall be installed below the glass, except for either condition listed in Section R308.6.5. All other panes in the multiple glazing may be of any type listed in Section R308.6.2.

R308.6.5 Screens not required. Screens shall not be required when fully tempered glass is used as single glazing or the inboard pane in multiple glazing and either of the following conditions are met:

1. Glass area 16 square feet (1.49 m²) or less. Highest point of glass not more than 12 feet (3658 mm) above

a walking surface or other accessible area, nominal glass thickness not more than $^3/_{16}$ inch (4.8 mm), and (for multiple glazing only) the other pane or panes fully tempered, laminated or wired glass.

2. Glass area greater than 16 square feet (1.49 m²). Glass sloped 30 degrees (0.52 rad) or less from vertical, and highest point of glass not more than 10 feet (3048 mm) above a walking surface or other accessible area.

R308.6.6 Glass in greenhouses. Any glazing material is permitted to be installed without screening in the sloped areas of greenhouses, provided the greenhouse height at the ridge does not exceed 20 feet (6096 mm) above grade.

R308.6.7 Screen characteristics. The screen and its fastenings shall be capable of supporting twice the weight of the glazing, be firmly and substantially fastened to the framing members, and have a mesh opening of no more than 1 inch by 1 inch (25 mm by 25 mm).

R308.6.8 Curbs for skylights. All unit skylights installed in a roof with a pitch flatter than three units vertical in 12 units horizontal (25-percent slope) shall be mounted on a curb extending at least 4 inches (102 mm) above the plane of the roof unless otherwise specified in the manufacturer's installation instructions.

R308.6.9 Testing and labeling. Unit skylights shall be tested by an approved independent laboratory, and bear a label identifying manufacturer, performance grade rating and approved inspection agency to indicate compliance with the requirements of AAMA/WDMA/CSA 101/I.S.2/A440.

SECTION R309
GARAGES AND CARPORTS

R309.1 Opening protection. Openings from a private garage directly into a room used for sleeping purposes shall not be permitted. Other openings between the garage and residence shall be equipped with solid wood doors not less than $1^3/_8$ inches (35 mm) in thickness, solid or honeycomb core steel doors not less than $1^3/_8$ inches (35 mm) thick, or 20-minute fire-rated doors.

R309.1.1 Duct penetration. Ducts in the garage and ducts penetrating the walls or ceilings separating the dwelling from the garage shall be constructed of a minimum No. 26 gage (0.48 mm) sheet steel or other approved material and shall have no openings into the garage.

R309.1.2 Other penetrations. Penetrations through the separation required in Section R309.2 shall be protected by filling the opening around the penetrating item with approved material to resist the free passage of flame and products of combustion.

R309.2 Separation required. The garage shall be separated from the residence and its attic area by not less than $^1/_2$-inch (12.7 mm) gypsum board applied to the garage side. Garages beneath habitable rooms shall be separated from all habitable rooms above by not less than $^5/_8$-inch (15.9 mm) Type X gypsum board or equivalent. Where the separation is a floor-ceiling assembly, the structure supporting the separation shall also be protected by not less than $^1/_2$-inch (12.7 mm) gypsum board or equivalent. Garages located less than 3 feet (914 mm) from a dwelling unit

on the same lot shall be protected with not less than $^1/_2$-inch (12.7 mm) gypsum board applied to the interior side of exterior walls that are within this area. Openings in these walls shall be regulated by Section R309.1. This provision does not apply to garage walls that are perpendicular to the adjacent dwelling unit wall.

R309.3 Floor surface. Garage floor surfaces shall be of approved noncombustible material.

The area of floor used for parking of automobiles or other vehicles shall be sloped to facilitate the movement of liquids to a drain or toward the main vehicle entry doorway.

R309.4 Carports. Carports shall be open on at least two sides. Carport floor surfaces shall be of approved noncombustible material. Carports not open on at least two sides shall be considered a garage and shall comply with the provisions of this section for garages.

Exception: Asphalt surfaces shall be permitted at ground level in carports.

The area of floor used for parking of automobiles or other vehicles shall be sloped to facilitate the movement of liquids to a drain or toward the main vehicle entry doorway.

R309.5 Flood hazard areas. For buildings located in flood hazard areas as established by Table R301.2(1), garage floors shall be:

1. Elevated to or above the design flood elevation as determined in Section R324; or

2. Located below the design flood elevation provided they are at or above grade on all sides, are used solely for parking, building access, or storage, meet the requirements of Section R324, and are otherwise constructed in accordance with this code.

R309.6 Automatic garage door openers. Automatic garage door openers, if provided, shall be listed in accordance with UL 325.

SECTION R310
EMERGENCY ESCAPE AND RESCUE OPENINGS

R310.1 Emergency escape and rescue required. Basements and every sleeping room shall have at least one operable emergency escape and rescue opening. Such opening shall open directly into a public street, public alley, yard or court. Where basements contain one or more sleeping rooms, emergency egress and rescue openings shall be required in each sleeping room, but shall not be required in adjoining areas of the basement. Where emergency escape and rescue openings are provided they shall have a sill height of not more than 44 inches (1118 mm) above the floor. Where a door opening having a threshold below the adjacent ground elevation serves as an emergency escape and rescue opening and is provided with a bulkhead enclosure, the bulkhead enclosure shall comply with Section R310.3. The net clear opening dimensions required by this section shall be obtained by the normal operation of the emergency escape and rescue opening from the inside. Emergency escape and rescue openings with a finished sill height below the adjacent ground elevation shall be provided with a window well in accordance with Section R310.2. Emergency escape and rescue openings shall open directly into a public way, or to a yard or court that opens to a public way.

> **Exception:** Basements used only to house mechanical equipment and not exceeding total floor area of 200 square feet (18.58 m²).

R310.1.1 Minimum opening area. All emergency escape and rescue openings shall have a minimum net clear opening of 5.7 square feet (0.530 m²).

> **Exception:** Grade floor openings shall have a minimum net clear opening of 5 square feet (0.465 m²).

R310.1.2 Minimum opening height. The minimum net clear opening height shall be 24 inches (610 mm).

R310.1.3 Minimum opening width. The minimum net clear opening width shall be 20 inches (508 mm).

R310.1.4 Operational constraints. Emergency escape and rescue openings shall be operational from the inside of the room without the use of keys, tools or special knowledge.

R310.2 Window wells. The minimum horizontal area of the window well shall be 9 square feet (0.9 m²), with a minimum horizontal projection and width of 36 inches (914 mm). The area of the window well shall allow the emergency escape and rescue opening to be fully opened.

> **Exception:** The ladder or steps required by Section R310.2.1 shall be permitted to encroach a maximum of 6 inches (152 mm) into the required dimensions of the window well.

R310.2.1 Ladder and steps. Window wells with a vertical depth greater than 44 inches (1118 mm) shall be equipped with a permanently affixed ladder or steps usable with the window in the fully open position. Ladders or steps required by this section shall not be required to comply with Sections R311.5 and R311.6. Ladders or rungs shall have an inside width of at least 12 inches (305 mm), shall project at least 3 inches (76 mm) from the wall and shall be spaced not more than 18 inches (457 mm) on center vertically for the full height of the window well.

R310.3 Bulkhead enclosures. Bulkhead enclosures shall provide direct access to the basement. The bulkhead enclosure with the door panels in the fully open position shall provide the minimum net clear opening required by Section R310.1.1. Bulkhead enclosures shall also comply with Section R311.5.8.2.

R310.4 Bars, grilles, covers and screens. Bars, grilles, covers, screens or similar devices are permitted to be placed over emergency escape and rescue openings, bulkhead enclosures, or window wells that serve such openings, provided the minimum net clear opening size complies with Sections R310.1.1 to R310.1.3, and such devices shall be releasable or removable from the inside without the use of a key, tool, special knowledge or force greater than that which is required for normal operation of the escape and rescue opening.

R310.5 Emergency escape windows under decks and porches. Emergency escape windows are allowed to be installed under decks and porches provided the location of the deck allows the emergency escape window to be fully opened and provides a path not less than 36 inches (914 mm) in height to a yard or court.

SECTION R311
MEANS OF EGRESS

R311.1 General. Stairways, ramps, exterior egress balconies, hallways and doors shall comply with this section.

R311.2 Construction.

R311.2.1 Attachment. Required exterior egress balconies, exterior exit stairways and similar means of egress components shall be positively anchored to the primary structure to resist both vertical and lateral forces. Such attachment shall not be accomplished by use of toenails or nails subject to withdrawal.

R311.2.2 Under stair protection. Enclosed accessible space under stairs shall have walls, under stair surface and any soffits protected on the enclosed side with ¹/₂-inch (13 mm) gypsum board.

R311.3 Hallways. The minimum width of a hallway shall be not less than 3 feet (914 mm).

R311.4 Doors.

R311.4.1 Exit door required. Not less than one exit door conforming to this section shall be provided for each dwelling unit. The required exit door shall provide for direct access from the habitable portions of the dwelling to the exterior without requiring travel through a garage. Access to habitable levels not having an exit in accordance with this section shall be by a ramp in accordance with Section R311.6 or a stairway in accordance with Section R311.5.

R311.4.2 Door type and size. The required exit door shall be a side-hinged door not less than 3 feet (914 mm) in width and 6 feet 8 inches (2032 mm) in height. Other doors shall not be required to comply with these minimum dimensions.

R311.4.3 Landings at doors. There shall be a floor or landing on each side of each exterior door. The floor or landing at the exterior door shall not be more than 1.5 inches (38 mm) lower than the top of the threshold. The landing shall be permitted to have a slope not to exceed 0.25 unit vertical in 12 units horizontal (2-percent).

> **Exceptions:**
> 1. Where a stairway of two or fewer risers is located on the exterior side of a door, other than the required exit door, a landing is not required for the exterior side of the door provided the door, other than an exterior storm or screen door does not swing over the stairway.
> 2. The exterior landing at an exterior doorway shall not be more than 7³/₄ inches (196 mm) below the top of the threshold, provided the door, other than an exterior storm or screen door does not swing over the landing.
> 3. The height of floors at exterior doors other than the exit door required by Section R311.4.1 shall not be more than 7³/₄ inches (186 mm) lower than the top of the threshold.

The width of each landing shall not be less than the door served. Every landing shall have a minimum dimension of 36 inches (914 mm) measured in the direction of travel.

R311.4.4 Type of lock or latch. All egress doors shall be readily openable from the side from which egress is to be made without the use of a key or special knowledge or effort.

R311.5 Stairways.

R311.5.1 Width. Stairways shall not be less than 36 inches (914 mm) in clear width at all points above the permitted handrail height and below the required headroom height. Handrails shall not project more than 4.5 inches (114 mm) on either side of the stairway and the minimum clear width of the stairway at and below the handrail height, including treads and landings, shall not be less than 31.5 inches (787 mm) where a handrail is installed on one side and 27 inches (698 mm) where handrails are provided on both sides.

Exception: The width of spiral stairways shall be in accordance with Section R311.5.8.

R311.5.2 Headroom. The minimum headroom in all parts of the stairway shall not be less than 6 feet 8 inches (2036 mm) measured vertically from the sloped plane adjoining the tread nosing or from the floor surface of the landing or platform.

R311.5.3 Stair treads and risers.

R311.5.3.1 Riser height. The maximum riser height shall be $7^3/_4$ inches (196 mm). The riser shall be measured vertically between leading edges of the adjacent treads. The greatest riser height within any flight of stairs shall not exceed the smallest by more than $^3/_8$ inch (9.5 mm).

R311.5.3.2 Tread depth. The minimum tread depth shall be 10 inches (254 mm). The tread depth shall be measured horizontally between the vertical planes of the foremost projection of adjacent treads and at a right angle to the tread's leading edge. The greatest tread depth within any flight of stairs shall not exceed the smallest by more than $^3/_8$ inch (9.5 mm). Winder treads shall have a minimum tread depth of 10 inches (254 mm) measured as above at a point 12 inches (305 mm) from the side where the treads are narrower. Winder treads shall have a minimum tread depth of 6 inches (152 mm) at any point. Within any flight of stairs, the largest winder tread depth at the 12 inch (305 mm) walk line shall not exceed the smallest by more than $^3/_8$ inch (9.5 mm).

R311.5.3.3 Profile. The radius of curvature at the leading edge of the tread shall be no greater than $^9/_{16}$ inch (14 mm). A nosing not less than $^3/_4$ inch (19 mm) but not more than $1^1/_4$ inch (32 mm) shall be provided on stairways with solid risers. The greatest nosing projection shall not exceed the smallest nosing projection by more than $^3/_8$ inch (9.5 mm) between two stories, including the nosing at the level of floors and landings. Beveling of nosing shall not exceed $^1/_2$ inch (12.7 mm). Risers shall be vertical or sloped from the underside of the leading edge of the tread above at an angle not more than 30 degrees (0.51 rad) from the vertical. Open risers are permitted, provided that the opening between treads does not permit the passage of a 4-inch diameter (102 mm) sphere.

Exceptions:

1. A nosing is not required where the tread depth is a minimum of 11 inches (279 mm).

2. The opening between adjacent treads is not limited on stairs with a total rise of 30 inches (762 mm) or less.

R311.5.4 Landings for stairways. There shall be a floor or landing at the top and bottom of each stairway. A flight of stairs shall not have a vertical rise larger than 12 feet (3658 mm) between floor levels or landings. The width of each landing shall not be less than the width of the stairway served. Every landing shall have a minimum dimension of 36 inches (914 mm) measured in the direction of travel.

Exception: A floor or landing is not required at the top of an interior flight of stairs, including stairs in an enclosed garage, provided that a door does not swing over the stairs.

R311.5.5 Stairway walking surface. The walking surface of treads and landings of stairways shall be sloped no steeper than one unit vertical in 48 inches horizontal (2-percent slope).

R311.5.6 Handrails. Handrails shall be provided on at least one side of each continuous run of treads or flight with four or more risers.

R311.5.6.1 Height. Handrail height, measured vertically from the sloped plane adjoining the tread nosing, or finish surface of ramp slope, shall be not less than 34 inches (864 mm) and not more than 38 inches (965 mm).

R311.5.6.2 Continuity. Handrails for stairways shall be continuous for the full length of the flight, from a point directly above the top riser of the flight to a point directly above the lowest riser of the flight. Handrail ends shall be returned or shall terminate in newel posts or safety terminals. Handrails adjacent to a wall shall have a space of not less than $1^1/_2$ inch (38 mm) between the wall and the handrails.

Exceptions:

1. Handrails shall be permitted to be interrupted by a newel post at the turn.

2. The use of a volute, turnout, starting easing or starting newel shall be allowed over the lowest tread.

R311.5.6.3 Handrail grip size. All required handrails shall be of one of the following types or provide equivalent graspability.

1. Type I. Handrails with a circular cross section shall have an outside diameter of at least $1^1/_4$ inches (32 mm) and not greater than 2 inches (51 mm). If the handrail is not circular it shall have a perimeter dimension of at least 4 inches (102 mm) and not greater than $6^1/_4$ inches (160 mm) with a maximum cross section of dimension of $2^1/_4$ inches (57 mm).

2. Type II. Handrails with a perimeter greater than $6^1/_4$ inches (160 mm) shall provide a graspable finger

recess area on both sides of the profile. The finger recess shall begin within a distance of $^3/_4$ inch (19 mm) measured vertically from the tallest portion of the profile and achieve a depth of at least $^5/_{16}$ inch (8 mm) within $^7/_8$ inch (22 mm) below the widest portion of the profile. This required depth shall continue for at least $^3/_8$ inch (10 mm) to a level that is not less than $1^3/_4$ inches (45 mm) below the tallest portion of the profile. The minimum width of the handrail above the recess shall be $1^1/_4$ inches (32 mm) to a maximum of $2^3/_4$ inches (70 mm). Edges shall have a minimum radius of 0.01 inch (0.25 mm).

R311.5.7 Illumination. All stairs shall be provided with illumination in accordance with Section R303.6.

R311.5.8 Special stairways. Spiral stairways and bulkhead enclosure stairways shall comply with all requirements of Section R311.5 except as specified below.

R311.5.8.1 Spiral stairways. Spiral stairways are permitted, provided the minimum width shall be 26 inches (660 mm) with each tread having a $7^1/_2$-inches (190 mm) minimum tread depth at 12 inches from the narrower edge. All treads shall be identical, and the rise shall be no more than $9^1/_2$ inches (241 mm). A minimum headroom of 6 feet 6 inches (1982 mm) shall be provided.

R311.5.8.2 Bulkhead enclosure stairways. Stairways serving bulkhead enclosures, not part of the required building egress, providing access from the outside grade level to the basement shall be exempt from the requirements of Sections R311.4.3 and R311.5 where the maximum height from the basement finished floor level to grade adjacent to the stairway does not exceed 8 feet (2438 mm), and the grade level opening to the stairway is covered by a bulkhead enclosure with hinged doors or other approved means.

R311.6 Ramps.

R311.6.1 Maximum slope. Ramps shall have a maximum slope of one unit vertical in twelve units horizontal (8.3-percent slope).

Exception: Where it is technically infeasible to comply because of site constraints, ramps may have a maximum slope of one unit vertical in eight horizontal (12.5 percent slope).

R311.6.2 Landings required. A minimum 3-foot-by-3-foot (914 mm by 914 mm) landing shall be provided:

1. At the top and bottom of ramps.

2. Where doors open onto ramps.

3. Where ramps change direction.

R311.6.3 Handrails required. Handrails shall be provided on at least one side of all ramps exceeding a slope of one unit vertical in 12 units horizontal (8.33-percent slope).

R311.6.3.1 Height. Handrail height, measured above the finished surface of the ramp slope, shall be not less than 34 inches (864 mm) and not more than 38 inches (965 mm).

R311.6.3.2 Handrail grip size. Handrails on ramps shall comply with Section R311.5.6.3.

R311.6.3.3 Continuity. Handrails where required on ramps shall be continuous for the full length of the ramp. Handrail ends shall be returned or shall terminate in newel posts or safety terminals. Handrails adjacent to a wall shall have a space of not less than 1.5 inches (38 mm) between the wall and the handrails.

SECTION R312
GUARDS

R312.1 Guards. Porches, balconies, ramps or raised floor surfaces located more than 30 inches (762 mm) above the floor or grade below shall have guards not less than 36 inches (914 mm) in height. Open sides of stairs with a total rise of more than 30 inches (762 mm) above the floor or grade below shall have guards not less than 34 inches (864 mm) in height measured vertically from the nosing of the treads.

Porches and decks which are enclosed with insect screening shall be equipped with guards where the walking surface is located more than 30 inches (762 mm) above the floor or grade below.

R312.2 Guard opening limitations. Required guards on open sides of stairways, raised floor areas, balconies and porches shall have intermediate rails or ornamental closures which do not allow passage of a sphere 4 inches (102mm) or more in diameter.

Exceptions:

1. The triangular openings formed by the riser, tread and bottom rail of a guard at the open side of a stairway are permitted to be of such a size that a sphere 6 inches (152 mm) cannot pass through.

2. Openings for required guards on the sides of stair treads shall not allow a sphere $4^3/_8$ inches (107 mm) to pass through.

SECTION R313
SMOKE ALARMS

R313.1 Smoke detection and notification. All smoke alarms shall be listed in accordance with UL 217 and installed in accordance with the provisions of this code and the household fire warning equipment provisions of NFPA 72.

Household fire alarm systems installed in accordance with NFPA 72 that include smoke alarms, or a combination of smoke detector and audible notification device installed as required by this section for smoke alarms, shall be permitted. The household fire alarm system shall provide the same level of smoke detection and alarm as required by this section for smoke alarms in the event the fire alarm panel is removed or the system is not connected to a central station.

R313.2 Location. Smoke alarms shall be installed in the following locations:

1. In each sleeping room.

2. Outside each separate sleeping area in the immediate vicinity of the bedrooms.

3. On each additional story of the dwelling, including basements but not including crawl spaces and uninhabitable attics. In dwellings or dwelling units with split levels and without an intervening door between the adjacent levels, a smoke alarm installed on the upper level shall suffice for the adjacent lower level provided that the lower level is less than one full story below the upper level.

When more than one smoke alarm is required to be installed within an individual dwelling unit the alarm devices shall be interconnected in such a manner that the actuation of one alarm will activate all of the alarms in the individual unit.

R313.2.1 Alterations, repairs and additions. When alterations, repairs or additions requiring a permit occur, or when one or more sleeping rooms are added or created in existing dwellings, the individual dwelling unit shall be equipped with smoke alarms located as required for new dwellings; the smoke alarms shall be interconnected and hard wired.

Exceptions:

1. Interconnection and hard-wiring of smoke alarms in existing areas shall not be required where the alterations or repairs do not result in the removal of interior wall or ceiling finishes exposing the structure, unless there is an attic, crawl space or basement available which could provide access for hard wiring and interconnection without the removal of interior finishes.

2. Work involving the exterior surfaces of dwellings, such as the replacement of roofing or siding, or the addition or replacement of windows or doors, or the addition of a porch or deck, are exempt from the requirements of this section.

R313.3 Power source. In new construction, the required smoke alarms shall receive their primary power from the building wiring when such wiring is served from a commercial source, and when primary power is interrupted, shall receive power from a battery. Wiring shall be permanent and without a disconnecting switch other than those required for overcurrent protection. Smoke alarms shall be permitted to be battery operated when installed in buildings without commercial power or in buildings that undergo alterations, repairs or additions regulated by Section R313.2.1.

SECTION R314
FOAM PLASTIC

R314.1 General. The provisions of this section shall govern the materials, design, application, construction and installation of foam plastic materials.

R314.2 Labeling and identification. Packages and containers of foam plastic insulation and foam plastic insulation components delivered to the job site shall bear the label of an approved agency showing the manufacturer's name, the product listing, product identification and information sufficient

to determine that the end use will comply with the requirements.

R314.3 Surface burning characteristics. Unless otherwise allowed in Section R314.5 or R314.6, all foam plastic or foam plastic cores used as a component in manufactured assemblies used in building construction shall have a flame spread index of not more than 75 and shall have a smoke-developed index of not more than 450 when tested in the maximum thickness intended for use in accordance with ASTM E 84. Loose-fill-type foam plastic insulation shall be tested as board stock for the flame spread index and smoke-developed index.

Exception: Foam plastic insulation more than 4 inches thick shall have a maximum flame spread index of 75 and a smoke-developed index of 450 where tested at a minimum thickness of 4 inches, provided the end use is approved in accordance with Section R314.6 using the thickness and density intended for use.

R314.4 Thermal barrier. Unless otherwise allowed in Section R314.5 or Section R314.6, foam plastic shall be separated from the interior of a building by an approved thermal barrier of minimum 0.5 inch (12.7 mm) gypsum wallboard or an approved finish material equivalent to a thermal barrier material that will limit the average temperature rise of the unexposed surface to no more than 250°F (139°C) after 15 minutes of fire exposure complying with the ASTM E 119 standard time temperature curve. The thermal barrier shall be installed in such a manner that it will remain in place for 15 minutes based on NFPA 286 with the acceptance criteria of Section R315.4, FM 4880, UL 1040 or UL 1715.

R314.5 Specific requirements. The following requirements shall apply to these uses of foam plastic unless specifically approved in accordance with Section R314.6 or by other sections of the code or the requirements of Sections R314.2 through R314.4 have been met.

R314.5.1 Masonry or concrete construction. The thermal barrier specified in Section R314.4 is not required in a masonry or concrete wall, floor or roof when the foam plastic insulation is separated from the interior of the building by a minimum 1-inch (25 mm) thickness of masonry or concrete.

R314.5.2 Roofing. The thermal barrier specified in Section R314.4 is not required when the foam plastic in a roof assembly or under a roof covering is installed in accordance with the code and the manufacturer's installation instructions and is separated from the interior of the building by tongue-and-groove wood planks or wood structural panel sheathing in accordance with Section R803, not less than $^{15}/_{32}$ inch (11.9 mm) thick bonded with exterior glue and identified as Exposure 1, with edges supported by blocking or tongue-and-groove joints or an equivalent material. The smoke-developed index for roof applications shall not be limited.

R314.5.3 Attics. The thermal barrier specified in Section 314.4 is not required where attic access is required by Section R807.1 and where the space is entered only for service of utilities and when the foam plastic insulation is protected against ignition using one of the following ignition barrier materials:

1. 1.5-inch-thick (38 mm) mineral fiber insulation;

2. 0.25-inch-thick (6.4 mm) wood structural panels;

3. 0.375-inch (9.5 mm) particleboard;

4. 0.25-inch (6.4 mm) hardboard;

5. 0.375-inch (9.5 mm) gypsum board; or

6. Corrosion-resistant steel having a base metal thickness of 0.016 inch (0.406 mm).

The above ignition barrier is not required where the foam plastic insulation has been tested in accordance with Section R314.6.

R314.5.4 Crawl spaces. The thermal barrier specified in Section R314.4 is not required where crawlspace access is required by Section R408.4 and where entry is made only for service of utilities and the foam plastic insulation is protected against ignition using one of the following ignition barrier materials:

1. 1.5-inch-thick (38 mm) mineral fiber insulation;

2. 0.25-inch-thick (6.4 mm) wood structural panels;

3. 0.375-inch (9.5 mm) particleboard;

4. 0.25-inch (6.4 mm) hardboard;

5. 0.375-inch (9.5 mm) gypsum board; or

6. Corrosion-resistant steel having a base metal thickness of 0.016 inch (0.41 mm).

The above ignition barrier is not required where the foam plastic insulation has been tested in accordance with Section R314.6.

R314.5.5 Foam-filled exterior doors. Foam-filled exterior doors are exempt from the requirements of Sections R314.3 and R314.4.

R314.5.6 Foam-filled garage doors. Foam-filled garage doors in attached or detached garages are exempt from the requirements of Sections R314.3 and R314.4.

R314.5.7 Foam backer board. The thermal barrier specified in Section R314.4 is not required where siding backer board foam plastic insulation has a maximum thickness of 0.5 inch (12.7 mm) and a potential heat of not more than 2000 Btu per square foot (22 720 kJ/m²) when tested in accordance with NFPA 259 provided that:

1. The foam plastic insulation is separated from the interior of the building by not less than 2 inches (51 mm) of mineral fiber insulation or

2. The foam plastic insulation is installed over existing exterior wall finish in conjunction with re-siding or

3. The foam plastic insulation has been tested in accordance with Section R314.6.

R314.5.8 Re-siding. The thermal barrier specified in Section R314.4 is not required where the foam plastic insulation is installed over existing exterior wall finish in conjunction with re-siding provided the foam plastic has a maximum thickness of 0.5 inch (12.7 mm) and a potential

heat of not more than 2000 Btu per square foot (22 720 kJ/m²) when tested in accordance with NFPA 259.

R314.5.9 Interior trim. The thermal barrier specified in Section R314.4 is not required for exposed foam plastic interior trim, provided all of the following are met:

1. The minimum density is 20 pounds per cubic foot (320 kg/m³).

2. The maximum thickness of the trim is 0.5 inch (12.7 mm) and the maximum width is 8 inches (204 mm).

3. The interior trim shall not constitute more than 10 percent of the aggregate wall and ceiling area of any room or space.

4. The flame spread index does not exceed 75 when tested per ASTM E 84. The smoke-developed index is not limited.

R314.5.10 Interior finish. Foam plastics shall be permitted as interior finish where approved in accordance with R314.6. Foam plastics that are used as interior finish shall also meet the flame spread and smoke-developed requirements of Section R315.

R314.5.11 Sill plates and headers. Foam plastic shall be permitted to be spray applied to a sill plate and header without the thermal barrier specified in Section R314.4 subject to all of the following:

1. The maximum thickness of the foam plastic shall be $3^1/_4$ inches (83 mm).

2. The density of the foam plastic shall be in the range of 1.5 to 2.0 pounds per cubic foot (24 to 32 kg/m³).

3. The foam plastic shall have a flame spread index of 25 or less and an accompanying smoke developed index of 450 or less when tested in accordance with ASTM E 84.

R314.5.12 Sheathing. Foam plastic insulation used as sheathing shall comply with Section R314.3 and Section R314.4. Where the foam plastic sheathing is exposed to the attic space at a gable or kneewall, the provisions of Section R314.5.3 shall apply.

R314.6 Specific approval. Foam plastic not meeting the requirements of Sections R314.3 through R314.5 shall be specifically approved on the basis of one of the following approved tests: NFPA 286 with the acceptance criteria of Section R315.4, FM4880, UL 1040 or UL 1715, or fire tests related to actual end-use configurations. The specific approval shall be based on the actual end use configuration and shall be performed on the finished foam plastic assembly in the maximum thickness intended for use. Assemblies tested shall include seams, joints and other typical details used in the installation of the assembly and shall be tested in the manner intended for use.

R314.7 Termite damage. The use of foam plastics in areas of "very heavy" termite infestation probability shall be in accordance with Section R320.5.

SECTION R315
FLAME SPREAD AND SMOKE DENSITY

R315.1 Wall and ceiling. Wall and ceiling finishes shall have a flame-spread classification of not greater than 200.

> **Exception:** Flame-spread requirements for finishes shall not apply to trim defined as picture molds, chair rails, baseboards and handrails; to doors and windows or their frames; or to materials that are less than $^{1}/_{28}$ inch (0.91 mm) in thickness cemented to the surface of walls or ceilings if these materials have a flame-spread characteristic no greater than paper of this thickness cemented to a noncombustible backing.

R315.2 Smoke-developed index. Wall and ceiling finishes shall have a smoke-developed index of not greater than 450.

R315.3 Testing. Tests shall be made in accordance with ASTM E 84.

R315.4 Alternate test method. As an alternate to having a flame-spread classification of not greater than 200 and a smoke developed index of not greater than 450 when tested in accordance with ASTM E 84, wall and ceiling finishes, other than textiles, shall be permitted to be tested in accordance with NFPA 286. Materials tested in accordance with NFPA 286 shall meet the following criteria:

During the 40 kW exposure, the interior finish shall comply with Item 1. During the 160 kW exposure, the interior finish shall comply with Item 2. During the entire test, the interior finish shall comply with Item 3.

1. During the 40 kW exposure, flames shall not spread to the ceiling.

2. During the 160 kW exposure, the interior finish shall comply with the following:

   2.1. Flame shall not spread to the outer extremity of the sample on any wall or ceiling.

   2.2. Flashover, as defined in NFPA 286, shall not occur.

3. The total smoke released throughout the NFPA 286 test shall not exceed 1,000 m².

SECTION R316
INSULATION

R316.1 Insulation. Insulation materials, including facings, such as vapor retarders or vapor permeable membranes installed within floor-ceiling assemblies, roof-ceiling assemblies, wall assemblies, crawl spaces and attics shall have a flame-spread index not to exceed 25 with an accompanying smoke-developed index not to exceed 450 when tested in accordance with ASTM E 84.

Exceptions:

1. When such materials are installed in concealed spaces, the flame-spread and smoke-developed limitations do not apply to the facings, provided that the facing is installed in substantial contact with the unexposed surface of the ceiling, floor or wall finish.

2. Cellulose loose-fill insulation, which is not spray applied, complying with the requirements of Section

R316.3, shall only be required to meet the smoke-developed index of not more than 450.

R316.2 Loose-fill insulation. Loose-fill insulation materials that cannot be mounted in the ASTM E 84 apparatus without a screen or artificial supports shall comply with the flame spread and smoke-developed limits of Sections R316.1 and R316.4 when tested in accordance with CAN/ULC S102.2.

> **Exception:** Cellulose loose-fill insulation shall not be required to comply with the flame spread index requirement of CAN/ULC S102.2, provided such insulation complies with the requirements of Section R316.3.

R316.3 Cellulose loose-fill insulation. Cellulose loose-fill insulation shall comply with CPSC 16 CFR, Parts 1209 and 1404. Each package of such insulating material shall be clearly labeled in accordance with CPSC 16 CFR, Parts 1209 and 1404.

R316.4 Exposed attic insulation. All exposed insulation materials installed on attic floors shall have a critical radiant flux not less than 0.12 watt per square centimeter.

R316.5 Testing. Tests for critical radiant flux shall be made in accordance with ASTM E 970.

SECTION R317
DWELLING UNIT SEPARATION

R317.1 Two-family dwellings. Dwelling units in two-family dwellings shall be separated from each other by wall and/or floor assemblies having not less than a 1-hour fire-resistance rating when tested in accordance with ASTM E 119. Fire-resistance-rated floor-ceiling and wall assemblies shall extend to and be tight against the exterior wall, and wall assemblies shall extend to the underside of the roof sheathing.

Exceptions:

1. A fire-resistance rating of $^{1}/_{2}$ hour shall be permitted in buildings equipped throughout with an automatic sprinkler system installed in accordance with NFPA 13.

2. Wall assemblies need not extend through attic spaces when the ceiling is protected by not less than $^{5}/_{8}$-inch (15.9 mm) Type X gypsum board and an attic draft stop constructed as specified in Section R502.12.1 is provided above and along the wall assembly separating the dwellings. The structural framing supporting the ceiling shall also be protected by not less than $^{1}/_{2}$-inch (12.7 mm) gypsum board or equivalent.

R317.1.1 Supporting construction. When floor assemblies are required to be fire-resistance-rated by Section R317.1, the supporting construction of such assemblies shall have an equal or greater fire-resistive rating.

R317.2 Townhouses. Each townhouse shall be considered a separate building and shall be separated by fire-resistance-rated wall assemblies meeting the requirements of Section R302 for exterior walls.

> **Exception:** A common 2-hour fire-resistance-rated wall is permitted for townhouses if such walls do not contain plumbing or mechanical equipment, ducts or vents in the

cavity of the common wall. Electrical installations shall be installed in accordance with Chapters 33 through 42. Penetrations of electrical outlet boxes shall be in accordance with Section R317.3.

R317.2.1 Continuity. The fire-resistance-rated wall or assembly separating townhouses shall be continuous from the foundation to the underside of the roof sheathing, deck or slab. The fire-resistance rating shall extend the full length of the wall or assembly, including wall extensions through and separating attached enclosed accessory structures.

R317.2.2 Parapets. Parapets constructed in accordance with Section R317.2.3 shall be constructed for townhouses as an extension of exterior walls or common walls in accordance with the following:

1. Where roof surfaces adjacent to the wall or walls are at the same elevation, the parapet shall extend not less than 30 inches (762 mm) above the roof surfaces.

2. Where roof surfaces adjacent to the wall or walls are at different elevations and the higher roof is not more than 30 inches (762 mm) above the lower roof, the parapet shall extend not less than 30 inches (762 mm) above the lower roof surface.

   Exception: A parapet is not required in the two cases above when the roof is covered with a minimum class C roof covering, and the roof decking or sheathing is of noncombustible materials or approved fire-retardant-treated wood for a distance of 4 feet (1219 mm) on each side of the wall or walls, or one layer of $^5/_8$-inch (15.9 mm) Type X gypsum board is installed directly beneath the roof decking or sheathing, supported by a minimum of nominal 2-inch (51 mm) ledgers attached to the sides of the roof framing members, for a minimum distance of 4 feet (1220 mm) on each side of the wall or walls.

3. A parapet is not required where roof surfaces adjacent to the wall or walls are at different elevations and the higher roof is more than 30 inches (762 mm) above the lower roof. The common wall construction from the lower roof to the underside of the higher roof deck shall have not less than a 1-hour fire-resistence rating. The wall shall be rated for exposure from both sides.

R317.2.3 Parapet construction. Parapets shall have the same fire-resistance rating as that required for the supporting wall or walls. On any side adjacent to a roof surface, the parapet shall have noncombustible faces for the uppermost 18 inches (457 mm), to include counterflashing and coping materials. Where the roof slopes toward a parapet at slopes greater than two units vertical in 12 units horizontal (16.7-percent slope), the parapet shall extend to the same height as any portion of the roof within a distance of 3 feet (914 mm), but in no case shall the height be less than 30 inches (762 mm).

R317.2.4 Structural independence. Each individual townhouse shall be structurally independent.

Exceptions:

1. Foundations supporting exterior walls or common walls.

2. Structural roof and wall sheathing from each unit may fasten to the common wall framing.

3. Nonstructural wall coverings.

4. Flashing at termination of roof covering over common wall.

5. Townhouses separated by a common 2-hour fire-resistance-rated wall as provided in Section R317.2.

R317.3 Rated penetrations. Penetrations of wall or floor/ceiling assemblies required to be fire-resistance rated in accordance with Section R317.1 or R317.2 shall be protected in accordance with this section.

R317.3.1 Through penetrations. Through penetrations of fire-resistance-rated wall or floor assemblies shall comply with Section R317.3.1.1 or R317.3.1.2.

Exception: Where the penetrating items are steel, ferrous or copper pipes, tubes or conduits, the annular space shall be protected as follows:

1. In concrete or masonry wall or floor assemblies where the penetrating item is a maximum 6 inches (152 mm) nominal diameter and the area of the opening through the wall does not exceed 144 square inches (92 900 mm²), concrete, grout or mortar is permitted where installed to the full thickness of the wall or floor assembly or the thickness required to maintain the fire-resistance rating.

2. The material used to fill the annular space shall prevent the passage of flame and hot gases sufficient to ignite cotton waste where subjected to ASTM E 119 time temperature fire conditions under a minimum positive pressure differential of 0.01 inch of water (3 Pa) at the location of the penetration for the time period equivalent to the fire resistance rating of the construction penetrated.

R317.3.1.1 Fire-resistance-rated assembly. Penetrations shall be installed as tested in the approved fire-resistance-rated assembly.

R317.3.1.2 Penetration firestop system. Penetrations shall be protected by an approved penetration firestop system installed as tested in accordance with ASTM E 814 or UL 1479, with a minimum positive pressure differential of 0.01 inch of water (3 Pa) and shall have an F rating of not less than the required fire-resistance rating of the wall or floor/ceiling assembly penetrated.

R317.3.2 Membrane penetrations. Membrane penetrations shall comply with Section R317.3.1. Where walls are required to have a fire-resistance rating, recessed fixtures shall be so installed such that the required fire resistance will not be reduced.

Exceptions:

1. Membrane penetrations of maximum 2-hour fire-resistance-rated walls and partitions by steel electrical boxes that do not exceed 16 square inches (0.0103 m²) in area provided the aggregate area of the openings through the membrane does not exceed 100 square inches (0.0645 m²) in any 100 square feet (9.29 m²) of wall area. The annular space between the wall membrane and the box shall not exceed $^1/_8$ inch (3.1 mm). Such boxes on opposite sides of the wall shall be separated as follows:

   1.1. By a horizontal distance of not less than 24 inches (610 mm) except at walls or partitions constructed using parallel rows of studs or staggered studs;

   1.2. By a horizontal distance of not less than the depth of the wall cavity when the wall cavity is filled with cellulose loose-fill, rockwool or slag mineral wool insulation;

   1.3. By solid fire blocking in accordance with Section R602.8.1;

   1.4. By protecting both boxes with listed putty pads; or

   1.5. By other listed materials and methods.

2. Membrane penetrations by listed electrical boxes of any materials provided the boxes have been tested for use in fire-resistance-rated assemblies and are installed in accordance with the instructions included in the listing. The annular space between the wall membrane and the box shall not exceed $^1/_8$ inch (3.1 mm) unless listed otherwise. Such boxes on opposite sides of the wall shall be separated as follows:

   2.1. By a horizontal distance of not less than 24 inches (610 mm) except at walls or partitions constructed using parallel rows of studs or staggered studs;

   2.2. By solid fire blocking in accordance with Section R602.8;

   2.3. By protecting both boxes with listed putty pads; or

   2.4. By other listed materials and methods.

3. The annular space created by the penetration of a fire sprinkler provided it is covered by a metal escutcheon plate.

SECTION R318
MOISTURE VAPOR RETARDERS

R318.1 Moisture control. In all framed walls, floors and roof/ceilings comprising elements of the building thermal envelope, a vapor retarder shall be installed on the warm-in-winter side of the insulation.

Exceptions:

1. In construction where moisture or freezing will not damage the materials.

2. Where the framed cavity or space is ventilated to allow moisture to escape.

3. In counties identified as in climate zones 1 through 4 in Table N1101.2.

SECTION R319
PROTECTION AGAINST DECAY

R319.1 Location required. Protection from decay shall be provided in the following locations by the use of naturally durable wood or wood that is preservative treated in accordance with AWPA U1 for the species, product, preservative and end use. Preservatives shall be listed in Section 4 of AWPA U1.

1. Wood joists or the bottom of a wood structural floor when closer than 18 inches (457 mm) or wood girders when closer than 12 inches (305 mm) to the exposed ground in crawl spaces or unexcavated area located within the periphery of the building foundation.

2. All wood framing members that rest on concrete or masonry exterior foundation walls and are less than 8 inches (203 mm) from the exposed ground.

3. Sills and sleepers on a concrete or masonry slab that is in direct contact with the ground unless separated from such slab by an impervious moisture barrier.

4. The ends of wood girders entering exterior masonry or concrete walls having clearances of less than 0.5 inch (12.7 mm) on tops, sides and ends.

5. Wood siding, sheathing and wall framing on the exterior of a building having a clearance of less than 6 inches (152 mm) from the ground.

6. Wood structural members supporting moisture-permeable floors or roofs that are exposed to the weather, such as concrete or masonry slabs, unless separated from such floors or roofs by an impervious moisture barrier.

7. Wood furring strips or other wood framing members attached directly to the interior of exterior masonry walls or concrete walls below grade except where an approved vapor retarder is applied between the wall and the furring strips or framing members.

R319.1.1 Field treatment. Field-cut ends, notches and drilled holes of preservative-treated wood shall be treated in the field in accordance with AWPA M4.

R319.1.2 Ground contact. All wood in contact with the ground, embedded in concrete in direct contact with the ground or embedded in concrete exposed to the weather that supports permanent structures intended for human occupancy shall be approved pressure-preservative-treated wood suitable for ground contact use, except untreated wood may be used where entirely below groundwater level or continuously submerged in fresh water.

R319.1.3 Geographical areas. In geographical areas where experience has demonstrated a specific need, approved naturally durable or pressure-preservative-treated wood shall be used for those portions of wood members that form the structural supports of buildings, balconies, porches or similar permanent building appurtenances when those members are exposed to the weather without adequate protection from a roof, eave, overhang or other covering that would prevent moisture or water accumulation on the surface or at joints between members. Depending on local experience, such members may include:

1. Horizontal members such as girders, joists and decking.

2. Vertical members such as posts, poles and columns.

3. Both horizontal and vertical members.

R319.1.4 Wood columns. Wood columns shall be approved wood of natural decay resistance or approved pressure-preservative-treated wood.

Exceptions:

1. Columns exposed to the weather or in basements when supported by concrete piers or metal pedestals projecting 1 inch (25.4 mm) above a concrete floor or 6 inches (152 mm) above exposed earth and the earth is covered by an approved impervious moisture barrier.

2. Columns in enclosed crawl spaces or unexcavated areas located within the periphery of the building when supported by a concrete pier or metal pedestal at a height more than 8 inches (203mm) from exposed earth and the earth is covered by an impervious moisture barrier.

R319.1.5 Exposed glued-laminated timbers. The portions of glued-laminated timbers that form the structural supports of a building or other structure and are exposed to weather and not properly protected by a roof, eave or similar covering shall be pressure treated with preservative, or be manufactured from naturally durable or preservative-treated wood.

R319.2 Quality mark. Lumber and plywood required to be pressure-preservative-treated in accordance with Section R319.1 shall bear the quality mark of an approved inspection agency that maintains continuing supervision, testing and inspection over the quality of the product and that has been approved by an accreditation body that complies with the requirements of the American Lumber Standard Committee treated wood program.

R319.2.1 Required information. The required quality mark on each piece of pressure-preservative-treated lumber or plywood shall contain the following information:

1. Identification of the treating plant.

2. Type of preservative.

3. The minimum preservative retention.

4. End use for which the product was treated.

5. Standard to which the product was treated.

6. Identity of the approved inspection agency.

7. The designation "Dry," if applicable.

Exception: Quality marks on lumber less than 1 inch (25.4 mm) nominal thickness, or lumber less than nominal 1 inch by 5 inches (25.4 mm by 127 mm) or 2 inches by 4 inches (51 mm by 102 mm) or lumber 36 inches (914 mm) or less in length shall be applied by stamping the faces of exterior pieces or by end labeling not less than 25 percent of the pieces of a bundled unit.

R319.3 Fasteners. Fasteners for pressure-preservative and fire-retardant-treated wood shall be of hot-dipped zinc-coated galvanized steel, stainless steel, silicon bronze or copper. The coating weights for zinc-coated fasteners shall be in accordance with ASTM A 153.

Exceptions:

1. One-half-inch (12.7 mm) diameter or larger steel bolts.

2. Fasteners other than nails and timber rivets shall be permitted to be of mechanically deposited zinc-coated steel with coating weights in accordance with ASTM B 695, Class 55, minimum.

SECTION R320
PROTECTION AGAINST
SUBTERRANEAN TERMITES

R320.1 Subterranean termite control methods. In areas subject to damage from termites as indicated by Table R301.2(1), methods of protection shall be one of the following methods or a combination of these methods:

1. Chemical termiticide treatment, as provided in Section R320.2.

2. Termite baiting system installed and maintained according to the label.

3. Pressure-preservative-treated wood in accordance with the AWPA standards listed in Section R319.1.

4. Naturally termite-resistant wood as provided in Section R320.3.

5. Physical barriers as provided in Section R320.4.

R320.1.1 Quality mark. Lumber and plywood required to be pressure-preservative-treated in accordance with Section R320.1 shall bear the quality mark of an approved inspection agency which maintains continuing supervision, testing and inspection over the quality of the product and which has been approved by an accreditation body which complies with the requirements of the American Lumber Standard Committee treated wood program.

R320.1.2 Field treatment. Field-cut ends, notches, and drilled holes of pressure-preservative-treated wood shall be retreated in the field in accordance with AWPA M4.

R320.2 Chemical termiticide treatment. Chemical termiticide treatment shall include soil treatment and/or field applied wood treatment. The concentration, rate of application and method of treatment of the chemical termiticide shall be in strict accordance with the termiticide label.

R320.3 Naturally resistant wood. Heartwood of redwood and eastern red cedar shall be considered termite resistant.

R320.4 Barriers. Approved physical barriers, such as metal or plastic sheeting or collars specifically designed for termite prevention, shall be installed in a manner to prevent termites from entering the structure. Shields placed on top of an exterior foundation wall are permitted to be used only if in combination with another method of protection.

R320.5 Foam plastic protection. In areas where the probability of termite infestation is "very heavy" as indicated in Figure R301.2(6), extruded and expanded polystyrene, polyisocyanurate and other foam plastics shall not be installed on the exterior face or under interior or exterior foundation walls or slab foundations located below grade. The clearance between foam plastics installed above grade and exposed earth shall be at least 6 inches (152 mm).

Exceptions:

1. Buildings where the structural members of walls, floors, ceilings and roofs are entirely of noncombustible materials or pressure-preservative-treated wood.

2. When in addition to the requirements of Section R320.1, an approved method of protecting the foam plastic and structure from subterranean termite damage is used.

3. On the interior side of basement walls.

SECTION R321
SITE ADDRESS

R321.1 Premises identification. Approved numbers or addresses shall be provided for all new buildings in such a position as to be plainly visible and legible from the street or road fronting the property.

SECTION R322
ACCESSIBILITY

R322.1 Scope. Where there are four or more dwelling units or sleeping units in a single structure, the provisions of Chapter 11 of the *International Building Code* for Group R-3 shall apply.

SECTION R323
ELEVATORS AND PLATFORM LIFTS

R323.1 Elevators. Where provided, passenger elevators, limited-use/limited-application elevators or private residence elevators shall comply with ASME A17.1.

R323.2 Platform lifts. Where provided, platform lifts shall comply with ASME A18.1.

R323.3 Accessibility. Elevators or platform lifts that are part of an accessible route required by Chapter 11 of the *International Building Code*, shall comply with ICC A117.1.

SECTION R324
FLOOD-RESISTANT CONSTRUCTION

R324.1 General. Buildings and structures constructed in whole or in part in flood hazard areas (including A or V Zones) as established in Table R301.2(1) shall be designed and constructed in accordance with the provisions contained in this section.

Exception: Buildings and structures located in whole or in part in identified floodways as established in Table R301.2(1) shall be designed and constructed as stipulated in the *International Building Code*.

R324.1.1 Structural systems. All structural systems of all buildings and structures shall be designed, connected and anchored to resist flotation, collapse or permanent lateral movement due to structural loads and stresses from flooding equal to the design flood elevation.

R324.1.2 Flood-resistant construction. All buildings and structures erected in areas prone to flooding shall be constructed by methods and practices that minimize flood damage.

R324.1.3 Establishing the design flood elevation. The design flood elevation shall be used to define areas prone to flooding, and shall describe, at a minimum, the base flood elevation at the depth of peak elevation of flooding (including wave height) which has a 1 percent (100-year flood) or greater chance of being equaled or exceeded in any given year.

R324.1.3.1 Determination of design flood elevations. If design flood elevations are not specified, the building official is authorized to require the applicant to:

1. Obtain and reasonably use data available from a federal, state or other source; or

2. Determine the design flood elevation in accordance with accepted hydrologic and hydraulic engineering practices used to define special flood hazard areas. Determinations shall be undertaken by a registered design professional who shall document that the technical methods used reflect currently accepted engineering practice. Studies, analyses and computations shall be submitted in sufficient detail to allow thorough review and approval.

R324.1.3.2 Determination of impacts. In riverine flood hazard areas where design flood elevations are specified but floodways have not been designated, the applicant shall demonstrate that the effect of the proposed buildings and structures on design flood elevations, including fill, when combined with all other existing and anticipated flood hazard area encroachments, will not increase the design flood elevation more than 1 foot (305 mm) at any point within the jurisdiction.

R324.1.4 Lowest floor. The lowest floor shall be the floor of the lowest enclosed area, including basement, but excluding any unfinished flood-resistant enclosure that is useable solely for vehicle parking, building access or limited storage provided that such enclosure is not built so as to render the building or structure in violation of this section.

R324.1.5 Protection of mechanical and electrical systems. Electrical systems, equipment and components, and heating, ventilating, air conditioning and plumbing appliances, plumbing fixtures, duct systems, and other service equipment shall be located at or above the design flood elevation. If replaced as part of a substantial improvement, electrical systems, equipment and components, and heating, ventilating, air conditioning, and plumbing appliances, plumbing fixtures, duct systems, and other service equipment shall meet the requirements of this section. Systems, fixtures, and equipment and components shall not be mounted on or penetrate through walls intended to break away under flood loads.

> **Exception:** Electrical systems, equipment and components, and heating, ventilating, air conditioning and plumbing appliances, plumbing fixtures, duct systems, and other service equipment are permitted to be located below the design flood elevation provided that they are designed and installed to prevent water from entering or accumulating within the components and to resist hydrostatic and hydrodynamic loads and stresses, including the effects of buoyancy, during the occurrence of flooding to the design flood elevation in compliance with the flood-resistant construction requirements of the *International Building Code*. Electrical wiring systems are permitted to be located below the design flood elevation provided they conform to the provisions of the electrical part of this code for wet locations.

R324.1.6 Protection of water supply and sanitary sewage systems. New and replacement water supply systems shall be designed to minimize or eliminate infiltration of flood waters into the systems in accordance with the plumbing provisions of this code. New and replacement sanitary sewage systems shall be designed to minimize or eliminate infiltration of floodwaters into systems and discharges from systems into floodwaters in accordance with the plumbing provisions of this code and Chapter 3 of the *International Private Sewage Disposal Code*.

R324.1.7 Flood-resistant materials. Building materials used below the design flood elevation shall comply with the following:

1. All wood, including floor sheathing, shall be pressure-preservative-treated in accordance with AWPA U1 for the species, product, preservative and end use or be the decay-resistant heartwood of redwood, black locust or cedars. Preservatives shall be listed in Section 4 of AWPA U1.

2. Materials and installation methods used for flooring and interior and exterior walls and wall coverings shall conform to the provisions of FEMA/FIA-TB-2.

R324.1.8 Manufactured housing. New or replacement manufactured housing shall be elevated in accordance with Section R324.2 and the anchor and tie-down requirements of Sections AE604 and AE605 of Appendix E shall apply. The foundation and anchorage of manufactured housing to be located in identified flood ways as established in Table R301.2(1) shall be designed and constructed in accordance

with the applicable provisions in the *International Building Code*.

R324.1.9 As-built elevation documentation. A registered design professional shall prepare and seal documentation of the elevations specified in Section R324.2 or R324.3.

R324.2 Flood hazard areas (including A Zones). Areas that have been determined to be prone to flooding but not subject to high velocity wave action shall be designated as flood hazard areas. All buildings and structures constructed in whole or in part in flood hazard areas shall be designed and constructed in accordance with Sections R324.2.1 through R324.2.3.

R324.2.1 Elevation requirements.

1. Buildings and structures shall have the lowest floors elevated to or above the design flood elevation.

2. In areas of shallow flooding (AO Zones), buildings and structures shall have the lowest floor (including basement) elevated at least as high above the highest adjacent grade as the depth number specified in feet (mm) on the FIRM, or at least 2 feet (610 mm) if a depth number is not specified.

3. Basement floors that are below grade on all sides shall be elevated to or above the design flood elevation.

> **Exception:** Enclosed areas below the design flood elevation, including basements whose floors are not below grade on all sides, shall meet the requirements of Section R324.2.2.

R324.2.2 Enclosed area below design flood elevation. Enclosed areas, including crawl spaces, that are below the design flood elevation shall:

1. Be used solely for parking of vehicles, building access or storage.

2. Be provided with flood openings that meet the following criteria:

   2.1. There shall be a minimum of two openings on different sides of each enclosed area; if a building has more than one enclosed area below the design flood elevation, each area shall have openings on exterior walls.

   2.2. The total net area of all openings shall be at least 1 square inch (645 mm²) for each square foot (0.093 m²) of enclosed area, or the openings shall be designed and the construction documents shall include a statement that the design and installation will provide for equalization of hydrostatic flood forces on exterior walls by allowing for the automatic entry and exit of floodwaters.

   2.3. The bottom of each opening shall be 1 foot (305 mm) or less above the adjacent ground level.

   2.4. Openings shall be at least 3 inches (76 mm) in diameter.

   2.5. Any louvers, screens or other opening covers shall allow the automatic flow of floodwaters into and out of the enclosed area.

2.6. Openings installed in doors and windows, that meet requirements 2.1 through 2.5, are acceptable; however, doors and windows without installed openings do not meet the requirements of this section.

R324.2.3 Foundation design and construction. Foundation walls for all buildings and structures erected in flood hazard areas shall meet the requirements of Chapter 4.

Exception: Unless designed in accordance with Section R404:

1. The unsupported height of 6-inch (152 mm) plain masonry walls shall be no more than 3 feet (914 mm).

2. The unsupported height of 8-inch (203 mm) plain masonry walls shall be no more than 4 feet (1219 mm).

3. The unsupported height of 8-inch (203 mm) reinforced masonry walls shall be no more than 8 feet (2438 mm).

For the purpose of this exception, unsupported height is the distance from the finished grade of the under-floor space and the top of the wall.

R324.3 Coastal high-hazard areas (including V Zones). Areas that have been determined to be subject to wave heights in excess of 3 feet (914 mm) or subject to high-velocity wave action or wave-induced erosion shall be designated as coastal high-hazard areas. Buildings and structures constructed in whole or in part in coastal high-hazard areas shall be designed and constructed in accordance with Sections R324.3.1 through R324.3.6.

R324.3.1 Location and site preparation.

1. Buildings and structures shall be located landward of the reach of mean high tide.

2. For any alteration of sand dunes and mangrove stands the building official shall require submission of an engineering analysis which demonstrates that the proposed alteration will not increase the potential for flood damage.

R324.3.2 Elevation requirements.

1. All buildings and structures erected within coastal high hazard areas shall be elevated so that the lowest portion of all structural members supporting the lowest floor, with the exception of mat or raft foundations, piling, pile caps, columns, grade beams and bracing, is located at or above the design flood elevation.

2. Basement floors that are below grade on all sides are prohibited.

3. The use of fill for structural support is prohibited.

4. The placement of fill beneath buildings and structures is prohibited.

Exception: Walls and partitions enclosing areas below the design flood elevation shall meet the requirements of Sections R324.3.4 and R324.3.5.

R324.3.3 Foundations. Buildings and structures erected in coastal high-hazard areas shall be supported on pilings or columns and shall be adequately anchored to those pilings or columns. Pilings shall have adequate soil penetrations to resist the combined wave and wind loads (lateral and uplift). Water loading values used shall be those associated with the design flood. Wind loading values shall be those required by this code. Pile embedment shall include consideration of decreased resistance capacity caused by scour of soil strata surrounding the piling. Pile systems design and installation shall be certified in accordance with Section R324.3.6. Mat, raft or other foundations that support columns shall not be permitted where soil investigations that are required in accordance with Section R401.4 indicate that soil material under the mat, raft or other foundation is subject to scour or erosion from wave-velocity flow conditions. Slabs, pools, pool decks and walkways shall be located and constructed to be structurally independent of buildings and structures and their foundations to prevent transfer of flood loads to the buildings and structures during conditions of flooding, scour or erosion from wave-velocity flow conditions, unless the buildings and structures and their foundation are designed to resist the additional flood load.

R324.3.4 Walls below design flood elevation. Walls and partitions are permitted below the elevated floor, provided that such walls and partitions are not part of the structural support of the building or structure and:

1. Electrical, mechanical, and plumbing system components are not to be mounted on or penetrate through walls that are designed to break away under flood loads; and

2. Are constructed with insect screening or open lattice; or

3. Are designed to break away or collapse without causing collapse, displacement or other structural damage to the elevated portion of the building or supporting foundation system. Such walls, framing and connections shall have a design safe loading resistance of not less than 10 (479 Pa) and no more than 20 pounds per square foot (958 Pa); or

4. Where wind loading values of this code exceed 20 pounds per square foot (958 Pa), the construction documents shall include documentation prepared and sealed by a registered design professional that:

   4.1. The walls and partitions below the design flood elevation have been designed to collapse from a water load less than that which would occur during the design flood.

   4.2. The elevated portion of the building and supporting foundation system have been designed to withstand the effects of wind and flood loads acting simultaneously on all building components (structural and nonstructural). Water loading values used shall be those associated with the design flood. Wind loading values shall be those required by this code.

R324.3.5 Enclosed areas below design flood elevation. Enclosed areas below the design flood elevation shall be used solely for parking of vehicles, building access or storage.

R324.3.6 Construction documents. The construction documents shall include documentation that is prepared and sealed by a registered design professional that the design and methods of construction to be used meet the applicable criteria of this section.

CHAPTER 4

FOUNDATIONS

SECTION R401
GENERAL

R401.1 Application. The provisions of this chapter shall control the design and construction of the foundation and foundation spaces for all buildings. In addition to the provisions of this chapter, the design and construction of foundations in areas prone to flooding as established by Table R301.2(1) shall meet the provisions of Section R324. Wood foundations shall be designed and installed in accordance with AF&PA Report No. 7.

Exception: The provisions of this chapter shall be permitted to be used for wood foundations only in the following situations:

1. In buildings that have no more than two floors and a roof.

2. When interior basement and foundation walls are constructed at intervals not exceeding 50 feet (15 240 mm).

Wood foundations in Seismic Design Category D_0, D_1 or D_2 shall be designed in accordance with accepted engineering practice.

R401.2 Requirements. Foundation construction shall be capable of accommodating all loads according to Section R301 and of transmitting the resulting loads to the supporting soil. Fill soils that support footings and foundations shall be designed, installed and tested in accordance with accepted engineering practice. Gravel fill used as footings for wood and precast concrete foundations shall comply with Section R403.

R401.3 Drainage. Surface drainage shall be diverted to a storm sewer conveyance or other approved point of collection so as to not create a hazard. Lots shall be graded to drain surface water away from foundation walls. The grade shall fall a minimum of 6 inches (152 mm) within the first 10 feet (3048 mm).

Exception: Where lot lines, walls, slopes or other physical barriers prohibit 6 inches (152 mm) of fall within 10 feet (3048 mm), the final grade shall slope away from the foundation at a minimum slope of 5 percent and the water shall be directed to drains or swales to ensure drainage away from the structure. Swales shall be sloped a minimum of 2 percent when located within 10 feet (3048 mm) of the building foundation. Impervious surfaces within 10 feet (3048 mm) of the building foundation shall be sloped a minimum of 2 percent away from the building.

R401.4 Soil tests. In areas likely to have expansive, compressible, shifting or other unknown soil characteristics, the building official shall determine whether to require a soil test to determine the soil's characteristics at a particular location. This test shall be made by an approved agency using an approved method.

R401.4.1 Geotechnical evaluation. In lieu of a complete geotechnical evaluation, the load-bearing values in Table R401.4.1 shall be assumed.

TABLE R401.4.1
PRESUMPTIVE LOAD–BEARING VALUES OF FOUNDATION MATERIALS[a]

CLASS OF MATERIAL	LOAD-BEARING PRESSURE (pounds per square foot)
Crystalline bedrock	12,000
Sedimentary and foliated rock	4,000
Sandy gravel and/or gravel (GW and GP)	3,000
Sand, silty sand, clayey sand, silty gravel and clayey gravel (SW, SP, SM, SC, GM and GC)	2,000
Clay, sandy clay, silty clay, clayey silt, silt and sandy silt (CL, ML, MH and CH)	1,500[b]

For SI: 1 pound per square foot = 0.0479 kPa.

a. When soil tests are required by Section R401.4, the allowable bearing capacities of the soil shall be part of the recommendations.

b. Where the building official determines that in-place soils with an allowable bearing capacity of less than 1,500 psf are likely to be present at the site, the allowable bearing capacity shall be determined by a soils investigation.

R401.4.2 Compressible or shifting soil. Instead of a complete geotechnical evaluation, when top or subsoils are compressible or shifting, they shall be removed to a depth and width sufficient to assure stable moisture content in each active zone and shall not be used as fill or stabilized within each active zone by chemical, dewatering or presaturation.

SECTION R402
MATERIALS

R402.1 Wood foundations. Wood foundation systems shall be designed and installed in accordance with the provisions of this code.

R402.1.1 Fasteners. Fasteners used below grade to attach plywood to the exterior side of exterior basement or crawlspace wall studs, or fasteners used in knee wall construction, shall be of Type 304 or 316 stainless steel. Fasteners used above grade to attach plywood and all lumber-to-lumber fasteners except those used in knee wall construction shall be of Type 304 or 316 stainless steel, silicon bronze, copper, hot-dipped galvanized (zinc coated) steel nails, or hot-tumbled galvanized (zinc coated) steel nails. Electro-galvanized steel nails and galvanized (zinc coated) steel staples shall not be permitted.

R402.1.2 Wood treatment. All lumber and plywood shall be pressure-preservative treated and dried after treatment in accordance with AWPA U1 (Commodity Specification A, Use Category 4B and Section 5.2), and shall bear the label of an accredited agency. Where lumber and/or plywood is cut or drilled after treatment, the treated surface shall be field treated with copper naphthenate, the concentration of which shall contain a minimum of 2 percent copper metal, by repeated brushing, dipping or soaking until the wood absorbs no more preservative.

R402.2 Concrete. Concrete shall have a minimum specified compressive strength of f'_c, as shown in Table R402.2. Concrete subject to moderate or severe weathering as indicated in Table R301.2(1) shall be air entrained as specified in Table R402.2. The maximum weight of fly ash, other pozzolans, silica fume, slag or blended cements that is included in concrete mixtures for garage floor slabs and for exterior porches, carport slabs and steps that will be exposed to deicing chemicals shall not exceed the percentages of the total weight of cementitious materials specified in Section 4.2.3 of ACI 318. Materials used to produce concrete and testing thereof shall comply with the applicable standards listed in Chapter 3 of ACI 318.

R402.3 Precast concrete. Approved precast concrete foundations shall be designed and installed in accordance with the provisions of this code and the manufacturer's installation instructions.

SECTION R403
FOOTINGS

R403.1 General. All exterior walls shall be supported on continuous solid or fully grouted masonry or concrete footings, wood foundations, or other approved structural systems which shall be of sufficient design to accommodate all loads according to Section R301 and to transmit the resulting loads to the soil within the limitations as determined from the character of the soil. Footings shall be supported on undisturbed natural soils or engineered fill.

TABLE R403.1
MINIMUM WIDTH OF CONCRETE OR MASONRY FOOTINGS
(inches)[a]

	LOAD-BEARING VALUE OF SOIL (psf)			
	1,500	2,000	3,000	≥4,000
Conventional light–frame construction				
1-story	12	12	12	12
2-story	15	12	12	12
3-story	23	17	12	12
4-inch brick veneer over light frame or 8-inch hollow concrete masonry				
1-story	12	12	12	12
2-story	21	16	12	12
3-story	32	24	16	12
8-inch solid or fully grouted masonry				
1-story	16	12	12	12
2-story	29	21	14	12
3-story	42	32	21	16

For SI: 1 inch = 25.4 mm, 1 pound per square foot = 0.0479 kPa.

a. Where minimum footing width is 12 inches, use of a single wythe of solid or fully grouted 12-inch nominal concrete masonry units is permitted.

R403.1.1 Minimum size. Minimum sizes for concrete and masonry footings shall be as set forth in Table R403.1 and Figure R403.1(1). The footing width, W, shall be based on the load-bearing value of the soil in accordance with Table R401.4.1. Spread footings shall be at least 6 inches (152 mm) thick. Footing projections, P, shall be at least 2 inches

TABLE R402.2
MINIMUM SPECIFIED COMPRESSIVE STRENGTH OF CONCRETE

	MINIMUM SPECIFIED COMPRESSIVE STRENGTH[a] (f'_c)		
	Weathering Potential[b]		
TYPE OR LOCATION OF CONCRETE CONSTRUCTION	Negligible	Moderate	Severe
Basement walls, foundations and other concrete not exposed to the weather	2,500	2,500	2,500[c]
Basement slabs and interior slabs on grade, except garage floor slabs	2,500	2,500	2,500[c]
Basement walls, foundation walls, exterior walls and other vertical concrete work exposed to the weather	2,500	3,000[d]	3,000[d]
Porches, carport slabs and steps exposed to the weather, and garage floor slabs	2,500	3,000[d,e,f]	3,500[d,e,f]

For SI: 1 pound per square inch = 6.895 kPa.

a. Strength at 28 days psi.

b. See Table R301.2(1) for weathering potential.

c. Concrete in these locations that may be subject to freezing and thawing during construction shall be air-entrained concrete in accordance with Footnote d.

d. Concrete shall be air-entrained. Total air content (percent by volume of concrete) shall be not less than 5 percent or more than 7 percent.

e. See Section R402.2 for maximum cementitious materials content.

f. For garage floors with a steel troweled finish, reduction of the total air content (percent by volume of concrete) to not less than 3 percent is permitted if the specified compressive strength of the concrete is increased to not less than 4,000 psi.

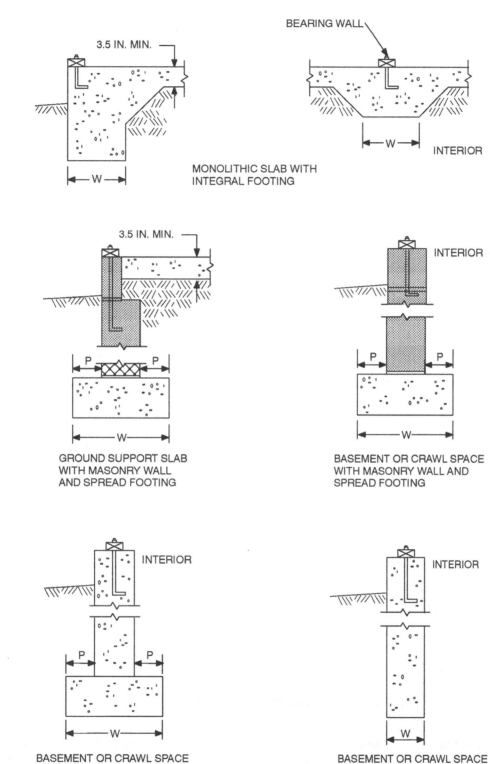

3.5 IN. MIN.

MONOLITHIC SLAB WITH
INTEGRAL FOOTING

BEARING WALL

W
INTERIOR

3.5 IN. MIN.

P   P

W

GROUND SUPPORT SLAB
WITH MASONRY WALL
AND SPREAD FOOTING

INTERIOR

P   P

W

BASEMENT OR CRAWL SPACE
WITH MASONRY WALL AND
SPREAD FOOTING

INTERIOR

P   P

W

BASEMENT OR CRAWL SPACE
WITH CONCRETE WALL AND
SPREAD FOOTING

INTERIOR

W

BASEMENT OR CRAWL SPACE
WITH FOUNDATION WALL
BEARING DIRECTLY ON SOIL

For SI: 1 inch = 25.4 mm.

FIGURE R403.1(1)
CONCRETE AND MASONRY FOUNDATION DETAILS

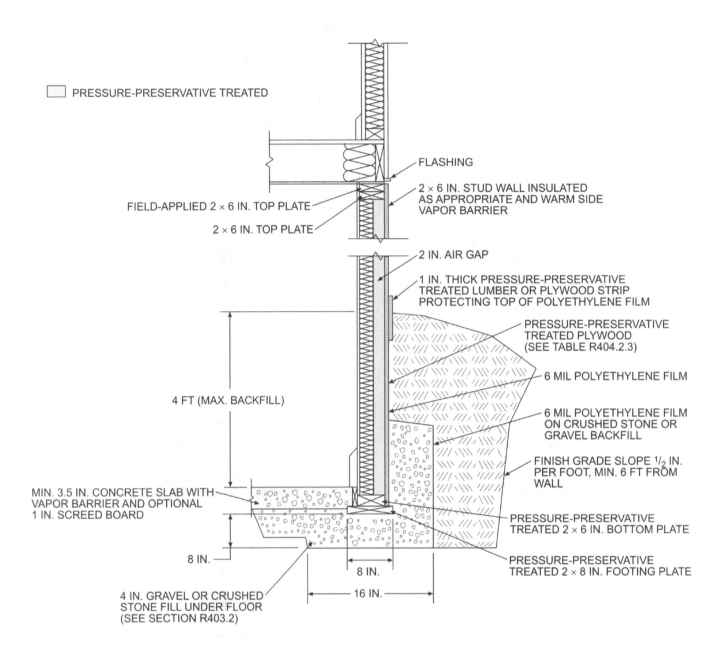

PRESSURE-PRESERVATIVE TREATED

FLASHING

2 × 6 IN. STUD WALL INSULATED AS APPROPRIATE AND WARM SIDE VAPOR BARRIER

FIELD-APPLIED 2 × 6 IN. TOP PLATE

2 × 6 IN. TOP PLATE

2 IN. AIR GAP

1 IN. THICK PRESSURE-PRESERVATIVE TREATED LUMBER OR PLYWOOD STRIP PROTECTING TOP OF POLYETHYLENE FILM

PRESSURE-PRESERVATIVE TREATED PLYWOOD (SEE TABLE R404.2.3)

6 MIL POLYETHYLENE FILM

4 FT (MAX. BACKFILL)

6 MIL POLYETHYLENE FILM ON CRUSHED STONE OR GRAVEL BACKFILL

FINISH GRADE SLOPE $1/2$ IN. PER FOOT, MIN. 6 FT FROM WALL

MIN. 3.5 IN. CONCRETE SLAB WITH VAPOR BARRIER AND OPTIONAL 1 IN. SCREED BOARD

PRESSURE-PRESERVATIVE TREATED 2 × 6 IN. BOTTOM PLATE

PRESSURE-PRESERVATIVE TREATED 2 × 8 IN. FOOTING PLATE

8 IN.

8 IN.

16 IN.

4 IN. GRAVEL OR CRUSHED STONE FILL UNDER FLOOR (SEE SECTION R403.2)

For SI: 1 inch 25.4 = mm, 1 foot =304.8, 1 mil =0.0254 mm.

FIGURE R403.1(2)
PERMANENT WOOD FOUNDATION BASEMENT WALL SECTION

(51 mm) and shall not exceed the thickness of the footing. The size of footings supporting piers and columns shall be based on the tributary load and allowable soil pressure in accordance with Table R401.4.1. Footings for wood foundations shall be in accordance with the details set forth in Section R403.2, and Figures R403.1(2) and R403.1(3).

R403.1.2 Continuous footing in Seismic Design Categories D_0, D_1 and D_2. The braced wall panels at exterior walls of buildings located in Seismic Design Categories D_0, D_1

and D_2 shall be supported by continuous footings. All required interior braced wall panels in buildings with plan dimensions greater than 50 feet (15 240 mm) shall also be supported by continuous footings.

R403.1.3 Seismic reinforcing. Concrete footings located in Seismic Design Categories D_0, D_1 and D_2, as established in Table R301.2(1), shall have minimum reinforcement. Bottom reinforcement shall be located a minimum of 3 inches (76 mm) clear from the bottom of the footing.

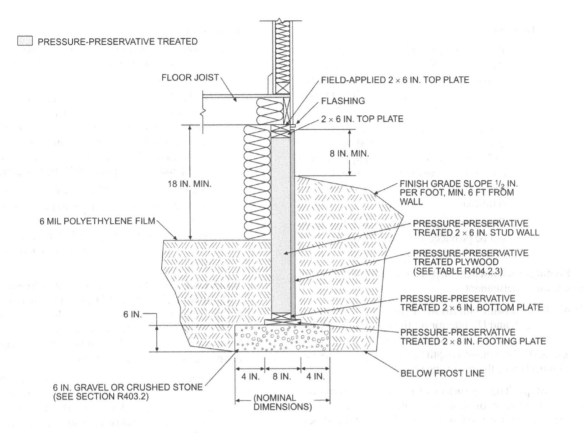

PRESSURE-PRESERVATIVE TREATED

FLOOR JOIST

FIELD-APPLIED 2 × 6 IN. TOP PLATE

FLASHING

2 × 6 IN. TOP PLATE

8 IN. MIN.

18 IN. MIN.

6 MIL POLYETHYLENE FILM

FINISH GRADE SLOPE 1/2 IN. PER FOOT, MIN. 6 FT FROM WALL

PRESSURE-PRESERVATIVE TREATED 2 × 6 IN. STUD WALL

PRESSURE-PRESERVATIVE TREATED PLYWOOD (SEE TABLE R404.2.3)

PRESSURE-PRESERVATIVE TREATED 2 × 6 IN. BOTTOM PLATE

6 IN.

PRESSURE-PRESERVATIVE TREATED 2 × 8 IN. FOOTING PLATE

BELOW FROST LINE

6 IN. GRAVEL OR CRUSHED STONE (SEE SECTION R403.2)

4 IN.   8 IN.   4 IN.

(NOMINAL DIMENSIONS)

For SI: 1 inch = 25.4 mm, 1 foot = 304.8 mm, 1 mil = 0.0254 mm.

FIGURE R403.1(3)
PERMANENT WOOD FOUNDATION CRAWL SPACE SECTION

In Seismic Design Categories D_0, D_1 and D_2 where a construction joint is created between a concrete footing and a stem wall, a minimum of one No. 4 bar shall be installed at not more than 4 feet (1219 mm) on center. The vertical bar shall extend to 3 inches (76 mm) clear of the bottom of the footing, have a standard hook and extend a minimum of 14 inches (357 mm) into the stem wall.

In Seismic Design Categories D_0, D_1 and D_2 where a grouted masonry stem wall is supported on a concrete footing and stem wall, a minimum of one No. 4 bar shall be installed at not more than 4 feet on center. The vertical bar shall extend to 3 inches (76 mm) clear of the bottom of the footing and have a standard hook.

In Seismic Design Categories D_0, D_1 and D_2 masonry stem walls without solid grout and vertical reinforcing are not permitted.

Exception: In detached one- and two-family dwellings which are three stories or less in height and constructed with stud bearing walls, plain concrete footings without longitudinal reinforcement supporting walls and isolated plain concrete footings supporting columns or pedestals are permitted.

R403.1.3.1 Foundations with stemwalls. Foundations with stem walls shall have installed a minimum of one No. 4 bar within 12 inches (305 mm) of the top of the wall

and one No. 4 bar located 3 inches (76 mm) to 4 inches (102 mm) from the bottom of the footing.

R403.1.3.2 Slabs-on-ground with turned-down footings. Slabs-on-ground with turned-down footings shall have a minimum of one No. 4 bar at the top and bottom of the footing.

Exception: For slabs-on-ground cast monolithically with a footing, one No. 5 bar or two No. 4 bars shall be located in the middle third of the footing depth.

R403.1.4 Minimum depth. All exterior footings shall be placed at least 12 inches (305 mm) below the undisturbed ground surface. Where applicable, the depth of footings shall also conform to Sections R403.1.4.1 through R403.1.4.2.

R403.1.4.1 Frost protection. Except where otherwise protected from frost, foundation walls, piers and other permanent supports of buildings and structures shall be protected from frost by one or more of the following methods:

1. Extended below the frost line specified in Table R301.2.(1);

2. Constructing in accordance with Section R403.3;

3. Constructing in accordance with ASCE 32; or

4. Erected on solid rock.

Exceptions:

1. Protection of freestanding accessory structures with an area of 600 square feet (56 m²) or less, of light-framed construction, with an eave height of 10 feet (3048 mm) or less shall not be required.

2. Protection of freestanding accessory structures with an area of 400 square feet (37 m²) or less, of other than light-framed construction, with an eave height of 10 feet (3048 mm) or less shall not be required.

3. Decks not supported by a dwelling need not be provided with footings that extend below the frost line.

Footings shall not bear on frozen soil unless the frozen condition is permanent.

R403.1.4.2 Seismic conditions. In Seismic Design Categories D_0, D_1 and D_2, interior footings supporting bearing or bracing walls and cast monolithically with a slab on grade shall extend to a depth of not less than 12 inches (305 mm) below the top of the slab.

R403.1.5 Slope. The top surface of footings shall be level. The bottom surface of footings shall not have a slope exceeding one unit vertical in 10 units horizontal (10-percent slope). Footings shall be stepped where it is necessary to change the elevation of the top surface of the footings or where the slope of the bottom surface of the footings will exceed one unit vertical in ten units horizontal (10-percent slope).

R403.1.6 Foundation anchorage. When braced wall panels are supported directly on continuous foundations, the wall wood sill plate or cold-formed steel bottom track shall be anchored to the foundation in accordance with this section.

The wood sole plate at exterior walls on monolithic slabs and wood sill plate shall be anchored to the foundation with anchor bolts spaced a maximum of 6 feet (1829 mm) on center. There shall be a minimum of two bolts per plate section with one bolt located not more than 12 inches (305 mm) or less than seven bolt diameters from each end of the plate section. In Seismic Design Categories D_0, D_1 and D_2, anchor bolts shall be spaced at 6 feet (1829 mm) on center and located within 12 inches (305 mm) of the ends of each plate section at interior braced wall lines when required by Section R602.10.9 to be supported on a continuous foundation. Bolts shall be at least $^1/_2$ inch (13 mm) in diameter and shall extend a minimum of 7 inches (178 mm) into masonry or concrete. Interior bearing wall sole plates on monolithic slab foundation shall be positively anchored with approved fasteners. A nut and washer shall be tightened on each bolt of the plate. Sills and sole plates shall be protected against decay and termites where required by Sections R319 and R320. Cold-formed steel framing systems shall be fastened to the wood sill plates or anchored directly to the foundation as required in Section R505.3.1 or R603.3.1.

Exceptions:

1. Foundation anchorage, spaced as required to provide equivalent anchorage to $^1/_2$-inch-diameter (13 mm) anchor bolts.

2. Walls 24 inches (610 mm) total length or shorter connecting offset braced wall panels shall be anchored to the foundation with a minimum of one anchor bolt located in the center third of the plate section and shall be attached to adjacent braced wall panels per Figure R602.10.5 at corners.

3. Walls 12 inches (305 mm) total length or shorter connecting offset braced wall panels shall be permitted to be connected to the foundation without anchor bolts. The wall shall be attached to adjacent braced wall panels per Figure R602.10.5 at corners.

R403.1.6.1 Foundation anchorage in Seismic Design Categories C, D_0, D_1 and D_2. In addition to the requirements of Section R403.1.6, the following requirements shall apply to wood light-frame structures in Seismic Design Categories D_0, D_1 and D_2 and wood light-frame townhouses in Seismic Design Category C.

1. Plate washers conforming to Section R602.11.1 shall be provided for all anchor bolts over the full length of required braced wall lines. Properly sized cut washers shall be permitted for anchor bolts in wall lines not containing braced wall panels.

2. Interior braced wall plates shall have anchor bolts spaced at not more than 6 feet (1829 mm) on center and located within 12 inches (305 mm) of the ends of each plate section when supported on a continuous foundation.

3. Interior bearing wall sole plates shall have anchor bolts spaced at not more than 6 feet (1829 mm) on center and located within 12 inches (305 mm) of the ends of each plate section when supported on a continuous foundation.

4. The maximum anchor bolt spacing shall be 4 feet (1219 mm) for buildings over two stories in height.

5. Stepped cripple walls shall conform to Section R602.11.3.

6. Where continuous wood foundations in accordance with Section R404.2 are used, the force transfer shall have a capacity equal to or greater than the connections required by Section R602.11.1 or the braced wall panel shall be connected to the wood foundations in accordance with the braced wall panel-to-floor fastening requirements of Table R602.3(1).

R403.1.7 Footings on or adjacent to slopes. The placement of buildings and structures on or adjacent to slopes steeper than 1 unit vertical in 3 units horizontal (33.3-percent slope) shall conform to Sections R403.1.7.1 through R403.1.7.4.

R403.1.7.1 Building clearances from ascending slopes. In general, buildings below slopes shall be set a

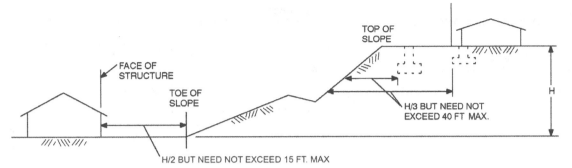

For SI: 1 foot = 304.8 mm.

FIGURE R403.1.7.1
FOUNDATION CLEARANCE FROM SLOPES

sufficient distance from the slope to provide protection from slope drainage, erosion and shallow failures. Except as provided in Section R403.1.7.4 and Figure R403.1.7.1, the following criteria will be assumed to provide this protection. Where the existing slope is steeper than one unit vertical in one unit horizontal (100-percent slope), the toe of the slope shall be assumed to be at the intersection of a horizontal plane drawn from the top of the foundation and a plane drawn tangent to the slope at an angle of 45 degrees (0.79 rad) to the horizontal. Where a retaining wall is constructed at the toe of the slope, the height of the slope shall be measured from the top of the wall to the top of the slope.

R403.1.7.2 Footing setback from descending slope surfaces. Footings on or adjacent to slope surfaces shall be founded in material with an embedment and setback from the slope surface sufficient to provide vertical and lateral support for the footing without detrimental settlement. Except as provided for in Section R403.1.7.4 and Figure R403.1.7.1, the following setback is deemed adequate to meet the criteria. Where the slope is steeper than one unit vertical in one unit horizontal (100-percent slope), the required setback shall be measured from an imaginary plane 45 degrees (0.79 rad) to the horizontal, projected upward from the toe of the slope.

R403.1.7.3 Foundation elevation. On graded sites, the top of any exterior foundation shall extend above the elevation of the street gutter at point of discharge or the inlet of an approved drainage device a minimum of 12 inches (305 mm) plus 2 percent. Alternate elevations are permitted subject to the approval of the building official, provided it can be demonstrated that required drainage to the point of discharge and away from the structure is provided at all locations on the site.

R403.1.7.4 Alternate setback and clearances. Alternate setbacks and clearances are permitted, subject to the approval of the building official. The building official is permitted to require an investigation and recommendation of a qualified engineer to demonstrate that the intent of this section has been satisfied. Such an investigation shall include consideration of material, height of slope, slope gradient, load intensity and erosion characteristics of slope material.

R403.1.8 Foundations on expansive soils. Foundation and floor slabs for buildings located on expansive soils shall be designed in accordance with Section 1805.8 of the *International Building Code.*

Exception: Slab-on-ground and other foundation systems which have performed adequately in soil conditions similar to those encountered at the building site are permitted subject to the approval of the building official.

R403.1.8.1 Expansive soils classifications. Soils meeting all four of the following provisions shall be considered expansive, except that tests to show compliance with Items 1, 2 and 3 shall not be required if the test prescribed in Item 4 is conducted:

1. Plasticity Index (PI) of 15 or greater, determined in accordance with ASTM D 4318.

2. More than 10 percent of the soil particles pass a No. 200 sieve (75 mm), determined in accordance with ASTM D 422.

3. More than 10 percent of the soil particles are less than 5 micrometers in size, determined in accordance with ASTM D 422.

4. Expansion Index greater than 20, determined in accordance with ASTM D 4829.

R403.2 Footings for wood foundations. Footings for wood foundations shall be in accordance with Figures R403.1(2) and R403.1(3). Gravel shall be washed and well graded. The maximum size stone shall not exceed $^{3}/_{4}$ inch (19.1 mm). Gravel shall be free from organic, clayey or silty soils. Sand shall be coarse, not smaller than $^{1}/_{16}$-inch (1.6 mm) grains and shall be free from organic, clayey or silty soils. Crushed stone shall have a maximum size of $^{1}/_{2}$ inch (12.7 mm).

R403.3 Frost protected shallow foundations. For buildings where the monthly mean temperature of the building is maintained at a minimum of 64°F (18°C), footings are not required to extend below the frost line when protected from frost by insulation in accordance with Figure R403.3(1) and Table R403.3. Foundations protected from frost in accordance with Figure R403.3(1) and Table R403.3 shall not be used for unheated spaces such as porches, utility rooms, garages and carports, and shall not be attached to basements or crawl spaces

TABLE R403.3
MINIMUM INSULATION REQUIREMENTS FOR FROST-PROTECTED FOOTINGS IN HEATED BUILDINGS[a]

AIR FREEZING INDEX (°F-days)[b]	VERTICAL INSULATION R-VALUE[c,d]	HORIZONTAL INSULATION R-VALUE[c,e]		HORIZONTAL INSULATION DIMENSIONS PER FIGURE R403.3(1) (inches)		
		Along walls	At corners	A	B	C
1,500 or less	4.5	Not required	Not required	Not required	Not required	Not required
2,000	5.6	Not required	Not required	Not required	Not required	Not required
2,500	6.7	1.7	4.9	12	24	40
3,000	7.8	6.5	8.6	12	24	40
3,500	9.0	8.0	11.2	24	30	60
4,000	10.1	10.5	13.1	24	36	60

a. Insulation requirements are for protection against frost damage in heated buildings. Greater values may be required to meet energy conservation standards. Interpolation between values is permissible.

b. See Figure R403.3(2) for Air Freezing Index values.

c. Insulation materials shall provide the stated minimum R–values under long-term exposure to moist, below-ground conditions in freezing climates. The following R–values shall be used to determine insulation thicknesses required for this application: Type II expanded polystyrene—2.4R per inch; Type IV extruded polystyrene—4.5R per inch; Type VI extruded polystyrene—4.5R per inch; Type IX expanded polystyrene—3.2R per inch; Type X extruded polystyrene—4.5R per inch.

d. Vertical insulation shall be expanded polystyrene insulation or extruded polystyrene insulation.

e. Horizontal insulation shall be extruded polystyrene insulation.

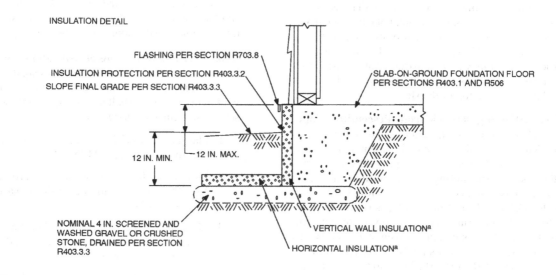

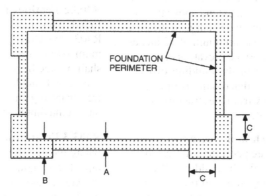

For SI: 1 inch = 25.4 mm.

a. See Table R403.3 for required dimensions and R-values for vertical and horizontal insulation.

FIGURE R403.3(1)
INSULATION PLACEMENT FOR FROST-PROTECTED FOOTINGS IN HEATED BUILDINGS

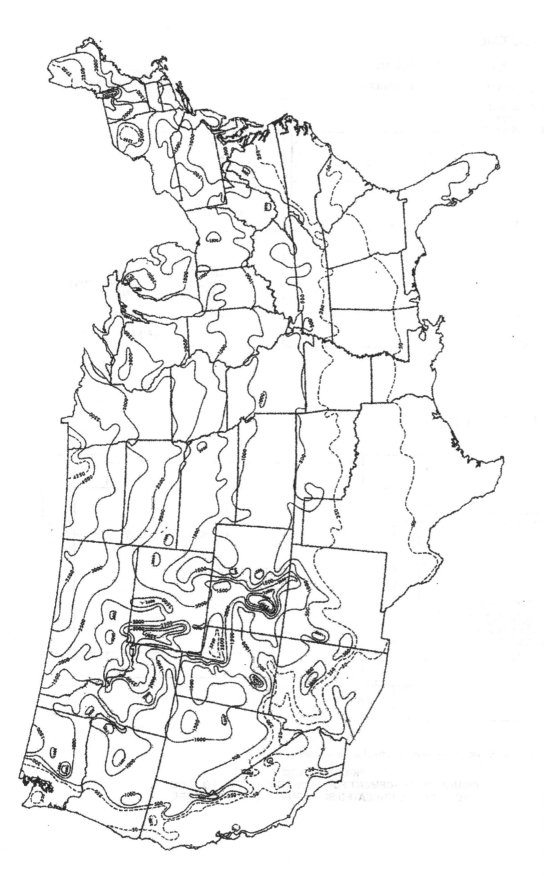

For SI: °C = [(°F)-32]/1.8.

NOTE: The air-freezing index is defined as cumulative degree days below 32°F. It is used as a measure of the combined magnitude and duration of air temperature below freezing. The index was computed over a 12-month period (July-June) for each of the 3,044 stations used in the above analysis. Data from the 1951-80 period were fitted to a Weibull probability distribution to produce an estimate of the 100-year return period.

FIGURE R403.3(2)
AIR–FREEZING INDEX
AN ESTIMATE OF THE 100-YEAR RETURN PERIOD

INSULATION DETAIL

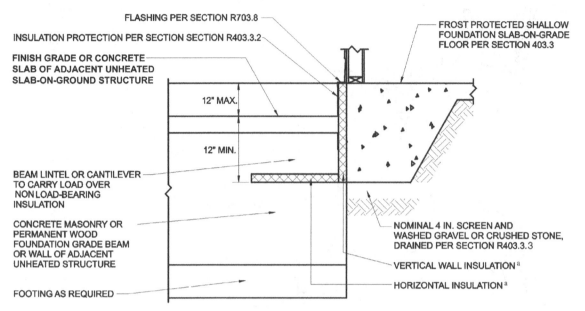

- FLASHING PER SECTION R703.8
- INSULATION PROTECTION PER SECTION SECTION R403.3.2
- FINISH GRADE OR CONCRETE SLAB OF ADJACENT UNHEATED SLAB-ON-GROUND STRUCTURE
- 12" MAX.
- 12" MIN.
- BEAM LINTEL OR CANTILEVER TO CARRY LOAD OVER NON LOAD-BEARING INSULATION
- CONCRETE MASONRY OR PERMANENT WOOD FOUNDATION GRADE BEAM OR WALL OF ADJACENT UNHEATED STRUCTURE
- FOOTING AS REQUIRED
- FROST PROTECTED SHALLOW FOUNDATION SLAB-ON-GRADE FLOOR PER SECTION 403.3
- NOMINAL 4 IN. SCREEN AND WASHED GRAVEL OR CRUSHED STONE, DRAINED PER SECTION R403.3.3
- VERTICAL WALL INSULATION [a]
- HORIZONTAL INSULATION [a]

HORIZONTAL INSULATION PLAN

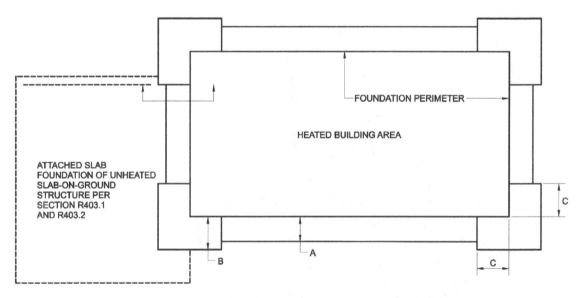

- FOUNDATION PERIMETER
- HEATED BUILDING AREA
- ATTACHED SLAB FOUNDATION OF UNHEATED SLAB-ON-GROUND STRUCTURE PER SECTION R403.1 AND R403.2
- C
- A
- B
- C

For SI: 1 inch = 25.4 mm.

a. See Table R403.3 for required dimensions and *R*-values for vertical and horizontal insulation.

FIGURE R403.3(3)
INSULATION PLACEMENT FOR FROST-PROTECTED FOOTINGS
ADJACENT TO UNHEATED SLAB-ON-GROUND STRUCTURE

that are not maintained at a minimum monthly mean temperature of 64°F (18°C).

Materials used below grade for the purpose of insulating footings against frost shall be labeled as complying with ASTM C 578.

R403.3.1 Foundations adjoining frost protected shallow foundations. Foundations that adjoin frost protected shallow foundations shall be protected from frost in accordance with Section R403.1.4.

R403.3.1.1 Attachment to unheated slab-on-ground structure. Vertical wall insulation and horizontal insulation of frost protected shallow foundations that adjoin a slab-on-ground foundation that does not have a monthly mean temperature maintained at a minimum of 64°F (18°C), shall be in accordance with Figure R403.3(3) and Table R403.3. Vertical wall insulation shall extend between the frost protected shallow foundation and the adjoining slab foundation. Required horizontal insulation shall be continuous under the adjoining slab foundation and through any foundation walls adjoining the frost protected shallow foundation. Where insulation passes through a foundation wall, it shall either be of a type complying with this section and having bearing capacity equal to or greater than the structural loads imposed by the building, or the building shall be designed and constructed using beams, lintels, cantilevers or other means of transferring building loads such that the structural loads of the building do not bear on the insulation.

R403.3.1.2 Attachment to heated structure. Where a frost protected shallow foundation abuts a structure that has a monthly mean temperature maintained at a mini-

mum of 64°F (18°C), horizontal insulation and vertical wall insulation shall not be required between the frost protected shallow foundation and the adjoining structure. Where the frost protected shallow foundation abuts the heated structure, the horizontal insulation and vertical wall insulation shall extend along the adjoining foundation in accordance with Figure R403.3(4) a distance of not less than Dimension A in Table R403.3.

> **Exception:** Where the frost protected shallow foundation abuts the heated structure to form an inside corner, vertical insulation extending along the adjoining foundation is not required.

R403.3.2 Protection of horizontal insulation below ground. Horizontal insulation placed less than 12 inches (305 mm) below the ground surface or that portion of horizontal insulation extending outward more than 24 inches (610 mm) from the foundation edge shall be protected against damage by use of a concrete slab or asphalt paving on the ground surface directly above the insulation or by cementitious board, plywood rated for below-ground use, or other approved materials placed below ground, directly above the top surface of the insulation.

R403.3.3 Drainage. Final grade shall be sloped in accordance with Section R401.3. In other than Group I Soils, as detailed in Table R405.1, gravel or crushed stone beneath horizontal insulation below ground shall drain to daylight or into an approved sewer system.

R403.3.4 Termite damage. The use of foam plastic in areas of "very heavy" termite infestation probability shall be in accordance with Section R320.5.

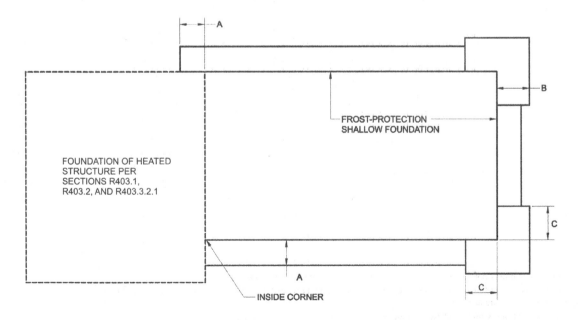

FIGURE R403.3(4)
INSULATION PLACEMENT FOR FROST-PROTECTED
FOOTINGS ADJACENT TO HEATED STRUCTURE

SECTION R404
FOUNDATION AND RETAINING WALLS

R404.1 Concrete and masonry foundation walls. Concrete and masonry foundation walls shall be selected and constructed in accordance with the provisions of Section R404 or in accordance with ACI 318, ACI 332, NCMA TR68–A or ACI 530/ASCE 5/TMS 402 or other approved structural standards. When ACI 318, ACI 332 or ACI 530/ASCE 5/TMS 402 or the provisions of Section R404 are used to design concrete or masonry foundation walls, project drawings, typical details and specifications are not required to bear the seal of the architect or engineer responsible for design, unless otherwise required by the state law of the jurisdiction having authority.

Foundation walls that meet all of the following shall be considered laterally supported:

1. Full basement floor shall be 3.5 inches (89 mm) thick concrete slab poured tight against the bottom of the foundation wall.
2. Floor joists and blocking shall be connected to the sill plate at the top of wall by the prescriptive method called out in Table R404.1(1), or; shall be connected with an approved connector with listed capacity meeting Table R404.1(1).
3. Bolt spacing for the sill plate shall be no greater than per Table R404.1(2).
4. Floor shall be blocked perpendicular to the floor joists. Blocking shall be full depth within two joist spaces of the

TABLE R404.1(1)
TOP REACTIONS AND PRESCRIPTIVE SUPPORT FOR FOUNDATION WALLS[a]

MAXIMUM WALL HEIGHT (feet)	MAXIMUM UNBALANCED BACKFILL HEIGHT (feet)	HORIZONTAL REACTION TO TOP (plf)		
		Soil Classes (Letter indicates connection types[b])		
		GW, GP, SW and SP soils	GM, GC, SM-SC and ML soils	SC, MH, ML-CL and inorganic CL soils
7	4	45.7 A	68.6 A	91.4 A
	5	89.3 A	133.9 B	178.6 B
	6	154.3 B	231.4 C	308.6 C
	7	245.0 C	367.5 C	490.0 D
8	4	40.0 A	60.0 A	80.0 A
	5	78.1 A	117.2 B	156.3 B
	6	135.0 B	202.5 B	270.0 C
	7	214.0 B	321.6 C	428.8 C
	8	320.0 C	480.0 C	640.0 D
9	4	35.6 A	53.3 A	71.1 A
	5	69.4 A	104.2 B	138.9 B
	6	120.0 B	180.0 B	240.0 C
	7	190.6 B	285.8 C	381.1 C
	8	284.4 C	426.7 C	568.9 D
	9	405.0 C	607.5 D	810.0 D

For SI: 1 foot = 304.8 mm, 1 pound = 0.454 kg, 1 plf = pounds per linear foot = 1.488 kg/m.

a. Loads are pounds per linear foot of wall. Prescriptive options are limited to maximum joist and blocking spacing of 24 inches on center.

b. Prescriptive Support Requirements:

Type	Joist/blocking Attachment Requirement
A	3-8d per joist per Table R602.3(1).
B	1-20 gage angle clip each joist with 5-8d per leg.
C	1-$\frac{1}{4}$-inch thick steel angle. Horizontal leg attached to sill bolt adjacent to joist/blocking, vertical leg attached to joist/blocking with $\frac{1}{2}$-inch minimum diameter bolt.
D	2-$\frac{1}{4}$-inch thick steel, angles, one on each side of joist/blocking. Attach each angle to adjacent sill bolt through horizontal leg. Bolt to joist/blocking with $\frac{1}{2}$-inch minimum diameter bolt common to both angles.

TABLE R404.1(2)
MAXIMUM PLATE ANCHOR-BOLT SPACING FOR SUPPORTED FOUNDATION WALL[a]

MAXIMUM WALL HEIGHT (feet)	MAXIMUM UNBALANCED BACKFILL HEIGHT (feet)	ANCHOR BOLT SPACING (inches)		
		Soil Classes		
		GW, GP, SW and SP soils	GM, GC, SM-SC and ML soils	SC, MH, ML-CL and inorganic CL soils
7	4	72	58	43
	5	44	30	22
	6	26	17	13
	7	16	11	8
8	4	72	66	50
	5	51	34	25
	6	29	20	15
	7	18	12	9
	8	12	8	6
9	4	72	72	56
	5	57	38	29
	6	33	22	17
	7	21	14	10
	8	14	9	7
	9	10	7	5

For SI: 1 inch = 25.4 mm, 1 foot = 304.8 mm.

a. Spacing is based on $^1/_2$-inch diameter anchor bolts. For $^5/_8$-inch diameter anchor bolts, spacing may be multiplied by 1.27, with a maximum spacing of 72 inches.

foundation wall, and be flat-blocked with minimum 2-inch by 4-inch (51 mm by 102 mm) blocking elsewhere.

5. Where foundation walls support unbalanced load on opposite sides of the building, such as a daylight basement, the building aspect ratio, L/W, shall not exceed the value specified in Table R404.1(3). For such foundation walls, the rim board shall be attached to the sill with a 20 gage metal angle clip at 24 inches (610 mm) on center, with five 8d nails per leg, or an approved connector supplying 230 pounds per linear foot (3.36 kN/m) capacity.

R404.1.1 Masonry foundation walls. Concrete masonry and clay masonry foundation walls shall be constructed as set forth in Table R404.1.1(1), R404.1.1(2), R404.1.1(3) or R404.1.1(4) and shall also comply with the provisions of Section R404 and the applicable provisions of Sections R606, R607 and R608. In Seismic Design Categories D_0, D_1 and D_2, concrete masonry and clay masonry foundation walls shall also comply with Section R404.1.4. Rubble stone masonry foundation walls shall be constructed in accordance with Sections R404.1.8 and R607.2.2. Rubble stone masonry walls shall not be used in Seismic Design Categories D_0, D_1 and D_2.

R404.1.2 Concrete foundation walls. Concrete foundation walls shall be constructed as set forth in Table R404.1.1(5) and shall also comply with the provisions of Section R404 and the applicable provisions of Section R402.2. In Seismic Design Categories D_0, D_1 and D_2, concrete foundation walls shall also comply with Section R404.1.4.

R404.1.3 Design required. Concrete or masonry foundation walls shall be designed in accordance with accepted engineering practice when either of the following conditions exists:

1. Walls are subject to hydrostatic pressure from groundwater.

2. Walls supporting more than 48 inches (1219 mm) of unbalanced backfill that do not have permanent lateral support at the top or bottom.

R404.1.4 Seismic Design Categories D_0, D_1 and D_2. In addition to the requirements of Tables R404.1.1(1) and R404.1.1(5), plain concrete and plain masonry foundation walls located in Seismic Design Categories D_0, D_1 and D_2, as established in Table R301.2(1), shall comply with the following.

1. Wall height shall not exceed 8 feet (2438 mm).

2. Unbalanced backfill height shall not exceed 4 feet (1219 mm).

3. Minimum reinforcement for plain concrete foundation walls shall consist of one No. 4 (No. 13) horizontal bar located in the upper 12 inches (305 mm) of the wall.

4. Minimum thickness for plain concrete foundation walls shall be 7.5 inches (191 mm) except that 6 inches (152 mm) is permitted when the maximum height is 4 feet, 6 inches (1372 mm).

5. Minimum nominal thickness for plain masonry foundation walls shall be 8 inches (203 mm).

6. Masonry stem walls shall have a minimum vertical reinforcement of one No. 3 (No. 10) bar located a maximum of 4 feet (1220 mm) on center in grouted cells. Vertical reinforcement shall be tied to the horizontal reinforcement in the footings.

Foundation walls located in Seismic Design Categories D_0, D_1 and D_2, as established in Table R301.2(1), supporting more than 4 feet (1219 mm) of unbalanced backfill or exceeding 8 feet (2438 mm) in height shall be constructed in accordance with Table R404.1.1(2), R404.1.1(3) or R404.1.1(4) for masonry, or Table R404.1.1(5) for con-

TABLE R404.1(3)
MAXIMUM ASPECT RATIO, L/W FOR UNBALANCED FOUNDATIONS

MAXIMUM WALL HEIGHT (feet)	MAXIMUM UNBALANCED BACKFILL HEIGHT (feet)	SOIL CLASSES		
		GW, GP, SW and SP soils	GM, GC, SM-SC and ML soils	SC, MH, ML-CL and inorganic CL soils
7	4	4.0	4.0	4.0
	5	4.0	3.4	2.6
	6	3.0	2.0	1.5
	7	1.9	1.2	0.9
8	4	4.0	4.0	4.0
	5	4.0	3.9	2.9
	6	3.4	2.3	1.7
	7	2.1	1.4	1.1
	8	1.4	1.0	0.7
9	4	4.0	4.0	4.0
	5	4.0	4.0	3.3
	6	3.8	2.6	1.9
	7	2.4	1.6	1.2
	8	1.6	1.1	0.8
	9	1.1	0.8	0.6

For SI: 1 foot = 304.8 mm.

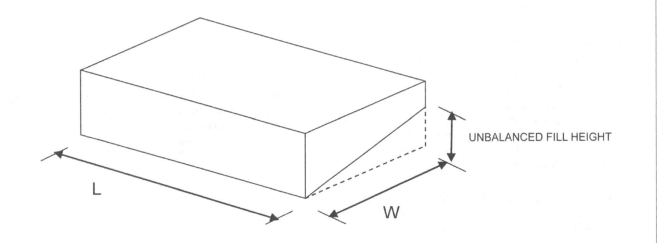

crete. Where Table R404.1.1(5) permits plain concrete walls, not less than No. 4 (No. 13) vertical bars at a spacing not exceeding 48 inches (1219 mm) shall be provided. Insulating concrete form foundation walls shall be reinforced as required in Table R404.4(1), R404.4(2), R404.4(3), R404.4(4) or R404.4(5). Where no vertical reinforcement is required by Table R404.4(2), R404.4(3) or R404.4(4) there shall be a minimum of one No. 4 (No. 13) bar at 48 inches (1220 mm) on center. All concrete and masonry foundation walls shall have two No. 4 (No. 13) horizontal bars located in the upper 12 inches (305 mm) of the wall.

R404.1.5 Foundation wall thickness based on walls supported. The thickness of concrete and masonry foundation walls shall not be less than the thickness of the wall supported, except that foundation walls of at least 8-inch (203 mm) nominal thickness shall be permitted under brick-veneered frame walls and under 10-inch-wide (254 mm) cavity walls where the total height of the wall supported, including gables, is not more than 20 feet (6096 mm), provided the requirements of Sections R404.1.1 and R404.1.2 are met.

R404.1.5.1 Pier and curtain wall foundations. Use of Pier and curtain wall foundations shall be permitted to support light-frame construction not more than two stories in height, provided the following requirements are met:

1. All load-bearing walls shall be placed on continuous concrete footings placed integrally with the exterior wall footings.

2. The minimum actual thickness of a load-bearing masonry wall shall be not less than 4 inches (102 mm) nominal or $3^3/_8$ inches (92 mm) actual thickness, and shall be bonded integrally with piers spaced in accordance with Section R606.9.

3. Piers shall be constructed in accordance with Section R606.6 and Section R606.6.1, and shall be bonded into the load-bearing masonry wall in accordance with Section R608.1.1 or Section R608.1.1.2.

4. The maximum height of a 4-inch (102 mm) load-bearing masonry foundation wall supporting

TABLE R404.1.1(1)
PLAIN MASONRY FOUNDATION WALLS

MAXIMUM WALL HEIGHT (feet)	MAXIMUM UNBALANCED BACKFILL HEIGHT[c] (feet)	PLAIN MASONRY[a] MINIMUM NOMINAL WALL THICKNESS (inches) Soil classes[b]		
		GW, GP, SW and SP	GM, GC, SM, SM-SC and ML	SC, MH, ML-CL and inorganic CL
5	4	6 solid[d] or 8	6 solid[d] or 8	6 solid[d] or 8
	5	6 solid[d] or 8	8	10
6	4	6 solid[d] or 8	6 solid[d] or 8	6 solid[d] or 8
	5	6 solid[d] or 8	8	10
	6	8	10	12
7	4	6 solid[d] or 8	8	8
	5	6 solid[d] or 8	10	10
	6	10	12	10 solid[d]
	7	12	10 solid[d]	12 solid[d]
8	4	6 solid[d] or 8	6 solid[d] or 8	8
	5	6 solid[d] or 8	10	12
	6	10	12	12 solid[d]
	7	12	12 solid[d]	Footnote e
	8	10 solid[d]	12 solid[d]	Footnote e
9	4	6 solid[d] or 8	6 solid[d] or 8	8
	5	8	10	12
	6	10	12	12 solid[d]
	7	12	12 solid[d]	Footnote e
	8	12 solid[d]	Footnote e	Footnote e
	9	Footnote e	Footnote e	Footnote e

For SI: 1 inch = 25.4 mm, 1 foot = 304.8 mm, 1 pound per square inch = 6.895 Pa.

a. Mortar shall be Type M or S and masonry shall be laid in running bond. Ungrouted hollow masonry units are permitted except where otherwise indicated.

b. Soil classes are in accordance with the Unified Soil Classification System. Refer to Table R405.1.

c. Unbalanced backfill height is the difference in height between the exterior finish ground level and the lower of the top of the concrete footing that supports the foundation wall or the interior finish ground level. Where an interior concrete slab-on-grade is provided and is in contact with the interior surface of the foundation wall, measurement of the unbalanced backfill height from the exterior finish ground level to the top of the interior concrete slab is permitted.

d. Solid grouted hollow units or solid masonry units.

e. Wall construction shall be in accordance with either Table R404.1.1(2), Table R404.1.1(3), Table R404.1.1(4), or a design shall be provided.

wood-frame walls and floors shall not be more than 4 feet (1219 mm).

5. Anchorage shall be in accordance with Section R403.1.6, Figure R404.1.5(1), or as specified by engineered design accepted by the building official.

6. The unbalanced fill for 4-inch (102 mm) foundation walls shall not exceed 24 inches (610 mm) for solid masonry or 12 inches (305 mm) for hollow masonry.

7. In Seismic Design Categories D_0, D_1 and D_2, prescriptive reinforcement shall be provided in the horizontal and vertical direction. Provide minimum horizontal joint reinforcement of two No.9 gage wires spaced not less than 6 inches (152 mm) or one $^1/_4$ inch (6.4 mm) diameter wire at 10 inches (254 mm) on center vertically. Provide minimum vertical reinforcement of one No. 4 bar at 48 inches (1220 mm) on center horizontally grouted in place.

R404.1.6 Height above finished grade. Concrete and masonry foundation walls shall extend above the finished grade adjacent to the foundation at all points a minimum of 4 inches (102 mm) where masonry veneer is used and a minimum of 6 inches (152 mm) elsewhere.

R404.1.7 Backfill placement. Backfill shall not be placed against the wall until the wall has sufficient strength and has been anchored to the floor above, or has been sufficiently braced to prevent damage by the backfill.

Exception: Bracing is not required for walls supporting less than 4 feet (1219 mm) of unbalanced backfill.

R404.1.8 Rubble stone masonry. Rubble stone masonry foundation walls shall have a minimum thickness of 16 inches (406 mm), shall not support an unbalanced backfill exceeding 8 feet (2438 mm) in height, shall not support a soil pressure greater than 30 pounds per square foot per foot (4.71 kPa/m), and shall not be constructed in Seismic Design Categories D_0, D_1, D_2 or townhouses in Seismic Design Category C, as established in Figure R301.2(2).

TABLE R404.1.1(2)
8-INCH MASONRY FOUNDATION WALLS WITH REINFORCING
WHERE d > 5 INCHES[a, c]

WALL HEIGHT	HEIGHT OF UNBALANCED BACKFILL[e]	MINIMUM VERTICAL REINFORCEMENT[b,c]		
		Soil classes and lateral soil load[d] (psf per foot below grade)		
		GW, GP, SW and SP soils 30	GM, GC, SM, SM-SC and ML soils 45	SC, ML-CL and inorganic CL soils 60
6 feet 8 inches	4 feet (or less)	#4 at 48″ o.c.	#4 at 48″ o.c.	#4 at 48″ o.c.
	5 feet	#4 at 48″ o.c.	#4 at 48″ o.c.	#4 at 48″ o.c.
	6 feet 8 inches	#4 at 48″ o.c.	#5 at 48″ o.c.	#6 at 48″ o.c.
7 feet 4 inches	4 feet (or less)	#4 at 48″ o.c.	#4 at 48″ o.c.	#4 at 48″ o.c.
	5 feet	#4 at 48″ o.c.	#4 at 48″ o.c.	#4 at 48″ o.c.
	6 feet	#4 at 48″ o.c.	#5 at 48″ o.c.	#5 at 48″ o.c.
	7 feet 4 inches	#5 at 48″ o.c.	#6 at 48″ o.c.	#6 at 40″ o.c.
8 feet	4 feet (or less)	#4 at 48″ o.c.	#4 at 48″ o.c.	#4 at 48″ o.c.
	5 feet	#4 at 48″ o.c.	#4 at 48″ o.c.	#4 at 48″ o.c.
	6 feet	#4 at 48″ o.c.	#5 at 48″ o.c.	#5 at 48″ o.c.
	7 feet	#5 at 48″ o.c.	#6 at 48″ o.c.	#6 at 40″ o.c.
	8 feet	#5 at 48″ o.c.	#6 at 48″ o.c.	#6 at 32″ o.c.
8 feet 8 inches	4 feet (or less)	#4 at 48″ o.c.	#4 at 48″ o.c.	#4 at 48″ o.c.
	5 feet	#4 at 48″ o.c.	#4 at 48″ o.c.	#5 at 48″ o.c.
	6 feet	#4 at 48″ o.c.	#5 at 48″ o.c.	#6 at 48″ o.c.
	7 feet	#5 at 48″ o.c.	#6 at 48″ o.c.	#6 at 40″ o.c.
	8 feet 8 inches	#6 at 48″ o.c.	#6 at 32″ o.c.	#6 at 24″ o.c.
9 feet 4 inches	4 feet (or less)	#4 at 48″ o.c.	#4 at 48″ o.c.	#4 at 48″ o.c.
	5 feet	#4 at 48″ o.c.	#4 at 48″ o.c.	#5 at 48″ o.c.
	6 feet	#4 at 48″ o.c.	#5 at 48″ o.c.	#6 at 48″ o.c.
	7 feet	#5 at 48″ o.c.	#6 at 48″ o.c.	#6 at 40″ o.c.
	8 feet	#6 at 48″ o.c.	#6 at 40″ o.c.	#6 at 24″ o.c.
	9 feet 4 inches	#6 at 40″ o.c.	#6 at 24″ o.c.	#6 at 16″ o.c.
10 feet	4 feet (or less)	#4 at 48″ o.c.	#4 at 48″ o.c.	#4 at 48″ o.c.
	5 feet	#4 at 48″ o.c.	#4 at 48″ o.c.	#5 at 48″ o.c.
	6 feet	#4 at 48″ o.c.	#5 at 48″ o.c.	#6 at 48″ o.c.
	7 feet	#5 at 48″ o.c.	#6 at 48″ o.c.	#6 at 32″ o.c.
	8 feet	#6 at 48″ o.c.	#6 at 32″ o.c.	#6 at 24″ o.c.
	9 feet	#6 at 40″ o.c.	#6 at 24″ o.c.	#6 at 16″ o.c.
	10 feet	#6 at 32″ o.c.	#6 at 16″ o.c.	#6 at 16″ o.c.

For SI: 1 inch = 25.4 mm, 1 foot = 304.8 mm, 1 pound per square foot per foot = 0.157 kPa/mm.

a. Mortar shall be Type M or S and masonry shall be laid in running bond.

b. Alternative reinforcing bar sizes and spacings having an equivalent cross-sectional area of reinforcement per lineal foot of wall shall be permitted provided the spacing of the reinforcement does not exceed 72 inches.

c. Vertical reinforcement shall be Grade 60 minimum. The distance (d) from the face of the soil side of the wall to the center of vertical reinforcement shall be at least 5 inches.

d. Soil classes are in accordance with the Unified Soil Classification System and design lateral soil loads are for moist conditions without hydrostatic pressure. Refer to Table R405.1.

e. Unbalanced backfill height is the difference in height between the exterior finish ground level and the lower of the top of the concrete footing that supports the foundation wall or the interior finish ground level. Where an interior concrete slab-on-grade is provided and is in contact with the interior surface of the foundation wall, measurement of the unbalanced backfill height from the exterior finish ground level to the top of the interior concrete slab is permitted.

R404.2 Wood foundation walls. Wood foundation walls shall be constructed in accordance with the provisions of Sections R404.2.1 through R404.2.6 and with the details shown in Figures R403.1(2) and R403.1(3).

R404.2.1 Identification. All load-bearing lumber shall be identified by the grade mark of a lumber grading or inspection agency which has been approved by an accreditation body that complies with DOC PS 20. In lieu of a grade mark, a certificate of inspection issued by a lumber grading or inspection agency meeting the requirements of this section shall be accepted. Wood structural panels shall conform to DOC PS 1 or DOC PS 2 and shall be identified by a grade mark or certificate of inspection issued by an approved agency.

R404.2.2 Stud size. The studs used in foundation walls shall be 2-inch by 6-inch (51 mm by 152 mm) members. When spaced 16 inches (406 mm) on center, a wood species with an F_b value of not less than 1,250 pounds per square inch (8612 kPa) as listed in AF&PA/NDS shall be used. When spaced 12 inches (305 mm) on center, an F_b of not less than 875 psi (6029 kPa) shall be required.

R404.2.3 Height of backfill. For wood foundations that are not designed and installed in accordance with AF&PA Report

No.7, the height of backfill against a foundation wall shall not exceed 4 feet (1219 mm). When the height of fill is more than 12 inches (305 mm) above the interior grade of a crawl space or floor of a basement, the thickness of the plywood sheathing shall meet the requirements of Table R404.2.3.

R404.2.4 Backfilling. Wood foundation walls shall not be backfilled until the basement floor and first floor have been constructed or the walls have been braced. For crawl space construction, backfill or bracing shall be installed on the interior of the walls prior to placing backfill on the exterior.

R404.2.5 Drainage and dampproofing. Wood foundation basements shall be drained and dampproofed in accordance with Sections R405 and R406, respectively.

R404.2.6 Fastening. Wood structural panel foundation wall sheathing shall be attached to framing in accordance with Table R602.3(1) and Section R402.1.1.

R404.3 Wood sill plates. Wood sill plates shall be a minimum of 2-inch by 4-inch (51 mm by 102 mm) nominal lumber. Sill plate anchorage shall be in accordance with Sections R403.1.6 and R602.11.

TABLE R404.1.1(3)
10-INCH MASONRY FOUNDATION WALLS WITH REINFORCING
WHERE d > 6.75 INCHES[a, c]

WALL HEIGHT	HEIGHT OF UNBALANCED BACKFILL[e]	MINIMUM VERTICAL REINFORCEMENT[b, c]		
		Soil classes and later soil load[d] (psf per foot below grade)		
		GW, GP, SW and SP soils 30	GM, GC, SM, SM-SC and ML soils 45	SC, ML-CL and inorganic CL soils 60
6 feet 8 inches	4 feet (or less)	#4 at 56″ o.c.	#4 at 56″ o.c.	#4 at 56″ o.c.
	5 feet	#4 at 56″ o.c.	#4 at 56″ o.c.	#4 at 56″ o.c.
	6 feet 8 inches	#4 at 56″ o.c.	#5 at 56″ o.c.	#5 at 56″ o.c.
7 feet 4 inches	4 feet (or less)	#4 at 56″ o.c.	#4 at 56″ o.c.	#4 at 56″ o.c.
	5 feet	#4 at 56″ o.c.	#4 at 56″ o.c.	#4 at 56″ o.c.
	6 feet	#4 at 56″ o.c.	#4 at 56″ o.c.	#5 at 56″ o.c.
	7 feet 4 inches	#4 at 56″ o.c.	#5 at 56″ o.c.	#6 at 56″ o.c.
8 feet	4 feet (or less)	#4 at 56″ o.c.	#4 at 56″ o.c.	#4 at 56″ o.c.
	5 feet	#4 at 56″ o.c.	#4 at 56″ o.c.	#4 at 56″ o.c.
	6 feet	#4 at 56″ o.c.	#4 at 56″ o.c.	#5 at 56″ o.c.
	7 feet	#4 at 56″ o.c.	#5 at 56″ o.c.	#6 at 56″ o.c.
	8 feet	#5 at 56″ o.c.	#6 at 56″ o.c.	#6 at 48″ o.c.
8 feet 8 inches	4 feet (or less)	#4 at 56″ o.c.	#4 at 56″ o.c.	#4 at 56″ o.c.
	5 feet	#4 at 56″ o.c.	#4 at 56″ o.c.	#4 at 56″ o.c.
	6 feet	#4 at 56″ o.c.	#4 at 56″ o.c.	#5 at 56″ o.c.
	7 feet	#4 at 56″ o.c.	#5 at 56″ o.c.	#6 at 56″ o.c.
	8 feet 8 inches	#5 at 56″ o.c.	#6 at 48″ o.c.	#6 at 32″ o.c.
9 feet 4 inches	4 feet (or less)	#4 at 56″ o.c.	#4 at 56″ o.c.	#4 at 56″ o.c.
	5 feet	#4 at 56″ o.c.	#4 at 56″ o.c.	#4 at 56″ o.c.
	6 feet	#4 at 56″ o.c.	#5 at 56″ o.c.	#5 at 56″ o.c.
	7 feet	#4 at 56″ o.c.	#5 at 56″ o.c.	#6 at 56″ o.c.
	8 feet	#5 at 56″ o.c.	#6 at 56″ o.c.	#6 at 40″ o.c.
	9 feet 4 inches	#6 at 56″ o.c.	#6 at 40″ o.c.	#6 at 24″ o.c.
10 feet	4 feet (or less)	#4 at 56″ o.c.	#4 at 56″ o.c.	#4 at 56″ o.c.
	5 feet	#4 at 56″ o.c.	#4 at 56″ o.c.	#4 at 56″ o.c.
	6 feet	#4 at 56″ o.c.	#5 at 56″ o.c.	#5 at 56″ o.c.
	7 feet	#5 at 56″ o.c.	#6 at 56″ o.c.	#6 at 48″ o.c.
	8 feet	#5 at 56″ o.c.	#6 at 48″ o.c.	#6 at 40″ o.c.
	9 feet	#6 at 56″ o.c.	#6 at 40″ o.c.	#6 at 24″ o.c.
	10 feet	#6 at 48″ o.c.	#6 at 32″ o.c.	#6 at 24″ o.c.

For SI: 1 inch = 25.4 mm, 1 foot = 304.8 mm, 1 pound per square foot per foot = 0.157 kPa/mm.

a. Mortar shall be Type M or S and masonry shall be laid in running bond.

b. Alternative reinforcing bar sizes and spacings having an equivalent cross-sectional area of reinforcement per lineal foot of wall shall be permitted provided the spacing of the reinforcement does not exceed 72 inches.

c. Vertical reinforcement shall be Grade 60 minimum. The distance, d, from the face of the soil side of the wall to the center of vertical reinforcement shall be at least 6.75 inches.

d. Soil classes are in accordance with the Unified Soil Classification System and design lateral soil loads are for moist conditions without hydrostatic pressure. Refer to Table R405.1.

e. Unbalanced backfill height is the difference in height between the exterior finish ground level and the lower of the top of the concrete footing that supports the foundation wall or the interior finish ground level. Where an interior concrete slab-on-grade is provided and is in contact with the interior surface of the foundation wall, measurement of the unbalanced backfill height from the exterior finish ground level to the top of the interior concrete slab is permitted.

TABLE R404.1.1(4)
12-INCH MASONRY FOUNDATION WALLS WITH REINFORCING
WHERE d > 8.75 INCHES[a, c]

WALL HEIGHT	HEIGHT OF UNBALANCED BACKFILL[e]	MINIMUM VERTICAL REINFORCEMENT[b, c]		
		Soil classes and lateral soil load[d] (psf per foot below grade)		
		GW, GP, SW and SP soils 30	GM, GC, SM, SM-SC and ML soils 45	SC, ML-CL and inorganic CL soils 60
6 feet 8 inches	4 feet (or less)	#4 at 72″ o.c.	#4 at 72″ o.c.	#4 at 72″ o.c.
	5 feet	#4 at 72″ o.c.	#4 at 72″ o.c.	#4 at 72″ o.c.
	6 feet 8 inches	#4 at 72″ o.c.	#4 at 72″ o.c.	#5 at 72″ o.c.
7 feet 4 inches	4 feet (or less)	#4 at 72″ o.c.	#4 at 72″ o.c.	#4 at 72″ o.c.
	5 feet	#4 at 72″ o.c.	#4 at 72″ o.c.	#4 at 72″ o.c.
	6 feet	#4 at 72″ o.c.	#4 at 72″ o.c.	#5 at 72″ o.c.
	7 feet 4 inches	#4 at 72″ o.c.	#5 at 72″ o.c.	#6 at 72″ o.c.
8 feet	4 feet (or less)	#4 at 72″ o.c.	#4 at 72″ o.c.	#4 at 72″ o.c.
	5 feet	#4 at 72″ o.c.	#4 at 72″ o.c.	#4 at 72″ o.c.
	6 feet	#4 at 72″ o.c.	#4 at 72″ o.c.	#5 at 72″ o.c.
	7 feet	#4 at 72″ o.c.	#5 at 72″ o.c.	#6 at 72″ o.c.
	8 feet	#5 at 72″ o.c.	#6 at 72″ o.c.	#6 at 64″ o.c.
8 feet 8 inches	4 feet (or less)	#4 at 72″ o.c.	#4 at 72″ o.c.	#4 at 72″ o.c.
	5 feet	#4 at 72″ o.c.	#4 at 72″ o.c.	#4 at 72″ o.c.
	6 feet	#4 at 72″ o.c.	#4 at 72″ o.c.	#5 at 72″ o.c.
	7 feet	#4 at 72″ o.c.	#5 at 72″ o.c.	#6 at 72″ o.c.
	8 feet 8 inches	#5 at 72″ o.c.	#7 at 72″ o.c.	#6 at 48″ o.c.
9 feet 4 inches	4 feet (or less)	#4 at 72″ o.c.	#4 at 72″ o.c.	#4 at 72″ o.c.
	5 feet	#4 at 72″ o.c.	#4 at 72″ o.c.	#4 at 72″ o.c.
	6 feet	#4 at 72″ o.c.	#5 at 72″ o.c.	#5 at 72″ o.c.
	7 feet	#4 at 72″ o.c.	#5 at 72″ o.c.	#6 at 72″ o.c.
	8 feet	#5 at 72″ o.c.	#6 at 72″ o.c.	#6 at 56″ o.c.
	9 feet 4 inches	#6 at 72″ o.c.	#6 at 48″ o.c.	#6 at 40″ o.c.
10 feet	4 feet (or less)	#4 at 72″ o.c.	#4 at 72″ o.c.	#4 at 72″ o.c.
	5 feet	#4 at 72″ o.c.	#4 at 72″ o.c.	#4 at 72″ o.c.
	6 feet	#4 at 72″ o.c.	#5 at 72″ o.c.	#5 at 72″ o.c.
	7 feet	#4 at 72″ o.c.	#6 at 72″ o.c.	#6 at 72″ o.c.
	8 feet	#5 at 72″ o.c.	#6 at 72″ o.c.	#6 at 48″ o.c.
	9 feet	#6 at 72″ o.c.	#6 at 56″ o.c.	#6 at 40″ o.c.
	10 feet	#6 at 64″ o.c.	#6 at 40″ o.c.	#6 at 32″ o.c.

For SI: 1 inch = 25.4 mm, 1 foot = 304.8 mm, 1 pound per square foot per foot = 0.157 kPa/mm.

a. Mortar shall be Type M or S and masonry shall be laid in running bond.

b. Alternative reinforcing bar sizes and spacings having an equivalent cross-sectional area of reinforcement per lineal foot of wall shall be permitted provided the spacing of the reinforcement does not exceed 72 inches.

c. Vertical reinforcement shall be Grade 60 minimum. The distance, d, from the face of the soil side of the wall to the center of vertical reinforcement shall be at least 8.75 inches.

d. Soil classes are in accordance with the Unified Soil Classification System and design lateral soil loads are for moist conditions without hydrostatic pressure. Refer to Table R405.1.

e. Unbalanced backfill height is the difference in height between the exterior finish ground level and the lower of the top of the concrete footing that supports the foundation wall or the interior finish ground levels. Where an interior concrete slab-on-grade is provided and in contact with the interior surface of the foundation wall, measurement of the unbalanced backfill height is permitted to be measured from the exterior finish ground level to the top of the interior concrete slab is permitted.

R404.4 Insulating concrete form foundation walls. Insulating concrete form (ICF) foundation walls shall be designed and constructed in accordance with the provisions of this section or in accordance with the provisions of ACI 318. When ACI 318 or the provisions of this section are used to design insulating concrete form foundation walls, project drawings, typical details and specifications are not required to bear the seal of the architect or engineer responsible for design unless otherwise required by the state law of the jurisdiction having authority.

R404.4.1 Applicability limits. The provisions of this section shall apply to the construction of insulating concrete form foundation walls for buildings not more than 60 feet (18 288 mm) in plan dimensions, and floors not more than 32 feet (9754 mm) or roofs not more than 40 feet (12 192 mm) in clear span. Buildings shall not exceed two stories in height above grade with each story not more than 10 feet (3048 mm) high. Foundation walls constructed in accordance with the provisions of this section shall be limited to buildings subjected to a maximum ground snow load of 70 psf (3.35 kPa) and located in Seismic Design Category A, B or C. In Seismic Design Categories D$_0$, D$_1$ and D$_2$, foundation walls shall comply with Section R404.1.4. Insulating concrete form foundation walls supporting above-grade concrete walls shall be reinforced as required for the above-

TABLE R404.1.1(5)
CONCRETE FOUNDATION WALLS[h, i, j, k]

MAXIMUM WALL HEIGHT (feet)	MAXIMUM UNBALANCED BACKFILL HEIGHT[b] (feet)	MINIMUM VERTICAL REINFORCEMENT SIZE AND SPACING[c, d, e, f, l]											
		Soil classes[a] and design lateral soil (psf per foot of depth)											
		GW, GP, SW and SP 30				GM, GC, SM, SM-SC and ML 45				SC, ML-CL and inorganic CL 60			
		Minimum wall thickness (inches)											
		5.5	7.5	9.5	11.5	5.5	7.5	9.5	11.5	5.5	7.5	9.5	11.5
5	4	PC	PC	PC	PC	PC	PC	PC	PC	PC	PC	PC	PC
	5	PC	PC	PC	PC	PC	PC	PC	PC	PC	PC	PC	PC
6	4	PC	PC	PC	PC	PC	PC	PC	PC	PC	PC	PC	PC
	5	PC	PC	PC	PC	PC	PC[g]	PC	PC	#4@35″	PC[g]	PC	PC
	6	PC	PC	PC	PC	#5@48″	PC	PC	PC	#5@36″	PC	PC	PC
7	4	PC	PC	PC	PC	PC	PC	PC	PC	PC	PC	PC	PC
	5	PC	PC	PC	PC	PC	PC	PC	PC	#5@47″	PC	PC	PC
	6	PC	PC	PC	PC	#5@42″	PC	PC	PC	#6@43″	#5@48″	PC[g]	PC
	7	#5@46″	PC	PC	PC	#6@42″	#5@46″	PC[g]	PC	#6@34″	#6@48″	PC	PC
8	4	PC	PC	PC	PC	PC	PC	PC	PC	PC	PC	PC	PC
	5	PC	PC	PC	PC	#4@38″	PC[g]	PC	PC	#5@43″	PC	PC	PC
	6	#4@37″	PC[g]	PC	PC	#5@37″	PC	PC	PC	#6@37″	#5@43″	PC[g]	PC
	7	#5@40″	PC	PC	PC	#6@37″	#5@41″	PC	PC	#6@34″	#6@43″	PC	PC
	8	#6@43″	#5@47″	PC[g]	PC	#6@34″	#6@43″	PC	PC	#6@27″	#6@32″	#6@44″	PC
9	4	PC	PC	PC	PC	PC	PC	PC	PC	PC	PC	PC	PC
	5	PC	PC	PC	PC	#4@35″	PC[g]	PC	PC	#5@40″	PC	PC[c]	PC
	6	#4@34″	PC[g]	PC	PC	#6@48″	PC	PC	PC	#6@36″	#5@39″	PC[g]	PC
	7	#5@36″	PC	PC	PC	#6@34″	#5@37″	PC	PC	#6@33″	#6@38″	#5@37″	PC[g]
	8	#6@38″	#5@41″	PC[g]	PC	#6@33″	#6@38″	#5@37″	PC[g]	#6@24″	#7@39″	#6@39″	#4@48″[h]
	9	#6@34″	#6@46″	PC	PC	#6@26″	#7@41″	#6@41″	PC	#6@19″	#7@31″	#7@41″	#6@39″
10	4	PC	PC	PC	PC	PC	PC	PC	PC	PC	PC	PC	PC
	5	PC	PC	PC	PC	#4@33″	PC[g]	PC	PC	#5@38″	PC	PC	PC
	6	#5@48″	PC[g]	PC	PC	#6@45″	PC	PC	PC	#6@34″	#5@37″	PC	PC
	7	#6@47″	PC	PC	PC	#6@34″	#6@48″	PC	PC	#6@30″	#6@35″	#6@48″	PC[g]
	8	#6@34″	#5@38″	PC	PC	#6@30″	#7@47″	#6@47″	PC[g]	#6@22″	#7@35″	#7@48″	#6@45″[h]
	9	#6@34″	#6@41″	#4@48″	PC[g]	#6@23″	#7@37″	#7@48″	#4@48″[h]	DR	#6@22″	#7@37″	#7@47″
	10	#6@28″	#7@45″	#6@45″	PC	DR	#7@31″	#7@40″	#6@38″	DR	#6@22″	#7@30″	#7@38″

For SI: 1 inch = 25.4 mm, 1 foot = 304.8 mm, 1 pound per square foot = 0.0479 kPa; 1 pound per square foot per foot = 0.157 kPa/mm.

a. Soil classes are in accordance with the United Soil Classification System. Refer to Table R405.1.

b. Unbalanced backfill height is the difference in height of the exterior and interior finish ground levels. Where there is an interior concrete slab, the unbalanced backfill height shall be measured from the exterior finish ground level to the top of the interior concrete slab.

c. The size and spacing of vertical reinforcement shown in the table is based on the use of reinforcement with a minimum yield strength of 60,000 psi. Vertical reinforcement with a minimum yield strength of 40,000 psi or 50,000 psi is permitted, provided the same size bar is used and the spacing shown in the table is reduced by multiplying the spacing by 0.67 or 0.83, respectively.

d. Vertical reinforcement, when required, shall be placed nearest the inside face of the wall a distance d from the outside face (soil side) of the wall. The distance d is equal to the wall thickness, t, minus 1.25 inches plus one-half the bar diameter, d_b ($d = t - (1.25 + d_b/2)$). The reinforcement shall be placed within a tolerance of $\pm \frac{3}{8}$ inch where d is less than or equal to 8 inches, or $\pm \frac{1}{2}$ inch where d is greater than 8 inches.

e. In lieu of the reinforcement shown, smaller reinforcing bar sizes and closer spacings resulting in an equivalent cross-sectional area of reinforcement per linear foot of wall are permitted.

f. Concrete cover for reinforcement measured from the inside face of the wall shall not be less than $\frac{3}{4}$ inch. Concrete cover for reinforcement measured from the outside face of the wall shall not be less than $1\frac{1}{2}$ inches for No. 5 bars and smaller, and not less than 2 inches for larger bars.

g. The minimum thickness is permitted to be reduced 2 inches, provided the minimum specified compressive strength of concrete f_c', is 4,000 psi.

(continued)

TABLE R404.1.1(5)—continued
CONCRETE FOUNDATION WALLS[h, i, j, k]

h. A plain concrete wall with a minimum thickness of 11.5 inches is permitted, provided minimum specified compressive strength of concrete, f_c', is 3,500 psi.

i. Concrete shall have a specified compressive strength of not less than 2,500 psi at 28 days, unless a higher strength is required by note g or h.

j. "DR" means design is required in accordance with ACI 318 or ACI 332.

k. "PC" means plain concrete.

l. Where vertical reinforcement is required, horizontal reinforcement shall be provided in accordance with the requirements of Section R404.4.6.2 for ICF foundation walls.

grade wall immediately above or the requirements in Tables R404.4(1), R404.4(2), R404.4(3), R404.4(4) or R404.4(5), whichever is greater.

R404.4.2 Flat insulating concrete form wall systems. Flat ICF wall systems shall comply with Figure R611.3, shall have a minimum concrete thickness of 5.5 inches (140 mm), and shall have reinforcement in accordance with Table R404.4(1), R404.4(2) or R404.4(3). Alternatively, for 7.5-inch (191 mm) and 9.5-inch (241 mm) flat ICF wall systems, use of Table R404.1.1(5) shall be permitted, provided the vertical reinforcement is of the grade and located within the wall as required by that table.

R404.4.3 Waffle-grid insulating concrete form wall systems. Waffle-grid wall systems shall have a minimum nominal concrete thickness of 6 inches (152 mm) for the horizontal and vertical concrete members (cores) and shall be reinforced in accordance with Table R404.4(4). The minimum core dimension shall comply with Table R611.2 and Figure R611.4.

R404.4.4 Screen-grid insulating concrete form wall systems. Screen-grid ICF wall systems shall have a minimum nominal concrete thickness of 6 inches (152 mm) for the horizontal and vertical concrete members (cores). The minimum core dimensions shall comply with Table R611.2 and Figure R611.5. Walls shall have reinforcement in accordance with Table R404.4(5).

R404.4.5 Concrete material. Ready-mixed concrete for insulating concrete form walls shall be in accordance with Section R402.2. Maximum slump shall not be greater than 6 inches (152 mm) as determined in accordance with ASTM C 143. Maximum aggregate size shall not be larger than $^3/_4$ inch (19.1 mm).

Exception: Concrete mixes conforming to the ICF manufacturer's recommendations.

R404.4.6 Reinforcing steel.

R404.4.6.1 General. Reinforcing steel shall meet the requirements of ASTM A 615, A 706 or A 996. The minimum yield strength of reinforcing steel shall be 40,000 psi (Grade 40) (276 MPa). Vertical and horizontal wall reinforcements shall be placed no closer to the outside face of the wall than one-half the wall thickness. Steel reinforcement for foundation walls shall have concrete cover in accordance with ACI 318.

Exception: Where insulated concrete forms are used and the form remains in place as cover for the concrete, the minimum concrete cover for the reinforcing steel is permitted to be reduced to $^3/_4$ inch (19.1 mm).

R404.4.6.2 Horizontal reinforcement. When vertical reinforcement is required, ICF foundation walls shall have horizontal reinforcement in accordance with this section. ICF foundation walls up to 8 feet (2438 mm) in height shall have a minimum of one continuous No. 4 horizontal reinforcing bar placed at 48 inches (1219 mm) on center with one bar located within 12 inches (305 mm) of the top of the wall story. ICF Foundation walls greater than 8 feet (2438 mm) in height shall have a minimum of one continuous No. 4 horizontal reinforcing bar placed at 36 inches (914 mm) on center with one bar located within 12 inches (305 mm) of the top of the wall story.

R404.4.6.3 Wall openings. Vertical wall reinforcement required by Section R404.4.2, R404.4.3 or R404.4.4 that is interrupted by wall openings shall have additional vertical reinforcement of the same size placed within 12 inches (305 mm) of each side of the opening.

R404.4.7 Foam plastic insulation. Foam plastic insulation in insulating concrete foam construction shall comply with this section.

R404.4.7.1 Material. Insulating concrete form material shall meet the surface burning characteristics of Section R314.3. A thermal barrier shall be provided on the building interior in accordance with Section R314.4.

R404.4.7.2 Termite hazards. In areas where hazard of termite damage is very heavy in accordance with Figure R301.2(6), foam plastic insulation shall be permitted below grade on foundation walls in accordance with one of the following conditions:

1. When in addition to the requirements in Section R320.1, an approved method of protecting the foam plastic and structure from subterranean termite damage is provided.

2. The structural members of walls, floors, ceilings and roofs are entirely of noncombustible materials or pressure preservatively treated wood.

3. On the interior side of basement walls.

R404.4.8 Foundation wall thickness based on walls supported. The thickness of ICF foundation walls shall not be less than the thickness of the wall supported above.

R404.4.9 Height above finished ground. ICF foundation walls shall extend above the finished ground adjacent to the foundation at all points a minimum of 4 inches (102 mm) where masonry veneer is used and a minimum of 6 inches (152 mm) elsewhere.

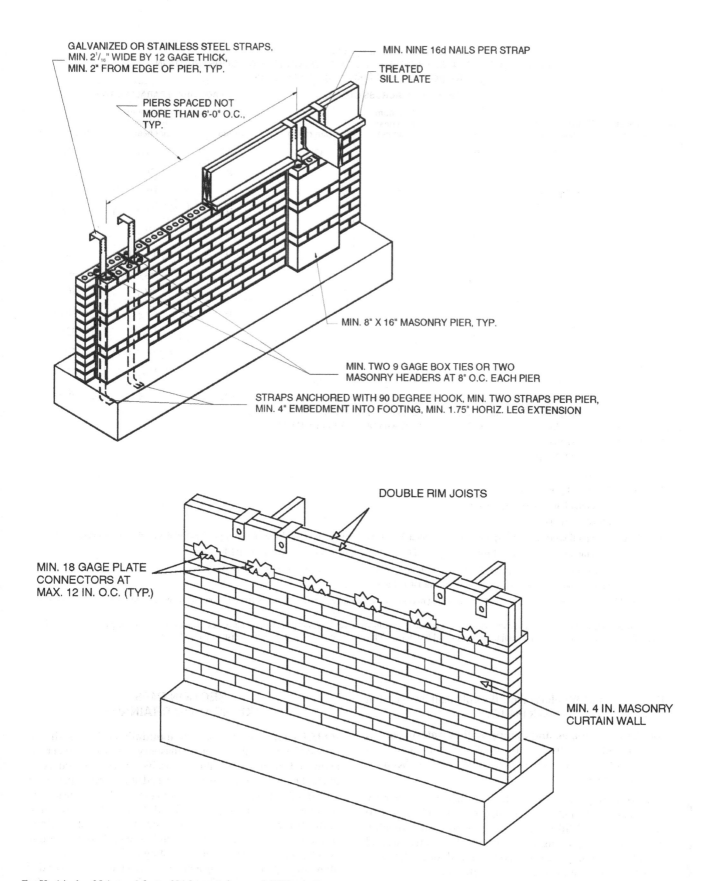

GALVANIZED OR STAINLESS STEEL STRAPS, MIN. 2¹/₁₆" WIDE BY 12 GAGE THICK, MIN. 2" FROM EDGE OF PIER, TYP.

MIN. NINE 16d NAILS PER STRAP

TREATED SILL PLATE

PIERS SPACED NOT MORE THAN 6'-0" O.C., TYP.

MIN. 8" X 16" MASONRY PIER, TYP.

MIN. TWO 9 GAGE BOX TIES OR TWO MASONRY HEADERS AT 8" O.C. EACH PIER

STRAPS ANCHORED WITH 90 DEGREE HOOK, MIN. TWO STRAPS PER PIER, MIN. 4" EMBEDMENT INTO FOOTING, MIN. 1.75" HORIZ. LEG EXTENSION

DOUBLE RIM JOISTS

MIN. 18 GAGE PLATE CONNECTORS AT MAX. 12 IN. O.C. (TYP.)

MIN. 4 IN. MASONRY CURTAIN WALL

For SI: 1 inch = 25.4 mm, 1 foot = 304.8 mm, 1 degree = 0.0157 rad.

FIGURE R404.1.5(1)
FOUNDATION WALL CLAY MASONRY CURTAIN WALL WITH CONCRETE MASONRY PIERS

TABLE R404.2.3
PLYWOOD GRADE AND THICKNESS FOR WOOD FOUNDATION CONSTRUCTION
(30 pcf equivalent-fluid weight soil pressure)

HEIGHT OF FILL (inches)	STUD SPACING (inches)	FACE GRAIN ACROSS STUDS			FACE GRAIN PARALLEL TO STUDS		
		Grade[a]	Minimum thickness (inches)	Span rating	Grade[a]	Minimum thickness (inches)[b,c]	Span rating
24	12	B	$^{15}/_{32}$	32/16	A	$^{15}/_{32}$	32/16
					B	$^{15}/_{32}{}^{c}$	32/16
	16	B	$^{15}/_{32}$	32/16	A	$^{15}/_{32}{}^{c}$	32/16
					B	$^{19}/_{32}{}^{c}$ (4, 5 ply)	40/20
36	12	B	$^{15}/_{32}$	32/16	A	$^{15}/_{32}$	32/16
					B	$^{15}/_{32}{}^{c}$ (4, 5 ply)	32/16
					B	$^{19}/_{32}$ (4, 5 ply)	40/20
	16	B	$^{15}/_{32}{}^{c}$	32/16	A	$^{19}/_{32}$	40/20
					B	$^{23}/_{32}$	48/24
48	12	B	$^{15}/_{32}$	32/16	A	$^{15}/_{32}{}^{c}$	32/16
					B	$^{19}/_{32}{}^{c}$ (4, 5 ply)	40/20
	16	B	$^{19}/_{32}$	40/20	A	$^{19}/_{32}{}^{c}$	40/20
					A	$^{23}/_{32}$	48/24

For SI: 1 inch = 25.4 mm, 1 foot = 304.8 mm, 1 pound per cubic foot = 0.1572 kN/m³.

a. Plywood shall be of the following minimum grades in accordance with DOC PS 1 or DOC PS 2:

   1. DOC PS 1 Plywood grades marked:

      1.1. Structural I C-D (Exposure 1)

      1.2. C-D (Exposure 1)

   2. DOC PS 2 Plywood grades marked:

      2.1. Structural I Sheathing (Exposure 1)

      2.2. Sheathing (Exposure 1)

   3. Where a major portion of the wall is exposed above ground and a better appearance is desired, the following plywood grades marked exterior are suitable:

      3.1. Structural I A-C, Structural I B-C or Structural I C-C (Plugged) in accordance with DOC PS 1

      3.2. A-C Group 1, B-C Group 1, C-C (Plugged) Group 1 or MDO Group 1 in accordance with DOC PS 1

      3.3. Single Floor in accordance with DOC PS 1 or DOC PS 2

b. Minimum thickness $^{15}/_{32}$ inch, except crawl space sheathing may be $^3/_8$ inch for face grain across studs 16 inches on center and maximum 2-foot depth of unequal fill.

c. For this fill height, thickness and grade combination, panels that are continuous over less than three spans (across less than three stud spacings) require blocking 16 inches above the bottom plate. Offset adjacent blocks and fasten through studs with two 16d corrosion-resistant nails at each end.

R404.4.10 Backfill placement. Backfill shall be placed in accordance with Section R404.1.7.

R404.4.11 Drainage and dampproofing/waterproofing. ICF foundation basements shall be drained and dampproofed/waterproofed in accordance with Sections R405 and R406.

R404.5 Retaining walls. Retaining walls that are not laterally supported at the top and that retain in excess of 24 inches (610 mm) of unbalanced fill shall be designed to ensure stability against overturning, sliding, excessive foundation pressure and water uplift. Retaining walls shall be designed for a safety factor of 1.5 against lateral sliding and overturning.

SECTION R405
FOUNDATION DRAINAGE

R405.1 Concrete or masonry foundations. Drains shall be provided around all concrete or masonry foundations that retain earth and enclose habitable or usable spaces located below grade. Drainage tiles, gravel or crushed stone drains, perforated pipe or other approved systems or materials shall be installed at or below the area to be protected and shall discharge by gravity or mechanical means into an approved drainage system. Gravel or crushed stone drains shall extend at least 1 foot (305 mm) beyond the outside edge of the footing and 6 inches (152 mm) above the top of the footing and be covered with an approved filter membrane material. The top of open joints of drain tiles shall be protected with strips of building paper, and the drainage tiles or perforated pipe shall be placed on a minimum of 2 inches (51 mm) of washed gravel or crushed rock at least one sieve size

TABLE R404.4(1)
5.5-INCH THICK FLAT ICF FOUNDATION WALLS[a, b, c, d]

HEIGHT OF BASEMENT WALL (feet)	MAXIMUM UNBALANCED BACKFILL HEIGHT[e] (feet)	MINIMUM VERTICAL REINFORCEMENT SIZE AND SPACING		
		Soil classes[f] and design lateral soil load (psf per foot of depth)		
		GW, GP, SW and SP 30	GM, GC, SM, SM-SC and ML 45	SC, ML-CL and inorganic CL 60
8	4	#4@48″	#4@48″	#4@48″
	5	#4@48″	#3@12″; #4@22″; #5@32″	#3@8″; #4@14″; #5@20″; #6@26″
	6	#3@12″; #4@22″; #5@30″	#3@8″; #4@14″; #5@20″; #6@24″	#3@6″; #4@10″: #5@14″; #6@20″
	7	#3@8″; #4@14″; #5@22″; #6@26″	#3@5″; #4@10″; #5@14″; #6@18″	#3@4″; #4@6″; #5@10″; #6@14″
9	4	#4@48″	#4@48″	#4@48″
	5	#4@48″	#3@12″; #4@20″; #5@28″; #6@36″	#3@8″; #4@14″; #5@20″; #6@22″
	6	#3@10″; #4@20″; #5@28″; #6@34″	#3@6″; #4@12″; #5@18″; #6@20″	#4@8″; #5@14″; #6@16″
	7	#3@8″; #4@14″; #5@20″; #6@22″	#4@8″; #5@12″; #6@16″	#4@6″; #5@10″; #6@12″
	8	#3@6″; #4@10″; #5@14″; #6@16″	#4@6″; #5@10″; #6@12″	#4@4″; #5@6″; #6@8″
10	4	#4@48″	#4@48″	#4@48″
	5	#4@48″	#3@10″; #4@18″; #5@26″; #6@30″	#3@6″; #4@14″; #5@18″; #6@20″
	6	#3@10″; #4@18″; #5@24″; #6@30″	#3@6″; #4@12″; #5@16″; #6@18″	#3@4″; #4@8″; #5@12″; #6@14″
	7	#3@6″; #4@12″; #5@16″; #6@18″	#3@4″; #4@8″; #5@12″	#4@6″; #5@8″; #6@10″
	8	#4@8″; #5@12″; #6@14″	#4@6″; #5@8″; #6@12″	#4@4″; #5@6″; #6@8″
	9	#4@6″; #5@10″; #6@12″	#4@4″; #5@6″; #6@8″	#5@4″; #6@6″

For SI: 1 inch = 25.4 mm, 1 foot = 304.8 mm, 1 pound per square inch = 6.895 kPa, 1 pound per square foot = 0.0479 kPa.

a. This table is based on concrete with a minimum specified concrete strength of 2500 psi, reinforcing steel with a minimum yield strength of 40,000 psi. When reinforcing steel with a minimum yield strength of 60,000 psi is used, the spacing of the reinforcement shall be increased to 1.5 times the spacing value in the table but in no case greater than 48 inches on center.

b. This table is not intended to prohibit the use of an ICF manufacturer's tables based on engineering analysis in accordance with ACI 318.

c. Deflection criteria: $L/240$.

d. Interpolation between rebar sizes and spacing is not permitted.

e. Unbalanced backfill height is the difference in height of the exterior and interior finished ground. Where an interior concrete slab is provided, the unbalanced backfill height shall be measured from the exterior finished ground level to the top of the interior concrete slab.

f. Soil classes are in accordance with the Unified Soil Classification System. Refer to Table R405.1.

larger than the tile joint opening or perforation and covered with not less than 6 inches (152 mm) of the same material.

Exception: A drainage system is not required when the foundation is installed on well-drained ground or sand-gravel mixture soils according to the Unified Soil Classification System, Group I Soils, as detailed in Table R405.1.

R405.2 Wood foundations. Wood foundations enclosing habitable or usable spaces located below grade shall be adequately drained in accordance with Sections R405.2.1 through R405.2.3.

R405.2.1 Base. A porous layer of gravel, crushed stone or coarse sand shall be placed to a minimum thickness of 4 inches (102 mm) under the basement floor. Provision shall be made for automatic draining of this layer and the gravel or crushed stone wall footings.

R405.2.2 Moisture barrier. A 6-mil-thick (0.15 mm) polyethylene moisture barrier shall be applied over the porous layer with the basement floor constructed over the polyethylene.

R405.2.3 Drainage system. In other than Group I soils, a sump shall be provided to drain the porous layer and footings.

TABLE R404.4(2)
7.5-INCH-THICK FLAT ICF FOUNDATION WALLS[a, b, c, d, e]

HEIGHT OF BASEMENT WALL (feet)	MAXIMUM UNBALANCED BACKFILL HEIGHT[f] (feet)	MINIMUM VERTICAL REINFORCEMENT SIZE AND SPACING		
		Soil classes[g] and design lateral soil load (psf per foot of depth)		
		GW, GP, SW and SP 30	GM, GC, SM, SM-SC and ML 45	SC, ML-CL and inorganic CL 60
8	6	N/R	N/R	#3@6"; #4@12"; #5@18"; #6@24"
	7	N/R	#3@8"; #4@14"; #5@20"; #6@28"	#3@6"; #4@10"; #5@16"; #6@20"
9	6	N/R	N/R	#3@8"; #4@14"; #5@20"; #6@28"
	7	N/R	#3@6"; #4@12"; #5@18"; #6@26"	#3@4"; #4@8"; #5@14"; #6@18"
	8	#3@8"; #4@14"; #5@22"; #6@28"	#3@4"; #4@8"; #5@14"; #6@18"	#3@4"; #4@6"; #5@10"; #6@14"
10	6	N/R	N/R	#3@6"; #4@12"; #5@18"; #6@26"
	7	N/R	#3@6"; #4@12"; #5@18"; #6@24"	#3@4"; #4@8"; #5@12"; #6@18"
	8	#3@6"; #4@12"; #5@20"; #6@26"	#3@4"; #4@8"; #5@12"; #6@16"	#3@4"; #4@6"; #5@8"; #6@12"
	9	#3@6"; #4@10"; #5@14"; #6@20"	#3@4"; #4@6"; #5@10"; #6@12"	#4@4"; #5@6"; #6@10"

For SI: 1 inch = 25.4 mm, 1 foot = 304.8 mm, 1 pound per square inch = 6.895 kPa, 1 pound per square foot = 0.0479 kPa.

a. This table is based on concrete with a minimum specified concrete strength of 2500 psi, reinforcing steel with a minimum yield strength of 40,000 psi. When reinforcing steel with a minimum yield strength of 60,000 psi is used, the spacing of the reinforcement shall be increased to 1.5 times the spacing value in the table.

b. This table is not intended to prohibit the use of an ICF manufacturer's tables based on engineering analysis in accordance with ACI 318.

c. N/R denotes "not required."

d. Deflection criteria: L/240.

e. Interpolation between rebar sizes and spacing is not permitted.

f. Unbalanced backfill height is the difference in height of the exterior and interior finished ground. Where an interior concrete slab is provided, the unbalanced backfill height shall be measured from the exterior finished ground level to the top of the interior concrete slab.

The sump shall be at least 24 inches (610 mm) in diameter or 20 inches square (0.0129 m²), shall extend at least 24 inches (610 mm) below the bottom of the basement floor and shall be capable of positive gravity or mechanical drainage to remove any accumulated water. The drainage system shall discharge into an approved sewer system or to daylight.

SECTION R406
FOUNDATION WATERPROOFING AND DAMPPROOFING

R406.1 Concrete and masonry foundation dampproofing.
Except where required by Section R406.2 to be waterproofed, foundation walls that retain earth and enclose interior spaces and floors below grade shall be dampproofed from the top of the footing to the finished grade. Masonry walls shall have not less than 3/8 inch (9.5 mm) portland cement parging applied to the exterior of the wall. The parging shall be dampproofed in accordance with one of the following:

1. Bituminous coating.

2. 3 pounds per square yard (1.63 kg/m²) of acrylic modified cement.

3. 1/8-inch (3.2 mm) coat of surface-bonding cement complying with ASTM C 887.

4. Any material permitted for waterproofing in Section R406.2.

5. Other approved methods or materials.

   Exception: Parging of unit masonry walls is not required where a material is approved for direct application to the masonry.

Concrete walls shall be dampproofed by applying any one of the above listed dampproofing materials or any one of the waterproofing materials listed in Section R406.2 to the exterior of the wall.

R406.2 Concrete and masonry foundation waterproofing.
In areas where a high water table or other severe soil-water conditions are known to exist, exterior foundation walls that retain earth and enclose interior spaces and floors below grade shall be waterproofed from the top of the footing to the finished grade. Walls shall be waterproofed in accordance with one of the following:

1. 2-ply hot-mopped felts.

2. 55 pound (25 kg) roll roofing.

3. 6-mil (0.15 mm) polyvinyl chloride.

4. 6-mil (0.15 mm) polyethylene.

5. 40-mil (1 mm) polymer-modified asphalt.

TABLE R404.4(3)
9.5-INCH-THICK FLAT ICF FOUNDATION WALLS[a, b, c, d, e]

HEIGHT OF BASEMENT WALL (feet)	MAXIMUM UNBALANCED BACKFILL HEIGHT[f] (feet)	MINIMUM VERTICAL REINFORCEMENT SIZE AND SPACING		
		Soil classes[g] and design lateral soil load (psf per foot of depth)		
		GW, GP, SW and SP 30	GM, GC, SM, SM-SC and ML 45	SC, ML-CL and inorganic CL 60
8	7	N/R	N/R	N/R
9	6	N/R	N/R	N/R
	7	N/R	N/R	#3@6"; #4@12"; #5@18"; #6@26"
	8	N/R	#3@6"; #4@12"; #5@18"; #6@26"	#3@4"; #4@8"; #5@14"; #6@18"
10	5	N/R	N/R	N/R
	6	N/R	N/R	N/R
	7	N/R	N/R	#3@6"; #4@10"; #5@18"; #6@24"
	8	N/R	#3@6"; #4@12"; #5@16"; #6@24"	#3@4"; #4@8"; #5@12"; #6@16"
	9	#3@4"; #4@10"; #5@14"; #6@20"	#3@4"; #4@8"; #5@12"; #6@18"	#3@4"; #4@6"; #5@10"; #6@12"

For SI: 1 inch = 25.4 mm, 1 foot = 304.8 mm, 1 pound per square inch = 6.895 kPa, 1 pound per square foot = 0.0479 kPa.

a. This table is based on concrete with a minimum specified concrete strength of 2500 psi, reinforcing steel with a minimum yield strength of 40,000 psi. When reinforcing steel with a minimum yield strength of 60,000 psi is used, the spacing of the reinforcement shall be increased to 1.5 times the spacing value in the table.

b. This table is not intended to prohibit the use of an ICF manufacturer's tables based on engineering analysis in accordance with ACI 318.

c. N/R denotes "not required."

d. Deflection criteria: L/240.

e. Interpolation between rebar sizes and spacing is not permitted.

f. Unbalanced backfill height is the difference in height of the exterior and interior finished ground. Where an interior concrete slab is provided, the unbalanced backfill height shall be measured from the exterior finished ground level to the top of the interior concrete slab.

g. Soil classes are in accordance with the Unified Soil Classification System. Refer to Table R405.1.

6. 60-mil (1.5 mm) flexible polymer cement.

7. $^{1}/_{8}$ inch (3 mm) cement-based, fiber-reinforced, waterproof coating.

8. 60-mil (0.22 mm) solvent-free liquid-applied synthetic rubber.

Exception: Organic-solvent-based products such as hydrocarbons, chlorinated hydrocarbons, ketones and esters shall not be used for ICF walls with expanded polystyrene form material. Use of plastic roofing cements, acrylic coatings, latex coatings, mortars and pargings to seal ICF walls is permitted. Cold-setting asphalt or hot asphalt shall conform to type C of ASTM D 449. Hot asphalt shall be applied at a temperature of less than 200°F (93°C).

All joints in membrane waterproofing shall be lapped and sealed with an adhesive compatible with the membrane.

R406.3 Dampproofing for wood foundations. Wood foundations enclosing habitable or usable spaces located below grade shall be dampproofed in accordance with Sections R406.3.1 through R406.3.4.

R406.3.1 Panel joint sealed. Plywood panel joints in the foundation walls shall be sealed full length with a caulking compound capable of producing a moisture-proof seal under the conditions of temperature and moisture content at which it will be applied and used.

R406.3.2 Below-grade moisture barrier. A 6-mil-thick (0.15 mm) polyethylene film shall be applied over the below-grade portion of exterior foundation walls prior to backfilling. Joints in the polyethylene film shall be lapped 6 inches (152 mm) and sealed with adhesive. The top edge of the polyethylene film shall be bonded to the sheathing to form a seal. Film areas at grade level shall be protected from mechanical damage and exposure by a pressure preservatively treated lumber or plywood strip attached to the wall several inches above finish grade level and extending approximately 9 inches (229 mm) below grade. The joint between the strip and the wall shall be caulked full length prior to fastening the strip to the wall. Other coverings appropriate to the architectural treatment may also be used. The polyethylene film shall extend down to the bottom of the wood footing plate but shall not overlap or extend into the gravel or crushed stone footing.

R406.3.3 Porous fill. The space between the excavation and the foundation wall shall be backfilled with the same material used for footings, up to a height of 1 foot (305 mm) above the footing for well-drained sites, or one-half the total back-fill height for poorly drained sites. The porous fill shall be covered with strips of 30-pound (13.6 kg) asphalt paper or 6-mil (0.15 mm) polyethylene to permit water seepage while avoiding infiltration of fine soils.

TABLE R404.4(4)
WAFFLE GRID ICF FOUNDATION WALLS[a, b, c, d, e]

MINIMUM NOMINAL WALL THICKNESS[f] (inches)	HEIGHT OF BASEMENT WALL (feet)	MAXIMUM UNBALANCED BACKFILL HEIGHT[g] (feet)	MINIMUM VERTICAL REINFORCEMENT SIZE AND SPACING		
			Soil classes[h] and design lateral soil load (psf per foot of depth)		
			GW, GP, SW and SP 30	GM, GC, SM, SM-SC and ML 45	SC, ML-CL and inorganic CL 60
6	8	4	#4@48″	#3@12″; #4@24″	#3@12″
		5	#3@12″; #5@24″	#4@12″	#7@12″
		6	#4@12″	Design required	Design required
		7	#7@12″	Design required	Design required
	9	4	#4@48″	#3@12″; #5@24″	#3@12″
		5	#3@12″	#4@12″	Design required
		6	#5@12″	Design required	Design required
		7	Design required	Design required	Design required
	10	4	#4@48″	#4@12″	#5@12″
		5	#3@12″	Design required	Design required
		6	Design required	Design required	Design required
		7	Design required	Design required	Design required
8	8	4	N/R	N/R	N/R
		5	N/R	#3@12″; #4@24″; #5@36″	#3@12″; #5@24″
		6	#3@12″; #4@24″; #5@36″	#4@12″; #5@24″	#4@12″
		7	#3@12″; #6@24″	#4@12″	#5@12″
	9	4	N/R	N/R	N/R
		5	N/R	#3@12″; #5@24″	#3@12″; #5@24″
		6	#3@12″; #4@24″	#4@12″	#4@12″
		7	#4@12″; #5@24″	#5@12″	#5@12″
		8	#4@12″	#5@12″	#8@12″
	10	4	N/R	#3@12″; #4@24″; #6@36″	#3@12″; #5@24″
		5	N/R	#3@12″; #4@24″; #6@36″	#4@12″; #5@24″
		6	#3@12″; #5@24″	#4@12″	#5@12″
		7	#4@12″	#5@12″	#6@12″
		8	#4@12″	#6@12″	Design required
		9	#5@12″	Design required	Design required

For SI: 1 inch = 25.4 mm, 1 foot = 304.8 mm, 1 pound per square inch = 6.895 kPa, 1 pound per square foot = 0.0479 kPa.

a. This table is based on concrete with a minimum specified concrete strength of 2500 psi, reinforcing steel with a minimum yield strength of 40,000 psi. When reinforcing steel with a minimum yield strength of 60,000 psi is used, the spacing of the reinforcement shall be increased 12 inches but in no case greater than 48 inches on center.

b. This table is not intended to prohibit the use of an ICF manufacturer's tables based on engineering analysis in accordance with ACI 318.

c. N/R denotes "not required."

d. Deflection criteria: $L/240$.

e. Interpolation between rebar sizes and spacing is not permitted.

f. Refer to Table R611.4(2) for wall dimensions.

g. Unbalanced backfill height is the difference in height of the exterior and interior finished ground. Where an interior concrete slab is provided, the unbalanced backfill height shall be measured from the exterior finished ground level to the top of the interior concrete slab.

h. Soil classes are in accordance with the Unified Soil Classification System. Refer to Table R405.1.

TABLE R404.4(5)
SCREEN-GRID ICF FOUNDATION WALLS[a, b c, d, e]

MINIMUM NOMINAL WALL THICKNESS[f] (inches)	HEIGHT OF BASEMENT WALL (feet)	MAXIMUM UNBALANCED BACKFILL HEIGHT[g] (feet)	MINIMUM VERTICAL REINFORCEMENT SIZE AND SPACING		
			Soil classes[h] and design lateral soil load (psf per foot of depth)		
			GW, GP, SW and SP 30	GM, GC, SM, SM-SC and ML 45	SC, ML-CL and inorganic CL 60
6	8	4	#4@48″	#3@12″; #4@24″; #5@36″	#3@12″; #5@24″
		5	#3@12″; #4@24″	#3@12″	#4@12″
		6	#4@12″	#5@12″	Design required
		7	#4@12″	Design required	Design required
	9	4	#4@48″	#3@12″; #4@24″	#3@12″; #6@24″
		5	#3@12″; #5@24″	#4@12″	#7@12″
		6	#4@12″	Design required	Design required
		7	Design required	Design required	Design required
		8	Design required	Design required	Design required
	10	4	#4@48″	#3@12″; #5@24″	#3@12″
		5	#3@12″	#4@12″	#7@12″
		6	#4@12″	Design required	Design required
		7	Design required	Design required	Design required
		8	Design required	Design required	Design required

For SI: 1 inch = 25.4 mm, 1 foot = 304.8 mm, 1 pound per square inch = 6.895 kPa, 1 pound per square foot = 0.0479 kPa.

a. This table is based on concrete with a minimum specified concrete strength of 2500 psi, reinforcing steel with a minimum yield strength of 40,000 psi. When reinforcing steel with a minimum yield strength of 60,000 psi is used, the spacing of the reinforcement in the shaded cells shall be increased 12 inches.

b. This table is not intended to prohibit the use of an ICF manufacturer's tables based on engineering analysis in accordance with ACI 318.

c. N/R denotes "not required."

d. Deflection criteria: L/240.

e. Interpolation between rebar sizes and spacing is not permitted.

f. Refer to Table R611.4(2) for wall dimensions.

g. Unbalanced backfill height is the difference in height of the exterior and interior finished ground. Where an interior concrete slab is provided, the unbalanced backfill height shall be measured from the exterior finished ground level to the top of the interior concrete slab.

h. Soil classes are in accordance with the Unified Soil Classification System. Refer to Table R405.1.

R406.3.4 Backfill. The remainder of the excavated area shall be backfilled with the same type of soil as was removed during the excavation.

SECTION R407
COLUMNS

R407.1 Wood column protection. Wood columns shall be protected against decay as set forth in Section R319.

R407.2 Steel column protection. All surfaces (inside and outside) of steel columns shall be given a shop coat of rust-inhibitive paint, except for corrosion-resistant steel and steel treated with coatings to provide corrosion resistance.

R407.3 Structural requirements. The columns shall be restrained to prevent lateral displacement at the bottom end. Wood columns shall not be less in nominal size than 4 inches by 4 inches (102 mm by 102 mm) and steel columns shall not be less than 3-inch-diameter (76 mm) standard pipe or approved equivalent.

Exception: In Seismic Design Categories A, B and C columns no more than 48 inches (1219 mm) in height on a pier or footing are exempt from the bottom end lateral displacement requirement within underfloor areas enclosed by a continuous foundation.

SECTION R408
UNDER-FLOOR SPACE

R408.1 Ventilation. The under-floor space between the bottom of the floor joists and the earth under any building (except space occupied by a basement) shall have ventilation openings through foundation walls or exterior walls. The minimum net area of ventilation openings shall not be less than 1 square foot (0.0929 m²) for each 150 square feet (14 m²) of under-floor space area. One such ventilating opening shall be within 3 feet (914 mm) of each corner of the building.

R408.2 Openings for under-floor ventilation. The minimum net area of ventilation openings shall not be less than 1 square foot (0.0929 m²) for each 150 square feet (14 m²) of under-floor area. One ventilating opening shall be within 3 feet (914 mm) of each corner of the building. Ventilation openings shall be covered for their height and width with any of the following

TABLE R405.1
PROPERTIES OF SOILS CLASSIFIED ACCORDING TO THE UNIFIED SOIL CLASSIFICATION SYSTEM

SOIL GROUP	UNIFIED SOIL CLASSIFICATION SYSTEM SYMBOL	SOIL DESCRIPTION	DRAINAGE CHARACTERISTICS[a]	FROST HEAVE POTENTIAL	VOLUME CHANGE POTENTIAL EXPANSION[b]
Group I	GW	Well-graded gravels, gravel sand mixtures, little or no fines	Good	Low	Low
	GP	Poorly graded gravels or gravel sand mixtures, little or no fines	Good	Low	Low
	SW	Well-graded sands, gravelly sands, little or no fines	Good	Low	Low
	SP	Poorly graded sands or gravelly sands, little or no fines	Good	Low	Low
	GM	Silty gravels, gravel-sand-silt mixtures	Good	Medium	Low
	SM	Silty sand, sand-silt mixtures	Good	Medium	Low
Group II	GC	Clayey gravels, gravel-sand-clay mixtures	Medium	Medium	Low
	SC	Clayey sands, sand-clay mixture	Medium	Medium	Low
	ML	Inorganic silts and very fine sands, rock flour, silty or clayey fine sands or clayey silts with slight plasticity	Medium	High	Low
	CL	Inorganic clays of low to medium plasticity, gravelly clays, sandy clays, silty clays, lean clays	Medium	Medium	Medium to Low
Group III	CH	Inorganic clays of high plasticity, fat clays	Poor	Medium	High
	MH	Inorganic silts, micaceous or diatomaceous fine sandy or silty soils, elastic silts	Poor	High	High
Group IV	OL	Organic silts and organic silty clays of low plasticity	Poor	Medium	Medium
	OH	Organic clays of medium to high plasticity, organic silts	Unsatisfactory	Medium	High
	Pt	Peat and other highly organic soils	Unsatisfactory	Medium	High

For SI: 1 inch = 25.4 mm.

a. The percolation rate for good drainage is over 4 inches per hour, medium drainage is 2 inches to 4 inches per hour, and poor is less than 2 inches per hour.

b. Soils with a low potential expansion typically have a plasticity index (PI) of 0 to 15, soils with a medium potential expansion have a PI of 10 to 35 and soils with a high potential expansion have a PI greater than 20.

materials provided that the least dimension of the covering shall not exceed $^1/_4$ inch (6.4 mm):

1. Perforated sheet metal plates not less than 0.070 inch (1.8 mm) thick.

2. Expanded sheet metal plates not less than 0.047 inch (1.2 mm) thick.

3. Cast-iron grill or grating.

4. Extruded load-bearing brick vents.

5. Hardware cloth of 0.035 inch (0.89 mm) wire or heavier.

6. Corrosion-resistant wire mesh, with the least dimension being $^1/_8$ inch (3.2 mm).

R408.3 Unvented crawl space. Ventilation openings in under-floor spaces specified in Sections R408.1 and R408.2 shall not be required where:

1. Exposed earth is covered with a continuous vapor retarder. Joints of the vapor retarder shall overlap by 6 inches (152 mm) and shall be sealed or taped. The edges of the vapor retarder shall extend at least 6 inches (152 mm) up the stem wall and shall be attached and sealed to the stem wall; and

2. One of the following is provided for the under-floor space:

    2.1. Continuously operated mechanical exhaust ventilation at a rate equal to 1 cfm (0.47 L/s) for each 50 ft² (4.7 m²) of crawlspace floor area, including an air pathway to the common area (such as a duct or transfer grille), and perimeter walls insulated in accordance with Section N1102.2.8;

    2.2. Conditioned air supply sized to deliver at a rate equal to 1 cfm (0.47 L/s) for each 50 ft² (4.7 m²) of under-floor area, including a return air pathway to the common area (such as a duct or transfer grille), and perimeter walls insulated in accordance with Section N1102.2.8;

2.3. Plenum complying with Section M1601.4, if under-floor space is used as a plenum.

R408.4 Access. Access shall be provided to all under-floor spaces. Access openings through the floor shall be a minimum of 18 inches by 24 inches (457 mm by 610 mm). Openings through a perimeter wall shall be not less than 16 inches by 24 inches (407 mm by 610 mm). When any portion of the through-wall access is below grade, an areaway not less than 16 inches by 24 inches (407 mm by 610 mm) shall be provided. The bottom of the areaway shall be below the threshold of the access opening. Through wall access openings shall not be located under a door to the residence. See Section M1305.1.4 for access requirements where mechanical equipment is located under floors.

R408.5 Removal of debris. The under-floor grade shall be cleaned of all vegetation and organic material. All wood forms used for placing concrete shall be removed before a building is occupied or used for any purpose. All construction materials shall be removed before a building is occupied or used for any purpose.

R408.6 Finished grade. The finished grade of under-floor surface may be located at the bottom of the footings; however, where there is evidence that the groundwater table can rise to within 6 inches (152 mm) of the finished floor at the building perimeter or where there is evidence that the surface water does not readily drain from the building site, the grade in the under-floor space shall be as high as the outside finished grade, unless an approved drainage system is provided.

R408.7 Flood resistance. For buildings located in areas prone to flooding as established in Table R301.2(1):

1. Walls enclosing the under-floor space shall be provided with flood openings in accordance with Section R324.2.2.

2. The finished ground level of the under-floor space shall be equal to or higher than the outside finished ground level.

   Exception: Under-floor spaces that meet the requirements of FEMA/FIA TB 11-1.

CHAPTER 5

FLOORS

SECTION R501
GENERAL

R501.1 Application. The provisions of this chapter shall control the design and construction of the floors for all buildings including the floors of attic spaces used to house mechanical or plumbing fixtures and equipment.

R501.2 Requirements. Floor construction shall be capable of accommodating all loads according to Section R301 and of transmitting the resulting loads to the supporting structural elements.

SECTION R502
WOOD FLOOR FRAMING

R502.1 Identification. Load-bearing dimension lumber for joists, beams and girders shall be identified by a grade mark of a lumber grading or inspection agency that has been approved by an accreditation body that complies with DOC PS 20. In lieu of a grade mark, a certificate of inspection issued by a lumber grading or inspection agency meeting the requirements of this section shall be accepted.

R502.1.1 Preservative-treated lumber. Preservative treated dimension lumber shall also be identified as required by Section R319.1.

R502.1.2 Blocking and subflooring. Blocking shall be a minimum of utility grade lumber. Subflooring may be a minimum of utility grade lumber or No. 4 common grade boards.

R502.1.3 End-jointed lumber. Approved end-jointed lumber identified by a grade mark conforming to Section R502.1 may be used interchangeably with solid-sawn members of the same species and grade.

R502.1.4 Prefabricated wood I-joists. Structural capacities and design provisions for prefabricated wood I-joists shall be established and monitored in accordance with ASTM D 5055.

R502.1.5 Structural glued laminated timbers. Glued laminated timbers shall be manufactured and identified as required in AITC A190.1 and ASTM D 3737.

R502.1.6 Structural log members. Stress grading of structural log members of nonrectangular shape, as typically used in log buildings, shall be in accordance with ASTM D 3957. Such structural log members shall be identified by the grade mark of an approved lumber grading or inspection agency. In lieu of a grade mark on the material, a certificate of inspection as to species and grade issued by a lumber-grading or inspection agency meeting the requirements of this section shall be permitted to be accepted.

R502.2 Design and construction. Floors shall be designed and constructed in accordance with the provisions of this chapter, Figure R502.2 and Sections R319 and R320 or in accordance with AF&PA/NDS.

R502.2.1 Framing at braced wall lines. A load path for lateral forces shall be provided between floor framing and braced wall panels located above or below a floor, as specified in Section R602.10.8.

R502.2.2 Decks. Where supported by attachment to an exterior wall, decks shall be positively anchored to the primary structure and designed for both vertical and lateral loads as applicable. Such attachment shall not be accomplished by the use of toenails or nails subject to withdrawal. Where positive connection to the primary building structure cannot be verified during inspection, decks shall be self-supporting. For decks with cantilevered framing members, connections to exterior walls or other framing members, shall be designed and constructed to resist uplift resulting from the full live load specified in Table R301.5 acting on the cantilevered portion of the deck.

R502.3 Allowable joist spans. Spans for floor joists shall be in accordance with Tables R502.3.1(1) and R502.3.1(2). For other grades and species and for other loading conditions, refer to the AF&PA Span Tables for Joists and Rafters.

R502.3.1 Sleeping areas and attic joists. Table R502.3.1(1) shall be used to determine the maximum allowable span of floor joists that support sleeping areas and attics that are accessed by means of a fixed stairway in accordance with Section R311.5 provided that the design live load does not exceed 30 psf (1.44 kPa) and the design dead load does not exceed 20 psf (0.96 kPa). The allowable span of ceiling joists that support attics used for limited storage or no storage shall be determined in accordance with Section R802.4.

R502.3.2 Other floor joists. Table R502.3.1(2) shall be used to determine the maximum allowable span of floor joists that support all other areas of the building, other than sleeping rooms and attics, provided that the design live load does not exceed 40 psf (1.92 kPa) and the design dead load does not exceed 20 psf (0.96 kPa).

R502.3.3 Floor cantilevers. Floor cantilever spans shall not exceed the nominal depth of the wood floor joist. Floor cantilevers constructed in accordance with Table R502.3.3(1) shall be permitted when supporting a light-frame bearing wall and roof only. Floor cantilevers supporting an exterior balcony are permitted to be constructed in accordance with Table R502.3.3(2).

R502.4 Joists under bearing partitions. Joists under parallel bearing partitions shall be of adequate size to support the load. Double joists, sized to adequately support the load, that are separated to permit the installation of piping or vents shall be full depth solid blocked with lumber not less than 2 inches (51 mm) in nominal thickness spaced not more than 4 feet (1219 mm) on center. Bearing partitions perpendicular to joists shall not be offset from supporting girders, walls or partitions more than the joist depth unless such joists are of sufficient size to carry the additional load.

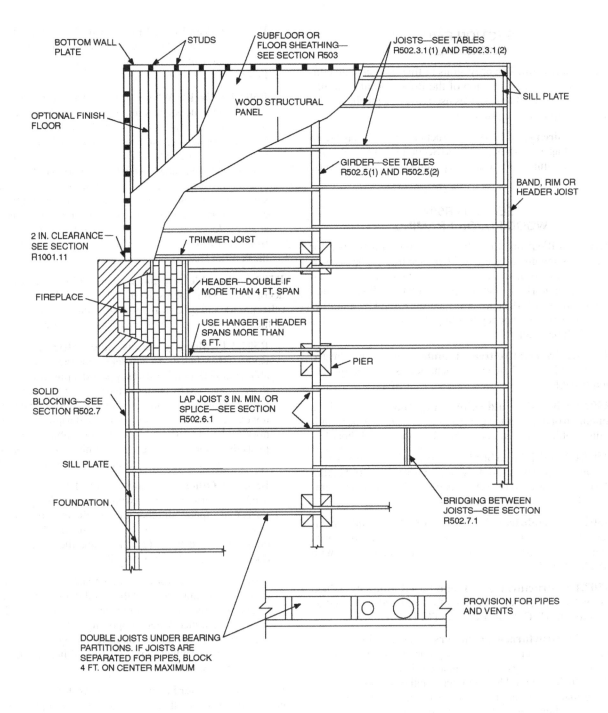

BOTTOM WALL PLATE

STUDS

SUBFLOOR OR FLOOR SHEATHING— SEE SECTION R503

JOISTS—SEE TABLES R502.3.1(1) AND R502.3.1(2)

SILL PLATE

OPTIONAL FINISH FLOOR

WOOD STRUCTURAL PANEL

GIRDER—SEE TABLES R502.5(1) AND R502.5(2)

BAND, RIM OR HEADER JOIST

2 IN. CLEARANCE— SEE SECTION R1001.11

TRIMMER JOIST

FIREPLACE

HEADER—DOUBLE IF MORE THAN 4 FT. SPAN

USE HANGER IF HEADER SPANS MORE THAN 6 FT.

PIER

SOLID BLOCKING—SEE SECTION R502.7

LAP JOIST 3 IN. MIN. OR SPLICE—SEE SECTION R502.6.1

SILL PLATE

FOUNDATION

BRIDGING BETWEEN JOISTS—SEE SECTION R502.7.1

PROVISION FOR PIPES AND VENTS

DOUBLE JOISTS UNDER BEARING PARTITIONS. IF JOISTS ARE SEPARATED FOR PIPES, BLOCK 4 FT. ON CENTER MAXIMUM

For SI: 1 inch = 25.4 mm, 1 foot = 304.8 mm.

FIGURE R502.2
FLOOR CONSTRUCTION

TABLE R502.3.1(1)
FLOOR JOIST SPANS FOR COMMON LUMBER SPECIES
(Residential sleeping areas, live load = 30 psf, L/Δ = 360)[a]

JOIST SPACING (inches)	SPECIES AND GRADE		DEAD LOAD = 10 psf				DEAD LOAD = 20 psf			
			2×6	2×8	2×10	2×12	2×6	2×8	2×10	2×12
						Maximum floor joist spans				
			(ft - in.)	(ft - in.)	(ft - in.)	(ft - in.)	(ft - in.)	(ft - in.)	(ft - in.)	(ft - in.)
12	Douglas fir-larch	SS	12- 6	16- 6	21- 0	25- 7	12- 6	16- 6	21- 0	25- 7
	Douglas fir-larch	#1	12- 0	15-10	20- 3	24- 8	12- 0	15- 7	19- 0	22- 0
	Douglas fir-larch	#2	11-10	15- 7	19-10	23- 0	11- 6	14- 7	17- 9	20- 7
	Douglas fir-larch	#3	9- 8	12- 4	15- 0	17- 5	8- 8	11- 0	13- 5	15- 7
	Hem-fir	SS	11-10	15- 7	19-10	24- 2	11-10	15- 7	19-10	24- 2
	Hem-fir	#1	11- 7	15- 3	19- 5	23- 7	11- 7	15- 2	18- 6	21- 6
	Hem-fir	#2	11- 0	14- 6	18- 6	22- 6	11- 0	14- 4	17- 6	20- 4
	Hem-fir	#3	9- 8	12- 4	15- 0	17- 5	8- 8	11- 0	13- 5	15- 7
	Southern pine	SS	12- 3	16- 2	20- 8	25- 1	12- 3	16- 2	20- 8	25- 1
	Southern pine	#1	12- 0	15-10	20- 3	24- 8	12- 0	15-10	20- 3	24- 8
	Southern pine	#2	11-10	15- 7	19-10	24- 2	11-10	15- 7	18- 7	21- 9
	Southern pine	#3	10- 5	13- 3	15- 8	18- 8	9- 4	11-11	14- 0	16- 8
	Spruce-pine-fir	SS	11- 7	15- 3	19- 5	23- 7	11- 7	15- 3	19- 5	23- 7
	Spruce-pine-fir	#1	11- 3	14-11	19- 0	23- 0	11- 3	14- 7	17- 9	20- 7
	Spruce-pine-fir	#2	11- 3	14-11	19- 0	23- 0	11- 3	14- 7	17- 9	20- 7
	Spruce-pine-fir	#3	9- 8	12- 4	15- 0	17- 5	8- 8	11- 0	13- 5	15- 7
16	Douglas fir-larch	SS	11- 4	15- 0	19- 1	23- 3	11- 4	15- 0	19- 1	23- 0
	Douglas fir-larch	#1	10-11	14- 5	18- 5	21- 4	10- 8	13- 6	16- 5	19- 1
	Douglas fir-larch	#2	10- 9	14- 1	17- 2	19-11	9-11	12- 7	15- 5	17-10
	Douglas fir-larch	#3	8- 5	10- 8	13- 0	15- 1	7- 6	9- 6	11- 8	13- 6
	Hem-fir	SS	10- 9	14- 2	18- 0	21-11	10- 9	14- 2	18- 0	21-11
	Hem-fir	#1	10- 6	13-10	17- 8	20- 9	10- 4	13- 1	16- 0	18- 7
	Hem-fir	#2	10- 0	13- 2	16-10	19- 8	9-10	12- 5	15- 2	17- 7
	Hem-fir	#3	8- 5	10- 8	13- 0	15- 1	7- 6	9- 6	11- 8	13- 6
	Southern pine	SS	11- 2	14- 8	18- 9	22-10	11- 2	14- 8	18- 9	22-10
	Southern pine	#1	10-11	14- 5	18- 5	22- 5	10-11	14- 5	17-11	21- 4
	Southern pine	#2	10- 9	14- 2	18- 0	21- 1	10- 5	13- 6	16- 1	18-10
	Southern pine	#3	9- 0	11- 6	13- 7	16- 2	8- 1	10- 3	12- 2	14- 6
	Spruce-pine-fir	SS	10- 6	13-10	17- 8	21- 6	10- 6	13-10	17- 8	21- 4
	Spruce-pine-fir	#1	10- 3	13- 6	17- 2	19-11	9-11	12- 7	15- 5	17-10
	Spruce-pine-fir	#2	10- 3	13- 6	17- 2	19-11	9-11	12- 7	15- 5	17-10
	Spruce-pine-fir	#3	8- 5	10- 8	13- 0	15- 1	7- 6	9- 6	11- 8	13- 6
19.2	Douglas fir-larch	SS	10- 8	14- 1	18- 0	21-10	10- 8	14- 1	18- 0	21- 0
	Douglas fir-larch	#1	10- 4	13- 7	16- 9	19- 6	9- 8	12- 4	15- 0	17- 5
	Douglas fir-larch	#2	10- 1	12-10	15- 8	18- 3	9- 1	11- 6	14- 1	16- 3
	Douglas fir-larch	#3	7- 8	9- 9	11-10	13- 9	6-10	8- 8	10- 7	12- 4
	Hem-fir	SS	10- 1	13- 4	17- 0	20- 8	10- 1	13- 4	17- 0	20- 7
	Hem-fir	#1	9-10	13- 0	16- 4	19- 0	9- 6	12- 0	14- 8	17- 0
	Hem-fir	#2	9- 5	12- 5	15- 6	17- 1	8-11	11- 4	13-10	16- 1
	Hem-fir	#3	7- 8	9- 9	11- 10	13- 9	6-10	8- 8	10- 7	12- 4
	Southern pine	SS	10- 6	13-10	17- 8	21- 6	10- 6	13-10	17- 8	21- 6
	Southern pine	#1	10- 4	13- 7	17- 4	21- 1	10- 4	13- 7	16- 4	19- 6
	Southern pine	#2	10- 1	13- 4	16- 5	19- 3	9- 6	12- 4	14- 8	17- 2
	Southern pine	#3	8- 3	10- 6	12- 5	14- 9	7- 4	9- 5	11- 1	13- 2
	Spruce-pine-fir	SS	9- 10	13- 0	16- 7	20- 2	9-10	13- 0	16- 7	19- 6
	Spruce-pine-fir	#1	9- 8	12- 9	15- 8	18- 3	9- 1	11- 6	14- 1	16- 3
	Spruce-pine-fir	#2	9- 8	12- 9	15- 8	18- 3	9- 1	11- 6	14- 1	16- 3
	Spruce-pine-fir	#3	7- 8	9- 9	11-10	13- 9	6-10	8- 8	10- 7	12- 4
24	Douglas fir-larch	SS	9-11	13- 1	16- 8	20- 3	9-11	13- 1	16- 2	18- 9
	Douglas fir-larch	#1	9- 7	12- 4	15- 0	17- 5	8- 8	11- 0	13- 5	15- 7
	Douglas fir-larch	#2	9- 1	11- 6	14- 1	16- 3	8- 1	10- 3	12- 7	14- 7
	Douglas fir-larch	#3	6-10	8- 8	10- 7	12- 4	6- 2	7- 9	9- 6	11- 0
	Hem-fir	SS	9- 4	12- 4	15- 9	19- 2	9- 4	12- 4	15- 9	18- 5
	Hem-fir	#1	9- 2	12- 0	14- 8	17- 0	8- 6	10- 9	13- 1	15- 2
	Hem-fir	#2	8- 9	11- 4	13-10	16- 1	8- 0	10- 2	12- 5	14- 4
	Hem-fir	#3	6-10	8- 8	10- 7	12- 4	6- 2	7- 9	9- 6	11- 0
	Southern pine	SS	9- 9	12-10	16- 5	19-11	9- 9	12-10	16- 5	19-11
	Southern pine	#1	9- 7	12- 7	16- 1	19- 6	9- 7	12- 4	14- 7	17- 5
	Southern pine	#2	9- 4	12- 4	14- 8	17- 2	8- 6	11- 0	13- 1	15- 5
	Southern pine	#3	7- 4	9- 5	11- 1	13- 2	6- 7	8- 5	9-11	11-10
	Spruce-pine-fir	SS	9- 2	12- 1	15- 5	18- 9	9- 2	12- 1	15- 0	17- 5
	Spruce-pine-fir	#1	8-11	11- 6	14- 1	16- 3	8- 1	10- 3	12- 7	14- 7
	Spruce-pine-fir	#2	8-11	11- 6	14- 1	16- 3	8- 1	10- 3	12- 7	14- 7
	Spruce-pine-fir	#3	6-10	8- 8	10- 7	12- 4	6- 2	7- 9	9- 6	11- 0

For SI: 1 inch = 25.4 mm, 1 foot = 304.8 mm, 1 pound per square foot = 0.0479 kPa.

NOTE: Check sources for availability of lumber in lengths greater than 20 feet.

a. Dead load limits for townhouses in Seismic Design Category C and all structures in Seismic Design Categories D_0, D_1 and D_2 shall be determined in accordance with Section R301.2.2.2.1.

TABLE R502.3.1(2)
FLOOR JOIST SPANS FOR COMMON LUMBER SPECIES
(Residential living areas, live load = 40 psf, L/Δ = 360)[b]

JOIST SPACING (inches)	SPECIES AND GRADE		DEAD LOAD = 10 psf				DEAD LOAD = 20 psf			
			2×6	2×8	2×10	2×12	2×6	2×8	2×10	2×12
						Maximum floor joist spans				
			(ft - in.)	(ft - in.)	(ft - in.)	(ft - in.)	(ft - in.)	(ft - in.)	(ft - in.)	(ft - in.)
12	Douglas fir-larch	SS	11- 4	15- 0	19- 1	23- 3	11- 4	15- 0	19- 1	23- 3
	Douglas fir-larch	#1	10-11	14- 5	18- 5	22- 0	10-11	14- 2	17- 4	20- 1
	Douglas fir-larch	#2	10- 9	14- 2	17- 9	20- 7	10- 6	13- 3	16- 3	18-10
	Douglas fir-larch	#3	8- 8	11- 0	13- 5	15- 7	7-11	10- 0	12- 3	14- 3
	Hem-fir	SS	10- 9	14- 2	18- 0	21-11	10- 9	14- 2	18- 0	21-11
	Hem-fir	#1	10- 6	13-10	17- 8	21- 6	10- 6	13-10	16-11	19- 7
	Hem-fir	#2	10- 0	13- 2	16-10	20- 4	10- 0	13- 1	16- 0	18- 6
	Hem-fir	#3	8- 8	11- 0	13- 5	15- 7	7-11	10- 0	12- 3	14- 3
	Southern pine	SS	11- 2	14- 8	18- 9	22-10	11- 2	14- 8	18- 9	22-10
	Southern pine	#1	10-11	14- 5	18- 5	22- 5	10-11	14- 5	18- 5	22- 5
	Southern pine	#2	10- 9	14- 2	18- 0	21- 9	10- 9	14- 2	16-11	19-10
	Southern pine	#3	9- 4	11-11	14- 0	16- 8	8- 6	10-10	12-10	15- 3
	Spruce-pine-fir	SS	10- 6	13-10	17- 8	21- 6	10- 6	13-10	17- 8	21- 6
	Spruce-pine-fir	#1	10- 3	13- 6	17- 3	20- 7	10- 3	13- 3	16- 3	18-10
	Spruce-pine-fir	#2	10- 3	13- 6	17- 3	20- 7	10- 3	13- 3	16- 3	18-10
	Spruce-pine-fir	#3	8- 8	11- 0	13- 5	15- 7	7-11	10- 0	12- 3	14- 3
16	Douglas fir-larch	SS	10- 4	13- 7	17- 4	21- 1	10- 4	13- 7	17- 4	21- 0
	Douglas fir-larch	#1	9-11	13- 1	16- 5	19- 1	9- 8	12- 4	15- 0	17- 5
	Douglas fir-larch	#2	9- 9	12- 7	15- 5	17-10	9- 1	11- 6	14- 1	16- 3
	Douglas fir-larch	#3	7- 6	9- 6	11- 8	13- 6	6-10	8- 8	10- 7	12- 4
	Hem-fir	SS	9- 9	12-10	16- 5	19-11	9- 9	12-10	16- 5	19-11
	Hem-fir	#1	9- 6	12- 7	16- 0	18- 7	9- 6	12- 0	14- 8	17- 0
	Hem-fir	#2	9- 1	12- 0	15- 2	17- 7	8-11	11- 4	13-10	16- 1
	Hem-fir	#3	7- 6	9- 6	11- 8	13- 6	6-10	8- 8	10- 7	12- 4
	Southern pine	SS	10- 2	13- 4	17- 0	20- 9	10- 2	13- 4	17- 0	20- 9
	Southern pine	#1	9-11	13- 1	16- 9	20- 4	9-11	13- 1	16- 4	19- 6
	Southern pine	#2	9- 9	12-10	16- 1	18-10	9- 6	12- 4	14- 8	17- 2
	Southern pine	#3	8- 1	10- 3	12- 2	14- 6	7- 4	9- 5	11- 1	13- 2
	Spruce-pine-fir	SS	9- 6	12- 7	16- 0	19- 6	9- 6	12- 7	16- 0	19- 6
	Spruce-pine-fir	#1	9- 4	12- 3	15- 5	17-10	9- 1	11- 6	14- 1	16- 3
	Spruce-pine-fir	#2	9- 4	12- 3	15- 5	17-10	9- 1	11- 6	14- 1	16- 3
	Spruce-pine-fir	#3	7- 6	9- 6	11- 8	13- 6	6-10	8- 8	10- 7	12- 4
19.2	Douglas fir-larch	SS	9- 8	12-10	16- 4	19-10	9- 8	12-10	16- 4	19- 2
	Douglas fir-larch	#1	9- 4	12- 4	15- 0	17- 5	8-10	11- 3	13- 8	15-11
	Douglas fir-larch	#2	9- 1	11- 6	14- 1	16- 3	8- 3	10- 6	12-10	14-10
	Douglas fir-larch	#3	6-10	8- 8	10- 7	12- 4	6- 3	7-11	9- 8	11- 3
	Hem-fir	SS	9- 2	12- 1	15- 5	18- 9	9- 2	12- 1	15- 5	18- 9
	Hem-fir	#1	9- 0	11-10	14- 8	17- 0	8- 8	10-11	13- 4	15- 6
	Hem-fir	#2	8- 7	11- 3	13-10	16- 1	8- 2	10- 4	12- 8	14- 8
	Hem-fir	#3	6-10	8- 8	10- 7	12- 4	6- 3	7-11	9- 8	11- 3
	Southern pine	SS	9- 6	12- 7	16- 0	19- 6	9- 6	12- 7	16- 0	19- 6
	Southern pine	#1	9- 4	12- 4	15- 9	19- 2	9- 4	12- 4	14-11	17- 9
	Southern pine	#2	9- 2	12- 1	14- 8	17- 2	8- 8	11- 3	13- 5	15- 8
	Southern pine	#3	7- 4	9- 5	11- 1	13- 2	6- 9	8- 7	10- 1	12- 1
	Spruce-pine-fir	SS	9- 0	11-10	15- 1	18- 4	9- 0	11-10	15- 1	17- 9
	Spruce-pine-fir	#	8- 9	11- 6	14- 1	16- 3	8- 3	10- 6	12-10	14-10
	Spruce-pine-fir	#2	8- 9	11- 6	14- 1	16- 3	8- 3	10- 6	12-10	14-10
	Spruce-pine-fir	#3	6-10	8- 8	10- 7	12- 4	6- 3	7-11	9- 8	11- 3
24	Douglas fir-larch	SS	9- 0	11-11	15- 2	18- 5	9- 0	11-11	14- 9	17- 1
	Douglas fir-larch	#1	8- 8	11- 0	13- 5	15- 7	7-11	10- 0	12- 3	14- 3
	Douglas fir-larch	#2	8- 1	10- 3	12- 7	14- 7	7- 5	9- 5	11- 6	13- 4
	Douglas fir-larch	#3	6- 2	7- 9	9- 6	11- 0	5- 7	7- 1	8- 8	10- 1
	Hem-fir	SS	8- 6	11- 3	14- 4	17- 5	8- 6	11- 3	14- 4	16-10[a]
	Hem-fir	#1	8- 4	10- 9	13- 1	15- 2	7- 9	9- 9	11-11	13-10
	Hem-fir	#2	7-11	10- 2	12- 5	14- 4	7- 4	9- 3	11- 4	13- 1
	Hem-fir	#3	6- 2	7- 9	9- 6	11- 0	5- 7	7- 1	8- 8	10- 1
	Southern pine	SS	8-10	11- 8	14-11	18- 1	8-10	11- 8	14-11	18- 1
	Southern pine	#1	8- 8	11- 5	14- 7	17- 5	8- 8	11- 3	13- 4	15-11
	Southern pine	#2	8- 6	11- 0	13- 1	15- 5	7- 9	10- 0	12- 0	14- 0
	Southern pine	#3	6- 7	8- 5	9-11	11-10	6- 0	7- 8	9- 1	10- 9
	Spruce-pine-fir	SS	8- 4	11- 0	14- 0	17- 0	8- 4	11- 0	13- 8	15-11
	Spruce-pine-fir	#1	8- 1	10- 3	12- 7	14- 7	7- 5	9- 5	11- 6	13- 4
	Spruce-pine-fir	#2	8- 1	10- 3	12- 7	14- 7	7- 5	9- 5	11- 6	13- 4
	Spruce-pine-fir	#3	6- 2	7- 9	9- 6	11- 0	5- 7	7- 1	8- 8	10- 1

For SI: 1 inch = 25.4 mm, 1 foot = 304.8 mm, 1 pound per square foot = 0.0479 kPa.

NOTE: Check sources for availability of lumber in lengths greater than 20 feet.

a. End bearing length shall be increased to 2 inches.

b. Dead load limits for townhouses in Seismic Design Category C and all structures in Seismic Design Categories D_0, D_1, and D_2 shall be determined in accordance with Section R301.2.2.2.1.

TABLE R502.3.3(1)
CANTILEVER SPANS FOR FLOOR JOISTS SUPPORTING LIGHT-FRAME EXTERIOR BEARING WALL AND ROOF ONLY[a, b, c, f, g, h]
(Floor Live Load ≤ 40 psf, Roof Live Load ≤ 20 psf)

Member & Spacing	Maximum Cantilever Span (Uplift Force at Backspan Support in Lbs.)[d, e]											
	Ground Snow Load											
	≤ 20 psf			30 psf			50 psf			70 psf		
	Roof Width			Roof Width			Roof Width			Roof Width		
	24 ft	32 ft	40 ft	24 ft	32 ft	40 ft	24 ft	32 ft	40 ft	24 ft	32 ft	40 ft
2 × 8 @ 12″	20″ (177)	15″ (227)	—	18″ (209)	—	—	—	—	—	—	—	—
2 × 10 @ 16″	29″ (228)	21″ (297)	16″ (364)	26″ (271)	18″ (354)	—	20″ (375)	—	—	—	—	—
2 × 10 @ 12″	36″ (166)	26″ (219)	20″ (270)	34″ (198)	22″ (263)	16″ (324)	26″ (277)	—	—	19″ (356)	—	—
2 × 12 @ 16″	—	32″ (287)	25″ (356)	36″ (263)	29″ (345)	21″ (428)	29″ (367)	20″ (484)	—	23″ (471)	—	—
2 × 12 @ 12″	—	42″ (209)	31″ (263)	—	37″ (253)	27″ (317)	36″ (271)	27″ (358)	17″ (447)	31″ (348)	19″ (462)	—
2 × 12 @ 8″	—	48″ (136)	45″ (169)	—	48″ (164)	38″ (206)	—	40″ (233)	26″ (294)	36″ (230)	29″ (304)	18″ (379)

For SI: 1 inch = 25.4 mm, 1 pound per square foot = 0.0479 kPa.

a. Tabulated values are for clear-span roof supported solely by exterior bearing walls.
b. Spans are based on No. 2 Grade lumber of Douglas fir-larch, hem-fir, southern pine, and spruce-pine-fir for repetitive (3 or more) members.
c. Ratio of backspan to cantilever span shall be at least 3:1.
d. Connections capable of resisting the indicated uplift force shall be provided at the backspan support.
e. Uplift force is for a backspan to cantilever span ratio of 3:1. Tabulated uplift values are permitted to be reduced by multiplying by a factor equal to 3 divided by the actual backspan ratio provided (3/backspan ratio).
f. See Section R301.2.2.2.2, Item 1, for additional limitations on cantilevered floor joists for detached one- and two-family dwellings in Seismic Design Category D_0, D_1, or D_2 and townhouses in Seismic Design Category C, D_0, D_1, or D_2.
g. A full-depth rim joist shall be provided at the cantilevered end of the joists. Solid blocking shall be provided at the cantilever support.
h. Linear interpolation shall be permitted for building widths and ground snow loads other than shown.

TABLE R502.3.3(2)
CANTILEVER SPANS FOR FLOOR JOISTS SUPPORTING EXTERIOR BALCONY[a, b, e, f]

Member Size	Spacing	Maximum Cantilever Span (Uplift Force at Backspan Support in lb)[c, d]		
		Ground Snow Load		
		≤ 30 psf	50 psf	70 psf
2 × 8	12″	42″ (139)	39″ (156)	34″ (165)
2 × 8	16″	36″ (151)	34″ (171)	29″ (180)
2 × 10	12″	61″ (164)	57″ (189)	49″ (201)
2 × 10	16″	53″ (180)	49″ (208)	42″ (220)
2 × 10	24″	43″ (212)	40″ (241)	34″ (255)
2 × 12	16″	72″ (228)	67″ (260)	57″ (268)
2 × 12	24″	58″ (279)	54″ (319)	47″ (330)

For SI: 1 inch = 25.4 mm, 1 pound per square foot = 0.0479 kPa.

a. Spans are based on No. 2 Grade lumber of Douglas fir-larch, hem-fir, southern pine, and spruce-pine-fir for repetitive (3 or more) members.
b. Ratio of backspan to cantilever span shall be at least 2:1.
c. Connections capable of resisting the indicated uplift force shall be provided at the backspan support.
d. Uplift force is for a backspan to cantilever span ratio of 2:1. Tabulated uplift values are permitted to be reduced by multiplying by a factor equal to 2 divided by the actual backspan ratio provided (2/backspan ratio).
e. A full-depth rim joist shall be provided at the cantilevered end of the joists. Solid blocking shall be provided at the cantilevered support.
f. Linear interpolation shall be permitted for ground snow loads other than shown.

TABLE R502.5(1)
GIRDER SPANS[a] AND HEADER SPANS[a] FOR EXTERIOR BEARING WALLS
(Maximum spans for Douglas fir-larch, hem-fir, southern pine and spruce-pine-fir[b] and required number of jack studs)

GIRDERS AND HEADERS SUPPORTING	SIZE	GROUND SNOW LOAD (psf)[e]																	
		30						50						70					
		Building width[c] (feet)																	
		20		28		36		20		28		36		20		28		36	
		Span	NJ[d]	Span	NJ[d]	Span	NJ[d]	Span	NJ[d]	Span	NJ[d]	Span	NJ[d]	Span	NJ[d]	Span	NJ[d]	Span	NJ[d]
Roof and ceiling	2-2×4	3-6	1	3-2	1	2-10	1	3-2	1	2-9	1	2-6	1	2-10	1	2-6	1	2-3	1
	2-2×6	5-5	1	4-8	1	4-2	1	4-8	1	4-1	1	3-8	2	4-2	1	3-8	2	3-3	2
	2-2×8	6-10	1	5-11	2	5-4	2	5-11	2	5-2	2	4-7	2	5-4	2	4-7	2	4-1	2
	2-2×10	8-5	2	7-3	2	6-6	2	7-3	2	6-3	2	5-7	2	6-6	2	5-7	2	5-0	2
	2-2×12	9-9	2	8-5	2	7-6	2	8-5	2	7-3	2	6-6	2	7-6	2	6-6	2	5-10	3
	3-2×8	8-4	1	7-5	1	6-8	1	7-5	1	6-5	2	5-9	2	6-8	1	5-9	2	5-2	2
	3-2×10	10-6	1	9-1	2	8-2	2	9-1	2	7-10	2	7-0	2	8-2	2	7-0	2	6-4	2
	3-2×12	12-2	2	10-7	2	9-5	2	10-7	2	9-2	2	8-2	2	9-5	2	8-2	2	7-4	2
	4-2×8	9-2	1	8-4	1	7-8	1	8-4	1	7-5	1	6-8	1	7-8	1	6-8	1	5-11	2
	4-2×10	11-8	1	10-6	1	9-5	2	10-6	1	9-1	2	8-2	2	9-5	2	8-2	2	7-3	2
	4-2×12	14-1	1	12-2	2	10-11	2	12-2	2	10-7	2	9-5	2	10-11	2	9-5	2	8-5	2
Roof, ceiling and one center-bearing floor	2-2×4	3-1	1	2-9	1	2-5	1	2-9	1	2-5	1	2-2	1	2-7	1	2-3	1	2-0	1
	2-2×6	4-6	1	4-0	1	3-7	2	4-1	1	3-7	2	3-3	2	3-9	2	3-3	2	2-11	2
	2-2×8	5-9	2	5-0	2	4-6	2	5-2	2	4-6	2	4-1	2	4-9	2	4-2	2	3-9	2
	2-2×10	7-0	2	6-2	2	5-6	2	6-4	2	5-6	2	5-0	2	5-9	2	5-1	2	4-7	3
	2-2×12	8-1	2	7-1	2	6-5	2	7-4	2	6-5	2	5-9	3	6-8	2	5-10	3	5-3	3
	3-2×8	7-2	1	6-3	2	5-8	2	6-5	2	5-8	2	5-1	2	5-11	2	5-2	2	4-8	2
	3-2×10	8-9	2	7-8	2	6-11	2	7-11	2	6-11	2	6-3	2	7-3	2	6-4	2	5-8	2
	3-2×12	10-2	2	8-11	2	8-0	2	9-2	2	8-0	2	7-3	2	8-5	2	7-4	2	6-7	2
	4-2×8	8-1	1	7-3	1	6-7	1	7-5	1	6-6	1	5-11	2	6-10	1	6-0	2	5-5	2
	4-2×10	10-1	1	8-10	2	8-0	2	9-1	2	8-0	2	7-2	2	8-4	2	7-4	2	6-7	2
	4-2×12	11-9	2	10-3	2	9-3	2	10-7	2	9-3	2	8-4	2	9-8	2	8-6	2	7-7	2
Roof, ceiling and one clear span floor	2-2×4	2-8	1	2-4	1	2-1	1	2-7	1	2-3	1	2-0	1	2-5	1	2-1	1	1-10	1
	2-2×6	3-11	1	3-5	2	3-0	2	3-10	2	3-4	2	3-0	2	3-6	2	3-1	2	2-9	2
	2-2×8	5-0	2	4-4	2	3-10	2	4-10	2	4-2	2	3-9	2	4-6	2	3-11	2	3-6	2
	2-2×10	6-1	2	5-3	2	4-8	2	5-11	2	5-1	2	4-7	3	5-6	2	4-9	2	4-3	3
	2-2×12	7-1	2	6-1	3	5-5	3	6-10	2	5-11	3	5-4	3	6-4	2	5-6	3	5-0	3
	3-2×8	6-3	2	5-5	2	4-10	2	6-1	2	5-3	2	4-8	2	5-7	2	4-11	2	4-5	2
	3-2×10	7-7	2	6-7	2	5-11	2	7-5	2	6-5	2	5-9	2	6-10	2	6-0	2	5-4	2
	3-2×12	8-10	2	7-8	2	6-10	2	8-7	2	7-5	2	6-8	2	7-11	2	6-11	2	6-3	2
	4-2×8	7-2	1	6-3	2	5-7	2	7-0	1	6-1	2	5-5	2	6-6	1	5-8	2	5-1	2
	4-2×10	8-9	2	7-7	2	6-10	2	8-7	2	7-5	2	6-7	2	7-11	2	6-11	2	6-2	2
	4-2×12	10-2	2	8-10	2	7-11	2	9-11	2	8-7	2	7-8	2	9-2	2	8-0	2	7-2	2
Roof, ceiling and two center-bearing floors	2-2×4	2-7	1	2-3	1	2-0	1	2-6	1	2-2	1	1-11	1	2-4	1	2-0	1	1-9	1
	2-2×6	3-9	2	3-3	2	2-11	2	3-8	2	3-2	2	2-10	2	3-5	2	3-0	2	2-8	2
	2-2×8	4-9	2	4-2	2	3-9	2	4-7	2	4-0	2	3-8	2	4-4	2	3-9	2	3-5	2
	2-2×10	5-9	2	5-1	2	4-7	3	5-8	2	4-11	2	4-5	3	5-3	2	4-7	3	4-2	3
	2-2×12	6-8	2	5-10	3	5-3	3	6-6	2	5-9	3	5-2	3	6-1	3	5-4	3	4-10	3
	3-2×8	5-11	2	5-2	2	4-8	2	5-9	2	5-1	2	4-7	2	5-5	2	4-9	2	4-3	2
	3-2×10	7-3	2	6-4	2	5-8	2	7-1	2	6-2	2	5-7	2	6-7	2	5-9	2	5-3	2
	3-2×12	8-5	2	7-4	2	6-7	2	8-2	2	7-2	2	6-5	3	7-8	2	6-9	2	6-1	3
	4-2×8	6-10	1	6-0	2	5-5	2	6-8	1	5-10	2	5-3	2	6-3	2	5-6	2	4-11	2
	4-2×10	8-4	2	7-4	2	6-7	2	8-2	2	7-2	2	6-5	2	7-7	2	6-8	2	6-0	2
	4-2×12	9-8	2	8-6	2	7-8	2	9-5	2	8-3	2	7-5	2	8-10	2	7-9	2	7-0	2

(continued)

TABLE R502.5(1)—continued
GIRDER SPANS[a] AND HEADER SPANS[a] FOR EXTERIOR BEARING WALLS
(Maximum spans for Douglas fir-larch, hem-fir, southern pine and spruce-pine-fir[b] and required number of jack studs)

GIRDERS AND HEADERS SUPPORTING	SIZE	GROUND SNOW LOAD (psf)[e]																
		30						50						70					
		Building width[c] (feet)																	
		20		28		36		20		28		36		20		28		36	
		Span	NJ[d]	Span	NJ[d]	Span	NJ[d]	Span	NJ[d]	Span	NJ[d]	Span	NJ[d]	Span	NJ[d]	Span	NJ[d]	Span	NJ[d]
Roof, ceiling, and two clear span floors	2-2×4	2-1	1	1-8	1	1-6	2	2-0	1	1-8	1	1-5	2	2-0	1	1-8	1	1-5	2
	2-2×6	3-1	2	2-8	2	2-4	2	3-0	2	2-7	2	2-3	2	2-11	2	2-7	2	2-3	2
	2-2×8	3-10	2	3-4	2	3-0	3	3-10	2	3-4	2	2-11	3	3-9	2	3-3	2	2-11	3
	2-2×10	4-9	2	4-1	3	3-8	3	4-8	2	4-0	3	3-7	3	4-7	3	4-0	3	3-6	3
	2-2×12	5-6	3	4-9	3	4-3	3	5-5	3	4-8	3	4-2	3	5-4	3	4-7	3	4-1	4
	3-2×8	4-10	2	4-2	2	3-9	2	4-9	2	4-1	2	3-8	2	4-8	2	4-1	2	3-8	2
	3-2×10	5-11	2	5-1	2	4-7	3	5-10	2	5-0	2	4-6	3	5-9	2	4-11	2	4-5	3
	3-2×12	6-10	2	5-11	3	5-4	3	6-9	2	5-10	3	5-3	3	6-8	2	5-9	3	5-2	3
	4-2×8	5-7	2	4-10	2	4-4	2	5-6	2	4-9	2	4-3	2	5-5	2	4-8	2	4-2	2
	4-2×10	6-10	2	5-11	2	5-3	2	6-9	2	5-10	2	5-2	2	6-7	2	5-9	2	5-1	2
	4-2×12	7-11	2	6-10	2	6-2	3	7-9	2	6-9	2	6-0	3	7-8	2	6-8	2	5-11	3

For SI: 1 inch = 25.4 mm, 1 pound per square foot = 0.0479 kPa.

a. Spans are given in feet and inches.

b. Tabulated values assume #2 grade lumber.

c. Building width is measured perpendicular to the ridge. For widths between those shown, spans are permitted to be interpolated.

d. NJ - Number of jack studs required to support each end. Where the number of required jack studs equals one, the header is permitted to be supported by an approved framing anchor attached to the full-height wall stud and to the header.

e. Use 30 psf ground snow load for cases in which ground snow load is less than 30 psf and the roof live load is equal to or less than 20 psf.

R502.5 Allowable girder spans. The allowable spans of girders fabricated of dimension lumber shall not exceed the values set forth in Tables R502.5(1) and R502.5(2).

R502.6 Bearing. The ends of each joist, beam or girder shall have not less than 1.5 inches (38 mm) of bearing on wood or metal and not less than 3 inches (76 mm) on masonry or concrete except where supported on a 1-inch-by-4-inch (25.4 mm by 102 mm) ribbon strip and nailed to the adjacent stud or by the use of approved joist hangers.

R502.6.1 Floor systems. Joists framing from opposite sides over a bearing support shall lap a minimum of 3 inches (76 mm) and shall be nailed together with a minimum three 10d face nails. A wood or metal splice with strength equal to or greater than that provided by the nailed lap is permitted.

R502.6.2 Joist framing. Joists framing into the side of a wood girder shall be supported by approved framing anchors or on ledger strips not less than nominal 2 inches by 2 inches (51 mm by 51 mm).

R502.7 Lateral restraint at supports. Joists shall be supported laterally at the ends by full-depth solid blocking not less than 2 inches (51 mm) nominal in thickness; or by attachment to a full-depth header, band or rim joist, or to an adjoining stud or shall be otherwise provided with lateral support to prevent rotation.

Exception: In Seismic Design Categories D_0, D_1 and D_2, lateral restraint shall also be provided at each intermediate support.

R502.7.1 Bridging. Joists exceeding a nominal 2 inches by 12 inches (51 mm by 305 mm) shall be supported laterally by solid blocking, diagonal bridging (wood or metal), or a continuous 1-inch-by-3-inch (25.4 mm by 76 mm) strip nailed across the bottom of joists perpendicular to joists at intervals not exceeding 8 feet (2438 mm).

R502.8 Drilling and notching. Structural floor members shall not be cut, bored or notched in excess of the limitations specified in this section. See Figure R502.8.

R502.8.1 Sawn lumber. Notches in solid lumber joists, rafters and beams shall not exceed one-sixth of the depth of the member, shall not be longer than one-third of the depth of the member and shall not be located in the middle one-third of the span. Notches at the ends of the member shall not exceed one-fourth the depth of the member. The tension side of members 4 inches (102 mm) or greater in nominal thickness shall not be notched except at the ends of the members. The diameter of holes bored or cut into members shall not exceed one-third the depth of the member. Holes shall not be closer than 2 inches (51 mm) to the top or bottom of the member, or to any other hole located in the member. Where the member is also notched, the hole shall not be closer than 2 inches (51 mm) to the notch.

R502.8.2 Engineered wood products. Cuts, notches and holes bored in trusses, structural composite lumber, structural glue-laminated members or I-joists are prohibited except where permitted by the manufacturer's recommendations or where the effects of such alterations are specifically considered in the design of the member by a registered design professional.

TABLE R502.5(2)
GIRDER SPANS[a] AND HEADER SPANS[a] FOR INTERIOR BEARING WALLS
(Maximum spans for Douglas fir-larch, hem-fir, southern pine and spruce-pine-fir[b] and required number of jack studs)

HEADERS AND GIRDERS SUPPORTING	SIZE	BUILDING WIDTH[c] (feet)					
		20		28		36	
		Span	NJ[d]	Span	NJ[d]	Span	NJ[d]
One floor only	2-2×4	3-1	1	2-8	1	2-5	1
	2-2×6	4-6	1	3-11	1	3-6	1
	2-2×8	5-9	1	5-0	2	4-5	2
	2-2×10	7-0	2	6-1	2	5-5	2
	2-2×12	8-1	2	7-0	2	6-3	2
	3-2×8	7-2	1	6-3	1	5-7	2
	3-2×10	8-9	1	7-7	2	6-9	2
	3-2×12	10-2	2	8-10	2	7-10	2
	4-2×8	9-0	1	7-8	1	6-9	1
	4-2×10	10-1	1	8-9	1	7-10	2
	4-2×12	11-9	1	10-2	2	9-1	2
Two floors	2-2×4	2-2	1	1-10	1	1-7	1
	2-2×6	3-2	2	2-9	2	2-5	2
	2-2×8	4-1	2	3-6	2	3-2	2
	2-2×10	4-11	2	4-3	2	3-10	3
	2-2×12	5-9	2	5-0	3	4-5	3
	3-2×8	5-1	2	4-5	2	3-11	2
	3-2×10	6-2	2	5-4	2	4-10	2
	3-2×12	7-2	2	6-3	2	5-7	3
	4-2×8	6-1	1	5-3	2	4-8	2
	4-2×10	7-2	2	6-2	2	5-6	2
	4-2×12	8-4	2	7-2	2	6-5	2

For SI: 1 inch = 25.4 mm, 1 foot = 304.8 mm.

a. Spans are given in feet and inches.

b. Tabulated values assume #2 grade lumber.

c. Building width is measured perpendicular to the ridge. For widths between those shown, spans are permitted to be interpolated.

d. NJ - Number of jack studs required to support each end. Where the number of required jack studs equals one, the header is permitted to be supported by an approved framing anchor attached to the full-height wall stud and to the header.

R502.9 Fastening. Floor framing shall be nailed in accordance with Table R602.3(1). Where posts and beam or girder construction is used to support floor framing, positive connections shall be provided to ensure against uplift and lateral displacement.

R502.10 Framing of openings. Openings in floor framing shall be framed with a header and trimmer joists. When the header joist span does not exceed 4 feet (1219 mm), the header joist may be a single member the same size as the floor joist. Single trimmer joists may be used to carry a single header joist that is located within 3 feet (914 mm) of the trimmer joist bearing. When the header joist span exceeds 4 feet (1219 mm), the trimmer joists and the header joist shall be doubled and of sufficient cross section to support the floor joists framing into the header. Approved hangers shall be used for the header joist to trimmer joist connections when the header joist span exceeds 6 feet (1829 mm). Tail joists over 12 feet (3658 mm) long shall be supported at the header by framing anchors or on ledger strips not less than 2 inches by 2 inches (51 mm by 51 mm).

R502.11 Wood trusses.

R502.11.1 Design. Wood trusses shall be designed in accordance with approved engineering practice. The design and manufacture of metal plate connected wood trusses shall comply with ANSI/TPI 1. The truss design drawings shall be prepared by a registered professional where required by the statutes of the jurisdiction in which the project is to be constructed in accordance with Section R106.1.

R502.11.2 Bracing. Trusses shall be braced to prevent rotation and provide lateral stability in accordance with the requirements specified in the construction documents for the building and on the individual truss design drawings. In the absence of specific bracing requirements, trusses shall be braced in accordance with the Building Component Safety Information (BCSI 1-03) Guide to Good Practice for Handling, Installing & Bracing of Metal Plate Connected Wood Trusses.

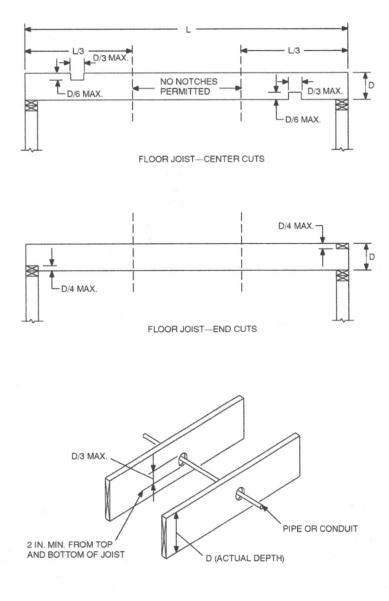

FLOOR JOIST—CENTER CUTS

FLOOR JOIST—END CUTS

For SI: 1 inch = 25.4 mm.

FIGURE R502.8
CUTTING, NOTCHING AND DRILLING

R502.11.3 Alterations to trusses. Truss members and components shall not be cut, notched, spliced or otherwise altered in any way without the approval of a registered design professional. Alterations resulting in the addition of load (e.g., HVAC equipment, water heater, etc.), that exceed the design load for the truss, shall not be permitted without verification that the truss is capable of supporting the additional loading.

R502.11.4 Truss design drawings. Truss design drawings, prepared in compliance with Section R502.11.1, shall be submitted to the building official and approved prior to installation. Truss design drawings shall be provided with the shipment of trusses delivered to the job site. Truss design drawings shall include, at a minimum, the information specified below:

1. Slope or depth, span and spacing.

2. Location of all joints.

3. Required bearing widths.

4. Design loads as applicable:

   4.1. Top chord live load;

   4.2. Top chord dead load;

   4.3. Bottom chord live load;

   4.4. Bottom chord dead load;

   4.5. Concentrated loads and their points of application; and

   4.6. Controlling wind and earthquake loads.

5. Adjustments to lumber and joint connector design values for conditions of use.

6. Each reaction force and direction.

7. Joint connector type and description, e.g., size, thickness or gauge, and the dimensioned location of each joint connector except where symmetrically located relative to the joint interface.

8. Lumber size, species and grade for each member.

9. Connection requirements for:

   9.1. Truss-to-girder-truss;

   9.2. Truss ply-to-ply; and

   9.3. Field splices.

10. Calculated deflection ratio and/or maximum description for live and total load.

11. Maximum axial compression forces in the truss members to enable the building designer to design the size, connections and anchorage of the permanent continuous lateral bracing. Forces shall be shown on the truss drawing or on supplemental documents.

12. Required permanent truss member bracing location.

R502.12 Draftstopping required. When there is usable space both above and below the concealed space of a floor/ceiling assembly, draftstops shall be installed so that the area of the concealed space does not exceed 1,000 square feet (92.9 m²). Draftstopping shall divide the concealed space into approximately equal areas. Where the assembly is enclosed by a floor membrane above and a ceiling membrane below draftstopping

shall be provided in floor/ceiling assemblies under the following circumstances:

1. Ceiling is suspended under the floor framing.

2. Floor framing is constructed of truss-type open-web or perforated members.

R502.12.1 Materials. Draftstopping materials shall not be less than $^1/_2$-inch (12.7 mm) gypsum board, $^3/_8$-inch (9.5 mm) wood structural panels, $^3/_8$-inch (9.5 mm) Type 2-M-W particleboard or other approved materials adequately supported. Draftstopping shall be installed parallel to the floor framing members unless otherwise approved by the building official. The integrity of all draftstops shall be maintained.

R502.13 Fireblocking required. Fireblocking shall be provided in accordance with Section R602.8.

SECTION R503
FLOOR SHEATHING

R503.1 Lumber sheathing. Maximum allowable spans for lumber used as floor sheathing shall conform to Tables R503.1, R503.2.1.1(1) and R503.2.1.1(2).

R503.1.1 End joints. End joints in lumber used as subflooring shall occur over supports unless end-matched lumber is used, in which case each piece shall bear on at least two joists. Subflooring may be omitted when joist spacing does not exceed 16 inches (406 mm) and a 1-inch (25.4 mm) nominal tongue-and-groove wood strip flooring is applied perpendicular to the joists.

TABLE R503.1
MINIMUM THICKNESS OF LUMBER FLOOR SHEATHING

JOIST OR BEAM SPACING (inches)	MINIMUM NET THICKNESS	
	Perpendicular to joist	Diagonal to joist
24	$^{11}/_{16}$	$^3/_4$
16	$^5/_8$	$^5/_8$
48[a]		
54[b]	$1^1/_2$ T & G	N/A
60[c]		

For SI: 1 inch = 25.4 mm, 1 pound per square inch = 6.895 kPa.

a. For this support spacing, lumber sheathing shall have a minimum F_b of 675 and minimum E of 1,100,000 (see AF&PA/NDS).

b. For this support spacing, lumber sheathing shall have a minimum F_b of 765 and minimum E of 1,400,000 (see AF&PA/NDS).

c. For this support spacing, lumber sheathing shall have a minimum F_b of 855 and minimum E of 1,700,000 (see AF&PA/NDS).

R503.2 Wood structural panel sheathing.

R503.2.1 Identification and grade. Wood structural panel sheathing used for structural purposes shall conform to DOC PS 1, DOC PS 2 or, when manufactured in Canada, CSA 0437 or CSA 0325. All panels shall be identified by a grade mark of certificate of inspection issued by an approved agency.

R503.2.1.1 Subfloor and combined subfloor underlayment. Where used as subflooring or combination subfloor underlayment, wood structural panels shall be of one of the grades specified in Table R503.2.1.1(1).

TABLE R503.2.1.1(1)
ALLOWABLE SPANS AND LOADS FOR WOOD STRUCTURAL PANELS FOR ROOF
AND SUBFLOOR SHEATHING AND COMBINATION SUBFLOOR UNDERLAYMENT[a, b, c]

SPAN RATING	MINIMUM NOMINAL PANEL THICKNESS (inch)	ALLOWABLE LIVE LOAD (psf)[h, i]		MAXIMUM SPAN (inches)		LOAD (pounds per square foot, at maximum span)		MAXIMUM SPAN (inches)
		SPAN @ 16 o.c.	SPAN @ 24 o.c.	With edge support[d]	Without edge support	Total load	Live load	
Sheathing[e]				Roof[f]				Subfloor[j]
12/0	$5/16$	—	—	12	12	40	30	0
16/0	$5/16$	30	—	16	16	40	30	0
20/0	$5/16$	50	—	20	20	40	30	0
24/0	$3/8$	100	30	24	20[g]	40	30	0
24/16	$7/16$	100	40	24	24	50	40	16
32/16	$15/32, 1/2$	180	70	32	28	40	30	16[h]
40/20	$19/32, 5/8$	305	130	40	32	40	30	20[h,i]
48/24	$23/32, 3/4$	—	175	48	36	45	35	24
60/32	$7/8$	—	305	60	48	45	35	32
Underlayment, C-C plugged, single floor[e]				Roof[f]				Combination subfloor underlayment[k]
16 o.c.	$19/32, 5/8$	100	40	24	24	50	40	16[i]
20 o.c.	$19/32, 5/8$	150	60	32	32	40	30	20[i,j]
24 o.c.	$23/32, 3/4$	240	100	48	36	35	25	24
32 o.c.	$7/8$	—	185	48	40	50	40	32
48 o.c.	$13/32, 11/8$	—	290	60	48	50	40	48

For SI: 1 inch = 25.4 mm, 1 pound per square foot = 0.0479 kPa.

a. The allowable total loads were determined using a dead load of 10 psf. If the dead load exceeds 10 psf, then the live load shall be reduced accordingly.

b. Panels continuous over two or more spans with long dimension perpendicular to supports. Spans shall be limited to values shown because of possible effect of concentrated loads.

c. Applies to panels 24 inches or wider.

d. Lumber blocking, panel edge clips (one midway between each support, except two equally spaced between supports when span is 48 inches), tongue-and-groove panel edges, or other approved type of edge support.

e. Includes Structural I panels in these grades.

f. Uniform load deflection limitation: $1/180$ of span under live load plus dead load, $1/240$ of span under live load only.

g. Maximum span 24 inches for $15/32$-and $1/2$-inch panels.

h. Maximum span 24 inches where $3/4$-inch wood finish flooring is installed at right angles to joists.

i. Maximum span 24 inches where 1.5 inches of lightweight concrete or approved cellular concrete is placed over the subfloor.

j. Unsupported edges shall have tongue-and-groove joints or shall be supported with blocking unless minimum nominal $1/4$-inch thick underlayment with end and edge joints offset at least 2 inches or 1.5 inches of lightweight concrete or approved cellular concrete is placed over the subfloor, or $3/4$-inch wood finish flooring is installed at right angles to the supports. Allowable uniform live load at maximum span, based on deflection of $1/360$ of span, is 100 psf.

k. Unsupported edges shall have tongue-and-groove joints or shall be supported by blocking unless nominal $1/4$-inch-thick underlayment with end and edge joints offset at least 2 inches or $3/4$-inch wood finish flooring is installed at right angles to the supports. Allowable uniform live load at maximum span, based on deflection of $1/360$ of span, is 100 psf, except panels with a span rating of 48 on center are limited to 65 psf total uniform load at maximum span.

l. Allowable live load values at spans of 16" o.c. and 24" o.c taken from reference standard APA E30, APA Engineered Wood Construction Guide. Refer to reference standard for allowable spans not listed in the table.

When sanded plywood is used as combination subfloor underlayment, the grade shall be as specified in Table R503.2.1.1(2).

TABLE R503.2.1.1(2)
ALLOWABLE SPANS FOR SANDED PLYWOOD
COMBINATION SUBFLOOR UNDERLAYMENT[a]

IDENTIFICATION	SPACING OF JOISTS (inches)		
	16	20	24
Species group[b]	—	—	—
1	$^1/_2$	$^5/_8$	$^3/_4$
2, 3	$^5/_8$	$^3/_4$	$^7/_8$
4	$^3/_4$	$^7/_8$	1

For SI: 1 inch = 25.4 mm, 1 pound per square foot = 0.0479 kPa.

a. Plywood continuous over two or more spans and face grain perpendicular to supports. Unsupported edges shall be tongue-and-groove or blocked except where nominal $^1/_4$-inch-thick underlayment or $^3/_4$-inch wood finish floor is used. Allowable uniform live load at maximum span based on deflection of $^1/_{360}$ of span is 100 psf.

b. Applicable to all grades of sanded exterior-type plywood.

R503.2.2 Allowable spans. The maximum allowable span for wood structural panels used as subfloor or combination subfloor underlayment shall be as set forth in Table R503.2.1.1(1), or APA E30. The maximum span for sanded plywood combination subfloor underlayment shall be as set forth in Table R503.2.1.1(2).

R503.2.3 Installation. Wood structural panels used as subfloor or combination subfloor underlayment shall be attached to wood framing in accordance with Table R602.3(1) and shall be attached to cold-formed steel framing in accordance with Table R505.3.1(2).

R503.3 Particleboard.

R503.3.1 Identification and grade. Particleboard shall conform to ANSI A208.1 and shall be so identified by a grade mark or certificate of inspection issued by an approved agency.

R503.3.2 Floor underlayment. Particleboard floor underlayment shall conform to Type PBU and shall not be less than $^1/_4$ inch (6.4 mm) in thickness.

R503.3.3 Installation. Particleboard underlayment shall be installed in accordance with the recommendations of the manufacturer and attached to framing in accordance with Table R602.3(1).

SECTION R504
PRESSURE PRESERVATIVELY TREATED-WOOD FLOORS (ON GROUND)

R504.1 General. Pressure preservatively treated-wood basement floors and floors on ground shall be designed to withstand axial forces and bending moments resulting from lateral soil pressures at the base of the exterior walls and floor live and dead loads. Floor framing shall be designed to meet joist deflection requirements in accordance with Section R301.

R504.1.1 Unbalanced soil loads. Unless special provision is made to resist sliding caused by unbalanced lateral soil loads, wood basement floors shall be limited to applications where the differential depth of fill on opposite exterior foundation walls is 2 feet (610 mm) or less.

R504.1.2 Construction. Joists in wood basement floors shall bear tightly against the narrow face of studs in the foundation wall or directly against a band joist that bears on the studs. Plywood subfloor shall be continuous over lapped joists or over butt joints between in-line joists. Sufficient blocking shall be provided between joists to transfer lateral forces at the base of the end walls into the floor system.

R504.1.3 Uplift and buckling. Where required, resistance to uplift or restraint against buckling shall be provided by interior bearing walls or properly designed stub walls anchored in the supporting soil below.

R504.2 Site preparation. The area within the foundation walls shall have all vegetation, topsoil and foreign material removed, and any fill material that is added shall be free of vegetation and foreign material. The fill shall be compacted to assure uniform support of the pressure preservatively treated-wood floor sleepers.

R504.2.1 Base. A minimum 4-inch-thick (102 mm) granular base of gravel having a maximum size of $^3/_4$ inch (19.1 mm) or crushed stone having a maximum size of $^1/_2$ inch (12.7 mm) shall be placed over the compacted earth.

R504.2.2 Moisture barrier. Polyethylene sheeting of minimum 6-mil (0.15 mm) thickness shall be placed over the granular base. Joints shall be lapped 6 inches (152 mm) and left unsealed. The polyethylene membrane shall be placed over the pressure preservatively treated-wood sleepers and shall not extend beneath the footing plates of the exterior walls.

R504.3 Materials. All framing materials, including sleepers, joists, blocking and plywood subflooring, shall be pressure-preservative treated and dried after treatment in accordance with AWPA U1 (Commodity Specification A, Use Category 4B and section 5.2), and shall bear the label of an accredited agency.

SECTION R505
STEEL FLOOR FRAMING

R505.1 Cold-formed steel floor framing. Elements shall be straight and free of any defects that would significantly affect structural performance. Cold-formed steel floor framing members shall comply with the requirements of this section.

R505.1.1 Applicability limits. The provisions of this section shall control the construction of steel floor framing for buildings not greater than 60 feet (18,288 mm) in length perpendicular to the joist span, not greater than 40 feet (12 192 mm) in width parallel to the joist span, and not greater than two stories in height. Steel floor framing constructed in accordance with the provisions of this section shall be limited to sites subjected to a maximum design wind speed of 110 miles per hour (49 m/s), Exposure A, B, or C, and a maximum ground snow load of 70 psf (3.35 kPa).

R505.1.2 In-line framing. When supported by steel-framed walls in accordance with Section R603, steel floor framing shall be constructed with floor joists located directly in-line with load-bearing studs located below the joists with a maximum tolerance of $^3/_4$ inch (19.1 mm) between the center lines of the joist and the stud.

R505.1.3 Floor trusses. The design, quality assurance, installation and testing of cold-formed steel trusses shall be in accordance with the AISI Standard for Cold-formed Steel Framing-Truss Design (COFS/Truss). Truss members shall not be notched, cut or altered in any manner without an approved design.

R505.2 Structural framing. Load-bearing floor framing members shall comply with Figure R505.2(1) and with the dimensional and minimum thickness requirements specified in Tables R505.2(1) and R505.2(2). Tracks shall comply with Figure R505.2(2) and shall have a minimum flange width of $1^1/_4$ inches (32 mm). The maximum inside bend radius for members shall be the larger of $^3/_{32}$ inch (2.4 mm) or twice the uncoated steel thickness. Holes in joist webs shall comply with all of the following conditions:

1. Holes shall conform to Figure R505.2(3);

2. Holes shall be permitted only along the centerline of the web of the framing member;

3. Holes shall have a center-to-center spacing of not less than 24 inches (610 mm);

4. Holes shall have a web hole width not greater than 0.5 times the member depth, or $2^1/_2$ inches (64.5 mm);

5. Holes shall have a web hole length not exceeding $4^1/_2$ inches (114 mm); and

6. Holes shall have a minimum distance between the edge of the bearing surface and the edge of the web hole of not less than 10 inches (254 mm).

Framing members with web holes not conforming to the above requirements shall be patched in accordance with Section R505.3.6 or designed in accordance with accepted engineering practices.

R505.2.1 Material. Load-bearing members used in steel floor construction shall be cold-formed to shape from structural quality sheet steel complying with the requirements of one of the following:

1. ASTM A 653: Grades 33, 37, 40 and 50 (Class 1 and 3).

2. ASTM A 792: Grades 33, 37, 40 and 50A.

3. ASTM A 875: Grades 33, 37, 40 and 50 (Class 1 and 3).

4. ASTM A 1003: Grades 33, 37, 40 and 50.

R505.2.2 Identification. Load-bearing steel framing members shall have a legible label, stencil, stamp or embossment with the following information as a minimum:

1. Manufacturer's identification.

2. Minimum uncoated steel thickness in inches (mm).

3. Minimum coating designation.

4. Minimum yield strength, in kips per square inch (ksi) (kPa).

R505.2.3 Corrosion protection. Load-bearing steel framing shall have a metallic coating complying with one of the following:

1. A minimum of G 60 in accordance with ASTM A 653.

2. A minimum of AZ 50 in accordance with ASTM A 792.

TABLE R505.2(1)
COLD-FORMED STEEL JOIST SIZES

MEMBER DESIGNATION[a]	WEB DEPTH (inches)	MINIMUM FLANGE WIDTH (inches)	MAXIMUM FLANGE WIDTH (inches)	MINIMUM LIP SIZE (inches)
550S162-t	5.5	1.625	2	0.5
800S162-t	8	1.625	2	0.5
1000S162-t	10	1.625	2	0.5
1200S162-t	12	1.625	2	0.5

For SI: 1 inch = 25.4 mm, 1 mil = 0.0254 mm.

a. The member designation is defined by the first number representing the member depth in 0.01 inch, the letter "S" representing a stud or joist member, the second number representing the flange width in 0.01 inch, and the letter "t" shall be a number representing the minimum base metal thickness in mils [See Table R505.2(2)].

TABLE R505.2(2)
MINIMUM THICKNESS OF COLD-FORMED STEEL MEMBERS

DESIGNATION (mils)	MINIMUM UNCOATED THICKNESS (inches)	REFERENCE GAGE NUMBER
33	0.033	20
43	0.043	18
54	0.054	16
68	0.068	14

For SI: 1 inch = 25.4 mm, 1 mil = 0.0254 mm.

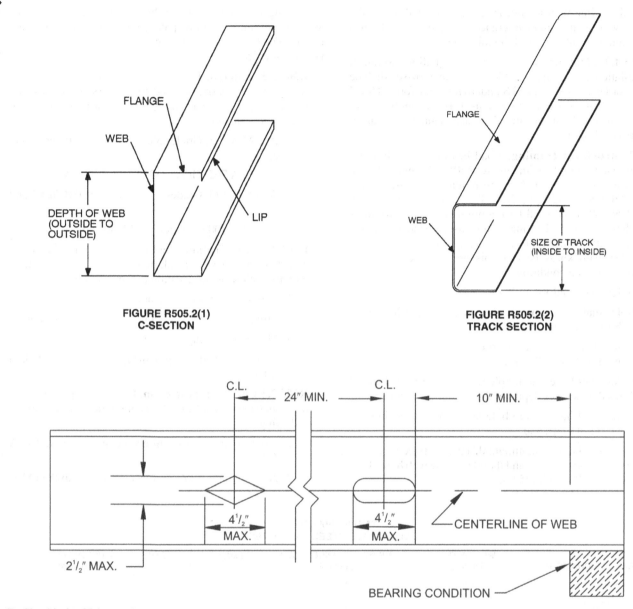

FIGURE R505.2(1)
C-SECTION

FIGURE R505.2(2)
TRACK SECTION

For SI: 1 inch = 25.4 mm.

FIGURE R505.2(3)
FLOOR JOIST WEB HOLES

3. A minimum of GF 60 in accordance with ASTM A 875.

R505.2.4 Fastening requirements. Screws for steel-to-steel connections shall be installed with a minimum edge distance and center-to-center spacing of 0.5 inch (12.7 mm), shall be self-drilling tapping, and shall conform to SAE J78. Floor sheathing shall be attached to steel joists with minimum No. 8 self-drilling tapping screws that conform to SAE J78. Screws attaching floor-sheathing-to-steel joists shall have a minimum head diameter of 0.292 inch (7.4 mm) with countersunk heads and shall be installed with a minimum edge distance of 0.375 inch (9.5 mm). Gypsum board ceilings shall be attached to steel joists with minimum No. 6 screws conforming to ASTM C 954 and shall be installed in accordance with Section R702. For all connections, screws shall extend through the steel a minimum of three exposed threads. All self-drilling tapping screws conforming to SAE J78 shall have a Type II coating in accordance with ASTM B 633.

Where No. 8 screws are specified in a steel to steel connection the required number of screws in the connection is permitted to be reduced in accordance with the reduction factors in Table R505.2.4 when larger screws are used or when one of the sheets of steel being connected is thicker than 33 mils (0.84 mm). When applying the reduction factor the resulting number of screws shall be rounded up.

TABLE R505.2.4
SCREW SUBSTITUTION FACTOR

SCREW SIZE	THINNEST CONNECTED STEEL SHEET (mils)	
	33	43
#8	1.0	0.67
#10	0.93	0.62
#12	0.86	0.56

For SI: 1 mil = 0.0254 mm.

R505.3 Floor construction. Cold-formed steel floors shall be constructed in accordance with this section and Figure R505.3.

R505.3.1 Floor to foundation or bearing wall connections. Cold-formed steel floors shall be anchored to foundations, wood sills or load-bearing walls in accordance with Table R505.3.1(1) and Figure R505.3.1(1), R505.3.1(2), R505.3.1(3), R505.3.1(4), R505.3.1(5) or R505.3.1(6). Continuous steel joists supported by interior load-bearing walls shall be constructed in accordance with Figure R505.3.1(7). Lapped steel joists shall be constructed in accordance with Figure R505.3.1(8). Fastening of steel joists to other framing members shall be in accordance with Table R505.3.1(2).

R505.3.2 Allowable joist spans. The clear span of cold-formed steel floor joists shall not exceed the limits set forth in Tables R505.3.2(1), R505.3.2(2), and R505.3.2(3). Floor joists shall have a minimum bearing length of 1.5 inches (38 mm). When continuous joists are used, the interior bearing supports shall be located within 2 feet (610 mm) of mid span of the steel joists, and the individual spans shall not exceed the span in Tables R505.3.2(2) and R505.3.2(3). Bearing stiffeners shall be installed at each bearing location in accordance with Section R505.3.4 and as shown in Figure R505.3.

Blocking is not required for continuous back-to-back floor joists at bearing supports. Blocking shall be installed between the joists for single continuous floor joists across bearing supports. Blocking shall be spaced at a maximum of 12 feet (3660 mm) on center. Blocking shall consist of C-shape or track section with a minimum thickness of 33 mils (0.84 mm). Blocking shall be fastened to each adjacent joist through a 33-mil (0.84 mm) clip angle, bent web of blocking or flanges of web stiffeners with two No. 8 screws on each side. The minimum depth of the blocking shall be equal to the depth of the joist minus 2 inches (51 mm). The minimum length of the angle shall be equal to the depth of the joist minus 2 inches (51 mm).

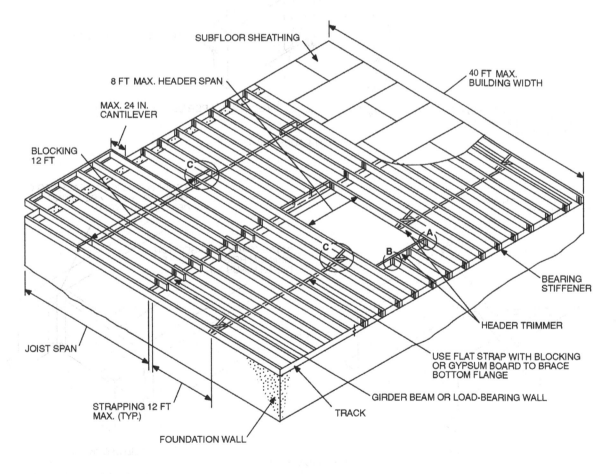

For SI: 1 inch = 25.4 mm, 1 foot = 304.8 mm.

FIGURE R505.3
STEEL FLOOR CONSTRUCTION

(continued)

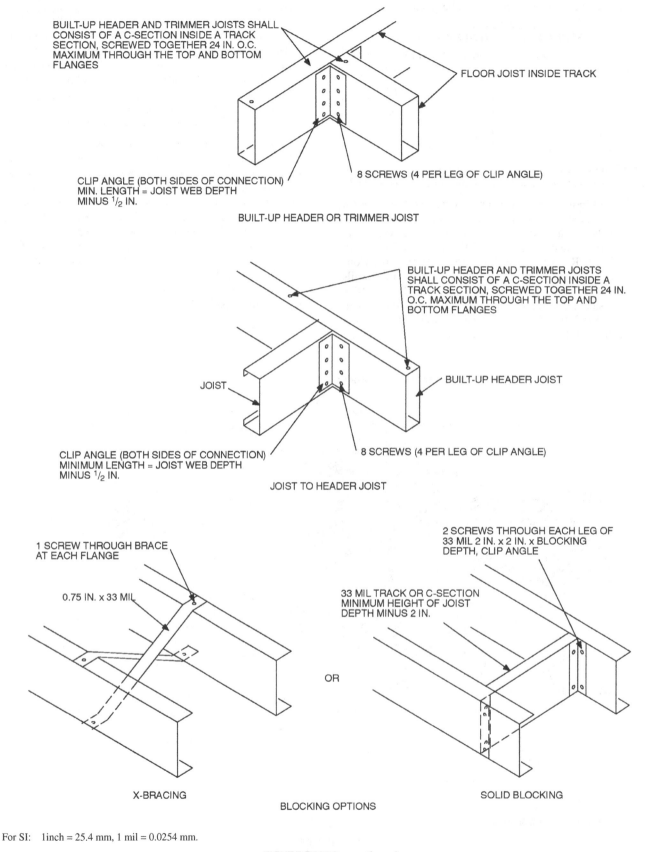

BUILT-UP HEADER AND TRIMMER JOISTS SHALL CONSIST OF A C-SECTION INSIDE A TRACK SECTION, SCREWED TOGETHER 24 IN. O.C. MAXIMUM THROUGH THE TOP AND BOTTOM FLANGES

FLOOR JOIST INSIDE TRACK

8 SCREWS (4 PER LEG OF CLIP ANGLE)

CLIP ANGLE (BOTH SIDES OF CONNECTION) MIN. LENGTH = JOIST WEB DEPTH MINUS $\frac{1}{2}$ IN.

BUILT-UP HEADER OR TRIMMER JOIST

BUILT-UP HEADER AND TRIMMER JOISTS SHALL CONSIST OF A C-SECTION INSIDE A TRACK SECTION, SCREWED TOGETHER 24 IN. O.C. MAXIMUM THROUGH THE TOP AND BOTTOM FLANGES

JOIST

BUILT-UP HEADER JOIST

8 SCREWS (4 PER LEG OF CLIP ANGLE)

CLIP ANGLE (BOTH SIDES OF CONNECTION) MINIMUM LENGTH = JOIST WEB DEPTH MINUS $\frac{1}{2}$ IN.

JOIST TO HEADER JOIST

1 SCREW THROUGH BRACE AT EACH FLANGE

0.75 IN. x 33 MIL

X-BRACING

2 SCREWS THROUGH EACH LEG OF 33 MIL 2 IN. x 2 IN. x BLOCKING DEPTH, CLIP ANGLE

33 MIL TRACK OR C-SECTION MINIMUM HEIGHT OF JOIST DEPTH MINUS 2 IN.

OR

SOLID BLOCKING

BLOCKING OPTIONS

For SI: 1inch = 25.4 mm, 1 mil = 0.0254 mm.

FIGURE R505.3—continued
STEEL FLOOR CONSTRUCTION

TABLE R505.3.1(1)
FLOOR TO FOUNDATION OR BEARING WALL CONNECTION REQUIREMENTS[a, b]

	WIND SPEED (mph) AND EXPOSURE	
FRAMING CONDITION	**Up to 110 A/B or 85 C or Seismic Design Categories A, B, C**	**Up to 110 C**
Floor joist to wall track of exterior steel load-bearing wall per Figure R505.3.1(1)	2-No. 8 screws	3-No. 8 screws
Floor joist track to wood sill per Figure R505.3.1(2)	Steel plate spaced at 3′ o.c., with 4-No. 8 screws and 4-10d or 6-8d common nails	Steel plate, spaced at 2′ o.c., with 4-No. 8 screws and 4-10d or 6-8d common nails
Floor joist track to foundation per Figure R505.3.1(3)	$1/_2''$ minimum diameter anchor bolt and clip angle spaced at 6′ o.c. with 8-No. 8 screws	$1/_2''$ minimum diameter anchor bolt and clip angle spaced at 4′ o.c. with 8-No. 8 screws
Joist cantilever to wall track per Figure R505.3.1(4)	2-No. 8 screws per stiffener or bent plate	3-No. 8 screws per stiffener or bent plate
Joist cantilever to wood sill per Figure R505.3.1(5)	Steel plate spaced at 3′ o.c., with 4-No. 8 screws and 4-10d or 6-8d common nails	Steel plate spaced at 2′ o.c., with 4-No. 8 screws and 4-10d or 6-8d common nails
Joist cantilever to foundation per Figure R505.3.1(6)	$1/_2''$ minimum diameter anchor bolt and clip angle spaced at 6′ o.c. with 8-No. 8 screws	$1/_2''$ minimum diameter anchor bolt and clip angle spaced at 4′ o.c. with 8-No. 8 screws

For SI: 1 inch = 25.4 mm, 1 foot = 304.8 mm, 1 mile per hour = 0.447 m/s.

a. Anchor bolts shall be located not more than 12 inches from corners or the termination of bottom tracks (e.g., at door openings). Bolts shall extend a minimum of 15 inches into masonry or 7 inches into concrete.

b. All screw sizes shown are minimum.

TABLE R505.3.1(2)
FLOOR FASTENING SCHEDULE[a]

DESCRIPTION OF BUILDING ELEMENTS	NUMBER AND SIZE OF FASTENERS	SPACING OF FASTENERS
Floor joist to track of an interior load-bearing wall per Figures R505.3.1(7) and R505.3.1(8)	2 No. 8 screws	Each joist
Floor joist to track at end of joist	2 No. 8 screws	One per flange or two per bearing stiffener
Subfloor to floor joists	No. 8 screws	6″ o.c. on edges and 12″ o.c. at intermediate supports

For SI: 1 inch = 25.4 mm.

a. All screw sizes shown are minimum.

R505.3.3 Joist bracing. The top flanges of steel joists shall be laterally braced by the application of floor sheathing fastened to the joists in accordance with Table R505.3.1(2). Floor joists with spans that exceed 12 feet (3658 mm) shall have the bottom flanges laterally braced in accordance with one of the following:

1. Gypsum board installed with minimum No. 6 screws in accordance with Section R702.

2. Continuous steel strapping installed in accordance with Figure R505.3. Steel straps shall be at least 1.5 inches (38 mm) in width and 33 mils (0.84 mm) in thickness. Straps shall be fastened to the bottom flange at each joist with at least one No. 8 screw and shall be fastened to blocking with at least two No. 8 screws. Blocking or bridging (X-bracing) shall be installed between joists in-line with straps at a maximum spacing of 12 feet (3658 mm) measured perpendicular to the joist run and at the termination of all straps.

R505.3.4 Bearing stiffeners. Bearing stiffeners shall be installed at all bearing locations for steel floor joists. A bearing stiffener shall be fabricated from a minimum 33 mil (0.84 mm) C-section or 43 mil (1.09 mm) track section. Each stiffener shall be fastened to the web of the joist with a minimum of four No. 8 screws equally spaced as shown in Figure R505.3.4. Stiffeners shall extend across the full depth of the web and shall be installed on either side of the web.

R505.3.5 Cutting and notching. Flanges and lips of load-bearing steel floor framing members shall not be cut or notched.

R505.3.6 Hole patching. Web holes not conforming to the requirements in Section R505.2 shall be designed in accordance with one of the following:

1. Framing members shall be replaced or designed in accordance with accepted engineering practices when web holes exceed the following size limits:

   1.1. The depth of the hole, measured across the web, exceeds 70 percent of the flat width of the web; or

   1.2. The length of the hole measured along the web, exceeds 10 inches (254 mm) or the depth of the web, whichever is greater.

2. Web holes not exceeding the dimensional requirements in Section R505.3.6, Item 1, shall be patched with a solid steel plate, stud section, or track section in accordance with Figure R505.3.6. The steel patch shall, as a minimum, be of the same thickness as the receiving member and shall extend at least 1 inch (25 mm) beyond all edges of the hole. The steel patch shall be fastened to the web of the receiving member with No.8 screws spaced no greater than 1 inch (25 mm) center-to-center along the edges of the patch with minimum edge distance of $1/_2$ inch (13 mm).

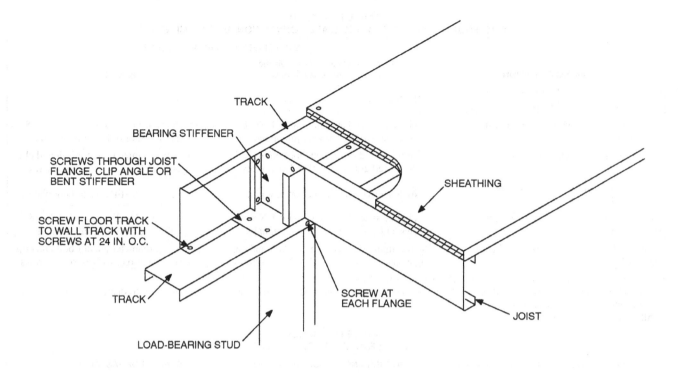

TRACK

BEARING STIFFENER

SCREWS THROUGH JOIST
FLANGE, CLIP ANGLE OR
BENT STIFFENER

SHEATHING

SCREW FLOOR TRACK
TO WALL TRACK WITH
SCREWS AT 24 IN. O.C.

TRACK

SCREW AT
EACH FLANGE

JOIST

LOAD-BEARING STUD

For SI: 1 inch = 25.4 mm, 1 mil = 0.0254 mm.

FIGURE R505.3.1(1)
FLOOR TO LOAD-BEARING WALL STUD CONNECTION

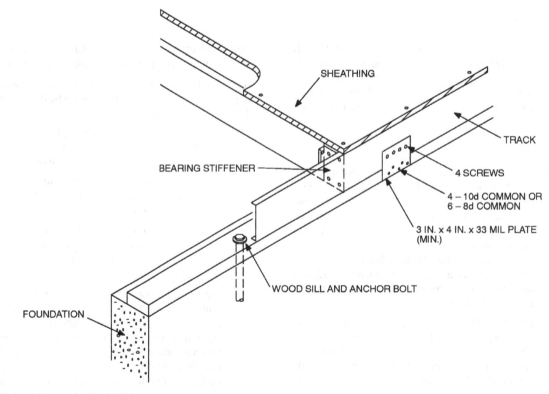

SHEATHING

TRACK

BEARING STIFFENER

4 SCREWS

4 – 10d COMMON OR
6 – 8d COMMON

3 IN. x 4 IN. x 33 MIL PLATE
(MIN.)

WOOD SILL AND ANCHOR BOLT

FOUNDATION

For SI: 1 inch = 25.4 mm, 1 mil = 0.0254 mm.

FIGURE R505.3.1(2)
FLOOR TO WOOD SILL CONNECTION

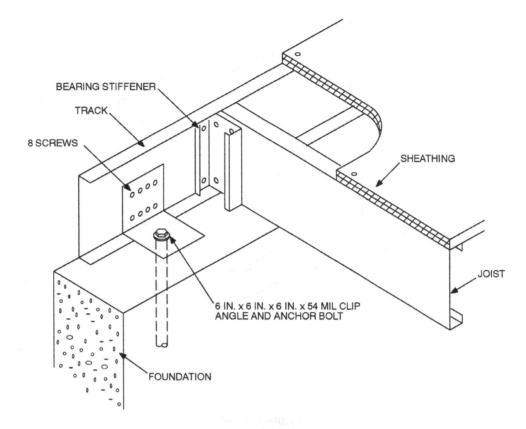

For SI: 1 inch = 25.4 mm, 1 mil = 0.0254 mm.

FIGURE R505.3.1(3)
FLOOR TO FOUNDATION CONNECTION

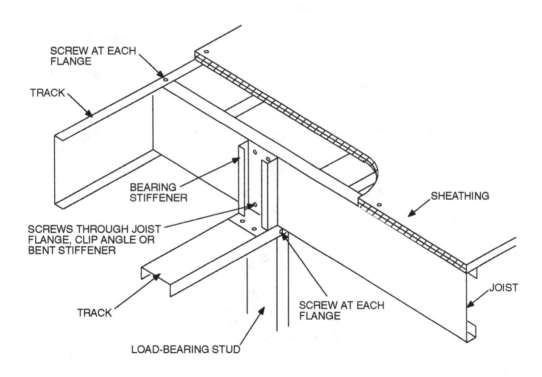

FIGURE R505.3.1(4)
FLOOR CANTILEVER TO LOAD-BEARING WALL CONNECTION

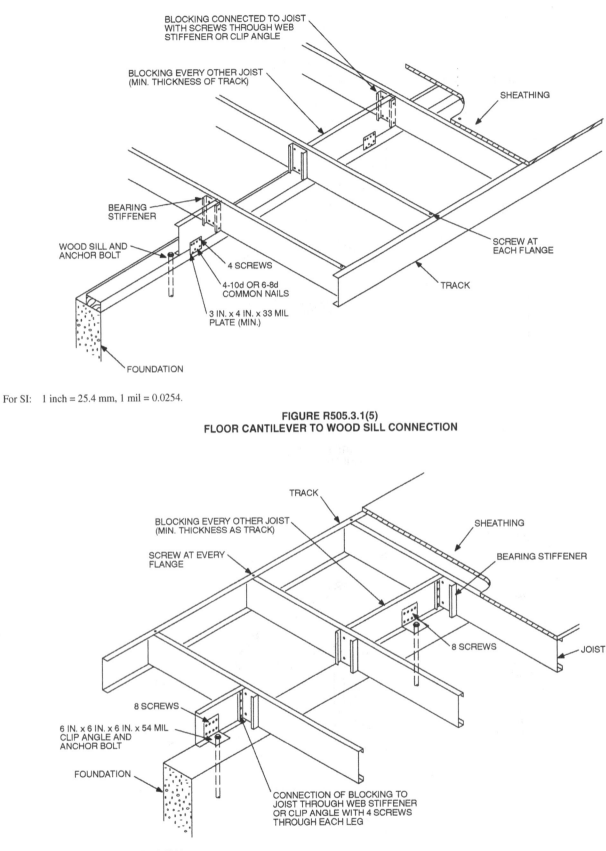

For SI: 1 inch = 25.4 mm, 1 mil = 0.0254.

FIGURE R505.3.1(5)
FLOOR CANTILEVER TO WOOD SILL CONNECTION

For SI: 1 inch = 25.4 mm, 1 mil = 0.0254.

FIGURE R505.3.1(6)
FLOOR CANTILEVER TO FOUNDATION CONNECTION

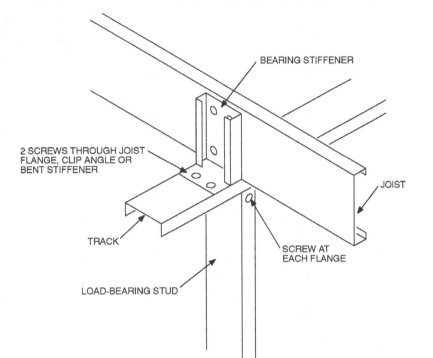

FIGURE R505.3.1(7)
CONTINUOUS JOIST SPAN SUPPORTED ON STUD

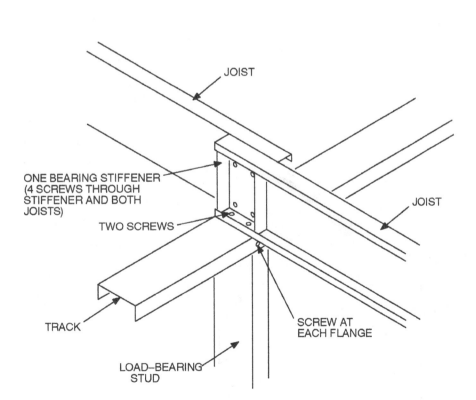

FIGURE R505.3.1(8)
LAPPED JOISTS SUPPORTED ON STUD

TABLE R505.3.2(1)
ALLOWABLE SPANS FOR COLD-FORMED STEEL JOISTS—SINGLE SPANS[a, b] 33 ksi STEEL

JOIST DESIGNATION	30 PSF LIVE LOAD				40 PSF LIVE LOAD			
	Spacing (inches)				Spacing (inches)			
	12	16	19.2	24	12	16	19.2	24
550S162-33	11'-7"	10'-7"	9'-6"	8'-6"	10'-7"	9'-3"	8'-6"	7'-6"
550S162-43	12'-8"	11'-6"	10'-10"	10'-2"	11'-6"	10'-5"	9'-10"	9'-1"
550S162-54	13'-7"	12'-4"	11'-7"	10'-9"	12'-4"	11'-2"	10'-6"	9'-9"
550S162-68	14'-7"	13'-3"	12'-6"	11'-7"	13'-3"	12'-0"	11'-4"	10'-6"
550S162-97	16'-2"	14'-9"	13'-10"	12'-10"	14'-9"	13'-4"	12'-7"	11'-8"
800S162-33	15'-8"	13'-11"	12'-9"	11'-5"	14'-3"	12'-5"	11'-3"	9'-0"
800S162-43	17'-1"	15'-6"	14'-7"	13'-7"	15'-6"	14'-1"	13'-3"	12'-4"
800S162-54	18'-4"	16'-8"	15'-8"	14'-7"	16'-8"	15'-2"	14'-3"	13'-3"
800S162-68	19'-9"	17'-11"	16'-10"	15'-8"	17'-11"	16'-3"	15'-4"	14'-2"
800S162-97	22'-0"	20'-0"	16'-10"	17'-5"	20'-0"	18'-2"	17'-1"	15'-10"
1000S162-43	20'-6"	18'-8"	17'-6"	15'-8"	18'-8"	16'-11"	15'-6"	13'-11"
1000S162-54	22'-1"	20'-0"	18'-10"	17'-6"	20'-0"	18'-2"	17'-2"	15'-11"
1000S162-68	23'-9"	21'-7"	20'-3"	18'-10"	21'-7"	19'-7"	18'-5"	17'-1"
1000S162-97	26'-6"	24'-1"	22'-8"	21'-0"	24'-1"	21'-10"	20'-7"	19'-1"
1200S162-43	23'-9"	20'-10"	19'-0"	16'-8"	21'-5"	18'-6"	16'-6"	13'-2"
1200S162-54	25'-9"	23'-4"	22'-0"	20'-1"	23'-4"	21'-3"	20'-0"	17'-10"
1200S162-68	27'-8"	25'-1"	23'-8"	21'-11"	25'-1"	22'-10"	21'-6"	21'-1"
1200S162-97	30'-11"	28'-1"	26'-5"	24'-6"	28'-1"	25'-6"	24'-0"	22'-3"

For SI: 1 inch = 25.4 mm, 1 foot = 304.8 mm, 1 pound per square foot = 0.0479 kPa.

a. Deflection criteria: L/480 for live loads, L/240 for total loads.

b. Floor dead load = 10 psf.

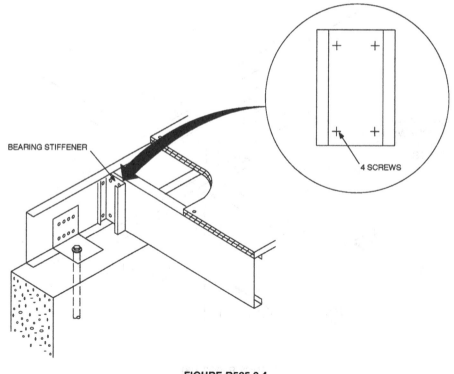

FIGURE R505.3.4
BEARING STIFFENER

TABLE R505.3.2(2)
ALLOWABLE SPANS FOR COLD-FORMED STEEL JOISTS—MULTIPLE SPANS[a, b] 33 ksi STEEL

JOIST DESIGNATION	30 PSF LIVE LOAD				40 PSF LIVE LOAD			
	Spacing (inches)				Spacing (inches)			
	12	16	19.2	24	12	16	19.2	24
550S162-33	12'-1"	10'-5"	9'-6"	8'-6"	10'-9"	9'-3"	8'-6"	7'-6"
550S162-43	14'-5"	12'-5"	11'-4"	10'-2"	12'-9"	11'-11"	10'-1"	9'-0"
550S162-54	16'-3"	14'-1"	12'-10"	11'-6"	14'-5"	12'-6"	11'-5"	10'-2"
550S162-68	19'-7"	17'-9"	16'-9"	15'-6"	17'-9"	16'-2"	15'-2"	14'-1"
550S162-97	21'-9"	19'-9"	18'-7"	17'-3"	19'-9"	17'-11"	16'-10"	15'-4"
800S162-33	14'-8"	11'-10"	10'-4"	8'-8"	12'-4"	9'-11"	8'-7"	7'-2"
800S162-43	20'-0"	17'-4"	15'-9"	14'-1"	17'-9"	15'-4"	14'-0"	12'-0"
800S162-54	23'-7"	20'-5"	18'-8"	16'-8"	21'-0"	18'-2"	16'-7"	14'-10"
800S162-68	26'-5"	23'-1"	21'-0"	18'-10"	23'-8"	20'-6"	18'-8"	16'-9"
800S162-97	29'-6"	26'-10"	25'-3"	22'-8"	26'-10"	24'-4"	22'-6"	20'-2"
1000S162-43	22'-2"	18'-3"	16'-0"	13'-7"	18'-11"	15'-5"	13'-6"	11'-5"
1000S162-54	26'-2"	22'-8"	20'-8"	18'-6"	23'-3"	20'-2"	18'-5"	16'-5"
1000S162-68	31'-5"	27'-2"	24'-10"	22'-2"	27'-11"	24'-2"	22'-1"	19'-9"
1000S162-97	35'-6"	32'-3"	29'-11"	26'-9"	32'-3"	29'-2"	26'-7"	23'-9"
1200S162-43	21'-8"	17'-6"	15'-3"	12'-10"	18'-3"	14'-8"	12'-8"	10'-6[2]
1200S162-54	28'-5"	24'-8"	22'-6"	19'-6"	25'-3"	21'-11"	19'-4"	16'-6"
1200S162-68	33'-7"	29'-1"	26'-6"	23'-9"	29'-10"	25'-10"	23'-7"	21'-1"
1200S162-97	41'-5"	37'-8"	34'-6"	30'-10"	37'-8"	33'-6"	30'-7"	27'-5"

For SI: 1 inch = 25.4 mm, 1 foot = 304.8 mm, 1 pound per square foot = 0.0479 kPa.

a. Deflection criteria: $L/480$ for live loads, $L/240$ for total loads.

b. Floor dead load = 10 psf.

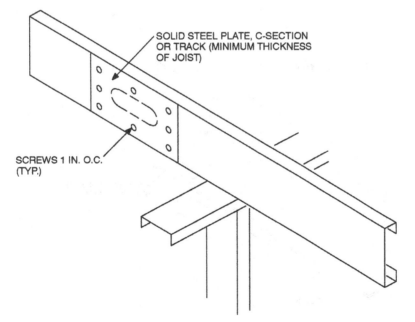

SOLID STEEL PLATE, C-SECTION
OR TRACK (MINIMUM THICKNESS
OF JOIST)

SCREWS 1 IN. O.C.
(TYP.)

For SI: 1 inch = 25.4 mm.

FIGURE R505.3.6
HOLE PATCH

TABLE R505.3.2(3)
ALLOWABLE SPANS FOR COLD-FORMED STEEL JOISTS—MULTIPLE SPANS[a, b] 50 ksi STEEL

JOIST DESIGNATION	30 PSF LIVE LOAD				40 PSF LIVE LOAD			
	Spacing (inches)				Spacing (inches)			
	12	16	19.2	24	12	16	19.2	24
550S162-33	13'-11"	12'-0"	11'-0"	9'-3"	12'-3"	10'-8"	9'-7"	8'-4"
550S162-43	16'-3"	14'-1"	12'-10"	11'-6"	14'-6"	12'-6"	11'-5"	10'-3"
550S162-54	18'-2"	16'-6"	15'-4"	13'-8"	16'-6"	14'-11"	13'-7"	12'-2"
550S162-68	19'-6"	17'-9"	16'-8"	15'-6"	17'-9"	16'-1"	15'-2"	14'-0"
550S162-97	21'-9"	19'-9"	18'-6"	17'-2"	19'-8"	17'-10"	16'-8"	15'-8"
800S162-33	15'-6"	12'-6"	10'-10"	9'-1"	13'-0"	10'-5"	8'-11"	6'-9"
800S162-43	22'-0"	19'-1"	17'-5"	15'-0"	19'-7"	16'-11"	14'-10"	12'-8"
800S162-54	24'-6"	22'-4"	20'-6"	17'-11"	22'-5"	19'-9"	17'-11"	15'-10"
800S162-68	26'-6"	24'-1"	22'-8"	21'-0"	24'-1"	21'-10"	20'-7"	19'-2"
800S162-97	29'-9"	26'-8"	25'-2"	23'-5"	26'-8"	24'-3"	22'-11"	21'-4"
1000S162-43	23'-6"	19'-2"	16'-9"	14'-2"	19'-11"	16'-2"	14'-0"	11'-9"
1000S162-54	28'-2"	23'-10"	21'-7"	18'-11"	24'-8"	20'-11"	18'-9"	18'-4"
1000S162-68	31'-10"	28'-11"	27'-2"	25'-3"	28'-11"	26'-3"	24'-9"	22'-9"
1000S162-97	35'-4"	32'-1"	30'-3"	28'-1"	32'-1"	29'-2"	27'-6"	25'-6"
1200S162-43	22'-11"	18'-5"	16'-0"	13'-4"	19'-2"	15'-4"	13'-2"	10'-6"
1200S162-54	32'-8"	28'-1"	24'-9"	21'-2"	29'-0"	23'-10"	20'-11"	17'-9"
1200S162-68	37'-1"	32'-5"	29'-4"	25'-10"	33'-4"	28'-6"	25'-9"	22'-7"
1200S162-97	41'-2"	37'-6"	35'-3"	32'-9"	37'-6"	34'-1"	32'-1"	29'-9"

For SI: 1 inch = 25.4 mm, 1 foot = 304.8 mm, 1 pound per square foot = 0.0479 kPa.
a. Deflection criteria: $L/480$ for live loads, $L/240$ for total loads.
b. Floor dead load = 10 psf.

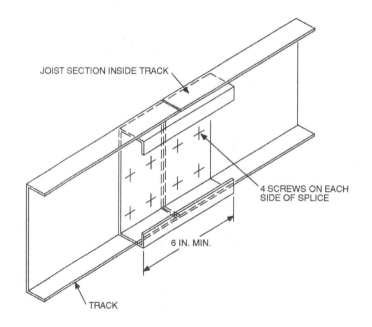

For SI: 1 inch = 25.4 mm.

FIGURE R505.3.8
TRACK SPLICE

R505.3.7 Floor cantilevers. Floor cantilevers shall not exceed 24 inches (610 mm) as illustrated in Figure R505.3. The cantilever back-span shall extend a minimum of 6 feet (1830 mm) within the building, and shall be fastened to a bearing condition in accordance with Section R505.3.1. Floor cantilevers shall be permitted only on the second floor of a two-story building or the first floor of a one-story building. Floor framing that is cantilevered and supports the cantilevered floor only shall consist of single joist members in accordance with Section R505.3.2. Floor framing that is cantilevered and supports the cantilevered floor and the roof framing load above shall consist of double joist members of the same size and material thickness as that for single joist members in accordance with Section R505.3.2, and shall be fastened web-to-web with minimum No. 8 screws at 24 inches (610 mm) maximum on-center spacing top and bottom. Built-up floor framing consisting of a C-section inside a track section, fastened at the top and bottom flanges by minimum No. 8 screws at 24 inches (610 mm) maximum on center spacing, is permitted in lieu of the web-to-web double joist method.

R505.3.8 Splicing. Joists and other structural members shall not be spliced. Splicing of tracks shall conform with Figure R505.3.8.

R505.3.9 Framing of openings. Openings in floor framing shall be framed with header and trimmer joists. Header joist spans shall not exceed 8 feet (2438 mm). Header and trimmer joists shall be fabricated from joist and track sections, which shall be of a minimum size and thickness as the adjacent floor joists and shall be installed in accordance with Figure R505.3. Each header joist shall be connected to trimmer joists with a minimum of four 2-inch-by-2-inch (51 mm by 51 mm) clip angles. Each clip angle shall be fastened to both the header and trimmer joists with four No. 8 screws, evenly spaced, through each leg of the clip angle. The clip angles shall have a steel thickness not less than that of the floor joist.

SECTION R506
CONCRETE FLOORS (ON GROUND)

R506.1 General. Concrete slab-on-ground floors shall be a minimum 3.5 inches (89 mm) thick (for expansive soils, see Section R403.1.8). The specified compressive strength of concrete shall be as set forth in Section R402.2.

R506.2 Site preparation. The area within the foundation walls shall have all vegetation, top soil and foreign material removed.

R506.2.1 Fill. Fill material shall be free of vegetation and foreign material. The fill shall be compacted to assure uniform support of the slab, and except where approved, the fill depths shall not exceed 24 inches (610 mm) for clean sand or gravel and 8 inches (203 mm) for earth.

R506.2.2 Base. A 4-inch-thick (102 mm) base course consisting of clean graded sand, gravel, crushed stone or crushed blast-furnace slag passing a 2-inch (51 mm) sieve shall be placed on the prepared subgrade when the slab is below grade.

> **Exception:** A base course is not required when the concrete slab is installed on well-drained or sand-gravel mixture soils classified as Group I according to the United Soil Classification System in accordance with Table R405.1.

R506.2.3 Vapor retarder. A 6 mil (0.006 inch; 152 μm) polyethylene or approved vapor retarder with joints lapped not less than 6 inches (152 mm) shall be placed between the concrete floor slab and the base course or the prepared subgrade where no base course exists.

> **Exception:** The vapor retarder may be omitted:
>
> 1. From garages, utility buildings and other unheated accessory structures.
>
> 2. From driveways, walks, patios and other flatwork not likely to be enclosed and heated at a later date.
>
> 3. Where approved by the building official, based on local site conditions.

R506.2.4 Reinforcement support. Where provided in slabs on ground, reinforcement shall be supported to remain in place from the center to upper one third of the slab for the duration of the concrete placement.

CHAPTER 6
WALL CONSTRUCTION

SECTION R601
GENERAL

R601.1 Application. The provisions of this chapter shall control the design and construction of all walls and partitions for all buildings.

R601.2 Requirements. Wall construction shall be capable of accommodating all loads imposed according to Section R301 and of transmitting the resulting loads to the supporting structural elements.

R601.2.1 Compressible floor-covering materials. Compressible floor-covering materials that compress more than $^1/_{32}$ inch (0.8 mm) when subjected to 50 pounds (23 kg) applied over 1 inch square (645 mm) of material and are greater than $^1/_8$ inch (3 mm) in thickness in the uncompressed state shall not extend beneath walls, partitions or columns, which are fastened to the floor.

SECTION R602
WOOD WALL FRAMING

R602.1 Identification. Load-bearing dimension lumber for studs, plates and headers shall be identified by a grade mark of a lumber grading or inspection agency that has been approved by an accreditation body that complies with DOC PS 20. In lieu of a grade mark, a certification of inspection issued by a lumber grading or inspection agency meeting the requirements of this section shall be accepted.

R602.1.1 End-jointed lumber. Approved end-jointed lumber identified by a grade mark conforming to Section R602.1 may be used interchangeably with solid-sawn members of the same species and grade.

R602.1.2 Structural glued laminated timbers. Glued laminated timbers shall be manufactured and identified as required in AITC A190.1 and ASTM D 3737.

R602.1.3 Structural log members. Stress grading of structural log members of nonrectangular shape, as typically used in log buildings, shall be in accordance with ASTM D 3957. Such structural log members shall be identified by the grade mark of an approved lumber grading or inspection agency. In lieu of a grade mark on the material, a certificate of inspection as to species and grade issued by a lumber-grading or inspection agency meeting the requirements of this section shall be permitted to be accepted.

R602.2 Grade. Studs shall be a minimum No. 3, standard or stud grade lumber.

Exception: Bearing studs not supporting floors and nonbearing studs may be utility grade lumber, provided the studs are spaced in accordance with Table R602.3(5).

R602.3 Design and construction. Exterior walls of wood-frame construction shall be designed and constructed in accordance with the provisions of this chapter and Figures R602.3(1) and R602.3(2) or in accordance with AF&PA's NDS. Components of exterior walls shall be fastened in accordance with Tables R602.3(1) through R602.3(4). Exterior walls covered with foam plastic sheathing shall be braced in accordance with Section R602.10. Structural sheathing shall be fastened directly to structural framing members.

R602.3.1 Stud size, height and spacing. The size, height and spacing of studs shall be in accordance with Table R602.3.(5).

Exceptions:

1. Utility grade studs shall not be spaced more than 16 inches (406 mm) on center, shall not support more than a roof and ceiling, and shall not exceed 8 feet (2438 mm) in height for exterior walls and load-bearing walls or 10 feet (3048 mm) for interior nonload-bearing walls.

2. Studs more than 10 feet (3048 mm) in height which are in accordance with Table R602.3.1.

R602.3.2 Top plate. Wood stud walls shall be capped with a double top plate installed to provide overlapping at corners and intersections with bearing partitions. End joints in top plates shall be offset at least 24 inches (610 mm). Joints in plates need not occur over studs. Plates shall be not less than 2-inches (51 mm) nominal thickness and have a width at least equal to the width of the studs.

Exception: A single top plate may be installed in stud walls, provided the plate is adequately tied at joints, corners and intersecting walls by a minimum 3-inch-by-6-inch by a 0.036-inch-thick (76 mm by 152 mm by 0.914 mm) galvanized steel plate that is nailed to each wall or segment of wall by six 8d nails on each side, provided the rafters or joists are centered over the studs with a tolerance of no more than 1 inch (25 mm). The top plate may be omitted over lintels that are adequately tied to adjacent wall sections with steel plates or equivalent as previously described.

R602.3.3 Bearing studs. Where joists, trusses or rafters are spaced more than 16 inches (406 mm) on center and the bearing studs below are spaced 24 inches (610 mm) on center, such members shall bear within 5 inches (127 mm) of the studs beneath.

Exceptions:

1. The top plates are two 2-inch by 6-inch (38 mm by 140 mm) or two 3-inch by 4-inch (64 mm by 89 mm) members.

2. A third top plate is installed.

3. Solid blocking equal in size to the studs is installed to reinforce the double top plate.

TABLE R602.3(1)
FASTENER SCHEDULE FOR STRUCTURAL MEMBERS

DESCRIPTION OF BUILDING ELEMENTS	NUMBER AND TYPE OF FASTENER[a,b,c]	SPACING OF FASTENERS
Joist to sill or girder, toe nail	3-8d ($2\text{-}^1/_2'' \times 0.113''$)	—
$1'' \times 6''$ subfloor or less to each joist, face nail	2-8d ($2^1/_2'' \times 0.113''$) 2 staples, $1^3/_4''$	—
$2''$ subfloor to joist or girder, blind and face nail	2-16d ($3^1/_2'' \times 0.135''$)	—
Sole plate to joist or blocking, face nail	16d ($3^1/_2'' \times 0.135''$)	$16''$ o.c.
Top or sole plate to stud, end nail	2-16d ($3^1/_2'' \times 0.135''$)	—
Stud to sole plate, toe nail	3-8d ($2^1/_2'' \times 0.113''$) or 2-16d ($3^1/_2'' \times 0.135''$)	—
Double studs, face nail	10d ($3'' \times 0.128''$)	$24''$ o.c.
Double top plates, face nail	10d ($3'' \times 0.128''$)	$24''$ o.c.
Sole plate to joist or blocking at braced wall panels	3-16d ($3^1/_2'' \times 0.135''$)	$16''$ o.c.
Double top plates, minimum 24-inch offset of end joints, face nail in lapped area	8-16d ($3^1/_2'' \times 0.135''$)	—
Blocking between joists or rafters to top plate, toe nail	3-8d ($2^1/_2'' \times 0.113''$)	—
Rim joist to top plate, toe nail	8d ($2^1/_2'' \times 0.113''$)	$6''$ o.c.
Top plates, laps at corners and intersections, face nail	2-10d ($3'' \times 0.128''$)	—
Built-up header, two pieces with $1/_2''$ spacer	16d ($3^1/_2'' \times 0.135''$)	$16''$ o.c. along each edge
Continued header, two pieces	16d ($3^1/_2'' \times 0.135''$)	$16''$ o.c. along each edge
Ceiling joists to plate, toe nail	3-8d ($2^1/_2'' \times 0.113''$)	—
Continuous header to stud, toe nail	4-8d ($2^1/_2'' \times 0.113''$)	—
Ceiling joist, laps over partitions, face nail	3-10d ($3'' \times 0.128''$)	—
Ceiling joist to parallel rafters, face nail	3-10d ($3'' \times 0.128''$)	—
Rafter to plate, toe nail	2-16d ($3^1/_2'' \times 0.135''$)	—
$1''$ brace to each stud and plate, face nail	2-8d ($2^1/_2'' \times 0.113''$) 2 staples, $1^3/_4''$	— —
$1'' \times 6''$ sheathing to each bearing, face nail	2-8d ($2^1/_2'' \times 0.113''$) 2 staples, $1^3/_4''$	— —
$1'' \times 8''$ sheathing to each bearing, face nail	2-8d ($2^1/_2'' \times 0.113''$) 3 staples, $1^3/_4''$	— —
Wider than $1'' \times 8''$ sheathing to each bearing, face nail	3-8d ($2^1/_2'' \times 0.113''$) 4 staples, $1^3/_4''$	— —
Built-up corner studs	10d ($3'' \times 0.128''$)	$24''$o.c.
Built-up girders and beams, 2-inch lumber layers	10d ($3'' \times 0.128''$)	Nail each layer as follows: $32''$ o.c. at top and bottom and staggered. Two nails at ends and at each splice.
$2''$ planks	2-16d ($3^1/_2'' \times 0.135''$)	At each bearing
Roof rafters to ridge, valley or hip rafters: toe nail face nail	 4-16d ($3^1/_2'' \times 0.135''$) 3-16d ($3^1/_2'' \times 0.135''$)	 — —
Rafter ties to rafters, face nail	3-8d ($2^1/_2'' \times 0.113''$)	—
Collar tie to rafter, face nail, or $1^1/_4'' \times 20$ gage ridge strap	3-10d ($3'' \times 0.128''$)	—

(continued)

TABLE R602.3(1)—continued
FASTENER SCHEDULE FOR STRUCTURAL MEMBERS

DESCRIPTION OF BUILDING MATERIALS	DESCRIPTION OF FASTENER[b, c, e]	SPACING OF FASTENERS	
		Edges (inches)[i]	Intermediate supports[c,e] (inches)
Wood structural panels, subfloor, roof and wall sheathing to framing, and particleboard wall sheathing to framing			
$^5/_{16}"$-$^1/_2"$	6d common ($2" \times 0.113"$) nail (subfloor, wall) 8d common ($2^1/_2" \times 0.131"$) nail (roof)[f]	6	12[g]
$^{19}/_{32}"$ -1"	8d common nail ($2^1/_2" \times 0.131"$)	6	12[g]
$1^1/_8"$-$1^1/_4"$	10d common ($3" \times 0.148"$) nail or 8d ($2^1/_2" \times 0.131"$) deformed nail	6	12
Other wall sheathing[h]			
$^1/_2"$ structural cellulosic fiberboard sheathing	$1^1/_2"$ galvanized roofing nail 8d common ($2^1/_2" \times 0.131"$) nail; staple 16 ga., $1^1/_2"$ long	3	6
$^{25}/_{32}"$ structural cellulosic fiberboard sheathing	$1^3/_4"$ galvanized roofing nail 8d common ($2^1/_2" \times 0.131"$) nail; staple 16 ga., $1^3/_4"$ long	3	6
$^1/_2"$ gypsum sheathing[d]	$1^1/_2"$ galvanized roofing nail; 6d common ($2" \times 0.131"$) nail; staple galvanized $1^1/_2"$ long; $1^1/_4"$ screws, Type W or S	4	8
$^5/_8"$ gypsum sheathing[d]	$1^3/_4"$ galvanized roofing nail; 8d common ($2^1/_2" \times 0.131"$) nail; staple galvanized $1^5/_8"$ long; $1^5/_8"$ screws, Type W or S	4	8
Wood structural panels, combination subfloor underlayment to framing			
$^3/_4"$ and less	6d deformed ($2" \times 0.120"$) nail or 8d common ($2^1/_2" \times 0.131"$) nail	6	12
$^7/_8"$-1"	8d common ($2^1/_2" \times 0.131"$) nail or 8d deformed ($2^1/_2" \times 0.120"$) nail	6	12
$1^1/_8"$-$1^1/_4"$	10d common ($3" \times 0.148"$) nail or 8d deformed ($2^1/_2" \times 0.120"$) nail	6	12

For SI: 1 inch = 25.4 mm, 1 foot = 304.8 mm, 1 mile per hour = 0.447 m/s; 1ksi = 6.895 MPa.

a. All nails are smooth-common, box or deformed shanks except where otherwise stated. Nails used for framing and sheathing connections shall have minimum average bending yield strengths as shown: 80 ksi for shank diameter of 0.192 inch (20d common nail), 90 ksi for shank diameters larger than 0.142 inch but not larger than 0.177 inch, and 100 ksi for shank diameters of 0.142 inch or less.

b. Staples are 16 gage wire and have a minimum $^7/_{16}$-inch on diameter crown width.

c. Nails shall be spaced at not more than 6 inches on center at all supports where spans are 48 inches or greater.

d. Four-foot-by-8-foot or 4-foot-by-9-foot panels shall be applied vertically.

e. Spacing of fasteners not included in this table shall be based on Table R602.3(2).

f. For regions having basic wind speed of 110 mph or greater, 8d deformed ($2^1/_2" \times 0.120$) nails shall be used for attaching plywood and wood structural panel roof sheathing to framing within minimum 48-inch distance from gable end walls, if mean roof height is more than 25 feet, up to 35 feet maximum.

g. For regions having basic wind speed of 100 mph or less, nails for attaching wood structural panel roof sheathing to gable end wall framing shall be spaced 6 inches on center. When basic wind speed is greater than 100 mph, nails for attaching panel roof sheathing to intermediate supports shall be spaced 6 inches on center for minimum 48-inch distance from ridges, eaves and gable end walls; and 4 inches on center to gable end wall framing.

h. Gypsum sheathing shall conform to ASTM C 79 and shall be installed in accordance with GA 253. Fiberboard sheathing shall conform to ASTM C 208.

i. Spacing of fasteners on floor sheathing panel edges applies to panel edges supported by framing members and required blocking and at all floor perimeters only. Spacing of fasteners on roof sheathing panel edges applies to panel edges supported by framing members and required blocking. Blocking of roof or floor sheathing panel edges perpendicular to the framing members need not be provided except as required by other provisions of this code. Floor perimeter shall be supported by framing members or solid blocking.

TABLE R602.3(2)
ALTERNATE ATTACHMENTS

NOMINAL MATERIAL THICKNESS (inches)	DESCRIPTION[a, b] OF FASTENER AND LENGTH (inches)	SPACING[c] OF FASTENERS	
		Edges (inches)	Intermediate supports (inches)
Wood structural panels subfloor, roof and wall sheathing to framing and particleboard wall sheathing to framing[f]			
up to $^1/_2$	Staple 15 ga. $1^3/_4$	4	8
	0.097 - 0.099 Nail $2^1/_4$	3	6
	Staple 16 ga. $1^3/_4$	3	6
$^{19}/_{32}$ and $^5/_8$	0.113 Nail 2	3	6
	Staple 15 and 16 ga. 2	4	8
	0.097 - 0.099 Nail $2^1/_4$	4	8
$^{23}/_{32}$ and $^3/_4$	Staple 14 ga. 2	4	8
	Staple 15 ga. $1^3/_4$	3	6
	0.097 - 0.099 Nail $2^1/_4$	4	8
	Staple 16 ga. 2	4	8
1	Staple 14 ga. $2^1/_4$	4	8
	0.113 Nail $2^1/_4$	3	6
	Staple 15 ga. $2^1/_4$	4	8
	0.097 - 0.099 Nail $2^1/_2$	4	8

NOMINAL MATERIAL THICKNESS (inches)	DESCRIPTION[a, b] OF FASTENER AND LENGTH (inches)	SPACING[c] OF FASTENERS	
		Edges (inches)	Body of panel[d] (inches)
Floor underlayment; plywood-hardboard-particleboard[f]			
Plywood			
$^1/_4$ and $^5/_{16}$	$1^1/_4$ ring or screw shank nail—minimum $12^1/_2$ ga. (0.099″) shank diameter	3	6
	Staple 18 ga., $^7/_8$, $^3/_{16}$ crown width	2	5
$^{11}/_{32}$, $^3/_8$, $^{15}/_{32}$, $^1/_2$ and $^{19}/_{32}$	$1^1/_4$ ring or screw shank nail—minimum $12^1/_2$ ga. (0.099″) shank diameter	6	8[e]
$^5/_8$, $^{23}/_{32}$ and $^3/_4$	$1^1/_2$ ring or screw shank nail—minimum $12^1/_2$ ga. (0.099″) shank diameter	6	8
	Staple 16 ga. $1^1/_2$	6	8
Hardboard[f]			
0.200	$1^1/_2$ long ring-grooved underlayment nail	6	6
	4d cement-coated sinker nail	6	6
	Staple 18 ga., $^7/_8$ long (plastic coated)	3	6
Particleboard			
$^1/_4$	4d ring-grooved underlayment nail	3	6
	Staple 18 ga., $^7/_8$ long, $^3/_{16}$ crown	3	6
$^3/_8$	6d ring-grooved underlayment nail	6	10
	Staple 16 ga., $1^1/_8$ long, $^3/_8$ crown	3	6
$^1/_2$, $^5/_8$	6d ring-grooved underlayment nail	6	10
	Staple 16 ga., $1^5/_8$ long, $^3/_8$ crown	3	6

For SI: 1 inch = 25.4 mm.

a. Nail is a general description and may be T-head, modified round head or round head.

b. Staples shall have a minimum crown width of $^7/_{16}$-inch on diameter except as noted.

c. Nails or staples shall be spaced at not more than 6 inches on center at all supports where spans are 48 inches or greater. Nails or staples shall be spaced at not more than 12 inches on center at intermediate supports for floors.

d. Fasteners shall be placed in a grid pattern throughout the body of the panel.

e. For 5-ply panels, intermediate nails shall be spaced not more than 12 inches on center each way.

f. Hardboard underlayment shall conform to ANSI/AHA A135.4.

TABLE R602.3(3)
WOOD STRUCTURAL PANEL WALL SHEATHING

PANEL SPAN RATING	PANEL NOMINAL THICKNESS (inch)	MAXIMUM STUD SPACING (inches)	
		Siding nailed to:[a]	
		Stud	Sheathing
12/0, 16/0, 20/0, or wall —16 o.c.	$^5/_{16}$, $^3/_8$	16	16[b]
24/0, 24/16, 32/16 or wall—24 o.c.	$^3/_8$, $^7/_{16}$, $^{15}/_{32}$, $^1/_2$	24	24[c]

For SI: 1 inch = 25.4 mm.

a. Blocking of horizontal joints shall not be required.

b. Plywood sheathing $^3/_8$-inch thick or less shall be applied with long dimension across studs.

c. Three-ply plywood panels shall be applied with long dimension across studs.

TABLE R602.3(4)
ALLOWABLE SPANS FOR PARTICLEBOARD WALL SHEATHING[a]

THICKNESS (inch)	GRADE	STUD SPACING (inches)	
		When siding is nailed to studs	When siding is nailed to sheathing
$^3/_8$	M-1 Exterior glue	16	—
$^1/_2$	M-2 Exterior glue	16	16

For SI: 1 inch = 25.4 mm.

a. Wall sheathing not exposed to the weather. If the panels are applied horizontally, the end joints of the panel shall be offset so that four panels corners will not meet. All panel edges must be supported. Leave a $^1/_{16}$-inch gap between panels and nail no closer than $^3/_8$ inch from panel edges.

TABLE R602.3(5)
SIZE, HEIGHT AND SPACING OF WOOD STUDS[a]

STUD SIZE (inches)	BEARING WALLS					NONBEARING WALLS	
	Laterally unsupported stud height[a] (feet)	Maximum spacing when supporting roof and ceiling only (inches)	Maximum spacing when supporting one floor, roof and ceiling (inches)	Maximum spacing when supporting two floors, roof and ceiling (inches)	Maximum spacing when supporting one floor only (inches)	Laterally unsupported stud height[a] (feet)	Maximum spacing (inches)
2 × 3[b]	—	—	—	—	—	10	16
2 × 4	10	24	16	—	24	14	24
3 × 4	10	24	24	16	24	14	24
2 × 5	10	24	24	—	24	16	24
2 × 6	10	24	24	16	24	20	24

For SI: 1 inch = 25.4 mm.

a. Listed heights are distances between points of lateral support placed perpendicular to the plane of the wall. Increases in unsupported height are permitted where justified by analysis.

b. Shall not be used in exterior walls.

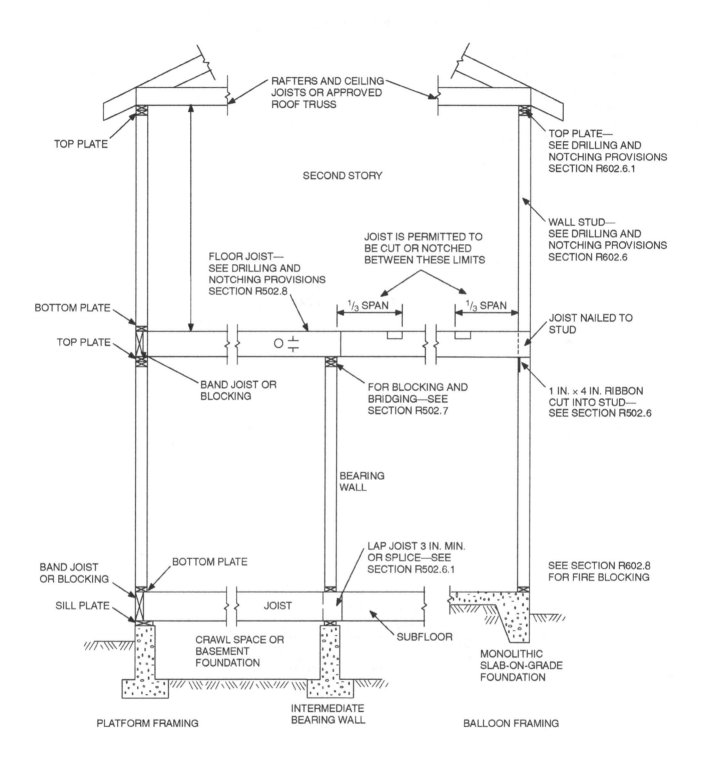

RAFTERS AND CEILING
JOISTS OR APPROVED
ROOF TRUSS

TOP PLATE

TOP PLATE—
SEE DRILLING AND
NOTCHING PROVISIONS
SECTION R602.6.1

SECOND STORY

WALL STUD—
SEE DRILLING AND
NOTCHING PROVISIONS
SECTION R602.6

FLOOR JOIST—
SEE DRILLING AND
NOTCHING PROVISIONS
SECTION R502.8

JOIST IS PERMITTED TO
BE CUT OR NOTCHED
BETWEEN THESE LIMITS

BOTTOM PLATE

$^1/_3$ SPAN $^1/_3$ SPAN

TOP PLATE

JOIST NAILED TO
STUD

BAND JOIST OR
BLOCKING

FOR BLOCKING AND
BRIDGING—SEE
SECTION R502.7

1 IN. × 4 IN. RIBBON
CUT INTO STUD—
SEE SECTION R502.6

BEARING
WALL

BAND JOIST
OR BLOCKING

BOTTOM PLATE

LAP JOIST 3 IN. MIN.
OR SPLICE—SEE
SECTION R502.6.1

SEE SECTION R602.8
FOR FIRE BLOCKING

SILL PLATE

JOIST

SUBFLOOR

CRAWL SPACE OR
BASEMENT
FOUNDATION

MONOLITHIC
SLAB-ON-GRADE
FOUNDATION

PLATFORM FRAMING

INTERMEDIATE
BEARING WALL

BALLOON FRAMING

For SI: 1 inch = 25.4 mm.

FIGURE R602.3(1)
TYPICAL WALL, FLOOR AND ROOF FRAMING

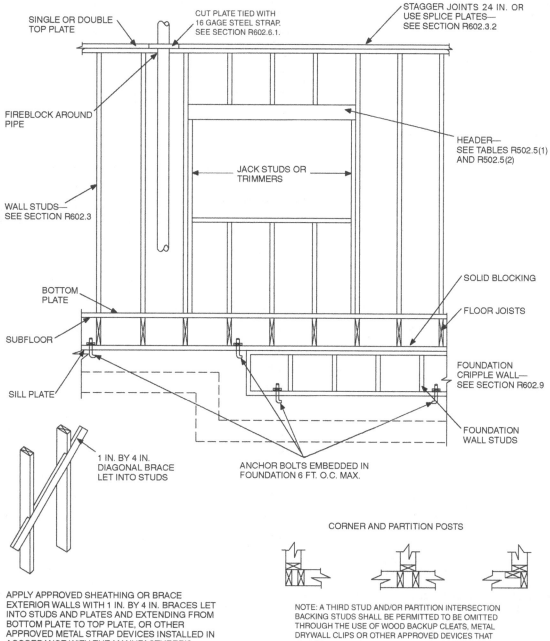

SINGLE OR DOUBLE TOP PLATE

CUT PLATE TIED WITH 16 GAGE STEEL STRAP. SEE SECTION R602.6.1.

STAGGER JOINTS 24 IN. OR USE SPLICE PLATES— SEE SECTION R602.3.2

FIREBLOCK AROUND PIPE

HEADER— SEE TABLES R502.5(1) AND R502.5(2)

JACK STUDS OR TRIMMERS

WALL STUDS— SEE SECTION R602.3

BOTTOM PLATE

SOLID BLOCKING

FLOOR JOISTS

SUBFLOOR

FOUNDATION CRIPPLE WALL— SEE SECTION R602.9

SILL PLATE

FOUNDATION WALL STUDS

1 IN. BY 4 IN. DIAGONAL BRACE LET INTO STUDS

ANCHOR BOLTS EMBEDDED IN FOUNDATION 6 FT. O.C. MAX.

CORNER AND PARTITION POSTS

APPLY APPROVED SHEATHING OR BRACE EXTERIOR WALLS WITH 1 IN. BY 4 IN. BRACES LET INTO STUDS AND PLATES AND EXTENDING FROM BOTTOM PLATE TO TOP PLATE, OR OTHER APPROVED METAL STRAP DEVICES INSTALLED IN ACCORDANCE WITH THE MANUFACTURER'S SPECIFICATIONS. SEE SECTION R602.10.

NOTE: A THIRD STUD AND/OR PARTITION INTERSECTION BACKING STUDS SHALL BE PERMITTED TO BE OMITTED THROUGH THE USE OF WOOD BACKUP CLEATS, METAL DRYWALL CLIPS OR OTHER APPROVED DEVICES THAT WILL SERVE AS ADEQUATE BACKING FOR THE FACING MATERIALS.

For SI: 1 inch = 25.4 mm, 1 foot = 304.8 mm.

FIGURE R602.3(2)
FRAMING DETAILS

TABLE R602.3.1
MAXIMUM ALLOWABLE LENGTH OF WOOD WALL STUDS EXPOSED TO WIND SPEEDS OF 100 mph OR LESS
IN SEISMIC DESIGN CATEGORIES A, B, C, D_0, D_1 AND D_2[b, c]

HEIGHT (feet)	ON-CENTER SPACING (inches)			
	24	16	12	8
Supporting a roof only				
>10	2 × 4	2 × 4	2 × 4	2 × 4
12	2 × 6	2 × 4	2 × 4	2 × 4
14	2 × 6	2 × 6	2 × 6	2 × 4
16	2 × 6	2 × 6	2 × 6	2 × 4
18	NA[a]	2 × 6	2 × 6	2 × 6
20	NA[a]	NA[a]	2 × 6	2 × 6
24	NA[a]	NA[a]	NA[a]	2 × 6
Supporting one floor and a roof				
>10	2 × 6	2 × 4	2 × 4	2 × 4
12	2 × 6	2 × 6	2 × 6	2 × 4
14	2 × 6	2 × 6	2 × 6	2 × 6
16	NA[a]	2 × 6	2 × 6	2 × 6
18	NA[a]	2 × 6	2 × 6	2 × 6
20	NA[a]	NA[a]	2 × 6	2 × 6
24	NA[a]	NA[a]	NA[a]	2 × 6
Supporting two floors and a roof				
>10	2 × 6	2 × 6	2 × 4	2 × 4
12	2 × 6	2 × 6	2 × 6	2 × 6
14	2 × 6	2 × 6	2 × 6	2 × 6
16	NA[a]	NA[a]	2 × 6	2 × 6
18	NA[a]	NA[a]	2 × 6	2 × 6
20	NA[a]	NA[a]	NA[a]	2 × 6
22	NA[a]	NA[a]	NA[a]	NA[a]
24	NA[a]	NA[a]	NA[a]	NA[a]

For SI: 1 inch = 25.4 mm, 1 foot = 304.8 mm, 1 pound per square foot = 0.0479 kPa,
1 pound per square inch = 6.895 kPa, 1 mile per hour = 0.447 m/s.

a. Design required.

b. Applicability of this table assumes the following: Snow load not exceeding 25 psf, f_b, not less than 1310 psi determined by multiplying the AF&PA NDS tabular base design value by the repetitive use factor, and by the size factor for all species except southern pine, E not less than 1.6×10^6 psi, tributary dimensions for floors and roofs not exceeding 6 feet, maximum span for floors and roof not exceeding 12 feet, eaves not over 2 feet in dimension and exterior sheathing. Where the conditions are not within these parameters, design is required.

c. Utility, standard, stud and No. 3 grade lumber of any species are not permitted.

(continued)

TABLE R602.3.1—continued
MAXIMUM ALLOWABLE LENGTH OF WOOD WALL STUDS EXPOSED TO WIND SPEEDS OF 100 mph OR LESS
IN SEISMIC DESIGN CATEGORIES A, B, C, D_0, D_1 and D_2

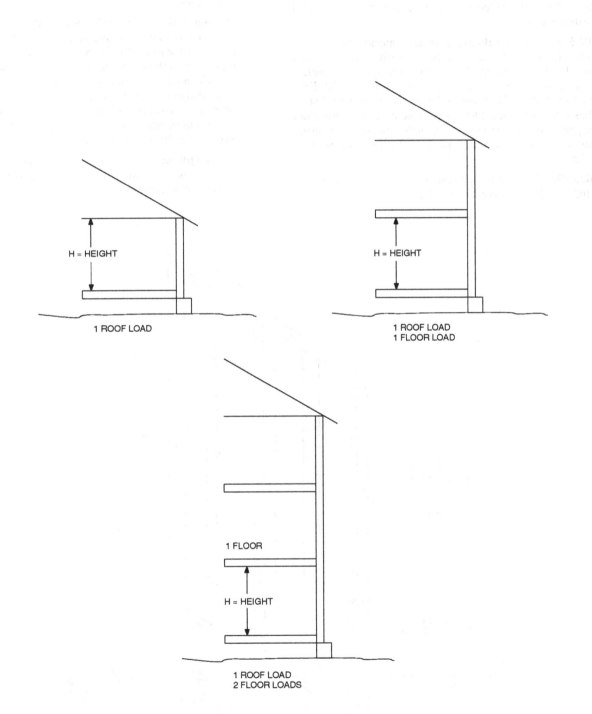

1 ROOF LOAD

1 ROOF LOAD
1 FLOOR LOAD

1 FLOOR

1 ROOF LOAD
2 FLOOR LOADS

R602.3.4 Bottom (sole) plate. Studs shall have full bearing on a nominal 2-by (38 mm) or larger plate or sill having a width at least equal to the width of the studs.

R602.4 Interior load-bearing walls. Interior load-bearing walls shall be constructed, framed and fireblocked as specified for exterior walls.

R602.5 Interior nonbearing walls. Interior nonbearing walls shall be permitted to be constructed with 2-inch-by-3-inch (51 mm by 76 mm) studs spaced 24 inches (610 mm) on center or, when not part of a braced wall line, 2-inch-by-4-inch (51 mm by 102 mm) flat studs spaced at 16 inches (406 mm) on center. Interior nonbearing walls shall be capped with at least a single top plate. Interior nonbearing walls shall be fireblocked in accordance with Section R602.8.

R602.6 Drilling and notching–studs. Drilling and notching of studs shall be in accordance with the following:

1. Notching. Any stud in an exterior wall or bearing partition may be cut or notched to a depth not exceeding 25 percent of its width. Studs in nonbearing partitions may be notched to a depth not to exceed 40 percent of a single stud width.

2. Drilling. Any stud may be bored or drilled, provided that the diameter of the resulting hole is no more than 60 percent of the stud width, the edge of the hole is no more than 5/8 inch (16 mm) to the edge of the stud, and the hole is not located in the same section as a cut or notch. Studs located in exterior walls or bearing partitions drilled over 40 percent and up to 60 percent shall also be doubled with no more than two successive doubled studs bored. See Figures R602.6(1) and R602.6(2).

   Exception: Use of approved stud shoes is permitted when they are installed in accordance with the manufacturer's recommendations.

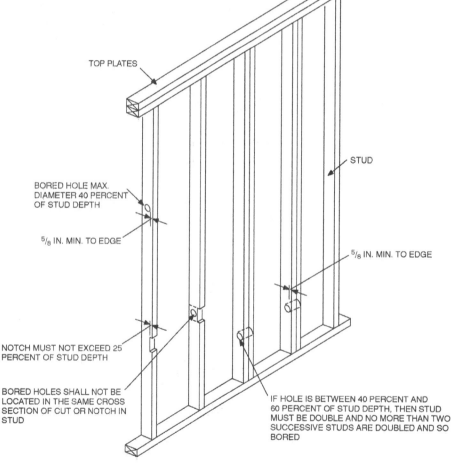

TOP PLATES

STUD

BORED HOLE MAX. DIAMETER 40 PERCENT OF STUD DEPTH

5/8 IN. MIN. TO EDGE

5/8 IN. MIN. TO EDGE

NOTCH MUST NOT EXCEED 25 PERCENT OF STUD DEPTH

BORED HOLES SHALL NOT BE LOCATED IN THE SAME CROSS SECTION OF CUT OR NOTCH IN STUD

IF HOLE IS BETWEEN 40 PERCENT AND 60 PERCENT OF STUD DEPTH, THEN STUD MUST BE DOUBLE AND NO MORE THAN TWO SUCCESSIVE STUDS ARE DOUBLED AND SO BORED

For SI: 1 inch = 25.4 mm.
NOTE: Condition for exterior and bearing walls.

FIGURE R602.6(1)
NOTCHING AND BORED HOLE LIMITATIONS FOR EXTERIOR WALLS AND BEARING WALLS

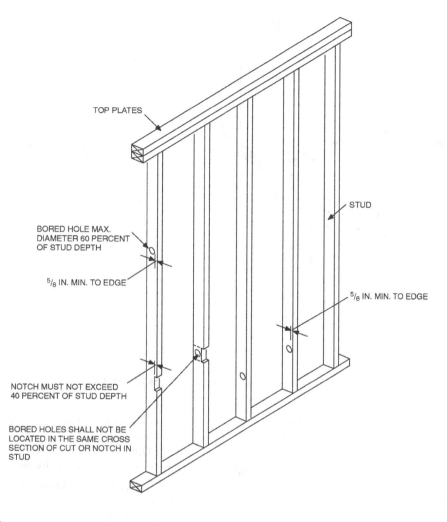

TOP PLATES

STUD

BORED HOLE MAX.
DIAMETER 60 PERCENT
OF STUD DEPTH

$^5/_8$ IN. MIN. TO EDGE

$^5/_8$ IN. MIN. TO EDGE

NOTCH MUST NOT EXCEED
40 PERCENT OF STUD DEPTH

BORED HOLES SHALL NOT BE
LOCATED IN THE SAME CROSS
SECTION OF CUT OR NOTCH IN
STUD

For SI: 1 inch = 25.4 mm.

FIGURE R602.6(2)
NOTCHING AND BORED HOLE LIMITATIONS FOR INTERIOR NONBEARING WALLS

R602.6.1 Drilling and notching of top plate. When piping or ductwork is placed in or partly in an exterior wall or interior load-bearing wall, necessitating cutting, drilling or notching of the top plate by more than 50 percent of its width, a galvanized metal tie of not less than 0.054 inch thick (1.37 mm) (16 ga) and 1 $^1/_2$ inches (38 mm) wide shall be fastened across and to the plate at each side of the opening with not less than eight 16d nails at each side or equivalent. See Figure R602.6.1.

Exception: When the entire side of the wall with the notch or cut is covered by wood structural panel sheathing.

R602.7 Headers. For header spans see Tables R502.5(1) and R502.5(2).

R602.7.1 Wood structural panel box headers. Wood structural panel box headers shall be constructed in accordance with Figure R602.7.2 and Table R602.7.2.

R602.7.2 Nonbearing walls. Load-bearing headers are not required in interior or exterior nonbearing walls. A single

flat 2-inch-by-4-inch (51 mm by 102 mm) member may be used as a header in interior or exterior nonbearing walls for openings up to 8 feet (2438 mm) in width if the vertical distance to the parallel nailing surface above is not more than 24 inches (610 mm). For such nonbearing headers, no cripples or blocking are required above the header.

R602.8 Fireblocking required. Fireblocking shall be provided to cut off all concealed draft openings (both vertical and horizontal) and to form an effective fire barrier between stories, and between a top story and the roof space. Fireblocking shall be provided in wood-frame construction in the following locations.

1. In concealed spaces of stud walls and partitions, including furred spaces and parallel rows of studs or staggered studs; as follows:

    1.1. Vertically at the ceiling and floor levels.

    1.2. Horizontally at intervals not exceeding 10 feet (3048 mm).

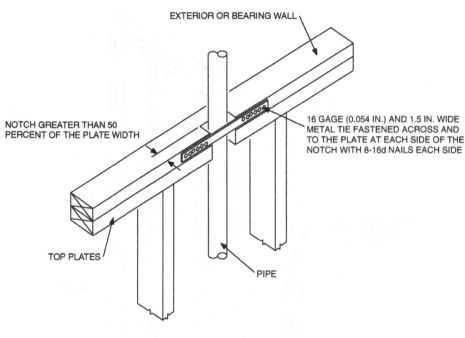

EXTERIOR OR BEARING WALL

NOTCH GREATER THAN 50
PERCENT OF THE PLATE WIDTH

16 GAGE (0.054 IN.) AND 1.5 IN. WIDE
METAL TIE FASTENED ACROSS AND
TO THE PLATE AT EACH SIDE OF THE
NOTCH WITH 8-16d NAILS EACH SIDE

TOP PLATES

PIPE

For SI: 1 inch = 25.4 mm.

FIGURE R602.6.1
TOP PLATE FRAMING TO ACCOMMODATE PIPING

2. At all interconnections between concealed vertical and horizontal spaces such as occur at soffits, drop ceilings and cove ceilings.

3. In concealed spaces between stair stringers at the top and bottom of the run. Enclosed spaces under stairs shall comply with Section R311.2.2.

4. At openings around vents, pipes, ducts, cables and wires at ceiling and floor level, with an approved material to resist the free passage of flame and products of combustion.

5. For the fireblocking of chimneys and fireplaces, see Section R1003.19.

6. Fireblocking of cornices of a two-family dwelling is required at the line of dwelling unit separation.

R602.8.1 Materials. Except as provided in Section R602.8, Item 4, fireblocking shall consist of 2-inch (51 mm) nominal lumber, or two thicknesses of 1-inch (25.4 mm) nominal lumber with broken lap joints, or one thickness of $^{23}/_{32}$-inch (18.3 mm) wood structural panels with joints backed by $^{23}/_{32}$-inch (18.3 mm) wood structural panels or one thickness of $^3/_4$-inch (19.1 mm) particleboard with joints backed by $^3/_4$-inch (19.1 mm) particleboard, $^1/_2$-inch (12.7 mm) gypsum board, or $^1/_4$-inch (6.4 mm) cement-based millboard. Batts or blankets of mineral wool or glass fiber or other approved materials installed in such a manner as to be securely retained in place shall be permitted as an acceptable fire block. Batts or blankets of mineral or glass fiber or other approved nonrigid materials shall be permitted for compliance with the 10 foot horizontal fireblocking in walls constructed using parallel rows of studs or staggered studs.

Loose-fill insulation material shall not be used as a fire block unless specifically tested in the form and manner intended for use to demonstrate its ability to remain in place and to retard the spread of fire and hot gases.

R602.8.1.1 Unfaced fiberglass. Unfaced fiberglass batt insulation used as fireblocking shall fill the entire cross section of the wall cavity to a minimum height of 16 inches (406 mm) measured vertically. When piping, conduit or similar obstructions are encountered, the insulation shall be packed tightly around the obstruction.

R602.8.1.2 Fireblocking integrity. The integrity of all fireblocks shall be maintained.

R602.9 Cripple walls. Foundation cripple walls shall be framed of studs not smaller than the studding above. When exceeding 4 feet (1219 mm) in height, such walls shall be framed of studs having the size required for an additional story.

Cripple walls with a stud height less than 14 inches (356 mm) shall be sheathed on at least one side with a wood structural panel that is fastened to both the top and bottom plates in accordance with Table R602.3(1), or the cripple walls shall be constructed of solid blocking. Cripple walls shall be supported on continuous foundations.

R602.10 Wall bracing. All exterior walls shall be braced in accordance with this section. In addition, interior braced wall lines shall be provided in accordance with Section R602.10.1.1. For buildings in Seismic Design Categories D$_0$, D$_1$ and D$_2$, walls shall be constructed in accordance with the additional requirements of Sections R602.10.9, R602.10.11, and R602.11.

TABLE R602.7.2
MAXIMUM SPANS FOR WOOD STRUCTURAL PANEL BOX HEADERS[a]

HEADER CONSTRUCTION[b]	HEADER DEPTH (inches)	HOUSE DEPTH (feet)				
		24	26	28	30	32
Wood structural panel—one side	9	4	4	3	3	—
	15	5	5	4	3	3
Wood structural panel—both sides	9	7	5	5	4	3
	15	8	8	7	7	6

For SI: 1 inch = 25.4 mm, 1 foot = 304.8 mm.

a. Spans are based on single story with clear-span trussed roof or two-story with floor and roof supported by interior-bearing walls.

b. See Figure R602.7.2 for construction details.

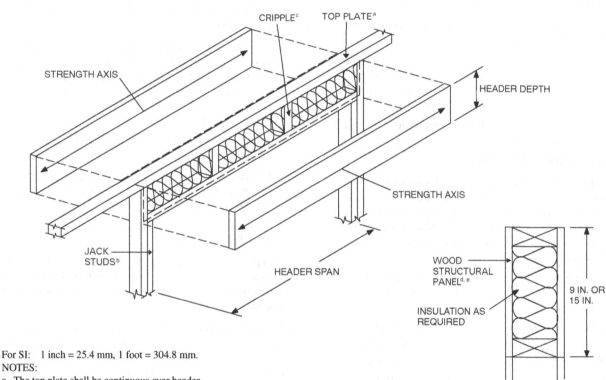

For SI: 1 inch = 25.4 mm, 1 foot = 304.8 mm.

NOTES:

a. The top plate shall be continuous over header.

b. Jack studs shall be used for spans over 4 feet.

c. Cripple spacing shall be the same as for studs.

d. Wood structural panel faces shall be single pieces of $^{15}/_{32}$-inch-thick Exposure 1 (exterior glue) or thicker, installed on the interior or exterior or both sides of the header.

e. Wood structural panel faces shall be nailed to framing and cripples with 8d common or galvanized box nails spaced 3 inches on center, staggering alternate nails $^{1}/_{2}$ inch. Galvanized nails shall be hot-dipped or tumbled.

FIGURE R602.7.2
TYPICAL WOOD STRUCTURAL PANEL BOX HEADER CONSTRUCTION

R602.10.1 Braced wall lines. Braced wall lines shall consist of braced wall panel construction in accordance with Section R602.10.3. The amount and location of bracing shall be in accordance with Table R602.10.1 and the amount of bracing shall be the greater of that required by the seismic design category or the design wind speed. Braced wall panels shall begin no more than 12.5 feet (3810 mm) from each end of a braced wall line. Braced wall panels that are counted as part of a braced wall line shall be in line, except that offsets out-of-plane of up to 4 feet (1219 mm) shall be permitted provided that the total out-to-out offset dimension in any braced wall line is not more than 8 feet (2438 mm).

R602.10.1.1 Spacing. Spacing of braced wall lines shall not exceed 35 feet (10 668 mm) on center in both the longitudinal and transverse directions in each story.

Exception: Spacing of braced wall lines not exceeding 50 feet shall be permitted where:

1. The wall bracing installed equals or exceeds the amount of bracing required by Table R602.10.1 multiplied by a factor equal to the braced wall line spacing divided by 35 feet and

2. The length-to-width ratio for the floor or roof diaphragm does not exceed 3:1.

R602.10.2 Cripple wall bracing.

R602.10.2.1 Seismic design categories other than D₂. In Seismic Design Categories other than D_2, cripple walls shall be braced with an amount and type of bracing as required for the wall above in accordance with Table R602.10.1 with the following modifications for cripple wall bracing:

1. The percent bracing amount as determined from Table R602.10.1 shall be increased by 15 percent and

2. The wall panel spacing shall be decreased to 18 feet (5486 mm) instead of 25 feet (7620 mm).

R602.10.2.2 Seismic Design Category D₂. In Seismic Design Category D_2, cripple walls shall be braced in accordance with Table R602.10.1.

R602.10.2.3 Redesignation of cripple walls. In any seismic design category, cripple walls are permitted to be redesignated as the first story walls for purposes of determining wall bracing requirements. If the cripple walls are redesignated, the stories above the redesignated story shall be counted as the second and third stories, respectively.

R602.10.3 Braced wall panel construction methods. The construction of braced wall panels shall be in accordance with one of the following methods:

1. Nominal 1-inch-by-4-inch (25 mm by 102 mm) continuous diagonal braces let in to the top and bottom plates and the intervening studs or approved metal strap devices installed in accordance with the manufacturer's specifications. The let-in bracing shall be placed at an angle not more than 60 degrees (1.06

rad) or less than 45 degrees (0.79 rad) from the horizontal.

2. Wood boards of ⁵/₈ inch (16 mm) net minimum thickness applied diagonally on studs spaced a maximum of 24 inches (610 mm). Diagonal boards shall be attached to studs in accordance with Table R602.3(1).

3. Wood structural panel sheathing with a thickness not less than ⁵/₁₆ inch (8 mm) for 16-inch (406 mm) stud spacing and not less than ³/₈ inch (9 mm) for 24-inch (610 mm) stud spacing. Wood structural panels shall be installed in accordance with Table R602.3(3).

4. One-half-inch (13 mm) or ²⁵/₃₂-inch (20 mm) thick structural fiberboard sheathing applied vertically or horizontally on studs spaced a maximum of 16 inches (406 mm) on center. Structural fiberboard sheathing shall be installed in accordance with Table R602.3(1).

5. Gypsum board with minimum ¹/₂-inch (13 mm) thickness placed on studs spaced a maximum of 24 inches (610 mm) on center and fastened at 7 inches (178 mm) on center with the size nails specified in Table R602.3(1) for sheathing and Table R702.3.5 for interior gypsum board.

6. Particleboard wall sheathing panels installed in accordance with Table R602.3(4).

7. Portland cement plaster on studs spaced a maximum of 16 inches (406 mm) on center and installed in accordance with Section R703.6.

8. Hardboard panel siding when installed in accordance with Table R703.4.

Exception: Alternate braced wall panels constructed in accordance with Section R602.10.6.1 or R602.10.6.2 shall be permitted to replace any of the above methods of braced wall panels.

R602.10.4 Length of braced panels. For Methods 2, 3, 4, 6, 7 and 8 above, each braced wall panel shall be at least 48 inches (1219 mm) in length, covering a minimum of three stud spaces where studs are spaced 16 inches (406 mm) on center and covering a minimum of two stud spaces where studs are spaced 24 inches (610 mm) on center. For Method 5 above, each braced wall panel shall be at least 96 inches (2438 mm) in length where applied to one face of a braced wall panel and at least 48 inches (1219 mm) where applied to both faces.

Exceptions:

1. Lengths of braced wall panels for continuous wood structural panel sheathing shall be in accordance with Section R602.10.5.

2. Lengths of alternate braced wall panels shall be in accordance with Section R602.10.6.1 or Section R602.10.6.2.

R602.10.5 Continuous wood structural panel sheathing. When continuous wood structural panel sheathing is provided in accordance with Method 3 of Section R602.10.3 on all sheathable areas of all exterior walls, and interior braced

TABLE R602.10.1
WALL BRACING

SEISMIC DESIGN CATEGORY OR WIND SPEED	CONDITION	TYPE OF BRACE[b, c]	AMOUNT OF BRACING[a, d, e]
Category A and B ($S_s \leq 0.35g$ and $S_{ds} \leq 0.33g$) or 100 mph or less	One story Top of two or three story	Methods 1, 2, 3, 4, 5, 6, 7 or 8	Located in accordance with Section R602.10 and at least every 25 feet on center but not less than 16% of braced wall line for Methods 2 through 8.
	First story of two story Second story of three story	Methods 1, 2, 3, 4, 5, 6, 7 or 8	Located in accordance with Section R602.10 and at least every 25 feet on center but not less than 16% of braced wall line for Method 3 or 25% of braced wall line for Methods 2, 4, 5, 6, 7 or 8.
	First story of three story	Methods 2, 3, 4, 5, 6, 7 or 8	Located in accordance with Section R602.10 and at least every 25 feet on center but not less than 25% of braced wall line for Method 3 or 35% of braced wall line for Methods 2, 4, 5, 6, 7 or 8.
Category C ($S_s \leq 0.6g$ and $S_{ds} \leq 0.50g$) or less than 110 mph	One story Top of two or three story	Methods 1, 2, 3, 4, 5, 6, 7 or 8	Located in accordance with Section R602.10 and at least every 25 feet on center but not less than 16% of braced wall line for Method 3 or 25% of braced wall line for Methods 2, 4, 5, 6, 7 or 8.
	First story of two story Second story of three story	Methods 2, 3, 4, 5, 6, 7 or 8	Located in accordance with Section R602.10 and at least every 25 feet on center but not less than 30% of braced wall line for Method 3 or 45% of braced wall line for Methods 2, 4, 5, 6, 7 or 8.
	First story of three story	Methods 2, 3, 4, 5, 6, 7 or 8	Located in accordance with Section R602.10 and at least every 25 feet on center but not less than 45% of braced wall line for Method 3 or 60% of braced wall line for Methods 2, 4, 5, 6, 7 or 8.
Categories D_0 and D_1 ($S_s \leq 1.25g$ and $S_{ds} \leq 0.83g$) or less than 110 mph	One story Top of two or three story	Methods 2, 3, 4, 5, 6, 7 or 8	Located in accordance with Section R602.10 and at least every 25 feet on center but not less than 20% of braced wall line for Method 3 or 30% of braced wall line for Methods 2, 4, 5, 6, 7 or 8.
	First story of two story Second story of three story	Methods 2, 3, 4, 5, 6, 7 or 8	Located in accordance with Section R602.10 and at least every 25 feet on center but not less than 45% of braced wall line for Method 3 or 60% of braced wall line for Methods 2, 4, 5, 6, 7 or 8.
	First story of three story	Methods 2, 3, 4, 5, 6, 7 or 8	Located in accordance with Section R602.10 and at least every 25 feet on center but not less than 60% of braced wall line for Method 3 or 85% of braced wall line for Methods 2, 4, 5, 6, 7 or 8.
Category D_2 or less than 110 mph	One story Top of two story	Methods 2, 3, 4, 5, 6, 7 or 8	Located in accordance with Section R602.10 and at least every 25 feet on center but not less than 25% of braced wall line for Method 3 or 40% of braced wall line for Methods 2, 4, 5, 6, 7 or 8.
	First story of two story	Methods 2, 3, 4, 5, 6, 7 or 8	Located in accordance with Section R602.10 and at least every 25 feet on center but not less than 55% of braced wall line for Method 3 or 75% of braced wall line for Methods 2, 4, 5, 6, 7 or 8.
	Cripple walls	Method 3	Located in accordance with Section R602.10 and at least every 25 feet on center but not less than 75% of braced wall line.

For SI: 1 inch = 25.4 mm, 1 foot = 304.8 mm, 1 pound per square foot = 0.0479kPa, 1 mile per hour = 0.477 m/s.

a. Wall bracing amounts are based on a soil site class "D." Interpolation of bracing amounts between the S_{ds} values associated with the seismic design categories shall be permitted when a site specific S_{ds} value is determined in accordance with Section 1613.5 of the *International Building Code.*

b. Foundation cripple wall panels shall be braced in accordance with Section R602.10.2.

c. Methods of bracing shall be as described in Section R602.10.3. The alternate braced wall panels described in Section R602.10.6.1 or R602.10.6.2 shall also be permitted.

d. The bracing amounts for Seismic Design Categories are based on a 15 psf wall dead load. For walls with a dead load of 8 psf or less, the bracing amounts shall be permitted to be multiplied by 0.85 provided that the adjusted bracing amount is not less than that required for the site's wind speed. The minimum length of braced panel shall not be less than required by Section R602.10.3.

e. When the dead load of the roof/ceiling exceeds 15 psf, the bracing amounts shall be increased in accordance with Section R301.2.2.2.1. Bracing required for a site's wind speed shall not be adjusted.

wall lines, where required, including areas above and below openings, bracing wall panel lengths shall be in accordance with Table R602.10.5. Wood structural panel sheathing shall be installed at corners in accordance with Figure R602.10.5. The bracing amounts in Table R602.10.1 for Method 3 shall be permitted to be multiplied by a factor of 0.9 for wall with a maximum opening height that does not exceed 85 percent of the wall height or a factor of 0.8 for walls with a maximum opening height that does not exceed 67 percent of the wall height.

R602.10.6 Alternate braced wall panel construction methods. Alternate braced wall panels shall be constructed in accordance with Sections R602.10.6.1 and R602.10.6.2.

R602.10.6.1 Alternate braced wall panels. Alternate braced wall lines constructed in accordance with one of the following provisions shall be permitted to replace each 4 feet (1219 mm) of braced wall panel as required by Section R602.10.4. The maximum height and minimum width of each panel shall be in accordance with Table R602.10.6:

1. In one-story buildings, each panel shall be sheathed on one face with $^3/_8$-inch-minimum-thickness (10 mm) wood structural panel sheathing nailed with 8d common or galvanized box nails in accordance with Table R602.3(1) and blocked at all wood structural panel sheathing edges. Two anchor bolts installed in accordance with Figure R403.1(1) shall be provided in each panel. Anchor bolts shall be placed at panel quarter points. Each panel end stud shall have a tie-down device fastened to the foundation, capable of providing an uplift capacity in accordance with Table R602.10.6. The tie down device shall be installed in accordance with the manufacturer's recommendations. The panels shall be supported directly on a foundation or on floor framing supported directly on a foundation which is continuous across the entire length of the braced wall line. This foundation shall be reinforced with not less than one No. 4 bar top and bottom. When the continuous foundation is required to have a depth greater than 12 inches (305 mm), a minimum 12-inch-by-12-inch (305 mm by 305 mm) continuous footing or turned down slab edge is permitted at door openings in the braced wall line. This continuous footing or turned down slab edge shall be reinforced with not less than one No. 4 bar top and bottom. This reinforcement shall be lapped 15 inches (381 mm) with the reinforcement required in the continuous foundation located directly under the braced wall line.

2. In the first story of two-story buildings, each braced wall panel shall be in accordance with Item 1 above, except that the wood structural panel sheathing shall be installed on both faces, sheathing edge nailing spacing shall not exceed 4 inches (102 mm) on center, at least three anchor bolts shall be placed at one-fifth points.

R602.10.6.2 Alternate braced wall panel adjacent to a door or window opening. Alternate braced wall panels constructed in accordance with one of the following provisions are also permitted to replace each 4 feet (1219 mm) of braced wall panel as required by Section R602.10.4 for use adjacent to a window or door opening with a full-length header:

1. In one-story buildings, each panel shall have a length of not less than 16 inches (406 mm) and a height of not more than 10 feet (3048 mm). Each panel shall be sheathed on one face with a single layer of $^3/_8$-inch-minimum-thickness (10 mm)

TABLE R602.10.5
LENGTH REQUIREMENTS FOR BRACED WALL PANELS IN A CONTINUOUSLY SHEATHED WALL[a, b, c]

MINIMUM LENGTH OF BRACED WALL PANEL (inches)			MAXIMUM OPENING HEIGHT NEXT TO THE BRACED WALL PANEL (% of wall height)
8-foot wall	**9-foot wall**	**10-foot wall**	
48	54	60	100
32	36	40	85
24	27	30	65

For SI: 1 inch = 25.4 mm, 1 foot = 305 mm, 1 pound per square foot = 0.0479 kPa.

a. Linear interpolation shall be permitted.

b. Full-height sheathed wall segments to either side of garage openings that support light frame roofs only, with roof covering dead loads of 3 psf or less shall be permitted to have a 4:1 aspect ratio.

c. Walls on either or both sides of openings in garages attached to fully sheathed dwellings shall be permitted to be built in accordance with Section R602.10.6.2 and Figure R602.10.6.2 except that a single bottom plate shall be permitted and two anchor bolts shall be placed at 1/3 points. In addition, tie-down devices shall not be required and the vertical wall segment shall have a maximum 6:1 height-to-width ratio (with height being measured from top of header to the bottom of the sill plate). This option shall be permitted for the first story of two-story applications in Seismic Design Categories A through C.

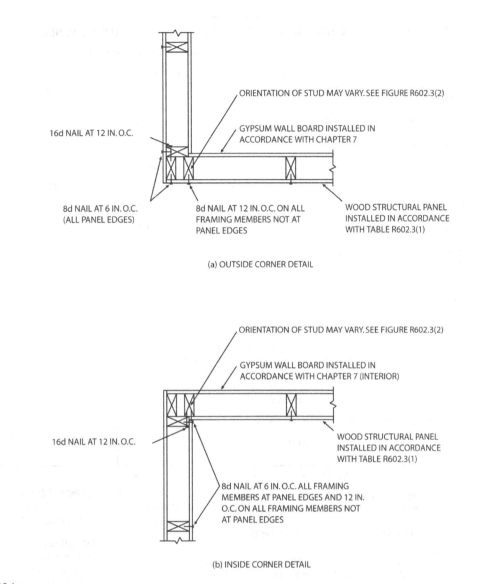

ORIENTATION OF STUD MAY VARY. SEE FIGURE R602.3(2)

GYPSUM WALL BOARD INSTALLED IN ACCORDANCE WITH CHAPTER 7

16d NAIL AT 12 IN. O.C.

8d NAIL AT 6 IN. O.C. (ALL PANEL EDGES)

8d NAIL AT 12 IN. O.C. ON ALL FRAMING MEMBERS NOT AT PANEL EDGES

WOOD STRUCTURAL PANEL INSTALLED IN ACCORDANCE WITH TABLE R602.3(1)

(a) OUTSIDE CORNER DETAIL

ORIENTATION OF STUD MAY VARY. SEE FIGURE R602.3(2)

GYPSUM WALL BOARD INSTALLED IN ACCORDANCE WITH CHAPTER 7 (INTERIOR)

16d NAIL AT 12 IN. O.C.

WOOD STRUCTURAL PANEL INSTALLED IN ACCORDANCE WITH TABLE R602.3(1)

8d NAIL AT 6 IN. O.C. ALL FRAMING MEMBERS AT PANEL EDGES AND 12 IN. O.C. ON ALL FRAMING MEMBERS NOT AT PANEL EDGES

(b) INSIDE CORNER DETAIL

For SI: 1 inch = 25.4 mm.
Gypsum board nails deleted for clarity.

FIGURE R602.10.5
TYPICAL EXTERIOR CORNER FRAMING FOR CONTINUOUS STRUCTURAL
PANEL SHEATHING; SHOWING REQUIRED STUD-TO-STUD NAILING

wood structural panel sheathing nailed with 8d common or galvanized box nails in accordance with Figure R602.10.6.2. The wood structural panel sheathing shall extend up over the solid sawn or glued-laminated header and shall be nailed in accordance with Figure R602.10.6.2. Use of a built-up header consisting of at least two 2 x 12s and fastened in accordance with Table R602.3(1) shall be permitted. A spacer, if used, shall be placed on the side of the built-up beam opposite the wood structural panel sheathing. The header shall extend between the inside faces of the first full-length outer studs of each panel. The clear span of the header between the inner studs of each panel shall be not less than 6 feet (1829 mm) and not more than 18 feet (5486 mm) in length. A strap

with an uplift capacity of not less than 1000 pounds (4448 N) shall fasten the header to the side of the inner studs opposite the sheathing. One anchor bolt not less than $^5/_8$-inch-diameter (16 mm) and installed in accordance with Section R403.1.6 shall be installed in the center of each sill plate. The studs at each end of the panel shall have a tie-down device fastened to the foundation with an uplift capacity of not less than 4,200 pounds (18 683 N).

Where a panel is located on one side of the opening, the header shall extend between the inside face of the first full-length stud of the panel and the bearing studs at the other end of the opening. A strap with an uplift capacity of not less than 1000 pounds (4448 N) shall fasten the header to the bearing

TABLE R602.10.6
MINIMUM WIDTHS AND TIE-DOWN FORCES OF ALTERNATE BRACED WALL PANELS

SEISMIC DESIGN CATEGORY AND WINDSPEED	TIE-DOWN FORCE (lb)	HEIGHT OF BRACED WALL PANEL				
		Sheathed Width				
		8 ft. 2' - 4"	9 ft. 2' - 8"	10 ft. 2' - 8"	11 ft. 3' - 2"	12 ft. 3' - 6"
SDC A, B, and C Windspeed < 110 mph	R602.10.6.1, Item 1	1800	1800	1800	2000	2200
	R602.10.6.1, Item 2	3000	3000	3000	3300	3600
		Sheathed Width				
		2' - 8"	2' - 8"	2' - 8"	Note a	Note a
SDC D$_0$, D$_1$ and D$_2$ Windspeed < 110 mph	R602.10.6.1, Item 1	1800	1800	1800	—	—
	R602.10.6.1, Item 2	3000	3000	3000	—	—

For SI: 1 inch = 25.4 mm, 1 foot = 304.8 mm.
a. Not permitted because maximum height is 10 feet.

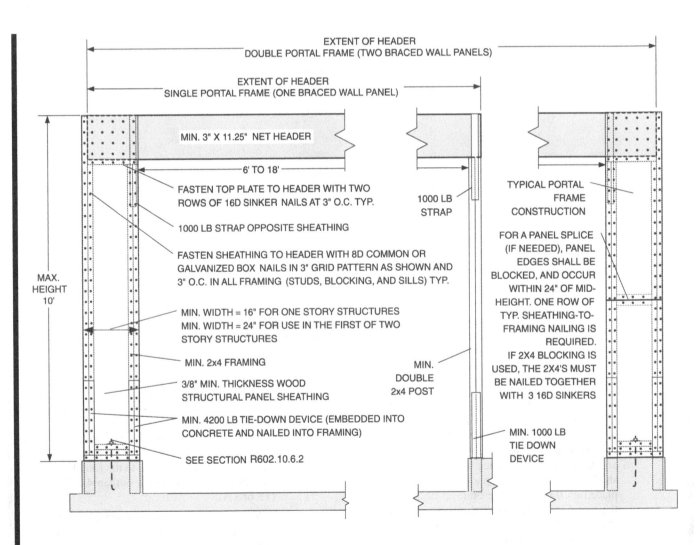

For SI: 1 inch = 25.4 mm, 1 foot = 304.8 mm, 1 pound = 0.454 kg.

FIGURE R602.10.6.2
ALTERNATE BRACED WALL PANEL ADJACENT TO A DOOR OR WINDOW OPENING

studs. The bearing studs shall also have a tie-down device fastened to the foundation with an uplift capacity of not less than 1000 pounds (4448 N).

The tie-down devices shall be an embedded-strap type, installed in accordance with the manufacturer's recommendations. The panels shall be supported directly on a foundation which is continuous across the entire length of the braced wall line. The foundation shall be reinforced with not less than one No. 4 bar top and bottom.

Where the continuous foundation is required to have a depth greater than 12 inches (305 mm), a minimum 12-inch-by-12-inch (305 mm by 305 mm) continuous footing or turned down slab edge is permitted at door openings in the braced wall line. This continuous footing or turned down slab edge shall be reinforced with not less than one No. 4 bar top and bottom. This reinforcement shall be lapped not less than 15 inches (381 mm) with the reinforcement required in the continuous foundation located directly under the braced wall line.

2. In the first story of two-story buildings, each wall panel shall be braced in accordance with Item 1 above, except that each panel shall have a length of not less than 24 inches (610 mm).

R602.10.7 Panel joints. All vertical joints of panel sheathing shall occur over, and be fastened to, common studs. Horizontal joints in braced wall panels shall occur over, and be fastened to, common blocking of a minimum $1^1/_2$ inch (38 mm) thickness.

Exception: Blocking is not required behind horizontal joints in Seismic Design Categories A and B and detached dwellings in Seismic Design Category C when constructed in accordance with Section R602.10.3, braced-wall-panel construction method 3 and Table R602.10.1, method 3, or where permitted by the manufacturer's installation requirements for the specific sheathing material.

R602.10.8 Connections. Braced wall line sole plates shall be fastened to the floor framing and top plates shall be connected to the framing above in accordance with Table R602.3(1). Sills shall be fastened to the foundation or slab in accordance with Sections R403.1.6 and R602.11. Where joists are perpendicular to the braced wall lines above, blocking shall be provided under and in line with the braced wall panels. Where joists are perpendicular to braced wall lines below, blocking shall be provided over and in line with the braced wall panels. Where joists are parallel to braced wall lines above or below, a rim joist or other parallel framing member shall be provided at the wall to permit fastening per Table R602.3(1).

R602.10.9 Interior braced wall support. In one-story buildings located in Seismic Design Category D_2, interior braced wall lines shall be supported on continuous foundations at intervals not exceeding 50 feet (15 240 mm). In two-story buildings located in Seismic Design Category D_2, all interior braced wall panels shall be supported on continuous foundations.

Exception: Two-story buildings shall be permitted to have interior braced wall lines supported on continuous foundations at intervals not exceeding 50 feet (15 240 mm) provided that:

1. The height of cripple walls does not exceed 4 feet (1219 mm).

2. First-floor braced wall panels are supported on doubled floor joists, continuous blocking or floor beams.

3. The distance between bracing lines does not exceed twice the building width measured parallel to the braced wall line.

R602.10.10 Design of structural elements. Where a building, or portion thereof, does not comply with one or more of the bracing requirements in this section, those portions shall be designed and constructed in accordance with accepted engineering practice.

R602.10.11 Bracing in Seismic Design Categories D_0, D_1 and D_2. Structures located in Seismic Design Categories D_0, D_1 and D_2 shall have exterior and interior braced wall lines.

R602.10.11.1 Braced wall line spacing. Spacing between braced wall lines in each story shall not exceed 25 feet (7620 mm) on center in both the longitudinal and transverse directions.

Exception: In one- and two-story buildings, spacing between two adjacent braced wall lines shall not exceed 35 feet (10 363 mm) on center in order to accommodate one single room not exceeding 900 square feet (84 m²) in each dwelling unit. Spacing between all other braced wall lines shall not exceed 25 feet (7620 mm).

R602.10.11.2 Braced wall panel location. Exterior braced wall lines shall have a braced wall panel at each end of the braced wall line.

Exception: For braced wall panel construction Method 3 of Section R602.10.3, the braced wall panel shall be permitted to begin no more than 8 feet (2438 mm) from each end of the braced wall line provided the following is satisfied:

1. A minimum 24-inch-wide (610 mm) panel is applied to each side of the building corner and the two 24-inch (610 mm) panels at the corner shall be attached to framing in accordance with Figure R602.10.5; or

2. The end of each braced wall panel closest to the corner shall have a tie-down device fastened to the stud at the edge of the braced wall panel closest to the corner and to the foundation or framing below. The tie-down device shall be capable of providing an uplift allowable design value of at least 1,800 pounds (8 kN). The tie-down device shall be installed in accordance with the manufacturer's recommendations.

R602.10.11.3 Collectors. A designed collector shall be provided if a braced wall panel is not located at each end of a braced wall line as indicated in Section R602.10.11.2, or, when using the Section R602.10.11.2 exception, if a braced wall panel is more than 8 feet (2438 mm) from each end of a braced wall line.

R602.10.11.4 Cripple wall bracing. In addition to the requirements of Section R602.10.2, where interior braced wall lines occur without a continuous foundation below, the length of parallel exterior cripple wall bracing shall be one and one-half times the length required by Table R602.10.1. Where cripple walls braced using Method 3 of Section R602.10.3 cannot provide this additional length, the capacity of the sheathing shall be increased by reducing the spacing of fasteners along the perimeter of each piece of sheathing to 4 inches (102 mm) on center.

R602.10.11.5 Sheathing attachment. Adhesive attachment of wall sheathing shall not be permitted in Seismic Design Categories C, D_0, D_1 and D_2.

R602.11 Framing and connections for Seismic Design Categories D_0, D_1 and D_2. The framing and connections details of buildings located in Seismic Design Categories D_0, D_1 and D_2 shall be in accordance with Sections R602.11.1 through R602.11.3.

R602.11.1 Wall anchorage. Braced wall line sills shall be anchored to concrete or masonry foundations in accordance with Sections R403.1.6 and R602.11. For all buildings in Seismic Design Categories D_0, D_1 and D_2 and townhouses in Seismic Design Category C, plate washers, a minimum of 0.229 inch by 3 inches by 3 inches (5.8 mm by 76 mm by 76 mm) in size, shall be installed between the foundation sill plate and the nut. The hole in the plate washer is permitted to be diagonally slotted with a width of up to $^3/_{16}$ inch (5 mm) larger than the bolt diameter and a slot length not to exceed $1^3/_4$ inches (44 mm), provided a standard cut washer is placed between the plate washer and the nut.

R602.11.2 Interior braced wall panel connections. Interior braced wall lines shall be fastened to floor and roof framing in accordance with Table R602.3(1), to required foundations in accordance with Section R602.11.1, and in accordance with the following requirements:

1. Floor joists parallel to the top plate shall be toe-nailed to the top plate with at least 8d nails spaced a maximum of 6 inches (152 mm) on center.

2. Top plate laps shall be face-nailed with at least eight 16d nails on each side of the splice.

R602.11.3 Stepped foundations. Where stepped foundations occur, the following requirements apply:

1. Where the height of a required braced wall panel that extends from foundation to floor above varies more than 4 feet (1220 mm), the braced wall panel shall be constructed in accordance with Figure R602.11.3.

2. Where the lowest floor framing rests directly on a sill bolted to a foundation not less than 8 feet (2440 mm) in length along a line of bracing, the line shall be considered as braced. The double plate of the cripple stud

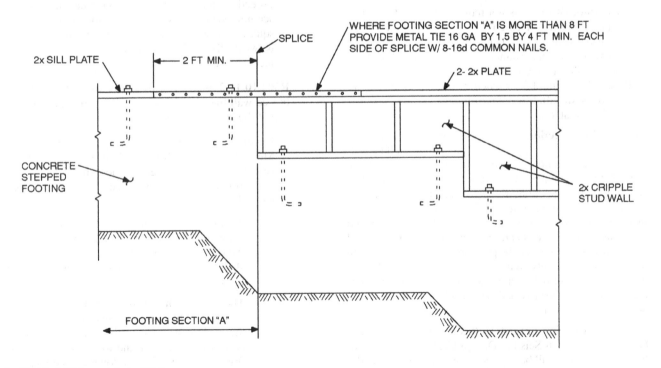

For SI: 1 inch 25.4 mm, 1 foot = 304.8 mm.

Note: Where footing Section "A" is less than 8 feet long in a 25-foot-long wall, install bracing at cripple stud wall.

FIGURE R602.11.3
STEPPED FOUNDATION CONSTRUCTION

wall beyond the segment of footing that extends to the lowest framed floor shall be spliced by extending the upper top plate a minimum of 4 feet (1219 mm) along the foundation. Anchor bolts shall be located a maximum of 1 foot and 3 feet (305 and 914 mm) from the step in the foundation.

3. Where cripple walls occur between the top of the foundation and the lowest floor framing, the bracing requirements for a story shall apply.

4. Where only the bottom of the foundation is stepped and the lowest floor framing rests directly on a sill bolted to the foundations, the requirements of Section R602.11.1 shall apply.

SECTION R603
STEEL WALL FRAMING

R603.1 General. Elements shall be straight and free of any defects that would significantly affect structural performance. Cold-formed steel wall framing members shall comply with the requirements of this section.

R603.1.1 Applicability limits. The provisions of this section shall control the construction of exterior steel wall framing and interior load-bearing steel wall framing for buildings not more than 60 feet (18 288 mm) long perpendicular to the joist or truss span, not more than 40 feet (12 192 mm) wide parallel to the joist or truss span, and not more than two stories in height. All exterior walls installed in accordance with the provisions of this section shall be considered as load-bearing walls. Steel walls constructed in accordance with the provisions of this section shall be limited to sites subjected to a maximum design wind speed of 110 miles per hour (49 m/s) Exposure A, B or C and a maximum ground snow load of 70 psf (3.35 kPa).

R603.1.2 In-line framing. Load-bearing steel studs constructed in accordance with Section R603 shall be located directly in-line with joists, trusses and rafters with a maximum tolerance of $^3/_4$ inch (19.1 mm) between their center lines. Interior load-bearing steel stud walls shall be supported on foundations or shall be located directly above load-bearing walls with a maximum tolerance of $^3/_4$ inch (19 mm) between the centerline of the studs.

R603.2 Structural framing. Load-bearing steel wall framing members shall comply with Figure R603.2(1) and with the dimensional and minimum thickness requirements specified in Tables R603.2(1) and R603.2(2). Tracks shall comply with Figure R603.2(2) and shall have a minimum flange width of $1^1/_4$ inches (32 mm). The maximum inside bend radius for members shall be the greater of $^3/_{32}$ inch (2.4 mm) or twice the uncoated steel thickness. Holes in wall studs and other structural members shall comply with all of the following conditions:

1. Holes shall conform to Figure R603.2(3);

2. Holes shall be permitted only along the centerline of the web of the framing member;

3. Holes shall have a center-to-center spacing of not less than 24 inches (610 mm);

4. Holes shall have a width not greater than 0.5 times the member depth, or $1^1/_2$ inches (38.1 mm);

5. Holes shall have a length not exceeding $4^1/_2$ inches (114 mm); and

6. Holes shall have a minimum distance between the edge of the bearing surface and the edge of the hole of not less than 10 inches (254 mm).

Framing members with web holes violating the above requirements shall be patched in accordance with Section R603.3.5 or designed in accordance with accepted engineering practices.

TABLE R603.2(1)
LOAD-BEARING COLD-FORMED STEEL STUD SIZES

MEMBER DESIGNATION[a]	WEB DEPTH (inches)	MINIMUM FLANGE WIDTH (inches)	MAXIMUM FLANGE WIDTH (inches)	MINIMUM LIP SIZE (inches)
350S162-t	3.5	1.625	2	0.5
550S162-t	5.5	1.625	2	0.5

For SI: 1 inch = 25.4 mm; 1 mil = 0.0254 mm.

a. The member designation is defined by the first number representing the member depth in hundredths of an inch "S" representing a stud or joist member, the second number representing the flange width in hundredths of an inch, and the letter "t" shall be a number representing the minimum base metal thickness in mils [See Table R603.2(2)].

TABLE R603.2(2)
MINIMUM THICKNESS OF COLD-FORMED STEEL STUDS

DESIGNATION (mils)	MINIMUM UNCOATED THICKNESS (inches)	REFERENCE GAGE NUMBER
33	0.033	20
43	0.043	18
54	0.054	16
68	0.068	14

For SI: 1 inch = 25.4 mm, 1 mil = 0.0254 mm.

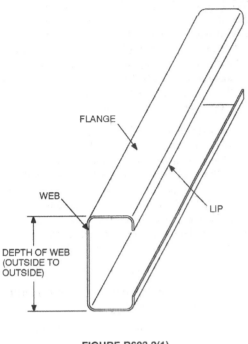

FIGURE R603.2(1)
C-SECTION

FIGURE R603.2(2)
TRACK SECTION

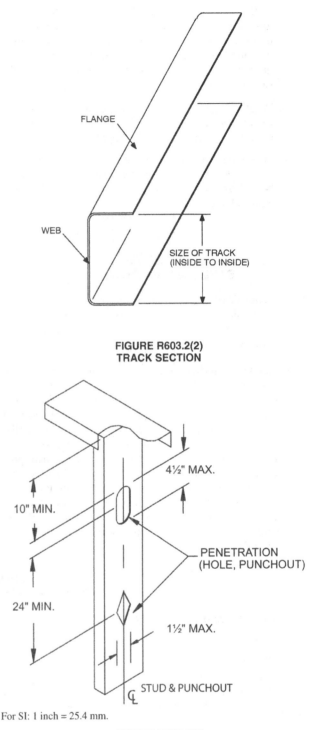

For SI: 1 inch = 25.4 mm.

FIGURE R603.2(3)
WEB HOLES

R603.2.1 Material. Load-bearing steel framing members shall be cold-formed to shape from structural quality sheet steel complying with the requirements of one of the following:

1. ASTM A 653: Grades 33, 37, 40 and 50 (Class 1 and 3).

2. ASTM A 792: Grades 33, 37, 40 and 50A.

3. ASTM A 875: Grades 33, 37, 40 and 50 (Class 1 and 3).

4. ASTM A 1003: Grades 33, 37, 40 and 50.

R603.2.2 Identification. Load-bearing steel framing members shall have a legible label, stencil, stamp or embossment with the following information as a minimum:

1. Manufacturer's identification.

2. Minimum uncoated steel thickness in inches (mm).

3. Minimum coating designation.

4. Minimum yield strength, in kips per square inch (ksi) (kN).

R603.2.3 Corrosion protection. Load-bearing steel framing shall have a metallic coating complying with one of the following:

1. A minimum of G 60 in accordance with ASTM A 653.

2. A minimum of AZ 50 in accordance with ASTM A 792.

3. A minimum of GF 60 in accordance with ASTM A 875.

R603.2.4 Fastening requirements. Screws for steel-to-steel connections shall be installed with a minimum edge distance and center-to-center spacing of $1/2$ inch (12.7 mm), shall be self-drilling tapping and shall conform to SAE J 78. Structural sheathing shall be attached to steel studs with minimum No. 8 self-drilling tapping screws that conform to SAE J 78. Screws for attaching structural sheathing to steel wall framing shall have a minimum head diameter of 0.292 inch (7.4 mm) with countersunk heads and shall be installed with a minimum edge distance of $3/8$ inch (9.5 mm). Gypsum board shall be attached to steel wall framing with minimum No. 6 screws conforming to ASTM C 954 and shall be installed in accordance with Section R702. For all connections, screws shall extend through the steel a minimum of three exposed threads. All self-drilling tapping screws conforming to SAE J 78 shall have a Type II coating in accordance with ASTM B 633.

Where No. 8 screws are specified in a steel-to-steel connection the required number of screws in the connection is permitted to be reduced in accordance with the reduction factors in Table R603.2.4, when larger screws are used or when one of the sheets of steel being connected is thicker than 33 mils (0.84 mm). When applying the reduction factor the resulting number of screws shall be rounded up.

TABLE R603.2.4
SCREW SUBSTITUTION FACTOR

SCREW SIZE	THINNEST CONNECTED STEEL SHEET (mils)	
	33	43
#8	1.0	0.67
#10	0.93	0.62
#12	0.86	0.56

For SI: 1 mil = 0.0254 mm.

R603.3 Wall construction. All exterior steel framed walls and interior load-bearing steel framed walls shall be constructed in accordance with the provisions of this section and Figure R603.3.

R603.3.1 Wall to foundation or floor connections. Steel framed walls shall be anchored to foundations or floors in accordance with Table R603.3.1 and Figure R603.3.1(1) or R603.3.1(2).

R603.3.2 Load-bearing walls. Steel studs shall comply with Tables R603.3.2(2) through R603.3.2(21). The tabulated stud thickness for structural walls shall be used when the attic load is 10 psf (0.48 kPa) or less. When an attic storage load is greater than 10 psf (0.48 kPa) but less than or equal to 20 psf (0.96 kPa), the next higher snow load column value from Tables R603.3.2(2) through R603.3.2(21) shall be used to select the stud size. The tabulated stud thickness for structural walls supporting one floor, roof and ceiling shall be used when the second floor live load is 30 psf (1.44 kPa). When the second floor live load is greater than 30 psf (1.44 kPa) but less than or equal to 40 psf (1.92 kPa) the design value in the next higher snow load column from Tables R603.2(12) through R603.3.2(21) shall be used to select the stud size.

Fastening requirements shall be in accordance with Section R603.2.4 and Table R603.3.2(1). Tracks shall have the same minimum thickness as the wall studs. Exterior walls with a minimum of $^1/_2$-inch (13 mm) gypsum board installed in accordance with Section R702 on the interior surface and wood structural panels of minimum $^7/_{16}$-inch thick (11mm) oriented-strand board or $^{15}/_{32}$-inch thick (12 mm) plywood installed in accordance with Table R603.3.2(1) on the outside surface shall be permitted to use the next thinner stud from Tables R603.3.2(2) through R603.3.2(13) but not less than 33 mils (0.84 mm). Interior load-bearing walls with a minimum $^1/_2$-inch (13 mm) gypsum board installed in accordance with Section R702 on both sides of the wall shall be permitted to use

the next thinner stud from Tables R603.3.2(2) through R603.3.2(13) but not less than 33 mils (0.84 mm).

R603.3.3 Stud bracing. The flanges of steel studs shall be laterally braced in accordance with one of the following:

1. Gypsum board installed with minimum No. 6 screws in accordance with Section R702 or structural sheathing installed in accordance with Table R603.3.2(1).

2. Horizontal steel strapping installed in accordance with Figure R603.3 at mid-height for 8-foot (2438 mm) walls, and one-third points for 9-foot and 10-foot (2743 mm and 3048 mm) walls. Steel straps shall be at least 1.5 inches in width and 33 mils in thickness (38 mm by 0.84 mm). Straps shall be attached to the flanges of studs with at least one No. 8 screw. In-line blocking shall be installed between studs at the termination of all straps. Straps shall be fastened to the blocking with at least two No. 8 screws.

3. Sheathing on one side and strapping on the other side. Sheathing shall be installed in accordance with Method #1 above. Steel straps shall be installed in accordance with Method #2 above.

R603.3.4 Cutting and notching. Flanges and lips of steel studs and headers shall not be cut or notched.

R603.3.5 Hole patching. Web holes violating the requirements in Section R603.2 shall be designed in accordance with one of the following:

1. Framing members shall be replaced or designed in accordance with accepted engineering practices when web holes exceed the following size limits:

    1.1. The depth of the hole, measured across the web, exceeds 70 percent of the flat width of the web; or

    1.2. The length of the hole measured along the web exceeds 10 inches (254 mm) or the depth of the web, whichever is greater.

2. Web holes not exceeding the dimensional requirements in R603.3.5(1) shall be patched with a solid steel plate, stud section, or track section in accordance with Figure R603.3.5. The steel patch shall be as a minimum the same thickness as the receiving member and shall extend at least 1 inch (25 mm) beyond all edges of the hole. The steel patch shall be fastened to the web of the receiving member with No. 8 screws spaced no more than 1 inch (25 mm) center-to-center along the edges of the patch with a minimum edge distance of $^1/_2$ inch (13 mm).

R603.3.6 Splicing. Steel studs and other structural members shall not be spliced. Tracks shall be spliced in accordance with Figure R603.3.6.

R603.4 Corner framing. Corner studs and the top tracks shall be installed in accordance with Figure R603.4.

R603.5 Exterior wall covering. The method of attachment of exterior wall covering materials to cold-formed steel stud wall framing shall conform to the manufacturer's installation instructions.

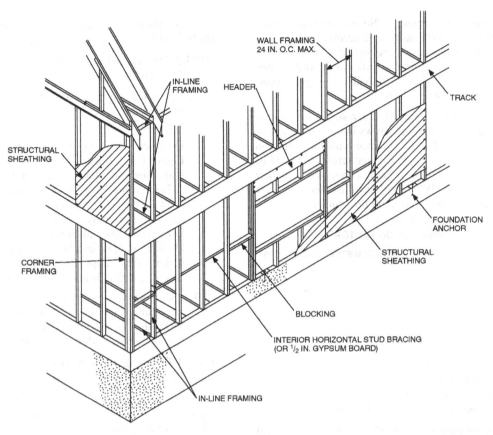

For SI: 1 inch = 25.4 mm.

FIGURE R603.3
STEEL WALL CONSTRUCTION

TABLE R603.3.1
WALL TO FOUNDATION OR FLOOR CONNECTION REQUIREMENTS[a,b,c]

FRAMING CONDITION	BASIC WIND SPEED (mph) AND EXPOSURE		
	85 A/B or Seismic Design Categories A, B and C	85 C or less than 110 A/B	Less than 110 C
Wall bottom track to floor joist or track	1-No. 8 screw at 12″ o.c.	1-No. 8 screw at 12″o.c.	2-No. 8 screw at 12″ o.c.
Wall bottom track to wood sill per Figure R603.3.1(2)	Steel plate spaced at 4′ o.c., with 4-No. 8 screws and 4-10d or 6-8d common nails	Steel plate spaced at 3′ o.c., with 4-No. 8 screws and 4-10d or 6-8d common nails	Steel plate spaced at 2′ o.c., with 4-No. 8 screws and 4-10d or 6-8d common nails
Wall bottom track to foundation per Figure R603.3.1(1)	$^1/_2$″ minimum diameter anchor bolt at 6′ o.c.	$^1/_2$″ minimum diameter anchor bolt at 6′ o.c.	$^1/_2$″ minimum diameter anchor bolt at 4′ o.c.
Wind uplift connector capacity for 16-inch stud spacing[c]	N/R	N/R	65 lb
Wind uplift connector capacity for 24-inch stud spacing[c]	N/R	N/R	100 lb

For SI: 1 inch = 25.4 mm, 1 foot = 304.8 mm, 1 mile per hour = 0.447 m/s, 1 pound = 4.4 N.

a. Anchor bolts shall be located not more than 12 inches from corners or the termination of bottom tracks (e.g., at door openings or corners). Bolts shall extend a minimum of 7 inches into concrete or masonry.

b. All screw sizes shown are minimum.

c. N/R = uplift connector not required. Uplift connectors are in addition to other connection requirements and shall be applied in accordance with Section R603.8.

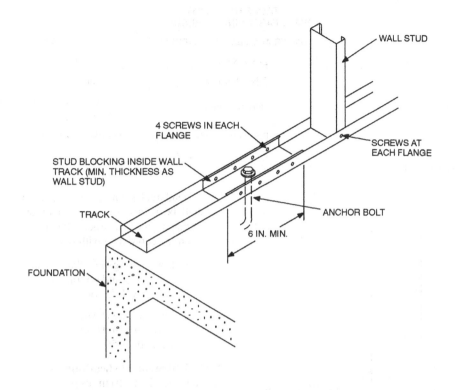

For SI: 1 inch = 25.4 mm.

**FIGURE R603.3.1(1)
WALL TO FOUNDATION CONNECTION**

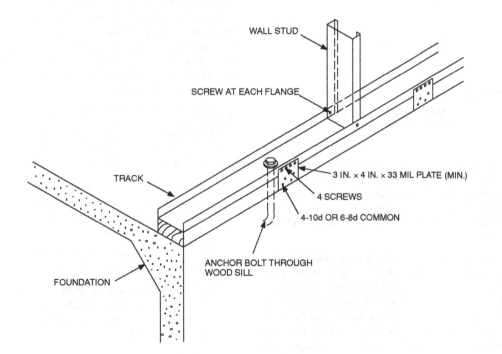

For SI: 1 inch = 25.4 mm.

**FIGURE R603.3.1(2)
WALL TO WOOD SILL CONNECTION**

DESCRIPTION OF BUILDING ELEMENT	NUMBER AND SIZE OF FASTENERS[a]	SPACING OF FASTENERS
Floor joist to track of load-bearing wall	2-No. 8 screws	Each joist
Wall stud to top or bottom track	2-No. 8 screws	Each end of stud, one per flange
Structural sheathing to wall studs	No. 8 screws	6″ o.c. on edges and 12″ o.c. at intermediate supports
Roof framing to wall	Approved design or tie down in accordance with Section R802.11	

For SI: 1 inch = 25.4 mm.

a. All screw sizes shown are minimum.

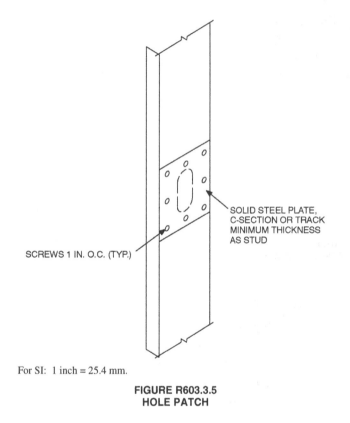

For SI: 1 inch = 25.4 mm.

**FIGURE R603.3.5
HOLE PATCH**

R603.6 Headers. Headers shall be installed above wall openings in all exterior walls and interior load-bearing walls in accordance with Figure R603.6 and Tables R603.6(1) through R603.6(8), or shall be designed in accordance with the AISI Standard for Cold-formed Steel Framing–Header Design (COFS/Header Design).

R603.6.1 Jack and king studs, and head track. The number of jack and king studs shall comply with Table R603.6(9). King and jack studs shall be of the same dimension and thickness as the adjacent wall studs. Headers constructed of C-shape framing members shall be connected to king studs in accordance with Table R603.6.(10). One-half the total number of screws shall be applied to the header and one-half to the king stud by use of a minimum 2-inch by 2-inch (51 mm by 51 mm) clip angle or 4-inch-wide (102 mm) steel plate. The clip angle or plate shall extend the depth of the header minus $^1/_2$ inch (13 mm) and shall have a minimum thickness of the header members or the wall studs, whichever is thicker.

Head track spans shall comply with Table R603.6(11) and shall be in accordance with Figures R603.3 and R603.6. Increasing the head track tabular value shall not be prohibited when in accordance with one of the following:

1. For openings less than 4 feet (1219 mm) in height that have a top and bottom head track, multiply the tabular value by 1.75; or

2. For openings less than 6 feet (1829 mm) in height that have a top and bottom head track, multiply the tabular value by 1.50.

R603.7 Structural sheathing. In areas where the basic wind speed is less than 110 miles per hour (49 m/s), wood structural panel sheathing shall be installed on all exterior walls of buildings in accordance with this section. Wood structural panel sheathing shall consist of minimum $^7/_{16}$-inch-thick (11 mm) oriented-strand board or $^{15}/_{32}$-inch-thick (12 mm) plywood and shall be installed on all exterior wall surfaces in accordance with Section R603.7.1 and Figure R603.3. The minimum length of full height sheathing on exterior walls shall be determined in accordance with Table R603.7, but shall not be less than 20 percent of the braced wall length in any case. The minimum percentage of full height sheathing in Table R603.7 shall include only those sheathed wall sections, uninterrupted by openings, which are a minimum of 48 inches (1120 mm) wide. The minimum percentage of full-height structural sheathing shall be multiplied by 1.10 for 9-foot-high (2743 mm) walls and multiplied by 1.20 for 10-foot-high (3048 mm) walls. In addition, structural sheathing shall:

1. Be installed with the long dimension parallel to the stud framing and shall cover the full vertical height of studs, from the bottom of the bottom track to the top of the top track of each story.

2. Be applied to each end (corners) of each of the exterior walls with a minimum 48-inch-wide (1219 mm) panel.

R603.7.1 Structural sheathing fastening. All edges and interior areas of wood structural panel sheathing shall be fastened to a framing member and tracks in accordance with Table R603.3.2(1).

R603.7.2 Hold-down requirements. Multiplying the percentage of structural sheathing required in Table R603.7 by 0.6 is permitted where a hold-down anchor with a capacity of 4,300 pounds (19 kN) is provided at each end of exterior walls. Installations of a single hold-down anchor at wall corners is permitted.

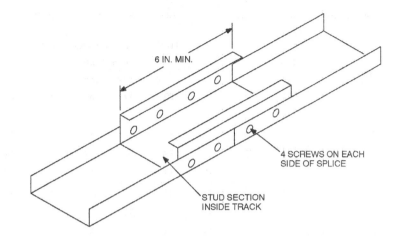

For SI: 1 inch = 25.4 mm.

FIGURE R603.3.6
TRACK SPLICE

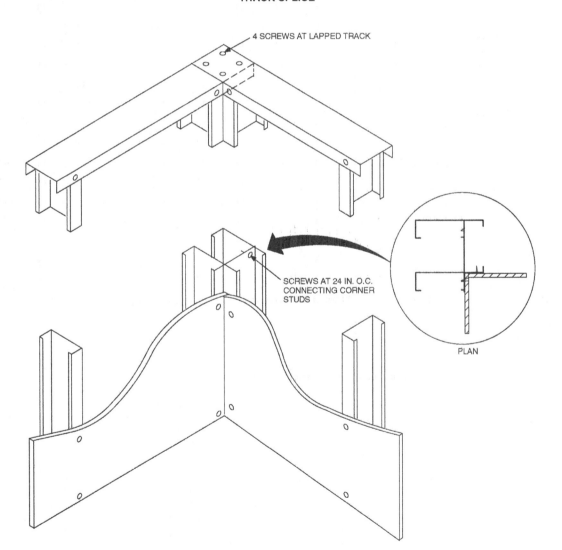

For SI: 1 inch = 25.4

FIGURE R603.4
CORNER FRAMING

TABLE R603.3.2(2)
24-FOOT-WIDE BUILDING SUPPORTING ROOF AND CEILING ONLY[a, b, c]
33 ksi STEEL

WIND SPEED		MEMBER SIZE	STUD SPACING (inches)	MINIMUM STUD THICKNESS (mils)											
				8-Foot Studs				9-Foot Studs				10-Foot Studs			
				Ground Snow Load (psf)											
Exp. A/B	Exp. C			20	30	50	70	20	30	50	70	20	30	50	70
85 mph	—	350S162	16	33	33	33	33	33	33	33	33	33	33	33	33
			24	33	33	33	33	33	33	33	33	33	33	33	43
		550S162	16	33	33	33	33	33	33	33	33	33	33	33	33
			24	33	33	33	33	33	33	33	33	33	33	33	33
90 mph	—	350S162	16	33	33	33	33	33	33	33	33	33	33	33	33
			24	33	33	33	33	33	33	33	33	33	33	33	43
		550S162	16	33	33	33	33	33	33	33	33	33	33	33	33
			24	33	33	33	33	33	33	33	33	33	33	33	33
100 mph	85 mph	350S162	16	33	33	33	33	33	33	33	33	33	33	33	33
			24	33	33	33	33	33	33	33	33	33	33	33	43
		550S162	16	33	33	33	33	33	33	33	33	33	33	33	33
			24	33	33	33	33	33	33	33	33	33	33	33	33
110 mph	90 mph	350S162	16	33	33	33	33	33	33	33	33	33	33	33	33
			24	33	33	33	33	33	33	33	43	33	33	33	43
		550S162	16	33	33	33	33	33	33	33	33	33	33	33	33
			24	33	33	33	33	33	33	33	33	33	33	33	33
—	100 mph	350S162	16	33	33	33	33	33	33	33	33	33	33	33	33
			24	33	33	33	43	33	33	33	43	43	43	43	43
		550S162	16	33	33	33	33	33	33	33	33	33	33	33	33
			24	33	33	33	33	33	33	33	33	33	33	33	33
—	110 mph	350S162	16	33	33	33	33	33	33	33	33	33	33	33	33
			24	33	33	33	43	43	43	43	43	54	54	54	54
		550S162	16	33	33	33	33	33	33	33	33	33	33	33	33
			24	33	33	33	33	33	33	33	33	33	33	33	33

For SI: 1 inch = 25.4 mm, 1 foot = 304.8 mm, 1 mil = 0.0254 mm, 1 mile per hour = 0.447 m/s, 1 pound per square foot = 0.0479 kPa, 1 ksi = 1000 psi = 6.895 MPa.

a. Deflection criterion: $L/240$.
b. Design load assumptions:
   Roof/ceiling dead load is 12 psf.
   Attic live load is 10 psf.
c. Building width is in the direction of horizontal framing members supported by the wall studs.

TABLE R603.3.2(3)
24-FOOT-WIDE BUILDING SUPPORTING ROOF AND CEILING ONLY[a, b, c]
50 ksi STEEL

WIND SPEED		MEMBER SIZE	STUD SPACING (inches)	MINIMUM STUD THICKNESS (mils)											
				8-Foot Studs				9-Foot Studs				10-Foot Studs			
				Ground Snow Load (psf)											
Exp. A/B	Exp. C			20	30	50	70	20	30	50	70	20	30	50	70
85 mph	—	350S162	16	33	33	33	33	33	33	33	33	33	33	33	33
			24	33	33	33	33	33	33	33	33	33	33	33	33
		550S162	16	33	33	33	33	33	33	33	33	33	33	33	33
			24	33	33	33	33	33	33	33	33	33	33	33	33
90 mph	—	350S162	16	33	33	33	33	33	33	33	33	33	33	33	33
			24	33	33	33	33	33	33	33	33	33	33	33	33
		550S162	16	33	33	33	33	33	33	33	33	33	33	33	33
			24	33	33	33	33	33	33	33	33	33	33	33	33
100 mph	85 mph	350S162	16	33	33	33	33	33	33	33	33	33	33	33	33
			24	33	33	33	33	33	33	33	33	33	33	33	33
		550S162	16	33	33	33	33	33	33	33	33	33	33	33	33
			24	33	33	33	33	33	33	33	33	33	33	33	33
110 mph	90 mph	350S162	16	33	33	33	33	33	33	33	33	33	33	33	33
			24	33	33	33	33	33	33	33	33	33	33	33	33
		550S162	16	33	33	33	33	33	33	33	33	33	33	33	33
			24	33	33	33	33	33	33	33	33	33	33	33	33
—	100 mph	350S162	16	33	33	33	33	33	33	33	33	33	33	33	33
			24	33	33	33	33	33	33	33	33	43	43	43	43
		550S162	16	33	33	33	33	33	33	33	33	33	33	33	33
			24	33	33	33	33	33	33	33	33	33	33	33	33
—	110 mph	350S162	16	33	33	33	33	33	33	33	33	33	33	33	33
			24	33	33	33	33	33	33	33	33	54	54	54	54
		550S162	16	33	33	33	33	33	33	33	33	33	33	33	33
			24	33	33	33	33	33	33	33	33	33	33	33	33

For SI: 1 inch = 25.4 mm, 1 foot = 304.8 mm, 1 mil = 0.0254 mm, 1 mile per hour = 0.447 m/s, 1 pound per square foot = 0.0479 kPa, 1 ksi = 1000 psi = 6.895 MPa.

a. Deflection criterion: $L/240$.
b. Design load assumptions:
   Roof/ceiling dead load is 12 psf.
   Attic live load is 10 psf.
c. Building width is in the direction of horizontal framing members supported by the wall studs.

TABLE R603.3.2(4)
28-FOOT-WIDE BUILDING SUPPORTING ROOF AND CEILING ONLY[a, b, c]
33 ksi STEEL

WIND SPEED		MEMBER SIZE	STUD SPACING (inches)	MINIMUM STUD THICKNESS (mils)											
				8-Foot Studs				9-Foot Studs				10-Foot Studs			
				Ground Snow Load (psf)											
Exp. A/B	Exp. C			20	30	50	70	20	30	50	70	20	30	50	70
85 mph	—	350S162	16	33	33	33	33	33	33	33	33	33	33	33	33
			24	33	33	33	43	33	33	33	43	33	33	33	43
		550S162	16	33	33	33	33	33	33	33	33	33	33	33	33
			24	33	33	33	33	33	33	33	33	33	33	33	33
90 mph	—	350S162	16	33	33	33	33	33	33	33	33	33	33	33	33
			24	33	33	33	43	33	33	33	43	33	33	33	43
		550S162	16	33	33	33	33	33	33	33	33	33	33	33	33
			24	33	33	33	33	33	33	33	33	33	33	33	33
100 mph	85 mph	350S162	16	33	33	33	33	33	33	33	33	33	33	33	33
			24	33	33	33	43	33	33	33	43	33	33	43	43
		550S162	16	33	33	33	33	33	33	33	33	33	33	33	33
			24	33	33	33	33	33	33	33	33	33	33	33	33
110 mph	90 mph	350S162	16	33	33	33	33	33	33	33	33	33	33	33	33
			24	33	33	33	43	33	33	33	43	33	33	43	43
		550S162	16	33	33	33	33	33	33	33	33	33	33	33	33
			24	33	33	33	33	33	33	33	33	33	33	33	33
—	100 mph	350S162	16	33	33	33	33	33	33	33	33	33	33	33	33
			24	33	33	33	43	33	33	43	43	43	43	43	43
		550S162	16	33	33	33	33	33	33	33	33	33	33	33	33
			24	33	33	33	33	33	33	33	33	33	33	33	33
—	110 mph	350S162	16	33	33	33	33	33	33	33	33	33	33	33	33
			24	33	33	33	43	43	43	43	43	54	54	54	54
		550S162	16	33	33	33	33	33	33	33	33	33	33	33	33
			24	33	33	33	33	33	33	33	33	33	33	33	33

For SI: 1 inch = 25.4 mm, 1 foot = 304.8 mm, 1 mil = 0.0254 mm, 1 mile per hour = 0.447 m/s, 1 pound per square foot = 0.0479 kPa, 1 ksi = 1000 psi = 6.895 MPa.

a. Deflection criterion: L/240.
b. Design load assumptions:
   Roof/ceiling dead load is 12 psf.
   Attic live load is 10 psf.
c. Building width is in the direction of horizontal framing members supported by the wall studs.

2006 INTERNATIONAL RESIDENTIAL CODE®

TABLE R603.3.2(5)
28-FOOT-WIDE BUILDING SUPPORTING ROOF AND CEILING ONLY[a, b, c]
50 ksi STEEL

WIND SPEED		MEMBER SIZE	STUD SPACING (inches)	MINIMUM STUD THICKNESS (mils)											
				8-Foot Studs				9-Foot Studs				10-Foot Studs			
				Ground Snow Load (psf)											
Exp. A/B	Exp. C			20	30	50	70	20	30	50	70	20	30	50	70
85 mph	—	350S162	16	33	33	33	33	33	33	33	33	33	33	33	33
			24	33	33	33	33	33	33	33	33	33	33	33	43
		550S162	16	33	33	33	33	33	33	33	33	33	33	33	33
			24	33	33	33	33	33	33	33	33	33	33	33	33
90 mph	—	350S162	16	33	33	33	33	33	33	33	33	33	33	33	33
			24	33	33	33	33	33	33	33	33	33	33	33	43
		550S162	16	33	33	33	33	33	33	33	33	33	33	33	33
			24	33	33	33	33	33	33	33	33	33	33	33	33
100 mph	85 mph	350S162	16	33	33	33	33	33	33	33	33	33	33	33	33
			24	33	33	33	33	33	33	33	33	33	33	33	43
		550S162	16	33	33	33	33	33	33	33	33	33	33	33	33
			24	33	33	33	33	33	33	33	33	33	33	33	33
110 mph	90 mph	350S162	16	33	33	33	33	33	33	33	33	33	33	33	33
			24	33	33	33	33	33	33	33	33	33	33	43	43
		550S162	16	33	33	33	33	33	33	33	33	33	33	33	33
			24	33	33	33	33	33	33	33	33	33	33	33	33
—	100 mph	350S162	16	33	33	33	33	33	33	33	33	33	33	33	33
			24	33	33	33	33	33	33	33	33	43	43	43	43
		550S162	16	33	33	33	33	33	33	33	33	33	33	33	33
			24	33	33	33	33	33	33	33	33	33	33	33	33
—	110 mph	350S162	16	33	33	33	33	33	33	33	33	33	33	33	33
			24	33	33	33	43	33	33	33	43	54	54	54	54
		550S162	16	33	33	33	33	33	33	33	33	33	33	33	33
			24	33	33	33	33	33	33	33	33	33	33	33	33

For SI: 1 inch = 25.4 mm, 1 foot = 304.8 mm, 1 mil = 0.0254 mm, 1 mile per hour = 0.447 m/s, 1 pound per square foot = 0.0479 kPa, 1 ksi = 1000 psi = 6.895 MPa.

a. Deflection criterion: $L/240$.
b. Design load assumptions:
   Roof/ceiling dead load is 12 psf.
   Attic live load is 10 psf.
c. Building width is in the direction of horizontal framing members supported by the wall studs.

TABLE R603.3.2(6)
32-FOOT-WIDE BUILDING SUPPORTING ROOF AND CEILING ONLY[a, b, c]
33 ksi STEEL

WIND SPEED		MEMBER SIZE	STUD SPACING (inches)	MINIMUM STUD THICKNESS (mils)											
				8-Foot Studs				9-Foot Studs				10-Foot Studs			
Exp. A/B	Exp. C			Ground Snow Load (psf)											
				20	30	50	70	20	30	50	70	20	30	50	70
85 mph	—	350S162	16	33	33	33	33	33	33	33	33	33	33	33	33
			24	33	33	33	43	33	33	33	43	33	33	43	43
		550S162	16	33	33	33	33	33	33	33	33	33	33	33	33
			24	33	33	33	33	33	33	33	33	33	33	33	33
90 mph	—	350S162	16	33	33	33	33	33	33	33	33	33	33	33	33
			24	33	33	33	43	33	33	33	43	33	33	43	43
		550S162	16	33	33	33	33	33	33	33	33	33	33	33	33
			24	33	33	33	33	33	33	33	33	33	33	33	33
100 mph	85 mph	350S162	16	33	33	33	33	33	33	33	33	33	33	33	33
			24	33	33	33	43	33	33	33	43	33	33	43	43
		550S162	16	33	33	33	33	33	33	33	33	33	33	33	33
			24	33	33	33	33	33	33	33	33	33	33	33	33
110 mph	90 mph	350S162	16	33	33	33	33	33	33	33	33	33	33	33	33
			24	33	33	33	43	33	33	33	43	33	33	43	43
		550S162	16	33	33	33	33	33	33	33	33	33	33	33	33
			24	33	33	33	33	33	33	33	33	33	33	33	33
—	100 mph	350S162	16	33	33	33	33	33	33	33	33	33	33	33	43
			24	33	33	43	43	33	43	43	43	43	43	43	54
		550S162	16	33	33	33	33	33	33	33	33	33	33	33	33
			24	33	33	33	33	33	33	33	33	33	33	33	43
—	110 mph	350S162	16	33	33	33	33	33	33	33	33	33	33	33	43
			24	33	33	43	43	33	43	43	43	54	54	54	54
		550S162	16	33	33	33	33	33	33	33	33	33	33	33	33
			24	33	33	33	33	33	33	33	33	33	33	33	43

For SI: 1 inch = 25.4 mm, 1 foot = 304.8 mm, 1 mil = 0.0254 mm, 1 mile per hour = 0.447 m/s, 1 pound per square foot = 0.0479 kPa, 1 ksi = 1000 psi = 6.895 MPa.

a. Deflection criterion: L/240.
b. Design load assumptions:
   Roof/ceiling dead load is 12 psf.
   Attic live load is 10 psf.
c. Building width is in the direction of horizontal framing members supported by the wall studs.

TABLE R603.3.2(7)
32-FOOT-WIDE BUILDING SUPPORTING ROOF AND CEILING ONLY[a, b, c]
50 ksi STEEL

WIND SPEED		MEMBER SIZE	STUD SPACING (inches)	MINIMUM STUD THICKNESS (mils)											
				8-Foot Studs				9-Foot Studs				10-Foot Studs			
				Ground Snow Load (psf)											
Exp. A/B	Exp. C			20	30	50	70	20	30	50	70	20	30	50	70
85 mph	—	350S162	16	33	33	33	33	33	33	33	33	33	33	33	33
			24	33	33	33	33	33	33	33	33	33	33	33	43
		550S162	16	33	33	33	33	33	33	33	33	33	33	33	33
			24	33	33	33	33	33	33	33	33	33	33	33	33
90 mph	—	350S162	16	33	33	33	33	33	33	33	33	33	33	33	33
			24	33	33	33	33	33	33	33	33	33	33	33	43
		550S162	16	33	33	33	33	33	33	33	33	33	33	33	33
			24	33	33	33	33	33	33	33	33	33	33	33	33
100 mph	85 mph	350S162	16	33	33	33	33	33	33	33	33	33	33	33	33
			24	33	33	33	43	33	33	33	33	33	33	33	43
		550S162	16	33	33	33	33	33	33	33	33	33	33	33	33
			24	33	33	33	33	33	33	33	33	33	33	33	33
110 mph	90 mph	350S162	16	33	33	33	33	33	33	33	33	33	33	33	33
			24	33	33	33	43	33	33	33	33	33	33	33	43
		550S162	16	33	33	33	33	33	33	33	33	33	33	33	33
			24	33	33	33	33	33	33	33	33	33	33	33	33
—	100 mph	350S162	16	33	33	33	33	33	33	33	33	33	33	33	33
			24	33	33	33	43	33	33	33	43	43	43	43	43
		550S162	16	33	33	33	33	33	33	33	33	33	33	33	33
			24	33	33	33	33	33	33	33	33	33	33	33	33
—	110 mph	350S162	16	33	33	33	33	33	33	33	33	33	33	33	33
			24	33	33	33	43	33	33	33	43	54	54	54	54
		550S162	16	33	33	33	33	33	33	33	33	33	33	33	33
			24	33	33	33	33	33	33	33	33	33	33	33	33

For SI: 1 inch = 25.4 mm, 1 foot = 304.8 mm, 1 mil = 0.0254 mm, 1 mile per hour = 0.447 m/s, 1 pound per square foot = 0.0479 kPa, 1 ksi = 1000 psi = 6.895 MPa.

a. Deflection criterion: $L/240$.
b. Design load assumptions:
   Roof/ceiling dead load is 12 psf.
   Attic live load is 10 psf.
c. Building width is in the direction of horizontal framing members supported by the wall studs.

TABLE R603.3.2(8)
36-FOOT-WIDE BUILDING SUPPORTING ROOF AND CEILING ONLY[a, b, c]
33 ksi STEEL

WIND SPEED		MEMBER SIZE	STUD SPACING (inches)	MINIMUM STUD THICKNESS (mils)											
				8-Foot Studs				9-Foot Studs				10-Foot Studs			
				Ground Snow Load (psf)											
Exp. A/B	Exp. C			20	30	50	70	20	30	50	70	20	30	50	70
85 mph	—	350S162	16	33	33	33	33	33	33	33	33	33	33	33	33
			24	33	33	43	43	33	33	33	43	33	33	43	43
		550S162	16	33	33	33	33	33	33	33	33	33	33	33	33
			24	33	33	33	33	33	33	33	33	33	33	33	33
90 mph	—	350S162	16	33	33	33	33	33	33	33	33	33	33	33	33
			24	33	33	43	43	33	33	33	43	33	33	43	43
		550S162	16	33	33	33	33	33	33	33	33	33	33	33	33
			24	33	33	33	33	33	33	33	33	33	33	33	33
100 mph	85 mph	350S162	16	33	33	33	33	33	33	33	33	33	33	33	33
			24	33	33	33	43	33	33	43	43	33	33	43	54
		550S162	16	33	33	33	33	33	33	33	33	33	33	33	33
			24	33	33	33	33	33	33	33	33	33	33	33	43
110 mph	90 mph	350S162	16	33	33	33	33	33	33	33	33	33	33	33	33
			24	33	33	43	43	33	33	43	43	33	43	43	54
		550S162	16	33	33	33	33	33	33	33	33	33	33	33	33
			24	33	33	33	33	33	33	33	33	33	33	33	43
—	100 mph	350S162	16	33	33	33	33	33	33	33	33	33	33	33	43
			24	33	33	43	43	43	43	43	43	43	43	43	54
		550S162	16	33	33	33	33	33	33	33	33	33	33	33	33
			24	33	33	33	43	33	33	33	43	33	33	33	43
—	110 mph	350S162	16	33	33	33	33	33	33	33	33	33	33	33	43
			24	33	33	43	54	43	43	43	54	54	54	54	54
		550S162	16	33	33	33	33	33	33	33	33	33	33	33	33
			24	33	33	33	43	33	33	33	43	33	33	33	33

For SI: 1 inch = 25.4 mm, 1 foot = 304.8 mm, 1 mil = 0.0254 mm, 1 mile per hour = 0.447 m/s, 1 pound per square foot = 0.0479 kPa,
1 ksi = 1000 psi = 6.895 MPa.

a. Deflection criterion: $L/240$.
b. Design load assumptions:
   Roof/ceiling dead load is 12 psf.
   Attic live load is 10 psf.
c. Building width is in the direction of horizontal framing members supported by the wall studs.

TABLE R603.3.2(9)
36-FOOT-WIDE BUILDING SUPPORTING ROOF AND CEILING ONLY[a, b, c]
50 ksi STEEL

WIND SPEED		MEMBER SIZE	STUD SPACING (inches)	MINIMUM STUD THICKNESS (mils)											
				8-Foot Studs				9-Foot Studs				10-Foot Studs			
				Ground Snow Load (psf)											
Exp. A/B	Exp. C			20	30	50	70	20	30	50	70	20	30	50	70
85 mph	—	350S162	16	33	33	33	33	33	33	33	33	33	33	33	33
			24	33	33	33	43	33	33	33	43	33	33	33	43
		550S162	16	33	33	33	33	33	33	33	33	33	33	33	33
			24	33	33	33	33	33	33	33	33	33	33	33	33
90 mph	—	350S162	16	33	33	33	33	33	33	33	33	33	33	33	33
			24	33	33	33	43	33	33	33	43	33	33	33	43
		550S162	16	33	33	33	33	33	33	33	33	33	33	33	33
			24	33	33	33	33	33	33	33	33	33	33	33	33
100 mph	85 mph	350S162	16	33	33	33	33	33	33	33	33	33	33	33	33
			24	33	33	33	43	33	33	33	43	33	33	33	43
		550S162	16	33	33	33	33	33	33	33	33	33	33	33	33
			24	33	33	33	33	33	33	33	33	33	33	33	33
110 mph	90 mph	350S162	16	33	33	33	33	33	33	33	33	33	33	33	33
			24	33	33	33	43	33	33	33	43	33	33	43	43
		550S162	16	33	33	33	33	33	33	33	33	33	33	33	33
			24	33	33	33	33	33	33	33	33	33	33	33	33
—	100 mph	350S162	16	33	33	33	33	33	33	33	33	33	33	33	33
			24	33	33	33	43	33	33	33	43	43	43	43	43
		550S162	16	33	33	33	33	33	33	33	33	33	33	33	33
			24	33	33	33	33	33	33	33	33	33	33	33	33
—	110 mph	350S162	16	33	33	33	33	33	33	33	33	33	33	33	33
			24	33	33	33	43	33	33	33	43	54	54	54	54
		550S162	16	33	33	33	33	33	33	33	33	33	33	33	33
			24	33	33	33	33	33	33	33	33	33	33	33	33

For SI: 1 inch = 25.4 mm, 1 foot = 304.8 mm, 1 mil = 0.0254 mm, 1 mile per hour = 0.447 m/s, 1 pound per square foot = 0.0479 kPa,
   1 ksi = 1000 psi = 6.895 MPa.

a. Deflection criterion: L/240.
b. Design load assumptions:
   Roof/ceiling dead load is 12 psf.
   Attic live load is 10 psf.
c. Building width is in the direction of horizontal framing members supported by the wall studs.

TABLE R603.3.2(10)
40-FOOT-WIDE BUILDING SUPPORTING ROOF AND CEILING ONLY[a, b, c]
33 ksi STEEL

WIND SPEED		MEMBER SIZE	STUD SPACING (inches)	MINIMUM STUD THICKNESS (mils)											
				8-Foot Studs				9-Foot Studs				10-Foot Studs			
				Ground Snow Load (psf)											
Exp. A/B	Exp. C			20	30	50	70	20	30	50	70	20	30	50	70
85 mph	—	350S162	16	33	33	33	33	33	33	33	33	33	33	33	43
			24	33	33	43	43	33	33	43	43	33	33	43	54
		550S162	16	33	33	33	33	33	33	33	33	33	33	33	33
			24	33	33	33	43	33	33	33	33	33	33	33	43
90 mph	—	350S162	16	33	33	33	33	33	33	33	33	33	33	33	43
			24	33	33	43	43	33	33	43	43	33	33	43	54
		550S162	16	33	33	33	33	33	33	33	33	33	33	33	33
			24	33	33	33	43	33	33	33	33	33	33	33	43
100 mph	85 mph	350S162	16	33	33	33	33	33	33	33	33	33	33	33	43
			24	33	33	43	43	33	33	43	43	33	33	43	54
		550S162	16	33	33	33	33	33	33	33	33	33	33	33	33
			24	33	33	33	43	33	33	33	43	33	33	33	33
110 mph	90 mph	350S162	16	33	33	33	33	33	33	33	33	33	33	33	43
			24	33	33	43	54	33	33	43	43	43	43	43	54
		550S162	16	33	33	33	33	33	33	33	33	33	33	33	33
			24	33	33	33	43	33	33	33	43	33	33	33	43
—	100 mph	350S162	16	33	33	33	43	33	33	33	43	33	33	33	43
			24	33	33	43	54	43	43	43	54	43	43	43	54
		550S162	16	33	33	33	33	33	33	33	33	33	33	33	33
			24	33	33	33	43	33	33	33	43	33	33	33	43
—	110 mph	350S162	16	33	33	33	43	33	33	33	43	33	33	43	43
			24	33	43	43	54	33	43	43	54	54	54	54	54
		550S162	16	33	33	33	33	33	33	33	33	33	33	33	33
			24	33	33	33	43	33	33	33	43	33	33	33	43

For SI: 1 inch = 25.4 mm, 1 foot = 304.8 mm, 1 mil = 0.0254 mm, 1 mile per hour = 0.447 m/s, 1 pound per square foot = 0.0479 kPa, 1 ksi = 1000 psi = 6.895 MPa.

a. Deflection criterion: L/240.
b. Design load assumptions:
   Roof/ceiling dead load is 12 psf.
   Attic live load is 10 psf.
c. Building width is in the direction of horizontal framing members supported by the wall studs.

TABLE R603.3.2(11)
40-FOOT-WIDE BUILDING SUPPORTING ROOF AND CEILING ONLY[a, b, c]
50 ksi STEEL

WIND SPEED		MEMBER SIZE	STUD SPACING (inches)	MINIMUM STUD THICKNESS (mils)											
				8-Foot Studs				9-Foot Studs				10-Foot Studs			
				Ground Snow Load (psf)											
Exp. A/B	Exp. C			20	30	50	70	20	30	50	70	20	30	50	70
85 mph	—	350S162	16	33	33	33	33	33	33	33	33	33	33	33	33
			24	33	33	33	43	33	33	33	43	33	33	43	43
		550S162	16	33	33	33	33	33	33	33	33	33	33	33	33
			24	33	33	33	33	33	33	33	33	33	33	33	33
90 mph	—	350S162	16	33	33	33	33	33	33	33	33	33	33	33	33
			24	33	33	33	43	33	33	33	43	33	33	43	43
		550S162	16	33	33	33	33	33	33	33	33	33	33	33	33
			24	33	33	33	33	33	33	33	33	33	33	33	33
100 mph	85 mph	350S162	16	33	33	33	33	33	33	33	33	33	33	33	33
			24	33	33	33	43	33	33	33	43	33	33	43	43
		550S162	16	33	33	33	33	33	33	33	33	33	33	33	33
			24	33	33	33	33	33	33	33	33	33	33	33	33
110 mph	90 mph	350S162	16	33	33	33	33	33	33	33	33	33	33	33	33
			24	33	33	33	43	33	33	33	43	33	33	43	43
		550S162	16	33	33	33	33	33	33	33	33	33	33	33	33
			24	33	33	33	33	33	33	33	33	33	33	33	33
—	100 mph	350S162	16	33	33	33	33	33	33	33	33	33	33	33	33
			24	33	33	43	43	33	33	33	43	43	43	43	54
		550S162	16	33	33	33	33	33	33	33	33	33	33	33	33
			24	33	33	33	33	33	33	33	33	33	33	33	33
—	110 mph	350S162	16	33	33	33	33	33	33	33	33	33	33	33	43
			24	33	33	43	43	33	33	43	43	54	54	54	54
		550S162	16	33	33	33	33	33	33	33	33	33	33	33	33
			24	33	33	33	33	33	33	33	33	33	33	33	33

For SI: 1 inch = 25.4 mm, 1 foot = 304.8 mm, 1 mil = 0.0254 mm, 1 mile per hour = 0.447 m/s, 1 pound per square foot = 0.0479 kPa,
1 ksi = 1000 psi = 6.895 MPa.

a. Deflection criterion: $L/240$.
b. Design load assumptions:
Roof/ceiling dead load is 12 psf.
Attic live load is 10 psf.
c. Building width is in the direction of horizontal framing members supported by the wall studs.

TABLE R603.3.2(12)
24-FOOT-WIDE BUILDING SUPPORTING ONE FLOOR, ROOF AND CEILING[a, b, c]
33 ksi STEEL

WIND SPEED		MEMBER SIZE	STUD SPACING (inches)	MINIMUM STUD THICKNESS (mils)											
				8-Foot Studs				9-Foot Studs				10-Foot Studs			
				Ground Snow Load (psf)											
Exp. A/B	Exp. C			20	30	50	70	20	30	50	70	20	30	50	70
85 mph	—	350S162	16	33	33	33	33	33	33	33	33	33	33	33	33
			24	33	33	33	43	33	33	33	43	43	43	43	43
		550S162	16	33	33	33	33	33	33	33	33	33	33	33	33
			24	33	33	33	33	33	33	33	33	33	33	33	33
90 mph	—	350S162	16	33	33	33	33	33	33	33	33	33	33	33	33
			24	33	33	33	43	33	33	33	43	43	43	43	43
		550S162	16	33	33	33	33	33	33	33	33	33	33	33	33
			24	33	33	33	33	33	33	33	33	33	33	33	33
100 mph	85 mph	350S162	16	33	33	33	33	33	33	33	33	33	33	33	33
			24	33	33	33	43	33	33	43	43	43	43	43	43
		550S162	16	33	33	33	33	33	33	33	33	33	33	33	33
			24	33	33	33	33	33	33	33	33	33	33	33	33
110 mph	90 mph	350S162	16	33	33	33	33	33	33	33	33	33	33	33	33
			24	33	33	43	43	43	43	43	43	43	43	43	43
		550S162	16	33	33	33	33	33	33	33	33	33	33	33	33
			24	33	33	33	33	33	33	33	33	33	33	33	33
—	100 mph	350S162	16	33	33	33	33	33	33	33	33	33	33	43	43
			24	43	43	43	43	43	43	43	43	43	43	54	54
		550S162	16	33	33	33	33	33	33	33	33	33	33	33	33
			24	33	33	33	33	33	33	33	33	33	33	33	33
—	110 mph	350S162	16	33	33	33	33	33	33	33	33	43	43	43	43
			24	43	43	43	43	43	43	43	43	54	54	54	54
		550S162	16	33	33	33	33	33	33	33	33	33	33	33	33
			24	33	33	33	33	33	33	33	33	33	33	33	43

For SI: 1 inch = 25.4 mm, 1 foot = 304.8 mm, 1 mil = 0.0254 mm, 1 mile per hour = 0.447 m/s, 1 pound per square foot = 0.0479 kPa, 1 ksi = 1000 psi = 6.895 MPa.

a. Deflection criterion: L/240.
b. Design load assumptions:
    Second floor dead load is 10 psf.
    Second floor live load is 30 psf.
    Roof/ceiling dead load is 12 psf.
    Attic live load is 10 psf.
c. Building width is in the direction of horizontal framing members supported by the wall studs.

TABLE R603.3.2(13)
24-FOOT-WIDE BUILDING SUPPORTING ONE FLOOR, ROOF AND CEILING[a, b, c]
50 ksi STEEL

WIND SPEED		MEMBER SIZE	STUD SPACING (inches)	MINIMUM STUD THICKNESS (mils)											
				8-Foot Studs				9-Foot Studs				10-Foot Studs			
Exp. A/B	Exp. C			Ground Snow Load (psf)											
				20	30	50	70	20	30	50	70	20	30	50	70
85 mph	—	350S162	16	33	33	33	33	33	33	33	33	33	33	33	33
			24	33	33	33	33	33	33	33	33	33	33	33	43
		550S162	16	33	33	33	33	33	33	33	33	33	33	33	33
			24	33	33	33	33	33	33	33	33	33	33	33	33
90 mph	—	350S162	16	33	33	33	33	33	33	33	33	33	33	33	33
			24	33	33	33	33	33	33	33	33	33	33	33	43
		550S162	16	33	33	33	33	33	33	33	33	33	33	33	33
			24	33	33	33	33	33	33	33	33	33	33	33	33
100 mph	85 mph	350S162	16	33	33	33	33	33	33	33	33	33	33	33	33
			24	33	33	33	33	33	33	33	33	33	33	33	43
		550S162	16	33	33	33	33	33	33	33	33	33	33	33	33
			24	33	33	33	33	33	33	33	33	33	33	33	33
110 mph	90 mph	350S162	16	33	33	33	33	33	33	33	33	33	33	33	33
			24	33	33	33	33	33	33	33	33	33	33	43	43
		550S162	16	33	33	33	33	33	33	33	33	33	33	33	33
			24	33	33	33	33	33	33	33	33	33	33	33	33
—	100 mph	350S162	16	33	33	33	33	33	33	33	33	33	33	33	33
			24	33	33	33	43	33	33	33	43	43	43	43	43
		550S162	16	33	33	33	33	33	33	33	33	33	33	33	33
			24	33	33	33	33	33	33	33	33	33	33	33	33
—	110 mph	350S162	16	33	33	33	33	33	33	33	33	33	33	33	33
			24	33	33	33	43	33	43	43	43	54	54	54	54
		550S162	16	33	33	33	33	33	33	33	33	33	33	33	33
			24	33	33	33	33	33	33	33	33	33	33	33	33

For SI: 1 inch = 25.4 mm, 1 foot = 304.8 mm, 1 mil = 0.0254 mm, 1 mile per hour = 0.447 m/s, 1 pound per square foot = 0.0479 kPa,
1 ksi = 1000 psi = 6.895 MPa.

a. Deflection criterion: $L/240$.
b. Design load assumptions:
   Second floor dead load is 10 psf.
   Second floor live load is 30 psf.
   Roof/ceiling dead load is 12 psf.
   Attic live load is 10 psf.
c. Building width is in the direction of horizontal framing members supported by the wall studs.

TABLE R603.3.2(14)
28-FOOT-WIDE BUILDING SUPPORTING ONE FLOOR, ROOF AND CEILING[a, b, c]
33 ksi STEEL

WIND SPEED		MEMBER SIZE	STUD SPACING (inches)	MINIMUM STUD THICKNESS (mils)											
				8-Foot Studs				9-Foot Studs				10-Foot Studs			
				Ground Snow Load (psf)											
Exp. A/B	Exp. C			20	30	50	70	20	30	50	70	20	30	50	70
85 mph	—	350S162	16	33	33	33	33	33	33	33	33	33	33	33	33
			24	33	43	43	43	33	33	43	43	43	43	43	43
		550S162	16	33	33	33	33	33	33	33	33	33	33	33	33
			24	33	33	33	33	33	33	33	33	33	33	33	33
90 mph	—	350S162	16	33	33	33	33	33	33	33	33	33	33	33	33
			24	33	43	43	43	33	33	43	43	43	43	43	43
		550S162	16	33	33	33	33	33	33	33	33	33	33	33	33
			24	33	33	33	33	33	33	33	33	33	33	33	33
100 mph	85 mph	350S162	16	33	33	33	33	33	33	33	33	33	33	33	33
			24	33	43	43	43	43	43	43	43	43	43	43	43
		550S162	16	33	33	33	33	33	33	33	33	33	33	33	33
			24	33	33	33	33	33	33	33	33	33	33	33	33
110 mph	90 mph	350S162	16	33	33	33	33	33	33	33	33	33	33	33	33
			24	43	43	43	43	43	43	43	43	43	43	43	54
		550S162	16	33	33	33	33	33	33	33	33	33	33	33	33
			24	33	33	33	33	33	33	33	33	33	33	33	33
—	100 mph	350S162	16	33	33	33	33	33	33	33	33	43	43	43	43
			24	43	43	43	43	43	43	43	43	54	54	54	54
		550S162	16	33	33	33	33	33	33	33	33	33	33	33	33
			24	33	33	33	33	33	33	33	33	33	33	33	43
—	110 mph	350S162	16	33	33	33	33	33	33	33	43	43	43	43	43
			24	43	43	43	43	43	43	54	54	54	54	54	54
		550S162	16	33	33	33	33	33	33	33	33	33	33	33	33
			24	33	33	33	33	33	33	33	33	33	43	43	43

For SI: 1 inch = 25.4 mm, 1 foot = 304.8 mm, 1 mil = 0.0254 mm, 1 mile per hour = 0.447 m/s, 1 pound per square foot = 0.0479 kPa, 1 ksi = 1000 psi = 6.895 MPa.

a. Deflection criterion: $L/240$.

b. Design load assumptions:
   Second floor dead load is 10 psf.
   Second floor live load is 30 psf.
   Roof/ceiling dead load is 12 psf.
   Attic live load is 10 psf.

c. Building width is in the direction of horizontal framing members supported by the wall studs.

TABLE R603.3.2(15)
28-FOOT-WIDE BUILDING SUPPORTING ONE FLOOR, ROOF AND CEILING[a, b, c]
50 ksi STEEL

WIND SPEED		MEMBER SIZE	STUD SPACING (inches)	MINIMUM STUD THICKNESS (mils)											
				8-Foot Studs				9-Foot Studs				10-Foot Studs			
Exp. A/B	Exp. C			Ground Snow Load (psf)											
				20	30	50	70	20	30	50	70	20	30	50	70
85 mph	—	350S162	16	33	33	33	33	33	33	33	33	33	33	33	33
			24	33	33	33	43	33	33	33	33	33	33	33	43
		550S162	16	33	33	33	33	33	33	33	33	33	33	33	33
			24	33	33	33	33	33	33	33	33	33	33	33	33
90 mph	—	350S162	16	33	33	33	33	33	33	33	33	33	33	33	33
			24	33	33	33	43	33	33	33	33	33	33	33	43
		550S162	16	33	33	33	33	33	33	33	33	33	33	33	33
			24	33	33	33	33	33	33	33	33	33	33	33	33
100 mph	85 mph	350S162	16	33	33	33	33	33	33	33	33	33	33	33	33
			24	33	33	33	43	33	33	33	33	33	33	43	43
		550S162	16	33	33	33	33	33	33	33	33	33	33	33	33
			24	33	33	33	33	33	33	33	33	33	33	33	33
110 mph	90 mph	350S162	16	33	33	33	33	33	33	33	33	33	33	33	33
			24	33	33	33	43	33	33	33	43	43	43	43	43
		550S162	16	33	33	33	33	33	33	33	33	33	33	33	33
			24	33	33	33	33	33	33	33	33	33	33	33	33
—	100 mph	350S162	16	33	33	33	33	33	33	33	33	33	33	33	33
			24	33	33	43	43	33	33	43	43	43	43	43	43
		550S162	16	33	33	33	33	33	33	33	33	33	33	33	33
			24	33	33	33	33	33	33	33	33	33	33	33	33
—	110 mph	350S162	16	33	33	33	33	33	33	33	33	33	33	33	43
			24	33	43	43	43	43	43	43	43	54	54	54	54
		550S162	16	33	33	33	33	33	33	33	33	33	33	33	33
			24	33	33	33	33	33	33	33	33	33	33	33	33

For SI: 1 inch = 25.4 mm, 1 foot = 304.8 mm, 1 mil = 0.0254 mm, 1 mile per hour = 0.447 m/s, 1 pound per square foot = 0.0479 kPa, 1 ksi = 1000 psi = 6.895 MPa.

a. Deflection criterion: L/240.
b. Design load assumptions:
   Second floor dead load is 10 psf.
   Second floor live load is 30 psf.
   Roof/ceiling dead load is 12 psf.
   Attic live load is 10 psf.
c. Building width is in the direction of horizontal framing members supported by the wall studs.

TABLE R603.3.2(16)
32-FOOT-WIDE BUILDING SUPPORTING ONE FLOOR, ROOF AND CEILING[a, b, c]
33 ksi STEEL

WIND SPEED		MEMBER SIZE	STUD SPACING (inches)	MINIMUM STUD THICKNESS (mils)											
				8-Foot Studs				9-Foot Studs				10-Foot Studs			
				Ground Snow Load (psf)											
Exp. A/B	Exp. C			20	30	50	70	20	30	50	70	20	30	50	70
85 mph	—	350S162	16	33	33	33	33	33	33	33	33	33	33	33	33
			24	43	43	43	43	43	43	43	43	43	43	43	54
		550S162	16	33	33	33	33	33	33	33	33	33	33	33	33
			24	33	33	33	43	33	33	33	33	33	33	33	43
90 mph	—	350S162	16	33	33	33	33	33	33	33	33	33	33	33	33
			24	43	43	43	43	43	43	43	43	43	43	43	54
		550S162	16	33	33	33	33	33	33	33	33	33	33	33	33
			24	33	33	33	43	33	33	33	33	33	33	33	43
100 mph	85 mph	350S162	16	33	33	33	33	33	33	33	33	33	33	33	43
			24	43	43	43	43	43	43	43	43	43	43	43	54
		550S162	16	33	33	33	33	33	33	33	33	33	33	33	33
			24	33	33	33	43	33	33	33	33	33	33	33	43
110 mph	90 mph	350S162	16	33	33	33	33	33	33	33	33	33	33	33	43
			24	43	43	43	43	43	43	43	43	43	43	43	54
		550S162	16	33	33	33	33	33	33	33	33	33	33	33	33
			24	33	33	33	43	33	33	33	33	33	33	33	43
—	100 mph	350S162	16	33	33	33	33	33	33	33	33	43	43	43	43
			24	43	43	43	54	43	43	43	54	54	54	54	54
		550S162	16	33	33	33	33	33	33	33	33	33	33	33	33
			24	33	33	33	43	33	33	33	43	33	33	43	43
—	110 mph	350S162	16	33	33	33	43	33	33	43	43	43	43	43	43
			24	43	43	43	54	43	54	54	54	54	54	54	68
		550S162	16	33	33	33	33	33	33	33	33	33	33	33	33
			24	33	33	33	43	33	33	33	43	43	43	43	43

For SI: 1 inch = 25.4 mm, 1 foot = 304.8 mm, 1 mil = 0.0254 mm, 1 mile per hour = 0.447 m/s, 1 pound per square foot = 0.0479 kPa, 1 ksi = 1000 psi = 6.895 MPa.

a. Deflection criterion: $L/240$.
b. Design load assumptions:
   Second floor dead load is 10 psf.
   Second floor live load is 30 psf.
   Roof/ceiling dead load is 12 psf.
   Attic live load is 10 psf.
c. Building width is in the direction of horizontal framing members supported by the wall studs.

TABLE R603.3.2(17)
32-FOOT-WIDE BUILDING SUPPORTING ONE FLOOR, ROOF AND CEILING[a, b, c]
50 ksi STEEL

WIND SPEED		MEMBER SIZE	STUD SPACING (inches)	MINIMUM STUD THICKNESS (mils)											
				8-Foot Studs				9-Foot Studs				10-Foot Studs			
				Ground Snow Load (psf)											
Exp. A/B	Exp. C			20	30	50	70	20	30	50	70	20	30	50	70
85 mph	—	350S162	16	33	33	33	33	33	33	33	33	33	33	33	33
			24	33	43	43	43	33	33	33	43	33	33	43	43
		550S162	16	33	33	33	33	33	33	33	33	33	33	33	33
			24	33	33	33	33	33	33	33	33	33	33	33	33
90 mph	—	350S162	16	33	33	33	33	33	33	33	33	33	33	33	33
			24	33	43	43	43	33	33	33	43	33	33	43	43
		550S162	16	33	33	33	33	33	33	33	33	33	33	33	33
			24	33	33	33	33	33	33	33	33	33	33	33	33
100 mph	85 mph	350S162	16	33	33	33	33	33	33	33	33	33	33	33	33
			24	33	43	43	43	33	33	33	43	43	43	43	43
		550S162	16	33	33	33	33	33	33	33	33	33	33	33	33
			24	33	33	33	33	33	33	33	33	33	33	33	33
110 mph	90 mph	350S162	16	33	33	33	33	33	33	33	33	33	33	33	33
			24	33	43	43	43	33	33	33	43	43	43	43	54
		550S162	16	33	33	33	33	33	33	33	33	33	33	33	33
			24	33	33	33	33	33	33	33	33	33	33	33	33
—	100 mph	350S162	16	33	33	33	33	33	33	33	33	33	33	33	33
			24	33	43	43	43	43	43	43	43	43	43	43	54
		550S162	16	33	33	33	33	33	33	33	33	33	33	33	33
			24	33	33	33	33	33	33	33	33	33	33	33	33
—	110 mph	350S162	16	33	33	33	33	33	33	33	33	33	33	33	43
			24	43	43	43	43	43	43	43	43	54	54	54	54
		550S162	16	33	33	33	33	33	33	33	33	33	33	33	33
			24	33	33	33	33	33	33	33	33	33	33	33	33

For SI: 1 inch = 25.4 mm, 1 foot = 304.8 mm, 1 mil = 0.0254 mm, 1 mile per hour = 0.447 m/s, 1 pound per square foot = 0.0479 kPa,
1 ksi = 1000 psi = 6.895 MPa.

a. Deflection criterion: L/240.
b. Design load assumptions:
    Second floor dead load is 10 psf.
    Second floor live load is 30 psf.
    Roof/ceiling dead load is 12 psf.
    Attic live load is 10 psf.
c. Building width is in the direction of horizontal framing members supported by the wall studs.

TABLE R603.3.2(18)
36-FOOT-WIDE BUILDING SUPPORTING ONE FLOOR, ROOF AND CEILING[a, b, c]
33 ksi STEEL

WIND SPEED		MEMBER SIZE	STUD SPACING (inches)	MINIMUM STUD THICKNESS (mils)											
				8-Foot Studs				9-Foot Studs				10-Foot Studs			
Exp. A/B	Exp. C			Ground Snow Load (psf)											
				20	30	50	70	20	30	50	70	20	30	50	70
85 mph	—	350S162	16	33	33	33	33	33	33	33	33	33	33	33	43
			24	43	43	43	54	43	43	43	43	43	43	43	54
		550S162	16	33	33	33	33	33	33	33	33	33	33	33	33
			24	43	43	43	43	33	33	33	43	33	33	33	43
90 mph	—	350S162	16	33	33	33	33	33	33	33	33	33	33	33	43
			24	43	43	43	54	43	43	43	43	43	43	43	54
		550S162	16	33	33	33	33	33	33	33	33	33	33	33	33
			24	43	43	43	43	33	33	33	43	33	33	43	43
100 mph	85 mph	350S162	16	33	33	33	33	33	33	33	33	33	33	33	43
			24	43	43	43	54	43	43	43	54	43	43	54	54
		550S162	16	33	33	33	33	33	33	33	33	33	33	33	33
			24	43	43	43	43	33	33	33	43	33	33	43	43
110 mph	90 mph	350S162	16	33	33	33	43	33	33	33	33	33	33	43	43
			24	43	43	43	54	43	43	43	54	43	43	54	54
		550S162	16	33	33	33	33	33	33	33	33	33	33	33	33
			24	43	43	43	43	33	33	33	43	33	33	43	43
—	100 mph	350S162	16	33	33	33	43	33	33	43	43	43	43	43	43
			24	43	43	43	54	43	43	54	54	54	54	54	54
		550S162	16	33	33	33	33	33	33	33	33	33	33	33	33
			24	43	43	43	43	33	33	33	43	43	43	43	43
—	110 mph	350S162	16	33	33	33	43	33	33	43	43	43	43	43	43
			24	43	43	43	54	43	54	54	54	54	54	54	68
		550S162	16	33	33	33	33	33	33	33	33	33	33	33	33
			24	33	33	33	43	33	33	33	43	43	43	43	43

For SI: 1 inch = 25.4 mm, 1 foot = 304.8 mm, 1 mil = 0.0254 mm, 1 mile per hour = 0.447 m/s, 1 pound per square foot = 0.0479 kPa, 1 ksi = 1000 psi = 6.895 MPa.

a. Deflection criterion: $L/240$.
b. Design load assumptions:
   Second floor dead load is 10 psf.
   Second floor live load is 30 psf.
   Roof/ceiling dead load is 12 psf.
   Attic live load is 10 psf.
c. Building width is in the direction of horizontal framing members supported by the wall studs.

TABLE R603.3.2(19)
36-FOOT-WIDE BUILDING SUPPORTING ONE FLOOR, ROOF AND CEILING[a, b, c]
50 ksi STEEL

WIND SPEED		MEMBER SIZE	STUD SPACING (inches)	MINIMUM STUD THICKNESS (mils)											
				8-Foot Studs				9-Foot Studs				10-Foot Studs			
Exp. A/B	Exp. C			Ground Snow Load (psf)											
				20	30	50	70	20	30	50	70	20	30	50	70
85 mph	—	350S162	16	33	33	33	33	33	33	33	33	33	33	33	33
			24	43	43	43	43	33	33	43	43	43	43	43	43
		550S162	16	33	33	33	33	33	33	33	33	33	33	33	33
			24	33	33	33	43	33	33	33	33	33	33	33	33
90 mph	—	350S162	16	33	33	33	33	33	33	33	33	33	33	33	33
			24	43	43	43	43	33	33	43	43	43	43	43	43
		550S162	16	33	33	33	33	33	33	33	33	33	33	33	33
			24	33	43	33	43	33	33	33	33	33	33	33	33
100 mph	85 mph	350S162	16	33	33	33	33	33	33	33	33	33	33	33	33
			24	43	43	43	43	33	33	43	43	43	43	43	43
		550S162	16	33	33	33	33	33	33	33	33	33	33	33	33
			24	33	33	33	43	33	33	33	33	33	33	33	33
110 mph	90 mph	350S162	16	33	33	33	33	33	33	33	33	33	33	33	33
			24	43	43	43	43	33	33	43	43	43	43	43	54
		550S162	16	33	33	33	33	33	33	33	33	33	33	33	33
			24	43	33	33	43	33	33	33	33	33	33	33	33
—	100 mph	350S162	16	33	33	33	33	33	33	33	33	33	33	33	43
			24	43	43	43	43	43	43	43	43	43	43	54	54
		550S162	16	33	33	33	33	33	33	33	33	33	33	33	33
			24	33	33	33	43	33	33	33	33	33	33	33	43
—	110 mph	350S162	16	33	33	33	33	33	33	33	33	33	33	43	43
			24	43	43	43	54	43	43	43	43	54	54	54	54
		550S162	16	33	33	33	33	33	33	33	33	33	33	33	33
			24	33	33	33	33	33	33	33	33	33	33	33	43

For SI: 1 inch = 25.4 mm, 1 foot = 304.8 mm, 1 mil = 0.0254 mm, 1 mile per hour = 0.447 m/s 1 pound per square foot = 0.0479 kPa,
1 ksi = 1000 psi = 6.895 MPa.

a. Deflection criterion: L/240.
b. Design load assumptions:
  Second floor dead load is 10 psf.
  Second floor live load is 30 psf.
  Roof/ceiling dead load is 12 psf.
  Attic live load is 10 psf.
c. Building width is in the direction of horizontal framing members supported by the wall studs.

TABLE R603.3.2(20)
40-FOOT-WIDE BUILDING SUPPORTING ONE FLOOR, ROOF AND CEILING[a, b, c]
33 ksi STEEL

WIND SPEED		MEMBER SIZE	STUD SPACING (inches)	MINIMUM STUD THICKNESS (mils)											
				8-Foot Studs				9-Foot Studs				10-Foot Studs			
Exp. A/B	Exp. C			Ground Snow Load (psf)											
				20	30	50	70	20	30	50	70	20	30	50	70
85 mph	—	350S162	16	33	33	33	43	33	33	33	43	33	33	33	43
			24	43	54	54	54	43	43	43	54	43	43	54	54
		550S162	16	33	33	33	33	33	33	33	33	33	33	33	33
			24	43	43	43	43	43	43	43	43	33	43	43	43
90 mph	—	350S162	16	33	33	33	43	33	33	33	43	33	33	33	43
			24	43	54	54	54	43	43	43	54	43	43	54	54
		550S162	16	33	33	33	33	33	33	33	33	33	33	33	33
			24	43	43	43	43	43	43	43	43	33	43	43	43
100 mph	85 mph	350S162	16	33	33	33	43	33	33	33	43	33	33	43	43
			24	43	54	54	54	43	43	43	54	43	43	54	54
		550S162	16	33	33	33	33	33	33	33	33	33	33	33	33
			24	43	43	43	43	43	43	43	43	33	43	43	43
110 mph	90 mph	350S162	16	33	33	33	43	33	33	33	43	33	43	43	43
			24	43	54	54	54	43	43	43	54	43	54	54	54
		550S162	16	33	33	33	33	33	33	33	33	33	33	33	33
			24	43	43	43	43	43	43	43	43	33	43	43	43
—	100 mph	350S162	16	33	33	33	43	33	33	43	43	33	43	43	43
			24	43	54	54	54	54	54	54	54	43	54	54	68
		550S162	16	33	33	33	33	33	33	33	33	33	33	33	33
			24	43	43	43	43	43	43	43	43	33	43	43	43
—	110 mph	350S162	16	33	33	43	43	43	43	43	43	33	43	43	43
			24	54	54	54	54	54	54	54	54	54	68	68	68
		550S162	16	33	33	33	33	33	33	33	33	33	33	33	33
			24	43	43	43	43	43	43	43	43	33	43	43	43

For SI: 1 inch = 25.4 mm, 1 foot = 304.8 mm, 1 mil = 0.0254 mm, 1 mile per hour = 0.447 m/s, 1 pound per square foot = 0.0479 kPa,
1 ksi = 1000 psi = 6.895 MPa.
a. Deflection criterion: $L/240$.
b. Design load assumptions:
   Second floor dead load is 10 psf.
   Second floor live load is 30 psf.
   Roof/ceiling dead load is 12 psf.
   Attic live load is 10 psf.
c. Building width is in the direction of horizontal framing members supported by the wall studs.

TABLE R603.3.2(21)
40-FOOT-WIDE BUILDING SUPPORTING ONE FLOOR, ROOF AND CEILING[a, b, c]
50 ksi STEEL

WIND SPEED		MEMBER SIZE	STUD SPACING (inches)	MINIMUM STUD THICKNESS (mils)											
				8-Foot Studs				9-Foot Studs				10-Foot Studs			
				Ground Snow Load (psf)											
Exp. A/B	Exp. C			20	30	50	70	20	30	50	70	20	30	50	70
85 mph	—	350S162	16	33	33	33	33	33	33	33	33	33	33	33	33
			24	43	43	43	54	43	43	43	43	43	43	43	54
		550S162	16	33	33	33	33	33	33	33	33	33	33	33	33
			24	33	33	43	43	33	33	33	33	33	33	33	43
90 mph	—	350S162	16	33	33	33	33	33	33	33	33	33	33	33	33
			24	43	43	43	54	43	43	43	43	43	43	43	54
		550S162	16	33	33	33	33	33	33	33	33	33	33	33	33
			24	33	33	43	43	33	33	33	33	33	33	33	43
100 mph	85 mph	350S162	16	33	33	33	33	33	33	33	33	33	33	33	33
			24	43	43	43	54	43	43	43	43	43	43	43	54
		550S162	16	33	33	33	33	33	33	33	33	33	33	33	33
			24	33	33	43	43	33	33	33	33	33	33	33	43
110 mph	90 mph	350S162	16	33	33	33	33	33	33	33	33	33	33	33	43
			24	43	43	43	54	43	43	43	43	43	43	43	54
		550S162	16	33	33	33	33	33	33	33	33	33	33	33	33
			24	33	33	43	43	33	33	33	33	33	33	33	43
—	100 mph	350S162	16	33	33	33	33	33	33	33	33	33	33	33	43
			24	43	43	43	54	43	43	43	43	43	54	54	54
		550S162	16	33	33	33	33	33	33	33	33	33	33	33	33
			24	33	33	43	43	33	33	33	33	33	33	33	43
—	110 mph	350S162	16	33	33	33	43	33	33	33	43	33	43	43	43
			24	43	43	43	54	43	43	43	54	54	54	54	54
		550S162	16	33	33	33	33	33	33	33	33	33	33	33	33
			24	33	33	43	43	33	33	43	43	33	33	43	43

For SI: 1 inch = 25.4 mm, 1 foot = 304.8 mm, 1 mil = 0.0254 mm, 1 mile per hour = 0.447 m/s, 1 pound per square foot = 0.0479 kPa,
1 ksi = 1000 psi = 6.895 MPa.

a. Deflection criterion: L/240.
b. Design load assumptions:
　　Second floor dead load is 10 psf.
　　Second floor live load is 30 psf.
　　Roof/ceiling dead load is 12 psf.
　　Attic live load is 10 psf.
c. Building width is in the direction of horizontal framing members supported by the wall studs.

TABLE R603.6(1)
BOX-BEAM HEADER SPANS
Headers supporting roof and ceiling only (33 ksi steel)[a, b, c]

MEMBER DESIGNATION	GROUND SNOW LOAD (20 psf) Building width[c]					GROUND SNOW LOAD (30 psf) Building width[c]				
	24'	28'	32'	36'	40'	24'	28'	32'	36'	40'
2-350S162-33	3'-10"	3'-5"	3'-0"	2'-6"	2'-2"	3'-3"	2'-9"	2'-4"	—	—
2-350S162-43	5'-1"	4'-8"	4'-4"	4'-0"	3'-7"	4'-6"	4'-2"	3'-8"	3'-4"	2'-11"
2-350S162-54	5'-9"	5'-4"	5'-0"	4'-9"	4'-5"	5'-3"	4'-10"	4'-6"	4'-2"	3'-10"
2-350S162-68	6'-7"	6'-1"	5'-9"	5'-5"	5'-1"	6'-0"	5'-6"	5'-2"	4'-10"	4'-7"
2-350S162-97	8'-0"	7'-5"	7'-0"	6'-6"	6'-3"	7'-3"	6'-9"	6'-4"	6'-0"	5'-7"
2-550S162-33	5'-8"	5'-0"	4'-5"	3'-11"	3'-4"	4'-9"	4'-1"	3'-6"	2'-11"	—
2-550S162-43	7'-2"	6'-8"	6'-3"	5'-8"	5'-2"	6'-6"	5'-11"	5'-3"	4'-9"	4'-3"
2-550S162-54	8'-2"	7'-7"	7'-2"	6'-9"	6'-5"	7'-5"	6'-11"	6'-6"	6'-0"	5'-6"
2-550S162-68	9'-3"	8'-7"	8'-0"	7'-8"	7'-3"	8'-5"	7'-10"	7'-4"	7'-0"	6'-7"
2-550S162-97	11'-2"	10'-5"	9'-10"	9'-3"	8'-11"	10'-2"	9'-6"	9'-1"	8'-5"	8'-0"
2-800S162-33	6'-9"	5'-11"	5'-2"	4'-6"	3'-10"	5'-6"	4'-6"	4'-0"	—	—
2-800S162-43	9'-0"	8'-5"	7'-8"	7'-0"	6'-4"	8'-1"	7'-3"	6'-6"	5'-9"	5'-2"
2-800S162-54	10'-9"	10'-0"	9'-5"	8'-11"	8'-4"	9'-9"	9'-1"	8'-6"	7'-9"	7'-1"
2-800S162-68	12'-2"	11'-4"	10'-8"	10'-2"	9'-7"	11'-1"	10'-4"	9'-9"	9'-3"	8'-9"
2-800S162-97	14'-9"	13'-9"	13'-0"	12'-3"	11'-7"	13'-5"	12'-6"	11'-10"	11'-2"	10'-7"
2-1000S162-43	10'-0"	9'-2"	8'-4"	7'-6"	6'-9"	8'-9"	7'-10"	7'-0"	6'-2"	5'-5"
2-1000S162-54	12'-0"	11'-2"	10'-6"	9'-11"	9'-2"	10'-11"	10'-2"	9'-3"	8'-6"	7'-9"
2-1000S162-68	14'-5"	13'-6"	12'-8"	12'-0"	11'-5"	13'-2"	12'-3"	11'-6"	11'-0"	10'-4"
2-1000S162-97	17'-5"	16'-4"	15'-4"	14'-6"	13'-11"	16'-0"	14'-11"	14'-0"	13'-3"	12'-7"
2-1200S162-43	10'-10"	9'-9"	8'-9"	7'-11"	7'-1"	9'-3"	8'-2"	7'-2"	6'-4"	5'-6"
2-1200S162-54	13'-0"	12'-2"	11'-6"	10'-7"	9'-9"	11'-11"	11'-0"	10'-0"	9'-0"	8'-2"
2-1200S162-68	15'-5"	14'-5"	13'-6"	12'-11"	12'-3"	14'-0"	13'-2"	12'-4"	11'-9"	10'-11"
2-1200S162-97	20'-1"	18'-9"	17'-9"	16'-9"	16'-0"	18'-4"	17'-2"	16'-2"	15'-3"	14'-7"

For SI: 1 inch = 25.4 mm, 1 foot = 304.8 mm, 1 pound per square foot = 0.0479 kPa, 1 pound per square inch = 6.895 kPa.

a. Deflection criteria: $L/360$ for live loads, $L/240$ for total loads.
b. Design load assumptions:
   Roof/Ceiling dead load is 12 psf.
   Attic dead load is 10 psf.
c. Building width is in the direction of horizontal framing members supported by the header.

TABLE R603.6(2)
BOX-BEAM HEADER SPANS
Headers supporting roof and ceiling only (33 ksi steel)[a, b, c]

MEMBER DESIGNATION	GROUND SNOW LOAD (50 psf)					GROUND SNOW LOAD (70 psf)				
	Building width[c]					Building width[c]				
	24′	28′	32′	36′	40′	24′	28′	32′	36′	40′
2-350S162-33	—	—	—	—	—	—	—	—	—	—
2-350S162-43	3′-2″	2′-7″	2′-2″	—	—	2′-0″	—	—	—	—
2-350S162-54	4′-1″	3′-6″	3′-1″	2′-8″	2′-3″	3′-0″	2′-6″	—	—	—
2-350S162-68	4′-9″	4′-5″	4′-0″	3′-7″	3′-3″	4′-0″	3′-4″	3′-0″	2′-6″	2′-1″
2-350S162-97	5′-10″	5′-5″	5′-1″	4′-9″	4′-6″	5′-0″	4′-7″	4′-4″	4′-0″	3′-9″
2-550S162-33	2′-9″	—	—	—	—	—	—	—	—	—
2-550S162-43	4′-7″	3′-11″	3′-3″	—	—	3′-2″	—	—	—	—
2-550S162-54	5′-10″	5′-2″	4′-6″	4′-0″	3′-6″	4′-5″	3′-9″	3′-1″	—	—
2-550S162-68	6′-10″	6′-4″	5′-9″	5′-3″	4′-9″	5′-7″	5′-0″	4′-4″	3′-9″	3′-3″
2-550S162-97	8′-4″	7′-9″	7′-3″	6′-10″	6′-6″	7′-2″	6′-8″	6′-3″	5′-11″	5′-7″
2-800S162-33	—	—	—	—	—	—	—	—	—	—
2-800S162-43	5′-7″	4′-9″	3′-11″	—	—	—	—	—	—	—
2-800S162-54	7′-7″	6′-8″	5′-11″	5′-2″	4′-6″	5′-9″	4′-10″	—	—	—
2-800S162-68	9′-1″	8′-4″	7′-6″	6′-10″	6′-3″	7′-4″	6′-6″	5′-9″	5′-0″	4′-4″
2-800S162-97	11′-0″	10′-4″	9′-8″	9′-2″	8′-9″	9′-6″	8′-11″	8′-4″	7′-11″	7′-6″
2-1000S162-43	6′-0″	4′-11″	—	—	—	—	—	—	—	—
2-1000S162-54	8′-4″	7′-4″	6′-4″	5′-7″	4′-9″	6′-3″	5′-2″	—	—	—
2-1000S162-68	10′-9″	9′-9″	8′-10″	8′-0″	7′-3″	8′-7″	7′-7″	6′-7″	5′-9″	5′-0″
2-1000S162-97	13′-1″	12′-3″	11′-6″	10′-11″	10′-4″	11′-4″	10′-7″	10′-0″	9′-5″	8′-11″
2-1200S162-43	6′-1″	—	—	—	—	—	—	—	—	—
2-1200S162-54	8′-9″	7′-8″	6′-7″	5′-9″	—	6′-6″	—	—	—	—
2-1200S162-68	11′-6″	10′-4″	9′-4″	8′-4″	7′-7″	9′-1″	8′-0″	6′-11″	6′-0″	—
2-1200S162-97	15′-1″	14′-1″	13′-3″	12′-7″	12′-0″	13′-1″	12′-3″	11′-6″	11′-0″	10′-2″

For SI: 1 inch = 25.4 mm, 1 foot = 304.8 mm, 1 pound per square foot = 0.0479 kPa, 1 pound per square inch = 6.895 kPa.

a. Deflection criteria: $L/360$ for live loads, $L/240$ for total loads.

b. Design load assumptions:
   Roof/Ceiling dead load is 12 psf.
   Attic dead load is 10 psf.

c. Building width is in the direction of horizontal framing members supported by the header

TABLE R603.6(3)
BOX-BEAM HEADER SPANS
Headers supporting one floor, roof and ceiling (33 ksi steel)[a, b, c]

MEMBER DESIGNATION	GROUND SNOW LOAD (20 psf)					GROUND SNOW LOAD (30 psf)				
	Building width[c]					Building width[c]				
	24'	28'	32'	36'	40'	24'	28'	32'	36'	40'
2-350S162-33	—	—	—	—	—	—	—	—	—	—
2-350S162-43	2'-6"	—	—	—	—	2'-5"	—	—	—	—
2-350S162-54	3'-6"	3'-0"	2'-6"	—	—	3'-4"	2'-10"	2'-4"	—	—
2-350S162-68	4'-4"	3'-11"	3'-5"	3'-0"	2'-7"	4'-3"	3'-9"	3'-3"	2'-10"	2'-6"
2-350S162-97	5'-4"	5'-0"	4'-7"	4'-4"	4'-1"	5'-4"	4'-11"	4'-6"	4'-3"	4'-0"
2-550S162-33	—	—	—	—	—	—	—	—	—	—
2-550S162-43	3'-9"	3'-0"	—	—	—	3'-7"	2'-11"	—	—	—
2-550S162-54	5'-0"	4'-4"	3'-9"	3'-2"	—	4'-10"	4'-2"	3'-6"	3'-0"	—
2-550S162-68	6'-3"	5'-6"	5'-0"	4'-5"	4'-0"	6'-1"	5'-5"	4'-9"	4'-3"	3'-9"
2-550S162-97	7'-8"	7'-2"	6'-8"	6'-4"	6'-0"	7'-6"	7'-0"	6'-6"	6'-2"	5'-10"
2-800S162-33	—	—	—	—	—	—	—	—	—	—
2-800S162-43	4'-6"	—	—	—	—	4'-4"	—	—	—	—
2-800S162-54	6'-6"	5'-7"	4'-10"	4'-1"	—	6'-4"	5'-5"	4'-7"	—	—
2-800S162-68	8'-2"	7'-3"	6'-6"	5'-10"	5'-2"	8'-0"	7'-0"	6'-4"	5'-6"	5'-0"
2-800S162-97	10'-1"	9'-6"	8'-11"	8'-6"	8'-0"	10'-0"	9'-4"	8'-9"	8'-3"	7'-11"
2-1000S162-43	4'-9"	—	—	—	—	—	—	—	—	—
2-1000S162-54	7'-1"	6'-0"	5'-2"	—	—	6'-10"	5'-10"	4'-11"	—	—
2-1000S162-68	9'-7"	8'-6"	7'-7"	6'-9"	6'-0"	9'-4"	8'-4"	7'-4"	6'-6"	5'-9"
2-1000S162-97	12'-0"	11'-3"	10'-7"	10'-0"	9'-6"	11'-11"	11'-1"	10'-5"	9'-11"	9'-5"
2-1200S162-43	—	—	—	—	—	—	—	—	—	—
2-1200S162-54	7'-6"	6'-4"	—	—	—	7'-2"	6'-0"	—	—	—
2-1200S162-68	10'-1"	9'-0"	8'-0"	7'-0"	6'-2"	9'-11"	8'-9"	7'-9"	6'-9"	6'-0"
2-1200S162-97	14'-0"	13'-0"	12'-3"	11'-7"	11'-0"	13'-9"	12'-10"	12'-0"	11'-6"	10'-11"

For SI: 1 inch = 25.4 mm, 1 foot = 304.8 mm, 1 pound per square foot = 0.0479 kPa, 1 pound per square inch = 6.895 kPa.

a. Deflection criteria: $L/360$ for live loads, $L/240$ for total loads.

b. Design load assumptions:
   Roof/Ceiling dead load is 12 psf.
   Attic dead load is 10 psf.

c. Building width is in the direction of horizontal framing members supported by the header.

TABLE R603.6(4)
BOX-BEAM HEADER SPANS
Headers supporting one floor, roof and ceiling (33 ksi steel)[a, b, c]

MEMBER DESIGNATION	GROUND SNOW LOAD (50 psf) Building width[c]					GROUND SNOW LOAD (70 psf) Building width[c]				
	24′	28′	32′	36′	40′	24′	28′	32′	36′	40′
2-350S162-33	—	—	—	—	—	—	—	—	—	—
2-350S162-43	—	—	—	—	—	—	—	—	—	—
2-350S162-54	2′-6″	2′-1″	—	—	—	—	—	—	—	—
2-350S162-68	3′-6″	3′-0″	2′-6″	2′-2″	—	2′-9″	2′-2″	—	—	—
2-350S162-97	4′-9″	4′-5″	4′-1″	3′-10″	3′-7″	4′-2″	3′-11″	3′-7″	3′-4″	2′-11″
2-550S162-33	—	—	—	—	—	—	—	—	—	—
2-550S162-43	—	—	—	—	—	—	—	—	—	—
2-550S162-54	3′-11″	3′-4″	—	—	—	2′-10″	—	—	—	—
2-550S162-68	5′-2″	4′-6″	3′-11″	3′-4″	2′-10″	4′-1″	3′-5″	2′-9″	—	—
2-550S162-97	6′-10″	6′-4″	6′-0″	5′-7″	5′-4″	6′-1″	5′-7″	5′-4″	4′-9″	4′-4″
2-800S162-33	—	—	—	—	—	—	—	—	—	—
2-800S162-43	—	—	—	—	—	—	—	—	—	—
2-800S162-54	5′-1″	4′-2″	—	—	—	—	—	—	—	—
2-800S162-68	6′-9″	6′-1″	5′-2″	4′-5″	—	5′-5″	4′-6″	—	—	—
2-800S162-97	9′-1″	8′-6″	8′-0″	7′-6″	7′-1″	8′-2″	7′-7″	7′-0″	6′-5″	5′-10″
2-1000S162-43	—	—	—	—	—	—	—	—	—	—
2-1000S162-54	5′-6″	—	—	—	—	—	—	—	—	—
2-1000S162-68	7′-10″	6′-11″	6′-0″	5′-2″	—	6′-4″	5′-4″	—	—	—
2-1000S162-97	10′-10″	10′-1″	9′-6″	9′-0″	8′-4″	9′-9″	9′-2″	8′-4″	7′-7″	7′-0″
2-1200S162-43	—	—	—	—	—	—	—	—	—	—
2-1200S162-54	5′-7″	—	—	—	—	—	—	—	—	—
2-1200S162-68	8′-4″	7′-2″	6′-2″	—	—	6′-6″	—	—	—	—
2-1200S162-97	12′-6″	11′-8″	11′-0″	10′-4″	9′-6″	11′-3″	10′-6″	9′-6″	8′-8″	8′-0″

For SI: 1 inch = 25.4 mm, 1 foot = 304.8 mm, 1 pound per square foot = 0.0479 kPa, 1 pound per square inch = 6.895 kPa.

a. Deflection criteria: $L/360$ for live loads, $L/240$ for total loads.

b. Design load assumptions:
   Roof/Ceiling dead load is 12 psf.
   Attic dead load is 10 psf.

c. Building width is in the direction of horizontal framing members supported by the header.

TABLE R603.6(5)
BACK-TO-BACK HEADER SPANS
Headers supporting roof and ceiling only (33 ksi steel)[a, b, c]

MEMBER DESIGNATION	GROUND SNOW LOAD (20 psf)					GROUND SNOW LOAD (30 psf)				
	Building width[c]					Building width[c]				
	24'	28'	32'	36'	40'	24'	28'	32'	36'	40'
2-350S162-33	3'-7"	3'-1"	2'-8"	2'-4"	—	2'-11"	2'-6"	—	—	—
2-350S162-43	5'-0"	4'-8"	4'-4"	3'-10"	3'-7"	4'-6"	4'-0"	3'-8"	3'-4"	2'-11"
2-350S162-54	5'-9"	5'-5"	5'-0"	4'-9"	4'-6"	5'-3"	4'-10"	4'-6"	4'-4"	3'-11"
2-350S162-68	6'-7"	6'-2"	5'-9"	5'-5"	5'-2"	5'-11"	5'-7"	5'-2"	4'-10"	4'-7"
2-350S162-97	7'-11"	7'-6"	6'-11"	6'-7"	6'-6"	7'-4"	6'-9"	6'-4"	5'-11"	5'-8"
2-550S162-33	5'-5"	4'-9"	4'-4"	3'-9"	3'-5"	4'-7"	3'-11"	3'-5"	2'-11"	—
2-550S162-43	7'-3"	6'-8"	6'-2"	5'-8"	5'-4"	6'-6"	5'-10"	5'-5"	4'-10"	4'-6"
2-550S162-54	8'-2"	7'-8"	7'-2"	6'-9"	6'-5"	7'-5"	6'-10"	6'-6"	6'-1"	5'-9"
2-550S162-68	9'-4"	8'-8"	8'-7"	7'-8"	7'-4"	8'-6"	7'-10"	7'-5"	6'-11"	6'-7"
2-550S162-97	11'-3"	10'-6"	9'-11"	9'-4"	8'-10"	10'-3"	9'-6"	8'-11"	8'-6"	8'-0"
2-800S162-33	6'-9"	5'-11"	5'-5"	4'-9"	4'-4"	6'-9"	5'-0"	4'-5"	3'-9"	—
2-800S162-43	9'-1"	8'-6"	7'-9"	7'-3"	6'-8"	8'-3"	7'-6"	6'-9"	6'-3"	5'-8"
2-800S162-54	10'-9"	10'-1"	9'-6"	8'-11"	8'-6"	9'-9"	9'-2"	8'-7"	8'-2"	7'-8"
2-800S162-68	12'-3"	11'-5"	10'-9"	10'-2"	9'-8"	11'-2"	10'-5"	9'-9"	9'-4"	8'-9"
2-800S162-97	14'-9"	13'-9"	13'-0"	12'-4"	11'-8"	13'-6"	12'-7"	11'-10"	11'-2"	10'-8"
2-1000S162-43	10'-1"	9'-5"	8'-8"	8'-0"	7'-6"	9'-1"	8'-4"	7'-7"	6'-11"	6'-5"
2-1000S162-54	12'-0"	11'-5"	10'-7"	10'-0"	9'-6"	11'-0"	10'-3"	9'-7"	9'-1"	8'-6"
2-1000S162-68	14'-6"	13'-6"	12'-8"	12'-0"	11'-6"	13'-2"	12'-4"	11'-7"	10'-11"	10'-6"
2-1000S162-97	17'-6"	16'-5"	15'-5"	14'-7"	13'-10"	16'-0"	14'-10"	14'-0"	13'-4"	12'-8"
2-1200S162-43	11'-0"	10'-4"	9'-6"	8'-9"	8'-2"	10'-0"	9'-1"	8'-4"	7'-7"	7'-0"
2-1200S162-54	13'-1"	12'-3"	11'-6"	10'-10"	10'-5"	11'-10"	11'-1"	10'-6"	9'-10"	9'-4"
2-1200S162-68	15'-6"	14'-6"	13'-7"	12'-10"	12'-3"	14'-1"	13'-2"	12'-5"	11'-9"	11'-2"
2-1200S162-97	20'-2"	18'-9"	17'-9"	16'-9"	16'-0"	18'-4"	17'-2"	16'-2"	15'-5"	14'-7"

For SI: 1 inch = 25.4 mm, 1 foot = 304.8 mm, 1 pound per square foot = 0.0479 kPa, 1 pound per square inch = 6.895 kPa.

a. Deflection criteria: $L/360$ for live loads, $L/240$ for total loads.
b. Design load assumptions:
   Roof/Ceiling dead load is 12 psf.
   Attic dead load is 10 psf.
c. Building width is in the direction of horizontal framing members supported by the header.

TABLE R603.6(6)
BACK-TO-BACK HEADER SPANS
Headers supporting roof and ceiling only (33 ksi steel)[a,b,c]

MEMBER DESIGNATION	GROUND SNOW LOAD (50 psf)					GROUND SNOW LOAD (70 psf)				
	Building width[c]					Building width[c]				
	24'	28'	32'	36'	40'	24'	28'	32'	36'	40'
2-350S162-33	—	—	—	—	—	—	—	—	—	—
2-350S162-43	3'-2"	2'-8"	2'-4"	—	—	2'-3"	—	—	—	—
2-350S162-54	4'-3"	3'-8"	3'-5"	2'-11"	2'-8"	3'-4"	2'-9"	2'-5"	2'-0"	1'-7"
2-350S162-68	4'-9"	4'-6"	4'-2"	3'-10"	3'-7"	4'-1"	3'-9"	3'-5"	3'-1"	2'-8"
2-350S162-97	5'-10"	5'-6"	5'-2"	4'-10"	4'-7"	5'-0"	4'-8"	4'-5"	4'-2"	3'-10"
2-550S162-33	2'-9"	—	—	—	—	—	—	—	—	—
2-550S162-43	4'-9"	4'-2"	3'-8"	3'-3"	2'-8"	3'-6"	2'-10"	—	—	—
2-550S162-54	6'-0"	5'-6"	4'-11"	4'-6"	4'-1"	4'-10"	4'-4"	3'-10"	3'-6"	2'-11"
2-550S162-68	6'-10"	6'-5"	5'-11"	5'-8"	5'-4"	5'-11"	5'-6"	4'-11"	4'-7"	4'-3"
2-550S162-97	8'-5"	7'-7"	4'-4"	6'-10"	6'-7"	7'-3"	6'-8"	6'-4"	5'-11"	5'-8"
2-800S162-33	3'-8"	—	—	—	—	—	—	—	—	—
2-800S162-43	6'-1"	5'-5"	4'-9"	4'-2"	3'-8"	4'-7"	3'-9"	—	—	—
2-800S162-54	8'-0"	7'-4"	6'-8"	6'-1"	5'-7"	6'-6"	5'-8"	5'-3"	4'-8"	4'-3"
2-800S162-68	9'-1"	8'-6"	7'-11"	7'-7"	7'-1"	7'-10"	7'-5"	6'-9"	6'-3"	5'-9"
2-800S162-97	11'-1"	10'-4"	9'-8"	9'-2"	8'-9"	9'-8"	8'-11"	8'-5"	7'-11"	7'-7"
2-1000S162-43	6'-9"	6'-0"	5'-5"	4'-9"	4'-2"	5'-2"	4'-5"	—	—	—
2-1000S162-54	8'-11"	8'-2"	7'-6"	6'-10"	6'-6"	7'-4"	6'-7"	5'-10"	5'-4"	4'-9"
2-1000S162-68	10'-9"	10'-1"	9'-6"	8'-11"	8'-6"	9'-5"	8'-9"	8'-1"	7'-6"	6'-10"
2-1000S162-97	13'-1"	12'-4"	11'-6"	10'-10"	10'-5"	11'-5"	10'-7"	9'-11"	9'-6"	8'-11"
2-1200S162-43	7'-6"	6'-7"	5'-10"	5'-2"	4'-7"	5'-8"	4'-10"	—	—	—
2-1200S162-54	9'-9"	8'-10"	8'-1"	7'-6"	6'-10"	7'-11"	7'-2"	6'-6"	5'-9"	5'-3"
2-1200S162-68	11'-7"	10'-9"	10'-2"	9'-7"	9'-1"	10'-2"	9'-6"	8'-7"	7'-11"	7'-5"
2-1200S162-97	15'-1"	14'-1"	13'-4"	12'-7"	12'-0"	13'-2"	12'-4"	11'-7"	10'-11"	10'-6"

For SI: 1 inch = 25.4 mm, 1 foot = 304.8 mm, 1 pound per square foot = 0.0479 kPa, 1 pound per square inch = 6.895 kPa.

a. Deflection criteria: $L/360$ for live loads, $L/240$ for total loads.

b. Design load assumptions:
   Roof/Ceiling dead load is 12 psf.
   Attic dead load is 10 psf.

c. Building width is in the direction of horizontal framing members supported by the header.

TABLE R603.6(7)
BACK-TO-BACK HEADER SPANS
Headers supporting one floor, roof and ceiling (33 ksi steel)[a, b, c]

MEMBER DESIGNATION	GROUND SNOW LOAD (20 psf)					GROUND SNOW LOAD (30 psf)				
	Building width[c]					Building width[c]				
	24′	28′	32′	36′	40′	24′	28′	32′	36′	40′
2-350S162-33	—	—	—	—	—	—	—	—	—	—
2-350S162-43	2′-7″	2′-1″	—	—	—	2′-6″	—	—	—	—
2-350S162-54	3′-8″	3′-3″	2′-9″	2′-6″	2′-1″	3′-7″	3′-1″	2′-8″	2′-5″	1′-11″
2-350S162-68	4′-5″	4′-0″	3′-9″	3′-6″	3′-1″	4′-4″	3′-11″	3′-8″	3′-5″	3′-0″
2-350S162-97	5′-5″	4′-11″	4′-8″	4′-5″	4′-1″	5′-4″	4′-10″	4′-7″	4′-4″	4′-0″
2-550S162-33	—	—	—	—	—	—	—	—	—	—
2-550S162-43	4′-1″	3′-6″	2′-10″	2′-5″	—	3′-11″	3′-5″	2′-9″	—	—
2-550S162-54	5′-5″	4′-9″	4′-4″	3′-10″	3′-6″	5′-3″	4′-8″	4′-3″	3′-9″	3′-5″
2-550S162-68	6′-6″	5′-10″	5′-6″	5′-1″	4′-8″	6′-2″	5′-9″	5′-5″	4′-11″	4′-7″
2-550S162-97	7′-8″	7′-2″	6′-8″	6′-5″	5′-11″	7′-7″	7′-0″	6′-7″	6′-3″	5′-10″
2-800S162-33	—	—	—	—	—	—	—	—	—	—
2-800S162-43	5′-4″	4′-7″	3′-10″	3′-4″	—	5′-1″	4′-5″	3′-9″	—	—
2-800S162-54	7′-3″	6′-6″	5′-10″	5′-4″	4′-9″	6′-11″	6′-4″	5′-8″	5′-3″	4′-8″
2-800S162-68	8′-5″	7′-9″	7′-5″	6′-9″	6′-5″	8′-4″	7′-8″	7′-3″	6′-8″	6′-3″
2-800S162-97	10′-2″	9′-6″	8′-11″	8′-6″	8′-1″	10′-0″	9′-5″	8′-9″	8′-5″	7′-10″
2-1000S162-43	5′-9″	5′-1″	4′-5″	3′-9″	—″	5′-8″	4′-11″	4′-4″	—	—
2-1000S162-54	7′-11″	7′-3″	6′-7″	5′-11″	5′-5″	7′-9″	7′-1″	6′-5″	5′-9″	5′-4″
2-1000S162-68	9′-11″	9′-4″	8′-9″	8′-1″	7′-7″	9′-10″	9′-2″	8′-7″	7′-11″	7′-5″
2-1000S162-97	12′-1″	11′-4″	10′-8″	10′-0″	9′-7″	11′-11″	11′-1″	10′-6″	9′-10″	9′-6″
2-1200S162-43	6′-6″	5′-8″	4′-10″	4′-2″	—	6′-4″	5′-6″	4′-8″	—	—
2-1200S162-54	8′-8″	7′-10″	7′-2″	6′-6″	5′-11″	8′-6″	7′-8″	6′-11″	6′-5″	5′-9″
2-1200S162-68	10′-8″	9′-11″	9′-5″	8′-8″	8′-1″	10′-6″	9′-9″	9′-3″	8′-6″	7′-11″
2-1200S162-97	13′-11″	13′-0″	12′-4″	11′-7″	11′-1″	13′-9″	12′-10″	12′-1″	11′-6″	10′-10″

For SI: 1 inch = 25.4 mm, 1 foot = 304.8 mm, 1 pound per square foot = 0.0479 kPa, 1 pound per square inch = 6.895 kPa.

a. Deflection criteria: $L/360$ for live loads, $L/240$ for total loads.

b. Design load assumptions:

    Second floor dead load is 10 psf.

    Roof/Ceiling dead load is 12 psf.

    Second floor live load is 30 psf.

    Roof/ceiling load is 12 psf.

    Attic dead load is 10 psf.

c. Building width is in the direction of horizontal framing members supported by the header.

TABLE R603.6(8)
BACK-TO-BACK HEADER SPANS
Headers supporting one floor, roof and ceiling (33 ksi steel)[a, b, c]

MEMBER DESIGNATION	GROUND SNOW LOAD (50 psf) Building width[c]					GROUND SNOW LOAD (70 psf) Building width[c]				
	24'	28'	32'	36'	40'	24'	28'	32'	36'	40'
2-350S162-33	—	—	—	—	—	—	—	—	—	—
2-350S162-43	—	—	—	—	—	—	—	—	—	—
2-350S162-54	2'-11"	2'-6"	2'-2"	—	—	2'-4"	—	—	—	—
2-350S162-68	3'-10"	3'-6"	3'-2"	2'-9"	2'-6"	3'-4"	2'-10"	2'-6"	2'-3"	1'-10"
2-350S162-97	4'-9"	4'-5"	4'-2"	3'-10"	3'-8"	4'-3"	3'-10"	3'-8"	3'-5"	3'-2"
2-550S162-33	—	—	—	—	—	—	—	—	—	—
2-550S162-43	3'-1"	2'-5"	—	—	—	—	—	—	—	—
2-550S162-54	4'-6"	3'-10"	3'-6"	3'-0"	2'-7"	3'-8"	3'-1"	2'-7"	2'-1"	—
2-550S162-68	5'-7"	5'-1"	4'-8"	4'-4"	3'-11"	4'-10"	4'-5"	3'-10"	3'-6"	3'-3"
2-550S162-97	6'-10"	6'-5"	5'-11"	5'-8"	5'-5"	6'-2"	5'-8"	5'-5"	3'-1"	4'-9"
2-800S162-33	—	—	—	—	—	—	—	—	—	—
2-800S162-43	4'-1"	3'-5"	—	—	—	—	—	—	—	—
2-800S162-54	6'-0"	5'-5"	4'-9"	4'-4"	3'-9"	4'-11"	4'-5"	3'-9"	3'-2"	—
2-800S162-68	7'-6"	6'-10"	6'-5"	5'-10"	5'-5"	6'-7"	5'-10"	5'-5"	4'-10"	4'-6"
2-800S162-97	9'-1"	8'-6"	8'-0"	7'-7"	7'-4"	8'-3"	7'-8"	7'-3"	6'-9"	6'-6"
2-1000S162-43	4'-8"	3'-10"	—	—	—	—	—	—	—	—
2-1000S162-54	6'-9"	6'-0"	5'-5"	4'-10"	4'-4"	5'-8"	4'-10"	4'-4"	3'-8"	—
2-1000S162-68	8'-10"	8'-3"	7'-7"	6'-11"	6'-6"	7'-9"	7'-1"	6'-6"	5'-10"	5'-5"
2-1000S162-97	10'-10"	10'-3"	9'-7"	9'-1"	8'-8"	9'-9"	9'-3"	8'-7"	8'-3"	7'-9"
2-1200S162-43	5'-1"	4'-4[2]	—	—	—	—	—	—	—	—
2-1200S162-54	7'-5"	6'-6"	5'-10"	5'-4"	4'-9"	6'-3"	5'-5"	4'-8"	4'-1"	—
2-1200S162-68	9'-7"	8'-9"	8'-1"	7'-6"	6'-11"	8'-8"	7'-7"	6'-11"	6'-4"	5'-9"
2-1200S162-97	12'-6"	11'-8"	11'-1"	10'-6"	9'-11"	11'-4"	10'-7"	10'-0"	9'-6"	9'-0"

For SI: 1 inch = 25.4 mm, 1 foot = 304.8 mm, 1 pound per square foot = 0.0479 kPa, 1 pound per square inch = 6.895 kPa.

a. Deflection criteria: $L/360$ for live loads, $L/240$ for total loads.

b. Design load assumptions:
   Second floor dead load is 10 psf.
   Roof/Ceiling dead load is 12 psf.
   Second floor live load is 30 psf.
   Roof/ceiling dead load is 12 psf.
   Attic dead load is 10 psf.

c. Building width is in the direction of horizontal framing members supported by the header.

TABLE R603.6(9)
TOTAL NUMBER OF JACK AND KING STUDS REQUIRED AT EACH END OF AN OPENING

SIZE OF OPENING (feet-inches)	24" O.C. STUD SPACING		16" O.C. STUD SPACING	
	No. of jack studs	No. of king studs	No. of jack studs	No. of king studs
Up to 3'-6"	1	1	1	1
> 3'-6" to 5'-0"	1	2	1	2
> 5'-0" to 5'-6"	1	2	2	2
> 5'-6" to 8'-0"	1	2	2	2
> 8'-0" to 10'-6"	2	2	2	3
> 10'-6" to 12'-0"	2	2	3	3
> 12'-0" to 13'-0"	2	3	3	3
> 13'-0" to 14'-0"	2	3	3	4
> 14'-0" to 16'-0"	2	3	3	4
> 16'-0" to 18'-0"	3	3	4	4

For SI: 1 inch = 25.4 mm, 1 foot = 304.8 mm.

TABLE R603.6(10)
HEADER TO KING STUD CONNECTION REQUIREMENTS[a, b, c, d]

HEADER SPAN (feet)	BASIC WIND SPEED (mph), EXPOSURE		
	85 A/B or Seismic Design Categories A, B, C, D_0, D_1 and D_2	85 C or less than 110 A/B	Less than 110 C
≤ 4'	4-No. 8 screws	4-No. 8 screws	6-No. 8 screws
> 4' to 8'	4-No. 8 screws	4-No. 8 screws	8-No. 8 screws
> 8' to 12'	4-No. 8 screws	6-No. 8 screws	10-No. 8 screws
> 12' to 16'	4-No. 8 screws	8-No. 8 screws	12-No. 8 screws

For SI: 1 inch = 25.4 mm, 1 foot = 304.8 mm, 1 mile per hour = 0.447 m/s, 1 pound = 4.448 N.

a. All screw sizes shown are minimum.

b. For headers located on the first floor of a two-story building, the total number of screws may be reduced by two screws, but the total number of screws shall be no less than four.

c. For roof slopes of 6:12 or greater, the required number of screws may be reduced by half, but the total number of screws shall be no less than four.

d. Screws can be replaced by an uplift connector which has a capacity of the number of screws multiplied by 164 pounds (e.g., 12-No. 8 screws can be replaced by an uplift connector whose capacity exceeds 12 × 164 pounds = 1,968 pounds).

TABLE R603.6(11)
HEAD TRACK SPAN (33 ksi Steel)

BASIC WIND SPEED (mph)		ALLOWABLE HEAD TRACK SPAN[a, b] (ft-in)					
Exposure		Track Designation					
A/B	C	350T125-33	350T125-43	350T125-54	550T125-33	550T125-43	550T125-54
85		5'-0"	5'-7"	6'-2"	5'-10"	6'-8"	7'-0"
90		4'-10"	5'-5"	6'-0"	5'-8"	6'-3"	6'-10"
100	85	4'-6"	5'-1"	5'-8"	5'-4"	5'-11"	6'-5"
110	90	4'-2"	4'-9"	5'-4"	5'-1"	5'-7"	6'-1"
	100	3'-11"	4'-6"	5'-0"	4'-10"	5'-4"	5'-10"
	110	3'-8"	4'-2"	4'-9"	4'-1"	5'-1"	5'-7"

For SI: 1 inch = 25.4 mm, 1 foot = 304.8 mm, 1 mile per hour = 0.447 m/s.

a. Deflection Limit: L/240

b. Head track spans are based on components and cladding wind speeds and a 49-inch tributary span.

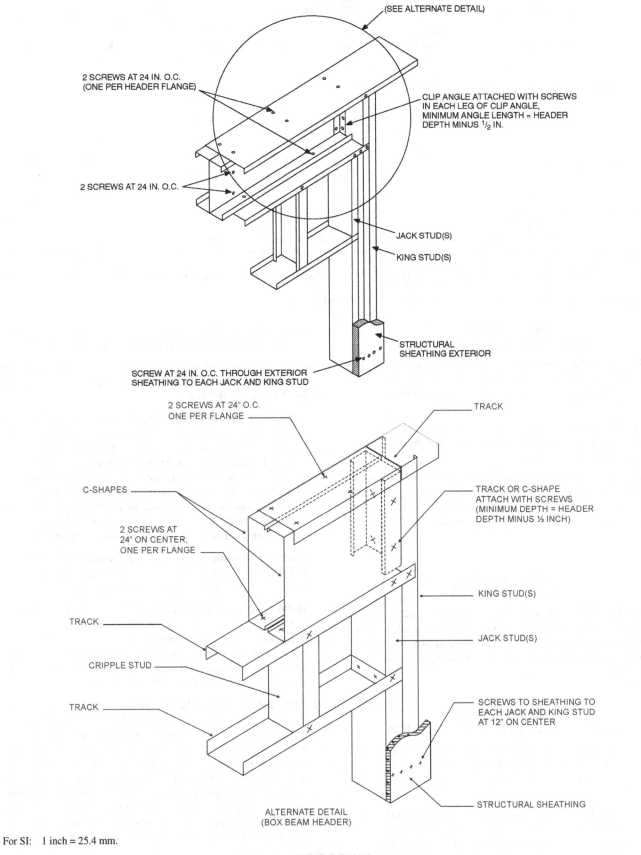

(SEE ALTERNATE DETAIL)

2 SCREWS AT 24 IN. O.C.
(ONE PER HEADER FLANGE)

CLIP ANGLE ATTACHED WITH SCREWS
IN EACH LEG OF CLIP ANGLE,
MINIMUM ANGLE LENGTH = HEADER
DEPTH MINUS $1/2$ IN.

2 SCREWS AT 24 IN. O.C.

JACK STUD(S)

KING STUD(S)

STRUCTURAL
SHEATHING EXTERIOR

SCREW AT 24 IN. O.C. THROUGH EXTERIOR
SHEATHING TO EACH JACK AND KING STUD

2 SCREWS AT 24" O.C.
ONE PER FLANGE

TRACK

C-SHAPES

TRACK OR C-SHAPE
ATTACH WITH SCREWS
(MINIMUM DEPTH = HEADER
DEPTH MINUS ½ INCH)

2 SCREWS AT
24" ON CENTER,
ONE PER FLANGE

KING STUD(S)

TRACK

JACK STUD(S)

CRIPPLE STUD

TRACK

SCREWS TO SHEATHING TO
EACH JACK AND KING STUD
AT 12" ON CENTER

STRUCTURAL SHEATHING

ALTERNATE DETAIL
(BOX BEAM HEADER)

For SI: 1 inch = 25.4 mm.

**FIGURE R603.6
HEADER DETAIL**

TABLE R603.7
MINIMUM PERCENTAGE OF FULL HEIGHT STRUCTURAL SHEATHING ON EXTERIOR WALLS[a, b, c, d, e]

WALL SUPPORTING	ROOF SLOPE	WIND SPEED (mph) AND EXPOSURE				
		85 A/B	100 A/B	110 A/B or 85 C	100 C	110 C
Roof and ceiling only	3:12	8	9	12	16	20
	6:12	12	15	20	26	35
	9:12	21	25	30	50	58
	12:12	30	35	40	66	75
One story, roof and ceiling	3:12	24	30	35	50	66
	6:12	25	30	40	58	74
	9:12	35	40	55	74	91
	12:12	40	50	65	100	115

For SI: 1 mile per hour = 0.447 m/s.

a. Linear interpolation shall be permitted.

b. Bracing amount shall not be less than 20 percent of the wall length after all applicable adjustments are made.

c. Minimum percentages are based on a building aspect ratio of 1:1. Minimum percentages for the shorter walls of a building shall be multiplied by a factor of 1.5 and 2.0 for building aspect ratios of 1.5:1 and 2:1 respectively.

d. For hip roofed homes with continuous structural sheathing, the amount of bracing shall be permitted to be multiplied by a factor of 0.95 for roof slopes not exceeding 7:12 and a factor of 0.9 for roof slopes greater than 7:12.

e. Sheathing percentages are permitted to be reduced in accordance with Section R603.7.2.

SECTION R604
WOOD STRUCTURAL PANELS

R604.1 Identification and grade. Wood structural panels shall conform to DOC PS 1 or DOC PS 2. All panels shall be identified by a grade mark or certificate of inspection issued by an approved agency.

R604.2 Allowable spans. The maximum allowable spans for wood structural panel wall sheathing shall not exceed the values set forth in Table R602.3(3).

R604.3 Installation. Wood structural panel wall sheathing shall be attached to framing in accordance with Table R602.3(1). Wood structural panels marked Exposure 1 or Exterior are considered water-repellent sheathing under the code.

SECTION R605
PARTICLEBOARD

R605.1 Identification and grade. Particleboard shall conform to ANSI A208.1 and shall be so identified by a grade mark or certificate of inspection issued by an approved agency. Particleboard shall comply with the grades specified in Table R602.3(4).

SECTION R606
GENERAL MASONRY CONSTRUCTION

R606.1 General. Masonry construction shall be designed and constructed in accordance with the provisions of this section or in accordance with the provisions of ACI 530/ASCE 5/TMS 402.

R606.1.1 Professional registration not required. When the empirical design provisions of ACI 530/ASCE 5/TMS 402 Chapter 5 or the provisions of this section are used to design masonry, project drawings, typical details and specifications are not required to bear the seal of the architect or engineer responsible for design, unless otherwise required by the state law of the jurisdiction having authority.

R606.2 Thickness of masonry. The nominal thickness of masonry walls shall conform to the requirements of Sections R606.2.1 through R606.2.4.

R606.2.1 Minimum thickness. The minimum thickness of masonry bearing walls more than one story high shall be 8 inches (203 mm). Solid masonry walls of one-story dwellings and garages shall not be less than 6 inches (152 mm) in thickness when not greater than 9 feet (2743 mm) in height, provided that when gable construction is used, an additional 6 feet (1829 mm) is permitted to the peak of the gable. Masonry walls shall be laterally supported in either the horizontal or vertical direction at intervals as required by Section R606.9.

R606.2.2 Rubble stone masonry wall. The minimum thickness of rough, random or coursed rubble stone masonry walls shall be 16 inches (406 mm).

R606.2.3 Change in thickness. Where walls of masonry of hollow units or masonry-bonded hollow walls are decreased in thickness, a course of solid masonry shall be constructed between the wall below and the thinner wall above, or special units or construction shall be used to transmit the loads from face shells or wythes above to those below.

R606.2.4 Parapet walls. Unreinforced solid masonry parapet walls shall not be less than 8 inches (203 mm) thick and their height shall not exceed four times their thickness. Unreinforced hollow unit masonry parapet walls shall be not less than 8 inches (203 mm) thick, and their height shall not exceed three times their thickness. Masonry parapet walls in areas subject to wind loads of 30 pounds per square foot (1.44 kPa) located in Seismic Design Category D_0, D_1 or D_2, or on townhouses in Seismic Design Category C shall be reinforced in accordance with Section R606.12.

R606.3 Corbeled masonry. Solid masonry units shall be used for corbeling. The maximum corbeled projection beyond the face of the wall shall not be more than one-half of the wall thickness or one-half the wythe thickness for hollow walls; the maximum projection of one unit shall not exceed one-half the height of the unit or one-third the thickness at right angles to the wall. When corbeled masonry is used to support floor or roof-framing members, the top course of the corbel shall be a header course or the top course bed joint shall have ties to the vertical wall. The hollow space behind the corbeled masonry shall be filled with mortar or grout.

R606.4 Support conditions. Bearing and support conditions shall be in accordance with Sections R606.4.1 and R606.4.2.

R606.4.1 Bearing on support. Each masonry wythe shall be supported by at least two-thirds of the wythe thickness.

R606.4.2 Support at foundation. Cavity wall or masonry veneer construction may be supported on an 8-inch (203 mm) foundation wall, provided the 8-inch (203 mm) wall is corbeled with solid masonry to the width of the wall system above. The total horizontal projection of the corbel shall not exceed 2 inches (51 mm) with individual corbels projecting not more than one-third the thickness of the unit or one-half the height of the unit.

R606.5 Allowable stresses. Allowable compressive stresses in masonry shall not exceed the values prescribed in Table R606.5. In determining the stresses in masonry, the effects of all loads and conditions of loading and the influence of all forces affecting the design and strength of the several parts shall be taken into account.

R606.5.1 Combined units. In walls or other structural members composed of different kinds or grades of units, materials or mortars, the maximum stress shall not exceed the allowable stress for the weakest of the combination of units, materials and mortars of which the member is composed. The net thickness of any facing unit that is used to resist stress shall not be less than 1.5 inches (38 mm).

R606.6 Piers. The unsupported height of masonry piers shall not exceed ten times their least dimension. When structural clay tile or hollow concrete masonry units are used for isolated piers to support beams and girders, the cellular spaces shall be filled solidly with concrete or Type M or S mortar, except that unfilled hollow piers may be used if their unsupported height is not more than four times their least dimension. Where hollow masonry units are solidly filled with concrete or Type M, S or N mortar, the allowable compressive stress shall be permitted to be increased as provided in Table R606.5.

R606.6.1 Pier cap. Hollow piers shall be capped with 4 inches (102 mm) of solid masonry or concrete or shall have cavities of the top course filled with concrete or grout or other approved methods.

R606.7 Chases. Chases and recesses in masonry walls shall not be deeper than one-third the wall thickness, and the maximum length of a horizontal chase or horizontal projection shall not exceed 4 feet (1219 mm), and shall have at least 8 inches (203 mm) of masonry in back of the chases and recesses and between adjacent chases or recesses and the jambs of openings. Chases and recesses in masonry walls shall be designed and constructed so as not to reduce the required strength or required fire resistance of the wall and in no case shall a chase or recess

be permitted within the required area of a pier. Masonry directly above chases or recesses wider than 12 inches (305 mm) shall be supported on noncombustible lintels.

TABLE R606.5
ALLOWABLE COMPRESSIVE STRESSES FOR EMPIRICAL DESIGN OF MASONRY

CONSTRUCTION; COMPRESSIVE STRENGTH OF UNIT, GROSS AREA	ALLOWABLE COMPRESSIVE STRESSES[a] GROSS CROSS-SECTIONAL AREA[b]	
	Type M or S mortar	Type N mortar
Solid masonry of brick and other solid units of clay or shale; sand-lime or concrete brick:		
8,000 + psi	350	300
4,500 psi	225	200
2,500 psi	160	140
1,500 psi	115	100
Grouted[c] masonry, of clay or shale; sand-lime or concrete:		
4,500+ psi	225	200
2,500 psi	160	140
1,500 psi	115	100
Solid masonry of solid concrete masonry units:		
3,000+ psi	225	200
2,000 psi	160	140
1,200 psi	115	100
Masonry of hollow load-bearing units:		
2,000+ psi	140	120
1,500 psi	115	100
1,000 psi	75	70
700 psi	60	55
Hollow walls (cavity or masonry bonded[d]) solid units:		
2,500+ psi	160	140
1,500 psi	115	100
Hollow units	75	70
Stone ashlar masonry:		
Granite	720	640
Limestone or marble	450	400
Sandstone or cast stone	360	320
Rubble stone masonry:		
Coarse, rough or random	120	100

For SI: 1 pound per square inch = 6.895 kPa.

a. Linear interpolation shall be used for determining allowable stresses for masonry units having compressive strengths that are intermediate between those given in the table.

b. Gross cross-sectional area shall be calculated on the actual rather than nominal dimensions.

c. See Section R608.

d. Where floor and roof loads are carried upon one wythe, the gross cross-sectional area is that of the wythe under load; if both wythes are loaded, the gross cross-sectional area is that of the wall minus the area of the cavity between the wythes. Walls bonded with metal ties shall be considered as cavity walls unless the collar joints are filled with mortar or grout.

R606.8 Stack bond. In unreinforced masonry where masonry units are laid in stack bond, longitudinal reinforcement consisting of not less than two continuous wires each with a minimum aggregate cross-sectional area of 0.017 square inch (11 mm²) shall be provided in horizontal bed joints spaced not more than 16 inches (406 mm) on center vertically.

R606.9 Lateral support. Masonry walls shall be laterally supported in either the horizontal or the vertical direction. The maximum spacing between lateral supports shall not exceed the distances in Table R606.9. Lateral support shall be provided by cross walls, pilasters, buttresses or structural frame members when the limiting distance is taken horizontally, or by floors or roofs when the limiting distance is taken vertically.

TABLE R606.9
SPACING OF LATERAL SUPPORT FOR MASONRY WALLS

CONSTRUCTION	MAXIMUM WALL LENGTH TO THICKNESS OR WALL HEIGHT TO THICKNESS[a,b]
Bearing walls:	
Solid or solid grouted	20
All other	18
Nonbearing walls:	
Exterior	18
Interior	36

For SI: 1 foot = 304.8 mm.

a. Except for cavity walls and cantilevered walls, the thickness of a wall shall be its nominal thickness measured perpendicular to the face of the wall. For cavity walls, the thickness shall be determined as the sum of the nominal thicknesses of the individual wythes. For cantilever walls, except for parapets, the ratio of height to nominal thickness shall not exceed 6 for solid masonry, or 4 for hollow masonry. For parapets, see Section R606.2.4.

b. An additional unsupported height of 6 feet is permitted for gable end walls.

R606.9.1 Horizontal lateral support. Lateral support in the horizontal direction provided by intersecting masonry walls shall be provided by one of the methods in Section R606.9.1.1 or Section R606.9.1.2.

R606.9.1.1 Bonding pattern. Fifty percent of the units at the intersection shall be laid in an overlapping masonry bonding pattern, with alternate units having a bearing of not less than 3 inches (76 mm) on the unit below.

R606.9.1.2 Metal reinforcement. Interior nonload- bearing walls shall be anchored at their intersections, at vertical intervals of not more than 16 inches (406 mm) with joint reinforcement of at least 9 gage [0.148 in. (4mm)], or $^1/_4$ inch (6 mm) galvanized mesh hardware cloth. Intersecting masonry walls, other than interior nonloadbearing walls, shall be anchored at vertical intervals of not more than 8 inches (203 mm) with joint reinforcement of at least 9 gage and shall extend at least 30 inches (762 mm) in each direction at the intersection. Other metal ties, joint reinforcement or anchors, if used, shall be spaced to provide equivalent area of anchorage to that required by this section.

R606.9.2 Vertical lateral support. Vertical lateral support of masonry walls in Seismic Design Category A, B or C shall be provided in accordance with one of the methods in Section R606.9.2.1 or Section R606.9.2.2.

R606.9.2.1 Roof structures. Masonry walls shall be anchored to roof structures with metal strap anchors spaced in accordance with the manufacturer's instructions, $^1/_2$-inch (13 mm) bolts spaced not more than 6 feet (1829 mm) on center, or other approved anchors. Anchors shall be embedded at least 16 inches (406 mm)

into the masonry, or be hooked or welded to bond beam reinforcement placed not less than 6 inches (152 mm) from the top of the wall.

R606.9.2.2 Floor diaphragms. Masonry walls shall be anchored to floor diaphragm framing by metal strap anchors spaced in accordance with the manufacturer's instructions, $^1/_2$-inch-diameter (13 mm) bolts spaced at intervals not to exceed 6 feet (1829 mm) and installed as shown in Figure R606.11(1), or by other approved methods.

R606.10 Lintels. Masonry over openings shall be supported by steel lintels, reinforced concrete or masonry lintels or masonry arches, designed to support load imposed.

R606.11 Anchorage. Masonry walls shall be anchored to floor and roof systems in accordance with the details shown in Figure R606.11(1), R606.11(2) or R606.11(3). Footings may be considered as points of lateral support.

R606.12 Seismic requirements. The seismic requirements of this section shall apply to the design of masonry and the construction of masonry building elements located in Seismic Design Category D₀, D₁ or D₂. Townhouses in Seismic Design Category C shall comply with the requirements of Section R606.12.2. These requirements shall not apply to glass unit masonry conforming to Section R610 or masonry veneer conforming to Section R703.7.

R606.12.1 General. Masonry structures and masonry elements shall comply with the requirements of Sections R606.12.2 through R606.12.4 based on the seismic design category established in Table R301.2(1). Masonry structures and masonry elements shall comply with the requirements of Section R606.12 and Figures R606.11(1), R606.11(2) and R606.11(3) or shall be designed in accordance with ACI 530/ASCE 5/TMS 402.

R606.12.1.1 Floor and roof diaphragm construction. Floor and roof diaphragms shall be constructed of wood structural panels attached to wood framing in accordance with Table R602.3(1) or to cold-formed steel floor framing in accordance with Table R505.3.1(2) or to cold-formed steel roof framing in accordance with Table R804.3. Additionally, sheathing panel edges perpendicular to framing members shall be backed by blocking, and sheathing shall be connected to the blocking with fasteners at the edge spacing. For Seismic Design Categories C, D₀, D₁ and D₂, where the width-to-thickness dimension of the diaphragm exceeds 2-to-1, edge spacing of fasteners shall be 4 inches (102 mm) on center.

R606.12.2 Seismic Design Category C. Townhouses located in Seismic Design Category C shall comply with the requirements of this section.

R606.12.2.1 Design of elements not part of the lateral force-resisting system.

R606.12.2.1.1 Load-bearing frames or columns. Elements not part of the lateral-force-resisting system shall be analyzed to determine their effect on the response of the system. The frames or columns shall be adequate for vertical load carrying capacity and induced moment caused by the design story drift.

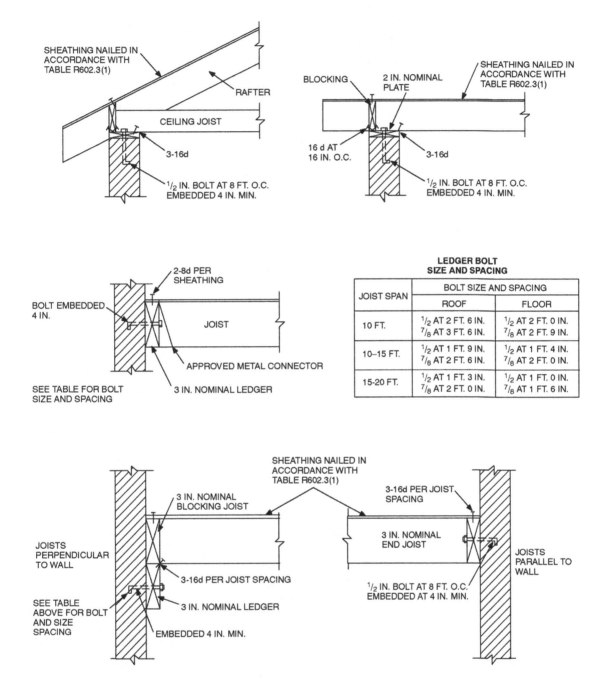

JOIST SPAN	BOLT SIZE AND SPACING	
	ROOF	FLOOR
10 FT.	$1/_2$ AT 2 FT. 6 IN. $7/_8$ AT 3 FT. 6 IN.	$1/_2$ AT 2 FT. 0 IN. $7/_8$ AT 2 FT. 9 IN.
10–15 FT.	$1/_2$ AT 1 FT. 9 IN. $7/_8$ AT 2 FT. 6 IN.	$1/_2$ AT 1 FT. 4 IN. $7/_8$ AT 2 FT. 0 IN.
15-20 FT.	$1/_2$ AT 1 FT. 3 IN. $7/_8$ AT 2 FT. 0 IN.	$1/_2$ AT 1 FT. 0 IN. $7/_8$ AT 1 FT. 6 IN.

LEDGER BOLT SIZE AND SPACING

NOTE: Where bolts are located in hollow masonry, the cells in the courses receiving the bolt shall be grouted solid.

For SI: 1 inch = 25.4 mm, 1 foot = 304.8 mm, 1 pound per square foot = 0.0479 kPa.

FIGURE R606.11(1)
ANCHORAGE REQUIREMENTS FOR MASONRY WALLS LOCATED IN SEISMIC DESIGN CATEGORY
A, B OR C AND WHERE WIND LOADS ARE LESS THAN 30 PSF

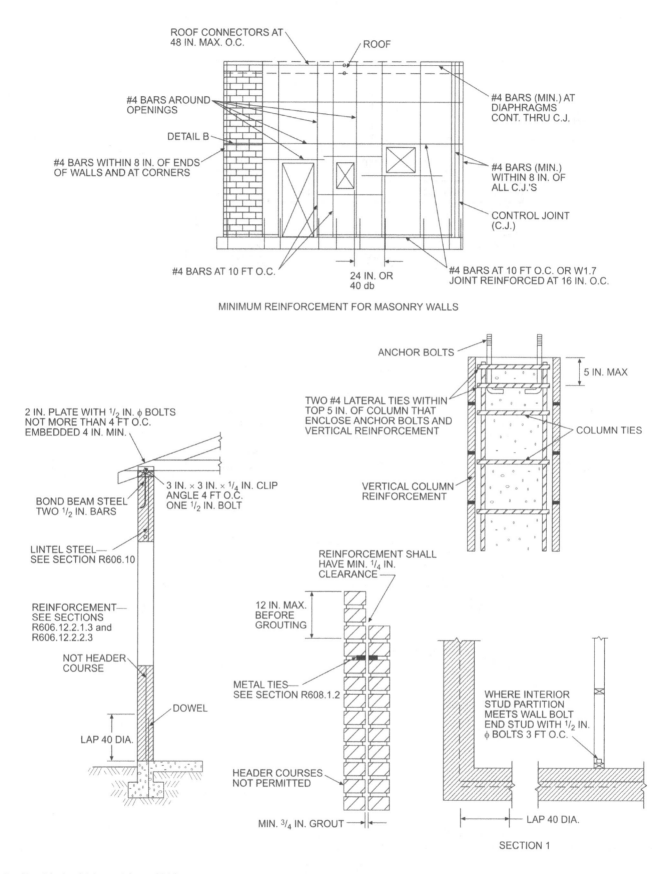

For SI: 1 inch = 25.4 mm, 1 foot = 304.8 mm.

FIGURE R606.11(2)
REQUIREMENTS FOR REINFORCED GROUTED MASONRY CONSTRUCTION IN SEISMIC DESIGN CATEGORY C

2006 INTERNATIONAL RESIDENTIAL CODE®

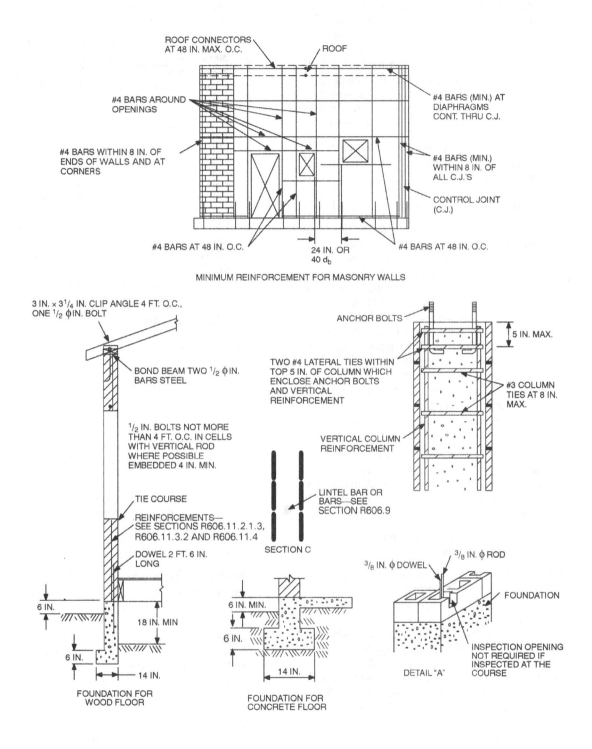

ROOF CONNECTORS AT 48 IN. MAX. O.C.

ROOF

#4 BARS AROUND OPENINGS

#4 BARS WITHIN 8 IN. OF ENDS OF WALLS AND AT CORNERS

#4 BARS (MIN.) AT DIAPHRAGMS CONT. THRU C.J.

#4 BARS (MIN.) WITHIN 8 IN. OF ALL C.J.'S

CONTROL JOINT (C.J.)

#4 BARS AT 48 IN. O.C.

24 IN. OR 40 d_b

#4 BARS AT 48 IN. O.C.

MINIMUM REINFORCEMENT FOR MASONRY WALLS

3 IN. × 3$\frac{1}{4}$ IN. CLIP ANGLE 4 FT. O.C., ONE $\frac{1}{2}$ φIN. BOLT

ANCHOR BOLTS

5 IN. MAX.

BOND BEAM TWO $\frac{1}{2}$ φIN. BARS STEEL

TWO #4 LATERAL TIES WITHIN TOP 5 IN. OF COLUMN WHICH ENCLOSE ANCHOR BOLTS AND VERTICAL REINFORCEMENT

#3 COLUMN TIES AT 8 IN. MAX.

$\frac{1}{2}$ IN. BOLTS NOT MORE THAN 4 FT. O.C. IN CELLS WITH VERTICAL ROD WHERE POSSIBLE EMBEDDED 4 IN. MIN.

VERTICAL COLUMN REINFORCEMENT

TIE COURSE

LINTEL BAR OR BARS—SEE SECTION R606.9

REINFORCEMENTS— SEE SECTIONS R606.11.2.1.3, R606.11.3.2 AND R606.11.4

DOWEL 2 FT. 6 IN. LONG

SECTION C

$\frac{3}{8}$ IN. φ ROD

$\frac{3}{8}$ IN. φ DOWEL

6 IN.

18 IN. MIN

6 IN. MIN.

6 IN.

FOUNDATION

6 IN.

14 IN.

FOUNDATION FOR WOOD FLOOR

14 IN.

FOUNDATION FOR CONCRETE FLOOR

DETAIL "A"

INSPECTION OPENING NOT REQUIRED IF INSPECTED AT THE COURSE

NOTE: A full bed joint must be provided. All cells containing vertical bars are to be filled to the top of wall and provide inspection opening as shown on detail "A." Horizontal bars are to be laid as shown on detail "B." Lintel bars are to be laid as shown on Section C.

For SI: 1 inch = 25.4 mm, 1 foot = 304.8 mm.

FIGURE R606.11(3)
REQUIREMENTS FOR REINFORCED MASONRY CONSTRUCTION IN SEISMIC DESIGN CATEGORY D$_0$, D$_1$ OR D$_2$

R606.12.2.1.2 Masonry partition walls. Masonry partition walls, masonry screen walls and other masonry elements that are not designed to resist vertical or lateral loads, other than those induced by their own weight, shall be isolated from the structure so that vertical and lateral forces are not imparted to these elements. Isolation joints and connectors between these elements and the structure shall be designed to accommodate the design story drift.

R606.12.2.1.3 Reinforcement requirements for masonry elements. Masonry elements listed in Section R606.12.2.1.2 shall be reinforced in either the horizontal or vertical direction as shown in Figure R606.11(2) and in accordance with the following:

1. Horizontal reinforcement. Horizontal joint reinforcement shall consist of at least two longitudinal W1.7 wires spaced not more than 16 inches (406 mm) for walls greater than 4 inches (102 mm) in width and at least one longitudinal W1.7 wire spaced not more than 16 inches (406 mm) for walls not exceeding 4 inches (102 mm) in width; or at least one No. 4 bar spaced not more than 48 inches (1219 mm). Where two longitudinal wires of joint reinforcement are used, the space between these wires shall be the widest that the mortar joint will accommodate. Horizontal reinforcement shall be provided within 16 inches (406 mm) of the top and bottom of these masonry elements.

2. Vertical reinforcement. Vertical reinforcement shall consist of at least one No. 4 bar spaced not more than 48 inches (1219 mm). Vertical reinforcement shall be located within 16 inches (406 mm) of the ends of masonry walls.

R606.12.2.2 Design of elements part of the lateral-force-resisting system.

R606.12.2.2.1 Connections to masonry shear walls. Connectors shall be provided to transfer forces between masonry walls and horizontal elements in accordance with the requirements of Section 2.1.8 of ACI 530/ASCE 5/TMS 402. Connectors shall be designed to transfer horizontal design forces acting either perpendicular or parallel to the wall, but not less than 200 pounds per linear foot (2919 N/m) of wall. The maximum spacing between connectors shall be 4 feet (1219 mm). Such anchorage mechanisms shall not induce tension stresses perpendicular to grain in ledgers or nailers.

R606.12.2.2.2 Connections to masonry columns. Connectors shall be provided to transfer forces between masonry columns and horizontal elements in accordance with the requirements of Section 2.1.8 of ACI 530/ASCE 5/TMS 402. Where anchor bolts are used to connect horizontal elements to the tops of columns, the bolts shall be placed within lateral ties. Lateral ties shall enclose both the vertical bars in the column and the anchor bolts. There shall be a minimum of two No. 4 lateral ties provided in the top 5 inches (127 mm) of the column.

R606.12.2.2.3 Minimum reinforcement requirements for masonry shear walls. Vertical reinforcement of at least one No. 4 bar shall be provided at corners, within 16 inches (406 mm) of each side of openings, within 8 inches (203 mm) of each side of movement joints, within 8 inches (203 mm) of the ends of walls, and at a maximum spacing of 10 feet (3048 mm).

Horizontal joint reinforcement shall consist of at least two wires of W1.7 spaced not more than 16 inches (406 mm); or bond beam reinforcement of at least one No. 4 bar spaced not more than 10 feet (3048 mm) shall be provided. Horizontal reinforcement shall also be provided at the bottom and top of wall openings and shall extend not less than 24 inches (610 mm) nor less than 40 bar diameters past the opening; continuously at structurally connected roof and floor levels; and within 16 inches (406 mm) of the top of walls.

R606.12.3 Seismic Design Category D_0 or D_1. Structures in Seismic Design Category D_0 or D_1 shall comply with the requirements of Seismic Design Category C and the additional requirements of this section.

R606.12.3.1 Design requirements. Masonry elements other than those covered by Section R606.12.2.1.2 shall be designed in accordance with the requirements of Chapter 1 and Sections 2.1 and 2.3 of ACI 530/ASCE 5/TMS 402 and shall meet the minimum reinforcement requirements contained in Sections R606.12.3.2 and R606.12.3.2.1.

Exception: Masonry walls limited to one story in height and 9 feet (2743 mm) between lateral supports need not be designed provided they comply with the minimum reinforcement requirements of Sections R606.12.3.2 and R606.12.3.2.1.

R606.12.3.2 Minimum reinforcement requirements for masonry walls. Masonry walls other than those covered by Section R606.12.2.1.3 shall be reinforced in both the vertical and horizontal direction. The sum of the cross-sectional area of horizontal and vertical reinforcement shall be at least 0.002 times the gross cross-sectional area of the wall, and the minimum cross-sectional area in each direction shall be not less than 0.0007 times the gross cross-sectional area of the wall. Reinforcement shall be uniformly distributed. Table R606.12.3.2 shows the minimum reinforcing bar sizes required for varying thicknesses of masonry walls. The maximum spacing of reinforcement shall be 48 inches (1219 mm) provided that the walls are solid grouted and constructed of hollow open-end units, hollow units laid with full head joints or two wythes of solid units. The maximum spacing of reinforcement shall be 24 inches (610 mm) for all other masonry.

R606.12.3.2.1 Shear wall reinforcement requirements. The maximum spacing of vertical and horizontal reinforcement shall be the smaller of one-third the length of the shear wall, one-third the height of the

TABLE R606.12.3.2
MINIMUM DISTRIBUTED WALL REINFORCEMENT FOR BUILDING ASSIGNED TO SEISMIC DESIGN CATEGORY D_0 or D_1

NOMINAL WALL THICKNESS (inches)	MINIMUM SUM OF THE VERTICAL AND HORIZONTAL REINFORCEMENT AREAS[a] (square inches per foot)	MINIMUM REINFORCEMENT AS DISTRIBUTED IN BOTH HORIZONTAL AND VERTICAL DIRECTIONS[b] (square inches per foot)	MINIMUM BAR SIZE FOR REINFORCEMENT SPACED AT 48 INCHES
6	0.135	0.047	#4
8	0.183	0.064	#5
10	0.231	0.081	#6
12	0.279	0.098	#6

For SI: 1 inch = 25.4 mm, 1 foot = 304.8 mm, 1 square inch per foot = 2064 mm^2/m.

a. Based on the minimum reinforcing ratio of 0.002 times the gross cross-sectional area of the wall.

b. Based on the minimum reinforcing ratio each direction of 0.0007 times the gross cross-sectional area of the wall.

shear wall, or 48 inches (1219 mm). The minimum cross-sectional area of vertical reinforcement shall be one-third of the required shear reinforcement. Shear reinforcement shall be anchored around vertical reinforcing bars with a standard hook.

R606.12.3.3 Minimum reinforcement for masonry columns. Lateral ties in masonry columns shall be spaced not more than 8 inches (203 mm) on center and shall be at least $3/_8$ inch (9.5 mm) diameter. Lateral ties shall be embedded in grout.

R606.12.3.4 Material restrictions. Type N mortar or masonry cement shall not be used as part of the lateral-force-resisting system.

R606.12.3.5 Lateral tie anchorage. Standard hooks for lateral tie anchorage shall be either a 135-degree (2.4 rad) standard hook or a 180-degree (3.2 rad) standard hook.

R606.12.4 Seismic Design Category D_2. All structures in Seismic Design Category D_2 shall comply with the requirements of Seismic Design Category D_1 and to the additional requirements of this section.

R606.12.4.1 Design of elements not part of the lateral-force-resisting system. Stack bond masonry that is not part of the lateral-force-resisting system shall have a horizontal cross-sectional area of reinforcement of at least 0.0015 times the gross cross-sectional area of masonry. Table R606.12.4.1 shows minimum reinforcing bar sizes for masonry walls. The maximum spacing of horizontal reinforcement shall be 24 inches (610 mm). These elements shall be solidly grouted and shall be constructed of hollow open-end units or two wythes of solid units.

TABLE R606.12.4.1
MINIMUM REINFORCING FOR STACKED BONDED MASONRY WALLS IN SEISMIC DESIGN CATEGORY D_2

NOMINAL WALL THICKNESS (inches)	MINIMUM BAR SIZE SPACED AT 24 INCHES
6	#4
8	#5
10	#5
12	#6

For SI: 1 inch = 25.4 mm.

R606.12.4.2 Design of elements part of the lateral-force-resisting system. Stack bond masonry that

is part of the lateral-force-resisting system shall have a horizontal cross-sectional area of reinforcement of at least 0.0025 times the gross cross-sectional area of masonry. Table R606.12.4.2 shows minimum reinforcing bar sizes for masonry walls. The maximum spacing of horizontal reinforcement shall be 16 inches (406 mm). These elements shall be solidly grouted and shall be constructed of hollow open-end units or two wythes of solid units.

TABLE R606.12.4.2
MINIMUM REINFORCING FOR STACKED BONDED MASONRY WALLS IN SEISMIC DESIGN CATEGORY D_2

NOMINAL WALL THICKNESS (inches)	MINIMUM BAR SIZE SPACED AT 16 INCHES
6	#4
8	#5
10	#5
12	#6

For SI: 1 inch = 25.4 mm.

R606.13 Protection for reinforcement. Bars shall be completely embedded in mortar or grout. Joint reinforcement embedded in horizontal mortar joints shall not have less than $5/_8$-inch (15.9 mm) mortar coverage from the exposed face. All other reinforcement shall have a minimum coverage of one bar diameter over all bars, but not less than $3/_4$ inch (19 mm), except where exposed to weather or soil, in which case the minimum coverage shall be 2 inches (51 mm).

R606.14 Beam supports. Beams, girders or other concentrated loads supported by a wall or column shall have a bearing of at least 3 inches (76 mm) in length measured parallel to the beam upon solid masonry not less than 4 inches (102 mm) in thickness, or upon a metal bearing plate of adequate design and dimensions to distribute the load safely, or upon a continuous reinforced masonry member projecting not less than 4 inches (102 mm) from the face of the wall.

R606.14.1 Joist bearing. Joists shall have a bearing of not less than $1^1/_2$ inches (38 mm), except as provided in Section R606.14, and shall be supported in accordance with Figure R606.11(1).

R606.15 Metal accessories. Joint reinforcement, anchors, ties and wire fabric shall conform to the following: ASTM A 82 for wire anchors and ties; ASTM A 36 for plate, headed and bent-bar anchors; ASTM A 510 for corrugated sheet metal anchors and ties; ASTM A 951 for joint reinforcement; ASTM B 227 for copper-clad steel wire ties; or ASTM A 167 for stainless steel hardware.

R606.15.1 Corrosion protection. Minimum corrosion protection of joint reinforcement, anchor ties and wire fabric for use in masonry wall construction shall conform to Table R606.15.1.

TABLE R606.15.1
MINIMUM CORROSION PROTECTION

MASONRY METAL ACCESSORY	STANDARD
Joint reinforcement, interior walls	ASTM A 641, Class 1
Wire ties or anchors in exterior walls completely embedded in mortar or grout	ASTM A 641, Class 3
Wire ties or anchors in exterior walls not completely embedded in mortar or grout	ASTM A 153, Class B-2
Joint reinforcement in exterior walls or interior walls exposed to moist environment	ASTM A 153, Class B-2
Sheet metal ties or anchors exposed to weather	ASTM A 153, Class B-2
Sheet metal ties or anchors completely embedded in mortar or grout	ASTM A 653, Coating Designation G60
Stainless steel hardware for any exposure	ASTM A 167, Type 304

SECTION R607
UNIT MASONRY

R607.1 Mortar. Mortar for use in masonry construction shall comply with ASTM C 270. The type of mortar shall be in accordance with Sections R607.1.1, R607.1.2 and R607.1.3 and shall meet the proportion specifications of Table R607.1 or the property specifications of ASTM C 270.

R607.1.1 Foundation walls. Masonry foundation walls constructed as set forth in Tables R404.1.1(1) through R404.1.1(4) and mortar shall be Type M or S.

R607.1.2 Masonry in Seismic Design Categories A, B and C. Mortar for masonry serving as the lateral-force-resisting system in Seismic Design Categories A, B and C shall be Type M, S or N mortar.

R607.1.3 Masonry in Seismic Design Categories D_0, D_1 and D_2. Mortar for masonry serving as the lateral-force-resisting system in Seismic Design Categories D_0, D_1 and D_2 shall be Type M or S portland cement-lime or mortar cement mortar.

R607.2 Placing mortar and masonry units.

R607.2.1 Bed and head joints. Unless otherwise required or indicated on the project drawings, head and bed joints shall be $^3/_8$ inch (10 mm) thick, except that the thickness of the bed joint of the starting course placed over foundations shall not be less than $^1/_4$ inch (7 mm) and not more than $^3/_4$ inch (19 mm).

R607.2.1.1 Mortar joint thickness tolerance. Mortar joint thickness shall be within the following tolerances from the specified dimensions:

1. Bed joint: + $^1/_8$ inch (3 mm).

2. Head joint: $^1/_4$ inch (7 mm), + $^3/_8$ inch (10 mm).

3. Collar joints: $^1/_4$ inch (7 mm), + $^3/_8$ inch (10 mm).

Exception: Nonload-bearing masonry elements and masonry veneers designed and constructed in accordance with Section R703.7 are not required to meet these tolerances.

R607.2.2 Masonry unit placement. The mortar shall be sufficiently plastic and units shall be placed with sufficient pressure to extrude mortar from the joint and produce a tight joint. Deep furrowing of bed joints that produces voids shall not be permitted. Any units disturbed to the extent that initial bond is broken after initial placement shall be removed and relaid in fresh mortar. Surfaces to be in contact with mortar shall be clean and free of deleterious materials.

R607.2.2.1 Solid masonry. Solid masonry units shall be laid with full head and bed joints and all interior vertical joints that are designed to receive mortar shall be filled.

R607.2.2.2 Hollow masonry. For hollow masonry units, head and bed joints shall be filled solidly with mortar for a distance in from the face of the unit not less than the thickness of the face shell.

R607.3 Installation of wall ties. The installation of wall ties shall be as follows:

1. The ends of wall ties shall be embedded in mortar joints. Wall tie ends shall engage outer face shells of hollow units by at least $^1/_2$ inch (13 mm). Wire wall ties shall be embedded at least $1^1/_2$ inches (38 mm) into the mortar bed of solid masonry units or solid grouted hollow units.

2. Wall ties shall not be bent after being embedded in grout or mortar.

SECTION R608
MULTIPLE WYTHE MASONRY

R608.1 General. The facing and backing of multiple wythe masonry walls shall be bonded in accordance with Section R608.1.1, R608.1.2 or R608.1.3. In cavity walls, neither the facing nor the backing shall be less than 3 inches (76 mm) nominal in thickness and the cavity shall not be more than 4 inches (102 mm) nominal in width. The backing shall be at least as thick as the facing.

Exception: Cavities shall be permitted to exceed the 4-inch (102 mm) nominal dimension provided tie size and tie spacing have been established by calculation.

R608.1.1 Bonding with masonry headers. Bonding with solid or hollow masonry headers shall comply with Sections R608.1.1.1 and R608.1.1.2.

R608.1.1.1 Solid units. Where the facing and backing (adjacent wythes) of solid masonry construction are bonded by means of masonry headers, no less than 4 percent of the wall surface of each face shall be composed of

TABLE R607.1
MORTAR PROPORTIONS[a, b]

MORTAR	TYPE	Portland cement or blended cement	Mortar cement			Masonry cement			Hydrated lime[c] or lime putty	Aggregate ratio (measured in damp, loose conditions)
			M	S	N	M	S	N		
Cement-lime	M	1	—	—	—	—	—	—	$1/4$	
	S	1	—	—	—	—	—	—	over $1/4$ to $1/2$	
	N	1	—	—	—	—	—	—	over $1/2$ to $1\,1/4$	
	O	1	—	—	—	—	—	—	over $1\,1/4$ to $2\,1/2$	
Mortar cement	M	1	—	—	1	—	—	—	—	Not less than $2\,1/4$ and not more than 3 times the sum of separate volumes of lime, if used, and cement
	M	—	1	—	—	—	—	—		
	S	$1/2$	—	—	1	—	—	—		
	S	—	—	1	—	—	—	—		
	N	—	—	—	1	—	—	—		
	O	—	—	—	1	—	—	—		
Masonry cement	M	1				—	—	1	—	
	M	—				1	—	—		
	S	$1/2$				—	—	1		
	S	—				—	1	—		
	N	—				—	—	1		
	O	—				—	—	1		

For SI: 1 cubic foot = 0.0283 m³, 1 pound = 0.454 kg.

a. For the purpose of these specifications, the weight of 1 cubic foot of the respective materials shall be considered to be as follows:

Portland Cement	94 pounds	Masonry Cement	Weight printed on bag
Mortar Cement	Weight printed on bag	Hydrated Lime	40 pounds
Lime Putty (Quicklime)	80 pounds	Sand, damp and loose	80 pounds of dry sand

b. Two air-entraining materials shall not be combined in mortar.

c. Hydrated lime conforming to the requirements of ASTM C 207.

headers extending not less than 3 inches (76 mm) into the backing. The distance between adjacent full-length headers shall not exceed 24 inches (610 mm) either vertically or horizontally. In walls in which a single header does not extend through the wall, headers from the opposite sides shall overlap at least 3 inches (76 mm), or headers from opposite sides shall be covered with another header course overlapping the header below at least 3 inches (76 mm).

R608.1.1.2 Hollow units. Where two or more hollow units are used to make up the thickness of a wall, the stretcher courses shall be bonded at vertical intervals not exceeding 34 inches (864 mm) by lapping at least 3 inches (76 mm) over the unit below, or by lapping at vertical intervals not exceeding 17 inches (432 mm) with units that are at least 50 percent thicker than the units below.

R608.1.2 Bonding with wall ties or joint reinforcement. Bonding with wall ties or joint reinforcement shall comply with Sections R608.1.2.1 through R608.1.2.3.

R608.1.2.1 Bonding with wall ties. Bonding with wall ties, except as required by Section R610, where the facing and backing (adjacent wythes) of masonry walls are bonded with $3/16$-inch-diameter (5 mm) wall ties embedded in the horizontal mortar joints, there shall be at least one metal tie for each 4.5 square feet (0.418 m²) of wall area. Ties in alternate courses shall be staggered. The maximum vertical distance between ties shall not exceed 24 inches (610 mm), and the maximum horizontal dis-

tance shall not exceed 36 inches (914 mm). Rods or ties bent to rectangular shape shall be used with hollow masonry units laid with the cells vertical. In other walls, the ends of ties shall be bent to 90-degree (0.79 rad) angles to provide hooks no less than 2 inches (51 mm) long. Additional bonding ties shall be provided at all openings, spaced not more than 3 feet (914 mm) apart around the perimeter and within 12 inches (305 mm) of the opening.

R608.1.2.2 Bonding with adjustable wall ties. Where the facing and backing (adjacent wythes) of masonry are bonded with adjustable wall ties, there shall be at least one tie for each 2.67 square feet (0.248 m²) of wall area. Neither the vertical nor the horizontal spacing of the adjustable wall ties shall exceed 24 inches (610 mm). The maximum vertical offset of bed joints from one wythe to the other shall be 1.25 inches (32 mm). The maximum clearance between connecting parts of the ties shall be $1/16$ inch (2 mm). When pintle legs are used, ties shall have at least two $3/16$-inch-diameter (5 mm) legs.

R608.1.2.3 Bonding with prefabricated joint reinforcement. Where the facing and backing (adjacent wythes) of masonry are bonded with prefabricated joint reinforcement, there shall be at least one cross wire serving as a tie for each 2.67 square feet (0.248 m²) of wall area. The vertical spacing of the joint reinforcement shall not exceed 16 inches (406 mm). Cross wires on prefabricated joint reinforcement shall not be smaller than No. 9 gage. The longitudinal wires shall be embedded in the mortar.

R608.1.3 Bonding with natural or cast stone. Bonding with natural and cast stone shall conform to Sections R608.1.3.1 and R608.1.3.2.

R608.1.3.1 Ashlar masonry. In ashlar masonry, bonder units, uniformly distributed, shall be provided to the extent of not less than 10 percent of the wall area. Such bonder units shall extend not less than 4 inches (102 mm) into the backing wall.

R608.1.3.2 Rubble stone masonry. Rubble stone masonry 24 inches (610 mm) or less in thickness shall have bonder units with a maximum spacing of 3 feet (914 mm) vertically and 3 feet (914 mm) horizontally, and if the masonry is of greater thickness than 24 inches (610 mm), shall have one bonder unit for each 6 square feet (0.557 m^2) of wall surface on both sides.

R608.2 Masonry bonding pattern. Masonry laid in running and stack bond shall conform to Sections R608.2.1 and R608.2.2.

R608.2.1 Masonry laid in running bond. In each wythe of masonry laid in running bond, head joints in successive courses shall be offset by not less than one-fourth the unit length, or the masonry walls shall be reinforced longitudinally as required in Section R608.2.2.

R608.2.2 Masonry laid in stack bond. Where unit masonry is laid with less head joint offset than in Section R608.2.1, the minimum area of horizontal reinforcement placed in mortar bed joints or in bond beams spaced not more than 48 inches (1219 mm) apart, shall be 0.0007 times the vertical cross-sectional area of the wall.

SECTION R609
GROUTED MASONRY

R609.1 General. Grouted multiple-wythe masonry is a form of construction in which the space between the wythes is solidly filled with grout. It is not necessary for the cores of masonry units to be filled with grout. Grouted hollow unit masonry is a form of construction in which certain cells of hollow units are continuously filled with grout.

R609.1.1 Grout. Grout shall consist of cementitious material and aggregate in accordance with ASTM C 476 and the proportion specifications of Table R609.1.1. Type M or Type S mortar to which sufficient water has been added to produce pouring consistency can be used as grout.

R609.1.2 Grouting requirements. Maximum pour heights and the minimum dimensions of spaces provided for grout placement shall conform to Table R609.1.2. If the work is stopped for one hour or longer, the horizontal construction joints shall be formed by stopping all tiers at the same elevation and with the grout 1 inch (25 mm) below the top.

TABLE R609.1.1
GROUT PROPORTIONS BY VOLUME FOR MASONRY CONSTRUCTION

TYPE	PORTLAND CEMENT OR BLENDED CEMENT SLAG CEMENT	HYDRATED LIME OR LIME PUTTY	AGGREGATE MEASURED IN A DAMP, LOOSE CONDITION	
			Fine	Coarse
Fine	1	0 to 1/10	$2^1/_4$ to 3 times the sum of the volume of the cementitious materials	—
Coarse	1	0 to 1/10	$2^1/_4$ to 3 times the sum of the volume of the cementitious materials	1 to 2 times the sum of the volumes of the cementitious materials

TABLE R609.1.2
GROUT SPACE DIMENSIONS AND POUR HEIGHTS

GROUT TYPE	GROUT POUR MAXIMUM HEIGHT (feet)	MINIMUM WIDTH OF GROUT SPACES[a,b] (inches)	MINIMUM GROUT[b,c] SPACE DIMENSIONS FOR GROUTING CELLS OF HOLLOW UNITS (inches x inches)
Fine	1	0.75	1.5 × 2
	5	2	2 × 3
	12	2.5	2.5 × 3
	24	3	3 × 3
Coarse	1	1.5	1.5 × 3
	5	2	2.5 × 3
	12	2.5	3 × 3
	24	3	3 × 4

For SI: 1 inch = 25.4 mm, 1 foot = 304.8 mm.

a. For grouting between masonry wythes.

b. Grout space dimension is the clear dimension between any masonry protrusion and shall be increased by the horizontal projection of the diameters of the horizontal bars within the cross section of the grout space.

c. Area of vertical reinforcement shall not exceed 6 percent of the area of the grout space.

R609.1.3 Grout space (cleaning). Provision shall be made for cleaning grout space. Mortar projections that project more than 0.5 inch (13 mm) into grout space and any other foreign matter shall be removed from grout space prior to inspection and grouting.

R609.1.4 Grout placement. Grout shall be a plastic mix suitable for pumping without segregation of the constituents and shall be mixed thoroughly. Grout shall be placed by pumping or by an approved alternate method and shall be placed before any initial set occurs and in no case more than $1^1/_2$ hours after water has been added. Grouting shall be done in a continuous pour, in lifts not exceeding 5 feet (1524 mm). It shall be consolidated by puddling or mechanical vibrating during placing and reconsolidated after excess moisture has been absorbed but before plasticity is lost.

R609.1.4.1 Grout pumped through aluminum pipes. Grout shall not be pumped through aluminum pipes.

R609.1.5 Cleanouts. Where required by the building official, cleanouts shall be provided as specified in this section. The cleanouts shall be sealed before grouting and after inspection.

R609.1.5.1 Grouted multiple-wythe masonry. Cleanouts shall be provided at the bottom course of the exterior wythe at each pour of grout where such pour exceeds 5 feet (1524 mm) in height.

R609.1.5.2 Grouted hollow unit masonry. Cleanouts shall be provided at the bottom course of each cell to be grouted at each pour of grout, where such pour exceeds 4 feet (1219 mm) in height.

R609.2 Grouted multiple-wythe masonry. Grouted multiple-wythe masonry shall conform to all the requirements specified in Section R609.1 and the requirements of this section.

R609.2.1 Bonding of backup wythe. Where all interior vertical spaces are filled with grout in multiple-wythe construction, masonry headers shall not be permitted. Metal wall ties shall be used in accordance with Section R608.1.2 to prevent spreading of the wythes and to maintain the vertical alignment of the wall. Wall ties shall be installed in accordance with Section R608.1.2 when the backup wythe in multiple-wythe construction is fully grouted.

R609.2.2 Grout spaces. Fine grout shall be used when interior vertical space to receive grout does not exceed 2 inches (51 mm) in thickness. Interior vertical spaces exceeding 2 inches (51 mm) in thickness shall use coarse or fine grout.

R609.2.3 Grout barriers. Vertical grout barriers or dams shall be built of solid masonry across the grout space the entire height of the wall to control the flow of the grout horizontally. Grout barriers shall not be more than 25 feet (7620 mm) apart. The grouting of any section of a wall between control barriers shall be completed in one day with no interruptions greater than one hour.

R609.3 Reinforced grouted multiple-wythe masonry. Reinforced grouted multiple-wythe masonry shall conform to all the requirements specified in Sections R609.1 and R609.2 and the requirements of this section.

R609.3.1 Construction. The thickness of grout or mortar between masonry units and reinforcement shall not be less than $^1/_4$ inch (7 mm), except that $^1/_4$-inch (7 mm) bars may be laid in horizontal mortar joints at least $^1/_2$ inch (13 mm) thick, and steel wire reinforcement may be laid in horizontal mortar joints at least twice the thickness of the wire diameter.

R609.4 Reinforced hollow unit masonry. Reinforced hollow unit masonry shall conform to all the requirements of Section R609.1 and the requirements of this section.

R609.4.1 Construction. Requirements for construction shall be as follows:

1. Reinforced hollow-unit masonry shall be built to preserve the unobstructed vertical continuity of the cells to be filled. Walls and cross webs forming cells to be filled shall be full-bedded in mortar to prevent leakage of grout. Head and end joints shall be solidly filled with mortar for a distance in from the face of the wall or unit not less than the thickness of the longitudinal face shells. Bond shall be provided by lapping units in successive vertical courses.

2. Cells to be filled shall have vertical alignment sufficient to maintain a clear, unobstructed continuous vertical cell of dimensions prescribed in Table R609.1.2.

3. Vertical reinforcement shall be held in position at top and bottom and at intervals not exceeding 200 diameters of the reinforcement.

4. Cells containing reinforcement shall be filled solidly with grout. Grout shall be poured in lifts of 8-foot (2438 mm) maximum height. When a total grout pour exceeds 8 feet (2438 mm) in height, the grout shall be placed in lifts not exceeding 5 feet (1524 mm) and special inspection during grouting shall be required.

5. Horizontal steel shall be fully embedded by grout in an uninterrupted pour.

SECTION R610
GLASS UNIT MASONRY

R610.1 General. Panels of glass unit masonry located in load-bearing and nonload-bearing exterior and interior walls shall be constructed in accordance with this section.

R610.2 Materials. Hollow glass units shall be partially evacuated and have a minimum average glass face thickness of $^3/_{16}$ inch (5 mm). The surface of units in contact with mortar shall be treated with a polyvinyl butyral coating or latex-based paint. The use of reclaimed units is prohibited.

R610.3 Units. Hollow or solid glass block units shall be standard or thin units.

R610.3.1 Standard units. The specified thickness of standard units shall be at least $3^7/_8$ inches (98 mm).

R610.3.2 Thin units. The specified thickness of thin units shall be at least $3^1/_8$ inches (79 mm) for hollow units and at least 3 inches (76 mm) for solid units.

R610.4 Isolated panels. Isolated panels of glass unit masonry shall conform to the requirements of this section.

R610.4.1 Exterior standard-unit panels. The maximum area of each individual standard-unit panel shall be 144 square feet (13.4 m²) when the design wind pressure is 20 psf (958 Pa). The maximum area of such panels subjected to design wind pressures other than 20 psf (958 Pa) shall be in accordance with Figure R610.4.1. The maximum panel dimension between structural supports shall be 25 feet (7620 mm) in width or 20 feet (6096 mm) in height.

R610.4.2 Exterior thin-unit panels. The maximum area of each individual thin-unit panel shall be 85 square feet (7.9 m²). The maximum dimension between structural supports shall be 15 feet (4572 mm) in width or 10 feet (3048 mm) in height. Thin units shall not be used in applications where the design wind pressure as stated in Table R301.2(1) exceeds 20 psf (958 Pa).

R610.4.3 Interior panels. The maximum area of each individual standard-unit panel shall be 250 square feet (23.2 m²). The maximum area of each thin-unit panel shall be 150 square feet (13.9 m²). The maximum dimension between structural supports shall be 25 feet (7620 mm) in width or 20 feet (6096 mm) in height.

R610.4.4 Curved panels. The width of curved panels shall conform to the requirements of Sections R610.4.1, R610.4.2 and R610.4.3, except additional structural supports shall be provided at locations where a curved section joins a straight section, and at inflection points in multicurved walls.

R610.5 Panel support. Glass unit masonry panels shall conform to the support requirements of this section.

R610.5.1 Deflection. The maximum total deflection of structural members that support glass unit masonry shall not exceed $^1/_{600}$.

R610.5.2 Lateral support. Glass unit masonry panels shall be laterally supported along the top and sides of the panel. Lateral supports for glass unit masonry panels shall be designed to resist a minimum of 200 pounds per lineal feet (2918 N/m) of panel, or the actual applied loads, whichever is greater. Except for single unit panels, lateral support shall be provided by panel anchors along the top and sides spaced a maximum of 16 inches (406 mm) on center or by channel-type restraints. Single unit panels shall be supported by channel-type restraints.

Exceptions:

1. Lateral support is not required at the top of panels that are one unit wide.

2. Lateral support is not required at the sides of panels that are one unit high.

R610.5.2.1 Panel anchor restraints. Panel anchors shall be spaced a maximum of 16 inches (406 mm) on center in both jambs and across the head. Panel anchors shall be embedded a minimum of 12 inches (305 mm) and shall be provided with two fasteners so as to resist the loads specified in Section R610.5.2.

R610.5.2.2 Channel-type restraints. Glass unit masonry panels shall be recessed at least 1 inch (25 mm) within channels and chases. Channel-type restraints shall be oversized to accommodate expansion material in the opening, packing and sealant between

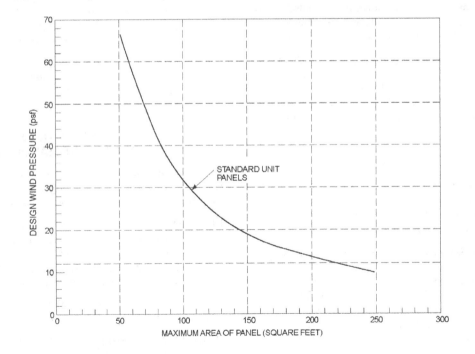

For SI: 1 square foot = 0.0929 m², 1 pound per square foot = 0.0479 kPa.

FIGURE R610.4.1
GLASS UNIT MASONRY DESIGN WIND LOAD RESISTANCE

the framing restraints, and the glass unit masonry perimeter units.

R610.6 Sills. Before bedding of glass units, the sill area shall be covered with a water base asphaltic emulsion coating. The coating shall shall be a minimum of $^1/_8$ inch (3 mm) thick.

R610.7 Expansion joints. Glass unit masonry panels shall be provided with expansion joints along the top and sides at all structural supports. Expansion joints shall be a minimum of $^3/_8$ inch (10 mm) in thickness and shall have sufficient thickness to accommodate displacements of the supporting structure. Expansion joints shall be entirely free of mortar and other debris and shall be filled with resilient material.

R610.8 Mortar. Glass unit masonry shall be laid with Type S or N mortar. Mortar shall not be retempered after initial set. Mortar unused within $1^1/_2$ hours after initial mixing shall be discarded.

R610.9 Reinforcement. Glass unit masonry panels shall have horizontal joint reinforcement spaced a maximum of 16 inches (406 mm) on center located in the mortar bed joint. Horizontal joint reinforcement shall extend the entire length of the panel but shall not extend across expansion joints. Longitudinal wires shall be lapped a minimum of 6 inches (152 mm) at splices. Joint reinforcement shall be placed in the bed joint immediately below and above openings in the panel. The reinforcement shall have not less than two parallel longitudinal wires of size W1.7 or greater, and have welded cross wires of size W1.7 or greater.

R610.10 Placement. Glass units shall be placed so head and bed joints are filled solidly. Mortar shall not be furrowed. Head and bed joints of glass unit masonry shall be $^1/_4$ inch (6.4 mm) thick, except that vertical joint thickness of radial panels shall not be less than $^1/_8$ inch (3 mm) or greater than $^5/_8$ inch (16 mm). The bed joint thickness tolerance shall be minus $^1/_{16}$ inch (1.6 mm) and plus $^1/_8$ inch (3 mm). The head joint thickness tolerance shall be plus or minus $^1/_8$ inch (3 mm).

SECTION R611
INSULATING CONCRETE FORM
WALL CONSTRUCTION

R611.1 General. Insulating Concrete Form (IFC) walls shall be designed and constructed in accordance with the provisions of this section or in accordance with the provisions of ACI 318. When ACI 318 or the provisions of this section are used to design insulating concrete form walls, project drawings, typical details and specifications are not required to bear the seal of the architect or engineer responsible for design, unless otherwise required by the state law of the jurisdiction having authority.

R611.2 Applicability limits. The provisions of this section shall apply to the construction of insulating concrete form walls for buildings not greater than 60 feet (18 288 mm) in plan dimensions, and floors not greater than 32 feet (9754 mm) or roofs not greater than 40 feet (12 192 mm) in clear span. Buildings shall not exceed two stories in height above-grade. ICF walls shall comply with the requirements in Table R611.2. Walls constructed in accordance with the provisions of this section shall be limited to buildings subjected to a maximum design wind speed of 150 miles per hour (67 m/s), and Seismic Design Categories A, B, C, D_0, D_1 and D_2. The provisions of this section shall not apply to the construction of ICF walls for buildings or portions of buildings considered irregular as defined in Section R301.2.2.2.2.

For townhouses in Seismic Design Category C and all buildings in Seismic Design Category D_0, D_1 or D_2, the provisions of this section shall apply only to buildings meeting the following requirements.

1. Rectangular buildings with a maximum building aspect ratio of 2:1. The building aspect ratio shall be determined by dividing the longest dimension of the building by the shortest dimension of the building.

2. Walls are aligned vertically with the walls below.

TABLE R611.2
REQUIREMENTS FOR ICF WALLS[b]

WALL TYPE AND NOMINAL SIZE	MAXIMUM WALL WEIGHT (psf)[c]	MINIMUM WIDTH OF VERTICAL CORE (inches)[a]	MINIMUM THICKNESS OF VERTICAL CORE (inches)[a]	MAXIMUM SPACING OF VERTICAL CORES (inches)	MAXIMUM SPACING OF HORIZONTAL CORES (inches)	MINIMUM WEB THICKNESS (inches)
3.5″ Flat[d]	44[d]	N/A	N/A	N/A	N/A	N/A
5.5″ Flat	69	N/A	N/A	N/A	N/A	N/A
7.5″ Flat	94	N/A	N/A	N/A	N/A	N/A
9.5″ Flat	119	N/A	N/A	N/A	N/A	N/A
6″ Waffle-Grid	56	6.25	5	12	16	2
8″ Waffle-Grid	76	7	7	12	16	2
6″ Screen-Grid	53	5.5	5.5	12	12	N/A

For SI: 1 inch = 25.4 mm; 1 pound per cubic foot = 16.018 kg/m³; 1 pound per square foot = 0.0479 kPa.

a. For width "W", thickness "T", spacing, and web thickness, refer to Figures R611.4 and R611.5.

b. N/A indicates not applicable.

c. Wall weight is based on a unit weight of concrete of 150 pcf. The tabulated values do not include any allowance for interior and exterior finishes.

d. For all buildings in Seismic Design Category A or B, and detached one- and two-family dwellings in Seismic Design Category C the actual wall thickness is permitted to be up to 1 inch thicker than shown and the maximum wall weight to be 56 psf. Construction requirements and other limitations within Section R611 for 3.5-inch flat ICF walls shall apply. Interpolation between provisions for 3.5-inch and 5.5-inch flat ICF walls is not permitted.

3. Cantilever and setback construction shall not be permitted.

4. The weight of interior and exterior finishes applied to ICF walls shall not exceed 8 psf (380 Pa).

5. The gable portion of ICF walls shall be constructed of light-frame construction.

R611.3 Flat insulating concrete form wall systems. Flat ICF wall systems shall comply with Figure R611.3 and shall have reinforcement in accordance with Tables R611.3(1) and R611.3(2) and Section R611.7.

R611.4 Waffle-grid insulating concrete form wall systems. Waffle-grid wall systems shall comply with Figure R611.4 and shall have reinforcement in accordance with Tables R611.3(1) and R611.4(1) and Section R611.7. The minimum core dimensions shall comply with Table R611.2.

R611.5 Screen-grid insulating concrete form wall systems. Screen-grid ICF wall systems shall comply with Figure R611.5 and shall have reinforcement in accordance with Tables R611.3(1) and R611.5 and Section R611.7. The minimum core dimensions shall comply with Table R611.2.

R611.6 Material. Insulating concrete form wall materials shall comply with this section.

R611.6.1 Concrete material. Ready-mixed concrete for insulating concrete form walls shall be in accordance with Section R402.2. Maximum slump shall not be greater than 6 inches (152 mm) as determined in accordance with ASTM

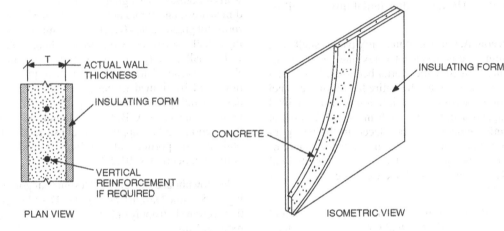

FIGURE R611.3
FLAT ICF WALL SYSTEM

TABLE R611.3(1)
DESIGN WIND PRESSURE FOR USE WITH TABLES R611.3(2), R611.4(1), AND R611.5 FOR ABOVE GRADE WALLS[a]

WIND SPEED (mph)[e]	DESIGN WIND PRESSURE (psf)					
	Enclosed[b]			Partially Enclosed[b]		
	Exposure[c]			Exposure[c]		
	B	C	D	B	C	D
85	18	24	29	23	31	37
90	20	27	32	25	35	41
100	24	34	39	31	43	51
110	29	41	48	38	52	61
120	35	48	57	45	62	73
130	41	56	66	53	73	85[d]
140	47	65	77	61	84[d]	99[d]
150	54	75	88[d]	70	96[d]	114[d]

For SI: 1 pound per square foot = 0.0479 kPa; 1 mile per hour = 0.447 m/s; 1 foot = 304.8 mm; 1 square foot = 0.0929 m².

a. This table is based on ASCE 7-98 components and cladding wind pressures using a mean roof height of 35 ft and a tributary area of 10 ft².

b. Buildings in wind-borne debris regions as defined in Section R202 shall be considered as "Partially Enclosed" unless glazed openings are protected in accordance with Section R301.2.1.2, in which case the building shall be considered as "Enclosed." All other buildings shall be classified as "Enclosed."

c. Exposure Categories shall be determined in accordance with Section R301.2.1.4.

d. For wind pressures greater than 80 psf, design is required in accordance with ACI 318 and approved manufacturer guidelines.

e. Interpolation is permitted between wind speeds.

C 143. Maximum aggregate size shall not be larger than $^3/_4$ inch (19 mm).

Exception: Concrete mixes conforming to the ICF manufacturer's recommendations.

In Seismic Design Categories D_0, D_1 and D_2, the minimum concrete compressive strength shall be 3,000 psi (20.5 MPa).

R611.6.2 Reinforcing steel. Reinforcing steel shall meet the requirements of ASTM A 615, A 706, or A 996. Except in Seismic Design Categories D_0, D_1 and D_2, the minimum yield strength of reinforcing steel shall be 40,000 psi (Grade 40) (276 MPa). In Seismic Design Categories D_0, D_1 and D_2, reinforcing steel shall meet the requirements of ASTM A 706 for low-alloy steel with a minimum yield strength of 60,000 psi (Grade 60) (414 Mpa).

TABLE R611.3(2)
MINIMUM VERTICAL WALL REINFORCEMENT FOR FLAT ICF ABOVE-GRADE WALLS[a, b, c, d]

Design Wind Pressure [Table R611.3(1)] (psf)	Maximum Unsupported Wall Height (feet)	Minimum Vertical Reinforcement[d, e, f]					
		Nonload-Bearing Wall or Supporting Roof		Supporting Light-Framed Second Story and Roof		Supporting ICF Second Story and Roof	
		Minimum Wall Thickness (inches)					
		3.5[g]	5.5	3.5[g]	5.5	3.5[g]	5.5
20	8	#4@48	#4@48	#4@48	#4@48	#4@48	#4@48
	9	#4@48	#4@48	#4@48	#4@48	#4@48	#4@48
	10	#4@38	#4@48	#4@40	#4@48	#4@42	#4@48
30	8	#4@42	#4@48	#4@46	#4@48	#4@48	#4@48
	9	#4@32; #5@48	#4@48	#4@34; #5@48	#4@48	#4@34; #5@48	#4@48
	10	Design Required	#4@48	Design Required	#4@48	Design Required	#4@48
40	8	#4@30; #5@48	#4@48	#4@30; #5@48	#4@48	#4@32; #5@48	#4@48
	9	Design Required	#4@42	Design Required	#4@46	Design Required	#4@48
	10	Design Required	#4@32; #5@48	Design Required	#4@34; #5@48	Design Required	#4@38
50	8	#4@20; #5@30	#4@42	#4@22; #5@34	#4@46	#4@24; #5@36	#4@48
	9	Design Required	#4@34; #5@48	Design Required	#4@34; #5@48	Design Required	#4@38
	10	Design Required	#4@26; #5@38	Design Required	#4@26; #5@38	Design Required	#4@28; #5@46
60	8	Design Required	#4@34; #5@48	Design Required	#4@36	Design Required	#4@40
	9	Design Required	#4@26; #5@38	Design Required	#4@28; #5@46	Design Required	#4@34; #5@48
	10	Design Required	#4@22; #5@34	Design Required	#4@22; #5@34	Design Required	#4@26; #5@38
70	8	Design Required	#4@28; #5@46	Design Required	#4@30; #5@48	Design Required	#4@34; #5@48
	9	Design Required	#4@22; #5@34	Design Required	#4@22; #5@34	Design Required	#4@24; #5@36
	10	Design Required	#4@16; #5@26	Design Required	#4@18; #5@28	Design Required	#4@20; #5@30
80	8	Design Required	#4@26; #5@38	Design Required	#4@26; #5@38	Design Required	#4@28; #5@46
	9	Design Required	#4@20; #5@30	Design Required	#4@20; #5@30	Design Required	#4@21; #5@34
	10	Design Required	#4@14; #5@24	Design Required	#4@14; #5@24	Design Required	#4@16; #5@26

For SI: 1 inch = 25.4 mm; 1 foot = 304.8 mm; 1 mile per hour = 0.447 m/s; 1 pound per square inch = 6.895 kPa.

a. This table is based on reinforcing bars with a minimum yield strength of 40,000 psi and concrete with a minimum specified compressive strength of 2,500 psi. For Seismic Design Categories D_0, D_1 and D_2, reinforcing bars shall have a minimum yield strength of 60,000 psi. See Section R611.6.2.

b. Deflection criterion is $L/240$, where L is the height of the wall story in inches.

c. Interpolation shall not be permitted.

d. Reinforcement spacing for 3.5 inch walls shall be permitted to be multiplied by 1.6 when reinforcing steel with a minimum yield strength of 60,000 psi is used. Reinforcement shall not be less than one #4 bar at 48 inches (1.2 m) on center.

e. Reinforcement spacing for 5.5 inch (139.7 mm) walls shall be permitted to be multiplied by 1.5 when reinforcing steel with a minimum yield strength of 60,000 psi is used. Reinforcement shall not be less than one #4 bar at 48 inches on center.

f. See Section R611.7.1.2 for limitations on maximum spacing of vertical reinforcement in Seismic Design Categories C, D_0, D_1 and D_2.

g. A 3.5-inch wall shall not be permitted if wood ledgers are used to support floor or roof loads. See Section R611.8.

TABLE R611.4(1)
MINIMUM VERTICAL WALL REINFORCEMENT FOR WAFFLE-GRID ICF ABOVE-GRADE WALLS[a, b, c]

Design Wind Pressure [Table R611.3(1)] (psf)	Maximum Unsupported Wall Height (feet)	MINIMUM VERTICAL REINFORCEMENT[d, e]					
		Nonload-Bearing Wall or Supporting Roof		Supporting Light-Framed Second Story and Roof		Supporting ICF Second Story and Roof	
		Minimum Wall Thickness (inches)					
		6	8	6	8	6	8
20	8	#4@48	#4@48	#4@48	#4@48	#4@48	#4@48
	9	#4@48	#4@48	#4@48	#4@48	#4@48	#4@48
	10	#4@48	#4@48	#4@48	#4@48	#4@48	#4@48
30	8	#4@48	#4@48	#4@48	#4@48	#4@48	#4@48
	9	#4@48	#4@48	#4@48	#4@48	#4@48	#4@48
	10	#4@36; #5@48	#4@48	#4@36; #5@48	#4@48	#4@36; #5@48	#4@48
40	8	#4@36; #5@48	#4@48	#4@48	#4@48	#4@48	#4@48
	9	#4@36; #5@48	#4@48	#4@36; #5@48	#4@48	#4@36; #5@48	#4@48
	10	#4@24; #5@36	#4@36; #5@48	#4@24; #5@36	#4@48	#4@24; #5@36	#4@48
50	8	#4@36; #5@48	#4@48	#4@36; #5@48	#4@48	#4@36; #5@48	#4@48
	9	#4@24; #5@36	#4@36; #5@48	#4@24; #5@36	#4@48	#4@24; #5@48	#4@48
	10	Design Required	#4@36; #5@48	Design Required	#4@36; #5@48	Design Required	#4@36; #5@48
60	8	#4@24; #5@36	#4@48	#4@24; #5@36	#4@48	#4@24; #5@48	#4@48
	9	Design Required	#4@36; #5@48	Design Required	#4@36; #5@48	Design Required	#4@36; #5@48
	10	Design Required	#4@24; #5@36	Design Required	#4@24; #5@36	Design Required	#4@24; #5@48
70	8	#4@24; #5@36	#4@36; #5@48	#4@24; #5@36	#4@36; #5@48	#4@24; #5@36	#4@48
	9	Design Required	#4@24; #5@36	Design Required	#4@24; #5@48	Design Required	#4@24; #5@48
	10	Design Required	#4@12; #5@36	Design Required	#4@24; #5@36	Design Required	#4@24; #5@36
80	8	#4@12; #5@24	#4@24; #5@48	#4@12; #5@24	#4@24; #5@48	#4@12; #5@24	#4@36; #5@48
	9	Design Required	#4@24; #5@36	Design Required	#4@24; #5@36	Design Required	#4@24; #5@36
	10	Design Required	#4@12; #5@24	Design Required	#4@12; #5@24	Design Required	#4@12; #5@24

For SI: 1 foot = 304.8 mm; 1 inch = 25.4 mm; 1 mile per hour = 0.447 m/s; 1 pound per square inch = 6.895 MPa.

a. This table is based on reinforcing bars with a minimum yield strength of 40,000 psi and concrete with a minimum specified compressive strength of 2,500 psi. For Seismic Design Categories D_0, D_1 and D_2, reinforcing bars shall have a minimum yield strength of 60,000 psi. See Section R611.6.2.

b. Deflection criterion is $L/240$, where L is the height of the wall story in inches.

c. Interpolation shall not be permitted.

d. Increasing reinforcement spacing by 12 inches shall be permitted when reinforcing steel with a minimum yield strength of 60,000 psi is used or substitution of No. 4 reinforcing bars for #5 bars shall be permitted when reinforcing steel with a minimum yield strength of 60,000 psi is used at the same spacing required for #5 bars. Reinforcement shall not be less than one #4 bar at 48 inches on center.

e. See Section R611.7.1.2 for limitations on maximum spacing of vertical reinforcement in Seismic Design Categories C, D_0, D_1 and D_2.

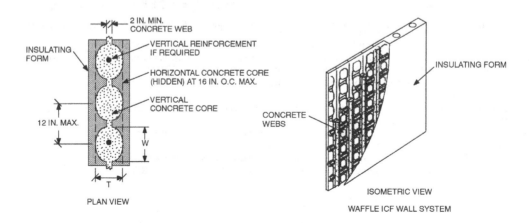

FIGURE R611.4
WAFFLE-GRID ICF WALL SYSTEM

TABLE R611.5
MINIMUM VERTICAL WALL REINFORCEMENT FOR SCREEN-GRID ICF ABOVE-GRADE WALLS[a, b, c]

DESIGN WIND PRESSURE [TABLE R611.3(1)] (psf)	MAXIMUM UNSUPPORTED WALL HEIGHT (feet)	MINIMUM VERTICAL REINFORCEMENT[d,e]		
		Nonload-Bearing Wall or Supporting Roof	Supporting Light-Framed Second Story and Roof	Supporting ICF Second Story and Roof
20	8	#4@48	#4@48	#4@48
	9	#4@48	#4@48	#4@48
	10	#4@48	#4@48	#4@48
30	8	#4@48	#4@48	#4@48
	9	#4@48	#4@48	#4@48
	10	#4@36; #5@48	#4@48	#4@48
40	8	#4@48	#4@48	#4@48
	9	#4@36; #5@48	#4@36; #5@48	#4@48
	10	#4@24; #5@48	#4@24; #5@48	#4@24; #5@48
50	8	#4@36; #5@48	#4@36; #5@48	#4@48
	9	#4@24; #5@48	#4@24; #5@48	#4@24; #5@48
	10	Design Required	Design Required	Design Required
60	8	#4@24; #5@48	#4@24; #5@48	#4@36; #5@48
	9	#4@24; #5@36	#4@24; #5@36	#4@24; #5@36
	10	Design Required	Design Required	Design Required
70	8	#4@24; #5@36	#4@24; #5@36	#4@24; #5@36
	9	Design Required	Design Required	Design Required
	10	Design Required	Design Required	Design Required
080	8	#4@12; #5@36	#4@24; #5@36	#4@24; #5@36
	9	Design Required	Design Required	Design Required
	10	Design Required	Design Required	Design Required

For SI: 1 inch = 25.4 mm, 1 foot = 304.8 mm, 1 mile per hour = 0.447 m/s; 1 pound per square inch = 6.895 kPa.

a. This table is based on reinforcing bars with a minimum yield strength of 40,000 psi and concrete with a minimum specified compressive strength of 2,500 psi. For Seismic Design Categories D_0, D_1 and D_2, reinforcing bars shall have a minimum yield strength of 60,000 psi. See Section R611.6.2.

b. Deflection criterion is $L/240$, where L is the height of the wall story in inches.

c. Interpolation shall not be permitted.

d. Increasing reinforcement spacing by 12 inches shall be permitted when reinforcing steel with a minimum yield strength of 60,000 psi is used. Reinforcement shall not be less than one #4 bar at 48 inches on center.

e. See Section R611.7.1.2 for limitations on maximum spacing of vertical reinforcement in Seismic Design Categories C, D_0, D_1 and D_2.

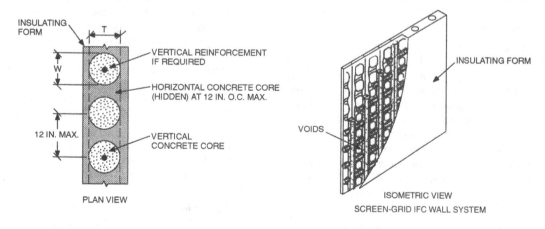

For SI: 1 inch = 25.4 mm.

FIGURE R611.5
SCREEN-GRID IFC WALL SYSTEM

R611.6.3 Insulation materials. Insulating concrete forms material shall meet the surface burning characteristics of Section R314.3. A thermal barrier shall be provided on the building interior in accordance with Section R314.4 or Section R702.3.4.

R611.7 Wall construction. Insulating concrete form walls shall be constructed in accordance with the provisions of this section and Figure R611.7(1).

R611.7.1 Reinforcement.

R611.7.1.1 Location. Vertical and horizontal wall reinforcement shall be placed within the middle third of the wall. Steel reinforcement shall have a minimum concrete cover in accordance with ACI 318.

> **Exception:** Where insulated concrete forms are used and the form remains in place as cover for the concrete, the minimum concrete cover for the reinforcing steel is permitted to be reduced to $^3/_4$ inch (19 mm).

R611.7.1.2 Vertical steel. Above-grade concrete walls shall have reinforcement in accordance with Sections R611.3, R611.4, or R611.5 and R611.7.2. Where the design wind pressure exceeds 40 psf (1.92 kPa) in accordance with Table R611.3(1) or for townhouses in Seismic Design Category C and all buildings in Seismic Design Categories D_0, D_1 and D_2, vertical wall reinforcement in the top-most ICF story shall terminate with a 90-degree (1.57 rad) standard hook in accordance with Section R611.7.1.5. The free end of the hook shall be within 4 inches (102 mm) of the top of the ICF wall and shall be oriented parallel to the horizontal steel in the top of the wall.

For townhouses in Seismic Design Category C, the minimum vertical reinforcement shall be one No. 5 bar at 24 inches (610 mm) on center or one No. 4 at 16 inches (407 mm) on center. For all buildings in Seismic Design Categories D_0, D_1 and D_2, the minimum vertical reinforcement shall be one No. 5 bar at 18 inches (457 mm) on center or one No. 4 at 12 inches (305 mm) on center.

Above-grade ICF walls shall be supported on concrete foundations reinforced as required for the above-grade wall immediately above, or in accordance with Tables R404.4(1) through R404.4(5), whichever requires the greater amount of reinforcement.

Vertical reinforcement shall be continuous from the bottom of the foundation wall to the roof. Lap splices, if required, shall comply with Section R611.7.1.4. Where vertical reinforcement in the above-grade wall is not continuous with the foundation wall reinforcement, dowel bars with a size and spacing to match the vertical ICF wall reinforcement shall be embedded 40 d_b into the foundation wall and shall be lap spliced with the above-grade wall reinforcement. Alternatively, for No. 6 and larger bars, the portion of the bar embedded in the foundation wall shall be embedded 24 inches in the foundation wall and shall have a standard hook.

R611.7.1.3 Horizontal reinforcement. Concrete walls with a minimum thickness of 4 inches (102 mm) shall have a minimum of one continuous No. 4 horizontal reinforcing bar placed at 32 inches (812 mm) on center with

one bar within 12 inches (305 mm) of the top of the wall story. Concrete walls 5.5 inches (140 mm) thick or more shall have a minimum of one continuous No. 4 horizontal reinforcing bar placed at 48 inches (1219 mm) on center with one bar located within 12 inches (305 mm) of the top of the wall story.

For townhouses in Seismic Design Category C, the minimum horizontal reinforcement shall be one No. 5 bar at 24 inches (610 mm) on center or one No. 4 at 16 inches (407 mm) on center. For all buildings in Seismic Design Categories D_0, D_1 and D_2, the minimum horizontal reinforcement shall be one No. 5 bar at 18 inches (457 mm) on center or one No. 4 at 12 inches (305 mm) on center.

Horizontal reinforcement shall be continuous around building corners using corner bars or by bending the bars. In either case, the minimum lap splice shall be 24 inches (610 mm). For townhouses in Seismic Design Category C and for all buildings in Seismic Design Categories D_0, D_1 and D_2, each end of all horizontal reinforcement shall terminate with a standard hook or lap splice.

R611.7.1.4 Lap splices. Where lap splicing of vertical or horizontal reinforcing steel is necessary, the lap splice shall be in accordance with Figure R611.7.1.4 and a minimum of 40 d_b, where d_b is the diameter of the smaller bar. The maximum distance between noncontact parallel bars at a lap splice shall not exceed 8 d_b.

R611.7.1.5 Standard hook. Where the free end of a reinforcing bar is required to have a standard hook, the hook shall be a 180-degree bend plus 4 d_b extension but not less than $2^1/_2$ inches, or a 90-degree bend plus 12 d_b extension.

R611.7.2 Wall openings. Wall openings shall have a minimum of 8 inches (203 mm) of depth of concrete for flat and waffle-grid ICF walls and 12 inches (305 mm) for screen-grid walls over the length of the opening. When the depth of concrete above the opening is less than 12 inches for flat or waffle-grid walls, lintels in accordance with Section R611.7.3 shall be provided. Reinforcement around openings shall be provided in accordance with Table R611.7(1) and Figure R611.7(2). Reinforcement placed horizontally above or below an opening shall extend a minimum of 24 inches (610 mm) beyond the limits of the opening. Wall opening reinforcement shall be provided in addition to the reinforcement required by Sections R611.3, R611.4, R611.5 and R611.7.1. The perimeter of all wall openings shall be framed with a minimum 2-inch by 4-inch plate, anchored to the wall with $^1/_2$-inch (13 mm) diameter anchor bolts spaced a maximum of 24 inches (610 mm) on center. The bolts shall be embedded into the concrete a minimum of 4 inches (102 mm) and have a minimum of $1^1/_2$ inches (38 mm) of concrete cover to the face of the wall.

> **Exception:** The 2-inch by 4-inch plate is not required where the wall is formed to provide solid concrete around the perimeter of the opening with a minimum depth of 4 inches (102 mm) for the full thickness of the wall.

TABLE R611.7(1)
MINIMUM WALL OPENING REINFORCEMENT REQUIREMENTS IN ICF WALLS[a]

WALL TYPE AND OPENING WIDTH (L) (feet)	MINIMUM HORIZONTAL OPENING REINFORCEMENT	MINIMUM VERTICAL OPENING REINFORCEMENT
Flat, Waffle-, and Screen-Grid: $L < 2$	None required	None required
Flat, Waffle-, and Screen-Grid: $L \geq 2$	Provide lintels in accordance with Section R611.7.3. Provide one No. 4 bar within 12 inches from the bottom of the opening. Top and bottom lintel reinforcement shall extend a minimum of 24 inches beyond the limits of the opening.	In locations with wind speeds less than or equal to 110 mph or in Seismic Design Categories A and B, provide one No. 4 bar for the full height of the wall story within 12 inches of each side of the opening. In locations with wind speeds greater than 110 mph, townhouses in Seismic Design Category C, or all buildings in Seismic Design Categories D_0, D_1 and D_2, provide two No. 4 bars or one No. 5 bar for the full height of the wall story within 12 inches of each side of the opening.

For SI: 1 inch = 25.4 mm, 1 foot = 304.8 mm, 1 mile per hour = 0.447 m/s; 1 pound per square inch = 6.895 kPa.

a. This table is based on concrete with a minimum specified compressive strength of 2,500 psi, reinforcing steel with a minimum yield strength of 40,000 psi and an assumed equivalent rectangular cross section. This table is not intended to prohibit the use of ICF manufacturer's tables based on engineering analysis in accordance with ACI 318.

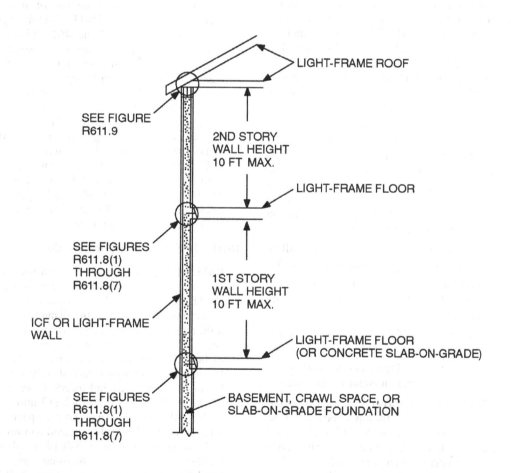

For SI: 1 foot = 304.8 mm.
NOTE: Section cut through flat wall or vertical core of waffle- or screen-grid walls.

FIGURE R611.7(1)
ICF WALL CONSTRUCTION

R611.7.3 Lintels.

R611.7.3.1 General requirements. Lintels shall be provided over all openings greater than or equal to 2 feet (610 mm) in width. Lintels for flat ICF walls shall be constructed in accordance with Figure R611.7(3) and Table R611.7(2) or R611.7(3). Lintels for waffle-grid ICF walls shall be constructed in accordance with Figure R611.7(4) or Figure R611.7(5) and Table R611.7(4) or R611.7(5). Lintels for screen-grid ICF walls shall be constructed in accordance with Figure R611.7(6) or Figure R611.7(7). Lintel construction in accordance with Figure R611.7(3) shall be permitted with waffle-grid and screen-grid ICF wall construction. Lintel depths are permitted to be increased by the height of the ICF wall located directly above the opening, provided that the lintel depth spans the entire length of the opening.

R611.7.3.2 Stirrups. Where required, No. 3 stirrups shall be installed in flat, waffle-grid and screen-grid wall lintels in accordance with the following:

1. For flat walls the stirrups shall be spaced at a maximum spacing of $d/2$ where d equals the depth of the lintel (D) minus the bottom cover of concrete as shown in Figure R611.7(3). Stirrups shall not be required in the middle portion of the span (A) per Figure R611.7(2), for flat walls for a length not to exceed the values shown in parenthesis in Tables R611.7(2) and R611.7(3) or for spans in accordance with Table R611.7(8).

2. For waffle-grid walls a minimum of two No. 3 stirrups shall be placed in each vertical core of waffle-grid lintels. Stirrups shall not be required in the middle portion of the span (A) per Figure R611.7(2), for waffle-grid walls for a length not to exceed the values shown in parenthesis in Tables R611.7(4) and R611.7(5) or for spans in accordance with Table R611.7(8).

3. For screen-grid walls one No. 3 stirrup shall be placed in each vertical core of screen-grid lintels.

    Exception: Stirrups are not required in screen-grid lintels meeting the following requirements:

    1. Lintel Depth (D) = 12 inches (305 mm) - spans less than or equal 3 feet 7 inches.
    2. Lintel Depth (D) = 24 inches (610 mm) - spans less than or equal 4 feet 4 inches.

R611.7.3.3 Horizontal reinforcement. One No. 4 horizontal bar shall be provided in the top of the lintel. Horizontal reinforcement placed within 12 inches (305 mm) of the top of the wall in accordance with Section R611.7.1.3 shall be permitted to serve as the top or bottom reinforcement in the lintel provided the reinforcement meets the location requirements in Figure R611.7(2), R611.7(3), R611.7(4), R611.7(5),

R611.7(6), or R611.7(7), and the size requirements in Tables R611.7(2), R611.7(3), R611.7(4), R611.7(5), R611.7(6), R611.7(7), or R611.7(8).

R611.7.3.4 Load-bearing walls. Lintels in flat ICF load-bearing walls shall comply with Table R611.7(2), Table RR611.7(3) or Table R611.7(8). Lintels in waffle-grid ICF load-bearing walls shall comply with Table R611.7(4), Table R611.7(5) or Table R611.7(8). Lintels in screen-grid ICF load-bearing walls shall comply with Table R611.7(6) or Table R611.7(7).

Where spans larger than those permitted in Table R611.7(2), Table R611.7(3), Table R611.7(4), Table R611.7(5), R611.7(6), R611.7(7) or R611.7(8) are required, the lintels shall comply with Table R611.7 (9).

R611.7.3.5 Nonload-bearing walls. Lintels in nonload-bearing flat, waffle-grid and screen-grid ICF walls shall comply with Table R611.7 (10). Stirrups are not required.

R611.7.4 Minimum length of wall without openings. The wind velocity pressures of Table R611.7.4 shall be used to determine the minimum amount of solid wall length in accordance with Tables R611.7(9A) through R611.7(10B) and Figure R611.7.4. Table R611.7(11) shall be used to determine the minimum amount of solid wall length for townhouses in Seismic Design Category C, and all buildings in Seismic Design Categories D_0, D_1 and D_2 for all types of ICF walls. The greater amount of solid wall length required by wind loading or seismic loading shall apply. The minimum percentage of solid wall length shall include only those solid wall segments that are a minimum of 24 inches (610 mm) in length. The maximum distance between wall segments included in determining solid wall length shall not exceed 18 feet (5486 mm). A minimum length of 24 inches (610 mm) of solid wall segment, extending the full height of each wall story, shall occur at all interior and exterior corners of exterior walls.

R611.8 ICF wall-to-floor connections.

R611.8.1 Top bearing. Floors bearing on the top of ICF foundation walls in accordance with Figure R611.8(1) shall have the wood sill plate anchored to the ICF wall with minimum $^1/_2$-inch (13 mm) diameter bolts embedded a minimum of 7 inches (178 mm) and placed at a maximum spacing of 6 feet (1829 mm) on center and not more than 12 inches (305 mm) from corners. Anchor bolts for waffle-grid and screen-grid walls shall be located in the cores. In conditions where wind speeds are in excess of 90 miles per hour (40 m/s), the $^1/_2$-inch (13 mm) diameter anchor bolts shall be placed at a maximum spacing of 4 feet (1219 mm) on center. Bolts shall extend a minimum of 7 inches (178 mm) into concrete. Sill plates shall be protected against decay where required by Section R319. Cold-formed steel framing systems shall be anchored to the concrete in accordance with Section R505.3.1 or Section R603.3.1.

TABLE R611.7(2)
MAXIMUM ALLOWABLE CLEAR SPANS FOR ICF LINTELS FOR FLAT LOAD-BEARING WALLS[a, b, c, d, f]
NO. 4 BOTTOM BAR SIZE

MINIMUM LINTEL THICKNESS, T (inches)	LINTEL DEPTH, D (inches)	MAXIMUM CLEAR SPAN, (feet-inches) (Number is Middle of Span, A)[e]					
		Supporting Roof Only		Supporting Light-Framed 2nd Story and Roof		Supporting ICF Second Story and Roof	
		Ground Snow Load					
		30 psf	70 psf	30 psf	70 psf	30 psf	70 psf
3.5	8	4-9 (1-2)	4-2 (0-9)	3-10 (0-8)	3-4 (0-6)	3-5 (0-6)	3-1 (0-5)
	12	6-8 (1-11)	5-5 (1-3)	5-0 (1-1)	4-5 (0-10)	4-6 (0-10)	4-0 (0-8)
	16	7-11 (2-9)	6-5 (1-9)	6-0 (1-6)	5-3 (1-2)	5-4 (1-2)	4-10 (1-0)
	20	8-11 (3-5)	7-4 (2-3)	6-9 (1-11)	6-0 (1-6)	6-1 (1-7)	5-6 (1-3)
	24	9-10 (4-1)	8-1 (2-9)	7-6 (2-4)	6-7 (1-10)	6-9 (1-11)	6-1 (1-6)
5.5	8	5-2 (1-10)	4-2 (1-2)	3-10 (1-0)	3-5 (0-9)	3-5 (0-10)	3-1 (0-8)
	12	6-8 (3-0)	5-5 (2-0)	5-0 (1-9)	4-5 (1-4)	4-6 (1-4)	4-1 (1-1)
	16	7-10 (4-1)	6-5 (2-9)	6-0 (2-5)	5-3 (1-10)	5-4 (1-11)	4-10 (1-7)
	20	8-10 (5-3)	7-3 (3-6)	6-9 (3-1)	6-0 (2-4)	6-1 (2-5)	5-6 (2-0)
	24	9-8 (6-3)	8-0 (4-3)	7-5 (3-8)	6-7 (2-11)	6-8 (3-0)	6-0 (2-5)
7.5	8	5-2 (2-6)	4-2 (1-8)	3-11 (1-5)	3-5 (1-1)	3-6 (1-1)	3-2 (0-11)
	12	6-7 (4-0)	5-5 (2-8)	5-0 (2-4)	4-5 (1-10)	4-6 (1-10)	4-1 (1-6)
	16	7-9 (5-5)	6-5 (3-8)	5-11 (3-3)	5-3 (2-6)	5-4 (2-7)	4-10 (2-2)
	20	8-8 (6-10)	7-2 (4-8)	6-8 (4-2)	5-11 (3-3)	6-0 (3-4)	5-5 (2-9)
	24	9-6 (8-2)	7-11 (5-8)	7-4 (5-1)	6-6 (3-11)	6-7 (4-1)	6-0 (3-4)
9.5	8	5-2 (3-1)	4-2 (2-1)	3-11 (1-9)	3-5 (1-5)	3-6 (1-5)	3-2 (1-2)
	12	6-7 (5-0)	5-5 (3-4)	5-0 (3-0)	4-5 (2-4)	4-6 (2-5)	4-1 (1-11)
	16	7-8 (6-9)	6-4 (4-7)	5-11 (4-2)	5-3 (3-3)	5-4 (3-4)	4-10 (2-8)

For SI: 1 inch = 25.4 mm, 1 foot = 304.8 mm, 1 pound per square inch = 6.895 kPa, 1 pound per square foot = 0.0479 kPa.

a. This table is based on concrete with a minimum specified compressive strength of 2,500 psi, reinforcing steel with a minimum yield strength of 40,000 psi and an assumed equivalent rectangular cross section. When reinforcement with a minimum yield strength of 60,000 psi is used, the span lengths in the shaded cells shall be increased by 1.2 times the table values.

b. This table is not intended to prohibit the use of ICF manufacturer's tables based on engineering analysis in accordance with ACI 318.

c. Deflection criterion: L/240.

d. Design load assumptions:

Floor dead load is 10 psf	Attic live load is 20 psf
Floor live load is 30 psf	Roof dead load is 15 psf
Building width is 32 feet	ICF wall dead load is 69 psf
Light-framed wall dead load is 10 psf	

e. No. 3 stirrups are required at d/2 spacing except no stirrups are required for the distance, (A), shown in the middle portion of the span in accordance with Figure R611.7(2) and Section R611.7.3.2.

f. Interpolation is permitted between ground snow loads and between lintel depths.

TABLE R611.7(3)
MAXIMUM ALLOWABLE CLEAR SPANS FOR ICF LINTELS FOR FLAT LOAD-BEARING WALLS[a, b, c, d, f]
NO. 5 BOTTOM BAR SIZE

MINIMUM LINTEL THICKNESS, T (inches)	LINTEL DEPTH, D (inches)	MAXIMUM CLEAR SPAN, (feet-inches) (Number is Middle of Span, A)[e]					
		Supporting Roof		Supporting Light-Framed 2nd Story and Roof		Supporting ICF Second Story and Roof	
		Ground Snow Load					
		30 psf	70 psf	30 psf	70 psf	30 psf	70 psf
3.5	8	4-9 (1-2)	4-2 (0-9)	3-11 (0-8)	3-7 (0-6)	3-7 (0-6)	3-5 (0-5)
	12	7-2 (1-11)	6-3 (1-3)	5-11 (1-1)	5-5 (0-10)	5-5 (0-10)	5-0 (0-8)
	16	9-6 (2-9)	8-0 (1-9)	7-4 (1-6)	6-6 (1-2)	6-7 (1-2)	5-11 (1-0)
	20	11-1 (3-5)	9-1 (2-3)	8-4 (1-11)	7-5 (1-6)	7-6 (1-7)	6-9 (1-3)
	24	12-2 (4-1)	10-0 (2-9)	9-3 (2-4)	8-2 (1-10)	8-4 (1-11)	7-6 (1-6)
5.5	8	5-6 (1-10)	4-10 (1-2)	4-7 (1-0)	4-2 (0-9)	4-2 (0-10)	3-10 (0-8)
	12	8-3 (3-0)	6-9 (2-0)	6-3 (1-9)	5-6 (1-4)	5-7 (1-4)	5-0 (1-1)
	16	9-9 (4-1)	8-0 (2-9)	7-5 (2-5)	6-6 (1-10)	6-7 (1-11)	6-0 (1-7)
	20	10-11 (5-3)	9-0 (3-6)	8-4 (3-1)	7-5 (2-4)	7-6 (2-5)	6-9 (2-0)
	24	12-0 (6-3)	9-11 (4-3)	9-3 (3-8)	8-2 (2-11)	8-3 (3-0)	7-6 (2-5)
7.5	8	6-1 (2-6)	5-2 (1-8)	4-9 (1-5)	4-3 (1-1)	4-3 (1-1)	3-10 (0-11)
	12	8-2 (4-0)	6-9 (2-8)	6-3 (2-4)	5-6 (1-10)	5-7 (1-10)	5-0 (1-6)
	16	9-7 (5-5)	7-11 (3-8)	7-4 (3-3)	6-6 (2-6)	6-7 (2-7)	6-0 (2-2)
	20	10-10 (6-10)	8-11 (4-8)	8-4 (4-2)	7-4 (3-3)	7-6 (3-4)	6-9 (2-9)
	24	11-10 (8-2)	9-10 (5-8)	9-2 (5-1)	8-1 (3-11)	8-3 (4-1)	7-5 (3-4)
9.5	8	6-4 (3-1)	5-2 (2-1)	4-10 (1-9)	4-3 (1-5)	4-4 (1-5)	3-11 (1-2)
	12	8-2 (5-0)	6-8 (3-4)	6-2 (3-0)	5-6 (2-4)	5-7 (2-5)	5-0 (1-11)
	16	9-6 (6-9)	7-11 (4-7)	7-4 (4-2)	6-6 (3-3)	6-7 (3-4)	5-11 (2-8)
	20	10-8 (8-4)	8-10 (5-10)	8-3 (5-4)	7-4 (4-2)	7-5 (4-3)	6-9 (3-6)
	24	11-7 (10-0)	9-9 (6-11)	9-0 (6-5)	8-1 (5-0)	8-2 (5-2)	7-5 (4-3)

For SI: 1 inch = 25.4 mm, 1 foot = 304.8 mm, 1 pound per square inch = 6.895 kPa, 1 pound per square foot = 0.0479 kPa.

a. This table is based on concrete with a minimum specified compressive strength of 2,500 psi, reinforcing steel with a minimum yield strength of 40,000 psi and an assumed equivalent rectangular cross section. When reinforcement with a minimum yield strength of 60,000 psi is used the span lengths in the shaded cells shall be increased by 1.2 times the table values.

b. This table is not intended to prohibit the use of ICF manufacturer's tables based on engineering analysis in accordance with ACI 318.

c. Deflection criterion: $L/240$.

d. Design load assumptions:

Floor dead load is 10 psf	Attic live load is 20 psf
Floor live load is 30 psf	Roof dead load is 15 psf
Building width is 32 feet	ICF wall dead load is 69 psf
Light-framed wall dead load is 10 psf	

e. No. 3 stirrups are required at $d/2$ spacing except no stirrups are required for the distance, (A), shown in the middle portion of the span in accordance with Figure R611.7(2) and Section R611.7.3.2.

f. Interpolation is permitted between ground snow loads and between lintel depths.

2006 INTERNATIONAL RESIDENTIAL CODE®

TABLE R611.7(4)
MAXIMUM ALLOWABLE CLEAR SPANS FOR WAFFLE-GRID ICF WALL LINTELS[a, b, c, d, f]
NO. 4 BOTTOM BAR SIZE

NOMINAL LINTEL THICKNESS $T^{g,h}$ (inches)	LINTEL DEPTH D (inches)	MAXIMUM CLEAR SPAN (feet-inches) (Number is Middle of Span, A)[e]					
		Supporting Roof		Supporting Light-Framed 2nd Story and Roof		Supporting ICF Second Story and Roof	
		Ground Snow Load					
		30 psf	70 psf	30 psf	70 psf	30 psf	70 psf
6	8	5-2 (0-10)	4-2 (0-7)	3-10 (0-6)	3-5 (0-4)	3-6 (0-5)	3-2 (0-4)
	12	6-8 (1-5)	5-5 (0-11)	5-0 (0-9)	4-5 (0-7)	4-7 (0-8)	4-2 (0-6)
	16	7-11 (1-11)	6-6 (1-4)	6-0 (1-1)	5-3 (0-10)	5-6 (0-11)	4-11 (0-9)
	20	8-11 (2-6)	7-4 (1-8)	6-9 (1-5)	6-0 (1-1)	6-3 (1-2)	5-7 (0-11)
	24	9-10 (3-0)	8-1 (2-0)	7-6 (1-9)	6-7 (1-4)	6-10 (1-5)	6-2 (1-2)
8	8	5-2 (0-10)	4-3 (0-7)	3-11 (0-6)	3-5 (0-4)	3-7 (0-5)	3-2 (0-4)
	12	6-8 (1-5)	5-5 (0-11)	5-1 (0-9)	4-5 (0-7)	4-8 (0-8)	4-2 (0-6)
	16	7-10 (1-11)	6-5 (1-4)	6-0 (1-1)	5-3 (0-10)	5-6 (0-11)	4-11 (0-9)
	20	8-10 (2-6)	7-3 (1-8)	6-9 (1-5)	6-0 (1-1)	6-2 (1-2)	5-7 (0-11)
	24	9-8 (3-0)	8-0 (2-0)	7-5 (1-9)	6-7 (1-4)	6-10 (1-5)	6-2 (1-2)

For SI: 1 inch = 25.4 mm, 1 foot = 304.8 mm, 1 psi = 6.895 kPa, 1 psf = 0.0479 kPa.

a. This table is based on concrete with a minimum specified compressive strength of 2,500 psi, reinforcing steel with a minimum yield strength of 40,000 psi and an assumed equivalent rectangular cross section. When reinforcement with a minimum yield strength of 60,000 psi is used the span lengths in the shaded cells shall be increased by 1.2 times the table values.

b. This table is not intended to prohibit the use of ICF manufacturer's tables based on engineering analysis in accordance with ACI 318.

c. Deflection criterion: $L/240$.

d. Design load assumptions:

    Floor dead load is 10 psf   Attic live load is 20 psf
    Floor live load is 30 psf   Roof dead load is 15 psf
    Building width is 32 feet   ICF wall dead load is 55 psf
    Light-framed wall dead load is 10 psf

e. No. 3 stirrups are required at $d/2$ spacing except no stirrups are required for the distance, (A), shown in the middle portion of the span in accordance with Figure R611.7(2) and Section R611.7.3.2.

f. Interpolation is permitted between ground snow loads and between lintel depths.

g. For actual wall lintel width, refer to Table R611.2.

h. Lintel width corresponds to the nominal waffle-grid ICF wall thickness with a minimum thickness of 2 inches.

TABLE R611.7(5)
MAXIMUM ALLOWABLE CLEAR SPANS FOR WAFFLE-GRID ICF WALL LINTELS[a, b, c, d, f]
NO. 5 BOTTOM BAR SIZE

NOMINAL LINTEL THICKNESS, T[g, h] (inches)	LINTEL DEPTH D (inches)	MAXIMUM CLEAR SPAN (feet-inches) (Number is Middle of Span, A)[e]					
		Supporting Roof		Supporting Light-Framed 2nd Story and Roof		Supporting ICF Second Story and Roof	
		Ground Snow Load					
		30 psf	70 psf	30 psf	70 psf	30 psf	70 psf
6	8	5-4 (0-10)	4-8 (0-7)	4-5 (0-6)	4-1 (0-4)	4-5 (0-5)	3-10 (0-4)
	12	8-0 (1-5)	6-9 (0-11)	6-3 (0-9)	5-6 (0-7)	6-3 (0-8)	5-1 (0-6)
	16	9-9 (1-11)	8-0 (1-4)	7-5 (1-1)	6-6 (0-10)	7-5 (0-11)	6-1 (0-9)
	20	11-0 (2-6)	9-1 (1-8)	8-5 (1-5)	7-5 (1-1)	8-5 (1-2)	6-11 (0-11)
	24	12-2 (3-0)	10-0 (2-0)	9-3 (1-9)	8-2 (1-4)	9-3 (1-5)	7-8 (1-2)
8	8	6-0 (0-10)	5-2 (0-7)	4-9 (0-6)	4-3 (0-4)	4-9 (0-5)	3-11 (0-4)
	12	8-3 (1-5)	6-9 (0-11)	6-3 (0-9)	5-6 (0-7)	6-3 (0-8)	5-2 (0-6)
	16	9-9 (1-11)	8-0 (1-4)	7-5 (1-1)	6-6 (0-10)	7-5 (0-11)	6-1 (0-9)
	20	10-11 (2-6)	9-0 (1-8)	8-4 (1-5)	7-5 (1-1)	8-4 (1-2)	6-11 (0-11)
	24	12-0 (3-0)	9-11 (2-0)	9-2 (1-9)	8-2 (1-4)	9-2 (1-5)	7-8 (1-2)

For SI: 1 inch = 25.4 mm, 1 foot = 304.8 mm, 1 psi = 6.895 kPa, 1 psf = 0.0479 kPa.

a. This table is based on concrete with a minimum specified compressive strength of 2,500 psi, reinforcing steel with a minimum yield strength of 40,000 psi and an assumed equivalent rectangular cross section. When reinforcement with a minimum yield strength of 60,000 psi is used the span lengths in the shaded cells shall be increased by 1.2 times the table values.

b. This table is not intended to prohibit the use of ICF manufacturer's tables based on engineering analysis in accordance with ACI 318.

c. Deflection criterion: $L/240$.

d. Design load assumptions:

Floor dead load is 10 psf Attic live load is 20 psf
Floor live load is 30 psf Roof dead load is 15 psf
Building width is 32 feet ICF wall dead load is 53 psf
Light-framed wall dead load is 10 psf

e. No. 3 stirrups are required at $d/2$ spacing except no stirrups are required for the distance, (A), shown in the middle portion of the span in accordance with Figure R611.7(2) and Section R611.7.3.2.

f. Interpolation is permitted between ground snow loads and between lintel depths.

g. For actual wall lintel width, refer to Table R611.2.

h. Lintel width corresponds to the nominal waffle-grid ICF wall thickness with a minimum thickness of 2 inches.

TABLE R611.7(6)
MAXIMUM ALLOWABLE CLEAR SPANS FOR SCREEN-GRID ICF LINTELS IN LOAD-BEARING WALLS[a, b, c, d, e, f, g]
NO. 4 BOTTOM BAR SIZE

MINIMUM LINTEL THICKNESS, T (inches)[h,i]	MINIMUM LINTEL DEPTH, D (inches)	MAXIMUM CLEAR SPAN (feet-inches)					
		Supporting Roof		Supporting Light-Framed Second Story and Roof		Supporting ICF Second Story and Roof	
		Maximum Ground Snow Load (psf)					
		30	70	30	70	30	70
6	12	3-7	2-10	2-5	2-0	2-0	NA
	24	9-10	8-1	7-6	6-7	6-11	6-2

For SI: 1 inch = 25.4 mm, 1 foot = 304.8 mm, 1 psi = 6.895 kPa, 1 psf = 0.0479 kPa.

a. This table is based on concrete with a minimum specified compressive strength of 2,500 psi, reinforcing steel with a minimum yield strength of 40,000 psi and an assumed equivalent rectangular cross section. When reinforcement with a minimum yield strength of 60,000 psi is used the span lengths in the shaded cells shall be increased by 1.2 times the table values.

b. This table is not intended to prohibit the use of ICF manufacturer's tables based on engineering analysis in accordance with ACI 318.

c. Deflection criterion: $L/240$.

d Design load assumptions:
Floor dead load is 10 psf   Attic live load is 20 psf
Floor live load is 30 psf   Roof dead load is 15 psf
Maximum floor clear span is 32 ft   ICF wall dead load is 53 psf
Light-frame wall dead load is 10 psf

e. Stirrup requirements:
Stirrups are not required for lintels 12 inches deep.
One No. 3 stirrup is required in each vertical core for lintels 24 inches deep.

f. Interpolation is permitted between ground snow loads.

g. Flat ICF lintels may be used in lieu of screen-grid lintels.

h. For actual wall lintel width, refer to Table R611.2.

i. Lintel width corresponds to the nominal screen-grid ICF wall thickness.

TABLE R611.7(7)
MAXIMUM ALLOWABLE CLEAR SPANS FOR SCREEN-GRID ICF LINTELS IN LOAD-BEARING WALLS[a, b, c, d, e, f, g]
NO. 5 BOTTOM BAR SIZE

MINIMUM LINTEL THICKNESS, T (inches)[h,i]	MINIMUM LINTEL DEPTH, D (inches)	MAXIMUM CLEAR SPAN (feet-inches)					
		Supporting Roof		Supporting Light-Framed Second Story and Roof		Supporting ICF Second Story and Roof	
		Maximum Ground Snow Load (psf)					
		30	70	30	70	30	70
6	12	3-7	2-10	2-5	2-0	2-0	NA
	24	12-3	10-0	9-3	8-3	8-7	7-8

For SI: 1 inch = 25.4 mm, 1 foot = 304.8 mm, 1 pound per square inch = 6.895 kPa, 1 pound per square foot = 0.0479 kPa.

a. This table is based on concrete with a minimum specified compressive strength of 2,500 psi, reinforcing steel with a minimum yield strength of 40,000 psi and an assumed equivalent rectangular cross section. When reinforcement with a minimum yield strength of 60,000 psi is used the span lengths in the shaded cells shall be increased by 1.2 times the table values.

b. This table is not intended to prohibit the use of ICF manufacturer's tables based on engineering analysis in accordance with ACI 318.

c. Deflection criterion: $L/240$.

d.   Design load assumptions:
Floor dead load is 10 psf   Attic live load is 20 psf
Floor live load is 30 psf   Roof dead load is 15 psf
Maximum floor clear span is 32 ft   ICF wall dead load is 53 psf
Light-frame wall dead load is 10 psf

e. Stirrup requirements:
Stirrups are not required for lintels 12 inches deep.
One No. 3 stirrup is required in each vertical core for lintels 24 inches deep.

f. Interpolation is permitted between ground snow loads.

g. Flat ICF lintels may be used in lieu of screen-grid lintels.

h. For actual wall lintel width, refer to Table R611.2.

i. Lintel width corresponds to the nominal screen-grid ICF wall thickness.

TABLE R611.7(8)
MAXIMUM ALLOWABLE CLEAR SPANS FOR ICF LINTELS WITHOUT STIRRUPS IN LOAD-BEARING WALLS[a, b, c, d, e, f, g, h]
(NO. 4 OR NO. 5) BOTTOM BAR SIZE

MINIMUM LINTEL THICKNESS, T (inches)	MINIMUM LINTEL DEPTH, D (inches)	MAXIMUM CLEAR SPAN (feet-inches)					
		Supporting Roof Only		Supporting Light-Framed Second Story and Roof		Supporting ICF Second Story and Roof	
		MAXIMUM GROUND SNOW LOAD (psf)					
		30	70	30	70	30	70
Flat ICF Lintel							
3.5	8	2-6	2-6	2-6	2-4	2-5	2-2
	12	4-2	4-2	4-1	3-10	3-10	3-7
	16	4-11	4-8	4-6	4-2	4-2	3-11
	20	6-3	5-3	4-11	4-6	4-6	4-3
	24	7-7	6-4	6-0	5-6	5-6	5-2
5.5	8	2-10	2-6	2-6	2-5	2-6	2-2
	12	4-8	4-4	4-3	3-11	3-10	3-7
	16	6-5	5-1	4-8	4-2	4-3	3-11
	20	8-2	6-6	6-0	5-4	5-5	5-0
	24	9-8	7-11	7-4	6-6	6-7	6-1
7.5	8	3-6	2-8	2-7	2-5	2-5	2-2
	12	5-9	4-5	4-4	4-0	3-10	3-7
	16	7-9	6-1	5-7	4-10	4-11	4-5
	20	8-8	7-2	6-8	5-11	6-0	5-5
	24	9-6	7-11	7-4	6-6	6-7	6-0
9.5	8	4-2	3-1	2-9	2-5	2-5	2-2
	12	6-7	5-1	4-7	3-11	4-0	3-7
	16	7-10	6-4	5-11	5-3	5-4	4-10
	20	8-7	7-2	6-8	5-11	6-0	5-5
	24	9-4	7-10	7-3	6-6	6-7	6-0
Waffle-Grid ICF Lintel							
6 or 8	8	2-6	2-6	2-6	2-4	2-4	2-2
	12	4-2	4-2	4-1	3-8	3-9	3-7
	16	5-9	5-8	5-7	5-1	5-2	4-8
	20	7-6	7-4	6-9	6-0	6-3	5-7
	24	9-2	8-1	7-6	6-7	6-10	6-2

For SI: 1 inch = 25.4 mm; 1 foot = 304.8 mm; 1 pound per square foot = 0.0479 kPa; 1 pound per square inch = 6.895 kPa.

a. Table values are based on tensile reinforcement with a minimum yield strength of 40,000 psi (276 MPa), concrete with a minimum specified compressive strength of 2,500 psi, and a building width (clear span) of 32 feet.

b. Spans located in shaded cells shall be permitted to be multiplied by 1.05 when concrete with a minimum compressive strength of 3,000 psi is used or by 1.1 when concrete with a minimum compressive strength of 4,000 psi is used.

c. Deflection criterion is $L/240$, where L is the clear span of the lintel in inches.

d. Linear interpolation shall be permitted between ground snow loads and between lintel depths.

e. Lintel depth, D, shall be permitted to include the available height of ICF wall located directly above the lintel, provided that the increased lintel depth spans the entire length of the opening.

f. Spans shall be permitted to be multiplied by 1.05 for a building width (clear span) of 28 feet.

g. Spans shall be permitted to be multiplied by 1.1 for a building width (clear span) of 24 feet or less.

h. ICF wall dead load is 69 psf.

TABLE R611.7(9)
MINIMUM BOTTOM BAR ICF LINTEL REINFORCEMENT FOR LARGE CLEAR SPANS IN LOAD-BEARING WALLS[a,b,c,d,e,f,h]

MINIMUM LINTEL THICKNESS, T[g] (inches)	MINIMUM LINTEL DEPTH, D (inches)	MINIMUM BOTTOM LINTEL REINFORCEMENT					
		Supporting Light-Frame Roof Only		Supporting Light-Framed Second Story and Roof		Supporting ICF Second Story and Light-Frame Roof	
		Maximum Ground Snow Load (psf)					
		30	70	30	70	30	70
Flat ICF Lintel, 12 feet- 3 inches Maximum Clear Span							
3.5	24	1 #5	1 #7	D/R	D/R	D/R	D/R
5.5	20	1 #6	1 #7	D/R	D/R	D/R	D/R
	24	1 #5	1 #7	1 #7	1 #8	1 #8	D/R
7.5	16	1 #7; 2 #5	D/R	D/R	D/R	D/R	D/R
	20	1 #6; 2 #4	1#7; 2 #5	1 #8; 2 #6	D/R	D/R	D/R
	24	1 #6; 2 #4	1 #7; 2 #5	1 #7; 2 #5	1 #8; 2 #6	1 #8; 2 #6	1 #8; 2 #6
9.5	16	1 #7; 2 #5	D/R	D/R	D/R	D/R	D/R
	20	1 #6; 2 #4	1 #7; 2 #5	1 #8; 2 #6	1 #8; 2 #6	1 #8; 2 #6	1 #9; 2 #6
	24	1 #6; 2 #4	1 #7; 2 #5	1 #7; 2 #5	1 #7; 2 #6	1 #8; 2 #6	1 #9; 2 #6
Flat ICF Lintel, 16 feet-3 inches Maximum Clear Span							
5.5	24	1 #7	D/R	D/R	D/R	D/R	D/R
7.5	24	1 #7; 2 #5	D/R	D/R	D/R	D/R	D/R
9.5	24	1 #7; 2 #5	1 #9; 2 #6	1 #9; 2 #6	D/R	D/R	D/R
Waffle-Grid ICF Lintel, 12 feet-3 inches Maximum Clear Span							
6	20	1 #6	D/R	D/R	D/R	D/R	D/R
	24	1 #5	1 #7; 2 #5	1 #7; 2 #5	1 #8; 2 #6	1 #8; 2 #6	D/R
8	16	1 #7; 2 #5	D/R	D/R	D/R	D/R	D/R
	20	1 #6; 2 #4	1 #7; 2 #5	1 #8; 2 #6	D/R	D/R	D/R
	24	1 #5	1 #7; 2 #5	1 #7; 2 #5	1 #8; 2 #6	1 #8; 2 #6	1 #8; 2 #6
Screen-Grid ICF Lintel, 12 feet-3 inches Maximum Clear Span							
6	24	1 #5	1 #7	D/R	D/R	D/R	D/R

For SI: 1 inch = 25.4 mm, 1 foot = 304.8 mm, 1 psi = 6.895 kPa, 1 psf = 0.0479 kPa.

a. This table is based on concrete with a minimum specified compressive strength of 2,500 psi, reinforcing steel with a minimum yield strength of 40,000 psi and an assumed equivalent rectangular cross section. When reinforcement with a minimum yield strength of 60,000 psi is used the span lengths in the shaded cells shall be increased by 1.2 times the table values.

b. This table is not intended to prohibit the use of ICF manufacturers tables based on engineering analysis in accordance with ACI 318.

c. D/R indicates design is required.

d. Deflection criterion: $L/240$.

e. Interpolation is permitted between ground snow loads and between lintel depths.

f. No. 3 stirrups are required a maximum d/2 spacing for spans greater than 4 feet.

g. Actual thickness is shown for flat lintels; nominal thickness is given for waffle-grid and screen-grid lintels. Lintel thickness corresponds to the nominal waffle-grid and screen-grid ICF wall thickness. Refer to Table R611.2 for actual wall thickness.

h. ICF wall dead load varies based on wall thickness using 150 pcf concrete density.

TABLE R611.7(9A)
MINIMUM SOLID END WALL LENGTH REQUIREMENTS FOR FLAT ICF WALLS (WIND PERPENDICULAR TO RIDGE)[a, b, c]

WALL CATEGORY	BUILDING SIDE WALL LENGTH, L (feet)	Roof Slope	WIND VELOCITY PRESSURE FROM TABLE R611.7.4 (psf)							
			20	25	30	35	40	45	50	60
			Minimum Solid Wall Length on Building End Wall (feet)							
One-Story or Top Story of Two-Story	16	≤ 1:12	4.00	4.00	4.00	4.00	4.00	4.00	4.00	4.00
		5:12	4.00	4.00	4.00	4.00	4.00	4.00	4.25	4.50
		7:12[d]	4.00	4.25	4.25	4.50	4.75	4.75	5.00	5.50
		12:12[d]	4.25	4.50	4.75	5.00	5.25	5.50	5.75	6.25
	24	≤ 1:12	4.00	4.00	4.00	4.00	4.00	4.00	4.25	4.50
		5:12	4.00	4.00	4.00	4.25	4.25	4.50	4.50	4.75
		7:12[d]	4.25	4.50	4.75	5.00	5.25	5.50	5.75	6.25
		12:12[d]	4.75	5.00	5.25	5.75	6.00	6.50	6.75	7.50
	32	≤ 1:12	4.00	4.00	4.00	4.00	4.25	4.25	4.50	4.75
		5:12	4.00	4.00	4.25	4.50	4.50	4.75	5.00	5.25
		7:12[d]	4.50	5.00	5.25	5.50	6.00	6.25	6.50	7.25
		12:12[d]	5.00	5.50	6.00	6.50	7.00	7.25	7.75	8.75
	40	≤ 1:12	4.00	4.00	4.25	4.25	4.50	4.50	4.75	5.00
		5:12	4.00	4.25	4.50	4.75	4.75	5.00	5.25	5.50
		7:12[d]	4.75	5.25	5.75	6.00	6.50	7.00	7.25	8.00
		12:12[d]	5.50	6.00	6.50	7.25	7.75	8.25	8.75	10.0
	50	≤ 1:12	4.00	4.25	4.25	4.50	4.75	4.75	5.00	5.50
		5:12	4.25	4.50	4.75	5.00	5.25	5.50	5.75	6.00
		7:12[d]	5.25	5.75	6.25	6.75	7.25	7.75	8.25	9.25
		12:12[d]	6.00	6.75	7.50	8.00	8.75	9.50	10.25	11.5
	60	≤ 1:12	4.00	4.25	4.50	4.75	5.00	5.25	5.25	5.75
		5:12	4.50	4.75	5.00	5.25	5.50	5.75	6.00	6.75
		7:12[d]	5.50	6.25	6.75	7.50	8.00	8.50	9.25	10.25
		12:12[d]	6.50	7.25	8.25	9.00	9.75	10.5	11.5	13.0
First Story of Two-Story	16	≤ 1:12	4.00	4.25	4.50	4.75	5.00	5.25	5.25	5.75
		5:12	4.50	4.75	5.00	5.25	5.50	5.75	6.00	6.75
		7:12[d]	4.50	5.00	5.25	5.75	6.00	6.25	6.75	7.25
		12:12[d]	5.00	5.25	5.75	6.25	6.50	7.00	7.25	8.25
	24	≤ 1:12	4.50	4.75	5.00	5.25	5.50	5.75	6.00	6.75
		5:12	4.75	5.25	5.50	6.00	6.25	6.75	7.00	7.75
		7:12[d]	5.25	5.75	6.25	6.75	7.00	7.50	8.00	9.00
		12:12[d]	5.50	6.25	6.75	7.25	8.00	8.50	9.00	10.25
	32	≤ 1:12	4.75	5.00	5.50	5.75	6.25	6.50	6.75	7.50
		5:12	5.25	5.75	6.25	6.75	7.25	7.50	8.00	9.00
		7:12[d]	5.75	6.50	7.00	7.75	8.25	9.00	9.50	10.75
		12:12[d]	6.25	7.00	7.75	8.50	9.25	10.0	10.75	12.25
	40	≤ 1:12	5.00	5.50	5.75	6.25	6.75	7.25	7.50	8.50
		5:12	5.50	6.25	6.75	7.25	8.00	8.50	9.00	10.25
		7:12[d]	6.25	7.00	7.75	8.75	9.50	10.25	11.0	12.5
		12:12[d]	7.00	8.00	8.75	9.75	10.75	11.5	12.5	14.25
	50	≤ 1:12	5.50	6.00	6.50	7.00	7.50	8.00	8.50	9.50
		5:12	6.00	6.75	7.50	8.25	9.00	9.75	10.5	11.75
		7:12[d]	7.00	8.00	9.00	10.0	10.75	11.75	12.75	14.5
		12:12[d]	7.75	9.00	10.0	11.25	12.25	13.50	14.75	17.0
	60	≤ 1:12	5.75	6.50	7.00	7.50	8.25	8.75	9.50	10.75
		5:12	6.75	7.50	8.25	9.25	10.0	10.75	11.75	13.25
		7:12[d]	7.75	9.00	10.0	11.0	12.25	13.25	14.5	16.75
		12:12[d]	8.75	10.0	11.5	12.75	14.0	15.5	16.75	19.5

(continued)

Footnotes to Table R611.7 (9A)

For SI: 1 foot = 304.8 mm; 1 inch = 25.4 mm; 1 pound per square foot = 0.0479 kPa.

a. Table values are based on a 3.5 in thick flat wall. For a 5.5 in thick flat wall, multiply the table values by 0.9. The adjusted values shall not result in solid wall lengths less than 4 ft.

b. Table values are based on a maximum unsupported wall height of 10 ft.

c. Linear interpolation shall be permitted.

d. The minimum solid wall lengths shown in the table are based on a building with an end wall length "W" of 60 feet and a roof slope of less than 7:12. For roof slopes of 7:12 or greater and end wall length "W" greater than 30 feet, the minimum solid wall length determined from the table shall be multiplied by:

$1 + 0.4[(W-30)/30]$.

TABLE R611.7(9B)
MINIMUM SOLID SIDEWALL LENGTH REQUIREMENTS FOR FLAT ICF WALLS (WIND PARALLEL TO RIDGE) [a, b, c, d]

WALL CATEGORY	BUILDING END WALL WIDTH, W (feet)	WIND VELOCITY PRESSURE FROM TABLE R611.7.4 (psf)							
		20	25	30	35	40	45	50	60
		Minimum Solid Wall Length on Building Side Wall (feet)							
One-Story or Top Story of Two-Story	16	4.00	4.00	4.00	4.00	4.25	4.25	4.50	4.75
	24	4.00	4.25	4.50	4.75	4.75	5.00	5.25	5.50
	32	4.50	4.75	5.00	5.25	5.50	6.00	6.25	6.75
	40	5.00	5.50	5.75	6.25	6.75	7.00	7.50	8.25
	50	5.75	6.25	7.00	7.50	8.25	8.75	9.50	10.75
	60	6.50	7.50	8.25	9.25	10.0	10.75	11.75	13.25
First Story of Two-Story	16	4.25	4.50	4.75	5.00	5.25	5.50	5.75	6.50
	24	4.75	5.25	5.50	6.00	6.25	6.75	7.00	8.00
	32	5.50	6.00	6.50	7.00	7.50	8.00	8.75	9.75
	40	6.25	7.00	7.50	8.25	9.00	9.75	10.5	12.0
	50	7.25	8.25	9.25	10.25	11.25	12.25	13.25	15.25
	60	8.50	9.75	11.0	12.25	13.5	15.0	16.25	18.75

For SI: 1 foot = 304.8 mm; 1 inch = 25.4 mm; 1 pound per square foot = 0.0479 kPa.

a. Table values are based on a 3.5 in thick flat wall. For a 5.5 in thick flat wall, multiply the table values by 0.9. The adjusted values shall not result in solid wall lengths less than 4 ft.

b. Table values are based on a maximum unsupported wall height of 10 ft.

c. Table values are based on a maximum 12:12 roof pitch.

d. Linear interpolation shall be permitted.

TABLE R611.7(10)
MAXIMUM ALLOWABLE CLEAR SPANS FOR ICF LINTELS IN NONLOAD-BEARING WALLS WITHOUT STIRRUPS[a,b,c,d]
NO. 4 BOTTOM BAR

MINIMUM LINTEL THICKNESS, T (inches)	MINIMUM LINTEL DEPTH, D (inches)	MAXIMUM CLEAR SPAN	
		Supporting Light-Framed Nonbearing Wall (feet-inches)	Supporting ICF Second Story and Nonbearing Wall (feet-inches)
Flat ICF Lintel			
3.5	8	11-1	3-1
	12	15-11	5-1
	16	16-3	6-11
	20	16-3	8-8
	24	16-3	10-5
5.5	8	16-3	4-4
	12	16-3	7-0
	16	16-3	9-7
	20	16-3	12-0
	24	16-3	14-3
7.5	8	16-3	5-6
	12	16-3	8-11
	16	16-3	12-2
	20	16-3	15-3
	24	16-3	16-3
9.5	8	16-3	6-9
	12	16-3	10-11
	16	16-3	14-10
	20	16-3	16-3
	24	16-3	16-3
Waffle-Grid ICF Lintel			
6 or 8	8	9-1	2-11
	12	13-4	4-10
	16	16-3	6-7
	20	16-3	8-4
	24	16-3	9-11
Screen-Grid Lintel			
6	12	5-8	4-1
	24	16-3	9-1

For SI: 1 foot = 304.8 mm; 1 inch = 25.4 mm; 1 pounds per square foot = 0.0479 kPa.

a. This table is based on concrete with a minimum specified compressive strength of 2,500 psi, reinforcing steel with a minimum yield strength of 40,000 psi and an assumed equivalent rectangular cross section.

b. This table is not intended to prohibit the use of ICF manufacturers tables based on engineering analysis in accordance with ACI 318.

c. Deflection criterion is $L/240$, where L is the clear span of the lintel in inches.

d. Linear interpolation is permitted between lintel depths.

TABLE R611.7(10A)
MINIMUM SOLID END WALL LENGTH REQUIREMENTS FOR WAFFLE AND
SCREEN-GRID ICF WALLS (WIND PERPENDICULAR TO RIDGE)[a, b, c]

WALL CATEGORY	BUILDING SIDE WALL LENGTH, L (feet)	ROOF SLOPE	WIND VELOCITY PRESSURE FROM TABLE R611.7.4							
			20	25	30	35	40	45	50	60
			Minimum Solid Wall Length on Building End Wall (feet)							
One-Story or Top Story of Two-Story	16	≤1:12	4.00	4.00	4.00	4.00	4.00	4.00	4.00	4.25
		5:12	4.00	4.00	4.00	4.00	4.00	4.25	4.25	4.50
		7:12[d]	4.00	4.25	4.50	4.75	5.00	5.25	5.50	6.00
		12:12[d]	4.25	4.75	5.00	5.50	5.75	6.00	6.50	7.00
	24	≤1:12	4.00	4.00	4.00	4.00	4.00	4.25	4.25	4.50
		5:12	4.00	4.00	4.00	4.25	4.50	4.50	4.75	5.00
		7:12[d]	4.50	4.75	5.00	5.50	5.75	6.25	6.50	7.25
		12:12[d]	5.00	5.50	6.00	6.50	7.00	7.25	7.75	8.75
	32	≤1:12	4.00	4.00	4.00	4.25	4.25	4.50	4.75	5.00
		5:12	4.00	4.00	4.25	4.50	4.75	5.00	5.25	5.75
		7:12[d]	4.75	5.25	5.75	6.25	6.50	7.00	7.50	8.50
		12:12[d]	5.50	6.25	6.75	7.50	8.00	8.75	9.25	10.5
	40	≤1:12	4.00	4.00	4.25	4.50	4.50	4.75	5.00	5.50
		5:12	4.00	4.25	4.50	5.00	5.25	5.50	5.75	6.25
		7:12[d]	5.25	5.75	6.25	7.00	7.50	8.00	8.50	9.75
		12:12[d]	6.00	6.75	7.75	8.50	9.25	10.0	10.75	12.25
	50	≤1:12	4.00	4.25	4.50	4.75	5.00	5.25	5.50	6.00
		5:12	4.25	4.75	5.00	5.25	5.50	6.00	6.25	7.00
		7:12[d]	5.75	6.50	7.00	7.75	8.50	9.25	9.75	11.25
		12:12[d]	6.75	7.75	8.75	9.75	10.75	11.5	12.5	14.5
	60	≤1:12	4.25	4.50	4.75	5.00	5.25	5.50	5.75	6.50
		5:12	4.50	5.00	5.25	5.75	6.00	6.50	6.75	7.75
		7:12[d]	6.25	7.00	8.00	8.75	9.50	10.25	11.25	12.75
		12:12[d]	7.50	8.75	9.75	11.0	12.0	13.25	14.25	16.5
First Story of Two-Story	16	≤1:12	4.25	4.50	4.75	5.00	5.25	5.50	5.75	6.50
		5:12	4.50	5.00	5.25	5.75	6.00	6.50	6.75	7.75
		7:12[d]	4.75	5.25	5.75	6.25	6.75	7.25	7.75	8.75
		12:12[d]	5.25	5.75	6.50	7.00	7.50	8.00	8.75	9.75
	24	≤1:12	4.50	5.00	5.25	5.75	6.25	6.50	7.00	7.75
		5:12	5.00	5.75	6.25	6.75	7.25	7.75	8.25	9.25
		7:12[d]	5.75	6.25	7.00	7.75	8.25	9.00	9.75	11.0
		12:12[d]	6.25	7.00	7.75	8.50	9.50	10.25	11.0	12.75
	32	≤1:12	5.00	5.50	6.00	6.50	7.00	7.50	8.00	9.00
		5:12	5.75	6.25	7.00	7.75	8.25	9.00	9.75	11.0
		7:12[d]	6.50	7.25	8.25	9.00	10.0	10.75	11.75	13.5
		12:12[d]	7.25	8.25	9.25	10.25	11.25	12.5	13.5	15.5
	40	≤1:12	5.50	6.00	6.50	7.25	7.75	8.50	9.00	10.25
		5:12	6.25	7.00	7.75	8.75	9.50	10.25	11.0	12.75
		7:12[d]	7.25	8.25	9.25	10.5	11.5	12.5	13.75	15.75
		12:12[d]	8.00	9.50	10.75	12.0	13.25	14.5	15.75	18.25

(continued)

TABLE R611.7(10A)—continued
MINIMUM SOLID END WALL LENGTH REQUIREMENTS FOR WAFFLE AND
SCREEN-GRID ICF WALLS (WIND PERPENDICULAR TO RIDGE)[a, b, c]

WALL CATEGORY	BUILDING SIDE WALL LENGTH, L (feet)	ROOF SLOPE	WIND VELOCITY PRESSURE FROM TABLE R611.7.4							
			20	25	30	35	40	45	50	60
			Minimum Solid Wall Length on Building End Wall (feet)							
First Story of Two-Story	50	≤ 1:12	6.00	6.75	7.50	8.00	8.75	9.50	10.25	11.75
		5:12	7.00	8.00	9.00	10.0	11.0	12.0	13.0	14.75
		7:12[d]	8.25	9.50	10.75	12.25	13.5	14.75	16.0	18.75
		12:12[d]	9.25	11.0	12.5	14.0	15.5	17.25	18.75	22.0
	60	≤ 1:12	6.50	7.25	8.25	9.00	10.0	10.75	11.75	13.25
		5:12	7.75	8.75	10.0	11.25	12.25	13.5	14.75	17.0
		7:12[d]	9.25	10.75	12.25	14.0	15.5	17.0	18.5	21.75
		12:12[d]	10.5	12.25	14.25	16.25	18.0	20.0	21.75	25.5

For SI: 1 foot = 304.8 mm; 1 inch = 25.4 mm; 1 pound per square foot = 0.0479kPa.

a. Table values are based on a 6 in (152.4 mm) thick nominal waffle-grid wall. For a 8 in thick nominal waffle-grid wall, multiply the table values by 0.90.

b. Table values are based on a maximum unsupported wall height of 10 ft.

c. Linear interpolation is permitted.

d. The minimum solid wall lengths shown in the table are based on a building with an end wall length "W" of 60 feet and a roof slope of less than 7:12. For roof slopes of 7:12 or greater and end wall length "W" greater than 30 feet, the minimum solid wall length determined from the table shall be multiplied by: 1 + 0.4 [(W-30)/30].

TABLE R611.7(10B)
MINIMUM SOLID SIDE WALL LENGTH REQUIREMENTS FOR 6-INCH WAFFLE AND
SCREEN-GRID ICF WALLS (WIND PARALLEL TO RIDGE)[a, b, c, d]

WALL CATEGORY	BUILDING END WALL WIDTH, W (feet)	WIND VELOCITY PRESSURE FROM TABLE R611.7.4 (psf)							
		20	25	30	35	40	45	50	60
		Minimum Solid Wall Length on Building Side Wall (feet)							
One-Story or Top Story of Two-Story	16	4.00	4.00	4.00	4.25	4.25	4.50	4.75	5.00
	24	4.00	4.25	4.50	5.00	5.25	5.50	5.75	6.25
	32	4.50	5.00	5.50	5.75	6.25	6.75	7.00	8.00
	40	5.25	6.00	6.50	7.00	7.75	8.25	8.75	10.0
	50	6.50	7.25	8.00	9.00	9.75	10.75	11.5	13.25
	60	7.75	8.75	10.0	11.25	12.25	13.5	14.5	17.0
First Story of Two-Story	16	4.50	4.75	5.25	5.50	5.75	6.25	6.50	7.25
	24	5.00	5.75	6.25	6.75	7.25	7.75	8.25	9.50
	32	6.00	6.75	7.50	8.25	9.00	9.75	10.5	12.0
	40	7.00	8.00	9.00	10.0	11.0	12.0	13.0	15.0
	50	8.50	9.75	11.25	12.5	14.0	15.25	16.75	19.5
	60	10.25	12.0	13.75	15.5	17.25	19.0	21.0	24.5

For SI: 1 foot = 304.8 mm; 1 inch = 25.4 mm; 1 pound per square foot = 0.0479kPa.

a. Table values are based on a 6 in thick nominal waffle-grid wall. For a 8 in thick nominal waffle-grid wall, multiply the table values by 0.90.

b. Table values are based on a maximum unsupported wall height of 10 ft.

c. Table values are based on a maximum 12:12 roof pitch.

d. Linear interpolation shall be permitted.

TABLE R611.7(11)
MINIMUM PERCENTAGE OF SOLID WALL LENGTH ALONG EXTERIOR WALL LINES FOR TOWNHOUSES IN SEISMIC DESIGN
CATEGORY C AND ALL BUILDINGS IN SEISMIC DESIGN CATEGORIES D_0, D_1 AND D_2[a, b]

SEISMIC DESIGN CATEGORY (SDC)	MINIMUM SOLID WALL LENGTH (percent)		
	One-Story or Top Story of Two-Story	Wall Supporting Light-Framed Second Story and Roof	Wall Supporting ICF Second Story and Roof
Townhouses in SDC C[c]	20 percent	25 percent	35 percent
D_0[d] or D_1[d]	25 percent	30 percent	40 percent
D_2[d]	30 percent	35 percent	45 percent

For SI: 1 inch = 25.4 mm; 1 mile per hour = 0.447 m/s.

a. Base percentages are applicable for maximum unsupported wall height of 10-feet, light-frame gable construction, and all ICF wall types. These percentages assume that the maximum weight of the interior and exterior wall finishes applied to ICF walls do not exceed 8 psf.

b. For all walls, the minimum required length of solid walls shall be based on the table percent value multiplied by the minimum dimension of a rectangle inscribing the overall building plan.

c. Walls shall be reinforced with a minimum No. 5 bar (Grade 40 or 60) spaced a maximum of 24 inches on center each way or a No. 4 bar spaced a maximum of 16 inches on center each way. (Grade 40 or 60) spaced at a maximum of 16 inches on center each way.

d. Walls shall be constructed with a minimum concrete compressive strength of 3,000 psi and reinforced with minimum #5 rebar (Grade 60 ASTM A 706) spaced a maximum of 18 inches on center each way or No. 4 rebar (Grade 60 ASTM A706) spaced at a maximum of 12 inches (304.8 mm) on center each way. The minimum thickness of flat ICF walls shall be 5.5 inches.

TABLE R611.7.4
WIND VELOCITY PRESSURE FOR DETERMINATION OF
MINIMUM SOLID WALL LENGTH[a]

WIND SPEED (mph)[d]	VELOCITY PRESSURE (psf)		
	Exposure[b]		
	B	C	D
85	14	19	23
90	16	21	25
100	19	26	31
110	23	32	37
120	27	38	44
130	32	44	52
140	37	51	60
150	43	59	69[c]

For SI: 1 pound per square foot = 0.0479 kPa; 1 mile per hour = 0.447 m/s.

a. Table values are based on ASCE 7-98 Figure 6-4 using a mean roof height of 35 ft.

b. Exposure Categories shall be determined in accordance with Section R301.2.1.4.

c. Design is required in accordance with ACI 318 and approved manufacturer guidelines.

d. Interpolation is permitted between wind speeds.

R611.8.1.1 Top bearing requirements for Seismic Design Categories C, D_0, D_1 and D_2. For townhouses in Seismic Design Category C, wood sill plates attached to ICF walls shall be anchored with Grade A 307, $^3/_8$-inch-diameter (10 mm) headed anchor bolts embedded a minimum of 7 inches (178 mm) and placed at a maximum spacing of 36 inches (914 mm) on center. For all buildings in Seismic Design Category D_0 or D_1, wood sill plates attached to ICF walls shall be anchored with ASTM A 307, Grade A, $^3/_8$-inch-diameter (10 mm) headed anchor bolts embedded a minimum of 7 inches (178 mm) and placed at a maximum spacing of 24 inches (610 mm) on center. For all buildings in Seismic Design Category D_2, wood sill plates attached to ICF walls shall

be anchored with ASTM A 307, Grade A, $^3/_8$-inch-diameter (10 mm) headed anchor bolts embedded a minimum of 7 inches (178 mm) and placed at a maximum spacing of 16 inches (406 mm) on center. Larger diameter bolts than specified herein shall not be used.

For townhouses in Seismic Design Category C, each floor joist perpendicular to an ICF wall shall be attached to the sill plate with an 18-gage [(0.0478 in.) (1.2 mm)] angle bracket using 3 - 8d common nails per leg in accordance with Figure R611.8(1). For all buildings in Seismic Design Category D_0 or D_1, each floor joist perpendicular to an ICF wall shall be attached to the sill plate with an 18-gage [(0.0478 in.) (1.2 mm)] angle bracket using 4 - 8d common nails per leg in accordance with Figure R611.8(1). For all buildings in Seismic Design Category D_2, each floor joist perpendicular to an ICF wall shall be attached to the sill plate with an 18-gage [(0.0478 in.) (1.2 mm)] angle bracket using 6 - 8d common nails per leg in accordance with Figure R611.8(1).

For ICF walls parallel to floor framing in townhouses in Seismic Design Category C, full depth blocking shall be placed at 24 inches (610 mm) on center and shall be attached to the sill plate with an 18-gage [(0.0478 in.) (1.2 mm)] angle bracket using 5 - 8d common nails per leg in accordance with Figure R611.8(6). For ICF walls parallel to floor framing for all buildings in Seismic Design Category D_0 or D_1, full depth blocking shall be placed at 24 inches (610 mm) on center and shall be attached to the sill plate with an 18-gage [(0.0478 in.) (1.2 mm)] angle bracket using 6 - 8d common nails per leg in accordance with Figure R611.8(6). For ICF walls parallel to floor framing for all buildings in Seismic Design Category D_2, full depth blocking shall be placed at 24 inches (610 mm) on center and shall be attached to the sill plate with an 18-gage [(0.0478 in.) (1.2 mm)] angle bracket using 9 - 8d common nails per leg in accordance with Figure R611.8(6).

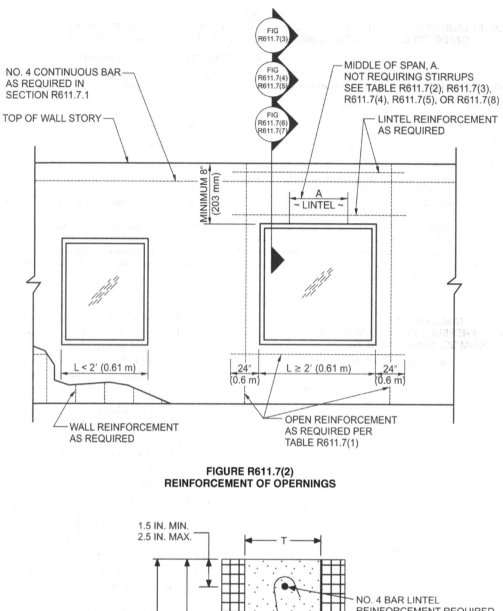

NO. 4 CONTINUOUS BAR
AS REQUIRED IN
SECTION R611.7.1

TOP OF WALL STORY

MIDDLE OF SPAN, A.
NOT REQUIRING STIRRUPS
SEE TABLE R611.7(2), R611.7(3),
R611.7(4), R611.7(5), OR R611.7(8)

LINTEL REINFORCEMENT
AS REQUIRED

FIG R611.7(3)
FIG R611.7(4) R611.7(5)
FIG R611.7(6) R611.7(7)

MINIMUM 8" (203 mm)

A ~ LINTEL ~

L < 2' (0.61 m)

24" (0.6 m)   L ≥ 2' (0.61 m)   24" (0.6 m)

WALL REINFORCEMENT
AS REQUIRED

OPEN REINFORCEMENT
AS REQUIRED PER
TABLE R611.7(1)

FIGURE R611.7(2)
REINFORCEMENT OF OPERNINGS

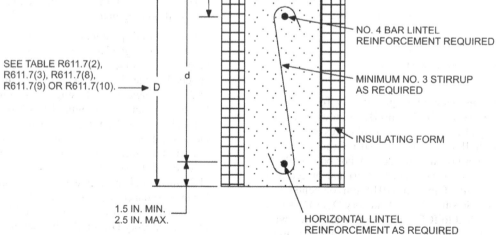

1.5 IN. MIN.
2.5 IN. MAX.

T

NO. 4 BAR LINTEL
REINFORCEMENT REQUIRED

SEE TABLE R611.7(2),
R611.7(3), R611.7(8),
R611.7(9) OR R611.7(10).

D   d

MINIMUM NO. 3 STIRRUP
AS REQUIRED

INSULATING FORM

1.5 IN. MIN.
2.5 IN. MAX.

HORIZONTAL LINTEL
REINFORCEMENT AS REQUIRED

For SI: 1 inch = 25.4 mm.
NOTE: Section cut through flat wall.

FIGURE R611.7(3)
ICF LINTELS FOR FLAT AND SCREEN-GRID WALLS

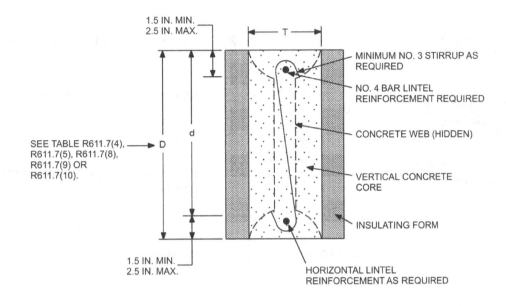

For SI: 1 inch = 25.4 mm.
NOTE: Section cut through vertical core of a waffle-grid lintel.

FIGURE R611.7(4)
SINGLE FORM HEIGHT WAFFLE-GRID LINTEL

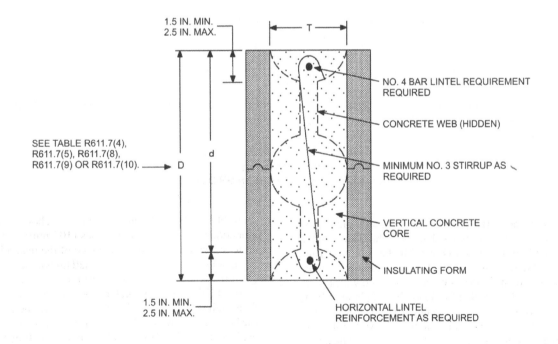

For SI: 1 inch = 25.4 mm.
NOTE: Section cut through vertical core of a waffle-grid lintel.

FIGURE R611.7(5)
DOUBLE FORM HEIGHT WAFFLE-GRID LINTEL

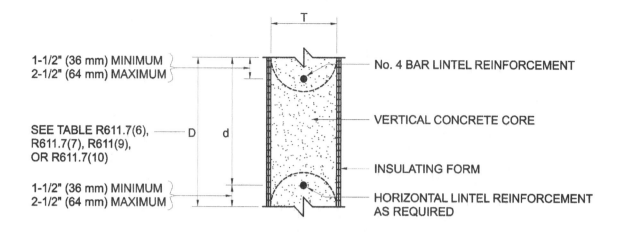

FIGURE R611.7(6)
SINGLE FORM HEIGHT SCREEN-GRID LINTEL

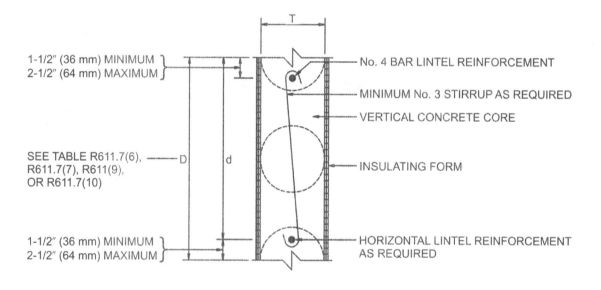

FIGURE R611.7(7)
DOUBLE FORM HEIGHT SCREEN-GRID LINTEL

R611.8.2 Ledger bearing. Wood ledger boards supporting bearing ends of joists or trusses shall be anchored to flat ICF walls with minimum thickness of 5.5 inches (140 mm) and to waffle- or screen-grid ICF walls with minimum nominal thickness of 6 inches (152 mm) in accordance with Figure R611.8(2), R611.8(3), R611.8(4) or R611.8(5) and Table R611.8(1). Wood ledger boards supporting bearing ends of joists or trusses shall be anchored to flat ICF walls with minimum thickness of 3.5 inches (140 mm) in accordance with Figure R611.8(5) and Table R611.8(1). The ledger shall be a minimum 2 by 8, No. 2 Southern Yellow Pine or No. 2 Douglas Fir. Ledgers anchored to nonload-bearing walls to support floor or roof sheathing shall be attached with $^{1}/_{2}$ inch (12.7 mm) diameter or headed anchor bolts spaced a maxi-

mum of 6 feet (1829 mm) on center. Anchor bolts shall be embedded a minimum of 4 inches (102 mm) into the concrete measured from the inside face of the insulating form. For insulating forms with a face shell thickness of 1.5 inches (38 mm) or less, the hole in the form shall be a minimum of 4 inches (102 mm) in diameter. For insulating forms with a face shell thicker than 1.5 inches (38 mm), the diameter of the hole in the form shall be increased by 1 inch (25 mm) for each $^{1}/_{2}$ inch (13 mm) of additional insulating form face shell thickness. The ledger board shall be in direct contact with the concrete at each bolt location.

R611.8.2.1 Ledger bearing requirements for Seismic Design Categories C, D_0, D_1 and D_2. Additional anchorage mechanisms connecting the wall to the floor system

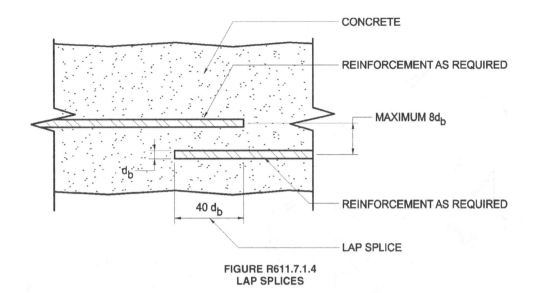

CONCRETE

REINFORCEMENT AS REQUIRED

MAXIMUM 8d$_b$

d_b

40 d$_b$

REINFORCEMENT AS REQUIRED

LAP SPLICE

**FIGURE R611.7.1.4
LAP SPLICES**

shall be installed at a maximum spacing of 6 feet (1829 mm) on center for townhouses in Seismic Design Category C and 4 feet (1220 mm) on center for all buildings in Seismic Design Categories D_0, D_1 and D_2. The additional anchorage mechanisms shall be attached to the ICF wall reinforcement and joist rafters or blocking in accordance with Figures R611.8(1) through R611.8(7). The additional anchorage shall be installed through an oversized hole in the ledger board that is $^1/_2$ inch (13 mm) larger than the anchorage mechanism diameter to prevent combined tension and shear in the mechanism. The blocking shall be attached to floor or roof sheathing in accordance with edge fastener spacing. Such additional anchorage shall not be accomplished by the use of toe nails or nails subject to withdrawal nor shall such anchorage mechanisms induce tension stresses perpendicular to grain in ledgers or nailers. The capacity of such anchors shall result in connections capable of resisting the design values listed in Table R611.8(2).The diaphragm sheathing fasteners applied directly to a ledger shall not be considered effective in providing the additional anchorage required by this section.

Where the additional anchorage mechanisms consist of threaded rods with hex nuts or headed bolts complying with ASTM A 307, Grade A or ASTM F 1554, Grade 36, the design tensile strengths shown in Table R611.9 shall be equal to or greater than the product of the design values listed in Table R611.8(2) and the spacing of the bolts in feet (mm). Anchor bolts shall be embedded as indicated in Table R611.9. Bolts with hooks shall not be used.

R611.8.3 Floor and roof diaphragm construction. Floor and roof diaphragms shall be constructed of wood structural panel sheathing attached to wood framing in accordance with Table R602.3(1) or Table R602.3(2) or to cold-formed steel floor framing in accordance with Table R505.3.1(2) or to cold-formed steel roof framing in accordance with Table R804.3.

R611.8.3.1 Floor and roof diaphragm construction requirements in Seismic Design Categories D_0, D_1 and D_2. The requirements of this section shall apply in addi-

tion to those required by Section R611.8.3. Edge spacing of fasteners in floor and roof sheathing shall be 4 inches (102 mm) on center for Seismic Design Category D_0 or D_1 and 3 inches (76 mm) on center for Seismic Design Category D_2. In Seismic Design Categories D_0, D_1 and D_2, all sheathing edges shall be attached to framing or blocking. Minimum sheathing fastener size shall be 0.113 inch (3 mm) diameter with a minimum penetration of $1^3/_8$-inches (35 mm) into framing members supporting the sheathing. Minimum wood structural panel thickness shall be $^7/_{16}$ inch (11 mm) for roof sheathing and $^{23}/_{32}$ inch (18 mm) for floor sheathing. Vertical offsets in floor framing shall not be permitted.

R611.9 ICF wall to top sill plate (roof) connections. Wood sill plates attaching roof framing to ICF walls shall be anchored with minimum $^1/_2$ inch (13 mm) diameter anchor bolt embedded a minimum of 7 inches (178 mm) and placed at 6 feet (1829 mm) on center in accordance with Figure R611.9. Anchor bolts shall be located in the cores of waffle-grid and screen-grid ICF walls. Roof assemblies subject to wind uplift pressure of 20 pounds per square foot (1.44 kPa) or greater as established in Table R301.2(2) shall have rafter or truss ties provided in accordance with Table R802.11.

R611.9.1 ICF wall to top sill plate (roof) connections for Seismic Design Categories C, D_0, D_1 and D_2. The requirements of this section shall apply in addition to those required by Section R611.9. The top of an ICF wall at a gable shall be attached to an attic floor in accordance with Section R611.8.1.1. For townhouses in Seismic Design Category C, attic floor diaphragms shall be constructed of structural wood sheathing panels attached to wood framing in accordance with Table R602.3(1) or Table R602.3(2). Edge spacing of fasteners in attic floor sheathing shall be 4 inches (102 mm) on center for Seismic Design Category D_0 or D_1 and 3 inches (76 mm) on center for Seismic Design Category D_2. In Seismic Design Categories D_0, D_1 and D_2, all sheathing edges shall be attached to framing or blocking. Minimum sheathing fastener size shall be 0.113 inch (2.8 mm) diameter with a

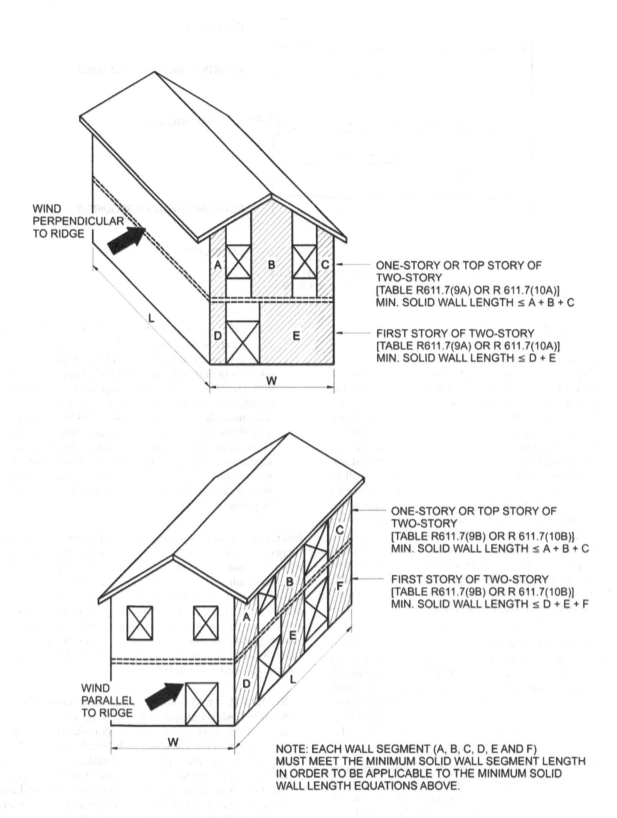

ONE-STORY OR TOP STORY OF
TWO-STORY
[TABLE R611.7(9A) OR R 611.7(10A)]
MIN. SOLID WALL LENGTH ≤ A + B + C

FIRST STORY OF TWO-STORY
[TABLE R611.7(9A) OR R 611.7(10A)]
MIN. SOLID WALL LENGTH ≤ D + E

ONE-STORY OR TOP STORY OF
TWO-STORY
[TABLE R611.7(9B) OR R 611.7(10B)]
MIN. SOLID WALL LENGTH ≤ A + B + C

FIRST STORY OF TWO-STORY
[TABLE R611.7(9B) OR R 611.7(10B)]
MIN. SOLID WALL LENGTH ≤ D + E + F

NOTE: EACH WALL SEGMENT (A, B, C, D, E AND F)
MUST MEET THE MINIMUM SOLID WALL SEGMENT LENGTH
IN ORDER TO BE APPLICABLE TO THE MINIMUM SOLID
WALL LENGTH EQUATIONS ABOVE.

FIGURE R611.7.4
MINIMUM SOLID WALL LENGTH

TABLE R611.8(1)
FLOOR LEDGER-ICF WALL CONNECTION (SIDE-BEARING CONNECTION) REQUIREMENTS[a, b, c]

MAXIMUM FLOOR CLEAR SPAN[d] (feet)	MAXIMUM ANCHOR BOLT SPACING[e] (inches)			
	Staggered $1/_2$-inch-diameter anchor bolts	Staggered $5/_8$-inch-diameter anchor bolts	Two $1/_2$-inch-diameter anchor bolts[f]	Two $5/_8$-inch-diameter anchor bolts[f]
8	18	20	36	40
10	16	18	32	36
12	14	18	28	36
14	12	16	24	32
16	10	14	20	28
18	9	13	18	26
20	8	11	16	22
22	7	10	14	20
24	7	9	14	18
26	6	9	12	18
28	6	8	12	16
30	5	8	10	16
32	5	7	10	14

For SI: 1 inch = 25.4 mm, 1 foot = 304.8 mm.

a. Minimum ledger board nominal depth shall be 8 inches. The thickness of the ledger board shall be a minimum of 2 inches. Thickness of ledger board is in nominal lumber dimensions. Ledger board shall be minimum No. 2 Grade.

b. Minimum edge distance shall be 2 inches for $1/_2$-inch-diameter anchor bolts and 2.5 inches for $5/_8$-inch-diameter anchor bolts.

c. Interpolation is permitted between floor spans.

d. Floor span corresponds to the clear span of the floor structure (i.e., joists or trusses) spanning between load-bearing walls or beams.

e. Anchor bolts shall extend through the ledger to the center of the flat ICF wall thickness or the center of the horizontal or vertical core thickness of the waffle-grid or screen-grid ICF wall system.

f. Minimum vertical distance between bolts shall be 1.5 inches for $1/_2$-inch-diameter anchor bolts and 2 inches for $5/_8$-inch-diameter anchor bolts.

minimum penetration of $1^3/_8$ inches (35 mm) into framing members supporting the sheathing. Minimum wood structural panel thickness shall be $7/_{16}$ inch (11 mm) for the attic floor sheathing. Where hipped roof construction is used, the use of a structural attic floor is not required.

For townhouses in Seismic Design Category C, wood sill plates attached to ICF walls shall be anchored with ASTM A 307, Grade A, $3/_8$-inch (10 mm) diameter anchor bolts embedded a minimum of 7 inches (178 mm) and placed at a maximum spacing of 36 inches (914 mm) on center. For all buildings in Seismic Design Category D_0 or D_1, wood sill plates attached to ICF walls shall be anchored with ASTM A 307, Grade A, $3/_8$-inch (10 mm) diameter anchor bolts embedded a minimum of 7 inches (178 mm) and placed at a maximum spacing of 16 inches (406 mm) on center. For all buildings in Seismic Design Category D_2, wood sill plates attached to ICF walls shall be anchored with ASTM A 307, Grade A, $3/_8$-inch (10 mm) diameter anchor bolts embedded a minimum of 7 inches (178 mm) and placed at a maximum spacing of 16 inches (406 mm) on center.

For townhouses in Seismic Design Category C, each floor joist shall be attached to the sill plate with an 18-gage [(0.0478 in.) (1.2 mm)] angle bracket using 3 - 8d common nails per leg in accordance with Figure R611.8(1). For all buildings in Seismic Design Category D_0 or D_1, each floor joist shall be attached to the sill plate with an 18-gage [(0.0478 in.) (1.2 mm)] angle bracket using 4 - 8d common nails per leg in accordance with Figure R611.8(1). For all buildings in Seismic Design Category D_2, each floor joist shall be attached to the sill plate with an 18-gage [(0.0478 in.) (1.2 mm)] angle bracket using 6-8d common nails per leg in accordance with Figure R611.8(1).

Where hipped roof construction is used without an attic floor, the following shall apply. For townhouses in Seismic Design Category C, each rafter shall be attached to the sill plate with an 18-gage [(0.0478 in.) (1.2 mm)] angle bracket using 3 - 8d common nails per leg in accordance with Figure R611.9. For all buildings in Seismic Design Category D_0 or D_1, each rafter shall be attached to the sill plate with an 18-gage [(0.0478 in.) (1.2 mm)] angle bracket using 4 - 8d common nails per leg in accordance with Figure R611.9. For all buildings in Seismic Design Category D_2, each rafter shall be attached to the sill plate with an 18-gage [(0.0478 in.) (1.2 mm)] angle bracket using 6-8d common nails per leg in accordance with Figure R611.9.

TABLE R611.8(2)
DESIGN VALUES (PLF) FOR FLOOR JOIST-TO-WALL ANCHORS REQUIRED FOR TOWNHOUSES
IN SEISMIC DESIGN CATEGORY C AND ALL BUILDINGS IN SEISMIC DESIGN CATEGORIES D_0, D_1, AND D_2[a, b]

WALL TYPE	SEISMIC DESIGN CATEGORY		
	C	D_0 or D_1	D_2
Flat 3.5	193	NP	NP
Flat 5.5	303	502	708
Flat 7.5	413	685	965
Flat 9.5	523	867	1,223
Waffle 6	246	409	577
Waffle 8	334	555	782
Screen 6	233	387	546

For SI: 1 pound per linear foot = 1.488 kg/m.

NP = Not Permitted

a. Table values are based on IBC Equation 16–64 using a tributary wall height of 11 feet. Table values shall be permitted to be reduced for tributary wall heights less than 11 feet by multiplying the table values by X/11, where X is the tributary wall height.

b. Values may be reduced by 30 percent when used for ASD.

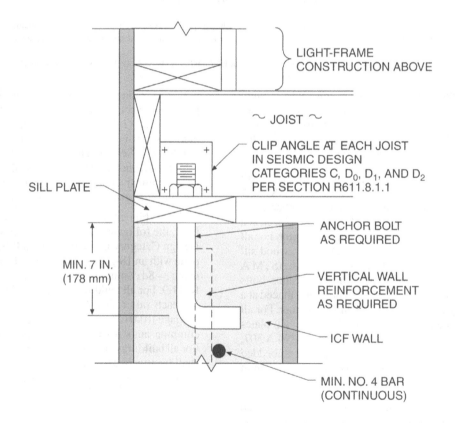

FIGURE R611.8(1)
SECTION CUT THROUGH FLAT WALL OR VERTICAL CORE
OF WAFFLE- OR SCREEN-GRID WALL

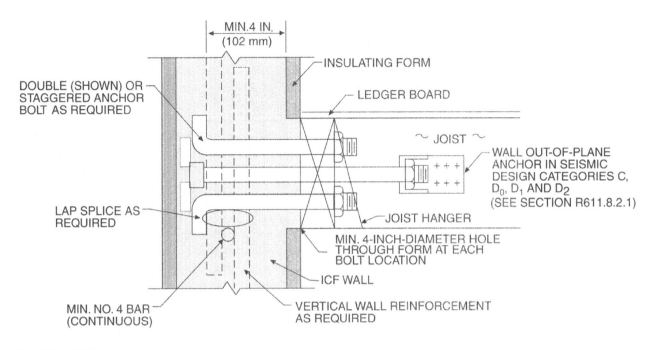

For SI: 1 inch = 25.4 mm.
NOTE: Section cut through flat wall or vertical core of a waffle- or screen-grid wall.

FIGURE R611.8(2)
FLOOR LEDGER—ICF WALL CONNECTION (SIDE-BEARING CONNECTION)

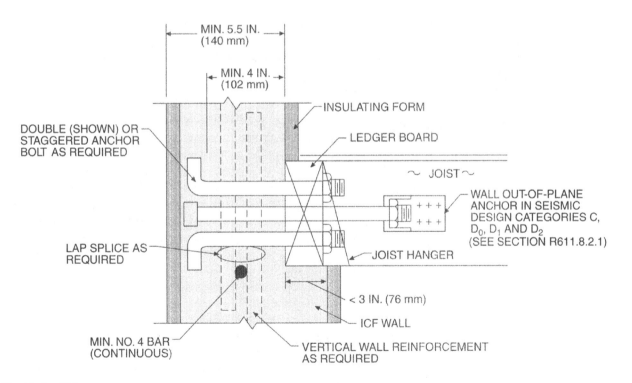

For SI: 1 inch = 25.4 mm.
NOTE: Section cut through flat wall or vertical core of a waffle- or screen-grid wall.

FIGURE R611.8(3)
FLOOR LEDGER—ICF WALL CONNECTION (LEDGE-BEARING CONNECTION)

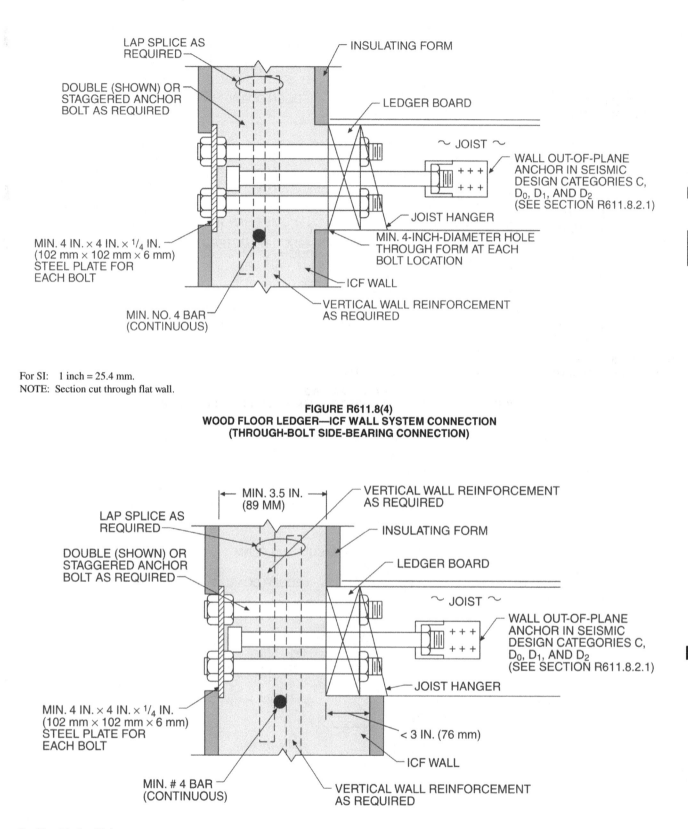

For SI: 1 inch = 25.4 mm.
NOTE: Section cut through flat wall.

FIGURE R611.8(4)
WOOD FLOOR LEDGER—ICF WALL SYSTEM CONNECTION
(THROUGH-BOLT SIDE-BEARING CONNECTION)

For SI: 1 inch = 25.4 mm.
NOTE: Section cut through flat wall.

FIGURE R611.8(5)
FLOOR LEDGER—ICF WALL CONNECTION

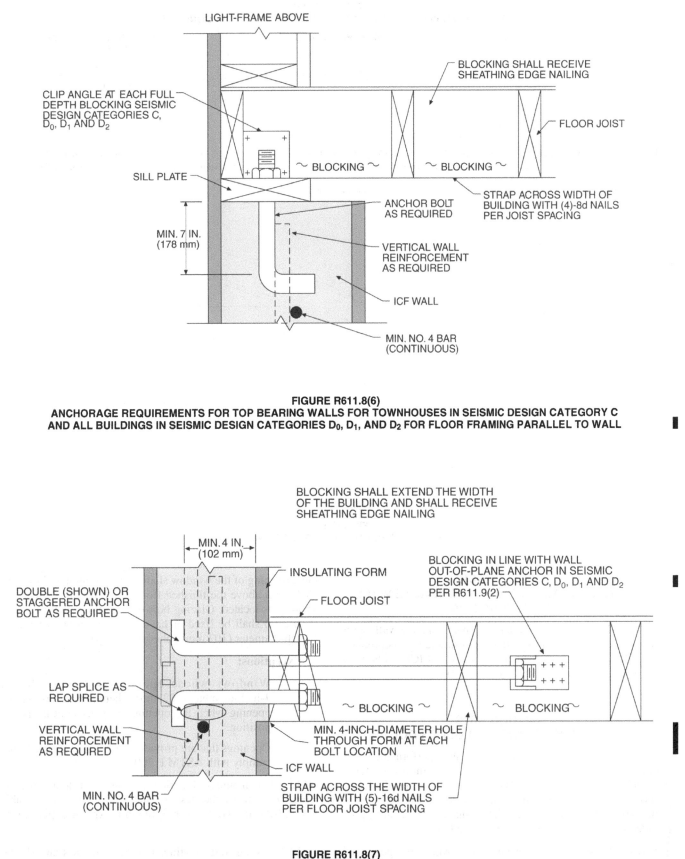

FIGURE R611.8(6)
ANCHORAGE REQUIREMENTS FOR TOP BEARING WALLS FOR TOWNHOUSES IN SEISMIC DESIGN CATEGORY C AND ALL BUILDINGS IN SEISMIC DESIGN CATEGORIES D$_0$, D$_1$, AND D$_2$ FOR FLOOR FRAMING PARALLEL TO WALL

FIGURE R611.8(7)
ANCHORAGE REQUIREMENTS FOR LEDGER BEARING WALLS FOR TOWNHOUSES IN SEISMIC DESIGN CATEGORY C AND ALL BUILDINGS IN SEISMIC DESIGN CATEGORIES D$_0$, D$_1$ AND D$_2$ FOR FLOOR FRAMING PARALLEL TO WALL

TABLE R611.9
DESIGN TENSILE STRENGTH OF HEADED BOLTS CAST IN CONCRETE[a]

DIAMETER OF BOLT (inches)	MINIMUM EMBEDMENT DEPTH (inches)	DESIGN TENSILE STRENGTH[b] (pounds)
$^1/_4$	2	1040
$^3/_8$ with washer[c]	$2^3/_4$[d]	2540
$^1/_2$ with washer[c]	4[d]	4630

For SI: 1 pound per square inch = 6.895 kPa.

a. Applicable to concrete of all strengths. See Notes (c) and (d).

b. Values are based on ASTM F 1554, Grade 36 bolts. Where ASTM A 307, Grade A headed bolts are used, the strength shall be increased by 1.034.

c. A hardened washer shall be installed at the nut embedded in the concrete or head of the bolt to increase the bearing area. The washer is not required where the concrete strength is 4000 psi or more.

d. Embedment depth shall be permitted to be reduced $^1/_4$-inch where 4000 psi concrete is used.

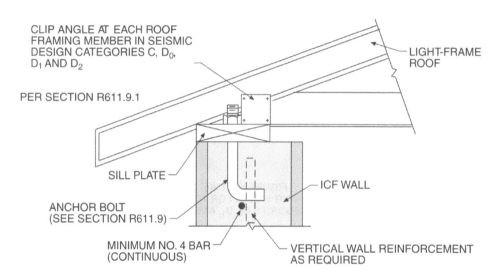

NOTE: Section cut through flat wall or vertical core of a waffle- or screen-grid wall.

FIGURE R611.9
ROOF SILL PLATE—ICF WALL CONNECTION

SECTION R612
CONVENTIONALLY FORMED CONCRETE WALL CONSTRUCTION

R612.1 General. Conventionally formed concrete walls with flat surfaces shall be designed and constructed in accordance with the provisions of Section R611 for Flat ICF walls or in accordance with the provisions of ACI 318.

SECTION R613
EXTERIOR WINDOWS AND GLASS DOORS

R613.1 General. This section prescribes performance and construction requirements for exterior window systems installed in wall systems. Windows shall be installed and flashed in accordance with the manufacturer's written installation instructions. Written installation instructions shall be provided by the manufacturer for each window.

R613.2 Window sills. In dwelling units, where the opening of an operable window is located more than 72 inches (1829 mm) above the finished grade or surface below, the lowest part of the clear opening of the window shall be a minimum of 24 inches (610 mm) above the finished floor of the room in which the window is located. Glazing between the floor and 24 inches (610 mm) shall be fixed or have openings through which a 4-inch-diameter (102 mm) sphere cannot pass.

Exceptions:

1. Windows whose openings will not allow a 4-inch-diameter (102 mm) sphere to pass through the opening when the opening is in its largest opened position.

2. Openings that are provided with window guards that comply with ASTM F 2006 or F 2090.

R613.3 Performance. Exterior windows and doors shall be designed to resist the design wind loads specified in Table R301.2(2) adjusted for height and exposure per Table R301.2(3).

R613.4 Testing and labeling. Exterior windows and sliding doors shall be tested by an approved independent laboratory, and bear a label identifying manufacturer, performance characteristics and approved inspection agency to indicate compliance with AAMA/WDMA/CSA 101/I.S.2/A440. Exterior side-hinged doors shall be tested and labeled as conforming to AAMA/WDMA/CSA 101/I.S.2/A440 or comply with Section R613.6.

Exception: Decorative glazed openings.

R613.4.1 Comparative analysis. Structural wind load design pressures for window and door units smaller than the size tested in accordance with Section R613.4 shall be permitted to be higher than the design value of the tested unit provided such higher pressures are determined by accepted engineering analysis. All components of the small unit shall be the same as those of the tested unit. Where such calculated design pressures are used, they shall be validated by an additional test of the window or door unit having the highest allowable design pressure.

R613.5 Vehicular access doors. Vehicular access doors shall be tested in accordance with either ASTM E 330 or ANSI/DASMA 108, and shall meet the acceptance criteria of ANSI/DASMA 108.

R613.6 Other exterior window and door assemblies. Exterior windows and door assemblies not included within the scope of Section R613.4 or Section R613.5 shall be tested in accordance with ASTM E 330. Glass in assemblies covered by this exception shall comply with Section R308.5.

R613.7 Wind-borne debris protection. Protection of exterior windows and glass doors in buildings located in wind-borne debris regions shall be in accordance with Section R301.2.1.2.

R613.7.1 Fenestration testing and labeling. Fenestration shall be tested by an approved independent laboratory, listed by an approved entity, and bear a label identifying manufacturer, performance characteristics, and approved inspection agency to indicate compliance with the requirements of the following specification:

1. ASTM E 1886 and ASTM E 1996; or

2. AAMA 506.

R613.8 Anchorage methods. The methods cited in this section apply only to anchorage of window and glass door assemblies to the main force-resisting system.

R613.8.1 Anchoring requirements. Window and glass door assemblies shall be anchored in accordance with the published manufacturer's recommendations to achieve the design pressure specified. Substitute anchoring systems used for substrates not specified by the fenestration manufacturer shall provide equal or greater anchoring performance as demonstrated by accepted engineering practice.

R613.8.2 Anchorage details. Products shall be anchored in accordance with the minimum requirements illustrated in Figures R613.8(1), R613.8(2), R613.8(3), R613.8(4), R613.8(5), R613.8(6), R613.8(7) and R613.8(8).

R613.8.2.1 Masonry, concrete or other structural substrate. Where the wood shim or buck thickness is less than $1^1/_2$ inches (38 mm), window and glass door assemblies shall be anchored through the jamb, or by jamb clip and anchors shall be embedded directly into the masonry, concrete or other substantial substrate material. Anchors shall adequately transfer load from the window or door frame into the rough opening substrate [see Figures R613.8(1) and R613.8(2).]

Where the wood shim or buck thickness is $1^1/_2$ inches (38 mm) or more, the buck is securely fastened to the masonry, concrete or other substantial substrate, and the buck extends beyond the interior face of the window or door frame, window and glass door assemblies shall be anchored through the jamb, or by jamb clip, or through the flange to the secured wood buck. Anchors shall be embedded into the secured wood buck to adequately transfer load from the window or door frame assembly [Figures R613.8(3), R613.8(4) and R613.8(5)].

R613.8.2.2 Wood or other approved framing material. Where the framing material is wood or other approved framing material, window and glass door assemblies shall be anchored through the frame, or by frame clip, or through the flange. Anchors shall be embedded into the frame construction to adequately

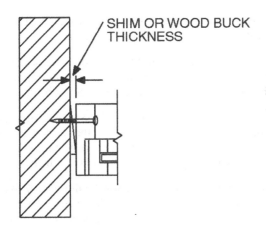

FIGURE R613.8(1)
THROUGH THE FRAME

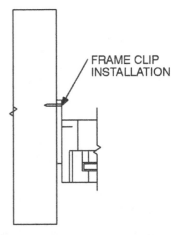

APPLY FRAME CLIP TO WINDOW OR DOOR IN ACCORDANCE WITH PUBLISHED MANUFACTURER'S RECOMMENDATIONS.

FIGURE R613.8(2)
FRAME CLIP

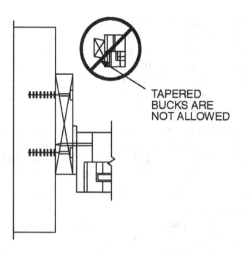

THROUGH THE FRAME ANCHORING METHOD. ANCHORS SHALL BE PROVIDED TO TRANSFER LOAD FROM THE WINDOW OR DOOR FRAME INTO THE ROUGH OPENING SUBSTRATE.

FIGURE R613.8(3)
THROUGH THE FRAME

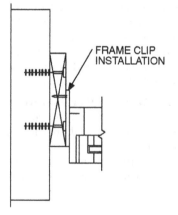
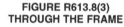

APPLY FRAME CLIP TO WINDOW OR DOOR FRAME IN ACCORDANCE WITH PUBLISHED MANUFACTURER'S RECOMMENDATIONS. ANCHORS SHALL BE PROVIDED TO TRANSFER LOAD FROM THE FRAME CLIP INTO THE ROUGH OPENING SUBSTRATE.

FIGURE R613.8(4)
FRAME CLIP

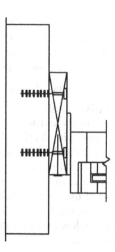

APPLY ANCHORS THROUGH FLANGE IN ACCORDANCE WITH PUBLISHED MANUFACTURER'S RECOMMENDATIONS.

FIGURE R613.8(5)
THROUGH THE FLANGE

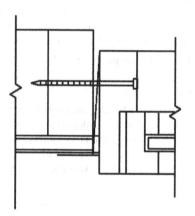

FIGURE R613.8(6)
THROUGH THE FRAME

transfer load [Figures R613.8(6), R613.8(7) and R613.8(8)].

R613.9 Mullions occurring between individual window and glass door assemblies.

R613.9.1 Mullions. Mullions shall be tested by an approved testing laboratory in accordance with AAMA 450, or be engineered in accordance with accepted engineering practice. Mullions tested as stand-alone units or qualified by engineering shall use performance criteria cited in Sections R613.9.2, R613.9.3 and R613.9.4. Mullions qualified by an actual test of an entire assembly shall comply with Sections R613.9.2 and R613.9.4.

R613.9.2 Load transfer. Mullions shall be designed to transfer the design pressure loads applied by the window and door assemblies to the rough opening substrate.

R613.9.3 Deflection. Mullions shall be capable of resisting the design pressure loads applied by the window and door assemblies to be supported without deflecting more than $L/175$, where L is the span of the mullion in inches.

R613.9.4 Structural safety factor. Mullions shall be capable of resisting a load of 1.5 times the design pressure loads applied by the window and door assemblies to be supported without exceeding the appropriate material stress levels. If tested by an approved laboratory, the 1.5 times the design pressure load shall be sustained for 10 seconds, and the permanent deformation shall not exceed 0.4 percent of the mullion span after the 1.5 times design pressure load is removed.

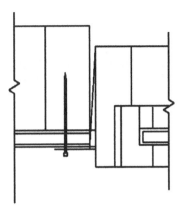

FIGURE R613.8(8)
THROUGH THE FLANGE

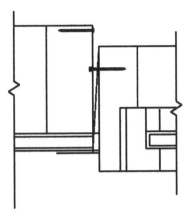

FIGURE R613.8(7)
FRAME CLIP

CHAPTER 7
WALL COVERING

SECTION R701
GENERAL

R701.1 Application. The provisions of this chapter shall control the design and construction of the interior and exterior wall covering for all buildings.

R701.2 Installation. Products sensitive to adverse weather shall not be installed until adequate weather protection for the installation is provided. Exterior sheathing shall be dry before applying exterior cover.

SECTION R702
INTERIOR COVERING

R702.1 General. Interior coverings or wall finishes shall be installed in accordance with this chapter and Table R702.1(1), Table R702.1(2), Table R702.1(3) and Table R702.3.5. Interior masonry veneer shall comply with the requirements of Section R703.7.1 for support and Section R703.7.4 for anchorage, except an air space is not required. Interior finishes and materials shall conform to the flame spread and smoke-density requirements of Section R315.

TABLE R702.1(1)
THICKNESS OF PLASTER

PLASTER BASE	FINISHED THICKNESS OF PLASTER FROM FACE OF LATH, MASONRY, CONCRETE (inches)	
	Gypsum plaster	Portland cement mortar
Expanded metal lath	$5/8$, minimum[a]	$5/8$, minimum[a]
Wire lath	$5/8$, minimum[a]	$3/4$, minimum (interior)[b] $7/8$, minimum (exterior)[b]
Gypsum lath[g]	$1/2$, minimum	$3/4$, minimum (interior)[b]
Masonry walls[c]	$1/2$, minimum	$1/2$, minimum
Monolithic concrete walls[c,d]	$5/8$, maximum	$7/8$, maximum
Monolithic concrete ceilings[c,d]	$3/8$, maximum[e]	$1/2$, maximum
Gypsum veneer base[f,g]	$1/16$, minimum	$3/4$, minimum (interior)[b]
Gypsum sheathing[g]	—	$3/4$, minimum (interior)[b] $7/8$, minimum (exterior)[b]

For SI: 1 inch = 25.4 mm.

a. When measured from back plane of expanded metal lath, exclusive of ribs, or self-furring lath, plaster thickness shall be $3/4$ inch minimum.

b. When measured from face of support or backing.

c. Because masonry and concrete surfaces may vary in plane, thickness of plaster need not be uniform.

d. When applied over a liquid bonding agent, finish coat may be applied directly to concrete surface.

e. Approved acoustical plaster may be applied directly to concrete or over base coat plaster, beyond the maximum plaster thickness shown.

f. Attachment shall be in accordance with Table R702.3.5.

g. Where gypsum board is used as a base for portland cement plaster, weather-resistant sheathing paper complying with Section R703.2 shall be provided.

TABLE R702.1(2)
GYPSUM PLASTER PROPORTIONS[a]

NUMBER	COAT	PLASTER BASE OR LATH	MAXIMUM VOLUME AGGREGATE PER 100 POUNDS NEAT PLASTER[b] (cubic feet)	
			Damp loose sand[a]	Perlite or vermiculite[c]
Two-coat work	Base coat	Gypsum lath	2.5	2
	Base coat	Masonry	3	3
Three-coat work	First coat	Lath	2[d]	2
	Second coat	Lath	3[d]	2[e]
	First and second coats	Masonry	3	3

For SI: 1 inch = 25.4 mm, 1 cubic foot = 0.0283 m³, 1 pound = 0.454 kg.

a. Wood-fibered gypsum plaster may be mixed in the proportions of 100 pounds of gypsum to not more than 1 cubic foot of sand where applied on masonry or concrete.

b. When determining the amount of aggregate in set plaster, a tolerance of 10 percent shall be allowed.

c. Combinations of sand and lightweight aggregate may be used, provided the volume and weight relationship of the combined aggregate to gypsum plaster is maintained.

d. If used for both first and second coats, the volume of aggregate may be 2.5 cubic feet.

e. Where plaster is 1 inch or more in total thickness, the proportions for the second coat may be increased to 3 cubic feet.

TABLE R702.1(3)
PORTLAND CEMENT PLASTER

	MAXIMUM VOLUME AGGREGATE PER VOLUME CEMENTITIOUS MATERIAL[a]					
	Portland cement plaster[b] maximum volume aggregate per volume cement	Portland cement-lime plaster[c]			MINIMUM PERIOD MOIST COATS	MINIMUM INTERVAL BETWEEN
Coat		Maximum volume lime per volume cement	Maximum volume sand per volume cement and lime	Approximate minimum thickness[d] curing (inches)		
First	4	$3/4$	4	$3/8$[e]	48 Hours[f]	48 Hours[g]
Second	5	$3/4$	5	First and second coats	48 Hours	7 Days[h]
Finish	3[i]	—	3[i]	$1/8$	—	Note h

For SI: 1 inch = 25.4 mm, 1 pound = 0.454 kg.

a. When determining the amount of aggregate in set plaster, a tolerance of 10 percent may be allowed.
b. From 10 to 20 pounds of dry hydrated lime (or an equivalent amount of lime putty) may be added as a plasticizing agent to each sack of Type I and Type II standard portland cement in base coat plaster.
c. No plasticizing agents shall be added.
d. See Table R702.1(1).
e. Measured from face of support or backing to crest of scored plaster.
f. Twenty-four-hour minimum period for moist curing of interior portland cement plaster.
g. Twenty-four hour minimum interval between coats of interior portland cement plaster.
h. Finish coat plaster may be applied to interior portland cement base coats after a 48-hour period.
i. For finish coat, plaster up to an equal part of dry hydrated lime by weight (or an equivalent volume of lime putty) may be added to Type I, Type II and Type III standard portland cement.

R702.2 Interior plaster. Gypsum plaster or portland cement plastering materials shall conform to ASTM C 5, C 28, C 35, C 37, C 59, C 61, C 587, C 588, C 631, C 847, C 897, C 933, C 1032 and C 1047, and shall be installed or applied in conformance with ASTM C 843, C 844 and C 1063. Plaster shall not be less than three coats when applied over metal lath and not less than two coats when applied over other bases permitted by this section, except that veneer plaster may be applied in one coat not to exceed $3/16$ inch (5 mm) thickness, provided the total thickness is as set forth in Table R702.1(1).

R702.2.1 Support. Support spacing for gypsum or metal lath on walls or ceilings shall not exceed 16 inches (406 mm) for $3/8$ inch thick (10 mm) or 24 inches (610 mm) for $1/2$-inch-thick (13 mm) plain gypsum lath. Gypsum lath shall be installed at right angles to support framing with end joints in adjacent courses staggered by at least one framing space.

R702.3 Gypsum board.

R702.3.1 Materials. All gypsum board materials and accessories shall conform to ASTM C 36, C 79, C 475, C 514, C 630, C 931, C 960, C 1002, C 1047, C 1177, C 1178, C 1278, C 1395 or C 1396 and shall be installed in accordance with the provisions of this section. Adhesives for the installation of gypsum board shall conform to ASTM C 557.

R702.3.2 Wood framing. Wood framing supporting gypsum board shall not be less than 2 inches (51 mm) nominal thickness in the least dimension except that wood furring strips not less than 1-inch-by-2 inch (25 mm by 51 mm) nominal dimension may be used over solid backing or framing spaced not more than 24 inches (610 mm) on center.

R702.3.3 Steel framing. Steel framing supporting gypsum board shall not be less than 1.25 inches (32 mm) wide in the least dimension. Light-gage nonload-bearing steel framing shall comply with ASTM C 645. Load-bearing steel framing and steel framing from 0.033 inch to 0.112 inch (1 mm to 3 mm) thick shall comply with ASTM C 955.

R702.3.4 Insulating concrete form walls. Foam plastics for insulating concrete form walls constructed in accordance with Sections R404.4 and R611 on the interior of habitable spaces shall be covered in accordance with Section R314.4. Use of adhesives in conjunction with mechanical fasteners is permitted. Adhesives used for interior and exterior finishes shall be compatible with the insulating form materials.

R702.3.5 Application. Maximum spacing of supports and the size and spacing of fasteners used to attach gypsum board shall comply with Table R702.3.5. Gypsum sheathing shall be attached to exterior walls in accordance with Table R602.3(1). Gypsum board shall be applied at right angles or parallel to framing members. All edges and ends of gypsum board shall occur on the framing members, except those edges and ends that are perpendicular to the framing members. Interior gypsum board shall not be installed where it is directly exposed to the weather or to water.

R702.3.6 Fastening. Screws for attaching gypsum board to wood framing shall be Type W or Type S in accordance with ASTM C 1002 and shall penetrate the wood not less than $5/8$ inch (16 mm). Screws for attaching gypsum board to light-gage steel framing shall be Type S in accordance with ASTM C 1002 and shall penetrate the steel not less than $3/8$ inch (10 mm). Screws for attaching gypsum board to steel framing 0.033 inch to 0.112 inch (1 mm to 3 mm) thick shall comply with ASTM C 954.

THICKNESS OF GYPSUM BOARD (inches)	APPLICATION	ORIENTATION OF GYPSUM BOARD TO FRAMING	MAXIMUM SPACING OF FRAMING MEMBERS (inches o.c.)	MAXIMUM SPACING OF FASTENERS (inches)		SIZE OF NAILS FOR APPLICATION TO WOOD FRAMING[c]
				Nails[a]	Screws[b]	
Application without adhesive						
$^3/_8$	Ceiling[d]	Perpendicular	16	7	12	13 gage, $1^1/_4''$ long, $^{19}/_{64}$ head; 0.098″ diameter, $1^1/_4''$ long, annular-ringed; or 4d cooler nail, 0.080″ diameter, $1^3/_8''$ long, $^7/_{32}''$ head.
	Wall	Either direction	16	8	16	
$^1/_2$	Ceiling	Either direction	16	7	12	13 gage, $1^3/_8''$ long, $^{19}/_{64}$ head; 0.098″ diameter, $1^1/_4''$ long, annular-ringed; 5d cooler nail, 0.086″ diameter, $1^5/_8''$ long, $^{15}/_{64}''$ head; or gypsum board nail, 0.086″ diameter, $1^5/_8''$ long, $^9/_{32}''$ head.
	Ceiling[d]	Perpendicular	24	7	12	
	Wall	Either direction	24	8	12	
	Wall	Either direction	16	8	16	
$^5/_8$	Ceiling	Either direction	16	7	12	13 gage, $1^5/_8''$ long, $^{19}/_{64}$ head; 0.098″ diameter, $1^3/_8''$ long, annular-ringed; 6d cooler nail, 0.092″ diameter, $1^7/_8''$ long, $^1/_4''$ head; or gypsum board nail, 0.0915″ diameter, $1^7/_8''$ long, $^{19}/_{64}''$ head.
	Ceiling[e]	Perpendicular	24	7	12	
	Wall	Either direction	24	8	12	
	Wall	Either direction	16	8	16	
Application with adhesive						
$^3/_8$	Ceiling[d]	Perpendicular	16	16	16	Same as above for $^3/_8''$ gypsum board
	Wall	Either direction	16	16	24	
$^1/_2$ or $^5/_8$	Ceiling	Either direction	16	16	16	Same as above for $^1/_2''$ and $^5/_8''$ gypsum board, respectively
	Ceiling[d]	Perpendicular	24	12	16	
	Wall	Either direction	24	16	24	
Two $^3/_8$ layers	Ceiling	Perpendicular	16	16	16	Base ply nailed as above for $^1/_2''$ gypsum board; face ply installed with adhesive
	Wall	Either direction	24	24	24	

For SI: 1 inch = 25.4 mm.

a. For application without adhesive, a pair of nails spaced not less than 2 inches apart or more than $2^1/_2$ inches apart may be used with the pair of nails spaced 12 inches on center.

b. Screws shall be Type S or W per ASTM C 1002 and shall be sufficiently long to penetrate wood framing not less than $^5/_8$ inch and metal framing not less than $^3/_8$ inch.

c. Where metal framing is used with a clinching design to receive nails by two edges of metal, the nails shall be not less than $^5/_8$ inch longer than the gypsum board thickness and shall have ringed shanks. Where the metal framing has a nailing groove formed to receive the nails, the nails shall have barbed shanks or be 5d, $13^1/_2$ gage, $1^5/_8$ inches long, $^{15}/_{64}$-inch head for $^1/_2$-inch gypsum board; and 6d, 13 gage, $1^7/_8$ inches long, $^{15}/_{64}$-inch head for $^5/_8$-inch gypsum board.

d. Three-eighths-inch-thick single-ply gypsum board shall not be used on a ceiling where a water-based textured finish is to be applied, or where it will be required to support insulation above a ceiling. On ceiling applications to receive a water-based texture material, either hand or spray applied, the gypsum board shall be applied perpendicular to framing. When applying a water-based texture material, the minimum gypsum board thickness shall be increased from $^3/_8$ inch to $^1/_2$ inch for 16-inch on center framing, and from $^1/_2$ inch to $^5/_8$ inch for 24-inch on center framing or $^1/_2$-inch sag-resistant gypsum ceiling board shall be used.

e. Type X gypsum board for garage ceilings beneath habitable rooms shall be installed perpendicular to the ceiling framing and shall be fastened at maximum 6 inches o.c. by minimum $1^7/_8$ inches 6d coated nails or equivalent drywall screws.

R702.3.7 Horizontal gypsum board diaphragm ceilings. Use of gypsum board shall be permitted on wood joists to create a horizontal diaphragm in accordance with Table R702.3.7. Gypsum board shall be installed perpendicular to ceiling framing members. End joints of adjacent courses of board shall not occur on the same joist. The maximum allowable diaphragm proportions shall be $1^1/_2$:1 between shear resisting elements. Rotation or cantilever conditions shall not be permitted. Gypsum board shall not be used in diaphragm ceilings to resist lateral forces imposed by masonry or concrete construction. All perimeter edges shall be blocked using wood members not less than 2-inch (51 mm) by 6-inch (152 mm) nominal dimension. Blocking material shall be installed flat over the top plate of the wall to provide a nailing surface not less than 2 inches (51 mm) in width for the attachment of the gypsum board.

R702.3.8 Water-resistant gypsum backing board. Gypsum board used as the base or backer for adhesive application of ceramic tile or other required nonabsorbent finish material shall conform to ASTM C 630 or C 1178. Use of water-resistant gypsum backing board shall be permitted on ceilings where framing spacing does not exceed 12 inches (305 mm) on center for $^1/_2$-inch-thick (13 mm) or 16 inches (406 mm) for $^5/_8$-inch-thick (16 mm) gypsum board. Water-resistant gypsum board shall not be installed over a vapor retarder in a shower or tub compartment. Cut or exposed edges, including those at wall intersections, shall be sealed as recommended by the manufacturer.

R702.3.8.1 Limitations. Water resistant gypsum backing board shall not be used where there will be direct exposure to water, or in areas subject to continuous high humidity.

TABLE R702.3.7
SHEAR CAPACITY FOR HORIZONTAL WOOD-FRAMED
GYPSUM BOARD DIAPHRAGM CEILING ASSEMBLIES

MATERIAL	THICKNESS OF MATERIAL (min.) (in.)	SPACING OF FRAMING MEMBERS (max.) (in.)	SHEAR VALUE[a, b] (plf of ceiling)	MINIMUM FASTENER SIZE[c, d]
Gypsum Board	$^1/_2$	16 o.c.	90	5d cooler or wallboard nail; $1^5/_8$-inch long; 0.086-inch shank; $^{15}/_{64}$-inch head
Gypsum Board	$^1/_2$	24 o.c.	70	5d cooler or wallboard nail; $1^5/_8$-inch long; 0.086-inch shank; $^{15}/_{64}$-inch head

For SI: 1 inch = 25.4 mm, 1 pound per linear foot = 1.488 kg/m.

a. Values are not cumulative with other horizontal diaphragm values and are for short-term loading caused by wind or seismic loading. Values shall be reduced 25 percent for normal loading.

b. Values shall be reduced 50 percent in Seismic Design Categories D_0, D_1, D_2 and E.

c. $1^1/_4$", #6 Type S or W screws may be substituted for the listed nails.

d. Fasteners shall be spaced not more than 7 inches on center at all supports, including perimeter blocking, and not less than $^3/_8$ inch from the edges and ends of the gypsum board.

R702.4 Ceramic tile.

R702.4.1 General. Ceramic tile surfaces shall be installed in accordance with ANSI A108.1, A108.4, A108.5, A108.6, A108.11, A118.1, A118.3, A136.1 and A137.1.

R702.4.2 Cement, fiber-cement and glass mat gypsum backers. Cement, fiber-cement or glass mat gypsum backers in compliance with ASTM C 1288, C 1325 or C 1178 and installed in accordance with manufacturers' recommendations shall be used as backers for wall tile in tub and shower areas and wall panels in shower areas.

R702.5 Other finishes. Wood veneer paneling and hardboard paneling shall be placed on wood or cold-formed steel framing spaced not more than 16 inches (406 mm) on center. Wood veneer and hard board paneling less than $^1/_4$ inch (6 mm) nominal thickness shall not have less than a $^3/_8$-inch (10 mm) gypsum board backer. Wood veneer paneling not less than $^1/_4$-inch (6 mm) nominal thickness shall conform to ANSI/HPVA HP-1. Hardboard paneling shall conform to ANSI/AHA A135.5.

R702.6 Wood shakes and shingles. Wood shakes and shingles shall conform to CSSB *Grading Rules for Wood Shakes and Shingles* and shall be permitted to be installed directly to the studs with maximum 24 inches (610 mm) on-center spacing.

R702.6.1 Attachment. Nails, staples or glue are permitted for attaching shakes or shingles to the wall, and attachment of the shakes or shingles directly to the surface shall be permitted provided the fasteners are appropriate for the type of wall surface material. When nails or staples are used, two fasteners shall be provided and shall be placed so that they are covered by the course above.

R702.6.2 Furring strips. Where furring strips are used, they shall be 1 inch by 2 inches or 1 inch by 3 inches (25 mm by 51 mm or 25 mm by 76 mm), spaced a distance on center equal to the desired exposure, and shall be attached to the wall by nailing through other wall material into the studs.

SECTION R703
EXTERIOR COVERING

R703.1 General. Exterior walls shall provide the building with a weather-resistant exterior wall envelope. The exterior wall envelope shall include flashing as described in Section R703.8. The exterior wall envelope shall be designed and constructed in a manner that prevents the accumulation of water within the wall assembly by providing a water-resistant barrier behind the exterior veneer as required by Section R703.2. and a means of draining water that enters the assembly to the exterior. Protection against condensation in the exterior wall assembly shall be provided in accordance with Chapter 11 of this code.

Exceptions:

1. A weather-resistant exterior wall envelope shall not be required over concrete or masonry walls designed in accordance with Chapter 6 and flashed according to Section R703.7 or R703.8.

2. Compliance with the requirements for a means of drainage, and the requirements of Section R703.2 and Section R703.8, shall not be required for an exterior wall envelope that has been demonstrated to resist wind-driven rain through testing of the exterior wall envelope, including joints, penetrations and intersections with dissimilar materials, in accordance with ASTM E 331 under the following conditions:

   2.1. Exterior wall envelope test assemblies shall include at least one opening, one control joint, one wall/eave interface and one wall sill. All tested openings and penetrations shall be representative of the intended end-use configuration.

   2.2. Exterior wall envelope test assemblies shall be at least 4 feet (1219 mm) by 8 feet (2438 mm) in size.

   2.3. Exterior wall assemblies shall be tested at a minimum differential pressure of 6.24 pounds per square foot (299 Pa).

2.4. Exterior wall envelope assemblies shall be subjected to a minimum test exposure duration of 2 hours.

The exterior wall envelope design shall be considered to resist wind-driven rain where the results of testing indicate that water did not penetrate: control joints in the exterior wall envelope; joints at the perimeter of openings penetration; or intersections of terminations with dissimilar materials.

R703.2 Water-resistive barrier. One layer of No. 15 asphalt felt, free from holes and breaks, complying with ASTM D 226 for Type 1 felt or other approved water-resistive barrier shall be applied over studs or sheathing of all exterior walls. Such felt or material shall be applied horizontally, with the upper layer lapped over the lower layer not less than 2 inches (51 mm). Where joints occur, felt shall be lapped not less than 6 inches (152 mm). The felt or other approved material shall be continuous to the top of walls and terminated at penetrations and building appendages in a manner to meet the requirements of the exterior wall envelope as described in Section R703.1.

Exception: Omission of the water-resistive barrier is permitted in the following situations:

1. In detached accessory buildings.
2. Under exterior wall finish materials as permitted in Table R703.4.
3. Under paperbacked stucco lath when the paper backing is an approved weather-resistive sheathing paper.

R703.3 Wood, hardboard and wood structural panel siding.

R703.3.1 Panel siding. Joints in wood, hardboard or wood structural panel siding shall be made as follows unless otherwise approved. Vertical joints in panel siding shall occur over framing members, unless wood or wood structural panel sheathing is used, and shall be shiplapped or covered with a batten. Horizontal joints in panel siding shall be lapped a minimum of 1 inch (25 mm) or shall be shiplapped or shall be flashed with Z-flashing and occur over solid blocking, wood or wood structural panel sheathing.

R703.3.2 Horizontal siding. Horizontal lap siding shall be lapped a minimum of 1 inch (25 mm), or 0.5 inch (13 mm) if rabbeted, and shall have the ends caulked, covered with a batten, or sealed and installed over a strip of flashing.

R703.4 Attachments. Unless specified otherwise, all wall coverings shall be securely fastened in accordance with Table R703.4 or with other approved aluminum, stainless steel, zinc-coated or other approved corrosion-resistant fasteners. Where the basic wind speed per Figure R301.2(4) is 110 miles per hour (49 m/s) or higher, the attachment of wall coverings shall be designed to resist the component and cladding loads specified in Table R301.2(2), adjusted for height and exposure in accordance with Table R301.2(3).

R703.5 Wood shakes and shingles. Wood shakes and shingles shall conform to CSSB *Grading Rules for Wood Shakes and Shingles.*

R703.5.1 Application. Wood shakes or shingles shall be applied either single-course or double-course over nominal $^1/_2$-inch (13 mm) wood-based sheathing or to furring strips over $^1/_2$-inch (13 mm) nominal nonwood sheathing . A permeable water-resistive barrier shall be provided over all sheathing, with horizontal overlaps in the membrane of not less than 2 inches (51mm) and vertical overlaps of not less than 6 inches (152 mm). Where furring strips are used, they shall be 1 inch by 3 inches or 1 inch by 4 inches (25 mm by 76 mm or 25 mm by 102 mm) and shall be fastened horizontally to the studs with 7d or 8d box nails and shall be spaced a distance on center equal to the actual weather exposure of the shakes or shingles, not to exceed the maximum exposure specified in Table R703.5.2. The spacing between adjacent shingles to allow for expansion shall not exceed $^1/_4$ inch (6 mm), and between adjacent shakes, it shall not exceed $^1/_2$ inch (13 mm). The offset spacing between joints in adjacent courses shall be a minimum of $1^1/_2$ inches (38 mm).

R703.5.2 Weather exposure. The maximum weather exposure for shakes and shingles shall not exceed that specified in Table R703.5.2.

R703.5.3 Attachment. Each shake or shingle shall be held in place by two hot-dipped zinc-coated, stainless steel, or aluminum nails or staples. The fasteners shall be long enough to penetrate the sheathing or furring strips by a minimum of $^1/_2$ inch (13 mm) and shall not be overdriven.

R703.5.3.1 Staple attachment. Staples shall not be less than 16 gage and shall have a crown width of not less than $^7/_{16}$ inch (11 mm), and the crown of the staples shall be parallel with the butt of the shake or shingle. In single-course application, the fasteners shall be concealed by the course above and shall be driven approximately 1 inch (25 mm) above the butt line of the succeeding course and $^3/_4$ inch (19 mm) from the edge. In double-course applications, the exposed shake or shingle shall be face-nailed with two casing nails, driven approximately 2 inches (51 mm) above the butt line and $^3/_4$ inch (19 mm) from each edge. In all applications, staples shall be concealed by the course above. With shingles wider than 8 inches (203 mm) two additional nails shall be required and shall be nailed approximately 1 inch (25 mm) apart near the center of the shingle.

R703.5.4 Bottom courses. The bottom courses shall be doubled.

R703.6 Exterior plaster. Installation of these materials shall be in compliance with ASTM C 926 and ASTM C 1063 and the provisions of this code.

R703.6.1 Lath. All lath and lath attachments shall be of corrosion-resistant materials. Expanded metal or woven wire lath shall be attached with $1^1/_2$-inch-long (38 mm), 11 gage nails having a $^7/_{16}$-inch (11.1 mm) head, or $^7/_8$-inch-long (22.2 mm), 16 gage staples, spaced at no more than 6 inches (152 mm), or as otherwise approved.

R703.6.2 Plaster. Plastering with portland cement plaster shall be not less than three coats when applied over metal lath or wire lath and shall be not less than two coats when applied over masonry, concrete, pressure-preservative treated wood or decay-resistant wood as specified in Section R319.1 or gypsum backing. If the plaster surface is completely covered by veneer or other facing material or is completely concealed, plaster application need be only two coats, provided the total thickness is as set forth in Table R702.1(1).

TABLE R703.4
WEATHER–RESISTANT SIDING ATTACHMENT AND MINIMUM THICKNESS

SIDING MATERIAL		NOMINAL THICKNES[a] (inches)	JOINT TREATMENT	WATER-RESISTIVE BARRIER REQUIRED	TYPE OF SUPPORTS FOR THE SIDING MATERIAL AND FASTENERS[b,c,d]					Number or spacing of fasteners
					Wood or wood structural panel sheathing	Fiberboard sheathing into stud	Gypsum sheathing into stud	Foam plastic sheathing into stud	Direct to studs	
Horizontal aluminum[e]	Without insulation	0.019[f]	Lap	Yes	0.120 nail 1½″ long	0.120 nail 2″ long	0.120 nail 2″ long	0.120 nail[y]	Not allowed	Same as stud spacing
		0.024	Lap	Yes	0.120 nail 1½″ long	0.120 nail 2″ long	0.120 nail 2″ long	0.120 nail[y]	Not allowed	
	With insulation	0.019	Lap	Yes	0.120 nail 1½″ long	0.120 nail 2½″ long	0.120 nail 2½″ long	0.120 nail[y]	0.120 nail 1½″ long	
Brick veneer[z] Concrete masonry veneer[z]		2 2	Section R703	Yes (Note l)	See Section R703 and Figure R703.7[g]					
Hardboard[k] Panel siding-vertical		7/16	—	Yes	Note n	Note n	Note n	Note n	Note n	6″ panel edges 12″ inter. sup.[o]
Hardboard[k] Lap-siding-horizontal		7/16	Note q	Yes	Note p	Note p	Note p	Note p	Note p	Same as stud spacing 2 per bearing
Steel[h]		29 ga.	Lap	Yes	0.113 nail 1¾″ Staple–1¾″	0.113 nail 2¾″ Staple–2½″	0.113 nail 2½″ Staple–2¼″	0.113 nail[y] Staple[y]	Not allowed	Same as stud spacing
Stone veneer		2	Section R703	Yes (Note l)	See Section R703 and Figure R703.7[g]					
Particleboard panels		3/8 – 1/2	—	Yes	6d box nail (2″ × 0.099″)	6d box nail (2″ × 0.099″)	6d box nail (2″ × 0.099″)	box nail[y]	6d box nail (2″ × 0.099″), 3/8 not allowed	6″ panel edge, 12″ inter. sup.
		5/8	—	Yes	6d box nail (2″ × 0.099″)	8d box nail (2½″ × 0.113″)	8d box nail (2½″ × 0.113″)	box nail[y]	6d box nail (2″ × 0.099″)	
Plywood panel[i] (exterior grade)		3/8	—	Yes	0.099 nail–2″	0.113 nail–2½″	0.099 nail–2″	0.113 nail[y]	0.099 nail–2″	6″ on edges, 12″ inter. sup.
Vinyl siding[m]		0.035	Lap	Yes	0.120 nail 1½″ Staple–1¾″	0.120 nail 2″ Staple–2½″	0.120 nail 2″ Staple–2½″	0.120 nail[y] Staple[y]	Not allowed	Same as stud spacing
Wood[j] rustic, drop		3/8 Min	Lap	Yes	Fastener penetration into stud–1″				0.113 nail–2½″ Staple–2″	Face nailing up to 6″ widths, 1 nail per bearing; 8″ widths and over, 2 nails per bearing
Shiplap		19/32 Average	Lap	Yes						
Bevel		7/16		Yes						
Butt tip		3/16	Lap	Yes						
Fiber cement panel siding[r]		5/16	Note s	Yes Note x	6d corrosion-resistant nail[t]	6d corrosion-resistant nail[t]	6d corrosion-resistant nail[t]	6d corrosion-resistant nail[t, y]	4d corrosion-resistant nail[u]	6″ o.c. on edges, 12″ o.c. on intermed. studs
Fiber cement lap siding[r]		5/16	Note v	Yes Note x	6d corrosion-resistant nail[t]	6d corrosion-resistant nail[t]	6d corrosion-resistant nail[t]	6d corrosion-resistant nail[t, y]	6d corrosion-resistant nail[w]	Note w

For SI: 1 inch = 25.4 mm.

a. Based on stud spacing of 16 inches on center where studs are spaced 24 inches, siding shall be applied to sheathing approved for that spacing.

b. Nail is a general description and shall be T-head, modified round head, or round head with smooth or deformed shanks.

c. Staples shall have a minimum crown width of 7/16-inch outside diameter and be manufactured of minimum 16 gage wire.

d. Nails or staples shall be aluminum, galvanized, or rust-preventative coated and shall be driven into the studs for fiberboard or gypsum backing.

e. Aluminum nails shall be used to attach aluminum siding.

f. Aluminum (0.019 inch) shall be unbacked only when the maximum panel width is 10 inches and the maximum flat area is 8 inches. The tolerance for aluminum siding shall be +0.002 inch of the nominal dimension.

g. All attachments shall be coated with a corrosion-resistant coating.

h. Shall be of approved type.

(continued)

Footnotes to Table R703.4—continued

i. Three-eighths-inch plywood shall not be applied directly to studs spaced more than 16 inches on center when long dimension is parallel to studs. Plywood $^1/_2$-inch or thinner shall not be applied directly to studs spaced more than 24 inches on center. The stud spacing shall not exceed the panel span rating provided by the manufacturer unless the panels are installed with the face grain perpendicular to the studs or over sheathing approved for that stud spacing.

j. Wood board sidings applied vertically shall be nailed to horizontal nailing strips or blocking set 24 inches on center. Nails shall penetrate $1^1/_2$ inches into studs, studs and wood sheathing combined, or blocking. A weather-resistive membrane shall be installed weatherboard fashion under the vertical siding unless the siding boards are lapped or battens are used.

k. Hardboard siding shall comply with AHA A135.6.

l. For masonry veneer, a weather-resistive sheathing paper is not required over a sheathing that performs as a weather-resistive barrier when a 1-inch air space is provided between the veneer and the sheathing. When the 1-inch space is filled with mortar, a weather-resistive sheathing paper is required over studs or sheathing.

m. Vinyl siding shall comply with ASTM D 3679.

n. Minimum shank diameter of 0.092 inch, minimum head diameter of 0.225 inch, and nail length must accommodate sheathing and penetrate framing $1^1/_2$ inches.

o. When used to resist shear forces, the spacing must be 4 inches at panel edges and 8 inches on interior supports.

p. Minimum shank diameter of 0.099 inch, minimum head diameter of 0.240 inch, and nail length must accommodate sheathing and penetrate framing $1^1/_2$ inches.

q. Vertical end joints shall occur at studs and shall be covered with a joint cover or shall be caulked.

r. Fiber cement siding shall comply with the requirements of ASTM C 1186.

s. See Section R703.10.1.

t. Minimum 0.102″ smooth shank, 0.255″ round head.

u. Minimum 0.099″ smooth shank, 0.250″ round head.

v. See Section R703.10.2.

w. Face nailing: 2 nails at each stud. Concealed nailing: one 11 gage $1^1/_2$ galv. roofing nail (0.371″ head diameter, 0.120″ shank) or 6d galv. box nail at each stud.

x. See Section R703.2 exceptions.

y. Minimum nail length must accommodate sheathing and penetrate framing $1^1/_2$ inches.

z. Adhered masonry veneer shall comply with the requirements in Sections 6.1 and 6.3 of ACI 530/ASCE 5/TMS-402.

TABLE R703.5.2
MAXIMUM WEATHER EXPOSURE FOR WOOD SHAKES AND SHINGLES ON EXTERIOR WALLS[a,b,c]
(Dimensions are in inches)

LENGTH	EXPOSURE FOR SINGLE COURSE	EXPOSURE FOR DOUBLE COURSE
Shingles[a]		
16	$7^1/_2$	12[b]
18	$8^1/_2$	14[c]
24	$11^1/_2$	16
Shakes[a]		
18	$8^1/_2$	14
24	$11^1/_2$	18

For SI: 1 inch = 25.4 mm.

a. Dimensions given are for No. 1 grade.

b. A maximum 10-inch exposure is permitted for No. 2 grade.

c. A maximum 11-inch exposure is permitted for No. 2 grade.

On wood-frame construction with an on-grade floor slab system, exterior plaster shall be applied to cover, but not extend below, lath, paper and screed.

The proportion of aggregate to cementitious materials shall be as set forth in Table R702.1(3).

R703.6.2.1 Weep screeds. A minimum 0.019-inch (0.5 mm) (No. 26 galvanized sheet gage), corrosion-resistant weep screed or plastic weep screed, with a minimum vertical attachment flange of $3^1/_2$ inches (89 mm) shall be provided at or below the foundation plate line on exterior stud walls in accordance with ASTM C 926. The weep screed shall be placed a minimum of 4 inches (102 mm) above the earth or 2 inches (51 mm) above paved areas and shall be of a type that will allow trapped water to drain to the exterior of the building. The weather-resistant barrier shall lap the attachment flange. The exterior lath shall cover and terminate on the attachment flange of the weep screed.

R703.6.3 Water-resistive barriers. Water-resistive barriers shall be installed as required in Section R703.2 and, where applied over wood-based sheathing, shall include a water-resistive vapor-permeable barrier with a performance at least equivalent to two layers of Grade D paper.

Exception: Where the water-resistive barrier that is applied over wood-based sheathing has a water resistance equal to or greater than that of 60 minute Grade D paper and is separated from the stucco by an intervening, substantially nonwater-absorbing layer or designed drainage space.

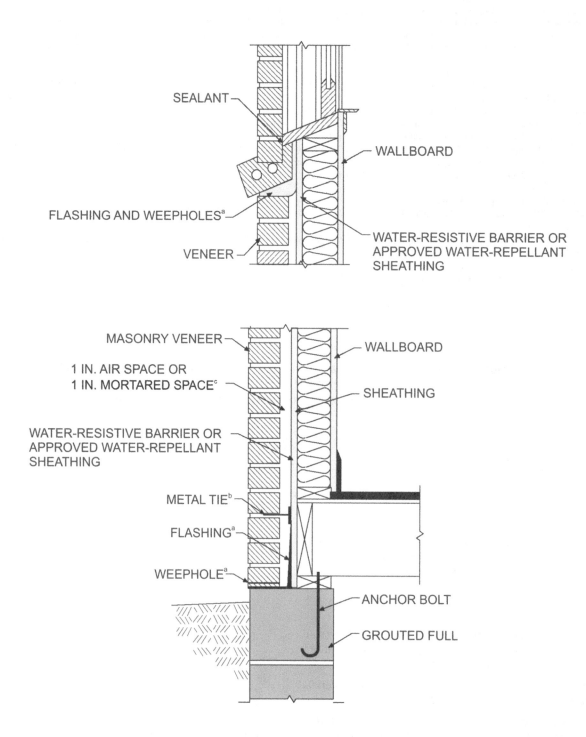

SEALANT

WALLBOARD

FLASHING AND WEEPHOLES[a]

WATER-RESISTIVE BARRIER OR
APPROVED WATER-REPELLANT
SHEATHING

VENEER

MASONRY VENEER

WALLBOARD

1 IN. AIR SPACE OR
1 IN. MORTARED SPACE[c]

SHEATHING

WATER-RESISTIVE BARRIER OR
APPROVED WATER-REPELLANT
SHEATHING

METAL TIE[b]

FLASHING[a]

WEEPHOLE[a]

ANCHOR BOLT

GROUTED FULL

For SI: 1 inch = 25.4 mm.

FIGURE R703.7
MASONRY VENEER WALL DETAILS

(continued)

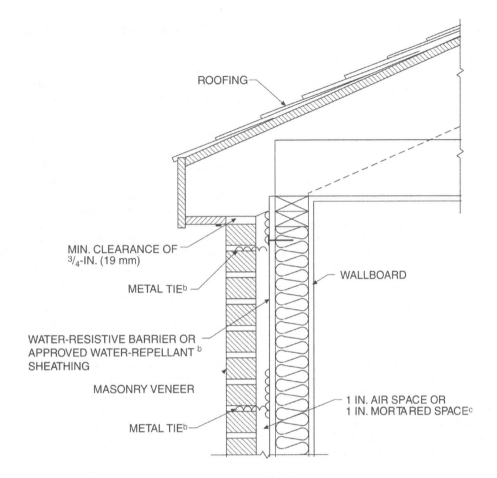

ROOFING

MIN. CLEARANCE OF
$^3/_4$-IN. (19 mm)

METAL TIE[b]

WATER-RESISTIVE BARRIER OR
APPROVED WATER-REPELLANT [b]
SHEATHING

MASONRY VENEER

METAL TIE[b]

WALLBOARD

1 IN. AIR SPACE OR
1 IN. MORTARED SPACE[c]

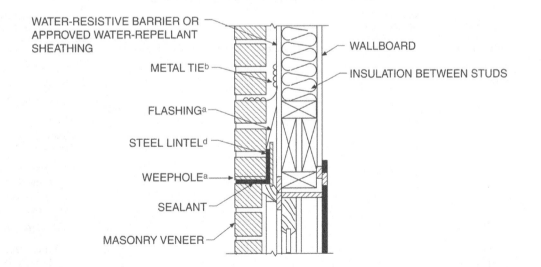

WATER-RESISTIVE BARRIER OR
APPROVED WATER-REPELLANT
SHEATHING

METAL TIE[b]

FLASHING[a]

STEEL LINTEL[d]

WEEPHOLE[a]

SEALANT

MASONRY VENEER

WALLBOARD

INSULATION BETWEEN STUDS

For SI: 1 inch = 25.4 mm.
a. See Sections R703.7.5, R703.7.6 and R703.8.
b. See Sections R703.2 and R703.7.4.
c. See Sections R703.7.4.2 and R703.7.4.3.
d. See Section R703.7.3.

**FIGURE R703.7—continued
MASONRY VENEER WALL DETAILS**

R703.7 Stone and masonry veneer, general. Stone and masonry veneer shall be installed in accordance with this chapter, Table R703.4 and Figure R703.7. These veneers installed over a backing of wood or cold-formed steel shall be limited to the first story above-grade and shall not exceed 5 inches (127 mm) in thickness.

Exceptions:

1. For all buildings in Seismic Design Categories A, B and C, exterior stone or masonry veneer, as specified in Table R703.7(1), with a backing of wood or steel framing shall be permitted to the height specified in Table R703.7(1) above a noncombustible foundation. Wall bracing at exterior and interior braced wall lines shall be in accordance with Section R602.10 or R603.7, and the additional requirements of Table R703.7(1).

2. For detached one- or two-family dwellings in Seismic Design Categories D_0, D_1 and D_2, exterior stone or masonry veneer, as specified in Table R703.7(2), with a backing of wood framing shall be permitted to the height specified in Table R703.7(2) above a noncombustible foundation. Wall bracing and hold downs at exterior and interior braced wall lines shall be in accordance with Sections R602.10 and R602.11 and the additional requirements of Table R703.7(2). In Seismic Design Categories D_0, D_1 and D_2, cripple walls shall not be permitted, and required interior braced wall lines shall be supported on continuous foundations.

R703.7.1 Interior veneer support. Veneers used as interior wall finishes shall be permitted to be supported on wood or cold-formed steel floors that are designed to support the loads imposed.

R703.7.2 Exterior veneer support. Except in Seismic Design Categories D_0, D_1 and D_2, exterior masonry veneers having an installed weight of 40 pounds per square foot (195 kg/m²) or less shall be permitted to be supported on wood or cold-formed steel construction. When masonry veneer supported by wood or cold-formed steel construction adjoins masonry veneer supported by the foundation, there shall be a movement joint between the veneer supported by the wood or cold-formed steel construction and the veneer supported by the foundation. The wood or cold-formed steel construction supporting the masonry veneer shall be designed to limit the deflection to $^1/_{600}$ of the span for the supporting members. The design of the wood or cold-formed steel construction shall consider the weight of the veneer and any other loads.

R703.7.2.1 Support by steel angle. A minimum 6 inches by 4 inches by $^5/_{16}$ inch (152 mm by 102 mm by 8 mm) steel angle, with the long leg placed vertically, shall be anchored to double 2 inches by 4 inches (51 mm by 102 mm) wood studs at a maximum on-center spacing of 16 inches (406 mm). Anchorage of the steel angle at every double stud spacing shall be a minimum of two $^7/_{16}$ inch

TABLE R703.7(1)
STONE OR MASONRY VENEER LIMITATIONS AND REQUIREMENTS, WOOD
OR STEEL FRAMING, SEISMIC DESIGN CATEGORIES A, B AND C

SEISMIC DESIGN CATEGORY	NUMBER OF WOOD OR STEEL FRAMED STORIES	MAXIMUM HEIGHT OF VENEER ABOVE NONCOMBUSTIBLE FOUNDATION[a] (feet)	MAXIMUM NOMINAL THICKNESS OF VENEER (inches)	MAXIMUM WEIGHT OF VENEER (psf)[b]	WOOD OR STEEL FRAMED STORY	MINIMUM SHEATHING AMOUNT (percent of braced wall line length)[c]
A or B	Steel: 1 or 2 Wood: 1, 2 or 3	30	5	50	all	Table R602.10.1 or Table R603.7
C	1	30	5	50	1 only	Table R602.10.1 or Table R603.7
	2	30	5	50	top	Table R602.10.1 or Table R603.7
					bottom	1.5 times length required by Table R602.10.1 or 1.5 times length required by Table R603.7
	Wood only: 3	30	5	50	top	Table R602.10.1
					middle	1.5 times length required by Table R602.10.1
					bottom	1.5 times length required by Table R602.10.1

For SI: 1 inch = 25.4 mm, 1 foot = 304.8 mm, 1 pound per square foot = 0.479 kPa.

a. An Additional 8 feet is permitted for gable end walls. See also story height limitations of Section R301.3.

b. Maximum weight is installed weight and includes weight of mortar, grout, lath and other materials used for installation. Where veneer is placed on both faces of a wall, the combined weight shall not exceed that specified in this table.

c. Applies to exterior and interior braced wall lines.

TABLE R703.7(2)
STONE OR MASONRY VENEER LIMITATIONS AND REQUIREMENTS, ONE- AND TWO-FAMILY DETACHED DWELLINGS, WOOD FRAMING, SEISMIC DESIGN CATEGORIES D₀, D₁ AND D₂

SEISMIC DESIGN CATEGORY	NUMBER OF WOOD FRAMED STORIES[a]	MAXIMUM HEIGHT OF VENEER ABOVE NONCOMBUSTIBLE FOUNDATION OR FOUNDATION WALL (feet)	MAXIMUM NOMINAL THICKNESS OF VENEER (inches)	MAXIMUM WEIGHT OF VENEER (psf)[b]	WOOD FRAMED STORY	MINIMUM SHEATHING AMOUNT (percent of braced wall line length)[c]	MINIMUM SHEATHING THICKNESS AND FASTENING	SINGLE STORY HOLD DOWN FORCE (lb)[d]	CUMULATIVE HOLD DOWN FORCE (lb)[e]
D₀	1	20[f]	4	40	1 only	35	$7/16$-inch wood structural panel sheathing with 8d common nails spaced at 4 inches on center at panel edges, 12 inches on center at intermediate supports. 8d common nails at 4 inches on center at braced wall panel end posts with hold down attached.	N/A	—
	2	20[f]	4	40	top	35		1900	—
					bottom	45		3200	5100
	3	30[g]	4	40	top	40		1900	—
					middle	45		3500	5400
					bottom	60		3500	8900
D₁	1	20[f]	4	40	1 only	45		2100	—
	2	20[f]	4	40	top	45		2100	—
					bottom	45		3700	5800
	3	20[f]	4	40	top	45		2100	—
					middle	45		3700	5800
					bottom	60		3700	9500
D₂	1	20[f]	3	30	1 only	55		2300	—
	2	20[f]	3	30	top	55		2300	—
					bottom	55		3900	6200

For SI: 1 inch = 25.4 mm, 1 foot = 304.8 mm, 1 pound per square foot = 0.479 kPa, 1 pound-force = 4.448 N.

a. Cripple walls are not permitted in Seismic Design Categories D₀, D₁ and D₂.

b. Maximum weight is installed weight and includes weight of mortar, grout and lath, and other materials used for installation.

c. Applies to exterior and interior braced wall lines.

d. Hold down force is minimum allowable stress design load for connector providing uplift tie from wall framing at end of braced wall panel at the noted story to wall framing at end of braced wall panel at the story below, or to foundation or foundation wall. Use single story hold down force where edges of braced wall panels do not align; a continuous load path to the foundation shall be maintained. [See Figure R703.7(1)(b)].

e. Where hold down connectors from stories above align with stories below, use cumulative hold down force to size middle and bottom story hold down connectors. [See Figure R703.7(1)(a)].

f. The veneer shall not exceed 20 feet in height above a noncombustible foundation, with an additional 8 feet permitted for gable end walls, or 30 feet in height with an additional 8 feet for gable end walls where the lower 10 feet has a backing of concrete or masonry wall. See also story height limitations of Section R301.3.

g. The veneer shall not exceed 30 feet in height above a noncombustible foundation, with an additional 8 feet permitted for gable end walls. See also story height limitations of Section R301.3.

(11 mm) diameter by 4 inch (102 mm) lag screws. The steel angle shall have a minimum clearance to underlying construction of $1/16$ inch (2 mm). A minimum of two-thirds the width of the masonry veneer thickness shall bear on the steel angle. Flashing and weep holes shall be located in the masonry veneer wythe in accordance with Figure R703.7.2.1. The maximum height of masonry veneer above the steel angle support shall be 12 feet, 8 inches (3861 mm). The air space separating the masonry veneer from the wood backing shall be in accordance with Sections R703.7.4 and R703.7.4.2. The method of support for the masonry veneer on wood construction shall be constructed in accordance with Figure R703.7.2.1.

The maximum slope of the roof construction without stops shall be 7:12. Roof construction with slopes greater than 7:12 but not more than 12:12 shall have stops of a minimum 3 inch × 3 inch × $1/4$ inch (76 mm × 76 mm × 6 mm) steel plate welded to the angle at 24 inches (610 mm) on center along the angle or as approved by the building official.

R703.7.2.2 Support by roof construction. A steel angle shall be placed directly on top of the roof construction. The roof supporting construction for the steel angle shall consist of a minimum of three 2-inch by 6-inch (51 mm by 152 mm) wood members. The wood member abutting the vertical wall stud construction shall be anchored with a minimum of three $5/8$-inch (16 mm) diameter by 5-inch (127 mm) lag screws to every wood stud spacing. Each additional roof member shall be anchored by the use of two 10d nails at every wood stud spacing. A minimum of two-thirds the width of the masonry veneer thickness shall bear on the steel angle. Flashing and weep holes shall be located in the masonry veneer wythe in accordance with Figure R703.7.2.2. The maximum height of the masonry veneer above the steel angle support shall be 12 feet, 8 inches (3861 mm). The air space separating the masonry veneer from the wood backing shall be in accordance with Sections R703.7.4 and R703.7.4.2. The support for the masonry veneer on wood construction shall be constructed in accordance with Figure R703.7.2.2.

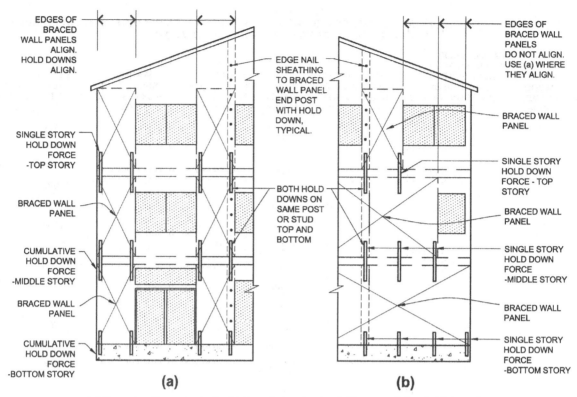

(a) Braced wall panels stacked (aligned story to story). Use cumulative hold down force.
(b) Braced wall panels not stacked. Use single story hold down force.

FIGURE R703.7(1)
HOLD DOWNS AT EXTERIOR AND INTERIOR BRACED WALL PANELS
WHEN USING STONE OR MASONRY VENEER

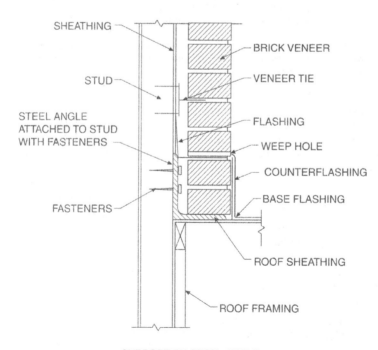

SUPPORT BY STEEL ANGLE

FIGURE R703.7.2.1
EXTERIOR MASONRY VENEER SUPPORT BY STEEL ANGLES

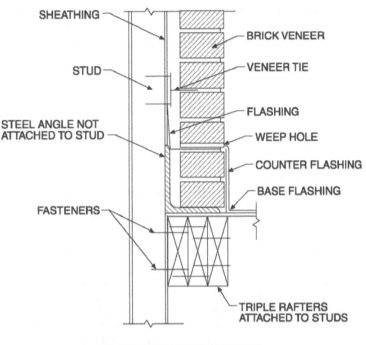

SHEATHING

STUD

STEEL ANGLE NOT
ATTACHED TO STUD

FASTENERS

BRICK VENEER

VENEER TIE

FLASHING

WEEP HOLE

COUNTER FLASHING

BASE FLASHING

TRIPLE RAFTERS
ATTACHED TO STUDS

SUPPORT BY ROOF MEMBERS

FIGURE R703.7.2.2
EXTERIOR MASONRY VENEER SUPPORT BY ROOF MEMBERS

The maximum slope of the roof construction without stops shall be 7:12. Roof construction with slopes greater than 7:12 but not more than 12:12 shall have stops of a minimum 3 inch × 3 inch × $^1/_4$ inch (76 mm × 76 mm × 6 mm) steel plate welded to the angle at 24 inches (610 mm) on center along the angle or as approved by the building official.

R703.7.3 Lintels. Masonry veneer shall not support any vertical load other than the dead load of the veneer above. Veneer above openings shall be supported on lintels of noncombustible materials and the allowable span shall not exceed the value set forth in Table R703.7.3. The lintels shall have a length of bearing not less than 4 inches (102 mm).

R703.7.4 Anchorage. Masonry veneer shall be anchored to the supporting wall with corrosion-resistant metal ties. Where veneer is anchored to wood backings by corrugated sheet metal ties, the distance separating the veneer from the sheathing material shall be a maximum of a nominal 1 inch (25 mm). Where the veneer is anchored to wood backings using metal strand wire ties, the distance separating the veneer from the sheathing material shall be a maximum of $4^1/_2$ inches (114 mm). Where the veneer is anchored to cold-formed steel backings, adjustable metal strand wire ties shall be used. Where veneer is anchored to cold-formed steel backings, the distance separating the veneer from the sheathing material shall be a maximum of $4^1/_2$ inches (114 mm).

R703.7.4.1 Size and spacing. Veneer ties, if strand wire, shall not be less in thickness than No. 9 U.S. gage [(0.148 in.) (4 mm)] wire and shall have a hook embedded in the mortar joint, or if sheet metal, shall be not less than No.

22 U.S. gage by [(0.0299 in.)(0.76 mm)] $^7/_8$ inch (22 mm) corrugated. Each tie shall be spaced not more than 24 inches (610 mm) on center horizontally and vertically and shall support not more than 2.67 square feet (0.25 m²) of wall area.

Exception: In Seismic Design Category D_0, D_1 or D_2 or townhouses in Seismic Design Category C or in wind areas of more than 30 pounds per square foot pressure (1.44 kPa), each tie shall support not more than 2 square feet (0.2 m²) of wall area.

R703.7.4.1.1 Veneer ties around wall openings. Veneer ties around wall openings. Additional metal ties shall be provided around all wall openings greater than 16 inches (406 mm) in either dimension. Metal ties around the perimeter of openings shall be spaced not more than 3 feet (9144 mm) on center and placed within 12 inches (305 mm) of the wall opening.

R703.7.4.2 Air space. The veneer shall be separated from the sheathing by an air space of a minimum of a nominal 1 inch (25 mm) but not more than $4^1/_2$ inches (114 mm).

R703.7.4. 3 Mortar or grout fill. As an alternate to the air space required by Section R703.7.4.2, mortar or grout shall be permitted to fill the air space . When the air space is filled with mortar, a water-resistive barrier is required over studs or sheathing. When filling the air space, replacing the sheathing and water-resistive barrier with a wire mesh and approved water-resistive barrier or an approved water-resistive barrier-backed reinforcement attached directly to the studs is permitted.

TABLE R703.7.3
ALLOWABLE SPANS FOR LINTELS SUPPORTING MASONRY VENEER[a, b, c]

SIZE OF STEEL ANGLE[a, c] (inches)	NO STORY ABOVE	ONE STORY ABOVE	TWO STORIES ABOVE	NO. OF $^1/_2$" OR EQUIVALENT REINFORCING BARS[b]
$3 \times 3 \times ^1/_4$	6'-0"	4'-6"	3'-0"	1
$4 \times 3 \times ^1/_4$	8'-0"	6'-0"	4'-6"	1
$5 \times 3^1/_2 \times ^5/_{16}$	10'-0"	8'-0"	6'-0"	2
$6 \times 3^1/_2 \times ^5/_{16}$	14'-0"	9'-6"	7'-0"	2
$2\text{-}6 \times 3^1/_2 \times ^5/_{16}$	20'-0"	12'-0"	9'-6"	4

For SI: 1 inch = 25.4 mm, 1 foot = 304.8 mm.

a. Long leg of the angle shall be placed in a vertical position.

b. Depth of reinforced lintels shall not be less than 8 inches and all cells of hollow masonry lintels shall be grouted solid. Reinforcing bars shall extend not less than 8 inches into the support.

c. Steel members indicated are adequate typical examples; other steel members meeting structural design requirements may be used.

R703.7.5 Flashing. Flashing shall be located beneath the first course of masonry above finished ground level above the foundation wall or slab and at other points of support, including structural floors, shelf angles and lintels when masonry veneers are designed in accordance with Section R703.7. See Section R703.8 for additional requirements.

R703.7.6 Weepholes. Weepholes shall be provided in the outside wythe of masonry walls at a maximum spacing of 33 inches (838 mm) on center. Weepholes shall not be less than $^3/_{16}$ inch (5 mm) in diameter. Weepholes shall be located immediately above the flashing.

R703.8 Flashing. Approved corrosion-resistant flashing shall be applied shingle-fashion in such a manner to prevent entry of water into the wall cavity or penetration of water to the building structural framing components. The flashing shall extend to the surface of the exterior wall finish. Approved corrosion-resistant flashings shall be installed at all of the following locations:

1. Exterior window and door openings. Flashing at exterior window and door openings shall extend to the surface of the exterior wall finish or to the water-resistive barrier for subsequent drainage.

2. At the intersection of chimneys or other masonry construction with frame or stucco walls, with projecting lips on both sides under stucco copings.

3. Under and at the ends of masonry, wood or metal copings and sills.

4. Continuously above all projecting wood trim.

5. Where exterior porches, decks or stairs attach to a wall or floor assembly of wood-frame construction.

6. At wall and roof intersections.

7. At built-in gutters.

R703.9 Exterior insulation finish systems, general. All Exterior Insulation Finish Systems (EIFS) shall be installed in accordance with the manufacturer's installation instructions and the requirements of this section. Decorative trim shall not be face nailed through the EIFS. The EIFS shall terminate not less than 6 inches (152 mm) above the finished ground level.

R703.9.1 Water-resistive barrier. All EIFS shall have a water-resistive barrier applied between the underlying water-sensitive building components and the exterior insulation, and a means of draining water to the exterior of the veneer. A water-resistive barrier shall be compliant with ASTM D 226 Type I asphalt saturated felt or equivalent, shall be applied horizontally with the upper layer lapped over the lower layer not less than 2 inches (51 mm), and shall have all vertical joints lapped not less than 6 inches (152 mm).

R703.9.2 Flashing, general. Flashing of EIFS shall be provided in accordance with the requirements of Section R703.8.

R703.10 Fiber cement siding.

R703.10.1 Panel siding. Panels shall be installed with the long dimension parallel to framing. Vertical joints shall occur over framing members and shall be sealed with caulking or covered with battens. Horizontal joints shall be flashed with Z-flashing and blocked with solid wood framing.

R703.10.2 Horizontal lap siding. Lap siding shall be lapped a minimum of $1^1/_4$ inches (32 mm) and shall have the ends sealed with caulking, covered with an H-section joint cover, or located over a strip of flashing. Lap siding courses may be installed with the fastener heads exposed or concealed, according to approved manufacturers' installation instructions.

R703.11 Vinyl siding. Vinyl siding shall be certified and labeled as conforming to the requirements of ASTM D 3679 by an approved quality control agency.

R703.11.1 Installation. Vinyl siding, soffit and accessories shall be installed in accordance with the manufacturer's installation instructions.

CHAPTER 8
ROOF-CEILING CONSTRUCTION

SECTION R801
GENERAL

R801.1 Application. The provisions of this chapter shall control the design and construction of the roof-ceiling system for all buildings.

R801.2 Requirements. Roof and ceiling construction shall be capable of accommodating all loads imposed according to Section R301 and of transmitting the resulting loads to the supporting structural elements.

R801.3 Roof drainage. In areas where expansive or collapsible soils are known to exist, all dwellings shall have a controlled method of water disposal from roofs that will collect and discharge roof drainage to the ground surface at least 5 feet (1524 mm) from foundation walls or to an approved drainage system.

SECTION R802
WOOD ROOF FRAMING

R802.1 Identification. Load-bearing dimension lumber for rafters, trusses and ceiling joists shall be identified by a grade mark of a lumber grading or inspection agency that has been approved by an accreditation body that complies with DOC PS 20. In lieu of a grade mark, a certificate of inspection issued by a lumber grading or inspection agency meeting the requirements of this section shall be accepted.

R802.1.1 Blocking. Blocking shall be a minimum of utility grade lumber.

R802.1.2 End-jointed lumber. Approved end-jointed lumber identified by a grade mark conforming to Section R802.1 may be used interchangeably with solid-sawn members of the same species and grade.

R802.1.3 Fire-retardant-treated wood. Fire-retardant-treated wood (FRTW) is any wood product which, when impregnated with chemicals by a pressure process or other means during manufacture, shall have, when tested in accordance with ASTM E 84, a listed flame spread index of 25 or less and shows no evidence of significant progressive combustion when the test is continued for an additional 20-minute period. In addition, the flame front shall not progress more than 10.5 feet (3200 mm) beyond the center line of the burners at any time during the test.

R802.1.3.1 Labeling. Fire-retardant-treated lumber and wood structural panels shall be labeled. The label shall contain:

1. The identification mark of an approved agency in accordance with Section 1703.5 of the *International Building Code*.

2. Identification of the treating manufacturer.

3. The name of the fire-retardant treatment.

4. The species of wood treated.

5. Flame spread and smoke-developed rating.

6. Method of drying after treatment.

7. Conformance to appropriate standards in accordance with Sections R802.1.3.2 through R802.1.3.5.

8. For FRTW exposed to weather, or a damp or wet location, the words "No increase in the listed classification when subjected to the Standard Rain Test" (ASTM D 2898).

R802.1.3.2 Strength adjustments. Design values for untreated lumber and wood structural panels as specified in Section R802.1 shall be adjusted for fire-retardant-treated wood. Adjustments to design values shall be based upon an approved method of investigation which takes into consideration the effects of the anticipated temperature and humidity to which the fire-retardant-treated wood will be subjected, the type of treatment and redrying procedures.

R802.1.3.2.1 Wood structural panels. The effect of treatment and the method of redrying after treatment, and exposure to high temperatures and high humidities on the flexure properties of fire-retardant-treated softwood plywood shall be determined in accordance with ASTM D 5516. The test data developed by ASTM D 5516 shall be used to develop adjustment factors, maximum loads and spans, or both for untreated plywood design values in accordance with ASTM D 6305. Each manufacturer shall publish the allowable maximum loads and spans for service as floor and roof sheathing for their treatment.

R802.1.3.2.2 Lumber. For each species of wood treated, the effect of the treatment and the method of redrying after treatment and exposure to high temperatures and high humidities on the allowable design properties of fire-retardant-treated lumber shall be determined in accordance with ASTM D 5664. The test data developed by ASTM D 5664 shall be used to develop modification factors for use at or near room temperature and at elevated temperatures and humidity in accordance with ASTM D 6841. Each manufacturer shall publish the modification factors for service at temperatures of not less than 80°F (27°C) and for roof framing. The roof framing modification factors shall take into consideration the climatological location.

R802.1.3.3 Exposure to weather. Where fire-retardant-treated wood is exposed to weather or damp or wet locations, it shall be identified as "Exterior" to indicate there is no increase in the listed flame spread index as defined in Section R802.1.3 when subjected to ASTM D 2898.

R802.1.3.4 Interior applications. Interior fire-retardant-treated wood shall have a moisture content of not over 28 percent when tested in accordance with ASTM D 3201 procedures at 92 percent relative humidity. Interior fire-retardant-treated wood shall be tested in accordance with Section R802.1.3.2.1 or R802.1.3.2.2. Interior fire-retardant-treated wood designated as Type A shall be tested in accordance with the provisions of this section.

R802.1.3.5 Moisture content. Fire-retardant-treated wood shall be dried to a moisture content of 19 percent or less for lumber and 15 percent or less for wood structural panels before use. For wood kiln dried after treatment (KDAT) the kiln temperatures shall not exceed those used in kiln drying the lumber and plywood submitted for the tests described in Section R802.1.3.2.1 for plywood and R802.1.3.2.2 for lumber.

R802.1.4 Structural glued laminated timbers. Glued laminated timbers shall be manufactured and identified as required in AITC A190.1 and ASTM D 3737.

R802.1.5 Structural log members. Stress grading of structural log members of nonrectangular shape, as typically used in log buildings, shall be in accordance with ASTM D 3957. Such structural log members shall be identified by the grade mark of an approved lumber grading or inspection agency. In lieu of a grade mark on the material, a certificate of inspection as to species and grade issued by a lumber-grading or inspection agency meeting the requirements of this section shall be permitted to be accepted.

R802.2 Design and construction. The framing details required in Section R802 apply to roofs having a minimum slope of three units vertical in 12 units horizontal (25-percent slope) or greater. Roof-ceilings shall be designed and constructed in accordance with the provisions of this chapter and Figures R606.11(1), R606.11(2) and R606.11(3) or in accordance with AFPA/NDS. Components of roof-ceilings shall be fastened in accordance with Table R602.3(1).

R802.3 Framing details. Rafters shall be framed to ridge board or to each other with a gusset plate as a tie. Ridge board shall be at least 1-inch (25 mm) nominal thickness and not less in depth than the cut end of the rafter. At all valleys and hips there shall be a valley or hip rafter not less than 2-inch (51 mm) nominal thickness and not less in depth than the cut end of the rafter. Hip and valley rafters shall be supported at the ridge by a brace to a bearing partition or be designed to carry and distribute the specific load at that point. Where the roof pitch is less than three units vertical in 12 units horizontal (25-percent slope), structural members that support rafters and ceiling joists, such as ridge beams, hips and valleys, shall be designed as beams.

R802.3.1 Ceiling joist and rafter connections. Ceiling joists and rafters shall be nailed to each other in accordance with Table R802.5.1(9), and the rafter shall be nailed to the top wall plate in accordance with Table R602.3(1). Ceiling joists shall be continuous or securely joined in accordance with Table R802.5.1(9) where they meet over interior partitions and are nailed to adjacent rafters to provide a continu-ous tie across the building when such joists are parallel to the rafters.

Where ceiling joists are not connected to the rafters at the top wall plate, joists connected higher in the attic shall be installed as rafter ties, or rafter ties shall be installed to provide a continuous tie. Where ceiling joists are not parallel to rafters, rafter ties shall be installed. Rafter ties shall be a minimum of 2-inch by 4-inch (51 mm by 102 mm) (nominal), installed in accordance with the connection requirements in Table R802.5.1(9), or connections of equivalent capacities shall be provided. Where ceiling joists or rafter ties are not provided, the ridge formed by these rafters shall be supported by a wall or girder designed in accordance with accepted engineering practice.

Collar ties or ridge straps to resist wind uplift shall be connected in the upper third of the attic space in accordance with Table R602.3(1).

Collar ties shall be a minimum of 1-inch by 4-inch (25 mm by 102 mm) (nominal), spaced not more than 4 feet (1219 mm) on center.

R802.3.2 Ceiling joists lapped. Ends of ceiling joists shall be lapped a minimum of 3 inches (76 mm) or butted over bearing partitions or beams and toenailed to the bearing member. When ceiling joists are used to provide resistance to rafter thrust, lapped joists shall be nailed together in accordance with Table R802.5.1(9) and butted joists shall be tied together in a manner to resist such thrust.

R802.4 Allowable ceiling joist spans. Spans for ceiling joists shall be in accordance with Tables R802.4(1) and R802.4(2). For other grades and species and for other loading conditions, refer to the AF&PA Span Tables for Joists and Rafters.

R802.5 Allowable rafter spans. Spans for rafters shall be in accordance with Tables R802.5.1(1) through R802.5.1(8). For other grades and species and for other loading conditions, refer to the AF&PA Span Tables for Joists and Rafters. The span of each rafter shall be measured along the horizontal projection of the rafter.

R802.5.1 Purlins. Installation of purlins to reduce the span of rafters is permitted as shown in Figure R802.5.1. Purlins shall be sized no less than the required size of the rafters that they support. Purlins shall be continuous and shall be supported by 2-inch by 4-inch (51 mm by 102 mm) braces installed to bearing walls at a slope not less than 45 degrees from the horizontal. The braces shall be spaced not more than 4 feet (1219 mm) on center and the unbraced length of braces shall not exceed 8 feet (2438 mm).

R802.6 Bearing. The ends of each rafter or ceiling joist shall have not less than $1^1/_2$ inches (38 mm) of bearing on wood or metal and not less than 3 inches (76 mm) on masonry or concrete.

R802.6.1 Finished ceiling material. If the finished ceiling material is installed on the ceiling prior to the attachment of the ceiling to the walls, such as in construction at a factory, a compression strip of the same thickness as the finish ceiling material shall be installed directly above the top plate of bearing walls if the compressive strength of the finish ceiling material is less than the loads it will be required to withstand. The compression strip shall cover the entire length of such top plate and shall be at least one-half the width of the top plate. It shall be of material capable of transmitting the loads transferred through it.

R802.7 Cutting and notching. Structural roof members shall not be cut, bored or notched in excess of the limitations specified in this section.

R802.7.1 Sawn lumber. Notches in solid lumber joists, rafters and beams shall not exceed one-sixth of the depth of the member, shall not be longer than one-third of the depth of the member and shall not be located in the middle one-third of the span. Notches at the ends of the member shall not exceed one-fourth the depth of the member. The tension side of members 4 inches (102 mm) or greater in nominal thickness shall not be notched except at the ends of the members. The diameter of the holes bored or cut into members shall not exceed one-third the depth of the member. Holes shall not be closer than 2 inches (51 mm) to the top or bottom of the member, or to any other hole located in the member. Where the member is also notched, the hole shall not be closer than 2 inches (51 mm) to the notch.

> **Exception:** Notches on cantilevered portions of rafters are permitted provided the dimension of the remaining portion of the rafter is not less than 4-inch nominal (102 mm) and the length of the cantilever does not exceed 24 inches (610 mm).

R802.7.2 Engineered wood products. Cuts, notches and holes bored in trusses, structural composite lumber, structural glue-laminated members or I-joists are prohibited except where permitted by the manufacturer's recommendations or where the effects of such alterations are specifically considered in the design of the member by a registered design professional.

R802.8 Lateral support. Rafters and ceiling joists having a depth-to-thickness ratio exceeding 5 to 1 based on nominal dimensions shall be provided with lateral support at points of bearing to prevent rotation.

R802.8.1 Bridging. Rafters and ceiling joists having a depth-to-thickness ratio exceeding 6 to 1 based on nominal dimensions shall be supported laterally by solid blocking, diagonal bridging (wood or metal) or a continuous 1-inch by 3-inch (25 mm by 76 mm) wood strip nailed across the rafters or ceiling joists at intervals not exceeding 8 feet (2438 mm).

R802.9 Framing of openings. Openings in roof and ceiling framing shall be framed with header and trimmer joists. When the header joist span does not exceed 4 feet (1219 mm), the header joist may be a single member the same size as the ceiling joist or rafter. Single trimmer joists may be used to carry a single header joist that is located within 3 feet (914 mm) of the trimmer joist bearing. When the header joist span exceeds 4 feet (1219 mm), the trimmer joists and the header joist shall be doubled and of sufficient cross section to support the ceiling joists or rafter framing into the header. Approved hangers shall be used for the header joist to trimmer joist connections when the header joist span exceeds 6 feet (1829 mm). Tail joists over 12 feet (3658 mm) long shall be supported at the header by

framing anchors or on ledger strips not less than 2 inches by 2 inches (51 mm by 51 mm).

R802.10 Wood trusses.

R802.10.1 Truss design drawings. Truss design drawings, prepared in conformance to Section R802.10.1, shall be provided to the building official and approved prior to installation. Truss design drawings shall include, at a minimum, the information specified below. Truss design drawing shall be provided with the shipment of trusses delivered to the jobsite.

1. Slope or depth, span and spacing.

2. Location of all joints.

3. Required bearing widths.

4. Design loads as applicable.

    4.1. Top chord live load (as determined from Section R301.6).

    4.2. Top chord dead load.

    4.3. Bottom chord live load.

    4.4. Bottom chord dead load.

    4.5. Concentrated loads and their points of application.

    4.6. Controlling wind and earthquake loads.

5. Adjustments to lumber and joint connector design values for conditions of use.

6. Each reaction force and direction.

7. Joint connector type and description (e.g., size, thickness or gage) and the dimensioned location of each joint connector except where symmetrically located relative to the joint interface.

8. Lumber size, species and grade for each member.

9. Connection requirements for:

    9.1. Truss to girder-truss.

    9.2. Truss ply to ply.

    9.3. Field splices.

10. Calculated deflection ratio and/or maximum description for live and total load.

11. Maximum axial compression forces in the truss members to enable the building designer to design the size, connections and anchorage of the permanent continuous lateral bracing. Forces shall be shown on the truss design drawing or on supplemental documents.

12. Required permanent truss member bracing location.

R802.10.2 Design. Wood trusses shall be designed in accordance with accepted engineering practice. The design and manufacture of metal-plate-connected wood trusses shall comply with ANSI/TPI 1. The truss design drawings shall be prepared by a registered professional where required by the statutes of the jurisdiction in which the project is to be constructed in accordance with Section R106.1.

TABLE R802.4(1)
CEILING JOIST SPANS FOR COMMON LUMBER SPECIES
(Uninhabitable attics without storage, live load = 10 psf, L/Δ = 240)

CEILING JOIST SPACING (inches)	SPECIES AND GRADE		DEAD LOAD = 5 psf			
			2 × 4	2 × 6	2 × 8	2 × 10
			Maximum ceiling joist spans			
			(feet - inches)	(feet - inches)	(feet - inches)	(feet - inches)
12	Douglas fir-larch	SS	13-2	20-8	Note a	Note a
	Douglas fir-larch	#1	12-8	19-11	Note a	Note a
	Douglas fir-larch	#2	12-5	19-6	25-8	Note a
	Douglas fir-larch	#3	10-10	15-10	20-1	24-6
	Hem-fir	SS	12-5	19-6	25-8	Note a
	Hem-fir	#1	12-2	19-1	25-2	Note a
	Hem-fir	#2	11-7	18-2	24-0	Note a
	Hem-fir	#3	10-10	15-10	20-1	24-6
	Southern pine	SS	12-11	20-3	Note a	Note a
	Southern pine	#1	12-8	19-11	Note a	Note a
	Southern pine	#2	12-5	19-6	25-8	Note a
	Southern pine	#3	11-6	17-0	21-8	25-7
	Spruce-pine-fir	SS	12-2	19-1	25-2	Note a
	Spruce-pine-fir	#1	11-10	18-8	24-7	Note a
	Spruce-pine-fir	#2	11-10	18-8	24-7	Note a
	Spruce-pine-fir	#3	10-10	15-10	20-1	24-6
16	Douglas fir-larch	SS	11-11	18-9	24-8	Note a
	Douglas fir-larch	#1	11-6	18-1	23-10	Note a
	Douglas fir-larch	#2	11-3	17-8	23-0	Note a
	Douglas fir-larch	#3	9-5	13-9	17-5	21-3
	Hem-fir	SS	11-3	17-8	23-4	Note a
	Hem-fir	#1	11-0	17-4	22-10	Note a
	Hem-fir	#2	10-6	16-6	21-9	Note a
	Hem-fir	#3	9-5	13-9	17-5	21-3
	Southern pine	SS	11-9	18-5	24-3	Note a
	Southern pine	#1	11-6	18-1	23-1	Note a
	Southern pine	#2	11-3	17-8	23-4	Note a
	Southern pine	#3	10-0	14-9	18-9	22-2
	Spruce-pine-fir	SS	11-0	17-4	22-10	Note a
	Spruce-pine-fir	#1	10-9	16-11	22-4	Note a
	Spruce-pine-fir	#2	10-9	16-11	22-4	Note a
	Spruce-pine-fir	#3	9-5	13-9	17-5	21-3
19.2	Douglas fir-larch	SS	11-3	17-8	23-3	Note a
	Douglas fir-larch	#1	10-10	17-0	22-5	Note a
	Douglas fir-larch	#2	10-7	16-7	21-0	25-8
	Douglas fir-larch	#3	8-7	12-6	15-10	19-5
	Hem-fir	SS	10-7	16-8	21-11	Note a
	Hem-fir	#1	10-4	16-4	21-6	Note a
	Hem-fir	#2	9-11	15-7	20-6	25-3
	Hem-fir	#3	8-7	12-6	15-10	19-5
	Southern -pine	SS	11-0	17-4	22-10	Note a
	Southern pine	#1	10-10	17-0	22-5	Note a
	Southern pine	#2	10-7	16-8	21-11	Note a
	Southern pine	#3	9-1	13-6	17-2	20-3
	Spruce-pine-fir	SS	10-4	16-4	21-6	Note a
	Spruce-pine-fir	#1	10-2	15-11	21-0	25-8
	Spruce-pine-fir	#2	10-2	15-11	21-0	25-8
	Spruce-pine-fir	#3	8-7	12-6	15-10	19-5

(continued)

TABLE R802.4(1)—continued
CEILING JOIST SPANS FOR COMMON LUMBER SPECIES
(Uninhabitable attics without storage, live load = 10 psf, L/Δ = 240)

CEILING JOIST SPACING (inches)	SPECIES AND GRADE		DEAD LOAD = 5 psf			
			2 × 4	2 × 6	2 × 8	2 × 10
			Maximum ceiling joist spans			
			(feet - inches)	(feet - inches)	(feet - inches)	(feet - inches)
24	Douglas fir-larch	SS	10-5	16-4	21-7	Note a
	Douglas fir-larch	#1	10-0	15-9	20-1	24-6
	Douglas fir-larch	#2	9-10	14-10	18-9	22-11
	Douglas fir-larch	#3	7-8	11-2	14-2	17-4
	Hem-fir	SS	9-10	15-6	20-5	Note a
	Hem-fir	#1	9-8	15-2	19-7	23-11
	Hem-fir	#2	9-2	14-5	18-6	22-7
	Hem-fir	#3	7-8	11-2	14-2	17-4
	Southern pine	SS	10-3	16-1	21-2	Note a
	Southern pine	#1	10-0	15-9	20-10	Note a
	Southern pine	#2	9-10	15-6	20-1	23-11
	Southern pine	#3	8-2	12-0	15-4	18-1
	Spruce-pine-fir	SS	9-8	15-2	19-11	25-5
	Spruce-pine-fir	#1	9-5	14-9	18-9	22-11
	Spruce-pine-fir	#2	9-5	14-9	18-9	22-11
	Spruce-pine-fir	#3	7-8	11-2	14-2	17-4

Check sources for availability of lumber in lengths greater than 20 feet.

For SI: 1 inch = 25.4 mm, 1 foot = 304.8 mm, 1 pound per square foot = 0.0479 kPa.

a. Span exceeds 26 feet in length.

TABLE R802.4(2)
CEILING JOIST SPANS FOR COMMON LUMBER SPECIES
(Uninhabitable attics with limited storage, live load = 20 psf, L/Δ = 240)

CEILING JOIST SPACING (inches)	SPECIES AND GRADE		DEAD LOAD = 10 psf			
			2 × 4	2 × 6	2 × 8	2 × 10
			Maximum ceiling joist spans			
			(feet - inches)	(feet - inches)	(feet - inches)	(feet - inches)
12	Douglas fir-larch	SS	10-5	16-4	21-7	Note a
	Douglas fir-larch	#1	10-0	15-9	20-1	24-6
	Douglas fir-larch	#2	9-10	14-10	18-9	22-11
	Douglas fir-larch	#3	7-8	11-2	14-2	17-4
	Hem-fir	SS	9-10	15-6	20-5	Note a
	Hem-fir	#1	9-8	15-2	19-7	23-11
	Hem-fir	#2	9-2	14-5	18-6	22-7
	Hem-fir	#3	7-8	11-2	14-2	17-4
	Southern pine	SS	10-3	16-1	21-2	Note a
	Southern pine	#1	10-0	15-9	20-10	Note a
	Southern pine	#2	9-10	15-6	20-1	23-11
	Southern pine	#3	8-2	12-0	15-4	18-1
	Spruce-pine-fir	SS	9-8	15-2	19-11	25-5
	Spruce-pine-fir	#1	9-5	14-9	18-9	22-11
	Spruce-pine-fir	#2	9-5	14-9	18-9	22-11
	Spruce-pine-fir	#3	7-8	11-2	14-2	17-4
16	Douglas fir-larch	SS	9-6	14-11	19-7	25-0
	Douglas fir-larch	#1	9-1	13-9	17-5	21-3
	Douglas fir-larch	#2	8-9	12-10	16-3	19-10
	Douglas fir-larch	#3	6-8	9-8	12-4	15-0
	Hem-fir	SS	8-11	14-1	18-6	23-8
	Hem-fir	#1	8-9	13-5	16-10	20-8
	Hem-fir	#2	8-4	12-8	16-0	19-7
	Hem-fir	#3	6-8	9-8	12-4	15-0
	Southern pine	SS	9-4	14-7	19-3	24-7
	Southern pine	#1	9-1	14-4	18-11	23-1
	Southern pine	#2	8-11	13-6	17-5	20-9
	Southern pine	#3	7-1	10-5	13-3	15-8
	Spruce-pine-fir	SS	8-9	13-9	18-1	23-1
	Spruce-pine-fir	#1	8-7	12-10	16-3	19-10
	Spruce-pine-fir	#2	8-7	12-10	16-3	19-10
	Spruce-pine-fir	#3	6-8	9-8	12-4	15-0
19.2	Douglas fir-larch	SS	8-11	14-0	18-5	23-4
	Douglas fir-larch	#1	8-7	12-6	15-10	19-5
	Douglas fir-larch	#2	8-0	11-9	14-10	18-2
	Douglas fir-larch	#3	6-1	8-10	11-3	13-8
	Hem-fir	SS	8-5	13-3	17-5	22-3
	Hem-fir	#1	8-3	12-3	15-6	18-11
	Hem-fir	#2	7-10	11-7	14-8	17-10
	Hem-fir	#3	6-1	8-10	11-3	13-8
	Southern pine	SS	8-9	13-9	18-1	23-1
	Southern pine	#1	8-7	13-6	17-9	21-1
	Southern pine	#2	8-5	12-3	15-10	18-11
	Southern pine	#3	6-5	9-6	12-1	14-4
	Spruce-pine-fir	SS	8-3	12-11	17-1	21-8
	Spruce-pine-fir	#1	8-0	11-9	14-10	18-2
	Spruce-pine-fir	#2	8-0	11-9	14-10	18-2
	Spruce-pine-fir	#3	6-1	8-10	11-3	13-8

(continued)

TABLE R802.4(2)—continued
CEILING JOIST SPANS FOR COMMON LUMBER SPECIES
(Uninhabitable attics with limited storage, live load = 20 psf, L/Δ = 240)

CEILING JOIST SPACING (inches)	SPECIES AND GRADE		DEAD LOAD = 10 psf			
			2 × 4	2 × 6	2 × 8	2 × 10
			Maximum Ceiling Joist Spans			
			(feet - inches)	(feet - inches)	(feet - inches)	(feet - inches)
24	Douglas fir-larch	SS	8-3	13-0	17-1	20-11
	Douglas fir-larch	#1	7-8	11-2	14-2	17-4
	Douglas fir-larch	#2	7-2	10-6	13-3	16-3
	Douglas fir-larch	#3	5-5	7-11	10-0	12-3
	Hem-fir	SS	7-10	12-3	16-2	20-6
	Hem-fir	#1	7-6	10-11	13-10	16-11
	Hem-fir	#2	7-1	10-4	13-1	16-0
	Hem-fir	#3	5-5	7-11	10-0	12-3
	Southern pine	SS	8-1	12-9	16-10	21-6
	Southern pine	#1	8-0	12-6	15-10	18-10
	Southern pine	#2	7-8	11-0	14-2	16-11
	Southern pine	#3	5-9	8-6	10-10	12-10
	Spruce-pine-fir	SS	7-8	12-0	15-10	19-5
	Spruce-pine-fir	#1	7-2	10-6	13-3	16-3
	Spruce-pine-fir	#2	7-2	10-6	13-3	16-3
	Spruce-pine-fir	#3	5-5	7-11	10-0	12-3

Check sources for availability of lumber in lengths greater than 20 feet.

For SI: 1 inch = 25.4 mm, 1 foot = 304.8 mm, 1 pound per square foot = 0.0479 kPa.

a. Span exceeds 26 feet in length.

TABLE R802.5.1(1)
RAFTER SPANS FOR COMMON LUMBER SPECIES
(Roof live load=20 psf, ceiling not attached to rafters, L/Δ = 180)

RAFTER SPACING (inches)	SPECIES AND GRADE		DEAD LOAD = 10 psf					DEAD LOAD = 20 psf				
			2 × 4	2 × 6	2 × 8	2 × 10	2 × 12	2 × 4	2 × 6	2 × 8	2 × 10	2 × 12
			Maximum rafter spans[a]									
			(feet - inches)	(feet - inches)	(feet - inches)	(feet - inches)	(feet - inches)	(feet - inches)	(feet - inches)	(feet - inches)	(feet - inches)	(feet - inches)
12	Douglas fir-larch	SS	11-6	18-0	23-9	Note b	Note b	11-6	18-0	23-5	Note b	Note b
	Douglas fir-larch	#1	11-1	17-4	22-5	Note b	Note b	10-6	15-4	19-5	23-9	Note b
	Douglas fir-larch	#2	10-10	16-7	21-0	25-8	Note b	9-10	14-4	18-2	22-3	25-9
	Douglas fir-larch	#3	8-7	12-6	15-10	19-5	22-6	7-5	10-10	13-9	16-9	19-6
	Hem-fir	SS	10-10	17-0	22-5	Note b	Note b	10-10	17-0	22-5	Note b	Note b
	Hem-fir	#1	10 -7	16-8	21-10	Note b	Note b	10-3	14-11	18-11	23-2	Note b
	Hem-fir	#2	10-1	15-11	20-8	25-3	Note b	9-8	14-2	17-11	21-11	25-5
	Hem-fir	#3	8-7	12-6	15-10	19-5	22-6	7-5	10-10	13-9	16-9	19-6
	Southern pine	SS	11-3	17-8	23-4	Note b	Note b	11-3	17-8	23-4	Note b	Note b
	Southern pine	#1	11-1	17-4	22-11	Note b	Note b	11-1	17-3	21-9	25-10	Note b
	Southern pine	#2	10-10	17-0	22-5	Note b	Note b	10-6	15-1	19-5	23-2	Note b
	Southern pine	#3	9-1	13-6	17-2	20-3	24-1	7-11	11-8	14-10	17-6	20-11
	Spruce-pine-fir	SS	10-7	16-8	21-11	Note b	Note b	10-7	16-8	21-9	Note b	Note b
	Spruce-pine-fir	#1	10-4	16-3	21-0	25-8	Note b	9-10	14-4	18-2	22-3	25-9
	Spruce-pine-fir	#2	10-4	16-3	21-0	25-8	Note b	9-10	14-4	18-2	22-3	25-9
	Spruce-pine-fir	#3	8-7	12-6	15-10	19-5	22-6	7-5	10-10	13-9	16-9	19-6
16	Douglas fir-larch	SS	10-5	16-4	21-7	Note b	Note b	10-5	16-0	20-3	24-9	Note b
	Douglas fir-larch	#1	10-0	15-4	19-5	23-9	Note b	9-1	13-3	16-10	20-7	23-10
	Douglas fir-larch	#2	9-10	14-4	18-2	22-3	25-9	8-6	12-5	15-9	19-3	22-4
	Douglas fir-larch	#3	7-5	10-10	13-9	16-9	19-6	6-5	9-5	11-11	14-6	16-10
	Hem-fir	SS	9-10	15-6	20-5	Note b	Note b	9-10	15-6	19-11	24-4	Note b
	Hem-fir	#1	9-8	14-11	18-11	23-2	Note b	8-10	12-11	16-5	20-0	23-3
	Hem-fir	#2	9-2	14-2	17-11	21-11	25-5	8-5	12-3	15-6	18-11	22-0
	Hem-fir	#3	7-5	10-10	13-9	16-9	19-6	6-5	9-5	11-11	14-6	16-10
	Southern pine	SS	10-3	16-1	21-2	Note b	Note b	10-3	16-1	21-2	Note b	Note b
	Southern pine	#1	10-0	15-9	20-10	25-10	Note b	10-0	15-0	18-10	22-4	Note b
	Southern pine	#2	9-10	15-1	19-5	23-2	Note b	9-1	13-0	16-10	20-1	23-7
	Southern pine	#3	7-11	11-8	14-10	17-6	20-11	6-10	10-1	12-10	15-2	18-1
	Spruce-pine-fir	SS	9-8	15-2	19-11	25-5	Note b	9-8	14-10	18-10	23-0	Note b
	Spruce-pine-fir	#1	9-5	14-4	18-2	22-3	25-9	8-6	12-5	15-9	19-3	22-4
	Spruce-pine-fir	#2	9-5	14-4	18-2	22-3	25-9	8-6	12-5	15-9	19-3	22-4
	Spruce-pine-fir	#3	7-5	10-10	13-9	16-9	19-6	6-5	9-5	11-11	14-6	16-10
19.2	Douglas fir-larch	SS	9-10	15-5	20-4	25-11	Note b	9-10	14-7	18-6	22-7	Note b
	Douglas fir-larch	#1	9-5	14-0	17-9	21-8	25-2	8-4	12-2	15-4	18-9	21-9
	Douglas fir-larch	#2	8-11	13-1	16-7	20-3	23-6	7-9	11-4	14-4	17-7	20-4
	Douglas fir-larch	#3	6-9	9-11	12-7	15-4	17-9	5-10	8-7	10-10	13-3	15-5
	Hem-fir	SS	9-3	14-7	19-2	24-6	Note b	9-3	14-4	18-2	22-3	25-9
	Hem-fir	#1	9-1	13-8	17-4	21-1	24-6	8-1	11-10	15-0	18-4	21-3
	Hem-fir	#2	8-8	12-11	16-4	20-0	23-2	7-8	11-2	14-2	17-4	20-1
	Hem-fir	#3	6-9	9-11	12-7	15-4	17-9	5-10	8-7	10-10	13-3	15-5
	Southern pine	SS	9-8	15-2	19-11	25-5	Note b	9-8	15-2	19-11	25-5	Note b
	Southern pine	#1	9-5	14-10	19-7	23-7	Note b	9-3	13-8	17-2	20-5	24-4
	Southern pine	#2	9-3	13-9	17-9	21-2	24-10	8-4	11-11	15-4	18-4	21-6
	Southern pine	#3	7-3	10-8	13-7	16-0	19-1	6-3	9-3	11-9	13-10	16-6
	Spruce-pine-fir	SS	9-1	14-3	18-9	23-11	Note b	9-1	13-7	17-2	21-0	24-4
	Spruce-pine-fir	#1	8-10	13-1	16-7	20-3	23-6	7-9	11-4	14-4	17-7	20-4
	Spruce-pine-fir	#2	8-10	13-1	16-7	20-3	23-6	7-9	11-4	14-4	17-7	20-4
	Spruce-pine-fir	#3	6-9	9-11	12-7	15-4	17-9	5-10	8-7	10-10	13-3	15-5

(continued)

TABLE R802.5.1(1)—continued
RAFTER SPANS FOR COMMON LUMBER SPECIES
(Roof live load=20 psf, ceiling not attached to rafters, L/Δ = 180)

RAFTER SPACING (inches)	SPECIES AND GRADE		DEAD LOAD = 10 psf					DEAD LOAD = 20 psf				
			2 × 4	2 × 6	2 × 8	2 × 10	2 × 12	2 × 4	2 × 6	2 × 8	2 × 10	2 × 12
			Maximum rafter spans[a]									
			(feet - inches)	(feet - inches)	(feet - inches)	(feet - inches)	(feet - inches)	(feet - inches)	(feet - inches)	(feet - inches)	(feet - inches)	(feet - inches)
24	Douglas fir-larch	SS	9-1	14-4	18-10	23-4	Note b	8-11	13-1	16-7	20-3	23-5
	Douglas fir-larch	#1	8-7	12-6	15-10	19-5	22-6	7-5	10-10	13-9	16-9	19-6
	Douglas fir-larch	#2	8-0	11-9	14-10	18-2	21-0	6-11	10-2	12-10	15-8	18-3
	Douglas fir-larch	#3	6-1	8-10	11-3	13-8	15-11	5-3	7-8	9-9	11-10	13-9
	Hem-fir	SS	8-7	13-6	17-10	22-9	Note b	8-7	12-10	16-3	19-10	23-0
	Hem-fir	#1	8-4	12-3	15-6	18-11	21-11	7-3	10-7	13-5	16-4	19-0
	Hem-fir	#2	7-11	11-7	14-8	17-10	20-9	6-10	10-0	12-8	15-6	17-11
	Hem-fir	#3	6-1	8-10	11-3	13-8	15-11	5-3	7-8	9-9	11-10	13-9
	Southern pine	SS	8-11	14-1	18-6	23-8	Note b	8-11	14-1	18-6	22-11	Note b
	Southern pine	#1	8-9	13-9	17-9	21-1	25-2	8-3	12-3	15-4	18-3	21-9
	Southern pine	#2	8-7	12-3	15-10	18-11	22-2	7-5	10-8	13-9	16-5	19-3
	Southern pine	#3	6-5	9-6	12-1	14-4	17-1	5-7	8-3	10-6	12-5	14-9
	Spruce-pine-fir	SS	8-5	13-3	17-5	21-8	25-2	8-4	12-2	15-4	18-9	21-9
	Spruce-pine-fir	#1	8-0	11-9	14-10	18-2	21-0	6-11	10-2	12-10	15-8	18-3
	Spruce-pine-fir	#2	8-0	11-9	14-10	18-2	21-0	6-11	10-2	12-10	15-8	18-3
	Spruce-pine-fir	#3	6-1	8-10	11-3	13-8	15-11	5-3	7-8	9-9	11-10	13-9

Check sources for availability of lumber in lengths greater than 20 feet.

For SI: 1 inch = 25.4 mm, 1 foot = 304.8 mm, 1 pound per square foot = 0.0479kPa.

a. The tabulated rafter spans assume that ceiling joists are located at the bottom of the attic space or that some other method of resisting the outward push of the rafters on the bearing walls, such as rafter ties, is provided at that location. When ceiling joists or rafter ties are located higher in the attic space, the rafter spans shall be multiplied by the factors given below:

H_C/H_R	Rafter Span Adjustment Factor
1/3	0.67
1/4	0.76
1/5	0.83
1/6	0.90
1/7.5 or less	1.00

where:

H_C = Height of ceiling joists or rafter ties measured vertically above the top of the rafter support walls.

H_R = Height of roof ridge measured vertically above the top of the rafter support walls.

b. Span exceeds 26 feet in length.

TABLE R802.5.1(2)
RAFTER SPANS FOR COMMON LUMBER SPECIES
(Roof live load=20 psf, ceiling attached to rafters, L/Δ = 240)

RAFTER SPACING (inches)	SPECIES AND GRADE		DEAD LOAD = 10 psf					DEAD LOAD = 20 psf				
			2 × 4	2 × 6	2 × 8	2 × 10	2 × 12	2 × 4	2 × 6	2 × 8	2 × 10	2 × 12
			Maximum rafter spans[a]									
			(feet-inches)	(feet-inches)	(feet-inches)	(feet-inches)	(feet-inches)	(feet-inches)	(feet-inches)	(feet-inches)	(feet-inches)	(feet-inches)
12	Douglas fir-larch	SS	10-5	16-4	21-7	Note b	Note b	10-5	16-4	21-7	Note b	Note b
	Douglas fir-larch	#1	10-0	15-9	20-10	Note b	Note b	10-0	15-4	19-5	23-9	Note b
	Douglas fir-larch	#2	9-10	15-6	20-5	25-8	Note b	9-10	14-4	18-2	22-3	25-9
	Douglas fir-larch	#3	8-7	12-6	15-10	19-5	22-6	7-5	10-10	13-9	16-9	19-6
	Hem-fir	SS	9-10	15-6	20-5	Note b	Note b	9-10	15-6	20-5	Note b	Note b
	Hem-fir	#1	9-8	15-2	19-11	25-5	Note b	9-8	14-11	18-11	23-2	Note b
	Hem-fir	#2	9-2	14-5	19-0	24-3	Note b	9-2	14-2	17-11	21-11	25-5
	Hem-fir	#3	8-7	12-6	15-10	19-5	22-6	7-5	10-10	13-9	16-9	19-6
	Southern pine	SS	10-3	16-1	21-2	Note b	Note b	10-3	16-1	21-2	Note b	Note b
	Southern pine	#1	10-0	15-9	20-10	Note b	Note b	10-0	15-9	20-10	25-10	Note b
	Southern pine	#2	9-10	15-6	20-5	Note b	Note b	9-10	15-1	19-5	23-2	Note b
	Southern pine	#3	9-1	13-6	17-2	20-3	24-1	7-11	11-8	14-10	17-6	20-11
	Spruce-pine-fir	SS	9-8	15-2	19-11	25-5	Note b	9-8	15-2	19-11	25-5	Note b
	Spruce-pine-fir	#1	9-5	14-9	19-6	24-10	Note b	9-5	14-4	18-2	22-3	25-9
	Spruce-pine-fir	#2	9-5	14-9	19-6	24-10	Note b	9-5	14-4	18-2	22-3	25-9
	Spruce-pine-fir	#3	8-7	12-6	15-10	19-5	22-6	7-5	10-10	13-9	16-9	19-6
16	Douglas fir-larch	SS	9-6	14-11	19-7	25-0	Note b	9-6	14-11	19-7	24-9	Note b
	Douglas fir-larch	#1	9-1	14-4	18-11	23-9	Note b	9-1	13-3	16-10	20-7	23-10
	Douglas fir-larch	#2	8-11	14-1	18-2	22-3	25-9	8-6	12-5	15-9	19-3	22-4
	Douglas fir-larch	#3	7-5	10-10	13-9	16-9	19-6	6-5	9-5	11-11	14-6	16-10
	Hem-fir	SS	8-11	14-1	18-6	23-8	Note b	8-11	14-1	18-6	23-8	Note b
	Hem-fir	#1	8-9	13-9	18-1	23-1	Note b	8-9	12-11	16-5	20-0	23-3
	Hem-fir	#2	8-4	13-1	17-3	21-11	25-5	8-4	12-3	15-6	18-11	22-0
	Hem-fir	#3	7-5	10-10	13-9	16-9	19-6	6-5	9-5	11-11	14-6	16-10
	Southern pine	SS	9-4	14-7	19-3	24-7	Note b	9-4	14-7	19-3	24-7	Note b
	Southern pine	#1	9-1	14-4	18-11	24-1	Note b	9-1	14-4	18-10	22-4	Note b
	Southern pine	#2	8-11	14-1	18-6	23-2	Note b	8-11	13-0	16-10	20-1	23-7
	Southern pine	#3	7-11	11-8	14-10	17-6	20-11	6-10	10-1	12-10	15-2	18-1
	Spruce-pine-fir	SS	8-9	13-9	18-1	23-1	Note b	8-9	13-9	18-1	23-0	Note b
	Spruce-pine-fir	#1	8-7	13-5	17-9	22-3	25-9	8-6	12-5	15-9	19-3	22-4
	Spruce-pine-fir	#2	8-7	13-5	17-9	22-3	25-9	8-6	12-5	15-9	19-3	22-4
	Spruce-pine-fir	#3	7-5	10-10	13-9	16-9	19-6	6-5	9-5	11-11	14-6	16-10
19.2	Douglas fir-larch	SS	8-11	14-0	18-5	23-7	Note b	8-11	14-0	18-5	22-7	Note b
	Douglas fir-larch	#1	8-7	13-6	17-9	21-8	25-2	8-4	12-2	15-4	18-9	21-9
	Douglas fir-larch	#2	8-5	13-1	16-7	20-3	23-6	7-9	11-4	14-4	17-7	20-4
	Douglas fir-larch	#3	6-9	9-11	12-7	15-4	17-9	5-10	8-7	10-10	13-3	15-5
	Hem-fir	SS	8-5	13-3	17-5	22-3	Note b	8-5	13-3	17-5	22-3	25-9
	Hem-fir	#1	8-3	12-11	17-1	21-1	24-6	8-1	11-10	15-0	18-4	21-3
	Hem-fir	#2	7-10	12-4	16-3	20-0	23-2	7-8	11-2	14-2	17-4	20-1
	Hem-fir	#3	6-9	9-11	12-7	15-4	17-9	5-10	8-7	10-10	13-3	15-5
	Southern pine	SS	8-9	13-9	18-1	23-1	Note b	8-9	13-9	18-1	23-1	Note b
	Southern pine	#1	8-7	13-6	17-9	22-8	Note b	8-7	13-6	17-2	20-5	24-4
	Southern pine	#2	8-5	13-3	17-5	21-2	24-10	8-4	11-11	15-4	18-4	21-6
	Southern pine	#3	7-3	10-8	13-7	16-0	19-1	6-3	9-3	11-9	13-10	16-6
	Spruce-pine-fir	SS	8-3	12-11	17-1	21-9	Note b	8-3	12-11	17-1	21-0	24-4
	Spruce-pine-fir	#1	8-1	12-8	16-7	20-3	23-6	7-9	11-4	14-4	17-7	20-4
	Spruce-pine-fir	#2	8-1	12-8	16-7	20-3	23-6	7-9	11-4	14-4	17-7	20-4
	Spruce-pine-fir	#3	6-9	9-11	12-7	15-4	17-9	5-10	8-7	10-10	13-3	15-5

(continued)

TABLE R802.5.1(2)—continued
RAFTER SPANS FOR COMMON LUMBER SPECIES
(Roof live load=20 psf, ceiling attached to rafters, L/Δ = 240)

RAFTER SPACING (inches)	SPECIES AND GRADE		DEAD LOAD = 10 psf					DEAD LOAD = 20 psf				
			2 × 4	2 × 6	2 × 8	2 × 10	2 × 12	2 × 4	2 × 6	2 × 8	2 × 10	2 × 12
			Maximum rafter spans[a]									
			(feet-inches)	(feet-inches)	(feet-inches)	(feet-inches)	(feet-inches)	(feet-inches)	(feet-inches)	(feet-inches)	(feet-inches)	(feet-inches)
24	Douglas fir-larch	SS	8-3	13-0	17-2	21-10	Note b	8-3	13-0	16-7	20-3	23-5
	Douglas fir-larch	#1	8-0	12-6	15-10	19-5	22-6	7-5	10-10	13-9	16-9	19-6
	Douglas fir-larch	#2	7-10	11-9	14-10	18-2	21-0	6-11	10-2	12-10	15-8	18-3
	Douglas fir-larch	#3	6-1	8-10	11-3	13-8	15-11	5-3	7-8	9-9	11-10	13-9
	Hem-fir	SS	7-10	12-3	16-2	20-8	25-1	7-10	12-3	16-2	19-10	23-0
	Hem-fir	#1	7-8	12-0	15-6	18-11	21-11	7-3	10-7	13-5	16-4	19-0
	Hem-fir	#2	7-3	11-5	14-8	17-10	20-9	6-10	10-0	12-8	15-6	17-11
	Hem-fir	#3	6-1	8-10	11-3	13-8	15-11	5-3	7-8	9-9	11-10	13-9
	Southern pine	SS	8-1	12-9	16-10	21-6	Note b	8-1	12-9	16-10	21-6	Note b
	Southern pine	#1	8-0	12-6	16-6	21-1	25-2	8-0	12-3	15-4	18-3	21-9
	Southern pine	#2	7-10	12-3	15-10	18-11	22-2	7-5	10-8	13-9	16-5	19-3
	Southern pine	#3	6-5	9-6	12-1	14-4	17-1	5-7	8-3	10-6	12-5	14-9
	Spruce-pine-fir	SS	7-8	12-0	15-10	20-2	24-7	7-8	12-0	15-4	18-9	21-9
	Spruce-pine-fir	#1	7-6	11-9	14-10	18-2	21-0	6-11	10-2	12-10	15-8	18-3
	Spruce-pine-fir	#2	7-6	11-9	14-10	18-2	21-0	6-11	10-2	12-10	15-8	18-3
	Spruce-pine-fir	#3	6-1	8-10	11-3	13-8	15-11	5-3	7-8	9-9	11-10	13-9

Check sources for availability of lumber in lengths greater than 20 feet.

For SI: 1 inch = 25.4 mm, 1 foot = 304.8 mm, 1 pound per square foot = 0.0479 kPa.

a. The tabulated rafter spans assume that ceiling joists are located at the bottom of the attic space or that some other method of resisting the outward push of the rafters on the bearing walls, such as rafter ties, is provided at that location. When ceiling joists or rafter ties are located higher in the attic space, the rafter spans shall be multiplied by the factors given below:

H_C/H_R	Rafter Span Adjustment Factor
1/3	0.67
1/4	0.76
1/5	0.83
1/6	0.90
1/7.5 or less	1.00

where:

H_C = Height of ceiling joists or rafter ties measured vertically above the top of the rafter support walls.

H_R = Height of roof ridge measured vertically above the top of the rafter support walls.

b. Span exceeds 26 feet in length.

TABLE R802.5.1(3)
RAFTER SPANS FOR COMMON LUMBER SPECIES
(Ground snow load=30 psf, ceiling not attached to rafters, L/Δ = 180)

RAFTER SPACING (inches)	SPECIES AND GRADE		DEAD LOAD = 10 psf					DEAD LOAD = 20 psf				
			2 × 4	2 × 6	2 × 8	2 × 10	2 × 12	2 × 4	2 × 6	2 × 8	2 × 10	2 × 12
			Maximum rafter spans[a]									
			(feet-inches)	(feet-inches)	(feet-inches)	(feet-inches)	(feet-inches)	(feet-inches)	(feet-inches)	(feet-inches)	(feet-inches)	(feet-inches)
12	Douglas fir-larch	SS	10-0	15-9	20-9	Note b	Note b	10-0	15-9	20-1	24-6	Note b
	Douglas fir-larch	#1	9-8	14-9	18-8	22-9	Note b	9-0	13-2	16-8	20-4	23-7
	Douglas fir-larch	#2	9-5	13-9	17-5	21-4	24-8	8-5	12-4	15-7	19-1	22-1
	Douglas fir-larch	#3	7-1	10-5	13-2	16-1	18-8	6-4	9-4	11-9	14-5	16-8
	Hem-fir	SS	9-6	14-10	19-7	25-0	Note b	9-6	14-10	19-7	24-1	Note b
	Hem-fir	#1	9-3	14-4	18-2	22-2	25-9	8-9	12-10	16-3	19-10	23-0
	Hem-fir	#2	8-10	13-7	17-2	21-0	24-4	8-4	12-2	15-4	18-9	21-9
	Hem-fir	#3	7-1	10-5	13-2	16-1	18-8	6-4	9-4	11-9	14-5	16-8
	Southern pine	SS	9-10	15-6	20-5	Note b	Note b	9-10	15-6	20-5	Note b	Note b
	Southern pine	#1	9-8	15-2	20-0	24-9	Note b	9-8	14-10	18-8	22-2	Note b
	Southern pine	#2	9-6	14-5	18-8	22-3	Note b	9-0	12-11	16-8	19-11	23-4
	Southern pine	#3	7-7	11-2	14-3	16-10	20-0	6-9	10-0	12-9	15-1	17-11
	Spruce-pine-fir	SS	9-3	14-7	19-2	24-6	Note b	9-3	14-7	18-8	22-9	Note b
	Spruce-pine-fir	#1	9-1	13-9	17-5	21-4	24-8	8-5	12-4	15-7	19-1	22-1
	Spruce-pine-fir	#2	9-1	13-9	17-5	21-4	24-8	8-5	12-4	15-7	19-1	22-1
	Spruce-pine-fir	#3	7-1	10-5	13-2	16-1	18-8	6-4	9-4	11-9	14-5	16-8
16	Douglas fir-larch	SS	9-1	14-4	18-10	23-9	Note b	9-1	13-9	17-5	21-3	24-8
	Douglas fir-larch	#1	8-9	12-9	16-2	19-9	22-10	7-10	11-5	14-5	17-8	20-5
	Douglas fir-larch	#2	8-2	11-11	15-1	18-5	21-5	7-3	10-8	13-6	16-6	19-2
	Douglas fir-larch	#3	6-2	9-0	11-5	13-11	16-2	5-6	8-1	10-3	12-6	14-6
	Hem-fir	SS	8-7	13-6	17-10	22-9	Note b	8-7	13-6	17-1	20-10	24-2
	Hem-fir	#1	8-5	12-5	15-9	19-3	22-3	7-7	11-1	14-1	17-2	19-11
	Hem-fir	#2	8-0	11-9	14-11	18-2	21-1	7-2	10-6	13-4	16-3	18-10
	Hem-fir	#3	6-2	9-0	11-5	13-11	16-2	5-6	8-1	10-3	12-6	14-6
	Southern pine	SS	8-11	14-1	18-6	23-8	Note b	8-11	14-1	18-6	23-8	Note b
	Southern pine	#1	8-9	13-9	18-1	21-5	25-7	8-8	12-10	16-2	19-2	22-10
	Southern pine	#2	8-7	12-6	16-2	19-3	22-7	7-10	11-2	14-5	17-3	20-2
	Southern pine	#3	6-7	9-8	12-4	14-7	17-4	5-10	8-8	11-0	13-0	15-6
	Spruce-pine-fir	SS	8-5	13-3	17-5	22-1	25-7	8-5	12-9	16-2	19-9	22-10
	Spruce-pine-fir	#1	8-2	11-11	15-1	18-5	21-5	7-3	10-8	13-6	16-6	19-2
	Spruce-pine-fir	#2	8-2	11-11	15-1	18-5	21-5	7-3	10-8	13-6	16-6	19-2
	Spruce-pine-fir	#3	6-2	9-0	11-5	13-11	16-2	5-6	8-1	10-3	12-6	14-6
19.2	Douglas fir-larch	SS	8-7	13-6	17-9	21-8	25-2	8-7	12-6	15-10	19-5	22-6
	Douglas fir-larch	#1	7-11	11-8	14-9	18-0	20-11	7-1	10-5	13-2	16-1	18-8
	Douglas fir-larch	#2	7-5	10-11	13-9	16-10	19-6	6-8	9-9	12-4	15-1	17-6
	Douglas fir-larch	#3	5-7	8-3	10-5	12-9	14-9	5-0	7-4	9-4	11-5	13-2
	Hem-fir	SS	8-1	12-9	16-9	21-4	24-8	8-1	12-4	15-7	19-1	22-1
	Hem-fir	#1	7-9	11-4	14-4	17-7	20-4	6-11	10-2	12-10	15-8	18-2
	Hem-fir	#2	7-4	10-9	13-7	16-7	19-3	6-7	9-7	12-2	14-10	17-3
	Hem-fir	#3	5-7	8-3	10-5	12-9	14-9	5-0	7-4	9-4	11-5	13-2
	Southern pine	SS	8-5	13-3	17-5	22-3	Note b	8-5	13-3	17-5	22-0	25-9
	Southern pine	#1	8-3	13-0	16-6	19-7	23-4	7-11	11-9	14-9	17-6	20-11
	Southern pine	#2	7-11	11-5	14-9	17-7	20-7	7-1	10-2	13-2	15-9	18-5
	Southern pine	#3	6-0	8-10	11-3	13-4	15-10	5-4	7-11	10-1	11-11	14-2
	Spruce-pine-fir	SS	7-11	12-5	16-5	20-2	23-4	7-11	11-8	14-9	18-0	20-11
	Spruce-pine-fir	#1	7-5	10-11	13-9	16-10	19-6	6-8	9-9	12-4	15-1	17-6
	Spruce-pine-fir	#2	7-5	10-11	13-9	16-10	19-6	6-8	9-9	12-4	15-1	17-6
	Spruce-pine-fir	#3	5-7	8-3	10-5	12-9	14-9	5-0	7-4	9-4	11-5	13-2

(continued)

TABLE R802.5.1(3)—continued
RAFTER SPANS FOR COMMON LUMBER SPECIES
(Ground snow load=30 psf, ceiling not attached to rafters, L/Δ = 180)

RAFTER SPACING (inches)	SPECIES AND GRADE		DEAD LOAD = 10 psf					DEAD LOAD = 20 psf				
			2 × 4	2 × 6	2 × 8	2 × 10	2 × 12	2 × 4	2 × 6	2 × 8	2 × 10	2 × 12
			Maximum rafter spans[a]									
			(feet - inches)	(feet - inches)	(feet - inches)	(feet - inches)	(feet - inches)	(feet - inches)	(feet - inches)	(feet - inches)	(feet - inches)	(feet - inches)
24	Douglas fir-larch	SS	7-11	12-6	15-10	19-5	22-6	7-8	11-3	14-2	17-4	20-1
	Douglas fir-larch	#1	7-1	10-5	13-2	16-1	18-8	6-4	9-4	11-9	14-5	16-8
	Douglas fir-larch	#2	6-8	9-9	12-4	15-1	17-6	5-11	8-8	11-0	13-6	15-7
	Douglas fir-larch	#3	5-0	7-4	9-4	11-5	13-2	4-6	6-7	8-4	10-2	11-10
	Hem-fir	SS	7-6	11-10	15-7	19-1	22-1	7-6	11-0	13-11	17-0	19-9
	Hem-fir	#1	6-11	10-2	12-10	15-8	18-2	6-2	9-1	11-6	14-0	16-3
	Hem-fir	#2	6-7	9-7	12-2	14-10	17-3	5-10	8-7	10-10	13-3	15-5
	Hem-fir	#3	5-0	7-4	9-4	11-5	13-2	4-6	6-7	8-4	10-2	11-10
	Southern pine	SS	7-10	12-3	16-2	20-8	25-1	7-10	12-3	16-2	19-8	23-0
	Southern pine	#1	7-8	11-9	14-9	17-6	20-11	7-1	10-6	13-2	15-8	18-8
	Southern pine	#2	7-1	10-2	13-2	15-9	18-5	6-4	9-2	11-9	14-1	16-6
	Southern pine	#3	5-4	7-11	10-1	11-11	14-2	4-9	7-1	9-0	10-8	12-8
	Spruce-pine-fir	SS	7-4	11-7	14-9	18-0	20-11	7-1	10-5	13-2	16-1	18-8
	Spruce-pine-fir	#1	6-8	9-9	12-4	15-1	17-6	5-11	8-8	11-0	13-6	15-7
	Spruce-pine-fir	#2	6-8	9-9	12-4	15-1	17-6	5-11	8-8	11-0	13-6	15-7
	Spruce-pine-fir	#3	5-0	7-4	9-4	11-5	13-2	4-6	6-7	8-4	10-2	11-10

Check sources for availability of lumber in lengths greater than 20 feet.

For SI: 1 inch = 25.4 mm, 1 foot = 304.8 mm, 1 pound per square foot = 0.0479 kPa.

a. The tabulated rafter spans assume that ceiling joists are located at the bottom of the attic space or that some other method of resisting the outward push of the rafters on the bearing walls, such as rafter ties, is provided at that location. When ceiling joists or rafter ties are located higher in the attic space, the rafter spans shall be multiplied by the factors given below:

H_C/H_R	Rafter Span Adjustment Factor
1/3	0.67
1/4	0.76
1/5	0.83
1/6	0.90
1/7.5 or less	1.00

where:

H_C = Height of ceiling joists or rafter ties measured vertically above the top of the rafter support walls.

H_R = Height of roof ridge measured vertically above the top of the rafter support walls.

b. Span exceeds 26 feet in length.

TABLE R802.5.1(4)
RAFTER SPANS FOR COMMON LUMBER SPECIES
(Ground snow load=50 psf, ceiling not attached to rafters, L/Δ = 180)

RAFTER SPACING (inches)	SPECIES AND GRADE		DEAD LOAD = 10 psf					DEAD LOAD = 20 psf				
			2 × 4	2 × 6	2 × 8	2 × 10	2 × 12	2 × 4	2 × 6	2 × 8	2 × 10	2 × 12
			Maximum rafter spans[a]									
			(feet - inches)	(feet - inches)	(feet - inches)	(feet - inches)	(feet - inches)	(feet - inches)	(feet - inches)	(feet - inches)	(feet - inches)	(feet - inches)
12	Douglas fir-larch	SS	8-5	13-3	17-6	22-4	26-0	8-5	13-3	17-0	20-9	24-0
	Douglas fir-larch	#1	8-2	12-0	15-3	18-7	21-7	7-7	11-2	14-1	17-3	20-0
	Douglas fir-larch	#2	7-8	11-3	14-3	17-5	20-2	7-1	10-5	13-2	16-1	18-8
	Douglas fir-larch	#3	5-10	8-6	10-9	13-2	15-3	5-5	7-10	10-0	12-2	14-1
	Hem-fir	SS	8-0	12-6	16-6	21-1	25-6	8-0	12-6	16-6	20-4	23-7
	Hem-fir	#1	7-10	11-9	14-10	18-1	21-0	7-5	10-10	13-9	16-9	19-5
	Hem-fir	#2	7-5	11-1	14-0	17-2	19-11	7-0	10-3	13-0	15-10	18-5
	Hem-fir	#3	5-10	8-6	10-9	13-2	15-3	5-5	7-10	10-0	12-2	14-1
	Southern pine	SS	8-4	13-0	17-2	21-11	Note b	8-4	13-0	17-2	21-11	Note b
	Southern pine	#1	8-2	12-10	16-10	20-3	24-1	8-2	12-6	15-9	18-9	22-4
	Southern pine	#2	8-0	11-9	15-3	18-2	21-3	7-7	10-11	14-1	16-10	19-9
	Southern pine	#3	6-2	9-2	11-8	13-9	16-4	5-9	8-5	10-9	12-9	15-2
	Spruce-pine-fir	SS	7-10	12-3	16-2	20-8	24-1	7-10	12-3	15-9	19-3	22-4
	Spruce-pine-fir	#1	7-8	11-3	14-3	17-5	20-2	7-1	10-5	13-2	16-1	18-8
	Spruce-pine-fir	#2	7-8	11-3	14-3	17-5	20-2	7-1	10-5	13-2	16-1	18-8
	Spruce-pine-fir	#3	5-10	8-6	10-9	13-2	15-3	5-5	7-10	10-0	12-2	14-1
16	Douglas fir-larch	SS	7-8	12-1	15-10	19-5	22-6	7-8	11-7	14-8	17-11	20-10
	Douglas fir-larch	#1	7-1	10-5	13-2	16-1	18-8	6-7	9-8	12-2	14-11	17-3
	Douglas fir-larch	#2	6-8	9-9	12-4	15-1	17-6	6-2	9-0	11-5	13-11	16-2
	Douglas fir-larch	#3	5-0	7-4	9-4	11-5	13-2	4-8	6-10	8-8	10-6	12-3
	Hem-fir	SS	7-3	11-5	15-0	19-1	22-1	7-3	11-5	14-5	17-8	20-5
	Hem-fir	#1	6-11	10-2	12-10	15-8	18-2	6-5	9-5	11-11	14-6	16-10
	Hem-fir	#2	6-7	9-7	12-2	14-10	17-3	6-1	8-11	11-3	13-9	15-11
	Hem-fir	#3	5-0	7-4	9-4	11-5	13-2	4-8	6-10	8-8	10-6	12-3
	Southern pine	SS	7-6	11-10	15-7	19-11	24-3	7-6	11-10	15-7	19-11	23-10
	Southern pine	#1	7-5	11-7	14-9	17-6	20-11	7-4	10-10	13-8	16-2	19-4
	Southern pine	#2	7-1	10-2	13-2	15-9	18-5	6-7	9-5	12-2	14-7	17-1
	Southern pine	#3	5-4	7-11	10-1	11-11	14-2	4-11	7-4	9-4	11-0	13-1
	Spruce-pine-fir	SS	7-1	11-2	14-8	18-0	20-11	7-1	10-9	13-8	15-11	19-4
	Spruce-pine-fir	#1	6-8	9-9	12-4	15-1	17-6	6-2	9-0	11-5	13-11	16-2
	Spruce-pine-fir	#2	6-8	9-9	12-4	15-1	17-6	6-2	9-0	11-5	13-11	16-2
	Spruce-pine-fir	#3	5-0	7-4	9-4	11-5	13-2	4-8	6-10	8-8	10-6	12-3
19.2	Douglas fir-larch	SS	7-3	11-4	14-6	17-8	20-6	7-3	10-7	13-5	16-5	19-0
	Douglas fir-larch	#1	6-6	9-6	12-0	14-8	17-1	6-0	8-10	11-2	13-7	15-9
	Douglas fir-larch	#2	6-1	8-11	11-3	13-9	15-11	5-7	8-3	10-5	12-9	14-9
	Douglas fir-larch	#3	4-7	6-9	8-6	10-5	12-1	4-3	6-3	7-11	9-7	11-2
	Hem-fir	SS	6-10	10-9	14-2	17-5	20-2	6-10	10-5	13-2	16-1	18-8
	Hem-fir	#1	6-4	9-3	11-9	14-4	16-7	5-10	8-7	10-10	13-3	15-5
	Hem-fir	#2	6-0	8-9	11-1	13-7	15-9	5-7	8-1	10-3	12-7	14-7
	Hem-fir	#3	4-7	6-9	8-6	10-5	12-1	4-3	6-3	7-11	9-7	11-2
	Southern pine	SS	7-1	11-2	14-8	18-9	22-10	7-1	11-2	14-8	18 7	21-9
	Southern pine	#1	7-0	10-8	13-5	16-0	19-1	6-8	9-11	12-5	14-10	17-8
	Southern pine	#2	6-6	9-4	12-0	14-4	16-10	6-0	8-8	11-2	13-4	15-7
	Southern pine	#3	4-11	7-3	9-2	10-10	12-11	4-6	6-8	8-6	10-1	12-0
	Spruce-pine-fir	SS	6-8	10-6	13-5	16-5	19-1	6-8	9-10	12-5	15-3	17-8
	Spruce-pine-fir	#1	6-1	8-11	11-3	13-9	15-11	5-7	8-3	10-5	12-9	14-9
	Spruce-pine-fir	#2	6-1	8-11	11-3	13-9	15-11	5-7	8-3	10-5	12-9	14-9
	Spruce-pine-fir	#3	4-7	6-9	8-6	10-5	12-1	4-3	6-3	7-11	9-7	11-2

(continued)

TABLE R802.5.1(4)—continued
RAFTER SPANS FOR COMMON LUMBER SPECIES
(Ground snow load=50 psf, ceiling not attached to rafters, L/Δ = 180)

RAFTER SPACING (inches)	SPECIES AND GRADE		DEAD LOAD = 10 psf					DEAD LOAD = 20 psf				
			2 × 4	2 × 6	2 × 8	2 × 10	2 × 12	2 × 4	2 × 6	2 × 8	2 × 10	2 × 12
			Maximum rafter spans[a]									
			(feet - inches)	(feet - inches)	(feet - inches)	(feet - inches)	(feet - inches)	(feet - inches)	(feet - inches)	(feet - inches)	(feet - inches)	(feet - inches)
24	Douglas fir-larch	SS	6-8	10-	13-0	15-10	18-4	6-6	9-6	12-0	14-8	17-0
	Douglas fir-larch	#1	5-10	8-6	10-9	13-2	15-3	5-5	7-10	10-0	12-2	14-1
	Douglas fir-larch	#2	5-5	7-11	10-1	12-4	14-3	5-0	7-4	9-4	11-5	13-2
	Douglas fir-larch	#3	4-1	6-0	7-7	9-4	10-9	3-10	5-7	7-1	8-7	10-0
	Hem-fir	SS	6-4	9-11	12-9	15-7	18-0	6-4	9-4	11-9	14-5	16-8
	Hem-fir	#1	5-8	8-3	10-6	12-10	14-10	5-3	7-8	9-9	11-10	13-9
	Hem-fir	#2	5-4	7-10	9-11	12-1	14-1	4-11	7-3	9-2	11-3	13-0
	Hem-fir	#3	4-1	6-0	7-7	9-4	10-9	3-10	5-7	7-1	8-7	10-0
	Southern pine	SS	6-7	10-4	13-8	17-5	21-0	6-7	10-4	13-8	16-7	19-5
	Southern pine	#1	6-5	9-7	12-0	14-4	17-1	6-0	8-10	11-2	13-3	15-9
	Southern pine	#2	5-10	8-4	10-9	12-10	15-1	5-5	7-9	10-0	11-11	13-11
	Southern pine	#3	4-4	6-5	8-3	9-9	11-7	4-1	6-0	7-7	9-0	10-8
	Spruce-pine-fir	SS	6-2	9-6	12-0	14-8	17-1	6-0	8-10	11-2	13-7	15-9
	Spruce-pine-fir	#1	5-5	7-11	10-1	12-4	14-3	5-0	7-4	9-4	11-5	13-2
	Spruce-pine-fir	#2	5-5	7-11	10-1	12-4	14-3	5-0	7-4	9-4	11-5	13-2
	Spruce-pine-fir	#3	4-1	6-0	7-7	9-4	10-9	3-10	5-7	7-1	8-7	10-0

Check sources for availability of lumber in lengths greater than 20 feet.

For SI: 1 inch = 25.4 mm, 1 foot = 304.8 mm, 1 pound per square foot = 0.0479 kPa.

a. The tabulated rafter spans assume that ceiling joists are located at the bottom of the attic space or that some other method of resisting the outward push of the rafters on the bearing walls, such as rafter ties, is provided at that location. When ceiling joists or rafter ties are located higher in the attic space, the rafter spans shall be multiplied by the factors given below:

H_C/H_R	Rafter Span Adjustment Factor
1/3	0.67
1/4	0.76
1/5	0.83
1/6	0.90
1/7.5 or less	1.00

where:

H_C = Height of ceiling joists or rafter ties measured vertically above the top of the rafter support walls.

H_R = Height of roof ridge measured vertically above the top of the rafter support walls.

b. Span exceeds 26 feet in length.

RAFTER SPANS FOR COMMON LUMBER SPECIES
(Ground snow load=30 psf, ceiling attached to rafters, L/Δ = 240)

RAFTER SPACING (inches)	SPECIES AND GRADE		DEAD LOAD = 10 psf					DEAD LOAD = 20 psf				
			2 × 4	2 × 6	2 × 8	2 × 10	2 × 12	2 × 4	2 × 6	2 × 8	2 × 10	2 × 12
			Maximum rafter spans[a]									
			(feet - inches)	(feet - inches)	(feet - inches)	(feet - inches)	(feet - inches)	(feet - inches)	(feet - inches)	(feet - inches)	(feet - inches)	(feet - inches)
12	Douglas fir-larch	SS	9-1	14-4	18-10	24-1	Note b	9-1	14-4	18-10	24-1	Note b
	Douglas fir-larch	#1	8-9	13-9	18-2	22-9	Note b	8-9	13-2	16-8	20-4	23-7
	Douglas fir-larch	#2	8-7	13-6	17-5	21-4	24-8	8-5	12-4	15-7	19-1	22-1
	Douglas fir-larch	#3	7-1	10-5	13-2	16-1	18-8	6-4	9-4	11-9	14-5	16-8
	Hem-fir	SS	8-7	13-6	17-10	22-9	Note b	8-7	13-6	17-10	22-9	Note b
	Hem-fir	#1	8-5	13-3	17-5	22-2	25-9	8-5	12-10	16-3	19-10	23-0
	Hem-fir	#2	8-0	12-7	16-7	21-0	24-4	8-0	12-2	15-4	18-9	21-9
	Hem-fir	#3	7-1	10-5	13-2	16-1	18-8	6-4	9-4	11-9	14-5	16-8
	Southern pine	SS	8-11	14-1	18-6	23-8	Note b	8-11	14-1	18-6	23-8	Note b
	Southern pine	#1	8-9	13-9	18-2	23-2	Note b	8-9	13-9	18-2	22-2	Note b
	Southern pine	#2	8-7	13-6	17-10	22-3	Note b	8-7	12-11	16-8	19-11	23-4
	Southern pine	#3	7-7	11-2	14-3	16-10	20-0	6-9	10-0	12-9	15-1	17-11
	Spruce-pine-fir	SS	8-5	13-3	17-5	22-3	Note b	8-5	13-3	17-5	22-3	Note b
	Spruce-pine-fir	#1	8-3	12-11	17-0	21-4	24-8	8-3	12-4	15-7	19-1	22-1
	Spruce-pine-fir	#2	8-3	12-11	17-0	21-4	24-8	8-3	12-4	15-7	19-1	22-1
	Spruce-pine-fir	#3	7-1	10-5	13-2	16-1	18-8	6-4	9-4	11-9	14-5	16-8
16	Douglas fir-larch	SS	8-3	13-0	17-2	21-10	Note b	8-3	13-0	17-2	21-3	24-8
	Douglas fir-larch	#1	8-0	12-6	16-2	19-9	22-10	7-10	11-5	14-5	17-8	20-5
	Douglas fir-larch	#2	7-10	11-11	15-1	18-5	21-5	7-3	10-8	13-6	16-6	19-2
	Douglas fir-larch	#3	6-2	9-0	11-5	13-11	16-2	5-6	8-1	10-3	12-6	14-6
	Hem-fir	SS	7-10	12-3	16-2	20-8	25-1	7-10	12-3	16-2	20-8	24-2
	Hem-fir	#1	7-8	12-0	15-9	19-3	22-3	7-7	11-1	14-1	17-2	19-11
	Hem-fir	#2	7-3	11-5	14-11	18-2	21-1	7-2	10-6	13-4	16-3	18-10
	Hem-fir	#3	6-2	9-0	11-5	13-11	16-2	5-6	8-1	10-3	12-6	14-6
	Southern pine	SS	8-1	12-9	16-10	21-6	Note b	8-1	12-9	16-10	21-6	Note b
	Southern pine	#1	8-0	12-6	16-6	21-1	25-7	8-0	12-6	16-2	19-2	22-10
	Southern pine	#2	7-10	12-3	16-2	19-3	22-7	7-10	11-2	14-5	17-3	20-2
	Southern pine	#3	6-7	9-8	12-4	14-7	17-4	5-10	8-8	11-0	13-0	15-6
	Spruce-pine-fir	SS	7-8	12-0	15-10	20-2	24-7	7-8	12-0	15-10	19-9	22-10
	Spruce-pine-fir	#1	7-6	11-9	15-1	18-5	21-5	7-3	10-8	13-6	16-6	19-2
	Spruce-pine-fir	#2	7-6	11-9	15-1	18-5	21-5	7-3	10-8	13-6	16-6	19-2
	Spruce-pine-fir	#3	6-2	9-0	11-5	13-11	16-2	5-6	8-1	10-3	12-6	14-6
19.2	Douglas fir-larch	SS	7-9	12-3	16-1	20-7	25-0	7-9	12-3	15-10	19-5	22-6
	Douglas fir-larch	#1	7-6	11-8	14-9	18-0	20-11	7-1	10-5	13-2	16-1	18-8
	Douglas fir-larch	#2	7-4	10-11	13-9	16-10	19-6	6-8	9-9	12-4	15-1	17-6
	Douglas fir-larch	#3	5-7	8-3	10-5	12-9	14-9	5-0	7-4	9-4	11-5	13-2
	Hem-fir	SS	7-4	11-7	15-3	19-5	23-7	7-4	11-7	15-3	19-1	22-1
	Hem-fir	#1	7-2	11-4	14-4	17-7	20-4	6-11	10-2	12-10	15-8	18-2
	Hem-fir	#2	6-10	10-9	13-7	16-7	19-3	6-7	9-7	12-2	14-10	17-3
	Hem-fir	#3	5-7	8-3	10-5	12-9	14-9	5-0	7-4	9-4	11-5	13-2
	Southern pine	SS	7-8	12-0	15-10	20-2	24-7	7-8	12-0	15-10	20-2	24-7
	Southern pine	#1	7-6	11-9	15-6	19-7	23-4	7-6	11-9	14-9	17-6	20-11
	Southern pine	#2	7-4	11-5	14-9	17-7	20-7	7-1	10-2	13-2	15-9	18-5
	Southern pine	#3	6-0	8-10	11-3	13-4	15-10	5-4	7-11	10-1	11-11	14-2
	Spruce-pine-fir	SS	7-2	11-4	14-11	19-0	23-1	7-2	11-4	14-9	18-0	20-11
	Spruce-pine-fir	#1	7-0	10-11	13-9	16-10	19-6	6-8	9-9	12-4	15-1	17-6
	Spruce-pine-fir	#2	7-0	10-11	13-9	16-10	19-6	6-8	9-9	12-4	15-1	17-6
	Spruce-pine-fir	#3	5-7	8-3	10-5	12-9	14-9	5-0	7-4	9-4	11-5	13-2

(continued)

TABLE R802.5.1(5)—continued
RAFTER SPANS FOR COMMON LUMBER SPECIES
(Ground snow load=30 psf, ceiling attached to rafters, L/Δ = 240)

RAFTER SPACING (inches)	SPECIES AND GRADE		DEAD LOAD = 10 psf					DEAD LOAD = 20 psf				
			2 × 4	2 × 6	2 × 8	2 × 10	2 × 12	2 × 4	2 × 6	2 × 8	2 × 10	2 × 12
			Maximum rafter spans[a]									
			(feet-inches)	(feet-inches)	(feet-inches)	(feet-inches)	(feet-inches)	(feet-inches)	(feet-inches)	(feet-inches)	(feet-inches)	(feet-inches)
24	Douglas fir-larch	SS	7-3	11-4	15-0	19-1	22-6	7-3	11-3	14-2	17-4	20-1
	Douglas fir-larch	#1	7-0	10-5	13-2	16-1	18-8	6-4	9-4	11-9	14-5	16-8
	Douglas fir-larch	#2	6-8	9-9	12-4	15-1	17-6	5-11	8-8	11-0	13-6	15-7
	Douglas fir-larch	#3	5-0	7-4	9-4	11-5	13-2	4-6	6-7	8-4	10-2	11-10
	Hem-fir	SS	6-10	10-9	14-2	18-0	21-11	6-10	10-9	13-11	17-0	19-9
	Hem-fir	#1	6-8	10-2	12-10	15-8	18-2	6-2	9-1	11-6	14-0	16-3
	Hem-fir	#2	6-4	9-7	12-2	14-10	17-3	5-10	8-7	10-10	13-3	15-5
	Hem-fir	#3	5-0	7-4	9-4	11-5	13-2	4-6	6-7	8-4	10-2	11-10
	Southern pine	SS	7-1	11-2	14-8	18-9	22-10	7-1	11-2	14-8	18-9	22-10
	Southern pine	#1	7-0	10-11	14-5	17-6	20-11	7-0	10-6	13-2	15-8	18-8
	Southern pine	#2	6-10	10-2	13-2	15-9	18-5	6-4	9-2	11-9	14-1	16-6
	Southern pine	#3	5-4	7-11	10-1	11-11	14-2	4-9	7-1	9-0	10-8	12-8
	Spruce-pine-fir	SS	6-8	10-6	13-10	17-8	20-11	6-8	10-5	13-2	16-1	18-8
	Spruce-pine-fir	#1	6-6	9-9	12-4	15-1	17-6	5-11	8-8	11-0	13-6	15-7
	Spruce-pine-fir	#2	6-6	9-9	12-4	15-1	17-6	5-11	8-8	11-0	13-6	15-7
	Spruce-pine-fir	#3	5-0	7-4	9-4	11-5	13-2	4-6	6-7	8-4	10-2	11-10

Check sources for availability of lumber in lengths greater than 20 feet.

For SI: 1 inch = 25.4 mm, 1 foot = 304.8 mm, 1 pound per square foot = 0.0479kPa.

a. The tabulated rafter spans assume that ceiling joists are located at the bottom of the attic space or that some other method of resisting the outward push of the rafters on the bearing walls, such as rafter ties, is provided at that location. When ceiling joists or rafter ties are located higher in the attic space, the rafter spans shall be multiplied by the factors given below:

H_C/H_R	Rafter Span Adjustment Factor
1/3	0.67
1/4	0.76
1/5	0.83
1/6	0.90
1/7.5 or less	1.00

where:

H_C = Height of ceiling joists or rafter ties measured vertically above the top of the rafter support walls.

H_R = Height of roof ridge measured vertically above the top of the rafter support walls.

b. Span exceeds 26 feet in length.

TABLE R802.5.1(6)
RAFTER SPANS FOR COMMON LUMBER SPECIES
(Ground snow load=50 psf, ceiling attached to rafters, L/Δ = 240)

RAFTER SPACING (inches)	SPECIES AND GRADE		DEAD LOAD = 10 psf					DEAD LOAD = 20 psf				
			2 × 4	2 × 6	2 × 8	2 × 10	2 × 12	2 × 4	2 × 6	2 × 8	2 × 10	2 × 12
			Maximum rafter spans[a]									
			(feet-inches)	(feet-inches)	(feet-inches)	(feet-inches)	(feet-inches)	(feet-inches)	(feet-inches)	(feet-inches)	(feet-inches)	(feet-inches)
12	Douglas fir-larch	SS	7-8	12-1	15-11	20-3	24-8	7-8	12-1	15-11	20-3	24-0
	Douglas fir-larch	#1	7-5	11-7	15-3	18-7	21-7	7-5	11-2	14-1	17-3	20-0
	Douglas fir-larch	#2	7-3	11-3	14-3	17-5	20-2	7-1	10-5	13-2	16-1	18-8
	Douglas fir-larch	#3	5-10	8-6	10-9	13-2	15-3	5-5	7-10	10-0	12-2	14-1
	Hem-fir	SS	7-3	11-5	15-0	19-2	23-4	7-3	11-5	15-0	19-2	23-4
	Hem-fir	#1	7-1	11-2	14-8	18-1	21-0	7-1	10-10	13-9	16-9	19-5
	Hem-fir	#2	6-9	10-8	14-0	17-2	19-11	6-9	10-3	13-0	15-10	18-5
	Hem-fir	#3	5-10	8-6	10-9	13-2	15-3	5-5	7-10	10-0	12-2	14-1
	Southern pine	SS	7-6	11-10	15-7	19-11	24-3	7-6	11-10	15-7	19-11	24-3
	Southern pine	#1	7-5	11-7	15-4	19-7	23-9	7-5	11-7	15-4	18-9	22-4
	Southern pine	#2	7-3	11-5	15-0	18-2	21-3	7-3	10-11	14-1	16-10	19-9
	Southern pine	#3	6-2	9-2	11-8	13-9	16-4	5-9	8-5	10-9	12-9	15-2
	Spruce-pine-fir	SS	7-1	11-2	14-8	18-9	22-10	7-1	11-2	14-8	18-9	22-4
	Spruce-pine-fir	#1	6-11	10-11	14-3	17-5	20-2	6-11	10-5	13-2	16-1	18-8
	Spruce-pine-fir	#2	6-11	10-11	14-3	17-5	20-2	6-11	10-5	13-2	16-1	18-8
	Spruce-pine-fir	#3	5-10	8-6	10-9	13-2	15-3	5-5	7-10	10-0	12-2	14-1
16	Douglas fir-larch	SS	7-0	11-0	14-5	18-5	22-5	7-0	11-0	14-5	17-11	20-10
	Douglas fir-larch	#1	6-9	10-5	13-2	16-1	18-8	6-7	9-8	12-2	14-11	17-3
	Douglas fir-larch	#2	6-7	9-9	12-4	15-1	17-6	6-2	9-0	11-5	13-11	16-2
	Douglas fir-larch	#3	5-0	7-4	9-4	11-5	13-2	4-8	6-10	8-8	10-6	12-3
	Hem-fir	SS	6-7	10-4	13-8	17-5	21-2	6-7	10-4	13-8	17-5	20-5
	Hem-fir	#1	6-5	10-2	12-10	15-8	18-2	6-5	9-5	11-11	14-6	16-10
	Hem-fir	#2	6-2	9-7	12-2	14-10	17-3	6-1	8-11	11-3	13-9	15-11
	Hem-fir	#3	5-0	7-4	9-4	11-5	13-2	4-8	6-10	8-8	10-6	12-3
	Southern pine	SS	6-10	10-9	14-2	18-1	22-0	6-10	10-9	14-2	18-1	22-0
	Southern pine	#1	6-9	10-7	13-11	17-6	20-11	6-9	10-7	13-8	16-2	19-4
	Southern pine	#2	6-7	10-2	13-2	15-9	18-5	6-7	9-5	12-2	14-7	17-1
	Southern pine	#3	5-4	7-11	10-1	11-11	14-2	4-11	7-4	9-4	11-0	13-1
	Spruce-pine-fir	SS	6-5	10-2	13-4	17-0	20-9	6-5	10-2	13-4	16-8	19-4
	Spruce-pine-fir	#1	6-4	9-9	12-4	15-1	17-6	6-2	9-0	11-5	13-11	16-2
	Spruce-pine-fir	#2	6-4	9-9	12-4	15-1	17-6	6-2	9-0	11-5	13-11	16-2
	Spruce-pine-fir	#3	5-0	7-4	9-4	11-5	13-2	4-8	6-10	8-8	10-6	12-3
19.2	Douglas fir-larch	SS	6-7	10-4	13-7	17-4	20-6	6-7	10-4	13-5	16-5	19-0
	Douglas fir-larch	#1	6-4	9-6	12-0	14-8	17-1	6-0	8-10	11-2	13-7	15-9
	Douglas fir-larch	#2	6-1	8-11	11-3	13-9	15-11	5-7	8-3	10-5	12-9	14-9
	Douglas fir-larch	#3	4-7	6-9	8-6	10-5	12-1	4-3	6-3	7-11	9-7	11-2
	Hem-fir	SS	6-2	9-9	12-10	16-5	19-11	6-2	9-9	12-10	16-1	18-8
	Hem-fir	#1	6-1	9-3	11-9	14-4	16-7	5-10	8-7	10-10	13-3	15-5
	Hem-fir	#2	5-9	8-9	11-1	13-7	15-9	5-7	8-1	10-3	12-7	14-7
	Hem-fir	#3	4-7	6-9	8-6	10-5	12-1	4-3	6-3	7-11	9-7	11-2
	Southern pine	SS	6-5	10-2	13-4	17-0	20-9	6-5	10-2	13-4	17-0	20-9
	Southern pine	#1	6-4	9-11	13-1	16-0	19-1	6-4	9-11	12-5	14-10	17-8
	Southern pine	#2	6-2	9-4	12-0	14-4	16-10	6-0	8-8	11-2	13-4	15-7
	Southern pine	#3	4-11	7-3	9-2	10-10	12-11	4-6	6-8	8-6	10-1	12-0
	Spruce-pine-fir	SS	6-1	9-6	12-7	16-0	19-1	6-1	9-6	12-5	15-3	17-8
	Spruce-pine-fir	#1	5-11	8-11	11-3	13-9	15-11	5-7	8-3	10-5	12-9	14-9
	Spruce-pine-fir	#2	5-11	8-11	11-3	13-9	15-11	5-7	8-3	10-5	12-9	14-9
	Spruce-pine-fir	#3	4-7	6-9	8-6	10-5	12-1	4-3	6-3	7-11	9-7	11-2

(continued)

TABLE R802.5.1(6)—continued
RAFTER SPANS FOR COMMON LUMBER SPECIES
(Ground snow load=50 psf, ceiling attached to rafters, L/Δ = 240)

RAFTER SPACING (inches)	SPECIES AND GRADE		DEAD LOAD = 10 psf					DEAD LOAD = 20 psf				
			2 × 4	2 × 6	2 × 8	2 × 10	2 × 12	2 × 4	2 × 6	2 × 8	2 × 10	2 × 12
			Maximum rafter spans[a]									
			(feet-inches)	(feet-inches)	(feet-inches)	(feet-inches)	(feet-inches)	(feet-inches)	(feet-inches)	(feet-inches)	(feet-inches)	(feet-inches)
24	Douglas fir-larch	SS	6-1	9-7	12-7	15-10	18-4	6-1	9-6	12-0	14-8	17-0
	Douglas fir-larch	#1	5-10	8-6	10-9	13-2	15-3	5-5	7-10	10-0	12-2	14-1
	Douglas fir-larch	#2	5-5	7-11	10-1	12-4	14-3	5-0	7-4	9-4	11-5	13-2
	Douglas fir-larch	#3	4-1	6-0	7-7	9-4	10-9	3-10	5-7	7-1	8-7	10-0
	Hem-fir	SS	5-9	9-1	11-11	15-2	18-0	5-9	9-1	11-9	14-5	15-11
	Hem-fir	#1	5-8	8-3	10-6	12-10	14-10	5-3	7-8	9-9	11-10	13-9
	Hem-fir	#2	5-4	7-10	9-11	12-1	14-1	4-11	7-3	9-2	11-3	13-0
	Hem-fir	#3	4-1	6-0	7-7	9-4	10-9	3-10	5-7	7-1	8-7	10-0
	Southern pine	SS	6-0	9-5	12-5	15-10	19-3	6-0	9-5	12-5	15-10	19-3
	Southern pine	#1	5-10	9-3	12-0	14-4	17-1	5-10	8-10	11-2	13-3	15-9
	Southern pine	#2	5-9	8-4	10-9	12-10	15-1	5-5	7-9	10-0	11-11	13-11
	Southern pine	#3	4-4	6-5	8-3	9-9	11-7	4-1	6-0	7-7	9-0	10-8
	Spruce-pine-fir	SS	5-8	8-10	11-8	14-8	17-1	5-8	8-10	11-2	13-7	15-9
	Spruce-pine-fir	#1	5-5	7-11	10-1	12-4	14-3	5-0	7-4	9-4	11-5	13-2
	Spruce-pine-fir	#2	5-5	7-11	10-1	12-4	14-3	5-0	7-4	9-4	11-5	13-2
	Spruce-pine-fir	#3	4-1	6-0	7-7	9-4	10-9	3-10	5-7	7-1	8-7	10-0

Check sources for availability of lumber in lengths greater than 20 feet.

For SI: 1 inch = 25.4 mm, 1 foot = 304.8 mm, 1 pound per square foot = 0.0479 kPa.

a. The tabulated rafter spans assume that ceiling joists are located at the bottom of the attic space or that some other method of resisting the outward push of the rafters on the bearing walls, such as rafter ties, is provided at that location. When ceiling joists or rafter ties are located higher in the attic space, the rafter spans shall be multiplied by the factors given below:

H_C/H_R	Rafter Span Adjustment Factor
1/3	0.67
1/4	0.76
1/5	0.83
1/6	0.90
1/7.5 or less	1.00

where:

H_C = Height of ceiling joists or rafter ties measured vertically above the top of the rafter support walls.

H_R = Height of roof ridge measured vertically above the top of the rafter support walls.

TABLE R802.5.1(7)
RAFTER SPANS FOR 70 PSF GROUND SNOW LOAD
(Ceiling not attached to rafters, L/Δ = 180)

RAFTER SPACING (inches)	SPECIES AND GRADE		DEAD LOAD = 10 psf					DEAD LOAD = 20 psf				
			2 × 4	2 × 6	2 × 8	2 × 10	2 × 12	2 × 4	2 × 6	2 × 8	2 × 10	2 × 12
			Maximum Rafter Spans[a]									
			(feet-inches)	(feet-inches)	(feet-inches)	(feet-inches)	(feet-inches)	(feet-inches)	(feet-inches)	(feet-inches)	(feet-inches)	(feet-inches)
12	Douglas fir-larch	SS	7-7	11-10	15-8	19-5	22-6	7-7	11-10	15-0	18-3	21-2
	Douglas fir-larch	#1	7-1	10-5	13-2	16-1	18-8	6-8	9-10	12-5	15-2	17-7
	Douglas fir-larch	#2	6-8	9-9	12-4	15-1	17-6	6-3	9-2	11-8	14-2	16-6
	Douglas fir-larch	#3	5-0	7-4	9-4	11-5	13-2	4-9	6-11	8-9	10-9	12-5
	Hem-fir	SS	7-2	11-3	14-9	18-10	22-1	7-2	11-3	14-8	18-0	20-10
	Hem-fir	#1	6-11	10-2	12-10	15-8	18-2	6-6	9-7	12-1	14-10	17-2
	Hem-fir	#2	6-7	9-7	12-2	14-10	17-3	6-2	9-1	11-5	14-0	16-3
	Hem-fir	#3	5-0	7-4	9-4	11-5	13-2	4-9	6-11	8-9	10-9	12-5
	Southern pine	SS	7-5	11-8	15-4	19-7	23-10	7-5	11-8	15-4	19-7	23-10
	Southern pine	#1	7-3	11-5	14-9	17-6	20-11	7-3	11-1	13-11	16-6	19-8
	Southern pine	#2	7-1	10-2	13-2	15-9	18-5	6-8	9-7	12-5	14-10	17-5
	Southern pine	#3	5-4	7-11	10-1	11-11	14-2	5-1	7-5	9-6	11-3	13-4
	Spruce-pine-fir	SS	7-0	11-0	14-6	18-0	20-11	7-0	11-0	13-11	17-0	19-8
	Spruce-pine-fir	#1	6-8	9-9	12-4	15-1	17-6	6-3	9-2	11-8	14-2	16-6
	Spruce-pine-fir	#2	6-8	9-9	12-4	15-1	17-6	6-3	9-2	11-8	14-2	16-6
	Spruce-pine-fir	#3	5-0	7-4	9-4	11-5	13-2	4-9	6-11	8-9	10-9	12-5
16	Douglas fir-larch	SS	6-10	10-9	13-9	16-10	19-6	6-10	10-3	13-0	15-10	18-4
	Douglas fir-larch	#1	6-2	9-0	11-5	13-11	16-2	5-10	8-6	10-9	13-2	15-3
	Douglas fir-larch	#2	5-9	8-5	10-8	13-1	15-2	5-5	7-11	10-1	12-4	14-3
	Douglas fir-larch	#3	4-4	6-4	8-1	9-10	11-5	4-1	6-0	7-7	9-4	10-9
	Hem-fir	SS	6-6	10-2	13-5	16-6	19-2	6-6	10-1	12-9	15-7	18-0
	Hem-fir	#1	6-0	8-9	11-2	13-7	15-9	5-8	8-3	10-6	12-10	14-10
	Hem-fir	#2	5-8	8-4	10-6	12-10	14-11	5-4	7-10	9-11	12-1	14-1
	Hem-fir	#3	4-4	6-4	8-1	9-10	11-5	4-1	6-0	7-7	9-4	10-9
	Southern pine	SS	6-9	10-7	14-0	17-10	21-8	6-9	10-7	14-0	17-10	21-0
	Southern pine	#1	6-7	10-2	12-9	15-2	18-1	6-5	9-7	12-0	14-4	17-1
	Southern pine	#2	6-2	8-10	11-5	13-7	16-0	5-10	8-4	10-9	12-10	15-1
	Southern pine	#3	4-8	6-10	8-9	10-4	12-3	4-4	6-5	8-3	9-9	11-7
	Spruce-pine-fir	SS	6-4	10-0	12-9	15-7	18-1	6-4	9-6	12-0	14-8	17-1
	Spruce-pine-fir	#1	5-9	8-5	10-8	13-1	15-2	5-5	7-11	10-1	12-4	14-3
	Spruce-pine-fir	#2	5-9	8-5	10-8	13-1	15-2	5-5	7-11	10-1	12-4	14-3
	Spruce-pine-fir	#3	4-4	6-4	8-1	9-10	11-5	4-1	6-0	7-7	9-4	10-9
19.2	Douglas fir-larch	SS	6-5	9-11	12-7	15-4	17-9	6-5	9-4	11-10	14-5	16-9
	Douglas fir-larch	#1	5-7	8-3	10-5	12-9	14-9	5-4	7-9	9-10	12-0	13-11
	Douglas fir-larch	#2	5-3	7-8	9-9	11-11	13-10	5-0	7-3	9-2	11-3	13-0
	Douglas fir-larch	#3	4-0	5-10	7-4	9-0	10-5	3-9	5-6	6-11	8-6	9-10
	Hem-fir	SS	6-1	9-7	12-4	15-1	17-4	6-1	9-2	11-8	14-2	15-5
	Hem-fir	#1	5-6	8-0	10-2	12-5	14-5	5-2	7-7	9-7	11-8	13-7
	Hem-fir	#2	5-2	7-7	9-7	11-9	13-7	4-11	7-2	9-1	11-1	12-10
	Hem-fir	#3	4-0	5-10	7-4	9-0	10-5	3-9	5-6	6-11	8-6	9-10
	Southern pine	SS	6-4	10-0	13-2	16-9	20-4	6-4	10-0	13-2	16-5	19-2
	Southern pine	#1	6-3	9-3	11-8	13-10	16-6	5-11	8-9	11-0	13-1	15-7
	Southern pine	#2	5-7	8-1	10-5	12-5	14-7	5-4	7-7	9-10	11-9	13-9
	Southern pine	#3	4-3	6-3	8-0	9-5	11-2	4-0	5-11	7-6	8-10	10-7
	Spruce-pine-fir	SS	6-0	9-2	11-8	14-3	16-6	5-11	8-8	11-0	13-5	15-7
	Spruce-pine-fir	#1	5-3	7-8	9-9	11-11	13-10	5-0	7-3	9-2	11-3	13-0
	Spruce-pine-fir	#2	5-3	7-8	9-9	11-11	13-10	5-0	7-3	9-2	11-3	13-0
	Spruce-pine-fir	#3	4-0	5-10	7-4	9-0	10-5	3-9	5-6	6-11	8-6	9-10

(continued)

TABLE R802.5.1(7)—continued
RAFTER SPANS FOR 70 PSF GROUND SNOW LOAD
(Ceiling not attached to rafters, L/Δ = 180)

RAFTER SPACING (inches)	SPECIES AND GRADE		DEAD LOAD = 10 psf					DEAD LOAD = 20 psf				
			2 × 4	2 × 6	2 × 8	2 × 10	2 × 12	2 × 4	2 × 6	2 × 8	2 × 10	2 × 12
			Maximum rafter spans[a]									
			(feet-inches)	(feet-inches)	(feet-inches)	(feet-inches)	(feet-inches)	(feet-inches)	(feet-inches)	(feet-inches)	(feet-inches)	(feet-inches)
24	Douglas fir-larch	SS	6-0	8-10	11-3	13-9	15-11	5-9	8-4	10-7	12-11	15-0
	Douglas fir-larch	#1	5-0	7-4	9-4	11-5	13-2	4-9	6-11	8-9	10-9	12-5
	Douglas fir-larch	#2	4-8	6-11	8-9	10-8	12-4	4-5	6-6	8-3	10-0	11-8
	Douglas fir-larch	#3	3-7	5-2	6-7	8-1	9-4	3-4	4-11	6-3	7-7	8-10
	Hem-fir	SS	5-8	8-8	11-0	13-6	13-11	5-7	8-3	10-5	12-4	12-4
	Hem-fir	#1	4-11	7-2	9-1	11-1	12-10	4-7	6-9	8-7	10-6	12-2
	Hem-fir	#2	4-8	6-9	8-7	10-6	12-2	4-4	6-5	8-1	9-11	11-6
	Hem-fir	#3	3-7	5-2	6-7	8-1	9-4	3-4	4-11	6-3	7-7	8-10
	Southern pine	SS	5-11	9-3	12-2	15-7	18-2	5-11	9-3	12-2	14-8	17-2
	Southern pine	#1	5-7	8-3	10-5	12-5	14-9	5-3	7-10	9-10	11-8	13-11
	Southern pine	#2	5-0	7-3	9-4	11-1	13-0	4-9	6-10	8-9	10-6	12-4
	Southern pine	#3	3-9	5-7	7-1	8-5	10-0	3-7	5-3	6-9	7-11	9-5
	Spruce-pine-fir	SS	5-6	8-3	10-5	12-9	14-9	5-4	7-9	9-10	12-0	12-11
	Spruce-pine-fir	#1	4-8	6-11	8-9	10-8	12-4	4-5	6-6	8-3	10-0	11-8
	Spruce-pine-fir	#2	4-8	6-11	8-9	10-8	12-4	4-5	6-6	8-3	10-0	11-8
	Spruce-pine-fir	#3	3-7	5-2	6-7	8-1	9-4	3-4	4-11	6-3	7-7	8-10

Check sources for availability of lumber in lengths greater than 20 feet.

For SI: 1 inch = 25.4 mm, 1 foot = 304.8 mm, 1 pound per square foot = 0.0479 kPa.

a. The tabulated rafter spans assume that ceiling joists are located at the bottom of the attic space or that some other method of resisting the outward push of the rafters on the bearing walls, such as rafter ties, is provided at that location. When ceiling joists or rafter ties are located higher in the attic space, the rafter spans shall be multiplied by the factors given below:

H_C/H_R	Rafter Span Adjustment Factor
1/3	0.67
1/4	0.76
1/5	0.83
1/6	0.90
1/7.5 or less	1.00

where:

H_C = Height of ceiling joists or rafter ties measured vertically above the top of the rafter support walls.

H_R = Height of roof ridge measured vertically above the top of the rafter support walls.

TABLE R802.5.1(8)
RAFTER SPANS FOR 70 PSF GROUND SNOW LOAD
(Ceiling attached to rafters, L/Δ = 240)

RAFTER SPACING (inches)	SPECIES AND GRADE		DEAD LOAD = 10 psf					DEAD LOAD = 20 psf				
			2 × 4	2 × 6	2 × 8	2 × 10	2 × 12	2 × 4	2 × 6	2 × 8	2 × 10	2 × 12
			Maximum rafter spans[a]									
			(feet - inches)	(feet - inches)	(feet - inches)	(feet - inches)	(feet - inches)	(feet - inches)	(feet - inches)	(feet - inches)	(feet - inches)	(feet - inches)
12	Douglas fir-larch	SS	6-10	10-9	14-3	18-2	22-1	6-10	10-9	14-3	18-2	21-2
	Douglas fir-larch	#1	6-7	10-5	13-2	16-1	18-8	6-7	9-10	12-5	15-2	17-7
	Douglas fir-larch	#2	6-6	9-9	12-4	15-1	17-6	6-3	9-2	11-8	14-2	16-6
	Douglas fir-larch	#3	5-0	7-4	9-4	11-5	13-2	4-9	6-11	8-9	10-9	12-5
	Hem-fir	SS	6-6	10-2	13-5	17-2	20-10	6-6	10-2	13-5	17-2	20-10
	Hem-fir	#1	6-4	10-0	12-10	15-8	18-2	6-4	9-7	12-1	14-10	17-2
	Hem-fir	#2	6-1	9-6	12-2	14-10	17-3	6-1	9-1	11-5	14-0	16-3
	Hem-fir	#3	5-0	7-4	9-4	11-5	13-2	4-9	6-11	8-9	10-9	12-5
	Southern pine	SS	6-9	10-7	14-0	17-10	21-8	6-9	10-7	14-0	17-10	21-8
	Southern pine	#1	6-7	10-5	13-8	17-6	20-11	6-7	10-5	13-8	16-6	19-8
	Southern pine	#2	6-6	10-2	13-2	15-9	18-5	6-6	9-7	12-5	14-10	17-5
	Southern pine	#3	5-4	7-11	10-1	11-11	14-2	5-1	7-5	9-6	11-3	13-4
	Spruce-pine-fir	SS	6-4	10-0	13-2	16-9	20-5	6-4	10-0	13-2	16-9	19-8
	Spruce-pine-fir	#1	6-2	9-9	12-4	15-1	17-6	6-2	9-2	11-8	14-2	16-6
	Spruce-pine-fir	#2	6-2	9-9	12-4	15-1	17-6	6-2	9-2	11-8	14-2	16-6
	Spruce-pine-fir	#3	5-0	7-4	9-4	11-5	13-2	4-9	6-11	8-9	10-9	12-5
16	Douglas fir-larch	SS	6-3	9-10	12-11	16-6	19-6	6-3	9-10	12-11	15-10	18-4
	Douglas fir-larch	#1	6-0	9-0	11-5	13-11	16-2	5-10	8-6	10-9	13-2	15-3
	Douglas fir-larch	#2	5-9	8-5	10-8	13-1	15-2	5-5	7-11	10-1	12-4	14-3
	Douglas fir-larch	#3	4-4	6-4	8-1	9-10	11-5	4-1	6-0	7-7	9-4	10-9
	Hem-fir	SS	5-11	9-3	12-2	15-7	18-11	5-11	9-3	12-2	15-7	18-0
	Hem-fir	#1	5-9	8-9	11-2	13-7	15-9	5-8	8-3	10-6	12-10	14-10
	Hem-fir	#2	5-6	8-4	10-6	12-10	14-11	5-4	7-10	9-11	12-1	14-1
	Hem-fir	#3	4-4	6-4	8-1	9-10	11-5	4-1	6-0	7-7	9-4	10-9
	Southern pine	SS	6-1	9-7	12-8	16-2	19-8	6-1	9-7	12-8	16-2	19-8
	Southern pine	#1	6-0	9-5	12-5	15-2	18-1	6-0	9-5	12-0	14-4	17-1
	Southern pine	#2	5-11	8-10	11-5	13-7	16-0	5-10	8-4	10-9	12-10	15-1
	Southern pine	#3	4-8	6-10	8-9	10-4	12-3	4-4	6-5	8-3	9-9	11-7
	Spruce-pine-fir	SS	5-9	9-1	11-11	15-3	18-1	5-9	9-1	11-11	14-8	17-1
	Spruce-pine-fir	#1	5-8	8-5	10-8	13-1	15-2	5-5	7-11	10-1	12-4	14-3
	Spruce-pine-fir	#2	5-8	8-5	10-8	13-1	15-2	5-5	7-11	10-1	12-4	14-3
	Spruce-pine-fir	#3	4-4	6-4	8-1	9-10	11-5	4-1	6-0	7-7	9-4	10-9
19.2	Douglas fir-larch	SS	5-10	9-3	12-2	15-4	17-9	5-10	9-3	11-10	14-5	16-9
	Douglas fir-larch	#1	5-7	8-3	10-5	12-9	14-9	5-4	7-9	9-10	12-0	13-11
	Douglas fir-larch	#2	5-3	7-8	9-9	11-11	13-10	5-0	7-3	9-2	11-3	13-0
	Douglas fir-larch	#3	4-0	5-10	7-4	9-0	10-5	3-9	5-6	6-11	8-6	9-10
	Hem-fir	SS	5-6	8-8	11-6	14-8	17-4	5-6	8-8	11-6	14-2	15-5
	Hem-fir	#1	5-5	8-0	10-2	12-5	14-5	5-2	7-7	9-7	11-8	13-7
	Hem-fir	#2	5-2	7-7	9-7	11-9	13-7	4-11	7-2	9-1	11-1	12-10
	Hem-fir	#3	4-0	5-10	7-4	9-0	10-5	3-9	5-6	6-11	8-6	9-10
	Southern pine	SS	5-9	9-1	11-11	15-3	18-6	5-9	9-1	11-11	15-3	18-6
	Southern pine	#1	5-8	8-11	11-8	13-10	16-6	5-8	8-9	11-0	13-1	15-7
	Southern pine	#2	5-6	8-1	10-5	12-5	14-7	5-4	7-7	9-10	11-9	13-9
	Southern pine	#3	4-3	6-3	8-0	9-5	11-2	4-0	5-11	7-6	8-10	10-7
	Spruce-pine-fir	SS	5-5	8-6	11-3	14-3	16-6	5-5	8-6	11-0	13-5	15-7
	Spruce-pine-fir	#1	5-3	7-8	9-9	11-11	13-10	5-0	7-3	9-2	11-3	13-0
	Spruce-pine-fir	#2	5-3	7-8	9-9	11-11	13-10	5-0	7-3	9-2	11-3	13-0
	Spruce-pine-fir	#3	4-0	5-10	7-4	9-0	10-5	3-9	5-6	6-11	8-6	9-10

(continued)

TABLE R802.5.1(8)—continued
RAFTER SPANS FOR 70 PSF GROUND SNOW LOAD[a]
(Ceiling attached to rafters, L/Δ = 240)

RAFTER SPACING (inches)	SPECIES AND GRADE		DEAD LOAD = 10 psf					DEAD LOAD = 20 psf				
			2 × 4	2 × 6	2 × 8	2 × 10	2 × 12	2 × 4	2 × 6	2 × 8	2 × 10	2 × 12
			Maximum rafter spans[a]									
			(feet - inches)	(feet - inches)	(feet - inches)	(feet - inches)	(feet - inches)	(feet - inches)	(feet - inches)	(feet - inches)	(feet - inches)	(feet - inches)
24	Douglas fir-larch	SS	5-5	8-7	11-3	13-9	15-11	5-5	8-4	10-7	12-11	15-0
	Douglas fir-larch	#1	5-0	7-4	9-4	11-5	13-2	4-9	6-11	8-9	10-9	12-5
	Douglas fir-larch	#2	4-8	6-11	8-9	10-8	12-4	4-5	6-6	8-3	10-0	11-8
	Douglas fir-larch	#3	3-7	5-2	6-7	8-1	9-4	3-4	4-11	6-3	7-7	8-10
	Hem-fir	SS	5-2	8-1	10-8	13-6	13-11	5-2	8-1	10-5	12-4	12-4
	Hem-fir	#1	4-11	7-2	9-1	11-1	12-10	4-7	6-9	8-7	10-6	12-2
	Hem-fir	#2	4-8	6-9	8-7	10-6	12-2	4-4	6-5	8-1	9-11	11-6
	Hem-fir	#3	3-7	5-2	6-7	8-1	9-4	3-4	4-11	6-3	7-7	8-10
	Southern pine	SS	5-4	8-5	11-1	14-2	17-2	5-4	8-5	11-1	14-2	17-2
	Southern pine	#1	5-3	8-3	10-5	12-5	14-9	5-3	7-10	9-10	11-8	13-11
	Southern pine	#2	5-0	7-3	9-4	11-1	13-0	4-9	6-10	8-9	10-6	12-4
	Southern pine	#3	3-9	5-7	7-1	8-5	10-0	3-7	5-3	6-9	7-11	9-5
	Spruce-pine-fir	SS	5-0	7-11	10-5	12-9	14-9	5-0	7-9	9-10	12-0	12-11
	Spruce-pine-fir	#1	4-8	6-11	8-9	10-8	12-4	4-5	6-6	8-3	10-0	11-8
	Spruce-pine-fir	#2	4-8	6-11	8-9	10-8	12-4	4-5	6-6	8-3	10-0	11-8
	Spruce-pine-fir	#3	3-7	5-2	6-7	8-1	9-4	3-4	4-11	6-3	7-7	8-10

Check sources for availability of lumber in lengths greater than 20 feet.

For SI: 1 inch = 25.4 mm, 1 foot = 304.8 mm, 1 pound per square foot = 0.0479 kPa.

a. The tabulated rafter spans assume that ceiling joists are located at the bottom of the attic space or that some other method of resisting the outward push of the rafters on the bearing walls, such as rafter ties, is provided at that location. When ceiling joists or rafter ties are located higher in the attic space, the rafter spans shall be multiplied by the factors given below:

H_C/H_R	Rafter Span Adjustment Factor
1/3	0.67
1/4	0.76
1/5	0.83
1/6	0.90
1/7.5 or less	1.00

where:

H_C = Height of ceiling joists or rafter ties measured vertically above the top of the rafter support walls.

H_R = Height of roof ridge measured vertically above the top of the rafter support walls.

TABLE R802.5.1(9)
RAFTER/CEILING JOIST HEEL JOINT CONNECTIONS[a, b, c, d, e, f, g]

RAFTER SLOPE	RAFTER SPACING (inches)	GROUND SNOW LOAD (psf)											
		30				50				70			
		Roof span (feet)											
		12	20	28	36	12	20	28	36	12	20	28	36
		Required number of 16d common nails[a,b] per heel joint splices[c,d,e,f]											
3:12	12	4	6	8	11	5	8	12	15	6	11	15	20
	16	5	8	11	14	6	11	15	20	8	14	20	26
	24	7	11	16	21	9	16	23	30	12	21	30	39
4:12	12	3	5	6	8	4	6	9	11	5	8	12	15
	16	4	6	8	11	5	8	12	15	6	11	15	20
	24	5	9	12	16	7	12	17	22	9	16	23	29
5:12	12	3	4	5	7	3	5	7	9	4	7	9	12
	16	3	5	7	9	4	7	9	12	5	9	12	16
	24	4	7	10	13	6	10	14	18	7	13	18	23
7:12	12	3	3	4	5	3	4	5	7	3	5	7	9
	16	3	4	5	6	3	5	7	9	4	6	9	11
	24	3	5	7	9	4	7	10	13	5	9	13	17
9:12	12	3	3	3	4	3	3	4	5	3	4	5	7
	16	3	3	4	5	3	4	5	7	3	5	7	9
	24	3	4	6	7	3	6	8	10	4	7	10	13
12:12	12	3	3	3	3	3	3	3	4	3	3	4	5
	16	3	3	3	4	3	3	4	5	3	4	5	7
	24	3	3	4	6	3	4	6	8	3	6	8	10

For SI: 1 inch = 25.4 mm, 1 foot = 304.8 mm, 1 pound per square foot = 0.0479kPa.

a. 40d box nails shall be permitted to be substituted for 16d common nails.

b. Nailing requirements shall be permitted to be reduced 25 percent if nails are clinched.

c. Heel joint connections are not required when the ridge is supported by a load-bearing wall, header or ridge beam.

d. When intermediate support of the rafter is provided by vertical struts or purlins to a loadbearing wall, the tabulated heel joint connection requirements shall be permitted to be reduced proportionally to the reduction in span.

e. Equivalent nailing patterns are required for ceiling joist to ceiling joist lap splices.

f. When rafter ties are substituted for ceiling joists, the heel joint connection requirement shall be taken as the tabulated heel joint connection requirement for two-thirds of the actual rafter-slope.

g. Tabulated heel joint connection requirements assume that ceiling joists or rafter ties are located at the bottom of the attic space. When ceiling joists or rafter ties are located higher in the attic, heel joint connection requirements shall be increased by the following factors:

H_C/H_R	Heel Joint Connection Adjustment Factor
1/3	1.5
1/4	1.33
1/5	1.25
1/6	1.2
1/10 or less	1.11

where:

H_C = Height of ceiling joists or rafter ties measured vertically above the top of the rafter support walls.

H_R = Height of roof ridge measured vertically above the top of the rafter support walls.

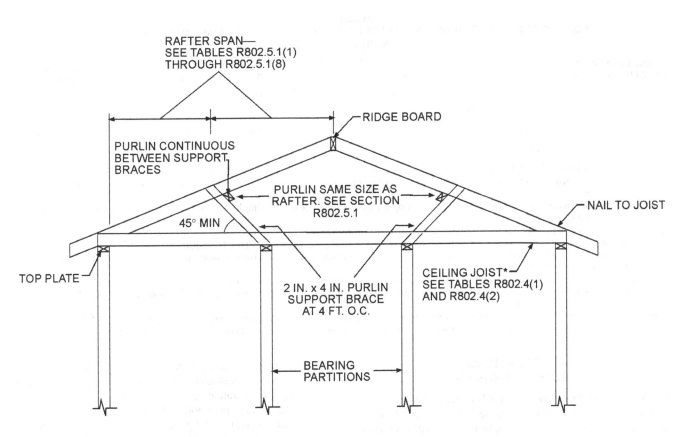

RAFTER SPAN—
SEE TABLES R802.5.1(1)
THROUGH R802.5.1(8)

RIDGE BOARD

PURLIN CONTINUOUS
BETWEEN SUPPORT
BRACES

PURLIN SAME SIZE AS
RAFTER. SEE SECTION
R802.5.1

45° MIN

NAIL TO JOIST

TOP PLATE

2 IN. x 4 IN. PURLIN
SUPPORT BRACE
AT 4 FT. O.C.

CEILING JOIST*
SEE TABLES R802.4(1)
AND R802.4(2)

BEARING
PARTITIONS

For SI: 1 inch = 25.4 mm, 1 foot = 304.8 mm, 1 degree = 0.018 rad.
NOTE: Where ceiling joints run perpendicular to the rafters, rafter ties shall be nailed to each rafter near the top of the ceiling joist.

FIGURE R802.5.1
BRACED RAFTER CONSTRUCTION

R802.10.2.1 Applicability limits. The provisions of this section shall control the design of truss roof framing when snow controls for buildings not greater than 60 feet (18 288 mm) in length perpendicular to the joist, rafter or truss span, not greater than 36 feet (10 973 mm) in width parallel to the joist span or truss, not greater than two stories in height with each story not greater than 10 feet (3048 mm) high, and roof slopes not smaller than 3:12 (25-percent slope) or greater than 12:12 (100-percent slope). Truss roof framing constructed in accordance with the provisions of this section shall be limited to sites subjected to a maximum design wind speed of 110 miles per hour (49 m/s), Exposure A, B or C, and a maximum ground snow load of 70 psf (3352 Pa). Roof snow load is to be computed as: $0.7\, p_g$.

R802.10.3 Bracing. Trusses shall be braced to prevent rotation and provide lateral stability in accordance with the requirements specified in the construction documents for the building and on the individual truss design drawings. In the absence of specific bracing requirements, trusses shall be braced in accordance with the Building Component Safety Information (BCSI 1-03) Guide to Good Practice for Handling, Installing & Bracing of Metal Plate Connected Wood Trusses.

R802.10.4 Alterations to trusses. Truss members shall not be cut, notched, drilled, spliced or otherwise altered in any way without the approval of a registered design professional. Alterations resulting in the addition of load (e.g., HVAC equipment, water heater) that exceeds the design load for the truss shall not be permitted without verification that the truss is capable of supporting such additional loading.

R802.10.5 Truss to wall connection. Trusses shall be connected to wall plates by the use of approved connectors having a resistance to uplift of not less than 175 pounds (779 N) and shall be installed in accordance with the manufacturer's specifications. For roof assemblies subject to wind uplift pressures of 20 pounds per square foot (960 Pa) or greater, as established in Table R301.2(2), adjusted for height and exposure per Table R301.2(3), see section R802.11.

R802.11 Roof tie-down.

R802.11.1 Uplift resistance. Roof assemblies which are subject to wind uplift pressures of 20 pounds per square foot (960 Pa) or greater shall have roof rafters or trusses attached to their supporting wall assemblies by connections capable of providing the resistance required in Table R802.11. Wind uplift pressures shall be determined using an effective wind area of 100 square feet (9.3 m2) and Zone 1 in Table R301.2(2), as adjusted for height and exposure per Table R301.2(3).

A continuous load path shall be designed to transmit the uplift forces from the rafter or truss ties to the foundation.

TABLE R802.11
REQUIRED STRENGTH OF TRUSS OR RAFTER CONNECTIONS TO RESIST WIND UPLIFT FORCES[a, b, c, e, f]
(Pounds per connection)

BASIC WIND SPEED (mph) (3–second gust)	ROOF SPAN (feet)							OVERHANGS[d] (pounds/foot)
	12	20	24	28	32	36	40	
85	-72	-120	-145	-169	-193	-217	-241	-38.55
90	-91	-151	-181	-212	-242	-272	-302	-43.22
100	-131	-218	-262	-305	-349	-393	-436	-53.36
110	-175	-292	-351	-409	-467	-526	-584	-64.56

For SI: 1 inch = 25.4 mm, 1 foot = 305 mm, 1 mph = 0.447 m/s, 1 pound/foot = 14.5939 N/m, 1 pound = 0.454 kg.

a. The uplift connection requirements are based on a 30 foot mean roof height located in Exposure B. For Exposures C and D and for other mean roof heights, multiply the above loads by the Adjustment Coefficients in Table R301.2(3).

b. The uplift connection requirements are based on the framing being spaced 24 inches on center. Multiply by 0.67 for framing spaced 16 inches on center and multiply by 0.5 for framing spaced 12 inches on center.

c. The uplift connection requirements include an allowance for 10 pounds of dead load.

d. The uplift connection requirements do not account for the effects of overhangs. The magnitude of the above loads shall be increased by adding the overhang loads found in the table. The overhang loads are also based on framing spaced 24 inches on center. The overhang loads given shall be multiplied by the overhang projection and added to the roof uplift value in the table.

e. The uplift connection requirements are based on wind loading on end zones as defined in Figure 6-2 of ASCE 7. Connection loads for connections located a distance of 20% of the least horizontal dimension of the building from the corner of the building are permitted to be reduced by multiplying the table connection value by 0.7 and multiplying the overhang load by 0.8.

f. For wall-to-wall and wall-to-foundation connections, the capacity of the uplift connector is permitted to be reduced by 100 pounds for each full wall above. (For example, if a 600-pound rated connector is used on the roof framing, a 500-pound rated connector is permitted at the next floor level down).

SECTION R803
ROOF SHEATHING

R803.1 Lumber sheathing. Allowable spans for lumber used as roof sheathing shall conform to Table R803.1. Spaced lumber sheathing for wood shingle and shake roofing shall conform to the requirements of Sections R905.7 and R905.8. Spaced lumber sheathing is not allowed in Seismic Design Category D_2.

TABLE R803.1
MINIMUM THICKNESS OF LUMBER ROOF SHEATHING

RAFTER OR BEAM SPACING (inches)	MINIMUM NET THICKNESS (inches)
24	$5/_8$
48[a]	
60[b]	$1^1/_2$ T & G
72[c]	

For SI: 1 inch = 25.4 mm.

a. Minimum 270 F_b, 340,000 E.

b. Minimum 420 F_b, 660,000 E.

c. Minimum 600 F_b, 1,150,000 E.

R803.2 Wood structural panel sheathing.

R803.2.1 Identification and grade. Wood structural panels shall conform to DOC PS 1, DOC PS 2 or, when manufactured in Canada, CSA 0437, and shall be identified by a grade mark or certificate of inspection issued by an approved agency. Wood structural panels shall comply with the grades specified in Table R503.2.1.1(1).

R803.2.1.1 Exposure durability. All wood structural panels, when designed to be permanently exposed in outdoor applications, shall be of an exterior exposure durability. Wood structural panel roof sheathing exposed to the underside may be of interior type bonded with exterior glue, identified as Exposure 1.

R803.2.1.2 Fire-retardant-treated plywood. The allowable unit stresses for fire-retardant-treated plywood, including fastener values, shall be developed from an approved method of investigation that considers the effects of anticipated temperature and humidity to which the fire-retardant-treated plywood will be subjected, the type of treatment and redrying process. The fire-retardant-treated plywood shall be graded by an approved agency.

R803.2.2 Allowable spans. The maximum allowable spans for wood structural panel roof sheathing shall not exceed the values set forth in Table R503.2.1.1(1), or APA E30.

R803.2.3 Installation. Wood structural panel used as roof sheathing shall be installed with joints staggered or not staggered in accordance with Table R602.3(1), or APA E30 for wood roof framing or with Table R804.3 for steel roof framing.

SECTION R804
STEEL ROOF FRAMING

R804.1 General. Elements shall be straight and free of any defects that would significantly affect their structural performance. Cold-formed steel roof framing members shall comply with the requirements of this section.

R804.1.1 Applicability limits. The provisions of this section shall control the construction of steel roof framing for buildings not greater than 60 feet (18 288 mm) perpendicular to the joist, rafter or truss span, not greater than 40 feet (12 192 mm) in width parallel to the joist span or truss, not greater than two stories in height and roof slopes not smaller than 3:12 (25-percent slope) or greater than 12:12 (100 percent slope). Steel roof framing constructed in accordance with the provisions of this section shall be limited to sites subjected to a maximum design wind speed of 110 miles per hour (49 m/s), Exposure A, B, or C, and a maximum ground snow load of 70 pounds per square foot (3350 Pa).

R804.1.2 In-line framing. Steel roof framing constructed in accordance with Section R804 shall be located directly in line with load-bearing studs below with a maximum tolerance of $^3/_4$ inch (19 mm) between the centerline of the stud and the roof joist/rafter.

R804.1.3 Roof trusses. The design, quality assurance, installation and testing of cold-formed steel trusses shall be in accordance with the AISI Standard for Cold-formed Steel Framing-Truss Design (COFS/Truss). Truss members shall not be notched, cut or altered in any manner without an approved design.

R804.2 Structural framing. Load-bearing steel roof framing members shall comply with Figure R804.2(1) and with the dimensional and minimum thickness requirements specified in Tables R804.2(1) and R804.2(2). Tracks shall comply with Figure R804.2(2) and shall have a minimum flange width of $1^1/_4$ inches (32 mm). The maximum inside bend radius for load-bearing members shall be the greater of $^3/_{32}$ inch (2.4 mm) or twice the uncoated steel thickness. Holes in roof framing members shall comply with all of the following conditions:

1. Holes shall conform to Figure R804.2(3);

2. Holes shall be permitted only along the centerline of the web of the framing member;

3. Holes shall have a center-to-center spacing of not less than 24 inches (610 mm);

4. Holes shall have a width not greater than 0.5 times the member depth, or $2^1/_2$ inches (64 mm);

5. Holes shall have a length not exceeding $4^1/_2$ inches (114 mm); and

6. Holes shall have a minimum distance between the edge of the bearing surface and the edge of the hole of not less than 10 inches (254 mm).

Framing members with web holes not conforming to these requirements shall be patched in accordance with Section R804.3.6 or designed in accordance with accepted engineering practices.

R804.2.1 Material. Load-bearing steel framing members shall be cold-formed to shape from structural quality sheet steel complying with the requirements of one of the following:

1. ASTM A 653: Grades 33, 37, 40 and 50 (Class 1 and 3).

2. ASTM A 792: Grades 33, 37, 40 and 50A.

3. ASTM A 875: Grades 33, 37, 40 and 50 (Class 1 and 3).

4. ASTM A 1003: Grades 33, 37, 40 and 50.

R804.2.2 Identification. Load-bearing steel framing members shall have a legible label, stencil, stamp or embossment with the following information as a minimum:

1. Manufacturer's identification.

2. Minimum uncoated steel thickness in inches (mm).

3. Minimum coating designation.

4. Minimum yield strength, in kips per square inch (ksi).

R804.2.3 Corrosion protection. Load-bearing steel framing shall have a metallic coating complying with one of the following:

1. A minimum of G 60 in accordance with ASTM A 653.

2. A minimum of AZ 50 in accordance with ASTM A 792.

3. A minimum of GF 60 in accordance with ASTM A 875.

TABLE R804.2(1)
LOAD-BEARING COLD-FORMED STEEL MEMBER SIZES

NOMINAL MEMBER SIZE MEMBER DESIGNATION[a]	WEB DEPTH (inches)	MINIMUM FLANGE WIDTH (inches)	MAXIMUM FLANGE WIDTH (inches)	MINIMUM LIP SIZE (inches)
350S162-t	3.5	1.625	2	0.5
550S162-t	5.5	1.625	2	0.5
800S162-t	8	1.625	2	0.5
1000S162-t	10	1.625	2	0.5
1200S162-t	12	1.625	2	0.5

For SI: 1 inch = 25.4 mm.

a. The member designation is defined by the first number representing the member depth in hundredths of an inch, the letter "s" representing a stud or joist member, the second number representing the flange width in hundredths of an inch, and the letter "t" shall be a number representing the minimum base metal thickness in mils [see Table R804.2(2)].

TABLE R804.2(2)
MINIMUM THICKNESS OF COLD-FORMED STEEL ROOF FRAMING MEMBERS

DESIGNATION (mils)	MINIMUM UNCOATED THICKNESS (inches)	REFERENCED GAGE NUMBER
33	0.033	20
43	0.043	18
54	0.054	16
68	0.068	14

For SI: 1 inch = 25.4 mm, 1 mil = 0.0254 mm.

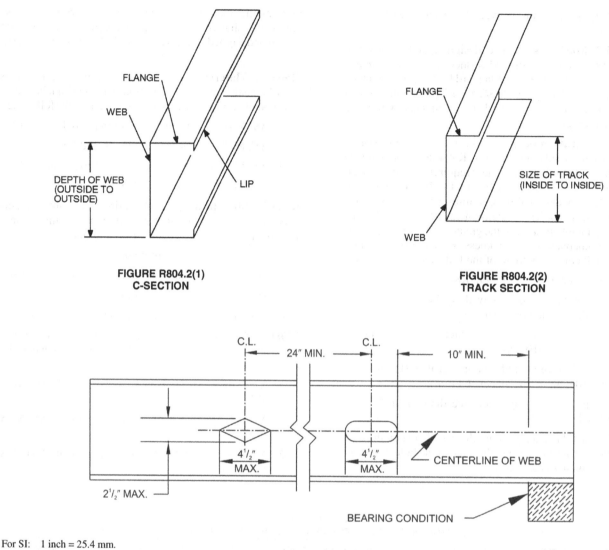

FIGURE R804.2(1)
C-SECTION

FIGURE R804.2(2)
TRACK SECTION

For SI: 1 inch = 25.4 mm.

FIGURE R804.2(3)
WEB HOLES

R804.2.4 Fastening requirements. Screws for steel-to-steel connections shall be installed with a minimum edge distance and center-to-center spacing of $^1/_2$ inch (13 mm), shall be self-drilling tapping, and shall conform to SAE J78. Structural sheathing shall be attached to roof rafters with minimum No. 8 self-drilling tapping screws that conform to SAE J78. Screws for attaching structural sheathing to steel roof framing shall have a minimum head diameter of 0.292 inch (7.4 mm) with countersunk heads and shall be installed with a minimum edge distance of $^3/_8$ inch (10 mm). Gypsum board ceilings shall be attached to steel joists with minimum No. 6 screws conforming to ASTM C 954 and shall be installed in accordance with Section R805. For all connections, screws shall extend through the steel a minimum of three exposed threads. All self-drilling tapping screws conforming to SAE J78 shall have a minimum Type II coating in accordance with ASTM B 633.

Where No. 8 screws are specified in a steel-to-steel connection, reduction of the required number of screws in the connection is permitted in accordance with the reduction factors in Table R804.2.4 when larger screws are used or when one of the sheets of steel being connected is thicker that 33 mils (0.84 mm). When applying the reduction factor, the resulting number of screws shall be rounded up.

TABLE R804.2.4
SCREW SUBSTITUTION FACTOR

SCREW SIZE	THINNEST CONNECTED STEEL SHEET (mils)	
	33	43
#8	1.0	0.67
#10	0.93	0.62
#12	0.86	0.56

For SI: 1 mil = 0.0254 mm.

R804.3 Roof construction. Steel roof systems constructed in accordance with the provisions of this section shall consist of both ceiling joists and rafters in accordance with Figure R804.3 and fastened in accordance with Table R804.3.

R804.3.1 Allowable ceiling joist spans. The clear span of cold-formed steel ceiling joists shall not exceed the limits set forth in Tables R804.3.1(1) through R804.3.1(8). Ceiling joists shall have a minimum bearing length of 1.5 inches (38 mm) and shall be connected to rafters (heel joint) in accordance with Figure R804.3.1(1) and Table R804.3.1. When continuous joists are framed across interior bearing supports, the interior bearing supports shall be located within 24 inches (610 mm) of midspan of the ceiling joist, and the individual spans shall not exceed the applicable spans in Tables R804.3.1(2), R804.3.1(4), R804.3.1(6), R804.3.1(8). Where required in Tables R804.3.1(1) through R804.3.1(8), bearing stiffeners shall be installed at each bearing location in accordance with Section R804.3.8 and Figure R804.3.8. When the attic is to be used as an occupied space, the ceiling joists shall be designed in accordance with Section R505.

TABLE R804.3
ROOF FRAMING FASTENING SCHEDULE[a,b]

DESCRIPTION OF BUILDING ELEMENTS	NUMBER AND SIZE OF FASTENERS	SPACING OF FASTENERS
Ceiling joist to top track of load-bearing wall	2 No. 10 screws	Each joist
Roof sheathing (oriented strand board or plywood) to rafters	No. 8 screws	6″ o.c. on edges and 12″ o.c. at interior supports. 6″ o.c. at gable end truss
Truss to bearing wall[a]	2 No. 10 screws	Each truss
Gable end truss to endwall top track	No. 10 screws	12″ o.c.
Rafter to ceiling joist	Minimum No. 10 screws, per Table R804.3.1	Evenly spaced, not less than $\frac{1}{2}$″ from all edges.

For SI: 1 inch = 25.4 mm, 1 foot = 304.8 mm, 1 pound per square foot = 0.0479kPa, 1 mil = 0.0254 mm.

a. Screws shall be applied through the flanges of the truss or ceiling joist or a 54 mil clip angle shall be used with two No. 10 screws in each leg. See Section R804.4 for additional requirements to resist uplift forces.

b. Spacing of fasteners on roof sheathing panel edges applies to panel edges supported by framing members and at all roof plane perimeters. Blocking of roof sheathing panel edges perpendicular to the framing members shall not be required except at the intersection of adjacent roof planes. Roof perimeter shall be supported by framing members or cold-formed blocking of the same depth and gage as the floor members.

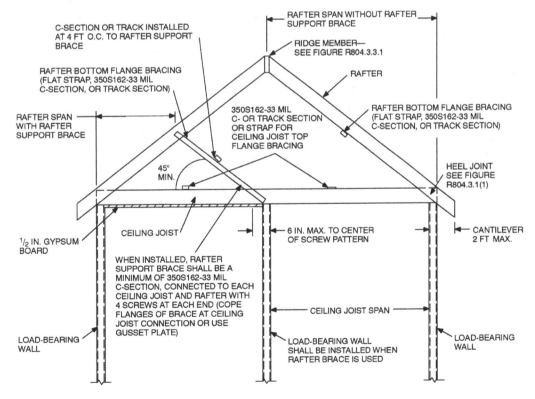

For SI: 1 inch = 25.4 mm, 1 foot = 304.8 mm, 1 mil = 0.0254 mm.

FIGURE R804.3
STEEL ROOF CONSTRUCTION

TABLE R804.3.1(1)
CEILING JOIST SPANS
SINGLE SPANS WITH BEARING STIFFENERS
10 lb per sq ft LIVE LOAD (NO ATTIC STORAGE)[a, b, c] 33 ksi STEEL

MEMBER DESIGNATION	ALLOWABLE SPAN (feet-inches)					
	Lateral Support of Top (Compression) Flange					
	Unbraced		Mid-Span Bracing		Third-Point Bracing	
	Ceiling Joist Spacing (inches)					
	16	24	16	24	16	24
350S162-33	9'-5"	8'-6"	12'-2"	10'-4"	12'-2"	10'-7"
350S162-43	10'-3"	9'-2"	12'-10"	11'-2"	12'-10"	11'-2"
350S162-54	11'-1"	9'-11"	13'-9"	12'-0"	13'-9"	12'-0"
350S162-68	12'-1"	10'-9"	14'-8"	12'-10"	14'-8"	12'-10"
350S162-97	14'-4"	12'-7"	16'-4"	14'-3"	16'-4"	14'-3"
550S162-33	10'-7"	9'-6"	14'-10"	12'-10"	15'-11"	13'-4"
550S162-43	11'-8"	10'-6"	16'-4"	14'-3"	17'-10"	15'-3"
550S162-54	12'-6"	11'-2"	17'-7"	15'-7"	19'-5"	16'-10"
550S162-68	13'-6"	12'-1"	19'-2"	17'-1"	21'-0"	18'-4"
550S162-97	15'-9"	13'-11"	21'-8"	19'-3"	23'-5"	20'-5"
800S162-33	12'-2"	10'-11"	17'-8"	15'-10"	19'-10"	17'-1"
800S162-43	13'-0"	11'-9"	18'-10"	17'-0"	21'-6"	19'-1"
800S162-54	13'-10"	12'-5"	20'-0"	18'-0"	22'-9"	20'-4"
800S162-68	14'-11"	13'-4"	21'-3"	19'-1"	24'-1"	21'-8"
800S162-97	17'-1"	15'-2"	23'-10"	21'-3"	26'-7"	23'-10"
1000S162-43	13'-11"	12'-6"	20'-2"	18'-3"	23'-1"	20'-9"
1000S162-54	14'-9"	13'-3"	21'-4"	19'-3"	24'-4"	22'-0"
1000S162-68	15'-10"	14'-2"	22'-8"	20'-5"	25'-9"	23'-2"
1000S162-97	18'-0"	16'-0"	25'-3"	22'-7"	28'-3"	25'-4"
1200S162-43	14'-8"	13'-3"	21'-4"	19'-3"	24'-5"	21'-8"
1200S162-54	15'-7"	14'-0"	22'-6"	20'-4"	25'-9"	23'-2"
1200S162-68	16'-8"	14'-11"	23'-11"	21'-6"	27'-2"	24'-6"
1200S162-97	18'-9"	16'-9"	26'-6"	23'-8"	29'-9"	26'-9"

For SI: 1 inch = 25.4 mm, 1 foot = 304.8 mm, 1 pound per square foot = 0.0479 kPa.

a. Deflection criterion: $L/240$ for total loads.

b. Ceiling dead load = 5 psf.

c. Bearing stiffeners are required at all bearing points and concentrated load locations.

TABLE R804.3.1(2)
CEILING JOIST SPANS
TWO EQUAL SPANS WITH BEARING STIFFENERS
10 lb per sq ft LIVE LOAD (NO ATTIC STORAGE)[a, b, c] 33 ksi STEEL

MEMBER DESIGNATION	ALLOWABLE SPAN (feet-inches)					
	Lateral Support of Top (Compression) Flange					
	Unbraced		Mid-Span Bracing		Third-Point Bracing	
	Ceiling Joist Spacing (inches)					
	16	24	16	24	16	24
350S162-33	12'-11"	10'-11"	13'-5"	10'-11"	13'-5"	10'-11"
350S162-43	14'-2"	12'-8"	15'-10"	12'-11"	15'-10"	12'-11"
350S162-54	15'-6"	13'-10"	17'-1"	14'-6"	17'-9"	14'-6"
350S162-68	17'-3"	15'-3"	18'-6"	16'-1"	19'-8"	16'-1"
350S162-97	20'-10"	18'-4"	21'-5"	18'-10"	21'-11"	18'-10"
550S162-33	14'-4"	12'-11"	16'-7"	14'-1"	17'-3"	14'-1"
550S162-43	16'-0"	14'-1"	17'-11"	16'-1"	20'-7"	16'-10"
550S162-54	17'-4"	15'-6"	19'-5"	17'-6"	23'-2"	19'-0"
550S162-68	19'-1"	16'-11"	20'-10"	18'-8"	25'-2"	21'-5"
550S162-97	22'-8"	19'-9"	23'-6"	20'-11"	27'-11"	25'-1"
800S162-33	16'-5"	14'-10"	19'-2"	17'-3"	23'-1"	18'-3"
800S162-43	17'-9"	15'-11"	20'-6"	18'-5"	25'-0"	22'-6"
800S162-54	19'-1"	17'-1"	21'-8"	19'-6"	26'-4"	23'-9"
800S162-68	20'-9"	18'-6"	23'-1"	20'-9"	28'-0"	25'-2"
800S162-97	24'-5"	21'-6"	26'-0"	23'-2"	31'-1"	27'-9"
1000S162-43	18'-11"	17'-0"	21'-11"	19'-9"	26'-8"	24'-1"
1000S162-54	20'-3"	18'-2"	23'-2"	20'-10"	28'-2"	25'-5"
1000S162-68	21'-11"	19'-7"	24'-7"	22'-2"	29'-10"	26'-11"
1000S162-97	25'-7"	22'-7"	27'-6"	24'-6"	33'-0"	29'-7"
1200S162-43	19'-11"	17'-11"	23'-1"	20'-10"	28'-3"	25'-6"
1200S162-54	21'-3"	19'-1"	24'-5"	22'-0"	29'-9"	26'-10"
1200S162-68	23'-0"	20'-7"	25'-11"	23'-4"	31'-6"	28'-4"
1200S162-97	26'-7"	23'-6"	28'-9"	25'-10"	34'-8"	31'-1"

For SI: 1 inch = 25.4 mm, 1 foot = 304.8 mm, 1 pound per square foot = 0.0479 kPa.

a. Deflection criterion: $L/240$ for total loads.

b. Ceiling dead load = 5 psf.

c. Bearing stiffeners are required at all bearing points and concentrated load locations.

TABLE R804.3.1(3)
CEILING JOIST SPANS
SINGLE SPANS WITH BEARING STIFFENERS
20 lb per sq ft LIVE LOAD (LIMITED ATTIC STORAGE)[a, b, c] 33 ksi STEEL

MEMBER DESIGNATION	ALLOWABLE SPAN (feet-inches)					
	Lateral Support of Top (Compression) Flange					
	Unbraced		Mid-Span Bracing		Third-Point Bracing	
	Ceiling Joist Spacing (inches)					
	16	24	16	24	16	24
350S162-33	8'-2"	7'-2"	9'-9"	8'-1"	9'-11"	8'-1"
350S162-43	8'-10"	7'-10"	11'-0"	9'-5"	11'-0"	9'-7"
350S162-54	9'-6"	8'-6"	11'-9"	10'-3"	11'-9"	10'-3"
350S162-68	10'-4"	9'-2"	12'-7"	11'-0"	12'-7"	11'-0"
350S162-97	12'-1"	10'-8"	14'-0"	12'-0"	14'-0"	12'-0"
550S162-33	9'-2"	8'-3"	12'-2"	10'-2"	12'-6"	10'-5"
550S162-43	10'-1"	9'-1"	13'-7"	11'-7"	14'-5"	12'-2"
550S162-54	10'-9"	9'-8"	14'-10"	12'-10"	15'-11"	13'-6"
550S162-68	11'-7"	10'-4"	16'-4"	14'-0"	17'-5"	14'-11"
550S162-97	13'-4"	11'-10"	18'-5"	16'-2"	20'-1"	17'-1"
800S162-33	10'-7"	9'-6"	15'-1"	13'-0"	16'-2"	13'-7"
800S162-43	11'-4"	10'-2"	16'-5"	14'-6"	18'-2"	15'-9"
800S162-54	12'-0"	10'-9"	17'-4"	15'-6"	19'-6"	17'-0"
800S162-68	12'-10"	11'-6"	18'-5"	16'-6"	20'-10"	18'-3"
800S162-97	14'-7"	12'-11"	20'-5"	18'-3"	22'-11"	20'-5"
1000S162-43	12'-1"	10'-11"	17'-7"	15'-10"	19'-11"	17'-3"
1000S162-54	12'-10"	11'-6"	18'-7"	16'-9"	21'-2"	18'-10"
1000S162-68	13'-8"	12'-3"	19'-8"	17'-8"	22'-4"	20'-1"
1000S162-97	15'-4"	13'-8"	21'-8"	19'-5"	24'-5"	21'-11"
1200S162-43	12'-9"	11'-6"	18'-7"	16'-6"	20'-9"	18'-2"
1200S162-54	13'-6"	12'-2"	19'-7"	17'-8"	22'-5"	20'-2"
1200S162-68	14'-4"	12'-11"	20'-9"	18'-8"	23'-7"	21'-3"
1200S162-97	16'-1"	14'-4"	22'-10"	20'-6"	25'-9"	23'-2"

For SI: 1 inch = 25.4 mm, 1 foot = 304.8 mm, 1 pound per square foot = 0.0479 kPa.

a. Deflection criterion: $L/240$ for total loads.
b. Ceiling dead load = 5 psf.
c. Bearing stiffeners are required at all bearing points and concentrated load locations.

TABLE R804.3.1(4)
CEILING JOIST SPANS
TWO EQUAL SPANS WITH BEARING STIFFENERS
20 lb per sq ft LIVE LOAD (LIMITED ATTIC STORAGE)[a, b, c] 33 ksi STEEL

MEMBER DESIGNATION	ALLOWABLE SPAN (feet-inches)					
	Lateral Support of Top (Compression) Flange					
	Unbraced		Mid-Span Bracing		Third-Point Bracing	
	Ceiling Joist Spacing (inches)					
	16	24	16	24	16	24
350S162-33	10'-2"	8'-4"	10'-2"	8'-4"	10'-2"	8'-4"
350S162-43	12'-1"	9'-10"	12'-1"	9'-10"	12'-1"	9'-10"
350S162-54	13'-3"	11'-0"	13'-6"	11'-0"	13'-6"	11'-0"
350S162-68	14'-7"	12'-3"	15'-0"	12'-3"	15'-0"	12'-3"
350S162-97	17'-6"	14'-3"	17'-6"	14'-3"	17'-6"	14'-3"
550S162-33	12'-5"	10'-9"	13'-2"	10'-9"	13'-2"	10'-9"
550S162-43	13'-7"	12'-1"	15'-6"	12'-9"	15'-8"	12'-9"
550S162-54	14'-11"	13'-4"	16'-10"	14'-5"	17'-9"	14'-5"
550S162-68	16'-3"	14'-5"	18'-0"	16'-1"	20'-0"	16'-4"
550S162-97	19'-1"	16'-10"	20'-3"	18'-0"	23'-10"	19'-5"
800S162-33	14'-3"	12'-4"	16'-7"	12'-4"	16'-7"	12'-4"
800S162-43	15'-4"	13'-10"	17'-9"	16'-0"	21'-8"	17'-9"
800S162-54	16'-5"	14'-9"	18'-10"	16'-11"	22'-11"	20'-6"
800S162-68	17'-9"	15'-11"	20'-0"	18'-0"	24'-3"	21'-10"
800S162-97	20'-8"	18'-3"	22'-3"	19'-11"	26'-9"	24'-0"
1000S162-43	16'-5"	14'-9"	19'-0"	17'-2"	23'-3"	18'-11"
1000S162-54	17'-6"	15'-8"	20'-1"	18'-1"	24'-6"	22'-1"
1000S162-68	18'-10"	16'-10"	21'-4"	19'-2"	25'-11"	23'-4"
1000S162-97	21'-8"	19'-3"	23'-7"	21'-2"	28'-5"	25'-6"
1200S162-43	17'-3"	15'-7"	20'-1"	18'-2"	24'-6"	18'-3"
1200S162-54	18'-5"	16'-6"	21'-3"	19'-2"	25'-11"	23'-5"
1200S162-68	19'-9"	17'-8"	22'-6"	20'-3"	27'-4"	24'-8"
1200S162-97	22'-7"	20'-1"	24'-10"	22'-3"	29'-11"	26'-11"

For SI: 1 inch = 25.4 mm, 1 foot = 304.8 mm, 1 pound per square foot = 0.0479 kPa.

a. Deflection criterion: L/240 for total loads.
b. Ceiling dead load = 5 psf.
c. Bearing stiffeners are required at all bearing points and concentrated load locations.

TABLE R804.3.1(5)
CEILING JOIST SPANS
SINGLE SPANS WITHOUT BEARING STIFFENERS
10 lb per sq ft LIVE LOAD (NO ATTIC STORAGE)[a, b] 33 ksi STEEL

MEMBER DESIGNATION	ALLOWABLE SPAN (feet-inches)					
	Lateral Support of Top (Compression) Flange					
	Unbraced		Mid-Span Bracing		Third-Point Bracing	
	Ceiling Joist Spacing (inches)					
	16	24	16	24	16	24
350S162-33	9'-5"	8'-6"	12'-2"	10'-4"	12'-2"	10'-7"
350S162-43	10'-3"	9'-12"	13'-2"	11'-6"	13'-2"	11'-6"
350S162-54	11'-1"	9'-11"	13'-9"	12'-0"	13'-9"	12'-0"
350S162-68	12'-1"	10'-9"	14'-8"	12'-10"	14'-8"	12'-10"
350S162-97	14'-4"	12'-7"	16'-10"	14'-3"	16'-4"	14'-3"
550S162-33	10'-7"	9'-6"	14'-10"	12'-10"	15'-11"	13'-4"
550S162-43	11'-8"	10'-6"	16'-4"	14'-3"	17'-10"	15'-3"
550S162-54	12'-6"	11'-2"	17'-7"	15'-7"	19'-5"	16'-10"
550S162-68	13'-6"	12'-1"	19'-2"	17'-0"	21'-0"	18'-4"
550S162-97	15'-9"	13'-11"	21'-8"	19'-3"	23'-5"	20'-5"
800S162-33	—	—	—	—	—	—
800S162-43	13'-0"	11'-9"	18'-10"	17'-0"	21'-6"	19'-0"
800S162-54	13'-10"	12'-5"	20'-0"	18'-0"	22'-9"	20'-4"
800S162-68	14'-11"	13'-4"	21'-3"	19'-1"	24'-1"	21'-8"
800S162-97	17'-1"	15'-2"	23'-10"	21'-3"	26'-7"	23'-10"
1000S162-43	—	—	—	—	—	—
1000S162-54	14'-9"	13'-3"	21'-4"	19'-3"	24'-4"	22'-0"
1000S162-68	15'-10"	14'-2"	22'-8"	20'-5"	25'-9"	23'-2"
1000S162-97	18'-0"	16'-0"	25'-3"	22'-7"	28'-3"	25'-4"
1200S162-43	—	—	—	—	—	—
1200S162-54	—	—	—	—	—	—
1200S162-68	16'-8"	14'-11"	23'-11"	21'-6"	27'-2"	24'-6"
1200S162-97	18'-9"	16'-9"	26'-6"	23'-8"	29'-9"	26'-9"

For SI: 1 inch = 25.4 mm, 1 foot = 304.8 mm, 1 pound per square foot = 0.0479 kPa.

a. Deflection criterion: $L/240$ for total loads.

b. Ceiling dead load = 5 psf.

TABLE R804.3.1(6)
CEILING JOIST SPANS
TWO EQUAL SPANS WITHOUT BEARING STIFFENERS
10 lb per sq ft LIVE LOAD (NO ATTIC STORAGE)[a, b] 33 ksi STEEL

MEMBER DESIGNATION	ALLOWABLE SPAN (feet-inches)					
	Lateral Support of Top (Compression) Flange					
	Unbraced		Mid-Span Bracing		Third-Point Bracing	
	Ceiling Joist Spacing (inches)					
	16	24	16	24	16	24
350S162-33	11'-9"	8'-11"	11'-9"	8'-11"	11'-9"	8'-11"
350S162-43	14'-2"	11'-7"	14'-11"	11'-7"	14'-11"	11'-7"
350S162-54	15'-6"	13'-10"	17'-1"	13'-10"	17'-7"	13'-10"
350S162-68	17'-3"	15'-3"	18'-6"	16'-1"	19'-8"	16'-1"
350S162-97	20'-10"	18'-4"	21'-5"	18'-9"	21'-11"	18'-9"
550S162-33	13'-4"	9'-11"	13'-4"	9'-11"	13'-4"	9'-11"
550S162-43	16'-0"	13'-6"	17'-9"	13'-6"	17'-9"	13'-6"
550S162-54	17'-4"	15'-6"	19'-5"	16'-10"	21'-9"	16'-10"
550S162-68	19'-1"	16'-11"	20'-10"	18'-8"	24'-11"	20'-6"
550S162-97	22'-8"	20'-0"	23'-9"	21'-1"	28'-2"	25'-1"
800S162-33	—	—	—	—	—	—
800S162-43	17'-9"	15'-7"	20'-6"	15'-7"	21'-0"	15'-7"
800S162-54	19'-1"	17'-1"	21'-8"	19'-6"	26'-4"	23'-10"
800S162-68	20'-9"	18'-6"	23'-1"	20'-9"	28'-0"	25'-2"
800S162-97	24'-5"	21'-6"	26'-0"	23'-2"	31'-1"	27'-9"
1000S162-43	—	—	—	—	—	—
1000S162-54	20'-3"	18'-2"	23'-2"	20'-10"	28'-2"	21'-2"
1000S162-68	21'-11"	19'-7"	24'-7"	22'-2"	29'-10"	26'-11"
1000S162-97	25'-7"	22'-7"	27'-6"	24'-6"	33'-0"	29'-7"
1200S162-43	—	—	—	—	—	—
1200S162-54	—	—	—	—	—	—
1200S162-68	23'-0"	20'-7"	25'-11"	23'-4"	31'-6"	28'-4"
1200S162-97	26'-7"	23'-6"	28'-9"	25'-10"	34'-8"	31'-1"

For SI: 1 inch = 25.4 mm, 1 foot = 304.8 mm, 1 pound per square foot = 0.0479 kPa.
a. Deflection criterion: L/240 for total loads.
b. Ceiling dead load = 5 psf.

TABLE R804.3.1(7)
CEILING JOIST SPANS
SINGLE SPANS WITHOUT BEARING STIFFENERS
20 lb per sq ft LIVE LOAD (LIMITED ATTIC STORAGE)[a, b] 33 ksi STEEL

MEMBER DESIGNATION	ALLOWABLE SPAN (feet-inches)					
	Lateral Support of Top (Compression) Flange					
	Unbraced		Mid-Span Bracing		Third-Point Bracing	
	Ceiling Joist Spacing (inches)					
	16	24	16	24	16	24
350S162-33	8'-2"	6'-10"	9'-9"	6'-10"	9'-11"	6'-10"
350S162-43	8'-10"	7'-10"	11'-0"	9'-5"	11'-0"	9'-7"
350S162-54	9'-6"	8'-6"	11'-9"	10'-3"	11'-9"	10'-3"
350S162-68	10'-4"	9'-2"	12'-7"	11'-0"	12'-7"	11'-0"
350S162-97	12'-10"	10'-8"	13'-9"	12'-0"	13'-9"	12'-0"
550S162-33	9'-2"	8'-3"	12'-2"	8'-5"	12'-6"	8'-5"
550S162-43	10'-1"	9'-1"	13'-7"	11'-8"	14'-5"	12'-2"
550S162-54	10'-9"	9'-8"	14'-10"	12'-10"	15'-11"	13'-6"
550S162-68	11'-7"	10'-4"	16'-4"	14'-0"	17'-5"	14'-11"
550S162-97	13'-4"	11'-10"	18'-5"	16'-2"	20'-1"	17'-4"
800S162-33	—	—	—	—	—	—
800S162-43	11'-4"	10'-1"	16'-5"	13'-6"	18'-1"	13'-6"
800S162-54	20'-0"	10'-9"	17'-4"	15'-6"	19'-6"	27'-0"
800S162-68	12'-10"	11'-6"	18'-5"	16'-6"	20'-10"	18'-3"
800S162-97	14'-7"	12'-11"	20'-5"	18'-3"	22'-11"	20'-5"
1000S162-43	—	—	—	—	—	—
1000S162-54	12'-10"	11'-6"	18'-7"	16'-9"	21'-2"	15'-5"
1000S162-68	13'-8"	12'-3"	19'-8"	17'-8"	22'-4"	20'-1"
1000S162-97	15'-4"	13'-8"	21'-8"	19'-5"	24'-5"	21'-11"
1200S162-43	—	—	—	—	—	—
1200S162-54	—	—	—	—	—	—
1200S162-68	14'-4"	12'-11"	20'-9"	18'-8"	23'-7"	21'-3"
1200S162-97	16'-1"	14'-4"	22'-10"	20'-6"	25'-9"	23'-2"

For SI: 1 inch = 25.4 mm, 1 foot = 304.8 mm, 1 pound per square foot = 0.0479 kPa.

a. Deflection criterion: $L/240$ for total loads.

b. Ceiling dead load = 5 psf.

TABLE R804.3.1(8)
CEILING JOIST SPANS
TWO EQUAL SPANS WITHOUT BEARING STIFFENERS
20 lb per sq ft LIVE LOAD (LIMITED ATTIC STORAGE)[a, b] 33 ksi STEEL

MEMBER DESIGNATION	ALLOWABLE SPAN (feet-inches)					
	Lateral Support of Top (Compression) Flange					
	Unbraced		Mid-Span Bracing		Third-Point Bracing	
	Ceiling Joist Spacing (inches)					
	16	24	16	24	16	24
350S162-33	8'-1"	6'-1"	8'-1"	6'-1"	8'-1"	6'-1"
350S162-43	10'-7"	8'-1"	10'-7"	8'-1"	10'-7"	8'-1"
350S162-54	12'-8"	9'-10"	12'-8"	9'-10"	12'-8"	9'-10"
350S162-68	14'-7"	11'-10"	14'-11"	11'-10"	14'-11"	11'-10"
350S162-97	17'-6"	14'-3"	17'-6"	14'-3"	17'-6"	14'-3"
550S162-33	8'-11"	6'-8"	8'-11"	6'-8"	8'-11"	6'-8"
550S162-43	12'-3"	9'-2"	12'-3"	9'-2"	12'-3"	9'-2"
550S162-54	14'-11"	11'-8"	15'-4"	11'-8"	15'-4"	11'-8"
550S162-68	16'-3"	14'-5"	18'-0"	15'-8"	18'-10"	14'-7"
550S162-97	19'-1"	16'-10"	20'-3"	18'-0"	23'-9"	19'-5"
800S162-33	—	—	—	—	—	—
800S162-43	13'-11"	9'-10"	13'-11"	9'-10"	13'-11"	9'-10"
800S162-54	16'-5"	13'-9"	18'-8"	13'-9"	18'-8"	13'-9"
800S162-68	17'-9"	15'-11"	20'-0"	18'-0"	24'-1"	18'-3"
800S162-97	20'-8"	18'-3"	22'-3"	19'-11"	26'-9"	24'-0"
1000S162-43	—	—	—	—	—	—
1000S162-54	17'-6"	13'-11"	19'-1"	13'-11"	19'-1"	13'-11"
1000S162-68	18'-10"	16'-10"	21'-4"	19'-2"	25'-11"	19'-7"
1000S162-97	21'-8"	19'-3"	23'-7"	21'-2"	28'-5"	25'-6"
1200S162-43	—	—	—	—	—	—
1200S162-54	—	—	—	—	—	—
1200S162-68	19'-9"	17'-8"	22'-6"	19'-8"	26'-8"	19'-8"
1200S162-97	22'-7"	20'-1"	24'-10"	22'-3"	29'-11"	26'-11"

For SI: 1 inch = 25.4 mm, 1 foot = 304.8 mm, 1 pound per square foot = 0.0479 kPa.
a. Deflection criterion: *L*/240 for total loads.
b. Ceiling dead load = 5 psf.

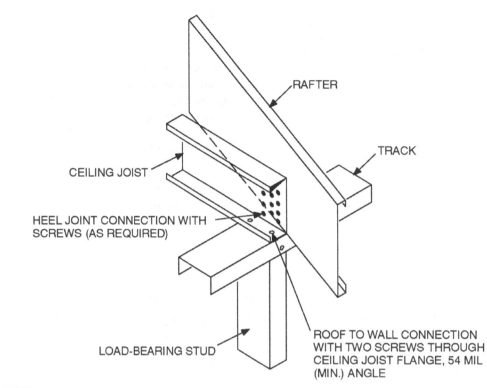

For SI: 1 mil = 0.0254 mm.

FIGURE R804.3.1(1)
JOIST TO RAFTER CONNECTION

TABLE R804.3.1
NUMBER OF SCREWS REQUIRED FOR CEILING JOIST TO RAFTER CONNECTION[a]

	NUMBER OF SCREWS																		
	Building width (feet)																		
	24				28				32				36				40			
	Ground snow load (psf)																		
ROOF SLOPE	20	30	50	70	20	30	50	70	20	30	50	70	20	30	50	70	20	30	50	70
3/12	5	6	9	11	5	7	10	13	6	8	11	15	7	8	13	17	8	9	14	19
4/12	4	5	7	9	4	5	8	10	5	6	9	12	5	7	10	13	6	7	11	14
5/12	3	4	6	7	4	4	6	8	4	5	7	10	5	5	8	11	5	6	9	12
6/12	3	3	5	6	3	4	6	7	4	4	6	8	4	5	7	9	4	5	8	10
7/12	3	3	4	6	3	3	5	7	3	4	6	7	4	4	6	8	4	5	7	9
8/12	2	3	4	5	3	3	5	6	3	4	5	7	3	4	6	8	4	4	6	8
9/12	2	3	4	5	3	3	4	6	3	3	5	6	3	4	5	7	3	4	6	8
10/12	2	2	4	5	2	3	4	5	3	3	5	6	3	3	5	7	3	4	6	7
11/12	2	2	3	4	2	3	4	5	3	3	4	6	3	3	5	6	3	4	5	7
12/12	2	2	3	4	2	3	4	5	2	3	4	5	3	3	5	6	3	4	5	7

For SI: 1 inch = 25.4 mm, 1 foot = 304.8 mm, 1 pound per square foot = 0.0479 kPa.
a. Screws shall be No. 10.

R804.3.2 Ceiling joist bracing. The bottom flanges of steel ceiling joists shall be laterally braced in accordance with Section R702. The top flanges of steel ceiling joists shall be laterally braced with a minimum of 33 mil (0.84 mm) C-section, 33 mil (0.84 mm) track section or $1^1/_2$ inch by 33 mil (38 mm by 0.84 mm) continuous steel strapping as required in Tables R804.3.1(1) through R804.3.1(8). Lateral bracing shall be installed in accordance with Figure R804.3. C-section, tracks or straps shall be fastened to the top flange at each joist with at least one No. 8 screw and shall be fastened to blocking with at least two No. 8 screws. Blocking or bridging (X-bracing) shall be installed between joists in line with strap bracing at a maximum spacing of 12 feet (3658 mm) measured perpendicular to the joists, and at the termination of all straps. The third-point bracing span values from Tables R804.3.1(1) through R804.3.1(8) shall be used for straps installed at closer spacings than third-point bracing, or when sheathing is applied to the top of the ceiling joists.

R804.3.3 Allowable rafter spans. The horizontal projection of the rafter span, as shown in Figure R804.3, shall not exceed the limits set forth in Table R804.3.3(1). Wind speeds shall be converted to equivalent ground snow loads in accordance with Table R804.3.3(2). Rafter spans shall be selected based on the higher of the ground snow load or the equivalent snow load converted from the wind speed. When required, a rafter support brace shall be a minimum of 350S162-33 C-section with maximum length of 8 feet (2438 mm) and shall be connected to a ceiling joist and rafter with four No. 10 screws at each end.

R804.3.3.1 Rafter framing. Rafters shall be connected to a parallel ceiling joist to form a continuous tie between exterior walls in accordance with Figures R804.3 and R804.3.1(1) and Table R804.3.1. Rafters shall be connected to a ridge member with a minimum 2-inch by 2-inch (51 mm by 51 mm) clip angle fastened with minimum No. 10 screws to the ridge member in accordance with Figure R804.3.3.1 and Table R804.3.3.1. The clip angle shall have a minimum steel thickness as the rafter member and shall extend the full depth of the rafter member. The ridge member shall be fabricated from a C-section and a track section, which shall be of a minimum size and steel thickness as the adjacent rafters and shall be installed in accordance with Figure R804.3.3.1.

R804.3.3.2 Roof cantilevers. Roof cantilevers shall not exceed 24 inches (610 mm) in accordance with Figure R804.3. Roof cantilevers shall be supported by a header in accordance with Section R603.6 or shall be supported by the floor framing in accordance with Section R505.3.7.

R804.3.4 Rafter bottom flange bracing. The bottom flanges of steel rafters shall be continuously braced with a minimum 33-mil (0.84 mm) C-section, 33-mil (0.84 mm) track section, or a $1^1/_2$-inch by 33-mil (38 mm by 0.84 mm) steel strapping at a maximum spacing of 8 feet (2438 mm) as measured parallel to the rafters. Bracing shall be installed in accordance with Figure R804.3. The C-section, track section, or straps shall be fastened to blocking with at least two No. 8 screws. Blocking or bridging (X-bracing) shall be

installed between rafters in-line with the continuous bracing at a maximum spacing of 12 feet (3658 mm) measured perpendicular to the rafters and at the termination of all straps. The ends of continuous bracing shall be fastened to blocking with at least two No. 8 screws.

R804.3.5 Cutting and notching. Flanges and lips of load-bearing steel roof framing members shall not be cut or notched. Holes in webs shall be in accordance with Section R804.2.

R804.3.6 Hole patching. Web holes not conforming to the requirements in Section R804.2 shall be designed in accordance with one of the following:

1. Framing members shall be replaced or designed in accordance with accepted engineering practices when web holes exceed the following size limits:

   1.1. The depth of the hole, measured across the web, exceeds 70 percent of the flat width of the web; or,

   1.2. The length of the hole, measured along the web, exceeds 10 inches (254 mm) or the depth of the web, whichever is greater.

2. Web holes not exceeding the dimensional requirements in Section R804.3.6, Item 1 shall be patched with a solid steel plate, stud section, or track section in accordance with Figure R804.3.6. The steel patch shall be of a minimum thickness as the receiving member and shall extend at least 1 inch (25 mm) beyond all edges of the hole. The steel patch shall be fastened to the web of the receiving member with No. 8 screws spaced no greater than 1 inch (25 mm) center-to-center along the edges of the patch with minimum edge distance of $^1/_2$ inch (13 mm).

R804.3.7 Splicing. Rafters and other structural members, except ceiling joists, shall not be spliced. Splices in ceiling joists shall only be permitted at interior bearing points and shall be constructed in accordance with Figure R804.3.7(1). Spliced ceiling joists shall be connected with the same number and size of screws on connection. Splicing of tracks shall conform to Figure R804.3.7(2).

R804.3.8 Bearing stiffener. A bearing stiffener shall be fabricated from a minimum 33-mil (0.84 mm) C-section or track section. Each stiffener shall be fastened to the web of the ceiling joist with a minimum of four No. 8 screws equally spaced as shown in Figure R804.3.8. Stiffeners shall extend across the full depth of the web and shall be installed on either side of the web.

R804.3.9 Headers. Roof-ceiling framing above wall openings shall be supported on headers. The allowable spans for headers in bearing walls shall not exceed the values set forth in Table R603.6(1).

R804.3.10 Framing of opening. Openings in roof and ceiling framing shall be framed with headers and trimmers between ceiling joists or rafters. Header joist spans shall not exceed 4 feet (1219 mm). Header and trimmer joists shall be fabricated from joist and track sections, which shall be of a minimum size and thickness in accordance with Figures R804.3.10(1) and R804.3.10(2). Each header joist shall be

TABLE R804.3.3(1)
ALLOWABLE HORIZONTAL RAFTER SPANS[a, b, c] 33 ksi STEEL

MEMBER DESIGNATION	ALLOWABLE SPAN MEASURED HORIZONTALLY (feet-inches)							
	Ground Snow Load							
	20 psf		30 psf		50 psf		70 psf	
	Rafter spacing (in)							
	16	24	16	24	16	24	16	24
550S162-33	14'-0"	11'-5"	11'-10"	9'-8"	9'-5"	7'-8"	8'-1"	6'-7"
550S162-43	16'-6"	13'-10"	14'-4"	11'-9"	11'-5"	9'-4"	9'-10"	8'-0"
550S162-54	17'-9"	15'-6"	15'-6"	13'-2"	12'-11"	10'-6"	11'-1"	9'-0"
550S162-68	19'-0"	16'-7"	16'-8"	14'-7"	14'-1"	11'-10"	12'-5"	10'-2"
550S162-97	21'-2"	18'-6"	18'-7"	16'-2"	15'-8"	13'-8"	14'-0"	12'-2"
800S162-33	17'-0"	13'-11"	14'-5"	11'-9"	11'-6"	7'-9"	8'-6"	5'-8"
800S162-43	21'-1"	17'-3"	17'-10"	14'-7"	14'-3"	11'-7"	12'-2"	9'-11"
800S162-54	23'-11"	20'-4"	21'-0"	17'-3"	16'-10"	13'-9"	14'-5"	11'-9"
800S162-68	25'-9"	22'-6"	22'-7"	19'-5"	19'-0"	15'-6"	16'-3"	13'-3"
800S162-97	28'-9"	25'-1"	25'-2"	22'-0"	21'-3"	18'-7"	19'-0"	16'-0"
1000S162-43	23'-4"	19'-1"	19'-9"	16'-2"	15'-9"	12'-11"	13'-6"	10'-0"
1000S162-54	27'-8"	22'-7"	23'-5"	19'-1"	18'-8"	15'-3"	16'-0"	13'-1"
1000S162-68	30'-11"	27'-0"	27'-2"	22'-11"	22'-5"	18'-3"	19'-2"	15'-8"
1000S162-97	34'-7"	30'-2"	30'-4"	26'-6"	25'-7"	22'-1"	22'-10"	18'-11"
1200S162-43	25'-5"	20'-9"	21'-6"	17'-6"	17'-1"	11'-5"	12'-6"	8'-6"
1200S162-54	30'-0"	24'-6"	25'-5"	20'-9"	20'-3"	16'-7"	17'-5"	14'-2"
1200S162-68	35'-5"	28'-11"	30'-0"	24'-6"	23'-11"	19'-6"	20'-6"	16'-9"
1200S162-97	40'-4"	35'-3"	35'-5"	30'-11"	29'-10"	25'-5"	26'-8"	21'-9"

For SI: 1 inch = 25.4 mm, 1 foot = 304.8 mm, 1 pound per square foot = 0.0479 kPa.

a. Table provides maximum horizontal rafter spans in feet and inches for slopes between 3:12 and 12:12.
b. Deflection criterion: $L/240$ for live loads and $L/180$ for total loads.
c. Roof dead load = 12 psf.

TABLE R804.3.3(2)
BASIC WIND SPEED TO EQUIVALENT SNOW LOAD CONVERSION

BASIC WIND SPEED AND EXPOSURE		EQUIVALENT GROUND SNOW LOAD (psf)									
		Roof slope									
Exp. A/B	Exp. C	3:12	4:12	5:12	6:12	7:12	8:12	9:12	10:12	11:12	12:12
85 mph	—	20	20	20	20	20	20	30	30	30	30
100 mph	85 mph	20	20	20	20	30	30	30	30	50	50
110 mph	100 mph	20	20	20	20	30	50	50	50	50	50
—	110 mph	30	30	30	50	50	50	70	70	70	—

For SI: 1 mile per hour = 0.447 m/s, 1 pound per square foot = 0.0479 kPa.

connected to a trimmer joist with a minimum of four 2-inch by 2-inch (51 by 51 mm) clip angles. Each clip angle shall be fastened to both the header and trimmer joists with four No. 8 screws, evenly spaced, through each leg of the clip angle. The clip angles shall have a steel thickness not less than that of the floor joist.

R804.4 Roof tie-down. Roof assemblies subject to wind uplift pressures of 20 pounds per square foot (0.96 kN/m²) or greater, as established in Table R301.2(2), shall have rafter-to-bearing wall ties provided in accordance with Table R802.11.

TABLE R804.3.3.1
NUMBER OF SCREWS REQUIRED AT EACH LEG OF CLIP
ANGLE FOR RAFTER TO RIDGE MEMBER CONNECTION[a]

BUILDING WIDTH (feet)	NUMBER OF SCREWS			
	Ground snow load (psf)			
	0 to 20	21 to 30	31 to 50	51 to 70
24	2	2	3	4
28	2	3	4	5
32	2	3	4	5
36	3	3	5	6
40	3	4	5	7

For SI: 1 inch = 25.4 mm, 1 foot = 304.8 mm, 1 pound per square foot = 0.0479 kPa.
a. Screws shall be No. 10 minimum.

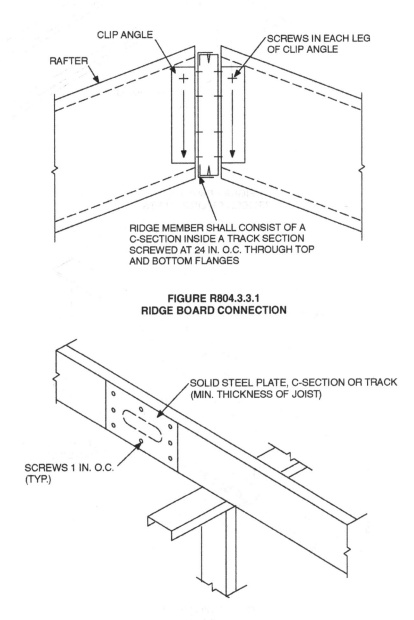

For SI: 1 inch = 25.4 mm.

FIGURE R804.3.3.1
RIDGE BOARD CONNECTION

For SI: 1 inch = 25.4 mm.

FIGURE R804.3.6
HOLE PATCHING

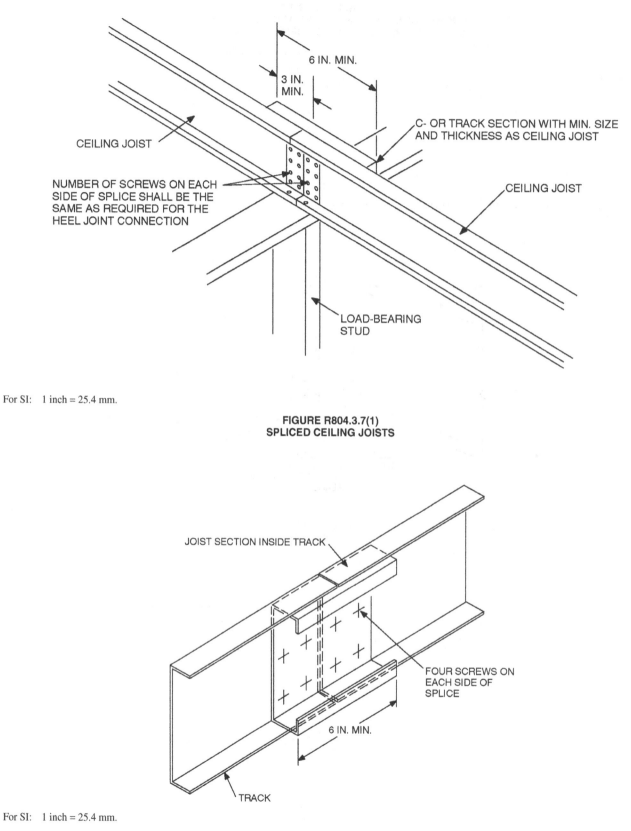

For SI: 1 inch = 25.4 mm.

**FIGURE R804.3.7(1)
SPLICED CEILING JOISTS**

For SI: 1 inch = 25.4 mm.

**FIGURE R804.3.7(2)
TRACK SPLICE**

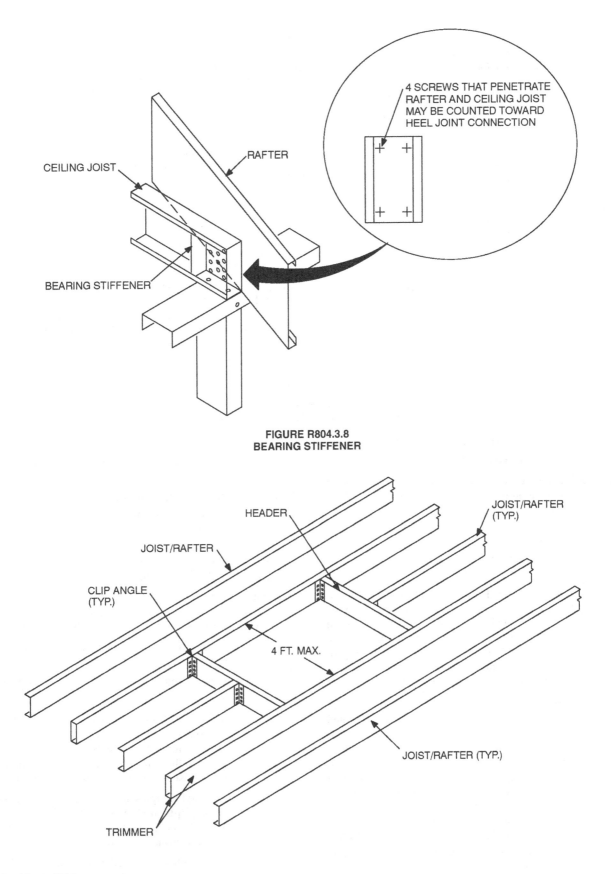

FIGURE R804.3.8
BEARING STIFFENER

4 SCREWS THAT PENETRATE RAFTER AND CEILING JOIST MAY BE COUNTED TOWARD HEEL JOINT CONNECTION

RAFTER

CEILING JOIST

BEARING STIFFENER

HEADER

JOIST/RAFTER (TYP.)

JOIST/RAFTER

CLIP ANGLE (TYP.)

4 FT. MAX.

JOIST/RAFTER (TYP.)

TRIMMER

For SI: 1 foot = 304.8 mm.

FIGURE R804.3.10(1)
ROOF OPENING

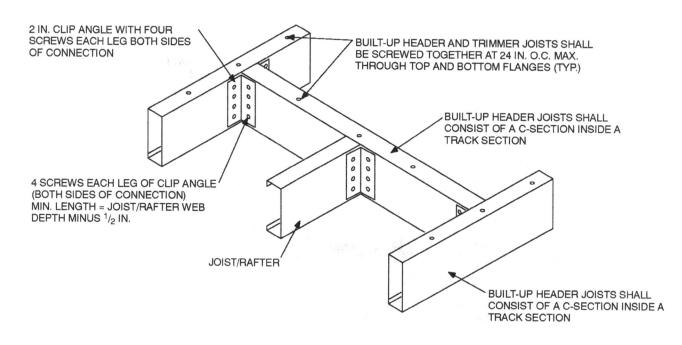

2 IN. CLIP ANGLE WITH FOUR SCREWS EACH LEG BOTH SIDES OF CONNECTION

BUILT-UP HEADER AND TRIMMER JOISTS SHALL BE SCREWED TOGETHER AT 24 IN. O.C. MAX. THROUGH TOP AND BOTTOM FLANGES (TYP.)

BUILT-UP HEADER JOISTS SHALL CONSIST OF A C-SECTION INSIDE A TRACK SECTION

4 SCREWS EACH LEG OF CLIP ANGLE (BOTH SIDES OF CONNECTION) MIN. LENGTH = JOIST/RAFTER WEB DEPTH MINUS $^1/_2$ IN.

JOIST/RAFTER

BUILT-UP HEADER JOISTS SHALL CONSIST OF A C-SECTION INSIDE A TRACK SECTION

For SI: 1 inch = 25.4 mm.

FIGURE R804.3.10(2)
HEADER TO TRIMMER CONNECTION

SECTION R805
CEILING FINISHES

R805.1 Ceiling installation. Ceilings shall be installed in accordance with the requirements for interior wall finishes as provided in Section R702.

SECTION R806
ROOF VENTILATION

R806.1 Ventilation required. Enclosed attics and enclosed rafter spaces formed where ceilings are applied directly to the underside of roof rafters shall have cross ventilation for each separate space by ventilating openings protected against the entrance of rain or snow. Ventilating openings shall be provided with corrosion-resistant wire mesh, with $^1/_8$ inch (3.2 mm) minimum to $^1/_4$ inch (6 mm) maximum openings.

R806.2 Minimum area. The total net free ventilating area shall not be less than $^1/_{150}$ of the area of the space ventilated except that reduction of the total area to $^1/_{300}$ is permitted, provided that at least 50 percent and not more than 80 percent of the required ventilating area is provided by ventilators located in the upper portion of the space to be ventilated at least 3 feet (914 mm) above the eave or cornice vents with the balance of the required ventilation provided by eave or cornice vents. As an alternative, the net free cross-ventilation area may be reduced to $^1/_{300}$ when a vapor barrier having a transmission rate not exceeding 1 perm (5.7×10^{-11} kg/s · m² · Pa) is installed on the warm-in-winter side of the ceiling.

R806.3 Vent and insulation clearance. Where eave or cornice vents are installed, insulation shall not block the free flow of air. A minimum of a 1-inch (25 mm) space shall be provided between the insulation and the roof sheathing and at the location of the vent.

R806.4 Conditioned attic assemblies. Unvented conditioned attic assemblies (spaces between the ceiling joists of the top story and the roof rafters) are permitted under the following conditions:

1. No interior vapor retarders are installed on the ceiling side (attic floor) of the unvented attic assembly.

2. An air-impermeable insulation is applied in direct contact to the underside/interior of the structural roof deck. "Air-impermeable" shall be defined by ASTM E 283.

    Exception: In Zones 2B and 3B, insulation is not required to be air impermeable.

3. In the warm humid locations as defined in Section N1101.2.1:

    3.1. For asphalt roofing shingles: A 1-perm (5.7×10^{-11} kg/s · m² · Pa) or less vapor retarder (determined using Procedure B of ASTM E 96) is placed to the exterior of the structural roof deck; that is, just above the roof structural sheathing.

    3.2. For wood shingles and shakes: a minimum continuous $^1/_4$-inch (6 mm) vented air space separates the shingles/shakes and the roofing felt placed over the structural sheathing.

4. In Zones 3 through 8 as defined in Section N1101.2, sufficient insulation is installed to maintain the monthly average temperature of the condensing surface above 45°F (7°C). The condensing surface is defined as either

the structural roof deck or the interior surface of an air-impermeable insulation applied in direct contact with the underside/interior of the structural roof deck. "Air-impermeable" is quantitatively defined by ASTM E 283. For calculation purposes, an interior temperature of 68°F (20°C) is assumed. The exterior temperature is assumed to be the monthly average outside temperature.

SECTION R807
ATTIC ACCESS

R807.1 Attic access. Buildings with combustible ceiling or roof construction shall have an attic access opening to attic areas that exceed 30 square feet (2.8 m²) and have a vertical height of 30 inches (762 mm) or more.

The rough-framed opening shall not be less than 22 inches by 30 inches (559 mm by 762 mm) and shall be located in a hallway or other readily accessible location. A 30-inch (762 mm) minimum unobstructed headroom in the attic space shall be provided at some point above the access opening. See Section M1305.1.3 for access requirements where mechanical equipment is located in attics.

SECTION R808
INSULATION CLEARANCE

R808.1 Combustible insulation. Combustible insulation shall be separated a minimum of 3 inches (76 mm) from recessed luminaires, fan motors and other heat-producing devices.

Exception: Where heat-producing devices are listed for lesser clearances, combustible insulation complying with the listing requirements shall be separated in accordance with the conditions stipulated in the listing.

Recessed luminaires installed in the building thermal envelope shall meet the requirements of Section N1102.4.3.

CHAPTER 9
ROOF ASSEMBLIES

SECTION R901
GENERAL

R901.1 Scope. The provisions of this chapter shall govern the design, materials, construction and quality of roof assemblies.

SECTION R902
ROOF CLASSIFICATION

R902.1 Roofing covering materials. Roofs shall be covered with materials as set forth in Sections R904 and R905. Class A, B or C roofing shall be installed in areas designated by law as requiring their use or when the edge of the roof is less than 3 feet (914 mm) from a property line. Classes A, B and C roofing required to be listed by this section shall be tested in accordance with UL 790 or ASTM E 108. Roof assemblies with coverings of brick, masonry, slate, clay or concrete roof tile, exposed concrete roof deck, ferrous or copper shingles or sheets, and metal sheets and shingles, shall be considered Class A roof coverings.

R902.2 Fire-retardant-treated shingles and shakes. Fire-retardant-treated wood shakes and shingles shall be treated by impregnation with chemicals by the full-cell vacuum-pressure process, in accordance with AWPA C1. Each bundle shall be marked to identify the manufactured unit and the manufacturer, and shall also be labeled to identify the classification of the material in accordance with the testing required in Section R902.1, the treating company and the quality control agency.

SECTION R903
WEATHER PROTECTION

R903.1 General. Roof decks shall be covered with approved roof coverings secured to the building or structure in accordance with the provisions of this chapter. Roof assemblies shall be designed and installed in accordance with this code and the approved manufacturer's installation instructions such that the roof assembly shall serve to protect the building or structure.

R903.2 Flashing. Flashings shall be installed in a manner that prevents moisture from entering the wall and roof through joints in copings, through moisture permeable materials and at intersections with parapet walls and other penetrations through the roof plane.

R903.2.1 Locations. Flashings shall be installed at wall and roof intersections, wherever there is a change in roof slope or direction and around roof openings. Where flashing is of metal, the metal shall be corrosion resistant with a thickness of not less than 0.019 inch (0.5 mm) (No. 26 galvanized sheet).

R903.3 Coping. Parapet walls shall be properly coped with noncombustible, weatherproof materials of a width no less than the thickness of the parapet wall.

R903.4 Roof drainage. Unless roofs are sloped to drain over roof edges, roof drains shall be installed at each low point of the roof. Where required for roof drainage, scuppers shall be placed level with the roof surface in a wall or parapet. The scupper shall be located as determined by the roof slope and contributing roof area.

R903.4.1 Overflow drains and scuppers. Where roof drains are required, overflow drains having the same size as the roof drains shall be installed with the inlet flow line located 2 inches (51 mm) above the low point of the roof, or overflow scuppers having three times the size of the roof drains and having a minimum opening height of 4 inches (102 mm) shall be installed in the adjacent parapet walls with the inlet flow located 2 inches (51 mm) above the low point of the roof served. The installation and sizing of overflow drains, leaders and conductors shall comply with the *International Plumbing Code.*

Overflow drains shall discharge to an approved location and shall not be connected to roof drain lines.

R903.5 Hail exposure. Hail exposure, as specified in Sections R903.5.1 and R903.5.2, shall be determined using Figure R903.5.

R903.5.1 Moderate hail exposure. One or more hail days with hail diameters larger than 1.5 inches (38 mm) in a 20-year period.

R903.5.2 Severe hail exposure. One or more hail days with hail diameters larger than or equal to 2.0 inches (51 mm) in a 20-year period.

SECTION R904
MATERIALS

R904.1 Scope. The requirements set forth in this section shall apply to the application of roof covering materials specified herein. Roof assemblies shall be applied in accordance with this chapter and the manufacturer's installation instructions. Installation of roof assemblies shall comply with the applicable provisions of Section R905.

R904.2 Compatibility of materials. Roof assemblies shall be of materials that are compatible with each other and with the building or structure to which the materials are applied.

R904.3 Material specifications and physical characteristics. Roof covering materials shall conform to the applicable standards listed in this chapter. In the absence of applicable standards or where materials are of questionable suitability, testing by an approved testing agency shall be required by the building official to determine the character, quality and limitations of application of the materials.

R904.4 Product identification. Roof covering materials shall be delivered in packages bearing the manufacturer's identifying marks and approved testing agency labels when required.

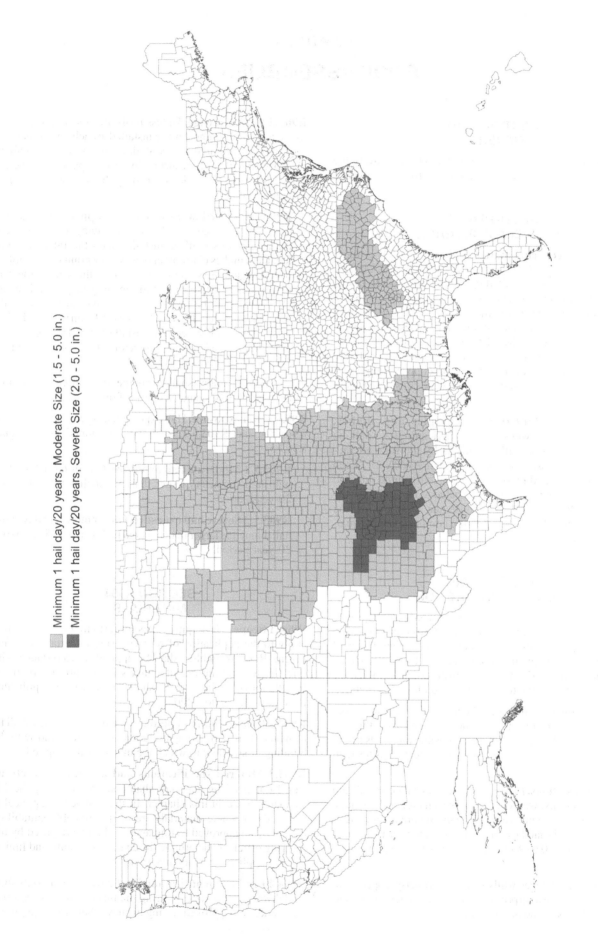

Minimum 1 hail day/20 years, Moderate Size (1.5 - 5.0 in.)

Minimum 1 hail day/20 years, Severe Size (2.0 - 5.0 in.)

FIGURE R903.5
HAIL EXPOSURE MAP

Bulk shipments of materials shall be accompanied by the same information issued in the form of a certificate or on a bill of lading by the manufacturer.

SECTION R905
REQUIREMENTS FOR ROOF COVERINGS

R905.1 Roof covering application. Roof coverings shall be applied in accordance with the applicable provisions of this section and the manufacturer's installation instructions. Unless otherwise specified in this section, roof coverings shall be installed to resist the component and cladding loads specified in Table R301.2(2), adjusted for height and exposure in accordance with Table R301.2(3).

R905.2 Asphalt shingles. The installation of asphalt shingles shall comply with the provisions of this section.

R905.2.1 Sheathing requirements. Asphalt shingles shall be fastened to solidly sheathed decks.

R905.2.2 Slope. Asphalt shingles shall be used only on roof slopes of two units vertical in 12 units horizontal (2:12) or greater. For roof slopes from two units vertical in 12 units horizontal (2:12) up to four units vertical in 12 units horizontal (4:12), double underlayment application is required in accordance with Section R905.2.7.

R905.2.3 Underlayment. Unless otherwise noted, required underlayment shall conform to ASTM D 226 Type I, ASTM D 4869 Type I, or ASTM D 6757.

Self-adhering polymer modified bitumen sheet shall comply with ASTM D 1970.

R905.2.4 Asphalt shingles. Asphalt shingles shall have self-seal strips or be interlocking, and comply with ASTM D 225 or D 3462.

R905.2.4.1 Wind resistance of asphalt shingles. Asphalt shingles shall be installed in accordance with Section R905.2.6. Shingles classified using ASTM D 3161 are acceptable for use in wind zones less than 110 mph (49 m/s). Shingles classified using ASTM D 3161, Class F, are acceptable for use in all cases where special fastening is required.

R905.2.5 Fasteners. Fasteners for asphalt shingles shall be galvanized steel, stainless steel, aluminum or copper roofing nails, minimum 12 gage [0.105 inch (3 mm)] shank with a minimum $^3/_8$-inch (10 mm) diameter head, ASTM F 1667, of a length to penetrate through the roofing materials and a minimum of $^3/_4$ inch (19 mm) into the roof sheathing. Where the roof sheathing is less than $^3/_4$ inch (19 mm) thick, the fasteners shall penetrate through the sheathing. Fasteners shall comply with ASTM F 1667.

R905.2.6 Attachment. Asphalt shingles shall have the minimum number of fasteners required by the manufacturer. For normal application, asphalt shingles shall be secured to the roof with not less than four fasteners per strip shingle or two fasteners per individual shingle. Where the roof slope exceeds 20 units vertical in 12 units horizontal (167 percent slope), special methods of fastening are required. For roofs located where the basic wind speed per Figure R301.2(4) is 110 mph (49 m/s) or higher, special methods of fastening are required. Special fastening methods shall be tested in accordance with ASTM D 3161, Class F. Asphalt shingle wrappers shall bear a label indicating compliance with ASTM D 3161, Class F.

R905.2.7 Underlayment application. For roof slopes from two units vertical in 12 units horizontal (17-percent slope), up to four units vertical in 12 units horizontal (33-percent slope), underlayment shall be two layers applied in the following manner. Apply a 19-inch (483 mm) strip of underlayment felt parallel to and starting at the eaves, fastened sufficiently to hold in place. Starting at the eave, apply 36-inch-wide (914 mm) sheets of underlayment, overlapping successive sheets 19 inches (483 mm), and fastened sufficiently to hold in place. Distortions in the underlayment shall not interfere with the ability of the shingles to seal. For roof slopes of four units vertical in 12 units horizontal (33-percent slope) or greater, underlayment shall be one layer applied in the following manner. Underlayment shall be applied shingle fashion, parallel to and starting from the eave and lapped 2 inches (51 mm), fastened sufficiently to hold in place. Distortions in the underlayment shall not interfere with the ability of the shingles to seal. End laps shall be offset by 6 feet (1829 mm).

R905.2.7.1 Ice barrier. In areas where there has been a history of ice forming along the eaves causing a backup of water as designated in Table R301.2(1), an ice barrier that consists of a least two layers of underlayment cemented together or of a self-adhering polymer modified bitumen sheet, shall be used in lieu of normal underlayment and extend from the lowest edges of all roof surfaces to a point at least 24 inches (610 mm) inside the exterior wall line of the building.

Exception: Detached accessory structures that contain no conditioned floor area.

R905.2.7.2 Underlayment and high wind. Underlayment applied in areas subject to high winds [above 110 mph (49 m/s) per Figure R301.2(4)] shall be applied with corrosion-resistant fasteners in accordance with manufacturer's installation instructions. Fasteners are to be applied along the overlap not farther apart than 36 inches (914 mm) on center.

R905.2.8 Flashing. Flashing for asphalt shingles shall comply with this section.

R905.2.8.1 Base and cap flashing. Base and cap flashing shall be installed in accordance with manufacturer's installation instructions. Base flashing shall be of either corrosion-resistant metal of minimum nominal 0.019-inch (0.5 mm) thickness or mineral surface roll roofing weighing a minimum of 77 pounds per 100 square feet (4 kg/m^2). Cap flashing shall be corrosion-resistant metal of minimum nominal 0.019-inch (0.5 mm) thickness.

R905.2.8.2 Valleys. Valley linings shall be installed in accordance with the manufacturer's installation instructions before applying shingles. Valley linings of the following types shall be permitted:

1. For open valley (valley lining exposed) lined with metal, the valley lining shall be at least 24 inches

(610 mm) wide and of any of the corrosion-resistant metals in Table R905.2.8.2.

2. For open valleys, valley lining of two plies of mineral surfaced roll roofing, complying with ASTM D 3909 or ASTM D 6380 Class M, shall be permitted. The bottom layer shall be 18 inches (457mm) and the top layer a minimum of 36 inches (914 mm) wide.

3. For closed valleys (valley covered with shingles), valley lining of one ply of smooth roll roofing complying with ASTM D 6380 Class S Type III, Class M Type II, or ASTM D 3909 and at least 36 inches wide (914 mm) or valley lining as described in Items 1 and 2 above shall be permitted. Specialty underlayment complying with ASTM D 1970 may be used in lieu of the lining material.

TABLE R905.2.8.2
VALLEY LINING MATERIAL

MATERIAL	MINIMUM THICKNESS (inches)	GAGE	WEIGHT (pounds)
Cold–rolled copper	0.0216 nominal	—	ASTM B 370, 16 oz. per square foot
Lead–coated copper	0.0216 nominal	—	ASTM B 101, 16 oz. per square foot
High–yield copper	0.0162 nominal	—	ASTM B 370, 12 oz. per square foot
Lead–coated high–yield copper	0.0162 nominal	—	ASTM B 101, 12 oz. per square foot
Aluminum	0.024	—	—
Stainless steel	—	28	—
Galvanized steel	0.0179	26 (zinc coated G90)	—
Zinc alloy	0.027	—	—
Lead	—	—	$2^{1}/_{2}$
Painted terne	—	—	20

For SI: 1 inch = 25.4 mm, 1 pound = 0.454 kg.

R905.2.8.3 Crickets and saddles. A cricket or saddle shall be installed on the ridge side of any chimney or penetration more than 30 inches (762 mm) wide as measured perpendicular to the slope. Cricket or saddle coverings shall be sheet metal or of the same material as the roof covering.

R905.2.8.4 Sidewall flashing. Flashing against a vertical sidewall shall be by the step-flashing method.

R905.2.8.5 Other flashing. Flashing against a vertical front wall, as well as soil stack, vent pipe and chimney flashing, shall be applied according to the asphalt shingle manufacturer's printed instructions.

R905.3 Clay and concrete tile. The installation of clay and concrete shall comply with the provisions of this section. Clay roof tile shall comply with ASTM C 1167.

R905.3.1 Deck requirements. Concrete and clay tile shall be installed only over solid sheathing or spaced structural sheathing boards.

R905.3.2 Deck slope. Clay and concrete roof tile shall be installed on roof slopes of two and one-half units vertical in 12 units horizontal ($2^{1}/_{2}$:12) or greater. For roof slopes from two and one-half units vertical in 12 units horizontal ($2^{1}/_{2}$:12) to four units vertical in 12 units horizontal (4:12), double underlayment application is required in accordance with Section R905.3.3.

R905.3.3 Underlayment. Unless otherwise noted, required underlayment shall conform to ASTM D 226 Type II; ASTM D 2626 Type I; or ASTM D 6380 Class M mineral surfaced roll roofing.

R905.3.3.1 Low slope roofs. For roof slopes from two and one-half units vertical in 12 units horizontal ($2^{1}/_{2}$:12), up to four units vertical in 12 units horizontal (4:12), underlayment shall be a minimum of two layers underlayment applied as follows:

1. Starting at the eave, a 19-inch (483 mm) strip of underlayment shall be applied parallel with the eave and fastened sufficiently in place.

2. Starting at the eave, 36-inch-wide (914 mm) strips of underlayment felt shall be applied, overlapping successive sheets 19 inches (483 mm), and fastened sufficiently in place.

R905.3.3.2 High slope roofs. For roof slopes of four units vertical in 12 units horizontal (4:12) or greater, underlayment shall be a minimum of one layer of underlayment felt applied shingle fashion, parallel to and starting from the eaves and lapped 2 inches (51 mm), fastened sufficiently in place.

R905.3.3.3 Underlayment and high wind. Underlayment applied in areas subject to high wind [over 110 miles per hour (49 m/s) per Figure R301.2(4)] shall be applied with corrosion-resistant fasteners in accordance with manufacturer's installation instructions. Fasteners are to be applied along the overlap not farther apart than 36 inches (914 mm) on center.

R905.3.4 Tile. Clay roof tile shall comply with ASTM C 1167.

R905.3.5 Concrete tile. Concrete roof tile shall comply with ASTM C 1492.

R905.3.6 Fasteners. Nails shall be corrosion resistant and not less than 11 gage, $^{5}/_{16}$-inch (11 mm) head, and of sufficient length to penetrate the deck a minimum of $^{3}/_{4}$ inch (19 mm) or through the thickness of the deck, whichever is less. Attaching wire for clay or concrete tile shall not be smaller than 0.083 inch (2 mm). Perimeter fastening areas include three tile courses but not less than 36 inches (914 mm) from either side of hips or ridges and edges of eaves and gable rakes.

R905.3.7 Application. Tile shall be applied in accordance with this chapter and the manufacturer's installation instructions, based on the following:

1. Climatic conditions.

2. Roof slope.

3. Underlayment system.

4. Type of tile being installed.

Clay and concrete roof tiles shall be fastened in accordance with this section and the manufacturer's installation instructions. Perimeter tiles shall be fastened with a minimum of one fastener per tile. Tiles with installed weight less than 9 pounds per square foot (0.4 kg/m²) require a minimum of one fastener per tile regardless of roof slope. Clay and concrete roof tile attachment shall be in accordance with the manufacturer's installation instructions where applied in areas where the wind speed exceeds 100 miles per hour (45 m/s) and on buildings where the roof is located more than 40 feet (12 192 mm) above grade. In areas subject to snow, a minimum of two fasteners per tile is required. In all other areas, clay and concrete roof tiles shall be attached in accordance with Table R905.3.7.

TABLE R905.3.7
CLAY AND CONCRETE TILE ATTACHMENT

SHEATHING	ROOF SLOPE	NUMBER OF FASTENERS
Solid without battens	All	One per tile
Spaced or solid with battens and slope < 5:12	Fasteners not required	—
Spaced sheathing without battens	5:12 ≤ slope < 12:12	One per tile/every other row
	12:12 ≤ slope < 24:12	One per tile

R905.3.8 Flashing. At the juncture of roof vertical surfaces, flashing and counterflashing shall be provided in accordance with this chapter and the manufacturer's installation instructions and, where of metal, shall not be less than 0.019 inch (0.5 mm) (No. 26 galvanized sheet gage) corrosion-resistant metal. The valley flashing shall extend at least 11 inches (279 mm) from the centerline each way and have a splash diverter rib not less than 1 inch (25 mm) high at the flow line formed as part of the flashing. Sections of flashing shall have an end lap of not less than 4 inches (102 mm). For roof slopes of three units vertical in 12 units horizontal (25-percent slope) and greater, valley flashing shall have a 36-inch-wide (914 mm) underlayment of one layer of Type I underlayment running the full length of the valley, in addition to other required underlayment. In areas where the average daily temperature in January is 25°F (-4°C) or less, metal valley flashing underlayment shall be solid-cemented to the roofing underlayment for slopes less than seven units vertical in 12 units horizontal (58-percent slope) or be of self-adhering polymer modified bitumen sheet.

R905.4 Metal roof shingles. The installation of metal roof shingles shall comply with the provisions of this section.

R905.4.1 Deck requirements. Metal roof shingles shall be applied to a solid or closely fitted deck, except where the roof covering is specifically designed to be applied to spaced sheathing.

R905.4.2 Deck slope. Metal roof shingles shall not be installed on roof slopes below three units vertical in 12 units horizontal (25-percent slope).

R905.4.3 Underlayment. Underlayment shall comply with ASTM D 226, Type I or ASTM D 4869, Type I or II.

R905.4.3.1 Ice barrier. In areas where there has been a history of ice forming along the eaves causing a backup of water as designated in Table R301.2(1), an ice barrier that consists of at least two layers of underlayment cemented together or a self-adhering polymer modified bitumen sheet shall be used in place of normal underlayment and extend from the lowest edges of all roof surfaces to a point at least 24 inches (610 mm) inside the exterior wall line of the building.

Exception: Detached accessory structures that contain no conditioned floor area.

R905.4.4 Material standards. Metal roof shingle roof coverings shall comply with Table R905.10.3(1). The materials used for metal roof shingle roof coverings shall be naturally corrosion resistant or be made corrosion resistant in accordance with the standards and minimum thicknesses listed in Table R905.10.3(2).

R905.4.5 Application. Metal roof shingles shall be secured to the roof in accordance with this chapter and the approved manufacturer's installation instructions.

R905.4.6 Flashing. Roof valley flashing shall be of corrosion-resistant metal of the same material as the roof covering or shall comply with the standards in Table R905.10.3(1). The valley flashing shall extend at least 8 inches (203 mm) from the center line each way and shall have a splash diverter rib not less than ³/₄ inch (19 mm) high at the flow line formed as part of the flashing. Sections of flashing shall have an end lap of not less than 4 inches (102 mm). The metal valley flashing shall have a 36-inch-wide (914 mm) underlayment directly under it consisting of one layer of underlayment running the full length of the valley, in addition to underlayment required for metal roof shingles. In areas where the average daily temperature in January is 25°F (-4°C) or less , the metal valley flashing underlayment shall be solid cemented to the roofing underlayment for roof slopes under seven units vertical in 12 units horizontal (58-percent slope) or self-adhering polymer modified bitumen sheet.

R905.5 Mineral-surfaced roll roofing. The installation of mineral-surfaced roll roofing shall comply with this section.

R905.5.1 Deck requirements. Mineral-surfaced roll roofing shall be fastened to solidly sheathed roofs.

R905.5.2 Deck slope. Mineral-surfaced roll roofing shall not be applied on roof slopes below one unit vertical in 12 units horizontal (8-percent slope).

R905.5.3 Underlayment. Underlayment shall comply with ASTM D 226, Type I or ASTM D 4869, Type I or II.

R905.5.3.1 Ice barrier. In areas where there has been a history of ice forming along the eaves causing a backup of water as designated in Table R301.2(1), an ice barrier that consists of at least two layers of underlayment cemented together or a self-adhering polymer modified bitumen sheet shall be used in place of normal underlayment and extend from the lowest edges of all roof surfaces to a point at least 24 inches (610 mm) inside the exterior wall line of the building.

> **Exception:** Detached accessory structures that contain no conditioned floor area.

R905.5.4 Material standards. Mineral-surfaced roll roofing shall conform to ASTM D 3909 or ASTM D 6380, Class M.

R905.5.5 Application. Mineral-surfaced roll roofing shall be installed in accordance with this chapter and the manufacturer's installation instructions.

R905.6 Slate and slate-type shingles. The installation of slate and slate-type shingles shall comply with the provisions of this section.

R905.6.1 Deck requirements. Slate shingles shall be fastened to solidly sheathed roofs.

R905.6.2 Deck slope. Slate shingles shall be used only on slopes of four units vertical in 12 units horizontal (33-percent slope) or greater.

R905.6.3 Underlayment. Underlayment shall comply with ASTM D 226, Type I or ASTM D 4869, Type I or II.

R905.6.3.1 Ice barrier. In areas where there has been a history of ice forming along the eaves causing a backup of water as designated in Table R301.2(1), an ice barrier that consists of at least two layers of underlayment cemented together or a self-adhering polymer modified bitumen sheet shall be used in lieu of normal underlayment and extend from the lowest edges of all roof surfaces to a point at least 24 inches (610 mm) inside the exterior wall line of the building.

> **Exception:** Detached accessory structures that contain no conditioned floor area.

R905.6.4 Material standards. Slate shingles shall comply with ASTM C 406.

R905.6.5 Application. Minimum headlap for slate shingles shall be in accordance with Table R905.6.5. Slate shingles shall be secured to the roof with two fasteners per slate. Slate shingles shall be installed in accordance with this chapter and the manufacturer's installation instructions.

TABLE R905.6.5
SLATE SHINGLE HEADLAP

SLOPE	HEADLAP (inches)
4:12 ≤ slope < 8:12	4
8:12 ≤ slope < 20:12	3
Slope ≤ 20:12	2

For SI: 1 inch = 25.4 mm.

R905.6.6 Flashing. Flashing and counterflashing shall be made with sheet metal. Valley flashing shall be a minimum of 15 inches (381 mm) wide. Valley and flashing metal shall be a minimum uncoated thickness of 0.0179-inch (0.5 mm) zinc coated G90. Chimneys, stucco or brick walls shall have a minimum of two plies of felt for a cap flashing consisting of a 4-inch-wide (102 mm) strip of felt set in plastic cement and extending 1 inch (25 mm) above the first felt and a top coating of plastic cement. The felt shall extend over the base flashing 2 inches (51 mm).

R905.7 Wood shingles. The installation of wood shingles shall comply with the provisions of this section.

R905.7.1 Deck requirements. Wood shingles shall be installed on solid or spaced sheathing. Where spaced sheathing is used, sheathing boards shall not be less than 1-inch by 4-inch (25.4 mm by 102 mm) nominal dimensions and shall be spaced on centers equal to the weather exposure to coincide with the placement of fasteners.

R905.7.1.1 Solid sheathing required. In areas where the average daily temperature in January is 25°F (-4°C) or less, solid sheathing is required on that portion of the roof requiring the application of an ice barrier.

R905.7.2 Deck slope. Wood shingles shall be installed on slopes of three units vertical in 12 units horizontal (25-percent slope) or greater.

R905.7.3 Underlayment. Underlayment shall comply with ASTM D 226, Type I or ASTM D 4869, Type I or II.

R905.7.3.1 Ice barrier. In areas where there has been a history of ice forming along the eaves causing a backup of water as designated in Table R301.2(1), an ice barrier that consists of at least two layers of underlayment cemented together or a self-adhering polymer modified bitumen sheet shall be used in lieu of normal underlayment and extend from the lowest edges of all roof surfaces to a point at least 24 inches (610 mm) inside the exterior wall line of the building.

> **Exception:** Detached accessory structures that contain no conditioned floor area.

R905.7.4 Material standards. Wood shingles shall be of naturally durable wood and comply with the requirements of Table R905.7.4.

TABLE R905.7.4
WOOD SHINGLE MATERIAL REQUIREMENTS

MATERIAL	MINIMUM GRADES	APPLICABLE GRADING RULES
Wood shingles of naturally durable wood	1, 2 or 3	Cedar Shake and Shingle Bureau

R905.7.5 Application. Wood shingles shall be installed according to this chapter and the manufacturer's installation instructions. Wood shingles shall be laid with a side lap not less than 1$^1/_2$ inches (38 mm) between joints in courses, and no two joints in any three adjacent courses shall be in direct alignment. Spacing between shingles shall not be less than $^1/_4$ inch to $^3/_8$ inch (6 mm to 10 mm). Weather exposure for wood shingles shall not exceed those set in Table R905.7.5. Fasteners for wood shingles shall be corrosion resistant with a minimum penetration of $^1/_2$ inch (13 mm) into the sheathing. For sheathing less than $^1/_2$ inch (13 mm) in thickness, the fasteners shall extend through the sheathing. Wood shingles shall be attached to the roof with two fasteners per shingle, positioned no more than $^3/_4$ inch (19 mm) from each edge and no more than 1 inch (25 mm) above the exposure line.

TABLE R905.7.5
WOOD SHINGLE WEATHER EXPOSURE AND ROOF SLOPE

ROOFING MATERIAL	LENGTH (inches)	GRADE	EXPOSURE (inches)	
			3:12 pitch to < 4:12	4:12 pitch or steeper
Shingles of naturally durable wood	16	No. 1	$3^3/_4$	5
		No. 2	$3^1/_2$	4
		No. 3	3	$3^1/_2$
	18	No. 1	$4^1/_4$	$5^1/_2$
		No. 2	4	$4^1/_2$
		No. 3	$3^1/_2$	4
	24	No. 1	$5^3/_4$	$7^1/_2$
		No. 2	$5^1/_2$	$6^1/_2$
		No. 3	5	$5^1/_2$

For SI: 1 inch = 25.4 mm.

R905.7.6 Valley flashing. Roof flashing shall be not less than No. 26 gage [0.019 inches (0.5 mm)] corrosion-resistant sheet metal and shall extend 10 inches (254 mm) from the centerline each way for roofs having slopes less than 12 units vertical in 12 units horizontal (100-percent slope), and 7 inches (178 mm) from the centerline each way for slopes of 12 units vertical in 12 units horizontal and greater. Sections of flashing shall have an end lap of not less than 4 inches (102 mm).

R905.7.7 Label required. Each bundle of shingles shall be identified by a label of an approved grading or inspection bureau or agency.

R905.8 Wood shakes. The installation of wood shakes shall comply with the provisions of this section.

R905.8.1 Deck requirements. Wood shakes shall be used only on solid or spaced sheathing. Where spaced sheathing is used, sheathing boards shall not be less than 1-inch by 4-inch (25 mm by 102 mm) nominal dimensions and shall be spaced on centers equal to the weather exposure to coincide with the placement of fasteners. Where 1-inch by 4-inch (25 mm by 102 mm) spaced sheathing is installed at 10 inches (254 mm) on center, additional 1-inch by 4-inch (25 mm by 102 mm) boards shall be installed between the sheathing boards.

R905.8.1.1 Solid sheathing required. In areas where the average daily temperature in January is 25°F (-4°C) or less, solid sheathing is required on that portion of the roof requiring an ice barrier.

R905.8.2 Deck slope. Wood shakes shall only be used on slopes of three units vertical in 12 units horizontal (25-percent slope) or greater.

R905.8.3 Underlayment. Underlayment shall comply with ASTM D 226, Type I or ASTM D 4869, Type I or II.

R905.8.3.1 Ice barrier. In areas where there has been a history of ice forming along the eaves causing a backup of water as designated in Table R301.2(1), an ice barrier that consists of at least two layers of underlayment cemented together or a self-adhering polymer modified bitumen sheet shall be used in place of normal underlayment and extend from the lowest edges of all roof surfaces to a point at least 24 inches (610 mm) inside the exterior wall line of the building.

Exception: Detached accessory structures that contain no conditioned floor area.

R905.8.4 Interlayment. Interlayment shall comply with ASTM D 226, Type I.

R905.8.5 Material standards. Wood shakes shall comply with the requirements of Table R905.8.5.

TABLE R905.8.5
WOOD SHAKE MATERIAL REQUIREMENTS

MATERIAL	MINIMUM GRADES	APPLICABLE GRADING RULES
Wood shakes of naturally durable wood	1	Cedar Shake and Shingle Bureau
Taper sawn shakes of naturally durable wood	1 or 2	Cedar Shake and Shingle Bureau
Preservative-treated shakes and shingles of naturally durable wood	1	Cedar Shake and Shingle Bureau
Fire-retardant-treated shakes and shingles of naturally durable wood	1	Cedar Shake and Shingle Bureau
Preservative-treated taper sawn shakes of Southern pine treated in accordance with AWPA Standard U1 (Commodity Specification A, Use Category 3B and Section 5.6)	1 or 2	Forest Products Laboratory of the Texas Forest Services

R905.8.6 Application. Wood shakes shall be installed according to this chapter and the manufacturer's installation instructions. Wood shakes shall be laid with a side lap not less than $1^1/_2$ inches (38 mm) between joints in adjacent courses. Spacing between shakes in the same course shall be $^1/_8$ inch to $^5/_8$ inch (3 mm to 16 mm) for shakes and tapersawn shakes of naturally durable wood and shall be $^1/_4$ inch to $^3/_8$ inch (6 mm to 10 mm) for preservative treated taper sawn shakes. Weather exposure for wood shakes shall not exceed those set forth in Table R905.8.6. Fasteners for wood shakes shall be corrosion-resistant, with a minimum penetration of $^1/_2$ inch (12.7 mm) into the sheathing. For sheathing less than $^1/_2$ inch (12.7 mm) in thickness, the fasteners shall extend through the sheathing. Wood shakes shall be attached to the roof with two fasteners per shake, positioned no more than 1 inch (25 mm) from each edge and no more than 2 inches (51 mm) above the exposure line.

R905.8.7 Shake placement. The starter course at the eaves shall be doubled and the bottom layer shall be either 15-inch (381 mm), 18-inch (457 mm) or 24-inch (610 mm) wood shakes or wood shingles. Fifteen-inch (381 mm) or 18-inch (457 mm) wood shakes may be used for the final course at the ridge. Shakes shall be interlaid with 18-inch-wide (457 mm) strips of not less than No. 30 felt shingled between each course in such a manner that no felt is exposed to the weather by positioning the lower edge of each felt strip above the butt end of the shake it covers a distance equal to twice the weather exposure.

TABLE R905.8.6
WOOD SHAKE WEATHER EXPOSURE AND ROOF SLOPE

ROOFING MATERIAL	LENGTH (inches)	GRADE	EXPOSURE (inches) 4:12 pitch or steeper
Shakes of naturally durable wood	18	No. 1	$7^1/_2$
	24	No. 1	10^a
Preservative-treated taper sawn shakes of Southern Yellow Pine	18	No. 1	$7^1/_2$
	24	No. 1	10
	18	No. 2	$5^1/_2$
	24	No. 2	$7^1/_2$
Taper-sawn shakes of naturally durable wood	18	No. 1	$7^1/_2$
	24	No. 1	10
	18	No. 2	$5^1/_2$
	24	No. 2	$7^1/_2$

For SI: 1 inch = 25.4 mm.
a. For 24-inch by $^3/_8$-inch handsplit shakes, the maximum exposure is $7^1/_2$ inches.

R905.8.8 Valley flashing. Roof valley flashing shall not be less than No. 26 gage [0.019 inch (0.5 mm)] corrosion-resistant sheet metal and shall extend at least 11 inches (279 mm) from the centerline each way. Sections of flashing shall have an end lap of not less than 4 inches (102 mm).

R905.8.9 Label required. Each bundle of shakes shall be identified by a label of an approved grading or inspection bureau or agency.

R905.9 Built-up roofs. The installation of built-up roofs shall comply with the provisions of this section.

R905.9.1 Slope. Built-up roofs shall have a design slope of a minimum of one-fourth unit vertical in 12 units horizontal (2-percent slope) for drainage, except for coal-tar built-up roofs, which shall have a design slope of a minimum one-eighth unit vertical in 12 units horizontal (1-percent slope).

R905.9.2 Material standards. Built-up roof covering materials shall comply with the standards in Table R905.9.2.

R905.9.3 Application. Built-up roofs shall be installed according to this chapter and the manufacturer's installation instructions.

R905.10 Metal roof panels. The installation of metal roof panels shall comply with the provisions of this section.

R905.10.1 Deck requirements. Metal roof panel roof coverings shall be applied to solid or spaced sheathing, except where the roof covering is specifically designed to be applied to spaced supports.

R905.10.2 Slope. Minimum slopes for metal roof panels shall comply with the following:

1. The minimum slope for lapped, nonsoldered-seam metal roofs without applied lap sealant shall be three units vertical in 12 units horizontal (25-percent slope).

2. The minimum slope for lapped, nonsoldered-seam metal roofs with applied lap sealant shall be one-half vertical unit in 12 units horizontal (4-percent slope). Lap sealants shall be applied in accordance with the approved manufacturer's installation instructions.

3. The minimum slope for standing-seam roof systems shall be one-quarter unit vertical in 12 units horizontal (2-percent slope).

R905.10.3 Material standards. Metal-sheet roof covering systems that incorporate supporting structural members shall be designed in accordance with the *International Building Code.* Metal-sheet roof coverings installed over structural decking shall comply with Table R905.10.3(1). The materials used for metal-sheet roof coverings shall be naturally corrosion resistant or provided with corrosion resistance in accordance with the standards and minimum thicknesses shown in Table R905.10.3(2).

R905.10.4 Attachment. Metal roof panels shall be secured to the supports in accordance with this chapter and the manufacturer's installation instructions. In the absence of manufacturer's installation instructions, the following fasteners shall be used:

1. Galvanized fasteners shall be used for steel roofs.

2. Three hundred series stainless steel fasteners shall be used for copper roofs.

3. Stainless steel fasteners are acceptable for metal roofs.

R905.11 Modified bitumen roofing. The installation of modified bitumen roofing shall comply with the provisions of this section.

TABLE R905.9.2
BUILT-UP ROOFING MATERIAL STANDARDS

MATERIAL STANDARD	STANDARD
Acrylic coatings used in roofing	ASTM D 6083
Aggregate surfacing	ASTM D 1863
Asphalt adhesive used in roofing	ASTM D 3747
Asphalt cements used in roofing	ASTM D 3019; D 2822; D 4586
Asphalt-coated glass fiber base sheet	ASTM D 4601
Asphalt coatings used in roofing	ASTM D 1227; D 2823; D 2824; D 4479
Asphalt glass felt	ASTM D 2178
Asphalt primer used in roofing	ASTM D 41
Asphalt-saturated and asphalt-coated organic felt base sheet	ASTM D 2626
Asphalt-saturated organic felt (perforated)	ASTM D 226
Asphalt used in roofing	ASTM D 312
Coal tar cements used in roofing	ASTM D 4022; D 5643
Coal-tar primer used in roofing, dampproofing and waterproofing	ASTM D 43
Coal-tar saturated organic felt	ASTM D 227
Coal-tar used in roofing	ASTM D 450, Types I or II
Glass mat, coal tar	ASTM D 4990
Glass mat, venting type	ASTM D 4897
Mineral-surfaced inorganic cap sheet	ASTM D 3909
Thermoplastic fabrics used in roofing	ASTM D 5665; D 5726

TABLE R905.10.3(1)
METAL ROOF COVERINGS STANDARDS

ROOF COVERING TYPE	STANDARD APPLICATION RATE/THICKNESS
Galvanized Steel	ASTM A 653 G90 Zinc Coated
Stainless Steel	ASTM A 240, 300 Series Alloys
Steel	ASTM A 924
Lead-coated Copper	ASTM B 101
Cold Rolled Copper	ASTM B 370 minimum 16 oz/square ft and 12 oz/square ft high yield copper for metal-sheet roof-covering systems; 12 oz/square ft for preformed metal shingle systems.
Hard Lead	2 lb/ sq ft
Soft Lead	3 lb/ sq ft
Aluminum	ASTM B 209, 0.024 minimum thickness for rollformed panels and 0.019 inch minimum thickness for pressformed shingles.
Terne (tin) and terne-coated stainless	Terne coating of 40 lb per double base box, field painted where applicable in accordance with manufacturer's installation instructions.
Zinc	0.027 inch minimum thickness: 99.995% electrolytic high grade zinc with alloy additives of copper (0.08 - 0.20%), titanium (0.07% - 0.12%) and aluminum (0.015%).

For SI: 1 ounce per square foot = 0.305 kg/m², 1 pound per square foot = 4.214 kg/m², 1 inch = 25.4 mm, 1 pound = 0.454 kg.

TABLE R905.10.3(2)
MINIMUM CORROSION RESISTANCE

55% Aluminum-zinc alloy coated steel	ASTM A 792 AZ 50
5% aluminum alloy-coated steel	ASTM A 875 GF60
Aluminum-coated steel	ASTM A 463 T2 65
Galvanized steel	ASTM A 653 G-90
Prepainted steel	ASTM A 755[a]

a. Paint systems in accordance with ASTM A 755 shall be applied over steel products with corrosion-resistant coatings complying with ASTM A 792, ASTM A 875, ASTM A 463, or ASTM A 653.

R905.11.1 Slope. Modified bitumen membrane roofs shall have a design slope of a minimum of one-fourth unit vertical in 12 units horizontal (2-percent slope) for drainage.

R905.11.2 Material standards. Modified bitumen roof coverings shall comply with the standards in Table R905.11.2.

TABLE R905.11.2
MODIFIED BITUMEN ROOFING MATERIAL STANDARDS

MATERIAL	STANDARD
Acrylic coating	ASTM D 6083
Asphalt adhesive	ASTM D 3747
Asphalt cement	ASTM D 3019
Asphalt coating	ASTM D 1227; D 2824
Asphalt primer	ASTM D 41
Modified bitumen roof membrane	ASTM D 6162; D 6163; D 6164; D 6222; D 6223; D 6298; CGSB 37–56M

R905.11.3 Application. Modified bitumen roofs shall be installed according to this chapter and the manufacturer's installation instructions.

R905.12 Thermoset single-ply roofing. The installation of thermoset single-ply roofing shall comply with the provisions of this section.

R905.12.1 Slope. Thermoset single-ply membrane roofs shall have a design slope of a minimum of one-fourth unit vertical in 12 units horizontal (2-percent slope) for drainage.

R905.12.2 Material standards. Thermoset single-ply roof coverings shall comply with ASTM D 4637, ASTM D 5019 or CGSB 37-GP-52M.

R905.12.3 Application. Thermoset single-ply roofs shall be installed according to this chapter and the manufacturer's installation instructions.

R905.13 Thermoplastic single-ply roofing. The installation of thermoplastic single-ply roofing shall comply with the provisions of this section.

R905.13.1 Slope. Thermoplastic single-ply membrane roofs shall have a design slope of a minimum of one-fourth unit vertical in 12 units horizontal (2-percent slope).

R905.13.2 Material standards. Thermoplastic single-ply roof coverings shall comply with ASTM D 4434, ASTM D 6754, ASTM D 6878, or CGSB CAN/CGSB 37.54.

R905.13.3 Application. Thermoplastic single-ply roofs shall be installed according to this chapter and the manufacturer's installation instructions.

R905.14 Sprayed polyurethane foam roofing. The installation of sprayed polyurethane foam roofing shall comply with the provisions of this section.

R905.14.1 Slope. Sprayed polyurethane foam roofs shall have a design slope of a minimum of one-fourth unit vertical in 12 units horizontal (2-percent slope) for drainage.

R905.14.2 Material standards. Spray-applied polyurethane foam insulation shall comply with ASTM C 1029.

R905.14.3 Application. Foamed-in-place roof insulation shall be installed in accordance with this chapter and the manufacturer's installation instructions. A liquid-applied protective coating that complies with Section R905.15 shall be applied no less than 2 hours nor more than 72 hours following the application of the foam.

R905.14.4 Foam plastics. Foam plastic materials and installation shall comply with Section R314.

R905.15 Liquid-applied coatings. The installation of liquid-applied coatings shall comply with the provisions of this section.

R905.15.1 Slope. Liquid-applied roofs shall have a design slope of a minimum of one-fourth unit vertical in 12 units horizontal (2-percent slope).

R905.15.2 Material standards. Liquid-applied roof coatings shall comply with ASTM C 836, C 957, D 1227, D 3468, D 6083 or D 6694.

R905.15.3 Application. Liquid-applied roof coatings shall be installed according to this chapter and the manufacturer's installation instructions.

SECTION R906
ROOF INSULATION

R906.1 General. The use of above-deck thermal insulation shall be permitted provided such insulation is covered with an approved roof covering and passes FM 4450 or UL 1256.

R906.2 Material standards. Above-deck thermal insulation board shall comply with the standards in Table R906.2.

TABLE R906.2
MATERIAL STANDARDS FOR ROOF INSULATION

Cellular glass board	ASTM C 552
Composite boards	ASTM C 1289, Type III, IV, V, or VI
Expanded polystyrene	ASTM C 578
Extruded polystyrene board	ASTM C 578
Perlite Board	ASTM C 728
Polyisocyanurate Board	ASTM C 1289, Type I or Type II
Wood fiberboard	ASTM C 208

SECTION R907
REROOFING

R907.1 General. Materials and methods of application used for re-covering or replacing an existing roof covering shall comply with the requirements of Chapter 9.

> **Exception:** Reroofing shall not be required to meet the minimum design slope requirement of one-quarter unit vertical in 12 units horizontal (2-percent slope) in Section R905 for roofs that provide positive roof drainage.

R907.2 Structural and construction loads. The structural roof components shall be capable of supporting the roof covering system and the material and equipment loads that will be encountered during installation of the roof covering system.

R907.3 Re-covering versus replacement. New roof coverings shall not be installed without first removing existing roof coverings where any of the following conditions occur:

1. Where the existing roof or roof covering is water-soaked or has deteriorated to the point that the existing roof or roof covering is not adequate as a base for additional roofing.

2. Where the existing roof covering is wood shake, slate, clay, cement or asbestos-cement tile.

3. Where the existing roof has two or more applications of any type of roof covering.

4. For asphalt shingles, when the building is located in an area subject to moderate or severe hail exposure according to Figure R903.5.

Exceptions:

1. Complete and separate roofing systems, such as standing-seam metal roof systems, that are designed to transmit the roof loads directly to the building's structural system and that do not rely on existing roofs and roof coverings for support, shall not require the removal of existing roof coverings.

2. Installation of metal panel, metal shingle, and concrete and clay tile roof coverings over existing wood shake roofs shall be permitted when the application is in accordance with Section R907.4.

3. The application of new protective coating over existing spray polyurethane foam roofing systems shall be permitted without tear-off of existing roof coverings.

R907.4 Roof recovering. Where the application of a new roof covering over wood shingle or shake roofs creates a combustible concealed space, the entire existing surface shall be covered with gypsum board, mineral fiber, glass fiber or other approved materials securely fastened in place.

R907.5 Reinstallation of materials. Existing slate, clay or cement tile shall be permitted for reinstallation, except that damaged, cracked or broken slate or tile shall not be reinstalled. Existing vent flashing, metal edgings, drain outlets, collars and metal counterflashings shall not be reinstalled where rusted, damaged or deteriorated. Aggregate surfacing materials shall not be reinstalled.

R907.6 Flashings. Flashings shall be reconstructed in accordance with approved manufacturer's installation instructions. Metal flashing to which bituminous materials are to be adhered shall be primed prior to installation.

CHAPTER 10

CHIMNEYS AND FIREPLACES

SECTION R1001
MASONRY FIREPLACES

R1001.1 General. Masonry fireplaces shall be constructed in accordance with this section and the applicable provisions of Chapters 3 and 4.

R1001.2 Footings and foundations. Footings for masonry fireplaces and their chimneys shall be constructed of concrete or solid masonry at least 12 inches (305 mm) thick and shall extend at least 6 inches (152 mm) beyond the face of the fireplace or foundation wall on all sides. Footings shall be founded on natural, undisturbed earth or engineered fill below frost depth. In areas not subjected to freezing, footings shall be at least 12 inches (305 mm) below finished grade.

R1001.2.1 Ash dump cleanout. Cleanout openings located within foundation walls below fireboxes, when provided, shall be equipped with ferrous metal or masonry doors and frames constructed to remain tightly closed except when in use. Cleanouts shall be accessible and located so that ash removal will not create a hazard to combustible materials.

R1001.3 Seismic reinforcing. Masonry or concrete chimneys in Seismic Design Category D_0, D_1 or D_2 shall be reinforced. Reinforcing shall conform to the requirements set forth in Table R1001.1 and Section R609, Grouted Masonry.

R1001.3.1 Vertical reinforcing. For chimneys up to 40 inches (1016 mm) wide, four No. 4 continuous vertical bars shall be placed between wythes of solid masonry or within the cells of hollow unit masonry and grouted in accordance with Section R609. Grout shall be prevented from bonding with the flue liner so that the flue liner is free to move with thermal expansion. For chimneys more than 40 inches (1016 mm) wide, two additional No. 4 vertical bars shall be provided for each additional flue incorporated into the chimney or for each additional 40 inches (1016 mm) in width or fraction thereof.

R1001.3.2 Horizontal reinforcing. Vertical reinforcement shall be placed within $^1/_4$-inch (6 mm) ties, or other reinforcing of equivalent net cross-sectional area, placed in the bed joints according to Section R607 at a minimum of every 18 inches (457 mm) of vertical height. Two such ties shall be installed at each bend in the vertical bars.

R1001.4 Seismic anchorage. Masonry or concrete chimneys in Seismic Design Categories D_0, D_1 or D_2 shall be anchored at each floor, ceiling or roof line more than 6 feet (1829 mm) above grade, except where constructed completely within the exterior walls. Anchorage shall conform to the requirements of Section R1001.4.1.

R1001.4.1 Anchorage. Two $^3/_{16}$-inch by 1-inch (5 mm by 25 mm) straps shall be embedded a minimum of 12 inches (305 mm) into the chimney. Straps shall be hooked around the outer bars and extend 6 inches (152 mm) beyond the bend. Each strap shall be fastened to a minimum of four floor ceiling or floor joists or rafters with two $^1/_2$-inch (13 mm) bolts.

R1001.5 Firebox walls. Masonry fireboxes shall be constructed of solid masonry units, hollow masonry units grouted solid, stone or concrete. When a lining of firebrick at least 2 inches (51 mm) thick or other approved lining is provided, the minimum thickness of back and side walls shall each be 8 inches (203 mm) of solid masonry, including the lining. The width of joints between firebricks shall not be greater than $^1/_4$ inch (6 mm). When no lining is provided, the total minimum thickness of back and side walls shall be 10 inches (254 mm) of solid masonry. Firebrick shall conform to ASTM C 27 or C 1261 and shall be laid with medium duty refractory mortar conforming to ASTM C 199.

R1001.5.1 Steel fireplace units. Installation of steel fireplace units with solid masonry to form a masonry fireplace is permitted when installed either according to the requirements of their listing or according to the requirements of this section. Steel fireplace units incorporating a steel firebox lining, shall be constructed with steel not less than $^1/_4$ inch (6 mm) thick, and an air circulating chamber which is ducted to the interior of the building. The firebox lining shall be encased with solid masonry to provide a total thickness at the back and sides of not less than 8 inches (203 mm), of which not less than 4 inches (102 mm) shall be of solid masonry or concrete. Circulating air ducts used with steel fireplace units shall be constructed of metal or masonry.

R1001.6 Firebox dimensions. The firebox of a concrete or masonry fireplace shall have a minimum depth of 20 inches (508 mm). The throat shall not be less than 8 inches (203 mm) above the fireplace opening. The throat opening shall not be less than 4 inches (102 mm) deep. The cross-sectional area of the passageway above the firebox, including the throat, damper and smoke chamber, shall not be less than the cross-sectional area of the flue.

Exception: Rumford fireplaces shall be permitted provided that the depth of the fireplace is at least 12 inches (305 mm) and at least one-third of the width of the fireplace opening, that the throat is at least 12 inches (305 mm) above the lintel and is at least $^1/_{20}$ the cross-sectional area of the fireplace opening.

R1001.7 Lintel and throat. Masonry over a fireplace opening shall be supported by a lintel of noncombustible material. The minimum required bearing length on each end of the fireplace opening shall be 4 inches (102 mm). The fireplace throat or damper shall be located a minimum of 8 inches (203 mm) above the lintel.

R1001.7.1 Damper. Masonry fireplaces shall be equipped with a ferrous metal damper located at least 8 inches (203 mm) above the top of the fireplace opening. Dampers shall be installed in the fireplace or the chimney venting the fireplace, and shall be operable from the room containing the fireplace.

TABLE R1001.1
SUMMARY OF REQUIREMENTS FOR MASONRY FIREPLACES AND CHIMNEYS

ITEM	LETTER[a]	REQUIREMENTS
Hearth slab thickness	A	4″
Hearth extension (each side of opening)	B	8″ fireplace opening < 6 square foot. 12″ fireplace opening ≥ 6 square foot.
Hearth extension (front of opening)	C	16″ fireplace opening < 6 square foot. 20″ fireplace opening ≥ 6 square foot.
Hearth slab reinforcing	D	Reinforced to carry its own weight and all imposed loads.
Thickness of wall of firebox	E	10″ solid brick or 8″ where a firebrick lining is used. Joints in firebrick $\frac{1}{4}$″ maximum.
Distance from top of opening to throat	F	8″
Smoke chamber wall thickness Unlined walls	G	6″ 8″
Chimney Vertical reinforcing[b]	H	Four No. 4 full-length bars for chimney up to 40″ wide. Add two No. 4 bars for each additional 40″ or fraction of width or each additional flue.
Horizontal reinforcing	J	$\frac{1}{4}$″ ties at 18″ and two ties at each bend in vertical steel.
Bond beams	K	No specified requirements.
Fireplace lintel	L	Noncombustible material.
Chimney walls with flue lining	M	Solid masonry units or hollow masonry units grouted solid with at least 4 inch nominal thickness.
Distances between adjacent flues	—	See Section R1003.13.
Effective flue area (based on area of fireplace opening)	P	See Section R1003.15.
Clearances: Combustible material Mantel and trim Above roof	R	See Sections R1001.11 and R1003.18. See Section R1001.11, Exception 4. 3′ at roofline and 2′ at 10′.
Anchorage[b] Strap Number Embedment into chimney Fasten to Bolts	S	$\frac{3}{16}$″ × 1″ Two 12″ hooked around outer bar with 6″ extension. 4 joists Two $\frac{1}{2}$″ diameter.
Footing Thickness Width	T	12″ min. 6″ each side of fireplace wall.

For SI: 1 inch = 25.4 mm, 1 foot = 304.8 mm, 1 square foot = 0.0929 m².

NOTE: This table provides a summary of major requirements for the construction of masonry chimneys and fireplaces. Letter references are to Figure R1001.1, which shows examples of typical construction. This table does not cover all requirements, nor does it cover all aspects of the indicated requirements. For the actual mandatory requirements of the code, see the indicated section of text.

a. The letters refer to Figure R1001.1.

b. Not required in Seismic Design Category A, B or C.

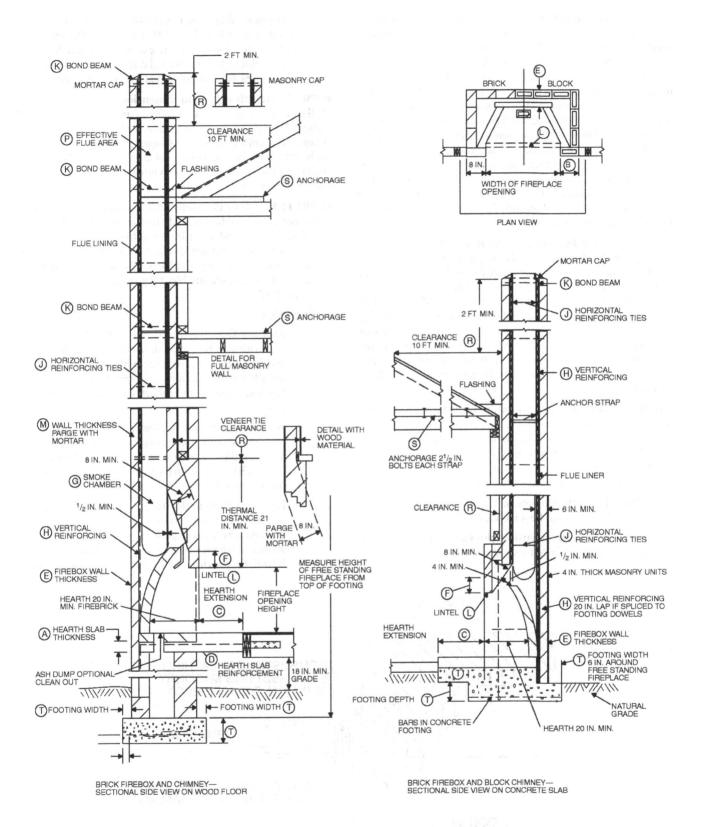

For SI: 1 inch = 25.4 mm, 1 foot = 304.8 mm.

FIGURE R1001.1
FIREPLACE AND CHIMNEY DETAILS

R1001.8 Smoke chamber. Smoke chamber walls shall be constructed of solid masonry units, hollow masonry units grouted solid, stone or concrete. Corbelling of masonry units shall not leave unit cores exposed to the inside of the smoke chamber. When a lining of firebrick at least 2 inches (51 mm) thick, or a lining of vitrified clay at least $^{5}/_{8}$ inch (16 mm) thick, is provided, the total minimum thickness of front, back and side walls shall be 6 inches (152 mm) of solid masonry, including the lining. Firebrick shall conform to ASTM C 27 or C 1261 and shall be laid with medium duty refractory mortar conforming to ASTM C 199. Where no lining is provided, the total minimum thickness of front, back and side walls shall be 8 inches (203 mm) of solid masonry. When the inside surface of the smoke chamber is formed by corbeled masonry, the inside surface shall be parged smooth.

R1001.8.1 Smoke chamber dimensions. The inside height of the smoke chamber from the fireplace throat to the beginning of the flue shall not be greater than the inside width of the fireplace opening. The inside surface of the smoke chamber shall not be inclined more than 45 degrees (0.79 rad) from vertical when prefabricated smoke chamber linings are used or when the smoke chamber walls are rolled or sloped rather than corbeled. When the inside surface of the smoke chamber is formed by corbeled masonry, the walls shall not be corbeled more than 30 degrees (0.52 rad) from vertical.

R1001.9 Hearth and hearth extension. Masonry fireplace hearths and hearth extensions shall be constructed of concrete or masonry, supported by noncombustible materials, and reinforced to carry their own weight and all imposed loads. No combustible material shall remain against the underside of hearths and hearth extensions after construction.

R1001.9.1 Hearth thickness. The minimum thickness of fireplace hearths shall be 4 inches (102 mm).

R1001.9.2 Hearth extension thickness. The minimum thickness of hearth extensions shall be 2 inches (51 mm).

Exception: When the bottom of the firebox opening is raised at least 8 inches (203 mm) above the top of the hearth extension, a hearth extension of not less than $^{3}/_{8}$-inch-thick (10 mm) brick, concrete, stone, tile or other approved noncombustible material is permitted.

R1001.10 Hearth extension dimensions. Hearth extensions shall extend at least 16 inches (406 mm) in front of and at least 8 inches (203 mm) beyond each side of the fireplace opening. Where the fireplace opening is 6 square feet (0.6 m²) or larger, the hearth extension shall extend at least 20 inches (508 mm) in front of and at least 12 inches (305 mm) beyond each side of the fireplace opening.

R1001.11 Fireplace clearance. All wood beams, joists, studs and other combustible material shall have a clearance of not less than 2 inches (51 mm) from the front faces and sides of masonry fireplaces and not less than 4 inches (102 mm) from the back faces of masonry fireplaces. The air space shall not be filled, except to provide fire blocking in accordance with Section R1001.12.

Exceptions:

1. Masonry fireplaces listed and labeled for use in contact with combustibles in accordance with UL 127 and installed in accordance with the manufacturer's installation instructions are permitted to have combustible material in contact with their exterior surfaces.

2. When masonry fireplaces are part of masonry or concrete walls, combustible materials shall not be in contact with the masonry or concrete walls less than 12 inches (306 mm) from the inside surface of the nearest firebox lining.

3. Exposed combustible trim and the edges of sheathing materials such as wood siding, flooring and drywall shall be permitted to abut the masonry fireplace side walls and hearth extension in accordance with Figure R1001.11, provided such combustible trim or sheath-

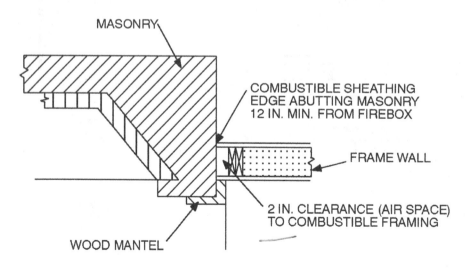

MASONRY

COMBUSTIBLE SHEATHING EDGE ABUTTING MASONRY 12 IN. MIN. FROM FIREBOX

FRAME WALL

2 IN. CLEARANCE (AIR SPACE) TO COMBUSTIBLE FRAMING

WOOD MANTEL

For SI: 1 inch = 25.4 mm.

FIGURE R1001.11
CLEARANCE FROM COMBUSTIBLES

ing is a minimum of 12 inches (305 mm) from the inside surface of the nearest firebox lining.

4. Exposed combustible mantels or trim may be placed directly on the masonry fireplace front surrounding the fireplace opening providing such combustible materials are not placed within 6 inches (152 mm) of a fireplace opening. Combustible material within 12 inches (306 mm) of the fireplace opening shall not project more than $^1/_8$ inch (3 mm) for each 1-inch (25 mm) distance from such an opening.

R1001.12 Fireplace fireblocking. Fireplace fireblocking shall comply with the provisions of Section R602.8.

SECTION R1002
MASONRY HEATERS

R1002.1 Definition. A masonry heater is a heating appliance constructed of concrete or solid masonry, hereinafter referred to as masonry, which is designed to absorb and store heat from a solid-fuel fire built in the firebox by routing the exhaust gases through internal heat exchange channels in which the flow path downstream of the firebox may include flow in a horizontal or downward direction before entering the chimney and which delivers heat by radiation from the masonry surface of the heater.

R1002.2 Installation. Masonry heaters shall be installed in accordance with this section and comply with one of the following:

1. Masonry heaters shall comply with the requirements of ASTM E 1602; or

2. Masonry heaters shall be listed and labeled in accordance with UL 1482 and installed in accordance with the manufacturer's installation instructions.

R1002.3 Footings and foundation. The firebox floor of a masonry heater shall be a minimum thickness of 4 inches (102 mm) of noncombustible material and be supported on a noncombustible footing and foundation in accordance with Section R1003.2.

R1002.4 Seismic reinforcing. In Seismic Design Categories D_0, D_1 and D_2, masonry heaters shall be anchored to the masonry foundation in accordance with Section R1003.3. Seismic reinforcing shall not be required within the body of a masonry heater whose height is equal to or less than 3.5 times it's body width and where the masonry chimney serving the heater is not supported by the body of the heater. Where the masonry chimney shares a common wall with the facing of the masonry heater, the chimney portion of the structure shall be reinforced in accordance with Section R1003.

R1002.5 Masonry heater clearance. Combustible materials shall not be placed within 36 inches (914 mm) of the outside surface of a masonry heater in accordance with NFPA 211 Section 8-7 (clearances for solid-fuel-burning appliances), and the required space between the heater and combustible material shall be fully vented to permit the free flow of air around all heater surfaces.

Exceptions:

1. When the masonry heater wall is at least 8 inches (203 mm) thick of solid masonry and the wall of the heat exchange channels is at least 5 inches (127 mm) thick of solid masonry, combustible materials shall not be placed within 4 inches (102 mm) of the outside surface of a masonry heater. A clearance of at least 8 inches (203 mm) shall be provided between the gas-tight capping slab of the heater and a combustible ceiling.

2. Masonry heaters tested and listed by an American National Standards Association (ANSI)-accredited laboratory to the requirements of UL1482 may be installed in accordance with the listing specifications and the manufacturer's written instructions.

SECTION R1003
MASONRY CHIMNEYS

R1003.1 Definition. A masonry chimney is a chimney constructed of concrete or masonry, hereinafter referred to as masonry. Masonry chimneys shall be constructed, anchored, supported and reinforced as required in this chapter.

R1003.2 Footings and foundations. Footings for masonry chimneys shall be constructed of concrete or solid masonry at least 12 inches (305 mm) thick and shall extend at least 6 inches (152 mm) beyond the face of the foundation or support wall on all sides. Footings shall be founded on natural undisturbed earth or engineered fill below frost depth. In areas not subjected to freezing, footings shall be at least 12 inches (305 mm) below finished grade.

R1003.3 Seismic reinforcing. Masonry or concrete chimneys shall be constructed, anchored, supported and reinforced as required in this chapter. In Seismic Design Category D_0, D_1 or D_2 masonry and concrete chimneys shall be reinforced and anchored as detailed in Section R1003.3.1, R1003.3.2 and R1003.4. In Seismic Design Category A, B or C, reinforcement and seismic anchorage is not required.

R1003.3.1 Vertical reinforcing. For chimneys up to 40 inches (1016 mm) wide, four No. 4 continuous vertical bars, anchored in the foundation, shall be placed in the concrete, or between wythes of solid masonry, or within the cells of hollow unit masonry, and grouted in accordance with Section R609.1.1. Grout shall be prevented from bonding with the flue liner so that the flue liner is free to move with thermal expansion. For chimneys more than 40 inches (1016 mm) wide, two additional No. 4 vertical bars shall be installed for each additional 40 inches (1016 mm) in width or fraction thereof.

R1003.3.2 Horizontal reinforcing. Vertical reinforcement shall be placed enclosed within $^1/_4$-inch (6 mm) ties, or other reinforcing of equivalent net cross-sectional area, spaced not to exceed 18 inches (457 mm) on center in concrete, or placed in the bed joints of unit masonry, at a minimum of every 18 inches (457 mm) of vertical height. Two such ties shall be installed at each bend in the vertical bars.

R1003.4 Seismic anchorage. Masonry and concrete chimneys and foundations in Seismic Design Category D_0, D_1 or D_2 shall be anchored at each floor, ceiling or roof line more than 6 feet (1829 mm) above grade, except where constructed completely within the exterior walls. Anchorage shall conform to the requirements in Section R1003.4.1.

R1003.4.1 Anchorage. Two $3/16$-inch by 1-inch (5 mm by 25 mm) straps shall be embedded a minimum of 12 inches (305 mm) into the chimney. Straps shall be hooked around the outer bars and extend 6 inches (152 mm) beyond the bend. Each strap shall be fastened to a minimum of four floor joists with two $1/2$-inch (13 mm) bolts.

R1003.5 Corbeling. Masonry chimneys shall not be corbeled more than one-half of the chimney's wall thickness from a wall or foundation, nor shall a chimney be corbeled from a wall or foundation that is less than 12 inches (305 mm) thick unless it projects equally on each side of the wall, except that on the second story of a two-story dwelling, corbeling of chimneys on the exterior of the enclosing walls may equal the wall thickness. The projection of a single course shall not exceed one-half the unit height or one-third of the unit bed depth, whichever is less.

R1003.6 Changes in dimension. The chimney wall or chimney flue lining shall not change in size or shape within 6 inches (152 mm) above or below where the chimney passes through floor components, ceiling components or roof components.

R1003.7 Offsets. Where a masonry chimney is constructed with a fireclay flue liner surrounded by one wythe of masonry, the maximum offset shall be such that the centerline of the flue above the offset does not extend beyond the center of the chimney wall below the offset. Where the chimney offset is supported by masonry below the offset in an approved manner, the maximum offset limitations shall not apply. Each individual corbeled masonry course of the offset shall not exceed the projection limitations specified in Section R1003.5.

R1003.8 Additional load. Chimneys shall not support loads other than their own weight unless they are designed and constructed to support the additional load. Construction of masonry chimneys as part of the masonry walls or reinforced concrete walls of the building shall be permitted.

R1003.9 Termination. Chimneys shall extend at least 2 feet (610 mm) higher than any portion of a building within 10 feet (3048 mm), but shall not be less than 3 feet (914 mm) above the highest point where the chimney passes through the roof.

R1003.9.1 Spark arrestors. Where a spark arrestor is installed on a masonry chimney, the spark arrestor shall meet all of the following requirements:

1. The net free area of the arrestor shall not be less than four times the net free area of the outlet of the chimney flue it serves.

2. The arrestor screen shall have heat and corrosion resistance equivalent to 19-gage galvanized steel or 24-gage stainless steel.

3. Openings shall not permit the passage of spheres having a diameter greater than $1/2$ inch (13 mm) nor block the passage of spheres having a diameter less than $3/8$ inch (10 mm).

4. The spark arrestor shall be accessible for cleaning and the screen or chimney cap shall be removable to allow for cleaning of the chimney flue.

R1003.10 Wall thickness. Masonry chimney walls shall be constructed of solid masonry units or hollow masonry units grouted solid with not less than a 4-inch (102 mm) nominal thickness.

R1003.10.1 Masonry veneer chimneys. Where masonry is used to veneer a frame chimney, through-flashing and weep holes shall be installed as required by Section R703.

R1003.11 Flue lining (material). Masonry chimneys shall be lined. The lining material shall be appropriate for the type of appliance connected, according to the terms of the appliance listing and manufacturer's instructions.

R1003.11.1 Residential-type appliances (general). Flue lining systems shall comply with one of the following:

1. Clay flue lining complying with the requirements of ASTM C 315 or equivalent.

2. Listed chimney lining systems complying with UL 1777.

3. Factory-built chimneys or chimney units listed for installation within masonry chimneys.

4. Other approved materials that will resist corrosion, erosion, softening or cracking from flue gases and condensate at temperatures up to 1,800°F (982°C).

R1003.11.2 Flue linings for specific appliances. Flue linings other than these covered in Section R1003.11.1, intended for use with specific types of appliances, shall comply with Sections R1003.11.3 through R1003.11.6.

R1003.11.3 Gas appliances. Flue lining systems for gas appliances shall be in accordance with Chapter 24.

R1003.11.4 Pellet fuel-burning appliances. Flue lining and vent systems for use in masonry chimneys with pellet fuel-burning appliances shall be limited to the following:

1. Flue lining systems complying with Section R1003.11.1.

2. Pellet vents listed for installation within masonry chimneys. (See Section R1003.11.6 for marking.)

R1003.11.5 Oil-fired appliances approved for use with Type L vent. Flue lining and vent systems for use in masonry chimneys with oil-fired appliances approved for use with Type L vent shall be limited to the following:

1. Flue lining systems complying with Section R1003.11.1.

2. Listed chimney liners complying with UL 641. (See Section R1003.11.6 for marking.)

R1003.11.6 Notice of usage. When a flue is relined with a material not complying with Section R1003.11.1, the chimney shall be plainly and permanently identified by a label attached to a wall, ceiling or other conspicuous location adjacent to where the connector enters the chimney. The label shall include the following message or equivalent language:

THIS CHIMNEY FLUE IS FOR USE ONLY WITH [TYPE OR CATEGORY OF APPLIANCE] APPLIANCES THAT BURN [TYPE OF FUEL]. DO NOT CONNECT OTHER TYPES OF APPLIANCES.

R1003.12 Clay flue lining (installation). Clay flue liners shall be installed in accordance with ASTM C 1283 and extend from a point not less than 8 inches (203 mm) below the lowest inlet or, in the case of fireplaces, from the top of the smoke chamber to a point above the enclosing walls. The lining shall be carried up vertically, with a maximum slope no greater than 30 degrees (0.52 rad) from the vertical.

Clay flue liners shall be laid in medium-duty refractory mortar conforming to ASTM C 199 with tight mortar joints left smooth on the inside and installed to maintain an air space or insulation not to exceed the thickness of the flue liner separating the flue liners from the interior face of the chimney masonry walls. Flue liners shall be supported on all sides. Only enough mortar shall be placed to make the joint and hold the liners in position.

R1003.12.1 Listed materials. Listed materials used as flue linings shall be installed in accordance with the terms of their listings and manufacturer's instructions.

R1003.12.2 Space around lining. The space surrounding a chimney lining system or vent installed within a masonry chimney shall not be used to vent any other appliance.

Exception: This shall not prevent the installation of a separate flue lining in accordance with the manufacturer's installation instructions.

R1003.13 Multiple flues. When two or more flues are located in the same chimney, masonry wythes shall be built between adjacent flue linings. The masonry wythes shall be at least 4 inches (102 mm) thick and bonded into the walls of the chimney.

Exception: When venting only one appliance, two flues may adjoin each other in the same chimney with only the flue lining separation between them. The joints of the adjacent flue linings shall be staggered at least 4 inches (102 mm).

R1003.14 Flue area (appliance). Chimney flues shall not be smaller in area than that of the area of the connector from the appliance [see Tables R1003.14(1) and R1003.14(2)]. The sizing of a chimney flue to which multiple appliance venting systems are connected shall be in accordance with Section M1805.3.

R1003.15 Flue area (masonry fireplace). Flue sizing for chimneys serving fireplaces shall be in accordance with Section R1003.15.1 or Section R1003.15.2.

R1003.15.1 Option 1. Round chimney flues shall have a minimum net cross-sectional area of at least $1/_{12}$ of the fireplace opening. Square chimney flues shall have a minimum net cross-sectional area of $1/_{10}$ of the fireplace opening. Rectangular chimney flues with an aspect ratio less than 2 to 1 shall have a minimum net cross-sectional area of $1/_{10}$ of the fireplace opening. Rectangular chimney flues with an aspect ratio of 2 to 1 or more shall have a minimum net cross-sectional area of $1/_{8}$ of the fireplace opening. Cross-sectional areas of clay flue linings are shown in Tables R1003.14(1)

and R1003.14(2) or as provided by the manufacturer or as measured in the field.

R1003.15.2 Option 2. The minimum net cross-sectional area of the chimney flue shall be determined in accordance with Figure R1003.15.2. A flue size providing at least the equivalent net cross-sectional area shall be used. Cross-sectional areas of clay flue linings are shown in Tables R1003.14(1) and R1003.14(2) or as provided by the manufacturer or as measured in the field. The height of the chimney shall be measured from the firebox floor to the top of the chimney flue.

TABLE R1003.14(1)
NET CROSS–SECTIONAL AREA OF ROUND FLUE SIZES[a]

FLUE SIZE, INSIDE DIAMETER (inches)	CROSS–SECTIONAL AREA (square inches)
6	28
7	38
8	50
10	78
$10^3/_4$	90
12	113
15	176
18	254

For SI: 1 inch = 25.4 mm, 1 square inch = 645.16 mm².

a. Flue sizes are based on ASTM C 315.

TABLE R1003.14(2)
NET CROSS–SECTIONAL AREA OF SQUARE AND RECTANGULAR FLUE SIZES

FLUE SIZE, OUTSIDE NOMINAL DIMENSIONS (inches)	CROSS–SECTIONAL AREA (square inches)
4.5 × 8.5	23
4.5 × 13	34
8 × 8	42
8.5 × 8.5	49
8 × 12	67
8.5 × 13	76
12 × 12	102
8.5 × 18	101
13 × 13	127
12 × 16	131
13 × 18	173
16 × 16	181
16 × 20	222
18 × 18	233
20 × 20	298
20 × 24	335
24 × 24	431

For SI: 1 inch = 25.4 mm, 1 square inch = 645.16 mm².

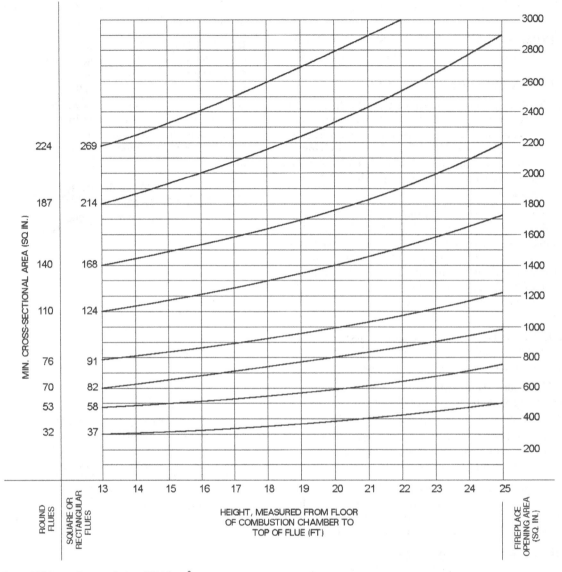

For SI: 1 foot = 304.8 mm, 1 square inch = 645.16 mm².

FIGURE R1003.15.2
FLUE SIZES FOR MASONRY CHIMNEYS

R1003.16 Inlet. Inlets to masonry chimneys shall enter from the side. Inlets shall have a thimble of fireclay, rigid refractory material or metal that will prevent the connector from pulling out of the inlet or from extending beyond the wall of the liner.

R1003.17 Masonry chimney cleanout openings. Cleanout openings shall be provided within 6 inches (152 mm) of the base of each flue within every masonry chimney. The upper edge of the cleanout shall be located at least 6 inches (152 mm) below the lowest chimney inlet opening. The height of the opening shall be at least 6 inches (152 mm). The cleanout shall be provided with a noncombustible cover.

Exception: Chimney flues serving masonry fireplaces where cleaning is possible through the fireplace opening.

R1003.18 Chimney clearances. Any portion of a masonry chimney located in the interior of the building or within the exterior wall of the building shall have a minimum air space clearance to combustibles of 2 inches (51 mm). Chimneys located entirely outside the exterior walls of the building, including chimneys that pass through the soffit or cornice, shall have a minimum air space clearance of 1 inch (25 mm). The air space shall not be filled, except to provide fire blocking in accordance with Section R1003.19.

Exceptions:

1. Masonry chimneys equipped with a chimney lining system listed and labeled for use in chimneys in contact with combustibles in accordance with UL 1777 and installed in accordance with the manufacturer's installation instructions are permitted to have combustible material in contact with their exterior surfaces.

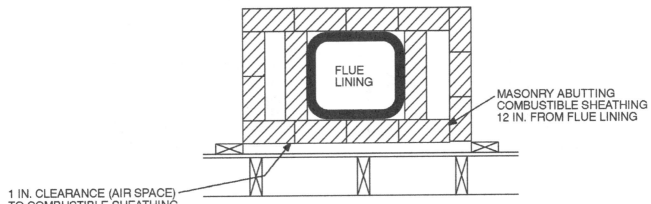

FIGURE R1003.18
CLEARANCE FROM COMBUSTIBLES

For SI: 1 inch = 25.4 mm.

2. When masonry chimneys are constructed as part of masonry or concrete walls, combustible materials shall not be in contact with the masonry or concrete wall less than 12 inches (305 mm) from the inside surface of the nearest flue lining.

3. Exposed combustible trim and the edges of sheathing materials, such as wood siding and flooring, shall be permitted to abut the masonry chimney side walls, in accordance with Figure R1003.18, provided such combustible trim or sheathing is a minimum of 12 inches (305 mm) from the inside surface of the nearest flue lining. Combustible material and trim shall not overlap the corners of the chimney by more than 1 inch (25 mm).

R1003.19 Chimney fireblocking. All spaces between chimneys and floors and ceilings through which chimneys pass shall be fireblocked with noncombustible material securely fastened in place. The fireblocking of spaces between chimneys and wood joists, beams or headers shall be self-supporting or be placed on strips of metal or metal lath laid across the spaces between combustible material and the chimney.

R1003.20 Chimney crickets. Chimneys shall be provided with crickets when the dimension parallel to the ridgeline is greater than 30 inches (762 mm) and does not intersect the ridgeline. The intersection of the cricket and the chimney shall be flashed and counterflashed in the same manner as normal roof-chimney intersections. Crickets shall be constructed in compliance with Figure R1003.20 and Table R1003.20.

TABLE R1003.20
CRICKET DIMENSIONS

ROOF SLOPE	H
12 - 12	$^{1}/_{2}$ of W
8 - 12	$^{1}/_{3}$ of W
6 - 12	$^{1}/_{4}$ of W
4 - 12	$^{1}/_{6}$ of W
3 - 12	$^{1}/_{8}$ of W

SECTION R1004
FACTORY-BUILT FIREPLACES

R1004.1 General. Factory-built fireplaces shall be listed and labeled and shall be installed in accordance with the conditions of the listing. Factory-built fireplaces shall be tested in accordance with UL 127.

R1004.2 Hearth extensions. Hearth extensions of approved factory-built fireplaces shall be installed in accordance with the listing of the fireplace. The hearth extension shall be readily distinguishable from the surrounding floor area.

R1004.3 Decorative shrouds. Decorative shrouds shall not be installed at the termination of chimneys for factory-built fireplaces except where the shrouds are listed and labeled for use with the specific factory-built fireplace system and installed in accordance with the manufacturer's installation instructions.

R1004.4 Unvented gas log heaters. An unvented gas log heater shall not be installed in a factory-built fireplace unless the fireplace system has been specifically tested, listed and labeled for such use in accordance with UL 127.

SECTION R1005
FACTORY-BUILT CHIMNEYS

R1005.1 Listing. Factory-built chimneys shall be listed and labeled and shall be installed and terminated in accordance with the manufacturer's installation instructions.

R1005.2 Decorative shrouds. Decorative shrouds shall not be installed at the termination of factory-built chimneys except where the shrouds are listed and labeled for use with the specific factory-built chimney system and installed in accordance with the manufacturer's installation instructions.

R1005.3 Solid-fuel appliances. Factory-built chimneys installed in dwelling units with solid-fuel-burning appliances shall comply with the Type HT requirements of UL 103 and shall be marked "Type HT and "Residential Type and Building Heating Appliance Chimney."

Exception: Chimneys for use with open combustion chamber fireplaces shall comply with the requirements of UL 103

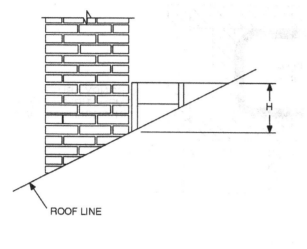

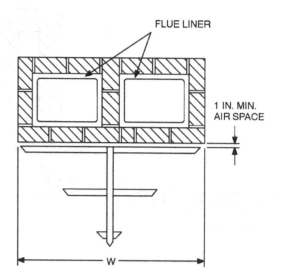

For SI: 1 inch = 25.4 mm.

FIGURE R1003.20
CHIMNEY CRICKET

and shall be marked "Residential Type and Building Heating Appliance Chimney."

Chimneys for use with open combustion chamber appliances installed in buildings other than dwelling units shall comply with the requirements of UL 103 and shall be marked "Building Heating Appliance Chimney" or "Residential Type and Building Heating Appliance Chimney."

R1005.4 Factory-built fireplaces. Chimneys for use with factory-built fireplaces shall comply with the requirements of UL 127.

R1005.5 Support. Where factory-built chimneys are supported by structural members, such as joists and rafters, those members shall be designed to support the additional load.

R1005.6 Medium-heat appliances. Factory-built chimneys for medium-heat appliances producing flue gases having a temperature above 1,000°F (538°C), measured at the entrance to the chimney shall comply with UL 959.

SECTION R1006
EXTERIOR AIR SUPPLY

R1006.1 Exterior air. Factory-built or masonry fireplaces covered in this chapter shall be equipped with an exterior air supply to assure proper fuel combustion unless the room is mechanically ventilated and controlled so that the indoor pressure is neutral or positive.

R1006.1.1 Factory-built fireplaces. Exterior combustion air ducts for factory-built fireplaces shall be a listed component of the fireplace and shall be installed according to the fireplace manufacturer's instructions.

R1006.1.2 Masonry fireplaces. Listed combustion air ducts for masonry fireplaces shall be installed according to the terms of their listing and the manufacturer's instructions.

R1006.2 Exterior air intake. The exterior air intake shall be capable of supplying all combustion air from the exterior of the dwelling or from spaces within the dwelling ventilated with outside air such as non-mechanically ventilated crawl or attic spaces. The exterior air intake shall not be located within the garage or basement of the dwelling nor shall the air intake be located at an elevation higher than the firebox. The exterior air intake shall be covered with a corrosion-resistant screen of $^1/_4$-inch (6 mm) mesh.

R1006.3 Clearance. Unlisted combustion air ducts shall be installed with a minimum 1-inch (25 mm) clearance to combustibles for all parts of the duct within 5 feet (1524 mm) of the duct outlet.

R1006.4 Passageway. The combustion air passageway shall be a minimum of 6 square inches (3870 mm²) and not more than 55 square inches (0.035 m²), except that combustion air systems for listed fireplaces shall be constructed according to the fireplace manufacturer's instructions.

R1006.5 Outlet. Locating the exterior air outlet in the back or sides of the firebox chamber or within 24 inches (610 mm) of the firebox opening on or near the floor is permitted. The outlet shall be closable and designed to prevent burning material from dropping into concealed combustible spaces.

Part IV — Energy Conservation

Use
MN 1322

CHAPTER 11

ENERGY EFFICIENCY

This chapter has been revised in its entirety; there will be no marginal markings

SECTION N1101
GENERAL

N1101.1 Scope. This chapter regulates the energy efficiency for the design and construction of buildings regulated by this code.

> **Exception:** Portions of the building envelope that do not enclose conditioned space.

N1101.2 Compliance. Compliance shall be demonstrated by either meeting the requirements of the *International Energy Conservation Code* or meeting the requirements of this chapter. Climate zones from Figure N1101.2 or Table N1101.2 shall be used in determining the applicable requirements from this chapter.

N1101.2.1 Warm humid counties. Warm humid counties are listed in Table N1101.2.1.

N1101.3 Identification. Materials, systems and equipment shall be identified in a manner that will allow a determination of compliance with the applicable provisions of this chapter.

N1101.4 Building thermal envelope insulation. An *R*-value identification mark shall be applied by the manufacturer to each piece of building thermal envelope insulation 12 inches (305 mm) or more wide. Alternately, the insulation installers shall provide a certification listing the type, manufacturer and *R*-value of insulation installed in each element of the building thermal envelope. For blown or sprayed insulation (fiberglass and cellulose), the initial installed thickness, settled thickness, settled *R*-value, installed density, coverage area and number of bags installed shall be listed on the certification. For sprayed polyurethane foam (SPF) insulation, the installed thickness of the area covered and *R*-value of installed thickness shall be listed on the certificate. The insulation installer shall sign, date and post the certificate in a conspicuous location on the job site.

N1101.4.1 Blown or sprayed roof/ceiling insulation. The thickness of blown in or sprayed roof/ceiling insulation (fiberglass or cellulose) shall be written in inches (mm) on markers that are installed at least one for every 300 ft^2 (28 m^2) throughout the attic space. The markers shall be affixed to the trusses or joists and marked with the minimum initial installed thickness with numbers a minimum of 1 inch (25 mm) high. Each marker shall face the attic access opening. Spray polyurethane foam thickness and installed *R*-value shall be listed on the certificate provided by the insulation installer.

N1101.4.2 Insulation mark installation. Insulating materials shall be installed such that the manufacturer's *R*-value mark is readily observable upon inspection.

N1101.5 Fenestration product rating. *U*-factors of fenestration products (windows, doors and skylights) shall be determined in accordance with NFRC 100 by an accredited, independent laboratory, and labeled and certified by the manufacturer. Products lacking such a labeled *U*-factor shall be assigned a default *U*-factor from Tables N1101.5(1) and N1101.5(2). The solar heat gain coefficient (SHGC) of glazed fenestration products (windows, glazed doors and skylights) shall be determined in accordance with NFRC 200 by an accredited, independent laboratory, and labeled and certified by the manufacturer. Products lacking such a labeled SHGC shall be assigned a default SHGC from Table N1101.5(3).

N1101.6 Installation. All materials, systems and equipment shall be installed in accordance with the manufacturer's installation instructions and the provisions of this code.

N1101.6.1 Protection of exposed foundation insulation. Insulation applied to the exterior of basement walls, crawl space walls, and the perimeter of slab-on-grade floors shall have a rigid, opaque and weather-resistant protective covering to prevent the degradation of the insulation's thermal performance. The protective covering shall cover the exposed exterior insulation and extend a minimum of 6 inches (152 mm) below grade.

N1101.7 Above code programs. The building official or other authority having jurisdiction shall be permitted to deem a national, state or local energy efficiency program to exceed the energy efficiency required by this chapter. Buildings approved in writing by such an energy efficiency program shall be considered in compliance with this chapter.

N1101.8 Certificate. A permanent certificate shall be posted on or in the electrical distribution panel. The certificate shall be completed by the builder or registered design professional. The certificate shall list the predominant *R*-values of insulation installed in or on ceiling/roof, walls, foundation (slab, basement wall, crawlspace wall and/or floor) and ducts outside conditioned spaces; *U*-factors for fenestration; and the solar heat gain coefficient (SHGC) of fenestration. Where there is more than one value for each component, the certificate shall list the value covering the largest area. The certificate shall list the type and efficiency of heating, cooling and service water heating equipment.

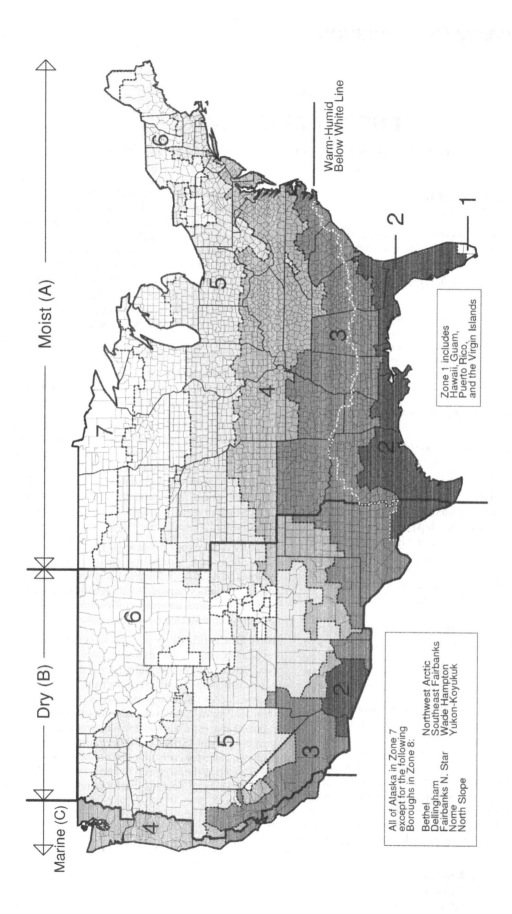

**FIGURE N1101.2
CLIMATE ZONES**

Moist (A)

Dry (B)

Marine (C)

Warm-Humid
Below White Line

Zone 1 includes
Hawaii, Guam,
Puerto Rico,
and the Virgin Islands

All of Alaska in Zone 7
except for the following
Boroughs in Zone 8:

Bethel   Northwest Arctic
Dellingham   Southeast Fairbanks
Fairbanks N. Star   Wade Hampton
Nome   Yukon-Koyukuk
North Slope

TABLE N1101.2
CLIMATE ZONES BY STATES AND COUNTIES

Alabama
Zone 3 except
Zone 2
Baldwin
Mobile

Alaska
Zone 7 except
Zone 8
Bethel
Dellingham
Fairbanks North Star
Nome
North Slope
Northwest Arctic
Southeast Fairbanks
Wade Hampton
Yukon-Koyukuk

Arizona
Zone 3 except
Zone 2
La Paz
Maricopa
Pima
Pinal
Yuma
Zone 4
Gila
Yavapai
Zone 5
Apache
Coconino
Navajo

Arkansas
Zone 3 except
Zone 4
Baxter
Benton
Boone
Carroll
Fulton
Izard
Madison
Marion
Newton
Searcy
Stone
Washington

California
Zone 3 Dry except
Zone 2
Imperial
Zone 3 Marine
Alameda
Marin

Mendocino
Monterey
Napa
San Benito
San Francisco
San Luis Obispo
San Mateo
Santa Barbara
Santa Clara
Santa Cruz
Sonoma
Ventura
Zone 4 Dry
Amador
Calaveras
El Dorado
Inyo
Lake
Mariposa
Trinity
Tuolumne
Zone 4 Marine
Del Norte
Humboldt
Zone 5
Lassen
Modoc
Nevada
Plumas
Sierra
Siskiyou
Zone 6
Alpine
Mono

Colorado
Zone 5 except
Zone 4
Baca
Las Animas
Otero
Zone 6
Alamosa
Archuleta
Chaffee
Conejos
Costilla
Custer
Dolores
Eagle
Moffat
Ouray
Rio Blanco
Saguache
San Miguel
Zone 7
Clear Creek
Grand

Gunnison
Hinsdale
Jackson
Lake
Mineral
Park
Pitkin
Rio Grande
Routt
San Juan
Summit

Connecticut
Zone 5

Delaware
Zone 4

Dist. of Columbia
Zone 4

Florida
Zone 2 except
Zone1
Broward
Dade
Monroe

Georgia
Zone 3 except
Zone 2
Appling
Atkinson
Bacon
Baker
Berrien
Brantley
Brooks
Bryan
Camden
Charlton
Chatham
Clinch
Colquitt
Cook
Decatur
Echols
Effingham
Evans
Glynn
Grady
Jeff Davis
Lanier
Liberty
Long
Lowndes
Mcintosh
Miller

Mitchell
Pierce
Seminole
Tattnall
Thomas
Toombs
Ware
Wayne
Zone 4
Banks
Catoosa
Dade
Dawson
Fannin
Floyd
Franklin
Gilmer
Gordon
Habersham
Hall
Lumpkin
Murray
Pickens
Rabun
Stephens
Towns
Union
Walker
White
Whitfield

Hawaii
Zone 1 Moist

Idaho
Zone 6 except
Zone 5
Ada
Benewah
Canyon
Cassia
Clearwater
Elmore
Gem
Gooding
Idaho
Jerome
Kootenai
Latah
Lewis
Lincoln
Minidoka
Nez Perce
Owyhee
Payette
Power
Shoshone
Twin Falls

Washington

Illinois
Zone 5 except
Zone 4
Alexander
Bond
Christian
Clay
Clinton
Crawford
Edwards
Effingham
Fayette
Franklin
Gallatin
Hamilton
Hardin
Jackson
Jasper
Jefferson
Johnson
Lawrence
Macoupin
Madison
Marion
Massac
Monroe
Montgomery
Perry
Pope
Pulaski
Randolph
Richland
Saline
Shelby
St clair
Union
Wabash
Washington
Wayne
White
Williamson

Indiana
Zone 5 except
Zone 4
Brown
Clark
Crawford
Daviess
Dearborn
Dubois
Floyd
Gibson
Greene
Harrison
Jackson

(continued)

TABLE N1101.2—continued
CLIMATE ZONES BY STATES AND COUNTIES

Jefferson	Webster	Morehouse	Oceana	Harrison
Jennings	Winnebago	Natchitoches	Ogemaw	Jackson
Knox	Winneshiek	Ouachita	Osceola	Pearl River
Lawrence	Worth	Red River	Oscoda	Stone
Martin	Wright	Richland	Otsego	
Monroe		Sabine	Presque Isle	**Missouri**
Ohio	**Kansas**	Tensas	Roscommon	**Zone 4 except**
Orange	**Zone 4 except**	Union	Sanilac	**Zone 5**
Perry	**Zone 5**	Vernon	Wexford	Adair
Pike	Cheyenne	Webster	**Zone 7**	Andrew
Posey	Cloud	West Carroll	Baraga	Atchison
Ripley	Decatur	Winn	Chippewa	Buchanan
Scott	Ellis		Gogebic	Caldwell
Spencer	Gove	**Maine**	Houghton	Chariton
Sullivan	Graham	**Zone 6 except**	Iron	Clark
Switzerland	Greeley	**Zone 7**	Keweenaw	Clinton
Vanderburgh	Hamilton	Aroostook	Luce	Daviess
Warrick	Jewell		Mackinac	De Kalb
Washington	Lane	**Maryland**	Ontonagon	Gentry
	Logan	**Zone 4 except**	Schoolcraft	Grundy
Iowa	Mitchell	**Zone 5**		Harrison
Zone 5 except	Ness	Garrett	**Minnesota**	Holt
Zone 6	Norton		**Zone 6 except**	Knox
Allamakee	Osborne	**Massachusetts**	**Zone 7**	Lewis
Black Hawk	Phillips	**Zone 5**	Aitkin	Linn
Bremer	Rawlins		Becker	Livingston
Buchanan	Republic	**Michigan**	Beltrami	Macon
Buena Vista	Rooks	**Zone 5 except**	Carlton	Marion
Butler	Scott	**Zone 6**	Cass	Mercer
Calhoun	Sheridan	Alcona	Clay	Nodaway
Cerro Gordo	Sherman	Alger	Clearwater	Pike
Cherokee	Smith	Alpena	Cook	Putnam
Chickasaw	Thomas	Antrim	Crow Wing	Ralls
Clay	Trego	Arenac	Grant	Schuyler
Clayton	Wallace	Benzie	Hubbard	Scotland
Delaware	Wichita	Charlevoix	Itasca	Shelby
Dickinson		Cheboygan	Kanabec	Sullivan
Emmet	**Kentucky**	Clare	Kittson	Worth
Fayette	**Zone 4**	Crawford	Koochiching	
Floyd		Delta	Lake of the Wood	**Montana**
Franklin	**Louisiana**	Dickinson	Mahnomen	**Zone 6**
Grundy	**Zone 2 except**	Emmet	Marshall	
Hamilton	**Zone 3**	Gladwin	Mille Lacs	**Nebraska**
Hancock	Bienville	Grand Traverse	Norman	**Zone 5**
Hardin	Bossier	Huron	Otter Tail	
Howard	Caddo	Iosco	Pennington	**Nevada**
Humboldt	Caldwell	Isabella	Pine	**Zone 5 except**
Ida	Catahoula	Kalkaska	Polk	**Zone 3**
Kossuth	Claiborne	Lake	Red Lake	Clark
Lyon	Concordia	Leelanau	Roseau	
Mitchell	De Soto	Manistee	St Louis	**New Hampshire**
O'Brien	East Carroll	Marquette	Wadena	**Zone 6 except**
Osceola	Franklin	Mason	Wilkin	**Zone 5**
Palo Alto	Grant	Mecosta		Cheshire
Plymouth	Jackson	Menominee	**Mississippi**	Hillsborough
Pocahontas	La Salle	Missaukee	**Zone 3 except**	Rockingham
Sac	Lincoln	Montmorency	**Zone 2**	Stafford
Sioux	Madison	Newaygo	Hancock	

(continued)

TABLE N1101.2—continued
CLIMATE ZONES BY STATES AND COUNTIES

New Jersey
Zone 4 except
Zone 5
Bergen
Hunterdon
Mercer
Morris
Passaic
Somerset
Sussex
Warren

New Mexico
Zone 4 except
Zone 3
Chaves
Dona Ana
Eddy
Hidalgo
Lea
Luna
Otero
Zone 5
Catron
Colfax
Harding
Los Alamos
McKinley
Mora
Rio Arriba
San Juan
San Miguel
Sandoval
Santa Fe
Taos
Torrance

New York
Zone 5 except
Zone 4
Bronx
Kings
Nassau
New York
Queens
Richmond
Suffolk
Westchester
Zone 6
Allegany
Broome
Cattaraugus
Chenango
Clinton
Delaware
Essex
Franklin
Fulton

Hamilton
Herkimer
Jefferson
Lewis
Madison
Montgomery
Oneida
Otsego
Schoharie
Schuyler
St Lawrence
Steuben
Sullivan
Tompkins
Ulster
Warren
Wyoming

North Carolina
Zone 3 except
Zone 4
Alamance
Alexander
Bertie
Buncombe
Burke
Calwell
Caswell
Catawba
Chatham
Cherokee
Clay
Cleveland
Davie
Durham
Forsyth
Franklin
Gates
Graham
Granville
Guilford
Halifax
Harnett
Haywood
Henderson
Hertford
Iredell
Jackson
Lee
Lincoln
Macon
Madison
McDowell
Nash
Northampton
Orange
Person
Polk

Rockingham
Rutherford
Stokes
Surry
Swain
Transylvania
Vance
Wake
Warren
Wilkes
Yadkin
Zone 5
Alleghany
Ashe
Avery
Mitchell
Watauga
Yancey

North Dakota
Zone 7 except
Zone 6
Adams
Billings
Bowman
Burleigh
Dickey
Dunn
Emmons
Golden Valley
Grant
Hettinger
La Moure
Logan
McIntosh
McKenzie
Mercer
Morton
Oliver
Ransom
Richland
Sargent
Sioux
Slope
Stark

Ohio
Zone 5 except
Zone 4
Adams
Brown
Clermont
Gallia
Hamilton
Lawrence
Pike
Scioto
Washington

Oklahoma
Zone 3 Moist except
Zone 4 Dry
Beaver
Cimarron
Texas

Oregon
Zone 4 Marine except
Zone 5 Dry
Baker
Crook
Deschutes
Gilliam
Grant
Harney
Hood River
Jefferson
Klamath
Lake
Malheur
Morrow
Sherman
Umatilla
Union
Wallowa
Wasco
Wheeler

Pennsylvania
Zone 5 except
Zone 4
Bucks
Chester
Delaware
Montgomery
Philadelphia
York
Zone 6
Cameron
Clearfield
Elk
McKean
Potter
Susquehanna
Tioga
Wayne

Rhode Island
Zone 5

South Carolina
Zone 3

South Dakota
Zone 6 except
Zone 5
Bennett

Bon Homme
Charles Mix
Clay
Douglas
Gregory
Hutchinson
Jackson
Mellette
Todd
Tripp
Union
Yankton

Tennessee
Zone 4 except
Zone 3
Chester
Crockett
Dyer
Fayette
Hardeman
Hardin
Haywood
Henderson
Lake
Lauderdale
Madison
McNairy
Shelby
Tipton

Texas
Zone 2 Moist except
Zone 2 Dry
Bandera
Dimmit
Edwards
Frio
Kinney
La Salle
Maverick
Medina
Real
Uvalde
Val Verde
Webb
Zapata
Zavala
Zone 3 Dry
Andrews
Baylor
Borden
Brewster
Callahan
Childress
Coke
Coleman
Collingsworth

(continued)

TABLE N1101.2—continued
CLIMATE ZONES BY STATES AND COUNTIES

Concho	Wheeler	Stephens	**Virginia**	Wyoming
Cottle	Wilbarger	Tarrant	**Zone 4**	
Crane	Winkler	Titus		**Wisconsin**
Crockett	**Zone 3 Moist**	Upshur	**Washington**	**Zone 6 except**
Crosby	Archer	Van Zandt	**Zone 4 Marine except**	**Zone 7**
Culberson	Bianco	Wichita	**Zone 5 Dry**	Ashland
Dawson	Bowie	Wise	Adams	Bayfield
Dickens	Brown	Wood	Asotin	Burnett
Ector	Burnet	Young	Benton	Douglass
El Paso	Camp	**Zone 4**	Chelan	Florence
Fisher	Cass	Armstrong	Columbia	Forest
Foard	Clay	Bailey	Douglas	Iron
Gaines	Collin	Briscoe	Franklin	Langlade
Garza	Comanche	Carson	Garfield	Lincoln
Glasscock	Cooke	Castro	Grant	Oneida
Hall	Dallas	Cochran	Kittitas	Price
Hardeman	Delta	Dallam	Klickitat	Sawyer
Haskell	Denton	Deaf Smith	Lincoln	Taylor
Hemphill	Eastland	Donley	Skamania	Vilas
Howard	Ellis	Floyd	Spokane	Washburn
Hudspeth	Erath	Gray	Walla Walla	
Irion	Fannin	Hale	Whitman	**Wyoming**
Jeff Davis	Franklin	Hansford	Yakima	**Zone 6 except**
Jones	Gillespie	Hartley	**Zone 6 Dry**	**Zone 5**
Kent	Grayson	Hockley	Ferry	Goshen
Kerr	Gregg	Hutchinson	Okanogan	Platte
Kimble	Hamilton	Lamb	Pend Oreille	**Zone 7**
King	Harrison	Lipscomb	Stevens	Lincoln
Knox	Henderson	Moore		Sublette
Loving	Hood	Ochiltree	**West Virginia**	Teto
Lubbock	Hopkins	Oldham	**Zone 5 except**	
Lynn	Hunt	Parmer	**Zone 4**	**American Samoa**
Martin	Jack	Potter	Berkely	**Zone 1 Moist**
Mason	Johnson	Randall	Boone	
McCulloch	Kaufman	Roberts	Braxton	**Guam**
Menard	Kendall	Sherman	Cabell	**Zone 1 Moist**
Midland	Lamar	Swisher	Calhoun	
Mitchell	Lampasas	Yoakum	Clay	**Northern Marianas**
Motley	Llano		Gilmer	**Zone 1 Moist**
Nolan	Marion	**Utah**	Jackson	
Pecos	Mills	**Zone 5 except**	Jefferson	**Puerto Rico**
Presidio	Montague	**Zone 3**	Kanawha	**Zone 1 Moist**
Reagan	Morris	Washington	Lincoln	
Reeves	Nacogdoches	**Zone 6**	Logan	**U.S. Virgin Islands**
Runnels	Navarro	Box Elder	Mason	**Zone 1 Moist**
Schleicher	Palo Pinto	Cache	McDowell	
Scurry	Panola	Carbon	Mercer	
Shackelford	Parker	Daggett	Mingo	
Sterling	Rains	Duchesne	Monroe	
Stonewall	Red River	Morgan	Morgan	
Sutton	Rockwall	Rich	Pleasants	
Taylor	Rusk	Summit	Putnam	
Terrell	Sabine	Uintah	Ritchie	
Terry	San Augustine	Wasatch	Roane	
Throckmorton	San Saba		Tyler	
Tom Green	Shelby	**Vermont**	Wayne	
Upton	Smith	**Zone 6**	Wirt	
Ward	Somervell		Wood	

TABLE N1101.2.1
WARM HUMID COUNTIES

Alabama

Autauga
Baldwin
Barbour
Bullock
Butler
Choctaw
Clarke
Coffee
Conecuh
Covington
Crenshaw
Dale
Dallas
Elmore
Escambia
Geneva
Henry
Houston
Lowndes
Macon
Marengo
Mobile
Monroe
Montgomery
Perry
Pike
Russell
Washington
Wilcox

Arkansas

Columbia
Hempstead
Lafayette
Little River
Miller
Sevier
Union

Florida

All

Georgia

All in Zone 2
  Plus
Ben Hill
Bleckley
Bulloch
Calhoun
Candler
Chattahoochee
Clay
Coffee
Crisp
Dodge
Dooly
Dougherty
Early
Emanuel

Houston
Irwin
Jenkins
Johnson
Laurens
Lee
Macon
Marion
Montgomery
Peach
Pulaski
Quitman
Randolph
Schley
Screven
Stewart
Sumter
Taylor
Telfair
Terrell
Tift
Treutlen
Turner
Twiggs
Webster
Wheeler
Wilcox
Worth

Hawaii

All

Louisiana

All in Zone 2
  Plus
Bienville
Bossier
Caddo
Caldwell
Catahoula
Claiborne
De Soto
Franklin
Grant
Jackson
La Salle
Lincoln
Madison
Natchitoches
Ouachita
Red River
Richland
Sabine
Tensas
Union
Vernon
Webster
Winn

Mississippi

All in Zone 2
  Plus
Adams
Amite
Claiborne
Copiah
Covington
Forrest
Franklin
George
Greene
Hinds
Jefferson
Jefferson Davis
Jones
Lamar
Lawrence
Lincoln
Marion
Perry
Pike
Rankin
Simpson
Smith
Walthall
Warren
Wayne
Wilkinson

North Carolina

Brunswick
Carteret
Columbus
New Hanover
Onslow
Pender

South Carolina

Allendale
Bamberg
Barnwell
Beaufort
Berkeley
Charleston
Colleton
Dorchester
Georgetown
Hampton
Horry
Jasper

Texas

All in Zone 2
  Plus
Blanco
Bowie
Brown
Burnet
Camp

Cass
Collin
Comanche
Dallas
Delta
Denton
Ellis
Erath
Franklin
Gillespie
Gregg
Hamilton
Harrison
Henderson
Hood
Hopkins
Hunt
Johnson
Haufman
Kendall
Lamar
Lampasas
Llano
Marion
Mills
Morris
Nacogdoches
Navarro
Palo Pinto
Panola
Parker
Rains
Red River
Rockwall
Rusk
Sabine
San Augustine
San Saba
Shelby
Smith
Somervell
Tarrant
Titus
Upshur
Van Zandt
Wood

American Samoa

All

Guam

All

Northern Marianas

All

Puerto Rico

All

U.S. Virgin Islands

All

TABLE N1101.5(1)
DEFAULT GLAZED FENESTRATION *U*-FACTORS

FRAME TYPE	SINGLE PANE	DOUBLE PANE	SKYLIGHT	
			Single	Double
Metal	1.2	0.8	2	1.3
Metal with thermal break	1.1	0.65	1.9	1.1
Nonmetal or metal clad	0.95	0.55	1.75	1.05
Glazed block	0.6			

TABLE N1101.5(2)
DEFAULT DOOR *U*-FACTORS

DOOR TYPE	*U*-FACTOR
Uninsulated metal	1.2
Insulated metal	0.6
Wood	0.5
Insulated, nonmetal edge, max 45% glazing, any glazing double pane	0.35

TABLE N1101.5(3)
DEFAULT GLAZED FENESTRATION SHGC

SINGLE GLAZED		DOUBLE GLAZED		GLAZED BLOCK
Clear	Tinted	Clear	Tinted	
0.8	0.7	0.7	0.6	0.6

SECTION N1102
BUILDING THERMAL ENVELOPE

N1102.1 Insulation and fenestration criteria. The building thermal envelope shall meet the requirements of Table N1102.1 based on the climate zone specified in Table N1101.2.

N1102.1.1 *R*-value computation. Insulation material used in layers, such as framing cavity insulation and insulating sheathing, shall be summed to compute the component *R*-value. The manufacturer's settled *R*-value shall be used for blown insulation. Computed *R*-values shall not include an *R*-value for other building materials or air films.

N1102.1.2 *U*-factor alternative. An assembly with a *U*-factor equal to or less than that specified in Table N1102.1.2 shall be permitted as an alternative to the *R*-value in Table N1102.1.

Exception: For mass walls not meeting the criterion for insulation location in Section N1102.2.3, the *U*-factor shall be permitted to be:

1. *U*-factor of 0.17 in Climate Zone 1

2. *U*-factor of 0.14 in Climate Zone 2

3. *U*-factor of 0.12 in Climate Zone 3

4. *U*-factor of 0.10 in Climate Zone 4 except Marine

5. *U*-factor of 0.082 in Climate Zone 5 and Marine 4

N1102.1.3 Total UA alternative. If the total building thermal envelope UA (sum of *U*-factor times assembly area) is less than or equal to the total UA resulting from using the *U*-factors in Table N1102.1.2, (multiplied by the same assembly area as in the proposed building), the building shall be considered in compliance with Table N1102.1. The UA calculation shall be done using a method consistent with the ASHRAE *Handbook of Fundamentals* and shall include the thermal bridging effects of framing materials. The SHGC requirements shall be met in addition to UA compliance.

N1102.2 Specific insulation requirements.

N1102.2.1 Ceilings with attic spaces. When Section N1102.1 would require R-38 in the ceiling, R-30 shall be deemed to satisfy the requirement for R-38 wherever the full height of uncompressed R-30 insulation extends over the wall top plate at the eaves. Similarly R-38 shall be deemed to satisfy the requirement for R-49 wherever the full height of uncompressed R-38 insulation extends over the wall top plate at the eaves.

N1102.2.2 Ceilings without attic spaces. Where Section N1102.1 would require insulation levels above R-30 and the design of the roof/ceiling assembly does not allow sufficient space for the required insulation, the minimum required insulation for such roof/ceiling assemblies shall be R-30. This reduction of insulation from the requirements of Section N1102.1 shall be limited to 500 ft² (46 m²) of ceiling area.

TABLE N1102.1
INSULATION AND FENESTRATION REQUIREMENTS BY COMPONENT[a]

CLIMATE ZONE	FENESTRATION U-FACTOR	SKYLIGHT[b] U-FACTOR	GLAZED FENESTRATION SHGC	CEILING R-VALUE	WOOD FRAME WALL R-VALUE	MASS WALL R-VALUE	FLOOR R-VALUE	BASEMENT[c] WALL R-VALUE	SLAB[d] R-VALUE AND DEPTH	CRAWL SPACE[c] WALL R-VALUE
1	1.2	0.75	0.40	30	13	3	13	0	0	0
2	0.75	0.75	0.40	30	13	4	13	0	0	0
3	0.65	0.65	0.40[e]	30	13	5	19	0	0	5/13
4 except Marine	0.40	0.60	NR	38	13	5	19	10/13	10, 2 ft	10/13
5 and Marine 4	0.35	0.60	NR	38	19 or 13 + 5[g]	13	30[f]	10/13	10, 2 ft	10/13
6	0.35	0.60	NR	49	19 or 13 + 5[g]	15	30[f]	10/13	10, 4 ft	10/13
7 and 8	0.35	0.60	NR	49	21	19	30[f]	10/13	10, 4 ft	10/13

a. *R*-values are minimums. *U*-factors and SHGC are maximums. R-19 insulation shall be permitted to be compressed into a 2 × 6 cavity.

b. The fenestration *U*-factor column excludes skylights. The solar heat gain coefficient (SHGC) column applies to all glazed fenestration.

c. The first *R*-value applies to continuous insulation, the second to framing cavity insulation; either insulation meets the requirement.

d. R-5 shall be added to the required slab edge *R*-values for heated slabs.

e. There are no solar heat gain coefficient (SHGC) requirements in the Marine Zone.

f. Or insulation sufficient to fill the framing cavity, R-19 minimum.

g. "13+5" means R-13 cavity insulation plus R-5 insulated sheathing. If structural sheathing covers 25% or less of the exterior, R-5 sheathing is not required where structural sheathing is used. If structural sheathing covers more than 25% of exterior, structural sheathing shall be supplemented with insulated sheathing of at least R-2.

TABLE N1102.1.2
EQUIVALENT U-FACTORS[a]

CLIMATE ZONE	FENESTRATION U-FACTOR	SKYLIGHT U-FACTOR	CEILING U-FACTOR	FRAME WALL U-FACTOR	MASS WALL U-FACTOR	FLOOR U-FACTOR	BASEMENT WALL U-FACTOR	CRAWL SPACE WALL U-FACTOR
1	1.20	0.75	0.035	0.082	0.197	0.064	0.360	0.477
2	0.75	0.75	0.035	0.082	0.165	0.064	0.360	0.477
3	0.65	0.65	0.035	0.082	0.141	0.047	0.360	0.136
4 except Marine	0.40	0.60	0.030	0.082	0.141	0.047	0.059	0.065
5 and Marine 4	0.35	0.60	0.030	0.060	0.082	0.033	0.059	0.065
6	0.35	0.60	0.026	0.060	0.06	0.033	0.059	0.065
7 and 8	0.35	0.60	0.026	0.057	0.057	0.033	0.059	0.065

a. Nonfenestration *U*-factors shall be obtained from measurement, calculation or an approved source.

N1102.2.3 Mass walls. Mass walls, for the purposes of this chapter, shall be considered walls of concrete block, concrete, insulated concrete form (ICF), masonry cavity, brick (other than brick veneer), earth (adobe, compressed earth block, rammed earth) and solid timber/logs. The provisions of Section N1102.1 for mass walls shall be applicable when at least 50 percent of the required insulation *R*-value is on the exterior of, or integral to, the wall. Walls that do not meet this criterion for insulation placement shall meet the wood frame wall insulation requirements of Section N1102.1.

Exception: For walls that do not meet this criterion for insulation placement, the minimum added insulation R-value shall be permitted to be:

1. *R*-value of 4 in Climate Zone 1

2. *R*-value of 6 in Climate Zone 2

3. *R*-value of 8 in Climate Zone 3

4. *R*-value of 10 in Climate Zone 4 except Marine

5. *R*-value of 13 in climate Zone 5 and Marine 4

N1102.2.4 Steel-frame ceilings, walls and floors. Steel-frame ceilings, walls and floors shall meet the insulation requirements of Table N1102.2.4 or shall meet the *U*-factor requirements in Table N1102.1.2. The calculation of the *U*-factor for a steel-frame envelope assembly shall use a series-parallel path calculation method.

TABLE N1102.2.4
STEEL-FRAME CEILING, WALL AND FLOOR INSULATION (*R*-VALUE)

WOOD FRAME *R*-VALUE REQUIREMENT	COLD-FORMED STEEL EQUIVALENT *R*-VALUE[a]
Steel Truss Ceilings[a]	
R-30	R-38 or R-30 + 3 or R-26 + 5
R-38	R-49 or R-38 + 3
R-49	R-38 + 5
Steel Joist Ceilings[b]	
R-30	R-38 in 2 × 4 or 2 × 6 or 2 × 8 R-49 in any framing
R-38	R-49 in 2 × 4 or 2 × 6 or 2 × 8 or 2 × 10
Steel Framed Wall	
R-13	R-13 + 5 or R-15 + 4 or R-21 + 3
R-19	R-13 + 9 or R-19 + 8 or R-25 + 7
R-21	R-13 +10 or R-19 + 9 or R-25 + 8
Steel Joist Floor	
R-13	R-19 in 2 × 6 R-19 + R-6 in 2 × 8 or 2 ×10
R-19	R-19 + R-6 in 2 × 6 R-19 + R-12 in 2 × 8 or 2 × 10

For SI: 1 inch = 25.4 mm.
a. Cavity insulation *R*-value is listed first, followed by continuous insulation *R*-value.
b. Insulation exceeding the height of the framing shall cover the framing.

N1102.2.5 Floors. Floor insulation shall be installed to maintain permanent contact with the underside of the subfloor decking.

N1102.2.6 Basement walls. Exterior walls associated with conditioned basements shall be insulated from the top of the basement wall down to 10 feet (3048 mm) below grade or to the basement floor, whichever is less. Walls associated with unconditioned basements shall meet this requirement unless the floor overhead is insulated in accordance with Sections N1102.1 and N1102.2.5.

N1102.2.7 Slab-on-grade floors. Slab-on-grade floors with a floor surface less than 12 inches below grade shall be insulated in accordance with Table N1102.1. The insulation shall extend downward from the top of the slab on the outside or inside of the foundation wall. Insulation located below grade shall be extended the distance provided in Table N1102.1 by any combination of vertical insulation, insulation extending under the slab or insulation extending out from the building. Insulation extending away from the building shall be protected by pavement or by a minimum of 10 inches (254 mm) of soil. The top edge of the insulation installed between the exterior wall and the edge of the interior slab shall be permitted to be cut at a 45-degree (0.79 rad) angle away from the exterior wall. Slab-edge insulation is not required in jurisdictions designated by the code official as having a very heavy termite infestation.

N1102.2.8 Crawl space walls. As an alternative to insulating floors over crawl spaces, insulation of crawl space walls when the crawl space is not vented to the outside is permitted.

Crawl space wall insulation shall be permanently fastened to the wall and extend downward from the floor to the finished grade level and then vertically and/or horizontally for at least an additional 24 inches (610 mm). Exposed earth in unvented crawl space foundations shall be covered with a continuous vapor retarder. All joints of the vapor retarder shall overlap by 6 inches (152 mm) and be sealed or taped. The edges of the vapor retarder shall extend at least 6 inches (152 mm) up the stem wall and shall be attached to the stem wall.

N1102.2.9 Masonry veneer. Insulation shall not be required on the horizontal portion of the foundation that supports a masonry veneer.

N1102.2.10 Thermally isolated sunroom insulation. The minimum ceiling insulation *R*-values shall be R-19 in zones 1 through 4 and R-24 in zones 5 though 8. The minimum wall *R*-value shall be R-13 in all zones. New wall(s) separating the sunroom from conditioned space shall meet the building thermal envelope requirements.

N1102.3 Fenestration.

N1102.3.1 *U*-factor. An area-weighted average of fenestration products shall be permitted to satisfy the *U*-factor requirements.

N1102.3.2 Glazed fenestration SHGC. An area-weighted average of fenestration products more than 50 percent glazed shall be permitted to satisfy the solar heat gain coefficient (SHGC) requirements.

N1102.3.3 Glazed fenestration exemption. Up to 15 square feet (1.4 m²) of glazed fenestration per dwelling unit shall be permitted to be exempt from *U*-factor and solar heat gain coefficient (SHGC) requirements in Section N1102.1.

N1102.3.4 Opaque door exemption. One opaque door assembly is exempted from the *U*-factor requirement in Section N1102.1.

N1102.3.5 Thermally isolated sunroom *U*-factor. For zones 4 through 8 the maximum fenestration *U*-factor shall be 0.50 and the maximum skylight *U*-factor shall be 0.75. New windows and doors separating the sunroom from conditioned space shall meet the building thermal envelope requirements.

N1102.3.6 Replacement fenestration. Where some or all of an existing fenestration unit is replaced with a new fenestration product, including sash and glazing, the replacement fenestration unit shall meet the applicable requirements for *U*-factor and solar heat gain coefficient (SHGC) in Table N1102.1.

N1102.4 Air leakage.

N1102.4.1 Building thermal envelope. The building thermal envelope shall be durably sealed to limit infiltration. The sealing methods between dissimilar materials shall allow for differential expansion and contraction. The following shall be caulked, gasketed, weatherstripped or otherwise sealed with an air barrier material, suitable film or solid material.

1. All joints, seams and penetrations.

2. Site-built windows, doors and skylights.

3. Openings between window and door assemblies and their respective jambs and framing.

4. Utility penetrations.

5. Dropped ceilings or chases adjacent to the thermal envelope.

6. Knee walls.

7. Walls and ceilings separating the garage from conditioned spaces.

8. Behind tubs and showers on exterior walls.

9. Common walls between dwelling units.

10. Other sources of infiltration.

N1102.4.2 Fenestration air leakage. Windows, skylights and sliding glass doors shall have an air infiltration rate of no more than 0.3 cubic foot per minute per square foot [1.5(L/s)/m^2], and swinging doors no more than 0.5 cubic foot per minute per square foot [2.5(L/s)/m^2], when tested according to NFRC 400 or AAMA/WDMA/CSA 101/I.S.2/A440 by an accredited, independent laboratory, and listed and labeled by the manufacturer.

Exception: Site-built windows, skylights and doors.

N1102.4.3 Recessed lighting. Recessed luminaires installed in the building thermal envelope shall be sealed to limit air leakage between conditioned and unconditioned spaces by being:

1. IC-rated and labeled with enclosures that are sealed or gasketed to prevent air leakage to the ceiling cavity or unconditioned space; or

2. IC-rated and labeled as meeting ASTM E 283 when tested at 1.57 pounds per square foot (75 Pa) pressure differential with no more than 2.0 cubic feet per minute (0.944 L/s) of air movement from the conditioned space to the ceiling cavity; or

3. Located inside an airtight sealed box with clearances of at least 0.5 inch (13 mm) from combustible material and 3 inches (76 mm) from insulation.

N1102.5 Moisture control. The building design shall not create conditions of accelerated deterioration from moisture condensation. Above-grade frame walls, floors and ceilings not ventilated to allow moisture to escape shall be provided with an approved vapor retarder. The vapor retarder shall be installed on the warm-in-winter side of the thermal insulation.

Exceptions:

1. In construction where moisture or its freezing will not damage the materials.

2. Frame walls, floors and ceilings in jurisdictions in Zones 1, 2, 3, 4A, and 4B. (Crawl space floor vapor retarders are not exempted.)

3. Where other approved means to avoid condensation are provided.

N1102.5.1 Maximum fenestration *U*-factor. The area weighted average maximum fenestration *U*-factor permitted using tradeoffs from Section N1102.1.3 in Zones 6 through 8 shall be 0.55.

To comply with this section, the maximum *U*-factor for skylights shall be 0.75 in zones 6 through 8.

SECTION N1103
SYSTEMS

N1103.1 Controls. At least one thermostat shall be installed for each separate heating and cooling system.

N1103.1.1 Heat pump supplementary heat. Heat pumps having supplementary electric-resistance heat shall have controls that, except during defrost, prevent supplemental heat operation when the heat pump compressor can meet the heating load.

N1103.2 Ducts.

N1103.2.1 Insulation. Supply and return ducts shall be insulated to a minimum of R-8. Ducts in floor trusses shall be insulated to a minimum of R-6.

Exception: Ducts or portions thereof located completely inside the building thermal envelope.

N1103.2.2 Sealing. Ducts, air handlers, filter boxes and building cavities used as ducts shall be sealed. Joints and seams shall comply with Section M1601.3.1.

N1103.2.3 Building cavities. Building framing cavities shall not be used as supply ducts.

N1103.3 Mechanical system piping insulation. Mechanical system piping capable of carrying fluids above 105°F (40°C) or below 55°F (13°C) shall be insulated to a minimum of R-2.

N1103.4 Circulating hot water systems. All circulating service hot water piping shall be insulated to at least R-2. Circulating hot water systems shall include an automatic or readily accessible manual switch that can turn off the hot water circulating pump when the system is not in use.

N1103.5 Mechanical ventilation. Outdoor air intakes and exhausts shall have automatic or gravity dampers that close when the ventilation system is not operating.

N1103.6 Equipment sizing. Heating and cooling equipment shall be sized as specified in Section M1401.3.

Part V — Mechanical

CHAPTER 12

MECHANICAL ADMINISTRATION

SECTION M1201
GENERAL

M1201.1 Scope. The provisions of Chapters 12 through 24 shall regulate the design, installation, maintenance, alteration and inspection of mechanical systems that are permanently installed and used to control environmental conditions within buildings. These chapters shall also regulate those mechanical systems, system components, equipment and appliances specifically addressed in this code.

M1201.2 Application. In addition to the general administration requirements of Chapter 1, the administrative provisions of this chapter shall also apply to the mechanical requirements of Chapters 13 through 24.

[EB] SECTION M1202
EXISTING MECHANICAL SYSTEMS

M1202.1 Additions, alterations or repairs. Additions, alterations, renovations or repairs to a mechanical system shall conform to the requirements for a new mechanical system without requiring the existing mechanical system to comply with all of the requirements of this code. Additions, alterations or repairs shall not cause an existing mechanical system to become unsafe, hazardous or overloaded. Minor additions, alterations or repairs to existing mechanical systems shall meet the provisions for new construction, unless such work is done in the same manner and arrangement as was in the existing system, is not hazardous, and is approved.

M1202.2 Existing installations. Except as otherwise provided for in this code, a provision in this code shall not require the removal, alteration or abandonment of, nor prevent the continued use and maintenance of, an existing mechanical system lawfully in existence at the time of the adoption of this code.

M1202.3 Maintenance. Mechanical systems, both existing and new, and parts thereof shall be maintained in proper operating condition in accordance with the original design and in a safe and sanitary condition. Devices or safeguards that are required by this code shall be maintained in compliance with the code edition under which installed. The owner or the owner's designated agent shall be responsible for maintenance of the mechanical systems. To determine compliance with this provision, the building official shall have the authority to require a mechanical system to be reinspected.

CHAPTER 13

GENERAL MECHANICAL SYSTEM REQUIREMENTS

SECTION M1301
GENERAL

M1301.1 Scope. The provisions of this chapter shall govern the installation of mechanical systems not specifically covered in other chapters applicable to mechanical systems. Installations of mechanical appliances, equipment and systems not addressed by this code shall comply with the applicable provisions of the *International Mechanical Code* and the *International Fuel Gas Code*.

M1301.1.1 Flood-resistant installation. In areas prone to flooding as established by Table R301.2(1), mechanical appliances, equipment and systems shall be located or installed in accordance with Section R324.1.5.

SECTION M1302
APPROVAL

M1302.1 Listed and labeled. Appliances regulated by this code shall be listed and labeled for the application in which they are installed and used, unless otherwise approved in accordance with Section R104.11.

SECTION M1303
LABELING OF APPLIANCES

M1303.1 Label information. A permanent factory-applied nameplate(s) shall be affixed to appliances on which shall appear, in legible lettering, the manufacturer's name or trademark, the model number, a serial number and the seal or mark of the testing agency. A label shall also include the following:

1. Electrical appliances. Electrical rating in volts, amperes and motor phase; identification of individual electrical components in volts, amperes or watts and motor phase; and in Btu/h (W) output and required clearances.

2. Absorption units. Hourly rating in Btu/h (W), minimum hourly rating for units having step or automatic modulating controls, type of fuel, type of refrigerant, cooling capacity in Btu/h (W) and required clearances.

3. Fuel-burning units. Hourly rating in Btu/h (W), type of fuel approved for use with the appliance and required clearances.

4. Electric comfort heating appliances. Name and trademark of the manufacturer; the model number or equivalent; the electric rating in volts, amperes and phase; Btu/h (W) output rating; individual marking for each electrical component in amperes or watts, volts and phase; required clearances from combustibles and a seal indicating approval of the appliance by an approved agency.

5. Maintenance instructions. Required regular maintenance actions and title or publication number for the operation and maintenance manual for that particular model and type of product.

SECTION M1304
TYPE OF FUEL

M1304.1 Fuel types. Fuel-fired appliances shall be designed for use with the type of fuel to which they will be connected and the altitude at which they are installed. Appliances that comprise parts of the building mechanical system shall not be converted for the use of a different fuel, except where approved and converted in accordance with the manufacturer's instructions. The fuel input rate shall not be increased or decreased beyond the limit rating for the altitude at which the appliance is installed.

SECTION M1305
APPLIANCE ACCESS

M1305.1 Appliance access for inspection service, repair and replacement. Appliances shall be accessible for inspection, service, repair and replacement without removing permanent construction, other appliances, or any other piping or ducts not connected to the appliance being inspected, serviced, repaired or replaced. A level working space at least 30 inches deep and 30 inches wide (762 mm by 762 mm) shall be provided in front of the control side to service an appliance. Installation of room heaters shall be permitted with at least an 18-inch (457 mm) working space. A platform shall not be required for room heaters.

M1305.1.1 Central furnaces. Central furnaces within compartments or alcoves shall have a minimum working space clearance of 3 inches (76 mm) along the sides, back and top with a total width of the enclosing space being at least 12 inches (305 mm) wider than the furnace. Furnaces having a firebox open to the atmosphere shall have at least a 6-inch (152 mm) working space along the front combustion chamber side. Combustion air openings at the rear or side of the compartment shall comply with the requirements of Chapter 17.

Exception: This section shall not apply to replacement appliances installed in existing compartments and alcoves where the working space clearances are in accordance with the equipment or appliance manufacturer's installation instructions.

M1305.1.2 Appliances in rooms. Appliances installed in a compartment, alcove, basement or similar space shall be accessed by an opening or door and an unobstructed passageway measuring not less than 24 inches (610 mm) wide and large enough to allow removal of the largest appliance in the space, provided there is a level service space of not less than 30 inches (762 mm) deep and the height of the appliance, but not less than 30 inches (762 mm), at the front or service side of the appliance with the door open.

M1305.1.3 Appliances in attics. Attics containing appliances requiring access shall be provided with an opening and a clear and unobstructed passageway large enough to allow removal of the largest appliance, but not less than 30 inches (762 mm) high and 22 inches (559 mm) wide and not more than 20 feet (6096 mm) long when measured along the centerline of the passageway from the opening to the appliance. The passageway shall have continuous solid flooring in accordance with Chapter 5 not less than 24 inches (610 mm) wide. A level service space at least 30 inches (762 mm) deep and 30 inches (762 mm) wide shall be present along all sides of the appliance where access is required. The clear access opening dimensions shall be a minimum of 20 inches by 30 inches (508 mm by 762 mm), where such dimensions are large enough to allow removal of the largest appliance.

Exceptions:

1. The passageway and level service space are not required where the appliance can be serviced and removed through the required opening.

2. Where the passageway is unobstructed and not less than 6 feet (1829 mm) high and 22 inches (559 mm) wide for its entire length, the passageway shall be not more than 50 feet (15 250 mm) long.

M1305.1.3.1 Electrical requirements. A luminaire controlled by a switch located at the required passageway opening and a receptacle outlet shall be installed at or near the appliance location in accordance with Chapter 38.

M1305.1.4 Appliances under floors. Underfloor spaces containing appliances requiring access shall have an unobstructed passageway large enough to remove the largest appliance, but not less than 30 inches (762 mm) high and 22 inches (559 mm) wide, nor more than 20 feet (6096 mm) long when measured along the centerline of the passageway from the opening to the appliance. A level service space at least 30 inches (762 mm) deep and 30 inches (762 mm) wide shall be present at the front or service side of the appliance. If the depth of the passageway or the service space exceeds 12 inches (305 mm) below the adjoining grade, the walls of the passageway shall be lined with concrete or masonry extending 4 inches (102 mm) above the adjoining grade in accordance with Chapter 4. The rough-framed access opening dimensions shall be a minimum of 22 inches by 30 inches (559 mm by 762 mm), where the dimensions are large enough to remove the largest appliance.

Exceptions:

1. The passageway is not required where the level service space is present when the access is open, and the appliance can be serviced and removed through the required opening.

2. Where the passageway is unobstructed and not less than 6 feet high (1929 mm) and 22 inches wide for its entire length, the passageway shall not be limited in length.

M1305.1.4.1 Ground clearance. Appliances supported from the ground shall be level and firmly supported on a concrete slab or other approved material extending above the adjoining ground. Appliances suspended from the floor shall have a clearance of not less than 6 inches (152 mm) from the ground.

M1305.1.4.2 Excavations. Excavations for appliance installations shall extend to a depth of 6 inches (152 mm) below the appliance and 12 inches (305 mm) on all sides, except that the control side shall have a clearance of 30 inches (762 mm).

M1305.1.4.3 Electrical requirements. A luminaire controlled by a switch located at the required passageway opening and a receptacle outlet shall be installed at or near the appliance location in accordance with Chapter 38.

SECTION M1306
CLEARANCES FROM COMBUSTIBLE CONSTRUCTION

M1306.1 Appliance clearance. Appliances shall be installed with the clearances from unprotected combustible materials as indicated on the appliance label and in the manufacturer's installation instructions.

M1306.2 Clearance reduction. Reduction of clearances shall be in accordance with the appliance manufacturer's instructions and Table M1306.2. Forms of protection with ventilated air space shall conform to the following requirements:

1. Not less than 1-inch (25 mm) air space shall be provided between the protection and combustible wall surface.

2. Air circulation shall be provided by having edges of the wall protection open at least 1 inch (25 mm).

3. If the wall protection is mounted on a single flat wall away from corners, air circulation shall be provided by having the bottom and top edges, or the side and top edges open at least 1 inch (25 mm).

4. Wall protection covering two walls in a corner shall be open at the bottom and top edges at least 1 inch (25 mm).

M1306.2.1 Solid fuel appliances. Table M1306.2 shall not be used to reduce the clearance required for solid-fuel appliances listed for installation with minimum clearances of 12 inches (305 mm) or less. For appliances listed for installation with minimum clearances greater than 12 inches (305 mm), Table M1306.2 shall not be used to reduce the clearance to less than 12 inches (305 mm).

TABLE M1306.2
REDUCTION OF CLEARANCES WITH SPECIFIED FORMS OF PROTECTION[a, b, c, d, e, f, g, h, i, j, k]

TYPE OF PROTECTION APPLIED TO AND COVERING ALL SURFACES OF COMBUSTIBLE MATERIAL WITHIN THE DISTANCE SPECIFIED AS THE REQUIRED CLEARANCE WITH NO PROTECTION [See Figures M1306.1 and M1306.2]	WHERE THE REQUIRED CLEARANCE WITH NO PROTECTION FROM APPLIANCE, VENT CONNECTOR, OR SINGLE WALL METAL PIPE IS:									
	36 inches		18 inches		12 inches		9 inches		6 inches	
	Allowable clearances with specified protection (Inches)[b]									
	Use column 1 for clearances above an appliance or horizontal connector. Use column 2 for clearances from an appliance, vertical connector and single-wall metal pipe.									
	Above column 1	Sides and rear column 2	Above column 1	Sides and rear column 2	Above column 1	Sides and rear column 2	Above column 1	Sides and rear column 2	Above column 1	Sides and rear column 2
3½-inch thick masonry wall without ventilated air space	—	24	—	12	—	9	—	6	—	5
½-in. insulation board over 1-inch glass fiber or mineral wool batts	24	18	12	9	9	6	6	5	4	3
24 gage sheet metal over 1-inch glass fiber or mineral wool batts reinforced with wire on rear face with ventilated air space	18	12	9	6	6	4	5	3	3	3
3½-inch thick masonry wall with ventilated air space	—	12	—	6	—	6	—	6	—	6
24 gage sheet metal with ventilated air space	18	12	9	6	6	4	5	3	3	2
½-inch thick insulation board with ventilated air space	18	12	9	6	6	4	5	3	3	3
24 gage sheet metal with ventilated air space over 24 gage sheet metal with ventilated air space	18	12	9	6	6	4	5	3	3	3
1-inch glass fiber or mineral wool batts sandwiched between two sheets 24 gage sheet metal with ventilated air space.	18	12	9	6	6	4	5	3	3	3

For SI: 1 inch = 25.4 mm, 1 pound per cubic foot = 16.019 kg/m^3, °C = [(°F)-32/1.8], 1 Btu/(h · ft^2 · °F/in.) = 0.001442299 (W/cm^2 · °C/cm).

a. Reduction of clearances from combustible materials shall not interfere with combustion air, draft hood clearance and relief, and accessibility of servicing.

b. Clearances shall be measured from the surface of the heat producing appliance or equipment to the outer surface of the combustible material or combustible assembly.

c. Spacers and ties shall be of noncombustible material. No spacer or tie shall be used directly opposite appliance or connector.

d. Where all clearance reduction systems use a ventilated air space, adequate provision for air circulation shall be provided as described. (See Figures M1306.1 and M1306.2.)

e. There shall be at least 1 inch between clearance reduction systems and combustible walls and ceilings for reduction systems using ventilated air space.

f. If a wall protector is mounted on a single flat wall away from corners, adequate air circulation shall be permitted to be provided by leaving only the bottom and top edges or only the side and top edges open with at least a 1-inch air gap.

g. Mineral wool and glass fiber batts (blanket or board) shall have a minimum density of 8 pounds per cubic foot and a minimum melting point of 1,500°F.

h. Insulation material used as part of a clearance reduction system shall have a thermal conductivity of 1.0 Btu inch per square foot per hour °F or less. Insulation board shall be formed of noncombustible material.

i. There shall be at least 1 inch between the appliance and the protector. In no case shall the clearance between the appliance and the combustible surface be reduced below that allowed in this table.

j. All clearances and thicknesses are minimum; larger clearances and thicknesses are acceptable.

k. Listed single-wall connectors shall be permitted to be installed in accordance with the terms of their listing and the manufacturer's instructions.

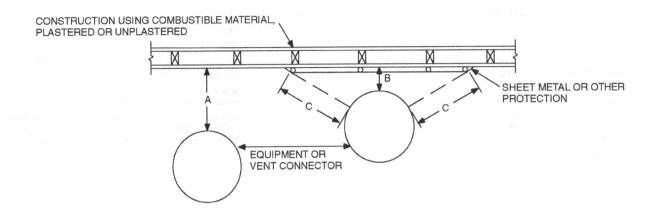

NOTE: "A" equals the required clearance with no protection. "B" equals the reduced clearance permitted in accordance with Table M1306.2. The protection applied to the construction using combustible material shall extend far enough in each direction to make "C" equal to "A."

FIGURE M1306.1
REDUCED CLEARANCE DIAGRAM

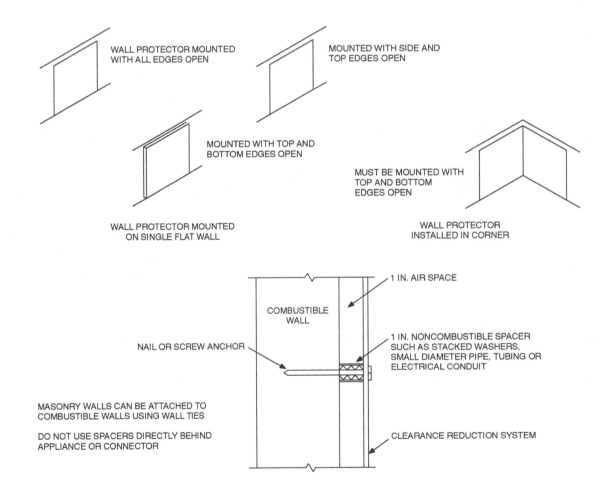

For SI: 1 inch = 25.4 mm.

FIGURE M1306.2
WALL PROTECTOR CLEARANCE REDUCTION SYSTEM

SECTION M1307
APPLIANCE INSTALLATION

M1307.1 General. Installation of appliances shall conform to the conditions of their listing and label and the manufacturer's installation instructions. The manufacturer's operating and installation instructions shall remain attached to the appliance.

M1307.2 Anchorage of appliances. Appliances designed to be fixed in position shall be fastened or anchored in an approved manner. In Seismic Design Categories D_1 and D_2, water heaters shall be anchored or strapped to resist horizontal displacement caused by earthquake motion. Strapping shall be at points within the upper one-third and lower one-third of the appliance's vertical dimensions. At the lower point, the strapping shall maintain a minimum distance of 4 inches (102 mm) above the controls.

M1307.3 Elevation of ignition source. Appliances having an ignition source shall be elevated such that the source of ignition is not less than 18 inches (457 mm) above the floor in garages. For the purpose of this section, rooms or spaces that are not part of the living space of a dwelling unit and that communicate with a private garage through openings shall be considered to be part of the garage.

M1307.3.1 Protection from impact. Appliances located in a garage or carport shall be protected from impact by automobiles.

M1307.4 Hydrogen generating and refueling operations. Ventilation shall be required in accordance with Section M1307.4.1, M1307.4.2 or M1307.4.3 in private garages that contain hydrogen-generating appliances or refueling systems. For the purpose of this section, rooms or spaces that are not part of the living space of a dwelling unit and that communicate directly with a private garage through openings shall be considered to be part of the private garage.

M1307.4.1 Natural ventilation. Indoor locations intended for hydrogen-generating or refueling operations shall be limited to a maximum floor area of 850 square feet (79 m²) and shall communicate with the outdoors in accordance with Sections M1307.4.1.1 and M1307.4.1.2. The maximum rated output capacity of hydrogen generating appliances shall not exceed 4 standard cubic feet per minute (1.9 L/s) of hydrogen for each 250 square feet (23 m²) of floor area in such spaces. The minimum cross-sectional dimension of air openings shall be 3 inches (76 mm). Where ducts are used, they shall be of the same cross-sectional area as the free area of the openings to which they connect. In those locations, equipment and appliances having an ignition source shall be located so that the source of ignition is not within 12 inches (305 mm) of the ceiling.

M1307.4.1.1 Two openings. Two permanent openings shall be constructed within the garage. The upper opening shall be located entirely within 12 inches (305 mm) of the ceiling of the garage. The lower opening shall be located entirely within 12 inches (305 mm) of the floor of the garage. Both openings shall be constructed in the same exterior wall. The openings shall communicate directly with the outdoors and shall have a minimum free area of ¹/₂ square foot per 1,000 cubic feet (1.7 m²/1000 m³) of garage volume.

M1307.4.1.2 Louvers and grilles. In calculating free area required by Section M1307.4.1, the required size of openings shall be based on the net free area of each opening. If the free area through a design of louver or grille is known, it shall be used in calculating the size opening required to provide the free area specified. If the design and free area are not known, it shall be assumed that wood louvers will have a 25-percent free area and metal louvers and grilles will have a 75-percent free area. Louvers and grilles shall be fixed in the open position.

M1307.4.2 Mechanical ventilation. Indoor locations intended for hydrogen-generating or refueling operations shall be ventilated in accordance with Section 502.16 of the *International Mechanical Code*. In these locations, equipment and appliances having an ignition source shall be located so that the source of ignition is below the mechanical ventilation outlet(s).

M1307.4.3 Specially engineered installations. As an alternative to the provisions of Sections M1307.4.1 and M1307.4.2, the necessary supply of air for ventilation and dilution of flammable gases shall be provided by an approved engineered system.

M1307.5 Electrical appliances. Electrical appliances shall be installed in accordance with Chapters 14, 15, 19, 20 and 33 through 42 of this code.

SECTION M1308
MECHANICAL SYSTEMS INSTALLATION

M1308.1 Drilling and notching. Wood-framed structural members shall be drilled, notched or altered in accordance with the provisions of Sections R502.8, R602.6, R602.6.1 and R802.7. Holes in cold-formed, steel-framed, load-bearing members shall be permitted only in accordance with Sections R505.2, R603.2 and R804.2. In accordance with the provisions of Sections R505.3.5, R603.3.4 and R804.3.5, cutting and notching of flanges and lips of cold-formed, steel-framed, load-bearing members shall not be permitted.

M1308.2 Protection against physical damage. In concealed locations where piping, other than cast-iron or galvanized steel, is installed through holes or notches in studs, joists, rafters or similar members less than 1.5 inches (38 mm) from the nearest edge of the member, the pipe shall be protected by shield plates. Protective shield plates shall be a minimum of 0.062-inch-thick (1.6 mm) steel, shall cover the area of the pipe where the member is notched or bored, and shall extend a minimum of 2 inches (51 mm) above sole plates and below top plates.

M1308.3 Foundations and supports. Foundations and supports for outdoor mechanical systems shall be raised at least 3 inches (76 mm) above the finished grade, and shall also conform to the manufacturer's installation instructions.

CHAPTER 14

HEATING AND COOLING EQUIPMENT

SECTION M1401
GENERAL

M1401.1 Installation. Heating and cooling equipment and appliances shall be installed in accordance with the manufacturer's installation instructions and the requirements of this code.

M1401.2 Access. Heating and cooling equipment shall be located with respect to building construction and other equipment to permit maintenance, servicing and replacement. Clearances shall be maintained to permit cleaning of heating and cooling surfaces; replacement of filters, blowers, motors, controls and vent connections; lubrication of moving parts; and adjustments.

M1401.3 Sizing. Heating and cooling equipment shall be sized based on building loads calculated in accordance with ACCA Manual J or other approved heating and cooling calculation methodologies.

M1401.4 Exterior installations. Equipment installed outdoors shall be listed and labeled for outdoor installation. Supports and foundations shall prevent excessive vibration, settlement or movement of the equipment. Supports and foundations shall be level and conform to the manufacturer's installation instructions.

M1401.5 Flood hazard. In areas prone to flooding as established by Table R301.2(1), heating and cooling equipment and appliances shall be located or installed in accordance with Section R324.1.5.

SECTION M1402
CENTRAL FURNACES

M1402.1 General. Oil-fired central furnaces shall conform to ANSI/UL 727. Electric furnaces shall conform to UL 1995.

M1402.2 Clearances. Clearances shall be provided in accordance with the listing and the manufacturer's installation instructions.

M1402.3 Combustion air. Combustion air shall be supplied in accordance with Chapter 17. Combustion air openings shall be unobstructed for a distance of not less than 6 inches (152 mm) in front of the openings.

SECTION M1403
HEAT PUMP EQUIPMENT

M1403.1 Heat pumps. The minimum unobstructed total area of the outside and return air ducts or openings to a heat pump shall be not less than 6 square inches per 1,000 Btu/h (13 208 mm^2/kW) output rating or as indicated by the conditions of the listing of the heat pump. Electric heat pumps shall conform to UL 1995.

M1403.2 Foundations and supports. Supports and foundations for the outdoor unit of a heat pump shall be raised at least 3 inches (76 mm) above the ground to permit free drainage of defrost water, and shall conform to the manufacturer's installation instructions.

SECTION M1404
REFRIGERATION COOLING EQUIPMENT

M1404.1 Compliance. Refrigeration cooling equipment shall comply with Section M1411.

SECTION M1405
BASEBOARD CONVECTORS

M1405.1 General. Electric baseboard convectors shall be installed in accordance with the manufacturer's installation instructions and Chapters 33 through 42 of this code.

SECTION M1406
RADIANT HEATING SYSTEMS

M1406.1 General. Electric radiant heating systems shall be installed in accordance with the manufacturer's installation instructions and Chapters 33 through 42 of this code.

M1406.2 Clearances. Clearances for radiant heating panels or elements to any wiring, outlet boxes and junction boxes used for installing electrical devices or mounting luminaires shall comply with Chapters 33 through 42 of this code.

M1406.3 Installation of radiant panels. Radiant panels installed on wood framing shall conform to the following requirements:

1. Heating panels shall be installed parallel to framing members and secured to the surface of framing members or mounted between framing members.

2. Panels shall be nailed or stapled only through the unheated portions provided for this purpose and shall not be fastened at any point closer than $^1/_4$ inch (7 mm) to an element.

3. Unless listed and labeled for field cutting, heating panels shall be installed as complete units.

M1406.4 Installation in concrete or masonry. Radiant heating systems installed in concrete or masonry shall conform to the following requirements:

1. Radiant heating systems shall be identified as being suitable for the installation, and shall be secured in place as specified in the manufacturer's installation instructions.

2. Radiant heating panels or radiant heating panel sets shall not be installed where they bridge expansion joints unless protected from expansion and contraction.

M1406.5 Gypsum panels. Where radiant heating systems are used on gypsum assemblies, operating temperatures shall not exceed 125°F (52°C).

M1406.6 Finish surfaces. Finish materials installed over radiant heating panels or systems shall be installed in accordance with the manufacturer's installation instructions. Surfaces shall be secured so that nails or other fastenings do not pierce the radiant heating elements.

SECTION M1407
DUCT HEATERS

M1407.1 General. Electric duct heaters shall be installed in accordance with the manufacturer's installation instructions and Chapters 33 through 42 of this code. Electric furnaces shall be tested in accordance with UL 1995.

M1407.2 Installation. Electric duct heaters shall be installed so that they will not create a fire hazard. Class 1 ducts, duct coverings and linings shall be interrupted at each heater to provide the clearances specified in the manufacturer's installation instructions. Such interruptions are not required for duct heaters listed and labeled for zero clearance to combustible materials. Insulation installed in the immediate area of each heater shall be classified for the maximum temperature produced on the duct surface.

M1407.3 Installation with heat pumps and air conditioners. Duct heaters located within 4 feet (1219 mm) of a heat pump or air conditioner shall be listed and labeled for such installations. The heat pump or air conditioner shall additionally be listed and labeled for such duct heater installations.

M1407.4 Access. Duct heaters shall be accessible for servicing, and clearance shall be maintained to permit adjustment, servicing and replacement of controls and heating elements.

M1407.5 Fan interlock. The fan circuit shall be provided with an interlock to prevent heater operation when the fan is not operating.

SECTION M1408
VENTED FLOOR FURNACES

M1408.1 General. Vented floor furnaces shall conform to UL 729 and be installed in accordance with their listing, the manufacturer's installation instructions and the requirements of this code.

M1408.2 Clearances. Vented floor furnaces shall be installed in accordance with their listing and the manufacturer's installation instructions.

M1408.3 Location. Location of floor furnaces shall conform to the following requirements:

1. Floor registers of floor furnaces shall be installed not less than 6 inches (152 mm) from a wall.

2. Wall registers of floor furnaces shall be installed not less than 6 inches (152 mm) from the adjoining wall at inside corners.

3. The furnace register shall be located not less than 12 inches (305 mm) from doors in any position, draperies or similar combustible objects.

4. The furnace register shall be located at least 5 feet (1524 mm) below any projecting combustible materials.

5. The floor furnace burner assembly shall not project into an occupied under-floor area.

6. The floor furnace shall not be installed in concrete floor construction built on grade.

7. The floor furnace shall not be installed where a door can swing within 12 inches (305 mm) of the grille opening.

M1408.4 Access. An opening in the foundation not less than 18 inches by 24 inches (457 mm by 610 mm), or a trap door not less than 22 inches by 30 inches (559 mm by 762 mm) shall be provided for access to a floor furnace. The opening and passageway shall be large enough to allow replacement of any part of the equipment.

M1408.5 Installation. Floor furnace installations shall conform to the following requirements:

1. Thermostats controlling floor furnaces shall be located in the room in which the register of the floor furnace is located.

2. Floor furnaces shall be supported independently of the furnace floor register.

3. Floor furnaces shall be installed not closer than 6 inches (152 mm) to the ground. Clearance may be reduced to 2 inches (51 mm), provided that the lower 6 inches (152 mm) of the furnace is sealed to prevent water entry.

4. Where excavation is required for a floor furnace installation, the excavation shall extend 30 inches (762 mm) beyond the control side of the floor furnace and 12 inches (305 mm) beyond the remaining sides. Excavations shall slope outward from the perimeter of the base of the excavation to the surrounding grade at an angle not exceeding 45 degrees (0.79 rad) from horizontal.

5. Floor furnaces shall not be supported from the ground.

SECTION M1409
VENTED WALL FURNACES

M1409.1 General. Vented wall furnaces shall conform to UL 730 and be installed in accordance with their listing, the manufacturer's installation instructions and the requirements of this code.

M1409.2 Location. The location of vented wall furnaces shall conform to the following requirements:

1. Vented wall furnaces shall be located where they will not cause a fire hazard to walls, floors, combustible furnishings or doors. Vented wall furnaces installed between bathrooms and adjoining rooms shall not circulate air from bathrooms to other parts of the building.

2. Vented wall furnaces shall not be located where a door can swing within 12 inches (305 mm) of the furnace air inlet or outlet measured at right angles to the opening.

Doorstops or door closers shall not be installed to obtain this clearance.

M1409.3 Installation. Vented wall furnace installations shall conform to the following requirements:

1. Required wall thicknesses shall be in accordance with the manufacturer's installation instructions.

2. Ducts shall not be attached to a wall furnace. Casing extensions or boots shall be installed only when listed as part of a listed and labeled appliance.

3. A manual shut off valve shall be installed ahead of all controls.

M1409.4 Access. Vented wall furnaces shall be provided with access for cleaning of heating surfaces; removal of burners; replacement of sections, motors, controls, filters and other working parts; and for adjustments and lubrication of parts requiring such attention. Panels, grilles and access doors that must be removed for normal servicing operations shall not be attached to the building construction.

SECTION M1410
VENTED ROOM HEATERS

M1410.1 General. Vented room heaters shall be tested in accordance with ASTM E 1509, UL 896 or UL 1482 and installed in accordance with their listing, the manufacturer's installation instructions and the requirements of this code.

M1410.2 Floor mounting. Room heaters shall be installed on noncombustible floors or approved assemblies constructed of noncombustible materials that extend at least 18 inches (457 mm) beyond the appliance on all sides.

Exceptions:

1. Listed room heaters shall be installed on noncombustible floors, assemblies constructed of noncombustible materials or listed floor protectors with materials and dimensions in accordance with the appliance manufacturer's instructions.

2. Room heaters listed for installation on combustible floors without floor protection shall be installed in accordance with the appliance manufacturer's instructions.

SECTION M1411
HEATING AND COOLING EQUIPMENT

M1411.1 Approved refrigerants. Refrigerants used in direct refrigerating systems shall conform to the applicable provisions of ANSI/ASHRAE 34.

M1411.2 Refrigeration coils in warm-air furnaces. Where a cooling coil is located in the supply plenum of a warm-air furnace, the furnace blower shall be rated at not less than 0.5-inch water column (124 Pa) static pressure unless the furnace is listed and labeled for use with a cooling coil. Cooling coils shall not be located upstream from heat exchangers unless listed and labeled for such use. Conversion of existing furnaces for use with cooling coils shall be permitted provided the fur-

nace will operate within the temperature rise specified for the furnace.

M1411.3 Condensate disposal. Condensate from all cooling coils or evaporators shall be conveyed from the drain pan outlet to an approved place of disposal. Condensate shall not discharge into a street, alley or other areas where it would cause a nuisance.

M1411.3.1 Auxiliary and secondary drain systems. In addition to the requirements of Section M1411.3, a secondary drain or auxiliary drain pan shall be required for each cooling or evaporator coil where damage to any building components will occur as a result of overflow from the equipment drain pan or stoppage in the condensate drain piping. Such piping shall maintain a minimum horizontal slope in the direction of discharge of not less than $^1/_8$ unit vertical in 12 units horizontal (1-percent slope). Drain piping shall be a minimum of $^3/_4$-inch (19 mm) nominal pipe size. One of the following methods shall be used:

1. An auxiliary drain pan with a separate drain shall be installed under the coils on which condensation will occur. The auxiliary pan drain shall discharge to a conspicuous point of disposal to alert occupants in the event of a stoppage of the primary drain. The pan shall have a minimum depth of 1.5 inches (38 mm), shall not be less than 3 inches (76 mm) larger than the unit or the coil dimensions in width and length and shall be constructed of corrosion-resistant material. Metallic pans shall have a minimum thickness of not less than 0.0276-inch (0.7 mm) galvanized sheet metal. Nonmetallic pans shall have a minimum thickness of not less than 0.0625 inch (1.6 mm).

2. A separate overflow drain line shall be connected to the drain pan provided with the equipment. This overflow drain shall discharge to a conspicuous point of disposal to alert occupants in the event of a stoppage of the primary drain. The overflow drain line shall connect to the drain pan at a higher level than the primary drain connection.

3. An auxiliary drain pan without a separate drain line shall be installed under the coils on which condensate will occur. This pan shall be equipped with a water level detection device conforming to UL 508 that will shut off the equipment served prior to overflow of the pan. The auxiliary drain pan shall be constructed in accordance with Item 1 of this section.

4. A water level detection device conforming to UL 508 shall be provided that will shut off the equipment served in the event that the primary drain is blocked. The device shall be installed in the primary drain line, the overflow drain line or the equipment-supplied drain pan, located at a point higher than the primary drain line connection and below the overflow rim of such pan.

M1411.3.1.1 Water level monitoring devices. On down-flow units and all other coils that have no secondary drain and no means to install an auxiliary drain pan, a water-level monitoring device shall be installed inside the primary drain pan. This device shall shut off the equipment served in the event that the primary drain becomes restricted. Externally installed devices and devices installed in the drain line shall not be permitted

M1411.3.2 Drain pipe materials and sizes. Components of the condensate disposal system shall be cast iron, galvanized steel, copper, polybutylene, polyethylene, ABS, CPVC or PVC pipe or tubing. All components shall be selected for the pressure and temperature rating of the installation. Condensate waste and drain line size shall be not less than $^3/_4$-inch (19 mm) internal diameter and shall not decrease in size from the drain pan connection to the place of condensate disposal. Where the drain pipes from more than one unit are manifolded together for condensate drainage, the pipe or tubing shall be sized in accordance with an approved method. All horizontal sections of drain piping shall be installed in uniform alignment at a uniform slope.

M1411.4 Auxiliary drain pan. Category IV condensing appliances shall have an auxiliary drain pan where damage to any building component will occur as a result of stoppage in the condensate drainage system. These pans shall be installed in accordance with the applicable provisions of Section M1411.3.

> **Exception:** Fuel-fired appliances that automatically shut down operation in the event of a stoppage in the condensate drainage system.

M1411.5 Insulation of refrigerant piping. Piping and fittings for refrigerant vapor (suction) lines shall be insulated with insulation having a thermal resistivity of at least R-4 and having external surface permeance not exceeding 0.05 perm [2.87 ng/(s · m^2 · Pa)] when tested in accordance with ASTM E 96.

SECTION M1412
ABSORPTION COOLING EQUIPMENT

M1412.1 Approval of equipment. Absorption systems shall be installed in accordance with the manufacturer's installation instructions.

M1412.2 Condensate disposal. Condensate from the cooling coil shall be disposed of as provided in Section M1411.3.

M1412.3 Insulation of piping. Refrigerant piping, brine piping and fittings within a building shall be insulated to prevent condensation from forming on piping.

M1412.4 Pressure-relief protection. Absorption systems shall be protected by a pressure-relief device. Discharge from the pressure-relief device shall be located where it will not create a hazard to persons or property.

SECTION M1413
EVAPORATIVE COOLING EQUIPMENT

M1413.1 General. Cooling equipment that uses evaporation of water for cooling shall be installed in accordance with the manufacturer's installation instructions. Evaporative coolers shall be installed on a level platform or base not less than 3 inches (76 mm) above the adjoining ground and secured to prevent displacement. Openings in exterior walls shall be flashed in accordance with Section R703.8.

M1413.2 Protection of potable water. The potable water system shall be protected from backflow in accordance with the provisions in Section P2902.

SECTION 1414
FIREPLACE STOVES

M1414.1 General. Fireplace stoves shall be listed, labeled and installed in accordance with the terms of the listing. Fireplace stoves shall be tested in accordance with UL 737.

M1414.2 Hearth extensions. Hearth extensions for fireplace stoves shall be installed in accordance with the listing of the fireplace stove. The supporting structure for a hearth extension for a fireplace stove shall be at the same level as the supporting structure for the fireplace unit. The hearth extension shall be readily distinguishable from the surrounding floor area.

SECTION M1415
MASONRY HEATERS

M1415.1 General. Masonry heaters shall be constructed in accordance with Section R1002.

CHAPTER 15

EXHAUST SYSTEMS

SECTION M1501
GENERAL

M1501.1 Outdoor discharge. The air removed by every mechanical exhaust system shall be discharged to the outdoors. Air shall not be exhausted into an attic, soffit, ridge vent or crawl space.

> **Exception:** Whole-house ventilation-type attic fans that discharge into the attic space of dwelling units having private attics shall be permitted.

SECTION M1502
CLOTHES DRYER EXHAUST

M1502.1 General. Dryer exhaust systems shall be independent of all other systems, and shall convey the moisture to the outdoors.

> **Exception:** This section shall not apply to listed and labeled condensing (ductless) clothes dryers.

M1502.2 Duct termination. Exhaust ducts shall terminate on the outside of the building. Exhaust duct terminations shall be in accordance with the dryer manufacturer's installation instructions. Exhaust ducts shall terminate not less than 3 feet (914 mm) in any direction from openings into buildings. Exhaust duct terminations shall be equipped with a backdraft damper. Screens shall not be installed at the duct termination.

M1502.3 Duct size. The diameter of the exhaust duct shall be as required by the clothes dryer's listing and the manufacturer's installation instructions.

M1502.4 Transition ducts. Transition ducts shall not be concealed within construction. Flexible transition ducts used to connect the dryer to the exhaust duct system shall be limited to single lengths, not to exceed 8 feet (2438 mm) and shall be listed and labeled in accordance with UL 2158A.

M1502.5 Duct construction. Exhaust ducts shall be constructed of minimum 0.016-inch-thick (0.4 mm) rigid metal ducts, having smooth interior surfaces with joints running in the direction of air flow. Exhaust ducts shall not be connected with sheet-metal screws or fastening means which extend into the duct.

M1502.6 Duct length. The maximum length of a clothes dryer exhaust duct shall not exceed 25 feet (7620 mm) from the dryer location to the wall or roof termination. The maximum length of the duct shall be reduced 2.5 feet (762 mm) for each 45-degree (0.8 rad) bend and 5 feet (1524 mm) for each 90-degree (1.6 rad) bend. The maximum length of the exhaust duct does not include the transition duct.

> **Exceptions:**
>
> 1. Where the make and model of the clothes dryer to be installed is known and the manufacturer's installation instructions for the dryer are provided to the building official, the maximum length of the exhaust duct, including any transition duct, shall be permitted to be in accordance with the dryer manufacturer's installation instructions.
>
> 2. Where large-radius 45-degree (0.8 rad) and 90-degree (1.6 rad) bends are installed, determination of the equivalent length of clothes dryer exhaust duct for each bend by engineering calculation in accordance with the ASHRAE Fundamentals Handbook shall be permitted.

SECTION M1503
RANGE HOODS

M1503.1 General. Range hoods shall discharge to the outdoors through a single-wall duct. The duct serving the hood shall have a smooth interior surface, shall be air tight and shall be equipped with a backdraft damper. Ducts serving range hoods shall not terminate in an attic or crawl space or areas inside the building.

> **Exception:** Where installed in accordance with the manufacturer's installation instructions, and where mechanical or natural ventilation is otherwise provided, listed and labeled ductless range hoods shall not be required to discharge to the outdoors.

M1503.2 Duct material. Single-wall ducts serving range hoods shall be constructed of galvanized steel, stainless steel or copper.

> **Exception:** Ducts for domestic kitchen cooking appliances equipped with down-draft exhaust systems shall be permitted to be constructed of schedule 40 PVC pipe provided that the installation complies with all of the following:
>
> 1. The duct shall be installed under a concrete slab poured on grade; and
>
> 2. The underfloor trench in which the duct is installed shall be completely backfilled with sand or gravel; and
>
> 3. The PVC duct shall extend not more than 1 inch (25 mm) above the indoor concrete floor surface; and
>
> 4. The PVC duct shall extend not more than 1 inch (25 mm) above grade outside of the building; and
>
> 5. The PVC ducts shall be solvent cemented.

M1503.3 Kitchen exhaust rates. Where domestic kitchen cooking appliances are equipped with ducted range hoods or down-draft exhaust systems, the fans shall be sized in accordance with Section M1507.3.

SECTION M1504
INSTALLATION OF MICROWAVE OVENS

M1504.1 Installation of microwave oven over a cooking appliance. The installation of a listed and labeled cooking appliance or microwave oven over a listed and labeled cooking appliance shall conform to the terms of the upper appliance's listing and label and the manufacturer's installation instructions. The microwave oven shall conform to UL 923.

SECTION M1505
OVERHEAD EXHAUST HOODS

M1505.1 General. Domestic open-top broiler units shall be provided with a metal exhaust hood, not less than 28 gage, with $^{1}/_{4}$ inch (6 mm) between the hood and the underside of combustible material or cabinets. A clearance of at least 24 inches (610 mm) shall be maintained between the cooking surface and the combustible material or cabinet. The hood shall be at least as wide as the broiler unit and shall extend over the entire unit. Such exhaust hood shall discharge to the outdoors and shall be equipped with a backdraft damper or other means to control infiltration/exfiltration when not in operation. Broiler units incorporating an integral exhaust system, and listed and labeled for use without an exhaust hood, need not be provided with an exhaust hood.

SECTION M1506
EXHAUST DUCTS

M1506.1 Ducts. Where exhaust duct construction is not specified in this chapter, such construction shall comply with Chapter 16.

SECTION M1507
MECHANICAL VENTILATION

M1507.1 General. Where toilet rooms and bathrooms are mechanically ventilated, the ventilation equipment shall be installed in accordance with this section.

M1507.2 Recirculation of air. Exhaust air from bathrooms and toilet rooms shall not be recirculated within a residence or to another dwelling unit and shall be exhausted directly to the outdoors. Exhaust air from bathrooms and toilet rooms shall not discharge into an attic, crawl space or other areas inside the building.

M1507.3 Ventilation rate. Ventilation systems shall be designed to have the capacity to exhaust the minimum air flow rate determined in accordance with Table M1507.3.

TABLE M1507.3
MINIMUM REQUIRED EXHAUST RATES FOR
ONE- AND TWO-FAMILY DWELLINGS

AREA TO BE VENTILATED	VENTILATION RATES
Kitchens	100 cfm intermittent or 25 cfm continuous
Bathrooms—Toilet Rooms	Mechanical exhaust capacity of 50 cfm intermittent or 20 cfm continuous

For SI: 1 cubic foot per minute = 0.0004719 m^3/s.

CHAPTER 16

DUCT SYSTEMS

SECTION M1601
DUCT CONSTRUCTION

M1601.1 Duct design. Duct systems serving heating, cooling and ventilation equipment shall be fabricated in accordance with the provisions of this section and ACCA Manual D or other approved methods.

M1601.1.1 Above-ground duct systems. Above-ground duct systems shall conform to the following:

1. Equipment connected to duct systems shall be designed to limit discharge air temperature to a maximum of 250°F (121°C).

2. Factory-made air ducts shall be constructed of Class 0 or Class 1 materials as designated in Table M1601.1.1(1).

3. Fibrous duct construction shall conform to the SMACNA *Fibrous Glass Duct Construction Standards* or NAIMA *Fibrous Glass Duct Construction Standards*.

4. Minimum thickness of metal duct material shall be as listed in Table M1601.1.1(2). Galvanized steel shall conform to ASTM A 653.

5. Use of gypsum products to construct return air ducts or plenums is permitted, provided that the air temperature does not exceed 125°F (52°C) and exposed surfaces are not subject to condensation.

6. Duct systems shall be constructed of materials having a flame spread index not greater than 200.

7. Stud wall cavities and the spaces between solid floor joists to be used as air plenums shall comply with the following conditions:

   7.1. These cavities or spaces shall not be used as a plenum for supply air.

   7.2. These cavities or spaces shall not be part of a required fire-resistance-rated assembly.

   7.3. Stud wall cavities shall not convey air from more than one floor level.

   7.4. Stud wall cavities and joist-space plenums shall be isolated from adjacent concealed spaces by tight-fitting fire blocking in accordance with Section R602.8.

TABLE M1601.1.1(1)
CLASSIFICATION OF FACTORY-MADE AIR DUCTS

DUCT CLASS	MAXIMUM FLAME-SPREAD RATING
0	0
1	25

M1601.1.2 Underground duct systems. Underground duct systems shall be constructed of approved concrete, clay, metal or plastic. The maximum duct temperature for plastic ducts shall not be greater than 150°F (66°C). Metal ducts shall be protected from corrosion in an approved manner or shall be completely encased in concrete not less than 2 inches (51 mm) thick. Nonmetallic ducts shall be installed in accordance with the manufacturer's installation instructions. Plastic pipe and fitting materials shall conform to cell classification 12454-B of ASTM D 1248 or ASTM D 1784 and external loading properties of ASTM D 2412. All ducts shall slope to an accessible point for drainage. Where encased in concrete, ducts shall be sealed and secured prior to any concrete being poured. Metallic ducts having an approved protective coating and nonmetallic ducts shall be installed in accordance with the manufacturer's installation instructions.

M1601.2 Factory-made ducts. Factory-made air ducts or duct material shall be approved for the use intended, and shall be installed in accordance with the manufacturer's installation instructions. Each portion of a factory-made air duct system shall bear a listing and label indicating compliance with UL 181 and UL 181A or UL 181B.

M1601.2.1 Duct insulation materials. Duct insulation materials shall conform to the following requirements:

1. Duct coverings and linings, including adhesives where used, shall have a flame spread index not higher than 25, and a smoke-developed index not over 50 when tested in accordance with ASTM E 84, using the speci-

TABLE M1601.1.1(2)
GAGES OF METAL DUCTS AND PLENUMS USED FOR HEATING OR COOLING

TYPE OF DUCT	SIZE (inches)	MINIMUM THICKNESS (inch)	EQUIVALENT GALVANIZED SHEET GAGE	APPROXIMATE ALUMINUM B & S GAGE
Round ducts and enclosed rectangular ducts	14 or less	0.013	30	26
	over 14	0.016	28	24
Exposed rectangular ducts	14 or less	0.016	28	24
	over 14	0.019	26	22

For SI: 1 inch = 25.4 mm.

men preparation and mounting procedures of ASTM E 2231.

2. Duct coverings and linings shall not flame, glow, smolder or smoke when tested in accordance with ASTM C 411 at the temperature to which they are exposed in service. The test temperature shall not fall below 250°F (121°C).

3. External duct insulation and factory-insulated flexible ducts shall be legibly printed or identified at intervals not longer than 36 inches (914 mm) with the name of the manufacturer; the thermal resistance *R*-value at the specified installed thickness; and the flame spread and smoke-developed indexes of the composite materials. All duct insulation product *R*-values shall be based on insulation only, excluding air films, vapor retarders or other duct components, and shall be based on tested C-values at 75°F (24°C) mean temperature at the installed thickness, in accordance with recognized industry procedures. The installed thickness of duct insulation used to determine its *R*-value shall be determined as follows:

    3.1. For duct board, duct liner and factory-made rigid ducts not normally subjected to compression, the nominal insulation thickness shall be used.

    3.2. For ductwrap, the installed thickness shall be assumed to be 75 percent (25-percent compression) of nominal thickness.

    3.3. For factory-made flexible air ducts, The installed thickness shall be determined by dividing the difference between the actual outside diameter and nominal inside diameter by two.

M1601.2.2 Vibration isolators. Vibration isolators installed between mechanical equipment and metal ducts shall be fabricated from approved materials and shall not exceed 10 inches (254 mm) in length.

M1601.3 Installation. Duct installation shall comply with Sections M1601.3.1 through M1601.3.6.

M1601.3.1 Joints and seams. Joints of duct systems shall be made substantially airtight by means of tapes, mastics, gasketing or other approved closure systems. Closure systems used with rigid fibrous glass ducts shall comply with UL 181A and shall be marked "181A-P" for pressure-sensitive tape, "181 A-M" for mastic or "181 A-H" for heat-sensitive tape. Closure systems used with flexible air ducts and flexible air connectors shall comply with UL 181B and shall be marked "181B-FX" for pressure-sensitive tape or "181B-M" for mastic. Duct connections to flanges of air distribution system equipment or sheet metal fittings shall be mechanically fastened. Mechanical fasteners for use with flexible nonmetallic air ducts shall comply with UL 181B and shall be marked 181B-C. Crimp joints for round metal ducts shall have a contact lap of at least 1¹/₂ inches (38 mm) and shall be mechanically fastened by means of at least three sheet-metal screws or rivets equally spaced around the joint.

M1601.3.2 Support. Metal ducts shall be supported by ¹/₂-inch (13 mm) wide 18-gage metal straps or 12-gage gal-

vanized wire at intervals not exceeding 10 feet (3048 mm) or other approved means. Nonmetallic ducts shall be supported in accordance with the manufacturer's installation instructions.

M1601.3.3 Fireblocking. Duct installations shall be fireblocked in accordance with Section R602.8.

M1601.3.4 Duct insulation. Duct insulation shall be installed in accordance with the following requirements:

1. A vapor retarder having a maximum permeance of 0.05 perm [(2.87 ng/(s m² Pa)] in accordance with ASTM E 96, or aluminum foil with a minimum thickness of 2 mils (0.05 mm), shall be installed on the exterior of insulation on cooling supply ducts that pass through nonconditioned spaces conducive to condensation.

2. Exterior duct systems shall be protected against the elements.

3. Duct coverings shall not penetrate a fireblocked wall or floor.

M1601.3.5 Factory-made air ducts. Factory-made air ducts shall not be installed in or on the ground, in tile or metal pipe, or within masonry or concrete.

M 1601.3.6 Duct separation. Ducts shall be installed with at least 4 inches (102 mm) separation from earth except where they meet the requirements of Section M1601.1.2.

M1601.3.7 Ducts located in garages. Ducts in garages shall comply with the requirements of Section R309.1.1.

M1601.3.8 Flood hazard areas. In areas prone to flooding as established by Table R301.2(1), duct systems shall be located or installed in accordance with Section R324.1.5.

M1601.4 Under-floor plenums. An under-floor space used as a supply plenum shall conform to the requirements of this section. Fuel gas lines and plumbing waste cleanouts shall not be located within the space.

M1601.4.1 General. The space shall be cleaned of loose combustible materials and scrap, and shall be tightly enclosed. The ground surface of the space shall be covered with a moisture barrier having a minimum thickness of 4 mils (0.1 mm).

M1601.4.2 Materials. The under-floor space, including the sidewall insulation, shall be formed by materials having flame-spread ratings not greater than 200 when tested in accordance with ASTM E 84.

M1601.4.3 Furnace connections. A duct shall extend from the furnace supply outlet to not less than 6 inches (152 mm) below the combustible framing. This duct shall comply with the provisions of Section M1601.1. A noncombustible receptacle shall be installed below any floor opening into the plenum in accordance with the following requirements:

1. The receptacle shall be securely suspended from the floor members and shall not be more than 18 inches (457 mm) below the floor opening.

2. The area of the receptacle shall extend 3 inches (76 mm) beyond the opening on all sides.

3. The perimeter of the receptacle shall have a vertical lip at least 1 inch (25 mm) high at the open sides.

M1601.4.4 Access. Access to an under-floor plenum shall be provided through an opening in the floor with minimum dimensions of 18 inches by 24 inches (457 mm by 610 mm).

M1601.4.5 Furnace controls. The furnace shall be equipped with an automatic control that will start the air-circulating fan when the air in the furnace bonnet reaches a temperature not higher than 150°F (66°C). The furnace shall additionally be equipped with an approved automatic control that limits the outlet air temperature to 200°F (93°C).

SECTION M1602
RETURN AIR

M1602.1 Return air. Return air shall be taken from inside the dwelling. Dilution of return air with outdoor air shall be permitted.

M1602.2 Prohibited sources. Outdoor and return air for a forced-air heating or cooling system shall not be taken from the following locations:

1. Closer than 10 feet (3048 mm) to an appliance vent outlet, a vent opening from a plumbing drainage system or the discharge outlet of an exhaust fan, unless the outlet is 3 feet (914 mm) above the outside air inlet.

2. Where flammable vapors are present; or where located less than 10 feet (3048 mm) above the surface of any abutting public way or driveway; or where located at grade level by a sidewalk, street, alley or driveway.

3. A room or space, the volume of which is less than 25 percent of the entire volume served by such system. Where connected by a permanent opening having an area sized in accordance with ACCA Manual D, adjoining rooms or spaces shall be considered as a single room or space for the purpose of determining the volume of such rooms or spaces.

   Exception: The minimum volume requirement shall not apply where the amount of return air taken from a room or space is less than or equal to the amount of supply air delivered to such room or space.

4. A closet, bathroom, toilet room, kitchen, garage, mechanical room, furnace room or other dwelling unit.

5. A room or space containing a fuel-burning appliance where such room or space serves as the sole source of return air.

   Exceptions:

   1. The fuel-burning appliance is a direct-vent appliance or an appliance not requiring a vent in accordance with Section M1801.1 or Chapter 24.

   2. The room or space complies with the following requirements:

      2.1. The return air shall be taken from a room or space having a volume exceed-

ing 1 cubic foot for each 10 Btu/h (9.6 L/W) of combined input rating of all fuel-burning appliances therein.

      2.2. The volume of supply air discharged back into the same space shall be approximately equal to the volume of return air taken from the space.

      2.3. Return-air inlets shall not be located within 10 feet (3048 mm) of any appliance firebox or draft hood in the same room or space.

3. Rooms or spaces containing solid-fuel burning appliances, provided that return-air inlets are located not less than 10 feet (3048 mm) from the firebox of such appliances.

M1602.3 Inlet opening protection. Outdoor air inlets shall be covered with screens having openings that are not less than $^1/_4$-inch (6 mm) and not greater than $^1/_2$-inch (12.7 mm).

CHAPTER 17

COMBUSTION AIR

SECTION M1701
GENERAL

M1701.1 Air supply. Liquid- and solid-fuel-burning appliances shall be provided with a supply of air for fuel combustion, draft hood dilution and ventilation of the space in which the appliance is installed, in accordance with Section M1702 or Section M1703. The methods of providing combustion air in this chapter do not apply to fireplaces, fireplace stoves and direct-vent appliances.

M1701.1.1 Buildings of unusually tight construction. In buildings of unusually tight construction, combustion air shall be obtained from outside the sealed thermal envelope. In buildings of ordinary tightness, insofar as infiltration is concerned, all or a portion of the combustion air for fuel-burning appliances may be obtained from infiltration when the room or space has a volume of 50 cubic feet per 1,000 Btu/h (4.83 L/W) input.

M1701.2 Exhaust and ventilation system. Air requirements for the operation of exhaust fans, kitchen ventilation systems, clothes dryers and fireplaces shall be considered in determining the adequacy of a space to provide combustion air.

M1701.3 Volume dampers prohibited. Volume dampers shall not be installed in combustion air openings.

M1701.4 Prohibited sources. Combustion air ducts and openings shall not connect appliance enclosures with space in which the operation of a fan may adversely affect the flow of combustion air. Combustion air shall not be obtained from an area in which flammable vapors present a hazard. Fuel-fired appliances shall not obtain combustion air from any of the following rooms or spaces:

1. Sleeping rooms.

2. Bathrooms.

3. Toilet rooms.

Exception: The following appliances shall be permitted to obtain combustion air from sleeping rooms, bathrooms and toilet rooms:

1. Solid fuel-fired appliances provided that the room is not a confined space and the building is not of unusually tight construction.

2. Appliances installed in an enclosure in which all combustion air is taken from the outdoors and the enclosure is equipped with a solid weatherstripped door and self-closing device.

M1701.5 Opening area. The free area of each opening shall be used for determining combustion air. Unless otherwise specified by the manufacturer or determined by actual measurement, the free area shall be considered 75 percent of the gross area for metal louvers and 25 percent of the gross area for wood louvers.

M1701.6 Opening location. In areas prone to flooding as established by Table R301.2(1), openings shall be located at or above the design flood elevation established in Section R324.1.5.

SECTION M1702
ALL AIR FROM INSIDE THE BUILDING

M1702.1 Required volume. Where the volume of the space in which fuel-burning appliances are installed is greater than 50 cubic feet per 1,000 Btu/h (4.83 L/W) of aggregate input rating in buildings of ordinary tightness, insofar as infiltration is concerned, normal infiltration shall be regarded as adequate to provide combustion air. Rooms communicating directly with the space in which the appliances are installed through openings not furnished with doors shall be considered part of the required volume.

M1702.2 Confined space. Where the space in which the appliance is located does not meet the criterion specified in Section M1702.1, two permanent openings to adjacent spaces shall be provided so that the combined volume of all spaces meets the criterion. One opening shall be within 12 inches (305 mm) of the top and one within 12 inches (305 mm) of the bottom of the space, as illustrated in Figure M1702.2. Each opening shall have a free area equal to a minimum of 1 square inch per 1,000 Btu/h (2201 mm²/kW) input rating of all appliances installed within the space, but not less than 100 square inches (64 415 mm²).

M1702.3 Unusually tight construction. Where the space is of adequate volume in accordance with Section M1702.1 or Section M1702.2, but is within a building sealed so tightly that infiltration air is not adequate for combustion, combustion air shall be obtained from outdoors or from spaces freely communicating with the outdoors in accordance with Section M1703.

SECTION M1703
ALL AIR FROM OUTDOORS

M1703.1 Outdoor air. Where the space in which fuel-burning appliances are located does not meet the criterion for indoor air specified in Section M1702, outside combustion air shall be supplied as specified in Section M1703.2.

M1703.2 Two openings or ducts. Outside combustion air shall be supplied through openings or ducts, as illustrated in Figures M1703.2(1), M1703.2(2), M1703.2(3) and M1703.2(4). One opening shall be within 12 inches (305 mm) of the top of the enclosure, and one within 12 inches (305 mm) of the bottom of the enclosure. Openings are permitted to connect to spaces directly communicating with the outdoors, such as ventilated crawl spaces or ventilated attic spaces. The same duct or opening shall not serve both combustion air openings. The duct serving the upper opening shall be level or extend upward from the appliance space.

M1703.2.1 Size of openings. Where directly communicating with the outdoors, or where communicating with the outdoors by means of vertical ducts, each opening shall have a free area of at least 1 square inch per 4,000 Btu/per hour (550 mm²/kW) of total input rating of all appliances in the space. Where horizontal ducts are used, each opening shall have a free area of at least 1 square inch per 2,000 Btu/per hour (1100 mm²/kW) of total input of all appliances in the space. Ducts shall be of the same minimum cross-sectional area as the required free area of the openings to which they connect. The minimum cross-sectional dimension of rectangular air ducts shall be 3 inches (76 mm).

M1703.3 Attic combustion air. Combustion air obtained from an attic area, as illustrated in Figure M1703.2(3), shall be in accordance with the following:

1. The attic ventilation shall be sufficient to provide the required volume of combustion air.

2. The combustion air opening shall be provided with a metal sleeve extending from the appliance enclosure to at least 6 inches (152 mm) above the top of the ceiling joists and ceiling insulation.

3. An inlet air duct within an outlet air duct shall be an acceptable means of supplying attic combustion air to an appliance room provided that the inlet duct extends at least 12 inches (305 mm) above the top of the outlet duct in the attic space, as illustrated in Figure M1703.3.

4. The end of ducts that terminate in an attic shall not be screened.

M1703.4 Under-floor combustion air. Combustion air obtained from under-floor areas, as illustrated in Figure M1703.2(4), shall have free opening areas to the outside equivalent to not less than twice the required combustion air opening.

M1703.5 Opening requirements. Outside combustion air openings shall be covered with corrosion-resistant screen or equivalent protection having not less than $^1/_4$-inch (6 mm) openings, and not greater than $^1/_2$-inch (13 mm) openings.

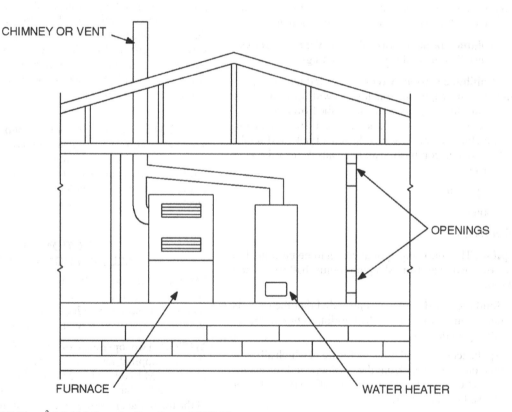

For SI: 1 square inch = 645.16 mm², 1 British thermal unit per hour = 0.2931 W.

NOTE: Each opening shall have a free area of not less than 1 square inch per 1,000 Btu/h of the total input rating of all appliances in the enclosure, but not less than 100 square inches.

FIGURE M1702.2
APPLIANCES LOCATED IN CONFINED SPACES—ALL AIR TAKEN FROM ADJACENT SPACES WITHIN THE BUILDING

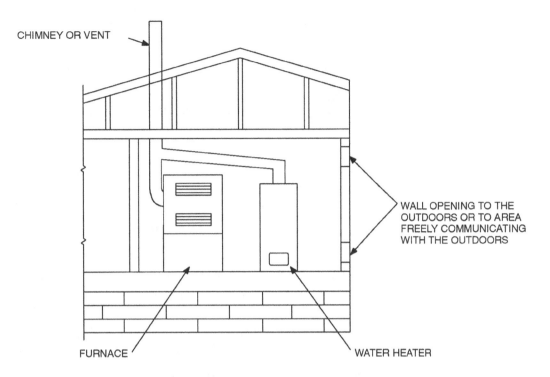

For SI: 1 square inch = 645.16 mm², 1 British thermal unit per hour = 0.2931 W.
NOTE: Each opening shall have a free area of not less than 1 square inch per 4,000 Btu/h of the total input rating of all appliances in the enclosure.

FIGURE M1703.2(1)
APPLIANCES LOCATED IN CONFINED SPACES—ALL AIR TAKEN
FROM OUTDOORS THROUGH TWO OPENINGS

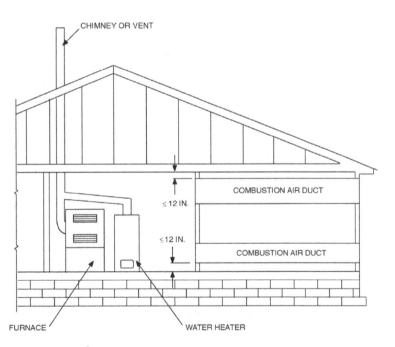

For SI: 1 inch = 25.4 mm, 1 square inch = 645.16 mm², 1 British thermal unit per hour = 0.2931 W.
NOTE: Each opening shall have a free area of at least 1 square inch per 2,000 Btu/h of the total input of all appliances in the space.

FIGURE M1703.2(2)
APPLIANCES LOCATED IN CONFINED SPACES—ALL AIR TAKEN
FROM OUTDOORS THROUGH HORIZONTAL DUCTS

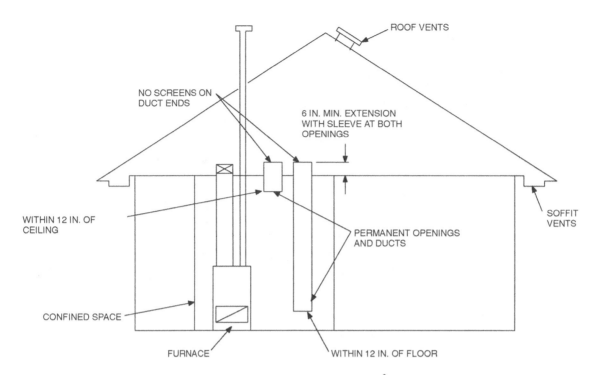

For SI: 1 inch = 25.4 mm, 1 British thermal unit per hour = 0.2931 W, 1 square inch = 645.16 mm².
NOTE: Each opening shall have a free area of at least 1 square inch per 4,000 Btu/h of the total input of all appliances in the space. The attic must be sufficiently vented for combustion air to be taken from the attic.

FIGURE M1703.2(3)
APPLIANCES LOCATED IN CONFINED SPACES—ALL AIR TAKEN FROM OUTDOORS THROUGH VENTILATED ATTIC

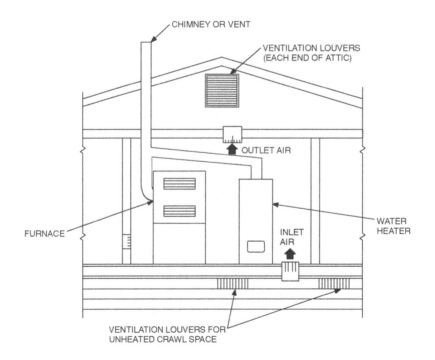

For SI: 1 square inch = 645.16 mm², 1 British thermal unit per hour = 0.2931 W.
NOTE: The inlet and outlet air openings shall have a free area of not less than 1 square inch per 4,000 Btu/h of the total input rating of all appliances in the enclosure.

FIGURE M1703.2(4)
APPLIANCES LOCATED IN CONFINED SPACES—INLET AIR TAKEN FROM
VENTILATED CRAWL SPACE AND OUTLET AIR TO VENTILATED ATTIC

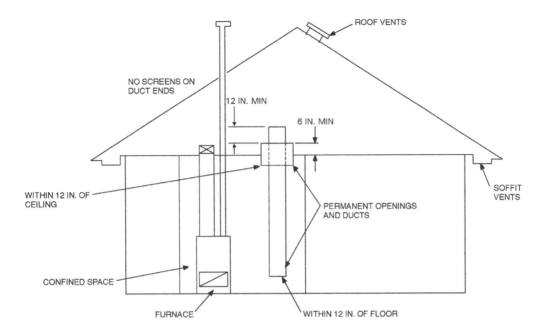

For SI: 1 inch = 25.4 mm, 1 British thermal unit per hour = 0.2931 W, 1 square inch = 645.16 mm².

NOTE: Each duct shall have a free area of at least 1 square inch per 4,000 Btu/h of the total input of all appliances in the space. The attic must be sufficiently ventilated to provide the required combustion air.

**FIGURE M1703.3
APPLIANCES LOCATED IN CONFINED SPACES—ALL AIR TAKEN
FROM OUTDOORS THROUGH VENTILATED ATTIC, INLET DUCT WITHIN OUTLET DUCT**

CHAPTER 18

CHIMNEYS AND VENTS

SECTION M1801
GENERAL

M1801.1 Venting required. Fuel-burning appliances shall be vented to the outdoors in accordance with their listing and label and manufacturer's installation instructions except appliances listed and labeled for unvented use. Venting systems shall consist of approved chimneys or vents, or venting assemblies that are integral parts of labeled appliances. Gas-fired appliances shall be vented in accordance with Chapter 24.

M1801.2 Draft requirements. A venting system shall satisfy the draft requirements of the appliance in accordance with the manufacturer's installation instructions, and shall be constructed and installed to develop a positive flow to convey combustion products to the outside atmosphere.

M1801.3 Existing chimneys and vents. Where an appliance is permanently disconnected from an existing chimney or vent, or where an appliance is connected to an existing chimney or vent during the process of a new installation, the chimney or vent shall comply with Sections M1801.3.1 through M1801.3.4.

M1801.3.1 Size. The chimney or vent shall be resized as necessary to control flue gas condensation in the interior of the chimney or vent and to provide the appliance, or appliances served, with the required draft. For the venting of oil-fired appliances to masonry chimneys, the resizing shall be done in accordance with NFPA 31.

M1801.3.2 Flue passageways. The flue gas passageway shall be free of obstructions and combustible deposits and shall be cleaned if previously used for venting a solid- or liquid-fuel-burning appliance or fireplace. The flue liner, chimney inner wall or vent inner wall shall be continuous and free of cracks, gaps, perforations, or other damage or deterioration that would allow the escape of combustion products, including gases, moisture and creosote.

M1801.3.3 Cleanout. Masonry chimneys shall be provided with a cleanout opening complying with Section R1003.17.

M1801.3.4 Clearances. Chimneys and vents shall have airspace clearance to combustibles in accordance with this code and the chimney or vent manufacturer's installation instructions.

Exception: Masonry chimneys equipped with a chimney lining system tested and listed for installation in chimneys in contact with combustibles in accordance with UL 1777, and installed in accordance with the manufacturer's instruction, shall not be required to have a clearance between combustible materials and exterior surfaces of the masonry chimney. Noncombustible firestopping shall be provided in accordance with this code.

M1801.4 Space around lining. The space surrounding a flue lining system or other vent installed within a masonry chimney shall not be used to vent any other appliance. This shall not prevent the installation of a separate flue lining in accordance with the manufacturer's installation instructions and this code.

M1801.5 Mechanical draft systems. A mechanical draft system shall be used only with appliances listed and labeled for such use. Provisions shall be made to prevent the flow of fuel to the equipment when the draft system is not operating. Forced draft systems and all portions of induced draft systems under positive pressure during operation shall be designed and installed to prevent leakage of flue gases into a building.

M1801.6 Direct-vent appliances. Direct-vent appliances shall be installed in accordance with the manufacturer's installation instructions.

M1801.7 Support. Venting systems shall be adequately supported for the weight of the material used.

M1801.8 Duct penetrations. Chimneys, vents and vent connectors shall not extend into or through supply and return air ducts or plenums.

M1801.9 Fireblocking. Vent and chimney installations shall be fireblocked in accordance with Section R602.8.

M1801.10 Unused openings. Unused openings in any venting system shall be closed or capped.

M1801.11 Multiple-appliance venting systems. Two or more listed and labeled appliances connected to a common natural draft venting system shall comply with the following requirements:

1. Appliances that are connected to common venting systems shall be located on the same floor of the dwelling.

   Exception: Engineered systems as provided for in Section G2427.

2. Inlets to common venting systems shall be offset such that no portion of an inlet is opposite another inlet.

3. Connectors serving appliances operating under a natural draft shall not be connected to any portion of a mechanical draft system operating under positive pressure.

M1801.12 Multiple solid fuel prohibited. A solid-fuel-burning appliance or fireplace shall not connect to a chimney passageway venting another appliance.

SECTION M1802
VENT COMPONENTS

M1802.1 Draft hoods. Draft hoods shall be located in the same room or space as the combustion air openings for the appliances.

M1802.2 Vent dampers. Vent dampers shall comply with Sections M1802.2.1 and M1802.2.2.

M1802.2.1 Manually operated. Manually operated dampers shall not be installed except in connectors or chimneys serving solid-fuel-burning appliances.

M1802.2.2 Automatically operated. Automatically operated dampers shall conform to UL 17 and be installed in accordance with the terms of their listing and label. The installation shall prevent firing of the burner when the damper is not opened to a safe position.

M1802.3 Draft regulators. Draft regulators shall be provided for oil-fired appliances that must be connected to a chimney. Draft regulators provided for solid-fuel-burning appliances to reduce draft intensity shall be installed and set in accordance with the manufacturer's installation instructions.

M1802.3.1 Location. Where required, draft regulators shall be installed in the same room or enclosure as the appliance so that no difference in pressure will exist between the air at the regulator and the combustion air supply.

SECTION M1803
CHIMNEY AND VENT CONNECTORS

M1803.1 General. Connectors shall be used to connect fuel-burning appliances to a vertical chimney or vent except where the chimney or vent is attached directly to the appliance.

M1803.2 Connectors for oil and solid fuel appliances. Connectors for oil and solid-fuel-burning appliances shall be constructed of factory-built chimney material, Type L vent material or single-wall metal pipe having resistance to corrosion and heat and thickness not less than that of galvanized steel as specified in Table M1803.2.

TABLE M1803.2
THICKNESS FOR SINGLE-WALL METAL PIPE CONNECTORS

DIAMETER OF CONNECTOR (inches)	GALVANIZED SHEET METAL GAGE NUMBER	MINIMUM THICKNESS (inch)
Less than 6	26	0.019
6 to 10	24	0.024
Over 10 through 16	22	0.029

For SI: 1 inch = 25.4 mm.

M1803.3 Installation. Vent and chimney connectors shall be installed in accordance with the manufacturer's installation instructions and within the space where the appliance is located. Appliances shall be located as close as practical to the vent or chimney. Connectors shall be as short and straight as possible and installed with a slope of not less than $^1/_4$ inch (6 mm) rise per foot of run. Connectors shall be securely supported and joints shall be fastened with sheet metal screws or rivets. Devices that obstruct the flow of flue gases shall not be installed in a connector unless listed and labeled or approved for such installations.

M1803.3.1 Floor, ceiling and wall penetrations. A chimney connector or vent connector shall not pass through any floor or ceiling. A chimney connector or vent connector shall not pass through a wall or partition unless the connector is listed and labeled for wall pass-through, or is routed through a device listed and labeled for wall pass-through and is installed in accordance with the conditions of its listing and label. Connectors for oil-fired appliances listed and labeled for Type L vents, passing through walls or partitions shall be in accordance with the following:

1. Type L vent material for oil appliances shall be installed with not less than listed and labeled clearances to combustible material.

2. Single-wall metal pipe shall be guarded by a ventilated metal thimble not less than 4 inches (102 mm) larger in diameter than the vent connector. A minimum 6 inches (152 mm) of clearance shall be maintained between the thimble and combustibles.

M1803.3.2 Length. The horizontal run of an uninsulated connector to a natural draft chimney shall not exceed 75 percent of the height of the vertical portion of the chimney above the connector. The horizontal run of a listed connector to a natural draft chimney shall not exceed 100 percent of the height of the vertical portion of the chimney above the connector.

M1803.3.3 Size. A connector shall not be smaller than the flue collar of the appliance.

Exception: Where installed in accordance with the appliance manufacturer's installation instructions.

M1803.3.4 Clearance. Connectors shall be installed with clearance to combustibles as set forth in Table M1803.3.4. Reduced clearances to combustible materials shall be in accordance with Table M1306.2 and Figure M1306.1.

TABLE M1803.3.4
CHIMNEY AND VENT CONNECTOR CLEARANCES TO COMBUSTIBLE MATERIALS[a]

TYPE OF CONNECTOR	MINIMUM CLEARANCE (inches)
Single–wall metal pipe connectors:	
Oil and solid-fuel appliances	18
Oil appliances listed for use with Type L vents	9
Type L vent piping connectors:	
Oil and solid-fuel appliances	9
Oil appliances listed for use with Type L vents	3[b]

For SI: 1 inch = 25.4 mm.

a. These minimum clearances apply to unlisted single-wall chimney and vent connectors. Reduction of required clearances is permitted as in Table M1306.2.

b. When listed Type L vent piping is used, the clearance shall be in accordance with the vent listing.

M1803.3.5 Access. The entire length of a connector shall be accessible for inspection, cleaning and replacement.

M1803.4 Connection to fireplace flue. Connection of appliances to chimney flues serving fireplaces shall comply with Sections M1803.4.1 through M1803.4.4.

M1803.4.1 Closure and accessibility. A noncombustible seal shall be provided below the point of connection to prevent entry of room air into the flue. Means shall be provided for access to the flue for inspection and cleaning.

M1803.4.2 Connection to factory-built fireplace flue. A different appliance shall not be connected to a flue serving a factory-built fireplace unless the appliance is specifically listed for such an installation. The connection shall be made in conformance with the appliance manufacturer's instructions.

M1803.4.3 Connection to masonry fireplace flue. A connector shall extend from the appliance to the flue serving a masonry fireplace to convey the flue gases directly into the flue. The connector shall be accessible or removable for inspection and cleaning of both the connector and the flue. Listed direct-connection devices shall be installed in accordance with their listing.

M1803.4.4 Size of flue. The size of the fireplace flue shall be in accordance with Section M1805.3.1.

SECTION M1804
VENTS

M1804.1 Type of vent required. Appliances shall be provided with a listed and labeled venting system as set forth in Table M1804.1.

TABLE M1804.1
VENT SELECTION CHART

VENT TYPES	APPLIANCE TYPES
Type L oil vents	Oil-burning appliances listed and labeled for venting with Type L vents
Pellet vents	Pellet fuel-burning appliances listed and labeled for use with pellet vents

M1804.2 Termination. Vent termination shall comply with Sections M1804.2.1 through M1804.2.6.

M1804.2.1 Through the roof. Vents passing through a roof shall extend through flashing and terminate in accordance with the manufacturer's installation requirements.

M1804.2.2 Decorative shrouds. Decorative shrouds shall not be installed at the termination of vents except where the shrouds are listed and labeled for use with the specific venting system and are installed in accordance with the manufacturer's installation instructions.

M1804.2.3 Natural draft appliances. Vents for natural draft appliances shall terminate at least 5 feet (1524 mm) above the highest connected appliance outlet, and natural draft gas vents serving wall furnaces shall terminate at an elevation at least 12 feet (3658 mm) above the bottom of the furnace.

M1804.2.4 Type L vent. Type L venting systems shall conform to UL 641 and shall terminate with a listed and labeled cap in accordance with the vent manufacturer's installation instructions not less than 2 feet (610 mm) above the roof and not less than 2 feet (610 mm) above any portion of the building within 10 feet (3048 mm).

M1804.2.5 Direct vent terminations. Vent terminals for direct-vent appliances shall be installed in accordance with the manufacturer's installation instructions.

M1804.2.6 Mechanical draft systems. Mechanical draft systems shall be installed in accordance with their listing,

the manufacturer's installation instructions and, except for direct vent appliances, the following requirements:

1. The vent terminal shall be located not less than 3 feet (914 mm) above a forced air inlet located within 10 feet (3048 mm).

2. The vent terminal shall be located not less than 4 feet (1219 mm) below, 4 feet (1219 mm) horizontally from, or 1 foot (305 mm) above any door, window or gravity air inlet into a dwelling.

3. The vent termination point shall not be located closer than 3 feet (914 mm) to an interior corner formed by two walls perpendicular to each other.

4. The bottom of the vent terminal shall be located at least 12 inches (305 mm) above finished ground level.

5. The vent termination shall not be mounted directly above or within 3 feet (914 mm) horizontally of an oil tank vent or gas meter.

6. Power exhauster terminations shall be located not less than 10 feet (3048 mm) from lot lines and adjacent buildings.

7. The discharge shall be directed away from the building.

M1804.3 Installation. Type L and pellet vents shall be installed in accordance with the terms of their listing and label and the manufacturer's installation instructions.

M1804.3.1 Size of single-appliance venting systems. An individual vent for a single appliance shall have a cross-sectional area equal to or greater than the area of the connector to the appliance, but not less than 7 square inches (4515 mm^2) except where the vent is an integral part of a listed and labeled appliance.

SECTION M1805
MASONRY AND FACTORY-BUILT CHIMNEYS

M1805.1 General. Masonry and factory-built chimneys shall be built and installed in accordance with Sections R1003 and R1005, respectively. Flue lining for masonry chimneys shall comply with Section R1003.11.

M1805.2 Masonry chimney connection. A chimney connector shall enter a masonry chimney not less than 6 inches (152 mm) above the bottom of the chimney. Where it is not possible to locate the connector entry at least 6 inches (152 mm) above the bottom of the chimney flue, a cleanout shall be provided by installing a capped tee in the connector next to the chimney. A connector entering a masonry chimney shall extend through, but not beyond, the wall and shall be flush with the inner face of the liner. Connectors, or thimbles where used, shall be firmly cemented into the masonry.

M1805.3 Size of chimney flues. The effective area of a natural draft chimney flue for one appliance shall be not less than the area of the connector to the appliance. The area of chimney flues connected to more than one appliance shall be not less than the area of the largest connector plus 50 percent of the areas of additional chimney connectors.

Exception: Chimney flues serving oil-fired appliances sized in accordance with NFPA 31.

M1805.3.1 Size of chimney flue for solid-fuel appliance. Except where otherwise specified in the manufacturer's installation instructions, the cross-sectional area of a flue connected to a solid-fuel-burning appliance shall be not less than the area of the flue collar or connector, and not larger than three times the area of the flue collar.

CHAPTER 19

SPECIAL FUEL-BURNING EQUIPMENT

SECTION M1901
RANGES AND OVENS

M1901.1 Clearances. Freestanding or built-in ranges shall have a vertical clearance above the cooking top of not less than 30 inches (762 mm) to unprotected combustible material. Reduced clearances are permitted in accordance with the listing and labeling of the range hoods or appliances.

M1901.2 Cooking appliances. Household cooking appliances shall be listed and labeled and shall be installed in accordance with the manufacturer's installation instructions. The installation shall not interfere with combustion air or access for operation and servicing.

SECTION M1902
SAUNA HEATERS

M1902.1 Locations and protection. Sauna heaters shall be protected from accidental contact by persons with a guard of material having a low thermal conductivity, such as wood. The guard shall have no substantial effect on the transfer of heat from the heater to the room.

M1902.2 Installation. Sauna heaters shall be installed in accordance with the manufacturer's installation instructions.

M1902.3 Combustion air. Combustion air and venting for a nondirect vent-type heater shall be provided in accordance with Chapters 17 and 18, respectively.

M1902.4 Controls. Sauna heaters shall be equipped with a thermostat that will limit room temperature to not greater than 194°F (90°C). Where the thermostat is not an integral part of the heater, the heat-sensing element shall be located within 6 inches (152 mm) of the ceiling.

SECTION M1903
STATIONARY FUEL CELL POWER PLANTS

M1903.1 General. Stationary fuel cell power plants having a power output not exceeding 1,000 kW, shall be tested in accordance with ANSI Z21.83 and shall be installed in accordance with the manufacturer's installation instructions and NFPA 853.

SECTION M1904
GASEOUS HYDROGEN SYSTEMS

M1904.1 Installation. Gaseous hydrogen systems shall be installed in accordance with the applicable requirements of Sections M1307.4 and M1903.1 and the *International Fuel Gas Code*, the *International Fire Code* and the *International Building Code*.

CHAPTER 20

BOILERS AND WATER HEATERS

SECTION M2001
BOILERS

M2001.1 Installation. In addition to the requirements of this code, the installation of boilers shall conform to the manufacturer's instructions. The manufacturer's rating data, the nameplate and operating instructions of a permanent type shall be attached to the boiler. Boilers shall have all controls set, adjusted and tested by the installer. A complete control diagram together with complete boiler operating instructions shall be furnished by the installer. Solid- and liquid-fuel-burning boilers shall be provided with combustion air as required by Chapter 17.

M2001.1.1 Standards. Oil-fired boilers and their control systems shall be listed and labeled in accordance with UL 726. Electric boilers and their control systems shall be listed in accordance with UL 834. Boilers shall be designed and constructed in accordance with the requirements of ASME CSD-1 and as applicable, the ASME *Boiler and Pressure Vessel Code*, Sections I and IV. Gas-fired boilers shall conform to the requirements listed in Chapter 24.

M2001.2 Clearance. Boilers shall be installed in accordance with their listing and label.

M2001.3 Valves. Every boiler or modular boiler shall have a shutoff valve in the supply and return piping. For multiple boiler or multiple modular boiler installations, each boiler or modular boiler shall have individual shutoff valves in the supply and return piping.

Exception: Shutoff valves are not required in a system having a single low-pressure steam boiler.

M2001.4 Flood-resistant installation. In areas prone to flooding as established in Table R301.2(1), boilers, water heaters and their control systems shall be located or installed in accordance with Section R324.1.5.

SECTION M2002
OPERATING AND SAFETY CONTROLS

M2002.1 Safety controls. Electrical and mechanical operating and safety controls for boilers shall be listed and labeled.

M2002.2 Hot water boiler gauges. Every hot water boiler shall have a pressure gauge and a temperature gauge, or combination pressure and temperature gauge. The gauges shall indicate the temperature and pressure within the normal range of the system's operation.

M2002.3 Steam boiler gauges. Every steam boiler shall have a water-gauge glass and a pressure gauge. The pressure gauge shall indicate the pressure within the normal range of the system's operation. The gauge glass shall be installed so that the midpoint is at the normal water level.

M2002.4 Pressure-relief valve. Boilers shall be equipped with pressure-relief valves with minimum rated capacities for the equipment served. Pressure-relief valves shall be set at the maximum rating of the boiler. Discharge shall be piped to drains by gravity to within 18 inches (457 mm) of the floor or to an open receptor.

M2002.5 Boiler low-water cutoff. All steam and hot water boilers shall be protected with a low-water cutoff control. The low-water cutoff shall automatically stop the combustion operation of the appliance when the water level drops below the lowest safe water level as established by the manufacturer.

SECTION M2003
EXPANSION TANKS

M2003.1 General. Hot water boilers shall be provided with expansion tanks. Nonpressurized expansion tanks shall be securely fastened to the structure or boiler and supported to carry twice the weight of the tank filled with water. Provisions shall be made for draining nonpressurized tanks without emptying the system.

M2003.1.1 Pressurized expansion tanks. Pressurized expansion tanks shall be consistent with the volume and capacity of the system. Tanks shall be capable of withstanding a hydrostatic test pressure of two and one-half times the allowable working pressure of the system.

M2003.2 Minimum capacity. The minimum capacity of expansion tanks shall be determined from Table M2003.2.

SECTION M2004
WATER HEATERS USED FOR SPACE HEATING

M2004.1 General. Water heaters used to supply both potable hot water and hot water for space heating shall be installed in accordance with this chapter, Chapter 24, Chapter 28 and the manufacturer's installation instructions.

SECTION M2005
WATER HEATERS

M2005.1 General. Water heaters shall be installed in accordance with the manufacturer's installation instructions and the requirements of this code. Water heaters installed in an attic shall conform to the requirements of Section M1305.1.3. Gas-fired water heaters shall conform to the requirements in Chapter 24. Domestic electric water heaters shall conform to UL 174 or UL 1453. Commercial electric water heaters shall conform to UL 1453. Oiled-fired water heaters shall conform to UL 732.

M2005.2 Prohibited locations. Fuel-fired water heaters shall not be installed in a room used as a storage closet. Water heaters located in a bedroom or bathroom shall be installed in a sealed enclosure so that combustion air will not be taken from the living space. Installation of direct-vent water heaters within an enclosure is not required.

TABLE M2003.2
EXPANSION TANK MINIMUM CAPACITY[a] FOR FORCED HOT-WATER SYSTEMS

SYSTEM VOLUME[b] (gallons)	PRESSURIZED DIAPHRAGM TYPE	NONPRESSURIZED TYPE
10	1.0	1.5
20	1.5	3.0
30	2.5	4.5
40	3.0	6.0
50	4.0	7.5
60	5.0	9.0
70	6.0	10.5
80	6.5	12.0
90	7.5	13.5
100	8.0	15.0

For SI: 1 gallon = 3.785 L, 1 pound per square inch gauge = 6.895 kPa, °C = [(°F)-32]/1.8.

a. Based on average water temperature of 195°F, fill pressure of 12 psig and a maximum operating pressure of 30 psig.

b. System volume includes volume of water in boiler, convectors and piping, not including the expansion tank.

M2005.2.1 Water heater access. Access to water heaters that are located in an attic or underfloor crawl space is permitted to be through a closet located in a sleeping room or bathroom where ventilation of those spaces is in accordance with this code.

M2005.3 Electric water heaters. Electric water heaters shall also be installed in accordance with the applicable provisions of Chapters 33 through 42.

M2005.4 Supplemental water-heating devices. Potable water heating devices that use refrigerant-to-water heat exchangers shall be approved and installed in accordance with the manufacturer's installation instructions.

SECTION M2006
POOL HEATERS

M2006.1 General. Pool and spa heaters shall be installed in accordance with the manufacturer's installation instructions. Oil-fired pool heaters shall be tested in accordance with UL 726. Electric pool and spa heaters shall be tested in accordance UL 1261.

M2006.2 Clearances. In no case shall the clearances interfere with combustion air, draft hood or flue terminal relief, or accessibility for servicing.

M2006.3 Temperature-limiting devices. Pool heaters shall have temperature-relief valves.

M2006.4 Bypass valves. Where an integral bypass system is not provided as a part of the pool heater, a bypass line and valve shall be installed between the inlet and outlet piping for use in adjusting the flow of water through the heater.

CHAPTER 21

HYDRONIC PIPING

SECTION M2101
HYDRONIC PIPING SYSTEMS INSTALLATION

M2101.1 General. Hydronic piping shall conform to Table M2101.1. Approved piping, valves, fittings and connections shall be installed in accordance with the manufacturer's installation instructions. Pipe and fittings shall be rated for use at the operating temperature and pressure of the hydronic system. Used pipe, fittings, valves or other materials shall be free of foreign materials.

M2101.2 System drain down. Hydronic piping systems shall be installed to permit draining the system. When the system drains to the plumbing drainage system, the installation shall conform to the requirements of Chapters 25 through 32 of this code.

M2101.3 Protection of potable water. The potable water system shall be protected from backflow in accordance with the provisions listed in Section P2902.

M2101.4 Pipe penetrations. Openings through concrete or masonry building elements shall be sleeved.

M2101.5 Contact with building material. A hydronic piping system shall not be in direct contact with any building material that causes the piping material to degrade or corrode.

M2101.6 Drilling and notching. Wood-framed structural members shall be drilled, notched or altered in accordance with the provisions of Sections R508.8, R602.6, R602.6.1 and R802.7. Holes in cold-formed, steel-framed, load-bearing members shall be permitted only in accordance with Sections R507.2, R603.2 and R804.2. In accordance with the provisions of Sections R505.3.5, R603.3.4 and R804.3.5, cutting and notching of flanges and lips of cold-formed, steel-framed, load-bearing members shall not be permitted.

M2101.7 Prohibited tee applications. Fluid in the supply side of a hydronic system shall not enter a tee fitting through the branch opening.

M2101.8 Expansion, contraction and settlement. Piping shall be installed so that piping, connections and equipment shall not be subjected to excessive strains or stresses. Provisions shall be made to compensate for expansion, contraction, shrinkage and structural settlement.

M2101.9 Piping support. Hangers and supports shall be of material of sufficient strength to support the piping, and shall be fabricated from materials compatible with the piping material. Piping shall be supported at intervals not exceeding the spacing specified in Table M2101.9.

M2101.10 Tests. Hydronic piping shall be tested hydrostatically at a pressure of not less than 100 pounds per square inch (psi) (690 kPa) for a duration of not less than 15 minutes.

SECTION M2102
BASEBOARD CONVECTORS

M2102.1 General. Baseboard convectors shall be installed in accordance with the manufacturer's installation instructions. Convectors shall be supported independently of the hydronic piping.

SECTION M2103
FLOOR HEATING SYSTEMS

M2103.1 Piping materials. Piping for embedment in concrete or gypsum materials shall be standard-weight steel pipe, copper tubing, cross-linked polyethylene/aluminum/cross-linked polyethylene (PEX-AL-PEX) pressure pipe, chlorinated polyvinyl chloride (CPVC), polybutylene, cross-linked polyethylene (PEX) tubing or polypropylene (PP) with a minimum rating of 100 psi at 180°F (690 kPa at 82°C).

M2103.2 Piping joints. Piping joints that are embedded shall be installed in accordance with the following requirements:

1. Steel pipe joints shall be welded.

2. Copper tubing shall be joined with brazing material having a melting point exceeding 1,000°F (538°C).

3. Polybutylene pipe and tubing joints shall be installed with socket-type heat-fused polybutylene fittings.

4. CPVC tubing shall be joined using solvent cement joints.

5. Polypropylene pipe and tubing joints shall be installed with socket-type heat-fused polypropylene fittings.

6. Cross-linked polyethylene (PEX) tubing shall be joined using cold expansion, insert or compression fittings.

M2103.3 Testing. Piping or tubing to be embedded shall be tested by applying a hydrostatic pressure of not less than 100 psi (690 kPa). The pressure shall be maintained for 30 minutes, during which all joints shall be visually inspected for leaks.

SECTION M2104
LOW TEMPERATURE PIPING

M2104.1 Piping materials. Low temperature piping for embedment in concrete or gypsum materials shall be as indicated in Table M2101.1.

M2104.2 Piping joints. Piping joints (other than those in Section M2103.2) that are embedded shall comply with the following requirements:

1. Cross-linked polyethylene (PEX) tubing shall be installed in accordance with the manufacturer's instructions.

2. Polyethylene tubing shall be installed with heat fusion joints.

3. Polypropylene (PP) tubing shall be installed in accordance with the manufacturer's instructions.

TABLE M2101.1
HYDRONIC PIPING MATERIALS

MATERIAL	USE CODE[a]	STANDARD[b]	JOINTS	NOTES
Brass pipe	1	ASTM B 43	Brazed, welded, threaded, mechanical and flanged fittings	
Brass tubing	1	ASTM B 135	Brazed, soldered and mechanical fittings	
Chlorinated poly (vinyl chloride) (CPVC) pipe and tubing	1, 2, 3	ASTM D 2846	Solvent cement joints, compression joints and threaded adapters	
Copper pipe	1	ASTM B 42, B 302	Brazed, soldered and mechanical fittings threaded, welded and flanged	
Copper tubing (type K, L or M)	1, 2	ASTM B 75, B 88, B 251, B 306	Brazed, soldered and flared mechanical fittings	Joints embedded in concrete
Cross-linked polyethylene (PEX)	1, 2, 3	ASTM F 876, F 877	(See PEX fittings)	Install in accordance with manufacturer's instructions.
Cross-linked polyethylene/aluminum/ cross-linked polyethylene-(PEX-AL-PEX) pressure pipe	1, 2	ASTM F 1281 or CAN/ CSA B137.10	Mechanical, crimp/insert	Install in accordance with manufacturer's instructions.
PEX Fittings		ASTM F 1807 ASTM F 1960 ASTM F 2098	Copper-crimp/insert fittings, cold expansion fittings, stainless steel clamp, insert fittings	Install in accordance with manufacturer's instructions
Plastic fittings PEX		ASTM F 1807		
Polybutylene (PB) pipe and tubing	1, 2, 3	ASTM D 3309	Heat-fusion, crimp/insert and compression	Joints in concrete shall be heat-fused.
Polyethylene (PE) pipe, tubing and fittings (for ground source heat pump loop systems)	1, 2, 4	ASTM D 2513; ASTM D 3350; ASTM D 2513; ASTM D 3035; ASTM D 2447; ASTM D 2683; ASTM F 1055; ASTM D 2837; ASTM D 3350; ASTM D 1693	Heat-fusion	
Polyproplylene (PP)	1, 2, 3	ISO 15874 ASTM F 2389	Heat-fusion joints, mechanical fittings, threaded adapters, compression joints	
Soldering fluxes	1	ASTM B 813	Copper tube joints	
Steel pipe	1, 2	ASTM A 53; A 106	Brazed, welded, threaded, flanged and mechanical fittings	Joints in concrete shall be welded. Galvanized pipe shall not be welded or brazed.
Steel tubing	1	ASTM A 254	Mechanical fittings, welded	

For SI: °C = [(°F)-32]/1.8.

a. Use code:
   1. Above ground.
   2. Embedded in radiant systems.
   3. Temperatures below 180°F only.
   4. Low temperature (below 130°F) applications only.

b. Standards as listed in Chapter 43.

TABLE M2101.9
HANGER SPACING INTERNALS

PIPING MATERIAL	MAXIMUM HORIZONTAL SPACING (feet)	MAXIMUM VERTICAL SPACING (feet)
ABS	4	10
CPVC ≤ 1 inch pipe or tubing	3	5
CPVC ≥ 1¼ inch	4	10
Copper or copper alloy pipe	12	10
Copper or copper alloy tubing	6	10
PB pipe or tubing	2.67	4
PE pipe or tubing	2.67	4
PEX tubing	2.67	4
PP < 1 inch pipe or tubing	2.67	4
PP > 1¼ inch	4	5
PVC	4	10
Steel pipe	12	15
Steel tubing	8	10

For SI: 1 inch = 25.4 mm, 1 foot = 304.8 mm.

M2104.2.1 Polyethylene plastic pipe and tubing for ground source heat pump loop systems. Joints between polyethylene plastic pipe and tubing or fittings for ground source heat pump loop systems shall be heat fusion joints conforming to Section M2104.2.1.1, electrofusion joints conforming to Section M2104.2.1.2 or stab-type insertion joints conforming to Section M2104.2.1.3.

M2104.2.1.1 Heat-fusion joints. Joints shall be of the socket-fusion, saddle-fusion or butt-fusion type, fabricated in accordance with the piping manufacturer's instructions. Joint surfaces shall be clean and free of moisture. Joint surfaces shall be heated to melt temperatures and joined. The joint shall be undisturbed until cool. Fittings shall be manufactured in accordance with ASTM D 2683.

M2104.2.1.2 Electrofusion joints. Joint surfaces shall be clean and free of moisture, and scoured to expose virgin resin. Joint surfaces shall be heated to melt temperatures for the period of time specified by the manufacturer. The joint shall be undisturbed until cool. Fittings shall be manufactured in accordance with ASTM F 1055.

M2104.2.1.3 Stab-type insert fittings. Joint surfaces shall be clean and free of moisture. Pipe ends shall be chamfered and inserted into the fitting to full depth. Fittings shall be manufactured in accordance with ASTM D 2513.

SECTION M2105
GROUND SOURCE HEAT PUMP SYSTEM LOOP PIPING

M2105.1 Testing. The assembled loop system shall be pressure tested with water at 100 psi (690 kPa) for 30 minutes with no observed leaks before connection (header) trenches are backfilled. Flow rates and pressure drops shall be compared to calculated values. If actual flow rate or pressure drop figures differ from calculated values by more than 10 percent, the problem shall be identified and corrected.

CHAPTER 22

SPECIAL PIPING AND STORAGE SYSTEMS

SECTION M2201
OIL TANKS

M2201.1 Materials. Supply tanks shall be listed and labeled and shall conform to UL 58 for underground tanks and UL 80 for indoor tanks.

M2201.2 Above-ground tanks. The maximum amount of fuel oil stored above ground or inside of a building shall be 660 gallons (2498 L). The supply tank shall be supported on rigid noncombustible supports to prevent settling or shifting.

M2201.2.1 Tanks within buildings. Supply tanks for use inside of buildings shall be of such size and shape to permit installation and removal from dwellings as whole units. Supply tanks larger than 10 gallons (38 L) shall be placed not less than 5 feet (1524 mm) from any fire or flame either within or external to any fuel-burning appliance.

M2201.2.2 Outside above-ground tanks. Tanks installed outside above ground shall be a minimum of 5 feet (1524 mm) from an adjoining property line. Such tanks shall be suitably protected from the weather and from physical damage.

M2201.3 Underground tanks. Excavations for underground tanks shall not undermine the foundations of existing structures. The clearance from the tank to the nearest wall of a basement, pit or property line shall not be less than 1 foot (305 mm). Tanks shall be set on and surrounded with noncorrosive inert materials such as clean earth, sand or gravel well tamped in place. Tanks shall be covered with not less than 1 foot (305 mm) of earth. Corrosion protection shall be provided in accordance with Section M2203.7.

M2201.4 Multiple tanks. Cross connection of two supply tanks shall be permitted in accordance with Section M2203.6.

M2201.5 Oil gauges. Inside tanks shall be provided with a device to indicate when the oil in the tank has reached a predetermined safe level. Glass gauges or a gauge subject to breakage that could result in the escape of oil from the tank shall not be used.

M2201.6 Flood-resistant installation. In areas prone to flooding as established by Table R301.2(1), tanks shall be installed at or above the design flood elevation established in Section R324 or shall be anchored to prevent flotation, collapse and lateral movement under conditions of the design flood.

M2201.7 Tanks abandoned or removed. Exterior above-grade fill piping shall be removed when tanks are abandoned or removed. Tank abandonment and removal shall be in accordance with the *International Fire Code*.

SECTION M2202
OIL PIPING, FITTING AND CONNECTIONS

M2202.1 Materials. Piping shall consist of steel pipe, copper tubing or steel tubing conforming to ASTM A 539. Aluminum tubing shall not be used between the fuel-oil tank and the burner units.

M2202.2 Joints and fittings. Piping shall be connected with standard fittings compatible with the piping material. Cast iron fittings shall not be used for oil piping. Unions requiring gaskets or packings, right or left couplings, and sweat fittings employing solder having a melting point less than 1,000°F (538°C) shall not be used for oil piping. Threaded joints and connections shall be made tight with a lubricant or pipe thread compound.

M2202.3 Flexible connectors. Flexible metal hose used where rigid connections are impractical or to reduce the effect of jarring and vibration shall be listed and labeled in accordance with UL 536 and shall be installed in compliance with its label and the manufacturer's installation instructions. Connectors made from combustible materials shall not be used inside of buildings or above ground outside of buildings.

SECTION M2203
INSTALLATION

M2203.1 General. Piping shall be installed in a manner to avoid placing stresses on the piping, and to accommodate expansion and contraction of the piping system.

M2203.2 Supply piping. Supply piping used in the installation of oil burners and appliances shall be not smaller than $^3/_8$-inch (9 mm) pipe or $^3/_8$-inch (9 mm) outside diameter tubing. Copper tubing and fittings shall be a minimum of Type L.

M2203.3 Fill piping. Fill piping shall terminate outside of buildings at a point at least 2 feet (610 mm) from any building opening at the same or lower level. Fill openings shall be equipped with a tight metal cover.

M2203.4 Vent piping. Vent piping shall be not smaller than $1^1/_4$-inch (32 mm) pipe. Vent piping shall be laid to drain toward the tank without sags or traps in which the liquid can collect. Vent pipes shall not be cross connected with fill pipes, lines from burners or overflow lines from auxiliary tanks. The lower end of a vent pipe shall enter the tank through the top and shall extend into the tank not more than 1 inch (25 mm).

M2203.5 Vent termination. Vent piping shall terminate outside of buildings at a point not less than 2 feet (610 mm), measured vertically or horizontally, from any building opening. Outer ends of vent piping shall terminate in a weather-proof cap or fitting having an unobstructed area at least equal to the cross-sectional area of the vent pipe, and shall be located sufficiently above the ground to avoid being obstructed by snow and ice.

M2203.6 Cross connection of tanks. Cross connection of two supply tanks, not exceeding 660 gallons (2498 L) aggregate capacity, with gravity flow from one tank to another, shall be acceptable providing that the two tanks are on the same horizontal plane.

M2203.7 Corrosion protection. Underground tanks and buried piping shall be protected by corrosion-resistant coatings or special alloys or fiberglass-reinforced plastic.

SECTION M2204
OIL PUMPS AND VALVES

M2204.1 Pumps. Oil pumps shall be positive displacement types that automatically shut off the oil supply when stopped. Automatic pumps shall be listed and labeled in accordance with UL 343 and shall be installed in accordance with their listing.

M2204.2 Shutoff valves. A readily accessible manual shutoff valve shall be installed between the oil supply tank and the burner. Where the shutoff valve is installed in the discharge line of an oil pump, a pressure-relief valve shall be incorporated to bypass or return surplus oil.

M2204.3 Maximum pressure. Pressure at the oil supply inlet to an appliance shall be not greater than 3 pounds per square inch (psi) (20.7 kPa).

M2204.4 Relief valves. Fuel-oil lines incorporating heaters shall be provided with relief valves that will discharge to a return line when excess pressure exists.

CHAPTER 23

SOLAR SYSTEMS

SECTION M2301
SOLAR ENERGY SYSTEMS

M2301.1 General. This section provides for the design, construction, installation, alteration and repair of equipment and systems using solar energy to provide space heating or cooling, hot water heating and swimming pool heating.

M2301.2 Installation. Installation of solar energy systems shall comply with Sections M2301.2.1 through M2301.2.9.

M2301.2.1 Access. Solar energy collectors, controls, dampers, fans, blowers and pumps shall be accessible for inspection, maintenance, repair and replacement.

M2301.2.2 Roof-mounted collectors. The roof shall be constructed to support the loads imposed by roof-mounted solar collectors. Roof-mounted solar collectors that serve as a roof covering shall conform to the requirements for roof coverings in Chapter 9 of this code. Where mounted on or above the roof coverings, the collectors and supporting structure shall be constructed of noncombustible materials or fire-retardant-treated wood equivalent to that required for the roof construction.

M2301.2.3 Pressure and temperature relief. System components containing fluids shall be protected with pressure- and temperature-relief valves. Relief devices shall be installed in sections of the system so that a section cannot be valved off or isolated from a relief device.

M2301.2.4 Vacuum relief. System components that might be subjected to pressure drops below atmospheric pressure during operation or shutdown shall be protected by a vacuum-relief valve.

M2301.2.5 Protection from freezing. System components shall be protected from damage resulting from freezing of heat-transfer liquids at the winter design temperature provided in Table R301.2(1). Freeze protection shall be provided by heating, insulation, thermal mass and heat transfer fluids with freeze points lower than the winter design temperature, heat tape or other approved methods, or combinations thereof.

> **Exception:** Where the winter design temperature is greater than 32°F (0°C).

M2301.2.6 Expansion tanks. Expansion tanks in solar energy systems shall be installed in accordance with Section M2003 in closed fluid loops that contain heat transfer fluid.

M2301.2.7 Roof and wall penetrations. Roof and wall penetrations shall be flashed and sealed in accordance with Chapter 9 of this code to prevent entry of water, rodents and insects.

M2301.2.8 Solar loop isolation. Valves shall be installed to allow the solar collectors to be isolated from the remainder of the system. Each isolation valve shall be labeled with the open and closed position.

M2301.2.9 Maximum temperature limitation. Systems shall be equipped with means to limit the maximum water temperature of the system fluid entering or exchanging heat with any pressurized vessel inside the dwelling to 180°F (82°C). This protection is in addition to the required temperature- and pressure-relief valves required by Section M2301.2.3.

M2301.3 Labeling. Labeling shall comply with Sections M2301.3.1 and M2301.3.2.

M2301.3.1 Collectors. Collectors shall be listed and labeled to show the manufacturer's name, model number, serial number, collector weight, collector maximum allowable temperatures and pressures, and the type of heat transfer fluids that are compatible with the collector. The label shall clarify that these specifications apply only to the collector.

M2301.3.2 Thermal storage units. Pressurized thermal storage units shall be listed and labeled to show the manufacturer's name, model number, serial number, storage unit maximum and minimum allowable operating temperatures and pressures, and the type of heat transfer fluids that are compatible with the storage unit. The label shall clarify that these specifications apply only to the thermal storage unit.

M2301.4 Prohibited heat transfer fluids. Flammable gases and liquids shall not be used as heat transfer fluids.

M2301.5 Backflow protection. Connections from the potable water supply to solar systems shall comply with Section P2902.5.5.

Part VI — Fuel Gas

CHAPTER 24
FUEL GAS

The text of this chapter is excerpted from the 2006 edition of the *International Fuel Gas Code* and has been modified where necessary to make such text conform to the scope of application of the *International Residential Code for One- and Two-Family Dwellings*. The section numbers appearing in parentheses after each section number represent the location of the corresponding text in the *International Fuel Gas Code*.

SECTION G2401 (101)
GENERAL

G2401.1 (101.2) Application. This chapter covers those fuel-gas piping systems, fuel-gas utilization equipment and related accessories, venting systems and combustion air configurations most commonly encountered in the construction of one- and two-family dwellings and structures regulated by this code.

Coverage of piping systems shall extend from the point of delivery to the outlet of the equipment shutoff valves (see "Point of delivery"). Piping systems requirements shall include design, materials, components, fabrication, assembly, installation, testing, inspection, operation and maintenance. Requirements for gas utilization equipment and related accessories shall include installation, combustion and ventilation air and venting and connections to piping systems.

The omission from this chapter of any material or method of installation provided for in the *International Fuel Gas Code* shall not be construed as prohibiting the use of such material or method of installation. Fuel-gas piping systems, fuel-gas utilization equipment and related accessories, venting systems and combustion air configurations not specifically covered in these chapters shall comply with the applicable provisions of the *International Fuel Gas Code*.

Gaseous hydrogen systems shall be regulated by Chapter 7 of the *International Fuel Gas Code*.

This chapter shall not apply to the following:

1. Liquified natural gas (LNG) installations.

2. Temporary LP-gas piping for buildings under construction or renovation that is not to become part of the permanent piping system.

3. Except as provided in Section G2412.1.1, gas piping, meters, gas pressure regulators, and other appurtenances used by the serving gas supplier in the distribution of gas, other than undiluted LP-gas.

4. Portable LP-gas equipment of all types that is not connected to a fixed fuel piping system.

5. Portable fuel cell appliances that are neither connected to a fixed piping system nor interconnected to a power grid.

6. Installation of hydrogen gas, LP-gas and compressed natural gas (CNG) systems on vehicles.

SECTION G2402 (201)
GENERAL

G2402.1 (201.1) Scope. Unless otherwise expressly stated, the following words and terms shall, for the purposes of this chapter, have the meanings indicated in this chapter.

G2402.2 (201.2) Interchangeability. Words used in the present tense include the future; words in the masculine gender include the feminine and neuter; the singular number includes the plural and the plural, the singular.

G2402.3 (201.3) Terms defined in other codes. Where terms are not defined in this code and are defined in the ICC *Electrical Code, International Building Code, International Fire Code, International Mechanical Code* or *International Plumbing Code*, such terms shall have meanings ascribed to them as in those codes.

SECTION G2403 (202)
GENERAL DEFINITIONS

AIR CONDITIONING, GAS FIRED. A gas-burning, automatically operated appliance for supplying cooled and/or dehumidified air or chilled liquid.

AIR, EXHAUST. Air being removed from any space or piece of equipment and conveyed directly to the atmosphere by means of openings or ducts.

AIR-HANDLING UNIT. A blower or fan used for the purpose of distributing supply air to a room, space or area.

AIR, MAKEUP. Air that is provided to replace air being exhausted.

ALTERATION. A change in a system that involves an extension, addition or change to the arrangement, type or purpose of the original installation.

ANODELESS RISER. A transition assembly in which plastic piping is installed and terminated above ground outside of a building.

APPLIANCE (EQUIPMENT). Any apparatus or equipment that utilizes gas as a fuel or raw material to produce light, heat, power, refrigeration or air conditioning.

APPLIANCE, FAN-ASSISTED COMBUSTION. An appliance equipped with an integral mechanical means to either draw or force products of combustion through the combustion chamber or heat exchanger.

APPLIANCE, AUTOMATICALLY CONTROLLED. Appliances equipped with an automatic burner ignition and safety shut-off device and other automatic devices, which accomplish complete turn-on and shut-off of the gas to the main burner or burners, and graduate the gas supply to the burner or burners, but do not affect complete shut-off of the gas.

APPLIANCE, UNVENTED. An appliance designed or installed in such a manner that the products of combustion are not conveyed by a vent or chimney directly to the outside atmosphere.

APPLIANCE, VENTED. An appliance designed and installed in such a manner that all of the products of combustion are conveyed directly from the appliance to the outside atmosphere through an approved chimney or vent system.

APPROVED. Acceptable to the code official or other authority having jurisdiction.

ATMOSPHERIC PRESSURE. The pressure of the weight of air and water vapor on the surface of the earth, approximately 14.7 pounds per square inch (psia) (101 kPa absolute) at sea level.

AUTOMATIC IGNITION. Ignition of gas at the burner(s) when the gas controlling device is turned on, including reignition if the flames on the burner(s) have been extinguished by means other than by the closing of the gas controlling device.

BAROMETRIC DRAFT REGULATOR. A balanced damper device attached to a chimney, vent connector, breeching or flue gas manifold to protect combustion equipment by controlling chimney draft. A double-acting barometric draft regulator is one whose balancing damper is free to move in either direction to protect combustion equipment from both excessive draft and backdraft.

BOILER, LOW-PRESSURE. A self-contained gas-fired appliance for supplying steam or hot water.

Hot water heating boiler. A boiler in which no steam is generated, from which hot water is circulated for heating purposes and then returned to the boiler, and that operates at water pressures not exceeding 160 psig (1100 kPa gauge) and at water temperatures not exceeding 250°F (121°C) at or near the boiler outlet.

Hot water supply boiler. A boiler, completely filled with water, which furnishes hot water to be used externally to itself, and that operates at water pressures not exceeding 160 psig (1100 kPa gauge) and at water temperatures not exceeding 250°F (121°C) at or near the boiler outlet.

Steam heating boiler. A boiler in which steam is generated and that operates at a steam pressure not exceeding 15 psig (100 kPa gauge).

BRAZING. A metal joining process wherein coalescence is produced by the use of a nonferrous filler metal having a melting point above 1,000°F (538°C), but lower than that of the base metal being joined. The filler material is distributed between the closely fitted surfaces of the joint by capillary action.

BTU. Abbreviation for British thermal unit, which is the quantity of heat required to raise the temperature of 1 pound (454 g) of water 1°F (0.56°C) (1 Btu = 1055 J).

BURNER. A device for the final conveyance of the gas, or a mixture of gas and air, to the combustion zone.

Induced-draft. A burner that depends on draft induced by a fan that is an integral part of the appliance and is located downstream from the burner.

Power. A burner in which gas, air or both are supplied at pressures exceeding, for gas, the line pressure, and for air, atmospheric pressure, with this added pressure being applied at the burner.

CHIMNEY. A primarily vertical structure containing one or more flues, for the purpose of carrying gaseous products of combustion and air from an appliance to the outside atmosphere.

Factory-built chimney. A listed and labeled chimney composed of factory-made components, assembled in the field in accordance with manufacturer's instructions and the conditions of the listing.

Masonry chimney. A field-constructed chimney composed of solid masonry units, bricks, stones or concrete.

CLEARANCE. The minimum distance through air measured between the heat-producing surface of the mechanical appliance, device or equipment and the surface of the combustible material or assembly.

CLOTHES DRYER. An appliance used to dry wet laundry by means of heated air.

Type 1. Factory-built package, multiple production. Primarily used in the family living environment. Usually the smallest unit physically and in function output.

CODE. These regulations, subsequent amendments thereto, or any emergency rule or regulation that the administrative authority having jurisdiction has lawfully adopted.

CODE OFFICIAL. The officer or other designated authority charged with the administration and enforcement of this code, or a duly authorized representative.

COMBUSTION. In the context of this code, refers to the rapid oxidation of fuel accompanied by the production of heat or heat and light.

COMBUSTION AIR. Air necessary for complete combustion of a fuel, including theoretical air and excess air.

COMBUSTION CHAMBER. The portion of an appliance within which combustion occurs.

COMBUSTION PRODUCTS. Constituents resulting from the combustion of a fuel with the oxygen of the air, including the inert gases, but excluding excess air.

CONCEALED LOCATION. A location that cannot be accessed without damaging permanent parts of the building structure or finish surface. Spaces above, below or behind readily removable panels or doors shall not be considered as concealed.

CONCEALED PIPING. Piping that is located in a concealed location (see "Concealed location").

CONDENSATE. The liquid that condenses from a gas (including flue gas) caused by a reduction in temperature or increase in pressure.

CONNECTOR, APPLIANCE (Fuel). Rigid metallic pipe and fittings, semirigid metallic tubing and fittings or a listed and labeled device that connects an appliance to the gas piping system.

CONNECTOR, CHIMNEY OR VENT. The pipe that connects an appliance to a chimney or vent.

CONTROL. A manual or automatic device designed to regulate the gas, air, water or electrical supply to, or operation of, a mechanical system.

CONVERSION BURNER. A unit consisting of a burner and its controls for installation in an appliance originally utilizing another fuel.

CUBIC FOOT. The amount of gas that occupies 1 cubic foot (0.02832 m³) when at a temperature of 60°F (16°C), saturated with water vapor and under a pressure equivalent to that of 30 inches of mercury (101 kPa).

DAMPER. A manually or automatically controlled device to regulate draft or the rate of flow of air or combustion gases.

DECORATIVE GAS APPLIANCE, VENTED. A vented appliance wherein the primary function lies in the aesthetic effect of the flames.

DECORATIVE GAS APPLIANCES FOR INSTALLATION IN VENTED FIREPLACES. A vented appliance designed for installation within the fire chamber of a vented fireplace, wherein the primary function lies in the aesthetic effect of the flames.

DEMAND. The maximum amount of gas input required per unit of time, usually expressed in cubic feet per hour, or Btu/h (1 Btu/h = 0.2931 W).

DESIGN FLOOD ELEVATION. The elevation of the "design flood," including wave height, relative to the datum specified on the community's legally designated flood hazard map.

DILUTION AIR. Air that is introduced into a draft hood and is mixed with the flue gases.

DIRECT-VENT APPLIANCES. Appliances that are constructed and installed so that all air for combustion is derived directly from the outside atmosphere and all flue gases are discharged directly to the outside atmosphere.

DRAFT. The pressure difference existing between the equipment or any component part and the atmosphere, that causes a continuous flow of air and products of combustion through the gas passages of the appliance to the atmosphere.

Mechanical or induced draft. The pressure difference created by the action of a fan, blower or ejector that is located between the appliance and the chimney or vent termination.

Natural draft. The pressure difference created by a vent or chimney because of its height, and the temperature difference between the flue gases and the atmosphere.

DRAFT HOOD. A nonadjustable device built into an appliance, or made as part of the vent connector from an appliance, that is designed to (1) provide for ready escape of the flue gases from the appliance in the event of no draft, backdraft, or stoppage beyond the draft hood, (2) prevent a backdraft from entering the appliance, and (3) neutralize the effect of stack action of the chimney or gas vent upon operation of the appliance.

DRAFT REGULATOR. A device that functions to maintain a desired draft in the appliance by automatically reducing the draft to the desired value.

DRIP. The container placed at a low point in a system of piping to collect condensate and from which the condensate is removable.

DUCT FURNACE. A warm-air furnace normally installed in an air-distribution duct to supply warm air for heating. This definition shall apply only to a warm-air heating appliance that depends for air circulation on a blower not furnished as part of the furnace.

DWELLING UNIT. A single unit providing complete, independent living facilities for one or more persons, including permanent provisions for living, sleeping, eating, cooking and sanitation.

EQUIPMENT. See "Appliance."

FIREPLACE. A fire chamber and hearth constructed of noncombustible material for use with solid fuels and provided with a chimney.

Masonry fireplace. A hearth and fire chamber of solid masonry units such as bricks, stones, listed masonry units or reinforced concrete, provided with a suitable chimney.

Factory-built fireplace. A fireplace composed of listed factory-built components assembled in accordance with the terms of listing to form the completed fireplace.

FLAME SAFEGUARD. A device that will automatically shut off the fuel supply to a main burner or group of burners when the means of ignition of such burners becomes inoperative, and when flame failure occurs on the burner or group of burners.

FLOOD HAZARD AREA. The greater of the following two areas:

1. The area within a floodplain subject to a 1 percent or greater chance of flooding in any given year.

2. This area designated as a flood hazard area on a community's flood hazard map, or otherwise legally designated.

FLOOR FURNACE. A completely self-contained furnace suspended from the floor of the space being heated, taking air for combustion from outside such space and with means for observing flames and lighting the appliance from such space.

FLUE, APPLIANCE. The passage(s) within an appliance through which combustion products pass from the combustion chamber of the appliance to the draft hood inlet opening on an appliance equipped with a draft hood or to the outlet of the appliance on an appliance not equipped with a draft hood.

FLUE COLLAR. That portion of an appliance designed for the attachment of a draft hood, vent connector or venting system.

FLUE GASES. Products of combustion plus excess air in appliance flues or heat exchangers.

FLUE LINER (LINING). A system or material used to form the inside surface of a flue in a chimney or vent, for the purpose of protecting the surrounding structure from the effects of combustion products and for conveying combustion products without leakage to the atmosphere.

FUEL GAS. A natural gas, manufactured gas, liquefied petroleum gas or mixtures of these gases.

FUEL GAS UTILIZATION EQUIPMENT. See "Appliance."

FURNACE. A completely self-contained heating unit that is designed to supply heated air to spaces remote from or adjacent to the appliance location.

FURNACE, CENTRAL FURNACE. A self-contained appliance for heating air by transfer of heat of combustion through metal to the air, and designed to supply heated air through ducts to spaces remote from or adjacent to the appliance location.

FURNACE PLENUM. An air compartment or chamber to which one or more ducts are connected and which forms part of an air distribution system.

GAS CONVENIENCE OUTLET. A permanently mounted, manually operated device that provides the means for connecting an appliance to, and disconnecting an appliance from, the gas supply piping. The device includes an integral, manually operated valve with a nondisplaceable valve member and is designed so that disconnection of an appliance only occurs when the manually operated valve is in the closed position.

GAS PIPING. An installation of pipe, valves or fittings installed on a premises or in a building and utilized to convey fuel gas.

GAS UTILIZATION EQUIPMENT. An appliance that utilizes gas as a fuel or raw material or both.

HAZARDOUS LOCATION. Any location considered to be a fire hazard for flammable vapors, dust, combustible fibers or other highly combustible substances. The location is not necessarily categorized in the *International Building Code* as a high-hazard use group classification.

HOUSE PIPING. See "Piping system."

IGNITION PILOT. A pilot that operates during the lighting cycle and discontinues during main burner operation.

IGNITION SOURCE. A flame spark or hot surface capable of igniting flammable vapors or fumes. Such sources include appliance burners, burner ignitors and electrical switching devices.

INFRARED RADIANT HEATER. A heater which directs a substantial amount of its energy output in the form of infrared radiant energy into the area to be heated. Such heaters are of either the vented or unvented type.

JOINT, FLARED. A metal-to-metal compression joint in which a conical spread is made on the end of a tube that is compressed by a flare nut against a mating flare.

JOINT, MECHANICAL. A general form of gas-tight joints obtained by the joining of metal parts through a positive-holding mechanical construction, such as flanged joint, threaded joint, flared joint or compression joint.

JOINT, PLASTIC ADHESIVE. A joint made in thermoset plastic piping by the use of an adhesive substance which forms a continuous bond between the mating surfaces without dissolving either one of them.

LIQUEFIED PETROLEUM GAS or LPG (LP-GAS). Liquefied petroleum gas composed predominately of propane, propylene, butanes or butylenes, or mixtures thereof that is gaseous under normal atmospheric conditions, but is capable of being liquefied under moderate pressure at normal temperatures.

LIVING SPACE. Space within a dwelling unit utilized for living, sleeping, eating, cooking, bathing, washing and sanitation purposes.

LOG LIGHTER, GAS-FIRED. A manually operated solid-fuel ignition appliance for installation in a vented solid-fuel-burning fireplace.

MAIN BURNER. A device or group of devices essentially forming an integral unit for the final conveyance of gas or a mixture of gas and air to the combustion zone, and on which combustion takes place to accomplish the function for which the appliance is designed.

METER. The instrument installed to measure the volume of gas delivered through it.

MODULATING. Modulating or throttling is the action of a control from its maximum to minimum position in either predetermined steps or increments of movement as caused by its actuating medium.

OFFSET (VENT). A combination of approved bends that make two changes in direction bringing one section of the vent out of line, but into a line parallel with the other section.

OUTLET. A threaded connection or bolted flange in a pipe system to which a gas-burning appliance is attached.

OXYGEN DEPLETION SAFETY SHUTOFF SYSTEM (ODS). A system designed to act to shut off the gas supply to the main and pilot burners if the oxygen in the surrounding atmosphere is reduced below a predetermined level.

PILOT. A small flame that is utilized to ignite the gas at the main burner or burners.

PIPING. Where used in this code, "piping" refers to either pipe or tubing, or both.

    Pipe. A rigid conduit of iron, steel, copper, brass or plastic.

    Tubing. Semirigid conduit of copper, aluminum, plastic or steel.

PIPING SYSTEM. All fuel piping, valves, and fittings from the outlet of the point of delivery to the outlets of the equipment shutoff valves.

PLASTIC, THERMOPLASTIC. A plastic that is capable of being repeatedly softened by increase of temperature and hardened by decrease of temperature.

POINT OF DELIVERY. For natural gas systems, the point of delivery is the outlet of the service meter assembly or the outlet of the service regulator or service shutoff valve where a meter is not provided. Where a valve is provided at the outlet of the service meter assembly, such valve shall be considered to be downstream of the point of delivery. For undiluted liquefied petroleum gas systems, the point of delivery shall be considered to be the outlet of the first regulator that reduces pressure to 2 psig (13.8 kPa) or less.

PRESSURE DROP. The loss in pressure due to friction or obstruction in pipes, valves, fittings, regulators and burners.

PRESSURE TEST. An operation performed to verify the gas-tight integrity of gas piping following its installation or modification.

READY ACCESS (TO). That which enables a device, appliance or equipment to be directly reached, without requiring the removal or movement of any panel, door or similar obstruction. (See "Access.")

REGULATOR. A device for controlling and maintaining a uniform gas supply pressure, either pounds-to-inches water column (MP regulator) or inches-to-inches water column (appliance regulator).

REGULATOR, GAS APPLIANCE. A pressure regulator for controlling pressure to the manifold of gas equipment.

REGULATOR, LINE GAS PRESSURE. A device placed in a gas line between the service pressure regulator and the equipment for controlling, maintaining or reducing the pressure in that portion of the piping system downstream of the device.

REGULATOR, MEDIUM-PRESSURE (MP Regulator). A line pressure regulator that reduces gas pressure from the range of greater than 0.5 psig (3.4 kPa) and less than or equal to 5 psig (34.5 kPa) to a lower pressure.

REGULATOR, PRESSURE. A device placed in a gas line for reducing, controlling and maintaining the pressure in that portion of the piping system downstream of the device.

REGULATOR, SERVICE PRESSURE. A device installed by the serving gas supplier to reduce and limit the service line gas pressure to delivery pressure.

RELIEF OPENING. The opening provided in a draft hood to permit the ready escape to the atmosphere of the flue products from the draft hood in the event of no draft, backdraft or stoppage beyond the draft hood, and to permit air into the draft hood in the event of a strong chimney updraft.

RELIEF VALVE (DEVICE). A safety valve designed to forestall the development of a dangerous condition by relieving either pressure, temperature or vacuum in the hot water supply system.

RELIEF VALVE, PRESSURE. An automatic valve which opens and closes a relief vent, depending on whether the pressure is above or below a predetermined value.

RELIEF VALVE, TEMPERATURE

Manual reset type. A valve which automatically opens a relief vent at a predetermined temperature and which must be manually returned to the closed position.

Reseating or self-closing type. An automatic valve which opens and closes a relief vent, depending on whether the temperature is above or below a predetermined value.

RELIEF VALVE, VACUUM. A valve that automatically opens and closes a vent for relieving a vacuum within the hot water supply system, depending on whether the vacuum is above or below a predetermined value.

RISER, GAS. A vertical pipe supplying fuel gas.

ROOM HEATER, UNVENTED. See "Unvented room heater."

ROOM HEATER, VENTED. A free-standing gas-fired heating unit used for direct heating of the space in and adjacent to that in which the unit is located. [See also "Vented room heater."]

SAFETY SHUTOFF DEVICE. See "Flame safeguard."

SHAFT. An enclosed space extending through one or more stories of a building, connecting vertical openings in successive floors, or floors and the roof.

SPECIFIC GRAVITY. As applied to gas, specific gravity is the ratio of the weight of a given volume to that of the same volume of air, both measured under the same condition.

THERMOSTAT

Electric switch type. A device that senses changes in temperature and controls electrically, by means of separate components, the flow of gas to the burner(s) to maintain selected temperatures.

Integral gas valve type. An automatic device, actuated by temperature changes, designed to control the gas supply to the burner(s) in order to maintain temperatures between predetermined limits, and in which the thermal actuating element is an integral part of the device.

1. Graduating thermostat. A thermostat in which the motion of the valve is approximately in direct proportion to the effective motion of the thermal element induced by temperature change.

2. Snap-acting thermostat. A thermostat in which the thermostatic valve travels instantly from the closed to the open position, and vice versa.

TRANSITION FITTINGS, PLASTIC TO STEEL. An adapter for joining plastic pipe to steel pipe. The purpose of this fitting is to provide a permanent, pressure-tight connection between two materials that cannot be joined directly one to another.

UNIT HEATER

High-static pressure type. A self-contained, automatically controlled, vented appliance having integral means for circulation of air against 0.2 inch (15 mm H_2O) or greater static pressure. Such appliance is equipped with provisions for attaching an outlet air duct and, where the appliance is for indoor installation remote from the space to be heated, is also equipped with provisions for attaching an inlet air duct.

Low-static pressure type. A self-contained, automatically controlled, vented appliance, intended for installation in the space to be heated without the use of ducts, having integral means for circulation of air. Such units are allowed to be equipped with louvers or face extensions made in accordance with the manufacturer's specifications.

UNVENTED ROOM HEATER. An unvented heating appliance designed for stationary installation and utilized to provide comfort heating. Such appliances provide radiant heat or convection heat by gravity or fan circulation directly from the heater and do not utilize ducts.

VALVE. A device used in piping to control the gas supply to any section of a system of piping or to an appliance.

Automatic. An automatic or semiautomatic device consisting essentially of a valve and operator that control the gas supply to the burner(s) during operation of an appliance. The operator shall be actuated by application of gas pressure on a flexible diaphragm, by electrical means, by mechanical means or by other approved means.

Automatic gas shutoff. A valve used in conjunction with an automatic gas shutoff device to shut off the gas supply to a water heating system. It shall be constructed integrally with the gas shutoff device or shall be a separate assembly.

Equipment shutoff. A valve located in the piping system, used to isolate individual equipment for purposes such as service or replacement.

Individual main burner. A valve that controls the gas supply to an individual main burner.

Main burner control. A valve that controls the gas supply to the main burner manifold.

Manual main gas-control. A manually operated valve in the gas line for the purpose of completely turning on or shutting off the gas supply to the appliance, except to pilot or pilots that are provided with independent shutoff.

Manual reset. An automatic shutoff valve installed in the gas supply piping and set to shut off when unsafe conditions occur. The device remains closed until manually reopened.

Service shutoff. A valve, installed by the serving gas supplier between the service meter or source of supply and the customer piping system, to shut off the entire piping system.

VENT. A pipe or other conduit composed of factory-made components, containing a passageway for conveying combustion products and air to the atmosphere, listed and labeled for use with a specific type or class of appliance.

Special gas vent. A vent listed and labeled for use with listed Category II, III and IV gas appliances.

Type B vent. A vent listed and labeled for use with appliances with draft hoods and other Category I appliances that are listed for use with Type B vents.

Type BW vent. A vent listed and labeled for use with wall furnaces.

Type L vent. A vent listed and labeled for use with appliances that are listed for use with Type L or Type B vents.

VENT CONNECTOR. See "Connector."

VENT PIPING

Breather. Piping run from a pressure-regulating device to the outdoors, designed to provide a reference to atmospheric pressure. If the device incorporates an integral pressure relief mechanism, a breather vent can also serve as a relief vent.

Relief. Piping run from a pressure-regulating or pressure-limiting device to the outdoors, designed to provide for the safe venting of gas in the event of excessive pressure in the gas piping system.

VENTED GAS APPLIANCE CATEGORIES. Appliances that are categorized for the purpose of vent selection are classified into the following four categories:

Category I. An appliance that operates with a nonpositive vent static pressure and with a vent gas temperature that avoids excessive condensate production in the vent.

Category II. An appliance that operates with a nonpositive vent static pressure and with a vent gas temperature that is capable of causing excessive condensate production in the vent.

Category III. An appliance that operates with a positive vent static pressure and with a vent gas temperature that avoids excessive condensate production in the vent.

Category IV. An appliance that operates with a positive vent static pressure and with a vent gas temperature that is capable of causing excessive condensate production in the vent.

VENTED ROOM HEATER. A vented self-contained, free-standing, nonrecessed appliance for furnishing warm air to the space in which it is installed, directly from the heater without duct connections.

VENTED WALL FURNACE. A self-contained vented appliance complete with grilles or equivalent, designed for incorporation in or permanent attachment to the structure of a building, mobile home or travel trailer, and furnishing heated air circulated by gravity or by a fan directly into the space to be heated through openings in the casing. This definition shall exclude floor furnaces, unit heaters and central furnaces as herein defined.

VENTING SYSTEM. A continuous open passageway from the flue collar or draft hood of an appliance to the outside atmosphere for the purpose of removing flue or vent gases. A venting system is usually composed of a vent or a chimney and vent connector, if used, assembled to form the open passageway.

WALL HEATER, UNVENTED TYPE. A room heater of the type designed for insertion in or attachment to a wall or partition. Such heater does not incorporate concealed venting arrangements in its construction and discharges all products of combustion through the front into the room being heated.

WATER HEATER. Any heating appliance or equipment that heats potable water and supplies such water to the potable hot water distribution system.

SECTION G2404 (301)
GENERAL

G2404.1 (301.1) Scope. This section shall govern the approval and installation of all equipment and appliances that comprise parts of the installations regulated by this code in accordance with Section G2401.

G2404.2 (301.1.1) Other fuels. The requirements for combustion and dilution air for gas-fired appliances shall be governed by Section G2407. The requirements for combustion and dilution air for appliances operating with fuels other than fuel gas shall be regulated by Chapter 17.

G2404.3 (301.3) Listed and labeled. Appliances regulated by this code shall be listed and labeled for the application in which they are used unless otherwise approved in accordance with Section R104.11. The approval of unlisted appliances in accordance with Section R104.11 shall be based upon approved engineering evaluation.

G2404.4 (301.8) Vibration isolation. Where means for isolation of vibration of an appliance is installed, an approved means for support and restraint of that appliance shall be provided.

G2404.5 (301.9) Repair. Defective material or parts shall be replaced or repaired in such a manner so as to preserve the original approval or listing.

G2404.6 (301.10) Wind resistance. Appliances and supports that are exposed to wind shall be designed and installed to resist the wind pressures determined in accordance with this code.

G2404.7 (301.11) Flood hazard. For structures located in flood hazard areas, the appliance, equipment and system installations regulated by this code shall be located at or above the design flood elevation and shall comply with the flood-resistant construction requirements of Section R324.

> **Exception:** The appliance, equipment and system installations regulated by this code are permitted to be located below the design flood elevation provided that they are designed and installed to prevent water from entering or accumulating within the components and to resist hydrostatic and hydrodynamic loads and stresses, including the effects of buoyancy, during the occurrence of flooding to the design flood elevation and shall comply with the flood-resistant construction requirements of Section R324.

G2404.8 (301.12) Seismic resistance. When earthquake loads are applicable in accordance with this code, the supports shall be designed and installed for the seismic forces in accordance with this code.

G2404.9 (301.14) Rodentproofing. Buildings or structures and the walls enclosing habitable or occupiable rooms and spaces in which persons live, sleep or work, or in which feed, food or foodstuffs are stored, prepared, processed, served or sold, shall be constructed to protect against the entry of rodents.

G2404.10 (307.5) Auxiliary drain pan. Category IV condensing appliances shall be provided with an auxiliary drain pan where damage to any building component will occur as a result of stoppage in the condensate drainage system. Such pan shall be installed in accordance with the applicable provisions of Section M1411.

> **Exception:** An auxiliary drain pan shall not be required for appliances that automatically shut down operation in the event of a stoppage in the condensate drainage system.

SECTION G2405 (302)
STRUCTURAL SAFETY

G2405.1 (302.1) Structural safety. The building shall not be weakened by the installation of any gas piping. In the process of installing or repairing any gas piping, the finished floors, walls, ceilings, tile work or any other part of the building or premises which are required to be changed or replaced shall be left in a safe structural condition in accordance with the requirements of this code.

G2405.2 (302.4) Alterations to trusses. Truss members and components shall not be cut, drilled, notched, spliced or otherwise altered in any way without the written concurrence and approval of a registered design professional. Alterations resulting in the addition of loads to any member (e.g., HVAC equipment, water heaters) shall not be permitted without verification that the truss is capable of supporting such additional loading.

G2405.3 (302.3.1) Engineered wood products. Cuts, notches and holes bored in trusses, structural composite lumber, structural glued-laminated members and I-joists are prohibited except where permitted by the manufacturer's recommendations or where the effects of such alterations are specifically considered in the design of the member by a registered design professional.

SECTION G2406 (303)
APPLIANCE LOCATION

G2406.1 (303.1) General. Appliances shall be located as required by this section, specific requirements elsewhere in this code and the conditions of the equipment and appliance listing.

G2406.2 (303.3) Prohibited locations. Appliances shall not be located in sleeping rooms, bathrooms, toilet rooms, storage closets or surgical rooms, or in a space that opens only into such rooms or spaces, except where the installation complies with one of the following:

1. The appliance is a direct-vent appliance installed in accordance with the conditions of the listing and the manufacturer's instructions.

2. Vented room heaters, wall furnaces, vented decorative appliances, vented gas fireplaces, vented gas fireplace heaters and decorative appliances for installation in vented solid fuel-burning fireplaces are installed in rooms that meet the required volume criteria of Section G2407.5.

3. A single wall-mounted unvented room heater is installed in a bathroom and such unvented room heater is equipped as specified in Section G2445.6 and has an input rating not greater than 6,000 Btu/h (1.76 kW). The

bathroom shall meet the required volume criteria of Section G2407.5.

4. A single wall-mounted unvented room heater is installed in a bedroom and such unvented room heater is equipped as specified in Section G2445.6 and has an input rating not greater than 10,000 Btu/h (2.93 kW). The bedroom shall meet the required volume criteria of Section G2407.5.

5. The appliance is installed in a room or space that opens only into a bedroom or bathroom, and such room or space is used for no other purpose and is provided with a solid weather-stripped door equipped with an approved self-closing device. All combustion air shall be taken directly from the outdoors in accordance with Section G2407.6.

G2406.3 (303.6) Outdoor locations. Equipment installed in outdoor locations shall be either listed for outdoor installation or provided with protection from outdoor environmental factors that influence the operability, durability and safety of the equipment.

SECTION G2407 (304)
COMBUSTION, VENTILATION AND DILUTION AIR

G2407.1 (304.1) General. Air for combustion, ventilation and dilution of flue gases for appliances installed in buildings shall be provided by application of one of the methods prescribed in Sections G2407.5 through G2407.9. Where the requirements of Section G2407.5 are not met, outdoor air shall be introduced in accordance with one of the methods prescribed in Sections G2407.6 through G2407.9. Direct-vent appliances, gas appliances of other than natural draft design and vented gas appliances other than Category I shall be provided with combustion, ventilation and dilution air in accordance with the appliance manufacturer's instructions.

Exception: Type 1 clothes dryers that are provided with makeup air in accordance with Section G2439.4.

G2407.2 (304.2) Appliance location. Appliances shall be located so as not to interfere with proper circulation of combustion, ventilation and dilution air.

G2407.3 (304.3) Draft hood/regulator location. Where used, a draft hood or a barometric draft regulator shall be installed in the same room or enclosure as the appliance served so as to prevent any difference in pressure between the hood or regulator and the combustion air supply.

G2407.4 (304.4) Makeup air provisions. Makeup air requirements for the operation of exhaust fans, kitchen ventilation systems, clothes dryers and fireplaces shall be considered in determining the adequacy of a space to provide combustion air requirements.

G2407.5 (304.5) Indoor combustion air. The required volume of indoor air shall be determined in accordance with Section G2407.5.1 or G2407.5.2, except that where the air infiltration rate is known to be less than 0.40 air changes per hour (ACH), Section G2407.5.2 shall be used. The total required volume shall be the sum of the required volume calculated for all appliances located within the space. Rooms communicating directly with the space in which the appliances are installed through openings not furnished with doors, and through combustion air openings sized and located in accordance with Section G2407.5.3, are considered to be part of the required volume.

G2407.5.1 (304.5.1) Standard method. The minimum required volume shall be 50 cubic feet per 1,000 Btu/h (4.8 m^3/kW).

G2407.5.2 (304.5.2) Known air-infiltration-rate method. Where the air infiltration rate of a structure is known, the minimum required volume shall be determined as follows:

For appliances other than fan assisted, calculate volume using Equation 24-1.

$$\text{Required Volume}_{other} \geq \frac{21\,\text{ft}^3}{ACH}\left(\frac{I_{other}}{1,000\,\text{Btu}/\text{hr}}\right)$$

(Equation 24-1)

For fan-assisted appliances, calculate volume using Equation 24-2.

$$\text{Required Volume}_{fan} \geq \frac{15\,\text{ft}^3}{ACH}\left(\frac{I_{fan}}{1,000\,\text{Btu}/\text{hr}}\right)$$

(Equation 24-2)

where:

I_{other} = All appliances other than fan assisted (input in Btu/h).

I_{fan} = Fan-assisted appliance (input in Btu/h).

ACH = Air change per hour (percent of volume of space exchanged per hour, expressed as a decimal).

For purposes of this calculation, an infiltration rate greater than 0.60 ACH shall not be used in Equations 24-1 and 24-2.

G2407.5.3 (304.5.3) Indoor opening size and location. Openings used to connect indoor spaces shall be sized and located in accordance with Sections G2407.5.3.1 and G2407.5.3.2 (see Figure G2407.5.3).

G2407.5.3.1 (304.5.3.1) Combining spaces on the same story. Each opening shall have a minimum free area of 1 square inch per 1,000 Btu/h (2,200 mm^2/kW) of the total input rating of all appliances in the space, but not less than 100 square inches (0.06 m^2). One opening shall commence within 12 inches (305 mm) of the top and one opening shall commence within 12 inches (305 mm) of the bottom of the enclosure. The minimum dimension of air openings shall be not less than 3 inches (76 mm).

G2407.5.3.2 (304.5.3.2) Combining spaces in different stories. The volumes of spaces in different stories shall be considered as communicating spaces where such spaces are connected by one or more openings in doors or floors having a total minimum free area of 2 square inches per 1,000 Btu/h (4402 mm^2/kW) of total input rating of all appliances.

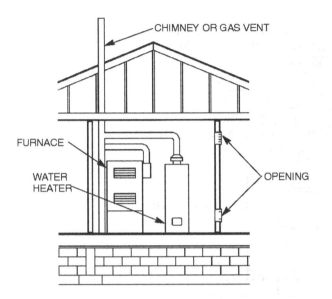

FIGURE G2407.5.3 (304.5.3)
ALL AIR FROM INSIDE THE BUILDING
(see Section G2407.5.3)

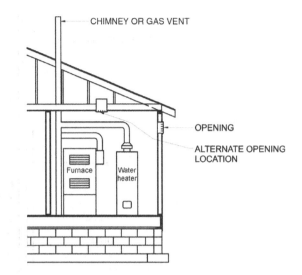

CHIMNEY OR GAS VENT

OPENING

ALTERNATE OPENING
LOCATION

Furnace   Water
           heater

FIGURE G2407.6.2 (304.6.2)
SINGLE COMBUSTION AIR OPENING,
ALL AIR FROM OUTDOORS
(see Section G2407.6.2)

G2407.6 (304.6) Outdoor combustion air. Outdoor combustion air shall be provided through opening(s) to the outdoors in accordance with Section G2407.6.1 or G2407.6.2. The minimum dimension of air openings shall be not less than 3 inches (76 mm).

G2407.6.1 (304.6.1) Two-permanent-openings method. Two permanent openings, one commencing within 12 inches (305 mm) of the top and one commencing within 12 inches (305 mm) of the bottom of the enclosure, shall be provided. The openings shall communicate directly, or by ducts, with the outdoors or spaces that freely communicate with the outdoors.

Where directly communicating with the outdoors, or where communicating with the outdoors through vertical ducts, each opening shall have a minimum free area of 1 square inch per 4,000 Btu/h (550 mm²/kW) of total input rating of all appliances in the enclosure [see Figures G2407.6.1(1) and G2407.6.1(2)].

Where communicating with the outdoors through horizontal ducts, each opening shall have a minimum free area of not less than 1 square inch per 2,000 Btu/h (1,100 mm²/kW) of total input rating of all appliances in the enclosure [see Figure G2407.6.1(3)].

G2407.6.2 (304.6.2) One-permanent-opening method. One permanent opening, commencing within 12 inches (305 mm) of the top of the enclosure, shall be provided. The appliance shall have clearances of at least 1 inch (25 mm) from the sides and back and 6 inches (152 mm) from the front of the appliance. The opening shall directly communicate with the outdoors or through a vertical or horizontal duct to the outdoors, or spaces that freely communicate with the outdoors (see Figure G2407.6.2) and shall have a minimum free area of 1 square inch per 3,000 Btu/h (734

mm²/kW) of the total input rating of all appliances located in the enclosure and not less than the sum of the areas of all vent connectors in the space.

G2407.7 (304.7) Combination indoor and outdoor combustion air. The use of a combination of indoor and outdoor combustion air shall be in accordance with Sections G2407.7.1 through G2407.7.3.

G2407.7.1 (304.7.1) Indoor openings. Where used, openings connecting the interior spaces shall comply with Section G2407.5.3.

G2407.7.2 (304.7.2) Outdoor opening location. Outdoor opening(s) shall be located in accordance with Section G2407.6.

G2407.7.3 (304.7.3) Outdoor opening(s) size. The outdoor opening(s) size shall be calculated in accordance with the following:

1. The ratio of interior spaces shall be the available volume of all communicating spaces divided by the required volume.

2. The outdoor size reduction factor shall be one minus the ratio of interior spaces.

3. The minimum size of outdoor opening(s) shall be the full size of outdoor opening(s) calculated in accordance with Section G2407.6, multiplied by the reduction factor. The minimum dimension of air openings shall be not less than 3 inches (76 mm).

G2407.8 (304.8) Engineered installations. Engineered combustion air installations shall provide an adequate supply of combustion, ventilation and dilution air and shall be approved.

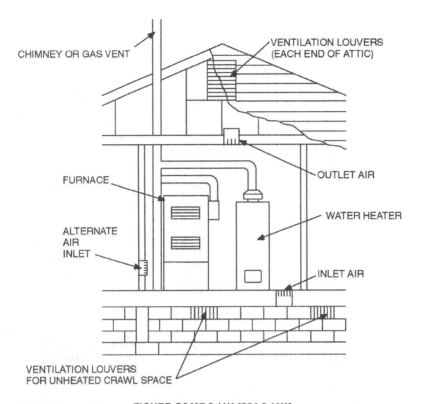

FIGURE G2407.6.1(1) [304.6.1(1)]
ALL AIR FROM OUTDOORS—INLET AIR FROM VENTILATED
CRAWL SPACE AND OUTLET AIR TO VENTILATED ATTIC (see Section G2407.6.1)

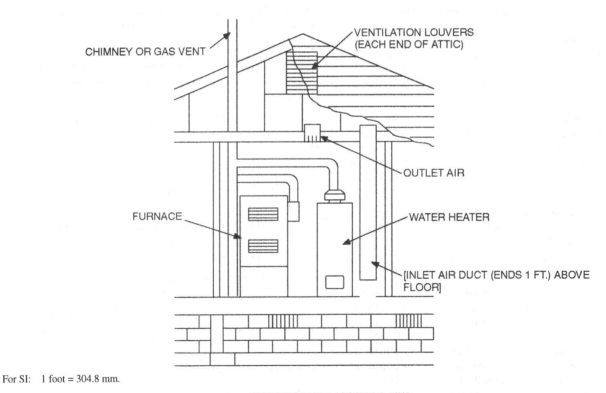

For SI: 1 foot = 304.8 mm.

FIGURE G2407.6.1(2) [304.6.1(2)]
ALL AIR FROM OUTDOORS THROUGH VENTILATED ATTIC
(see Section G2407.6.1)

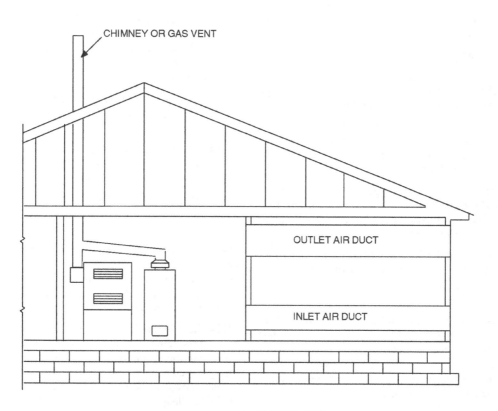

FIGURE G2407.6.1(3) [304.6.1(3)]
ALL AIR FROM OUTDOORS
(see Section G2407.6.1)

G2407.9 (304.9) Mechanical combustion air supply. Where all combustion air is provided by a mechanical air supply system, the combustion air shall be supplied from the outdoors at a rate not less than 0.35 cubic feet per minute per 1,000 Btu/h (0.034 m³/min per kW) of total input rating of all appliances located within the space.

G2407.9.1 (304.9.1) Makeup air. Where exhaust fans are installed, makeup air shall be provided to replace the exhausted air.

G2407.9.2 (304.9.2) Appliance interlock. Each of the appliances served shall be interlocked with the mechanical air supply system to prevent main burner operation when the mechanical air supply system is not in operation.

G2407.9.3 (304.9.3) Combined combustion air and ventilation air system. Where combustion air is provided by the building's mechanical ventilation system, the system shall provide the specified combustion air rate in addition to the required ventilation air.

G2407.10 (304.10) Louvers and grilles. The required size of openings for combustion, ventilation and dilution air shall be based on the net free area of each opening. Where the free area through a design of louver, grille or screen is known, it shall be used in calculating the size opening required to provide the free area specified. Where the design and free area of louvers and grilles are not known, it shall be assumed that wood louvers will have 25-percent free area and metal louvers and grilles will have

75-percent free area. Screens shall have a mesh size not smaller than 1/4 inch (6.4 mm). Nonmotorized louvers and grilles shall be fixed in the open position. Motorized louvers shall be interlocked with the appliance so that they are proven to be in the full open position prior to main burner ignition and during main burner operation. Means shall be provided to prevent the main burner from igniting if the louvers fail to open during burner start-up and to shut down the main burner if the louvers close during operation.

G2407.11 (304.11) Combustion air ducts. Combustion air ducts shall comply with all of the following:

1. Ducts shall be constructed of galvanized steel complying with Chapter 16 or of a material having equivalent corrosion resistance, strength and rigidity.

   Exception: Within dwellings units, unobstructed stud and joist spaces shall not be prohibited from conveying combustion air, provided that not more than one required fireblock is removed.

2. Ducts shall terminate in an unobstructed space allowing free movement of combustion air to the appliances.

3. Ducts shall serve a single enclosure.

4. Ducts shall not serve both upper and lower combustion air openings where both such openings are used. The separation between ducts serving upper and lower com-

bustion air openings shall be maintained to the source of combustion air.

5. Ducts shall not be screened where terminating in an attic space.

6. Horizontal upper combustion air ducts shall not slope downward toward the source of combustion air.

7. The remaining space surrounding a chimney liner, gas vent, special gas vent or plastic piping installed within a masonry, metal or factory-built chimney shall not be used to supply combustion air.

> **Exception:** Direct-vent gas-fired appliances designed for installation in a solid fuel-burning fireplace where installed in accordance with the manufacturer's instructions.

8. Combustion air intake openings located on the exterior of a building shall have the lowest side of such openings located not less than 12 inches (305 mm) vertically from the adjoining grade level.

G2407.12 (304.12) Protection from fumes and gases. Where corrosive or flammable process fumes or gases, other than products of combustion, are present, means for the disposal of such fumes or gases shall be provided. Such fumes or gases include carbon monoxide, hydrogen sulfide, ammonia, chlorine and halogenated hydrocarbons.

In barbershops, beauty shops and other facilities where chemicals that generate corrosive or flammable products, such as aerosol sprays, are routinely used, nondirect vent-type appliances shall be located in a mechanical room separated or partitioned off from other areas with provisions for combustion air and dilution air from the outdoors. Direct-vent appliances shall be installed in accordance with the appliance manufacturer's installation instructions.

SECTION G2408 (305)
INSTALLATION

G2408.1 (305.1) General. Equipment and appliances shall be installed as required by the terms of their approval, in accordance with the conditions of listing, the manufacturer's instructions and this code. Manufacturers' installation instructions shall be available on the job site at the time of inspection. Where a code provision is less restrictive than the conditions of the listing of the equipment or appliance or the manufacturer's installation instructions, the conditions of the listing and the manufacturer's installation instructions shall apply.

Unlisted appliances approved in accordance with Section 2404.3 shall be limited to uses recommended by the manufacturer and shall be installed in accordance with the manufacturer's instructions, the provisions of this code and the requirements determined by the code official.

G2408.2 (305.3) Elevation of ignition source. Equipment and appliances having an ignition source shall be elevated such that the source of ignition is not less than 18 inches (457 mm) above the floor in hazardous locations and public garages, private

garages, repair garages, motor fuel-dispensing facilities and parking garages. For the purpose of this section, rooms or spaces that are not part of the living space of a dwelling unit and that communicate directly with a private garage through openings shall be considered to be part of the private garage.

> **Exception:** Elevation of the ignition source is not required for appliances that are listed as flammable vapor ignition resistant.

G2408.3 (305.5) Private garages. Appliances located in private garages shall be installed with a minimum clearance of 6 feet (1829 mm) above the floor.

> **Exception:** The requirements of this section shall not apply where the appliances are protected from motor vehicle impact and installed in accordance with Section G2408.2.

G2408.4 (305.7) Clearances from grade. Equipment and appliances installed at grade level shall be supported on a level concrete slab or other approved material extending above adjoining grade or shall be suspended a minimum of 6 inches (152 mm) above adjoining grade.

G2408.5 (305.8) Clearances to combustible construction. Heat-producing equipment and appliances shall be installed to maintain the required clearances to combustible construction as specified in the listing and manufacturer's instructions. Such clearances shall be reduced only in accordance with Section G2409. Clearances to combustibles shall include such considerations as door swing, drawer pull, overhead projections or shelving and window swing. Devices, such as door stops or limits and closers, shall not be used to provide the required clearances.

SECTION G2409 (308)
CLEARANCE REDUCTION

G2409.1 (308.1) Scope. This section shall govern the reduction in required clearances to combustible materials and combustible assemblies for chimneys, vents, appliances, devices and equipment.

G2409.2 (308.2) Reduction table. The allowable clearance reduction shall be based on one of the methods specified in Table G2409.2 or shall utilize an assembly listed for such application. Where required clearances are not listed in Table G2409.2, the reduced clearances shall be determined by linear interpolation between the distances listed in the table. Reduced clearances shall not be derived by extrapolation below the range of the table. The reduction of the required clearances to combustibles for listed and labeled appliances and equipment shall be in accordance with the requirements of this section except that such clearances shall not be reduced where reduction is specifically prohibited by the terms of the appliance or equipment listing [see Figures G2409.2(1), G2409.2(2) and G2409.2(3)].

G2409.3 (308.3) Clearances for indoor air-conditioning appliances. Clearance requirements for indoor air-conditioning appliances shall comply with Sections G2409.3.1 through G2409.3.5.

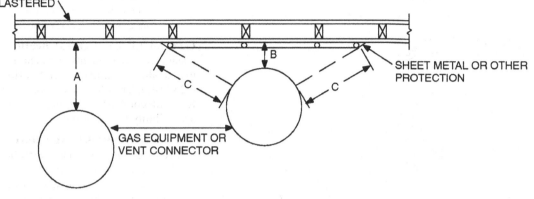

CONSTRUCTION USING COMBUSTIBLE MATERIAL,
PLASTERED OR UNPLASTERED

SHEET METAL OR OTHER
PROTECTION

GAS EQUIPMENT OR
VENT CONNECTOR

NOTES:

"A" equals the clearance with no protection.

"B" equals the reduced clearance permitted in accordance with Table G2409.2. The protection applied to the construction using combustible material shall extend far enough in each direction to make "C" equal to "A."

FIGURE G2409.2(1) [308.2(1)]
EXTENT OF PROTECTION NECESSARY TO REDUCE CLEARANCES
FROM GAS EQUIPMENT OR VENT CONNECTORS

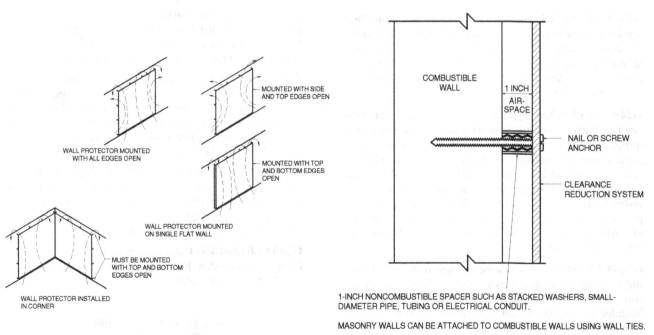

WALL PROTECTOR MOUNTED
WITH ALL EDGES OPEN

MOUNTED WITH SIDE
AND TOP EDGES OPEN

MOUNTED WITH TOP
AND BOTTOM EDGES
OPEN

WALL PROTECTOR MOUNTED
ON SINGLE FLAT WALL

MUST BE MOUNTED
WITH TOP AND BOTTOM
EDGES OPEN

WALL PROTECTOR INSTALLED
IN CORNER

COMBUSTIBLE
WALL

1 INCH
AIR-
SPACE

NAIL OR SCREW
ANCHOR

CLEARANCE
REDUCTION SYSTEM

1-INCH NONCOMBUSTIBLE SPACER SUCH AS STACKED WASHERS, SMALL-DIAMETER PIPE, TUBING OR ELECTRICAL CONDUIT.

MASONRY WALLS CAN BE ATTACHED TO COMBUSTIBLE WALLS USING WALL TIES.

DO NOT USE SPACERS DIRECTLY BEHIND APPLIANCE OR CONNECTOR.

For SI: 1 inch = 25.4 mm.

FIGURE G2409.2(2) [308.2(2)]
WALL PROTECTOR CLEARANCE REDUCTION SYSTEM

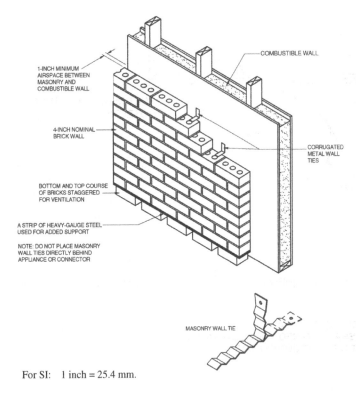

For SI: 1 inch = 25.4 mm.

FIGURE G2409.2(3) [308.2(3)]
MASONRY CLEARANCE REDUCTION SYSTEM

G2409.3.1 (308.3.1) Appliances installed in rooms that are large in comparison with the size of the appliances. Air-conditioning appliances installed in rooms that are large in comparison with the size of the appliance shall be installed with clearances in accordance with the manufacturer's instructions.

G2409.3.2 (308.3.2) Appliances installed in rooms that are not large in comparison with the size of the appliances. Air-conditioning appliances installed in rooms that are not large in comparison with the size of the appliance, such as alcoves and closets, shall be listed for such installations and installed in accordance with the manufacturer's instructions. Listed clearances shall not be reduced by the protection methods described in Table G2409.2, regardless of whether the enclosure is of combustible or noncombustible material.

G2409.3.3 (308.3.3) Clearance reduction. Air-conditioning appliances installed in rooms that are large in comparison with the size of the appliance shall be permitted to be installed with reduced clearances to combustible material, provided the combustible material or appliance is protected as described in Table G2409.2.

G2409.3.4 (308.3.4) Plenum clearances. Where the furnace plenum is adjacent to plaster on metal lath or noncombustible material attached to combustible material, the clearance shall be measured to the surface of the plaster or other noncombustible finish where the clearance specified is 2 inches (51 mm) or less.

G2409.3.5 (308.3.5) Clearance from supply ducts. Air-conditioning appliances shall have the clearance from supply ducts within 3 feet (914 mm) of the furnace plenum be not less than that specified from the furnace plenum. Clearance is not necessary beyond this distance.

G2409.4 (308.4) Central heating boilers and furnaces. Clearance requirements for central-heating boilers and furnaces shall comply with Sections G2409.4.1 through G2409.4.6. The clearance to these appliances shall not interfere with combustion air; draft hood clearance and relief; and accessibility for servicing.

G2409.4.1 (308.4.1) Appliances installed in rooms that are large in comparison with the size of the appliances. Central-heating furnaces and low-pressure boilers installed in rooms large in comparison with the size of the appliance shall be installed with clearances in accordance with the manufacturer's instructions.

G2409.4.2 (308.4.2) Appliances installed in rooms that are not large in comparison with the size of the appliances. Central-heating furnaces and low-pressure boilers installed in rooms that are not large in comparison with the size of the appliance, such as alcoves and closets, shall be listed for such installations. Listed clearances shall not be reduced by the protection methods described in Table G2409.2 and illustrated in Figures G2409.2(1) through G2409.2(3), regardless of whether the enclosure is of combustible or noncombustible material.

G2409.4.3 (308.4.3) Clearance reduction. Central heating furnaces and low-pressure boilers installed in rooms that are large in comparison with the size of the equipment shall be permitted to be installed with reduced clearances to combustible material provided the combustible material or equipment is protected as described in Table G2409.2.

G2409.4.4 (308.4.5) Plenum clearances. Where the furnace plenum is adjacent to plaster on metal lath or noncombustible material attached to combustible material, the clearance shall be measured to the surface of the plaster or other noncombustible finish where the clearance specified is 2 inches (51 mm) or less.

G2409.4.5 (308.4.6) Clearance from supply ducts. Central-heating furnaces shall have the clearance from supply ducts within 3 feet (914 mm) of the furnace plenum be not less than that specified from the furnace plenum. No clearance is necessary beyond this distance.

G2409.4.6 (308.4.4) Clearance for servicing appliances. Front clearance shall be sufficient for servicing the burner and the furnace or boiler.

SECTION G2410 (309)
ELECTRICAL

G2410.1 (309.1) Grounding. Gas piping shall not be used as a grounding electrode.

G2410.2 (309.2) Connections. Electrical connections between gas utilization equipment and the building wiring, including the grounding of the equipment, shall conform to Chapters 33 through 42.

TABLE G2409.2 (308.2)[a through k]
REDUCTION OF CLEARANCES WITH SPECIFIED FORMS OF PROTECTION

TYPE OF PROTECTION APPLIED TO AND COVERING ALL SURFACES OF COMBUSTIBLE MATERIAL WITHIN THE DISTANCE SPECIFIED AS THE REQUIRED CLEARANCE WITH NO PROTECTION [see Figures G2409.2(1), G2409.2(2), and G2409.2(3)]	WHERE THE REQUIRED CLEARANCE WITH NO PROTECTION FROM APPLIANCE, VENT CONNECTOR, OR SINGLE-WALL METAL PIPE IS: (inches)									
	36		18		12		9		6	
	Allowable clearances with specified protection (inches)									
	Use Column 1 for clearances above appliance or horizontal connector. Use Column 2 for clearances from appliance, vertical connector, and single-wall metal pipe.									
	Above Col. 1	Sides and rear Col. 2	Above Col. 1	Sides and rear Col. 2	Above Col. 1	Sides and rear Col. 2	Above Col. 1	Sides and rear Col. 2	Above Col. 1	Sides and rear Col. 2
1. 3½-inch-thick masonry wall without ventilated airspace	—	24	—	12	—	9	—	6	—	5
2. ½-inch insulation board over 1-inch glass fiber or mineral wool batts	24	18	12	9	9	6	6	5	4	3
3. 0.024-inch (nominal 24 gage) sheet metal over 1-inch glass fiber or mineral wool batts reinforced with wire on rear face with ventilated airspace	18	12	9	6	6	4	5	3	3	3
4. 3½-inch-thick masonry wall with ventilated airspace	—	12	—	6	—	6	—	6	—	6
5. 0.024-inch (nominal 24 gage) sheet metal with ventilated airspace	18	12	9	6	6	4	5	3	3	2
6. ½-inch-thick insulation board with ventilated airspace	18	12	9	6	6	4	5	3	3	3
7. 0.024-inch (nominal 24 gage) sheet metal with ventilated airspace over 0.024-inch (nominal 24 gage) sheet metal with ventilated airspace	18	12	9	6	6	4	5	3	3	3
8. 1-inch glass fiber or mineral wool batts sandwiched between two sheets 0.024-inch (nominal 24 gage) sheet metal with ventilated airspace	18	12	9	6	6	4	5	3	3	3

For SI: 1 inch = 25.4 mm, °C = [(°F - 32)/1.8], 1 pound per cubic foot = 16.02 kg/m³, 1 Btu per inch per square foot per hour per °F = 0.144 W/m² × K.

a. Reduction of clearances from combustible materials shall not interfere with combustion air, draft hood clearance and relief, and accessibility of servicing.

b. All clearances shall be measured from the outer surface of the combustible material to the nearest point on the surface of the appliance, disregarding any intervening protection applied to the combustible material.

c. Spacers and ties shall be of noncombustible material. No spacer or tie shall be used directly opposite an appliance or connector.

d. For all clearance reduction systems using a ventilated airspace, adequate provision for air circulation shall be provided as described [see Figures G2409.2(2) and G2409.2(3)].

e. There shall be at least 1 inch between clearance reduction systems and combustible walls and ceilings for reduction systems using ventilated airspace.

f. Where a wall protector is mounted on a single flat wall away from corners, it shall have a minimum 1-inch air gap. To provide air circulation, the bottom and top edges, or only the side and top edges, or all edges shall be left open.

g. Mineral wool batts (blanket or board) shall have a minimum density of 8 pounds per cubic foot and a minimum melting point of 1500°F.

h. Insulation material used as part of a clearance reduction system shall have a thermal conductivity of 1.0 Btu per inch per square foot per hour per °F or less.

i. There shall be at least 1 inch between the appliance and the protector. In no case shall the clearance between the appliance and the combustible surface be reduced below that allowed in this table.

j. All clearances and thicknesses are minimum; larger clearances and thicknesses are acceptable.

k. Listed single-wall connectors shall be installed in accordance with the manufacturer's installation instructions.

SECTION G2411 (310)
ELECTRICAL BONDING

G2411.1 (310.1) Gas pipe bonding. Each above-ground portion of a gas piping system that is likely to become energized shall be electrically continuous and bonded to an effective ground-fault current path. Gas piping shall be considered to be bonded where it is connected to appliances that are connected to the equipment grounding conductor of the circuit supplying that appliance.

SECTION G2412 (401)
GENERAL

G2412.1 (401.1) Scope. This section shall govern the design, installation, modification and maintenance of piping systems. The applicability of this code to piping systems extends from the point of delivery to the connections with the equipment and includes the design, materials, components, fabrication, assembly, installation, testing, inspection, operation and maintenance of such piping systems.

G2412.1.1 (401.1.1) Utility piping systems located within buildings. Utility service piping located within buildings shall be installed in accordance with the structural safety and fire protection provisions of this code.

G2412.2 (401.2) Liquefied petroleum gas storage. The storage system for liquefied petroleum gas shall be designed and installed in accordance with the *International Fire Code* and NFPA 58.

G2412.3 (401.3) Modifications to existing systems. In modifying or adding to existing piping systems, sizes shall be maintained in accordance with this chapter.

G2412.4 (401.4) Additional appliances. Where an additional appliance is to be served, the existing piping shall be checked to determine if it has adequate capacity for all appliances served. If inadequate, the existing system shall be enlarged as required or separate piping of adequate capacity shall be provided.

G2412.5 (401.5) Identification. For other than steel pipe, exposed piping shall be identified by a yellow label marked "Gas" in black letters. The marking shall be spaced at intervals not exceeding 5 feet (1524 mm). The marking shall not be required on pipe located in the same room as the equipment served.

G2412.6 (401.6) Interconnections. Where two or more meters are installed on the same premises, but supply separate consumers, the piping systems shall not be interconnected on the outlet side of the meters.

G2412.7 (401.7) Piping meter identification. Piping from multiple meter installations shall be marked with an approved permanent identification by the installer so that the piping system supplied by each meter is readily identifiable.

G2412.8 (401.8) Minimum sizes. All pipe utilized for the installation, extension and alteration of any piping system shall be sized to supply the full number of outlets for the intended purpose and shall be sized in accordance with Section G2413.

SECTION G2413 (402)
PIPE SIZING

G2413.1 (402.1) General considerations. Piping systems shall be of such size and so installed as to provide a supply of gas sufficient to meet the maximum demand without undue loss of pressure between the point of delivery and the appliance.

G2413.2 (402.2) Maximum gas demand. The volume of gas to be provided, in cubic feet per hour, shall be determined directly from the manufacturer's input ratings of the appliances served. Where an input rating is not indicated, the gas supplier, appliance manufacturer or a qualified agency shall be contacted, or the rating from Table G2413.2 shall be used for estimating the volume of gas to be supplied.

The total connected hourly load shall be used as the basis for pipe sizing, assuming that all appliances could be operating at full capacity simultaneously. Where a diversity of load can be established, pipe sizing shall be permitted to be based on such loads.

TABLE G2413.2 (402.2)
APPROXIMATE GAS INPUT FOR TYPICAL APPLIANCES

APPLIANCE	INPUT BTU/H (Approx.)
Space Heating Units	
Hydronic boiler	
Single family	100,000
Multifamily, per unit	60,000
Warm-air furnace	
Single family	100,000
Multifamily, per unit	60,000
Space and Water Heating Units	
Hydronic boiler	
Single family	120,000
Multifamily, per unit	75,000
Water Heating Appliances	
Water heater, automatic instantaneous	
Capacity at 2 gal./minute	142,800
Capacity at 4 gal./minute	285,000
Capacity at 6 gal./minute	428,400
Water heater, automatic storage, 30- to 40-gal. tank	35,000
Water heater, automatic storage, 50-gal. tank	50,000
Water heater, domestic, circulating or side-arm	35,000
Cooking Appliances	
Built-in oven or broiler unit, domestic	25,000
Built-in top unit, domestic	40,000
Range, free-standing, domestic	65,000
Other Appliances	
Barbecue	40,000
Clothes dryer, Type 1 (domestic)	35,000
Gas fireplace, direct vent	40,000
Gas light	2,500
Gas log	80,000
Refrigerator	3,000

For SI: 1 British thermal unit per hour = 0.293 W, 1 gallon = 3.785 L, 1 gallon per minute = 3.785 L/m.

G2413.3 (402.3) Sizing. Gas piping shall be sized in accordance with one of the following:

1. Pipe sizing tables or sizing equations in accordance with Section G2413.4.

2. The sizing tables included in a listed piping system's manufacturer's installation instructions.

3. Other approved engineering methods.

G2413.4 (402.4) Sizing tables and equations. Where Tables G2413.4(1) through G2413.4(8) are used to size piping or tubing, the pipe length shall be determined in accordance with Section G2413.4.1, G2413.4.2 or G2413.4.3.

Where Equations 24-3 and 24-4 are used to size piping or tubing, the pipe or tubing shall have smooth inside walls and the pipe length shall be determined in accordance with Section G2413.4.1, G2413.4.2 or G2413.4.3.

1. Low-pressure gas equation [Less than 1.5 pounds per square inch (psi) (10.3 kPa)]:

$$D = \frac{Q^{0.381}}{19.17 \left(\dfrac{\Delta H}{C_r \times L} \right)^{0.206}} \qquad \textbf{(Equation 24-3)}$$

2. High-pressure gas equation [1.5 psi (10.3 kPa) and above]:

$$D = \frac{Q^{0.381}}{18.93 \left[\dfrac{\left(P_1{}^2 - P_2{}^2 \right) \times Y}{C_r \times L} \right]^{0.206}} \qquad \textbf{(Equation 24-4)}$$

where:

D = Inside diameter of pipe, inches (mm).

Q = Input rate appliance(s), cubic feet per hour at 60°F (16°C) and 30-inch mercury column.

P_1 = Upstream pressure, psia (P_1 + 14.7).

P_2 = Downstream pressure, psia (P_2 + 14.7).

L = Equivalent length of pipe, feet.

ΔH = Pressure drop, inch water column (27.7 inch water column = 1 psi).

TABLE G2413.4 (402.4)
C_r AND Y VALUES FOR NATURAL GAS AND
UNDILUTED PROPANE AT STANDARD CONDITIONS

GAS	EQUATION FACTORS	
	C_r	Y
Natural gas	0.6094	0.9992
Undiluted propane	1.2462	0.9910

For SI: 1 cubic foot = 0.028 m³, 1 foot = 305 mm, 1 inch water column = 0.249 kPa, 1 pound per square inch = 6.895 kPa, 1 British thermal unit per hour = 0.293 W.

G2413.4.1 (402.4.1) Longest length method. The pipe size of each section of gas piping shall be determined using the longest length of piping from the point of delivery to the most remote outlet and the load of the section.

G2413.4.2 (402.4.2) Branch length method. Pipe shall be sized as follows:

1. Pipe size of each section of the longest pipe run from the point of delivery to the most remote outlet shall be determined using the longest run of piping and the load of the section.

2. The pipe size of each section of branch piping not previously sized shall be determined using the length of piping from the point of delivery to the most remote outlet in each branch and the load of the section.

G2413.4.3 (402.4.3) Hybrid pressure. The pipe size for each section of higher pressure gas piping shall be determined using the longest length of piping from the point of delivery to the most remote line pressure regulator. The pipe size from the line pressure regulator to each outlet shall be determined using the length of piping from the regulator to the most remote outlet served by the regulator.

G2413.5 (402.5) Allowable pressure drop. The design pressure loss in any piping system under maximum probable flow conditions, from the point of delivery to the inlet connection of the appliance, shall be such that the supply pressure at the appliance is greater than the minimum pressure required for proper appliance operation.

G2413.6 (402.6) Maximum design operating pressure. The maximum design operating pressure for piping systems located inside buildings shall not exceed 5 pounds per square inch gauge (psig) (34 kPa gauge) except where one or more of the following conditions are met:

1. The piping system is welded.

2. The piping is located in a ventilated chase or otherwise enclosed for protection against accidental gas accumulation.

3. The piping is a temporary installation for buildings under construction.

G2413.6.1 (402.6.1) Liquefied petroleum gas systems. The operating pressure for undiluted LP-gas systems shall not exceed 20 psig (140 kPa gauge). Buildings having systems designed to operate below -5°F (-21°C) or with butane or a propane-butane mix shall be designed to either accommodate liquid LP-gas or prevent LP-gas vapor from condensing into a liquid.

SECTION G2414 (403)
PIPING MATERIALS

G2414.1 (403.1) General. Materials used for piping systems shall comply with the requirements of this chapter or shall be approved.

G2414.2 (403.2) Used materials. Pipe, fittings, valves or other materials shall not be used again unless they are free of foreign materials and have been ascertained to be adequate for the service intended.

G2414.3 (403.3) Other materials. Material not covered by the standards specifications listed herein shall be investigated and tested to determine that it is safe and suitable for the proposed service, and, in addition, shall be recommended for that service by the manufacturer and shall be approved by the code official.

G2414.4 (403.4) Metallic pipe. Metallic pipe shall comply with Sections G2414.4.1 and G2414.4.2.

TABLE G2413.4(1) [402.4(2)]
SCHEDULE 40 METALLIC PIPE

Gas	Natural
Inlet Pressure	Less than 2 psi
Pressure Drop	0.5 in. w.c.
Specific Gravity	0.60

PIPE SIZE (inch)														
Nominal	$^1/_2$	$^3/_4$	1	$1^1/_4$	$1^1/_2$	2	$2^1/_2$	3	4	5	6	8	10	12
Actual ID	0.622	0.824	1.049	1.380	1.610	2.067	2.469	3.068	4.026	5.047	6.065	7.981	10.020	11.938
Length (ft)	Capacity in Cubic Feet of Gas per Hour													
10	172	360	678	1,390	2,090	4,020	6,400	11,300	23,100	41,800	67,600	139,000	252,000	399,000
20	118	247	466	957	1,430	2,760	4,400	7,780	15,900	28,700	46,500	95,500	173,000	275,000
30	95	199	374	768	1,150	2,220	3,530	6,250	12,700	23,000	37,300	76,700	139,000	220,000
40	81	170	320	657	985	1,900	3,020	5,350	10,900	19,700	31,900	65,600	119,000	189,000
50	72	151	284	583	873	1,680	2,680	4,740	9,660	17,500	28,300	58,200	106,000	167,000
60	65	137	257	528	791	1,520	2,430	4,290	8,760	15,800	25,600	52,700	95,700	152,000
70	60	126	237	486	728	1,400	2,230	3,950	8,050	14,600	23,600	48,500	88,100	139,000
80	56	117	220	452	677	1,300	2,080	3,670	7,490	13,600	22,000	45,100	81,900	130,000
90	52	110	207	424	635	1,220	1,950	3,450	7,030	12,700	20,600	42,300	76,900	122,000
100	50	104	195	400	600	1,160	1,840	3,260	6,640	12,000	19,500	40,000	72,600	115,000
125	44	92	173	355	532	1,020	1,630	2,890	5,890	10,600	17,200	35,400	64,300	102,000
150	40	83	157	322	482	928	1,480	2,610	5,330	9,650	15,600	32,100	58,300	92,300
175	37	77	144	296	443	854	1,360	2,410	4,910	8,880	14,400	29,500	53,600	84,900
200	34	71	134	275	412	794	1,270	2,240	4,560	8,260	13,400	27,500	49,900	79,000
250	30	63	119	244	366	704	1,120	1,980	4,050	7,320	11,900	24,300	44,200	70,000
300	27	57	108	221	331	638	1,020	1,800	3,670	6,630	10,700	22,100	40,100	63,400
350	25	53	99	203	305	587	935	1,650	3,370	6,100	9,880	20,300	36,900	58,400
400	23	49	92	189	283	546	870	1,540	3,140	5,680	9,190	18,900	34,300	54,300
450	22	46	86	177	266	512	816	1,440	2,940	5,330	8,620	17,700	32,200	50,900
500	21	43	82	168	251	484	771	1,360	2,780	5,030	8,150	16,700	30,400	48,100
550	20	41	78	159	239	459	732	1,290	2,640	4,780	7,740	15,900	28,900	45,700
600	19	39	74	152	228	438	699	1,240	2,520	4,560	7,380	15,200	27,500	43,600
650	18	38	71	145	218	420	669	1,180	2,410	4,360	7,070	14,500	26,400	41,800
700	17	36	68	140	209	403	643	1,140	2,320	4,190	6,790	14,000	25,300	40,100
750	17	35	66	135	202	389	619	1,090	2,230	4,040	6,540	13,400	24,400	38,600
800	16	34	63	130	195	375	598	1,060	2,160	3,900	6,320	13,000	23,600	37,300
850	16	33	61	126	189	363	579	1,020	2,090	3,780	6,110	12,600	22,800	36,100
900	15	32	59	122	183	352	561	992	2,020	3,660	5,930	12,200	22,100	35,000
950	15	31	58	118	178	342	545	963	1,960	3,550	5,760	11,800	21,500	34,000
1,000	14	30	56	115	173	333	530	937	1,910	3,460	5,600	11,500	20,900	33,100
1,100	14	28	53	109	164	316	503	890	1,810	3,280	5,320	10,900	19,800	31,400
1,200	13	27	51	104	156	301	480	849	1,730	3,130	5,070	10,400	18,900	30,000
1,300	12	26	49	100	150	289	460	813	1,660	3,000	4,860	9,980	18,100	28,700
1,400	12	25	47	96	144	277	442	781	1,590	2,880	4,670	9,590	17,400	27,600
1,500	11	24	45	93	139	267	426	752	1,530	2,780	4,500	9,240	16,800	26,600
1,600	11	23	44	89	134	258	411	727	1,480	2,680	4,340	8,920	16,200	25,600
1,700	11	22	42	86	130	250	398	703	1,430	2,590	4,200	8,630	15,700	24,800
1,800	10	22	41	84	126	242	386	682	1,390	2,520	4,070	8,370	15,200	24,100
1,900	10	21	40	81	122	235	375	662	1,350	2,440	3,960	8,130	14,800	23,400
2,000	NA	20	39	79	119	229	364	644	1,310	2,380	3,850	7,910	14,400	22,700

For SI: 1 inch = 25.4 mm, 1 foot = 304.8 mm, 1 pound per square inch = 6.895 kPa, 1-inch water column = 0.2488 kPa, 1 British thermal unit per hour = 0.2931 W, 1 cubic foot per hour = 0.0283 m^3/h, 1 degree = 0.01745 rad.

Notes:
1. NA means a flow of less than 10 cfh.
2. All table entries have been rounded to three significant digits.

**TABLE G2413.4(2) [402.4(3)]
SCHEDULE 40 METALLIC PIPE**

Gas	Natural
Inlet Pressure	2.0 psi
Pressure Drop	1.0 psi
Specific Gravity	0.60

PIPE SIZE (inch)									
Nominal	1/2	3/4	1	1 1/4	1 1/2	2	2 1/2	3	4
Actual ID	0.622	0.824	1.049	1.380	1.610	2.067	2.469	3.068	4.026
Length (ft)	Capacity in Cubic Feet of Gas per Hour								
10	1,510	3,040	5,560	11,400	17,100	32,900	52,500	92,800	189,000
20	1,070	2,150	3,930	8,070	12,100	23,300	37,100	65,600	134,000
30	869	1,760	3,210	6,590	9,880	19,000	30,300	53,600	109,000
40	753	1,520	2,780	5,710	8,550	16,500	26,300	46,400	94,700
50	673	1,360	2,490	5,110	7,650	14,700	23,500	41,500	84,700
60	615	1,240	2,270	4,660	6,980	13,500	21,400	37,900	77,300
70	569	1,150	2,100	4,320	6,470	12,500	19,900	35,100	71,600
80	532	1,080	1,970	4,040	6,050	11,700	18,600	32,800	67,000
90	502	1,010	1,850	3,810	5,700	11,000	17,500	30,900	63,100
100	462	934	1,710	3,510	5,260	10,100	16,100	28,500	58,200
125	414	836	1,530	3,140	4,700	9,060	14,400	25,500	52,100
150	372	751	1,370	2,820	4,220	8,130	13,000	22,900	46,700
175	344	695	1,270	2,601	3,910	7,530	12,000	21,200	43,300
200	318	642	1,170	2,410	3,610	6,960	11,100	19,600	40,000
250	279	583	1,040	2,140	3,210	6,180	9,850	17,400	35,500
300	253	528	945	1,940	2,910	5,600	8,920	15,800	32,200
350	232	486	869	1,790	2,670	5,150	8,210	14,500	29,600
400	216	452	809	1,660	2,490	4,790	7,640	13,500	27,500
450	203	424	759	1,560	2,330	4,500	7,170	12,700	25,800
500	192	401	717	1,470	2,210	4,250	6,770	12,000	24,400
550	182	381	681	1,400	2,090	4,030	6,430	11,400	23,200
600	174	363	650	1,330	2,000	3,850	6,130	10,800	22,100
650	166	348	622	1,280	1,910	3,680	5,870	10,400	21,200
700	160	334	598	1,230	1,840	3,540	5,640	9,970	20,300
750	154	322	576	1,180	1,770	3,410	5,440	9,610	19,600
800	149	311	556	1,140	1,710	3,290	5,250	9,280	18,900
850	144	301	538	1,100	1,650	3,190	5,080	8,980	18,300
900	139	292	522	1,070	1,600	3,090	4,930	8,710	17,800
950	135	283	507	1,040	1,560	3,000	4,780	8,460	17,200
1,000	132	275	493	1,010	1,520	2,920	4,650	8,220	16,800
1,100	125	262	468	960	1,440	2,770	4,420	7,810	15,900
1,200	119	250	446	917	1,370	2,640	4,220	7,450	15,200
1,300	114	239	427	878	1,320	2,530	4,040	7,140	14,600
1,400	110	230	411	843	1,260	2,430	3,880	6,860	14,000
1,500	106	221	396	812	1,220	2,340	3,740	6,600	13,500
1,600	102	214	382	784	1,180	2,260	3,610	6,380	13,000
1,700	99	207	370	759	1,140	2,190	3,490	6,170	12,600
1,800	96	200	358	736	1,100	2,120	3,390	5,980	12,200
1,900	93	195	348	715	1,070	2,060	3,290	5,810	11,900
2,000	91	189	339	695	1,040	2,010	3,200	5,650	11,500

For SI: 1 inch = 25.4 mm, 1 foot = 304.8 mm, 1 pound per square inch = 6.895 kPa, 1-inch water column = 0.2488 kPa,
1 British thermal unit per hour = 0.2931 W, 1 cubic foot per hour = 0.0283 m³/h, 1 degree = 0.01745 rad.

Note: All table entries have been rounded to three significant digits.

TABLE G2413.4(3) [402.4(7)]
SEMIRIGID COPPER TUBING

	Gas	Natural
	Inlet Pressure	Less than 2 psi
	Pressure Drop	0.5 in. w.c.
	Specific Gravity	0.60

Nominal	K & L	$^1/_4$	$^3/_8$	$^1/_2$	$^5/_8$	$^3/_4$	1	$1^1/_4$	$1^1/_2$	2
	ACR	$^3/_8$	$^1/_2$	$^5/_8$	$^3/_4$	$^7/_8$	$1^1/_8$	$1^3/_8$	—	—
Outside		0.375	0.500	0.625	0.750	0.875	1.125	1.375	1.625	2.125
Inside		0.305	0.402	0.527	0.652	0.745	0.995	1.245	1.481	1.959
Length (ft)		Capacity in Cubic Feet of Gas per Hour								
10		27	55	111	195	276	590	1,060	1,680	3,490
20		18	38	77	134	190	406	730	1,150	2,400
30		15	30	61	107	152	326	586	925	1,930
40		13	26	53	92	131	279	502	791	1,650
50		11	23	47	82	116	247	445	701	1,460
60		10	21	42	74	105	224	403	635	1,320
70		NA	19	39	68	96	206	371	585	1,220
80		NA	18	36	63	90	192	345	544	1,130
90		NA	17	34	59	84	180	324	510	1,060
100		NA	16	32	56	79	170	306	482	1,000
125		NA	14	28	50	70	151	271	427	890
150		NA	13	26	45	64	136	245	387	806
175		NA	12	24	41	59	125	226	356	742
200		NA	11	22	39	55	117	210	331	690
250		NA	NA	20	34	48	103	186	294	612
300		NA	NA	18	31	44	94	169	266	554
350		NA	NA	16	28	40	86	155	245	510
400		NA	NA	15	26	38	80	144	228	474
450		NA	NA	14	25	35	75	135	214	445
500		NA	NA	13	23	33	71	128	202	420
550		NA	NA	13	22	32	68	122	192	399
600		NA	NA	12	21	30	64	116	183	381
650		NA	NA	12	20	29	62	111	175	365
700		NA	NA	11	20	28	59	107	168	350
750		NA	NA	11	19	27	57	103	162	338
800		NA	NA	10	18	26	55	99	156	326
850		NA	NA	10	18	25	53	96	151	315
900		NA	NA	NA	17	24	52	93	147	306
950		NA	NA	NA	17	24	50	90	143	297
1,000		NA	NA	NA	16	23	49	88	139	289
1,100		NA	NA	NA	15	22	46	84	132	274
1,200		NA	NA	NA	15	21	44	80	126	262
1,300		NA	NA	NA	14	20	42	76	120	251
1,400		NA	NA	NA	13	19	41	73	116	241
1,500		NA	NA	NA	13	18	39	71	111	232
1,600		NA	NA	NA	13	18	38	68	108	224
1,700		NA	NA	NA	12	17	37	66	104	217
1,800		NA	NA	NA	12	17	36	64	101	210
1,900		NA	NA	NA	11	16	35	62	98	204
2,000		NA	NA	NA	11	16	34	60	95	199

For SI: 1 inch = 25.4 mm, 1 foot = 304.8 mm, 1 pound per square inch = 6.895 kPa, 1-inch water column = 0.2488 kPa,
1 British thermal unit per hour = 0.2931 W, 1 cubic foot per hour = 0.0283 m³/h, 1 degree = 0.01745 rad.

Notes:

1. Table capacities are based on Type K copper tubing inside diameter (shown), which has the smallest inside diameter of the copper tubing products.

2. NA means a flow of less than 10 cfh.

3. All table entries have been rounded to three significant digits.

**TABLE G2413.4(4) [402.4(10)]
SEMIRIGID COPPER TUBING**

Gas	Natural
Inlet Pressure	2.0 psi
Pressure Drop	1.0 psi
Specific Gravity	0.60

					TUBE SIZE (inch)					
Nominal	**K & L**	$^1/_4$	$^3/_8$	$^1/_2$	$^5/_8$	$^3/_4$	1	$1^1/_4$	$1^1/_2$	2
	ACR	$^3/_8$	$^1/_2$	$^5/_8$	$^3/_4$	$^7/_8$	$1^1/_8$	$1^3/_8$	—	—
Outside		0.375	0.500	0.625	0.750	0.875	1.125	1.375	1.625	2.125
Inside		0.305	0.402	0.527	0.652	0.745	0.995	1.245	1.481	1.959
Length (ft)		Capacity in Cubic Feet of Gas per Hour								
10		245	506	1,030	1,800	2,550	5,450	9,820	15,500	32,200
20		169	348	708	1,240	1,760	3,750	6,750	10,600	22,200
30		135	279	568	993	1,410	3,010	5,420	8,550	17,800
40		116	239	486	850	1,210	2,580	4,640	7,310	15,200
50		103	212	431	754	1,070	2,280	4,110	6,480	13,500
60		93	192	391	683	969	2,070	3,730	5,870	12,200
70		86	177	359	628	891	1,900	3,430	5,400	11,300
80		80	164	334	584	829	1,770	3,190	5,030	10,500
90		75	154	314	548	778	1,660	2,990	4,720	9,820
100		71	146	296	518	735	1,570	2,830	4,450	9,280
125		63	129	263	459	651	1,390	2,500	3,950	8,220
150		57	117	238	416	590	1,260	2,270	3,580	7,450
175		52	108	219	383	543	1,160	2,090	3,290	6,850
200		49	100	204	356	505	1,080	1,940	3,060	6,380
250		43	89	181	315	448	956	1,720	2,710	5,650
300		39	80	164	286	406	866	1,560	2,460	5,120
350		36	74	150	263	373	797	1,430	2,260	4,710
400		33	69	140	245	347	741	1,330	2,100	4,380
450		31	65	131	230	326	696	1,250	1,970	4,110
500		30	61	124	217	308	657	1,180	1,870	3,880
550		28	58	118	206	292	624	1,120	1,770	3,690
600		27	55	112	196	279	595	1,070	1,690	3,520
650		26	53	108	188	267	570	1,030	1,620	3,370
700		25	51	103	181	256	548	986	1,550	3,240
750		24	49	100	174	247	528	950	1,500	3,120
800		23	47	96	168	239	510	917	1,450	3,010
850		22	46	93	163	231	493	888	1,400	2,920
900		22	44	90	158	224	478	861	1,360	2,830
950		21	43	88	153	217	464	836	1,320	2,740
1,000		20	42	85	149	211	452	813	1,280	2,670
1,100		19	40	81	142	201	429	772	1,220	2,540
1,200		18	38	77	135	192	409	737	1,160	2,420
1,300		18	36	74	129	183	392	705	1,110	2,320
1,400		17	35	71	124	176	376	678	1,070	2,230
1,500		16	34	68	120	170	363	653	1,030	2,140
1,600		16	33	66	116	164	350	630	994	2,070
1,700		15	31	64	112	159	339	610	962	2,000
1,800		15	30	62	108	154	329	592	933	1,940
1,900		14	30	60	105	149	319	575	906	1,890
2,000		14	29	59	102	145	310	559	881	1,830

For SI: 1 inch = 25.4 mm, 1 foot = 304.8 mm, 1 pound per square inch = 6.895 kPa, 1-inch water column = 0.2488 kPa,
1 British thermal unit per hour = 0.2931 W, 1 cubic foot per hour = 0.0283 m³/h, 1 degree = 0.01745 rad.

Notes:
1. Table capacities are based on Type K copper tubing inside diameter (shown), which has the smallest inside diameter of the copper tubing products.
2. All table entries have been rounded to three significant digits.

TABLE G2413.4(5) [402.4(13)]
CORRUGATED STAINLESS STEEL TUBING (CSST)

	Gas	Natural
	Inlet Pressure	Less than 2 psi
	Pressure Drop	0.5 in. w.c.
	Specific Gravity	0.60

	TUBE SIZE (EHD)												
Flow Designation	13	15	18	19	23	25	30	31	37	46	48	60	62
Length (ft)	Capacity in Cubic Feet of Gas per Hour												
5	46	63	115	134	225	270	471	546	895	1,790	2,070	3,660	4,140
10	32	44	82	95	161	192	330	383	639	1,260	1,470	2,600	2,930
15	25	35	66	77	132	157	267	310	524	1,030	1,200	2,140	2,400
20	22	31	58	67	116	137	231	269	456	888	1,050	1,850	2,080
25	19	27	52	60	104	122	206	240	409	793	936	1,660	1,860
30	18	25	47	55	96	112	188	218	374	723	856	1,520	1,700
40	15	21	41	47	83	97	162	188	325	625	742	1,320	1,470
50	13	19	37	42	75	87	144	168	292	559	665	1,180	1,320
60	12	17	34	38	68	80	131	153	267	509	608	1,080	1,200
70	11	16	31	36	63	74	121	141	248	471	563	1,000	1,110
80	10	15	29	33	60	69	113	132	232	440	527	940	1,040
90	10	14	28	32	57	65	107	125	219	415	498	887	983
100	9	13	26	30	54	62	101	118	208	393	472	843	933
150	7	10	20	23	42	48	78	91	171	320	387	691	762
200	6	9	18	21	38	44	71	82	148	277	336	600	661
250	5	8	16	19	34	39	63	74	133	247	301	538	591
300	5	7	15	17	32	36	57	67	95	226	275	492	540

For SI: 1 inch = 25.4 mm, 1 foot = 304.8 mm, 1 pound per square inch = 6.895 kPa, 1-inch water column = 0.2488 kPa, 1 British thermal unit per hour = 0.2931 W, 1 cubic foot per hour = 0.0283 m^3/h, 1 degree = 0.01745 rad.

Notes:

1. Table includes losses for four 90-degree bends and two end fittings. Tubing runs with larger numbers of bends and/or fittings shall be increased by an equivalent length of tubing to the following equation: $L = 1.3n$, where L is additional length (feet) of tubing and n is the number of additional fittings and/or bends.
2. EHD—Equivalent Hydraulic Diameter, which is a measure of the relative hydraulic efficiency between different tubing sizes. The greater the value of EHD, the greater the gas capacity of the tubing.
3. All table entries have been rounded to three significant digits.

TABLE G2413.4(6) [402.4(16)]
CORRUGATED STAINLESS STEEL TUBING (CSST)

Gas	Natural
Inlet Pressure	2.0 psi
Pressure Drop	1.0 psi
Specific Gravity	0.60

	TUBE SIZE (EHD)												
Flow Designation	13	15	18	19	23	25	30	31	37	46	48	60	62
Length (ft)	Capacity in Cubic Feet of Gas per Hour												
10	270	353	587	700	1,100	1,370	2,590	2,990	4,510	9,600	10,700	18,600	21,600
25	166	220	374	444	709	876	1,620	1,870	2,890	6,040	6,780	11,900	13,700
30	151	200	342	405	650	801	1,480	1,700	2,640	5,510	6,200	10,900	12,500
40	129	172	297	351	567	696	1,270	1,470	2,300	4,760	5,380	9,440	10,900
50	115	154	266	314	510	624	1,140	1,310	2,060	4,260	4,820	8,470	9,720
75	93	124	218	257	420	512	922	1,070	1,690	3,470	3,950	6,940	7,940
80	89	120	211	249	407	496	892	1,030	1,640	3,360	3,820	6,730	7,690
100	79	107	189	222	366	445	795	920	1,470	3,000	3,420	6,030	6,880
150	64	87	155	182	302	364	646	748	1,210	2,440	2,800	4,940	5,620
200	55	75	135	157	263	317	557	645	1,050	2,110	2,430	4,290	4,870
250	49	67	121	141	236	284	497	576	941	1,890	2,180	3,850	4,360
300	44	61	110	129	217	260	453	525	862	1,720	1,990	3,520	3,980
400	38	52	96	111	189	225	390	453	749	1,490	1,730	3,060	3,450
500	34	46	86	100	170	202	348	404	552	1,330	1,550	2,740	3,090

For SI: 1 inch = 25.4 mm, 1 foot = 304.8 mm, 1 pound per square inch = 6.895 kPa, 1-inch water column = 0.2488 kPa,
1 British thermal unit per hour = 0.2931 W, 1 cubic foot per hour = 0.0283 m³/h, 1 degree = 0.01745 rad.

Notes:

1. Table does not include effect of pressure drop across the line regulator. Where regulator loss exceeds $^3/_4$ psi, DO NOT USE THIS TABLE. Consult with the regulator manufacturer for pressure drops and capacity factors. Pressure drops across a regulator can vary with flow rate.

2. CAUTION: Capacities shown in the table might exceed maximum capacity for a selected regulator. Consult with the regulator or tubing manufacturer for guidance.

3. Table includes losses for four 90-degree bends and two end fittings. Tubing runs with larger numbers of bends and/or fittings shall be increased by an equivalent length of tubing to the following equation: $L = 1.3n$ where L is additional length (feet) of tubing and n is the number of additional fittings and/or bends.

4. EHD—Equivalent Hydraulic Diameter, which is a measure of the relative hydraulic efficiency between different tubing sizes. The greater the value of EHD, the greater the gas capacity of the tubing.

5. All table entries have been rounded to three significant digits.

TABLE G2413.4(7) [402.4(19)]
POLYETHYLENE PLASTIC PIPE

Gas	Natural
Inlet Pressure	Less than 2 psi
Pressure Drop	0.5 in. w.c.
Specific Gravity	0.60

PIPE SIZE (in.)

Nominal OD	$^1/_2$	$^3/_4$	1	$1^1/_4$	$1^1/_2$	2
Designation	SDR 9.33	SDR 11.0	SDR 11.00	SDR 10.00	SDR 11.00	SDR 11.00
Actual ID	0.660	0.860	1.077	1.328	1.554	1.943
Length (ft)	Capacity in Cubic Feet of Gas per Hour					
10	201	403	726	1,260	1,900	3,410
20	138	277	499	865	1,310	2,350
30	111	222	401	695	1,050	1,880
40	95	190	343	594	898	1,610
50	84	169	304	527	796	1,430
60	76	153	276	477	721	1,300
70	70	140	254	439	663	1,190
80	65	131	236	409	617	1,110
90	61	123	221	383	579	1,040
100	58	116	209	362	547	983
125	51	103	185	321	485	871
150	46	93	168	291	439	789
175	43	86	154	268	404	726
200	40	80	144	249	376	675
250	35	71	127	221	333	598
300	32	64	115	200	302	542
350	29	59	106	184	278	499
400	27	55	99	171	258	464
450	26	51	93	160	242	435
500	24	48	88	152	229	411

For SI: 1 inch = 25.4 mm, 1 foot = 304.8 mm, 1 pound per square inch = 6.895 kPa, 1-inch water column = 0.2488 kPa,
1 British thermal unit per hour = 0.2931 W, 1 cubic foot per hour = 0.0283 m³/h, 1 degree = 0.01745 rad.

Note: All table entries have been rounded to three significant digits.

TABLE G2413.4(8) [402.4(20)]
POLYETHYLENE PLASTIC PIPE

	Gas	Natural
	Inlet Pressure	2.0 psi
	Pressure Drop	1.0 psi
	Specific Gravity	0.60

PIPE SIZE (in.)						
Nominal OD	$^1/_2$	$^3/_4$	1	$1^1/_4$	$1^1/_2$	2
Designation	SDR 9.33	SDR 11.0	SDR 11.00	SDR 10.00	SDR 11.00	SDR 11.00
Actual ID	0.660	0.860	1.077	1.328	1.554	1.943
Length (ft)	Capacity in Cubic Feet of Gas per Hour					
10	1,860	3,720	6,710	11,600	17,600	31,600
20	1,280	2,560	4,610	7,990	12,100	21,700
30	1,030	2,050	3,710	6,420	9,690	17,400
40	878	1,760	3,170	5,490	8,300	14,900
50	778	1,560	2,810	4,870	7,350	13,200
60	705	1,410	2,550	4,410	6,660	12,000
70	649	1,300	2,340	4,060	6,130	11,000
80	603	1,210	2,180	3,780	5,700	10,200
90	566	1,130	2,050	3,540	5,350	9,610
100	535	1,070	1,930	3,350	5,050	9,080
125	474	949	1,710	2,970	4,480	8,050
150	429	860	1,550	2,690	4,060	7,290
175	395	791	1,430	2,470	3,730	6,710
200	368	736	1,330	2,300	3,470	6,240
250	326	652	1,180	2,040	3,080	5,530
300	295	591	1,070	1,850	2,790	5,010
350	272	544	981	1,700	2,570	4,610
400	253	506	913	1,580	2,390	4,290
450	237	475	856	1,480	2,240	4,020
500	224	448	809	1,400	2,120	3,800
550	213	426	768	1,330	2,010	3,610
600	203	406	733	1,270	1,920	3,440
650	194	389	702	1,220	1,840	3,300
700	187	374	674	1,170	1,760	3,170
750	180	360	649	1,130	1,700	3,050
800	174	348	627	1,090	1,640	2,950
850	168	336	607	1,050	1,590	2,850
900	163	326	588	1,020	1,540	2,770
950	158	317	572	990	1,500	2,690
1,000	154	308	556	963	1,450	2,610
1,100	146	293	528	915	1,380	2,480
1,200	139	279	504	873	1,320	2,370
1,300	134	267	482	836	1,260	2,270
1,400	128	257	463	803	1,210	2,180
1,500	124	247	446	773	1,170	2,100
1,600	119	239	431	747	1,130	2,030
1,700	115	231	417	723	1,090	1,960
1,800	112	224	404	701	1,060	1,900
1,900	109	218	393	680	1,030	1,850
2,000	106	212	382	662	1,000	1,800

For SI: 1 inch = 25.4 mm, 1 foot = 304.8 mm, 1 pound per square inch = 6.895 kPa, 1-inch water column = 0.2488 kPa,
    1 British thermal unit per hour = 0.2931 W, 1 cubic foot per hour = 0.0283 m³/h, 1 degree = 0.01745 rad.
Note: All table entries have been rounded to three significant digits.

G2414.4.1 (403.4.1) Cast iron. Cast-iron pipe shall not be used.

G2414.4.2 (403.4.2) Steel. Steel and wrought-iron pipe shall be at least of standard weight (Schedule 40) and shall comply with one of the following:

1. ASME B 36.10, 10M;

2. ASTM A 53; or

3. ASTM A 106.

G2414.5 (403.5) Metallic tubing. Seamless copper, aluminum alloy or steel tubing shall be permitted to be used with gases not corrosive to such material.

G2414.5.1 (403.5.1) Steel tubing. Steel tubing shall comply with ASTM A 539 or ASTM A 254.

G2414.5.2 (403.5.2) Copper tubing. Copper tubing shall comply with standard Type K or L of ASTM B 88 or ASTM B 280.

Copper and brass tubing shall not be used if the gas contains more than an average of 0.3 grains of hydrogen sulfide per 100 standard cubic feet of gas (0.7 milligrams per 100 liters).

G2414.5.3 (403.5.4) Corrugated stainless steel tubing. Corrugated stainless steel tubing shall be listed in accordance with ANSI LC 1/CSA 6.26.

G2414.6 (403.6) Plastic pipe, tubing and fittings. Plastic pipe, tubing and fittings used to supply fuel gas shall be used outdoors, underground, only, and shall conform to ASTM D 2513. Pipe shall be marked "Gas" and "ASTM D 2513."

G2414.6.1 (403.6.1) Anodeless risers. Anodeless risers shall comply with the following:

1. Factory-assembled anodeless risers shall be recommended by the manufacturer for the gas used and shall be leak-tested by the manufacturer in accordance with written procedures.

2. Service head adapters and field-assembled anodeless risers incorporating service head adapters shall be recommended by the manufacturer for the gas used by the manufacturer and shall be designed certified to meet the requirements of Category I of ASTM D 2513, and U.S. Department of Transportation, Code of Federal Regulations, Title 49, Part 192.281(e). The manufacturer shall provide the user qualified installation instructions as prescribed by the U.S. Department of Transportation, Code of Federal Regulations, Title 49, Part 192.283(b).

G2414.6.2 (403.6.2) LP-gas systems. The use of plastic pipe, tubing and fittings in undiluted liquefied petroleum gas piping systems shall be in accordance with NFPA 58.

G2414.6.3 (403.6.3) Regulator vent piping. Plastic pipe, tubing and fittings used to connect regulator vents to remote vent terminations shall be of PVC conforming to UL 651. PVC vent piping shall not be installed indoors.

G2414.7 (403.7) Workmanship and defects. Pipe or tubing and fittings shall be clear and free from cutting burrs and defects in structure or threading, and shall be thoroughly brushed, and chip and scale blown.

Defects in pipe or tubing or fittings shall not be repaired. Defective pipe, tubing or fittings shall be replaced. (See Section G2417.1.2.)

G2414.8 (403.8) Protective coating. Where in contact with material or atmosphere exerting a corrosive action, metallic piping and fittings coated with a corrosion-resistant material shall be used. External or internal coatings or linings used on piping or components shall not be considered as adding strength.

G2414.9 (403.9) Metallic pipe threads. Metallic pipe and fitting threads shall be taper pipe threads and shall comply with ASME B1.20.1.

G2414.9.1 (403.9.1) Damaged threads. Pipe with threads that are stripped, chipped, corroded or otherwise damaged shall not be used. If a weld opens during the operation of cutting or threading, that portion of the pipe shall not be used.

G2414.9.2 (403.9.2) Number of threads. Field threading of metallic pipe shall be in accordance with Table G2414.9.2.

TABLE G2414.9.2 (403.9.2)
SPECIFICATIONS FOR THREADING METALLIC PIPE

IRON PIPE SIZE (inches)	APPROXIMATE LENGTH OF THREADED PORTION (inches)	APPROXIMATE NO. OF THREADS TO BE CUT
$^1/_2$	$^3/_4$	10
$^3/_4$	$^3/_4$	10
1	$^7/_8$	10
$1^1/_4$	1	11
$1^1/_2$	1	11

For SI: 1 inch = 25.4 mm.

G2414.9.3 (403.9.3) Thread compounds. Thread (joint) compounds (pipe dope) shall be resistant to the action of liquefied petroleum gas or to any other chemical constituents of the gases to be conducted through the piping.

G2414.10 (403.10) Metallic piping joints and fittings. The type of piping joint used shall be suitable for the pressure-temperature conditions and shall be selected giving consideration to joint tightness and mechanical strength under the service conditions. The joint shall be able to sustain the maximum end force due to the internal pressure and any additional forces due to temperature expansion or contraction, vibration, fatigue, or to the weight of the pipe and its contents.

G2414.10.1 (403.10.1) Pipe joints. Pipe joints shall be threaded, flanged, brazed or welded. Where nonferrous pipe is brazed, the brazing materials shall have a melting point in excess of 1,000°F (538°C). Brazing alloys shall not contain more than 0.05-percent phosphorus.

G2414.10.2 (403.10.2) Tubing joints. Tubing joints shall either be made with approved gas tubing fittings or be brazed with a material having a melting point in excess of

1,000°F (538°C). Brazing alloys shall not contain more than 0.05-percent phosphorus.

G2414.10.3 (403.10.3) Flared joints. Flared joints shall be used only in systems constructed from nonferrous pipe and tubing where experience or tests have demonstrated that the joint is suitable for the conditions and where provisions are made in the design to prevent separation of the joints.

G2414.10.4 (403.10.4) Metallic fittings. Metallic fittings, including valves, strainers and filters shall comply with the following:

1. Fittings used with steel or wrought-iron pipe shall be steel, brass, bronze, malleable iron, ductile iron or cast iron.

2. Fittings used with copper or brass pipe shall be copper, brass or bronze.

3. Cast-iron bushings shall be prohibited.

4. Special fittings. Fittings such as couplings, proprietary-type joints, saddle tees, gland-type compression fittings, and flared, flareless or compression-type tubing fittings shall be: used within the fitting manufacturer's pressure-temperature recommendations; used within the service conditions anticipated with respect to vibration, fatigue, thermal expansion or contraction; installed or braced to prevent separation of the joint by gas pressure or external physical damage; and shall be approved.

G2414.11 (403.11) Plastic piping, joints and fittings. Plastic pipe, tubing and fittings shall be joined in accordance with the manufacturers' instructions. Such joints shall comply with the following:

1. The joints shall be designed and installed so that the longitudinal pull-out resistance of the joints will be at least equal to the tensile strength of the plastic piping material.

2. Heat-fusion joints shall be made in accordance with qualified procedures that have been established and proven by test to produce gas-tight joints at least as strong as the pipe or tubing being joined. Joints shall be made with the joining method recommended by the pipe manufacturer. Heat fusion fittings shall be marked "ASTM D 2513."

3. Where compression-type mechanical joints are used, the gasket material in the fitting shall be compatible with the plastic piping and with the gas distributed by the system. An internal tubular rigid stiffener shall be used in conjunction with the fitting. The stiffener shall be flush with the end of the pipe or tubing and shall extend at least to the outside end of the compression fitting when installed. The stiffener shall be free of rough or sharp edges and shall not be a force fit in the plastic. Split tubular stiffeners shall not be used.

4. Plastic piping joints and fittings for use in liquefied petroleum gas piping systems shall be in accordance with NFPA 58.

SECTION G2415 (404)
PIPING SYSTEM INSTALLATION

G2415.1 (404.1) Prohibited locations. Piping shall not be installed in or through a circulating air duct, clothes chute, chimney or gas vent, ventilating duct, dumbwaiter or elevator shaft. Piping installed downstream of the point of delivery shall not extend through any townhouse unit other than the unit served by such piping.

G2415.2 (404.2) Piping in solid partitions and walls. Concealed piping shall not be located in solid partitions and solid walls, unless installed in a chase or casing.

G2415.3 (404.3) Piping in concealed locations. Portions of a piping system installed in concealed locations shall not have unions, tubing fittings, right and left couplings, bushings, compression couplings, and swing joints made by combinations of fittings.

Exceptions:

1. Tubing joined by brazing.

2. Fittings listed for use in concealed locations.

G2415.4 (404.4) Piping through foundation wall. Underground piping, where installed below grade through the outer foundation or basement wall of a building, shall be encased in a protective pipe sleeve. The annular space between the gas piping and the sleeve shall be sealed.

G2415.5 (404.5) Protection against physical damage. In concealed locations, where piping other than black or galvanized steel is installed through holes or notches in wood studs, joists, rafters or similar members less than 1.5 inches (38 mm) from the nearest edge of the member, the pipe shall be protected by shield plates. Shield plates shall be a minimum of $^1/_{16}$-inch-thick (1.6 mm) steel, shall cover the area of the pipe where the member is notched or bored and shall extend a minimum of 4 inches (102 mm) above sole plates, below top plates and to each side of a stud, joist or rafter.

G2415.6 (404.6) Piping in solid floors. Piping in solid floors shall be laid in channels in the floor and covered in a manner that will allow access to the piping with a minimum amount of damage to the building. Where such piping is subject to exposure to excessive moisture or corrosive substances, the piping shall be protected in an approved manner. As an alternative to installation in channels, the piping shall be installed in a conduit of Schedule 40 steel, wrought iron, PVC or ABS pipe with tightly sealed ends and joints. Both ends of such conduit shall extend not less than 2 inches (51 mm) beyond the point where the pipe emerges from the floor. The conduit shall be vented above grade to the outdoors and shall be installed so as to prevent the entry of water and insects.

G2415.7 (404.7) Above-ground piping outdoors. All piping installed outdoors shall be elevated not less than $3^1/_2$ inches (152 mm) above ground and where installed across roof surfaces, shall be elevated not less than $3^1/_2$ inches (152 mm) above the roof surface. Piping installed above ground, outdoors, and installed across the surface of roofs shall be securely supported and located where it will be protected from physical damage. Where passing through an outside wall, the piping shall also be protected against corrosion by coating or wrapping with an inert material. Where piping is encased in a protective pipe sleeve, the annular space between the piping and the sleeve shall be sealed.

G2415.8 (404.8) Protection against corrosion. Metallic pipe or tubing exposed to corrosive action, such as soil condition or moisture, shall be protected in an approved manner. Zinc coatings (galvanizing) shall not be deemed adequate protection for gas piping underground. Ferrous metal exposed in exterior locations shall be protected from corrosion in a manner satisfactory to the code official. Where dissimilar metals are joined underground, an insulating coupling or fitting shall be used. Piping shall not be laid in contact with cinders.

G2415.8.1 (404.8.1) Prohibited use. Uncoated threaded or socket welded joints shall not be used in piping in contact with soil or where internal or external crevice corrosion is known to occur.

G2415.8.2 (404.8.2) Protective coatings and wrapping. Pipe protective coatings and wrappings shall be approved for the application and shall be factory applied.

Exception: Where installed in accordance with the manufacturer's installation instructions, field application of coatings and wrappings shall be permitted for pipe nipples, fittings and locations where the factory coating or wrapping has been damaged or necessarily removed at joints.

G2415.9 (404.9) Minimum burial depth. Underground piping systems shall be installed a minimum depth of 12 inches (305 mm) below grade, except as provided for in Section G2415.9.1.

G2415.9.1 (404.9.1) Individual outside appliances. Individual lines to outside lights, grills or other appliances shall be installed a minimum of 8 inches (203 mm) below finished grade, provided that such installation is approved and is installed in locations not susceptible to physical damage.

G2415.10 (404.10) Trenches. The trench shall be graded so that the pipe has a firm, substantially continuous bearing on the bottom of the trench.

G2415.11 (404.11) Piping underground beneath buildings. Piping installed underground beneath buildings is prohibited except where the piping is encased in a conduit of wrought iron, plastic pipe, or steel pipe designed to withstand the superimposed loads. Such conduit shall extend into an occupiable portion of the building and, at the point where the conduit terminates in the building, the space between the conduit and the gas piping shall be sealed to prevent the possible entrance of any gas leakage. If the end sealing is capable of withstanding the full pressure of the gas pipe, the conduit shall be designed for the same pressure as the pipe. Such conduit shall extend not less than 4 inches (102 mm) outside the building, shall be vented above grade to the outdoors, and shall be installed so as prevent the entrance of water and insects. The conduit shall be protected from corrosion in accordance with Section G2415.8.

G2415.12 (404.12) Outlet closures. Gas outlets that do not connect to appliances shall be capped gas tight.

Exception: Listed and labeled flush-mounted-type quick-disconnect devices and listed and labeled gas conve-

nience outlets shall be installed in accordance with the manufacturer's installation instructions.

G2415.13 (404.13) Location of outlets. The unthreaded portion of piping outlets shall extend not less than 1 inch (25 mm) through finished ceilings and walls and where extending through floors, outdoor patios and slabs, shall not be less than 2 inches (51 mm) above them. The outlet fitting or piping shall be securely supported. Outlets shall not be placed behind doors. Outlets shall be located in the room or space where the appliance is installed.

Exception: Listed and labeled flush-mounted-type quick-disconnect devices and listed and labeled gas convenience outlets shall be installed in accordance with the manufacturer's installation instructions.

G2415.14 (404.14) Plastic pipe. The installation of plastic pipe shall comply with Sections G2415.14.1 through G2415.14.3.

G2415.14.1 (404.14.1) Limitations. Plastic pipe shall be installed outside underground only. Plastic pipe shall not be used within or under any building or slab or be operated at pressures greater than 100 psig (689 kPa) for natural gas or 30 psig (207 kPa) for LP gas.

Exceptions:

1. Plastic pipe shall be permitted to terminate above ground outside of buildings where installed in premanufactured anodeless risers or service head adapter risers that are installed in accordance with that manufacturer's installation instructions.

2. Plastic pipe shall be permitted to terminate with a wall head adapter within buildings where the plastic pipe is inserted in a piping material for fuel gas use in buildings.

G2415.14.2 (404.14.2) Connections. Connections made outside and underground between metallic and plastic piping shall be made only with transition fittings categorized as Category I in accordance with ASTM D 2513.

G2415.14.3 (404.14.3) Tracer. A yellow insulated copper tracer wire or other approved conductor shall be installed adjacent to underground nonmetallic piping. Access shall be provided to the tracer wire or the tracer wire shall terminate above ground at each end of the nonmetallic piping. The tracer wire size shall not be less than 18 AWG and the insulation type shall be suitable for direct burial.

G2415.15 (404.15) Prohibited devices. A device shall not be placed inside the piping or fittings that will reduce the cross-sectional area or otherwise obstruct the free flow of gas.

Exception: Approved gas filters.

G2415.16 (404.16) Testing of piping. Before any system of piping is put in service or concealed, it shall be tested to ensure that it is gas tight. Testing, inspection and purging of piping systems shall comply with Section G2417.

SECTION G2416 (405)
PIPING BENDS AND CHANGES IN DIRECTION

G2416.1 (405.1) General. Changes in direction of pipe shall be permitted to be made by the use of fittings, factory bends or field bends.

G2416.2 (405.2) Metallic pipe. Metallic pipe bends shall comply with the following:

1. Bends shall be made only with bending tools and procedures intended for that purpose.

2. All bends shall be smooth and free from buckling, cracks or other evidence of mechanical damage.

3. The longitudinal weld of the pipe shall be near the neutral axis of the bend.

4. Pipe shall not be bent through an arc of more than 90 degrees (1.6 rad).

5. The inside radius of a bend shall be not less than six times the outside diameter of the pipe.

G2416.3 (405.3) Plastic pipe. Plastic pipe bends shall comply with the following:

1. The pipe shall not be damaged and the internal diameter of the pipe shall not be effectively reduced.

2. Joints shall not be located in pipe bends.

3. The radius of the inner curve of such bends shall not be less than 25 times the inside diameter of the pipe.

4. Where the piping manufacturer specifies the use of special bending tools or procedures, such tools or procedures shall be used.

SECTION G2417 (406)
INSPECTION, TESTING AND PURGING

G2417.1 (406.1) General. Prior to acceptance and initial operation, all piping installations shall be inspected and pressure tested to determine that the materials, design, fabrication, and installation practices comply with the requirements of this code.

G2417.1.1 (406.1.1) Inspections. Inspection shall consist of visual examination, during or after manufacture, fabrication, assembly or pressure tests as appropriate.

G2417.1.2 (406.1.2) Repairs and additions. In the event repairs or additions are made after the pressure test, the affected piping shall be tested.

Minor repairs and additions are not required to be pressure tested provided that the work is inspected and connections are tested with a noncorrosive leak-detecting fluid or other approved leak-detecting methods.

G2417.1.3 (406.1.3) New branches. Where new branches are installed to new appliances, only the newly installed branches shall be required to be pressure tested. Connections between the new piping and the existing piping shall be tested with a noncorrosive leak-detecting fluid or other approved leak-detecting methods.

G2417.1.4 (406.1.4) Section testing. A piping system shall be permitted to be tested as a complete unit or in sections.

Under no circumstances shall a valve in a line be used as a bulkhead between gas in one section of the piping system and test medium in an adjacent section, unless two valves are installed in series with a valved "tell-tale" located between these valves. A valve shall not be subjected to the test pressure unless it can be determined that the valve, including the valve closing mechanism, is designed to safely withstand the test pressure.

G2417.1.5 (406.1.5) Regulators and valve assemblies. Regulator and valve assemblies fabricated independently of the piping system in which they are to be installed shall be permitted to be tested with inert gas or air at the time of fabrication.

G2417.2 (406.2) Test medium. The test medium shall be air, nitrogen, carbon dioxide or an inert gas. Oxygen shall not be used.

G2417.3 (406.3) Test preparation. Pipe joints, including welds, shall be left exposed for examination during the test.

Exception: Covered or concealed pipe end joints that have been previously tested in accordance with this code.

G2417.3.1 (406.3.1) Expansion joints. Expansion joints shall be provided with temporary restraints, if required, for the additional thrust load under test.

G2417.3.2 (406.3.2) Equipment isolation. Equipment that is not to be included in the test shall be either disconnected from the piping or isolated by blanks, blind flanges or caps.

G2417.3.3 (406.3.3) Appliance and equipment disconnection. Where the piping system is connected to appliances or equipment designed for operating pressures of less than the test pressure, such appliances or equipment shall be isolated from the piping system by disconnecting them and capping the outlet(s).

G2417.3.4 (406.3.4) Valve isolation. Where the piping system is connected to appliances or equipment designed for operating pressures equal to or greater than the test pressure, such appliances or equipment shall be isolated from the piping system by closing the individual appliance or equipment shutoff valve(s).

G2417.3.5 (406.3.5) Testing precautions. All testing of piping systems shall be done with due regard for the safety of employees and the public during the test. Prior to testing, the interior of the pipe shall be cleared of all foreign material.

G2417.4 (406.4) Test pressure measurement. Test pressure shall be measured with a manometer or with a pressure-measuring device designed and calibrated to read, record, or indicate a pressure loss caused by leakage during the pressure test period. The source of pressure shall be isolated before the pressure tests are made. Mechanical gauges used to measure test pressures shall have a range such that the highest end of the scale is not greater than five times the test pressure.

G2417.4.1 (406.4.1) Test pressure. The test pressure to be used shall be not less than one and one-half times the proposed maximum working pressure, but not less than 3 psig (20 kPa gauge), irrespective of design pressure. Where the test pressure exceeds 125 psig (862 kPa gauge), the test pressure shall not exceed a value that produces a hoop stress in the piping greater than 50 percent of the specified minimum yield strength of the pipe.

G2417.4.2 (406.4.2) Test duration. The test duration shall be not less than 10 minutes.

G2417.5 (406.5) Detection of leaks and defects. The piping system shall withstand the test pressure specified without showing any evidence of leakage or other defects. Any reduction of test pressures as indicated by pressure gauges shall be deemed to indicate the presence of a leak unless such reduction can be readily attributed to some other cause.

G2417.5.1 (406.5.1) Detection methods. The leakage shall be located by means of an approved combustible gas detector, a noncorrosive leak detection fluid or an equivalent nonflammable solution. Matches, candles, open flames or other methods that could provide a source of ignition shall not be used.

G2417.5.2 (406.5.2) Corrections. Where leakage or other defects are located, the affected portion of the piping system shall be repaired or replaced and retested.

G2417.6 (406.6) Piping system, appliance and equipment leakage check. Leakage checking of systems and equipment shall be in accordance with Sections G2417.6.1 through G2417.6.4.

G2417.6.1 (406.6.1) Test gases. Fuel gas shall be permitted to be used for leak checks in piping systems that have been tested in accordance with Section G2417.

G2417.6.2 (406.6.2) Before turning gas on. Before gas is introduced into a system of new gas piping, the entire system shall be inspected to determine that there are no open fittings or ends and that all valves at unused outlets are closed and plugged or capped.

G2417.6.3 (406.6.3) Leak check. Immediately after the gas is turned on into a new system or into a system that has been initially restored after an interruption of service, the piping system shall be checked for leakage. Where leakage is indicated, the gas supply shall be shut off until the necessary repairs have been made.

G2417.6.4 (406.6.4) Placing appliances and equipment in operation. Appliances and equipment shall be permitted to be placed in operation after the piping system has been checked for leakage and determined to be free of leakage and purged in accordance with Section G2417.7.2.

G2417.7 (406.7) Purging. Purging of piping shall comply with Sections G2417.7.1 through G2417.7.4.

G2417.7.1 (406.7.1) Removal from service. When gas piping is to be opened for servicing, addition or modification, the section to be worked on shall be turned off from the gas supply at the nearest convenient point, and the line pressure vented to the outdoors, or to ventilated areas of sufficient size to prevent accumulation of flammable mixtures.

G2417.7.2 (406.7.2) Placing in operation. When piping full of air is placed in operation, the air in the piping shall be displaced with fuel gas. The air can be safely displaced with fuel gas provided that a moderately rapid and continuous flow of fuel gas is introduced at one end of the line and air is vented out at the other end. The fuel gas flow should be continued without interruption until the vented gas is free of air. The point of discharge shall not be left unattended during purging. After purging, the vent shall then be closed.

G2417.7.3 (406.7.3) Discharge of purged gases. The open end of piping systems being purged shall not discharge into confined spaces or areas where there are sources of ignition unless precautions are taken to perform this operation in a safe manner by ventilation of the space, control or purging rate, and elimination of all hazardous conditions.

G2417.7.4 (406.7.4) Placing appliances and equipment in operation. After the piping system has been placed in operation, all appliances and equipment shall be purged and then placed in operation, as necessary.

SECTION G2418 (407)
PIPING SUPPORT

G2418.1 (407.1) General. Piping shall be provided with support in accordance with Section G2418.2.

G2418.2 (407.2) Design and installation. Piping shall be supported with pipe hooks, metal pipe straps, bands, brackets or hangers suitable for the size of piping, of adequate strength and quality, and located at intervals so as to prevent or damp out excessive vibration. Piping shall be anchored to prevent undue strains on connected equipment and shall not be supported by other piping. Pipe hangers and supports shall conform to the requirements of MSS SP-58 and shall be spaced in accordance with Section G2424. Supports, hangers, and anchors shall be installed so as not to interfere with the free expansion and contraction of the piping between anchors. All parts of the supporting equipment shall be designed and installed so they will not be disengaged by movement of the supported piping.

SECTION G2419 (408)
DRIPS AND SLOPED PIPING

G2419.1 (408.1) Slopes. Piping for other than dry gas conditions shall be sloped not less than 0.25 inch in 15 feet (6.4 mm in 4572 mm) to prevent traps.

G2419.2 (408.2) Drips. Where wet gas exists, a drip shall be provided at any point in the line of pipe where condensate could collect. A drip shall also be provided at the outlet of the meter and shall be installed so as to constitute a trap wherein an accumulation of condensate will shut off the flow of gas before the condensate will run back into the meter.

G2419.3 (408.3) Location of drips. Drips shall be provided with ready access to permit cleaning or emptying. A drip shall not be located where the condensate is subject to freezing.

G2419.4 (408.4) Sediment trap. Where a sediment trap is not incorporated as part of the gas utilization equipment, a sediment trap shall be installed downstream of the equipment shutoff valve as close to the inlet of the equipment as practical. The sediment trap shall be either a tee fitting with a capped nipple in the bottom opening of the run of the tee or other device approved as an effective sediment trap. Illuminating appliances, ranges, clothes dryers and outdoor grills need not be so equipped.

SECTION G2420 (409)
GAS SHUTOFF VALVES

G2420.1 (409.1) General. Piping systems shall be provided with shutoff valves in accordance with this section.

G2420.1.1 (409.1.1) Valve approval. Shutoff valves shall be of an approved type; shall be constructed of materials compatible with the piping; and shall comply with the standard that is applicable for the pressure and application, in accordance with Table G2420.1.1.

G2420.1.2 (409.1.2) Prohibited locations. Shutoff valves shall be prohibited in concealed locations and furnace plenums.

G2420.1.3 (409.1.3) Access to shutoff valves. Shutoff valves shall be located in places so as to provide access for operation and shall be installed so as to be protected from damage.

G2420.2 (409.2) Meter valve. Every meter shall be equipped with a shutoff valve located on the supply side of the meter.

G2420.3 (409.3.2) Individual buildings. In a common system serving more than one building, shutoff valves shall be installed outdoors at each building.

G2420.4 (409.4) MP regulator valves. A listed shutoff valve shall be installed immediately ahead of each MP regulator.

G2420.5 (409.5) Equipment shutoff valve. Each appliance shall be provided with a shutoff valve separate from the appliance. The shutoff valve shall be located in the same room as the appliance, not further than 6 feet (1829 mm) from the appliance, and shall be installed upstream from the union, connector or quick disconnect device it serves. Such shutoff valves shall be provided with access.

Exception: Shutoff valves for vented decorative appliances and decorative appliances for installation in vented fireplaces shall not be prohibited from being installed in an area remote from the appliance where such valves are provided with ready access. Such valves shall be permanently identified and shall serve no other equipment. Piping from the shutoff valve to within 3 feet (914 mm) of the appliance connection shall be sized in accordance with Section 402.

G2420.5.1 (409.5.1) Shutoff valve in fireplace. Equipment shutoff valves located in the firebox of a fireplace shall be installed in accordance with the appliance manufacturer's instructions.

SECTION G2421 (410)
FLOW CONTROLS

G2421.1 (410.1) Pressure regulators. A line pressure regulator shall be installed where the appliance is designed to operate at a lower pressure than the supply pressure. Line gas pressure regulators shall be listed as complying with ANSI Z21.80. Access shall be provided to pressure regulators. Pressure regulators shall be protected from physical damage. Regulators installed on the exterior of the building shall be approved for outdoor installation.

G2421.2 (410.2) MP regulators. MP pressure regulators shall comply with the following:

1. The MP regulator shall be approved and shall be suitable for the inlet and outlet gas pressures for the application.

2. The MP regulator shall maintain a reduced outlet pressure under lockup (no-flow) conditions.

3. The capacity of the MP regulator, determined by published ratings of its manufacturer, shall be adequate to supply the appliances served.

4. The MP pressure regulator shall be provided with access. Where located indoors, the regulator shall be vented to the outdoors or shall be equipped with a leak-limiting device, in either case complying with Section G2421.3.

5. A tee fitting with one opening capped or plugged shall be installed between the MP regulator and its upstream shutoff valve. Such tee fitting shall be positioned to allow connection of a pressure measuring instrument and to serve as a sediment trap.

6. A tee fitting with one opening capped or plugged shall be installed not less than 10 pipe diameters downstream of the MP regulator outlet. Such tee fitting shall be positioned to allow connection of a pressure measuring instrument.

G2421.3 (410.3) Venting of regulators. Pressure regulators that require a vent shall be vented directly to the outdoors. The vent shall be designed to prevent the entry of insects, water and foreign objects.

TABLE G2420.1.1
MANUAL GAS VALVE STANDARDS

VALVE STANDARDS	APPLIANCE SHUTOFF VALVE APPLICATION UP TO $^{1}/_{2}$ psig PRESSURE	OTHER VALVE APPLICATIONS			
		UP TO $^{1}/_{2}$ psig PRESSURE	UP TO 2 psig PRESSURE	UP TO 5 psig PRESSURE	UP TO 125 psig PRESSURE
ANSI Z21.15	X	—	—	—	—
CSA Requirement 3-88	X	X	X[a]	X[b]	—
ASME B16.44	X	X	X[a]	X[b]	—
ASME B16.33	X	X	X	X	X

For SI: 1 pound per square inch gauge = 6.895 kPa.
a. If labeled 2G.
b. If labeled 5G.

Exception: A vent to the outdoors is not required for regulators equipped with and labeled for utilization with an approved vent-limiting device installed in accordance with the manufacturer's instructions.

G2421.3.1 (410.3.1) Vent piping. Vent piping shall be not smaller than the vent connection on the pressure regulating device. Vent piping serving relief vents and combination relief and breather vents shall be run independently to the outdoors and shall serve only a single device vent. Vent piping serving only breather vents is permitted to be connected in a manifold arrangement where sized in accordance with an approved design that minimizes back pressure in the event of diaphragm rupture.

SECTION G2422 (411)
APPLIANCE CONNECTIONS

G2422.1 (411.1) Connecting appliances. Appliances shall be connected to the piping system by one of the following:

1. Rigid metallic pipe and fittings.

2. Corrugated stainless steel tubing (CSST) where installed in accordance with the manufacturer's instructions.

3. Listed and labeled appliance connectors in compliance with ANSI Z21.24 and installed in accordance with the manufacturer's installation instructions and located entirely in the same room as the appliance.

4. Listed and labeled quick-disconnect devices used in conjunction with listed and labeled appliance connectors.

5. Listed and labeled convenience outlets used in conjunction with listed and labeled appliance connectors.

6. Listed and labeled outdoor appliance connectors in compliance with ANSI Z21.75/CSA 6.27 and installed in accordance with the manufacturer's installation instructions.

G2422.1.1 (411.1.2) Protection from damage. Connectors and tubing shall be installed so as to be protected against physical damage.

G2422.1.2 (411.1.3) Connector installation. Appliance fuel connectors shall be installed in accordance with the manufacturer's instructions and Sections G24221.2.1 through G2422.1.2.4.

G2422.1.2.1 (411.1.3.1) Maximum length. Connectors shall have an overall length not to exceed 3 feet (914 mm), except for range and domestic clothes dryer connectors, which shall not exceed 6 feet (1829 mm) in overall length. Measurement shall be made along the centerline of the connector. Only one connector shall be used for each appliance.

Exception: Rigid metallic piping used to connect an appliance to the piping system shall be permitted to have a total length greater than 3 feet (914 mm), provided that the connecting pipe is sized as part of the piping system in accordance with Section G2413 and the location of the equipment shutoff valve complies with Section G2420.5.

G2422.1.2.2 (411.1.3.2) Minimum size. Connectors shall have the capacity for the total demand of the connected appliance.

G2422.1.2.3 (411.1.3.3) Prohibited locations and penetrations. Connectors shall not be concealed within, or extended through, walls, floors, partitions, ceilings or appliance housings.

Exception: Fireplace inserts that are factory equipped with grommets, sleeves or other means of protection in accordance with the listing of the appliance.

G2422.1.2.4 (411.1.3.4) Shutoff valve. A shutoff valve not less than the nominal size of the connector shall be installed ahead of the connector in accordance with Section G2420.5.

G2422.1.3 (411.1.4) Movable appliances. Where appliances are equipped with casters or are otherwise subject to periodic movement or relocation for purposes such as routine cleaning and maintenance, such appliances shall be connected to the supply system piping by means of an approved flexible connector designed and labeled for the application. Such flexible connectors shall be installed and protected against physical damage in accordance with the manufacturer's installation instructions.

SECTION G2423 (413)
CNG GAS-DISPENSING SYSTEMS

G2423.1 (413.1) General. Motor fuel-dispensing facilities for CNG fuel shall be in accordance with Section 413 of the *International Fuel Gas Code.*

SECTION G2424 (415)
PIPING SUPPORT INTERVALS

G2424.1 (415.1) Interval of support. Piping shall be supported at intervals not exceeding the spacing specified in Table G2424.1. Spacing of supports for CSST shall be in accordance with the CSST manufacturer's instructions.

TABLE G2424.1
SUPPORT OF PIPING

STEEL PIPE, NOMINAL SIZE OF PIPE (inches)	SPACING OF SUPPORTS (feet)	NOMINAL SIZE OF TUBING SMOOTH-WALL (inch O.D.)	SPACING OF SUPPORTS (feet)
$^1/_2$	6	$^1/_2$	4
$^3/_4$ or 1	8	$^5/_8$ or $^3/_4$	6
$1^1/_4$ or larger (horizontal)	10	$^7/_8$ or 1 (horizontal)	8
$1^1/_4$ or larger (vertical)	Every floor level	1 or Larger (vertical)	Every floor level

For SI: 1 inch = 25.4 mm, 1 foot = 304.8 mm.

SECTION G2425 (501)
GENERAL

G2425.1 (501.1) Scope. This section shall govern the installation, maintenance, repair and approval of factory-built and masonry chimneys, chimney liners, vents and connectors serving gas-fired appliances.

G2425.2 (501.2) General. Every appliance shall discharge the products of combustion to the outdoors, except for appliances exempted by Section G2425.8.

G2425.3 (501.3) Masonry chimneys. Masonry chimneys shall be constructed in accordance with Section G2427.5 and Chapter 10.

G2425.4 (501.4) Minimum size of chimney or vent. Chimneys and vents shall be sized in accordance with Section G2427.

G2425.5 (501.5) Abandoned inlet openings. Abandoned inlet openings in chimneys and vents shall be closed by an approved method.

G2425.6 (501.6) Positive pressure. Where an appliance equipped with a mechanical forced draft system creates a positive pressure in the venting system, the venting system shall be designed for positive pressure applications.

G2425.7 (501.7) Connection to fireplace. Connection of appliances to chimney flues serving fireplaces shall be in accordance with Sections G2425.7.1 through G2425.7.3.

G2425.7.1 (501.7.1) Closure and access. A noncombustible seal shall be provided below the point of connection to prevent entry of room air into the flue. Means shall be provided for access to the flue for inspection and cleaning.

G2425.7.2 (501.7.2) Connection to factory-built fireplace flue. An appliance shall not be connected to a flue serving a factory-built fireplace unless the appliance is specifically listed for such installation. The connection shall be made in accordance with the appliance manufacturer's installation instructions.

G2425.7.3 (501.7.3) Connection to masonry fireplace flue. A connector shall extend from the appliance to the flue serving a masonry fireplace such that the flue gases are exhausted directly into the flue. The connector shall be accessible or removable for inspection and cleaning of both the connector and the flue. Listed direct connection devices shall be installed in accordance with their listing.

G2425.8 (501.8) Equipment not required to be vented. The following appliances shall not be required to be vented:

1. Ranges.

2. Built-in domestic cooking units listed and marked for optional venting.

3. Hot plates and laundry stoves.

4. Type 1 clothes dryers (Type 1 clothes dryers shall be exhausted in accordance with the requirements of Section G2439).

5. Refrigerators.

6. Counter appliances.

7. Room heaters listed for unvented use.

Where the appliances and equipment listed in Items 5 through 7 above are installed so that the aggregate input rating exceeds 20 Btu per hour per cubic foot (207 watts per m³) of volume of the room or space in which such appliances and equipment are installed, one or more shall be provided with venting systems or other approved means for conveying the vent gases to the outdoor atmosphere so that the aggregate input rating of the remaining unvented appliances and equipment does not exceed the 20 Btu per hour per cubic foot (207 watts per m³) figure. Where the room or space in which the equipment is installed is directly connected to another room or space by a doorway, archway or other opening of comparable size that cannot be closed, the volume of such adjacent room or space shall be permitted to be included in the calculations.

G2425.9 (501.9) Chimney entrance. Connectors shall connect to a masonry chimney flue at a point not less than 12 inches (305 mm) above the lowest portion of the interior of the chimney flue.

G2425.10 (501.10) Connections to exhauster. Appliance connections to a chimney or vent equipped with a power exhauster shall be made on the inlet side of the exhauster. Joints on the positive pressure side of the exhauster shall be sealed to prevent flue-gas leakage as specified by the manufacturer's installation instructions for the exhauster.

G2425.11 (501.11) Masonry chimneys. Masonry chimneys utilized to vent appliances shall be located, constructed and sized as specified in the manufacturer's installation instructions for the appliances being vented and Section G2427.

G2425.12 (501.12) Residential and low-heat appliances flue lining systems. Flue lining systems for use with residential-type and low-heat appliances shall be limited to the following:

1. Clay flue lining complying with the requirements of ASTM C 315 or equivalent. Clay flue lining shall be installed in accordance with Chapter 10.

2. Listed chimney lining systems complying with UL 1777.

3. Other approved materials that will resist, without cracking, softening or corrosion, flue gases and condensate at temperatures up to 1,800°F (982°C).

G2425.13 (501.13) Category I appliance flue lining systems. Flue lining systems for use with Category I appliances shall be limited to the following:

1. Flue lining systems complying with Section G2425.12.

2. Chimney lining systems listed and labeled for use with appliances with draft hoods and other Category I gas appliances listed and labeled for use with Type B vents.

G2425.14 (501.14) Category II, III and IV appliance venting systems. The design, sizing and installation of vents for Category II, III and IV appliances shall be in accordance with the appliance manufacturer's installation instructions.

G2425.15 (501.15) Existing chimneys and vents. Where an appliance is permanently disconnected from an existing chimney or vent, or where an appliance is connected to an existing chimney or vent during the process of a new installation, the chimney or vent shall comply with Sections G2425.15.1 through G2425.15.4.

G2425.15.1 (501.15.1) Size. The chimney or vent shall be resized as necessary to control flue gas condensation in the interior of the chimney or vent and to provide the appliance or appliances served with the required draft. For Category I appliances, the resizing shall be in accordance with Section G2426.

G2425.15.2 (501.15.2) Flue passageways. The flue gas passageway shall be free of obstructions and combustible deposits and shall be cleaned if previously used for venting a solid or liquid fuel-burning appliance or fireplace. The flue liner, chimney inner wall or vent inner wall shall be continuous and shall be free of cracks, gaps, perforations, or other damage or deterioration that would allow the escape of combustion products, including gases, moisture and creosote.

G2425.15.3 (501.15.3) Cleanout. Masonry chimney flues shall be provided with a cleanout opening having a minimum height of 6 inches (152 mm). The upper edge of the opening shall be located not less than 6 inches (152 mm) below the lowest chimney inlet opening. The cleanout shall be provided with a tight-fitting, noncombustible cover.

G2425.15.4 (501.15.4) Clearances. Chimneys and vents shall have airspace clearance to combustibles in accordance with Chapter 10 and the chimney or vent manufacturer's installation instructions. Noncombustible firestopping or fireblocking shall be provided in accordance with Chapter 10.

Exception: Masonry chimneys equipped with a chimney lining system tested and listed for installation in chimneys in contact with combustibles in accordance with UL 1777, and installed in accordance with the manufacturer's instructions, shall not be required to have clearance between combustible materials and exterior surfaces of the masonry chimney.

SECTION G2426 (502)
VENTS

G2426.1 (502.1) General. All vents, except as provided in Section G2427.7, shall be listed and labeled. Type B and BW vents shall be tested in accordance with UL 441. Type L vents shall be tested in accordance with UL 641. Vents for Category II and III appliances shall be tested in accordance with UL 1738. Plastic vents for Category IV appliances shall not be required to be listed and labeled where such vents are as specified by the appliance manufacturer and are installed in accordance with the appliance manufacturer's installation instructions.

G2426.2 (502.2) Connectors required. Connectors shall be used to connect appliances to the vertical chimney or vent, except where the chimney or vent is attached directly to the appliance. Vent connector size, material, construction and installation shall be in accordance with Section G2427.

G2426.3 (502.3) Vent application. The application of vents shall be in accordance with Table G2427.4.

G2426.4 (502.4) Insulation shield. Where vents pass through insulated assemblies, an insulation shield constructed of not less than 26 gage sheet (0.016 inch) (0.4 mm) metal shall be installed to provide clearance between the vent and the insulation material. The clearance shall not be less than the clearance to combustibles specified by the vent manufacturer's installation instructions. Where vents pass through attic space, the shield shall terminate not less than 2 inches (51 mm) above the insulation materials and shall be secured in place to prevent displacement. Insulation shields provided as part of a listed vent system shall be installed in accordance with the manufacturer's installation instructions.

G2426.5 (502.5) Installation. Vent systems shall be sized, installed and terminated in accordance with the vent and appliance manufacturer's installation instructions and Section G2427.

G2426.6 (502.6) Support of vents. All portions of vents shall be adequately supported for the design and weight of the materials employed.

G2426.7 (502.7) Protection against physical damage. In concealed locations, where a vent is installed through holes or notches in studs, joists, rafters or similar members less than 1.5 inches (38 mm) from the nearest edge of the member, the vent shall be protected by shield plates. Shield plates shall be a minimum of $^1/_{16}$-inch-thick (1.6 mm) steel, shall cover the area of the vent where the member is notched or bored and shall extend a minimum of 4 inches (102 mm) above sole plates, below top plates and to each side of a stud, joist or rafter.

SECTION G2427 (503)
VENTING OF EQUIPMENT

G2427.1 (503.1) General. This section recognizes that the choice of venting materials and the methods of installation of venting systems are dependent on the operating characteristics of the appliance being vented. The operating characteristics of vented appliances can be categorized with respect to: (1) positive or negative pressure within the venting system; and (2) whether or not the appliance generates flue or vent gases that might condense in the venting system. See Section G2403 for the definitions of these vented appliance categories.

G2427.2 (503.2) Venting systems required. Except as permitted in Sections G2427.2.1, G2427.2.2 and G2425.8, all appliances shall be connected to venting systems.

G2427.2.1 (503.2.3) Direct-vent appliances. Listed direct-vent appliances shall be installed in accordance with the manufacturer's instructions and Section G2427.8, Item 3.

G2427.2.2 (503.2.4) Appliances with integral vents. Appliances incorporating integral venting means shall be considered properly vented where installed in accordance with the manufacturer's instructions and Section G2427.8, Items 1 and 2.

G2427.3 (503.3) Design and construction. A venting system shall be designed and constructed so as to develop a positive flow adequate to convey flue or vent gases to the outdoors.

G2427.3.1 (503.3.1) Appliance draft requirements. A venting system shall satisfy the draft requirements of the appliance in accordance with the manufacturer's instructions.

G2427.3.2 (503.3.2) Design and construction. Appliances required to be vented shall be connected to a venting system designed and installed in accordance with the provisions of Sections G2427.4 through G2427.15.

G2427.3.3 (503.3.3) Mechanical draft systems. Mechanical draft systems shall comply with the following:

1. Mechanical draft systems shall be listed and shall be installed in accordance with the manufacturer's installation instructions for both the appliance and the mechanical draft system.

2. Appliances, except incinerators, requiring venting shall be permitted to be vented by means of mechanical draft systems of either forced or induced draft design.

3. Forced draft systems and all portions of induced draft systems under positive pressure during operation shall be designed and installed so as to prevent leakage of flue or vent gases into a building.

4. Vent connectors serving appliances vented by natural draft shall not be connected into any portion of mechanical draft systems operating under positive pressure.

5. Where a mechanical draft system is employed, provisions shall be made to prevent the flow of gas to the main burners when the draft system is not performing so as to satisfy the operating requirements of the appliance for safe performance.

6. The exit terminals of mechanical draft systems shall be not less than 7 feet (2134 mm) above grade where located adjacent to public walkways and shall be located as specified in Section 503.8, Items 1 and 2.

G2427.3.4 (503.3.5) Circulating air ducts and furnace plenums. No portion of a venting system shall extend into or pass through any circulating air duct or furnace plenum.

G2427.3.5 (503.3.6) Above-ceiling air-handling spaces. Where a venting system passes through an above-ceiling air-handling space or other nonducted portion of an air-handling system, the venting system shall conform to one of the following requirements:

1. The venting system shall be a listed special gas vent; other venting system serving a Category III or Category IV appliance; or other positive pressure vent, with joints sealed in accordance with the appliance or vent manufacturer's instructions.

2. The venting system shall be installed such that fittings and joints between sections are not installed in the above-ceiling space.

3. The venting system shall be installed in a conduit or enclosure with sealed joints separating the interior of the conduit or enclosure from the ceiling space.

G2427.4 (503.4) Type of venting system to be used. The type of venting system to be used shall be in accordance with Table G2427.4.

G2427.4.1 (503.4.1) Plastic piping. Plastic piping used for venting appliances listed for use with such venting materials shall be approved.

G2427.4.2 (503.4.2) Special gas vent. Special gas vent shall be listed and installed in accordance with the special gas vent manufacturer's installation instructions.

TABLE G2427.4
TYPE OF VENTING SYSTEM TO BE USED

APPLIANCES	TYPE OF VENTING SYSTEM
Listed Category I appliances Listed appliances equipped with draft hood Appliances listed for use with Type B gas vent	Type B gas vent (Section G2427.6) Chimney (Section G2427.5) Single-wall metal pipe (Section G2427.7) Listed chimney lining system for gas venting (Section G2427.5.2) Special gas vent listed for these appliances (Section G2427.4.2)
Listed vented wall furnaces	Type B-W gas vent (Sections G2427.6, G2436)
Category II appliances	As specified or furnished by manufacturers of listed appliances (Sections G2427.4.1, G2427.4.2)
Category III appliances	As specified or furnished by manufacturers of listed appliances (Sections G2427.4.1, G2427.4.2)
Category IV appliances	As specified or furnished by manufacturers of listed appliances (Sections G2427.4.1, G2427.4.2)
Unlisted appliances	Chimney (Section G2427.5)
Decorative appliances in vented fireplaces	Chimney
Direct-vent appliances	See Section G2427.2.1
Appliances with integral vent	See Section G2427.2.2

G2427.5 (503.5) Masonry, metal, and factory-built chimneys. Masonry, metal and factory-built chimneys shall comply with Sections G2427.5.1 through G2427.5.9.

G2427.5.1 (503.5.1) Factory-built chimneys. Factory-built chimneys shall be installed in accordance with the manufacturer's installation instructions. Factory-built chimneys used to vent appliances that operate at a positive vent pressure shall be listed for such application.

G2427.5.2 (503.5.3) Masonry chimneys. Masonry chimneys shall be built and installed in accordance with NFPA 211 and shall be lined with approved clay flue lining, a listed chimney lining system or other approved material that will resist corrosion, erosion, softening or cracking from vent gases at temperatures up to 1,800°F (982°C).

Exception: Masonry chimney flues serving listed gas appliances with draft hoods, Category I appliances and other gas appliances listed for use with Type B vents shall be permitted to be lined with a chimney lining system specifically listed for use only with such appliances. The liner shall be installed in accordance with the liner manufacturer's installation instructions. A permanent identifying label shall be attached at the point where the connection is to be made to the liner. The label shall read: "This chimney liner is for appliances that burn gas only. Do not connect to solid or liquid fuel-burning appliances or incinerators."

G2427.5.3 (503.5.4) Chimney termination. Chimneys for residential-type or low-heat appliances shall extend at least 3 feet (914 mm) above the highest point where they pass through a roof of a building and at least 2 feet (610 mm) higher than any portion of a building within a horizontal distance of 10 feet (3048 mm) (see Figure G2427.5.3). Chimneys for medium-heat appliances shall extend at least 10 feet (3048 mm) higher than any portion of any building within 25 feet (7620 mm). Chimneys shall extend at least 5 feet (1524 mm) above the highest connected appliance draft hood outlet or flue collar. Decorative shrouds shall not be installed at the termination of factory-built chimneys except where such shrouds are listed and labeled for use with the specific factory-built chimney system and are installed in accordance with the manufacturer's installation instructions.

G2427.5.4 (503.5.5) Size of chimneys. The effective area of a chimney venting system serving listed appliances with

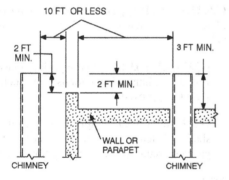

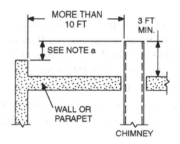

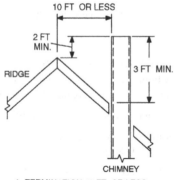

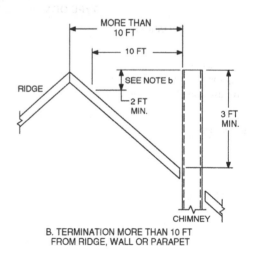

A. TERMINATION 10 FT OR LESS FROM RIDGE, WALL OR PARAPET

B. TERMINATION MORE THAN 10 FT FROM RIDGE, WALL OR PARAPET

For SI: 1 inch = 25.4 mm, 1 foot = 304.8 mm.
NOTES:
a. No height above parapet required when distance from walls or parapet is more than 10 feet.
b. Height above any roof surface within 10 feet horizontally.

FIGURE G2427.5.3 (503.5.4)
TYPICAL TERMINATION LOCATIONS FOR CHIMNEYS AND SINGLE-WALL METAL PIPES

draft hoods, Category I appliances, and other appliances listed for use with Type B vents shall be determined in accordance with one of the following methods:

1. The provisions of Section G2428.

2. For sizing an individual chimney venting system for a single appliance with a draft hood, the effective areas of the vent connector and chimney flue shall be not less than the area of the appliance flue collar or draft hood outlet, nor greater than seven times the draft hood outlet area.

3. For sizing a chimney venting system connected to two appliances with draft hoods, the effective area of the chimney flue shall be not less than the area of the larger draft hood outlet plus 50 percent of the area of the smaller draft hood outlet, nor greater than seven times the smallest draft hood outlet area.

4. Chimney venting systems using mechanical draft shall be sized in accordance with approved engineering methods.

5. Other approved engineering methods.

G2427.5.5 (503.5.6) Inspection of chimneys. Before replacing an existing appliance or connecting a vent connector to a chimney, the chimney passageway shall be examined to ascertain that it is clear and free of obstructions and it shall be cleaned if previously used for venting solid or liquid fuel-burning appliances or fireplaces.

G2427.5.5.1 (503.5.6.1) Chimney lining. Chimneys shall be lined in accordance with Chapter 10.

Exception: Existing chimneys shall be permitted to have their use continued when an appliance is replaced by an appliance of similar type, input rating and efficiency.

G2427.5.5.2 (503.5.6.2) Cleanouts. Cleanouts shall be examined to determine they will remain tightly closed when not in use.

G2427.5.5.3 (503.5.6.3) Unsafe chimneys. Where inspection reveals that an existing chimney is not safe for the intended application, it shall be repaired, rebuilt, lined, relined or replaced with a vent or chimney to conform to NFPA 211 and it shall be suitable for the appliances to be vented.

G2427.5.6 (503.5.7) Chimneys serving equipment burning other fuels. Chimneys serving equipment burning other fuels shall comply with Sections G2427.5.6.1 through G2427.5.6.4.

G2427.5.6.1 (503.5.7.1) Solid fuel-burning appliances. An appliance shall not be connected to a chimney flue serving a separate appliance designed to burn solid fuel.

G2427.5.6.2 (503.5.7.2) Liquid fuel-burning appliances. Where one chimney flue serves gas appliances and liquid fuel-burning appliances, the appliances shall be connected through separate openings or shall be connected through a single opening where joined by a suitable fitting located as close as practical to the chimney.

Where two or more openings are provided into one chimney flue, they shall be at different levels. Where the appliances are automatically controlled, they shall be equipped with safety shutoff devices.

G2427.5.6.3 (503.5.7.3) Combination gas- and solid fuel-burning appliances. A combination gas- and solid fuel-burning appliance equipped with a manual reset device to shut off gas to the main burner in the event of sustained backdraft or flue gas spillage shall be permitted to be connected to a single chimney flue. The chimney flue shall be sized to properly vent the appliance.

G2427.5.6.4 (503.5.7.4) Combination gas- and oil fuel-burning appliances. A listed combination gas- and oil fuel-burning appliance shall be permitted to be connected to a single chimney flue. The chimney flue shall be sized to properly vent the appliance.

G2427.5.7 (503.5.8) Support of chimneys. All portions of chimneys shall be supported for the design and weight of the materials employed. Factory-built chimneys shall be supported and spaced in accordance with the manufacturer's installation instructions.

G2427.5.8 (503.5.9) Cleanouts. Where a chimney that formerly carried flue products from liquid or solid fuel-burning appliances is used with an appliance using fuel gas, an accessible cleanout shall be provided. The cleanout shall have a tight-fitting cover and be installed so its upper edge is at least 6 inches (152 mm) below the lower edge of the lowest chimney inlet opening.

G2427.5.9 (503.5.10) Space surrounding lining or vent. The remaining space surrounding a chimney liner, gas vent, special gas vent or plastic piping installed within a masonry chimney flue shall not be used to vent another appliance. The insertion of another liner or vent within the chimney as provided in this code and the liner or vent manufacturer's instructions shall not be prohibited.

The remaining space surrounding a chimney liner, gas vent, special gas vent or plastic piping installed within a masonry, metal or factory-built chimney shall not be used to supply combustion air. Such space shall not be prohibited from supplying combustion air to direct-vent appliances designed for installation in a solid fuel-burning fireplace and installed in accordance with the manufacturer's installation instructions.

G2427.6 (503.6) Gas vents. Gas vents shall comply with Sections G2427.6.1 through G2427.6.10. (See Section G2403, Definitions.)

G2427.6.1 (503.6.1) Installation, general. Gas vents shall be installed in accordance with the terms of their listings and the manufacturer's instructions.

G2427.6.2 (503.6.2) Type B-W vent capacity. A Type B-W gas vent shall have a listed capacity not less than that of the listed vented wall furnace to which it is connected.

G2427.6.3 (503.6.4) Gas vent termination. A gas vent shall terminate in accordance with one of the following:

1. Gas vents that are 12 inches (305 mm) or less in size and located not less than 8 feet (2438 mm) from a ver-

tical wall or similar obstruction shall terminate above the roof in accordance with Figure G2427.6.3.

2. Gas vents that are over 12 inches (305 mm) in size or are located less than 8 feet (2438 mm) from a vertical wall or similar obstruction shall terminate not less than 2 feet (610 mm) above the highest point where they pass through the roof and not less than 2 feet (610 mm) above any portion of a building within 10 feet (3048 mm) horizontally.

3. As provided for direct-vent systems in Section G2427.2.1.

4. As provided for appliances with integral vents in Section G2427.2.2.

5. As provided for mechanical draft systems in Section G2427.3.3.

G2427.6.3.1 (503.6.4.1) Decorative shrouds. Decorative shrouds shall not be installed at the termination of gas vents except where such shrouds are listed for use with the specific gas venting system and are installed in accordance with manufacturer's installation instructions.

G2427.6.4 (503.6.5) Minimum height. A Type B or L gas vent shall terminate at least 5 feet (1524 mm) in vertical height above the highest connected appliance draft hood or flue collar. A Type B-W gas vent shall terminate at least 12 feet (3658 mm) in vertical height above the bottom of the wall furnace.

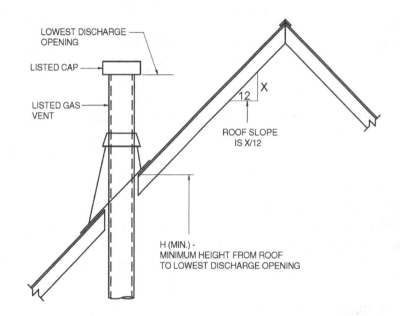

ROOF SLOPE	H (minimum) ft
Flat to $^6/_{12}$	1.0
Over $^6/_{12}$ to $^7/_{12}$	1.25
Over $^7/_{12}$ to $^8/_{12}$	1.5
Over $^8/_{12}$ to $^9/_{12}$	2.0
Over $^9/_{12}$ to $^{10}/_{12}$	2.5
Over $^{10}/_{12}$ to $^{11}/_{12}$	3.25
Over $^{11}/_{12}$ to $^{12}/_{12}$	4.0
Over $^{12}/_{12}$ to $^{14}/_{12}$	5.0
Over $^{14}/_{12}$ to $^{16}/_{12}$	6.0
Over $^{16}/_{12}$ to $^{18}/_{12}$	7.0
Over $^{18}/_{12}$ to $^{20}/_{12}$	7.5
Over $^{20}/_{12}$ to $^{21}/_{12}$	8.0

For SI: 1 foot = 304.8 mm.

FIGURE G2427.6.3 (503.6.4)
GAS VENT TERMINATION LOCATIONS FOR LISTED CAPS 12 INCHES
OR LESS IN SIZE AT LEAST 8 FEET FROM A VERTICAL WALL

G2427.6.5 (503.6.6) Roof terminations. Gas vents shall extend through the roof flashing, roof jack or roof thimble and terminate with a listed cap or listed roof assembly.

G2427.6.6 (503.6.7) Forced air inlets. Gas vents shall terminate not less than 3 feet (914 mm) above any forced air inlet located within 10 feet (3048 mm).

G2427.6.7 (503.6.8) Exterior wall penetrations. A gas vent extending through an exterior wall shall not terminate adjacent to the wall or below eaves or parapets, except as provided in Sections G2427.2.1 and G2427.3.3.

G2427.6.8 (503.6.9) Size of gas vents. Venting systems shall be sized and constructed in accordance with Section G2428 or other approved engineering methods and the gas vent and appliance manufacturer's installation instructions.

G2427.6.8.1 (503.6.9.1) Category I appliances. The sizing of natural draft venting systems serving one or more listed appliances equipped with a draft hood or appliances listed for use with Type B gas vent, installed in a single story of a building, shall be in accordance with one of the following methods:

1. The provisions of Section G2428.

2. For sizing an individual gas vent for a single, draft-hood-equipped appliance, the effective area of the vent connector and the gas vent shall be not less than the area of the appliance draft hood outlet, nor greater than seven times the draft hood outlet area.

3. For sizing a gas vent connected to two appliances with draft hoods, the effective area of the vent shall be not less than the area of the larger draft hood outlet plus 50 percent of the area of the smaller draft hood outlet, nor greater than seven times the smaller draft hood outlet area.

4. Approved engineering practices.

G2427.6.8.2 (503.6.9.2) Vent offsets. Type B and L vents sized in accordance with Item 2 or 3 of Section G2427.6.8.1 shall extend in a generally vertical direction with offsets not exceeding 45 degrees (0.79 rad), except that a vent system having not more than one 60-degree (1.04 rad) offset shall be permitted. Any angle greater than 45 degrees (0.79 rad) from the vertical is considered horizontal. The total horizontal distance of a vent plus the horizontal vent connector serving draft hood-equipped appliances shall be not greater than 75 percent of the vertical height of the vent.

G2427.6.8.3 (503.6.9.3) Category II, III and IV appliances. The sizing of gas vents for Category II, III and IV appliances shall be in accordance with the appliance manufacturer's instructions.

G2427.6.8.4 (503.6.9.4) Mechanical draft. Chimney venting systems using mechanical draft shall be sized in accordance with approved engineering methods.

G2427.6.9 (503.6.11) Support of gas vents. Gas vents shall be supported and spaced in accordance with the manufacturer's installation instructions.

G2427.6.10 (503.6.12) Marking. In those localities where solid and liquid fuels are used extensively, gas vents shall be permanently identified by a label attached to the wall or ceiling at a point where the vent connector enters the gas vent. The determination of where such localities exist shall be made by the code official. The label shall read:

"This gas vent is for appliances that burn gas. Do not connect to solid or liquid fuel-burning appliances or incinerators."

G2427.7 (503.7) Single-wall metal pipe. Single-wall metal pipe vents shall comply with Sections G2427.7.1 through G2427.7.12.

G2427.7.1 (503.7.1) Construction. Single-wall metal pipe shall be constructed of galvanized sheet steel not less than 0.0304 inch (0.7 mm) thick, or other approved, noncombustible, corrosion-resistant material.

G2427.7.2 (503.7.2) Cold climate. Uninsulated single-wall metal pipe shall not be used outdoors for venting appliances in regions where the 99-percent winter design temperature is below 32°F (0°C).

G2427.7.3 (503.7.3) Termination. Single-wall metal pipe shall terminate at least 5 feet (1524 mm) in vertical height above the highest connected appliance draft hood outlet or flue collar. Single-wall metal pipe shall extend at least 2 feet (610 mm) above the highest point where it passes through a roof of a building and at least 2 feet (610 mm) higher than any portion of a building within a horizontal distance of 10 feet (3048 mm) (see Figure G2427.6.4). An approved cap or roof assembly shall be attached to the terminus of a single-wall metal pipe (see also Section G2427.7.8, Item 3).

G2427.7.4 (503.7.4) Limitations of use. Single-wall metal pipe shall be used only for runs directly from the space in which the appliance is located through the roof or exterior wall to the outdoor atmosphere.

G2427.7.5 (503.7.5) Roof penetrations. A pipe passing through a roof shall extend without interruption through the roof flashing, roof jack, or roof thimble. Where a single-wall metal pipe passes through a roof constructed of combustible material, a noncombustible, nonventilating thimble shall be used at the point of passage. The thimble shall extend at least 18 inches (457 mm) above and 6 inches (152 mm) below the roof with the annular space open at the bottom and closed only at the top. The thimble shall be sized in accordance with Section G2427.10.15.

G2427.7.6 (503.7.6) Installation. Single-wall metal pipe shall not originate in any unoccupied attic or concealed space and shall not pass through any attic, inside wall, concealed space, or floor. The installation of a single-wall metal pipe through an exterior combustible wall shall comply with Section G2427.10.15. Single-wall metal pipe used for venting an incinerator shall be exposed and readily examinable for its full length and shall have suitable clearances maintained.

G2427.7.7 (503.7.7) Clearances. Minimum clearances from single-wall metal pipe to combustible material shall be in accordance with Table G2427.7.7. The clearance from single-wall metal pipe to combustible material shall be permitted to be reduced where the combustible material is protected as specified for vent connectors in Table G2409.2.

G2427.7.8 (503.7.8) Size of single-wall metal pipe. A venting system constructed of single-wall metal pipe shall be sized in accordance with one of the following methods and the appliance manufacturer's instructions:

1. For a draft-hood-equipped appliance, in accordance with Section G2428.

2. For a venting system for a single appliance with a draft hood, the areas of the connector and the pipe each shall be not less than the area of the appliance flue collar or draft hood outlet, whichever is smaller. The vent area shall not be greater than seven times the draft hood outlet area.

3. Other approved engineering methods.

G2427.7.9 (503.7.9) Pipe geometry. Any shaped single-wall metal pipe shall be permitted to be used, provided that its equivalent effective area is equal to the effective area of the round pipe for which it is substituted, and provided that the minimum internal dimension of the pipe is not less than 2 inches (51 mm).

G2427.7.10 (503.7.10) Termination capacity. The vent cap or a roof assembly shall have a venting capacity not less than that of the pipe to which it is attached.

G2427.7.11 (503.7.11) Support of single-wall metal pipe. All portions of single-wall metal pipe shall be supported for the design and weight of the material employed.

G2427.7.12 (503.7.12) Marking. Single-wall metal pipe shall comply with the marking provisions of Section G2427.6.10.

G2427.8 (503.8) Venting system termination location. The location of venting system terminations shall comply with the following (see Appendix C):

1. A mechanical draft venting system shall terminate at least 3 feet (914 mm) above any forced-air inlet located within 10 feet (3048 mm).

   Exceptions:

   1. This provision shall not apply to the combustion air intake of a direct-vent appliance.

   2. This provision shall not apply to the separation of the integral outdoor air inlet and flue gas discharge of listed outdoor appliances.

2. A mechanical draft venting system, excluding direct-vent appliances, shall terminate at least 4 feet (1219 mm) below, 4 feet (1219 mm) horizontally from, or 1 foot (305 mm) above any door, operable window, or gravity air inlet into any building. The bottom of the vent terminal shall be located at least 12 inches (305 mm) above grade.

3. The vent terminal of a direct-vent appliance with an input of 10,000 Btu per hour (3 kW) or less shall be located at least 6 inches (152 mm) from any air opening into a building, and such an appliance with an input over 10,000 Btu per hour (3 kW) but not over 50,000 Btu per hour (14.7 kW) shall be installed with a 9-inch (230 mm) vent termination clearance, and an appliance with an input over 50,000 Btu/h (14.7 kW) shall have at least a 12-inch (305 mm) vent termination clearance. The bottom of the vent terminal and the air intake shall be located at least 12 inches (305 mm) above grade.

4. Through-the-wall vents for Category II and IV appliances and noncategorized condensing appliances shall not terminate over public walkways or over an area where condensate or vapor could create a nuisance or hazard or could be detrimental to the operation of regulators, relief valves, or other equipment. Where local expe-

TABLE G2427.7.7 (503.8.7)[a]
CLEARANCES FOR CONNECTORS

APPLIANCE	MINIMUM DISTANCE FROM COMBUSTIBLE MATERIAL			
	Listed Type B gas vent material	Listed Type L vent material	Single-wall metal pipe	Factory-built chimney sections
Listed appliances with draft hoods and appliances listed for use with Type B gas vents	As listed	As listed	6 inches	As listed
Residential boilers and furnaces with listed gas conversion burner and with draft hood	6 inches	6 inches	9 inches	As listed
Residential appliances listed for use with Type L vents	Not permitted	As listed	9 inches	As listed
Listed gas-fired toilets	Not permitted	As listed	As listed	As listed
Unlisted residential appliances with draft hood	Not permitted	6 inches	9 inches	As listed
Residential and low-heat appliances other than above	Not permitted	9 inches	18 inches	As listed
Medium-heat appliances	Not permitted	Not permitted	36 inches	As listed

For SI: 1 inch = 25.4 mm.

a. These clearances shall apply unless the manufacturer's installation instructions for a listed appliance or connector specify different clearances, in which case the listed clearances shall apply.

rience indicates that condensate is a problem with Category I and III appliances, this provision shall also apply.

G2427.9 (503.9) Condensation drainage. Provisions shall be made to collect and dispose of condensate from venting systems serving Category II and IV appliances and noncategorized condensing appliances in accordance with Section 503.8, Item 4. Where local experience indicates that condensation is a problem, provision shall be made to drain off and dispose of condensate from venting systems serving Category I and III appliances in accordance with Section G2427.8, Item 4.

G2427.10 (503.10) Vent connectors for Category I equipment. Vent connectors for Category I equipment shall comply with Sections G2427.10.1 through G2427.10.15.

G2427.10.1 (503.10.1) Where required. A vent connector shall be used to connect an appliance to a gas vent, chimney or single-wall metal pipe, except where the gas vent, chimney or single-wall metal pipe is directly connected to the appliance.

G2427.10.2 (503.10.2) Materials. Vent connectors shall be constructed in accordance with Sections G2427.10.2.1 through G2427.10.2.4.

G2427.10.2.1 (503.10.2.1) General. A vent connector shall be made of noncombustible corrosion-resistant material capable of withstanding the vent gas temperature produced by the appliance and of sufficient thickness to withstand physical damage.

G2427.10.2.2 (503.10.2.2) Vent connectors located in unconditioned areas. Where the vent connector used for an appliance having a draft hood or a Category I appliance is located in or passes through attics, crawl spaces or other unconditioned spaces, that portion of the vent connector shall be listed Type B, Type L or listed vent material having equivalent insulation properties.

Exception: Single-wall metal pipe located within the exterior walls of the building in areas having a local 99-percent winter design temperature of 5°F (-15°C) or higher shall be permitted to be used in unconditioned spaces other than attics and crawl spaces.

G2427.10.2.3 (503.10.2.3) Residential-type appliance connectors. Where vent connectors for residential-type appliances are not installed in attics or other unconditioned spaces, connectors for listed appliances having draft hoods, appliances having draft hoods and equipped with listed conversion burners and Category I appliances shall be one of the following:

1. Type B or L vent material;

2. Galvanized sheet steel not less than 0.018 inch (0.46 mm) thick;

3. Aluminum (1100 or 3003 alloy or equivalent) sheet not less than 0.027 inch (0.69 mm) thick;

4. Stainless steel sheet not less than 0.012 inch (0.31 mm) thick;

5. Smooth interior wall metal pipe having resistance to heat and corrosion equal to or greater than that of Item 2, 3 or 4 above; or

6. A listed vent connector.

Vent connectors shall not be covered with insulation.

Exception: Listed insulated vent connectors shall be installed according to the terms of their listing.

G2427.10.2.4 (503.10.2.4) Low-heat equipment. A vent connector for a nonresidential, low-heat appliance shall be a factory-built chimney section or steel pipe having resistance to heat and corrosion equivalent to that for the appropriate galvanized pipe as specified in Table G2427.10.2.4. Factory-built chimney sections shall be joined together in accordance with the chimney manufacturer's instructions.

TABLE G2427.10.2.4 (503.10.2.4)
MINIMUM THICKNESS FOR GALVANIZED STEEL VENT CONNECTORS FOR LOW-HEAT APPLIANCES

DIAMETER OF CONNECTOR (inches)	MINIMUM THICKNESS (inch)
Less than 6	0.019
6 to less than 10	0.023
10 to 12 inclusive	0.029
14 to 16 inclusive	0.034
Over 16	0.056

For SI: 1 inch = 25.4 mm.

G2427.10.3 (503.10.3) Size of vent connector. Vent connectors shall be sized in accordance with Sections G2427.10.3.1 through G2427.3.5.

G2427.10.3.1 (503.10.3.1) Single draft hood and fan-assisted. A vent connector for an appliance with a single draft hood or for a Category I fan-assisted combustion system appliance shall be sized and installed in accordance with Section G2428 or other approved engineering methods.

G2427.10.3.2 (503.10.3.2) Multiple draft hood. For a single appliance having more than one draft hood outlet or flue collar, the manifold shall be constructed according to the instructions of the appliance manufacturer. Where there are no instructions, the manifold shall be designed and constructed in accordance with approved engineering practices. As an alternate method, the effective area of the manifold shall equal the combined area of the flue collars or draft hood outlets and the vent connectors shall have a minimum 1-foot (305 mm) rise.

G2427.10.3.3 (503.10.3.3) Multiple appliances. Where two or more appliances are connected to a common vent or chimney, each vent connector shall be sized in accordance with Section G2428 or other approved engineering methods.

As an alternative method applicable only when all of the appliances are draft hood equipped, each vent connector shall have an effective area not less than the area of the draft hood outlet of the appliance to which it is connected.

G2427.10.3.4 (503.10.3.4) Common connector/manifold. Where two or more appliances are vented through a common vent connector or vent manifold, the common vent connector or vent manifold shall be located at the highest level consistent with available headroom and the required clearance to combustible materials and shall be sized in accordance with Section G2428 or other approved engineering methods.

As an alternate method applicable only where there are two draft hood-equipped appliances, the effective area of the common vent connector or vent manifold and all junction fittings shall be not less than the area of the larger vent connector plus 50 percent of the area of the smaller flue collar outlet.

G2427.10.3.5 (503.10.3.5) Size increase. Where the size of a vent connector is increased to overcome installation limitations and obtain connector capacity equal to the appliance input, the size increase shall be made at the appliance draft hood outlet.

G2427.10.4 (503.10.4) Two or more appliances connected to a single vent. Where two or more vent connectors enter a common gas vent, chimney flue, or single-wall metal pipe, the smaller connector shall enter at the highest level consistent with the available headroom or clearance to combustible material. Vent connectors serving Category I appliances shall not be connected to any portion of a mechanical draft system operating under positive static pressure, such as those serving Category III or IV appliances.

G2427.10.5 (503.10.5) Clearance. Minimum clearances from vent connectors to combustible material shall be in accordance with Table G2427.7.7.

> **Exception:** The clearance between a vent connector and combustible material shall be permitted to be reduced where the combustible material is protected as specified for vent connectors in Table G2409.2.

G2427.10.6 (503.10.6) Flow resistance. A vent connector shall be installed so as to avoid turns or other construction features that create excessive resistance to flow of vent gases.

G2427.10.7 (503.10.7) Joints. Joints between sections of connector piping and connections to flue collars and draft hood outlets shall be fastened by one of the following methods:

1. Sheet metal screws.

2. Vent connectors of listed vent material assembled and connected to flue collars or draft hood outlets in accordance with the manufacturers' instructions.

3. Other approved means.

G2427.10.8 (503.10.8) Slope. A vent connector shall be installed without dips or sags and shall slope upward toward the vent or chimney at least $^1/_4$ inch per foot (21 mm/m).

> **Exception:** Vent connectors attached to a mechanical draft system installed in accordance with the manufacturers' instructions.

G2427.10.9 (503.10.9) Length of vent connector. A vent connector shall be as short as practical and the appliance located as close as practical to the chimney or vent. The maximum horizontal length of a single-wall connector shall be 75 percent of the height of the chimney or vent except for engineered systems. The maximum horizontal length of a Type B double-wall connector shall be 100 percent of the height of the chimney or vent except for engineered systems. For a chimney or vent system serving multiple appliances, the maximum length of an individual connector, from the appliance outlet to the junction with the common vent or another connector, shall be 100 percent of the height of the chimney or vent.

G2427.10.10 (503.10.10) Support. A vent connector shall be supported for the design and weight of the material employed to maintain clearances and prevent physical damage and separation of joints.

G2427.10.11 (503.10.11) Chimney connection. Where entering a flue in a masonry or metal chimney, the vent connector shall be installed above the extreme bottom to avoid stoppage. Where a thimble or slip joint is used to facilitate removal of the connector, the connector shall be firmly attached to or inserted into the thimble or slip joint to prevent the connector from falling out. Means shall be employed to prevent the connector from entering so far as to restrict the space between its end and the opposite wall of the chimney flue (see Section G2425.9).

G2427.10.12 (503.10.12) Inspection. The entire length of a vent connector shall be provided with ready access for inspection, cleaning, and replacement.

G2427.10.13 (503.10.13) Fireplaces. A vent connector shall not be connected to a chimney flue serving a fireplace unless the fireplace flue opening is permanently sealed.

G2427.10.14 (503.10.14) Passage through ceilings, floors or walls. Single-wall metal pipe connectors shall not pass through any wall, floor or ceiling except as permitted by Sections G2427.7.4 and G2427.10.15.

G2427.10.15 (503.10.15) Single-wall connector penetrations of combustible walls. A vent connector made of a single-wall metal pipe shall not pass through a combustible exterior wall unless guarded at the point of passage by a ventilated metal thimble not smaller than the following:

1. For listed appliances equipped with draft hoods and appliances listed for use with Type B gas vents, the thimble shall be not less than 4 inches (102 mm) larger in diameter than the vent connector. Where there is a run of not less than 6 feet (1829 mm) of vent connector in the open between the draft hood outlet and the thimble, the thimble shall be permitted to be not less than 2 inches (51 mm) larger in diameter than the vent connector.

2. For unlisted appliances having draft hoods, the thimble shall be not less than 6 inches (152 mm) larger in diameter than the vent connector.

3. For residential and low-heat appliances, the thimble shall be not less than 12 inches (305 mm) larger in diameter than the vent connector.

Exception: In lieu of thimble protection, all combustible material in the wall shall be removed from the vent connector a sufficient distance to provide the specified clearance from such vent connector to combustible material. Any material used to close up such opening shall be noncombustible.

G2427.11 (503.11) Vent connectors for Category II, III and IV appliances. Vent connectors for Category II, III and IV appliances shall be as specified for the venting systems in accordance with Section G2427.4.

G2427.12 (503.12) Draft hoods and draft controls. The installation of draft hoods and draft controls shall comply with Sections G2427.12.1 through G2427.12.7.

G2427.12.1 (503.12.1) Appliances requiring draft hoods. Vented appliances shall be installed with draft hoods.

Exception: Dual oven-type combination ranges; incinerators; direct-vent appliances; fan-assisted combustion system appliances; appliances requiring chimney draft for operation; single firebox boilers equipped with conversion burners with inputs greater than 400,000 Btu per hour (117 kW); appliances equipped with blast, power or pressure burners that are not listed for use with draft hoods; and appliances designed for forced venting.

G2427.12.2 (503.12.2) Installation. A draft hood supplied with or forming a part of a listed vented appliance shall be installed without alteration, exactly as furnished and specified by the appliance manufacturer.

G2427.12.2.1 (503.12.2.1) Draft hood required. If a draft hood is not supplied by the appliance manufacturer where one is required, a draft hood shall be installed, shall be of a listed or approved type and, in the absence of other instructions, shall be of the same size as the appliance flue collar. Where a draft hood is required with a conversion burner, it shall be of a listed or approved type.

G2427.12.2.2 (503.12.2.2) Special design draft hood. Where it is determined that a draft hood of special design is needed or preferable for a particular installation, the installation shall be in accordance with the recommendations of the appliance manufacturer and shall be approved.

G2427.12.3 (503.12.3) Draft control devices. Where a draft control device is part of the appliance or is supplied by the appliance manufacturer, it shall be installed in accordance with the manufacturer's instructions. In the absence of manufacturer's instructions, the device shall be attached to the flue collar of the appliance or as near to the appliance as practical.

G2427.12.4 (503.12.4) Additional devices. Appliances (except incinerators) requiring a controlled chimney draft shall be permitted to be equipped with a listed double-acting barometric-draft regulator installed and adjusted in accordance with the manufacturer's instructions.

G2427.12.5 (503.12.5) Location. Draft hoods and barometric draft regulators shall be installed in the same room or enclosure as the appliance in such a manner as to prevent any difference in pressure between the hood or regulator and the combustion air supply.

G2427.12.6 (503.12.6) Positioning. Draft hoods and draft regulators shall be installed in the position for which they were designed with reference to the horizontal and vertical planes and shall be located so that the relief opening is not obstructed by any part of the appliance or adjacent construction. The appliance and its draft hood shall be located so that the relief opening is accessible for checking vent operation.

G2427.12.7 (503.12.7) Clearance. A draft hood shall be located so its relief opening is not less than 6 inches (152 mm) from any surface except that of the appliance it serves and the venting system to which the draft hood is connected. Where a greater or lesser clearance is indicated on the appliance label, the clearance shall be not less than that specified on the label. Such clearances shall not be reduced.

G2427.13 (503.13) Manually operated dampers. A manually operated damper shall not be placed in the vent connector for any appliance. Fixed baffles shall not be classified as manually operated dampers.

G2427.14 (503.14) Automatically operated vent dampers. An automatically operated vent damper shall be of a listed type.

G2427.15 (503.15) Obstructions. Devices that retard the flow of vent gases shall not be installed in a vent connector, chimney, or vent. The following shall not be considered as obstructions:

1. Draft regulators and safety controls specifically listed for installation in venting systems and installed in accordance with the manufacturer's installation instructions.

2. Approved draft regulators and safety controls that are designed and installed in accordance with approved engineering methods.

3. Listed heat reclaimers and automatically operated vent dampers installed in accordance with the manufacturer's installation instructions.

4. Approved economizers, heat reclaimers, and recuperators installed in venting systems of equipment not required to be equipped with draft hoods, provided that the appliance manufacturer's instructions cover the installation of such a device in the venting system and performance in accordance with Sections G2427.3 and G2427.3.1 is obtained.

5. Vent dampers serving listed appliances installed in accordance with Sections G2428.2.1 and G2428.3.1 or other approved engineering methods.

SECTION G2428 (504)
SIZING OF CATEGORY I APPLIANCE
VENTING SYSTEMS

G2428.1 (504.1) Definitions. The following definitions apply to tables in this section.

APPLIANCE CATEGORIZED VENT DIAMETER/AREA. The minimum vent area/diameter permissible for Category I appliances to maintain a nonpositive vent static pressure when tested in accordance with nationally recognized standards.

FAN-ASSISTED COMBUSTION SYSTEM. An appliance equipped with an integral mechanical means to either draw or force products of combustion through the combustion chamber or heat exchanger.

FAN MIN. The minimum input rating of a Category I fan-assisted appliance attached to a vent or connector.

FAN MAX. The maximum input rating of a Category I fan-assisted appliance attached to a vent or connector.

NAT MAX. The maximum input rating of a Category I draft-hood-equipped appliance attached to a vent or connector.

FAN + FAN. The maximum combined appliance input rating of two or more Category I fan-assisted appliances attached to the common vent.

FAN + NAT. The maximum combined appliance input rating of one or more Category I fan-assisted appliances and one or more Category I draft-hood-equipped appliances attached to the common vent.

NA. Vent configuration is not permitted due to potential for condensate formation or pressurization of the venting system, or not applicable due to physical or geometric restraints.

NAT + NAT. The maximum combined appliance input rating of two or more Category I draft-hood-equipped appliances attached to the common vent.

G2428.2 (504.2) Application of single appliance vent Tables G2428.2(1) and G2428.2(2). The application of Tables G2428.2(1) and G2428.2(2) shall be subject to the requirements of Sections G2428.2.1 through G2428.2.15.

G2428.2.1 (504.2.1) Vent obstructions. These venting tables shall not be used where obstructions, as described in Section G2427.15, are installed in the venting system. The installation of vents serving listed appliances with vent dampers shall be in accordance with the appliance manufacturer's instructions or in accordance with the following:

1. The maximum capacity of the vent system shall be determined using the "NAT Max" column.

2. The minimum capacity shall be determined as if the appliance were a fan-assisted appliance, using the "FAN Min" column to determine the minimum

capacity of the vent system. Where the corresponding "FAN Min" is "NA," the vent configuration shall not be permitted and an alternative venting configuration shall be utilized.

G2428.2.2 (504.2.2) Minimum size. Where the vent size determined from the tables is smaller than the appliance draft hood outlet or flue collar, the smaller size shall be permitted to be used provided all of the following are met:

1. The total vent height (H) is at least 10 feet (3048 mm).

2. Vents for appliance draft hood outlets or flue collars 12 inches (305 mm) in diameter or smaller are not reduced more than one table size.

3. Vents for appliance draft hood outlets or flue collars larger than 12 inches (305 mm) in diameter are not reduced more than two table sizes.

4. The maximum capacity listed in the tables for a fan-assisted appliance is reduced by 10 percent (0.90 by maximum table capacity).

5. The draft hood outlet is greater than 4 inches (102 mm) in diameter. Do not connect a 3-inch-diameter (76 mm) vent to a 4-inch-diameter (102 mm) draft hood outlet. This provision shall not apply to fan-assisted appliances.

G2428.2.3 (504.2.3) Vent offsets. Single-appliance venting configurations with zero (0) lateral lengths in Tables G2428.2(1) and G2428.2(2) shall not have elbows in the venting system. Single-appliance venting configurations with lateral lengths include two 90-degree (1.57 rad) elbows. For each additional elbow up to and including 45 degrees (0.79 rad), the maximum capacity listed in the venting tables shall be reduced by 5 percent. For each additional elbow greater than 45 degrees (0.79 rad) up to and including 90 degrees (1.57 rad), the maximum capacity listed in the venting tables shall be reduced by 10 percent.

G2428.2.4 (504.2.4) Zero lateral. Zero (0) lateral (L) shall apply only to a straight vertical vent attached to a top outlet draft hood or flue collar.

G2428.2.5 (504.2.5) High altitude installations. Sea level input ratings shall be used when determining maximum capacity for high altitude installation. Actual input, derated for altitude, shall be used for determining minimum capacity for high altitude installation.

G2428.2.6 (504.2.6) Multiple input rate appliances. For appliances with more than one input rate, the minimum vent capacity (FAN Min) determined from the tables shall be less than the lowest appliance input rating, and the maximum vent capacity (FAN Max/NAT Max) determined from the tables shall be greater than the highest appliance rating input.

TABLE G2428.2(1) [504.2(1)]
TYPE B DOUBLE-WALL GAS VENT

Number of Appliances	Single
Appliance Type	Category I
Appliance Vent Connection	Connected directly to vent

HEIGHT (H) (feet)	LATERAL (L) (feet)	3 FAN Min	3 FAN Max	3 NAT Max	4 FAN Min	4 FAN Max	4 NAT Max	5 FAN Min	5 FAN Max	5 NAT Max	6 FAN Min	6 FAN Max	6 NAT Max	7 FAN Min	7 FAN Max	7 NAT Max	8 FAN Min	8 FAN Max	8 NAT Max	9 FAN Min	9 FAN Max	9 NAT Max
6	0	0	78	46	0	152	86	0	251	141	0	375	205	0	524	285	0	698	370	0	897	470
	2	13	51	36	18	97	67	27	157	105	32	232	157	44	321	217	53	425	285	63	543	370
	4	21	49	34	30	94	64	39	153	103	50	227	153	66	316	211	79	419	279	93	536	362
	6	25	46	32	36	91	61	47	149	100	59	223	149	78	310	205	93	413	273	110	530	354
8	0	0	84	50	0	165	94	0	276	155	0	415	235	0	583	320	0	780	415	0	1,006	537
	2	12	57	40	16	109	75	25	178	120	28	263	180	42	365	247	50	483	322	60	619	418
	5	23	53	38	32	103	71	42	171	115	53	255	173	70	356	237	83	473	313	99	607	407
	8	28	49	35	39	98	66	51	164	109	64	247	165	84	347	227	99	463	303	117	596	396
10	0	0	88	53	0	175	100	0	295	166	0	447	255	0	631	345	0	847	450	0	1,096	585
	2	12	61	42	17	118	81	23	194	129	26	289	195	40	402	273	48	533	355	57	684	457
	5	23	57	40	32	113	77	41	187	124	52	280	188	68	392	263	81	522	346	95	671	446
	10	30	51	36	41	104	70	54	176	115	67	267	175	88	376	245	104	504	330	122	651	427
15	0	0	94	58	0	191	112	0	327	187	0	502	285	0	716	390	0	970	525	0	1,263	682
	2	11	69	48	15	136	93	20	226	150	22	339	225	38	475	316	45	633	414	53	815	544
	5	22	65	45	30	130	87	39	219	142	49	330	217	64	463	300	76	620	403	90	800	529
	10	29	59	41	40	121	82	51	206	135	64	315	208	84	445	288	99	600	386	116	777	507
	15	35	53	37	48	112	76	61	195	128	76	301	198	98	429	275	115	580	373	134	755	491
20	0	0	97	61	0	202	119	0	349	202	0	540	307	0	776	430	0	1,057	575	0	1,384	752
	2	10	75	51	14	149	100	18	250	166	20	377	249	33	531	346	41	711	470	50	917	612
	5	21	71	48	29	143	96	38	242	160	47	367	241	62	519	337	73	697	460	86	902	599
	10	28	64	44	38	133	89	50	229	150	62	351	228	81	499	321	95	675	443	112	877	576
	15	34	58	40	46	124	84	59	217	142	73	337	217	94	481	308	111	654	427	129	853	557
	20	48	52	35	55	116	78	69	206	134	84	322	206	107	464	295	125	634	410	145	830	537

VENT DIAMETER—(D) inches
APPLIANCE INPUT RATING IN THOUSANDS OF BTU/H

(continued)

TABLE G2428.2(1) [504.2(1)]—continued
TYPE B DOUBLE-WALL GAS VENT

Number of Appliances	Single
Appliance Type	Category I
Appliance Vent Connection	Connected directly to vent

VENT DIAMETER—(D) inches
APPLIANCE INPUT RATING IN THOUSANDS OF BTU/H

HEIGHT (H) (feet)	LATERAL (L) (feet)	3 FAN Min	3 FAN Max	3 NAT Max	4 FAN Min	4 FAN Max	4 NAT Max	5 FAN Min	5 FAN Max	5 NAT Max	6 FAN Min	6 FAN Max	6 NAT Max	7 FAN Min	7 FAN Max	7 NAT Max	8 FAN Min	8 FAN Max	8 NAT Max	9 FAN Min	9 FAN Max	9 NAT Max
30	0	0	100	64	0	213	128	0	374	220	0	587	336	0	853	475	0	1,173	650	0	1,548	855
	2	9	81	56	13	166	112	14	283	185	18	432	280	27	613	394	33	826	535	42	1,072	700
	5	21	77	54	28	160	108	36	275	176	45	421	273	58	600	385	69	811	524	82	1,055	688
	10	27	70	50	37	150	102	48	262	171	59	405	261	77	580	371	91	788	507	107	1,028	668
	15	33	64	NA	44	141	96	57	249	163	70	389	249	90	560	357	105	765	490	124	1,002	648
	20	56	58	NA	53	132	90	66	237	154	80	374	237	102	542	343	119	743	473	139	977	628
	30	NA	NA	NA	73	113	NA	88	214	NA	104	346	219	131	507	321	149	702	444	171	929	594
50	0	0	101	67	0	216	134	0	397	232	0	633	363	0	932	518	0	1,297	708	0	1,730	952
	2	8	86	61	11	183	122	14	320	206	15	497	314	22	715	445	26	975	615	33	1,276	813
	5	20	82	NA	27	177	119	35	312	200	43	487	308	55	702	438	65	960	605	77	1,259	798
	10	26	76	NA	35	168	114	45	299	190	56	471	298	73	681	426	86	935	589	101	1,230	773
	15	59	70	NA	42	158	NA	54	287	180	66	455	288	85	662	413	100	911	572	117	1,203	747
	20	NA	NA	NA	50	149	NA	63	275	169	76	440	278	97	642	401	113	888	556	131	1,176	722
	30	NA	NA	NA	69	131	NA	84	250	NA	99	410	259	123	605	376	141	844	522	161	1,125	670

For SI: 1 inch = 25.4 mm, 1 foot = 304.8 mm, 1 British thermal unit per hour = 0.2931 W.

TABLE G2428.2(2) [504.2(2)]
TYPE B DOUBLE-WALL GAS VENT

Number of Appliances	Single
Appliance Type	Category I
Appliance Vent Connection	Single-wall metal connector

APPLIANCE INPUT RATING IN THOUSANDS OF BTU/H

HEIGHT (H) (feet)	LATERAL (L) (feet)	3 FAN Min	3 FAN Max	3 NAT Max	4 FAN Min	4 FAN Max	4 NAT Max	5 FAN Min	5 FAN Max	5 NAT Max	6 FAN Min	6 FAN Max	6 NAT Max	7 FAN Min	7 FAN Max	7 NAT Max	8 FAN Min	8 FAN Max	8 NAT Max	9 FAN Min	9 FAN Max	9 NAT Max	10 FAN Min	10 FAN Max	10 NAT Max	12 FAN Min	12 FAN Max	12 NAT Max
6	0	38	77	45	59	151	85	85	249	140	126	373	204	165	522	284	211	695	369	267	894	469	371	1,118	569	537	1,639	849
	2	39	51	36	60	96	66	85	156	104	123	231	156	159	320	213	201	423	284	251	541	368	347	673	453	498	979	648
	4	NA	NA	33	74	92	63	102	152	102	146	225	152	187	313	208	237	416	277	295	533	360	409	664	443	584	971	638
	6	NA	NA	31	83	89	60	114	147	99	163	220	148	207	307	203	263	409	271	327	526	352	449	656	433	638	962	627
8	0	37	83	50	58	164	93	83	273	154	123	412	234	161	580	319	206	777	414	258	1,002	536	360	1,257	658	521	1,852	967
	2	39	56	39	59	108	75	83	176	119	121	261	179	155	363	246	197	482	321	246	617	417	339	768	513	486	1,120	743
	5	NA	NA	37	77	102	69	107	168	114	151	252	171	193	352	235	245	470	311	305	604	404	418	754	500	598	1,104	730
	8	NA	NA	33	90	95	64	122	161	107	175	243	163	223	342	225	280	458	300	344	591	392	470	740	486	665	1,089	715
10	0	37	87	53	57	174	99	82	293	165	120	444	254	158	628	344	202	844	449	253	1,093	584	351	1,373	718	507	2,031	1,057
	2	39	61	41	59	117	80	82	193	128	119	287	194	153	400	272	193	531	354	242	681	456	332	849	559	475	1,242	848
	5	52	56	39	76	111	76	105	185	122	148	277	186	190	388	261	241	518	344	299	667	443	409	834	544	584	1,224	825
	10	NA	NA	34	97	100	68	132	171	112	188	261	171	237	369	241	296	497	325	363	643	423	492	808	520	688	1,194	788
15	0	36	93	57	56	190	111	80	325	186	116	499	283	153	713	388	195	966	523	244	1,259	681	336	1,591	838	488	2,374	1,237
	2	38	69	47	57	136	93	80	225	149	115	337	224	148	473	314	187	631	413	232	812	543	319	1,015	673	457	1,491	983
	5	51	63	44	75	128	86	102	216	140	144	326	217	182	459	298	231	616	400	287	795	526	392	997	657	562	1,469	963
	10	NA	NA	39	95	116	79	128	201	131	182	308	203	228	438	284	284	592	381	349	768	501	470	966	628	664	1,433	928
	15	NA	NA	NA	NA	NA	72	158	186	124	220	290	192	272	418	269	334	568	367	404	742	484	540	937	601	750	1,399	894
20	0	35	96	60	54	200	118	78	346	201	114	537	306	149	772	428	190	1,053	573	238	1,379	750	326	1,751	927	473	2,631	1,346
	2	37	74	50	56	148	99	78	248	165	113	375	248	144	528	344	182	708	468	227	914	611	309	1,146	754	443	1,689	1,098
	5	50	68	47	73	140	94	100	239	158	141	363	239	178	514	334	224	692	457	279	896	596	381	1,126	734	547	1,665	1,074
	10	NA	NA	41	93	129	86	125	223	146	177	344	224	222	491	316	277	666	437	339	866	570	457	1,092	702	646	1,626	1,037
	15	NA	NA	NA	NA	NA	80	155	208	136	216	325	210	264	469	301	325	640	419	393	838	549	526	1,060	677	730	1,587	1,005
	20	NA	NA	NA	NA	NA	NA	186	192	126	254	306	196	309	448	285	374	616	400	448	810	526	592	1,028	651	808	1,550	973

(continued)

TABLE G2428.2(2) [504.2(2)]—continued
TYPE B DOUBLE-WALL GAS VENT

		Number of Appliances	Single
		Appliance Type	Category I
		Appliance Vent Connection	Single-wall metal connector

VENT DIAMETER—(D) inches

APPLIANCE INPUT RATING IN THOUSANDS OF BTU/H

HEIGHT (H) (feet)	LATERAL (L) (feet)	3 FAN Min	3 FAN Max	3 NAT Max	4 FAN Min	4 FAN Max	4 NAT Max	5 FAN Min	5 FAN Max	5 NAT Max	6 FAN Min	6 FAN Max	6 NAT Max	7 FAN Min	7 FAN Max	7 NAT Max	8 FAN Min	8 FAN Max	8 NAT Max	9 FAN Min	9 FAN Max	9 NAT Max	10 FAN Min	10 FAN Max	10 NAT Max	12 FAN Min	12 FAN Max	12 NAT Max
30	0	34	99	63	53	211	127	76	372	219	110	584	334	144	849	472	184	1,168	647	229	1,542	852	312	1,971	1,056	454	2,996	1,545
	2	37	80	56	55	164	111	76	281	183	109	429	279	139	610	392	175	823	533	219	1,069	698	296	1,346	863	424	1,999	1,308
	5	49	74	52	72	157	106	98	271	173	136	417	271	171	595	382	215	806	521	269	1,049	684	366	1,324	846	524	1,971	1,283
	10	NA	NA	NA	91	144	98	122	255	168	171	397	257	213	570	367	265	777	501	327	1,017	662	440	1,287	821	620	1,927	1,234
	15	NA	NA	NA	115	131	NA	151	239	157	208	377	242	255	547	349	312	750	481	379	985	638	507	1,251	794	702	1,884	1,205
	20	NA	NA	NA	NA	NA	NA	181	223	NA	246	357	228	298	524	333	360	723	461	433	955	615	570	1,216	768	780	1,841	1,166
	30	NA	NA	NA	NA	NA	NA	NA	NA	NA	NA	NA	NA	389	477	305	461	670	426	541	895	574	704	1,147	720	937	1,759	1,101
50	0	33	99	66	51	213	133	73	394	230	105	629	361	138	928	515	176	1,292	704	220	1,724	948	295	2,223	1,189	428	3,432	1,818
	2	36	84	61	53	181	121	73	318	205	104	495	312	133	712	443	168	971	613	209	1,273	811	280	1,615	1,007	401	2,426	1,509
	5	48	80	NA	70	174	117	94	308	198	131	482	305	164	696	435	204	953	602	257	1,252	795	347	1,591	991	496	2,396	1,490
	10	NA	NA	NA	89	160	NA	118	292	186	162	461	292	203	671	420	253	923	583	313	1,217	765	418	1,551	963	589	2,347	1,455
	15	NA	NA	NA	112	148	NA	145	275	174	199	441	280	244	646	405	299	894	562	363	1,183	736	481	1,512	934	668	2,299	1,421
	20	NA	NA	NA	NA	NA	NA	176	257	NA	236	420	267	285	622	389	345	866	543	415	1,150	708	544	1,473	906	741	2,251	1,387
	30	NA	NA	NA	NA	NA	NA	NA	NA	NA	315	376	NA	373	573	NA	442	809	502	521	1,086	649	674	1,399	848	892	2,159	1,318

For SI: 1 inch = 25.4 mm, 1 foot = 304.8 mm, 1 British thermal unit per hour = 0.2931 W.

G2428.2.7 (504.2.7) Liner system sizing and connections. Listed corrugated metallic chimney liner systems in masonry chimneys shall be sized by using Table G2428.2(1) or G2428.2(2) for Type B vents with the maximum capacity reduced by 20 percent (0.80 × maximum capacity) and the minimum capacity as shown in Table G2428.2(1) or G2428.2(2). Corrugated metallic liner systems installed with bends or offsets shall have their maximum capacity further reduced in accordance with Section G2428.2.3. The 20-percent reduction for corrugated metallic chimney liner systems includes an allowance for one long-radius 90-degree (1.57 rad) turn at the bottom of the liner.

Connections between chimney liners and listed double-wall connectors shall be made with listed adapters designed for such purpose.

G2428.2.8 (504.2.8) Vent area and diameter. Where the vertical vent has a larger diameter than the vent connector, the vertical vent diameter shall be used to determine the minimum vent capacity, and the connector diameter shall be used to determine the maximum vent capacity. The flow area of the vertical vent shall not exceed seven times the flow area of the listed appliance categorized vent area, flue collar area, or draft hood outlet area unless designed in accordance with approved engineering methods.

G2428.2.9 (504.2.9) Chimney and vent locations. Tables G2428.2(1) and G2428.2(2) shall only be used for chimneys and vents not exposed to the outdoors below the roof line. A Type B vent or listed chimney lining system passing through an unused masonry chimney flue shall not be considered to be exposed to the outdoors. A Type B vent shall not be considered to be exposed to the outdoors where it passes through an unventilated enclosure or chase insulated to a value of not less than R-8.

G2428.2.10 (504.2.10) Corrugated vent connector size. Corrugated vent connectors shall be not smaller than the listed appliance categorized vent diameter, flue collar diameter, or draft hood outlet diameter.

G2428.2.11 (504.2.11) Vent connector size limitation. Vent connectors shall not be increased in size more than two sizes greater than the listed appliance categorized vent diameter, flue collar diameter or draft hood outlet diameter.

G2428.2.12 (504.2.12) Component commingling. In a single run of vent or vent connector, different diameters and types of vent and connector components shall be permitted to be used, provided that all such sizes and types are permitted by the tables.

G2428.2.13 (504.2.13) Draft hood conversion accessories. Draft hood conversion accessories for use with masonry chimneys venting listed Category I fan-assisted appliances shall be listed and installed in accordance with the manufacturer's installation instructions for such listed accessories.

G2428.2.14 (504.2.14) Table interpolation. Interpolation shall be permitted in calculating capacities for vent dimensions that fall between the table entries (see Example 3, Appendix B).

G2428.2.15 (504.2.15) Extrapolation prohibited. Extrapolation beyond the table entries shall not be permitted.

G2428.2.16 (504.2.16) Engineering calculations. For vent heights less than 6 feet (1829 mm) and greater than shown in the tables, engineering methods shall be used to calculate vent capacities.

G2428.3 (504.3) Application of multiple appliance vent Tables G2428.3(1) through G2428.3(4). The application of Tables G2428.3(1) through G2428.3(4) shall be subject to the requirements of Sections G2428.3.1 through G2428.3.22.

G2428.3.1 (504.3.1) Vent obstructions. These venting tables shall not be used where obstructions, as described in Section G2427.15, are installed in the venting system. The installation of vents serving listed appliances with vent dampers shall be in accordance with the appliance manufacturer's instructions or in accordance with the following:

1. The maximum capacity of the vent connector shall be determined using the NAT Max column.

2. The maximum capacity of the vertical vent or chimney shall be determined using the FAN+NAT column when the second appliance is a fan-assisted appliance, or the NAT+NAT column when the second appliance is equipped with a draft hood.

3. The minimum capacity shall be determined as if the appliance were a fan-assisted appliance.

    3.1. The minimum capacity of the vent connector shall be determined using the FAN Min column.

    3.2. The FAN+FAN column shall be used when the second appliance is a fan-assisted appliance, and the FAN+NAT column shall be used when the second appliance is equipped with a draft hood, to determine whether the vertical vent or chimney configuration is not permitted (NA). Where the vent configuration is NA, the vent configuration shall not be permitted and an alternative venting configuration shall be utilized.

G2428.3.2 (504.3.2) Connector length limit. The vent connector shall be routed to the vent utilizing the shortest possible route. Except as provided in Section G2428.3.3, the maximum vent connector horizontal length shall be 1.5 feet for each inch (18 mm per mm) of connector diameter as shown in Table G2428.3.2.

TABLE G2428.3(1) [504.3(1)]
TYPE B DOUBLE-WALL VENT

Number of Appliances	Two or more
Appliance Type	Category I
Appliance Vent Connection	Type B double-wall connector

VENT CONNECTOR CAPACITY

		TYPE B DOUBLE-WALL VENT AND CONNECTOR DIAMETER—(D) inches																						
		3			**4**			**5**			**6**			**7**			**8**			**9**			**10**		
VENT HEIGHT (H) (feet)	CONNECTOR RISE (R) (feet)	APPLIANCE INPUT RATING LIMITS IN THOUSANDS OF BTU/H																							
		FAN		NAT	FAN		NAT	FAN		NAT	FAN		NAT	FAN		NAT	FAN		NAT	FAN		NAT	FAN		NAT
		Min	Max	Max	Min	Max	Max	Min	Max	Max	Min	Max	Max	Min	Max	Max	Min	Max	Max	Min	Max	Max	Min	Max	Max
6	1	22	37	26	35	66	46	46	106	72	58	164	104	77	225	142	92	296	185	109	376	237	128	466	289
	2	23	41	31	37	75	55	48	121	86	60	183	124	79	253	168	95	333	220	112	424	282	131	526	345
	3	24	44	35	38	81	62	49	132	96	62	199	139	82	275	189	97	363	248	114	463	317	134	575	386
8	1	22	40	27	35	72	48	49	114	76	64	176	109	84	243	148	100	320	194	118	408	248	138	507	303
	2	23	44	32	36	80	57	51	128	90	66	195	129	86	269	175	103	356	230	121	454	294	141	564	358
	3	24	47	36	37	87	64	53	139	101	67	210	145	88	290	198	105	384	258	123	492	330	143	612	402
10	1	22	43	28	34	78	50	49	123	78	65	189	113	89	257	154	106	341	200	125	436	257	146	542	314
	2	23	47	33	36	86	59	51	136	93	67	206	134	91	282	182	109	374	238	128	479	305	149	596	372
	3	24	50	37	37	92	67	52	146	104	69	220	150	94	303	205	111	402	268	131	515	342	152	642	417
15	1	21	50	30	33	89	53	47	142	83	64	220	120	88	298	163	110	389	214	134	493	273	162	609	333
	2	22	53	35	35	96	63	49	153	99	66	235	142	91	320	193	112	419	253	137	532	323	165	658	394
	3	24	55	40	36	102	71	51	163	111	68	248	160	93	339	218	115	445	286	140	565	365	167	700	444
20	1	21	54	31	33	99	56	46	157	87	62	246	125	86	334	171	107	436	224	131	552	285	158	681	347
	2	22	57	37	34	105	66	48	167	104	64	259	149	89	354	202	110	463	265	134	587	339	161	725	414
	3	23	60	42	35	110	74	50	176	116	66	271	168	91	371	228	113	486	300	137	618	383	164	764	466
30	1	20	62	33	31	113	59	45	181	93	60	288	134	83	391	182	103	512	238	125	649	305	151	802	372
	2	21	64	39	33	118	70	47	190	110	62	299	158	85	408	215	105	535	282	129	679	360	155	840	439
	3	22	66	44	34	123	79	48	198	124	64	309	178	88	423	242	108	555	317	132	706	405	158	874	494

COMMON VENT CAPACITY

	TYPE B DOUBLE-WALL COMMON VENT DIAMETER (D)—inches																				
	4			**5**			**6**			**7**			**8**			**9**			**10**		
VENT HEIGHT (H) (feet)	COMBINED APPLIANCE INPUT RATING IN THOUSANDS OF BTU/H																				
	FAN +FAN	FAN +NAT	NAT +NAT	FAN +FAN	FAN +NAT	NAT +NAT	FAN +FAN	FAN +NAT	NAT +NAT	FAN +FAN	FAN +NAT	NAT +NAT	FAN +FAN	FAN +NAT	NAT +NAT	FAN +FAN	FAN +NAT	NAT +NAT	FAN +FAN	FAN +NAT	NAT +NAT
6	92	81	65	140	116	103	204	161	147	309	248	200	404	314	260	547	434	335	672	520	410
8	101	90	73	155	129	114	224	178	163	339	275	223	444	348	290	602	480	378	740	577	465
10	110	97	79	169	141	124	243	194	178	367	299	242	477	377	315	649	522	405	800	627	495
15	125	112	91	195	164	144	283	228	206	427	352	280	556	444	365	753	612	465	924	733	565
20	136	123	102	215	183	160	314	255	229	475	394	310	621	499	405	842	688	523	1,035	826	640
30	152	138	118	244	210	185	361	297	266	547	459	360	720	585	470	979	808	605	1,209	975	740
50	167	153	134	279	244	214	421	353	310	641	547	423	854	706	550	1,164	977	705	1,451	1,188	860

For SI: 1 inch = 25.4 mm, 1 foot = 304.8 mm, 1 British thermal unit per hour = 0.2931 W.

TABLE G2428.3(2) [504.3(2)]
TYPE B DOUBLE-WALL VENT

Number of Appliances	Two or more
Appliance Type	Category I
Appliance Vent Connection	Single-wall metal connector

VENT CONNECTOR CAPACITY

		SINGLE-WALL METAL VENT CONNECTOR DIAMETER—(D) inches																						
		3			4			5			6			7			8			9			10		
VENT HEIGHT (H) (feet)	CONNECTOR RISE (R) (feet)	APPLIANCE INPUT RATING LIMITS IN THOUSANDS OF BTU/H																						
		FAN		NAT	FAN		NAT	FAN		NAT	FAN		NAT	FAN		NAT	FAN		NAT	FAN		NAT	FAN		NAT
		Min	Max	Max	Min	Max	Max	Min	Max	Max	Min	Max	Max	Min	Max	Max	Min	Max	Max	Min	Max	Max	Min	Max	Max
6	1	NA	NA	26	NA	NA	46	NA	NA	71	NA	NA	102	207	223	140	262	293	183	325	373	234	447	463	286
	2	NA	NA	31	NA	NA	55	NA	NA	85	168	182	123	215	251	167	271	331	219	334	422	281	458	524	344
	3	NA	NA	34	NA	NA	62	121	131	95	175	198	138	222	273	188	279	361	247	344	462	316	468	574	385
8	1	NA	NA	27	NA	NA	48	NA	NA	75	NA	NA	106	226	240	145	285	316	191	352	403	244	481	502	299
	2	NA	NA	32	NA	NA	57	125	126	89	184	193	127	234	266	173	293	353	228	360	450	292	492	560	355
	3	NA	NA	35	NA	NA	64	130	138	100	191	208	144	241	287	197	302	381	256	370	489	328	501	609	400
10	1	NA	NA	28	NA	NA	50	119	121	77	182	186	110	240	253	150	302	335	196	372	429	252	506	534	308
	2	NA	NA	33	84	85	59	124	134	91	189	203	132	248	278	183	311	369	235	381	473	302	517	589	368
	3	NA	NA	36	89	91	67	129	144	102	197	217	148	257	299	203	320	398	265	391	511	339	528	637	413
15	1	NA	NA	29	79	87	52	116	138	81	177	214	116	238	291	158	312	380	208	397	482	266	556	596	324
	2	NA	NA	34	83	94	62	121	150	97	185	230	138	246	314	189	321	411	248	407	522	317	568	646	387
	3	NA	NA	39	87	100	70	127	160	109	193	243	157	255	333	215	331	438	281	418	557	360	579	690	437
20	1	49	56	30	78	97	54	115	152	84	175	238	120	233	325	165	306	425	217	390	538	276	546	664	336
	2	52	59	36	82	103	64	120	163	101	182	252	144	243	346	197	317	453	259	400	574	331	558	709	403
	3	55	62	40	87	107	72	125	172	113	190	264	164	252	363	223	326	476	294	412	607	375	570	750	457
30	1	47	60	31	77	110	57	112	175	89	169	278	129	226	380	175	296	497	230	378	630	294	528	779	358
	2	51	62	37	81	115	67	117	185	106	177	290	152	236	397	208	307	521	274	389	662	349	541	819	425
	3	54	64	42	85	119	76	122	193	120	185	300	172	244	412	235	316	542	309	400	690	394	555	855	482

COMMON VENT CAPACITY

	TYPE B DOUBLE-WALL COMMON VENT DIAMETER— (D) inches																				
	4			5			6			7			8			9			10		
VENT HEIGHT (H) (feet)	COMBINED APPLIANCE INPUT RATING IN THOUSANDS OF BTU/H																				
	FAN +FAN	FAN +NAT	NAT +NAT	FAN +FAN	FAN +NAT	NAT +NAT	FAN +FAN	FAN +NAT	NAT +NAT	FAN +FAN	FAN +NAT	NAT +NAT	FAN +FAN	FAN +NAT	NAT +NAT	FAN +FAN	FAN +NAT	NAT +NAT	FAN +FAN	FAN +NAT	NAT +NAT
6	NA	78	64	NA	113	99	200	158	144	304	244	196	398	310	257	541	429	332	665	515	407
8	NA	87	71	NA	126	111	218	173	159	331	269	218	436	342	285	592	473	373	730	569	460
10	NA	94	76	163	137	120	237	189	174	357	292	236	467	369	309	638	512	398	787	617	487
15	121	108	88	189	159	140	275	221	200	416	343	274	544	434	357	738	599	456	905	718	553
20	131	118	98	208	177	156	305	247	223	463	383	302	606	487	395	824	673	512	1,013	808	626
30	145	132	113	236	202	180	350	286	257	533	446	349	703	570	459	958	790	593	1,183	952	723
50	159	145	128	268	233	208	406	337	296	622	529	410	833	686	535	1,139	954	689	1,418	1,157	838

For SI: 1 inch = 25.4 mm, 1 foot = 304.8 mm, 1 British thermal unit per hour = 0.2931 W.

TABLE G2428.3(3) [504.3(3)]
MASONRY CHIMNEY

Number of Appliances	Two or more
Appliance Type	Category I
Appliance Vent Connection	Type B double-wall connector

VENT CONNECTOR CAPACITY

		TYPE B DOUBLE-WALL VENT CONNECTOR DIAMETER—(D) inches																							
		3			4			5			6			7			8			9			10		
VENT HEIGHT (H) (feet)	CONNECTOR RISE (R) (feet)	APPLIANCE INPUT RATING LIMITS IN THOUSANDS OF BTU/H																							
		FAN		NAT	FAN		NAT	FAN		NAT	FAN		NAT	FAN		NAT	FAN		NAT	FAN		NAT	FAN		NAT
		Min	Max	Max	Min	Max	Max	Min	Max	Max	Min	Max	Max	Min	Max	Max	Min	Max	Max	Min	Max	Max	Min	Max	Max
6	1	24	33	21	39	62	40	52	106	67	65	194	101	87	274	141	104	370	201	124	479	253	145	599	319
	2	26	43	28	41	79	52	53	133	85	67	230	124	89	324	173	107	436	232	127	562	300	148	694	378
	3	27	49	34	42	92	61	55	155	97	69	262	143	91	369	203	109	491	270	129	633	349	151	795	439
8	1	24	39	22	39	72	41	55	117	69	71	213	105	94	304	148	113	414	210	134	539	267	156	682	335
	2	26	47	29	40	87	53	57	140	86	73	246	127	97	350	179	116	473	240	137	615	311	160	776	394
	3	27	52	34	42	97	62	59	159	98	75	269	145	99	383	206	119	517	276	139	672	358	163	848	452
10	1	24	42	22	38	80	42	55	130	71	74	232	108	101	324	153	120	444	216	142	582	277	165	739	348
	2	26	50	29	40	93	54	57	153	87	76	261	129	103	366	184	123	498	247	145	652	321	168	825	407
	3	27	55	35	41	105	63	58	170	100	78	284	148	106	397	209	126	540	281	147	705	366	171	893	463
15	1	24	48	23	38	93	44	54	154	74	72	277	114	100	384	164	125	511	229	153	658	297	184	824	375
	2	25	55	31	39	105	55	56	174	89	74	299	134	103	419	192	128	558	260	156	718	339	187	900	432
	3	26	59	35	41	115	64	57	189	102	76	319	153	105	448	215	131	597	292	159	760	382	190	960	486
20	1	24	52	24	37	102	46	53	172	77	71	313	119	98	437	173	123	584	239	150	752	312	180	943	397
	2	25	58	31	39	114	56	55	190	91	73	335	138	101	467	199	126	625	270	153	805	354	184	1,011	452
	3	26	63	35	40	123	65	57	204	104	75	353	157	104	493	222	129	661	301	156	851	396	187	1,067	505

COMMON VENT CAPACITY

	MINIMUM INTERNAL AREA OF MASONRY CHIMNEY FLUE (square inches)																							
	12			19			28			38			50			63			78			113		
VENT HEIGHT (H) (feet)	COMBINED APPLIANCE INPUT RATING IN THOUSANDS OF BTU/H																							
	FAN +FAN	FAN +NAT	NAT +NAT	FAN +FAN	FAN +NAT	NAT +NAT	FAN +FAN	FAN +NAT	NAT +NAT	FAN +FAN	FAN +NAT	NAT +NAT	FAN +FAN	FAN +NAT	NAT +NAT	FAN +FAN	FAN +NAT	NAT +NAT	FAN +FAN	FAN +NAT	NAT +NAT	FAN +FAN	FAN +NAT	NAT +NAT
6	NA	74	25	NA	119	46	NA	178	71	NA	257	103	NA	351	143	NA	458	188	NA	582	246	1,041	853	NA
8	NA	80	28	NA	130	53	NA	193	82	NA	279	119	NA	384	163	NA	501	218	724	636	278	1,144	937	408
10	NA	84	31	NA	138	56	NA	207	90	NA	299	131	NA	409	177	606	538	236	776	686	302	1,226	1,010	454
15	NA	NA	36	NA	152	67	NA	233	106	NA	334	152	523	467	212	682	611	283	874	781	365	1,374	1,156	546
20	NA	NA	41	NA	NA	75	NA	250	122	NA	368	172	565	508	243	742	668	325	955	858	419	1,513	1,286	648
30	NA	NA	NA	NA	NA	NA	NA	270	137	NA	404	198	615	564	278	816	747	381	1,062	969	496	1,702	1,473	749
50	NA	NA	NA	NA	NA	NA	NA	NA	NA	NA	NA	NA	NA	620	328	879	831	461	1,165	1,089	606	1,905	1,692	922

For SI: 1 inch = 25.4 mm, 1 square inch = 645.16 mm², 1 foot = 304.8 mm, 1 British thermal unit per hour = 0.2931 W.

**TABLE G2428.3(4) [504.3(4)]
MASONRY CHIMNEY**

Number of Appliances	Two or more
Appliance Type	Category I
Appliance Vent Connection	Single-wall metal connector

VENT CONNECTOR CAPACITY

		SINGLE-WALL METAL VENT CONNECTOR DIAMETER (D)—inches																						
		3			4			5			6			7			8			9			10		
		APPLIANCE INPUT RATING LIMITS IN THOUSANDS OF BTU/H																						
VENT HEIGHT (H) (feet)	CONNECTOR RISE (R) (feet)	FAN		NAT	FAN		NAT	FAN		NAT	FAN		NAT	FAN		NAT	FAN		NAT	FAN		NAT	FAN		NAT
		Min	Max	Max	Min	Max	Max	Min	Max	Max	Min	Max	Max	Min	Max	Max	Min	Max	Max	Min	Max	Max	Min	Max	Max
6	1	NA	NA	21	NA	NA	39	NA	NA	66	179	191	100	231	271	140	292	366	200	362	474	252	499	594	316
	2	NA	NA	28	NA	NA	52	NA	NA	84	186	227	123	239	321	172	301	432	231	373	557	299	509	696	376
	3	NA	NA	34	NA	NA	61	134	153	97	193	258	142	247	365	202	309	491	269	381	634	348	519	793	437
8	1	NA	NA	21	NA	NA	40	NA	NA	68	195	208	103	250	298	146	313	407	207	387	530	263	529	672	331
	2	NA	NA	28	NA	NA	52	137	139	85	202	240	125	258	343	177	323	465	238	397	607	309	540	766	391
	3	NA	NA	34	NA	NA	62	143	156	98	210	264	145	266	376	205	332	509	274	407	663	356	551	838	450
10	1	NA	NA	22	NA	NA	41	130	151	70	202	225	106	267	316	151	333	434	213	410	571	273	558	727	343
	2	NA	NA	29	NA	NA	53	136	150	86	210	255	128	276	358	181	343	489	244	420	640	317	569	813	403
	3	NA	NA	34	97	102	62	143	166	99	217	277	147	284	389	207	352	530	279	430	694	363	580	880	459
15	1	NA	NA	23	NA	NA	43	129	151	73	199	271	112	268	376	161	349	502	225	445	646	291	623	808	366
	2	NA	NA	30	92	103	54	135	170	88	207	295	132	277	411	189	359	548	256	456	706	334	634	884	424
	3	NA	NA	34	96	112	63	141	185	101	215	315	151	286	439	213	368	586	289	466	755	378	646	945	479
20	1	NA	NA	23	87	99	45	128	167	76	197	303	117	265	425	169	345	569	235	439	734	306	614	921	347
	2	NA	NA	30	91	111	55	134	185	90	205	325	136	274	455	195	355	610	266	450	787	348	627	986	443
	3	NA	NA	35	96	119	64	140	199	103	213	343	154	282	481	219	365	644	298	461	831	391	639	1,042	496

COMMON VENT CAPACITY

	MINIMUM INTERNAL AREA OF MASONRY CHIMNEY FLUE (square inches)																							
	12			19			28			38			50			63			78			113		
VENT HEIGHT (H) (feet)	COMBINED APPLIANCE INPUT RATING IN THOUSANDS OF BTU/H																							
	FAN +FAN	FAN +NAT	NAT +NAT	FAN +FAN	FAN +NAT	NAT +NAT	FAN +FAN	FAN +NAT	NAT +NAT	FAN +FAN	FAN +NAT	NAT +NAT	FAN +FAN	FAN +NAT	NAT +NAT	FAN +FAN	FAN +NAT	NAT +NAT	FAN +FAN	FAN +NAT	NAT +NAT	FAN +FAN	FAN +NAT	NAT +NAT
6	NA	NA	25	NA	118	45	NA	176	71	NA	255	102	NA	348	142	NA	455	187	NA	579	245	NA	846	NA
8	NA	NA	28	NA	128	52	NA	190	81	NA	276	118	NA	380	162	NA	497	217	NA	633	277	1,136	928	405
10	NA	NA	31	NA	136	56	NA	205	89	NA	295	129	NA	405	175	NA	532	234	171	680	300	1,216	1,000	450
15	NA	NA	36	NA	NA	66	NA	230	105	NA	335	150	NA	400	210	677	602	280	866	772	360	1,359	1,139	540
20	NA	NA	NA	NA	NA	74	NA	247	120	NA	362	170	NA	503	240	765	661	321	947	849	415	1,495	1,264	640
30	NA	NA	NA	NA	NA	NA	NA	NA	135	NA	398	195	NA	558	275	808	739	377	1,052	957	490	1,682	1,447	740
50	NA	NA	NA	NA	NA	NA	NA	NA	NA	NA	NA	NA	NA	612	325	NA	821	456	1,152	1,076	600	1,879	1,672	910

For SI: 1 inch = 25.4 mm, 1 square inch = 645.16 mm^2, 1 foot = 304.8 mm, 1 British thermal unit per hour = 0.2931 W.

TABLE G2428.3.2 (504.3.2)
MAXIMUM VENT CONNECTOR LENGTH

CONNECTOR DIAMETER	CONNECTOR HORIZONTAL
Maximum (inches)	Length (feet)
3	4.5
4	6
5	7.5
6	9
7	10.5
8	12
9	13.5

For SI: 1 inch = 25.4 mm, 1 foot = 304.8 mm.

G2428.3.3 (504.3.3) Connectors with longer lengths. Connectors with longer horizontal lengths than those listed in Section G2428.3.2 are permitted under the following conditions:

1. The maximum capacity (FAN Max or NAT Max) of the vent connector shall be reduced 10 percent for each additional multiple of the length listed above. For example, the maximum length listed above for a 4-inch (102 mm) connector is 6 feet (1829 mm). With a connector length greater than 6 feet (1829 mm), but not exceeding 12 feet (3658 mm), the maximum capacity must be reduced by 10 percent (0.90 × maximum vent connector capacity). With a connector length greater than 12 feet (3658 mm), but not exceeding 18 feet (5486 mm), the maximum capacity must be reduced by 20 percent (0.80 × maximum vent capacity).

2. For a connector serving a fan-assisted appliance, the minimum capacity (FAN Min) of the connector shall be determined by referring to the corresponding single appliance table. For Type B double-wall connectors, Table G2428.2(1) shall be used. For single-wall connectors, Table G2428.2(2) shall be used. The height (H) and lateral (L) shall be measured according to the procedures for a single appliance vent, as if the other appliances were not present.

G2428.3.4 (504.3.4) Vent connector manifold. Where the vent connectors are combined prior to entering the vertical portion of the common vent to form a common vent manifold, the size of the common vent manifold and the common vent shall be determined by applying a 10-percent reduction (0.90 × maximum common vent capacity) to the common vent capacity part of the common vent tables. The length of the common vent connector manifold (L_M) shall not exceed 1.5 feet for each inch (18 mm per mm) of common vent connector manifold diameter (D) (see Appendix B Figure B-11).

G2428.3.5 (504.3.5) Common vertical vent offset. Where the common vertical vent is offset, the maximum capacity of the common vent shall be reduced in accordance with Section G2428.3.6. The horizontal length of the common vent

offset (L_o) shall not exceed 1.5 feet for each inch (18 mm per mm) of common vent diameter.

G2428.3.6 (504.3.6) Elbows in vents. For each elbow up to and including 45 degrees (0.79 rad) in the common vent, the maximum common vent capacity listed in the venting tables shall be reduced by 5 percent. For each elbow greater than 45 degrees (0.79 rad) up to and including 90 degrees (1.57 rad), the maximum common vent capacity listed in the venting tables shall be reduced by 10 percent.

G2428.3.7 (504.3.7) Elbows in connectors. The vent connector capacities listed in the common vent sizing tables include allowance for two 90-degree (1.57 rad) elbows. For each additional elbow up to and including 45 degrees (0.79 rad), the maximum vent connector capacity listed in the venting tables shall be reduced by 5 percent. For each elbow greater than 45 degrees (0.79 rad) up to and including 90 degrees (1.57 rad), the maximum vent connector capacity listed in the venting tables shall be reduced by 10 percent.

G2428.3.8 (504.3.8) Common vent minimum size. The cross-sectional area of the common vent shall be equal to or greater than the cross-sectional area of the largest connector.

G2428.3.9 (504.3.9) Common vent fittings. At the point where tee or wye fittings connect to a common vent, the opening size of the fitting shall be equal to the size of the common vent. Such fittings shall not be prohibited from having reduced-size openings at the point of connection of appliance vent connectors.

G2428.3.9.1 (504.3.9.1) Tee and wye fittings. Tee and wye fittings connected to a common vent shall be considered as part of the common vent and shall be constructed of materials consistent with that of the common vent.

G2428.3.10 (504.3.10) High altitude installations. Sea-level input ratings shall be used when determining maximum capacity for high altitude installation. Actual input, derated for altitude, shall be used for determining minimum capacity for high altitude installation.

G2428.3.11 (504.3.11) Connector rise measurement. Connector rise (R) for each appliance connector shall be measured from the draft hood outlet or flue collar to the centerline where the vent gas streams come together.

G2428.3.12 (504.3.12) Vent height measurement. For multiple appliances all located on one floor, available total height (H) shall be measured from the highest draft hood outlet or flue collar up to the level of the outlet of the common vent.

G2428.3.13 (504.3.17) Vertical vent maximum size. Where two or more appliances are connected to a vertical vent or chimney, the flow area of the largest section of vertical vent or chimney shall not exceed seven times the smallest listed appliance categorized vent areas, flue collar area, or draft hood outlet area unless designed in accordance with approved engineering methods.

G2428.3.14 (504.3.18) Multiple input rate appliances. For appliances with more than one input rate, the minimum vent connector capacity (FAN Min) determined from the tables shall be less than the lowest appliance input rating, and the maximum vent connector capacity (FAN Max or NAT Max) determined from the tables shall be greater than the highest appliance input rating.

G2428.3.15 (504.3.19) Liner system sizing and connections. Listed, corrugated metallic chimney liner systems in masonry chimneys shall be sized by using Table G2428.3(1) or G2428.3(2) for Type B vents, with the maximum capacity reduced by 20 percent (0.80 × maximum capacity) and the minimum capacity as shown in Table G2428.3(1) or G2428.3(2). Corrugated metallic liner systems installed with bends or offsets shall have their maximum capacity further reduced in accordance with Sections G2428.3.5 and G2428.3.6. The 20-percent reduction for corrugated metallic chimney liner systems includes an allowance for one long-radius 90-degree (1.57 rad) turn at the bottom of the liner. Where double-wall connectors are required, tee and wye fittings used to connect to the common vent chimney liner shall be listed double-wall fittings. Connections between chimney liners and listed double-wall fittings shall be made with listed adapter fittings designed for such purpose.

G2428.3.16 (504.3.20) Chimney and vent location. Tables G2428.3(1), G2428.3(2), G2428.3(3) and G2428.3(4) shall only be used for chimneys and vents not exposed to the outdoors below the roof line. A Type B vent or listed chimney lining system passing through an unused masonry chimney flue shall not be considered to be exposed to the outdoors. A Type B vent shall not be considered to be exposed to the outdoors where it passes through an unventilated enclosure or chase insulated to a value of not less than R-8.

G2428.3.17 (504.3.21) Connector maximum and minimum size. Vent connectors shall not be increased in size more than two sizes greater than the listed appliance categorized vent diameter, flue collar diameter, or draft hood outlet diameter. Vent connectors for draft-hood-equipped appliances shall not be smaller than the draft hood outlet diameter. Where a vent connector size(s) determined from the tables for a fan-assisted appliance(s) is smaller than the flue collar diameter, the use of the smaller size(s) shall be permitted provided that the installation complies with all of the following conditions:

1. Vent connectors for fan-assisted appliance flue collars 12 inches (305 mm) in diameter or smaller are not reduced by more than one table size [e.g., 12 inches to 10 inches (305 mm to 254 mm) is a one-size reduction] and those larger than 12 inches (305 mm) in diameter are not reduced more than two table sizes [e.g., 24 inches to 20 inches (610 mm to 508 mm) is a two-size reduction].

2. The fan-assisted appliance(s) is common vented with a draft-hood-equipped appliances(s).

3. The vent connector has a smooth interior wall.

G2428.3.18 (504.3.22) Component commingling. All combinations of pipe sizes, single-wall, and double-wall metal pipe shall be allowed within any connector run(s) or within the common vent, provided all of the appropriate tables permit all of the desired sizes and types of pipe, as if they were used for the entire length of the subject connector or vent. Where single-wall and Type B double-wall metal pipes are used for vent connectors within the same venting system, the common vent must be sized using Table G2428.3(2) or G2428.3(4), as appropriate.

G2428.3.19 (504.3.23) Draft hood conversion accessories. Draft hood conversion accessories for use with masonry chimneys venting listed Category I fan-assisted appliances shall be listed and installed in accordance with the manufacturer's installation instructions for such listed accessories.

G2428.3.20 (504.3.24) Multiple sizes permitted. Where a table permits more than one diameter of pipe to be used for a connector or vent, all the permitted sizes shall be permitted to be used.

G2428.3.21 (504.3.25) Table interpolation. Interpolation shall be permitted in calculating capacities for vent dimensions that fall between table entries. (See Example 3, Appendix B.)

G2428.3.22 (504.3.26) Extrapolation prohibited. Extrapolation beyond the table entries shall not be permitted.

G2428.3.23 (504.3.27) Engineering calculations. For vent heights less than 6 feet (1829 mm) and greater than shown in the tables, engineering methods shall be used to calculate vent capacities.

SECTION G2429 (505)
DIRECT-VENT, INTEGRAL VENT, MECHANICAL VENT AND VENTILATION/EXHAUST HOOD VENTING

G2429.1 (505.1) General. The installation of direct-vent and integral vent appliances shall be in accordance with Section G2427. Mechanical venting systems shall be designed and installed in accordance with Section G2427.

SECTION G2430 (506)
FACTORY-BUILT CHIMNEYS

G2430.1 (506.1) Listing. Factory-built chimneys for building heating appliances producing flue gases having a temperature not greater than 1,000°F (538°C), measured at the entrance to the chimney, shall be listed and labeled in accordance with UL 103 and shall be installed and terminated in accordance with the manufacturer's installation instructions.

G2430.2 (506.2) Support. Where factory-built chimneys are supported by structural members, such as joists and rafters, such members shall be designed to support the additional load.

SECTION G2431 (601)
GENERAL

G2431.1 (601.1) Scope. Sections G2432 through G2453 shall govern the approval, design, installation, construction, maintenance, alteration and repair of the appliances and equipment specifically identified herein.

SECTION G2432 (602)
DECORATIVE APPLIANCES FOR INSTALLATION IN FIREPLACES

G2432.1 (602.1) General. Decorative appliances for installation in approved solid fuel burning fireplaces shall be tested in accordance with ANSI Z21.60 and shall be installed in accordance with the manufacturer's installation instructions. Manually lighted natural gas decorative appliances shall be tested in accordance with ANSI Z21.84.

G2432.2 (602.2) Flame safeguard device. Decorative appliances for installation in approved solid fuel-burning fireplaces, with the exception of those tested in accordance with ANSI Z21.84, shall utilize a direct ignition device, an ignitor or a pilot flame to ignite the fuel at the main burner, and shall be equipped with a flame safeguard device. The flame safeguard device shall automatically shut off the fuel supply to a main burner or group of burners when the means of ignition of such burners becomes inoperative.

G2432.3 (602.3) Prohibited installations. Decorative appliances for installation in fireplaces shall not be installed where prohibited by Section G2406.2.

SECTION G2433 (603)
LOG LIGHTERS

G2433.1 (603.1) General. Log lighters shall be tested in accordance with CSA 8 and shall be installed in accordance with the manufacturer's installation instructions.

SECTION G2434 (604)
VENTED GAS FIREPLACES
(DECORATIVE APPLIANCES)

G2434.1 (604.1) General. Vented gas fireplaces shall be tested in accordance with ANSI Z21.50, shall be installed in accordance with the manufacturer's installation instructions and shall be designed and equipped as specified in Section G2432.2.

G2434.2 (604.2) Access. Panels, grilles, and access doors that are required to be removed for normal servicing operations shall not be attached to the building.

SECTION G2435 (605)
VENTED GAS FIREPLACE HEATERS

G2435.1 (605.1) General. Vented gas fireplace heaters shall be installed in accordance with the manufacturer's installation instructions, shall be tested in accordance with ANSI Z21.88 and shall be designed and equipped as specified in Section G2432.2.

SECTION G2436 (608)
VENTED WALL FURNACES

G2436.1 (608.1) General. Vented wall furnaces shall be tested in accordance with ANSI Z21.86/CSA 2.32 and shall be installed in accordance with the manufacturer's installation instructions.

G2436.2 (608.2) Venting. Vented wall furnaces shall be vented in accordance with Section G2427.

G2436.3 (608.3) Location. Vented wall furnaces shall be located so as not to cause a fire hazard to walls, floors, combustible furnishings or doors. Vented wall furnaces installed between bathrooms and adjoining rooms shall not circulate air from bathrooms to other parts of the building.

G2436.4 (608.4) Door swing. Vented wall furnaces shall be located so that a door cannot swing within 12 inches (305 mm) of an air inlet or air outlet of such furnace measured at right angles to the opening. Doorstops or door closers shall not be installed to obtain this clearance.

G2436.5 (608.5) Ducts prohibited. Ducts shall not be attached to wall furnaces. Casing extension boots shall not be installed unless listed as part of the appliance.

G2436.6 (608.6) Access. Vented wall furnaces shall be provided with access for cleaning of heating surfaces, removal of burners, replacement of sections, motors, controls, filters and other working parts, and for adjustments and lubrication of parts requiring such attention. Panels, grilles and access doors that are required to be removed for normal servicing operations shall not be attached to the building construction.

SECTION G2437 (609)
FLOOR FURNACES

G2437.1 (609.1) General. Floor furnaces shall be tested in accordance with ANSI Z21.86/CSA 2.32 and shall be installed in accordance with the manufacturer's installation instructions.

G2437.2 (609.2) Placement. The following provisions apply to floor furnaces:

1. Floors. Floor furnaces shall not be installed in the floor of any doorway, stairway landing, aisle or passageway of any enclosure, public or private, or in an exitway from any such room or space.

2. Walls and corners. The register of a floor furnace with a horizontal warm air outlet shall not be placed closer than 6 inches (152 mm) to the nearest wall. A distance of at least 18 inches (457 mm) from two adjoining sides of the floor furnace register to walls shall be provided to eliminate the necessity of occupants walking over the warm air discharge. The remaining sides shall be permitted to be placed not closer than 6 inches (152 mm) to a wall. Wall-register models shall not be placed closer than 6 inches (152 mm) to a corner.

3. Draperies. The furnace shall be placed so that a door, drapery, or similar object cannot be nearer than 12 inches (305 mm) to any portion of the register of the furnace.

4. Floor construction. Floor furnaces shall not be installed in concrete floor construction built on grade.

5. Thermostat. The controlling thermostat for a floor furnace shall be located within the same room or space as the floor furnace or shall be located in an adjacent room or space that is permanently open to the room or space containing the floor furnace.

G2437.3 (609.3) Bracing. The floor around the furnace shall be braced and headed with a support framework designed in accordance with Chapter 5.

G2437.4 (609.4) Clearance. The lowest portion of the floor furnace shall have not less than a 6-inch (152 mm) clearance from the grade level; except where the lower 6-inch (152 mm) portion of the floor furnace is sealed by the manufacturer to prevent entrance of water, the minimum clearance shall be reduced to not less than 2 inches (51 mm). Where these clearances cannot be provided, the ground below and to the sides shall be excavated to form a pit under the furnace so that the required clearance is provided beneath the lowest portion of the furnace. A 12-inch (305 mm) minimum clearance shall be provided on all sides except the control side, which shall have an 18-inch (457 mm) minimum clearance.

G2437.5 (609.5) First floor installation. Where the basement story level below the floor in which a floor furnace is installed is utilized as habitable space, such floor furnaces shall be enclosed as specified in Section G2437.6 and shall project into a nonhabitable space.

G2437.6 (609.6) Upper floor installations. Floor furnaces installed in upper stories of buildings shall project below into nonhabitable space and shall be separated from the nonhabitable space by an enclosure constructed of noncombustible materials. The floor furnace shall be provided with access, clearance to all sides and bottom of not less than 6 inches (152 mm) and combustion air in accordance with Section G2407.

SECTION G2438 (613)
CLOTHES DRYERS

G2438.1 (613.1) General. Clothes dryers shall be tested in accordance with ANSI Z21.5.1 and shall be installed in accordance with the manufacturer's installation instructions.

SECTION G2439 (614)
CLOTHES DRYER EXHAUST

G2439.1 (614.1) Installation. Clothes dryers shall be exhausted in accordance with the manufacturer's instructions. Dryer exhaust systems shall be independent of all other systems and shall convey the moisture and any products of combustion to the outside of the building.

G2439.2 (614.2) Duct penetrations. Ducts that exhaust clothes dryers shall not penetrate or be located within any fireblocking, draftstopping or any wall, floor/ceiling or other assembly required by this code to be fire-resistance rated, unless such duct is constructed of galvanized steel or aluminum of the thickness specified in the mechanical provisions of this code and the fire-resistance rating is maintained in accordance with this code. Fire dampers shall not be installed in clothes dryer exhaust duct systems.

G2439.3 (614.4) Exhaust installation. Dryer exhaust ducts for clothes dryers shall terminate on the outside of the building and shall be equipped with a backdraft damper. Screens shall not be installed at the duct termination. Ducts shall not be connected or installed with sheet metal screws or other fasteners that will obstruct the flow. Clothes dryer exhaust ducts shall not be connected to a vent connector, vent or chimney. Clothes dryer exhaust ducts shall not extend into or through ducts or plenums.

G2439.4 (614.5) Makeup air. Installations exhausting more than 200 cfm (0.09 m³/s) shall be provided with makeup air. Where a closet is designed for the installation of a clothes dryer, an opening having an area of not less than 100 square inches (0.0645 m²) for makeup air shall be provided in the closet enclosure, or makeup air shall be provided by other approved means.

G2439.5 (614.6) Clothes dryer ducts. Exhaust ducts for domestic clothes dryers shall be constructed of metal and shall have a smooth interior finish. The exhaust duct shall be a minimum nominal size of 4 inches (102 mm) in diameter. The entire exhaust system shall be supported and secured in place. The male end of the duct at overlapped duct joints shall extend in the direction of airflow. Clothes dryer transition ducts used to connect the appliance to the exhaust duct system shall be metal and limited to a single length not to exceed 8 feet (2438 mm) in length and shall be listed and labeled for the application. Transition ducts shall not be concealed within construction.

G2439.5.1 (614.6.1) Maximum length. The maximum length of a clothes dryer exhaust duct shall not exceed 25 feet (7620 mm) from the dryer location to the outlet terminal. The maximum length of the duct shall be reduced 2$^{1}/_{2}$ feet (762 mm) for each 45 degree (0.79 rad) bend and 5 feet (1524 mm) for each 90 degree (1.6 rad) bend. The maximum length of the exhaust duct does not include the transition duct.

> **Exception:** Where the make and model of the clothes dryer to be installed is known and the manufacturer's installation instructions for such dryer are provided to the code official, the maximum length of the exhaust duct, including any transition duct, shall be permitted to be in accordance with the dryer manufacturer's installation instructions.

G2439.5.2 (614.6.2) Rough-in-required. Where a compartment or space for a clothes dryer is provided, an exhaust duct system shall be installed.

SECTION G2440 (615)
SAUNA HEATERS

G2440.1 (615.1) General. Sauna heaters shall be installed in accordance with the manufacturer's installation instructions.

G2440.2 (615.2) Location and protection. Sauna heaters shall be located so as to minimize the possibility of accidental contact by a person in the room.

G2440.2.1 (615.2.1) Guards. Sauna heaters shall be protected from accidental contact by an approved guard or barrier of material having a low coefficient of thermal conductivity. The guard shall not substantially affect the transfer of heat from the heater to the room.

G2440.3 (615.3) Access. Panels, grilles and access doors that are required to be removed for normal servicing operations, shall not be attached to the building.

G2440.4 (615.4) Combustion and dilution air intakes. Sauna heaters of other than the direct-vent type shall be installed with the draft hood and combustion air intake located outside the sauna room. Where the combustion air inlet and the draft hood are in a dressing room adjacent to the sauna room, there shall be provisions to prevent physically blocking the combustion air inlet and the draft hood inlet, and to prevent physical contact with the draft hood and vent assembly, or warning notices shall be posted to avoid such contact. Any warning notice shall be easily readable, shall contrast with its background, and the wording shall be in letters not less than 0.25 inch (6.4 mm) high.

G2440.5 (615.5) Combustion and ventilation air. Combustion air shall not be taken from inside the sauna room. Combustion and ventilation air for a sauna heater not of the direct-vent type shall be provided to the area in which the combustion air inlet and draft hood are located in accordance with Section G2407.

G2440.6 (615.6) Heat and time controls. Sauna heaters shall be equipped with a thermostat which will limit room temperature to 194°F (90°C). If the thermostat is not an integral part of the sauna heater, the heat-sensing element shall be located within 6 inches (152 mm) of the ceiling. If the heat-sensing element is a capillary tube and bulb, the assembly shall be attached to the wall or other support, and shall be protected against physical damage.

G2440.6.1 (615.6.1) Timers. A timer, if provided to control main burner operation, shall have a maximum operating time of 1 hour. The control for the timer shall be located outside the sauna room.

G2440.7 (615.7) Sauna room. A ventilation opening into the sauna room shall be provided. The opening shall be not less than 4 inches by 8 inches (102 mm by 203 mm) located near the top of the door into the sauna room.

SECTION G2441 (617)
POOL AND SPA HEATERS

G2441.1 (617.1) General. Pool and spa heaters shall be tested in accordance with ANSI Z21.56 and shall be installed in accordance with the manufacturer's installation instructions.

SECTION G2442 (618)
FORCED-AIR WARM-AIR FURNACES

G2442.1 (618.1) General. Forced-air warm-air furnaces shall be tested in accordance with ANSI Z21.47 or UL 795 and shall be installed in accordance with the manufacturer's installation instructions.

G2442.2 (618.2) Forced-air furnaces. The minimum unobstructed total area of the outside and return air ducts or openings to a forced-air warm-air furnace shall be not less than 2 square inches for each 1,000 Btu/h (4402 mm²/W) output rating capacity of the furnace and not less than that specified in the furnace manufacturer's installation instructions. The minimum unobstructed total area of supply ducts from a forced-air warm-air furnace shall be not less than 2 square inches for each 1,000 Btu/h (4402 mm²/W) output rating capacity of the furnace and not less than that specified in the furnace manufacturer's installation instructions.

> **Exception:** The total area of the supply air ducts and outside and return air ducts shall not be required to be larger than the minimum size required by the furnace manufacturer's installation instructions.

G2442.3 (618.3) Dampers. Volume dampers shall not be placed in the air inlet to a furnace in a manner that will reduce the required air to the furnace.

G2442.4 (618.4) Circulating air ducts for forced-air warm-air furnaces. Circulating air for forced-air-type, warm-air furnaces shall be conducted into the blower housing from outside the furnace enclosure by continuous air-tight ducts.

G2442.5 (618.5) Prohibited sources. Outside or return air for a forced-air heating system shall not be taken from the following locations:

1. Closer than 10 feet (3048 mm) from an appliance vent outlet, a vent opening from a plumbing drainage system or the discharge outlet of an exhaust fan, unless the outlet is 3 feet (914 mm) above the outside air inlet.

2. Where there is the presence of objectionable odors, fumes or flammable vapors; or where located less than 10 feet (3048 mm) above the surface of any abutting public way or driveway; or where located at grade level by a sidewalk, street, alley or driveway.

3. A room or space, the volume of which is less than 25 percent of the entire volume served by such system. Where connected by a permanent opening having an area sized in accordance with Section G2442.2, adjoining rooms or spaces shall be considered as a single room or space for the purpose of determining the volume of such rooms or spaces.

   > **Exception:** The minimum volume requirement shall not apply where the amount of return air taken from a room or space is less than or equal to the amount of supply air delivered to such room or space.

4. A room or space containing an appliance where such a room or space serves as the sole source of return air.

   > **Exception:** This shall not apply where:
   >
   > 1. The appliance is a direct-vent appliance or an appliance not requiring a vent in accordance with Section G2425.8.
   >
   > 2. The room or space complies with the following requirements:
   >
   >   2.1. The return air shall be taken from a room or space having a volume exceeding 1 cubic foot (28 316.85 mm³) for each 10 Btu/h (9.6 L/W) of combined input rating of all fuel-burning appliances therein.

2.2. The volume of supply air discharged back into the same space shall be approximately equal to the volume of return air taken from the space.

2.3. Return-air inlets shall not be located within 10 feet (3048 mm) of any appliance firebox or draft hood in the same room or space.

3. Rooms or spaces containing solid-fuel burning appliances, provided that return-air inlets are located not less than 10 feet (3048 mm) from the firebox of such appliances.

5. A closet, bathroom, toilet room, kitchen, garage, mechanical room, boiler room or furnace room.

G2442.6 (618.6) Screen. Required outdoor air inlets shall be covered with a screen having $^1/_4$-inch (6.4 mm) openings. Required outdoor air inlets serving a nonresidential portion of a building shall be covered with screen having openings larger than $^1/_4$ inch (6.4 mm) and not larger than 1 inch (25 mm).

G2442.7 (618.7) Return-air limitation. Return air from one dwelling unit shall not be discharged into another dwelling unit.

SECTION G2443 (619)
CONVERSION BURNERS

G2443.1 (619.1) Conversion burners. The installation of conversion burners shall conform to ANSI Z21.8.

SECTION G2444 (620)
UNIT HEATERS

G2444.1 (620.1) General. Unit heaters shall be tested in accordance with ANSI Z83.8 and shall be installed in accordance with the manufacturer's installation instructions.

G2444.2 (620.2) Support. Suspended-type unit heaters shall be supported by elements that are designed and constructed to accommodate the weight and dynamic loads. Hangers and brackets shall be of noncombustible material.

G2444.3 (620.3) Ductwork. Ducts shall not be connected to a unit heater unless the heater is listed for such installation.

G2444.4 (620.4) Clearance. Suspended-type unit heaters shall be installed with clearances to combustible materials of not less than 18 inches (457 mm) at the sides, 12 inches (305 mm) at the bottom and 6 inches (152 mm) above the top where the unit heater has an internal draft hood or 1 inch (25 mm) above the top of the sloping side of the vertical draft hood.

Floor-mounted-type unit heaters shall be installed with clearances to combustible materials at the back and one side only of not less than 6 inches (152 mm). Where the flue gases are vented horizontally, the 6-inch (152 mm) clearance shall be measured from the draft hood or vent instead of the rear wall of the unit heater. Floor-mounted-type unit heaters shall not be installed on combustible floors unless listed for such installation.

Clearance for servicing all unit heaters shall be in accordance with the manufacturer's installation instructions.

Exception: Unit heaters listed for reduced clearance shall be permitted to be installed with such clearances in accordance with their listing and the manufacturer's instructions.

SECTION G2445 (621)
UNVENTED ROOM HEATERS

G2445.1 (621.1) General. Unvented room heaters shall be tested in accordance with ANSI Z21.11.2 and shall be installed in accordance with the conditions of the listing and the manufacturer's installation instructions.

G2445.2 (621.2) Prohibited use. One or more unvented room heaters shall not be used as the sole source of comfort heating in a dwelling unit.

G2445.3 (621.3) Input rating. Unvented room heaters shall not have an input rating in excess of 40,000 Btu/h (11.7 kW).

G2445.4 (621.4) Prohibited locations. The location of unvented room heaters shall comply with Section G2406.2.

G2445.5 (621.5) Room or space volume. The aggregate input rating of all unvented appliances installed in a room or space shall not exceed 20 Btu/h per cubic foot (0.21 kW/m^3) of volume of such room or space. Where the room or space in which the equipment is installed is directly connected to another room or space by a doorway, archway or other opening of comparable size that cannot be closed, the volume of such adjacent room or space shall be permitted to be included in the calculations.

G2445.6 (621.6) Oxygen-depletion safety system. Unvented room heaters shall be equipped with an oxygen-depletion-sensitive safety shutoff system. The system shall shut off the gas supply to the main and pilot burners when the oxygen in the surrounding atmosphere is depleted to the percent concentration specified by the manufacturer, but not lower than 18 percent. The system shall not incorporate field adjustment means capable of changing the set point at which the system acts to shut off the gas supply to the room heater.

G2445.7 (621.7) Unvented decorative room heaters. An unvented decorative room heater shall not be installed in a factory-built fireplace unless the fireplace system has been specifically tested, listed and labeled for such use in accordance with UL 127.

G2445.7.1 (621.7.1) Ventless firebox enclosures. Ventless firebox enclosures used with unvented decorative room heaters shall be listed as complying with ANSI Z21.91.

SECTION G2446 (622)
VENTED ROOM HEATERS

G2446.1 (622.1) General. Vented room heaters shall be tested in accordance with ANSI Z21.86/CSA 2.32, shall be designed and equipped as specified in Section G2432.2 and shall be installed in accordance with the manufacturer's installation instructions.

SECTION G2447 (623)
COOKING APPLIANCES

G2447.1 (623.1) Cooking appliances. Cooking appliances that are designed for permanent installation, including ranges, ovens, stoves, broilers, grills, fryers, griddles, hot plates and barbecues, shall be tested in accordance with ANSI Z21.1 or ANSI Z21.58 and shall be installed in accordance with the manufacturer's installation instructions.

G2447.2 (623.2) Prohibited location. Cooking appliances designed, tested, listed and labeled for use in commercial occupancies shall not be installed within dwelling units or within any area where domestic cooking operations occur.

G2447.3 (623.3) Domestic appliances. Cooking appliances installed within dwelling units and within areas where domestic cooking operations occur shall be listed and labeled as household-type appliances for domestic use.

G2447.4 (623.4) Range installation. Ranges installed on combustible floors shall be set on their own bases or legs and shall be installed with clearances of not less than that shown on the label.

SECTION G2448 (624)
WATER HEATERS

G2448.1 (624.1) General. Water heaters shall be tested in accordance with ANSI Z 21.10.1 and ANSI Z 21.10.3 and shall be installed in accordance with the manufacturer's installation instructions.

G2448.1.1 (624.1.1) Installation requirements. The requirements for water heaters relative to sizing, relief valves, drain pans and scald protection shall be in accordance with this code.

G2448.2 (624.2) Water heaters utilized for space heating. Water heaters utilized both to supply potable hot water and provide hot water for space-heating applications shall be listed and labeled for such applications by the manufacturer and shall be installed in accordance with the manufacturer's installation instructions and this code.

SECTION G2449 (627)
AIR CONDITIONING EQUIPMENT

G2449.1 (627.1) General. Air conditioning equipment shall be tested in accordance with ANSI Z21.40.1 or ANSI Z21.40.2 and shall be installed in accordance with the manufacturer's installation instructions.

G2449.2 (627.2) Independent piping. Gas piping serving heating equipment shall be permitted to also serve cooling equipment where such heating and cooling equipment cannot be operated simultaneously. (See Section G2413.)

G2449.3 (627.3) Connection of gas engine-powered air conditioners. To protect against the effects of normal vibration in service, gas engines shall not be rigidly connected to the gas supply piping.

G2449.4 (627.6) Installation. Air conditioning equipment shall be installed in accordance with the manufacturer's instructions. Unless the equipment is listed for installation on a combustible surface such as a floor or roof, or unless the surface is protected in an approved manner, equipment shall be installed on a surface of noncombustible construction with noncombustible material and surface finish and with no combustible material against the underside thereof.

SECTION G2450 (628)
ILLUMINATING APPLIANCES

G2450.1 (628.1) General. Illuminating appliances shall be tested in accordance with ANSI Z21.42 and shall be installed in accordance with the manufacturer's installation instructions.

G2450.2 (628.2) Mounting on buildings. Illuminating appliances designed for wall or ceiling mounting shall be securely attached to substantial structures in such a manner that they are not dependent on the gas piping for support.

G2450.3 (628.3) Mounting on posts. Illuminating appliances designed for post mounting shall be securely and rigidly attached to a post. Posts shall be rigidly mounted. The strength and rigidity of posts greater than 3 feet (914 mm) in height shall be at least equivalent to that of a 2.5-inch-diameter (64 mm) post constructed of 0.064-inch-thick (1.6 mm) steel or a 1-inch (25 mm) Schedule 40 steel pipe. Posts 3 feet (914 mm) or less in height shall not be smaller than $^3/_4$-inch (19.1 mm) Schedule 40 steel pipe. Drain openings shall be provided near the base of posts where there is a possibility of water collecting inside them.

G2450.4 (628.4) Appliance pressure regulators. Where an appliance pressure regulator is not supplied with an illuminating appliance and the service line is not equipped with a service pressure regulator, an appliance pressure regulator shall be installed in the line to the illuminating appliance. For multiple installations, one regulator of adequate capacity shall be permitted to serve more than one illuminating appliance.

SECTION G2451 (630)
INFRARED RADIANT HEATERS

G2451.1 (630.1) General. Infrared radiant heaters shall be tested in accordance with ANSI Z 83.6 and shall be installed in accordance with the manufacturer's installation instructions.

G2451.2 (630.2) Support. Infrared radiant heaters shall be fixed in a position independent of gas and electric supply lines. Hangers and brackets shall be of noncombustible material.

SECTION G2452 (631)
BOILERS

G2452.1 (631.1) Standards. Boilers shall be listed in accordance with the requirements of ANSI Z21.13 or UL 795. If applicable, the boiler shall be designed and constructed in accordance with the requirements of ASME CSD-1 and as applicable, the ASME *Boiler and Pressure Vessel Code*, Sections I, II, IV, V and IX and NFPA 85.

G2452.2 (631.2) Installation. In addition to the requirements of this code, the installation of boilers shall be in accordance with the manufacturer's instructions and this code. Operating instructions of a permanent type shall be attached to the boiler. Boilers shall have all controls set, adjusted and tested by the installer. A complete control diagram together with complete boiler operating instructions shall be furnished by the installer. The manufacturer's rating data and the nameplate shall be attached to the boiler.

G2452.3 (631.3) Clearance to combustible material. Clearances to combustible materials shall be in accordance with Section G2409.4.

SECTION G2453 (634)
CHIMNEY DAMPER OPENING AREA

G2453.1 (634.1) Free opening area of chimney dampers. Where an unlisted decorative appliance for installation in a vented fireplace is installed, the fireplace damper shall have a permanent free opening equal to or greater than specified in Table G2453.1.

TABLE G2453.1 (634.1)
FREE OPENING AREA OF CHIMNEY DAMPER FOR VENTING FLUE GASES
FROM UNLISTED DECORATIVE APPLIANCES FOR INSTALLATION IN VENTED FIREPLACES

CHIMNEY HEIGHT (feet)	MINIMUM PERMANENT FREE OPENING (square inches)[a]						
	8	13	20	29	39	51	64
	Appliance input rating (Btu per hour)						
6	7,800	14,000	23,200	34,000	46,400	62,400	80,000
8	8,400	15,200	25,200	37,000	50,400	68,000	86,000
10	9,000	16,800	27,600	40,400	55,800	74,400	96,400
15	9,800	18,200	30,200	44,600	62,400	84,000	108,800
20	10,600	20,200	32,600	50,400	68,400	94,000	122,200
30	11,200	21,600	36,600	55,200	76,800	105,800	138,600

For SI: 1 foot = 304.8 mm, 1 square inch = 645.16 mm^2, 1,000 Btu per hour = 0.293 kW.

a. The first six minimum permanent free openings (8 square inches to 51 square inches) correspond approximately to the cross-sectional areas of chimneys having diameters of 3 inches through 8 inches, respectively. The 64-square inch opening corresponds to the cross-sectional area of standard 8-inch by 8-inch chimney tile.

CHAPTER 25

PLUMBING ADMINISTRATION

SECTION P2501
GENERAL

P2501.1 Scope. The provisions of this chapter shall establish the general administrative requirements applicable to plumbing systems and inspection requirements of this code.

P2501.2 Application. In addition to the general administration requirements of Chapter 1, the administrative provisions of this chapter shall also apply to the plumbing requirements of Chapters 25 through 32.

SECTION P2502
EXISTING PLUMBING SYSTEMS

P2502.1 Existing building sewers and drains. Existing building sewers and drains shall be used in connection with new systems when found by examination and/or test to conform to the requirements prescribed by this document.

P2502.2 Additions, alterations or repairs. Additions, alterations, renovations or repairs to any plumbing system shall conform to that required for a new plumbing system without requiring the existing plumbing system to comply with all the requirements of this code. Additions, alterations or repairs shall not cause an existing system to become unsafe, insanitary or overloaded.

Minor additions, alterations, renovations and repairs to existing plumbing systems shall be permitted in the same manner and arrangement as in the existing system, provided that such repairs or replacement are not hazardous and are approved.

SECTION P2503
INSPECTION AND TESTS

P2503.1 Inspection required. New plumbing work and parts of existing systems affected by new work or alterations shall be inspected by the building official to ensure compliance with the requirements of this code.

P2503.2 Concealment. A plumbing or drainage system, or part thereof, shall not be covered, concealed or put into use until it has been tested, inspected and approved by the building official.

P2503.3 Responsibility of permittee. Test equipment, materials and labor shall be furnished by the permittee.

P2503.4 Building sewer testing. The building sewer shall be tested by insertion of a test plug at the point of connection with the public sewer and filling the building sewer with water, testing with not less than a 10-foot (3048 mm) head of water and be able to maintain such pressure for 15 minutes.

P2503.5 DWV systems testing. Rough and finished plumbing installations shall be tested in accordance with Sections P2503.5.1 and P2503.5.2.

P2503.5.1 Rough plumbing. DWV systems shall be tested on completion of the rough piping installation by water or air with no evidence of leakage. Either test shall be applied to the drainage system in its entirety or in sections after rough piping has been installed, as follows:

1. Water test. Each section shall be filled with water to a point not less than 10 feet (3048 mm) above the highest fitting connection in that section, or to the highest point in the completed system. Water shall be held in the section under test for a period of 15 minutes. The system shall prove leak free by visual inspection.

2. Air test. The portion under test shall be maintained at a gauge pressure of 5 pounds per square inch (psi) (34 kPa) or 10 inches of mercury column (34 kPa). This pressure shall be held without introduction of additional air for a period of 15 minutes.

P2503.5.2 Finished plumbing. After the plumbing fixtures have been set and their traps filled with water, their connections shall be tested and proved gas tight and/or water tight as follows:

1. Water tightness. Each fixture shall be filled and then drained. Traps and fixture connections shall be proven water tight by visual inspection.

2. Gas tightness. When required by the local administrative authority, a final test for gas tightness of the DWV system shall be made by the smoke or peppermint test as follows:

   2.1. Smoke test. Introduce a pungent, thick smoke into the system. When the smoke appears at vent terminals, such terminals shall be sealed and a pressure equivalent to a 1-inch water column (249 Pa) shall be applied and maintained for a test period of not less than 15 minutes.

   2.2. Peppermint test. Introduce 2 ounces (59 mL) of oil of peppermint into the system. Add 10 quarts (9464 mL) of hot water and seal all vent terminals. The odor of peppermint shall not be detected at any trap or other point in the system.

P2503.6 Water-supply system testing. Upon completion of the water-supply system or a section of it, the system or portion completed shall be tested and proved tight under a water pressure of not less than the working pressure of the system or, for piping systems other than plastic, by an air test of not less than

50 psi (345 kPa). This pressure shall be held for not less than 15 minutes. The water used for tests shall be obtained from a potable water source.

P2503.7 Inspection and testing of backflow prevention devices. Inspection and testing of backflow prevention devices shall comply with Sections P2503.7.1 and P2503.7.2.

P2503.7.1 Inspections. Inspections shall be made of all backflow prevention assemblies to determine whether they are operable.

P2503.7.2 Testing. Reduced pressure principle backflow preventers, double check valve assemblies, double-detector check valve assemblies and pressure vacuum breaker assemblies shall be tested at the time of installation, immediately after repairs or relocation and at least annually.

P2503.8 Test gauges. Gauges used for testing shall be as follows:

1. Tests requiring a pressure of 10 psi or less shall utilize a testing gauge having increments of 0.10 psi (0.69 kPa) or less.

2. Tests requiring a pressure higher than 10 psi (0.69 kPa) but less than or equal to 100 psi (690 kPa) shall use a testing gauge having increments of 1 psi (6.9 kPa) or less.

3. Tests requiring a pressure higher than 100 psi (690 kPa) shall use a testing gauge having increments of 2 psi (14 kPa) or less.

CHAPTER 26

GENERAL PLUMBING REQUIREMENTS

SECTION P2601
GENERAL

P2601.1 Scope. The provisions of this chapter shall govern the installation of plumbing not specifically covered in other chapters applicable to plumbing systems. The installation of plumbing, appliances, equipment and systems not addressed by this code shall comply with the applicable provisions of the *International Plumbing Code*.

P2601.2 Connection. Plumbing fixtures, drains and appliances used to receive or discharge liquid wastes or sewage shall be connected to the sanitary drainage system of the building or premises in accordance with the requirements of this code. This section shall not be construed to prevent indirect waste systems.

P2601.3 Flood hazard area. In areas prone to flooding as established by Table R301.2(1), plumbing fixtures, drains, and appliances shall be located or installed in accordance with Section R324.1.5.

SECTION P2602
INDIVIDUAL WATER SUPPLY AND SEWAGE DISPOSAL

P2602.1 General. The water-distribution and drainage system of any building or premises where plumbing fixtures are installed shall be connected to a public water supply or sewer system, respectively, if available. When either a public water-supply or sewer system, or both, are not available, or connection to them is not feasible, an individual water supply or individual (private) sewage-disposal system, or both, shall be provided.

P2602.2 Flood-resistant installation. In areas prone to flooding as established by Table R301.2(1):

1. Water supply systems shall be designed and constructed to prevent infiltration of floodwaters.

2. Pipes for sewage disposal systems shall be designed and constructed to prevent infiltration of floodwaters into the systems and discharges from the systems into floodwaters.

SECTION P2603
STRUCTURAL AND PIPING PROTECTION

P2603.1 General. In the process of installing or repairing any part of a plumbing and drainage installation, the finished floors, walls, ceilings, tile work or any other part of the building or premises that must be changed or replaced shall be left in a safe structural condition in accordance with the requirements of the building portion of this code.

P2603.2 Drilling and notching. Wood-framed structural members shall not be drilled, notched or altered in any manner except as provided in Sections R502.8, R602.5, R602.6,

R802.7 and R802.7.1. Holes in cold-formed steel-framed load-bearing members shall be permitted only in accordance with Sections R505.2, R603.2 and R804.2. In accordance with the provisions of Sections R603.3.4 and R804.3.5 cutting and notching of flanges and lips of cold-formed steel-framed load-bearing members shall not be permitted.

P2603.2.1 Protection against physical damage. In concealed locations, where piping, other than cast-iron or galvanized steel, is installed through holes or notches in studs, joists, rafters or similar members less than 1.5 inches (38 mm) from the nearest edge of the member, the pipe shall be protected by shield plates. Protective shield plates shall be a minimum of 0.062-inch-thick (1.6 mm) steel, shall cover the area of the pipe where the member is notched or bored and shall extend a minimum of 2 inches (51 mm) above sole plates and below top plates.

P2603.3 Breakage and corrosion. Pipes passing through or under walls shall be protected from breakage. Pipes passing through concrete or cinder walls and floors, cold-formed steel framing or other corrosive material shall be protected against external corrosion by a protective sheathing or wrapping or other means that will withstand any reaction from lime and acid of concrete, cinder or other corrosive material. Sheathing or wrapping shall allow for expansion and contraction of piping to prevent any rubbing action. Minimum wall thickness of material shall be 0.025 inch (0.64 mm).

P2603.4 Sleeves. Annular spaces between sleeves and pipes shall be filled or tightly caulked as approved by the building official. Annular spaces between sleeves and pipes in fire-rated assemblies shall be filled or tightly caulked in accordance with the building portion of this code.

P2603.5 Pipes through footings or foundation walls. Any pipe that passes under a footing or through a foundation wall shall be provided with a relieving arch; or there shall be built into the masonry wall a pipe sleeve two pipe sizes greater than the pipe passing through.

P2603.6 Freezing. In localities having a winter design temperature of 32°F (0°C) or lower as shown in Table R301.2(1) of this code, a water, soil or waste pipe shall not be installed outside of a building, in exterior walls, in attics or crawl spaces, or in any other place subjected to freezing temperature unless adequate provision is made to protect it from freezing by insulation or heat or both. Water service pipe shall be installed not less than 12 inches (305 mm) deep and not less than 6 inches (152 mm) below the frost line.

P2603.6.1 Sewer depth. Building sewers that connect to private sewage disposal systems shall be a minimum of [NUMBER] inches (mm) below finished grade at the point of septic tank connection. Building sewers shall be a minimum of [NUMBER] inches (mm) below grade.

SECTION P2604
TRENCHING AND BACKFILLING

P2604.1 Trenching and bedding. Where trenches are excavated such that the bottom of the trench forms the bed for the pipe, solid and continuous load-bearing support shall be provided between joints. Where over-excavated, the trench shall be backfilled to the proper grade with compacted earth, sand, fine gravel or similar granular material. Piping shall not be supported on rocks or blocks at any point. Rocky or unstable soil shall be over-excavated by two or more pipe diameters and brought to the proper grade with suitable compacted granular material.

P2604.2 Common trench. See Section P2904.4.2.

P2604.3 Backfilling. Backfill shall be free from discarded construction material and debris. Backfill shall be free from rocks, broken concrete and frozen chunks until the pipe is covered by at least 12 inches (305 mm) of tamped earth. Backfill shall be placed evenly on both sides of the pipe and tamped to retain proper alignment. Loose earth shall be carefully placed in the trench in 6-inch (152 mm) layers and tamped in place.

P2604.4 Protection of footings. Trenching installed parallel to footings shall not extend below the 45-degree (0.79 rad) bearing plane of the bottom edge of a wall or footing (see Figure P2604.4).

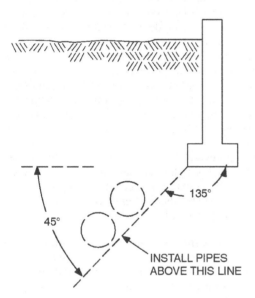

For SI: 1 degree = 0.018 rad.

FIGURE P2604.4
PIPE LOCATION WITH RESPECT TO FOOTINGS

SECTION P2605
SUPPORT

P2605.1 General. Piping shall be supported in accordance with the following:

1. Piping shall be supported to ensure alignment and prevent sagging, and allow movement associated with the expansion and contraction of the piping system.

2. Piping in the ground shall be laid on a firm bed for its entire length, except where support is otherwise provided.

3. Hangers and anchors shall be of sufficient strength to maintain their proportional share of the weight of pipe and contents and of sufficient width to prevent distortion to the pipe. Hangers and strapping shall be of approved material that will not promote galvanic action. Rigid support sway bracing shall be provided at changes in direction greater than 45 degrees (0.79 rad) for pipe sizes 4 inches (102 mm) and larger.

4. Piping shall be supported at distances not to exceed those indicated in Table P2605.1.

SECTION P2606
WATERPROOFING OF OPENINGS

P2606.1 General. Roof and exterior wall penetrations shall be made water tight. Joints at the roof, around vent pipes, shall be made water tight by the use of lead, copper or galvanized iron flashings or an approved elastomeric material. Counterflashing shall not restrict the required internal cross-sectional area of any vent.

SECTION P2607
WORKMANSHIP

P2607.1 General. Valves, pipes and fittings shall be installed in correct relationship to the direction of the flow. Burred ends shall be reamed to the full bore of the pipe.

SECTION P2608
MATERIALS EVALUATION AND LISTING

P2608.1 Identification. Each length of pipe and each pipe fitting, trap, fixture, material and device used in a plumbing system shall bear the identification of the manufacturer.

P2608.2 Installation of materials. All materials used shall be installed in strict accordance with the standards under which the materials are accepted and approved. In the absence of such installation procedures, the manufacturer's installation instructions shall be followed. Where the requirements of referenced standards or manufacturer's installation instructions do not conform to the minimum provisions of this code, the provisions of this code shall apply.

P2608.3 Plastic pipe, fittings and components. All plastic pipe, fittings and components shall be third-party certified as conforming to NSF 14.

P2608.4 Third-party testing and certification. All plumbing products and materials shall comply with the referenced standards, specifications and performance criteria of this code and shall be identified in accordance with Section P2608.1. Where required by Table P2608.4, plumbing products and materials shall either be tested by an approved third-party testing agency or certified by an approved third-party certification agency.

P2608.5 Water supply systems. Water service pipes, water distribution pipes and the necessary connecting pipes, fittings, control valves, faucets and all appurtenances used to dispense water intended for human ingestion shall be evaluated and listed as conforming to the requirements of NSF 61.

TABLE P2605.1
PIPING SUPPORT

PIPING MATERIAL	MAXIMUM HORIZONTAL SPACING (feet)	MAXIMUM VERTICAL SPACING
ABS pipe	4	10[b]
Aluminum tubing	10	15
Brass pipe	10	10
Cast-iron pipe	5[a]	15
Copper or copper alloy pipe	12	10
Copper or copper alloy tubing ($1^1/_4$ inch diameter and smaller)	6	10
Copper or copper alloy tubing ($1^1/_2$ inch diameter and larger)	10	10
Cross-linked polyethylene (PEX) pipe	2.67 (32 inches)	10[b]
Cross-linked polyethylene/aluminum/cross-linked polyethylene (PEX-AL-PEX) pipe	2.67 (32 inches)	4[b]
CPVC pipe or tubing (1 inch in diameter and smaller)	3	10[b]
CPVC pipe or tubing ($1^1/_4$ inch in diameter and larger)	4	10[b]
Lead pipe	Continuous	4
PB pipe or tubing	2.67 (32 inches)	4
Polyethylene/aluminum/polyethylene (PE-AL-PE) pipe	2.67 (32 inches)	4[b]
Polypropylene (PP) pipe or tubing 1 inch and smaller	2.67 (32 inches)	10[b]
Polypropylene (PP) pipe or tubing, $1^1/_4$ inches and larger	4	10[b]
PVC pipe	4	10[b]
Stainless steel drainage systems	10	10[b]
Steel pipe	12	15

For SI: 1 inch = 25.4 mm, 1 foot = 304.8 mm.

a. The maximum horizontal spacing of cast-iron pipe hangers shall be increased to 10 feet where 10-foot lengths of pipe are installed.

b. Midstory guide for sizes 2 inches and smaller.

TABLE P2608.4
PRODUCTS AND MATERIALS REQUIRING THIRD-PARTY TESTING AND THIRD-PARTY CERTIFICATION

PRODUCT OR MATERIAL	THIRD-PARTY CERTIFIED	THIRD-PARTY TESTED
Backflow prevention devices	Required	—
Plumbing appliance	Required	—
Plumbing fixtures	—	Required
Potable water supply system components and potable water fixture fittings	Required	—
Sanitary drainage and vent system components	Plastic pipe, fittings, and pipe related components	All others
Special waste system components	—	Required
Storm drainage system components	Plastic pipe, fittings, and pipe related components	All others
Subsoil drainage system components	—	Required
Waste fixture fittings	Plastic pipe, fittings, and pipe related components	All others
Water distribution system safety devices	Required	—

CHAPTER 27

PLUMBING FIXTURES

SECTION P2701
FIXTURES, FAUCETS AND FIXTURE FITTINGS

P2701.1 Quality of fixtures. Plumbing fixtures, faucets and fixture fittings shall be constructed of approved materials, shall have smooth impervious surfaces, shall be free from defects and concealed fouling surfaces, and shall conform to the standards cited in this code. Plumbing fixtures shall be provided with an adequate supply of potable water to flush and keep the fixtures in a clean and sanitary condition without danger of backflow or cross connection.

SECTION P2702
FIXTURE ACCESSORIES

P2702.1 Plumbing fixtures. Plumbing fixtures, other than water closets, shall be provided with approved strainers.

P2702.2 Waste fittings. Waste fittings shall conform to ASME A112.18.2, ASTM F 409, CSA B125 or to one of the standards listed in Table P3002.1(1) for above-ground drainage and vent pipe and fittings.

P2702.3 Plastic tubular fittings. Plastic tubular fittings shall conform to ASTM F 409 listed in Table P2701.1.

P2702.4 Carriers for wall-hung water closets. Carriers for wall-hung water closets shall conform to ASME A112.6.1 or ASME A112.6.2.

SECTION P2703
TAIL PIECES

P2703.1 Minimum size. Fixture tail pieces shall be not less than $1^1/_2$ inches (38 mm) in diameter for sinks, dishwashers, laundry tubs, bathtubs and similar fixtures, and not less than $1^1/_4$ inches (32 mm) in diameter for bidets, lavatories and similar fixtures.

SECTION P2704
ACCESS TO CONNECTIONS

P2704.1 General. Slip joints shall be made with an approved elastomeric gasket and shall be installed only on the trap outlet, trap inlet and within the trap seal. Fixtures with concealed slip-joint connections shall be provided with an access panel or utility space at least 12 inches (305 mm) in its smallest dimension or other approved arrangement so as to provide access to the slip connections for inspection and repair.

SECTION P2705
INSTALLATION

P2705.1 General. The installation of fixtures shall conform to the following:

1. Floor-outlet or floor-mounted fixtures shall be secured to the drainage connection and to the floor, where so designed, by screws, bolts, washers, nuts and similar fas-

teners of copper, brass or other corrosion-resistant material.

2. Wall-hung fixtures shall be rigidly supported so that strain is not transmitted to the plumbing system.

3. Where fixtures come in contact with walls and floors, the contact area shall be water tight.

4. Plumbing fixtures shall be usable.

5. The centerline of water closets or bidets shall not be less than 15 inches (381 mm) from adjacent walls or partitions or not less than 15 inches (381 mm) from the centerline of a bidet to the outermost rim of an adjacent water closet. There shall be at least 21 inches (533 mm) clearance in front of the water closet, bidet or lavatory to any wall, fixture or door.

6. The location of piping, fixtures or equipment shall not interfere with the operation of windows or doors.

7. In areas prone to flooding as established by Table R301.2(1), plumbing fixtures shall be located or installed in accordance with Section R324.1.5.

8. Integral fixture-fitting mounting surfaces on manufactured plumbing fixtures or plumbing fixtures constructed on site, shall meet the design requirements of ASME A112.19.2 or ASME A112.19.3.

SECTION P2706
WASTE RECEPTORS

P2706.1 General. Every waste receptor shall be of an approved type. Plumbing fixtures or other receptors receiving the discharge of indirect waste pipes shall be shaped and have a capacity to prevent splashing or flooding and shall be readily accessible for inspection and cleaning. Waste receptors and standpipes shall be trapped and vented and shall connect to the building drainage system. A removable strainer or basket shall cover the waste outlet of waste receptors. Waste receptors shall be installed in ventilated spaces. Waste receptors shall not be installed in bathrooms or in any inaccessible or unventilated space such as a closet. Ready access shall be provided to waste receptors.

Exception: Open hub waste receptors shall be permitted in the form of a hub or pipe extending not less than 1 inch (25 mm) above a water-impervious floor, and are not required to have a strainer.

P2706.2 Standpipes. Standpipes shall extend a minimum of 18 inches (457 mm) and a maximum of 42 inches (1067 mm) above the trap weir. Access shall be provided to all standpipe traps and drains for rodding.

P2706.2.1 Laundry tray connection. A laundry tray waste line is permitted to connect into a standpipe for the automatic clothes washer drain. The standpipes shall not be less than 30 inches (762 mm) as measured from the crown weir. The outlet of the laundry tray shall be a maximum horizontal distance of 30 inches (762 mm) from the standpipe trap.

TABLE P2701.1
PLUMBING FIXTURES, FAUCETS AND FIXTURE FITTINGS

MATERIAL	STANDARD
Air gap fittings for use with plumbing fixtures, appliances and appurtenances	ASME A112.1.3
Bathtub/whirlpool pressure-sealed doors	ASME A112.19.15
Diverters for faucets with hose spray anti-syphon type, residential application	ASSE 1025
Enameled cast-iron plumbing fixtures	ASME A112.19.1M, CSA B45.2
Floor drains	ASME A112.6.3
Floor-affixed supports for off-the-floor plumbing fixtures for public use	ASME A112.6.1M
Framing-affixed supports for off-the-floor water closets with concealed tanks	ASME A112.6.2
Handheld showers	ASSE 1014
Home laundry equipment	ASSE 1007
Hose connection vacuum breaker	ASSE 1052
Hot water dispensers, household storage type, electrical	ASSE 1023
Household dishwashing machines	ASSE 1006
Household disposers	ASSE 1008
Hydraulic performance for water closets and urinals	ASME A112.19.2
Individual pressure balancing valves for individual fixture fittings	ASSE 1066
Individual shower control valves anti-scald	ASSE 1016, CSA B125
Macerating toilet systems and related components	ASME A112.3.4
Nonvitreous ceramic plumbing fixtures	ASME A112.19.9M, CSA B45.1
Plastic bathtub units	ANSI Z124.1, CSA B45.1
Plastic lavatories	ANSI Z124.3, CSA B45.5
Plastic shower receptors and shower stall	ANSI Z124.2, CSA B45.5
Plastic sinks	ANSI Z124.6, CSA B45.5
Plastic water closet bowls and tanks	ANSI Z124.4, CSA B45.5
Plumbing fixture fittings	ASME A112.18.1M, CSA B125
Plumbing fixture waste fittings	ASME A112.18.2, ASTM F 409, CSA B125
Porcelain-enameled formed steel plumbing fixtures	ASME A112.19.4M, CSA B45.3
Pressurized flushing devices for plumbing fixtures	ASSE 1037
Specification for copper sheet and strip for building construction	ASTM B 370
Stainless steel plumbing fixtures (residential)	ASME A112.19.3M, CSA B45.4
Suction fittings for use in swimming pools, wading pools, spas, hot tubs and whirlpool bathtub appliances	ASME A112.19.8M
Temperature-actuated, flow reduction valves to individual fixture fittings	ASSE 1062
Thermoplastic accessible and replaceable plastic tube and tubular fittings	ASTM F 409
Trench drains	ASME A112.6.3
Trim for water closet bowls, tanks and urinals	ASME A112.19.5
Vacuum breaker wall hydrant—frost-resistant, automatic-draining type	ASSE 1019
Vitreous china plumbing fixtures	ASME A112.19.2M
Wall-mounted and pedestal-mounted, adjustable and pivoting lavatory and sink carrier systems	ASME A112.19.12
Water closet flush tank fill valves	ASSE 1002, CSA B125
Whirlpool bathtub appliances	ASME A112.19.7M

P2706.3 Prohibited waste receptors. Plumbing fixtures that are used for domestic or culinary purposes shall not be used to receive the discharge of indirect waste piping.

Exceptions:

1. A kitchen sink trap is acceptable for use as a receptor for a dishwasher.

2. A laundry tray is acceptable for use as a receptor for a clothes washing machine.

SECTION P2707
DIRECTIONAL FITTINGS

P2707.1 Directional fitting required. Approved directional-type branch fittings shall be installed in fixture tailpieces receiving the discharge from food waste disposal units or dishwashers.

SECTION P2708
SHOWERS

P2708.1 General. Shower compartments shall have at least 900 square inches (0.6 m²) of interior cross-sectional area. Shower compartments shall be not less than 30 inches (762 mm) in minimum dimension measured from the finished interior dimension of the shower compartment, exclusive of fixture valves, shower heads, soap dishes, and safety grab bars or rails. The minimum required area and dimension shall be measured from the finished interior dimension at a height equal to the top of the threshold and at a point tangent to its centerline and shall be continued to a height of not less than 70 inches (1778 mm) above the shower drain outlet. Hinged shower doors shall open outward. The wall area above built-in tubs having installed shower heads and in shower compartments shall be constructed in accordance with Section R702.4. Such walls shall form a water-tight joint with each other and with either the tub, receptor or shower floor.

Exceptions:

1. Fold-down seats shall be permitted in the shower, provided the required 900-square-inch (0.6 m²) dimension is maintained when the seat is in the folded-up position.

2. Shower compartments having not less than 25 inches (635 mm) in minimum dimension measured from the finished interior dimension of the compartment provided that the shower compartment has a minimum of 1,300 square inches (0.838 m²) of cross-sectional area.

P2708.1.1 Access. The shower compartment access and egress opening shall have a minimum clear and unobstructed finished width of 22 inches (559 mm).

P2708.2 Water-supply riser. The water supply riser from the shower valve to the shower head outlet shall be secured to the permanent structure.

P2708.3 Shower control valves. Individual shower and tub/shower combination valves shall be equipped with control valves of the pressure-balance, thermostatic-mixing or combination pressure-balance/thermostatic-mixing valve types with a high limit stop in accordance with ASSE 1016 or CSA B125. The high limit stop shall be set to limit water temperature to a maximum of 120°F (49°C). In-line thermostatic valves shall not be used for compliance with this section.

P2708.4 Hand showers. Hand-held showers shall conform to ASME A112.18.1 or CSA B125.1. Hand-held showers shall be provide backflow protection in accordance with ASME A112.18.1 or CSA B125.1 or shall be protected against backflow by a device complying with ASME A112.18.3.

SECTION P2709
SHOWER RECEPTORS

P2709.1 Construction. Shower receptors shall have a finished curb threshold not less than 1 inch (25 mm) below the sides and back of the receptor. The curb shall be not less than 2 inches (51 mm) and not more than 9 inches (229 mm) deep when measured from the top of the curb to the top of the drain. The finished floor shall slope uniformly toward the drain not less than $^1/_4$ unit vertical in 12 units horizontal (2-percent slope) nor more than $^1/_2$ inch (13 mm), and floor drains shall be flanged to provide a water-tight joint in the floor.

P2709.2 Lining required. The adjoining walls and floor framing enclosing on-site built-up shower receptors shall be lined with sheet lead, copper or a plastic liner material that complies with ASTM D 4068 or ASTM D 4551. The lining material shall extend not less than 3 inches (76 mm) beyond or around the rough jambs and not less than 3 inches (76 mm) above finished thresholds. Hot mopping shall be permitted in accordance with Section P2709.2.3.

P2709.2.1 PVC sheets. Plasticized polyvinyl chloride (PVC) sheets shall be a minimum of 0.040 inch (1 mm) thick, and shall meet the requirements of ASTM D 4551. Sheets shall be joined by solvent welding in accordance with the manufacturer's installation instructions.

P2709.2.2 Chlorinated polyethylene (CPE) sheets. Non-plasticized chlorinated polyethylene sheet shall be a minimum of 0.040 inch (1 mm) thick, and shall meet the requirements of ASTM D 4068. The liner shall be joined in accordance with the manufacturer's installation instructions.

P2709.2.3 Hot-mopping. Shower receptors lined by hot mopping shall be built-up with not less than three layers of standard grade Type 15 asphalt-impregnated roofing felt. The bottom layer shall be fitted to the formed subbase and each succeeding layer thoroughly hot-mopped to that below. All corners shall be carefully fitted and shall be made strong and water tight by folding or lapping, and each corner shall be reinforced with suitable webbing hot-mopped in place. All folds, laps and reinforcing webbing shall extend at least 4 inches (102 mm) in all directions from the corner and all webbing shall be of approved type and mesh, producing a tensile strength of not less than 50 pounds per inch (893 kg/m) in either direction.

P2709.3 Installation. Lining materials shall be pitched one-fourth unit vertical in 12 units horizontal (2-percent slope) to weep holes in the subdrain by means of a smooth, solidly formed subbase, shall be properly recessed and fastened to approved backing so as not to occupy the space required for the wall covering, and shall not be nailed or perforated at any point less than 1 inch (25.4 mm) above the finished threshold.

P2709.3.1 Materials. Lead and copper linings shall be insulated from conducting substances other than the connecting drain by 15-pound (6.80 kg) asphalt felt or its equivalent. Sheet lead liners shall weigh not less than 4 pounds per square foot (19.5 kg/m^2). Sheet copper liners shall weigh not less than 12 ounces per square foot (3.7 kg/m^2). Joints in lead and copper pans or liners shall be burned or silver brazed, respectively. Joints in plastic liner materials shall be jointed per the manufacturer's recommendations.

P2709.4 Receptor drains. An approved flanged drain shall be installed with shower subpans or linings. The flange shall be placed flush with the subbase and be equipped with a clamping ring or other device to make a water-tight connection between the lining and the drain. The flange shall have weep holes into the drain.

SECTION P2710
SHOWER WALLS

P2710.1 Bathtub and shower spaces. Shower walls shall be finished in accordance with Section R307.2.

SECTION P2711
LAVATORIES

P2711.1 Approval. Lavatories shall conform to ANSI Z124.3, ASME A112.19.1, ASME A112.19.2, ASME A112.19.3, ASME A112.19.4, ASME A112.19.9, CSA B45.1, CSA B45.2, CSA B45.3 or CSA B45.4.

P2711.2 Cultured marble lavatories. Cultured marble vanity tops with an integral lavatory shall conform to ANSI Z124.3 or CSA B45.5.

P2711.3 Lavatory waste outlets. Lavatories shall have waste outlets not less than 1$^1/_4$ inch (32 mm) in diameter. A strainer, pop-up stopper, crossbar or other device shall be provided to restrict the clear opening of the waste outlet.

P2711.4 Movable lavatory systems. Movable lavatory systems shall comply with ASME A112.19.12.

SECTION P2712
WATER CLOSETS

P2712.1 Approval. Water closets shall conform to the water consumption requirements of Section P2903.2 and shall conform to ANSI Z124.4, ASME A112.19.2, CSA B45.1, CSA B45.4 or CSA B45.5. Water closets shall conform to the hydraulic performance requirements of ASME A112.19.6. Water closets tanks shall conform to ANSI Z124.4, ASME A112.19.2, ASME A112.19.9, CSA B45.1, CSA B45.4 or CSA B45.5. Water closets that have an invisible seal and unventilated space or walls that are not thoroughly washed at each discharge shall be prohibited. Water closets that permit backflow of the contents of the bowl into the flush tank shall be prohibited.

P2712.2 Flushing devices required. Water closets shall be provided with a flush tank, flushometer tank or flushometer valve designed and installed to supply water in sufficient quantity and flow to flush the contents of the fixture, to cleanse the fixture and refill the fixture trap in accordance with ASME A112.19.2 and ASME A112.19.6.

P2712.3 Water supply for flushing devices. An adequate quantity of water shall be provided to flush and clean the fixture served. The water supply to flushing devices equipped for manual flushing shall be controlled by a float valve or other automatic device designed to refill the tank after each discharge and to completely shut off the water flow to the tank when the tank is filled to operational capacity. Provision shall be made to automatically supply water to the fixture so as to refill the trap after each flushing.

P2712.4 Flush valves in flush tanks. Flush valve seats in tanks for flushing water closets shall be at least 1 inch (25 mm) above the flood-level rim of the bowl connected thereto, except an approved water closet and flush tank combination designed so that when the tank is flushed and the fixture is clogged or partially clogged, the flush valve will close tightly so that water will not spill continuously over the rim of the bowl or backflow from the bowl to the tank.

P2712.5 Overflows in flush tanks. Flush tanks shall be provided with overflows discharging to the water closet connected thereto and such overflow shall be of sufficient size to prevent flooding the tank at the maximum rate at which the tanks are supplied with water according to the manufacturer's design conditions.

P2712.6 Access. All parts in a flush tank shall be accessible for repair and replacement.

P2712.7 Water closet seats. Water closets shall be equipped with seats of smooth, nonabsorbent material and shall be properly sized for the water closet bowl type.

P2712.8 Flush tank lining. Sheet copper used for flush tank linings shall have a minimum weight of 10 ounces per square foot (3 kg/m^2).

P2712.9 Electro-hydraulic water closets. Electro-hydraulic water closets shall conform to ASME A112.19.13.

SECTION P2713
BATHTUBS

P2713.1 Bathtub waste outlets and overflows. Bathtubs shall have outlets and overflows at least 1$^1/_2$ inches (38 mm) in diameter, and the waste outlet shall be equipped with an approved stopper.

P2713.2 Bathtub enclosures. Doors within a bathtub enclosure shall conform to ASME A112.19.15.

P2713.3 Bathtub and whirlpool bathtub valves. The hot water supplied to bathtubs and whirlpool bathtubs shall be limited to a maximum temperature of 120°F (49°C) by a water-temperature-limiting device that conforms to ASSE 1070, except where such protection is otherwise provided by a combination tub/shower valve in accordance with Section P2708.3.

SECTION P2714
SINKS

P2714.1 Sink waste outlets. Sinks shall be provided with waste outlets not less than $1^1/_2$ inches (38 mm) in diameter. A strainer, crossbar or other device shall be provided to restrict the clear opening of the waste outlet.

P2714.2 Movable sink systems. Movable sink systems shall comply with ASME A112.19.12.

SECTION P2715
LAUNDRY TUBS

P2715.1 Laundry tub waste outlet. Each compartment of a laundry tub shall be provided with a waste outlet not less than $1^1/_2$ inches (38 mm) in diameter and a strainer or crossbar to restrict the clear opening of the waste outlet.

SECTION P2716
FOOD WASTE GRINDER

P2716.1 Food waste grinder waste outlets. Food waste grinders shall be connected to a drain of not less than $1^1/_2$ inches (38 mm) in diameter.

P2716.2 Water supply required. Food waste grinders shall be provided with an adequate supply of water at a sufficient flow rate to ensure proper functioning of the unit.

SECTION P2717
DISHWASHING MACHINES

P2717.1 Protection of water supply. The water supply for dishwashers shall be protected by an air gap or integral backflow preventer.

P2717.2 Sink and dishwasher. A sink and dishwasher are permitted to discharge through a single $1^1/_2$-inch (38 mm) trap. The discharge pipe from the dishwasher shall be increased to a minimum of $^3/_4$ inch (19 mm) in diameter and shall be connected with a wye fitting to the sink tailpiece. The dishwasher waste line shall rise and be securely fastened to the underside of the counter before connecting to the sink tailpiece.

P2717.3 Sink, dishwasher and food grinder. The combined discharge from a sink, dishwasher, and waste grinder is permitted to discharge through a single $1^1/_2$ inch (38 mm) trap. The discharge pipe from the dishwasher shall be increased to a minimum of $^3/_4$ inch (19 mm) in diameter and shall connect with a wye fitting between the discharge of the food-waste grinder and the trap inlet or to the head of the food grinder. The dishwasher waste line shall rise and be securely fastened to the underside of the counter before connecting to the sink tail piece or the food grinder.

SECTION P2718
CLOTHES WASHING MACHINE

P2718.1 Waste connection. The discharge from a clothes washing machine shall be through an air break.

SECTION P2719
FLOOR DRAINS

P2719.1 Floor drains. Floor drains shall have waste outlets not less than 2 inches (51 mm) in diameter and shall be provided with a removable strainer. The floor drain shall be constructed so that the drain is capable of being cleaned. Access shall be provided to the drain inlet.

SECTION P2720
WHIRLPOOL BATHTUBS

P2720.1 Access to pump. Access shall be provided to circulation pumps in accordance with the fixture manufacturer's installation instructions. Where the manufacturer's instructions do not specify the location and minimum size of field fabricated access openings, a 12-inch by 12-inch (304 mm by 304 mm) minimum size opening shall be installed to provide access to the circulation pump. Where pumps are located more than 2 feet (609 mm) from the access opening, an 18-inch by 18-inch (457 mm by 457 mm) minimum size opening shall be installed. A door or panel shall be permitted to close the opening. In all cases, the access opening shall be unobstructed and be of the size necessary to permit the removal and replacement of the circulation pump.

P2720.2 Piping drainage. The circulation pump shall be accessibly located above the crown weir of the trap. The pump drain line shall be properly graded to ensure minimum water retention in the volute after fixture use. The circulation piping shall be installed to be self-draining.

P2720.3 Leak testing. Leak testing and pump operation shall be performed in accordance with the manufacturer's installation instructions.

P2720.4 Manufacturer's instructions. The product shall be installed in accordance with the manufacturer's installation instructions.

SECTION P2721
BIDET INSTALLATIONS

P2721.1 Water supply. The bidet shall be equipped with either an air-gap-type or vacuum-breaker-type fixture supply fitting.

P2721.2 Bidet water temperature. The discharge water temperature from a bidet fitting shall be limited to a maximum temperature of 110°F (43°C) by a water-temperature-limiting device conforming to ASSE 1070.

SECTION P2722
FIXTURE FITTING

P2722.1 General. Fixture supply valves and faucets shall comply with ASME A112.18.1 or CSA B125 as listed in Table P2701.1. Faucets and fixture fittings that supply drinking water for human ingestion shall conform to the requirements of NSF 61, Section 9. Flexible water connectors shall conform to the requirements of Section P2904.7.

P2722.2 Hot water. Fixture fittings and faucets that are supplied with both hot and cold water shall be installed and adjusted so that the left-hand side of the water temperature control represents the flow of hot water when facing the outlet.

Exception: Shower and tub/shower mixing valves conforming to ASSE 1016 or CSA B125, where the water temperature control corresponds to the markings on the device.

P2722.3 Hose-connected outlets. Faucets and fixture fittings with hose-connected outlets shall conform to ASME A112.18.3 or CSA B125.

P2722.4 Individual pressure-balancing in-line valves for individual fixture fittings. Where individual pressure-balancing in-line valves for individual fixture fittings are installed, the valves shall comply with ASSE 1066. Such valves shall be installed in an accessible location and shall not be used alone as a substitute for the balanced pressure, thermostatic or combination shower valves required in Section P2708.3.

SECTION P2723
MACERATING TOILET SYSTEMS

P2723.1 General. Macerating toilet systems shall be installed in accordance with manufacturer's installation instructions.

P2723.2 Drain. The minimum size of the drain from the macerating toilet system shall be $^3/_4$ inch (19 mm) in diameter.

SECTION P2724
SPECIALTY TEMPERATURE CONTROL
DEVICES AND VALVES

P2724.1 Temperature-actuated, flow-reduction devices for individual fixtures. Temperature-actuated, flow-reduction devices, where installed for individual fixture fittings, shall conform to ASSE 1062. Such valves shall not be used alone as a substitute for the balanced pressure, thermostatic or combination shower valves required for showers in Section P2708.3.

CHAPTER 28

WATER HEATERS

SECTION P2801
GENERAL

P2801.1 Required. Each dwelling shall have an approved automatic water heater or other type of domestic water-heating system sufficient to supply hot water to plumbing fixtures and appliances intended for bathing, washing or culinary purposes. Storage tanks shall be constructed of noncorrosive metal or shall be lined with noncorrosive material.

P2801.2 Installation. Water heaters shall be installed in accordance with this chapter and Chapters 20 and 24.

P2801.3 Location. Water heaters and storage tanks shall be located and connected to provide access for observation, maintenance, servicing and replacement.

P2801.4 Prohibited locations. Water heaters shall be located in accordance with Chapter 20.

P2801.5 Required pan. Where water heaters or hot water storage tanks are installed in locations where leakage of the tanks or connections will cause damage, the tank or water heater shall be installed in a galvanized steel pan having a minimum thickness of 24 gage (0.016 inch) (0.4 mm) or other pans for such use. Listed pans shall comply with CSA LC3.

P2801.5.1 Pan size and drain. The pan shall be not less than $1^1/_2$ inches (38 mm) deep and shall be of sufficient size and shape to receive all dripping or condensate from the tank or water heater. The pan shall be drained by an indirect waste pipe having a minimum diameter of $^3/_4$ inch (19 mm). Piping for safety pan drains shall be of those materials listed in Table P2904.5.

P2801.5.2 Pan drain termination. The pan drain shall extend full-size and terminate over a suitably located indirect waste receptor or shall extend to the exterior of the building and terminate not less than 6 inches (152 mm) and not more than 24 inches (610 mm) above the adjacent ground surface.

P2801.6 Water heaters installed in garages. Water heaters having an ignition source shall be elevated such that the source of ignition is not less than 18 inches (457 mm) above the garage floor.

P2801.7 Water heater seismic bracing. In Seismic Design Categories D_0, D_1 and D_2 and townhouses in Seismic Design Category C, water heaters shall be anchored or strapped in the upper one-third and in the lower one-third of the appliance to resist a horizontal force equal to one-third of the operating weight of the water heater, acting in any horizontal direction, or in accordance with the appliance manufacturer's recommendations.

SECTION P2802
WATER HEATERS USED FOR SPACE HEATING

P2802.1 Protection of potable water. Piping and components connected to a water heater for space heating applications shall be suitable for use with potable water in accordance with Chapter 29. Water heaters that will be used to supply potable water shall not be connected to a heating system or components previously used with nonpotable-water heating appliances. Chemicals for boiler treatment shall not be introduced into the water heater.

P2802.2 Temperature control. Where a combination water heater-space heating system requires water for space heating at temperatures exceeding 140°F (60°C), a master thermostatic mixing valve complying with ASSE 1017 shall be installed to temper the water to a temperature of 140°F (60°C) or less for domestic uses.

SECTION P2803
RELIEF VALVES

P2803.1 Relief valves required. Appliances and equipment used for heating water or storing hot water shall be protected by:

1. A separate pressure-relief valve and a separate temperature-relief valve; or

2. A combination pressure- and temperature-relief valve.

P2803.2 Rating. Relief valves shall have a minimum rated capacity for the equipment served and shall conform to ANSI Z 21.22.

P2803.3 Pressure relief valves. Pressure-relief valves shall have a relief rating adequate to meet the pressure conditions for the appliances or equipment protected. In tanks, they shall be installed directly into a tank tapping or in a water line close to the tank. They shall be set to open at least 25 psi (172 kPa) above the system pressure but not over 150 psi (1034 kPa). The relief-valve setting shall not exceed the tanks rated working pressure.

P2803.4 Temperature relief valves. Temperature-relief valves shall have a relief rating compatible with the temperature conditions of the appliances or equipment protected. The valves shall be installed such that the temperature-sensing element monitors the water within the top 6 inches (152 mm) of the tank. The valve shall be set to open at a maximum temperature of 210°F (99°C).

P2803.5 Combination pressure-/temperature-relief valves. Combination pressure-/temperature-relief valves shall comply with all the requirements for separate pressure- and temperature-relief valves.

P2803.6 Installation of relief valves. A check or shutoff valve shall not be installed in the following locations:

1. Between a relief valve and the termination point of the relief valve discharge pipe;

2. Between a relief valve and a tank; or

3. Between a relief valve and heating appliances or equipment.

P2803.6.1 Requirements for discharge pipe. The discharge piping serving a pressure-relief valve, temperature-relief valve or combination valve shall:

1. Not be directly connected to the drainage system.

2. Discharge through an air gap located in the same room as the water heater.

3. Not be smaller than the diameter of the outlet of the valve served and shall discharge full size to the air gap.

4. Serve a single relief device and shall not connect to piping serving any other relief device or equipment.

5. Discharge to the floor, to an indirect waste receptor or to the outdoors. Where discharging to the outdoors in areas subject to freezing, discharge piping shall be first piped to an indirect waste receptor through an air gap located in a conditioned area.

6. Discharge in a manner that does not cause personal injury or structural damage.

7. Discharge to a termination point that is readily observable by the building occupants.

8. Not be trapped.

9. Be installed to flow by gravity.

10. Not terminate more than 6 inches (152 mm) above the floor or waste receptor.

11. Not have a threaded connection at the end of the piping.

12. Not have valves or tee fittings.

13. Be constructed of those materials listed in Section P2904.5 or materials tested, rated and approved for such use in accordance with ASME A112.4.1.

CHAPTER 29

WATER SUPPLY AND DISTRIBUTION

SECTION P2901
GENERAL

P2901.1 Potable water required. Dwelling units shall be supplied with potable water in the amounts and pressures specified in this chapter. In a building where both a potable and nonpotable water-distribution system are installed, each system shall be identified by color marking, metal tag or other appropriate method. Any nonpotable outlet that could inadvertently be used for drinking or domestic purposes shall be posted.

SECTION P2902
PROTECTION OF POTABLE WATER SUPPLY

P2902.1 General. A potable water supply system shall be designed and installed as to prevent contamination from nonpotable liquids, solids or gases being introduced into the potable water supply. Connections shall not be made to a potable water supply in a manner that could contaminate the water supply or provide a cross-connection between the supply and a source of contamination unless an approved backflow-prevention device is provided. Cross-connections between an individual water supply and a potable public water supply shall be prohibited.

P2902.2 Plumbing fixtures. The supply lines and fittings for every plumbing fixture shall be installed to prevent backflow. Plumbing fixture fittings shall provide backflow protection in accordance with ASME A112.18.1.

P2902.3 Backflow protection. A means of protection against backflow shall be provided in accordance with Sections P2902.3.1 through P2902.3.6. Backflow prevention applications shall conform to Table P2902.3, except as specifically stated in Sections P2902.4 through P2902.5.5.

P2902.3.1 Air gaps. Air gaps shall comply with ASME A112.1.2 and air gap fittings shall comply with ASME A112.1.3. The minimum air gap shall be measured vertically from the lowest end of a water supply outlet to the flood level rim of the fixture or receptor into which such potable water outlets discharge. The minimum required air gap shall be twice the diameter of the effective opening of the outlet, but in no case less than the values specified in Table P2902.3.1. An air gap is required at the discharge point of a relief valve or piping. Air gap devices shall be incorporated in dishwashing and clothes washing appliances.

P2902.3.2 Atmospheric-type vacuum breakers. Pipe-applied atmospheric-type vacuum breakers shall conform to ASSE 1001 or CSA B64.1.1. Hose-connection vacuum breakers shall conform to ASSE 1011, ASSE 1019, ASSE 1035, ASSE 1052, CSA B64.2, CSA B64.2.1, CSA B64.2.1.1, CSA B64.2.2 or CSA B64.7. These devices shall operate under normal atmospheric pressure when the critical level is installed at the required height.

P2902.3.3 Backflow preventer with intermediate atmospheric vent. Backflow preventers with intermediate atmospheric vents shall conform to ASSE 1012 or CSA CAN/CSA B64.3. These devices shall be permitted to be installed where subject to continuous pressure conditions. The relief opening shall discharge by air gap and shall be prevented from being submerged.

P2902.3.4 Pressure-type vacuum breakers. Pressure-type vacuum breakers shall conform to ASSE 1020 or CSA B64.1.2 and spillproof vacuum breakers shall comply with ASSE 1056. These devices are designed for installation under continuous pressure conditions when the critical level is installed at the required height. Pressure-type vacuum breakers shall not be installed in locations where spillage could cause damage to the structure.

P2902.3.5 Reduced pressure principle backflow preventers. Reduced pressure principle backflow preventers shall conform to ASSE 1013, AWWA C511, CSA B64.4 or CSA B64.4.1. Reduced pressure detector assembly backflow preventers shall conform to ASSE 1047. These devices shall be permitted to be installed where subject to continuous pressure conditions. The relief opening shall discharge by air gap and shall be prevented from being submerged.

P2902.3.6 Double check-valve assemblies. Double check-valve assemblies shall conform to ASSE 1015, CSA B64.5, CSA B64.5.1 or AWWA C510. Double-detector check-valve assemblies shall conform to ASSE 1048. These devices shall be capable of operating under continuous pressure conditions.

P2902.4 Protection of potable water outlets. Potable water openings and outlets shall be protected by an air gap, reduced pressure principle backflow preventer with atmospheric vent, atmospheric-type vacuum breaker, pressure-type vacuum breaker or hose connection backflow preventer.

P2902.4.1 Fill valves. Flush tanks shall be equipped with an antisiphon fill valve conforming to ASSE 1002 or CSA B125. The fill valve backflow preventer shall be located at least 1 inch (25 mm) above the full opening of the overflow pipe.

P2902.4.2 Deck-mounted and integral vacuum breakers. Approved deck-mounted vacuum breakers and faucets with integral atmospheric or spill-proof vacuum breakers shall be installed in accordance with the manufacturer's installation instructions and the requirements for labeling with the critical level not less than 1 inch (25 mm) above the flood level rim.

P2902.4.3 Hose connection. Sillcocks, hose bibbs, wall hydrants and other openings with a hose connection shall be protected by an atmospheric-type or pressure-type vacuum breaker or a permanently attached hose connection vacuum breaker.

TABLE P2902.3
APPLICATION FOR BACKFLOW PREVENTERS

DEVICE	DEGREE OF HAZARD[a]	APPLICATION[b]	APPLICABLE STANDARDS
Air gap	High or low hazard	Backsiphonage or backpressure	ASME A112.1.2
Air gap fittings for use with plumbing fixtures, appliances and appurtenances	High or low hazard	Backsiphonage or backpressure	ASME A112.1.3
Antisiphon-type fill valves for gravity water closet flush tanks	High hazard	Backsiphonage only	ASSE 1002 CSA CAN/CSA B125
Backflow preventer with intermediate atmospheric vents	Low hazard	Backpressure or backsiphonage Sizes $^1/_4'' - ^3/_4''$	ASSE 1012 CSA B64.3
Double check backflow prevention assembly and double check fire protection backflow prevention assembly	Low hazard	Backpressure or backsiphonage Sizes $^3/_8'' - 16''$	ASSE 1015, AWWA C510 CSA B64.5, CSA B64.5.1
Double check detector fire protection backflow prevention assemblies	Low hazard	Backpressure or backsiphonage (Fire sprinkler systems) Sizes $2'' - 16''$	ASSE 1048
Dual-check-valve-type backflow preventer	Low hazard	Backpressure or backsiphonage Sizes $^1/_4'' - 1''$	ASSE 1024, CSA B64.6
Hose connection backflow preventer	High or low hazard	Low head backpressure, rated working pressure backpressure or backsiphonage, Sizes $^1/_2'' - 1''$	ASSE 1052, CSA B64.2.1.1
Hose-connection vacuum breaker	High or low hazard	Low head backpressure or backsiphonage Sizes $^1/_2'', ^3/_4'', 1''$	ASSE 1011, CSA B64.2, CSA B64.2.1
Laboratory faucet backflow preventer	High or low hazard	Low head backpressure and backsiphonage	ASSE 1035, CSA B64.7
Pipe-applied atmospheric-type vacuum breaker	High or low hazard	Backsiphonage only Sizes $^1/_4'' - 4''$	ASSE 1001 CSA B64.1.1
Pressure vacuum breaker assembly	High or low hazard	Backsiphonage only Sizes $^1/_2'' - 2''$	ASSE 1020, CSA B64.1.2
Reduced pressure detector fire protection backflow prevention assemblies	High or low hazard	Backsiphonage or backpressure (Fire sprinkler systems)	ASSE 1047
Reduced pressure principle backflow preventer and reduced pressure principle fire protection backflow preventer	High or low hazard	Backpressure or backsiphonage Sizes $^3/_8'' - 16''$	ASSE` 1013, AWWA C511 CSA B64.4, CSA B64.4.1
Spillproof vacuum breaker	High or low hazard	Backsiphonage only Sizes $^1/_4'' - 2''$	ASSE 1056
Vacuum breaker wall hydrants, frost-resistant, automatic draining type	High or low hazard	Low head backpressure or backsiphonage Sizes $^3/_4'' - 1''$	ASSE 1019, CSA B64.2.2

For SI: 1 inch = 25.4 mm.

a. Low hazard—See Pollution (Section 202). High hazard—See Contamination (Section 202).

b. See Backpressure (Section 202). See Backpressure, Low Head (Section 202). See Backsiphonage (Section 202).

Exceptions:

1. This section shall not apply to water heater and boiler drain valves that are provided with hose connection threads and that are intended only for tank or vessel draining.

2. This section shall not apply to water supply valves intended for connection of clothes washing machines where backflow prevention is otherwise provided or is integral with the machine.

P2902.5 Protection of potable water connections. Connections to the potable water shall conform to Sections P2902.5.1 through P2902.5.5.

P2902.5.1 Connections to boilers. The potable supply to the boiler shall be equipped with a backflow preventer with an intermediate atmospheric vent complying with ASSE 1012 or CSA B64.3. Where conditioning chemicals are introduced into the system, the potable water connection shall be protected by an air gap or a reduced pressure principle backflow preventer complying with ASSE 1013, CSA B64.4 or AWWA C511.

TABLE P2902.3.1
MINIMUM AIR GAPS

FIXTURE	MINIMUM AIR GAP	
	Away from a wall[a] (inches)	Close to a wall (inches)
Effective openings greater than 1 inch	Two times the diameter of the effective opening	Three times the diameter of the effective opening
Lavatories and other fixtures with effective opening not greater than $^1/_2$ inch in diameter	1	1.5
Over-rim bath fillers and other fixtures with effective openings not greater than 1 inch in diameter	2	3
Sink, laundry trays, gooseneck back faucets and other fixtures with effective openings not greater than $^3/_4$ inch in diameter	1.5	2.5

For SI: 1 inch = 25.4 mm.

a. Applicable where walls or obstructions are spaced from the nearest inside edge of the spout opening a distance greater than three times the diameter of the effective opening for a single wall, or a distance greater than four times the diameter of the effective opening for two intersecting walls.

P2902.5.2 Heat exchangers. Heat exchangers using an essentially toxic transfer fluid shall be separated from the potable water by double-wall construction. An air gap open to the atmosphere shall be provided between the two walls. Heat exchangers utilizing an essentially nontoxic transfer fluid shall be permitted to be of single-wall construction.

P2902.5.3 Lawn irrigation systems. The potable water supply to lawn irrigation systems shall be protected against backflow by an atmospheric-type vacuum breaker, a pressure-type vacuum breaker or a reduced pressure principle backflow preventer. A valve shall not be installed downstream from an atmospheric vacuum breaker. Where chemicals are introduced into the system, the potable water supply shall be protected against backflow by a reduced pressure principle backflow preventer.

P2902.5.4 Connections to automatic fire sprinkler systems. The potable water supply to automatic fire sprinkler systems shall be protected against backflow by a double check-valve assembly or a reduced pressure principle backflow preventer.

> **Exception:** Where systems are installed as a portion of the water distribution system in accordance with the requirements of this code and are not provided with a fire department connection, isolation of the water supply system shall not be required.

P2902.5.4.1 Additives or nonpotable source. Where systems contain chemical additives or antifreeze, or where systems are connected to a nonpotable secondary water supply, the potable water supply shall be protected against backflow by a reduced pressure principle backflow preventer. Where chemical additives or antifreeze is added to only a portion of an automatic fire sprinkler or standpipe system, the reduced pressure principle backflow preventer shall be permitted to be located so as to isolate that portion of the system.

P2902.5.5 Solar systems. The potable water supply to a solar system shall be equipped with a backflow preventer with intermediate atmospheric vent complying with ASSE 1012 or a reduced pressure principle backflow preventer complying with ASSE 1013. Where chemicals are used, the potable water supply shall be protected by a reduced pressure principle backflow preventer.

> **Exception:** Where all solar system piping is a part of the potable water distribution system, in accordance with the requirements of the *International Plumbing Code*, and all components of the piping system are listed for potable water use, cross-connection protection measures shall not be required.

P2902.6 Access. Backflow prevention devices shall be accessible for inspection and servicing.

SECTION P2903
WATER-SUPPLY SYSTEM

P2903.1 Water supply system design criteria. The water service and water distribution systems shall be designed and pipe sizes shall be selected such that under conditions of peak demand, the capacities at the point of outlet discharge shall not be less than shown in Table P2903.1.

TABLE P2903.1
REQUIRED CAPACITIES AT
POINT OF OUTLET DISCHARGE

FIXTURE AT POINT OF OUTLET	FLOW RATE (gpm)	FLOW PRESSURE (psi)
Bathtub	4	8
Bidet	2	4
Dishwasher	2.75	8
Laundry tub	4	8
Lavatory	2	8
Shower	3	8
Shower, temperature controlled	3	20
Sillcock, hose bibb	5	8
Sink	2.5	8
Water closet, flushometer tank	1.6	15
Water closet, tank, close coupled	3	8
Water closet, tank, one-piece	6	20

For SI: 1 gallon per minute = 3.785 L/m,
1 pound per square inch = 6.895 kPa.

P2903.2 Maximum flow and water consumption. The maximum water consumption flow rates and quantities for all plumbing fixtures and fixture fittings shall be in accordance with Table P2903.2.

TABLE P2903.2
MAXIMUM FLOW RATES AND CONSUMPTION FOR
PLUMBING FIXTURES AND FIXTURE FITTINGS[b]

PLUMBING FIXTURE OR FIXTURE FITTING	PLUMBING FIXTURE OR FIXTURE FITTING
Lavatory faucet	2.2 gpm at 60 psi
Shower head[a]	2.5 gpm at 80 psi
Sink faucet	2.2 gpm at 60 psi
Water closet	1.6 gallons per flushing cycle

For SI: 1 gallon per minute = 3.785 L/m,
1 pound per square inch = 6.895 kPa.
a. A handheld shower spray is also a shower head.
b. Consumption tolerances shall be determined from referenced standards.

P2903.3 Minimum pressure. Minimum static pressure (as determined by the local water authority) at the building entrance for either public or private water service shall be 40 psi (276 kPa).

P2903.3.1 Maximum pressure. Maximum static pressure shall be 80 psi (551 kPa). When main pressure exceeds 80 psi (551 kPa), an approved pressure-reducing valve conforming to ASSE 1003 shall be installed on the domestic water branch main or riser at the connection to the water-service pipe.

P2903.4 Thermal expansion control. A means for controlling increased pressure caused by thermal expansion shall be installed where required in accordance with Sections P2903.4.1 and P2903.4.2.

P2903.4.1 Pressure-reducing valve. For water service system sizes up to and including 2 inches (51 mm), a device for controlling pressure shall be installed where, because of thermal expansion, the pressure on the downstream side of a pressure-reducing valve exceeds the pressure-reducing valve setting.

P2903.4.2 Backflow prevention device or check valve. Where a backflow prevention device, check valve or other device is installed on a water supply system using storage water heating equipment such that thermal expansion causes an increase in pressure, a device for controlling pressure shall be installed.

P2903.5 Water hammer. The flow velocity of the water distribution system shall be controlled to reduce the possibility of water hammer. A water-hammer arrestor shall be installed where quick-closing valves are used. Water-hammer arrestors shall be installed in accordance with manufacturers' specifications. Water-hammer arrestors shall conform to ASSE 1010.

P2903.6 Determining water-supply fixture units. Supply loads in the building water-distribution system shall be determined by total load on the pipe being sized, in terms of water-supply fixture units (w.s.f.u.), as shown in Table P2903.6, and gallon per minute (gpm) flow rates [see Table P2903.6(1)]. For fixtures not listed, choose a w.s.f.u. value of a fixture with similar flow characteristics.

P2903.7 Size of water-service mains, branch mains and risers. The minimum size water service pipe shall be $^3/_4$ inch (19 mm). The size of water service mains, branch mains and risers shall be determined according to water supply demand [gpm (L/m)], available water pressure [psi (kPa)] and friction loss caused by the water meter and developed length of pipe [feet (m)], including equivalent length of fittings. The size of each water distribution system shall be determined according to the procedure outlined in this section or by other design methods conforming to acceptable engineering practice and approved by the administrative authority:

1. Obtain the minimum daily static service pressure [psi (kPa)] available (as determined by the local water authority) at the water meter or other source of supply at the installation location. Adjust this minimum daily static pressure [psi (kPa)] for the following conditions:

    1.1. Determine the difference in elevation between the source of supply and the highest water supply outlet. Where the highest water supply outlet is located above the source of supply, deduct 0.5 psi (3.4 kPa) for each foot (305 mm) of difference in elevation. Where the highest water supply outlet is located below the source of supply, add 0.5 psi (3.4 kPa) for each foot (305 mm) of difference in elevation.

    1.2. Where a water pressure reducing valve is installed in the water distribution system, the minimum daily static water pressure available is 80 percent of the minimum daily static water pressure at the source of supply or the set pressure downstream of the pressure reducing valve, whichever is smaller.

1.3. Deduct all pressure losses caused by special equipment such as a backflow preventer, water filter or water softener. Pressure loss data for each piece of equipment shall be obtained from the manufacturer of such devices.

1.4. Deduct the pressure in excess of 8 psi (55 kPa) caused by installation of special plumbing fixtures, such as temperature controlled showers and flushometer tank water closets.

Using the resulting minimum available pressure, find the corresponding pressure range in Table P2903.7.

2. The maximum developed length for water piping is the actual length of pipe between the source of supply and the most remote fixture, including either hot (through the water heater) or cold water branches multiplied by a factor of 1.2 to compensate for pressure loss through fittings.

Select the appropriate column in Table P2903.7 equal to or greater than the calculated maximum developed length.

3. To determine the size of water service pipe, meter and main distribution pipe to the building using the appropriate table, follow down the selected "maximum devel-

oped length" column to a fixture unit equal to, or greater than the total installation demand calculated by using the "combined" water supply fixture unit column of Table P2903.6. Read the water service pipe and meter sizes in the first left-hand column and the main distribution pipe to the building in the second left-hand column on the same row.

4. To determine the size of each water distribution pipe, start at the most remote outlet on each branch (either hot or cold branch) and, working back toward the main distribution pipe to the building, add up the water supply fixture unit demand passing through each segment of the distribution system using the related hot or cold column of Table P2903.6. Knowing demand, the size of each segment shall be read from the second left-hand column of the same table and a maximum developed length column selected in Steps 1 and 2, under the same or next smaller size meter row. In no case does the size of any branch or main need to be larger that the size of the main distribution pipe to the building established in Step 3.

P2903.8 Gridded and parallel water distribution system manifolds. Hot water and cold water manifolds installed with gridded or parallel-connected individual distribution lines to each fixture or fixture fittings shall be designed in accordance with Sections P2903.8.1 through P2903.8.6.

TABLE P2903.6
WATER-SUPPLY FIXTURE-UNIT VALUES FOR VARIOUS PLUMBING FIXTURES AND FIXTURE GROUPS

TYPE OF FIXTURES OR GROUP OF FIXTURES	WATER-SUPPLY FIXTURE-UNIT VALUE (w.s.f.u.)		
	Hot	Cold	Combined
Bathtub (with/without overhead shower head)	1.0	1.0	1.4
Clothes washer	1.0	1.0	1.4
Dishwasher	1.4	—	1.4
Full-bath group with bathtub (with/without shower head) or shower stall	1.5	2.7	3.6
Half-bath group (water closet and lavatory)	0.5	2.5	2.6
Hose bibb (sillcock)[a]	—	2.5	2.5
Kitchen group (dishwasher and sink with/without garbage grinder)	1.9	1.0	2.5
Kitchen sink	1.0	1.0	1.4
Laundry group (clothes washer standpipe and laundry tub)	1.8	1.8	2.5
Laundry tub	1.0	1.0	1.4
Lavatory	0.5	0.5	0.7
Shower stall	1.0	1.0	1.4
Water closet (tank type)	—	2.2	2.2

For SI: 1 gallon per minute = 3.785 L/m.

a. The fixture unit value 2.5 assumes a flow demand of 2.5 gpm, such as for an individual lawn sprinkler device. If a hose bibb/sill cock will be required to furnish a greater flow, the equivalent fixture-unit value may be obtained from this table or Table P2903.6(1).

P2903.8.1 Sizing of manifolds. Manifolds shall be sized in accordance with Table P2903.8.1. Total gallons per minute is the demand for all outlets.

P2903.8.2 Minimum size. Where the developed length of the distribution line is 60 feet (18 288 mm) or less, and the available pressure at the meter is a minimum of 40 pounds per square inch (276 kPa), the minimum size of individual distribution lines shall be $^3/_8$ inch (10 mm). Certain fixtures such as one-piece water closets and whirlpool bathtubs shall require a larger size where specified by the manufacturer. If a water heater is fed from the end of a cold water manifold, the manifold shall be one size larger than the water heater feed.

P2903.8.3 Orientation. Manifolds shall be permitted to be installed in a horizontal or vertical position.

P2903.8.4 Support and protection. Plastic piping bundles shall be secured in accordance with the manufacturer's installation instructions and supported in accordance with Section P2605. Bundles that have a change in direction equal to or greater than 45 degrees (0.79 rad) shall be protected from chafing at the point of contact with framing members by sleeving or wrapping.

TABLE P2903.6(1)
CONVERSIONS FROM WATER SUPPLY FIXTURE UNIT TO GALLON PER MINUTE FLOW RATES

SUPPLY SYSTEMS PREDOMINANTLY FOR FLUSH TANKS			SUPPLY SYSTEM PREDOMINANTLY FOR FLUSH VALVES		
Load	Demand		Load	Demand	
(Water supply fixture units)	(Gallons per minute)	(Cubic feet per minute)	(Water supply fixture units)	(Gallons per minute)	(Cubic feet per minute)
1	3.0	0.04104	—	—	—
2	5.0	0.0684	—	—	—
3	6.5	0.86892	—	—	—
4	8.0	1.06944	—	—	—
5	9.4	1.256592	5	15.0	2.0052
6	10.7	1.430376	6	17.4	2.326032
7	11.8	1.577424	7	19.8	2.646364
8	12.8	1.711104	8	22.2	2.967696
9	13.7	1.831416	9	24.6	3.288528
10	14.6	1.951728	10	27.0	3.60936
11	15.4	2.058672	11	27.8	3.716304
12	16.0	2.13888	12	28.6	3.823248
13	16.5	2.20572	13	29.4	3.930192
14	17.0	2.27256	14	30.2	4.037136
15	17.5	2.3394	15	31.0	4.14408
16	18.0	2.90624	16	31.8	4.241024
17	18.4	2.459712	17	32.6	4.357968
18	18.8	2.513184	18	33.4	4.464912
19	19.2	2.566656	19	34.2	4.571856
20	19.6	2.620128	20	35.0	4.6788
25	21.5	2.87412	25	38.0	5.07984
30	23.3	3.114744	30	42.0	5.61356
35	24.9	3.328632	35	44.0	5.88192
40	26.3	3.515784	40	46.0	6.14928
45	27.7	3.702936	45	48.0	6.41664
50	29.1	3.890088	50	50.0	6.684

For SI: 1 gallon per minute = 3.785 L/m, 1 cubic foot per minute = 0.4719 L/s.

TABLE P2903.7
MINIMUM SIZE OF WATER METERS, MAINS AND DISTRIBUTION PIPING
BASED ON WATER SUPPLY FIXTURE UNIT VALUES

Pressure Range—30 to 39 psi

METER AND SERVICE PIPE (inches)	DISTRIBUTION PIPE (inches)	MAXIMUM DEVELOPMENT LENGTH (feet)									
		40	60	80	100	150	200	250	300	400	500
$^3/_4$	$^1/_2$ a	2.5	2	1.5	1.5	1	1	.5	.5	0	0
$^3/_4$	$^3/_4$	9.5	7.5	6	5.5	4	3.5	3	2.5	2	1.5
$^3/_4$	1	32	25	20	16.5	11	9	7.5	6.5	5.5	4.5
1	1	32	32	27	21	13.5	10	8	7	5.5	5
$^3/_4$	$1^1/_4$	32	32	32	32	30	24	20	17	13	10.5
1	$1^1/_4$	80	80	70	61	45	34	27	22	16	12
$1^1/_2$	$1^1/_4$	80	80	80	75	54	40	31	25	17.5	13
1	$1^1/_2$	87	87	87	87	84	73	74	56	45	36
$1^1/_2$	$1^1/_2$	151	151	151	151	117	92	79	69	54	43

Pressure Range—40 to 49 psi

METER AND SERVICE PIPE (inches)	DISTRIBUTION PIPE (inches)	MAXIMUM DEVELOPMENT LENGTH (feet)									
		40	60	80	100	150	200	250	300	400	500
$^3/_4$	$^1/_2$ a	3	2.5	2	1.5	1.5	1	1	.5	.5	.5
$^3/_4$	$^3/_4$	9.5	9.5	8.5	7	5.5	4.5	3.5	3	2.5	2
$^3/_4$	1	32	32	32	26	18	13.5	10.5	9	7.5	6
1	1	32	32	32	32	21	15	11.5	9.5	7.5	6.5
$^3/_4$	$1^1/_4$	32	32	32	32	32	32	32	27	21	16.5
1	$1^1/_4$	80	80	80	80	65	52	42	35	26	20
$1^1/_2$	$1^1/_4$	80	80	80	80	75	59	48	39	28	21
1	$1^1/_2$	87	87	87	87	87	87	87	78	65	55
$1^1/_2$	$1^1/_2$	151	151	151	151	151	130	109	93	75	63

Pressure Range—50 to 60 psi

METER AND SERVICE PIPE (inches)	DISTRIBUTION PIPE (inches)	MAXIMUM DEVELOPMENT LENGTH (feet)									
		40	60	80	100	150	200	250	300	400	500
$^3/_4$	$^1/_2$ a	3	3	2.5	2	1.5	1	1	1	.5	.5
$^3/_4$	$^3/_4$	9.5	9.5	9.5	8.5	6.5	5	4.5	4	3	2.5
$^3/_4$	1	32	32	32	32	25	18.5	14.5	12	9.5	8
1	1	32	32	32	32	30	22	16.5	13	10	8
$^3/_4$	$1^1/_4$	32	32	32	32	32	32	32	32	29	24
1	$1^1/_4$	80	80	80	80	80	68	57	48	35	28
$1^1/_2$	$1^1/_4$	80	80	80	80	80	75	63	53	39	29
1	$1^1/_2$	87	87	87	87	87	87	87	87	82	70
$1^1/_2$	$1^1/_2$	151	151	151	151	151	151	139	120	94	79

(continued)

TABLE P2903.7—continued
MINIMUM SIZE OF WATER METERS, MAINS AND DISTRIBUTION PIPING
BASED ON WATER SUPPLY FIXTURE UNIT VALUES

Pressure Range—greater than 60 psi

METER AND SERVICE PIPE (inches)	DISTRIBUTION PIPE (inches)	MAXIMUM DEVELOPMENT LENGTH (feet)									
		40	60	80	100	150	200	250	300	400	500
$^3/_4$	$^1/_2$[a]	3	3	3	2.5	2	1.5	1.5	1	1	.5
$^3/_4$	$^3/_4$	9.5	9.5	9.5	9.5	7.5	6	5	4.5	3.5	3
$^3/_4$	1	32	32	32	32	32	24	19.5	15.5	11.5	9.5
1	1	32	32	32	32	32	28	22	17	12	9.5
$^3/_4$	$1^1/_4$	32	32	32	32	32	32	32	32	32	30
1	$1^1/_4$	80	80	80	80	80	80	69	60	46	36
$1^1/_2$	$1^1/_4$	80	80	80	80	80	80	76	65	50	38
1	$1^1/_2$	87	87	87	87	87	87	87	87	87	84
$1^1/_2$	$1^1/_2$	151	151	151	151	151	151	151	144	114	94

For SI: 1 inch = 25.4 mm, 1 foot = 304.8 mm, 1 pound per square inch = 6.895kPa.
a. Minimum size for building supply is $^3/_4$-inch pipe.

TABLE P2903.8.1
MANIFOLD SIZING

PLASTIC		METALLIC	
Nominal Size ID (inches)	Maximum[a] gpm	Nominal Size ID (inches)	Maximum[a] gpm
$^3/_4$	17	$^3/_4$	11
1	29	1	20
$1^1/_4$	46	$1^1/_4$	31
$1^1/_2$	66	$1^1/_2$	44

For SI: 1 inch = 25.4 mm, 1 gallon per minute = 3.785 L/m, 1 foot per second = 0.3048 m/s.
NOTE: See Table P2903.6 for w.s.f.u and Table P2903.6(1) for gallon-per-minute (gpm) flow rates.
a. Based on velocity limitation: plastic—12 fps; metal—8 fps.

P2903.8.5 Valving. Fixture valves, when installed, shall be located either at the fixture or at the manifold. If valves are installed at the manifold, they shall be labeled indicating the fixture served.

P2903.8.6 Hose bibb bleed. A readily accessible air bleed shall be installed in hose bibb supplies at the manifold or at the hose bibb exit point.

P2903.9 Valves. Valves shall be installed in accordance with Sections P2903.9.1 through P2903.9.3.

P2903.9.1 Service valve. Each dwelling unit shall be provided with an accessible main shutoff valve near the entrance of the water service. The valve shall be of a full-open type having nominal restriction to flow, with provision for drainage such as a bleed orifice or installation of a separate drain valve. Additionally, the water service shall be valved at the curb or property line in accordance with local requirements.

P2903.9.2 Water heater valve. A readily accessible full-open valve shall be installed in the cold-water supply pipe to each water heater at or near the water heater.

P2903.9.3 Fixture valves and access. Valves serving individual fixtures, appliances, risers and branches shall be provided with access. An individual shutoff valve shall be required on the fixture supply pipe to each plumbing fixture other than bathtubs and showers.

P2903.9.4 Valve requirements. Valves shall be of an approved type and compatible with the type of piping material installed in the system. Ball valves, gate valves, globe valves and plug valves intended to supply drinking water shall meet the requirements of NSF 61.

P2903.10 Hose bibb. Hose bibbs subject to freezing, including the "frost-proof" type, shall be equipped with an accessible stop-and-waste-type valve inside the building so that they can be controlled and/or drained during cold periods.

Exception: Frostproof hose bibbs installed such that the stem extends through the building insulation into an open heated or semiconditioned space need not be separately valved (see Figure P2903.10).

SECTION P2904
MATERIALS, JOINTS AND CONNECTIONS

P2904.1 Soil and groundwater. The installation of water service pipe, water distribution pipe, fittings, valves, appurtenances and gaskets shall be prohibited in soil and groundwater that is contaminated with solvents, fuels, organic compounds or other detrimental materials that cause permeation, corrosion, degradation or structural failure of the water service or water distribution piping material.

P2904.1.1 Investigation required. Where detrimental conditions are suspected by or brought to the attention of the building official, a chemical analysis of the soil and groundwater conditions shall be required to ascertain the acceptability of the water service material for the specific installation.

P2904.1.2 Detrimental condition. When a detrimental condition exists, approved alternate materials or alternate routing shall be required.

P2904.2 Lead content. Pipe and fittings used in the water-supply system shall have a maximum of 8 percent lead.

P2904.3 Polyethylene plastic piping installation. Polyethylene pipe shall be cut square using a cutter designed for plastic pipe. Except where joined by heat fusion, pipe ends shall be chamfered to remove sharp edges. Pipe that has been kinked shall not be installed. For bends, the installed radius of pipe curvature shall be greater than 30 pipe diameters or the coil radius when bending with the coil. Coiled pipe shall not be bent beyond straight. Bends shall not be permitted within 10 pipe diameters of any fitting or valve. Joints between polyethylene plastic pipe and fittings shall comply with Sections P2904.3.1 and P2904.3.2.

P2904.3.1 Heat-fusion joints. Joint surfaces shall be clean and free from moisture. Joint surfaces shall be heated to melting temperature and joined. The joint shall be undisturbed until cool. Joints shall be made in accordance with ASTM D 2657.

P2904.3.2 Mechanical joints. Mechanical joints shall be installed in accordance with the manufacturer's installation instructions.

P2904.4 Water service pipe. Water service pipe shall conform to NSF 61 and shall conform to one of the standards listed in Table P2904.4. Water service pipe or tubing, installed underground and outside of the structure, shall have a minimum working pressure rating of 160 pounds per square inch at 73°F (1103 kPa at 23°C). Where the water pressure exceeds 160 pounds per square inch (1103 kPa), piping material shall have a rated working pressure equal to or greater than the highest available pressure. Water service piping materials not third-party certified for water distribution shall terminate at or before the full open valve located at the entrance to the structure. Ductile iron water service piping shall be cement mortar lined in accordance with AWWA C104.

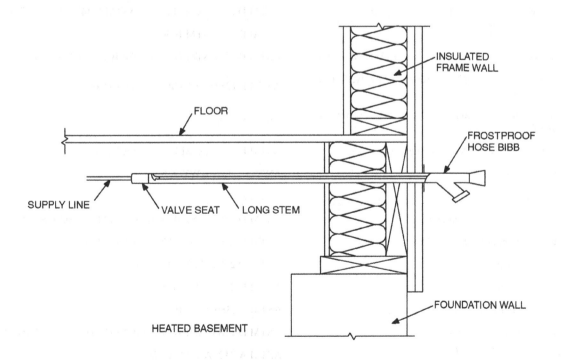

FIGURE P2903.10
TYPICAL FROSTPROOF HOSE BIBB INSTALLATION NOT REQUIRING SEPARATE VALUE

P2904.4.1 Dual check-valve-type backflow preventer. Where a dual check-valve backflow preventer is installed on the water supply system, it shall comply with ASSE 1024 or CSA B64.6.

P2904.4.2 Water service installation. Trenching, pipe installation and backfilling shall be in accordance with Section P2604. Water-service pipe is permitted to be located in the same trench with a building sewer provided such sewer is constructed of materials listed for underground use within a building in Section P3002.1. If the building sewer is not constructed of materials listed in Section P3002.1, the water-service pipe shall be separated from the building sewer by a minimum of 5 feet (1524 mm), measured horizontally, of undisturbed or compacted earth or placed on a solid ledge at least 12 inches (305 mm) above and to one side of the highest point in the sewer line.

> **Exception:** The required separation distance shall not apply where a water service pipe crosses a sewer pipe, provided that the water service pipe is sleeved to at least 5 feet (1524 mm), horizontally from the sewer pipe centerline, on both sides of the crossing with pipe materials listed in Tables P2904.4, P3002.1(1), P3002.1(2) or P3002.2.

P2904.5 Water-distribution pipe. Water-distribution piping within dwelling units shall conform to NSF 61 and shall conform to one of the standards listed in Table P2904.5. All hot-water-distribution pipe and tubing shall have a minimum pressure rating of 100 psi at 180°F (689 kPa at 82°C).

P2904.5.1 Under concrete slabs. Inaccessible water distribution piping under slabs shall be copper water tube minimum Type M, brass, ductile iron pressure pipe, cross-linked polyethylene/aluminum/cross-linked polyethylene (PEX-AL-PEX) pressure pipe, polyethylene/aluminum/polyethylene (PE-AL-PE) pressure pipe, chlorinated polyvinyl chloride (CPVC), polybutylene (PB), cross-linked polyethylene (PEX) plastic pipe or tubing or polypropylene (PP) pipe or tubing, all to be installed with approved fittings or bends. The minimum pressure rating for plastic pipe or tubing installed under slabs shall be 100 pounds per square inch at 180°F (689 kPa at 82°C).

P2904.6 Fittings. Pipe fittings shall be approved for installation with the piping material installed, and shall conform to the respective pipe standards listed in Table P2904.6. Pipe fittings used in the water supply system shall also conform to NSF 61.

TABLE P2904.4
WATER SERVICE PIPE

MATERIAL	STANDARD
Acrylonitrile butadiene styrene (ABS) plastic pipe	ASTM D 1527; ASTM D 2282
Asbestos-cement pipe	ASTM C 296
Brass pipe	ASTM B 43
Chlorinated polyvinyl chloride (CPVC) plastic pipe	ASTM D 2846; ASTM F 441; ASTM F 442; CSA B137.6
Copper or copper-alloy pipe	ASTM B 42; ASTM B 302
Copper or copper-alloy tubing (Type K, WK, L, WL, M or WM)	ASTM B 75; ASTM B 88; ASTM B 251; ASTM B 447
Cross-linked polyethylene/aluminum/cross-linked polyethylene (PEX-AL-PEX) pipe	ASTM F 1281; CSA CAN/CSA B137.10
Cross-linked polyethylene/aluminum/high-density polyethylene (PEX-AL-HDPE)	ASTM F 1986
Cross-linked polyethylene (PEX) plastic tubing	ASTM F 876; ASTM F 877; CSA B137.5
Ductile iron water pipe	AWWA C151; AWWA C115
Galvanized steel pipe	ASTM A 53
Polybutylene (PB) plastic pipe and tubing	ASTM D 2662; ASTM D 2666; ASTM D 3309; CSA B137.8M
Polyethylene/aluminum/polyethylene (PE-AL-PE) pipe	ASTM F 1282; CSA CAN/CSA-B137.9M
Polyethylene (PE) plastic pipe	ASTM D 2104; ASTM D 2239; CSA-B137.1
Polyethylene (PE) plastic tubing	ASTM D 2737; CSA B137.1
Polypropylene (PP) plastic pipe or tubing	ASTM F 2389; CSA B137.11
Polyvinyl chloride (PVC) plastic pipe	ASTM D 1785; ASTM D 2241; ASTM D 2672; CSA B137.3
Stainless steel (Type 304/304L) pipe	ASTM A 312; ASTM A 778
Stainless steel (Type 316/316L) pipe	ASTM A 312; ASTM A 778

TABLE P2904.5
WATER DISTRIBUTION PIPE

MATERIAL	STANDARD
Brass pipe	ASTM B 43
Chlorinated polyvinyl chloride (CPVC) plastic pipe and tubing	ASTM D 2846; ASTM F 441; ASTM F 442; CSA B137.6
Copper or copper-alloy pipe	ASTM B 42; ASTM B 302
Copper or copper-alloy tubing (Type K, WK, L, WL, M or WM)	ASTM B 75; ASTM B 88; ASTM B 251; ASTM B 447
Cross-linked polyethylene (PEX) plastic tubing	ASTM F 877; CSA B137.5
Cross-linked polyethylene/aluminum/cross-linked polyethylene (PEX-AL-PEX) pipe	ASTM F 1281; CSACAN/CSA-B137.10
Cross-linked polyethylene/aluminum/high-density polyethylene (PEX-AL-HDPE)	ASTM F 1986
Galvanized steel pipe	ASTM A 53
Polybutylene (PB) plastic pipe and tubing	ASTM D 3309; CSA CAN3-B137.8
Polyethylene/aluminum/polyethylene (PE-AL-PE) composite pipe	ASTM F 1282
Polypropylene (PP) plastic pipe or tubing	ASTM F 2389; CSA B137.11
Stainless steel (Type 304/304L) pipe	ASTM A 312; ASTM A 778
Stainless steel (Type 316/316L) pipe	ASTM A 312; ASTM A 778

TABLE P2904.6
PIPE FITTINGS

MATERIAL	STANDARD
Acrylonitrile butadiene styrene (ABS) plastic	ASTM D 2468
Brass	ASTM F1974
Cast-iron	ASME B16.4; ASME B16.12
Chlorinated polyvinyl chloride (CPVC) plastic	ASTM F 437; ASTM F 438; ASTM F 439; CSA B137.6
Copper or copper alloy	ASME B16.15; ASME B16.18; ASME B16.22; ASME B16.23; ASME B16.26; ASME B16.29
Cross-linked polyethylene/aluminum/high-density polyethylene (PEX-AL-HDPE)	ASTM F 1986
Fittings for cross-linked polyethylene (PEX) plastic tubing	ASTM F 877; ASTM F 1807; ASTM F 1960; ASTM F 2080; ASTM F 2159; CSA B137.5
Gray iron and ductile iron	AWWA C110; AWWA C153
Malleable iron	ASME B16.3
Polybutylene (PB) plastic	CSA B137.8
Polyethylene (PE) plastic	ASTM D 2609; CSA B137.1
Polypropylene (PP) plastic pipe or tubing	ASTM F 2389; CSA B137.11
Polyvinyl chloride (PVC) plastic	ASTM D 2464; ASTM D 2466; ASTM D 2467; CSA B137.2
Stainless steel (Type 304/304L) pipe	ASTM A 312; ASTM A 778
Stainless steel (Type 316/316L) pipe	ASTM A 312; ASTM A 778
Steel	ASME B16.9; ASME B16.11; ASME B16.28

P2904.7 Flexible water connectors. Flexible water connectors, exposed to continuous pressure, shall conform to ASME A112.18.6. Access shall be provided to all flexible water connectors.

P2904.8 Joint and connection tightness. Joints and connections in the plumbing system shall be gas tight and water tight for the intended use or required test pressure.

P2904.9 Plastic pipe joints. Joints in plastic piping shall be made with approved fittings by solvent cementing, heat fusion, corrosion-resistant metal clamps with insert fittings or compression connections. Flared joints for polyethylene pipe are permitted in accordance with Section P2904.3.

P2904.9.1 Solvent cementing. Solvent-cemented joints shall comply with Sections P2904.9.1.1 through P2904.9.1.3.

P2904.9.1.1 ABS plastic pipe. Solvent cement for ABS plastic pipe conforming to ASTM D 2235 shall be applied to all joint surfaces.

P2904.9.1.2 CPVC plastic pipe. Joint surfaces shall be clean and free from moisture and an approved primer shall be applied. Solvent cement for CPVC plastic pipe, orange in color and conforming to ASTM F 493, shall be applied to all joint surfaces. The parts shall be joined while the cement is wet and in accordance with ASTM D 2846 or ASTM F 493. Solvent-cement joints shall be permitted above or below ground.

Exception: A primer is not required where all of the following conditions apply:

1. The solvent cement used is third-party certified as conforming to ASTM F 493.

2. The solvent cement used is yellow in color.

3. The solvent cement is used only for joining $^1/_2$-inch (13 mm) through 2-inch (51 mm) diameter CPVC pipe and fittings.

4. The CPVC pipe and fittings are manufactured in accordance with ASTM D 2846.

P2904.9.1.3 PVC plastic pipe. A purple primer that conforms to ASTM F 656 shall be applied to PVC solvent cemented joints. Solvent cement for PVC plastic pipe conforming to ASTM D 2564 shall be applied to all joint surfaces.

P2904.9.1.4 Cross-linked polyethylene plastic (PEX). Joints between cross-linked polyethylene plastic tubing or fittings shall comply with Section P2904.9.1.4.1 or Section P2904.9.1.4.2.

P2904.9.1.4.1 Flared joints. Flared pipe ends shall be made by a tool designed for that operation.

P2904.9.1.4.2 Mechanical joints. Mechanical joints shall be installed in accordance with the manufacturer's instructions. Fittings for cross-linked polyethylene (PEX) plastic tubing as described in ASTM F 877, ASTM F 1807, ASTM F 1960, and ASTM F 2080 shall be installed in accordance with the manufacturer's installation instructions.

P2904.10 Polypropylene (PP) plastic. Joints between PP plastic pipe and fittings shall comply with Section P2904.10.1 or P2904.10.2.

P2904.10.1 Heat-fusion joints. Heat fusion joints for polypropylene pipe and tubing joints shall be installed with socket-type heat-fused polypropylene fittings, butt-fusion polypropylene fittings or electrofusion polypropylene fittings. Joint surfaces shall be clean and free from moisture. The joint shall be undisturbed until cool. Joints shall be made in accordance with ASTM F 2389.

P2904.10.2 Mechanical and compression sleeve joints. Mechanical and compression sleeve joints shall be installed in accordance with the manufacturer's installation instructions.

P2904.11 Stainless steel. Joints between stainless steel pipe and fittings shall comply with Sections P2904.11.1 and P2904.11.2.

P2904.11.1 Mechanical joints. Mechanical joints shall be installed in accordance with the manufacturer's instructions.

P2904.11.2 Welded joints. Joint surfaces shall be cleaned. The joint shall be welded autogenously or with an approved filler metal in accordance with ASTM A 312.

P2904.12 Threaded pipe joints. Threaded joints shall conform to American National Taper Pipe Thread specifications. Pipe ends shall be deburred and chips removed. Pipe joint compound shall be used only on male threads.

P2904.13 Soldered joints. Soldered joints in tubing shall be made with fittings approved for water piping and shall conform to ASTM B 828. Surfaces to be soldered shall be cleaned bright. The joints shall be properly fluxed and made with approved solder. Solders and fluxes used in potable water-supply systems shall have a maximum of 0.2 percent lead. Fluxes shall conform to ASTM B 813.

P2904.14 Flared joints. Flared joints in water tubing shall be made with approved fittings. The tubing shall be reamed and then expanded with a flaring tool.

P2904.15 Underground joints. Joints in polybutylene (PB) plastic pipe or tubing underground or under a concrete floor slab shall be installed using heat fusion, in accordance with the manufacturer's installation instructions. Joints in copper pipe or tube installed in a concrete floor slab or under a concrete floor slab on grade shall be installed using wrought-copper fittings and brazed joints.

P2904.16 Above-ground joints. Joints within the building between copper pipe, polybutylene tubing or CPVC tubing, in any combination with compatible outside diameters, are permitted to be made with the use of approved push-in mechanical fittings of a pressure-lock design.

P2904.17 Joints between different materials. Joints between different piping materials shall be made in accordance with Sections P2904.17.1, P2904.17.2 and P2904.17.3 or with a mechanical joint of the compression or mechanical sealing type having an elastomeric seal conforming to ASTM D 1869 or ASTM F 477. Joints shall be installed in accordance with the manufacturer's instructions.

P2904.17.1 Copper or copper-alloy tubing to galvanized steel pipe. Joints between copper or copper-alloy tubing and galvanized steel pipe shall be made with a brass fitting or dielectric fitting. The copper tubing shall be joined to the fitting in an approved manner, and the fitting shall be screwed to the threaded pipe.

P2904.17.2 Plastic pipe or tubing to other piping material. Joints between different grades of plastic pipe or between plastic pipe and other piping material shall be made with an approved adapter fitting. Joints between plastic pipe and cast-iron hub pipe shall be made by a caulked joint or a mechanical compression joint.

P2904.17.3 Stainless steel. Joints between stainless steel and different piping materials shall be made with a mechanical joint of the compression or mechanical-sealing type or a dielectric fitting.

P2904.18 Press joints. Press-type mechanical joints in copper tubing shall be made in accordance with the manufacturer's instructions using approved tools which affix the copper fitting with integral O-ring to the tubing.

SECTION P2905
CHANGES IN DIRECTION

P2905.1 Bends. Changes in direction in copper tubing are permitted to be made with bends having a radius of not less than four diameters of the tube, providing such bends are made by use of forming equipment that does not deform or create loss in cross-sectional area of the tube.

SECTION P2906
SUPPORT

P2906.1 General. Pipe and tubing support shall conform to Section P2605.

SECTION P2907
DRINKING WATER TREATMENT UNITS

P2907.1 Design. Drinking water treatment units shall meet the requirements of NSF 42, NSF 44 or NSF 53.

P2907.2 Reverse osmosis drinking water treatment units. Point-of-use reverse osmosis drinking water treatment units, designed for residential use, shall meet the requirements of NSF 58. Waste or discharge from reverse osmosis drinking water treatment units shall enter the drainage system through an air gap or an air gap device that meets the requirements of NSF 58.

P2907.3 Connection tubing. The tubing to and from drinking water treatment units shall be of a size and material as recommended by the manufacturer. The tubing shall comply with NSF 14, NSF 42, NSF 44, NSF 53, NSF 58 or NSF 61.

CHAPTER 30

SANITARY DRAINAGE

SECTION P3001
GENERAL

P3001.1 Scope. The provisions of this chapter shall govern the materials, design, construction and installation of sanitary drainage systems. Plumbing materials shall conform to the requirements of this chapter. The drainage, waste and vent (DWV) system shall consist of all piping for conveying wastes from plumbing fixtures, appliances and appurtenances, including fixture traps; above-grade drainage piping; below-grade drains within the building (building drain); below- and above-grade venting systems; and piping to the public sewer or private septic system.

P3001.2 Protection from freezing. No portion of the above grade DWV system other than vent terminals shall be located outside of a building, in attics or crawl spaces, concealed in outside walls, or in any other place subjected to freezing temperatures unless adequate provision is made to protect them from freezing by insulation or heat or both, except in localities having a winter design temperature above 32°F (0°C) (ASHRAE 97.5 percent column, winter, see Chapter 3).

P3001.3 Flood-resistant installation. In areas prone to flooding as established by Table R301.2(1), drainage, waste and vent systems shall be located and installed to prevent infiltration of floodwaters into the systems and discharges from the systems into floodwaters.

SECTION P3002
MATERIALS

P3002.1 Piping within buildings. Drain, waste and vent (DWV) piping in buildings shall be as shown in Tables P3002.1(1) and P3002.1(2) except that galvanized wrought-iron or galvanized steel pipe shall not be used underground and shall be maintained not less than 6 inches (152 mm) above ground. Allowance shall be made for the thermal expansion and contraction of plastic piping.

P3002.2 Building sewer. Building sewer piping shall be as shown in Table P3002.2. Forced main sewer piping shall conform to one of the standards for ABS plastic pipe, copper or copper-alloy tubing, PVC plastic pipe or pressure-rated pipe listed in Table P3002.2.

P3002.3 Fittings. Fittings shall be approved and compatible with the type of piping being used and shall be of a sanitary or DWV design for drainage and venting as shown in Table P3002.3. Water pipe fittings shall be permitted in engineer-designed systems where the design indicates compliance with Section P3101.2.1.

P3002.3.1 Drainage. Drainage fittings shall have a smooth interior waterway of the same diameter as the piping served. All fittings shall conform to the type of pipe used. Drainage fittings shall have no ledges, shoulders or reductions which can retard or obstruct drainage flow in the piping. Threaded drainage pipe fittings shall be of the recessed drainage type, black or galvanized. Drainage fittings shall be designed to maintain one-fourth unit vertical in 12 units horizontal (2-percent slope) grade.

P3002.4 Other materials. Sheet lead, lead bends, lead traps and sheet copper shall comply with Sections P3002.4.1 through P3002.4.3.

P3002.4.1 Sheet lead. Sheet lead for the following uses shall weigh not less than indicated below:

1. Flashing of vent terminals, 3 psf (15 kg/m²).

2. Prefabricated flashing for vent pipes, $2^1/_2$ psf (12 kg/m²).

P3002.4.2 Lead bends and traps. Lead bends and lead traps shall not be less than $^1/_8$-inch (3 mm) wall thickness.

P3002.4.3 Sheet copper. Sheet copper for the following uses shall weigh not less than indicated below:

1. General use, 12 ounces per square feet (4 kg/m²).

2. Flashing for vent pipes, 8 ounces per square feet (2.5 kg/m²).

SECTION P3003
JOINTS AND CONNECTIONS

P3003.1 Tightness. Joints and connections in the DWV system shall be gas tight and water tight for the intended use or pressure required by test.

P3003.2 Prohibited joints. Running threads and bands shall not be used in the drainage system. Drainage and vent piping shall not be drilled, tapped, burned or welded.

The following types of joints and connections shall be prohibited:

1. Cement or concrete.

2. Mastic or hot-pour bituminous joints.

3. Joints made with fittings not approved for the specific installation.

4. Joints between different diameter pipes made with elastomeric rolling O-rings.

5. Solvent-cement joints between different types of plastic pipe.

6. Saddle-type fittings.

P3003.3 ABS plastic. Joints between ABS plastic pipe or fittings shall comply with Sections P3003.3.1 through P3003.3.3.

P3003.3.1 Mechanical joints. Mechanical joints on drainage pipes shall be made with an elastomeric seal conforming to ASTM C 1173, ASTM D 3212 or CSA B602. Mechanical joints shall be installed only in underground systems unless otherwise approved. Joints shall be installed in accordance with the manufacturer's installation instructions.

TABLE P3002.1(1)
ABOVE-GROUND DRAINAGE AND VENT PIPE

MATERIAL	STANDARD
Acrylonitrile butadiene styrene (ABS) plastic pipe	ASTM D 2661; ASTM F 628; CSA B181.1
Brass pipe	ASTM B 43
Cast-iron pipe	ASTM A 74; CISPI 301; ASTM A 888
Coextruded composite ABS DWV schedule 40 IPS pipe (solid)	ASTM F 1488
Coextruded composite ABS DWV schedule 40 IPS pipe (cellular core)	ASTM F 1488
Coextruded composite PVC DWV schedule 40 IPS pipe (solid)	ASTM F 1488
Coextruded composite PVC DWV schedule 40 IPS pipe (cellular core)	ASTM F 1488; ASTM F 891
Coextruded composite PVC IPS-DR, PS140, PS200 DWV	ASTM F 1488
Copper or copper-alloy pipe	ASTM B 42; ASTM B 302
Copper or copper-alloy tubing (Type K, L, M or DWV)	ASTM B 75; ASTM B 88; ASTM B 251; ASTM B 306
Galvanized steel pipe	ASTM A 53
Polyolefin pipe	CSA B181.3
Polyvinyl chloride (PVC) plastic pipe (Type DWV)	ASTM D 2665; ASTM D 2949; CSA B181.2; ASTM F 1488
Stainless steel drainage systems, Types 304 and 316L	ASME A112.3.1

TABLE P3002.1(2)
UNDERGROUND BUILDING DRAINAGE AND VENT PIPE

MATERIAL	STANDARD
Acrylonitrile butadiene styrene (ABS) plastic pipe	ASTM D 2661; ASTM F 628; CSA B181.1
Asbestos-cement pipe	ASTM C 428
Cast-iron pipe	ASTM A 74; CISPI 301; ASTM A 888
Coextruded composite ABS DWV schedule 40 IPS pipe (solid)	ASTM F 1488
Coextruded composite ABS DWV schedule 40 IPS pipe (cellular core)	ASTM F 1488
Coextruded composite PVC DWV schedule 40 IPS pipe (solid)	ASTM F 1488
Coextruded composite PVC DWV schedule 40 IPS pipe (cellular core)	ASTM F 891; ASTM F 1488
Coextruded composite PVC IPS-DR, PS140, PS200 DWV	ASTM F 1488
Copper or copper alloy tubing (Type K, L, M or DWV)	ASTM B 75; ASTM B 88; ASTM B 251; ASTM B 306
Polyolefin pipe	ASTM F 1412; CSA B181.3
Polyvinyl chloride (PVC) plastic pipe (Type DWV)	ASTM D 2665; ASTM D 2949; CSA B181.2
Stainless steel drainage systems, Type 316L	ASME A112.3.1

P3003.3.2 Solvent cementing. Joint surfaces shall be clean and free from moisture. Solvent cement that conforms to ASTM D 2235 or CSA B181.1 shall be applied to all joint surfaces. The joint shall be made while the cement is wet. Joints shall be made in accordance with ASTM D 2235, ASTM D 2661, ASTM F 628 or CSA B181.1. Solvent-cement joints shall be permitted above or below ground.

P3003.3.3 Threaded joints. Threads shall conform to ASME B1.20.1. Schedule 80 or heavier pipe shall be permitted to be threaded with dies specifically designed for plastic pipe. Approved thread lubricant or tape shall be applied on the male threads only.

P3003.4 Asbestos-cement. Joints between asbestos-cement pipe or fittings shall be made with a sleeve coupling of the same composition as the pipe, sealed with an elastomeric ring conforming to ASTM D 1869.

P3003.5 Brass. Joints between brass pipe or fittings shall comply with Sections P3003.5.1 through P3003.5.3.

P3003.5.1 Brazed joints. All joint surfaces shall be cleaned. An approved flux shall be applied where required. The joint shall be brazed with a filler metal conforming to AWS A5.8.

P3003.5.2 Mechanical joints. Mechanical joints shall be installed in accordance with the manufacturer's installation instructions.

TABLE P3002.2
BUILDING SEWER PIPE

MATERIAL	STANDARD
Acrylonitrile butadiene styrene (ABS) plastic pipe	ASTM D 2661; ASTM D 2751; ASTM F 628
Asbestos-cement pipe	ASTM C 428
Cast-iron pipe	ASTM A 74; ASTM A 888; CISPI 301
Coextruded composite ABS DWV schedule 40 IPS pipe (solid)	ASTM F 1488
Coextruded composite ABS DWV schedule 40 IPS pipe (cellular core)	ASTM F 1488
Coextruded composite PVC DWV schedule 40 IPS pipe (solid)	ASTM F 1488
Coextruded composite PVC DWV schedule 40 IPS pipe (cellular core)	ASTM F 1488; ASTM F 891
Coextruded composite PVC IPS-DR-PS DWV, PS140, PS200	ASTM F 1488
Coextruded composite ABS sewer and drain DR-PS in PS35, PS50, PS100, PS140, PS200	ASTM F 1488
Coextruded composite PVC sewer and drain DR-PS in PS35, PS50, PS100, PS140, PS200	ASTM F 1488
Coextruded composite PVC sewer and drain PS 25, PS 50, PS 100 (cellular core)	ASTM F 891
Concrete pipe	ASTM C 14; ASTM C 76; CSA A257.1M; CSA A257.2M
Copper or copper-alloy tubing (Type K or L)	ASTM B 75; ASTM B 88; ASTM B 251
Polyethylene (PE) plastic pipe (SDR-PR)	ASTM F 714
Polyolefin pipe	ASTM F 1412; CSA B181.3
Polyvinyl chloride (PVC) plastic pipe (Type DWV, SDR 26, SDR 35, SDR41, PS50 or PS100)	ASTM D 2665; ASTM D 2949; ASTM D 3034; ASTM F 1412; CSA B182.2; CSA B182.4
Stainless steel drainage systems, Types 304 and 316L	ASME A112.3.1
Vitrified clay pipe	ASTM C 425; ASTM C 700

TABLE P3002.3
PIPE FITTINGS

MATERIAL	STANDARD
Acrylonitrile butadiene styrene (ABS) plastic pipe	ASTM D 3311; CSA B181.1; ASTM D 2661
Cast-iron pipe	ASME B 16.12; ASTM A 74; ASTM A 888; CISPI 301
Coextruded composite ABS DWV schedule 40 IPS pipe (solid or cellular core)	ASTM D 2661; ASTM D 3311; ASTM F 628
Coextruded composite ABS DWV schedule 40 IPS-DR, PS 140, PS 200 (solid or cellular core)	ASTM D 2665; ASTM D 3311; ASTM F 891
Coextruded composite ABS sewer and drain DR-PS in PS35, PS50, PS100, PS140, PS200	ASTM D 2751
Coextruded composite PVC DWV schedule 40 IPS-DR, PS 140, PS200 (solid and cellular core)	ASTM D 2665; ASTM D 3311; ASTM F 891
Coextruded composite PVC sewer and drain DR-PS in PS35, PS50, PS100, PS140, PS200	ASTM D 3034
Copper or copper alloy	ASME B 16.23; ASME B 16.29
Gray iron and ductile iron	AWWA C 110
Polyolefin	ASTM F 1412; CSA B181.3
Polyvinyl chloride (PVC) plastic pipe	ASTM D 3311; ASTM D 2665; ASTM F 1412; ASTM F 1866; CSA B 181.2; CSA B 182.4
Stainless steel drainage systems, Types 304 and 316L	ASME A112.3.1

P3003.5.3 Threaded joints. Threads shall conform to ASME B1.20.1. Pipe-joint compound or tape shall be applied on the male threads only.

P3003.6 Cast iron. Joints between cast-iron pipe or fittings shall comply with Sections P3003.6.1 through P3003.6.3.

P3003.6.1 Caulked joints. Joints for hub and spigot pipe shall be firmly packed with oakum or hemp. Molten lead shall be poured in one operation to a depth of not less than 1 inch (25 mm). The lead shall not recede more than $^{1}/_{8}$ inch (3 mm) below the rim of the hub and shall be caulked tight. Paint, varnish or other coatings shall not be permitted on the jointing material until after the joint has been tested and approved. Lead shall be run in one pouring and shall be caulked tight. Acid-resistant rope and acidproof cement shall be permitted.

P3003.6.2 Compression gasket joints. Compression gaskets for hub and spigot pipe and fittings shall conform to ASTM C 564. Gaskets shall be compressed when the pipe is fully inserted.

P3003.6.3 Mechanical joint coupling. Mechanical joint couplings for hubless pipe and fittings shall comply with CISPI 310 or ASTM C 1277. The elastomeric sealing sleeve shall conform to ASTM C 564 or CSA B602 and shall have a center stop. Mechanical joint couplings shall be installed in accordance with the manufacturer's installation instructions.

P3003.7 Concrete joints. Joints between concrete pipe and fittings shall be made with an elastomeric seal conforming to ASTM C 443, ASTM C 1173, CSA A257.3M or CSA B602.

P3003.8 Coextruded composite ABS pipe. Joints between coextruded composite pipe with an ABS outer layer or ABS fittings shall comply with Sections P3003.8.1 and P3003.8.2.

P3003.8.1 Mechanical joints. Mechanical joints on drainage pipe shall be made with an elastomeric seal conforming to ASTM C 1173, ASTM D 3212 or CSA B602. Mechanical joints shall not be installed in above-ground systems, unless otherwise approved. Joints shall be installed in accordance with the manufacturer's installation instructions.

P3003.8.2 Solvent cementing. Joint surfaces shall be clean and free from moisture. Solvent cement that conforms to ASTM D 2235 or CSA B181.1 shall be applied to all joint surfaces. The joint shall be made while the cement is wet. Joints shall be made in accordance with ASTM D 2235, ASTM D 2661, ASTM F 628 or CSA B181.1. Solvent-cement joints shall be permitted above or below ground.

P3003.9 Coextruded composite PVC pipe. Joints between coextruded composite pipe with a PVC outer layer or PVC fittings shall comply with Sections P3003.9.1 and P3003.9.2.

P3003.9.1 Mechanical joints. Mechanical joints on drainage pipe shall be made with an elastomeric seal conforming to ASTM D 3212. Mechanical joints shall not be installed in above-ground systems, unless otherwise approved. Joints shall be installed in accordance with the manufacturer's installation instructions.

P3003.9.2 Solvent cementing. Joint surfaces shall be clean and free from moisture. A purple primer that conforms to ASTM F 656 shall be applied. Solvent cement not purple in color and conforming to ASTM D 2564, CSA B137.3 or CSA B181.2 shall be applied to all joint surfaces. The joint shall be made while the cement is wet, and shall be in accordance with ASTM D 2855. Solvent-cement joints shall be permitted above or below ground.

P3003.10 Copper pipe. Joints between copper or copper-alloy pipe or fittings shall comply with Sections P3003.10.1 through P3003.10.4.

P3003.10.1 Brazed joints. All joint surfaces shall be cleaned. An approved flux shall be applied where required. The joint shall be brazed with a filler metal conforming to AWS A5.8.

P3003.10.2 Mechanical joints. Mechanical joints shall be installed in accordance with the manufacturer's installation instructions.

P3003.10.3 Soldered joints. Solder joints shall be made in accordance with the methods of ASTM B 828. All cut tube ends shall be reamed to the full inside diameter of the tube end. All joint surfaces shall be cleaned. A flux conforming to ASTM B 813 shall be applied. The joint shall be soldered with a solder conforming to ASTM B 32.

P3003.10.4 Threaded joints. Threads shall conform to ASME B1.20.1. Pipe-joint compound or tape shall be applied on the male threads only.

P3003.11 Copper tubing. Joints between copper or copper-alloy tubing or fittings shall comply with Sections P3003.11.1 through P3003.11.3.

P3003.11.1 Brazed joints. All joint surfaces shall be cleaned. An approved flux shall be applied where required. The joint shall be brazed with a filler metal conforming to AWS A5.8.

P3003.11.2 Mechanical joints. Mechanical joints shall be installed in accordance with the manufacturer's installation instructions.

P3003.11.3 Soldered joints. Solder joints shall be made in accordance with the methods of ASTM B 828. Cut tube ends shall be reamed to the full inside diameter of the tube end. All joint surfaces shall be cleaned. A flux conforming to ASTM B 813 shall be applied. The joint shall be soldered with a solder conforming to ASTM B 32.

P3003.12 Steel. Joints between galvanized steel pipe or fittings shall comply with Sections P3003.12.1 and P3003.12.2.

P3003.12.1 Threaded joints. Threads shall conform to ASME B1.20.1. Pipe-joint compound or tape shall be applied on the male threads only.

P3003.12.2 Mechanical joints. Joints shall be made with an approved elastomeric seal. Mechanical joints shall be installed in accordance with the manufacturer's installation instructions.

P3003.13 Lead. Joints between lead pipe or fittings shall comply with Sections P3003.13.1 and P3003.13.2.

P3003.13.1 Burned. Burned joints shall be uniformly fused together into one continuous piece. The thickness of the joint shall be at least as thick as the lead being joined. The filler metal shall be of the same material as the pipe.

P3003.13.2 Wiped. Joints shall be fully wiped, with an exposed surface on each side of the joint not less than $^3/_4$ inch (19 mm). The joint shall be at least $^3/_8$ inch (9.5 mm) thick at the thickest point.

P3003.14 PVC plastic. Joints between PVC plastic pipe or fittings shall comply with Sections P3003.14.1 through P3003.14.3.

P3003.14.1 Mechanical joints. Mechanical joints on drainage pipe shall be made with an elastomeric seal conforming to ASTM C 1173, ASTM D 3212 or CSA B602. Mechanical joints shall not be installed in above-ground systems, unless otherwise approved. Joints shall be installed in accordance with the manufacturer's installation instructions.

P3003.14.2 Solvent cementing. Joint surfaces shall be clean and free from moisture. A purple primer that conforms to ASTM F 656 shall be applied. Solvent cement not purple in color and conforming to ASTM D 2564, CSA B137.3 or CSA B181.2 shall be applied to all joint surfaces. The joint shall be made while the cement is wet, and shall be in accordance with ASTM D 2855. Solvent-cement joints shall be permitted above or below ground.

P3003.14.3 Threaded joints. Threads shall conform to ASME B1.20.1. Schedule 80 or heavier pipe shall be permitted to be threaded with dies specifically designed for plastic pipe. Approved thread lubricant or tape shall be applied on the male threads only.

P3003.15 Vitrified clay. Joints between vitrified clay pipe or fittings shall be made with an elastomeric seal conforming to ASTM C 425, ASTM C 1173 or CSA B602.

P3003.16 Polyolefin plastic. Joints between polyolefin plastic pipe and fittings shall comply with Sections P3003.16.1 and P3003.16.2.

P3003.16.1 Heat-fusion joints. Heat-fusion joints for polyolefin pipe and tubing joints shall be installed with socket-type heat-fused polyolefin fittings or electrofusion polyolefin fittings. Joint surfaces shall be clean and free from moisture. The joint shall be undisturbed until cool. Joints shall be made in accordance with ASTM F 1412 or CSA B181.3.

P3003.16.2 Mechanical and compression sleeve joints. Mechanical and compression sleeve joints shall be installed in accordance with the manufacturer's installation instructions.

P3003.17 Polyethylene plastic pipe. Joints between polyethylene plastic pipe and fittings shall be underground and shall comply with Section P3003.17.1 or P3003.17.2.

P3003.17.1 Heat fusion joints. Joint surfaces shall be clean and free from moisture. All joint surfaces shall be cut, heated to melting temperature and joined using tools specifically designed for the operation. Joints shall be undisturbed until cool. Joints shall be made in accordance with ASTM D 2657 and the manufacturer's installation instructions.

P3003.17.2 Mechanical joints. Mechanical joints in drainage piping shall be made with an elastomeric seal conforming to ASTM C 1173, ASTM D 3212 or CSA B602. Mechanical joints shall be installed in accordance with the manufacturer's installation instructions.

P3003.18 Joints between different materials. Joints between different piping materials shall be made with a mechanical joint of the compression or mechanical-sealing type conforming to ASTM C 1173, ASTM C 1460 or ASTM C 1461. Connectors and adapters shall be approved for the application and such joints shall have an elastomeric seal conforming to ASTM C 425, ASTM C 443, ASTM C 564, ASTM C 1440, ASTM D 1869, ASTM F 477, CSA A257.3M or CSA B602, or as required in Sections P3003.18.1 through P3003.18.6. Joints between glass pipe and other types of materials shall be made with adapters having a TFE seal. Joints shall be installed in accordance with the manufacturer's installation instructions.

P3003.18.1 Copper or copper-alloy tubing to cast-iron hub pipe. Joints between copper or copper-alloy tubing and cast-iron hub pipe shall be made with a brass ferrule or compression joint. The copper or copper-alloy tubing shall be soldered to the ferrule in an approved manner, and the ferrule shall be joined to the cast-iron hub by a caulked joint or a mechanical compression joint.

P3003.18.2 Copper or copper-alloy tubing to galvanized steel pipe. Joints between copper or copper-alloy tubing and galvanized steel pipe shall be made with a brass converter fitting or dielectric fitting. The copper tubing shall be soldered to the fitting in an approved manner, and the fitting shall be screwed to the threaded pipe.

P3003.18.3 Cast-iron pipe to galvanized steel or brass pipe. Joints between cast-iron and galvanized steel or brass pipe shall be made by either caulked or threaded joints or with an approved adapter fitting.

P3003.18.4 Plastic pipe or tubing to other piping material. Joints between different types of plastic pipe or between plastic pipe and other piping material shall be made with an approved adapter fitting. Joints between plastic pipe and cast-iron hub pipe shall be made by a caulked joint or a mechanical compression joint.

P3003.18.5 Lead pipe to other piping material. Joints between lead pipe and other piping material shall be made by a wiped joint to a caulking ferrule, soldering nipple, or bushing or shall be made with an approved adapter fitting.

P3003.18.6 Stainless steel drainage systems to other materials. Joints between stainless steel drainage systems and other piping materials shall be made with approved mechanical couplings.

P3003.19 Joints between drainage piping and water closets. Joints between drainage piping and water closets or similar fixtures shall be made by means of a closet flange compatible with the drainage system material, securely fastened to a structurally firm base. The inside diameter of the drainage pipe shall not be used as a socket fitting for a four by three closet flange. The joint shall be bolted, with an approved gasket, flange to fixture connection complying with ASME A112.4.3 or setting compound between the fixture and the closet flange.

SECTION P3004
DETERMINING DRAINAGE FIXTURE UNITS

P3004.1 DWV system load. The load on DWV-system piping shall be computed in terms of drainage fixture unit (d.f.u.) values in accordance with Table P3004.1.

SECTION P3005
DRAINAGE SYSTEM

P3005.1 Drainage fittings and connections. Changes in direction in drainage piping shall be made by the appropriate use of sanitary tees, wyes, sweeps, bends or by a combination of these drainage fittings in accordance with Table P3005.1. Change in direction by combination fittings, heel or side inlets or increasers shall be installed in accordance with Table P3005.1 and Sections P3005.1.1 through P3005.1.4. based on the pattern of flow created by the fitting.

TABLE P3004.1
DRAINAGE FIXTURE UNIT (d.f.u.) VALUES FOR VARIOUS PLUMBING FIXTURES

TYPE OF FIXTURE OR GROUP OF FIXTURES	DRAINAGE FIXTURE UNIT VALUE (d.f.u.)[a]
Bar sink	1
Bathtub (with or without shower head and/or whirlpool attachments)	2
Bidet	1
Clothes washer standpipe	2
Dishwasher	2
Floor drain[b]	0
Kitchen sink	2
Lavatory	1
Laundry tub	2
Shower stall	2
Water closet (1.6 gallons per flush)	3
Water closet (greater than 1.6 gallons per flush)	4
Full-bath group with bathtub (with 1.6 gallon per flush water closet, and with or without shower head and/or whirlpool attachment on the bathtub or shower stall)	5
Full-bath group with bathtub (water closet greater than 1.6 gallon per flush, and with or without shower head and/or whirlpool attachment on the bathtub or shower stall)	6
Half-bath group (1.6 gallon per flush water closet plus lavatory)	4
Half-bath group (water closet greater than 1.6 gallon per flush plus lavatory)	5
Kitchen group (dishwasher and sink with or without garbage grinder)	2
Laundry group (clothes washer standpipe and laundry tub)	3
Multiple-bath groups[c]: 　1.5 baths 　2 baths 　2.5 baths 　3 baths 　3.5 baths	7 8 9 10 11

For SI: 1 gallon = 3.785 L.

a. For a continuous or semicontinuous flow into a drainage system, such as from a pump or similar device, 1.5 fixture units shall be allowed per gpm of flow. For a fixture not listed, use the highest d.f.u. value for a similar listed fixture.

b. A floor drain itself adds no hydraulic load. However, where used as a receptor, the fixture unit value of the fixture discharging into the receptor shall be applicable.

c. Add 2 d.f.u. for each additional full bath.

TABLE P3005.1
FITTINGS FOR CHANGE IN DIRECTION

TYPE OF FITTING PATTERN	CHANGE IN DIRECTION		
	Horizontal to vertical[c]	Vertical to horizontal	Horizontal to horizontal
Sixteenth bend	X	X	X
Eighth bend	X	X	X
Sixth bend	X	X	X
Quarter bend	X	X[a]	X[a]
Short sweep	X	X[a,b]	X[a]
Long sweep	X	X	X
Sanitary tee	X[c]	—	—
Wye	X	X	X
Combination wye and eighth bend	X	X	X

For SI: 1 inch = 25.4 mm.

a. The fittings shall only be permitted for a 2-inch or smaller fixture drain.

b. Three inches and larger.

c. For a limitation on multiple connection fittings, see Section P3005.1.1.

P3005.1.1 Horizontal to vertical (multiple connection fittings). Double fittings such as double sanitary tees and tee-wyes or approved multiple connection fittings and back-to-back fixture arrangements that connect two or more branches at the same level shall be permitted as long as directly opposing connections are the same size and the discharge into directly opposing connections is from similar fixture types or fixture groups. Double sanitary tee patterns shall not receive the discharge of back-to-back water closets and fixtures or appliances with pumping action discharge.

Exception: Back-to-back water closet connections to double sanitary tee patterns shall be permitted where the horizontal developed length between the outlet of the water closet and the connection to the double sanitary tee is 18 inches (457 mm) or greater.

P3005.1.2 Heel- or side-inlet quarter bends, drainage. Heel-inlet quarter bends shall be an acceptable means of connection, except where the quarter bends serves a water closet. A low-heel inlet shall not be used as a wet-vented connection. Side-inlet quarter bends shall be an acceptable means of connection for both drainage, wet venting and stack venting arrangements.

P3005.1.3 Heel- or side-inlet quarter bends, venting. Heel-inlet or side-inlet quarter bends, or any arrangement of pipe and fittings producing a similar effect, shall be acceptable as a dry vent where the inlet is placed in a vertical position. The inlet is permitted to be placed in a horizontal position only where the entire fitting is part of a dry vent arrangement.

P3005.1.4 Water closet connection between flange and pipe. One-quarter bends 3 inches (76 mm) in diameter shall be acceptable for water closet or similar connections, provided a 4-inch by 3-inch (102 mm by 76 mm) flange is installed to receive the closet fixture horn. Alternately, a 4-inch by 3-inch (102 mm by 76 mm) elbow shall be acceptable with a 4-inch (102 mm) flange.

P3005.1.5 Dead ends. Dead ends shall be prohibited except where necessary to extend a cleanout or as an approved part of a rough-in more than 2 feet (610 mm) in length.

P3005.1.6 Provisions for future fixtures. Where drainage has been roughed-in for future fixtures, the drainage unit values of the future fixtures shall be considered in determining the required drain sizes. Such future installations shall be terminated with an accessible permanent plug or cap fitting.

P3005.1.7 Change in size. The size of the drainage piping shall not be reduced in size in the direction of the flow. A 4-inch by 3-inch (102 mm by 76 mm) water closet connection shall not be considered as a reduction in size.

P3005.2 Drainage pipe cleanouts. Drainage pipe cleanouts shall comply with Sections P3005.2.1 through P3005.2.11.

Exception: These provisions shall not apply to pressurized building drains and building sewers that convey the discharge of automatic pumping equipment to a gravity drainage system.

P3005.2.1 Materials. Cleanouts shall be liquid and gas tight. Cleanout plugs shall be brass or plastic.

P3005.2.2 Spacing. Cleanouts shall be installed not more than 100 feet (30 480 mm) apart in horizontal drainage lines measured from the upstream entrance of the cleanout.

P3005.2.3 Underground drainage cleanouts. When installed in underground drains, cleanouts shall be extended vertically to or above finished grade either inside or outside the building.

P3005.2.4 Change of direction. Cleanouts shall be installed at each fitting with a change of direction more than 45 degrees (0.79 rad) in the building sewer, building drain and horizontal waste or soil lines. Where more than one change of direction occurs in a run of piping, only one cleanout shall be required in each 40 feet (12 192 mm) of developed length of the drainage piping.

P3005.2.5 Accessibility. Cleanouts shall be accessible. Minimum clearance in front of cleanouts shall be 18 inches (457 mm) on 3-inch (76 mm) and larger pipes, and 12 inches (305 mm) on smaller pipes. Concealed cleanouts shall be provided with access of sufficient size to permit removal of the cleanout plug and rodding of the system. Cleanout plugs shall not be concealed by permanent finishing material.

P3005.2.6 Base of stacks. Accessible cleanouts shall be provided near the base of each vertical waste or soil stack. Alternatively, such cleanouts shall be installed outside the building within 3 feet (914 mm) of the building wall.

P3005.2.7 Building drain and building sewer junction. There shall be a cleanout near the junction of the building drain and building sewer. This cleanout shall be either inside or outside the building wall, provided that it is brought up to finish grade or to the lowest floor level. An approved two-way cleanout shall be permitted to serve as the required cleanout for both the building drain and the building sewer. The cleanout at the junction of the building drain and building sewer shall not be required where a cleanout on a 3-inch (76 mm) or larger diameter soil stack is located within a developed length of 10 feet (3048 mm) of the building drain and building sewer junction.

P3005.2.8 Direction of flow. Cleanouts shall be installed so that the cleanout opens to allow cleaning in the direction of the flow of the drainage line.

P3005.2.9 Cleanout size. Cleanouts shall be the same nominal size as the pipe they serve up to 4 inches (102 mm). For pipes larger than 4 inches (102 mm) nominal size, the minimum size of the cleanout shall be 4 inches (102 mm).

Exceptions:

1. "P" trap connections with slip joints or ground joint connections, or stack cleanouts that are not more than one pipe diameter smaller than the drain served, shall be permitted.

2. Cast-iron cleanouts sized in accordance with the referenced standards in Table P3002.3, ASTM A 74 for hub and spigot fittings or ASTM A 888 or CISPI 301 for hubless fittings.

P3005.2.10 Cleanout equivalent. A fixture trap or a fixture with integral trap, readily removable without disturbing concealed piping shall be acceptable as a cleanout equivalent.

P3005.2.11 Connections to cleanouts prohibited. Cleanout openings shall not be used for the installation of new fixtures except where approved and an acceptable alternate cleanout is provided.

P3005.3 Horizontal drainage piping slope. Horizontal drainage piping shall be installed in uniform alignment at uniform slopes not less than $1/4$ unit vertical in 12 units horizontal (2-percent slope) for $2^1/_2$-inch (64 mm) diameter and less, and not less than $1/_8$ unit vertical in 12 units horizontal (1-percent slope) for diameters of 3 inches (76 mm) or more.

P3005.4 Drain pipe sizing. Drain pipes shall be sized according to drainage fixture unit (d.f.u.) loads. The size of the drainage piping shall not be reduced in size in the direction of flow. The following general procedure is permitted to be used:

1. Draw an isometric layout or riser diagram denoting fixtures on the layout.

2. Assign d.f.u. values to each fixture group plus individual fixtures using Table P3004.1.

3. Starting with the top floor or most remote fixtures, work downstream toward the building drain accumulating d.f.u. values for fixture groups plus individual fixtures for each branch. Where multiple bath groups are being added, use the reduced d.f.u. values in Table P3004.1, which take into account probability factors of simultaneous use.

4. Size branches and stacks by equating the assigned d.f.u. values to pipe sizes shown in Table P3005.4.1.

5. Determine the pipe diameter and slope of the building drain and building sewer based on the accumulated d.f.u. values, using Table P3005.4.2.

P3005.4.1 Fixture branch and stack sizing.

1. Branches and stacks shall be sized according to Table P3005.4.1. Below grade drain pipes shall not be less than $1^1/_2$ inches (38 mm) in diameter.

2. Minimum stack size. Drain stacks shall not be smaller than the largest horizontal branch connected.

Exceptions:

1. A 4-inch by 3-inch (102 mm by 76 mm) closet bend or flange.

2. A 4-inch (102 mm) closet bend into a 3-inch (76 mm) stack tee shall be acceptable (see Section P3005.1.4).

TABLE P3005.4.1
MAXIMUM FIXTURE UNITS ALLOWED TO BE CONNECTED TO BRANCHES AND STACKS

NOMINAL PIPE SIZE (inches)	ANY HORIZONTAL FIXTURE BRANCH	ANY ONE VERTICAL STACK OR DRAIN
$1^1/_4$[a]	—	—
$1^1/_2$[b]	3	4
2[b]	6	10
$2^1/_2$[b]	12	20
3	20	48
4	160	240

For SI: 1 inch = 25.4 mm.

a. $1^1/_4$-inch pipe size limited to a single–fixture drain or trap arm. See Table P3201.7.

b. No water closets.

P3005.4.2 Building drain and sewer size and slope. Pipe sizes and slope shall be determined from Table P3005.4.2 on the basis of drainage load in fixture units (d.f.u.) computed from Table P3004.1.

TABLE P3005.4.2
MAXIMUM NUMBER OF FIXTURE UNITS ALLOWED TO BE CONNECTED TO THE BUILDING DRAIN, BUILDING DRAIN BRANCHES OR THE BUILDING SEWER

DIAMETER OF PIPE (inches)	SLOPE PER FOOT		
	$1/_8$ inch	$1/_4$ inch	$1/_2$ inch
$1^1/_2$[a,b]	—	Note a	Note a
2[b]	—	21	27
$2^1/_2$[b]	—	24	31
3	36	42	50
4	180	216	250

For SI: 1 inch = 25.4 mm, 1 foot = 304.8 mm.

a. $1^1/_2$-inch pipe size limited to a building drain branch serving not more than two waste fixtures, or not more than one waste fixture if serving a pumped discharge fixture or garbage grinder discharge.

b. No water closets.

P3005.5 Connections to offsets and bases of stacks. Horizontal branches shall connect to the bases of stacks at a point located not less than 10 times the diameter of the drainage stack downstream from the stack. Horizontal branches shall connect to horizontal stack offsets at a point located not less than 10 times the diameter of the drainage stack downstream from the upper stack.

SECTION P3006
SIZING OF DRAIN PIPE OFFSETS

P3006.1 Vertical offsets. An offset in a vertical drain, with a change of direction of 45 degrees (0.79 rad) or less from the vertical, shall be sized as a straight vertical drain.

P3006.2 Horizontal offsets above the lowest branch. A stack with an offset of more than 45 degrees (0.79 rad) from the vertical shall be sized as follows:

1. The portion of the stack above the offset shall be sized as for a regular stack based on the total number of fixture units above the offset.

2. The offset shall be sized as for a building drain in accordance with Table P3005.4.2.

3. The portion of the stack below the offset shall be sized as for the offset or based on the total number of fixture units on the entire stack, whichever is larger.

P3006.3 Horizontal offsets below the lowest branch. In soil or waste stacks below the lowest horizontal branch, there shall be no change in diameter required if the offset is made at an angle not greater than 45 degrees (0.79 rad) from the vertical. If an offset greater than 45 degrees (0.79 rad) from the vertical is made, the offset and stack below it shall be sized as a building drain (see Table P3005.4.2).

SECTION P3007
SUMPS AND EJECTORS

P3007.1 Sewage ejectors or sewage pumps. A sewage ejector, sewage pump, or grinder pump receiving discharge from a water closet shall have minimum discharge velocity of 1.9 feet per second (0.579 m/s) throughout the discharge piping to the point of connection with a gravity building drain, gravity sewer or pressure sewer system. A nongrinding pump or ejector shall be capable of passing a $1^1/_2$-inch-diameter (38 mm) solid ball, and the discharge piping shall be not less than 2 inches (51 mm) in diameter. The discharge piping of grinding pumps shall be not less than $1^1/_4$ inches (32 mm) in diameter. A check valve and a gate valve located on the discharge side of the check valve shall be installed in the pump or ejector discharge piping between the pump or ejector and the drainage system. Access shall be provided to such valves. Such valves shall be located above the sump cover or, where the discharge pipe from the ejector is below grade, the valves shall be accessibly located outside the sump below grade in an access pit with a removeable access cover.

Exception: Macerating toilet systems shall be permitted to have the discharge pipe sized in accordance with manufacturer's instructions, but not less than 0.75 inch (19 mm) in diameter.

P3007.2 Building drains below sewer (building subdrains). Building drains which cannot be discharged to the sewer by gravity flow shall be discharged into a tightly covered and vented sump from which the contents shall be lifted and discharged into the building gravity drainage system by automatic pumping equipment.

P3007.2.1 Drainage piping. The system of drainage piping below the sewer level shall be installed and vented in a manner similar to that of the gravity system. Only such drains that must be lifted for discharge shall be discharged into sumps. All other drains shall be discharged by gravity.

Exception: Macerating toilet systems shall be permitted as an alternate to the sewage pump or ejector system. The macerating toilet shall comply with ASME A112.3.4 or CSA B45.9 and shall be installed in accordance with the manufacturers' instructions.

SECTION P3008
BACKWATER VALVES

P3008.1 General. Fixtures that have flood level rims located below the elevation of the next upstream manhole cover of the public sewer serving such fixtures shall be protected from backflow of sewage by installing an approved backwater valve. Fixtures having flood level rims above the elevation of the next upstream manhole shall not discharge through the backwater valve. Backwater valves shall be provided with access.

P3008.2 Construction. Backwater valves shall have noncorrosive bearings, seats and self-aligning discs, and shall be constructed to ensure a positive mechanical seal. Valve access covers shall be water tight.

CHAPTER 31

VENTS

SECTION P3101
VENT SYSTEMS

P3101.1 General. This chapter shall govern the selection and installation of piping, tubing and fittings for vent systems. This chapter shall control the minimum diameter of vent pipes, circuit vents, branch vents and individual vents, and the size and length of vents and various aspects of vent stacks and stack vents. Additionally, this chapter regulates vent grades and connections, height above fixtures and relief vents for stacks and fixture traps, and the venting of sumps and sewers.

P3101.2 Trap seal protection. The plumbing system shall be provided with a system of vent piping that will permit the admission or emission of air so that the seal of any fixture trap shall not be subjected to a pneumatic pressure differential of more than 1 inch of water column (249 Pa).

> **P3101.2.1 Venting required.** Every trap and trapped fixture shall be vented in accordance with one of the venting methods specified in this chapter.

P3101.3 Use limitations. The plumbing vent system shall not be used for purposes other than the venting of the plumbing system.

P3101.4 Extension outside a structure. In climates where the 97.5-percent value for outside design temperature is 0°F (-18°C) or less (ASHRAE 97.5-percent column, winter, see Chapter 3), vent pipes installed on the exterior of the structure shall be protected against freezing by insulation, heat or both. Vent terminals shall be protected from frost closure in accordance with Section P3103.2.

P3101.5 Flood resistance. In areas prone to floodings as established by Table R301.2(1), vents shall be located at or above the design flood elevation established in Section R324.1.

SECTION P3102
VENT STACKS AND STACK VENTS

P3102.1 Required vent extension. The vent system serving each building drain shall have at least one vent pipe that extends to the outdoors.

P3102.2 Installation. The required vent shall be a dry vent that connects to the building drain or an extension of a drain that connects to the building drain. Such vent shall not be an island fixture vent as permitted by Section P3112.

P3102.3 Size. The required vent shall be sized in accordance with Section P3113.1 based on the required size of the building drain.

SECTION P3103
VENT TERMINALS

P3103.1 Roof extension. Open vent pipes that extend through a roof shall be terminated at least 6 inches (152 mm) above the roof or 6 inches (152 mm) above the anticipated snow accumulation, whichever is greater, except that where a roof is to be used for any purpose other than weather protection, the vent extension shall be run at least 7 feet (2134 mm) above the roof.

P3103.2 Frost closure. Where the 97.5-percent value for outside design temperature is 0°F (-18°C) or less, every vent extension through a roof or wall shall be a minimum of 3 inches (76 mm) in diameter. Any increase in the size of the vent shall be made inside the structure a minimum of 1 foot (305 mm) below the roof or inside the wall.

P3103.3 Flashings and sealing. The juncture of each vent pipe with the roof line shall be made water tight by an approved flashing. Vent extensions in walls and soffits shall be made weather tight by caulking.

P3103.4 Prohibited use. Vent terminals shall not be used as a flag pole or to support flag poles, TV aerials, or similar items, except when the piping has been anchored in an approved manner.

P3103.5 Location of vent terminal. An open vent terminal from a drainage system shall not be located less than 4 feet (1219 mm) directly beneath any door, openable window, or other air intake opening of the building or of an adjacent building, nor shall any such vent terminal be within 10 feet (3048 mm) horizontally of such an opening unless it is at least 2 feet (610 mm) above the top of such opening.

P3103.6 Extension through the wall. Vent terminals extending through the wall shall terminate a minimum of 10 feet (3048 mm) from the lot line and 10 feet (3048 mm) above the highest adjacent grade within 10 feet (3048 mm) horizontally of the vent terminal. Vent terminals shall not terminate under the overhang of a structure with soffit vents. Side wall vent terminals shall be protected to prevent birds or rodents from entering or blocking the vent opening.

SECTION P3104
VENT CONNECTIONS AND GRADES

3104.1 Connection. All individual branch and circuit vents shall connect to a vent stack, stack vent or extend to the open air.

> **Exception:** Individual, branch and circuit vents shall be permitted to terminate at an air admittance valve in accordance with Section P3114.

P3104.2 Grade. Vent and branch vent pipes shall be graded, connected and supported to allow moisture and condensate to drain back to the soil or waste pipe by gravity.

P3104.3 Vent connection to drainage system. Every dry vent connecting to a horizontal drain shall connect above the centerline of the horizontal drain pipe.

P3104.4 Vertical rise of vent. Every dry vent shall rise vertically to a minimum of 6 inches (152 mm) above the flood level rim of the highest trap or trapped fixture being vented.

P3104.5 Height above fixtures. A connection between a vent pipe and a vent stack or stack vent shall be made at least 6 inches (152 mm) above the flood level rim of the highest fixture served by the vent. Horizontal vent pipes forming branch vents shall be at least 6 inches (152 mm) above the flood level rim of the highest fixture served.

P3104.6 Vent for future fixtures. Where the drainage piping has been roughed-in for future fixtures, a rough-in connection for a vent shall be installed a minimum of one-half the diameter of the drain. The vent rough-in shall connect to the vent system or shall be vented by other means as provided in this chapter. The connection shall be identified to indicate that the connection is a vent.

SECTION P3105
FIXTURE VENTS

P3105.1 Distance of trap from vent. Each fixture trap shall have a protecting vent located so that the slope and the developed length in the fixture drain from the trap weir to the vent fitting are within the requirements set forth in Table P3105.1.

Exception: The developed length of the fixture drain from the trap weir to the vent fitting for self-siphoning fixtures, such as water closets, shall not be limited.

TABLE P3105.1
MAXIMUM DISTANCE OF FIXTURE TRAP FROM VENT

SIZE OF TRAP (inches)	SLOPE (inch per foot)	DISTANCE FROM TRAP (feet)
$1^1/_4$	$^1/_4$	5
$1^1/_2$	$^1/_4$	6
2	$^1/_4$	8
3	$^1/_8$	12
4	$^1/_8$	16

For SI: 1 inch = 25.4 mm, 1 foot = 304.8 mm,
1 inch per foot = 83.3 mm/m.

P3105.2 Fixture drains. The total fall in a fixture drain resulting from pipe slope shall not exceed one pipe diameter, nor shall the vent pipe connection to a fixture drain, except for water closets, be below the weir of the trap.

P3105.3 Crown vent. A vent shall not be installed within two pipe diameters of the trap weir.

SECTION P3106
INDIVIDUAL VENT

P3106.1 Individual vent permitted. Each trap and trapped fixture is permitted to be provided with an individual vent. The individual vent shall connect to the fixture drain of the trap or trapped fixture being vented.

SECTION P3107
COMMON VENT

P3107.1 Individual vent as common vent. An individual vent is permitted to vent two traps or trapped fixtures as a common vent. The traps or trapped fixtures being common vented shall be located on the same floor level.

P3107.2 Connection at the same level. Where the fixture drains being common vented connect at the same level, the vent connection shall be at the interconnection of the fixture drains or downstream of the interconnection.

P3107.3 Connection at different levels. Where the fixture drains connect at different levels, the vent shall connect as a vertical extension of the vertical drain. The vertical drain pipe connecting the two fixture drains shall be considered the vent for the lower fixture drain, and shall be sized in accordance with Table P3107.3. The upper fixture shall not be a water closet.

TABLE P3107.3
COMMON VENT SIZES

PIPE SIZE (inches)	MAXIMUM DISCHARGE FROM UPPER FIXTURE DRAIN (d.f.u.)
$1^1/_2$	1
2	4
$2^1/_2$ to 3	6

For SI: 1 inch = 25.4 mm.

SECTION P3108
WET VENTING

P3108.1 Horizontal wet vent permitted. Any combination of fixtures within two bathroom groups located on the same floor level are permitted to be vented by a horizontal wet vent. The wet vent shall be considered the vent for the fixtures and shall extend from the connection of the dry vent along the direction of the flow in the drain pipe to the most downstream fixture drain connection. Each fixture drain shall connect horizontally to the horizontal branch being wet vented or shall have a dry vent. Only the fixtures within the bathroom groups shall connect to the wet-vented horizontal branch drain. Any additional fixtures shall discharge downstream of the horizontal wet vent.

P3108.2 Vent connections. The dry vent connection to the wet vent shall be an individual vent or common vent to the lavatory, bidet, shower or bathtub. In vertical wet vent systems, the most upstream fixture drain connection shall be a dry-vented fixture drain connection. In horizontal wet-vent systems, not more than one wet-vented fixture drain shall discharge upstream of the dry-vented fixture drain connection.

P3108.3 Size. Horizontal and vertical wet vents shall be of a minimum size as specified in Table P3108.3, based on the fixture unit discharge to the wet vent. The dry vent serving the wet vent shall be sized based on the largest required diameter of pipe within the wet-vent system served by the dry vent.

TABLE P3108.3
WET VENT SIZE

TABLE P3108.3
WET VENT SIZE

WET VENT PIPE SIZE (inches)	FIXTURE UNIT LOAD (d.f.u.)
$1^1/_2$	1
2	4
$2^1/_2$	6
3	12
4	32

For SI: 1 inch = 25.4 mm.

P3108.4 Vertical wet vent permitted. A combination of fixtures located on the same floor level are permitted to be vented by a vertical wet vent. The vertical wet vent shall be considered the vent for the fixtures and shall extend from the connection of the dry vent down to the lowest fixture drain connection. Each wet-vented fixture shall connect independently to the vertical wet vent. All water closet drains shall connect at the same elevation. Other fixture drains shall connect above or at the same elevation as the water closet fixture drains. The dry vent connection to the vertical wet vent shall be an individual or common vent serving one or two fixtures.

P3108.5 Trap weir to wet vent distances. The maximum developed length of wet-vented fixture drains shall comply with Table P3105.1.

SECTION P3109
WASTE STACK VENT

P3109.1 Waste stack vent permitted. A waste stack shall be considered a vent for all of the fixtures discharging to the stack where installed in accordance with the requirements of this section.

P3109.2 Stack installation. The waste stack shall be vertical, and both horizontal and vertical offsets shall be prohibited between the lowest fixture drain connection and the highest fixture drain connection to the stack. Every fixture drain shall connect separately to the waste stack. The stack shall not receive the discharge of water closets or urinals.

P3109.3 Stack vent. A stack vent shall be installed for the waste stack. The size of the stack vent shall be not less than the size of the waste stack. Offsets shall be permitted in the stack vent and shall be located at least 6 inches (152 mm) above the flood level of the highest fixture, and shall be in accordance with Section P3104.5. The stack vent shall be permitted to connect with other stack vents and vent stacks in accordance with Section P3113.3.

P3109.4 Waste stack size. The waste stack shall be sized based on the total discharge to the stack and the discharge within a branch interval in accordance with Table P3109.4. The waste stack shall be the same size throughout the length of the waste stack.

TABLE P3109.4
WASTE STACK VENT SIZE

STACK SIZE (inches)	MAXIMUM NUMBER OF FIXTURE UNITS (d.f.u.)	
	Total discharge into one branch interval	Total discharge for stack
$1^1/_2$	1	2
2	2	4
$2^1/_2$	No limit	8
3	No limit	24
4	No limit	50

For SI: 1 inch = 25.4 mm.

SECTION P3110
CIRCUIT VENTING

P3110.1 Circuit vent permitted. A maximum of eight fixtures connected to a horizontal branch drain shall be permitted to be circuit vented. Each fixture drain shall connect horizontally to the horizontal branch being circuit vented. The horizontal branch drain shall be classified as a vent from the most downstream fixture drain connection to the most upstream fixture drain connection to the horizontal branch.

P3110.2 Vent connection. The circuit vent connection shall be located between the two most upstream fixture drains. The vent shall connect to the horizontal branch and shall be installed in accordance with Section P3104. The circuit vent pipe shall not receive the discharge of any soil or waste.

P3110.3 Slope and size of horizontal branch. The maximum slope of the vent section of the horizontal branch drain shall be one unit vertical in 12 units horizontal (8-percent slope). The entire length of the vent section of the horizontal branch drain shall be sized for the total drainage discharge to the branch in accordance with Table P3005.4.1.

P3110.4 Additional fixtures. Fixtures, other than the circuit vented fixtures are permitted to discharge, to the horizontal branch drain. Such fixtures shall be located on the same floor as the circuit vented fixtures and shall be either individually or common vented.

SECTION P3111
COMBINATION WASTE AND VENT SYSTEM

P3111.1 Type of fixtures. A combination waste and vent system shall not serve fixtures other than floor drains, standpipes, sinks, lavatories and drinking fountains. A combination waste and vent system shall not receive the discharge of a food waste grinder.

P3111.2 Installation. The only vertical pipe of a combination drain and vent system shall be the connection between the fixture drain and the horizontal combination waste and vent pipe. The maximum vertical distance shall be 8 feet (2438 mm).

P3111.2.1 Slope. The horizontal combination waste and vent pipe shall have a maximum slope of $^1/_2$ unit vertical in 12 units horizontal (4-percent slope). The minimum slope shall be in accordance with Section P3005.3.

P3111.2.2 Connection. The combination waste and vent pipe shall connect to a horizontal drain that is vented or a vent shall connect to the combination waste and vent. The vent connecting to the combination waste and vent pipe shall extend vertically a minimum of 6 inches (152 mm) above the flood level rim of the highest fixture being vented before offsetting horizontally.

P3111.2.3 Vent size. The vent shall be sized for the total fixture unit load in accordance with Section P3113.1.

P3111.2.4 Fixture branch or drain. The fixture branch or fixture drain shall connect to the combination waste and vent within a distance specified in Table P3105.1. The combination waste and vent pipe shall be considered the vent for the fixture.

P3111.3 Size. The minimum size of a combination waste and vent pipe shall be in accordance with Table P3111.3.

TABLE P3111.3
SIZE OF COMBINATION WASTE AND VENT PIPE

DIAMETER PIPE (inches)	MAXIMUM NUMBER OF FIXTURE UNITS (d.f.u.)	
	Connecting to a horizontal branch or stack	Connecting to a building drain or building subdrain
2	3	4
$2^1/_2$	6	26
3	12	31
4	20	50

For SI: 1 inch = 25.4 mm.

SECTION P3112
ISLAND FIXTURE VENTING

P3112.1 Limitation. Island fixture venting shall not be permitted for fixtures other than sinks and lavatories. Kitchen sinks with a dishwasher waste connection, a food waste grinder, or both, in combination with the kitchen sink waste, shall be permitted to be vented in accordance with this section.

P3112.2 Vent connection. The island fixture vent shall connect to the fixture drain as required for an individual or common vent. The vent shall rise vertically to above the drainage outlet of the fixture being vented before offsetting horizontally or vertically downward. The vent or branch vent for multiple island fixture vents shall extend to a minimum of 6 inches (152 mm) above the highest island fixture being vented before connecting to the outside vent terminal.

P3112.3 Vent installation below the fixture flood level rim. The vent located below the flood level rim of the fixture being vented shall be installed as required for drainage piping in accordance with Chapter 30, except for sizing. The vent shall be sized in accordance with Section P3113.1. The lowest point of the island fixture vent shall connect full size to the drainage system. The connection shall be to a vertical drain pipe or to the

top half of a horizontal drain pipe. Cleanouts shall be provided in the island fixture vent to permit rodding of all vent piping located below the flood level rim of the fixtures. Rodding in both directions shall be permitted through a cleanout.

SECTION P3113
VENT PIPE SIZING

P3113.1 Size of vents. The minimum required diameter of individual vents, branch vents, circuit vents, vent stacks and stack vents shall be at least one-half the required diameter of the drain served. The required size of the drain shall be determined in accordance with Chapter 30. Vent pipes shall be not less than $1^1/_4$ inches (32 mm) in diameter. Vents exceeding 40 feet (12 192 mm) in developed length shall be increased by one nominal pipe size for the entire developed length of the vent pipe.

P3113.2 Developed length. The developed length of individual, branch, and circuit vents shall be measured from the farthest point of vent connection to the drainage system, to the point of connection to the vent stack, stack vent or termination outside of the building.

P3113.3 Branch vents. Where branch vents are connected to a common branch vent, the common branch vent shall be sized in accordance with this section, based on the size of the common horizontal drainage branch that is or would be required to serve the total drainage fixture unit (dfu) load being vented.

P3113.4 Sump vents. Sump vent sizes shall be determined in accordance with Sections P3113.4.1 and P3113.4.2.

P3113.4.1 Sewage pumps and sewage ejectors other than pneumatic. Drainage piping below sewer level shall be vented in a manner similar to that of a gravity system. Building sump vent sizes for sumps with sewage pumps or sewage ejectors, other than pneumatic, shall be determined in accordance with Table P3113.4.1.

P3113.4.2 Pneumatic sewage ejectors. The air pressure relief pipe from a pneumatic sewage ejector shall be connected to an independent vent stack terminating as required for vent extensions through the roof. The relief pipe shall be sized to relieve air pressure inside the ejector to atmospheric pressure, but shall not be less than $1^1/_4$ inches (32 mm) in size.

SECTION P3114
AIR ADMITTANCE VALVES

P3114.1 General. Vent systems using air admittance valves shall comply with this section. Individual and branch-type air admittance valves shall conform to ASSE 1051. Stack-type air admittance valves shall conform to ASSE 1050.

P3114.2 Installation. The valves shall be installed in accordance with the requirements of this section and the manufacturer's installation instructions. Air admittance valves shall be installed after the DWV testing required by Section P2503.5.1 or P2503.5.2 has been performed.

P3114.3 Where permitted. Individual vents, branch vents, circuit vents and stack vents shall be permitted to terminate with a connection to an air admittance valve.

P3114.4 Location. Individual and branch air admittance valves shall be located a minimum of 4 inches (102 mm) above the horizontal branch drain or fixture drain being vented. Stack-type air admittance valves shall be located a minimum of 6 inches (152 mm) above the flood level rim of the highest fixture being vented. The air admittance valve shall be located within the maximum developed length permitted for the vent. The air admittance valve shall be installed a minimum of 6 inches (152 mm) above insulation materials where installed in attics.

P3114.5 Access and ventilation. Access shall be provided to all air admittance valves. The valve shall be located within a ventilated space that allows air to enter the valve.

P3114.6 Size. The air admittance valve shall be rated for the size of the vent to which the valve is connected.

P3114.7 Vent required. Within each plumbing system, a minimum of one stack vent or a vent stack shall extend outdoors to the open air.

TABLE P3113.4.1
SIZE AND LENGTH OF SUMP VENTS

DISCHARGE CAPACITY OF PUMP (gpm)	MAXIMUM DEVELOPED LENGTH OF VENT (feet)[a]				
	Diameter of vent (inches)				
	1¼	1½	2	2½	3
10	No limit[b]	No limit	No limit	No limit	No limit
20	270	No limit	No limit	No limit	No limit
40	72	160	No limit	No limit	No limit
60	31	75	270	No limit	No limit

For SI: 1 inch = 25.4 mm, 1 foot = 304.8 mm, 1 gallon per minute (gpm) = 3.785 L/m.

a. Developed length plus an appropriate allowance for entrance losses and friction caused by fittings, changes in direction and diameter. Suggested allowances shall be obtained from NBS Monograph 31 or other approved sources. An allowance of 50 percent of the developed length shall be assumed if a more precise value is not available.

b. Actual values greater than 500 feet.

CHAPTER 32

TRAPS

SECTION P3201
FIXTURE TRAPS

P3201.1 Design of traps. Traps shall be of standard design, shall have smooth uniform internal waterways, shall be self-cleaning and shall not have interior partitions except where integral with the fixture. Traps shall be constructed of lead, cast iron, cast or drawn brass or approved plastic. Tubular brass traps shall be not less than No. 20 gage (0.8 mm) thickness. Solid connections, slip joints and couplings are permitted to be used on the trap inlet, trap outlet, or within the trap seal. Slip joints shall be accessible.

P3201.2 Trap seals and trap seal protection. Traps shall have a liquid seal not less than 2 inches (51 mm) and not more than 4 inches (102 mm). Traps for floor drains shall be fitted with a trap primer or shall be of the deep seal design.

P3201.3 Trap setting and protection. Traps shall be set level with respect to their water seals and shall be protected from freezing. Trap seals shall be protected from siphonage, aspiration or back pressure by an approved system of venting (see Section P3101).

P3201.4 Building traps. Building traps shall not be installed, except in special cases where sewer gases are extremely corrosive or noxious, as directed by the building official.

P3201.5 Prohibited trap designs. The following types of traps are prohibited:

1. Bell traps.

2. Separate fixture traps with interior partitions, except those lavatory traps made of plastic, stainless steel or other corrosion-resistant material.

3. "S" traps.

4. Drum traps.

5. Trap designs with moving parts.

P3201.6 Number of fixtures per trap. Each plumbing fixture shall be separately trapped by a water seal trap. The vertical distance from the fixture outlet to the trap weir shall not exceed 24 inches (610 mm) and the horizontal distance shall not exceed 30 inches (762 mm) measured from the center line of the fixture outlet to the centerline of the inlet of the trap. The height of a clothes washer standpipe above a trap shall conform to Section P2706.2. Fixtures shall not be double trapped.

Exceptions:

1. Fixtures that have integral traps.

2. A single trap shall be permitted to serve two or three like fixtures limited to kitchen sinks, laundry tubs and lavatories. Such fixtures shall be adjacent to each other and located in the same room with a continuous waste arrangement. The trap shall be installed at the center fixture where three fixtures are installed. Com-

mon trapped fixture outlets shall be not more than 30 inches (762 mm) apart.

3. Connection of a laundry tray waste line into a standpipe for the automatic clothes-washer drain is permitted in accordance with Section P2706.2.1.

P3201.7 Size of fixture traps. Fixture trap size shall be sufficient to drain the fixture rapidly and not less than the size indicated in Table P3201.7. A trap shall not be larger than the drainage pipe into which the trap discharges.

TABLE P3201.7
SIZE OF TRAPS AND TRAP ARMS FOR PLUMBING FIXTURES

PLUMBING FIXTURE	TRAP SIZE MINIMUM (inches)
Bathtub (with or without shower head and/or whirlpool attachments)	$1^1/_2$
Bidet	$1^1/_4$
Clothes washer standpipe	2
Dishwasher (on separate trap)	$1^1/_2$
Floor drain	2
Kitchen sink (one or two traps, with or without dishwasher and garbage grinder)	$1^1/_2$
Laundry tub (one or more compartments)	$1^1/_2$
Lavatory	$1^1/_4$
Shower (based on the total flow rate through showerheads and bodysprays) Flow rate: 5.7 gpm and less / More than 5.7 gpm up to 12.3 gpm / More than 12.3 gpm up to 25.8 gpm / More than 25.8 gpm up to 55.6 gpm	$1^1/_2$ 2 3 4
Water closet	Note a

For SI: 1 inch = 25.4 mm.

a. Consult fixture standards for trap dimensions of specific bowls.

Part VIII — Electrical

CHAPTER 33

GENERAL REQUIREMENTS

This Electrical Part (Chapters 33 through 42) is produced and copyrighted by the National Fire Protection Association (NFPA) and is based on the 2005 *National Electrical Code*® (NEC®) (NFPA 70-2005), copyright 2005 National Fire Protection Association, all rights reserved. Use of the Electrical Part is pursuant to license with the NFPA.

The title *National Electrical Code*® and the acronym NEC® are registered trademarks of the National Fire Protection Association, Quincy, Massachusetts. See Appendix Q, *International Residential Code* Electrical Provisions/National Electrical Code Cross Reference.

SECTION E3301
GENERAL

E3301.1 Applicability. The provisions of Chapters 33 through 42 shall establish the general scope of the electrical system and equipment requirements of this code. Chapters 33 through 42 cover those wiring methods and materials most commonly encountered in the construction of one- and two-family dwellings and structures regulated by this code. Other wiring methods, materials and subject matter covered in the NFPA 70 are also allowed by this code.

E3301.2 Scope. Chapters 33 through 42 shall cover the installation of electrical systems, equipment and components indoors and outdoors that are within the scope of this code, including services, power distribution systems, fixtures, appliances, devices and appurtenances. Services within the scope of this code shall be limited to 120/240-volt, 0- to 400-ampere, single-phase systems. These chapters specifically cover the equipment, fixtures, appliances, wiring methods and materials that are most commonly used in the construction or alteration of one- and two-family dwellings and accessory structures regulated by this code. The omission from these chapters of any material or method of construction provided for in the referenced standard NFPA 70 shall not be construed as prohibiting the use of such material or method of construction. Electrical systems, equipment or components not specifically covered in these chapters shall comply with the applicable provisions of the NFPA 70.

E3301.3 Not covered. Chapters 33 through 42 do not cover the following:

1. Installations, including associated lighting, under the exclusive control of communications utilities and electric utilities.

2. Services over 400 amperes.

E3301.4 Additions and alterations. Any addition or alteration to an existing electrical system shall be made in conformity with the provisions of Chapters 33 through 42. Where additions subject portions of existing systems to loads exceeding those permitted herein, such portions shall be made to comply with Chapters 33 through 42.

SECTION E3302
BUILDING STRUCTURE PROTECTION

E3302.1 Drilling and notching. Wood-framed structural members shall not be drilled, notched or altered in any manner except as provided for in this code.

E3302.2 Penetrations of fire-resistance-rated assemblies. Electrical installations in hollow spaces, vertical shafts and ventilation or air-handling ducts shall be made so that the possible spread of fire or products of combustion will not be substantially increased. Electrical penetrations through fire-resistance-rated walls, partitions, floors or ceilings shall be protected by approved methods to maintain the fire-resistance rating of the element penetrated. Penetrations of fire-resistance-rated walls shall be limited as specified in Section R317.3.

E3302.3 Penetrations of firestops and draftstops. Penetrations through fire blocking and draftstopping shall be protected in an approved manner to maintain the integrity of the element penetrated.

SECTION E3303
INSPECTION AND APPROVAL

E3303.1 Approval. Electrical materials, components and equipment shall be approved.

E3303.2 Inspection required. New electrical work and parts of existing systems affected by new work or alterations shall be inspected by the building official to ensure compliance with the requirements of Chapters 33 through 42.

E3303.3 Listing and labeling. Electrical materials, components, devices, fixtures and equipment shall be listed for the application, shall bear the label of an approved agency and shall be installed, and used, or both, in accordance with the manufacturer's installation instructions.

SECTION E3304
GENERAL EQUIPMENT REQUIREMENTS

E3304.1 Voltages. Throughout Chapters 33 through 42, the voltage considered shall be that at which the circuit operates.

E3304.2 Interrupting rating. Equipment intended to interrupt current at fault levels shall have a minimum interrupting rating of 10,000 amperes. Equipment intended to interrupt current at levels other than fault levels shall have an interrupting rating at nominal circuit voltage sufficient for the current that must be interrupted.

E3304.3 Circuit characteristics. The overcurrent protective devices, total impedance, component short-circuit current ratings and other characteristics of the circuit to be protected shall be so selected and coordinated as to permit the circuit protective devices that are used to clear a fault to do so without extensive damage to the electrical components of the circuit. This fault shall be assumed to be either between two or more of the circuit conductors or between any circuit conductor and the grounding conductor or enclosing metal raceway. Listed products applied in accordance with their listing shall be considered to meet the requirements of this section.

E3304.4 Protection of equipment. Equipment identified only as "dry locations," "Type 1," or "indoor use only" shall be protected against permanent damage from the weather during building construction.

E3304.5 Unused openings. Unused cable or raceway openings in boxes, cabinets, meter socket enclosures, equipment cases or housings shall be effectively closed to afford protection substantially equivalent to the wall of the equipment. Where metallic plugs or plates are used with nonmetallic enclosures they shall be recessed at least $^1/_4$ inch (6 mm) from the outer surface of the enclosure.

E3304.6 Integrity of electrical equipment. Internal parts of electrical equipment, including busbars, wiring terminals, insulators and other surfaces, shall not be damaged or contaminated by foreign materials such as paint, plaster, cleaners or abrasives, and corrosive residues. There shall not be any damaged parts that might adversely affect safe operation or mechanical strength of the equipment such as parts that are broken; bent; cut; deteriorated by corrosion, chemical action, or overheating. Foreign debris shall be removed from equipment.

E3304.7 Mounting. Electrical equipment shall be firmly secured to the surface on which it is mounted. Wooden plugs driven into masonry, concrete, plaster, or similar materials shall not be used.

E3304.8 Energized parts guarded against accidental contact. Approved enclosures shall guard energized parts that are operating at 50 volts or more against accidental contact.

E3304.9 Prevent physical damage. In locations where electrical equipment is likely to be exposed to physical damage, enclosures or guards shall be so arranged and of such strength as to prevent such damage.

E3304.10 Equipment identification. The manufacturer's name, trademark or other descriptive marking by which the organization responsible for the product can be identified shall be placed on all electric equipment. Other markings shall be provided that indicate voltage, current, wattage or other ratings as specified elsewhere in Chapters 33 through 42. The marking shall have the durability to withstand the environment involved.

E3304.11 Identification of disconnecting means. Each disconnecting means shall be legibly marked to indicate its purpose, except where located and arranged so that the purpose is evident. The marking shall have the durability to withstand the environment involved.

SECTION E3305
EQUIPMENT LOCATION AND CLEARANCES

E3305.1 Working space and clearances. Sufficient access and working space shall be provided and maintained around all electrical equipment to permit ready and safe operation and maintenance of such equipment in accordance with this section and Figure E3305.1.

E3305.2 Working clearances for energized equipment and panelboards. Except as otherwise specified in Chapters 33 through 42, the dimension of the working space in the direction of access to panelboards and live parts likely to require examination, adjustment, servicing or maintenance while energized shall be not less than 36 inches (914 mm) in depth. Distances shall be measured from the energized parts where such parts are exposed or from the enclosure front or opening where such parts are enclosed. In addition to the 36-inch dimension (914 mm), the work space shall not be less than 30 inches (762 mm) wide in front of the electrical equipment and not less than the width of such equipment. The work space shall be clear and shall extend from the floor or platform to a height of 6.5 feet (1981 mm). In all cases, the work space shall allow at least a 90-degree opening of equipment doors or hinged panels. Equipment associated with the electrical installation located above or below the electrical equipment shall be permitted to extend not more than 6 inches (152 mm) beyond the front of the electrical equipment.

E3305.3 Dedicated panelboard space. The space equal to the width and depth of the panelboard and extending from the floor to a height of 6 feet (1829 mm) above the panelboard, or to the structural ceiling, whichever is lower, shall be dedicated to the electrical installation. Piping, ducts, leak protection apparatus and other equipment foreign to the electrical installation shall not be installed in such dedicated space. The area above the dedicated space shall be permitted to contain foreign systems, provided that protection is installed to avoid damage to the electrical equipment from condensation, leaks and breaks in such foreign systems (see Figure E3305.1).

> **Exception:** Suspended ceilings with removable panels shall be permitted within the 6-foot (1.8 m) dedicated space.

E3305.4 Location of working spaces and equipment. Required working space shall not be designated for storage. Panelboards and overcurrent protection devices shall not be located in clothes closets or bathrooms.

E3305.5 Access and entrance to working space. Access shall be provided to the required working space.

E3305.6 Illumination. Artificial illumination shall be provided for all working spaces for service equipment and panelboards installed indoors.

E3305.7 Headroom. The minimum headroom for working spaces for service equipment and panelboards shall be 6.5 feet (1981 mm).

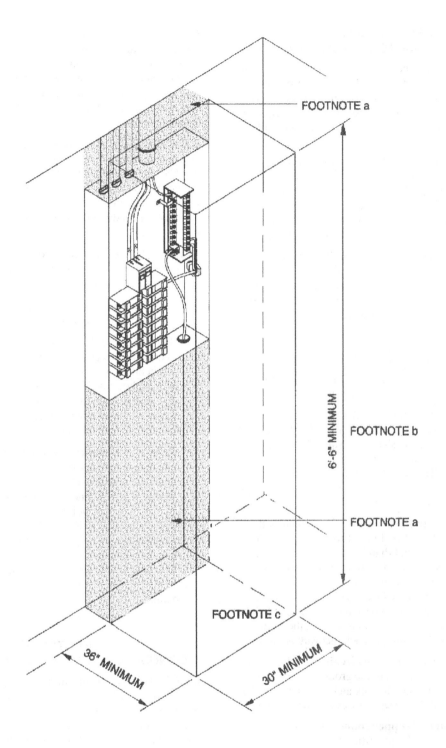

FIGURE E3305.1[a, b, c, d, e]
WORKING SPACE AND CLEARANCES

For SI: 1 inch = 25.4 mm, 1 foot = 304.8 mm.

a. Equipment, piping and ducts foreign to the electrical installation shall not be placed in the shaded areas extending from the floor to a height of 6 feet above the panelboard enclosure, or to the structural ceiling, whichever is lower.

b. The working space shall be clear and unobstructed from the floor to a height of 6.5 feet.

c. The working space shall not be designated for storage.

d. Panelboards, service equipment and similar enclosures shall not be located in bathrooms, toilet rooms and clothes closets.

e. Such work spaces shall be provided with artificial lighting where located indoors.

SECTION E3306
ELECTRICAL CONDUCTORS AND CONNECTIONS

E3306.1 General. This section provides general requirements for conductors, connections and splices. These requirements do not apply to conductors that form an integral part of equipment, such as motors, appliances and similar equipment, or to conductors specifically provided for elsewhere in Chapters 33 through 42.

E3306.2 Conductor material. Conductors used to conduct current shall be of copper except as otherwise provided in Chapters 33 through 42. Where the conductor material is not specified, the material and the sizes given in these chapters shall apply to copper conductors. Where other materials are used, the conductor sizes shall be changed accordingly.

E3306.3 Minimum size of conductors. The minimum size of conductors for feeders and branch circuits shall be 14 AWG copper and 12 AWG aluminum. The minimum size of service conductors shall be as specified in Chapter 35. The minimum size of class 2 remote control, signaling and power-limited circuits conductors shall be as specified in Chapter 42.

E3306.4 Stranded conductors. Where installed in raceways, conductors of size 8 AWG and larger shall be stranded. A solid 8 AWG conductor shall be permitted to be installed in a raceway only to meet the requirements of Section E4104.

E3306.5 Individual conductor insulation. Except where otherwise permitted in Sections E3505.1 and E3808.9, and E4203, current-carrying conductors shall be insulated. Insulated conductors shall have insulation types identified as RH, RHW, RHW-2, THHN, THHW, THW, THW-2, THWN, THWN-2, TW, UF, USE, USE-2, XHHW or XHHW-2. Insulation types shall be approved for the application.

E3306.6 Conductors in parallel. Circuit conductors that are connected in parallel shall be limited to sizes 1/0 AWG and larger. Conductors in parallel shall be of the same length, same conductor material, same circular mil area and same insulation type. Conductors in parallel shall be terminated in the same manner. Where run in separate raceways or cables, the raceway or cables shall have the same physical characteristics. Where conductors are in separate raceways or cables, the same number of conductors shall be used in each raceway or cable.

E3306.7 Conductors of the same circuit. All conductors of the same circuit and, where used, the grounded conductor and all equipment grounding conductors and bonding conductors shall be contained within the same raceway, cable or cord.

E3306.8 Aluminum and copper connections. Terminals and splicing connectors shall be identified for the material of the conductors joined. Conductors of dissimilar metals shall not be joined in a terminal or splicing connector where physical contact occurs between dissimilar conductors such as copper and aluminum, copper and copper-clad aluminum, or aluminum and copper-clad aluminum, except where the device is listed for the purpose and conditions of application. Materials such as inhibitors and compounds shall be suitable for the application and shall be of a type that will not adversely affect the conductors, installation or equipment.

E3306.9 Terminals. Connection of conductors to terminal parts shall be made without damaging the conductors and shall be made by means of pressure connectors, including set-screw type, by means of splices to flexible leads, or for conductor sizes of 10 AWG and smaller, by means of wire binding screws or studs and nuts having upturned lugs or the equivalent. Terminals for more than one conductor and terminals for connecting aluminum conductors shall be identified for the application.

E3306.10 Splices. Conductors shall be spliced or joined with splicing devices listed for the purpose. Splices and joints and the free ends of conductors shall be covered with an insulation equivalent to that of the conductors or with an insulating device listed for the purpose. Wire connectors or splicing means installed on conductors for direct burial shall be listed for such use.

E3306.10.1 Continuity. Conductors in raceways shall be continuous between outlets, boxes, and devices and shall be without splices or taps in the raceway.

Exception: Splices shall be permitted within surface-mounted raceways that have a removable cover.

E3306.10.2 Device connections. The continuity of a grounded conductor in multiwire branch circuits shall not be dependent on connection to devices such as receptacles and lampholders. The arrangement of grounding connections shall be such that the disconnection or the removal of a receptacle, luminaire or other device fed from the box does not interfere with or interrupt the grounding continuity.

E3306.10.3 Length of conductor for splice or termination. Where conductors are to be spliced, terminated or connected to fixtures or devices, a minimum length of 6 inches (150 mm) of free conductor shall be provided at each outlet, junction or switch point. The required length shall be measured from the point in the box where the conductor emerges from its raceway or cable sheath. Where the opening to an outlet, junction or switch point is less than 8 inches (200 mm) in any dimension, each conductor shall be long enough to extend at least 3 inches (75 mm) outside of such opening.

SECTION E3307
CONDUCTOR AND TERMINAL IDENTIFICATION

E3307.1 Grounded conductors. Insulated grounded conductors of sizes 6 AWG or smaller shall be identified by a continuous white or gray outer finish or by three continuous white stripes on other than green insulation along the entire length of the conductors. Conductors of sizes larger than 6 AWG shall be identified either by a continuous white or gray outer finish or by three continuous white stripes on other than green insulation along its entire length or at the time of installation by a distinctive white or gray marking at its terminations. This marking shall encircle the conductor or insulation.

E3307.2 Equipment grounding conductors. Equipment grounding conductors of sizes 6 AWG and smaller shall be identified by a continuous green color or a continuous green color with one or more yellow stripes on the insulation or covering, except where bare. Conductors with insulation or individual covering that is green, green with one or more yellow stripes, or otherwise identified as permitted by this section shall not be used for ungrounded or grounded circuit conductors.

Equipment grounding conductors larger than 6 AWG that are not identified as required for conductors of sizes 6 AWG and smaller shall, at the time of installation, be permanently identified as an equipment grounding conductor at each end and at every point where the conductor is accessible, except where such conductors are bare.

The required identification for conductors larger than 6 AWG shall encircle the conductor and shall be accomplished by one of the following:

1. Stripping the insulation or covering from the entire exposed length.

2. Coloring the exposed insulation or covering green.

3. Marking the exposed insulation or covering with green tape or green adhesive labels.

Exception: Conductors larger than 6 AWG shall not be required to be identified in conduit bodies that do not contain splices or unused hubs.

E3307.3 Ungrounded conductors. Insulation on the ungrounded conductors shall be a continuous color other than white, gray and green.

Exceptions:

1. An insulated conductor that is part of a cable or flexible cord assembly and that has a white or gray finish or a finish marking with three continuous white stripes shall be permitted to be used as an ungrounded conductor where it is permanently reidentified to indicate its use as an ungrounded conductor at all terminations and at each location where the conductor is visible and accessible. Identification shall encircle the insulation and shall be a color other than white, gray, and green.

2. Where a cable assembly contains an insulated conductor for single-pole, 3-way or 4-way switch loops and the conductor with white or gray insulation or a marking of three continuous white stripes is used for the supply to the switch but not as a return conductor from the switch to the switched outlet. In these applications, the conductor with white or gray insulation or with three continuous white stripes shall be permanently reidentified to indicate its use by painting or other effective means at its terminations and at each location where the conductor is visible and accessible.

E3307.4 Identification of terminals. Terminals for attachment to conductors shall be identified in accordance with Sections E3307.4.1 and E3307.4.2.

E3307.4.1 Device terminals. All devices excluding panelboards, provided with terminals for the attachment of conductors and intended for connection to more than one side of the circuit shall have terminals properly marked for identification, except where the terminal intended to be connected to the grounded conductor is clearly evident.

Exception: Terminal identification shall not be required for devices that have a normal current rating of over 30 amperes, other than polarized attachment caps and polarized receptacles for attachment caps as required in Section E3307.4.2.

E3307.4.2 Receptacles, plugs, and connectors. Receptacles, polarized attachment plugs and cord connectors for plugs and polarized plugs shall have the terminal intended for connection to the grounded (white) conductor identified. Identification shall be by a metal or metal coating substantially white in color or by the word "white" or the letter "W" located adjacent to the identified terminal. Where the terminal is not visible, the conductor entrance hole for the connection shall be colored white or marked with the word "white" or the letter "W."

CHAPTER 34

ELECTRICAL DEFINITIONS

SECTION E3401
GENERAL

E3401.1 Scope. This chapter contains definitions that shall apply only to the electrical requirements of Chapters 33 through 42. Unless otherwise expressly stated, the following terms shall, for the purpose of this code, have the meanings indicated in this chapter. Words used in the present tense include the future; the singular number includes the plural and the plural the singular. Where terms are not defined in this section and are defined in Section R202 of this code, such terms shall have the meanings ascribed to them in that section. Where terms are not defined in these sections, they shall have their ordinarily accepted meanings or such as the context implies.

ACCESSIBLE. (As applied to equipment.) Admitting close approach; not guarded by locked doors, elevation or other effective means.

ACCESSIBLE. (As applied to wiring methods.) Capable of being removed or exposed without damaging the building structure or finish, or not permanently closed in by the structure or finish of the building.

ACCESSIBLE, READILY. Capable of being reached quickly for operation, renewal or inspections, without requiring those to whom ready access is requisite to climb over or remove obstacles or to resort to portable ladders, etc.

AMPACITY. The current in amperes that a conductor can carry continuously under the conditions of use without exceeding its temperature rating.

APPLIANCE. Utilization equipment, normally built in standardized sizes or types, that is installed or connected as a unit to perform one or more functions such as clothes washing, air conditioning, food mixing, deep frying, etc.

APPROVED. Acceptable to the authority having jurisdiction.

ARC-FAULT CIRCUIT INTERRUPTER. A device intended to provide protection from the effects of arc-faults by recognizing characteristics unique to arcing and by functioning to de-energize the circuit when an arc-fault is detected.

ATTACHMENT PLUG (PLUG CAP) (PLUG). A device that, by insertion into a receptacle, establishes connection between the conductors of the attached flexible cord and the conductors connected permanently to the receptacle.

AUTOMATIC. Self-acting, operating by its own mechanism when actuated by some impersonal influence, as, for example, a change in current, pressure, temperature or mechanical configuration.

BATHROOM. An area, including a basin, with one or more of the following: a toilet, a tub or a shower.

BONDING. The permanent joining of metallic parts to form an electrically conductive path that will ensure electrical continuity and the capacity to conduct safely any current likely to be imposed.

BONDING JUMPER. A reliable conductor to ensure the required electrical conductivity between metal parts required to be electrically connected.

BONDING JUMPER (EQUIPMENT). The connection between two or more portions of the equipment grounding conductor.

BONDING JUMPER, MAIN. The connection between the grounded circuit conductor and the equipment grounding conductor at the service.

BRANCH CIRCUIT. The circuit conductors between the final overcurrent device protecting the circuit and the outlet(s).

BRANCH CIRCUIT, APPLIANCE. A branch circuit that supplies energy to one or more outlets to which appliances are to be connected, and that has no permanently connected luminaires that are not a part of an appliance.

BRANCH CIRCUIT, GENERAL PURPOSE. A branch circuit that supplies two or more receptacle outlets or outlets for lighting and appliances.

BRANCH CIRCUIT, INDIVIDUAL. A branch circuit that supplies only one utilization equipment.

BRANCH CIRCUIT, MULTIWIRE. A branch circuit consisting of two or more ungrounded conductors having voltage difference between them, and a grounded conductor having equal voltage difference between it and each ungrounded conductor of the circuit, and that is connected to the neutral or grounded conductor of the system.

CABINET. An enclosure designed either for surface or flush mounting and provided with a frame, mat or trim in which a swinging door or doors are or may be hung.

CIRCUIT BREAKER. A device designed to open and close a circuit by nonautomatic means and to open the circuit automatically on a predetermined overcurrent without damage to itself when properly applied within its rating.

CONCEALED. Rendered inaccessible by the structure or finish of the building. Wires in concealed raceways are considered concealed, even though they may become accessible by withdrawing them [see "Accessible (As applied to wiring methods)"].

CONDUCTOR

> **Bare.** A conductor having no covering or electrical insulation whatsoever.

> **Covered.** A conductor encased within material of composition or thickness that is not recognized by this code as electrical insulation.

> **Insulated.** A conductor encased within material of composition and thickness that is recognized by this code as electrical insulation.

CONDUIT BODY. A separate portion of a conduit or tubing system that provides access through a removable cover(s) to the interior of the system at a junction of two or more sections of the system or at a terminal point of the system. Boxes such as FS and FD or larger cast or sheet metal boxes are not classified as conduit bodies.

CONNECTOR, PRESSURE (SOLDERLESS). A device that establishes a connection between two or more conductors or between one or more conductors and a terminal by means of mechanical pressure and without the use of solder.

CONTINUOUS LOAD. A load where the maximum current is expected to continue for 3 hours or more.

COOKING UNIT, COUNTER-MOUNTED. A cooking appliance designed for mounting in or on a counter and consisting of one or more heating elements, internal wiring and built-in or separately mountable controls.

COPPER-CLAD ALUMINUM CONDUCTORS. Conductors drawn from a copper-clad aluminum rod with the copper metallurgically bonded to an aluminum core. The copper forms a minimum of 10 percent of the cross-sectional area of a solid conductor or each strand of a stranded conductor.

CUTOUT BOX. An enclosure designed for surface mounting and having swinging doors or covers secured directly to and telescoping with the walls of the box proper (see "Cabinet").

DEAD FRONT. Without live parts exposed to a person on the operating side of the equipment.

DEMAND FACTOR. The ratio of the maximum demand of a system, or part of a system, to the total connected load of a system or the part of the system under consideration.

DEVICE. A unit of an electrical system that is intended to carry or control but not utilize electric energy.

DISCONNECTING MEANS. A device, or group of devices, or other means by which the conductors of a circuit can be disconnected from their source of supply.

DWELLING

> **Dwelling unit.** A single unit, providing complete and independent living facilities for one or more persons, including permanent provisions for living, sleeping, cooking and sanitation.

> **One-family dwelling.** A building consisting solely of one dwelling unit.

> **Two-family dwelling.** A building consisting solely of two dwelling units.

ENCLOSED. Surrounded by a case, housing, fence or walls that will prevent persons from accidentally contacting energized parts.

ENCLOSURE. The case or housing of apparatus, or the fence or walls surrounding an installation, to prevent personnel from accidentally contacting energized parts or to protect the equipment from physical damage.

ENERGIZED. Electrically connected to, or is, a source of voltage.

EQUIPMENT. A general term including material, fittings, devices, appliances, luminaires, apparatus and the like used as a part of, or in connection with, an electrical installation.

EXPOSED. (As applied to live parts.) Capable of being inadvertently touched or approached nearer than a safe distance by a person. It is applied to parts not suitably guarded, isolated or insulated.

EXPOSED. (As applied to wiring methods.) On or attached to the surface or behind panels designed to allow access.

EXTERNALLY OPERABLE. Capable of being operated without exposing the operator to contact with live parts.

FEEDER. All circuit conductors between the service equipment, or the source of a separately derived system, or other power supply source and the final branch-circuit overcurrent device.

FITTING. An accessory such as a locknut, bushing or other part of a wiring system that is intended primarily to perform a mechanical rather than an electrical function.

GROUND. A conducting connection, whether intentional or accidental, between an electrical circuit or equipment and the earth, or to some conducting body that serves in place of the earth.

GROUNDED. Connected to earth or to some conducting body that serves in place of the earth.

GROUNDED, EFFECTIVELY. Intentionally connected to earth through a ground connection or connections of sufficiently low impedance and having sufficient current-carrying capacity to prevent the buildup of voltages that may result in undue hazards to connected equipment or to persons.

GROUNDED CONDUCTOR. A system or circuit conductor that is intentionally grounded.

GROUNDING CONDUCTOR. A conductor used to connect equipment or the grounded circuit of a wiring system to a grounding electrode or electrodes.

GROUNDING CONDUCTOR, EQUIPMENT. The conductor used to connect the noncurrent-carrying metal parts of equipment, raceways and other enclosures to the system grounded conductor, the grounding electrode conductor or both, at the service equipment, or at the source of a separately derived system.

GROUNDING ELECTRODE. A device that establishes an electrical connection to earth.

GROUNDING ELECTRODE CONDUCTOR. The conductor used to connect the grounding electrode(s) to the equipment grounding conductor, to the grounded conductor, or to both, at the service, at each building or structure where supplied by a feeder(s) or branch circuit(s), or at the source of a separately derived system.

GROUND-FAULT CIRCUIT-INTERRUPTER. A device intended for the protection of personnel that functions to de-energize a circuit or portion thereof within an established period of time when a current to ground exceeds the value for a Class A device.

GUARDED. Covered, shielded, fenced, enclosed or otherwise protected by means of suitable covers, casings, barriers, rails, screens, mats or platforms to remove the likelihood of approach or contact by persons or objects to a point of danger.

IDENTIFIED. (As applied to equipment.) Recognizable as suitable for the specific purpose, function, use, environment, application, etc., where described in a particular code requirement.

INTERRUPTING RATING. The highest current at rated voltage that a device is intended to interrupt under standard test conditions.

ISOLATED. (As applied to location.) Not readily accessible to persons unless special means for access are used.

LABELED. Equipment or materials to which has been attached a label, symbol or other identifying mark of an organization acceptable to the authority having jurisdiction and concerned with product evaluation that maintains periodic inspection of production of labeled equipment or materials and by whose labeling the manufacturer indicates compliance with appropriate standards or performance in a specified manner.

LIGHTING OUTLET. An outlet intended for the direct connection of a lampholder, a luminaire (lighting fixture) or a pendant cord terminating in a lampholder.

LISTED. Equipment, materials or services included in a list published by an organization that is acceptable to the authority having jurisdiction and concerned with evaluation of products or services, that maintains periodic inspection of production of listed equipment or materials or periodic evaluation of services, and whose listing states either that the equipment, material or services meets identified standards or has been tested and found suitable for a specified purpose.

LIVE PARTS. Energized conductive components.

LOCATION, DAMP. Location protected from weather and not subject to saturation with water or other liquids but subject to moderate degrees of moisture. Examples of such locations include partially protected locations under canopies, marquees, roofed open porches and like locations, and interior locations subject to moderate degrees of moisture, such as some basements, some barns and some cold-storage warehouses.

LOCATION, DRY. A location not normally subject to dampness or wetness. A location classified as dry may be temporarily subject to dampness or wetness, as in the case of a building under construction.

LOCATION, WET. Installations underground or in concrete slabs or masonry in direct contact with the earth and locations subject to saturation with water or other liquids, such as vehicle-washing areas, and locations exposed to weather.

LUMINAIRE. A complete lighting unit (lighting fixture) consisting of a lamp or lamps together with parts designed to distribute the light, to position and protect the lamps and ballast, where applicable, and to connect the lamps to the power supply.

MULTIOUTLET ASSEMBLY. A type of surface, or flush, or freestanding raceway; designed to hold conductors and receptacles, assembled in the field or at the factory.

OUTLET. A point on the wiring system at which current is taken to supply utilization equipment.

OVERCURRENT. Any current in excess of the rated current of equipment or the ampacity of a conductor. Such current might result from overload, short circuit or ground fault.

OVERLOAD. Operation of equipment in excess of normal, full-load rating, or of a conductor in excess of rated ampacity that, when it persists for a sufficient length of time, would cause damage or dangerous overheating. A fault, such as a short circuit or ground fault, is not an overload.

PANELBOARD. A single panel or group of panel units designed for assembly in the form of a single panel, including buses and automatic overcurrent devices, and equipped with or without switches for the control of light, heat or power circuits, designed to be placed in a cabinet or cutout box placed in or against a wall, partition or other support and accessible only from the front.

PLENUM. A compartment or chamber to which one or more air ducts are connected and that forms part of the air distribution system.

POWER OUTLET. An enclosed assembly that may include receptacles, circuit breakers, fuseholders, fused switches, buses and watt-hour meter mounting means, intended to supply and control power to mobile homes, recreational vehicles or boats, or to serve as a means for distributing power required to operate mobile or temporarily installed equipment.

PREMISES WIRING (SYSTEM). That interior and exterior wiring, including power, lighting, control and signal circuit wiring together with all of their associated hardware, fittings and wiring devices, both permanently and temporarily installed, that extends from the service point or source of power such as a battery, a solar photovoltaic system, or a generator, transformer, or converter winding, to the outlet(s). Such wiring does not include wiring internal to appliances, luminaires (fixtures), motors, controllers, and similar equipment.

QUALIFIED PERSON. One who has the skills and knowledge related to the construction and operation of the electrical equipment and installations and has received safety training on the hazards involved.

RACEWAY. An enclosed channel of metal or nonmetallic materials designed expressly for holding wires, cables, or busbars, with additional functions as permitted in this code. Raceways include, but are not limited to, rigid metal conduit, rigid nonmetallic conduit, intermediate metal conduit, liquid-tight flexible conduit, flexible metallic tubing, flexible metal conduit, electrical nonmetallic tubing, electrical metallic tubing, underfloor raceways, cellular concrete floor raceways, cellular metal floor raceways, surface raceways, wireways and busways.

RAINPROOF. Constructed, protected or treated so as to prevent rain from interfering with the successful operation of the apparatus under specified test conditions.

RAIN TIGHT. Constructed or protected so that exposure to a beating rain will not result in the entrance of water under specified test conditions.

RECEPTACLE. A receptacle is a contact device installed at the outlet for the connection of an attachment plug. A single receptacle is a single contact device with no other contact device on the same yoke. A multiple receptacle is two or more contact devices on the same yoke.

RECEPTACLE OUTLET. An outlet where one or more receptacles are installed.

SERVICE. The conductors and equipment for delivering energy from the serving utility to the wiring system of the premises served.

SERVICE CABLE. Service conductors made up in the form of a cable.

SERVICE CONDUCTORS. The conductors from the service point to the service disconnecting means.

SERVICE DROP. The overhead service conductors from the last pole or other aerial support to and including the splices, if any, connecting to the service-entrance conductors at the building or other structure.

SERVICE-ENTRANCE CONDUCTORS, OVERHEAD SYSTEM. The service conductors between the terminals of the service equipment and a point usually outside the building, clear of building walls, where joined by tap or splice to the service drop.

SERVICE-ENTRANCE CONDUCTORS, UNDERGROUND SYSTEM. The service conductors between the terminals of the service equipment and the point of connection to the service lateral.

SERVICE EQUIPMENT. The necessary equipment, usually consisting of a circuit breaker(s) or switch(es) and fuse(s), and their accessories, connected to the load end of the service conductors to a building or other structure, or an otherwise designated area, and intended to constitute the main control and cutoff of the supply.

SERVICE LATERAL. The underground service conductors between the street main, including any risers at a pole or other structure or from transformers, and the first point of connection to the service-entrance conductors in a terminal box or meter or other enclosure, inside or outside the building wall. Where there is no terminal box, meter or other enclosure with adequate space, the point of connection shall be considered to be the point of entrance of the service conductors into the building.

SERVICE POINT. Service point is the point of connection between the facilities of the serving utility and the premises wiring.

STRUCTURE. That which is built or constructed.

SWITCHES

General-use switch. A switch intended for use in general distribution and branch circuits. It is rated in amperes and is capable of interrupting its rated current at its rated voltage.

General-use snap switch. A form of general-use switch constructed so that it can be installed in device boxes or on box covers or otherwise used in conjunction with wiring systems recognized by this code.

Isolating switch. A switch intended for isolating an electric circuit from the source of power. It has no interrupting rating and is intended to be operated only after the circuit has been opened by some other means.

Motor-circuit switch. A switch, rated in horsepower that is capable of interrupting the maximum operating overload current of a motor of the same horsepower rating as the switch at the rated voltage.

UTILIZATION EQUIPMENT. Equipment that utilizes electric energy for electronic, electromechanical, chemical, heating, lighting or similar purposes.

VENTILATED. Provided with a means to permit circulation of air sufficient to remove an excess of heat, fumes or vapors.

VOLTAGE (OF A CIRCUIT). The greatest root-mean-square (rms) (effective) difference of potential between any two conductors of the circuit concerned.

VOLTAGE, NOMINAL. A nominal value assigned to a circuit or system for the purpose of conveniently designating its voltage class (e.g., 120/240). The actual voltage at which a circuit operates can vary from the nominal within a range that permits satisfactory operation of equipment.

VOLTAGE TO GROUND. For grounded circuits, the voltage between the given conductor and that point or conductor of the circuit that is grounded. For ungrounded circuits, the greatest voltage between the given conductor and any other conductor of the circuit.

WATERTIGHT. So constructed that moisture will not enter the enclosure under specified test conditions.

WEATHERPROOF. So constructed or protected that exposure to the weather will not interfere with successful operation.

CHAPTER 35

SERVICES

SECTION E3501
GENERAL SERVICES

E3501.1 Scope. This chapter covers service conductors and equipment for the control and protection of services and their installation requirements.

E3501.2 Number of services. One- and two-family dwellings shall be supplied by only one service.

E3501.3 One building or other structure not to be supplied through another. Service conductors supplying a building or other structure shall not pass through the interior of another building or other structure.

E3501.4 Other conductors in raceway or cable. Conductors other than service conductors shall not be installed in the same service raceway or service cable.

Exceptions:

1. Grounding conductors and bonding jumpers.

2. Load management control conductors having over-current protection.

E3501.5 Raceway seal. Where a service raceway enters from an underground distribution system, it shall be sealed in accordance with Section E3703.6.

E3501.6 Service disconnect required. Means shall be provided to disconnect all conductors in a building or other structure from the service entrance conductors.

E3501.6.1 Marking of service equipment and disconnects. Service disconnects shall be permanently marked as a service disconnect. Service equipment shall be listed for the purpose. Individual meter socket enclosures shall not be considered service equipment.

E3501.6.2 Service disconnect location. The service disconnecting means shall be installed at a readily accessible location either outside of a building or inside nearest the point of entrance of the service conductors. Service disconnecting means shall not be installed in bathrooms. Each occupant shall have access to the disconnect serving the dwelling unit in which they reside.

E3501.7 Maximum number of disconnects. The service disconnecting means shall consist of not more than six switches or six circuit breakers mounted in a single enclosure or in a group of separate enclosures.

SECTION E3502
SERVICE SIZE AND RATING

E3502.1 Ampacity of ungrounded conductors. Ungrounded service conductors shall have an ampacity of not less than the load served. For one-family dwellings, the ampacity of the ungrounded conductors shall be not less than 100 amperes, 3 wire. For all other installations, the ampacity of the ungrounded conductors shall be not less than 60 amperes.

E3502.2 Service load. The minimum load for ungrounded service conductors and service devices that serve 100 percent of the dwelling unit load shall be computed in accordance with Table E3502.2. Ungrounded service conductors and service devices that serve less than 100 percent of the dwelling unit load shall be computed as required for feeders in accordance with Chapter 36.

TABLE E3502.2
MINIMUM SERVICE LOAD CALCULATION

LOADS AND PROCEDURE
3 volt-amperes per square foot of floor area for general lighting and general use receptacle outlets.
Plus
1,500 volt-amperes × total number of 20-ampere-rated small appliance and laundry circuits.
Plus
The nameplate volt-ampere rating of all fastened-in-place, permanently connected or dedicated circuit-supplied appliances such as ranges, ovens, cooking units, clothes dryers and water heaters.
Apply the following demand factors to the above subtotal:
The minimum subtotal for the loads above shall be 100 percent of the first 10,000 volt-amperes of the sum of the above loads plus 40 percent of any portion of the sum that is in excess of 10,000 volt-amperes.
Plus the largest of the following:
Nameplate rating(s) of the air-conditioning and cooling equipment.
Nameplate rating(s) of the heating where a heat pump is used without any supplemental electric heating.
Nameplate rating of the electric thermal storage and other heating systems where the usual load is expected to be continuous at the full nameplate value. Systems qualifying under this selection shall not be figured under any other category in this table.
One-hundred percent of nameplate rating of the heat pump compressor and sixty-five percent of the supplemental electric heating load for central electric space-heating systems. If the heat pump compressor is prevented from operating at the same time as the supplementary heat, the compressor load does not need to be added to the supplementary heat load for the total central electric space-heating load.
Sixty-five percent of nameplate rating(s) of electric space-heating units if less than four separately controlled units.
Forty percent of nameplate rating(s) of electric space-heating units of four or more separately controlled units.
The minimum total load in amperes shall be the volt-ampere sum calculated above divided by 240 volts.

E3502.2.1 Services under 100 amperes. Services that are not required to be 100 amperes shall be sized in accordance with Chapter 36.

E3502.3 Rating of service disconnect. The combined rating of all individual service disconnects serving a single dwelling unit shall not be less than the load determined from Table E3502.2 and shall not be less than as specified in Section E3502.1.

E3502.4 Voltage rating. Systems shall be three-wire, 120/240-volt, single-phase with a grounded neutral.

SECTION E3503
SERVICE, FEEDER AND GROUNDING ELECTRODE CONDUCTOR SIZING

E3503.1 Grounded and ungrounded service conductor size. Conductors used as ungrounded service entrance conductors, service lateral conductors, and feeder conductors that serve as the main power feeder to a dwelling unit shall be those listed in Table E3503.1. The main power feeder shall be the feeder(s) between the main disconnect and the lighting and appliance branch-circuit panelboard(s). Ungrounded service conductors shall have a minimum size in accordance with Table E3503.1. The grounded conductor ampacity shall be not less than the maximum unbalance of the load and its size shall be not smaller

than the required minimum grounding electrode conductor size specified in Table E3503.1.

E3503.2 Ungrounded service conductors for accessory buildings and structures. Ungrounded conductors for other than dwelling units shall have an ampacity of not less than 60 amperes and shall be sized as required for feeders in Chapter 36.

Exceptions:

1. For limited loads of a single branch circuit, the service conductors shall have an ampacity of not less than 15 amperes.

2. For loads consisting of not more than two two-wire branch circuits, the service conductors shall have an ampacity of not less than 30 amperes.

E3503.3 Overload protection. Each ungrounded service conductor shall have overload protection.

E3503.3.1 Ungrounded conductor. Overload protection shall be provided by an overcurrent device installed in series with each ungrounded service conductor. The overcurrent device shall have a rating or setting not higher than the allowable ampacity specified in Table E3503.1. A set of

TABLE E3503.1
SERVICE CONDUCTOR AND GROUNDING ELECTRODE CONDUCTOR SIZING

CONDUCTOR TYPES AND SIZES—THHN, THHW, THW, THWN, USE, XHHW, THW-2, THWN-2, XHHW-2, SE, USE-2 (Parallel sets of 1/0 and larger conductors are permitted in either a single raceway or in separate raceways)		ALLOWABLE AMPACITY	MINIMUM GROUNDING ELECTRODE CONDUCTOR SIZE[a]	
Copper (AWG)	Aluminum and copper–clad aluminum (AWG)	Maximum load (amps)	Copper (AWG)	Aluminum (AWG)
4	2	100	8[b]	6[c]
3	1	110	8[b]	6[c]
2	1/0	125	8[b]	6[c]
1	2/0	150	6[c]	4
1/0	3/0	175	6[c]	4
2/0	4/0 or two sets of 1/0	200	4[d]	2[d]
3/0	250 kcmil or two sets of 2/0	225	4[d]	2[d]
4/0 or two sets of 1/0	300 kcmil or two sets of 3/0	250	2[d]	1/0[d]
250 kcmil or two sets of 2/0	350 kcmil or two sets of 4/0	300	2[d]	1/0[d]
350 kcmil or two sets of 3/0	500 kcmil or two sets of 250 kcmil	350	2[d]	1/0[d]
400 kcmil or two sets of 4/0	600 kcmil or two sets of 300 kcmil	400	1/0[d]	3/0[d]

For SI: 1 inch = 25.4 mm.

a. Where protected by a ferrous metal raceway, grounding electrode conductors shall be electrically bonded to the ferrous metal raceway at both ends.

b. Eight AWG grounding electrode conductors shall be protected with metal conduit or nonmetallic conduit.

c. Where not protected, 6 AWG grounding electrode conductors shall closely follow a structural surface for physical protection. The supports shall be spaced not more than 24 inches on center and shall be within 12 inches of any enclosure or termination.

d. Where the sole grounding electrode system is a ground rod or pipe as covered in Section E3508.2, the grounding electrode conductor shall not be required to be larger than 6 AWG copper or 4 AWG aluminum. Where the sole grounding electrode system is the footing steel as covered in Section E3508.1.2, the grounding electrode conductor shall not be required to be larger than 4 AWG copper conductor.

fuses shall be considered all the fuses required to protect all of the ungrounded conductors of a circuit. Single pole circuit breakers, grouped in accordance with Section E3501.7, shall be considered as one protective device.

Exception: Two to six circuit breakers or sets of fuses shall be permitted as the overcurrent device to provide the overload protection. The sum of the ratings of the circuit breakers or fuses shall be permitted to exceed the ampacity of the service conductors, provided that the calculated load does not exceed the ampacity of the service conductors.

E3503.3.2 Not in grounded conductor. Overcurrent devices shall not be connected in series with a grounded service conductor except where a circuit breaker is used that simultaneously opens all conductors of the circuit.

E3503.3.3 Location. The service overcurrent device shall be an integral part of the service disconnecting means or shall be located immediately adjacent thereto.

E3503.4 Grounding electrode conductor size. The grounding electrode conductors shall be sized based on the size of the service entrance conductors as required in Table E3503.1.

E3503.5 Temperature limitations. Except where the equipment is marked otherwise, conductor ampacities used in determining equipment termination provisions shall be based on Table E3503.1.

SECTION E3504
OVERHEAD SERVICE-DROP AND SERVICE CONDUCTOR INSTALLATION

E3504.1 Clearances on buildings. Open conductors and multiconductor cables without an overall outer jacket shall have a clearance of not less than 3 feet (914 mm) from the sides of doors, porches, decks, stairs, ladders, fire escapes and balconies, and from the sides and bottom of windows that open. See Figure E3504.1.

E3504.2 Vertical clearances. Service-drop conductors shall not have ready access and shall comply with Sections E3504.2.1 and E3504.2.2.

E3504.2.1 Above roofs. Conductors shall have a vertical clearance of not less than 8 feet (2438 mm) above the roof surface. The vertical clearance above the roof level shall be maintained for a distance of not less than 3 feet (914 mm) in all directions from the edge of the roof. See Figure E3504.2.1.

Exceptions:

1. Conductors above a roof surface subject to pedestrian traffic shall have a vertical clearance from the roof surface in accordance with Section E3504.2.2.

2. Where the roof has a slope of 4 inches (102 mm) in 12 inches (305 mm), or greater, the minimum clearance shall be 3 feet (914 mm).

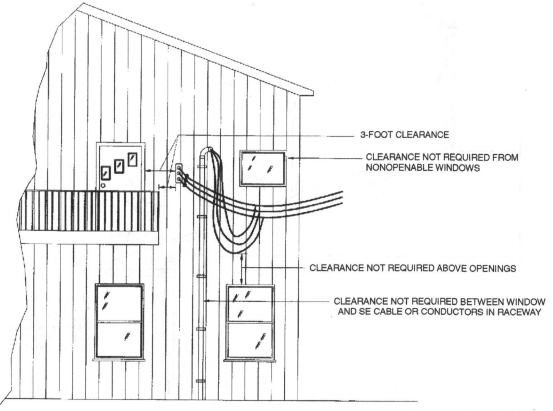

3-FOOT CLEARANCE

CLEARANCE NOT REQUIRED FROM NONOPENABLE WINDOWS

CLEARANCE NOT REQUIRED ABOVE OPENINGS

CLEARANCE NOT REQUIRED BETWEEN WINDOW AND SE CABLE OR CONDUCTORS IN RACEWAY

For SI: 1 foot = 304.8 mm.

FIGURE E3504.1
CLEARANCES FROM BUILDING OPENINGS

3. The minimum clearance above only the overhanging portion of the roof shall not be less than 18 inches (457 mm) where not more than 6 feet (1829 mm) of conductor length passes over 4 feet (1219 mm) or less of roof surface measured horizontally and such conductors are terminated at a through-the-roof raceway or approved support.

4. The requirement for maintaining the vertical clearance for a distance of 3 feet (914 mm) from the edge of the roof shall not apply to the final conductor span where the service drop is attached to the side of a building.

E3504.2.2 Vertical clearance from grade. Service-drop conductors shall have the following minimum clearances from final grade:

1. For service-drop cables supported on and cabled together with a grounded bare messenger wire, the minimum vertical clearance shall be 10 feet (3048 mm) at the electric service entrance to buildings, at the lowest point of the drip loop of the building electric entrance, and above areas or sidewalks accessed by pedestrians only. Such clearance shall be measured from final grade or other accessible surfaces.

2. Twelve feet (3658 mm)—over residential property and driveways.

3. Eighteen feet (5486 mm)—over public streets, alleys, roads or parking areas subject to truck traffic.

E3504.3 Point of attachment. The point of attachment of the service-drop conductors to a building or other structure shall provide the minimum clearances as specified in Sections E3504.1 through E3504.2.2. In no case shall the point of attachment be less than 10 feet (3048 mm) above finished grade.

E3504.4 Means of attachment. Multiconductor cables used for service drops shall be attached to buildings or other structures by fittings approved for the purpose.

E3504.5 Service masts as supports. Where a service mast is used for the support of service-drop conductors, it shall be of adequate strength or be supported by braces or guys to withstand the strain imposed by the service drop. Where raceway-type service masts are used, all equipment shall be approved. Only power service drop conductors shall be permitted to be attached to a service mast.

E3504.6 Supports over buildings. Service-drop conductors passing over a roof shall be securely supported. Where practicable, such supports shall be independent of the building.

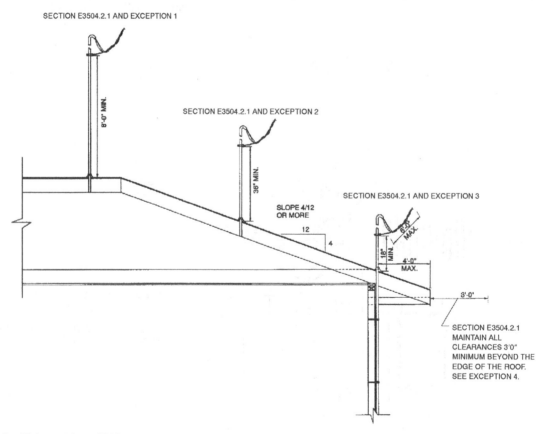

For SI: 1 inch = 25.4 mm, 1 foot = 304.8 mm.

FIGURE E3504.2.1
CLEARANCES FROM ROOFS

SECTION E3505
SERVICE-ENTRANCE CONDUCTORS

E3505.1 Insulation of service-entrance conductors. Service-entrance conductors entering or on the exterior of buildings or other structures shall be insulated in accordance with Section E3306.5.

Exceptions:

1. A copper grounded conductor shall not be required to be insulated where it is:

   1.1. In a raceway or part of a service cable assembly,

   1.2. Directly buried in soil of suitable condition, or

   1.3. Part of a cable assembly listed for direct burial without regard to soil conditions.

2. An aluminum or copper-clad aluminum grounded conductor shall not be required to be insulated where part of a cable or where identified for direct burial or utilization in underground raceways.

E3505.2 Wiring methods for services. Service-entrance wiring methods shall be installed in accordance with the applicable requirements in Chapter 37.

E3505.3 Spliced conductors. Service-entrance conductors shall be permitted to be spliced or tapped. Splices shall be made in enclosures or, if directly buried, with listed underground splice kits. Conductor splices shall be made in accordance with Chapters 33, 36, 37 and 38.

E3505.4 Protection against physical damage. Underground service-entrance conductors shall be protected against physical damage in accordance with Chapter 37.

E3505.5 Protection of service cables against damage. Above-ground service-entrance cables, where subject to physical damage, shall be protected by one or more of the following: rigid metal conduit, intermediate metal conduit, Schedule 80 rigid nonmetallic conduit, electrical metallic tubing or other approved means.

E3505.6 Locations exposed to direct sunlight. Insulated conductors and cables used where exposed to direct rays of the sun shall comply with one of the following:

1. The cables are listed, or listed and marked, as being sunlight resistant.

2. The conductors are listed, or listed and marked, as being sunlight resistant.

3. The conductors and cables are covered with insulating material, such as tape or sleeving, that is listed, or listed and marked, as being sunlight resistant.

E3505.7 Mounting supports. Service cables shall be supported by straps or other approved means within 12 inches (305 mm) of every service head, gooseneck or connection to a raceway or enclosure and at intervals not exceeding 30 inches (762 mm).

E3505.8 Raceways to drain. Where exposed to the weather, raceways enclosing service-entrance conductors shall be raintight and arranged to drain. Where embedded in masonry, raceways shall be arranged to drain.

E3505.9 Overhead service locations. Connections at service heads shall be in accordance with Sections E3505.9.1 through E3505.9.7.

E3505.9.1 Rain-tight service head. Service raceways shall be equipped with a rain-tight service head at the point of connection to service-drop conductors.

E3505.9.2 Service cable, service head or gooseneck. Service cable shall be equipped with a rain-tight service head or shall be formed into a gooseneck in an approved manner.

E3505.9.3 Service head location. Service heads, and goosenecks in service-entrance cables, shall be located above the point of attachment of the service-drop conductors to the building or other structure.

Exception: Where it is impracticable to locate the service head or gooseneck above the point of attachment, the service head or gooseneck location shall be not more than 24 inches (610 mm) from the point of attachment.

E3505.9.4 Separately bushed openings. Service heads shall have conductors of different potential brought out through separately bushed openings.

E3505.9.5 Drip loops. Drip loops shall be formed on individual conductors. To prevent the entrance of moisture, service-entrance conductors shall be connected to the service-drop conductors either below the level of the service head or below the level of the termination of the service-entrance cable sheath.

E3505.9.6 Conductor arrangement. Service-drop conductors and service-entrance conductors shall be arranged so that water will not enter service raceways or equipment.

E3505.9.7 Secured. Service cables shall be held securely in place.

SECTION E3506
SERVICE EQUIPMENT—GENERAL

E3506.1 Service equipment enclosures. Energized parts of service equipment shall be enclosed.

E3506.2 Working space. In no case shall the working space in the vicinity of service equipment be less than that specified in Chapter 33.

E3506.3 Available short-circuit current. Service equipment shall be suitable for the maximum fault current available at its supply terminals, but not less than 10,000 amperes.

E3506.4 Marking. Service equipment shall be marked to identify it as being suitable for use as service equipment. Individual meter socket enclosures shall not be considered service equipment.

SECTION E3507
SYSTEM GROUNDING

E3507.1 System service ground. The premises wiring system shall be grounded at the service with a grounding electrode conductor connected to a grounding electrode system as required by this code. Grounding electrode conductors shall be sized in accordance with Table E3503.1.

E3507.2 Location of grounding electrode conductor connection. The grounding electrode conductor shall be connected to the grounded service conductor at any accessible point from the load end of the service drop or service lateral to and including the terminal or bus to which the grounded service conductor is connected at the service disconnecting means. A grounding connection shall not be made to any grounded circuit conductor on the load side of the service disconnecting means, except as provided in Section E3507.3.

E3507.3 Buildings or structures supplied by feeder(s) or branch circuit(s). Buildings or structures supplied by feeder(s) or branch circuit(s) shall have a grounding electrode or grounding electrode system installed in accordance with Section E3508. The grounding electrode conductor(s) shall be connected in a manner specified in Section E3507.3.1 or E3507.3.2. Where there is no existing grounding electrode, the grounding electrode(s) required in Section E3508 shall be installed.

Exception: A grounding electrode shall not be required where only one branch circuit supplies the building or structure and the branch circuit includes an equipment grounding conductor for grounding the noncurrent-carrying parts of all equipment. For the purposes of this section, a multiwire branch circuit shall be considered as a single branch circuit.

E3507.3.1 Equipment grounding conductor. An equipment grounding conductor as described in Section E3808 shall be run with the supply conductors and connected to the building or structure disconnecting means and to the grounding electrode(s). The equipment grounding conductor shall be used for grounding or bonding of equipment, structures or frames required to be grounded or bonded. The equipment grounding conductor shall be sized in accordance with Section E3808.12. Any installed grounded conductor shall not be connected to the equipment grounding conductor or to the grounding electrode(s).

E3507.3.2 Grounded conductor. Where an equipment grounding conductor is not run with the supply conductors to the building or structure, and there are no continuous metallic paths bonded to the grounding system in both buildings or structures involved, and ground-fault protection of equipment has not been installed on the common service, the grounded circuit conductor run with the supply conductors to the building or structure shall be connected to the building or structure disconnecting means and to the grounding electrode(s) and shall be used for grounding or bonding of equipment, structures, or frames required to be grounded or bonded. The size of the grounded conductor shall be not smaller than the larger of:

1. That required by Section E3604.3.

2. That required by Section E3808.12.

E3507.4 Grounding electrode conductor. A grounding electrode conductor shall be used to connect the equipment grounding conductors, the service equipment enclosures, and the grounded service conductor to the grounding electrode(s).

E3507.5 Main bonding jumper. An unspliced main bonding jumper shall be used to connect the equipment grounding conductor(s) and the service-disconnect enclosure to the grounded conductor of the system within the enclosure for each service disconnect.

E3507.6 Common grounding electrode. Where an ac system is connected to a grounding electrode in or at a building or structure, the same electrode shall be used to ground conductor enclosures and equipment in or on that building or structure. Where separate services, feeders or branch circuits supply a building and are required to be connected to a grounding electrode(s), the same grounding electrode(s) shall be used. Two or more grounding electrodes that are effectively bonded together shall be considered as a single grounding electrode system.

SECTION E3508
GROUNDING ELECTRODE SYSTEM

E3508.1 Grounding electrode system. All electrodes specified in Sections E3508.1.1, E3508.1.2, E3508.1.3, E3508.1.4 and E3508.1.5 that are present at each building or structure served shall be bonded together to form the grounding electrode system. Where none of these electrodes are available, one or more of the electrodes specified in Sections E3508.1.3, E3508.1.4 and E3508.1.5 shall be installed and used.

Exception: Concrete-encased electrodes of existing buildings or structures shall not be required to be part of the grounding electrode system where the steel reinforcing bars or rods are not accessible for use without disturbing the concrete.

E3508.1.1 Metal underground water pipe. A metal underground water pipe that is in direct contact with the earth for 10 feet (3048 mm) or more, including any well casing effectively bonded to the pipe and that is electrically continuous, or made electrically continuous by bonding around insulating joints or insulating pipe to the points of connection of the grounding electrode conductor and the bonding conductors, shall be considered as a grounding electrode (see Section E3508.1). Interior metal water piping located more than 5 feet (1524 mm) from the entrance to the building shall not be used as part of the grounding electrode system or as a conductor to interconnect electrodes that are part of the grounding electrode system.

E3508.1.1.1 Installation. Continuity of the grounding path or the bonding connection to interior piping shall not rely on water meters, filtering devices and similar equipment. A metal underground water pipe shall be supplemented by an additional electrode of a type specified in Sections E3508.1.2 through E3508.1.5. The supplemental electrode shall be bonded to the grounding electrode conductor, the grounded service entrance conductor, a nonflexible grounded service raceway or any grounded service enclosure.

Where the supplemental electrode is a rod, pipe or plate electrode in accordance with Sections E3508.1.4 and E3508.1.5, that portion of the bonding jumper that is the sole connection to the supplemental grounding electrode shall not be required to be larger than 6 AWG copper or 4 AWG aluminum wire.

E3508.1.2 Concrete-encased electrode. An electrode encased by at least 2 inches (51 mm) of concrete, located within and near the bottom of a concrete foundation or footing that is in direct contact with the earth, consisting of at least 20 feet (6096 mm) of one or more bare or zinc-galvanized or other electrically conductive coated steel reinforcing bars or rods of not less than $^1/_2$ inch (12.7 mm) diameter, or consisting of at least 20 feet (6096 mm) of bare copper conductor not smaller than 4 AWG shall be considered as a grounding electrode. Reinforcing bars shall be permitted to be bonded together by the usual steel tie wires or other effective means.

E3508.1.3 Ground rings. A ground ring encircling the building or structure, in direct contact with the earth at a depth below the earth's surface of not less than 30 inches (762 mm), consisting of at least 20 feet (6096 mm) of bare copper conductor not smaller than 2 AWG shall be considered as a grounding electrode.

E3508.1.4 Rod and pipe electrodes. Rod and pipe electrodes not less than 8 feet (2438 mm) in length and consisting of the following materials shall be considered as a grounding electrode:

1. Electrodes of pipe or conduit shall not be smaller than trade size $^3/_4$ (metric designator 21) and, where of iron or steel, shall have the outer surface galvanized or otherwise metal-coated for corrosion protection.

2. Electrodes of rods of iron or steel shall be at least $^5/_8$ inch (15.9 mm) in diameter. Stainless steel rods less than $^5/_8$ inch (15.9 mm) in diameter, nonferrous rods or their equivalent shall be listed and shall be not less than $^1/_2$ inch (12.7 mm) in diameter.

E3508.1.4.1 Installation. The rod and pipe electrodes shall be installed such that at least 8 feet (2438 mm) of length is in contact with the soil. They shall be driven to a depth of not less than 8 feet (2438 mm) except that, where rock bottom is encountered, electrodes shall be driven at an oblique angle not to exceed 45 degrees from the vertical or shall be buried in a trench that is at least 30 inches (762 mm) deep. The upper end of the electrodes shall be flush with or below ground level except where the aboveground end and the grounding electrode conductor attachment are protected against physical damage.

E3508.1.5 Plate electrodes. A plate electrode that exposes not less than 2 square feet (0.186 m^2) of surface to exterior soil shall be considered as a grounding electrode. Electrodes of iron or steel plates shall be at least $^1/_4$ inch (6.4 mm) in thickness. Electrodes of nonferrous metal shall be at least 0.06 inch (1.5 mm) in thickness. Plate electrodes shall be installed not less than 30 inches (762 mm) below the surface of the earth.

E3508.2 Bonding jumper. The bonding jumper(s) used to connect the grounding electrodes together to form the grounding electrode system shall be installed in accordance with Sections E3510.2, and E3510.3, shall be sized in accordance with

Section E3503.4, and shall be connected in the manner specified in Section E3511.1.

E3508.3 Rod, pipe and plate electrode requirements. Where practicable, rod, pipe and plate electrodes shall be embedded below permanent moisture level. Such electrodes shall be free from nonconductive coatings such as paint or enamel. Where more than one such electrode is used, each electrode of one grounding system shall be not less than 6 feet (1829 mm) from any other electrode of another grounding system. Two or more grounding electrodes that are effectively bonded together shall be considered as a single grounding electrode system. That portion of a bonding jumper that is the sole connection to a rod, pipe or plate electrode shall not be required to be larger than 6 AWG copper or 4 AWG aluminum wire.

E3508.4 Resistance of rod, pipe and plate electrodes. A single electrode consisting of a rod, pipe or plate that does not have a resistance to ground of 25 ohms or less shall be augmented by one additional electrode of any of the types specified in Sections E3508.1.2 through E3508.1.5. Where multiple rod, pipe or plate electrodes are installed to meet the requirements of this section, they shall be not less than 6 feet (1829 mm) apart.

E3508.5 Aluminum electrodes. Aluminum electrodes shall not be permitted.

E3508.6 Metal underground gas piping system. A metal underground gas piping system shall not be used as a grounding electrode.

SECTION E3509
BONDING

E3509.1 General. Bonding shall be provided where necessary to ensure electrical continuity and the capacity to conduct safely any fault current likely to be imposed.

E3509.2 Bonding of services. The noncurrent-carrying metal parts of the following equipment shall be effectively bonded together:

1. The service raceways or service cable armor.

2. All service enclosures containing service conductors, including meter fittings, and boxes, interposed in the service raceway or armor.

3. Any metallic raceways or armor enclosing a grounding electrode conductor. Bonding shall apply at each end and to all intervening raceways, boxes and enclosures between the service equipment and the grounding electrode.

E3509.3 Bonding to other systems. An accessible means external to enclosures for connecting intersystem bonding and grounding electrode conductors shall be provided at the service equipment and at the disconnecting means for any additional buildings or structures by at least one of the following means:

1. Exposed nonflexible metallic service raceways.

2. Exposed grounding electrode conductor.

3. Approved means for the external connection of a copper or other corrosion-resistant bonding or grounding conductor to the service raceway or equipment.

E3509.4 Method of bonding at the service. Electrical continuity at service equipment, service raceways and service conductor enclosures shall be ensured by one or more of the methods specified in Sections E3509.4.1 through E3509.4.4.

Bonding jumpers meeting the other requirements of this code shall be used around concentric or eccentric knockouts that are punched or otherwise formed so as to impair the electrical connection to ground. Standard locknuts or bushings shall not be the sole means for the bonding required by this section.

E3509.4.1 Grounded service conductor. Equipment shall be bonded to the grounded service conductor in a manner provided in this code.

E3509.4.2 Threaded connections. Equipment shall be bonded by connections using threaded couplings or threaded bosses on enclosures. Such connections shall be made wrench tight.

E3509.4.3 Threadless couplings and connectors. Equipment shall be bonded by threadless couplings and connectors for metal raceways and metal-clad cables. Such couplings and connectors shall be made wrench tight. Standard locknuts or bushings shall not be used for the bonding required by this section.

E3509.4.4 Other devices. Equipment shall be bonded by other listed devices, such as bonding-type locknuts, bushings and bushings with bonding jumpers.

E3509.5 Sizing bonding jumper on supply side of service and main bonding jumper. The bonding jumper shall not be smaller than the sizes shown in Table E3503.1 for grounding electrode conductors. Where the service-entrance conductors are paralleled in two or more raceways or cables, the equipment bonding jumper, where routed with the raceways or cables, shall be run in parallel. The size of the bonding jumper for each raceway or cable shall be based on the size of the service-entrance conductors in each raceway or cable.

E3509.6 Metal water piping bonding. The metal water piping system shall be bonded to the service equipment enclosure, the grounded conductor at the service, the grounding electrode conductor where of sufficient size, or to the one or more grounding electrodes used. The bonding jumper shall be sized in accordance with Table E3503.1. The points of attachment of the bonding jumper(s) shall be accessible.

E3509.7 Bonding other metal piping. Where installed in or attached to a building or structure, metal piping systems, including gas piping, capable of becoming energized shall be bonded to the service equipment enclosure, the grounded conductor at the service, the grounding electrode conductor where of sufficient size, or to the one or more grounding electrodes used. The bonding jumper shall be sized in accordance with Table E3808.12 using the rating of the circuit capable of energizing the piping. The equipment grounding conductor for the circuit that is capable of energizing the piping shall be permitted to serve as the bonding means. The points of attachment of the bonding jumper(s) shall be accessible.

SECTION E3510
GROUNDING ELECTRODE CONDUCTORS

E3510.1 Continuous. The unspliced grounding electrode conductor shall run to any convenient grounding electrode available in the grounding electrode system, or to one or more grounding electrode(s) individually. The grounding electrode conductor shall be sized for the largest grounding electrode conductor required among all of the electrodes connected to it.

E3510.2 Securing and protection against physical damage. Where exposed, a grounding electrode conductor or its enclosure shall be securely fastened to the surface on which it is carried. A 4 AWG or larger conductor shall be protected where exposed to physical damage. A 6 AWG grounding conductor that is free from exposure to physical damage shall be permitted to be run along the surface of the building construction without metal covering or protection where it is securely fastened to the construction; otherwise, it shall be in rigid metal conduit, intermediate metal conduit, rigid nonmetallic conduit, electrical metallic tubing or cable armor. Grounding electrode conductors smaller than 6 AWG shall be in rigid metal conduit, intermediate metal conduit, rigid nonmetallic conduit, electrical metallic tubing or cable armor.

Bare aluminum or copper-clad aluminum grounding conductors shall not be used where in direct contact with masonry or the earth or where subject to corrosive conditions. Where used outside, aluminum or copper-clad aluminum grounding conductors shall not be installed within 18 inches (457 mm) of the earth.

E3510.3 Enclosures for grounding electrode conductors. Ferrous metal enclosures for grounding electrode conductors shall be electrically continuous from the point of attachment to cabinets or equipment to the grounding electrode, and shall be securely fastened to the ground clamp or fitting. Nonferrous metal enclosures shall not be required to be electrically continuous. Ferrous metal enclosures that are not physically continuous from cabinet or equipment to the grounding electrode shall be made electrically continuous by bonding each end to the grounding conductor. The bonding jumper for a grounding electrode conductor raceway shall be the same size or larger than the required enclosed grounding electrode conductor.

Where a raceway is used as protection for a grounding conductor, the installation shall comply with the requirements of Chapter 37.

SECTION E3511
GROUNDING ELECTRODE CONDUCTOR
CONNECTION TO THE GROUNDING ELECTRODES

E3511.1 Methods of grounding conductor connection to electrodes. The grounding or bonding conductor shall be connected to the grounding electrode by exothermic welding, listed lugs, listed pressure connectors, listed clamps or other listed means. Connections depending on solder shall not be used. Ground clamps shall be listed for the materials of the grounding electrode and the grounding electrode conductor and, where used on pipe, rod or other buried electrodes, shall also be listed for direct soil burial or concrete encasement. Not more than one conductor shall be connected to the grounding electrode by a single clamp or fitting unless the clamp or fitting is listed for multiple conductors. One of the methods indicated in the following items shall be used:

1. A pipe fitting, pipe plug or other approved device screwed into a pipe or pipe fitting.

2. A listed bolted clamp of cast bronze or brass, or plain or malleable iron.

3. For indoor telecommunications purposes only, a listed sheet metal strap-type ground clamp having a rigid metal base that seats on the electrode and having a strap of such material and dimensions that it is not likely to stretch during or after installation.

4. Other equally substantial approved means.

E3511.2 Accessibility. The connection of the grounding electrode conductor or bonding jumper to the grounding electrodes that are not buried or concrete encased shall be accessible.

E3511.3 Effective grounding path. The connection of the grounding electrode conductor or bonding jumper shall be made in a manner that will ensure a permanent and effective grounding path. Where necessary to ensure effective grounding for a metal piping system used as a grounding electrode, effective bonding shall be provided around insulated joints and sections and around any equipment that is likely to be disconnected for repairs or replacement. Bonding conductors shall be of sufficient length to permit removal of such equipment while retaining the integrity of the bond.

E3511.4 Protection of ground clamps and fittings. Ground clamps or other fittings shall be approved for applications without protection or shall be protected from physical damage by installing them where they are not likely to be damaged or by enclosing them in metal, wood or equivalent protective coverings.

E3511.5 Clean surfaces. Nonconductive coatings (such as paint, enamel and lacquer) on equipment to be grounded shall be removed from threads and other contact surfaces to ensure good electrical continuity or shall be connected by fittings that make such removal unnecessary.

CHAPTER 36

BRANCH CIRCUIT AND FEEDER REQUIREMENTS

SECTION E3601
GENERAL

E3601.1 Scope. This chapter covers branch circuits and feeders and specifies the minimum required branch circuits, the allowable loads and the required overcurrent protection for branch circuits and feeders that serve less than 100 percent of the total dwelling unit load. Feeder circuits that serve 100 percent of the dwelling unit load shall be sized in accordance with the procedures in Chapter 35.

E3601.2 Branch-circuit and feeder ampacity. Branch-circuit and feeder conductors shall have ampacities not less than the maximum load to be served. Where a branch circuit or a feeder supplies continuous loads or any combination of continuous and noncontinuous loads, the minimum branch-circuit or feeder conductor size, before the application of any adjustment or correction factors, shall have an allowable ampacity equal to or greater than the noncontinuous load plus 125 percent of the continuous load.

E3601.3 Selection of ampacity. Where more than one calculated or tabulated ampacity could apply for a given circuit length, the lowest value shall be used.

> **Exception:** Where two different ampacities apply to adjacent portions of a circuit, the higher ampacity shall be permitted to be used beyond the point of transition, a distance equal to 10 feet (3048 mm) or 10 percent of the circuit length figured at the higher ampacity, whichever is less.

E3601.4 Multi-outlet branch circuits. Conductors of multi-outlet branch circuits supplying more than one receptacle for cord-and-plug-connected portable loads shall have ampacities of not less than the rating of the branch circuit.

E3601.5 Multiwire branch circuits. All conductors for multiwire branch circuits shall originate from the same panelboard or similar distribution equipment. Where two or more devices on the same yoke or strap are supplied by a multiwire branch circuit, a means shall be provided at the point where the circuit originates to simultaneously disconnect all ungrounded conductors of the multiwire circuit. Except where all ungrounded conductors are opened simultaneously by the branch-circuit overcurrent device, multiwire branch circuits shall supply only line-to-neutral loads or only one appliance.

SECTION E3602
BRANCH CIRCUIT RATINGS

E3602.1 Branch-circuit voltage limitations. The voltage ratings of branch circuits that supply luminaires or receptacles for cord-and-plug-connected loads of up to 1,440 volt-amperes or of less than $^1/_4$ horsepower shall be limited to a maximum rating of 120 volts, nominal, between conductors.

Branch circuits that supply cord-and-plug-connected or permanently connected utilization equipment and appliances rated at over 1,440 volt-amperes or $^1/_4$ horsepower and greater shall be rated at 120 volts or 240 volts, nominal.

E3602.2 Branch-circuit ampere rating. Branch circuits shall be rated in accordance with the maximum allowable ampere rating or setting of the overcurrent protection device. The rating for other than individual branch circuits shall be 15, 20, 30, 40 and 50 amperes. Where conductors of higher ampacity are used, the ampere rating or setting of the specified over-current device shall determine the circuit rating.

E3602.3 Fifteen- and 20-ampere branch circuits. A 15- or 20-ampere branch circuit shall be permitted to supply lighting units, or other utilization equipment, or a combination of both. The rating of any one cord-and-plug-connected utilization equipment not fastened in place shall not exceed 80 percent of the branch-circuit ampere rating. The total rating of utilization equipment fastened in place, other than lighting fixtures, shall not exceed 50 percent of the branch-circuit ampere rating where lighting units, cord-and-plug-connected utilization equipment not fastened in place, or both, are also supplied.

E3602.4 Thirty-ampere branch circuits. A 30-ampere branch circuit shall be permitted to supply fixed utilization equipment. A rating of any one cord-and-plug-connected utilization equipment shall not exceed 80 percent of the branch-circuit ampere rating.

E3602.5 Branch circuits serving multiple loads or outlets. General-purpose branch circuits shall supply lighting outlets, appliances, equipment or receptacle outlets, and combinations of such. The rating of a fastened-in-place appliance or equipment, where used in combination on the same branch circuit with light fixtures, receptacles, and/or other appliances or equipment not fastened in place, shall not exceed 50 percent of the branch-circuit rating. Multi-outlet branch circuits serving lighting or receptacles shall be limited to a maximum branch-circuit rating of 20 amperes.

E3602.6 Branch circuits serving a single motor. Branch-circuit conductors supplying a single motor shall have an ampacity not less than 125 percent of the motor full-load current rating.

E3602.7 Branch circuits serving motor-operated and combination loads. For circuits supplying loads consisting of motor-operated utilization equipment that is fastened in place and that has a motor larger than $^1/_8$ horsepower in combination with other loads, the total calculated load shall be based on 125 percent of the largest motor load plus the sum of the other loads.

E3602.8 Branch-circuit inductive lighting loads. For circuits supplying luminaires having ballasts, the calculated load shall be based on the total ampere ratings of such units and not on the total watts of the lamps.

E3602.9 Branch-circuit load for ranges and cooking appliances. It shall be permissible to calculate the branch-circuit load for one range in accordance with Table E3604.2(2). The branch-circuit load for one wall-mounted oven or one counter-mounted cooking unit shall be the nameplate rating of the appliance. The branch-circuit load for a counter-mounted cooking unit and not more than two wall-mounted ovens all supplied from a single branch circuit and located in the same room shall be calculated by adding the nameplate ratings of the individual appliances and treating the total as equivalent to one range.

E3602.9.1 Minimum branch circuit for ranges. Ranges with a rating of 8.75 kVA or more shall be supplied by a branch circuit having a minimum rating of 40 amperes.

E3602.10 Branch circuits serving heating loads. Electric space-heating and water-heating appliances shall be considered continuous loads. Branch circuits supplying two or more outlets for fixed electric space-heating equipment shall be rated 15, 20, 25 or 30 amperes.

E3602.11 Branch circuits for air-conditioning and heat pump equipment. The ampacity of the conductors supplying multimotor and combination load equipment shall not be less than the minimum circuit ampacity marked on the equipment. The branch-circuit overcurrent device rating shall be the size and type marked on the appliance and shall be listed for the specific purpose.

E3602.12 Branch circuits serving room air conditioners. A room air conditioner shall be considered as a single motor unit in determining its branch-circuit requirements where all the following conditions are met:

1. It is cord- and attachment plug-connected.

2. The rating is not more than 40 amperes and 250 volts; single phase.

3. Total rated-load current is shown on the room air-conditioner nameplate rather than individual motor currents.

4. The rating of the branch-circuit short-circuit and ground-fault protective device does not exceed the ampacity of the branch-circuit conductors, or the rating of the branch-circuit conductors, or the rating of the receptacle, whichever is less.

E3602.12.1 Where no other loads are supplied. The total marked rating of a cord- and attachment plug-connected room air conditioner shall not exceed 80 percent of the rating of a branch circuit where no other appliances are also supplied.

E3602.12.2 Where lighting units or other appliances are also supplied. The total marked rating of a cord- and attachment plug-connected room air conditioner shall not exceed 50 percent of the rating of a branch circuit where lighting or other appliances are also supplied. Where the circuitry is interlocked to prevent simultaneous operation of the room air conditioner and energization of other outlets on the same branch circuit, a cord- and attachment-plug-connected room air conditioner shall not exceed 80 percent of the branch-circuit rating.

E3602.13 Branch-circuit requirement—summary. The requirements for circuits having two or more outlets, or recep-

tacles, other than the receptacle circuits of Section E3603.2, are summarized in Table E3602.13. Branch circuits in dwelling units shall supply only loads within that dwelling unit or loads associated only with that dwelling unit. Branch circuits required for the purpose of lighting, central alarm, signal, communications or other needs for public or common areas of a two-family dwelling shall not be supplied from equipment that supplies an individual dwelling unit.

TABLE E3602.13
BRANCH-CIRCUIT REQUIREMENTS—SUMMARY[a,b]

	CIRCUIT RATING		
	15 amp	20 amp	30 amp
Conductors: Minimum size (AWG) circuit conductors	14	12	10
Maximum overcurrent-protection device rating Ampere rating	15	20	30
Outlet devices: Lampholders permitted Receptacle rating (amperes)	Any type 15 maximum	Any type 15 or 20	N/A 30
Maximum load (amperes)	15	20	30

a. These gages are for copper conductors.

b. N/A means not allowed.

SECTION E3603
REQUIRED BRANCH CIRCUITS

E3603.1 Branch circuits for heating. Central heating equipment other than fixed electric space heating shall be supplied by an individual branch circuit. Permanently connected air-conditioning equipment, and auxiliary equipment directly associated with the central heating equipment such as pumps, motorized valves, humidifiers and electrostatic air cleaners, shall not be prohibited from connecting to the same branch circuit as the central heating equipment.

E3603.2 Kitchen and dining area receptacles. A minimum of two 20-ampere-rated branch circuits shall be provided to serve all wall and floor receptacle outlets located in the kitchen, pantry, breakfast area, dining area or similar area of a dwelling. The kitchen countertop receptacles shall be served by a minimum of two 20-ampere-rated branch circuits, either or both of which shall also be permitted to supply other receptacle outlets in the kitchen, pantry, breakfast and dining area including receptacle outlets for refrigeration appliances.

Exception: The receptacle outlet for refrigeration appliances shall be permitted to be supplied from an individual branch circuit rated 15 amperes or greater.

E3603.3 Laundry circuit. A minimum of one 20-ampere-rated branch circuit shall be provided for receptacles located in the laundry area and shall serve only receptacle outlets located in the laundry area.

E3603.4 Bathroom branch circuits. A minimum of one 20-ampere branch circuit shall be provided to supply bathroom receptacle outlet(s). Such circuits shall have no other outlets.

Exception: Where the 20-ampere circuit supplies a single bathroom, outlets for other equipment within the same bathroom shall be permitted to be supplied in accordance with Section E3602.

E3603.5 Number of branch circuits. The minimum number of branch circuits shall be determined from the total calculated load and the size or rating of the circuits used. The number of circuits shall be sufficient to supply the load served. In no case shall the load on any circuit exceed the maximum specified by Section E3602.

E3603.6 Branch-circuit load proportioning. Where the branch-circuit load is calculated on a volt-amperes-per-square-foot (m²) basis, the wiring system, up to and including the branch-circuit panelboard(s), shall have the capacity to serve not less than the calculated load. This load shall be evenly proportioned among multioutlet branch circuits within the panelboard(s). Branch-circuit overcurrent devices and circuits shall only be required to be installed to serve the connected load.

SECTION E3604
FEEDER REQUIREMENTS

E3604.1 Conductor size. Feeder conductors that do not serve 100 percent of the dwelling unit load and branch-circuit conductors shall be of a size sufficient to carry the load as determined by this chapter. Feeder conductors shall not be required to be larger than the service-entrance conductors that supply the dwelling unit. The load for feeder conductors that serve as the main power feeder to a dwelling unit shall be determined as specified in Chapter 35 for services.

E3604.2 Feeder loads. The minimum load in volt-amperes shall be calculated in accordance with the load calculation procedure prescribed in Table E3604.2(1). The associated table demand factors shall be applied to the actual load to determine the minimum load for feeders.

E3604.3 Feeder neutral load. The feeder neutral load shall be the maximum unbalance of the load determined in accordance with this chapter. The maximum unbalanced load shall be the maximum net calculated load between the neutral and any one ungrounded conductor. For a feeder or service supplying electric ranges, wall-mounted ovens, counter-mounted cooking units and electric dryers, the maximum unbalanced load shall be considered as 70 percent of the load on the ungrounded conductors.

E3604.4 Lighting and general use receptacle load. A unit load of not less than 3 volt-amperes shall constitute the minimum lighting and general use receptacle load for each square foot of floor area (33 VA for each square meter of floor area). The floor area for each floor shall be calculated from the outside dimensions of the building. The calculated floor area shall not include open porches, garages, or unused or unfinished spaces not adaptable for future use.

E3604.5 Ampacity and calculated loads. The calculated load of a feeder shall be not less than the sum of the loads on the branch circuits supplied, as determined by Section E3604, after any applicable demand factors permitted by Section E3604 have been applied.

Feeder conductors shall have sufficient ampacity to supply the load served. In no case shall the calculated load of a feeder be less than the sum of the loads on the branch circuits supplied as determined by this chapter after any permitted demand factors have been applied.

TABLE E3604.2(1)
FEEDER LOAD CALCULATION

LOAD CALCULATION PROCEDURE	APPLIED DEMAND FACTOR
Lighting and receptacles: A unit load of not less than 3 VA per square foot of total floor area shall constitute the lighting and 120-volt, 15- and 20-ampere general use receptacle load. 1,500 VA shall be added for each 20-ampere branch circuit serving receptacles in the kitchen, dining room, pantry, breakfast area and laundry area.	100 percent of first 3,000 VA or less and 35 percent of that in excess of 3,000 VA.
Plus	
Appliances and motors: The nameplate rating load of all fastened-in-place appliances other than dryers, ranges, air-conditioning and space-heating equipment.	100 percent of load for three or less appliances. 75 percent of load for four or more appliances.
Plus	
Fixed motors: Full-load current of motors plus 25 percent of the full load current of the largest motor.	
Plus	
Electric clothes dryer: The dryer load shall be 5,000 VA for each dryer circuit or the nameplate rating load of each dryer, whichever is greater.	
Plus	
Cooking appliances: The nameplate rating of ranges, wall-mounted ovens, counter-mounted cooking units and other cooking appliances rated in excess of 1.75 kVA shall be summed.	Demand factors shall be as allowed by Table E3604.2(2).
Plus the largest of either the heating or cooling load	

Largest of the following two selections:
  1. 100 percent of the nameplate rating(s) of the air conditioning and cooling, including heat pump compressors.
  2. 100 percent of the fixed electric space heating.	

For SI: 1 square foot = 0.0929 m².

TABLE E3604.2(2)
DEMAND LOADS FOR ELECTRIC RANGES, WALL-MOUNTED OVENS, COUNTER-MOUNTED
COOKING UNITS AND OTHER COOKING APPLIANCES OVER 1³/₄ kVA RATING[a,b]

NUMBER OF APPLIANCES	MAXIMUM DEMAND[b,c] Column A maximum 12 kVA rating	DEMAND FACTORS (percent)[d] Column B less than 3¹/₂ kVA rating	Column C 3¹/₂ to 8³/₄ kVA rating
1	8 kVA	80	80
2	11 kVA	75	65

a. Column A shall be used in all cases except as provided for in Footnote d.

b. For ranges all having the same rating and individually rated more than 12 kVA but not more than 27 kVA, the maximum demand in Column A shall be increased 5 percent for each additional kVA of rating or major fraction thereof by which the rating of individual ranges exceeds 12 kVA.

c. For ranges of unequal ratings and individually rated more than 8.75 kVA, but none exceeding 27 kVA, an average value of rating shall be computed by adding together the ratings of all ranges to obtain the total connected load (using 12 kVA for any ranges rated less than 12 kVA) and dividing by the total number of ranges; and then the maximum demand in Column A shall be increased 5 percent for each kVA or major fraction thereof by which this average value exceeds 12 kVA.

d. Over 1.75 kVA through 8.75 kVA. As an alternative to the method provided in Column A, the nameplate ratings of all ranges rated more than 1.75 kVA but not more than 8.75 kVA shall be added and the sum shall be multiplied by the demand factor specified in Column B or C for the given number of appliances.

SECTION E3605
CONDUCTOR SIZING
AND OVERCURRENT PROTECTION

E3605.1 General. Ampacities for conductors shall be determined based in accordance with Table E3605.1 and Sections E3605.2 and E3605.3.

E3605.2 Correction factor for ambient temperatures. For ambient temperatures other than 30°C (86°F), multiply the allowable ampacities specified in Table E3605.1 by the appropriate correction factor shown in Table E3605.2.

E3605.3 Adjustment factor for conductor proximity. Where the number of current-carrying conductors in a raceway or cable exceeds three, or where single conductors or multiconductor cables are stacked or bundled for distances greater than 24 inches (610 mm) without maintaining spacing and are not installed in raceways, the allowable ampacity of each conductor shall be reduced as shown in Table E3605.3.

Exceptions:

1. Adjustment factors shall not apply to conductors in nipples having a length not exceeding 24 inches (610 mm).

2. Adjustment factors shall not apply to underground conductors entering or leaving an outdoor trench if those conductors have physical protection in the form of rigid metal conduit, intermediate metal conduit, or rigid nonmetallic conduit having a length not exceeding 10 feet (3048 mm) and the number of conductors does not exceed four.

TABLE E3605.1
ALLOWABLE AMPACITIES

CONDUCTOR SIZE AWG kcmil	CONDUCTOR TEMPERATURE RATING						CONDUCTOR SIZE AWG kcmil
	60°C	75°C	90°C	60°C	75°C	90°C	
	Types TW, UF	Types RHW, THHW, THW, THWN, USE, XHHW	Types RHW-2, THHN, THHW, THW-2, THWN-2, XHHW, XHHW-2, USE-2	Types TW, UF	Types RHW, THHW, THW, THWN, USE, XHHW	Types RHW-2, THHN, THHW, THW-2, THWN-2, XHHW, XHHW-2, USE-2	
	Copper			Aluminum or copper-clad aluminum			
18	—	—	14	—	—	—	—
16	—	—	18	—	—	—	—
14	20	20	25	—	—	—	—
12	25	25	30	20	20	25	12
10	30	35	40	25	30	35	10
8	40	50	55	30	40	45	8
6	55	65	75	40	50	60	6
4	70	85	95	55	65	75	4
3	85	100	110	65	75	85	3
2	95	115	130	75	90	100	2
1	110	130	150	85	100	115	1
1/0	125	150	170	100	120	135	1/0
2/0	145	175	195	115	135	150	2/0
3/0	165	200	225	130	155	175	3/0
4/0	195	230	260	150	180	205	4/0

For SI: °C = [(°F)-32]/1.8.

TABLE E3605.2
AMBIENT TEMPERATURE CORRECTION FACTORS

AMBIENT TEMP. °C	FOR AMBIENT TEMPERATURES OTHER THAN 30°C (86°F), MULTIPLY THE ALLOWABLE AMPACITIES SPECIFIED IN TABLE E3605.1 BY THE APPROPRIATE FACTOR SHOWN BELOW						AMBIENT TEMP. °F
	CONDUCTOR TEMPERATURE RATING						
	60°C	75°C	90°C	60°C	75°C	90°C	
	Types TW, UF	Types RHW, THHW, THW, THWN, USE, XHHW	Types RHW-2, THHN, THHW, THW-2, THWN-2, XHHW, XHHW-2, USE-2	Types TW, UF	Types RHW, THHW, THW, THWN, USE, XHHW	Types RHW-2, THHN, THHW, THW-2, THWN-2, XHHW, XHHW-2, USE-2	
	Copper			Aluminum or copper-clad aluminum			
21-25	1.08	1.05	1.04	1.08	1.05	1.04	70-77
26-30	1.00	1.00	1.00	1.00	1.00	1.00	78-86
31-35	0.91	0.94	0.96	0.91	0.94	0.96	87-95
36-40	0.82	0.88	0.91	0.82	0.88	0.91	96-104
41-45	0.71	0.82	0.87	0.71	0.82	0.87	105-113
46-50	0.58	0.75	0.82	0.58	0.75	0.82	114-122
51-55	0.41	0.67	0.76	0.41	0.67	0.76	123-131
56-60	—	0.58	0.71	—	0.58	0.71	132-140
61-70	—	0.33	0.58	—	0.33	0.58	141-158
71-80	—	—	0.41	—	—	0.41	159-176

For SI: °C = [(°F)-32]/1.8.

3. Adjustment factors shall not apply to type AC cable or to type MC cable without an overall outer jacket meeting all of the following conditions:

3.1. Each cable has not more than three current-carrying conductors.

3.2. The conductors are 12 AWG copper.

3.3. Not more than 20 current-carrying conductors are bundled, stacked or supported on bridle rings. A 60 percent adjustment factor shall be applied where the current-carrying conductors in such cables exceed 20 and the cables are stacked or bundled for distances greater than 24 inches (610 mm) without maintaining spacing.

TABLE E3605.3
CONDUCTOR PROXIMITY ADJUSTMENT FACTORS

NUMBER OF CURRENT-CARRYING CONDUCTORS IN CABLE OR RACEWAY	PERCENT OF VALUES IN TABLE E3605.1
4-6	80
7-9	70
10-20	50
21-30	45
31-40	40
41 and above	35

E3605.4 Temperature limitations. The temperature rating associated with the ampacity of a conductor shall be so selected and coordinated to not exceed the lowest temperature rating of any connected termination, conductor or device. Conductors with temperature ratings higher than specified for terminations shall be permitted to be used for ampacity adjustment, correc-tion, or both. Except where the equipment is marked otherwise, conductor ampacities used in determining equipment termination provisions shall be based on Table E3605.1

E3605.4.1 Conductors rated 60°C. Except where the equipment is marked otherwise, termination provisions of equipment for circuits rated 100 amperes or less, or marked for 14 AWG through 1 AWG conductors, shall be used only for one of the following:

1. Conductors rated 60°C (140°F);

2. Conductors with higher temperature ratings, provided that the ampacity of such conductors is determined based on the 60°C (140°F) ampacity of the conductor size used;

3. Conductors with higher temperature ratings where the equipment is listed and identified for use with such conductors; or

4. For motors marked with design letters B, C, or D conductors having an insulation rating of 75°C (167°F) or higher shall be permitted to be used provided that the ampacity of such conductors does not exceed the 75°C (167°F) ampacity.

E3605.4.2 Conductors rated 75°C. Termination provisions of equipment for circuits rated over 100 amperes, or marked for conductors larger than 1 AWG, shall be used only for:

1. Conductors rated 75°C (167°F).

2. Conductors with higher temperature ratings provided that the ampacity of such conductors does not exceed the 75°C (167°F) ampacity of the conductor size

used, or provided that the equipment is listed and identified for use with such conductors.

E3605.4.3 Separately installed pressure connectors. Separately installed pressure connectors shall be used with conductors at the ampacities not exceeding the ampacity at the listed and identified temperature rating of the connector.

E3605.4.4 Conductors of Type NM cable. Conductors in NM cable assemblies shall be rated at 90°C (194°F). Types NM, NMC, and NMS cable identified by the markings NM-B, NMC-B, and NMS-B meet this requirement. The ampacity of Types NM, NMC, and NMS cable shall be at 60°C (140°F) conductors and shall comply with Section E3605.1 and Table E3605.5.3. The 90°C (194°F) rating shall be permitted to be used for ampacity correction and adjustment purposes provided that the final corrected or adjusted ampacity does not exceed that for a 60°C (140°F) rated conductor. Where more than two NM cables containing two or more current-carrying conductors are bundled together and pass through wood framing that is to be fire- or draft-stopped using thermal insulation or sealing foam, the allowable ampacity of each conductor shall be adjusted in accordance with Table E3605.3.

E3605.5 Overcurrent protection required. All ungrounded branch-circuit and feeder conductors shall be protected against overcurrent by an overcurrent device installed at the point where the conductors receive their supply. Overcurrent devices shall not be connected in series with a grounded conductor. Overcurrent protection and allowable loads for branch circuits and feeders that do not serve as the main power feeder to the dwelling unit load shall be in accordance with this chapter.

Branch-circuit conductors and equipment shall be protected by overcurrent protective devices having a rating or setting not exceeding the allowable ampacity specified in Table E3605.1 and Sections E3605.2, E3605.3 and E3605.4 except where otherwise permitted or required in Sections E3605.5.1 through E3605.5.3.

E3605.5.1 Cords. Cords shall be protected in accordance with Section E3809.2.

E3605.5.2 Overcurrent devices of the next higher rating. The next higher standard overcurrent device rating, above the ampacity of the conductors being protected, shall be permitted to be used, provided that all of the following conditions are met:

1. The conductors being protected are not part of a multioutlet branch circuit supplying receptacles for cord- and plug-connected portable loads.

2. The ampacity of conductors does not correspond with the standard ampere rating of a fuse or a circuit breaker without overload trip adjustments above its rating (but that shall be permitted to have other trip or rating adjustments).

3. The next higher standard device rating does not exceed 400 amperes.

E3605.5.3 Small conductors. Except as specifically permitted by Section E3605.5.4, the rating of overcurrent pro-

tection devices shall not exceed the ratings shown in Table E3605.5.3 for the conductors specified therein.

E3605.5.4 Air-conditioning and heat pump equipment. Air-conditioning and heat pump equipment circuit conductors shall be permitted to be protected against overcurrent in accordance with Section E3602.11.

E3605.6 Fuses and fixed trip circuit breakers. The standard ampere ratings for fuses and inverse time circuit breakers shall be considered 15, 20, 25, 30, 35, 40, 45, 50, 60, 70, 80, 90, 100, 110, 125, 150, 175, 200, 225, 250, 300, 350 and 400 amperes.

TABLE E3605.5.3
OVERCURRENT-PROTECTION RATING

COPPER		ALUMINUM OR COPPER-CLAD ALUMINUM	
Size (AWG)	Maximum overcurrent-protection-device rating[a] (amps)	Size (AWG)	Maximum overcurrent-protection-device rating[a] (amps)
14	15	12	15
12	20	10	25
10	30	8	30

a. The maximum overcurrent-protection-device rating shall not exceed the conductor allowable ampacity determined by the application of the correction and adjustment factors in accordance with Sections E3605.2 and E3605.3.

E3605.7 Location of overcurrent devices in or on premises. Overcurrent devices shall:

1. Be readily accessible.

2. Not be located where they will be exposed to physical damage.

3. Not be located where they will be in the vicinity of easily ignitible material such as in clothes closets.

4. Not be located in bathrooms.

5. Be installed so that the center of the grip of the operating handle of the switch or circuit breaker, when in its highest position, is not more than 6 feet 7 inches (2.0 m) above the floor or working platform.

Exceptions:

1. This section shall not apply to supplementary overcurrent protection that is integral to utilization equipment.

2. Overcurrent devices installed adjacent to the utilization equipment that they supply shall be permitted to be accessible by portable means.

E3605.8 Ready access for occupants. Each occupant shall have ready access to all overcurrent devices protecting the conductors supplying that occupancy.

E3605.9 Enclosures for overcurrent devices. Overcurrent devices shall be enclosed in cabinets or cutout boxes except where an overcurrent device is part of an assembly that provides equivalent protection. The operating handle of a circuit breaker shall be permitted to be accessible without opening a door or cover.

SECTION E3606
PANELBOARDS

E3606.1 Panelboard rating. All panelboards shall have a rating not less than that of the minimum service entrance or feeder capacity required for the calculated load.

E3606.2 Panelboard circuit identification. All circuits and circuit modifications shall be legibly identified as to their clear, evident, and specific purpose or use. The identification shall include sufficient detail to allow each circuit to be distinguished from all others. The identification shall be included in a circuit directory located on the face of the panelboard enclosure or inside the panel door.

E3606.3 Panelboard overcurrent protection. Panelboards shall be protected on the supply side by not more than two main circuit breakers or two sets of fuses having a combined rating not greater than that of the panelboard.

> **Exception:** Individual protection for a panelboard shall not be required if the panelboard feeder has overcurrent protection not greater than the rating of the panelboard.

E3606.4 Grounded conductor terminations. Each grounded conductor shall terminate within the panelboard on an individual terminal that is not also used for another conductor, except that grounded conductors of circuits with parallel conductors shall be permitted to terminate on a single terminal where the terminal is identified for connection of more than one conductor.

E3606.5 Back-fed devices. Plug-in-type overcurrent protection devices or plug-in-type main lug assemblies that are back-fed and used to terminate field-installed ungrounded supply conductors shall be secured in place by an additional fastener that requires other than a pull to release the device from the mounting means on the panel.

CHAPTER 37
WIRING METHODS

SECTION E3701
GENERAL REQUIREMENTS

E3701.1 Scope. This chapter covers the wiring methods for services, feeders and branch circuits for electrical power and distribution.

E3701.2 Allowable wiring methods. The allowable wiring methods for electrical installations shall be those listed in Table E3701.2. Single conductors shall be used only where part of one of the recognized wiring methods listed in Table E3701.2. As used in this code, abbreviations of the wiring-method types shall be as indicated in Table E3701.2.

TABLE E3701.2
ALLOWABLE WIRING METHODS

ALLOWABLE WIRING METHOD	DESIGNATED ABBREVIATION
Armored cable	AC
Electrical metallic tubing	EMT
Electrical nonmetallic tubing	ENT
Flexible metal conduit	FMC
Intermediate metal conduit	IMC
Liquidtight flexible conduit	LFC
Metal-clad cable	MC
Nonmetallic sheathed cable	NM
Rigid nonmetallic conduit	RNC
Rigid metallic conduit	RMC
Service entrance cable	SE
Surface raceways	SR
Underground feeder cable	UF
Underground service cable	USE

E3701.3 Circuit conductors. All conductors of a circuit, including equipment grounding conductors and bonding conductors, shall be contained in the same raceway, trench, cable or cord.

E3701.4 Wiring method applications. Wiring methods shall be applied in accordance with Table E3701.4.

SECTION E3702
ABOVE-GROUND INSTALLATION REQUIREMENTS

E3702.1 Installation and support requirements. Wiring methods shall be installed and supported in accordance with Table E3702.1.

E3702.2 Cables in accessible attics. Cables in attics or roof spaces provided with access shall be installed as specified in Sections E3702.2.1 and E3702.2.2.

E3702.2.1 Across structural members. Where run across the top of floor joists, or run within 7 feet (2134 mm) of floor or floor joists across the face of rafters or studding, in attics and roof spaces that are provided with access, the cable shall be protected by substantial guard strips that are at least as high as the cable. Where such spaces are not provided with access by permanent stairs or ladders, protection shall only be required within 6 feet (1829 mm) of the nearest edge of the attic entrance.

E3702.2.2 Cable installed through or parallel to framing members. Where cables are installed through or parallel to the sides of rafters, studs or floor joists, guard strips and running boards shall not be required, and the installation shall comply with Table E3702.1.

E3702.3 Exposed cable. In exposed work, except as provided for in Sections E3702.2 and E3702.4, cable assemblies shall be installed as specified in Sections E3702.3.1 and E3702.3.2.

E3702.3.1 Surface installation. Cables shall closely follow the surface of the building finish or running boards.

E3702.3.2 Protection from physical damage. Where subject to physical damage, cables shall be protected by rigid metal conduit, intermediate metal conduit, electrical metallic tubing, Schedule 80 PVC rigid nonmetallic conduit, or other approved means. Where passing through a floor, the cable shall be enclosed in rigid metal conduit, intermediate metal conduit, electrical metallic tubing, Schedule 80 PVC rigid nonmetallic conduit or other approved means extending not less than 6 inches (152 mm) above the floor.

E3702.3.3 Locations exposed to direct sunlight. Insulated conductors and cables used where exposed to direct rays of the sun shall be of a type listed for sunlight resistance, or of a type listed and marked "sunlight resistant," or shall be covered with insulating material, such as tape or sleeving, that is listed or listed and marked as being "sunlight resistant."

E3702.4 In unfinished basements. Where type SE or NM cable is run at angles with joists in unfinished basements, cable assemblies containing two or more conductors of sizes 6 AWG and larger and assemblies containing three or more conductors of sizes 8 AWG and larger shall not require additional protection where attached directly to the bottom of the joists. Smaller cables shall be run either through bored holes in joists or on running boards. NM cable used on a wall of an unfinished basement shall be permitted to be installed in a listed conduit or tubing. Such conduit or tubing shall be provided with a nonmetallic bushing or adapter at the point the where cable enters the raceway.

E3702.5 Bends. Bends shall be made so as not to damage the wiring method or reduce the internal diameter of raceways.

For types NM and SE cable, bends shall be so made, and other handling shall be such that the cable will not be damaged and the radius of the curve of the inner edge of any bend shall be not less than five times the diameter of the cable.

TABLE E3701.4
ALLOWABLE APPLICATIONS FOR WIRING METHODS[a, b, c, d, e, f, g, h, i]

ALLOWABLE APPLICATIONS (application allowed where marked with an "A")	AC	EMT	ENT	FMC	IMC RMC RNC	LFC[a]	MC	NM	SR	SE	UF	USE
Services	—	A	A[h]	A[i]	A	A[i]	A	—	—	A	—	A
Feeders	A	A	A	A	A	A	A	A	—	A[b]	A	A[b]
Branch circuits	A	A	A	A	A	A	A	A	A	A[c]	A	—
Inside a building	A	A	A	A	A	A	A	A	A	A	A	—
Wet locations exposed to sunlight	—	A	A[h]	A[d]	A	A	A	—	—	A	A[e]	A[e]
Damp locations	—	A	A	A[d]	A	A	A	—	—	A	A	A
Embedded in noncinder concrete in dry location	—	A	A	—	A	—	—	—	—	—	—	—
In noncinder concrete in contact with grade	—	A[f]	A	—	A[f]	—	—	—	—	—	—	—
Embedded in plaster not exposed to dampness	A	A	A	A	A	A	A	—	—	A	A	—
Embedded in masonry	—	A	A	—	A[f]	A	A	—	—	—	—	—
In masonry voids and cells exposed to dampness or below grade line	—	A[f]	A	A[d]	A[f]	A	A	—	—	A	A	—
Fished in masonry voids	A	—	—	A	—	A	A	A	—	A	A	—
In masonry voids and cells not exposed to dampness	A	A	A	A	A	A	A	A	—	A	A	—
Run exposed	A	A	A	A	A	A	A	A	A	A	A	A
Run exposed and subject to physical damage	—	—	—	—	A[g]	—	—	—	—	—	—	—
For direct burial	—	A[f]	—	—	A[f]	A	A[f]	—	—	—	A	A

For SI: 1 foot = 304.8 mm.

a. Liquid-tight flexible nonmetallic conduit without integral reinforcement within the conduit wall shall not exceed 6 feet in length.

b. The grounded conductor shall be insulated except where used to supply other buildings on the same premises. Type USE cable shall not be used inside buildings.

c. The grounded conductor shall be insulated.

d. Conductors shall be a type approved for wet locations and the installation shall prevent water from entering other raceways.

e. Shall be listed as "Sunlight Resistant."

f. Metal raceways shall be protected from corrosion and approved for the application.

g. RNC shall be Schedule 80.

h. Shall be listed as "Sunlight Resistant" where exposed to the direct rays of the sun.

i. Conduit shall not exceed 6 feet in length.

TABLE E3702.1
GENERAL INSTALLATION AND SUPPORT REQUIREMENTS FOR WIRING METHODS[a, b, c, d, e, f, g, h, i, j, k]

INSTALLATION REQUIREMENTS (Requirement applicable only to wiring methods marked "A")	AC MC	EMT IMC RMC	ENT	FMC LFC	NM UF	RNC	SE	SR[a]	USE
Where run parallel with the framing member or furring strip, the wiring shall be not less than $1^1/_4$ inches from the edge of a furring strip or a framing member such as a joist, rafter or stud or shall be physically protected.	A	—	A	A	A	—	A	—	—
Bored holes in framing members for wiring shall be located not less than $1^1/_4$ inches from the edge of the framing member or shall be protected with a minimum 0.0625-inch steel plate or sleeve, a listed steel plate or other physical protection.	A[k]	—	A[k]	A[k]	A[k]	—	A[k]	—	—
Where installed in grooves, to be covered by wallboard, siding, paneling, carpeting, or similar finish, wiring methods shall be protected by 0.0625-inch-thick steel plate, sleeve, or equivalent, a listed steel plate or by not less than $1^1/_4$-inch free space for the full length of the groove in which the cable or raceway is installed.	A	—	A	A	A	—	A	A	A
Securely fastened bushings or grommets shall be provided to protect wiring run through openings in metal framing members.	—	—	A[j]	—	A[j]	—	A[j]	—	—
The maximum number of 90-degree bends shall not exceed four between junction boxes.	—	A	A	A	—	A	—	—	—
Bushings shall be provided where entering a box, fitting or enclosure unless the box or fitting is designed to afford equivalent protection.	A	A	A	A	—	A	—	A	—
Ends of raceways shall be reamed to remove rough edges.	—	A	A	A	—	A	—	A	—
Maximum allowable on center support spacing for the wiring method in feet.	4.5[b, c]	10	3[b]	4.5b	4.5[i]	3[d]	2.5[e]	—	2.5[e]
Maximum support distance in inches from box or other terminations.	12[b, f]	36	36	12[b,g]	12[h,i]	36	12	—	12

For SI: 1 inch = 25.4 mm, 1 foot = 304.8 mm, 1 degree = 0.009 rad.

a. Installed in accordance with listing requirements.
b. Supports not required in accessible ceiling spaces between light fixtures where lengths do not exceed 6 feet.
c. Six feet for MC cable.
d. Five feet for trade sizes greater than 1 inch.
e. Two and one-half feet where used for service or outdoor feeder and 4.5 feet where used for branch circuit or indoor feeder.
f. Twenty-four inches where flexibility is necessary.
g. Thirty-six inches where flexibility is necessary.
h. Within 8 inches of boxes without cable clamps.
i. Flat cables shall not be stapled on edge.
j. Bushings and grommets shall remain in place and shall be listed for the purpose of cable protection.
k. See Sections R502.8 and R802.7 for additional limitations on the location of bored holes in horizontal framing members.

E3702.6 Raceways exposed to different temperatures. Where portions of a cable, raceway or sleeve are known to be subjected to different temperatures and where condensation is known to be a problem, as in cold storage areas of buildings or where passing from the interior to the exterior of a building, the raceway or sleeve shall be filled with an approved material to prevent the circulation of warm air to a colder section of the raceway or sleeve.

SECTION E3703
UNDERGROUND INSTALLATION REQUIREMENTS

E3703.1 Minimum cover requirements. Direct buried cable or raceways shall be installed in accordance with the minimum cover requirements of Table E3703.1.

E3703.2 Warning ribbon. Underground service conductors that are not encased in concrete and that are buried 18 inches (457 mm) or more below grade shall have their location identi-fied by a warning ribbon that is placed in the trench not less than 12 inches (300 mm) above the underground installation.

E3703.3 Protection from damage. Direct buried conductors and cables emerging from the ground shall be protected by enclosures or raceways extending from the minimum cover distance below grade required by Section E3703.1 to a point at least 8 feet (2438 mm) above finished grade. In no case shall the protection be required to exceed 18 inches (457 mm) below finished grade. Conductors entering a building shall be protected to the point of entrance. Where the enclosure or raceway is subject to physical damage, the conductors shall be installed in rigid metal conduit, intermediate metal conduit, Schedule 80 rigid nonmetallic conduit or the equivalent.

E3703.4 Splices and taps. Direct buried conductors or cables shall be permitted to be spliced or tapped without the use of splice boxes. The splices or taps shall be made by approved methods with materials listed for the application.

TABLE E3703.1
MINIMUM COVER REQUIREMENTS, BURIAL IN INCHES[a, b, c, d, e]

LOCATION OF WIRING METHOD OR CIRCUIT	TYPE OF WIRING METHOD OR CIRCUIT				
	1 Direct burial cables or conductors	2 Rigid metal conduit or intermediate metal conduit	3 Nonmetallic raceways listed for direct burial without concrete encasement or other approved raceways	4 Residential branch circuits rated 120 volts or less with GFCI protection and maximum overcurrent protection of 20 amperes	5 Circuits for control of irrigation and landscape lighting limited to not more than 30 volts and installed with type UF or in other identified cable or raceway
All locations not specified below	24	6	18	12	6
In trench below 2-inch-thick concrete or equivalent	18	6	12	6	6
Under a building	0 (In raceway only)	0	0	0 (In raceway only)	0 (In raceway only)
Under minimum of 4-inch-thick concrete exterior slab with no vehicular traffic and the slab extending not less than 6 inches beyond the underground installation	18	4	4	6 (Direct burial) 4 (In raceway)	6 (Direct burial) 4 (In raceway)
Under streets, highways, roads, alleys, driveways and parking lots	24	24	24	24	24
One- and two-family dwelling driveways and outdoor parking areas, and used only for dwelling-related purposes	18	18	18	12	18
In solid rock where covered by minimum of 2 inches concrete extending down to rock	2 (In raceway only)	2	2	2 (In raceway only)	2 (In raceway only)

For SI: 1 inch = 25.4 mm.

a. Raceways approved for burial only where encased concrete shall require concrete envelope not less than 2 inches thick.

b. Lesser depths shall be permitted where cables and conductors rise for terminations or splices or where access is otherwise required.

c. Where one of the wiring method types listed in columns 1 to 3 is combined with one of the circuit types in columns 4 and 5, the shallower depth of burial shall be permitted.

d. Where solid rock prevents compliance with the cover depths specified in this table, the wiring shall be installed in metal or nonmetallic raceway permitted for direct burial. The raceways shall be covered by a minimum of 2 inches of concrete extending down to the rock.

e. Cover is defined as the shortest distance in inches (millimeters) measured between a point on the top surface of any direct-buried conductor, cable, conduit or other raceway and the top surface of finished grade, concrete, or similar cover.

E3703.5 Backfill. Backfill containing large rock, paving materials, cinders, large or sharply angular substances, or corrosive material shall not be placed in an excavation where such materials cause damage to raceways, cables or other substructures or prevent adequate compaction of fill or contribute to corrosion of raceways, cables or other substructures. Where necessary to prevent physical damage to the raceway or cable, protection shall be provided in the form of granular or selected material, suitable boards, suitable sleeves or other approved means.

E3703.6 Raceway seals. Conduits or raceways shall be sealed or plugged at either or both ends where moisture will enter and contact live parts.

E3703.7 Bushing. A bushing, or terminal fitting, with an integral bushed opening shall be installed on the end of a conduit or other raceway that terminates underground where the conductors or cables emerge as a direct burial wiring method. A seal incorporating the physical protection characteristics of a bushing shall be considered equivalent to a bushing.

E3703.8 Single conductors. All conductors of the same circuit and, where present, the grounded conductor and all equipment grounding conductors shall be installed in the same raceway or shall be installed in close proximity in the same trench.

> **Exception:** Where conductors are installed in parallel in raceways, each raceway shall contain all conductors of the same circuit including grounding conductors.

E3703.9 Ground movement. Where direct buried conductors, raceways or cables are subject to movement by settlement or frost, direct buried conductors, raceways or cables shall be arranged to prevent damage to the enclosed conductors or to equipment connected to the raceways.

CHAPTER 38
POWER AND LIGHTING DISTRIBUTION

SECTION E3801
RECEPTACLE OUTLETS

E3801.1 General. Outlets for receptacles rated at 125 volts, 15- and 20-amperes shall be provided in accordance with Sections E3801.2 through E3801.11. Receptacle outlets required by this section shall be in addition to any receptacle that is part of a luminaire or appliance, that is located within cabinets or cupboards, or that is located over 5.5 feet (1676 mm) above the floor.

Permanently installed electric baseboard heaters equipped with factory-installed receptacle outlets, or outlets provided as a separate assembly by the baseboard manufacturer shall be permitted as the required outlet or outlets for the wall space utilized by such permanently installed heaters. Such receptacle outlets shall not be connected to the heater circuits.

E3801.2 General purpose receptacle distribution. In every kitchen, family room, dining room, living room, parlor, library, den, sun room, bedroom, recreation room, or similar room or area of dwelling units, receptacle outlets shall be installed in accordance with the general provisions specified in Sections E3801.2.1 through E3801.2.3 (see Figure E3801.2).

E3801.2.1 Spacing. Receptacles shall be installed so that no point measured horizontally along the floor line in any wall space is more than 6 feet (1829 mm), from a receptacle outlet.

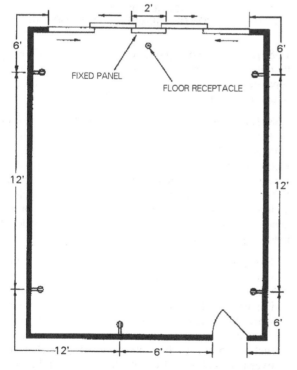

For SI: 1 foot = 304.8 mm.

FIGURE E3801.2
GENERAL USE RECEPTACLE DISTRIBUTION

E3801.2.2 Wall space. As used in this section, a wall space shall include the following:

1. Any space that is 2 feet (610 mm) or more in width, (including space measured around corners), and that is unbroken along the floor line by doorways, fireplaces, and similar openings.

2. The space occupied by fixed panels in exterior walls, excluding sliding panels.

3. The space created by fixed room dividers such as railings and freestanding bar-type counters.

E3801.2.3 Floor receptacles. Receptacle outlets in floors shall not be counted as part of the required number of receptacle outlets except where located within 18 inches (457 mm) of the wall.

E3801.3 Small appliance receptacles. In the kitchen, pantry, breakfast room, dining room, or similar area of a dwelling unit, the two or more 20-ampere small-appliance branch circuits required by Section E3603.2, shall serve all wall and floor receptacle outlets covered by Sections E3801.2 and E3801.4 and those receptacle outlets provided for refrigeration appliances.

Exceptions:

1. In addition to the required receptacles specified by Sections E3801.1 and E3801.2, switched receptacles supplied from a general-purpose branch circuit as defined in Section E3803.2, Exception 1 shall be permitted.

2. The receptacle outlet for refrigeration appliances shall be permitted to be supplied from an individual branch circuit rated at 15 amperes or greater.

E3801.3.1 Other outlets prohibited. The two or more small-appliance branch circuits specified in Section E3801.3 shall serve no other outlets.

Exceptions:

1. A receptacle installed solely for the electrical supply to and support of an electric clock in any of the rooms specified in Section E3801.3.

2. Receptacles installed to provide power for supplemental equipment and lighting on gas-fired ranges, ovens, and counter-mounted cooking units.

E3801.3.2 Limitations. Receptacles installed in a kitchen to serve countertop surfaces shall be supplied by not less than two small-appliance branch circuits, either or both of which shall also be permitted to supply receptacle outlets in the same kitchen and in other rooms specified in Section E3801.3. Additional small-appliance branch circuits shall be permitted to supply receptacle outlets in the kitchen and other rooms specified in Section E3801.3. A small-appliance branch circuit shall not serve more than one kitchen.

E3801.4 Countertop receptacles. In kitchens and dining rooms of dwelling units, receptacle outlets for counter spaces shall be installed in accordance with Sections E3801.4.1 through E3801.4.5 (see Figure E3801.4).

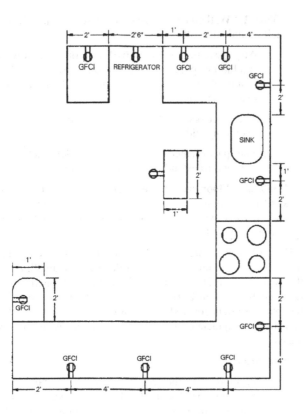

For SI: 1 foot = 304.8 mm.

FIGURE E3801.4
COUNTERTOP RECEPTACLES

E3801.4.1 Wall counter space. A receptacle outlet shall be installed at each wall counter space 12 inches (305 mm) or wider. Receptacle outlets shall be installed so that no point along the wall line is more than 24 inches (610 mm), measured horizontally from a receptacle outlet in that space.

> **Exception:** Receptacle outlets shall not be required on a wall directly behind a range or sink in the installation described in Figure E3801.4.1.

E3801.4.2 Island counter spaces. At least one receptacle outlet shall be installed at each island counter space with a long dimension of 24 inches (610 mm) or greater and a short dimension of 12 inches (305 mm) or greater. Where a rangetop or sink is installed in an island counter and the width of the counter behind the rangetop or sink is less than 12 inches (300 mm), the rangetop or sink has divided the island into two separate countertop spaces as defined in Section E3801.4.4.

E3801.4.3 Peninsular counter space. At least one receptacle outlet shall be installed at each peninsular counter space with a long dimension of 24 inches (610 mm) or greater and a short dimension of 12 inches (305 mm) or greater. A peninsular countertop is measured from the connecting edge.

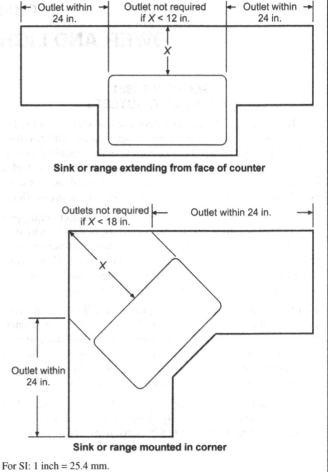

For SI: 1 inch = 25.4 mm.

FIGURE E3801.4.1
DETERMINATION OF AREA BEHIND SINK OR RANGE

E3801.4.4 Separate spaces. Countertop spaces separated by range tops, refrigerators, or sinks shall be considered as separate countertop spaces in applying the requirements of Sections E3801.4.1, E3801.4.2 and E3801.4.3.

E3801.4.5 Receptacle outlet location. Receptacle outlets shall be located not more than 20 inches (508 mm) above the countertop. Receptacle outlets shall not be installed in a face-up position in the work surfaces or countertops. Receptacle outlets rendered not readily accessible by appliances fastened in place, appliance garages, sinks or rangetops as addressed in the exception to Section E3801.4.1, or appliances occupying dedicated space shall not be considered as these required outlets.

> **Exception:** Receptacle outlets shall be permitted to be mounted not more than 12 inches (305 mm) below the countertop in construction designed for the physically impaired and for island and peninsular countertops where the countertop is flat across its entire surface and there are no means to mount a receptacle within 20 inches (457 mm) above the countertop, such as in an overhead cabinet. Receptacles mounted below the countertop in accordance with this exception shall not be located where the countertop extends more than 6 inches (152 mm) beyond its support base.

E3801.5 Appliance outlets. Appliance receptacle outlets installed for specific appliances, such as laundry equipment, shall be installed within 6 feet (1829 mm) of the intended location of the appliance.

E3801.6 Bathroom. At least one wall receptacle outlet shall be installed in bathrooms and such outlet shall be located within 36 inches (914 mm) of the outside edge of each lavatory basin. The receptacle outlet shall be located on a wall that is adjacent to the lavatory basin location.

Receptacle outlets shall not be installed in a face-up position in the work surfaces or countertops in a bathroom basin location.

> **Exception:** The receptacle shall not be required to be mounted on the wall or partition where it is installed on the side or face of the basin cabinet not more than 12 inches (300 mm) below the countertop.

E3801.7 Outdoor outlets. At least one receptacle outlet accessible at grade level and not more than 6 feet, 6 inches (1981 mm) above grade, shall be installed outdoors at the front and back of each dwelling unit having direct access to grade.

E3801.8 Laundry areas. At least one receptacle outlet shall be installed to serve laundry appliances.

E3801.9 Basements and garages. At least one receptacle outlet, in addition to any provided for laundry equipment, shall be installed in each basement and in each attached garage, and in each detached garage that is provided with electrical power. Where a portion of the basement is finished into one or more habitable room(s), each separate unfinished portion shall have a receptacle outlet installed in accordance with this section.

E3801.10 Hallways. Hallways of 10 feet (3048 mm) or more in length shall have at least one receptacle outlet. The hall length shall be considered the length measured along the centerline of the hall without passing through a doorway.

E3801.11 HVAC outlet. A 125-volt, single-phase, 15- or 20-ampere-rated receptacle outlet shall be installed at an accessible location for the servicing of heating, air-conditioning and refrigeration equipment. The receptacle shall be located on the same level and within 25 feet (7620 mm) of the heating, air-conditioning and refrigeration equipment. The receptacle outlet shall not be connected to the load side of the HVAC equipment disconnecting means.

> **Exception:** A receptacle outlet shall not be required for the servicing of evaporative coolers.

SECTION E3802
GROUND-FAULT AND ARC-FAULT
CIRCUIT-INTERRUPTER PROTECTION

E3802.1 Bathroom receptacles. All 125-volt, single-phase, 15- and 20-ampere receptacles installed in bathrooms shall have ground-fault circuit-interrupter protection for personnel.

E3802.2 Garage and accessory building receptacles. All 125-volt, single-phase, 15- or 20-ampere receptacles installed in garages and grade-level portions of unfinished accessory buildings used for storage or work areas shall have ground-fault circuit-interrupter protection for personnel (see Section E3802.11).

Exceptions:

1. Receptacles that are not readily accessible.

2. A single receptacle or a duplex receptacle for two appliances located within dedicated space for each appliance that in normal use is not easily moved from one place to another, and that is cord- and plug-connected.

E3802.3 Outdoor receptacles. All 125-volt, single-phase, 15- and 20-ampere receptacles installed outdoors shall have ground-fault circuit-interrupter protection for personnel.

Exception: Receptacles as covered in Section E4001.7.

E3802.4 Crawl space receptacles. Where a crawl space is at or below grade level, all 125-volt, single-phase, 15- and 20-ampere receptacles installed in such spaces shall have ground-fault circuit-interrupter protection for personnel.

E3802.5 Unfinished basement receptacles. All 125-volt, single-phase, 15- and 20-ampere receptacles installed in unfinished basements shall have ground-fault circuit-interrupter protection for personnel. For purposes of this section, unfinished basements are defined as portions or areas of the basement not intended as habitable rooms and limited to storage areas, work areas, and the like (see Section E3802.11).

Exceptions:

1. Receptacles that are not readily accessible.

2. A single receptacle or duplex receptacle for two appliances located within dedicated space for each appliance that in normal use is not easily moved from one place to another, and that is cord- and plug-connected.

3. A receptacle supplying only a permanently installed fire alarm or burglar alarm system.

E3802.6 Kitchen receptacles. All 125-volt, single-phase, 15- and 20-ampere receptacles that serve countertop surfaces shall have ground-fault circuit-interrupter protection for personnel.

E3802.7 Laundry, utility, and bar sink receptacles. All 125-volt, single-phase, 15- and 20-ampere receptacles that are located within 6 feet (1829 mm) of the outside edge of a laundry, utility or wet bar sink shall have ground-fault circuit-interrupter protection for personnel. Receptacle outlets shall not be installed in a face-up position in the work surfaces or countertops.

E3802.8 Boathouse receptacles. All 125-volt, single-phase, 15- or 20-ampere receptacles installed in boathouses used for storage or work areas shall have ground-fault circuit-interrupter protection for personnel.

E3802.9 Boat hoists. Ground-fault circuit-interrupter protection for personnel shall be provided for outlets that supply boat hoists supplied by 125-volt, 15- and 20-ampere branch circuits.

E3802.10 Electrically heated floors. Ground-fault circuit-interrupter protection for personnel shall be provided for electrically heated floors in bathrooms, and in hydromassage bathtub, spa and hot tub locations.

E3802.11 Exempt receptacles. Receptacles installed under exceptions to Sections E3802.2 and E3802.5 shall not be considered as meeting the requirements of Section E3801.9.

E3802.12 Arc-fault protection of bedroom outlets. All branch circuits that supply 120-volt, single-phase, 15- and

20-ampere outlets installed in bedrooms shall be protected by a combination type or branch/feeder type arc-fault circuit interrupter installed to provide protection of the entire branch circuit. Effective January 1, 2008, such arc-fault circuit interrupter devices shall be combination type.

Exception: The location of the arc-fault circuit interrupter shall be permitted to be at other than the origination of the branch circuit provided that:

1. The arc-fault circuit interrupter is installed within 6 feet (1.8 m) of the branch circuit overcurrent device as measured along the branch circuit conductors and

2. The circuit conductors between the branch circuit overcurrent device and the arc-fault circuit interrupter are installed in a metal raceway or a cable with a metallic sheath.

SECTION E3803
LIGHTING OUTLETS

E3803.1 General. Lighting outlets shall be provided in accordance with Sections E3803.2 through E3803.4.

E3803.2 Habitable rooms. At least one wall switch-controlled lighting outlet shall be installed in every habitable room and bathroom.

Exceptions:

1. In other than kitchens and bathrooms, one or more receptacles controlled by a wall switch shall be considered equivalent to the required lighting outlet.

2. Lighting outlets shall be permitted to be controlled by occupancy sensors that are in addition to wall switches, or that are located at a customary wall switch location and equipped with a manual override that will allow the sensor to function as a wall switch.

E3803.3 Additional locations. At least one wall-switch-controlled lighting outlet shall be installed in hallways, stairways, attached garages, and detached garages with electric power. At least one wall-switch-controlled lighting outlet shall be installed to provide illumination on the exterior side of each outdoor egress door having grade level access, including outdoor egress doors for attached garages and detached garages with electric power. A vehicle door in a garage shall not be considered as an outdoor egress door. Where one or more lighting outlets are installed for interior stairways, there shall be a wall switch at each floor level and landing level that includes an entryway to control the lighting outlets where the stairway between floor levels has six or more risers.

Exception: In hallways, stairways, and at outdoor egress doors, remote, central, or automatic control of lighting shall be permitted.

E3803.4 Storage or equipment spaces. In attics, under-floor spaces, utility rooms and basements, at least one lighting outlet shall be installed where these spaces are used for storage or contain equipment requiring servicing. Such lighting outlet shall be controlled by a wall switch or shall have an integral switch. At least one point of control shall be at the usual point of entry to these spaces. The lighting outlet shall be provided at or near the equipment requiring servicing.

SECTION E3804
GENERAL INSTALLATION REQUIREMENTS

E3804.1 Electrical continuity of metal raceways and enclosures. Metal raceways, cable armor and other metal enclosures for conductors shall be mechanically joined together into a continuous electric conductor and shall be connected to all boxes, fittings and cabinets so as to provide effective electrical continuity. Raceways and cable assemblies shall be mechanically secured to boxes, fittings cabinets and other enclosures.

Exception: Short sections of raceway used to provide cable assemblies with support or protection against physical damage.

E3804.2 Mechanical continuity—raceways and cables. Metal or nonmetallic raceways, cable armors and cable sheaths shall be continuous between cabinets, boxes, fittings or other enclosures or outlets.

Exception: Short sections of raceway used to provide cable assemblies with support or protection against physical damage.

E3804.3 Securing and supporting. Raceways, cable assemblies, boxes, cabinets and fittings shall be securely fastened in place.

E3804.3.1 Prohibited means of support. Cable wiring methods shall not be used as a means of support for other cables, raceways and nonelectrical equipment.

E3804.4 Raceways as means of support. Raceways shall be used as a means of support for other raceways, cables or nonelectric equipment only under the following conditions:

1. Where the raceway or means of support is identified for the purpose; or

2. Where the raceway contains power supply conductors for electrically controlled equipment and is used to support Class 2 circuit conductors or cables that are solely for the purpose of connection to the control circuits of the equipment served by such raceway; or

3. Where the raceway is used to support boxes or conduit bodies in accordance with Sections E3806.8.4 and E3806.8.5.

E3804.5 Raceway installations. Raceways shall be installed complete between outlet, junction or splicing points prior to the installation of conductors.

Exception: Short sections of raceways used to contain conductors or cable assemblies for protection from physical damage shall not be required to be installed complete between outlet, junction, or splicing points.

E3804.6 Conduit and tubing fill. The maximum number of conductors installed in conduit or tubing shall be in accordance with Tables E3804.6(1) through E3804.6(10).

E3804.7 Air handling—stud cavity and joist spaces. Where wiring methods having a nonmetallic covering pass through stud cavities and joist spaces used for air handling, such wiring shall pass through such spaces perpendicular to the long dimension of the spaces.

TABLE E3804.6(1)
MAXIMUM NUMBER OF CONDUCTORS IN ELECTRICAL METALLIC TUBING (EMT)[a]

TYPE LETTERS	CONDUCTOR SIZE AWG/kcmil	TRADE SIZES (inches)					
		$^1/_2$	$^3/_4$	1	$1^1/_4$	$1^1/_2$	2
RHW, RHW-2	14	4	7	11	20	27	46
	12	3	6	9	17	23	38
	10	2	5	8	13	18	30
	8	1	2	4	7	9	16
	6	1	1	3	5	8	13
	4	1	1	2	4	6	10
	3	1	1	1	4	5	9
	2	1	1	1	3	4	7
	1	0	1	1	1	3	5
	1/0	0	1	1	1	2	4
	2/0	0	1	1	1	2	4
	3/0	0	0	1	1	1	3
	4/0	0	0	1	1	1	3
TW	14	8	15	25	43	58	96
	12	6	11	19	33	45	74
	10	5	8	14	24	33	55
	8	2	5	8	13	18	30
RHW[a], RHW-2[a], THHW, THW, THW-2	14	6	10	16	28	39	64
	12	4	8	13	23	31	51
	10	3	6	10	18	24	40
	8	1	4	6	10	14	24
RHW[a], RHW-2[a], TW, THW, THHW, THW-2	6	1	3	4	8	11	18
	4	1	1	3	6	8	13
	3	1	1	3	5	7	12
	2	1	1	2	4	6	10
	1	1	1	1	3	4	7
	1/0	0	1	1	2	3	6
	2/0	0	1	1	1	3	5
	3/0	0	1	1	1	2	4
	4/0	0	0	1	1	1	3
THHN, THWN, THWN-2	14	12	22	35	61	84	138
	12	9	16	26	45	61	101
	10	5	10	16	28	38	63
	8	3	6	9	16	22	36
	6	2	4	7	12	16	26
	4	1	2	4	7	10	16
	3	1	1	3	6	8	13
	2	1	1	3	5	7	11
	1	1	1	1	4	5	8
	1/0	1	1	1	3	4	7
	2/0	0	1	1	2	3	6
	3/0	0	1	1	1	3	5
	4/0	0	1	1	1	2	4
XHHW, XHHW-2	14	8	15	25	43	58	96
	12	6	11	19	33	45	74
	10	5	8	14	24	33	55
	8	2	5	8	13	18	30
	6	1	3	6	10	14	22
	4	1	2	4	7	10	16
	3	1	1	3	6	8	14
	2	1	1	3	5	7	11
	1	1	1	1	4	5	8
	1/0	1	1	1	3	4	7
	2/0	0	1	1	2	3	6
	3/0	0	1	1	1	3	5
	4/0	0	1	1	1	2	4

For SI: 1 inch = 25.4 mm.

a. Types RHW, and RHW-2 without outer covering.

TABLE E3804.6(2)
MAXIMUM NUMBER OF CONDUCTORS IN ELECTRICAL NONMETALLIC TUBING (ENT)[a]

TYPE LETTERS	CONDUCTOR SIZE AWG/kcmil	TRADE SIZES (inches)					
		1/2	3/4	1	1 1/4	1 1/2	2
RHW, RHW-2	14	3	6	10	19	26	43
	12	2	5	9	16	22	36
	10	1	4	7	13	17	29
	8	1	1	3	6	9	15
	6	1	1	3	5	7	12
	4	1	1	2	4	6	9
	3	1	1	1	3	5	8
	2	0	1	1	3	4	7
	1	0	1	1	1	3	5
	1/0	0	0	1	1	2	4
	2/0	0	0	1	1	1	3
	3/0	0	0	1	1	1	3
	4/0	0	0	1	1	1	2
TW	14	7	13	22	40	55	92
	12	5	10	17	31	42	71
	10	4	7	13	23	32	52
	8	1	4	7	13	17	29
RHW[a], RHW-2[a], THHW, THW, THW-2	14	4	8	15	27	37	61
	12	3	7	12	21	29	49
	10	3	5	9	17	23	38
	8	1	3	5	10	14	23
RHW[a], RHW-2[a], TW, THW, THHW, THW-2	6	1	2	4	7	10	17
	4	1	1	3	5	8	13
	3	1	1	2	5	7	11
	2	1	1	2	4	6	9
	1	0	1	1	3	4	6
	1/0	0	1	1	2	3	5
	2/0	0	1	1	1	3	5
	3/0	0	0	1	1	2	4
	4/0	0	0	1	1	1	3
THHN, THWN, THWN-2	14	10	18	32	58	80	132
	12	7	13	23	42	58	96
	10	4	8	15	26	36	60
	8	2	5	8	15	21	35
	6	1	3	6	11	15	25
	4	1	1	4	7	9	15
	3	1	1	3	5	8	13
	2	1	1	2	5	6	11
	1	1	1	1	3	5	8
	1/0	0	1	1	3	4	7
	2/0	0	1	1	2	3	5
	3/0	0	1	1	1	3	4
	4/0	0	0	1	1	2	4
XHHW, XHHW-2	14	7	13	22	40	55	92
	12	5	10	17	31	42	71
	10	4	7	13	23	32	52
	8	1	4	7	13	17	29
	6	1	3	5	9	13	21
	4	1	1	4	7	9	15
	3	1	1	3	6	8	13
	2	1	1	2	5	6	11
	1	1	1	1	3	5	8
	1/0	0	1	1	3	4	7
	2/0	0	1	1	2	3	6
	3/0	0	1	1	1	3	5
	4/0	0	0	1	1	2	4

For SI: 1 inch = 25.4 mm.

a. Types RHW, and RHW-2 without outer covering.

TABLE E3804.6(3)
MAXIMUM NUMBER OF CONDUCTORS IN FLEXIBLE METALLIC CONDUIT (FMC)[a]

TYPE LETTERS	CONDUCTOR SIZE AWG/kcmil	TRADE SIZES (inches)					
		1/2	3/4	1	1 1/4	1 1/2	2
RHW, RHW-2	14	4	7	11	17	25	44
	12	3	6	9	14	21	37
	10	3	5	7	11	17	30
	8	1	2	4	6	9	15
	6	1	1	3	5	7	12
	4	1	1	2	4	5	10
	3	1	1	1	3	5	7
	2	1	1	1	3	4	7
	1	0	1	1	1	2	5
	1/0	0	1	1	1	2	4
	2/0	0	1	1	1	1	3
	3/0	0	0	1	1	1	3
TW	14	9	15	23	36	53	94
	12	7	11	18	28	41	72
	10	5	8	13	21	30	54
	8	3	5	7	11	17	30
RHW[a], RHW-2[a], THHW, THW, THW-2	14	6	10	15	24	35	62
	12	5	8	12	19	28	50
	10	4	6	10	15	22	39
	8	1	4	6	9	13	23
RHW[a], RHW-2[a], TW, THW, THHW, THW-2	6	1	3	4	7	10	18
	4	1	1	3	5	7	13
	3	1	1	3	4	6	11
	2	1	1	2	4	5	10
	1	1	1	1	2	4	7
	1/0	0	1	1	1	3	6
	2/0	0	1	1	1	3	5
	3/0	0	1	1	1	2	4
	4/0	0	0	1	1	1	3
	4/0	0	0	1	1	1	2
THHN, THWN, THWN-2	14	13	22	33	52	76	134
	12	9	16	24	38	56	98
	10	6	10	15	24	35	62
	8	3	6	9	14	20	35
	6	2	4	6	10	14	25
	4	1	2	4	6	9	16
	3	1	1	3	5	7	13
	2	1	1	3	4	6	11
	1	1	1	1	3	4	8
	1/0	1	1	1	2	4	7
	2/0	0	1	1	1	3	6
	3/0	0	1	1	1	2	5
	4/0	0	1	1	1	1	4
XHHW, XHHW-2	14	9	15	23	36	53	94
	12	7	11	18	28	41	72
	10	5	8	13	21	30	54
	8	3	5	7	11	17	30
	6	1	3	5	8	12	22
	4	1	2	4	6	9	16
	3	1	1	3	5	7	13
	2	1	1	3	4	6	11
	1	1	1	1	3	5	8
	1/0	1	1	1	2	4	7
	2/0	0	1	1	2	3	6
	3/0	0	1	1	1	3	5
	4/0	0	1	1	1	2	4

For SI: 1 inch = 25.4 mm.

a. Types RHW, and RHW-2 without outer covering.

TABLE E3804.6(4)
MAXIMUM NUMBER OF CONDUCTORS IN INTERMEDIATE METALLIC CONDUIT (IMC)[a]

TYPE LETTERS	CONDUCTOR SIZE AWG/kcmil	TRADE SIZES (inches)					
		$^1/_2$	$^3/_4$	1	$1^1/_4$	$1^1/_2$	2
RHW, RHW-2	14	4	8	13	22	30	49
	12	4	6	11	18	25	41
	10	3	5	8	15	20	33
	8	1	3	4	8	10	17
	6	1	1	3	6	8	14
	4	1	1	3	5	6	11
	3	1	1	2	4	6	9
	2	1	1	1	3	5	8
	1	0	1	1	2	3	5
	1/0	0	1	1	1	3	4
	2/0	0	1	1	1	2	4
	3/0	0	0	1	1	1	3
	4/0	0	0	1	1	1	3
TW	14	10	17	27	47	64	104
	12	7	13	21	36	49	80
	10	5	9	15	27	36	59
	8	3	5	8	15	20	33
RHW[a], RHW-2[a], THHW, THW, THW-2	14	6	11	18	31	42	69
	12	5	9	14	25	34	56
	10	4	7	11	19	26	43
	8	2	4	7	12	16	26
RHW[a], RHW-2[a], TW, THW, THHW, THW-2	6	1	3	5	9	12	20
	4	1	2	4	6	9	15
	3	1	1	3	6	8	13
	2	1	1	3	5	6	11
	1	1	1	1	3	4	7
	1/0	1	1	1	3	4	6
	2/0	0	1	1	2	3	5
	3/0	0	1	1	1	3	4
	4/0	0	1	1	1	2	4
THHN, THWN, THWN-2	14	14	24	39	68	91	149
	12	10	17	29	49	67	109
	10	6	11	18	31	42	68
	8	3	6	10	18	24	39
	6	2	4	7	13	17	28
	4	1	3	4	8	10	17
	3	1	2	4	6	9	15
	2	1	1	3	5	7	12
	1	1	1	2	4	5	9
	1/0	1	1	1	3	4	8
	2/0	1	1	1	3	4	6
	3/0	0	1	1	2	3	5
	4/0	0	1	1	1	2	4
XHHW, XHHW-2	14	10	17	27	47	64	104
	12	7	13	21	36	49	80
	10	5	9	15	27	36	59
	8	3	5	8	15	20	33
	6	1	4	6	11	15	24
	4	1	3	4	8	11	18
	3	1	2	4	7	9	15
	2	1	1	3	5	7	12
	1	1	1	2	4	5	9
	1/0	1	1	1	3	5	8
	2/0	1	1	1	3	4	6
	3/0	0	1	1	2	3	5
	4/0	0	1	1	1	2	4

For SI: 1 inch = 25.4 mm.

a. Types RHW, and RHW-2 without outer covering.

TABLE E3804.6(5)
MAXIMUM NUMBER OF CONDUCTORS IN LIQUID-TIGHT FLEXIBLE NONMETALLIC CONDUIT (FNMC-B)[a]

TYPE LETTERS	CONDUCTOR SIZE AWG/kcmil	TRADE SIZES (inches)						
		3/8	1/2	3/4	1	1 1/4	1 1/2	2
RHW, RHW-2	14	2	4	7	12	21	27	44
	12	1	3	6	10	17	22	36
	10	1	3	5	8	14	18	29
	8	1	1	2	4	7	9	15
	6	1	1	1	3	6	7	12
	4	0	1	1	2	4	6	9
	3	0	1	1	1	4	5	8
	2	0	1	1	1	3	4	7
	1	0	0	1	1	1	3	5
	1/0	0	0	1	1	1	2	4
	2/0	0	0	1	1	1	1	3
	3/0	0	0	0	1	1	1	3
	4/0	0	0	0	1	1	1	2
TW	14	5	9	15	25	44	57	93
	12	4	7	12	19	33	43	71
	10	3	5	9	14	25	32	53
	8	1	3	5	8	14	18	29
RHW[a], RGW-2[a], THHW, THW, THW-2	14	3	6	10	16	29	38	62
	12	3	5	8	13	23	30	50
	10	1	3	6	10	18	23	39
	8	1	1	4	6	11	14	23
RHW[a], RHW-2[a], TW, THW, THHW, THW-2	6	1	1	3	5	8	11	18
	4	1	1	1	3	6	8	13
	3	1	1	1	3	5	7	11
	2	0	1	1	2	4	6	9
	1	0	1	1	1	3	4	7
	1/0	0	0	1	1	2	3	6
	2/0	0	0	1	1	2	3	5
	3/0	0	0	1	1	1	2	4
	4/0	0	0	0	1	1	1	3
THHN, THWN, THWN-2	14	8	13	22	36	63	81	133
	12	5	9	16	26	46	59	97
	10	3	6	10	16	29	37	61
	8	1	3	6	9	16	21	35
	6	1	2	4	7	12	15	25
	4	1	1	2	4	7	9	15
	3	1	1	1	3	6	8	13
	2	1	1	1	3	5	7	11
	1	0	1	1	1	4	5	8
	1/0	0	1	1	1	3	4	7
	2/0	0	0	1	1	2	3	6
	3/0	0	0	1	1	1	3	5
	4/0	0	0	1	1	1	2	4
XHHW, XHHW-2	14	5	9	15	25	44	57	93
	12	4	7	12	19	33	43	71
	10	3	5	9	14	25	32	53
	8	1	3	5	8	14	18	29
	6	1	1	3	6	10	13	22
	4	1	1	2	4	7	9	16
	3	1	1	1	3	6	8	13
	2	1	1	1	3	5	7	11
	1	0	1	1	1	4	5	8
	1/0	0	1	1	1	3	4	7
	2/0	0	0	1	1	2	3	6
	3/0	0	0	1	1	1	3	5
	4/0	0	0	1	1	1	2	4

For SI: 1 inch = 25.4 mm.

a. Types RHW, and RHW-2 without outer covering.

TABLE E3804.6(6)
MAXIMUM NUMBER OF CONDUCTORS IN LIQUID-TIGHT FLEXIBLE NONMETALLIC CONDUIT (FNMC-A)[a]

TYPE LETTERS	CONDUCTOR SIZE AWG/kcmil	TRADE SIZES (inches)						
		3/8	1/2	3/4	1	1 1/4	1 1/2	2
RHW, RHW-2	14	2	4	7	11	20	27	45
	12	1	3	6	9	17	23	38
	10	1	3	5	8	13	18	30
	8	1	1	2	4	7	9	16
	6	1	1	1	3	5	7	13
	4	0	1	1	2	4	6	10
	3	0	1	1	1	4	5	8
	2	0	1	1	1	3	4	7
	1	0	0	1	1	1	3	5
	1/0	0	0	1	1	1	2	4
	2/0	0	0	1	1	1	1	4
	3/0	0	0	0	1	1	1	3
	4/0	0	0	0	1	1	1	3
TW	14	5	9	15	24	43	58	96
	12	4	7	12	19	33	44	74
	10	3	5	9	14	24	33	55
	8	1	3	5	8	13	18	30
RHW[a], RHW-2[a], THHW, THW, THW-2	14	3	6	10	16	28	38	64
	12	3	4	8	13	23	31	51
	10	1	3	6	10	18	24	40
	8	1	1	4	6	10	14	24
RHW[a], RHW-2[a], TW, THW, THHW, THW-2	6	1	1	3	4	8	11	18
	4	1	1	1	3	6	8	13
	3	1	1	1	3	5	7	11
	2	0	1	1	2	4	6	10
	1	0	1	1	1	3	4	7
	1/0	0	0	1	1	2	3	6
	2/0	0	0	1	1	1	3	5
	3/0	0	0	1	1	1	2	4
	4/0	0	0	0	1	1	1	3
THHN, THWN, THWN-2	14	8	13	22	35	62	83	137
	12	5	9	16	25	45	60	100
	10	3	6	10	16	28	38	63
	8	1	3	6	9	16	22	36
	6	1	2	4	6	12	16	26
	4	1	1	2	4	7	9	16
	3	1	1	1	3	6	8	13
	2	1	1	1	3	5	7	11
	1	0	1	1	1	4	5	8
	1/0	0	1	1	1	3	4	7
	2/0	0	0	1	1	2	3	6
	3/0	0	0	1	1	1	3	5
	4/0	0	0	1	1	1	2	4
XHHW, XHHW-2	14	5	9	15	24	43	58	96
	12	4	7	12	19	33	44	74
	10	3	5	9	14	24	33	55
	8	1	3	5	8	13	18	30
	6	1	1	3	5	10	13	22
	4	1	1	2	4	7	10	16
	3	1	1	1	3	6	8	14
	2	1	1	1	3	5	7	11
	1	0	1	1	1	4	5	8
	1/0	0	1	1	1	3	4	7
	2/0	0	0	1	1	2	3	6
	3/0	0	0	1	1	1	3	5
	4/0	0	0	1	1	1	2	4

For SI: 1 inch = 25.4 mm.

a. Types RHW, and RHW-2 without outer covering.

TABLE E3804.6(7)
MAXIMUM NUMBER OF CONDUCTORS IN LIQUID-TIGHT FLEXIBLE METAL CONDUIT (LFMC)[a]

TYPE LETTERS	CONDUCTOR SIZE AWG/kcmil	TRADE SIZES (inches)					
		1/2	3/4	1	1 1/4	1 1/2	2
RHW, RHW-2	14	4	7	12	21	27	44
	12	3	6	10	17	22	36
	10	3	5	8	14	18	29
	8	1	2	4	7	9	15
	6	1	1	3	6	7	12
	4	1	1	2	4	6	9
	3	1	1	1	4	5	8
	2	1	1	1	3	4	7
	1	0	1	1	1	3	5
	1/0	0	1	1	1	2	4
	2/0	0	1	1	1	1	3
	3/0	0	0	1	1	1	3
	4/0	0	0	1	1	1	2
TW	14	9	15	25	44	57	93
	12	7	12	19	33	43	71
	10	5	9	14	25	32	53
	8	3	5	8	14	18	29
RHW[a], RHW-2[a], THHW, THW, THW-2	14	6	10	16	29	38	62
	12	5	8	13	23	30	50
	10	3	6	10	18	23	39
	8	1	4	6	11	14	23
RHW[a], RHW-2[a], TW, THW, THHW, THW-2	6	1	3	5	8	11	18
	4	1	1	3	6	8	13
	3	1	1	3	5	7	11
	2	1	1	2	4	6	9
	1	1	1	1	3	4	7
	1/0	0	1	1	2	3	6
	2/0	0	1	1	2	3	5
	3/0	0	1	1	1	2	4
	4/0	0	0	1	1	1	3
THHN, THWN, THWN-2	14	13	22	36	63	81	133
	12	9	16	26	46	59	97
	10	6	10	16	29	37	61
	8	3	6	9	16	21	35
	6	2	4	7	12	15	25
	4	1	2	4	7	9	15
	3	1	1	3	6	8	13
	2	1	1	3	5	7	11
	1	1	1	1	4	5	8
	1/0	1	1	1	3	4	7
	2/0	0	1	1	2	3	6
	3/0	0	1	1	1	3	5
	4/0	0	1	1	1	2	4
XHHW, XHHW-2	14	9	15	25	44	57	93
	12	7	12	19	33	43	71
	10	5	9	14	25	32	53
	8	3	5	8	14	18	29
	6	1	3	6	10	13	22
	4	1	2	4	7	9	16
	3	1	1	3	6	8	13
	2	1	1	3	5	7	11
	1	1	1	1	4	5	8
	1/0	1	1	1	3	4	7
	2/0	0	1	1	2	3	6
	3/0	0	1	1	1	3	5
	4/0	0	1	1	1	2	4

For SI: 1 inch = 25.4 mm.

a. Types RHW, and RHW-2 without outer covering.

TABLE E3804.6(8)
MAXIMUM NUMBER OF CONDUCTORS IN RIGID METAL CONDUIT (RMC)[a]

TYPE LETTERS	CONDUCTOR SIZE AWG/kcmil	TRADE SIZES (inches)					
		1/2	3/4	1	1 1/4	1 1/2	2
RHW, RHW-2	14	4	7	12	21	28	46
	12	3	6	10	17	23	38
	10	3	5	8	14	19	31
	8	1	2	4	7	10	16
	6	1	1	3	6	8	13
	4	1	1	2	4	6	10
	3	1	1	2	4	5	9
	2	1	1	1	3	4	7
	1	0	1	1	1	3	5
	1/0	0	1	1	1	2	4
	2/0	0	1	1	1	2	4
	3/0	0	0	1	1	1	3
	4/0	0	0	1	1	1	3
TW	14	9	15	25	44	59	98
	12	7	12	19	33	45	75
	10	5	9	14	25	34	56
	8	3	5	8	14	19	31
RHW[a], RHW-2[a], THHW, THW, THW-2	14	6	10	17	29	39	65
	12	5	8	13	23	32	52
	10	3	6	10	18	25	41
	8	1	4	6	11	15	24
RHW[a], RHW-2[a], TW, THW, THHW, THW-2	6	1	3	5	8	11	18
	4	1	1	3	6	8	14
	3	1	1	3	5	7	12
	2	1	1	2	4	6	10
	1	1	1	1	3	4	7
	1/0	0	1	1	2	3	6
	2/0	0	1	1	2	3	5
	3/0	0	1	1	1	2	4
	4/0	0	0	1	1	1	3
THHN, THWN, THWN-2	14	13	22	36	63	85	140
	12	9	16	26	46	62	102
	10	6	10	17	29	39	64
	8	3	6	9	16	22	37
	6	2	4	7	12	16	27
	4	1	2	4	7	10	16
	3	1	1	3	6	8	14
	2	1	1	3	5	7	11
	1	1	1	1	4	5	8
	1/0	1	1	1	3	4	7
	2/0	0	1	1	2	3	6
	3/0	0	1	1	1	3	5
	4/0	0	1	1	1	2	4
XHHW, XHHW-2	14	9	15	25	44	59	98
	12	7	12	19	33	45	75
	10	5	9	14	25	34	56
	8	3	5	8	14	19	31
	6	1	3	6	10	14	23
	4	1	2	4	7	10	16
	3	1	1	3	6	8	14
	2	1	1	3	5	7	12
	1	1	1	1	4	5	9
	1/0	1	1	1	3	4	7
	2/0	0	1	1	2	3	6
	3/0	0	1	1	1	3	5
	4/0	0	1	1	1	2	4

For SI: 1 inch = 25.4 mm.

a. Types RHW, and RHW-2 without outer covering.

TABLE E3804.6(9)
MAXIMUM NUMBER OF CONDUCTORS IN RIGID PVC CONDUIT, SCHEDULE 80 (PVC-80)[a]

TYPE LETTERS	CONDUCTOR SIZE AWG/kcmil	TRADE SIZES (inches)					
		$^1/_2$	$^3/_4$	1	$1^1/_4$	$1^1/_2$	2
RHW, RHW-2	14	3	5	9	17	23	39
	12	2	4	7	14	19	32
	10	1	3	6	11	15	26
	8	1	1	3	6	8	13
	6	1	1	2	4	6	11
	4	1	1	1	3	5	8
	3	0	1	1	3	4	7
	2	0	1	1	3	4	6
	1	0	1	1	1	2	4
	1/0	0	0	1	1	1	3
	2/0	0	0	1	1	1	3
	3/0	0	0	1	1	1	3
	4/0	0	0	0	1	1	2
TW	14	6	11	20	35	49	82
	12	5	9	15	27	38	63
	10	3	6	11	20	28	47
	8	1	3	6	11	15	26
RHW[a], RHW-2[a], THHW, THW, THW-2	14	4	8	13	23	32	55
	12	3	6	10	19	26	44
	10	2	5	8	15	20	34
	8	1	3	5	9	12	20
RHW[a], RHW-2[a], TW, THW, THHW, THW-2	6	1	1	3	7	9	16
	4	1	1	3	5	7	12
	3	1	1	2	4	6	10
	2	1	1	1	3	5	8
	1	0	1	1	2	3	6
	1/0	0	1	1	1	3	5
	2/0	0	1	1	1	2	4
	3/0	0	0	1	1	1	3
	4/0	0	0	1	1	1	3
THHN, THWN, THWN-2	14	9	17	28	51	70	118
	12	6	12	20	37	51	86
	10	4	7	13	23	32	54
	8	2	4	7	13	18	31
	6	1	3	5	9	13	22
	4	1	1	3	6	8	14
	3	1	1	3	5	7	12
	2	1	1	2	4	6	10
	1	0	1	1	3	4	7
	1/0	0	1	1	2	3	6
	2/0	0	1	1	1	3	5
	3/0	0	1	1	1	2	4
	4/0	0	0	1	1	1	3
XHHW, XHHW-2	14	6	11	20	35	49	82
	12	5	9	15	27	38	63
	10	3	6	11	20	28	47
	8	1	3	6	11	15	26
	6	1	2	4	8	11	19
	4	1	1	3	6	8	14
	3	1	1	3	5	7	12
	2	1	1	2	4	6	10
	1	0	1	1	3	4	7
	1/0	0	1	1	2	3	6
	2/0	0	1	1	1	3	5
	3/0	0	1	1	1	2	4
	4/0	0	0	1	1	1	3

For SI: 1 inch = 25.4 mm.

a. Types RHW, and RHW-2 without outer covering.

TABLE E3804.6(10)
MAXIMUM NUMBER OF CONDUCTORS IN RIGID PVC CONDUIT SCHEDULE 40 (PVC-40)[a]

TYPE LETTERS	CONDUCTOR SIZE AWG/kcmil	TRADE SIZES (inches)					
		1/2	3/4	1	1 1/4	1 1/2	2
RHW, RHW-2	14	4	7	11	20	27	45
	12	3	5	9	16	22	37
	10	2	4	7	13	18	30
	8	1	2	4	7	9	15
	6	1	1	3	5	7	12
	4	1	1	2	4	6	10
	3	1	1	1	4	5	8
	2	1	1	1	3	4	7
	1	0	1	1	1	3	5
	1/0	0	1	1	1	2	4
	2/0	0	0	1	1	1	3
	3/0	0	0	1	1	1	3
	4/0	0	0	1	1	1	2
TW	14	8	14	24	42	57	94
	12	6	11	18	32	44	72
	10	4	8	13	24	32	54
	8	2	4	7	13	18	30
RHW[a], RHW-2[a], THHW, THW, THW-2	14	5	9	16	28	38	63
	12	4	8	12	22	30	50
	10	3	6	10	17	24	39
	8	1	3	6	10	14	23
RHW[a], RHW-2[a], TW, THW, THHW, THW-2	6	1	2	4	8	11	18
	4	1	1	3	6	8	13
	3	1	1	3	5	7	11
	2	1	1	2	4	6	10
	1	0	1	1	3	4	7
	1/0	0	1	1	2	3	6
	2/0	0	1	1	1	3	5
	3/0	0	1	1	1	2	4
	4/0	0	0	1	1	1	3
THHN, THWN, THWN-2	14	11	21	34	60	82	135
	12	8	15	25	43	59	99
	10	5	9	15	27	37	62
	8	3	5	9	16	21	36
	6	1	4	6	11	15	26
	4	1	2	4	7	9	16
	3	1	1	3	6	8	13
	2	1	1	3	5	7	11
	1	1	1	1	3	5	8
	1/0	1	1	1	3	4	7
	2/0	0	1	1	2	3	6
	3/0	0	1	1	1	3	5
	4/0	0	1	1	1	2	4
XHHW, XHHW-2	14	8	14	24	42	57	94
	12	6	11	18	32	44	72
	10	4	8	13	24	32	54
	8	2	4	7	13	18	30
	6	1	3	5	10	13	22
	4	1	2	4	7	9	16
	3	1	1	3	6	8	13
	2	1	1	3	5	7	11
	1	1	1	1	3	5	8
	1/0	1	1	1	3	4	7
	2/0	0	1	1	2	3	6
	3/0	0	1	1	1	3	5
	4/0	0	1	1	1	2	4

For SI: 1 inch = 25.4 mm.

a. Types RHW, and RHW-2 without outer covering.

SECTION E3805
BOXES, CONDUIT BODIES AND FITTINGS

E3805.1 Box, conduit body or fitting—where required. A box or conduit body shall be installed at each conductor splice point, outlet, switch point, junction point and pull point except as otherwise permitted in Sections E3805.1.1 through E3805.1.7.

Fittings and connectors shall be used only with the specific wiring methods for which they are designed and listed.

E3805.1.1 Equipment. An integral junction box or wiring compartment that is part of listed equipment shall be permitted to serve as a box or conduit body.

E3805.1.2 Protection. A box or conduit body shall not be required where cables enter or exit from conduit or tubing that is used to provide cable support or protection against physical damage. A fitting shall be provided on the end(s) of the conduit or tubing to protect the cable from abrasion.

E3805.1.3 Integral enclosure. A wiring device with integral enclosure identified for the use, having brackets that securely fasten the device to walls or ceilings of conventional on-site frame construction, for use with nonmetallic-sheathed cable, shall be permitted in lieu of a box or conduit body.

E3805.1.4 Fitting. A fitting identified for the use shall be permitted in lieu of a box or conduit body where such fitting is accessible after installation and does not contain spliced or terminated conductors.

E3805.1.5 Buried conductors. Splices and taps in buried conductors and cables shall not be required to be enclosed in a box or conduit body where installed in accordance with Section E3703.4.

E3805.1.6 Luminaires. Where a luminaire is listed to be used as a raceway, a box or conduit body shall not be required for wiring installed therein.

E3805.1.7 Closed loop. Where a device identified and listed as suitable for installation without a box is used with a closed-loop power-distribution system, a box or conduit body shall not be required.

E3805.2 Metal boxes. All metal boxes shall be grounded.

E3805.3 Nonmetallic boxes. Nonmetallic boxes shall be used only with nonmetallic-sheathed cable, cabled wiring methods, flexible cords and nonmetallic raceways.

Exceptions:

1. Where internal bonding means are provided between all entries, nonmetallic boxes shall be permitted to be used with metal raceways and metal-armored cables.

2. Where integral bonding means with a provision for attaching an equipment grounding jumper inside the box are provided between all threaded entries in nonmetallic boxes listed for the purpose, nonmetallic boxes shall be permitted to be used with metal raceways and metal-armored cables.

E3805.3.1 Nonmetallic-sheathed cable and nonmetallic boxes. Where nonmetallic-sheathed cable is used, the cable assembly, including the sheath, shall extend into the box not less than $1/4$ inch (6.4 mm) through a nonmetallic-sheathed cable knockout opening.

E3805.3.2 Securing to box. All permitted wiring methods shall be secured to the boxes.

Exception: Where nonmetallic-sheathed cable is used with boxes not larger than a nominal size of $2^1/_4$ inches by 4 inches (57 mm by 102 mm) mounted in walls or ceilings, and where the cable is fastened within 8 inches (203 mm) of the box measured along the sheath, and where the sheath extends through a cable knockout not less than $1/4$ inch (6.4 mm), securing the cable to the box shall not be required.

E3805.3.3 Conductor rating. Nonmetallic boxes shall be suitable for the lowest temperature-rated conductor entering the box.

E3805.4 Minimum depth of outlet boxes. Boxes shall have an internal depth of not less than 0.5 inch (12.7 mm). Boxes designed to enclose flush devices shall have an internal depth of not less than 0.938 inch (24 mm).

E3805.5 Boxes enclosing flush-mounted devices. Boxes enclosing flush-mounted devices shall be of such design that the devices are completely enclosed at the back and all sides and shall provide support for the devices. Screws for supporting the box shall not be used for attachment of the device contained therein.

E3805.6 Boxes at luminaire outlets. Boxes for luminare outlets shall be designed for the purpose. At every outlet used exclusively for lighting, the box shall be designed or installed so that a luminaire may be attached.

Exception: A wall-mounted luminaire weighing not more than 6 lb (3 kg) shall be permitted to be supported on other boxes or plaster rings that are secured to other boxes, provided the luminaire or its supporting yoke is secured to the box with no fewer than two No. 6 or larger screws.

E3805.7 Maximum luminaire weight. Outlet boxes or fittings installed as required by Section E3804.3 shall be permitted to support luminaires weighing 50 lb (23 kg) or less. A luminaire that weighs more than 50 lb (23 kg) shall be supported independently of the outlet box unless the outlet box is listed for the weight to be supported.

E3805.8 Floor boxes. Where outlet boxes for receptacles are installed in the floor, such boxes shall be listed specifically for that application.

E3805.9 Boxes at fan outlets. Outlet boxes and outlet box systems used as the sole support of ceiling-suspended fans (paddle) shall be marked by their manufacturer as suitable for this purpose and shall not support ceiling-suspended fans (paddle) that weigh more than 70 lb (32 kg). For outlet boxes and outlet box systems designed to support ceiling-suspended fans (paddle) that weigh more than 35 lb (16 kg), the required marking shall include the maximum weight to be supported.

E3805.10 Conduit bodies and junction, pull and outlet boxes to be accessible. Conduit bodies and junction, pull and outlet boxes shall be installed so that the wiring therein can be accessed without removing any part of the building or, in underground circuits, without excavating sidewalks, paving, earth or other substance used to establish the finished grade.

Exception: Boxes covered by gravel, light aggregate or noncohesive granulated soil shall be listed for the application, and the box locations shall be effectively identified and access shall be provided for excavation.

E3805.11 Damp or wet locations. In damp or wet locations, boxes, conduit bodies and fittings shall be placed or equipped so as to prevent moisture from entering or accumulating within the box, conduit body or fitting. Boxes, conduit bodies and fittings installed in wet locations shall be listed for use in wet locations.

E3805.12 Number of conductors in outlet, device, and junction boxes, and conduit bodies. Boxes and conduit bodies shall be of sufficient size to provide free space for all enclosed conductors. In no case shall the volume of the box, as calculated in Section E3805.12.1, be less than the box fill calculation as calculated in Section E3805.12.2. The minimum volume for conduit bodies shall be as calculated in Section E3805.12.3. The provisions of this section shall not apply to terminal housings supplied with motors.

E3805.12.1 Box volume calculations. The volume of a wiring enclosure (box) shall be the total volume of the assembled sections, and, where used, the space provided by plaster rings, domed covers, extension rings, etc., that are marked with their volume in cubic inches or are made from boxes the dimensions of which are listed in Table E3805.12.1.

E3805.12.1.1 Standard boxes. The volumes of standard boxes that are not marked with a cubic-inch capacity shall be as given in Table E3805.12.1.

E3805.12.1.2 Other boxes. Boxes 100 cubic inches (1640 cm³) or less, other than those described in Table E3805.12.1, and nonmetallic boxes shall be durably and legibly marked by the manufacturer with their cubic-inch capacity. Boxes described in Table E3805.12.1 that have a larger cubic inch capacity than is designated in the table shall be permitted to have their cubic-inch capacity marked as required by this section.

E3805.12.2 Box fill calculations. The volumes in Section E3805.12.2.1 through Section E3805.12.2.5, as applicable, shall be added together. No allowance shall be required for small fittings such as locknuts and bushings.

TABLE E3805.12.1
MAXIMUM NUMBER OF CONDUCTORS IN METAL BOXES[a]

BOX DIMENSIONS (inches trade size and type)	MAXIMUM CAPACITY (cubic inches)	MAXIMUM NUMBER OF CONDUCTORS[a]						
		No. 18	No. 16	No. 14	No. 12	No. 10	No. 8	No. 6
$4 \times 1^1/_4$ round or octagonal	12.5	8	7	6	5	5	4	2
$4 \times 1^1/_2$ round or octagonal	15.5	10	8	7	6	6	5	3
$4 \times 2^1/_8$ round or octagonal	21.5	14	12	10	9	8	7	4
$4 \times 1^1/_4$ square	18.0	12	10	9	8	7	6	3
$4 \times 1^1/_2$ square	21.0	14	12	10	9	8	7	4
$4 \times 2^1/_8$ square	30.3	20	17	15	13	12	10	6
$4^{11}/_{16} \times 1^1/_4$ square	25.5	17	14	12	11	10	8	5
$4^{11}/_{16} \times 1^1/_2$ square	29.5	19	16	14	13	11	9	5
$4^{11}/_{16} \times 2^1/_8$ square	42.0	28	24	21	18	16	14	8
$3 \times 2 \times 1^1/_2$ device	7.5	5	4	3	3	3	2	1
$3 \times 2 \times 2$ device	10.0	6	5	5	4	4	3	2
$3 \times 2 \times 2^1/_4$ device	10.5	7	6	5	4	4	3	2
$3 \times 2 \times 2^1/_2$ device	12.5	8	7	6	5	5	4	2
$3 \times 2 \times 2^3/_4$ device	14.0	9	8	7	6	5	4	2
$3 \times 2 \times 3^1/_2$ device	18.0	12	10	9	8	7	6	3
$4 \times 2^1/_8 \times 1^1/_2$ device	10.3	6	5	5	4	4	3	2
$4 \times 2^1/_8 \times 1^7/_8$ device	13.0	8	7	6	5	5	4	2
$4 \times 2^1/_8 \times 2^1/_8$ device	14.5	9	8	7	6	5	4	2
$3^1/_4 \times 2 \times 2^1/_2$ masonry box/gang	14.0	9	8	7	6	5	4	2
$3^3/_4 \times 2 \times 3^1/_2$ masonry box/gang	21.0	14	12	10	9	8	7	4

For SI: 1 inch = 25.4 mm, 1 cubic inch = 16.4 cm³.
a. Where no volume allowances are required by Sections E3805.12.2.2 through E3805.12.2.5.

E3805.12.2.1 Conductor fill. Each conductor that originates outside the box and terminates or is spliced within the box shall be counted once, and each conductor that passes through the box without splice or termination shall be counted once. A looped, unbroken conductor having a length equal to or greater than twice that required for free conductors by Section E3306.10.3, shall be counted twice. The conductor fill, in cubic inches, shall be computed using Table E3805.12.2.1. A conductor, no part of which leaves the box, shall not be counted.

> **Exception:** An equipment grounding conductor or not more than four fixture wires smaller than No. 14, or both, shall be permitted to be omitted from the calculations where such conductors enter a box from a domed fixture or similar canopy and terminate within that box.

E3805.12.2.2 Clamp fill. Where one or more internal cable clamps, whether factory or field supplied, are present in the box, a single volume allowance in accordance with Table E3805.12.2.1 shall be made based on the largest conductor present in the box. No allowance shall be required for a cable connector with its clamping mechanism outside the box.

E3805.12.2.3 Support fittings fill. Where one or more fixture studs or hickeys are present in the box, a single volume allowance in accordance with Table E3805.12.2.1 shall be made for each type of fitting based on the largest conductor present in the box.

TABLE E3805.12.2.1
VOLUME ALLOWANCE REQUIRED PER CONDUCTOR

SIZE OF CONDUCTOR (AWG)	FREE SPACE WITHIN BOX FOR EACH CONDUCTOR (cubic inches)
No. 18	1.50
No. 16	1.75
No. 14	2.00
No. 12	2.25
No. 10	2.50
No. 8	3.00
No. 6	5.00

For SI: 1 cubic inch = 16.4 cm^3.

E3805.12.2.4 Device or equipment fill. For each yoke or strap containing one or more devices or equipment, a double volume allowance in accordance with Table E3805.12.2.1 shall be made for each yoke or strap based on the largest conductor connected to a device(s) or equipment supported by that yoke or strap.

E3805.12.2.5 Equipment grounding conductor fill. Where one or more equipment grounding conductors or equipment bonding jumpers enters a box, a single volume allowance in accordance with Table E3805.12.2.1 shall be made based on the largest equipment grounding conductor or equipment bonding jumper present in the box.

E3805.12.3 Conduit bodies. Conduit bodies enclosing 6 AWG conductors or smaller, other than short radius conduit bodies, shall have a cross-sectional area not less than twice the cross-sectional area of the largest conduit or tubing to which it is attached. The maximum number of conductors permitted shall be the maximum number permitted by Table E3804.6 for the conduit to which it is attached.

E3805.12.3.1 Splices, taps or devices. Only those conduit bodies that are durably and legibly marked by the manufacturer with their cubic inch capacity shall be permitted to contain splices, taps or devices. The maximum number of conductors shall be calculated using the same procedure for similar conductors in other than standard boxes.

SECTION E3806
INSTALLATION OF BOXES, CONDUIT BODIES AND FITTINGS

E3806.1 Conductors entering boxes, conduit bodies or fittings. Conductors entering boxes, conduit bodies or fittings shall be protected from abrasion.

E3806.1.1 Insulated fittings. Where raceways containing ungrounded conductors 4 AWG or larger enter a cabinet, box enclosure, or raceway, the conductors shall be protected by a substantial fitting providing a smoothly rounded insulating surface, unless the conductors are separated from the fitting or raceway by substantial insulating material securely fastened in place.

> **Exception:** Where threaded hubs or bosses that are an integral part of a cabinet, box enclosure, or raceway provide a smoothly rounded or flared entry for conductors.

Conduit bushings constructed wholly of insulating material shall not be used to secure a fitting or raceway. The insulating fitting or insulating material shall have a temperature rating not less than the insulation temperature rating of the installed conductors.

E3806.2 Openings. Openings through which conductors enter shall be adequately closed.

E3806.3 Metal boxes, conduit bodies and fittings. Where raceway or cable is installed with metal boxes, or conduit bodies, the raceway or cable shall be secured to such boxes and conduit bodies.

E3806.4 Unused openings. Unused cable or raceway openings in boxes and conduit bodies shall be effectively closed to afford protection substantially equivalent to that of the wall of the box or conduit body. Metal plugs or plates used with nonmetallic boxes or conduit bodies shall be recessed at least 0.25 inch (6.4 mm) from the outer surface of the box or conduit body.

E3806.5 In wall or ceiling. In walls or ceilings of concrete, tile or other noncombustible material, boxes employing a flush-type cover or faceplate shall be installed so that the front edge of the box, plaster ring, extension ring, or listed extender will not be set back from the finished surface more than $1/4$ inch (6.4 mm). In walls and ceilings constructed of wood or other combustible material, boxes, plaster rings, extension rings and listed extenders shall be flush with the finished surface or project therefrom.

E3806.6 Plaster, gypsum board and plasterboard. Openings in plaster, gypsum board or plasterboard surfaces that accommodate boxes employing a flush-type cover or faceplate shall be made so that there are no gaps or open spaces greater than $1/8$ inch (3.2 mm) around the edge of the box.

E3806.7 Exposed surface extensions. Surface extensions from a flush-mounted box shall be made by mounting and mechanically securing a box or extension ring over the flush box.

> **Exception:** A surface extension shall be permitted to be made from the cover of a flush-mounted box where the cover is designed so it is unlikely to fall off, or be removed if its securing means becomes loose. The wiring method shall be flexible for a length sufficient to permit removal of the cover and provide access to the box interior and arranged so that any bonding or grounding continuity is independent of the connection between the box and cover.

E3806.8 Supports. Boxes and enclosures shall be supported in accordance with one or more of the provisions in Sections E3806.8.1 through E3806.8.6.

> **E3806.8.1 Surface mounting.** An enclosure mounted on a building or other surface shall be rigidly and securely fastened in place. If the surface does not provide rigid and secure support, additional support in accordance with other provisions of Section E3806.8 shall be provided.

> **E3806.8.2 Structural mounting.** An enclosure supported from a structural member of a building or from grade shall be rigidly supported either directly, or by using a metal, polymeric or wood brace.

>> **E3806.8.2.1 Nails and screws.** Nails and screws, where used as a fastening means, shall be attached by using brackets on the outside of the enclosure, or they shall pass through the interior within $1/4$ inch (6.4 mm) of the back or ends of the enclosure. Screws shall not be permitted to pass through the box except where exposed threads in the box are protected by an approved means to avoid abrasion of conductor insulation.

>> **E3806.8.2.2 Braces.** Metal braces shall be protected against corrosion and formed from metal that is not less than 0.020 inch (.508 mm) thick uncoated. Wood braces shall have a cross section not less than nominal 1 inch by 2 inches (25.4 mm by 51 mm). Wood braces in wet locations shall be treated for the conditions. Polymeric braces shall be identified as being suitable for the use.

> **E3806.8.3 Mounting in finished surfaces.** An enclosure mounted in a finished surface shall be rigidly secured thereto by clamps, anchors, or fittings identified for the application.

> **E3806.8.4 Raceway supported enclosures without devices or fixtures.** An enclosure that does not contain a device(s), other than splicing devices, or support a luminaire, lampholder or other equipment, and that is supported by entering raceways shall not exceed 100 cubic inches (1640 cm³) in size. The enclosure shall have threaded entries or have hubs identified for the purpose. The enclosure shall be supported by two or more conduits threaded wrenchtight into the enclosure or hubs. Each conduit shall be secured within 3 feet (914 mm) of the enclosure, or within 18 inches (457 mm) of the enclosure if all entries are on the same side of the enclosure.

>> **Exception:** Rigid metal, intermediate metal, or rigid nonmetallic conduit or electrical metallic tubing shall be permitted to support a conduit body of any size, provided that the conduit body is not larger in trade size than the largest trade size of the supporting conduit or electrical metallic tubing.

> **E3806.8.5 Raceway supported enclosures, with devices or luminaire.** An enclosure that contains a device(s), other than splicing devices, or supports a luminaire, lampholder or other equipment and is supported by entering raceways shall not exceed 100 cubic inches (1640 cm³) in size. The enclosure shall have threaded entries or have hubs identified for the purpose. The enclosure shall be supported by two or more conduits threaded wrench-tight into the enclosure or hubs. Each conduit shall be secured within 18 inches (457 mm) of the enclosure.

>> **Exceptions:**
>>
>> 1. Rigid metal or intermediate metal conduit shall be permitted to support a conduit body of any size, provided that the conduit bodies are not larger in trade size than the largest trade size of the supporting conduit.
>>
>> 2. An unbroken length(s) of rigid or intermediate metal conduit shall be permitted to support a box used for luminaire or lampholder support, or to support a wiring enclosure that is an integral part of a luminaire and used in lieu of a box in accordance with Section E3805.1.1, where all of the following conditions are met:
>>
>>   2.1. The conduit is securely fastened at a point so that the length of conduit beyond the last point of conduit support does not exceed 3 feet (914 mm).
>>
>>   2.2. The unbroken conduit length before the last point of conduit support is 12 inches (305 mm) or greater, and that portion of the conduit is securely fastened at some point not less than 12 inches (305 mm) from its last point of support.
>>
>>   2.3. Where accessible to unqualified persons, the luminaire or lampholder, measured to its lowest point, is not less than 8 feet (2438 mm) above grade or standing area and at least 3 feet (914 mm) measured horizontally to the 8-foot (2438 mm) elevation from windows, doors, porches, fire escapes, or similar locations.

2.4. A luminaire supported by a single conduit does not exceed 12 inches (305 mm) in any direction from the point of conduit entry.

2.5. The weight supported by any single conduit does not exceed 20 pounds (9.1 kg).

2.6. At the luminaire or lampholder end, the conduit(s) is threaded wrenchtight into the box, conduit body, or integral wiring enclosure, or into hubs identified for the purpose. Where a box or conduit body is used for support, the luminaire shall be secured directly to the box or conduit body, or through a threaded conduit nipple not over 3 inches (76 mm) long.

E3806.8.6 Enclosures in concrete or masonry. An enclosure supported by embedment shall be identified as being suitably protected from corrosion and shall be securely embedded in concrete or masonry.

E3806.9 Covers and canopies. Outlet boxes shall be effectively closed with a cover, faceplate or fixture canopy.

E3806.10 Metal covers and plates. Metal covers and plates shall be grounded.

E3806.11 Exposed combustible finish. Combustible wall or ceiling finish exposed between the edge of a fixture canopy or pan and the outlet box shall be covered with noncombustible material.

SECTION E3807
CABINETS AND PANELBOARDS

E3807.1 Enclosures for switches or overcurrent devices. Enclosures for switches or overcurrent devices shall not be used as junction boxes, auxiliary gutters, or raceways for conductors feeding through or tapping off to other switches or overcurrent devices, except where adequate space for this purpose is provided. The conductors shall not fill the wiring space at any cross section to more than 40 percent of the cross-sectional area of the space, and the conductors, splices, and taps shall not fill the wiring space at any cross section to more than 75 percent of the cross-sectional area of that space.

E3807.2 Damp or wet locations. In damp or wet locations, cabinets and panelboards of the surface type shall be placed or equipped so as to prevent moisture or water from entering and accumulating within the cabinet, and shall be mounted to provide an airspace not less than $^1/_4$ inch (6.4 mm) between the enclosure and the wall or other supporting surface. Cabinets installed in wet locations shall be weatherproof. For enclosures in wet locations, raceways and cables entering above the level of uninsulated live parts shall be installed with fittings listed for wet locations.

E3807.3 Position in wall. In walls of concrete, tile or other noncombustible material, cabinets and panelboards shall be installed so that the front edge of the cabinet will not set back of the finished surface more than $^1/_4$ inch (6.4 mm). In walls constructed of wood or other combustible material, cabinets

shall be flush with the finished surface or shall project therefrom.

E3807.4 Repairing plaster, drywall and plasterboard. Plaster, drywall, and plasterboard surfaces that are broken or incomplete shall be repaired so that there will not be gaps or open spaces greater than $^1/_8$ inch (3.2 mm) at the edge of the cabinet or cutout box employing a flush-type cover.

E3807.5 Unused openings. Unused cable and raceway openings in cabinets and panelboards shall be effectively closed to afford protection equivalent to that of the wall of the cabinet. Metal plugs and plates used with nonmetallic cabinets shall be recessed at least $^1/_4$ inch (6.4 mm) from the outer surface. Unused openings for circuit breakers and switches shall be closed using identified closures, or other approved means that provide protection substantially equivalent to the wall of the enclosure.

E3807.6 Conductors entering cabinets. Conductors entering cabinets and panelboards shall be protected from abrasion and shall comply with Section E3806.1.1.

E3807.7 Openings to be closed. Openings through which conductors enter cabinets, panelboards and meter sockets shall be adequately closed.

E3807.8 Cables. Where cables are used, each cable shall be secured to the cabinet, panelboard, cutout box, or meter socket enclosure.

Exception: Cables with entirely nonmetallic sheaths shall be permitted to enter the top of a surface-mounted enclosure through one or more sections of rigid raceway not less than 18 inches (457 mm) nor more than 10 feet (3048 mm) in length, provided all the following conditions are met:

1. Each cable is fastened within 12 inches (305 mm), measured along the sheath, of the outer end of the raceway.

2. The raceway extends directly above the enclosure and does not penetrate a structural ceiling.

3. A fitting is provided on each end of the raceway to protect the cable(s) from abrasion and the fittings remain accessible after installation.

4. The raceway is sealed or plugged at the outer end using approved means so as to prevent access to the enclosure through the raceway.

5. The cable sheath is continuous through the raceway and extends into the enclosure beyond the fitting not less than $^1/_4$ inch (6.4 mm).

6. The raceway is fastened at its outer end and at other points in accordance with Section E3702.1.

7. The allowable cable fill shall not exceed that permitted by Table E3807.8. A multiconductor cable having two or more conductors shall be treated as a single conductor for calculating the percentage of conduit fill area. For cables that have elliptical cross sections, the cross-sectional area calculation shall be based on the major diameter of the ellipse as a circle diameter.

TABLE E3807.8
PERCENT OF CROSS SECTION
OF CONDUIT AND TUBING FOR CONDUCTORS

NUMBER OF CONDUCTORS	MAXIMUM PERCENT OF CONDUIT AND TUBING AREA FILLED BY CONDUCTORS
1	53
2	31
Over 2	40

SECTION E3808
GROUNDING

E3808.1 Metal enclosures. Metal enclosures of conductors, devices and equipment shall be grounded.

Exceptions:

1. Short sections of metal enclosures or raceways used to provide cable assemblies with support or protection against physical damage.

2. A metal elbow that is installed in an underground installation of rigid nonmetallic conduit and is isolated from possible contact by a minimum cover of 18 inches (457 mm) to any part of the elbow or that is encased in not less than 2 inches (50 mm) of concrete.

E3808.2 Equipment fastened in place or connected by permanent wiring methods (fixed). Exposed noncurrent-carrying metal parts of fixed equipment likely to become energized shall be grounded where any of the following conditions apply:

1. Where within 8 feet (2438 mm) vertically or 5 feet (1524 mm) horizontally of earth or grounded metal objects and subject to contact by persons;

2. Where located in a wet or damp location and not isolated; or

3. Where in electrical contact with metal.

E3808.3 Specific equipment fastened in place or connected by permanent wiring methods. Exposed noncurrent-carrying metal parts of the following equipment and enclosures shall be grounded:

1. Luminaires as provided in Chapter 39.

2. Motor-operated water pumps, including submersible types. Where a submersible pump is used in a metal well casing, the well casing shall be bonded to the pump circuit equipment grounding conductor.

E3808.4 Effective ground-fault current path. Electrical equipment and wiring and other electrically conductive material likely to become energized shall be installed in a manner that creates a permanent, low-impedance circuit facilitating the operation of the overcurrent device. Such circuit shall be capable of safely carrying the maximum ground-fault current likely to be imposed on it from any point on the wiring system where a ground fault to the electrical supply source might occur.

E3808.5 Earth as a ground-fault current path. The earth shall not be considered as an effective ground-fault current path.

E3808.6 Load-side neutral. A grounding connection shall not be made to any grounded circuit conductor on the load side of the service disconnecting means.

Exception: A grounding conductor connection shall be made at each separate building where required by Section E3507.3.

E3808.7 Load-side equipment. A grounded circuit conductor shall not be used for grounding noncurrent-carrying metal parts of equipment on the load side of the service disconnecting means.

Exception: For separate buildings, in accordance with Section E3507.3.2

E3808.8 Types of equipment grounding conductors. The equipment grounding conductor run with or enclosing the circuit conductors shall be one or more or a combination of the following:

1. A copper, aluminum or copper-clad conductor. This conductor shall be solid or stranded; insulated, covered or bare; and in the form of a wire or a busbar of any shape.

2. Rigid metal conduit.

3. Intermediate metal conduit.

4. Electrical metallic tubing.

5. Armor of Type AC cable in accordance with Section E3808.4.

6. The combined metallic sheath and grounding conductor of interlocked metal tape-type MC cable where listed and identified for grounding.

7. The metallic sheath or the combined metallic sheath and grounding conductors of the smooth or corrugated tube type MC cable where listed and identified for grounding.

8. Other electrically continuous metal raceways and auxiliary gutters.

9. Surface metal raceways listed for grounding.

E3808.8.1 Flexible metal conduit. Flexible metal conduit shall be permitted as an equipment grounding conductor where all of the following conditions are met:

1. The conduit is terminated in fittings listed for grounding.

2. The circuit conductors contained in the conduit are protected by overcurrent devices rated at 20 amperes or less.

3. The combined length of flexible metal conduit and flexible metallic tubing and liquid-tight flexible metal conduit in the same ground return path does not exceed 6 feet (1829 mm).

4. An equipment grounding conductor shall be installed where the conduit is used to connect equipment where flexibility is necessary after installation.

E3808.8.2 Liquid-tight flexible metal conduit. Liquid-tight flexible metal conduit shall be permitted as an equipment grounding conductor where all of the following conditions are met:

1. The conduit is terminated in fittings listed for grounding.

2. For trade sizes $^3/_8$ through $^1/_2$ (metric designator 12 through 16), the circuit conductors contained in the conduit are protected by overcurrent devices rated at 20 amperes or less.

3. For trade sizes $^3/_4$ through $1^1/_4$ (metric designator 21 through 35), the circuit conductors contained in the conduit are protected by overcurrent devices rated at not more than 60 amperes and there is no flexible metal conduit, flexible metallic tubing, or liquid-tight flexible metal conduit in trade sizes $^3/_8$ inch or $^1/_2$ inch (9.5 mm through 12.7 mm) in the grounding path.

4. The combined length of flexible metal conduit and flexible metallic tubing and liquid tight flexible metal conduit in the same ground return path does not exceed 6 feet (1829 mm).

5. An equipment grounding conductor shall be installed where the conduit is used to connect equipment where flexibility is necessary after installation.

E3808.8.3 Nonmetallic sheathed cable (Type NM). In addition to the insulated conductors, the cable shall have an insulated or bare conductor for equipment grounding purposes only. Equipment grounding conductors shall be sized in accordance with Table E3808.12.

E3808.9 Equipment fastened in place or connected by permanent wiring methods. Noncurrent-carrying metal parts of equipment, raceways and other enclosures, where required to be grounded, shall be grounded by one of the following methods:

1. By any of the equipment grounding conductors permitted by Sections E3808.8 through E3808.8.3.

2. By an equipment grounding conductor contained within the same raceway, cable or cord, or otherwise run with the circuit conductors. Equipment grounding conductors shall be identified in accordance with Section E3307.2.

E3808.10 Methods of equipment grounding. Fixtures and equipment shall be considered grounded where mechanically connected to an equipment grounding conductor as specified in Sections E3808.8 through E3808.8.3. Wire type equipment grounding conductors shall be sized in accordance with Section E3808.12.

E3808.11 Equipment grounding conductor installation. Where an equipment grounding conductor consists of a raceway, cable armor or cable sheath or where such conductor is a wire within a raceway or cable, it shall be installed in accordance with the provisions of this chapter and Chapters 33 and 37 using fittings for joints and terminations approved for installation with the type of raceway or cable used. All connections, joints and fittings shall be made tight using suitable tools.

E3808.12 Equipment grounding conductor size. Copper, aluminum and copper-clad aluminum equipment grounding conductors of the wire type shall be not smaller than shown in Table E3808.12, but shall not be required to be larger than the circuit conductors supplying the equipment. Where a raceway or a cable armor or sheath is used as the equipment grounding conductor, as provided in Section E3808.8, it shall comply with Section E3808.4. Where ungrounded connectors are increased in size, equipment grounding conductors shall be increased proportionally according to the circular mil area of the ungrounded conductors.

TABLE E3808.12
EQUIPMENT GROUNDING CONDUCTOR SIZING

RATING OR SETTING OF AUTOMATIC OVERCURRENT DEVICE IN CIRCUIT AHEAD OF EQUIPMENT, CONDUIT, ETC., NOT EXCEEDING THE FOLLOWING RATINGS (amperes)	MINIMUM SIZE	
	Copper wire No. (AWG)	Aluminum or copper-clad aluminum wire No. (AWG)
15	14	12
20	12	10
30	10	8
40	10	8
60	10	8
100	8	6
200	6	4
300	4	2
400	3	1

E3808.12.1 Multiple circuits. Where a single equipment grounding conductor is run with multiple circuits in the same raceway or cable, it shall be sized for the largest overcurrent device protecting conductors in the raceway or cable.

E3808.13 Continuity and attachment of equipment grounding conductors to boxes. Where circuit conductors are spliced within a box or terminated on equipment within or supported by a box, any equipment grounding conductors associated with the circuit conductors shall be spliced or joined within the box or to the box with devices suitable for the use. Connections depending solely on solder shall not be used. Splices shall be made in accordance with Section E3306.10 except that insulation shall not be required. The arrangement of grounding connections shall be such that the disconnection or removal of a receptacle, luminaire or other device fed from the box will not interfere with or interrupt the grounding continuity.

E3808.14 Connecting receptacle grounding terminal to box. An equipment bonding jumper shall be used to connect the grounding terminal of a grounding-type receptacle to a grounded box except where grounded in accordance with one of the following:

1. Surface mounted box. Where the box is mounted on the surface, direct metal-to-metal contact between the device yoke and the box shall be permitted to ground the receptacle to the box. At least one of the insulating washers shall be removed from receptacles that do not have a contact yoke or device designed and listed to be used in conjunction with the supporting screws to establish the

grounding circuit between the device yoke and flush-type boxes. This provision shall not apply to cover-mounted receptacles except where the box and cover combination are listed as providing satisfactory ground continuity between the box and the receptacle.

2. Contact devices or yokes. Contact devices or yokes designed and listed for the purpose shall be permitted in conjunction with the supporting screws to establish the grounding circuit between the device yoke and flush-type boxes.

3. Floor boxes. The receptacle is installed in a floor box designed for and listed as providing satisfactory ground continuity between the box and the device.

E3808.15 Metal boxes. A connection shall be made between the one or more equipment grounding conductors and a metal box by means of a grounding screw that shall be used for no other purpose, or by means of a listed grounding device. Sheet-metal screws shall not be used to connect grounding conductors or connection devices to boxes.

E3808.16 Nonmetallic boxes. One or more equipment grounding conductors brought into a nonmetallic outlet box shall be arranged to allow connection to fittings or devices installed in that box.

E3808.17 Clean surfaces. Nonconductive coatings such as paint, lacquer and enamel on equipment to be grounded shall be removed from threads and other contact surfaces to ensure electrical continuity or the equipment shall be connected by means of fittings designed so as to make such removal unnecessary.

E3808.18 Bonding other enclosures. Metal raceways, cable armor, cable sheath, enclosures, frames, fittings and other metal noncurrent-carrying parts that serve as grounding conductors, with or without the use of supplementary equipment grounding conductors, shall be effectively bonded where necessary to ensure electrical continuity and the capacity to conduct safely any fault current likely to be imposed on them. Any nonconductive paint, enamel and similar coating shall be removed at threads, contact points and contact surfaces, or connections shall be made by means of fittings designed so as to make such removal unnecessary.

E3808.19 Size of equipment bonding jumper on load side of service. The equipment bonding jumper on the load side of the service overcurrent devices shall be sized, as a minimum, in accordance with Table E3808.12, but shall not be required to be larger than the circuit conductors supplying the equipment. An equipment bonding conductor shall be not smaller than No. 14 AWG.

A single common continuous equipment bonding jumper shall be permitted to bond two or more raceways or cables where the bonding jumper is sized in accordance with Table E3808.12 for the largest overcurrent device supplying circuits therein.

E3808.20 Installation—equipment bonding jumper. The equipment bonding jumper shall be permitted to be installed inside or outside of a raceway or enclosure. Where installed on the outside, the length of the equipment bonding jumper shall not exceed 6 feet (1829 mm) and shall be routed with the raceway or enclosure. Where installed inside of a raceway, the equipment bonding jumper shall comply with the requirements of Sections E3808.9, Item 2; E3808.13; E3808.15; and E3808.16.

SECTION E3809
FLEXIBLE CORDS

E3809.1 Where permitted. Flexible cords shall be used only for the connection of appliances where the fastening means and mechanical connections of such appliances are designed to permit ready removal for maintenance, repair or frequent interchange and the appliance is listed for flexible cord connection. Flexible cords shall not be installed as a substitute for the fixed wiring of a structure; shall not be run through holes in walls, structural ceilings, suspended ceilings, dropped ceilings or floors; shall not be concealed behind walls, floors, ceilings or located above suspended or dropped ceilings.

E3809.2 Loading and protection. The ampere load of flexible cords serving fixed appliances shall be in accordance with Table E3809.2. This table shall be used in conjunction with applicable end use product standards to ensure selection of the proper size and type. Where flexible cord is approved for and used with a specific listed appliance, it shall be considered to be protected where applied within the appliance listing requirements.

E3809.3 Splices. Flexible cord shall be used only in continuous lengths without splices or taps.

E3809.4 Attachment plugs. Where used in accordance with Section E3809.1, each flexible cord shall be equipped with an attachment plug and shall be energized from a receptacle outlet.

TABLE E3809.2
MAXIMUM AMPERE LOAD FOR FLEXIBLE CORDS

CORD SIZE (AWG)	CORD TYPES S, SE, SEO, SJ, SJE, SJEO, SJO, SJOO, SJT, SJTO, SJTOO, SO, SOO, SRD, SRDE, SRDT, ST, STD, SV, SVO, SVOO, SVTO, SVTOO	
	Maximum ampere load	
	Three current-carrying conductors	Two current-carrying conductors
18	7	10
16	10	13
14	15	18
12	20	25

CHAPTER 39

DEVICES AND LUMINAIRES

SECTION E3901
SWITCHES

E3901.1 Rating and application of snap switches. General-use snap switches shall be used within their ratings and shall control only the following loads:

1. Resistive and inductive loads, including electric-discharge lamps, not exceeding the ampere rating of the switch at the voltage involved.

2. Tungsten-filament lamp loads not exceeding the ampere rating of the switch at 120 volts.

3. Motor loads not exceeding 80 percent of the ampere rating of the switch at its rated voltage.

E3901.2 CO/ALR snap switches. Snap switches rated 20 amperes or less directly connected to aluminum conductors shall be marked CO/ALR.

E3901.3 Indicating. General-use and motor-circuit switches and circuit breakers shall clearly indicate whether they are in the open OFF or closed ON position. Where single-throw switches or circuit breaker handles are operated vertically rather than rotationally or horizontally, the up position of the handle shall be the ON position.

E3901.4 Time switches and similar devices. Time switches and similar devices shall be of the enclosed type or shall be mounted in cabinets or boxes or equipment enclosures. A barrier shall be used around energized parts to prevent operator exposure when making manual adjustments or switching.

E3901.5 Grounding of enclosures. Metal enclosures for switches or circuit breakers shall be grounded. Where nonmetallic enclosures are used with metal raceways or metal-armored cables, provisions shall be made for maintaining grounding continuity.

Metal boxes for switches shall be effectively grounded. Nonmetallic boxes for switches shall be installed with a wiring method that provides or includes an equipment grounding conductor.

E3901.6 Access. All switches and circuit breakers used as switches shall be located to allow operation from a readily accessible location. Such devices shall be installed so that the center of the grip of the operating handle of the switch or circuit breaker, when in its highest position, will not be more than 6 feet 7 inches (2007 mm) above the floor or working platform.

E3901.7 Wet locations. A switch or circuit breaker located in a wet location or outside of a building shall be enclosed in a weatherproof enclosure or cabinet. Switches shall not be installed within wet locations in tub or shower spaces unless installed as part of a listed tub or shower assembly.

E3901.8 Grounded conductors. Switches or circuit breakers shall not disconnect the grounded conductor of a circuit except where the switch or circuit breaker simultaneously disconnects all conductors of the circuit.

E3901.9 Switch connections. Three- and four-way switches shall be wired so that all switching occurs only in the ungrounded circuit conductor. Color coding of switch connection conductors shall comply with Section E3307.3. Where in metal raceways or metal-jacketed cables, wiring between switches and outlets shall be in accordance with Section E3306.7.

Exception: Switch loops do not require a grounded conductor.

E3901.10 Box mounted. Flush-type snap switches mounted in boxes that are recessed from the finished wall surfaces as covered in Section E3806.5 shall be installed so that the extension plaster ears are seated against the surface of the wall. Flush-type snap switches mounted in boxes that are flush with the finished wall surface or project therefrom shall be installed so that the mounting yoke or strap of the switch is seated against the box.

E3901.11 Snap switch faceplates. Faceplates provided for snap switches mounted in boxes and other enclosures shall be installed so as to completely cover the opening and, where the switch is flush mounted, seat against the finished surface.

E3901.11.1 Faceplate grounding. Snap switches, including dimmer and similar control switches, shall be effectively grounded and shall provide a means to ground metal face plates, whether or not a metal faceplate is installed. Snap switches shall be considered effectively grounded where either of the following conditions is met:

1. The switch is mounted with metal screws to a metal box or to a nonmetallic box with integral means for grounding devices.

2. An equipment grounding conductor or equipment bonding jumper is connected to an equipment grounding termination of the snap switch.

   Exception: Where a grounding means does not exist within the snap-switch enclosure or where the wiring method does not include or provide an equipment ground, a snap switch without a grounding connection shall be permitted for replacement purposes only. A snap switch wired under the provisions of this exception and located within reach of earth, grade, conducting floors, or other conducting surfaces shall be provided with a faceplate of nonconducting, noncombustible material or shall be protected by a ground-fault circuit interrupter.

E3901.12 Dimmer switches. General-use dimmer switches shall be used only to control permanently installed incandescent luminaires (lighting fixtures) except where listed for the control of other loads and installed accordingly.

SECTION E3902
RECEPTACLES

E3902.1 Rating and type. Receptacles and cord connectors shall be rated at not less than 15 amperes, 125 volts, or 15 amperes, 250 volts, and shall not be a lampholder type. Receptacles shall be rated in accordance with this section.

E3902.1.1 Single receptacle. A single receptacle installed on an individual branch circuit shall have an ampere rating not less than that of the branch circuit.

E3902.1.2 Two or more receptacles. Where connected to a branch circuit supplying two or more receptacles or outlets, receptacles shall conform to the values listed in Table E3902.1.2.

TABLE E3902.1.2
RECEPTACLE RATINGS FOR VARIOUS SIZE
MULTI-OUTLET CIRCUITS

CIRCUIT RATING (amperes)	RECEPTACLE RATING (amperes)
15	15
20	15 or 20
30	30
40	40 or 50
50	50

E3902.2 Grounding type. Receptacles installed on 15- and 20-ampere-rated branch circuits shall be of the grounding type.

E3902.3 CO/ALR receptacles. Receptacles rated at 20 amperes or less and directly connected to aluminum conductors shall be marked CO/ALR.

E3902.4 Faceplates. Metal face plates shall be grounded.

E3902.5 Position of receptacle faces. After installation, receptacle faces shall be flush with or project from face plates of insulating material and shall project a minimum of 0.015 inch (0.381 mm) from metal face plates. Faceplates shall be installed so as to completely cover the opening and seat against the mounting surface.

Exceptions:

1. Listed kits or assemblies encompassing receptacles and nonmetallic faceplates that cover the receptacle face, where the plate cannot be installed on any other receptacle, shall be permitted.
2. Listed nonmetallic faceplates that cover the receptacle face to a maximum thickness of 0.040 inches (1 mm) shall be permitted.

E3902.6 Receptacle mounted in boxes. Receptacles mounted in boxes that are set back from the finished wall surface as permitted by Section E3806.5 shall be installed so that the mounting yoke or strap of the receptacle is held rigidly at the finished surface of the wall. Receptacles mounted in boxes that are flush with the wall surface or project therefrom shall be so installed that the mounting yoke or strap is seated against the box or raised cover.

E3902.7 Receptacles mounted on covers. Receptacles mounted to and supported by a cover shall be held rigidly against the cover by more than one screw or shall be a device

assembly or box cover listed and identified for securing by a single screw.

E3902.8 Damp locations. A receptacle installed outdoors in a location protected from the weather or in other damp locations shall have an enclosure for the receptacle that is weatherproof when the receptacle cover(s) is closed and an attachment plug cap is not inserted. An installation suitable for wet locations shall also be considered suitable for damp locations. A receptacle shall be considered to be in a location protected from the weather where located under roofed open porches, canopies and similar structures and not subject to rain or water runoff.

E3902.9 Fifteen- and 20-ampere receptacles in wet locations. Where installed in a wet location, 15- and 20-ampere, 125- and 250-volt receptacles shall have an enclosure that is weatherproof whether or not the attachment plug cap is inserted.

E3902.10 Other receptacles in wet locations. Where a receptacle other than a 15- or 20-amp, 125- or 250-volt receptacle is installed in a wet location and where the product intended to be plugged into it is not attended while in use, the receptacle shall have an enclosure that is weatherproof both when the attachment plug cap is inserted and when it is removed. Where such receptacle is installed in a wet location and where the product intended to be plugged into it will be attended while in use, the receptacle shall have an enclosure that is weatherproof when the attachment plug cap is removed.

E3902.11 Bathtub and shower space. A receptacle shall not be installed within or directly over a bathtub or shower stall.

E3902.12 Flush mounting with faceplate. In damp or wet locations, the enclosure for a receptacle installed in an outlet box flush-mounted in a finished surface shall be made weatherproof by means of a weatherproof faceplate assembly that provides a water-tight connection between the plate and the finished surface.

SECTION E3903
FIXTURES

E3903.1 Energized parts. Luminaires, lampholders, lamps and receptacles shall not have energized parts normally exposed to contact.

E3903.2 Luminaires near combustible material. Luminaires shall be installed so that combustible material will not be subjected to temperatures in excess of 90°C (194°F).

E3903.3 Exposed conductive parts. The exposed metal parts of luminaires shall be grounded or insulated from ground and other conducting surfaces. Lamp tie wires, mounting screws, clips and decorative bands on glass spaced at least 1.5 inches (38 mm) from lamp terminals shall not be required to be grounded.

E3903.4 Screw-shell type. Lampholders of the screw-shell type shall be installed for use as lampholders only.

E3903.5 Recessed incandescent luminaires. Recessed incandescent luminaires shall have thermal protection and shall be listed as thermally protected.

Exceptions:

1. Thermal protection shall not be required in recessed luminaires listed for the purpose and installed in poured concrete.

2. Thermal protection shall not be required in recessed luminaires having design, construction, and thermal performance characteristics equivalent to that of thermally protected luminaires, and such luminaires are identified as inherently protected.

E3903.6 Thermal protection. The ballast of a fluorescent luminaire installed indoors shall have integral thermal protection. Replacement ballasts shall also have thermal protection integral with the ballast. A simple reactance ballast in a fluorescent luminaire with straight tubular lamps shall not be required to be thermally protected.

E3903.7 High-intensity discharge luminaires. Recessed high-intensity luminaires designed to be installed in wall or ceiling cavities shall have thermal protection and be identified as thermally protected. Thermal protection shall not be required in recessed high-intensity luminaires having design, construction and thermal performance characteristics equivalent to that of thermally protected luminaires, and such luminaires are identified as inherently protected. Thermal protection shall not be required in recessed high-intensity discharge luminaires installed in and identified for use in poured concrete. A recessed remote ballast for a high-intensity discharge luminaire shall have thermal protection that is integral with the ballast and shall be identified as thermally protected.

E3903.8 Wet or damp locations. Luminaires installed in wet or damp locations shall be installed so that water cannot enter or accumulate in wiring compartments, lampholders or other electrical parts. All luminaires installed in wet locations shall be marked SUITABLE FOR WET LOCATIONS. All luminaires installed in damp locations shall be marked SUITABLE FOR WET LOCATIONS or SUITABLE FOR DAMP LOCATIONS.

E3903.9 Lampholders in wet or damp locations. Lampholders installed in wet or damp locations shall be of the weatherproof type.

E3903.10 Bathtub and shower areas. Cord-connected luminaires, chain-, cable-, or cord-suspended-luminaires, lighting track, pendants, and ceiling-suspended (paddle) fans shall not have any parts located within a zone measured 3 feet (914 mm) horizontally and 8 feet (2438 mm) vertically from the top of a bathtub rim or shower stall threshold. This zone is all encompassing and includes the zone directly over the tub or shower. Luminaires located in this zone shall be listed for damp locations and where subject to shower spray, shall be listed for wet locations.

E3903.11 Luminaires in clothes closets. For the purposes of this section, storage space shall be defined as a volume bounded by the sides and back closet walls and planes extending from the closet floor vertically to a height of 6 feet(1829 mm) or the highest clothes-hanging rod and parallel to the walls at a horizontal distance of 24 inches (610 mm) from the sides and back of the closet walls respectively, and continuing vertically to the closet ceiling parallel to the walls at a horizon-tal distance of 12 inches (305 mm) or the width of the shelf, whichever is greater. For a closet that permits access to both sides of a hanging rod, the storage space shall include the volume below the highest rod extending 12 inches (305 mm) on either side of the rod on a plane horizontal to the floor extending the entire length of the rod (see Figure E3903.11).

The types of luminaires installed in clothes closets shall be limited to surface-mounted or recessed incandescent luminaires with completely enclosed lamps, and surface-mounted or recessed fluorescent luminaires. Incandescent luminaires with open or partially enclosed lamps and pendant luminaires or lamp-holders shall be prohibited. Luminaire installations shall be in accordance with one or more of the following:

1. Surface-mounted incandescent luminaires shall be installed on the wall above the door or on the ceiling, provided there is a minimum clearance of 12 inches (305 mm) between the fixture and the nearest point of a storage space.

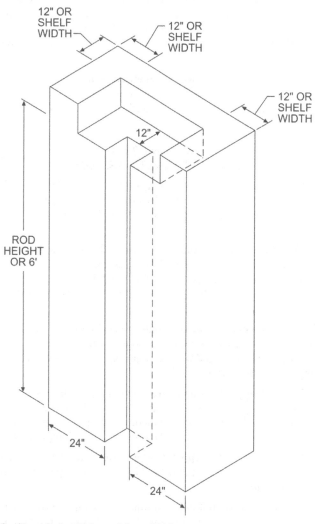

For SI: 1 inch = 25.4 mm, 1 foot = 304.8 mm.

FIGURE E3903.11
CLOSET STORAGE SPACE

2. Surface-mounted fluorescent luminaires shall be installed on the wall above the door or on the ceiling, provided there is a minimum clearance of 6 inches (152 mm) between the fixture and the nearest point of a storage space.

3. Recessed incandescent luminaires with a completely enclosed lamp shall be installed in the wall or the ceiling provided there is a minimum clearance of 6 inches (152 mm) between the luminaire and the nearest point of a storage space.

4. Recessed fluorescent luminaires shall be installed in the wall or on the ceiling provided there is a minimum clearance of 6 inches (152 mm) between the fixture and the nearest point of a storage space.

E3903.12 Luminaire wiring—general. Wiring on or within luminaires shall be neatly arranged and shall not be exposed to physical damage. Excess wiring shall be avoided. Conductors shall be arranged so that they are not subjected to temperatures above those for which the conductors are rated.

E3903.12.1 Polarization of luminaires. Luminaires shall be wired so that the screw shells of lampholders will be connected to the same luminaire or circuit conductor or terminal. The grounded conductor shall be connected to the screw shell.

E3903.12.2 Luminaires as raceways. Luminaires shall not be used as raceways for circuit conductors except where such luminaires are listed and marked for use as a raceway.

SECTION E3904
LUMINAIRE INSTALLATION

E3904.1 Outlet box covers. In a completed installation, each outlet box shall be provided with a cover except where covered by means of a luminaire canopy, lampholder or device with a faceplate.

E3904.2 Combustible material at outlet boxes. Combustible wall or ceiling finish exposed between the inside edge of a luminaire canopy or pan and the outlet box to which the luminaire connects shall be covered with a noncombustible material.

E3904.3 Access. Luminaires shall be installed so that the connections between the luminaire conductors and the circuit conductors can be accessed without requiring the disconnection of any part of the wiring.

E3904.4 Supports. Luminaires and lampholders shall be securely supported. A luminaire that weighs more than 6 pounds (2.72 kg) or exceeds 16 inches (406 mm) in any dimension shall not be supported by the screw shell of a lampholder.

E3904.5 Means of support. Outlet boxes or fittings installed as required by Sections E3805 and E3806 shall be permitted to support luminaires.

E3904.6 Exposed ballasts. Luminaires having exposed ballasts or transformers shall be installed so that such ballasts or transformers are not in contact with combustible material.

E3904.7 Combustible low-density cellulose fiberboard. Where a surface-mounted luminaire containing a ballast is installed on combustible low-density cellulose fiberboard, the luminaire shall be listed for this purpose or it shall be spaced not less than 1.5 inches (38 mm) from the surface of the fiberboard. Where such luminaires are partially or wholly recessed, the provisions of Sections E3904.8 and E3904.9 shall apply.

E3904.8 Recessed luminaire clearance. A recessed luminaire that is not identified for contact with insulation shall have all recessed parts spaced at least 0.5 inch (12.7 mm) from combustible materials. The points of support and the finish trim parts at the opening in the ceiling or wall surface shall be permitted to be in contact with combustible materials. A recessed luminaire that is identified for contact with insulation, Type IC, shall be permitted to be in contact with combustible materials at recessed parts, points of support, and portions passing through the building structure and at finish trim parts at the opening in the ceiling or wall.

E3904.9 Recessed luminaire installation. Thermal insulation shall not be installed above a recessed luminaire or within 3 inches (76 mm) of the recessed luminaire's enclosure, wiring compartment or ballast except where such luminaire is identified for contact with insulation, Type IC.

SECTION E3905
TRACK LIGHTING

E3905.1 Installation. Lighting track shall be permanently installed and permanently connected to a branch circuit having a rating not more than that of the track.

E3905.2 Fittings. Fittings identified for use on lighting track shall be designed specifically for the track on which they are to be installed. Fittings shall be securely fastened to the track, shall maintain polarization and grounding, and shall be designed to be suspended directly from the track. Only lighting track fittings shall be installed on lighting track. Lighting track fittings shall not be equipped with general-purpose receptacles.

E3905.3 Connected load. The connected load on lighting track shall not exceed the rating of the track.

E3905.4 Prohibited locations. Lighting track shall not be installed in the following locations:

1. Where likely to be subjected to physical damage.

2. In wet or damp locations.

3. Where subject to corrosive vapors.

4. In storage battery rooms.

5. In hazardous (classified) locations.

6. Where concealed.

7. Where extended through walls or partitions.

8. Less than 5 feet (1524 mm) above the finished floor except where protected from physical damage or the track operates at less than 30 volts rms open-circuit voltage.

9. Where prohibited by Section E3903.10.

E3905.5 Fastening. Lighting track shall be securely mounted so that each fastening will be suitable for supporting the maximum weight of luminaires that can be installed. Except where identified for supports at greater intervals, a single section 4 feet (1219 mm) or shorter in length shall have two supports and, where installed in a continuous row, each individual section of not more than 4 feet (1219 mm) in length shall have one additional support.

E3905.6 Grounding. Lighting track shall be grounded in accordance with Chapter 38, and the track sections shall be securely coupled to maintain continuity of the circuitry, polarization and grounding throughout.

CHAPTER 40

APPLIANCE INSTALLATION

SECTION E4001
GENERAL

E4001.1 Scope. This section covers installation requirements for appliances and fixed heating equipment.

E4001.2 Installation. Appliances and equipment shall be installed in accordance with the manufacturer's installation instructions. Electrically heated appliances and equipment shall be installed with the required clearances to combustible materials.

E4001.3 Flexible cords. Cord-and-plug-connected appliances shall use cords suitable for the environment and physical conditions likely to be encountered. Flexible cords shall be used only where the appliance is listed to be connected with a flexible cord. The cord shall be identified as suitable for the purpose in the installation instructions of the appliance manufacturer. Receptacles for cord-and-plug-connected appliances shall be accessible and shall be located to avoid physical damage to the flexible cord. Except for a listed appliance marked to indicate that it is protected by a system of double-insulation, the flexible cord supplying an appliance shall terminate in a grounding-type attachment plug. A receptacle for a cord-and-plug-connected range hood shall be supplied by an individual branch circuit. Specific appliances have additional requirements as specified in Table E4001.3 (see Section E3809).

TABLE E4001.3
FLEXIBLE CORD LENGTH

APPLIANCE	MINIMUM CORD LENGTH (inches)	MAXIMUM CORD LENGTH (inches)
Kitchen waste disposal	18	36
Built-in dishwasher	36	48
Trash compactor	36	48
Range hoods	18	36

For SI: 1 inch = 25.4 mm.

E4001.4 Overcurrent protection. Each appliance shall be protected against overcurrent in accordance with the rating of the appliance and its listing.

E4001.4.1 Single nonmotor-operated appliance. The overcurrent protection for a branch circuit that supplies a single nonmotor-operated appliance shall not exceed that marked on the appliance. Where the overcurrent protection rating is not marked and the appliance is rated at over 13.3 amperes, the overcurrent protection shall not exceed 150 percent of the appliance rated current. Where 150 percent of the appliance rating does not correspond to a standard overcurrent device ampere rating, the next higher standard rating shall be permitted. Where the overcurrent protection rating is not marked and the appliance is rated at 13.3 amperes or less, the overcurrent protection shall not exceed 20 amperes.

E4001.5 Disconnecting means. Each appliance shall be provided with a means to disconnect all ungrounded supply conductors. For fixed electric space-heating equipment, means shall be provided to disconnect the heater and any motor controller(s) and supplementary overcurrent-protective devices. Switches and circuit breakers used as a disconnecting means shall be of the indicating type. Disconnecting means shall be as set forth in Table E4001.5.

E4001.6 Support of ceiling-suspended paddle fans. Ceiling-suspended fans (paddle) shall be supported independently of an outlet box or by a listed outlet box or outlet box system identified for the use and installed in accordance with Section E3805.9.

E4001.7 Snow-melting and deicing equipment protection. Outdoor receptacles that are not readily accessible and are supplied from a dedicated branch circuit for electric snow-melting or deicing equipment shall be permitted to be installed without ground-fault circuit-interrupter protection for personnel. However, ground-fault protection of equipment shall be provided for fixed outdoor electric deicing and snow-melting equipment.

TABLE E4001.5
DISCONNECTING MEANS

DESCRIPTION	ALLOWED DISCONNECTING MEANS
Permanently connected appliance rated at not over 300 volt-amperes or $\frac{1}{8}$ horsepower.	Branch-circuit overcurrent device.
Permanently connected appliances rated in excess of 300 volt-amperes or $\frac{1}{8}$ horsepower.	Branch circuit breaker or switch located within sight of appliance or such devices in any location that are capable of being locked in the open position. The provision for locking or adding a lock to the disconnecting means shall be installed on or at the switch or circuit breaker used as the disconnecting means and shall remain in place with or without the lock installed.
Appliances listed for cord-and-plug connection.	A separable connector or attachment plug and receptacle provided with access.
Permanently installed heating equipment with motors rated at not over $\frac{1}{8}$ horsepower with supplementary overcurrent protection.	Disconnect, on the supply side of fuses, in sight from the supplementary overcurrent device, and in sight of the heating equipment or, in any location, if capable of being locked in the open position.
Heating equipment containing motors rated over $\frac{1}{8}$ horsepower with supplementary overcurrent protection.	Disconnect permitted to serve as required disconnect for both the heating equipment and the controller where, on the supply side of fuses, and in sight from the supplementary overcurrent devices, if the disconnecting means is also in sight from the controller, or is capable of being locked off.
Heating equipment containing no motor rated over $\frac{1}{8}$ horsepower without supplementary overcurrent protection.	Branch-circuit switch or circuit breaker where within sight from the heating equipment or capable of being locked off.
Heating equipment containing motors rated over $\frac{1}{8}$ horsepower without supplementary overcurrent protection.	Disconnecting means in sight from motor controller or as provided for heating equipment with motor rated over $\frac{1}{8}$ horsepower with supplementary overcurrent protection.
Air-conditioning condensing units and heat pump units.	A readily accessible disconnect within sight from unit as the only allowable means.[a]
Appliances and fixed heating equipment with unit switches having a marked OFF position.	Unit switch where an additional individual switch or circuit breaker serves as a redundant disconnecting means.
Thermostatically controlled fixed heating equipment.	Thermostats with a marked OFF position that directly open all ungrounded conductors, which when manually placed in the OFF position are designed so that the circuit cannot be energized automatically and that are located within sight of the equipment controlled.

a. The disconnecting means shall be permitted to be installed on or within the unit. It shall not be located on panels designed to allow access to the unit.

CHAPTER 41

SWIMMING POOLS

SECTION E4101
GENERAL

E4101.1 Scope. The provisions of this chapter shall apply to the construction and installation of electric wiring and equipment associated with all swimming pools, wading pools, decorative pools, fountains, hot tubs and spas, and hydromassage bathtubs, whether permanently installed or storable, and shall apply to metallic auxiliary equipment, such as pumps, filters and similar equipment. Sections E4102 through E4106 provide general rules for permanent pools, spas and hot tubs. Section E4107 provides specific rules for storable pools. Section E4108 provides specific rules for spas and hot tubs. Section E4109 provides specific rules for hydromassage bathtubs.

E4101.2 Definitions.

CORD-AND-PLUG-CONNECTED LIGHTING ASSEMBLY. A lighting assembly consisting of a cord-and-plug-connected transformer and a luminaire intended for installation in the wall of a spa, hot tub, or storable pool.

DRY-NICHE LUMINAIRE. A luminaire intended for installation in the wall of a pool or fountain in a niche that is sealed against the entry of pool water.

FORMING SHELL. A structure designed to support a wet-niche luminaire assembly and intended for mounting in a pool or fountain structure.

FOUNTAIN. Fountains, ornamental pools, display pools, and reflection pools. The definition does not include drinking fountains.

HYDROMASSAGE BATHTUB. A permanently installed bathtub equipped with a recirculating piping system, pump, and associated equipment. It is designed so it can accept, circulate and discharge water upon each use.

MAXIMUM WATER LEVEL. The highest level that water can reach before it spills out.

NO-NICHE LUMINAIRE. A luminaire intended for installation above or below the water without a niche.

PACKAGED SPA OR HOT TUB EQUIPMENT ASSEMBLY. A factory-fabricated unit consisting of water-circulating, heating and control equipment mounted on a common base, intended to operate a spa or hot tub. Equipment may include pumps, air blowers, heaters, luminaires, controls and sanitizer generators.

PERMANENTLY INSTALLED SWIMMING AND WADING POOLS. Those that are constructed in the ground or partially in the ground, and all others capable of holding water with a depth greater than 42 inches (1067 mm), and all pools installed inside of a building, regardless of water depth, whether or not served by electrical circuits of any nature.

POOL COVER, ELECTRICALLY OPERATED. Motor-driven equipment designed to cover and uncover the water surface of a pool by means of a flexible sheet or rigid frame.

SELF-CONTAINED SPA OR HOT TUB. A factory-fabricated unit consisting of a spa or hot tub vessel with all water-circulating, heating and control equipment integral to the unit. Equipment may include pumps, air blowers, heaters, luminaires, controls and sanitizer generators.

SPA OR HOT TUB. A hydromassage pool, or tub for recreational or therapeutic use, not located in health care facilities, designed for immersion of users, and usually having a filter, heater, and motor-driven blower. They are installed indoors or outdoors, on the ground or supporting structure, or in the ground or supporting structure. Generally, a spa or hot tub is not designed or intended to have its contents drained or discharged after each use.

STORABLE SWIMMING OR WADING POOL. Those that are constructed on or above the ground and are capable of holding water with a maximum depth of 42 inches (1067 mm), or a pool with nonmetallic, molded polymeric walls or inflatable fabric walls regardless of dimension.

THROUGH-WALL LIGHTING ASSEMBLY. A lighting assembly intended for installation above grade, on or through the wall of a pool, consisting of two interconnected groups of components separated by the pool wall.

WET-NICHE LUMINAIRE. A luminaire intended for installation in a forming shell mounted in a pool or fountain structure where the luminaire will be completely surrounded by water.

SECTION E4102
WIRING METHODS FOR POOLS, SPAS, HOT TUBS AND HYDROMASSAGE BATHTUBS

E4102.1 General. Wiring methods used in conjunction with permanently installed swimming pools, spas, hot tubs or hydromassage bathtubs shall be installed in accordance with Table E4102.1 and Chapter 37 except as otherwise stated in this section. Storable swimming pools shall comply with Section E4107.

E4102.2 Flexible cords. Flexible cords used in conjunction with a pool, spa, hot tub or hydromassage bathtub shall be installed in accordance with the following:

1. For other than underwater luminaires, fixed or stationary equipment, rated at 20 amperes or less shall be permitted to be connected with a flexible cord to facilitate the removal or disconnection for maintenance or repair. For other than storable pools, the flexible cord shall not exceed 3 feet (914 mm) in length. Cords that supply swimming pool equipment, shall have a copper equipment grounding conductor not smaller than 12 AWG and shall be provided with a grounding-type attachment plug.

2. Flexible cord that is supplied as part of a listed underwater swimming pool lighting luminaire shall be permitted to be installed in any of the permitted wiring methods from the luminaire to a deck box or other enclosure. Splices shall not be made within a raceway. The equipment grounding conductor shall be an insulated copper conductor that is not smaller than the supply conductors and not smaller than 16 AWG.

3. A listed packaged spa or hot tub installed outdoors that is GFCI protected, shall be permitted to be cord and plug connected provided that such cord does not exceed 15 feet (4572 mm) in length.

4. A listed packaged spa or hot tub rated at 20 amperes or less and installed indoors shall be permitted to be cord and plug connected to facilitate maintenance and repair.

5. For other than underwater and storable pool lighting luminaire, the requirements of Item 1 shall apply to any cord equipped luminaire that is located within 16 feet (4877 mm) radially from any point on the water surface.

E4102.3 Double insulated pool pumps. A listed cord-and-plug-connected pool pump incorporating an approved system of double insulation that provides a means for grounding only the internal and nonaccessible, noncurrent-carrying metal parts of the pump shall be connected to any wiring method recognized in Chapter 37 that is suitable for the location. Where the bonding grid is connected to the equipment grounding conductor of the motor circuit in accordance with Section E4104.2, Item 4, the branch circuit wiring shall comply with Sections E4102.1 and E4105.5.

TABLE E4102.1
ALLOWABLE APPLICATIONS FOR WIRING METHODS[a, b, c, d, e, f, g, h]

WIRING LOCATION OR PURPOSE (Application allowed where marked with an "A")	AC, FMC, NM, SR, SE	EMT	ENT	IMC, RMC, RNC	LFMC	LFNMC	UF	MC	Flex Cord
Panelboard(s) that supply pool equipment: from service equipment to panelboard	A[b, e]	A[c]	—	A	—	A	A[e]	A[e]	—
Wet-niche and no-niche luminaires: from branch circuit OCPD to deck or junction box	—	A[c]	A[b]	A	—	A	—	A[b]	—
Wet-niche and no-niche luminaires: from deck or junction box to forming shell	—	—	—	A[d]	—	A	—	—	A[h]
Dry niche: from branch circuit OCPD to luminaires	—	A[c]	A[b]	A	—	A	—	A[b]	—
Pool-associated motors: from branch circuit OCPD to motor	A[b]	A[c]	A[b]	A	A[f]	A	A[b]	A	A[h]
Packaged or self-contained outdoor spas and hot tubs with underwater luminaire: from branch circuit OCPD to spa or hot tub	—	A[c]	A[b]	A	A[f]	A[f]	—	—	A[h]
Packaged or self-contained outdoor spas and hot tubs without underwater luminaire: from branch circuit OCPD to spa or hot tub	A[b]	A[c]	A[b]	A	A[f]	A[f]	A[b]	A	A[h]
Indoor spa and hot tubs, hydromassage bathtubs, and other pool, spa or hot tub associated equipment: from branch circuit OCPD to equipment	A[b]	A[c]	A[b]	A	A	A	A	A	A[h]
Connection at pool lighting transformers	—	A[c]	A[b]	A	A[g]	A[g]	—	A[b]	

For SI: 1 foot = 304.8 mm.

a. For all wiring methods, see Section E4105 for equipment grounding conductor requirements.

b. Limited to use within buildings.

c. Limited to use on or within buildings.

d. Metal conduit shall be constructed of brass or other approved corrosion resistant metal.

e. Permitted only for existing installations in accordance with the exception to Section E4105.6.

f. Limited to use at pool, spa or hot tub equipment where flexibility is necessary. For spas and hot tubs, the maximum length shall be 6 feet (1.8 m).

g. Limited to use in individual lengths not to exceed 6 feet (1.8 m). The total length of all individual runs of LFMC and LFNMC shall not exceed 10 feet (3 m). LFNMC Type B shall be limited to lengths not exceeding 10 feet (3 m).

h. Flexible cord shall be installed in accordance with Section E4102.2.

SECTION E4103
EQUIPMENT LOCATION AND CLEARANCES

E4103.1 Receptacle outlets. Receptacles outlets shall be installed and located in accordance with Sections E4103.1.1 through E4103.1.6. Distances shall be measured as the shortest path that an appliance supply cord connected to the receptacle would follow without penetrating a floor, wall, ceiling, doorway with hinged or sliding door, window opening, or other effective permanent barrier.

E4103.1.1 Location. Receptacles that provide power for water-pump motors or other loads directly related to the circulation and sanitation system shall be permitted to be located between 5 feet and 10 feet (1524 mm and 3048 mm) from the inside walls of pools and outdoor spas and hot tubs, and, where so located, shall be single and of the locking and grounding type and shall be protected by ground-fault circuit interrupters.

Other receptacles on the property shall be located not less than 10 feet (3048 m) from the inside walls of pools and outdoor spas and hot tubs except where permitted by Section E4103.1.3.

E4103.1.2 Where required. At least one 125-volt, 15- or 20-ampere receptacle supplied by a general-purpose branch circuit shall be located a minimum of 10 feet (3048 mm) from and not more than 20 feet (6096 mm) from the inside wall of pools and outdoor spas and hot tubs except as permitted by Section E4103.1.3. This receptacle shall be located not more than 6 feet, 6 inches (1981 mm) above the floor, platform or grade level serving the pool, spa or hot tub.

E4103.1.3 Restricted space. Where a pool is within 10 feet (3.0 m) of a dwelling and the dimensions of the lot preclude meeting the required distances of Sections E4103.1.1 and E4103.1.2, not more than one receptacle outlet shall be permitted provided that such outlet is not less than 5 feet (1.5 m) measured horizontally from the inside wall of the pool.

E4103.1.4 GFCI protection. All 15- and 20-ampere, single phase, 125-volt receptacles located within 20 feet (6096 mm) of the inside walls of pools and outdoor spas and hot tubs shall be protected by a ground-fault circuit-interrupter. Receptacles that supply pool pump motors and that are rated 15 or 20 amperes, 125 volts through 250 volts, single phase, shall be provided with GFCI protection.

E4103.1.5 Indoor locations. Receptacles shall be located not less than 5 feet (1524 mm) from the inside walls of indoor spas and hot tubs. A minimum of one 125-volt receptacle shall be located between 5 feet (1524 mm) and 10 feet (3048 mm) from the inside walls of indoor spas or hot tubs.

E4103.1.6 Indoor GFCI protection. All 125-volt receptacles rated 30 amperes or less and located within 10 feet (3048 mm) of the inside walls of spas and hot tubs installed indoors, shall be protected by ground-fault circuit-interrupters.

E4103.2 Switching devices. Switching devices shall be located not less than 5 feet (1.5 m) horizontally from the inside walls of pools, spas and hot tubs except where separated from the pool, spa or hot tub by a solid fence, wall, or other permanent barrier. Switching devices located in a room or area containing a hydromassage bathtub shall be located in accordance with the general requirements of this code.

E4103.3 Disconnecting means. One or more means to disconnect all ungrounded conductors for all utilization equipment, other than lighting, shall be provided. Each of such means shall be readily accessible and within sight from the equipment it serves.

E4103.4 Luminaires and ceiling fans. Lighting outlets, luminaires, and ceiling-suspended paddle fans shall be installed and located in accordance with Sections E4103.4.1 through E4103.4.5.

E4103.4.1 Outdoor location. In outdoor pool, outdoor spas and outdoor hot tubs areas, luminaires, lighting outlets, and ceiling-suspended paddle fans shall not be installed over the pool or over the area extending 5 feet (1524 mm) horizontally from the inside walls of a pool except where no part of the luminaire or ceiling-suspended paddle fan is less than 12 feet (3658 mm) above the maximum water level.

E4103.4.2 Indoor locations. In indoor pool areas, the limitations of Section E4103.4.1 shall apply except where the luminaires, lighting outlets and ceiling-suspended paddle fans comply with all of the following conditions:

1. The luminaires are of a totally enclosed type;

2. A ground-fault circuit interrupter is installed in the branch circuit supplying the luminaires or ceiling-suspended (paddle) fans; and

3. The distance from the bottom of the luminaire or ceiling-suspended (paddle) fan to the maximum water level is not less than 7 feet, 6 inches (2286 mm).

E4103.4.3 Existing lighting outlets and luminaires. Existing lighting outlets and luminaires that are located within 5 feet (1524 mm) horizontally from the inside walls of pools and outdoor spas and hot tubs shall be permitted to be located not less than 5 feet (1524 mm) vertically above the maximum water level, provided that such luminaires and outlets are rigidly attached to the existing structure and are protected by a ground-fault circuit-interrupter.

E4103.4.4 Indoor spas and hot tubs.

1. Luminaires, lighting outlets, and ceiling-suspended paddle fans located over the spa or hot tub or within 5 feet (1524 mm) from the inside walls of the spa or hot tub shall be a minimum of 7 feet, 6 inches (2286 mm) above the maximum water level and shall be protected by a ground-fault circuit interrupter.

   Luminaires, lighting outlets, and ceiling-suspended paddle fans that are located 12 feet (3658 mm) or more above the maximum water level shall not require ground-fault circuit interrupter protection.

2. Luminaires protected by a ground-fault circuit interrupter and complying with Item 2.1 or 2.2 shall be permitted to be installed less than 7 feet, 6 inches (2286 mm) over a spa or hot tub.

   2.1. Recessed luminaires shall have a glass or plastic lens and nonmetallic or electrically isolated metal trim, and shall be suitable for use in damp locations.

2.2. Surface-mounted luminaires shall have a glass or plastic globe and a nonmetallic body or a metallic body isolated from contact. Such luminaires shall be suitable for use in damp locations.

E4103.4.5 GFCI protection in adjacent areas. Luminaires and outlets that are installed in the area extending between 5 feet (1524 mm) and 10 feet (3048 mm) from the inside walls of pools and outdoor spas and hot tubs shall be protected by ground-fault circuit-interrupters except where such fixtures and outlets are installed not less than 5 feet (1524 mm) above the maximum water level and are rigidly attached to the structure.

E4103.5 Overhead conductor clearances. Except where installed with the clearances specified in Table E4103.5, the following parts of pools and outdoor spas and hot tubs shall not be placed under existing service-drop conductors or any other open overhead wiring; nor shall such wiring be installed above the following:

1. Pools and the areas extending 10 feet (3048 mm) horizontally from the inside of the walls of the pool;

2. Diving structures; or

3. Observation stands, towers, and platforms.

Utility-owned, -operated and -maintained communications conductors, community antenna system coaxial cables and the supporting messengers shall be permitted at a height of not less than 10 feet (3048 mm) above swimming and wading pools, diving structures, and observation stands, towers, and platforms.

E4103.6 Underground wiring. Underground wiring shall not be installed under or within the area extending 5 feet(1524 mm) horizontally from the inside walls of pools and outdoor hot tubs and spas except where the wiring is installed to supply pool, spa or hot tub equipment or where space limitations prevent wiring from being routed 5 feet (1524 mm) or more horizontally from the inside walls. Where installed within 5 feet (1524 mm) of the inside walls, the wiring method shall be rigid metal conduit, intermediate metal conduit or a nonmetallic raceway system. Metal conduit shall be corrosion resistant and suitable for the location. The minimum raceway burial depth shall be in accordance with Table E4103.6.

SECTION E4104
BONDING

E4104.1 Performance. The equipotential bonding required by this section shall be installed to eliminate voltage gradients in the pool area as prescribed.

E4104.2 Bonded parts. The following parts shall be bonded together:

1. All metallic parts of pool, spa and hot tub structures, including the reinforcing metal of pool, spa and hot tub shells, coping stones, and decks. The usual steel tie wires shall be considered suitable for bonding the reinforcing steel together, and welding or special clamping shall not be required. Such tie wires shall be made tight. Where reinforcing steel is effectively insulated by a listed encapsulating nonconductive compound, at the time of manufacture and installation, it shall not be required to be bonded. Where reinforcing steel is encapsulated with

TABLE E4103.5
OVERHEAD CONDUCTOR CLEARANCES

	INSULATED SUPPLY OR SERVICE DROP CABLES, 0-750 VOLTS TO GROUND, SUPPORTED ON AND CABLED TOGETHER WITH AN EFFECTIVELY GROUNDED BARE MESSENGER OR EFFECTIVELY GROUNDED NEUTRAL CONDUCTOR (feet)	ALL OTHER SUPPLY OR SERVICE DROP CONDUCTORS (feet)	
		Voltage to ground	
		0-15 kV	Greater than 15 to 50 kV
A. Clearance in any direction to the water level, edge of water surface, base of diving platform, or permanently-anchored raft	22.5	25	27
B. Clearance in any direction to the diving platform	14.5	17	18

For SI: 1 foot = 304.8 mm.

TABLE E4103.6
MINIMUM BURIAL DEPTHS

WIRING METHOD	UNDERGROUND WIRING (inches)
Rigid metal conduit	6
Intermediate metal conduit	6
Nonmetallic raceways listed for direct burial without concrete encasement	18
Other approved raceways[a]	18

For SI: 1 inch = 25.4 mm.

a. Raceways approved for burial only where concrete-encased shall require a concrete envelope not less than 2 inches in thickness.

a nonconductive compound or another conductive material is not available, provisions shall be made for an alternate means to eliminate voltage gradients that would otherwise be provided by unencapsulated bonded reinforcing steel.

2. All metal forming shells and mounting brackets of no-niche luminaires except where a listed low-voltage lighting system with a nonmetallic forming shell is used that does not require bonding.

3. All metal fittings within or attached to pool, spa and hot tub structures. Isolated parts that are not over 4 inches (102 mm) in any dimension and do not penetrate into the pool structure more than 1 inch (25.4 mm) shall not require bonding. The metal bands or hoops used to secure wooden staves for a hot tub or spa shall not be required to be bonded.

4. Metal parts of electrical equipment associated with pool, spa and hot tub water circulating systems, including pump motors and metal parts of equipment associated with pool covers, including electric motors. Accessible metal parts of listed equipment incorporating an approved system of double insulation and providing a means for grounding internal nonaccessible, noncurrent-carrying metal parts shall not be bonded by a direct connection to the equipotential bonding grid. The means for grounding internal nonaccessible, noncurrent carrying metal parts shall be an equipment grounding conductor run with the power-supply conductors in the case of motors supplied with a flexible cord, or a grounding terminal in the case of motors intended for permanent connection. Where a double-insulated water-pump motor is installed under the provisions of this section, a solid 8 AWG copper conductor that is of sufficient length to make a bonding connection to a replacement motor shall be extended from the bonding grid to an accessible point in the motor vicinity. Where there is no connection between the swimming pool bonding grid and the equipment grounding system for the premises, this bonding conductor shall be connected to the equipment grounding conductor of the motor circuit.

5. Electrical devices and controls not associated with pools, spas or hot tubs and located within 5 feet (1.5 m) of such units.

6. Metal-sheathed cables and raceways, metal piping and all fixed metal parts that are within 5 feet (1524 mm) horizontally of the inside walls of the pool, spa or hot tub and that are within 12 feet (3658 mm) above the maximum water level of the pool or any observation stands, towers or platforms, or from any diving structures, and that are not separated from the pool by a permanent barrier.

7. For pool water heaters rated at more than 50 amperes and having specific instructions regarding bonding and grounding, only those parts designated to be bonded shall be bonded and only those parts designated to be grounded shall be grounded.

E4104.3 Parts not required to be bonded. Small conductive surfaces not likely to become energized, such as towel bars, mirror frames, and air and water jets and drain fittings that are not connected to metallic piping, and similar equipment installed on or within indoor spas and hot tubs shall not be required to be bonded.

E4104.4 Methods of bonding. It shall not be the intent to require that the 8 AWG or larger solid copper bonding conductor be extended or attached to any remote panelboard, service equipment, or any electrode, but only that it shall be employed to eliminate voltage gradients in the pool area as prescribed. Bonding shall be accomplished by one or more of the following methods:

1. Equipotential bonding grid. The parts specified in Section E4104.2 above shall be connected to an equipotential bonding grid with a solid copper conductor, insulated, covered, or bare, not smaller than 8 AWG or rigid metal conduit of brass or other identified corrosion resistant metal conduit. Connection shall be made by exothermic welding or by listed pressure connectors or clamps that are labeled as being suitable for the purpose and that are made of stainless steel, brass, copper or copper alloy.

The equipotential bonding grid shall conform to the contours of the pool and shall extend within or under paved walking surfaces for 3 feet (1 m) horizontally beyond the inside walls of the pool and shall be permitted to be any of the following:

Exception: The equipotential bonding grid shall not be required to be installed under the bottom of or vertically along the walls of vinyl lined polymer wall, fiberglass composite, or other pools constructed of nonconductive materials. Any metal parts of the pool, including metal structural supports, shall be bonded in accordance with Section E4104.1. For the purposes of this section, poured concrete, pneumatically applied (sprayed) concrete, and concrete block, with painted or plastered coatings, shall be considered as conductive materials.

1.1. The structural reinforcing steel of a concrete pool or deck where the reinforcing rods are bonded together by the usual steel tie wires made up tight or the equivalent. Where deck reinforcing steel is not an integral part of the pool, the deck reinforcing steel shall be bonded to the other parts of the bonding grid using a solid conductor not smaller than 8 AWG. Connections shall be in accordance with Item 1.4.

1.2. The wall of a bolted or welded metal pool.

1.3. As an alternative means, the system shall be constructed as specified in Items 1.3.1 through 1.3.3:

1.3.1. Materials and connections. The equipotential bonding grid shall be constructed of bare solid copper conductors not smaller than 8 AWG. Such conductors shall be bonded to each other at all points of crossing. Connections shall be made as required by Item 1.4.

1.3.2. Grid structure. The equipotential bonding grid shall cover the contour of the pool and the pool deck extending 3 feet (1 m) horizontally from the inside walls of the pool. The equipotential bonding grid shall be arranged in a 12 inch (300 mm) by 12 inch (300 mm) network of conductors in a uniformly spaced perpendicular grid pattern with tolerance of 4 inches (100 mm).

1.3.3. Securing. The below-grade grid shall be secured within or under the pool and deck media.

1.4. Connections. Where structural reinforcing steel or the walls of bolted or welded metal pool structures are used as an equipotential bonding grid for nonelectrical parts, the connections shall be connected by exothermic welding, listed pressure connectors, listed clamps, or other listed means. Connection devices or fittings that depend solely on solder shall not be used. Sheet metal screws shall not be used to connect bonding conductors or connection devices.

2. For indoor hot tubs and spas, metal to metal mounting on a common frame or base.

3. For indoor hot tubs and spas the interconnection of threaded metal piping and fittings.

4. For indoor hot tubs and spas the provision of a solid copper bonding jumper, insulated, covered, or bare, not smaller than 8 AWG.

SECTION E4105
GROUNDING

E4105.1 Equipment to be grounded. The following equipment shall be grounded:

1. Through-wall lighting assemblies and underwater luminaires other than those low-voltage systems listed for the application without a grounding conductor.

2. All electrical equipment located within 5 feet (1524 mm) of the inside wall of the pool, spa or hot tub.

3. All electrical equipment associated with the recirculating system of the pool, spa or hot tub.

4. Junction boxes.

5. Transformer enclosures.

6. Ground-fault circuit-interrupters.

7. Panelboards that are not part of the service equipment and that supply any electrical equipment associated with the pool, spa or hot tub.

E4105.2 Luminaires and related equipment. Through-wall lighting assemblies, wet-niche, dry-niche, or no-niche luminaires shall be connected to an insulated copper equipment grounding conductor sized in accordance with Table E3808.12 but not smaller than 12 AWG. The equipment

grounding conductor between the wiring chamber of the secondary winding of a transformer and a junction box shall be sized in accordance with the overcurrent device in such circuit. The junction box, transformer enclosure, or other enclosure in the supply circuit to a wet-niche or no-niche luminaire and the field-wiring chamber of a dry-niche luminaire shall be grounded to the equipment grounding terminal of the panelboard. The equipment grounding terminal shall be directly connected to the panelboard enclosure. The equipment grounding conductor shall be installed without joint or splice.

Exceptions:

1. Where more than one underwater luminaire is supplied by the same branch circuit, the equipment grounding conductor, installed between the junction boxes, transformer enclosures, or other enclosures in the supply circuit to wet-niche luminaires, or between the field-wiring compartments of dry-niche luminaires, shall be permitted to be terminated on grounding terminals.

2. Where an underwater luminaire is supplied from a transformer, ground-fault circuit-interrupter, clock-operated switch, or a manual snap switch that is located between the panelboard and a junction box connected to the conduit that extends directly to the underwater luminaire, the equipment grounding conductor shall be permitted to terminate on grounding terminals on the transformer, ground-fault circuit-interrupter, clock-operated switch enclosure, or an outlet box used to enclose a snap switch.

E4105.3 Nonmetallic conduit. Where a nonmetallic conduit is installed between a forming shell and a junction box, transformer enclosure, or other enclosure, a 8 AWG insulated copper bonding jumper shall be installed in this conduit except where a listed low-voltage lighting system not requiring grounding is used. The bonding jumper shall be terminated in the forming shell, junction box or transformer enclosure, or ground-fault circuit-interrupter enclosure. The termination of the 8 AWG bonding jumper in the forming shell shall be covered with, or encapsulated in, a listed potting compound to protect such connection from the possible deteriorating effect of pool water.

E4105.4 Flexible cords. Wet-niche luminaires that are supplied by a flexible cord or cable shall have all exposed noncurrent-carrying metal parts grounded by an insulated copper equipment grounding conductor that is an integral part of the cord or cable. This grounding conductor shall be connected to a grounding terminal in the supply junction box, transformer enclosure, or other enclosure. The grounding conductor shall not be smaller than the supply conductors and not smaller than 16 AWG.

E4105.5 Motors. Pool-associated motors shall be connected to an insulated copper equipment grounding conductor sized in accordance with Table E3808.12, but not smaller than 12 AWG. Where the branch circuit supplying the motor is installed in the interior of a one-family dwelling or in the interior of accessory buildings associated with a one-family dwelling, using a cable wiring method permitted by Table E4102.1, an uninsulated equipment grounding conductor shall be permitted provided that it is enclosed within the outer sheath of the cable assembly.

E4105.6 Panelboards. A panelboard that is not part of the service equipment, or source of a separately derived system shall have an equipment grounding conductor installed between its grounding terminal and the grounding terminal of the applicable service equipment or source of a separately derived system. The equipment grounding conductor shall be insulated, shall be sized in accordance with Table E3808.12, and shall be not smaller than 12 AWG.

> **Exception:** An existing feeder between an existing remote panelboard and service equipment shall be permitted to run in flexible metal conduit or an approved cable assembly that includes an equipment grounding conductor within its outer sheath. The equipment grounding conductor shall not be connected to the grounded conductor in the remote panelboard.

> **E4105.6.1 Separate buildings.** A feeder to a separate building or structure shall be permitted to supply swimming pool equipment branch circuits, or feeders supplying swimming pool equipment branch circuits, provided that the grounding arrangements in the separate building meet the requirements of Section E3507.3. Where installed in other than existing feeders covered in the exception to Section E4105.6, a separate equipment grounding conductor shall be an insulated conductor.

E4105.7 Cord-connected equipment. Where fixed or stationary equipment is connected with a flexible cord to facilitate removal or disconnection for maintenance, repair, or storage, as provided in Section E4102.2, the equipment grounding conductors shall be connected to a fixed metal part of the assembly. The removable part shall be mounted on or bonded to the fixed metal part.

E4105.8 Other equipment. Other electrical equipment shall be grounded in accordance with Section E3808.

SECTION E4106
EQUIPMENT INSTALLATION

E4106.1 Transformers. Transformers used for the supply of underwater luminaires, together with the transformer enclosure, shall be listed as a swimming pool and spa transformer. Such transformers shall be of an isolated winding type with an ungrounded secondary that has a grounded metal barrier between the primary and secondary windings.

E4106.2 Ground-fault circuit-interrupters. Ground-fault circuit-interrupters shall be self-contained units, circuit-breaker types, receptacle types or other approved types.

E4106.3 Wiring on load side of ground-fault circuit-interrupters and transformers. For other than grounding conductors, conductors installed on the load side of a ground-fault circuit-interrupter or transformer used to comply with the provisions of Section E4106.4, shall not occupy raceways, boxes, or enclosures containing other conductors except where the other conductors are protected by ground-fault circuit interrupters or are grounding conductors. Supply conductors to a feed-through type ground-fault circuit interrupter shall be permitted in the same enclosure. Ground-fault circuit interrupters shall be permitted in a panelboard that contains circuits protected by other than ground-fault circuit interrupters.

E4106.4 Underwater luminaires. The design of an underwater luminaire supplied from a branch circuit either directly or by way of a transformer meeting the requirements of Section E4106.1, shall be such that, where the fixture is properly installed without a ground-fault circuit-interrupter, there is no shock hazard with any likely combination of fault conditions during normal use (not relamping). In addition, a ground-fault circuit-interrupter shall be installed in the branch circuit supplying luminaires operating at more than 15 volts, so that there is no shock hazard during relamping. The installation of the ground-fault circuit-interrupter shall be such that there is no shock hazard with any likely fault-condition combination that involves a person in a conductive path from any ungrounded part of the branch circuit or the luminaire to ground. Compliance with this requirement shall be obtained by the use of a listed underwater luminaire and by installation of a listed ground-fault circuit-interrupter in the branch circuit. Luminaires that depend on submersion for safe operation shall be inherently protected against the hazards of overheating when not submerged.

> **E4106.4.1 Maximum voltage.** Luminaires shall not be installed for operation on supply circuits over 150 volts between conductors.

> **E4106.4.2 Luminaire location.** Luminaires mounted in walls shall be installed with the top of the fixture lens not less than 18 inches (457 mm) below the normal water level of the pool, except where the luminaire is listed and identified for use at a depth of not less than 4 inches (102 mm) below the normal water level of the pool. A luminaire facing upward shall have the lens adequately guarded to prevent contact by any person.

E4106.5 Wet-niche luminaires. Forming shells shall be installed for the mounting of all wet-niche underwater luminaires and shall be equipped with provisions for conduit entries. Conduit shall extend from the forming shell to a suitable junction box or other enclosure located as provided in Section E4106.9. Metal parts of the luminaire and forming shell in contact with the pool water shall be of brass or other approved corrosion-resistant metal.

The end of flexible-cord jackets and flexible-cord conductor terminations within a luminaire shall be covered with, or encapsulated in, a suitable potting compound to prevent the entry of water into the luminaire through the cord or its conductors. In addition, the grounding connection within a luminaire shall be similarly treated to protect such connection from the deteriorating effect of pool water in the event of water entry into the luminaire.

Luminaires shall be bonded to and secured to the forming shell by a positive locking device that ensures a low-resistance contact and requires a tool to remove the luminaire from the forming shell.

> **E4106.5.1 Servicing.** All luminaires shall be removable from the water for relamping or normal maintenance. Luminaires shall be installed in such a manner that personnel

can reach the luminaire for relamping, maintenance, or inspection while on the deck or equivalently dry location.

E4106.6 Dry-niche luminaires. Dry-niche luminaires shall be provided with provisions for drainage of water and means for accommodating one equipment grounding conductor for each conduit entry. Junction boxes shall not be required but, if used, shall not be required to be elevated or located as specified in Section E4106.9 if the luminaire is specifically identified for the purpose.

E4106.7 No-niche luminaires. No-niche luminaires shall be listed for the purpose and shall be installed in accordance with the requirements of Section E4106.5. Where connection to a forming shell is specified, the connection shall be to the mounting bracket.

E4106.8 Through-wall lighting assembly. A through-wall lighting assembly shall be equipped with a threaded entry or hub, or a nonmetallic hub, for the purpose of accommodating the termination of the supply conduit. A through-wall lighting assembly shall meet the construction requirements of Section E4105.4 and be installed in accordance with the requirements of Section E4106.5 Where connection to a forming shell is specified, the connection shall be to the conduit termination point.

E4106.9 Junction boxes and enclosures for transformers or ground-fault circuit interrupters. Junction boxes for underwater luminaires and enclosures for transformers and ground-fault circuit-interrupters that supply underwater luminaires shall comply with the following:

E4106.9.1 Junction boxes. A junction box connected to a conduit that extends directly to a forming shell or mounting bracket of a no-niche luminaire shall be:

1. Listed as a swimming pool junction box;

2. Equipped with threaded entries or hubs or a nonmetallic hub;

3. Constructed of copper, brass, suitable plastic, or other approved corrosion-resistant material;

4. Provided with electrical continuity between every connected metal conduit and the grounding terminals by means of copper, brass, or other approved corrosion-resistant metal that is integral with the box; and

5. Located not less than 4 inches (102 mm), measured from the inside of the bottom of the box, above the ground level, or pool deck, or not less than 8 inches (203 mm) above the maximum pool water level, whichever provides the greatest elevation, and shall be located not less than 4 feet (1219 mm) from the inside wall of the pool, unless separated from the pool by a solid fence, wall or other permanent barrier. Where used on a lighting system operating at 15 volts or less, a flush deck box shall be permitted provided that an approved potting compound is used to fill the box to prevent the entrance of moisture; and the flush deck box is located not less than 4 feet (1219 mm) from the inside wall of the pool.

E4106.9.2 Other enclosures. An enclosure for a transformer, ground-fault circuit-interrupter or a similar device connected to a conduit that extends directly to a forming shell or mounting bracket of a no-niche luminaire shall be:

1. Listed and labeled for the purpose, comprised of copper, brass, suitable plastic, or other approved corrosion-resistant material;

2. Equipped with threaded entries or hubs or a nonmetallic hub;

3. Provided with an approved seal, such as duct seal at the conduit connection, that prevents circulation of air between the conduit and the enclosures;

4. Provided with electrical continuity between every connected metal conduit and the grounding terminals by means of copper, brass or other approved corrosion-resistant metal that is integral with the enclosures; and

5. Located not less than 4 inches (102 mm), measured from the inside bottom of the enclosure, above the ground level or pool deck, or not less than 8 inches (203 mm) above the maximum pool water level, whichever provides the greater elevation, and shall be located not less than 4 feet (1219 mm) from the inside wall of the pool, except where separated from the pool by a solid fence, wall or other permanent barrier.

E4106.9.3 Protection of junction boxes and enclosures. Junction boxes and enclosures mounted above the grade of the finished walkway around the pool shall not be located in the walkway unless afforded additional protection, such as by location under diving boards or adjacent to fixed structures.

E4106.9.4 Grounding terminals. Junction boxes, transformer enclosures, and ground-fault circuit-interrupter enclosures connected to a conduit that extends directly to a forming shell or mounting bracket of a no-niche luminaire shall be provided with grounding terminals in a quantity not less than the number of conduit entries plus one.

E4106.9.5 Strain relief. The termination of a flexible cord of an underwater luminaire within a junction box, transformer enclosure, ground-fault circuit-interrupter, or other enclosure shall be provided with a strain relief.

E4106.10 Underwater audio equipment. Underwater audio equipment shall be identified for the purpose.

E4106.10.1 Speakers. Each speaker shall be mounted in an approved metal forming shell, the front of which is enclosed by a captive metal screen, or equivalent, that is bonded to and secured to the forming shell by a positive locking device that ensures a low-resistance contact and requires a tool to open for installation or servicing of the speaker. The forming shell shall be installed in a recess in the wall or floor of the pool.

E4106.10.2 Wiring methods. Rigid metal conduit or intermediate metal conduit of brass or other identified corrosion-resistant metal, rigid nonmetallic conduit, or liquid tight flexible nonmetallic conduit (LFNC-B) shall extend from the forming shell to a suitable junction box or other enclosure as provided in Section E4106.9. Where rigid nonmetallic conduit or liquid tight flexible nonmetallic conduit is used, an 8 AWG solid or stranded insulated copper bonding jumper shall be installed in this conduit with provisions for terminating in the forming shell and the junction box. The termination of the 8 AWG bonding jumper in the forming shell shall be covered with, or encapsulated in, a suitable potting compound to protect such connection from the possible deteriorating effect of pool water.

E4106.10.3 Forming shell and metal screen. The forming shell and metal screen shall be of brass or other approved corrosion-resistant metal. All forming shells shall include provisions for terminating an 8 AWG copper conductor.

E4106.11 Electrically operated pool covers. The electric motors, controllers, and wiring for pool covers shall be located not less than 5 feet (1524 mm) from the inside wall of the pool except where separated from the pool by a wall, cover, or other permanent barrier. Electric motors installed below grade level shall be of the totally enclosed type. The electric motor and controller shall be connected to a circuit protected by a ground-fault circuit-interrupter. The device that controls the operation of the motor for an electrically operated pool cover shall be located so that the operator has full view of the pool.

E4106.12 Electric pool water heaters. All electric pool water heaters shall have the heating elements subdivided into loads not exceeding 48 amperes and protected at not more than 60 amperes. The ampacity of the branch-circuit conductors and the rating or setting of overcurrent protective devices shall be not less than 125 percent of the total nameplate load rating.

E4106.13 Pool area heating. The provisions of Sections E4106.13.1 through E4106.13.3 shall apply to all pool deck areas, including a covered pool, where electrically operated comfort heating units are installed within 20 feet (6096 mm) of the inside wall of the pool.

E4106.13.1 Unit heaters. Unit heaters shall be rigidly mounted to the structure and shall be of the totally enclosed or guarded types. Unit heaters shall not be mounted over the pool or within the area extending 5 feet (1524 mm) horizontally from the inside walls of a pool.

E4106.13.2 Permanently wired radiant heaters. Electric radiant heaters shall be suitably guarded and securely fastened to their mounting devices. Heaters shall not be installed over a pool or within the area extending 5 feet (1524 mm) horizontally from the inside walls of the pool and shall be mounted not less than 12 feet (3658 mm) vertically above the pool deck.

E4106.13.3 Radiant heating cables prohibited. Radiant heating cables embedded in or below the deck shall be prohibited.

E4106.14 Double insulated pool pumps. A listed cord-and-plug-connected pool pump incorporating an approved system of double insulation that provides a means for grounding only the internal and nonaccessible, non-current-carrying metal parts of the pump shall be permitted to be used with permanently installed swimming pools. Branch circuit wiring to the pump shall comply with Section E4102.3.

SECTION E4107
STORABLE SWIMMING POOLS

E4107.1 Pumps. A cord-connected pool filter pump for use with storable pools shall incorporate an approved system of double insulation or its equivalent and shall be provided with means for grounding only the internal and nonaccessible noncurrent-carrying metal parts of the appliance.

The means for grounding shall be an equipment grounding conductor run with the power-supply conductors in a flexible cord that is properly terminated in a grounding-type attachment plug having a fixed grounding contact.

E4107.2 Ground-fault circuit-interrupters required. Electrical equipment, including power-supply cords, used with storable pools shall be protected by ground-fault circuit-interrupters. All 125-volt receptacles located within 20 feet (6.0 m) of the inside walls of a storable pool shall be protected by a ground-fault circuit interrupter. In determining these dimensions, the distance to be measured shall be the shortest path that the supply cord of an appliance connected to the receptacle would follow without passing through a floor, wall, ceiling, doorway with hinged or sliding door, window opening, or other effective permanent barrier.

E4107.3 Luminaires. Luminaires for storable pools shall not have exposed metal parts and shall be listed for the purpose as an assembly. In addition, luminaires for storable pools shall comply with the requirements of Section E4107.3.1 or E4107.3.2.

E4107.3.1 Fifteen (15) volts or less. A luminaire installed in or on the wall of a storable pool shall be part of a cord- and plug-connected lighting assembly. The assembly shall:

1. Have a luminaire lamp that operates at 15 volts or less;

2. Have an impact-resistant polymeric lens, luminaire body, and transformer enclosure;

3. Have a transformer meeting the requirements of section E4106.1 with a primary rating not over 150 volts; and

4. Have no exposed metal parts.

E4107.3.2 Not over 150 volts. A lighting assembly without a transformer, and with the luminaire lamp(s) operating at not over 150 volts, shall be permitted to be cord- and plug-connected where the assembly is listed as an assembly for the purpose and complies with all of the following:

1. It has an impact-resistant polymeric lens and luminaire body.

2. A ground-fault circuit interrupter with open neutral protection is provided as an integral part of the assembly.

3. The luminaire lamp is permanently connected to the ground-fault circuit interrupter with open-neutral protection.

4. It complies with the requirements of Section E4106.4.

5. It has no exposed metal parts.

E4107.4 Receptacle locations. Receptacles shall be not less than 10 feet (3.0 m) from the inside walls of a pool. In determining these dimensions, the distance to be measured shall be the shortest path that the supply cord of an appliance connected to the receptacle would follow without passing through a floor, wall, ceiling, doorway with hinged or sliding door, window opening, or other effective permanent barrier.

SECTION E4108
SPAS AND HOT TUBS

E4108.1 Ground-fault circuit-interrupters. The outlet(s) that supplies a self-contained spa or hot tub, or a packaged spa or hot tub equipment assembly, or a field-assembled spa or hot tub with a heater load of 50 amperes or less, shall be protected by a ground-fault circuit-interrupter.

A listed self-contained unit or listed packaged equipment assembly marked to indicate that integral ground-fault circuit-interrupter protection is provided for all electrical parts within the unit or assembly, including pumps, air blowers, heaters, luminaires, controls, sanitizer generators and wiring, shall not require that the outlet supply be protected by a ground-fault circuit interrupter.

A combination pool/hot tub or spa assembly commonly bonded need not be protected by a ground-fault circuit interrupter.

E4108.2 Electric water heaters. Electric spa and hot tub water heaters shall be listed and shall have the heating elements subdivided into loads not exceeding 48 amperes and protected at not more than 60 amperes. The ampacity of the branch-circuit conductors, and the rating or setting of overcurrent protective devices, shall be not less than 125 percent of the total nameplate load rating.

E4108.3 Underwater audio equipment. Underwater audio equipment used with spas and hot tubs shall comply with the provisions of Section E4106.10.

E4108.4 Emergency switch for spas and hot tubs. A clearly labeled emergency shutoff or control switch for the purpose of stopping the motor(s) that provides power to the recirculation system and jet system shall be installed at a point that is readily accessible to the users, adjacent to and within sight of the spa or hot tub and not less than 5 feet (1.5 m) away from the spa or hot tub. This requirement shall not apply to single-family dwellings.

SECTION E4109
HYDROMASSAGE BATHTUBS

E4109.1 Ground-fault circuit-interrupters. Hydromassage bathtubs and their associated electrical components shall be protected in accordance with Section E4108. All 125-volt, single-phase receptacles not exceeding 30 amperes and located within 5 feet (1524 mm) measured horizontally of the inside walls of a hydromassage tub shall be protected by a ground-fault circuit interrupter(s).

E4109.2 Other electric equipment. Luminaires, switches, receptacles, and other electrical equipment located in the same room, and not directly associated with a hydromassage bathtub, shall be installed in accordance with the requirements of this code relative to the installation of electrical equipment in bathrooms.

E4109.3 Accessibility. Hydromassage bathtub electrical equipment shall be accessible without damaging the building structure or building finish.

E4109.4 Bonding. All metal piping systems and all grounded metal parts in contact with the circulating water shall be bonded together using a solid copper bonding jumper, insulated, covered, or bare, not smaller than 8 AWG.

CHAPTER 42

CLASS 2 REMOTE-CONTROL, SIGNALING
AND POWER-LIMITED CIRCUITS

SECTION E4201
GENERAL

E4201.1 Scope. This chapter contains requirements for power supplies and wiring methods associated with Class 2 remote-control, signaling, and power-limited circuits that are not an integral part of a device or appliance. Other classes of remote-control, signaling and power-limited conductors shall comply with Article 725 of NFPA 70.

E4201.2 Definitions.

CLASS 2 CIRCUIT. That portion of the wiring system between the load side of a Class 2 power source and the connected equipment. Due to its power limitations, a Class 2 circuit considers safety from a fire initiation standpoint and provides acceptable protection from electric shock.

REMOTE-CONTROL CIRCUIT. Any electrical circuit that controls any other circuit through a relay or an equivalent device.

SIGNALING CIRCUIT. Any electrical circuit that energizes signaling equipment.

SECTION E4202
POWER SOURCES

E4202.1 Power sources for Class 2 circuits. The power source for a Class 2 circuit shall be one of the following:

1. A listed Class 2 transformer.

2. A listed Class 2 power supply.

3. Other listed equipment marked to identify the Class 2 power source.

4. Listed information technology (computer) equipment limited power circuits.

5. A dry cell battery provided that the voltage is 30 volts or less and the capacity is equal to or less than that available from series connected No. 6 carbon zinc cells.

E4202.2 Interconnection of power sources. A Class 2 power source shall not have its output connections paralleled or otherwise interconnected with another Class 2 power source except where listed for such interconnection.

SECTION E4203
WIRING METHODS

E4203.1 Wiring methods on supply side of Class 2 power source. Conductors and equipment on the supply side of the power source shall be installed in accordance with the appropriate requirements of Chapters 33 through 40. Transformers or other devices supplied from electric light or power circuits shall be protected by an over current device rated at not over 20 amperes. The input leads of a transformer or other power

source supplying Class 2 circuits shall be permitted to be smaller than 14 AWG, if not over 12 inches (305 mm) long and if the conductor insulation is rated at not less than 600 volts. In no case shall such leads be smaller than 18 AWG.

E4203.2 Wiring methods and materials on load side of the Class 2 power source. Class 2 cables installed as wiring within buildings shall be listed as being resistant to the spread of fire and listed as meeting the criteria specified in Sections E4203.2.1 through E4203.2.3. Cables shall be marked in accordance with Section E4203.2.4. Cable substitutions as described in Table E4203.2 and wiring methods covered in Chapter 37 shall also be permitted.

TABLE E4203.2
CABLE USES AND PERMITTED SUBSTITUTIONS

CABLE TYPE	USE	PERMITTED SUBSTITUTIONS[a]
CL2P	Class 2 Plenum Cable	CMP, CL3P
CL2	Class 2 Cable	CMP, CL3P, CL2P, CMR, CL3R, CL2R CMG, CM, CL3
CL2X	Class 2 Cable, Limited Use	CMP, CL3P CL2P, CMR, CL3R, CL2R, CMG, CM, CL3, CL2, CMX, CL3X

a. For identification of cables other than Class 2 cables, see NFPA 70.

E4203.2.1 Type CL2P cables. Cables installed in ducts, plenums and other spaces used to convey environmental air shall be Type CL2P cables listed as being suitable for the use and listed as having adequate fire-resistant and low smoke-producing characteristics.

E4203.2.2 Type CL2 cables. Cables for general-purpose use, shall be listed as being resistant to the spread of fire and listed for the use.

E4203.2.3 Type CL2X cables. Type CL2X limited-use cable shall be listed as being suitable for use in dwellings and for the use and in raceways and shall also be listed as being flame retardant. Cables with a diameter of less than 0.25 inches (6.4 mm) shall be permitted to be installed without a raceway.

E4203.2.4 Marking. Cables shall be marked in accordance with Table E4203.2. Voltage ratings shall not be marked on cables.

SECTION E4204
INSTALLATION REQUIREMENTS

E4204.1 Separation from other conductors. In cables, compartments, enclosures, outlet boxes, device boxes, and raceways, conductors of Class 2 circuits shall not be placed in any cable, compartment, enclosure, outlet box, device box, raceway, or similar fitting with conductors of electric light, power, Class 1 and nonpower-limited fire alarm circuits.

Exceptions:

1. Where the conductors of the electric light, power, Class 1 and nonpower-limited fire alarm circuits are separated by a barrier from the Class 2 circuits. In enclosures, Class 2 circuits shall be permitted to be installed in a raceway within the enclosure to separate them from Class 1, electric light, power and nonpower-limited fire alarm circuits.

2. Class 2 conductors in compartments, enclosures, device boxes, outlet boxes and similar fittings where electric light, power, Class 1 or nonpower-limited fire alarm circuit conductors are introduced solely to connect to the equipment connected to the Class 2 circuits. The electric light, power, Class 1 and nonpower-limited fire alarm circuit conductors shall be routed to maintain a minimum of $^1/_4$ inch (6.4 mm) separation from the conductors and cables of the Class 2 circuits; or the electric light power, Class 1 and nonpower-limited fire alarm circuit conductors operate at 150 volts or less to ground and the Class 2 circuits are installed using Types CL3, CL3R, or CL3P or permitted substitute cables, and provided that these Class 3 cable conductors extending beyond their jacket are separated by a minimum of $^1/_4$ inch (6.4 mm) or by a nonconductive sleeve or nonconductive barrier from all other conductors.

E4204.2 Other applications. Conductors of Class 2 circuits shall be separated by not less than 2 inches (51 mm) from conductors of any electric light, power, Class 1 or nonpower-limited fire alarm circuits except where one of the following conditions is met:

1. All of the electric light, power, Class 1 and nonpower-limited fire alarm circuit conductors are in raceways or in metal-sheathed, metal-clad, nonmetallic-sheathed or Type UF cables.

2. All of the Class 2 circuit conductors are in raceways or in metal-sheathed, metal-clad, nonmetallic-sheathed or Type UF cables.

E4204.3 Class 2 circuits with communications circuits. Where Class 2 circuit conductors are in the same cable as communications circuits, the Class 2 circuits shall be classified as communications circuits and shall meet the requirements of Article 800 of NFPA 70. The cables shall be listed as communications cables or multipurpose cables.

Cables constructed of individually listed Class 2 and communications cables under a common jacket shall be permitted to be classified as communications cables. The fire-resistance rating of the composite cable shall be determined by the performance of the composite cable.

E4204.4 Class 2 cables with other circuit cables. Jacketed cables of Class 2 circuits shall be permitted in the same enclosure or raceway with jacketed cables of any of the following:

1. Power-limited fire alarm systems in compliance with Article 760 of NFPA 70.

2. Nonconductive and conductive optical fiber cables in compliance with Article 770 of NFPA 70.

3. Communications circuits in compliance with Article 800 of NFPA 70.

4. Community antenna television and radio distribution systems in compliance with Article 820 of NFPA 70.

5. Low-power, network-powered broadband communications in compliance with Article 830 of NFPA 70.

E4204.5 Installation of conductors and cables. Cables and conductors installed exposed on the surface of ceilings and sidewalls shall be supported by the building structure in such a manner that they will not be damaged by normal building use. Such cables shall be supported by straps, staples, hangers, or similar fittings designed so as to not damage the cable. The installation shall comply with Table E3702.1 regarding cables run parallel with framing members and furring strips. The installation of wires and cables shall not prevent access to equipment nor prevent removal of panels, including suspended ceiling panels. Raceways shall not be used as a means of support for Class 2 circuit conductors, except where the supporting raceway contains conductors supplying power to the functionally associated equipment controlled by the Class 2 conductors.

Part IX—Referenced Standards

CHAPTER 43

REFERENCED STANDARDS

This chapter lists the standards that are referenced in various sections of this document. The standards are listed herein by the promulgating agency of the standard, the standard identification, the effective date and title, and the section or sections of this document that reference the standard. The application of the referenced standard shall be as specified in Section R102.4.

AAMA

American Architectural Manufacturers Association
1827 Walden Office Square, Suite 550
Schaumburg, IL 60173

Standard reference number	Title	Referenced in code section number
101/I.S.2/A440—05	Specifications for Windows, Doors and Unit Skylights	R308.6.9, R613.4, N1102.4.2
450—00	Voluntary Performance Rating Method for Mulled Fenestration Assemblies	R613.9.1
506—00	Voluntary Specifications for Hurricane Impact and Cycle Testing of Fenestration Products	R613.7.1

ACI

American Concrete Institute
38800 Country Club Drive
Farmington Hills, MI 48333

Standard reference number	Title	Referenced in code section number
318—05	Building Code Requirements for Structural Concrete.	R402.2, R404.1, Table R404.1.1(5), R404.4, R404.4.6.1, Table R404.4(1), Table R404.4(2), Table R404.4(3), Table R404.4(4), Table R404.4(5), R611.1, Table R611.3(1), Table R611.7(1), Table R611.7(2), Table R611.7(3), Table R611.7(4), Table R611.7(5), Table R611.7(6), Table R611.7(7), Table R611.7(9), Table R611.7(10), R611.7.1.1, Table R611.7.4, R612.1
332—04	Requirements for Residential Concrete Construction.	R404.1, Table R404.1.1(5)
530—05	Building Code Requirements for Masonry Structures	R404.1, R606.1, R606.1.1, R606.12.1, R606.12.2.2.1, R606.12.2.2.2, R606.12.3.1
530.1—05	Specifications for Masonry Structures	R404.1, R606.1, R606.1.1, R606.12.1, R606.12.2.2.1, R606.12.2.2.2, R606.12.3.1

ACCA

Air Conditioning Contractors of America
2800 Shirlington Road, Suite 300
Arlington, VA 22206

Standard reference number	Title	Referenced in code section number
Manual D—95	Residential Duct Systems	M1601.1, M1602.2
Manual J—02	Residential Load Calculation—Eighth Edition	M1401.3

AFPA

American Forest and Paper Association
111 19th Street, NW, Suite 800
Washington, DC 20036

Standard reference number	Title	Referenced in code section number
NDS—05	National Design Specification (NDS) for Wood Construction—with 2005 Supplement	R404.2.2, R502.2, Table R503.1, R602.3, R802.2
WFCM—2001	Wood Frame Construction Manual for One- and Two-family Dwellings	R301.2.1.1

| AFPA—93 | Span Tables for Joists and Rafters | R502.3, R802.4, R802.5 |
| T.R. No. 7—87 | Basic Requirements for Permanent Wood Foundation System | R401.1 |

AHA

American Hardboard Association
1210 West Northwest Highway
Palatine, IL 60067

Standard reference number	Title	Referenced in code section number
A135.4—04	Basic Hardboard	Table R602.3(1)
A135.5—04	Prefinished Hardboard Paneling	R702.5
A135.6—98	Hardboard Siding	Table R703.4

AISI

American Iron and Steel Institute
1140 Connecticut Ave, Suite 705
Washington, DC 20036

Standard reference number	Title	Referenced in code section number
Header—04	Standard for Cold-formed Steel Framing-Header Design	R603.6
PM—2001	Standard for Cold-formed Steel Framing-Prescriptive Method for One- and Two-family Dwellings (including 2004 Supplement)	R301.1.1, R301.2.1.1(4), R301.2.2.4.1, R301.2.2.4.5
Truss—04	Standard for Cold-formed Steel Framing-Truss Design	R804.1.3

AITC

American Institute of Timber Construction
7012 S. Revere Parkway, Suite 140
Englewood, CO 80112

Standard reference number	Title	Referenced in code section number
AITC A 190.1—02	Structural Glued Laminated Timber	R502.1.5, R602.1.2, R802.1.4

ANSI

American National Standards Institute
25 West 43rd Street, Fourth Floor
New York, NY 10036

Standard reference number	Title	Referenced in code section number
A108.1A—99	Installation of Ceramic Tile in the Wet-set Method, with Portland Cement Mortar	R702.4.1
A108.1B—99	Installation of Ceramic Tile, Quarry Tile on a Cured Portland Cement Mortar Setting Bed with Dry-set or Latex-Portland Mortar	R702.4.1
A108.4—99	Installation of Ceramic Tile with Organic Adhesives or Water Cleanable Tile-setting Epoxy Adhesive	R702.4.1
A108.5—99	Installation of Ceramic Tile with Dry-set Portland Cement Mortar or Latex-Portland Cement Mortar	R702.4.1
A108.6—99	Installation of Ceramic Tile with Chemical Resistant, Water Cleanable Tile-setting and -grouting Epoxy	R702.4.1
A108.11—99	Interior Installation of Cementitious Backer Units	R702.4.1
A118.1—99	American National Standard Specifications for Dry-set Portland Cement Mortar	R702.4.1
A118.3—99	American National Standard Specifications for Chemical Resistant, Water Cleanable Tile-setting and Grouting Epoxy and Water Cleanable Tile-setting Epoxy Adhesive	R702.4.1
A136.1—99	American National Standard Specifications for Organic Adhesives for Installation of Ceramic Tile	R702.4.1
A137.1—88	American National Standard Specifications for Ceramic Tile	R702.4.1
A208.1—99	Particleboard	R503.3.1, R605.1
LC1—97	Interior Fuel Gas Piping Systems Using Corrugated Stainless Steel Tubing —with Addenda LC 1a-1999 and LC 1b-2001	G2414.5.3
Z21.1—03	Household Cooking Gas Appliances—with Addenda Z21.1a-2003 and Z21.1b-2003	G2447.1

ANSI—continued

Z21.5.1—02	Gas Clothes Dryers—Volume I—Type I Clothes Dryers—with Addenda Z21.5.1a-2003	G2438.1
Z21.8—94(R2002)	Installation of Domestic Gas Conversion Burners	G2443.1
Z21.10.1—04	Gas Water Heaters—Volume I—Storage, Water Heaters with Input Ratings of 75,000 Btu per hour or Less	G2448.1
Z21.10.3—01	Gas Water Heaters—Volume III—Storage, Water Heaters with Input Ratings above 75,000 Btu per hour, Circulating and Instantaneous Water Heaters—with Addenda Z21.10.3a-2003 and Z21.10.3b-2004	G2448.1
Z21.11.2—02	Gas-fired Room Heaters—Volume II—Unvented Room Heaters—with Addenda Z21.11.2a-2003	G2445.1
Z21.13—04	Gas-fired Low-Pressure Steam and Hot Water Boilers	G2452.1
Z21.15—97(R2003)	Manually Operated Gas Valves for Appliances, Appliance Connector Valves and Hose End Valves —with Addenda Z21.15a-2001 (R2003)	G2420.1.1
Z21.22—99(R2003)	Relief Valves for Hot Water Supply Systems—with Addenda Z21.22a-2000 (R2003) and 21.22b-2001 (R2003)	P2803.2
Z21.24-97	Connectors for Gas Appliances	G2422.1
Z21.40.1—96(R2002)	Gas-fired Heat Activated Air Conditioning and Heat Pump Appliances—with Z21.40.1a-97 (R2002)	G2449.1
Z21.40.2—96(R2002)	Gas-fired Work Activated Air Conditioning and Heat Pump Appliances (Internal Combustion) —with Addenda Z21.40.2a-1997 (R2002)	G2449.1
Z21.42—93(R2002)	Gas-fired Illuminating Appliances	G2450.1
Z21.47—03	Gas-fired Central Furnaces	R617.1, G2442.1
Z21.50—03	Vented Gas Fireplaces—with Addenda Z21.50a-2003	G2434.1
Z21.56—01	Gas-fired Pool Heaters—with Addenda Z21.56a-2004 and Z21.56b—2004	R616.1, G2441.1
Z21.58—95(R2002)	Outdoor Cooking Gas Appliances—with Addenda Z21.58a-1998 (R2002) and Z21.58b-2002	R622.1, G2447.1
Z21.60—03	Decorative Gas Appliances for Installation in Solid Fuel Burning Fireplaces—with Addenda Z21.60a-2003	G2432.1
Z21.69—02	Connectors for Movable Gas Appliances with Addenda Z21.69a—2003	G2422.1
Z21.75/CSA 6.27—01	Connectors for Outdoor Gas Appliances	G2422.1
Z21.80—03	Line Pressure Regulators	G2421.1
Z21.84—02	Manually-listed, Natural Gas Decorative Gas Appliances for Installation in Solid Fuel Burning Fireplaces —with Addenda Z21.84a -2003	G2432.1, G2432.2
Z21.86—04	Gas-fired Vented Space Heating Appliances	G2436.1, G2437.1, G2446.1
Z21.88—02	Vented Gas Fireplace Heaters—with Addenda A21.88a-2003 and Z21.88b—2004	G2435.1
Z21.91—01	Ventless Firebox Enclosures for Gas-fired Unvented Decorative Room Heaters	G2445.7.1
Z83.6—90(R1998)	Gas-fired Infrared Heaters	G2442.1, G2449.1, G2451.1
Z83.8—02	Gas-fired Unit Heaters and Gas-fired Duct Furnaces—with Addenda Z83.8a-2003	G2444.1
Z124.1—95	Plastic Bathtub Units	Table P2701.1
Z124.2—95	Plastic Shower Receptors and Shower Stalls	Table P2701.1
Z124.3—95	Plastic Lavatories	Table P2701.1, P2711.1, P2711.2
Z124.4—96	Plastic Water Closet Bowls and Tanks	Table P2701.1, P2712.1
Z124.6—97	Plastic Sinks	Table P2701.1

APA

APA–The Engineered Wood Association
P. O. Box 11700
Tacoma, WA 98411-0700

Standard reference number	Title	Referenced in code section number
APA E30—03	Engineered Wood Construction Guide	R503.2.2, R803.2.2, R803.2.3

ASCE

American Society of Civil Engineers
1801 Alexander Bell Drive
Reston, VA 20191

Standard reference number	Title	Referenced in code section number
5—05	Building Code Requirements for Masonry Structures	R404.1, R606.1, R606.1.1, R606.12.1, R606.12.2.2.1, R606.12.2.2.2, R606.12.3.1
6—05	Specifications for Masonry Structures	R404.1, R606.1, R606.1.1, R606.12.1, R606.12.2.2.1, R606.12.2.2.2, R606.12.3.1
7—05	Minimum Design Loads for Buildings and Other Structures	R301.2.1.1
32—01	Design and Construction of Frost Protected Shallow Foundations	R403.1.4.1

ASHRAE

American Society of Heating, Refrigerating
and Air-Conditioning Engineers, Inc.
1791 Tullie Circle, NE
Atlanta, GA 30329

Standard reference number	Title	Referenced in code section number
34—2004	Designation and Safety Classification of Refrigerants	M1411.1
ASHRAE—2001	ASHRAE Fundamentals Handbook—2001	N1102.1.3, M1502.6, P3001.2, P3002.3, P3101.4, P3103.2

ASME

American Society of Mechanical Engineers
Three Park Avenue
New York, NY 10016-5990

Standard reference number	Title	Referenced in code section number
A17.1—2004	Safety Code for Elevators	R323.1
A18.1—2003	Safety Standard for Platforms and Stairway Chair Lifts	R323.2
A112.1.2—1991(R2002)	Air Gaps in Plumbing Systems	P2902.3.1, Table P2902.3
A112.1.3—2000	Air Gap Fittings for Use with Plumbing Fixtures, Appliances, and Appurtenances	Table P2701.1, P2902.3.1
A112.3.1—93	Performance Standard and Installation Procedures for Stainless Steel Drainage Systems for Sanitary, Storm and Chemical Applications Above and Below Ground	Table P3002.1(1), Table P3002.1(2), P3002.2, P3002.3
A112.3.4—2000	Macerating Toilet Systems and Related Components	Table P2701.1, P3007.2.1
A112.4.1—1993(R2002)	Water Heater Relief Valve Drain Tubes	P2803.6.2
A112.4.3—1999	Plastic Fittings for Connecting Water Closets to the Sanitary Drainage System	P3003.19, P3003.4.5
A112.6.1M—1997(R2002)	Floor Affixed Supports for Off-the-floor Plumbing Fixtures for Public Use	P2702.4, Table P2704.1
A112.6.2—2000	Framing-Affixed Supports for Off-the-floor Water Closets with Concealed Tanks	Table P2701.1, P2702.4
A112.6.3—2001	Floor and Trench Drains	Table P2701.1
A112.18.1—2003	Plumbing Fixture Fittings	Table P2701.1, P2708.4, P2722.1, P2902.2
A112.18.2—2002	Plumbing Fixture Waste Fittings	Table P2701.1, P2702.2
A112.18.3M—2002	Performance Requirements for Backflow Protection Devices and Systems in Plumbing Fixture Fittings	P2708.4, P2722.3
A112.18.6—2003	Flexible Water Connectors	P2904.7
A112.19.1M—1994(R1999)	Enameled Cast Iron Plumbing Fixtures—with 1998 and 2000 Supplements	Table P2701.1, P2711.1
A112.19.2—2003	Vitreous China Plumbing Fixtures—and Hydraulic Requirements for Water Closets and Urinals	Table P2701.1, P2705.1, P2711.1, P2712.1, P2712.2
A112.19.3M—2000	Stainless Steel Plumbing Fixtures (Designed for Residential Use)—with 2002 Supplement	Table P2701.1, P2705.1, P2711.1
A112.19.4M—1994(R1999)	Porcelain Enameled Formed Steel Plumbing Fixtures—with 1998 and 2000 Supplements	Table P2701.1, P2711.1
A112.19.5—1999	Trim for Water-closet Bowls, Tanks, and Urinals	Table P2701.1
A112.19.6—1995	Hydraulic Performance Requirements for Water Closets and Urinals	Table P2701.1, P2712.1, P2712.2
A112.19.7M—1995	Whirlpool Bathtub Appliances	Table P2701.1
A112.19.8M—1987(R1996)	Suction Fittings for Use in Swimming Pools, Wading Pools, Spas, Hot Tubs, and Whirlpool Bathtub Appliances	Table P2701.1
A112.19.9M—1991(R2002)	Non-vitreous Ceramic Plumbing Fixtures—with 2002 Supplement	Table P2701.1, P27.11.1, P2712.1
A112.19.12—2000	Wall Mounted and Pedestal Mounted, Adjustable and Pivoting Lavatory and Sink Carrier Systems	Table P2701.1, P2711.4, P2714.2
A112.19.13—2001	Electrohydraulic Water Closets	P2712.9
A112.19.15—2001	Bathtub/Whirlpool Bathtubs with Pressure Sealed Doors	Table P2701.1, P2713.2
B1.20.1—1983(R2001)	Pipe Threads, General Purpose (Inch)	G2414.9, P3003.3.3, P3003.5.3, P3003.10.4, P3003.12.1, P3003.14.3
B16.3—1998	Malleable Iron Threaded Fittings Classes 150 and 300	Table P2904.6
B16.4—1998	Gray-iron Threaded Fittings Classes 125 and 250	Table P2904.6
B16.9—2003	Factory-made Wrought Steel Buttwelding Fittings	Table P2904.6
B16.11—2001	Forged Fittings, Socket-welding and Threaded	Table P2904.6
B16.12—1998	Cast Iron Threaded Drainage Fittings	Table P2904.6
B16.15—1985(R1994)	Cast Bronze Threaded Fittings	Table P2904.6
B16.18—2001	Cast Copper Alloy Solder Joint Pressure Fittings	Table P2904.6
B16.22—2001	Wrought Copper and Copper Alloy Solder Joint Pressure Fittings	Table P2904.6
B16.23—2002	Cast Copper Alloy Solder Joint Drainage Fittings (DWV)	Table P2904.6, Table P3002.3
B16.26—1988	Cast Copper Alloy Fittings for Flared Copper Tubes	Table P2904.6
B16.28—1994	Wrought Steel Buttwelding Short Radius Elbows and Returns	Table P2904.6
B16.29—2001	Wrought Copper and Wrought Copper Alloy Solder Joint Drainage Fittings-DWV	Table P2904.6 Table P3002.3
B16.33—2002	Manually Operated Metallic Gas Valves for Use in Gas Piping Systems up to 125 psig (Sizes $\frac{1}{2}$ through 2)	G2420.1.1
B16.44-01	Manually Operated Metallic Gas Valves For Use in House Piping Systems	Table G2420.1.1
B36.10M—2000	Welded and Seamless Wrought-steel Pipe	G2414.4.2

| BPVC—2001 | ASME Boiler and Pressure Vessel Code (2001 Edition) (Sections I, II, IV, V, VI & IX) .... M2001.1.1, G2452.1 |
| CSD-1—2002 | Controls and Safety Devices for Automatically Fired Boilers................................... M2001.1.1, G2452.1 |

ASSE

American Society of Sanitary Engineering
28901 Clemens Road, Suite A
Westlake, OH 44145

Standard reference number	Title	Referenced in code section number
1001—02	Performance Requirements for Atmospheric Type Vacuum Breakers .... Table P2902.3, P2902.3.2	
1002—99	Performance Requirements for Antisiphen Fill Valves (Ballcocks) for Gravity Water Closet Flush Tank .... Table P2701.1, Table P2902.3, P2902.4.1	
1003—01	Performance Requirements for Water Pressure Reducing Valves.. P2903.3.1	
1006—89	Performance Requirements for Residential Use Dishwashers .... Table P2701.1	
1007—92	Performance Requirements for Home Laundry Equipment....................................... Table P2701.1	
1008—89	Performance Requirements for Household Food Waste Disposer Units .... Table P2701.1	
1010—96	Performance Requirements for Water Hammer Arresters P2903.5	
1011—93	Performance Requirements for Hose Connection Vacuum Breakers........................ Table P2902.3, P2902.3.2	
1012—02	Performance Requirements for Backflow Preventers with Intermediate Atmospheric Vent .... Table P2902.3 P2902.3.3, P2902.5.1, P2902.5.5	
1013—99	Performance Requirements for Reduced Pressure Principle Backflow Preventers and Reduced Pressure Fire Protection Principle Backflow Preventers......... Table P2902.3, P2902.5.5, P2902.5.1, P2902.5.5	
1014—90	Performance Requirements for Hand-held Shower .... Table P2701.1	
1015—99	Performance Requirements For Double Check Backflow Prevention Assemblies and Double Check Fire Protection Backflow Prevention Assemblies .... Table P2902.3, P2902.3.6	
1016—96	Performance Requirements for Individual Thermostatic, Pressure Balancing and Combination Control Valves for Bathing Facilities .... Table P2701.1, P2708.3, P2722.2	
1017—99	Performance Requirements for Temperature Actuated Mixing Valves for Hot Water Distribution Systems P2802.2	
1019—97	Performance Requirements for Wall Hydrants, Freezeless, Automatic Automatic Draining, Anti-backflow Types .... Table P2701.1, Table P2902.3	
1020—98	Performance Requirements for Pressure Vacuum Breaker Assembly .... Table P2902.3, P2902.3.4	
1023—79	Performance Requirements for Hot Water Dispensers Household Storage Type Electrical .... Table P2701.1	
1024—04	Performance Requirements for Dual Check Valve Type Backflow Preventers .... Table P2902.3	
1025—78	Performance Requirements for Diverters for Plumbing Faucets with Hose Spray, Anti-siphon Type, Residential Applications .... Table P2701.1	
1035—02	Performance Requirements for Laboratory Faucet Backflow Preventers .... Table P2902.3, P2902.3.2	
1037—90	Performance Requirements for Pressurized Flushing Devices for Plumbing Fixtures...................... Table P2701.1	
1047—99	Performance Requirements for Reduced Pressure Detector Fire Protection Backflow Prevention Assemblies .... P2902.3.5, Table P2902.3	
1048—99	Performance Requirements for Double Check Detector Fire Protection Backflow Prevention Assemblies .... P2902.3.6, Table P2902.3	
1050—02	Performance Requirements for Stack Air Admittance Valves for Sanitary Drainage Systems.................... P3114.1	
1051—02	Performance Requirements for Individual and Branch Type Air Admittance Valves for Plumbing Drainage Systems P3114.1	
1052—93	Performance Requirements for Hose Connection Backflow Preventers........... Table P2701.1, Table P2902.3, P2902.3.2	
1056—01	Performance Requirements for Spill Resistant Vacuum....................... Table P2902.3, P2902.3.4	
1062—97	Performance Requirements for Temperature Actuated, Flow Reduction Valves to Individual Fixture Fittings .... Table P2701.1, P2724.1	
1066—97	Performance Requirements for Individual Pressure Balancing Valves for Individual Fixture Fittings .. Table P2701.1, P2722.4	
1070—04	Performance Requirements for Water Temperature Limiting Devices P2713.3, P2721.2	

ASTM

ASTM International
100 Barr Harbor Drive
West Conshohocken, PA 19428

Standard reference number	Title	Referenced in code section number
A 36/A 36M—04	Specification for Carbon Structural Steel... R606.15	
A 53/A 53M—02	Specification for Pipe, Steel, Black and Hot-dipped, Zinc-coated Welded and Seamless .... Table M2101.1, G2414.4.2, Table P2904.4.1, Table P2904.5, Table P3002.1(1)	
A 74—04	Specification for Cast Iron Soil Pipe and Fittings................... Table P3002.1(1), Table P3002.1(2), Table P3002.2 Table P3002.3, P3005.2.9	

ASTM—continued

2003 INTERNATIONAL RESIDENTIAL CODE®

ASTM—continued

B 828—02	Practice for Making Capillary Joints by Soldering of Copper and Copper Alloy Tube and Fittings	P2904.13, P3003.10.3, P3003.11.3
C 5—03	Specification for Quicklime for Structural Purposes	R702.2
C 14—03	Specification for Concrete Sewer, Storm Drain and Culvert Pipe	Table P3002.2
C 27—98(2002)	Specification for Standard Classification of Fireclay and High-alumina Refractory Brick	R1001.5, R1001.8
C 28/C28M—00e01	Specification for Gypsum Plasters	R702.2
C 34—03	Specification for Structural Clay Load-Bearing Wall Tile	Table R301.2(1)
C 35—95(2001)	Specification for Inorganic Aggregates for Use in Gypsum Plaster	R702.2
C 36/C 0036M—03	Specification for Gypsum Wallboard	R702.3.1
C 37/C 0037M—01	Specification for Gypsum Lath	R702.2
C 55—03	Specification for Concrete Brick	R202, Table R301.2(1)
C 59/C 0059M—00	Specification for Gypsum Casting and Molding Plaster	R702.2
C 61/C 0061M—00	Specification for Gypsum Keene's Cement	R702.2
C 62—04	Specification for Building Brick (Solid Masonry Units Made from Clay or Shale)	R202, Table R301.2(1)
C 67-03ae01	Test Methods of Sampling and Testing Brick and Structural Clay Tile	R905.3.5
C 73-99a	Specification for Calcium Silicate Face Brick (Sand Lime Brick)	R202, Table R301.2(1)
C 76—04a	Specification for Reinforced Concrete Culvert, storm Drain, and Sewer Pipe	Table P3002.2
C79/C 79—04a	Specification for Treated Core and Nontreated Core Gypsum Sheathing Board	Table R602.3(1), R702.3.1
C 90—03	Specification for Load-bearing Concrete Masonry Units	Table R301.2(1)
C 129—03	Specification for Nonload-bearing Concrete Masonry Units	Table R301.2(1)
C140—03	Test Methods of Sampling and Testing Concrete Masonry Units and Related Units	R905.3.5
C 143/C 0143M—03	Test Method for Slump or Hydraulic Cement Concrete	R404.4.5, R611.6.1
C 145—85	Specification for Solid Load-bearing Concrete Masonry Units	R202, Table R301.2(1)
C 199—84(2000)	Test Method for Pier Test for Refractory Mortar	R1001.9, R1003.5, R1003.8
C 207—04	Specification for Hydrated Lime for Masonry Purposes	Table R607.1
C 208—95(2001)	Specification for Cellulosic Fiber Insulating Board	Table R602.3(1)
C 216—04a	Specification for Facing Brick (Solid Masonry Units Made from Clay or Shale)	R202, Table R301.2(1)
C 270—04	Specification for Mortar for Unit Masonry	R607.1
C 296—00	Specification for Asbestos Cement Pressure Pipe	Table P2904.4
C 315—02	Specification for Clay Flue Linings	Table R1001.11(1), Table R1001.11(2), R1001.8.1, G2425.12
C 406—00	Specifications for Roofing Slate	R905.6.4
C 411—97	Test Method for Hot-surface Performance of High-temperature Thermal Insulation	M1601.2.1
C 425—04	Specification for Compression Joints for Vitrified Clay Pipe and Fittings	Table P3002.2, P3003.15, P3003.18
C428—97(2002)	Specification for Asbestos-Cement Nonpressure Sewer Pipe	Table P3002.2
C 443—03	Specification for Joints for Concrete Pipe and Manholes, Using Rubber Gaskets	P3003.7, P3003.18
C 475/C 475M—02	Specification for Joint Compound and Joint Tape for Finishing Gypsum Wallboard	R702.3.1
C 476—02	Specification for Grout for Masonry	R609.1.1
C 514—01	Specification for Nails for the Application of Gypsum Wallboard	R702.3.1
C552—03	Standard Specification for Cellular Glass Thermal Insulation	Table R906.2
C 557—03	Specification for Adhesives for Fastening Gypsum Wallboard to Wood Framing	R702.3.1
C564—04a	Specification for Rubber Gaskets for Cast Iron Soil Pipe and Fittings	P3003.6.2, P3003.6.3, P3003.18
C 578—04	Specification for Rigid, Cellular Polystyrene Thermal Insulation	R403.3, Table R906.2
C 587—02	Specification for Gypsum Veneer Plaster	R702.2
C 588/C 588M—01	Specification for Gypsum Base for Veneer Plasters	R702.2
C 630/0630M—03	Specification for Water-resistant Gypsum Backing Board	R702.3.1, R702.3.8
C 631—95a (2000)	Specification for Bonding Compounds for Interior Gypsum Plastering	R702.2
C 645—04	Specification for Nonstructural Steel Framing Members	R702.3.3
C 652—04a	Specification for Hollow Brick (Hollow Masonry Units Made from Clay or Shale)	R202, Table R301.2(1)
C 700—02	Specification for Vitrified Clay Pipe, Extra Strength, Standard Strength, and Perforated	Table P3002.2
C 728—97[EI]	Standard Specification for Perlite Thermal Insulation Board	Table R906.2
C 836—03	Specification for High Solids Content, Cold Liquid-Applied Elastomeric Waterproofing Membrane for Use with Separate Wearing Course	R905.15.2
C 843—99e01	Specification for Application of Gypsum Veneer Plaster	R702.2
C 844—99	Specification for Application of Gypsum Base to Receive Gypsum Veneer Plaster	R702.2
C 847—(2000)	Specification for Metal Lath	R702.2
C 887—79(2001)	Specification for Packaged, Dry, Combined Materials for Surface Bonding Mortar	R406.1
C 897—00	Specification for Aggregate for Job-mixed Portland Cement-based Plasters	R702.2
C926—98a	Specification for Application of Portland Cement Based-Plaster	R703.6
C 931/C 931M—04	Specification for Exterior Gypsum Soffit Board	R702.3.1

ASTM—continued

ASTM—continued

D 2104—03	Specification for Polyethylene (PE) Plastic Pipe, Schedule 40	Table P2904.4
D 2178—97a	Specification for Asphalt Glass Felt Used in Roofing and Waterproofing	Table R905.9.2
D 2235—01	Specification for Solvent Cement for Acrylonitrile-Butadiene-Styrene (ABS) Plastic Pipe and Fittings	P2904.9.1.1, Table P3002.1, Table P3002.2 P3003.3.2, P3003.8.2
D 2239—03	Specification for Polyethylene (PE) Plastic Pipe (SIDR-PR) Based on Controlled Inside Diameter	Table P2904.4
D 2241—04a	Specification for Poly (Vinyl Chloride) (PVC) Pressure-rated Pipe (SDR-Series)	Table P2904.4
D 2282—99e01	Specification for Acrylonitrile-Butadiene-Styrene (ABS) Plastic Pipe (SDR-PR)	Table P2904.4
D 2412—02	Test Method for Determination of External Loading Characteristics of Plastic Pipe by Parallel-plate Loading	M1601.1.2
D 2447—03	Specification for Polyethylene (PE) Plastic Pipe Schedules 40 and 80, Based on Outside Diameter	Table M2101.1
D 2464—99	Specification for Threaded Poly (Vinyl Chloride) (PVC) Plastic Pipe Fittings, Schedule 80	Table P2904.6
D 2466—02	Specification for Poly(Vinyl Chloride) (PVC) Plastic Pipe Fittings, Schedule 40	Table P2904.6
D 2467—04	Specification for Poly (Vinyl Chloride) (PVC) Plastic Pipe Fittings, Schedule 80	Table P2904.6
D 2468—96a	Specification for Acrylonitrile-Butadiene-Styrene (ABS) Plastic Pipe Fittings, Schedule 40	Table P2904.6
D 2513—04a	Specification for Thermoplastic Gas Pressure Pipe, Tubing, and Fittings	Table M2101.1, M2104.2.1.3, G2414.6, G2414.6.1, G2414.11, G2415.14.3
D 2564—02	Specification for Solvent Cements for Poly (Vinyl Chloride) (PVC) Plastic Piping Systems	P2904.9.1.3, Table P3002.1, Table P3002.2 P3003.9.2, P3003.14.2
D 2609—02	Specification for Plastic Insert Fittings for Polyethylene (PE) Plastic Pipe	Table P2904.6
D 2626—04	Specification for Asphalt-Saturated and Coated Organic Felt Base Sheet Used in Roofing	R905.3.3, Table R905.9.2
D 2657—97	Standard Practice for Heat Fusion-joining of Polyolefin Pipe Fittings	P2904.3.1, P3003.17.1
D 2661—02	Specification for Acrylonitrile-Butadiene-Styrene (ABS) Schedule 40 Plastic Drain, Waste, and Vent Pipe and Fittings	Table P3002.1(1), Table P3002.1(2), Table P3002.2, Table P3002.3, Table P3002.4, P3003.3.2, P3003.8.2
D 2662—96a	Specification for Polybutylene (PB) Plastic Pipe (SDR-PR) Based on Controlled Inside Diameter	Table P2904.4
D 2665—04ae01	Specification for Poly (Vinyl Chloride) (PVC) Plastic Drain, Waste, and Vent Pipe and Fittings	Table P3002.1(1) Table P3002.1(2), Table P3002.2, P3002.3, Table P3002.4
D 2666—96a(2003)	Specification for Polybutylene (PB) Plastic Tubing	Table P2904.4
D 2672—03	Specification for Joints for IPS PVC Pipe Using Solvent Cement	Table P3002.1, Table, P3002.2, Table P2904.4
D 2683—98	Specification for Socket-Type Polyethylene Fittings for Outside Diameter-controlled Polyethylene Pipe and Tubing	Table M2101.1, M2104.2.1.1
D 2737—03	Specification for Polyethylene (PE) Plastic Tubing	Table P2904.4
D 2751—96a	Specification for Acrylonitrile-Butadiene-Styrene (ABS) Sewer Pipe and Fittings	Table P3002.2, Table P3002.3
D 2822—91(1997)e01	Specification for Asphalt Roof Cement	Table R905.9.2, Table R905.11.2
D 2823—90(1997)e[1]	Specification for Asphalt Roof Coatings	Table R905.9.2, Table R905.11.2
D 2824—04	Specification for Aluminum-Pigmented Asphalt Roof Coatings, Non-fibered, Asbestos Fibered, and Fibered without Asbestos	Table R905.9.2, Table R905.11.2
D 2837—04	Test Method for Obtaining Hydrostatic Design Basis for Thermoplastic Pipe Materials	Table M2101.1
D 2846/D 2846M—99	Specification for Chlorinated Poly (Vinyl Chloride) (CPVC) Plastic Hot- and Cold-water Distribution Systems	Table M2101.1, Table P2904.4, Table P2904.5, P2904.9.1.2
D 2855-96(2002)	Standard Practice for Making Solvent-Cemented Joints with Poly (Vinyl Chloride) (PVC) Pipe and Fittings	P3003.9.2, P3003.14.2
D 2898—94(1999)	Test Methods for Accelerated Weathering of Fire-retardant-treated Wood for Fire Testing	R802.1.3.3, R902.2
D 2949—01a	Specification for 3.25-in. Outside Diameter Poly (Vinyl Chloride) (PVC) Plastic Drain, Waste, and Vent Pipe and Fittings	Table P3002.1(1), Table P3002.1(2), Table P3002.2, Table P3002.3
D 3019—e01	Specification for Lap Cement Used with Asphalt Roll Roofing, Non-fibered, Asbestos Fibered, and Non-asbestos Fibered	Table R905.9.2, Table R905.11.2
D 3034—04	Specification for Type PSM Poly (Vinyl Chloride) (PVC) Sewer Pipe and Fittings	Table P3002.2, P3002.3, Table P3002.4
D 3035—03a	Specification for Polyethylene (PE) Plastic Pipe (DR-PR) Based On Controlled Outside Diameter	Table M2101.1
D 3161—03b	Test Method for Wind Resistance of Asphalt Shingles (Fan Induced Method)	R905.2.4.1, R905.2.6
D 3201—94(2003)	Test Method for Hygroscopic Properties of Fire-retardant Wood and Wood-base Products	R802.1.3.4
D 3212—96a(2003)	Specification for Joints for Drain and Sewer Plastic Pipes Using Flexible Elastomeric Seals	Table P3002.2, P3003.3.1, P3003.8.1, P3003.9.1, P3003.14.1, P3003.17.2
D 3309—96a(2002)	Specification for Polybutylene (PB) Plastic Hot- and Cold-water Distribution Systems	Table M2101.1, Table P2904.4, Table P2904.5
D 3311—02	Specification for Drain, Waste, and Vent (DWV) Plastic Fittings Patters	P3002.3
D 3350—02a	Specification for Polyethylene Plastic Pipe and Fitting Materials	Table M2101.1
D 3462—04	Specification for Asphalt Shingles Made From Glass Felt and Surfaced with Mineral Granules	R905.2.4
D 3468—99	Specification for Liquid-applied Neoprene and Chlorosulfanated Polyethylene Used in Roofing and Waterproofing	Table R905.15.2
D 3679—04	Specification for Rigid Poly (Vinyl Chloride) (PVC) Siding	Table R703.4

D 3737—03	Practice for Establishing Allowable Properties for Structural Glued Laminated Timber (Glulam)	R502.1.5, R602.1.2, R802.1.4
D 3747—79(2000)e01	Specification for Emulsified Asphalt Adhesive for Adhering Roof Insulation	Table R905.9.2, Table R905.11.2
D 3909—97b	Specification for Asphalt Roll Roofing (Glass Felt) Surfaced with Mineral Granules	R905.2.8.2, R905.3.3, R905.5.4, Table R905.9.2, Table R906.3.2,
D 3957—03	Standard Practices for Establishing Stress Grades for Structural Members Used in Log Buildings	R502.1.6, R602.1.3, R802.1.5
D 4022—94(2000)e01	Specification for Coal Tar Roof Cement, Asbestos Containing	Table R905.9.2
D 4068—01	Specification for Chlorinated Polyethylene (CPE) Sheeting for Concealed Water Containment Membrane	P2709.2.2
D 4318—00	Test Methods for Liquid Limit, Plastic Limit and Plasticity Index of Soils	R403.1.7.5.1
D 4434—04	Specification for Poly (Vinyl Chloride) Sheet Roofing	R905.13.2
D 4479—00	Specification for Asphalt Roof Coatings-Asbestos-free	Table R905.9.2
D 4551—96(2001)	Specification for Poly (Vinyl) Chloride (PVC) Plastic Flexible Concealed Water-containment Membrane	P2709.2.1
D 4586—00	Specification for Asphalt Roof Cement-Asbestos-free	Table R905.9.2
D 4601—98	Specification for Asphalt-coated Glass Fiber Base Sheet Used in Roofing	Table R905.9.2
D 4637—04	Specification for EPDM Sheet Used in Single-ply Roof Membrane	R905.12.2
D 4829—03	Test Method for Expansion Index of Soils	R403.1.8.1
D 4869—04	Specification for Asphalt-Saturated (Organic Felt) Underlayment Used in Steep Slope Roofing	R905.2.3, R905.4.3, R905.5.3, R905.6.3, R905.7.3, R905.8.3
D 4897—01	Specification for Asphalt Coated Glass-fiber Venting Base Sheet Used in Roofing	Table R905.9.2
D 4990—97a	Specification for Coal Tar Glass Felt Used in Roofing and Waterproofing	Table R905.9.2
D 5019-96 e01	Specification for Reinforced Non-Vulcanized Polymeric Sheet Used in Roofing Membrane	R905.12.2
D 5055—04	Specification for Establishing and Monitoring Structural Capacities of Prefabricated Wood I-Joists	R502.1.4
D 5516—03	Test Method for Evaluating the Flexural Properties of Fire-Retardant-treated Softwood Plywood Exposed to the Elevated Temperatures	R802.1.3.2.1
D 5643—94(2000)e01	Specification for Coal Tar Roof Cement Asbestos-free	Table R905.9.2
D 5664—02	Test Methods For Evaluating the Effects of Fire-Retardant Treatments and Elevated Temperatures on Strength Properties of Fire-retardant-treated Lumber	R802.1.3.2.2
D 5665—99a	Specification for Thermoplastic Fabrics Used in Cold-applied Roofing and Waterproofing	Table R905.9.2
D 5726—98	Specification for Thermoplastic Fabrics Used in Hot-applied Roofing and Waterproofing	Table R905.9.2
D 6083—97a	Specification for Liquid Applied Acrylic Coating Used in Roofing	Table R905.9.2, Table R905.11.2, Table R905.15.2
D 6162—00a	Specification for Styrene Butadiene Styrene (SBS) Modified Bituminous Sheet Materials Using a Combination of Polyester and Glass Fiber Reinforcements	Table R905.11.2
D 6163—00e01	Specification for Styrene Butadiene Styrene (SBS) Modified Bituminous Sheet Materials Using Glass Fiber Reinforcements	Table R905.11.2
D 6164—00	Specification for Styrene Butadiene Styrene (SBS) Modified Bituminous Sheet Materials Using Polyester Reinforcements	Table R905.11.2
D 6221—00	Specification for Reinforced Bituminous Flashing Sheets for Roofing and Waterproofing	Table R905.11.2
D 6222—02	Specification for Atactic Polypropelene (APP) Modified Bituminous Sheet Materials Using Polyester Reinforcement	Table R905.11.2
D 6223—02	Specification for Atactic Polypropelene (APP) Modified Bituminous Sheet Materials Using a Combination of Polyester and Glass Fiber Reinforcement	Table R905.11.2
D 6298—00	Specification for Fiberglass Reinforced Styrene-Butadiene-Styrene (SBS) Modified Bituminous Sheets with a Factory Applied Metal Surface	Table R905.11.2
D 6305—02e01	Practice for Calculating Bending Strength Design Adjustment Factors for Fire-Retardant-Treated Plywood Roof Sheathing	R802.1.3.2.1
D 6380—01[E1]	Standard Specification for Asphalt Roll Roofing (Organic Felt)	R905.2.8.2
D 6694—01	Standard Specification Liquid-Applied Silicone Coating Used in spray Polurethane Foam roofing[1]	R905.15.2
D 6754—02	Standard Specification for Ketone Ethylene ester Based Sheet Roofing[1]	R905.13.2
D 6757—02	Standard Specification for Inorganic Underlayment for Use with Steep Slope Roofing Products[1]	R905.2.3
D 6841—03	Standard Practice for Calculating Design Value Treatment Adjustment Factors for Fire-retardant-treated Lumber	R802.1.3.2.2
D 6878—03	Standard Specification for Thermoplastic Polyolefin Based Sheet Roofing[1]	R905.13.2
E 84—04	Test Method for Surface Burning Characteristics of Building Materials	R202, R314.1.1, R314.2.6, R314.3, R315.3, R316.1, R316.2, R802.1.3, M1601.2.1, M1601.4.2
E 96—00e01	Test Method for Water Vapor Transmission of Materials	R202, R806.4, M1411.4, M1601.3.4
E 108—04	Test Methods for Fire Tests of Roof Coverings	R902.1, R902.2
E 119—00	Test Methods for Fire Tests of Building Construction and Materials	R314.1.2, R317.1, R317.3.1
E 136—99e01	Test Method for Behavior of Materials in a Vertical Tube Furnace at 750°C	R202
E 283—04	Test Method for determining the Rate of Air Leakage Through Exterior Windows, Curtain Walls, and Doors Under specified Pressure Differences Across the Specimen	R806.4

ASTM—continued

E 330—02	Test Method for Structural Performance of Exterior Windows, Curtain Walls, and Doors by Uniform Static Air Pressure Difference	R613.3
E 331-00	Test Method for Water Penetration of Exterior Windows, Skylights, Doors and Curtain Walls by uniform Static Air Pressure Difference	R703.1
E 814—02	Test Method for Fire Tests of Through-Penetration Firestops	R317.3.1.2
E 970—00	Test Method for Critical Radiant Flux of Exposed Attic Floor Insulation Using a Radiant Heat Energy Source	R316.5
E 1509—04	Standard Specification for Room Heaters, Pellet Fuel-burning Type	M1410.1
E 1602—03	Guide for Construction of Solid Fuel Burning Masonry Heaters	R1002.2
E 1886—04	Test Method for Performance of Exterior Windows, Curtain Walls, Doors and Storm Shutters Impacted by Missles and Exposed to Cyclic Pressure Differentials	R301.2.1.2, R613.7.1
E 1996—04	Specification for Performance of Exterior Windows, Curtain Walls, Doors and Storm Shutters Impacted by Windborne Debris in Hurricanes	R301.2.1.2, R613.7.1
E 2231—04	Standard Practice for Specimen Preparation and Mounting of Pipe and Duct Insulation Materials to Assess Surface Burning Characteristics	M1601.2.1
F 409—02	Specification for Thermoplastic Accessible and Replaceable Plastic Tube and Tubular Fittings	Table P2701.1, P2702.2, P2702.3
F 437—99	Specification for Threaded Chlorinated Poly (Vinyl Chloride) (CPVC) Plastic Pipe Fittings, Schedule 80	Table P2904.6
F 438—04	Specification for Socket-type Chlorinated Poly (Vinyl Chloride) (CPVC) Plastic Pipe Fittings, Schedule 40	Table P2904.6
F 439—02e01	Specification for Socket-type Chlorinated Poly (Vinyl Chloride) (CPVC) Plastic Pipe Fittings, Schedule 80	Table P2904.6
F 441/F 441M—02	Specification for Chlorinated Poly (Vinyl Chloride) (CPVC) Plastic Pipe, Schedules 40 and 80	Table P2904.4, Table P2904.5
F 442/F 442M—99	Specification for Chlorinated Poly (Vinyl Chloride) (CPVC) Plastic Pipe (SDR-PR)	Table P2904.4, Table P2904.5
F 477—02e01	Specification for Elastomeric Seals (Gaskets) for joining Plastic Pipe	P2904.17, P3003.18
F 493—04	Specification for Solvent Cements for Chlorinated Poly (Vinyl Chloride) (CPVC) Plastic Pipe and Fittings	P2904.9.1.2
F 628—01	Specification for Acrylonitrile-Butadiene-Styrene (ABS) Schedule 40 Plastic Drain, Waste, and Vent Pipe with a Cellular Core	Table 3002.1(1), Table P3002.1(2), Table P3002.2, Table P3002.3, P3003.3.2, P3003.8.2
F 656—02	Specification for Primers for Use in Solvent Cement Joints of Poly (Vinyl Chloride) (PVC) Plastic Pipe and Fittings	P2904.9.1.3, Table P3002.1, Table P3002.2, P3003.9.2, P3003.14.2
F 714—03	Specification for Polyethylene (PE) Plastic Pipe (SDR-PR) Based on Outside Diameter	Table P3002.2
F 789—95a	Specification for Type PS-46 and Type PS-115 Poly (Vinyl Chloride) (PVC) Plastic Gravity Flow Sewer Pipe and Fittings	Table P3002.2
F 876—04	Specification for Cross-linked Polyethylene (PEX) Tubing	Table M2101.1, Table P2904.4
F 877—02e01	Specification for Cross-linked Polyethylene (PEX) Plastic Hot- and Cold-water Distribution Systems	Table M2101.1, Table P2904.4, P2904.9.1.4.2, Table P2904.5, Table 2904.6
F 891—00e01	Specification for Coextruded Poly (Vinyl Chloride) (PVC) Plastic Pipe with a Cellular Core	Table P3002.1(1), Table P3002.1(2), Table P3002.2, Table 3002.3, Table P3002.4
F 1055—98e01	Specification for Electrofusion Type Polyethylene Fittings for Outside Diameter Controlled Polyethylene Pipe and Fittings	Table M2101.1, M2104.2.1.2
F 1281—03	Specification for Crosslinked Polyethylene/Aluminum/Crosslinked Polyethylene (PEX-AL-PEX) Pressure Pipe	Table M2101.1, Table P2904.4, Table P2904.5
F 1282—03	Specification for Polyethylene/Aluminum/Polyethylene (PE-AL-PE) Composite Pressure Pipe	Table P2904.4, Table P2904.5
F 1412—01	Specification for Polyolefin Pipe and Fittings for Corrosive Waste Drainage	Table P3002.2, Table P3002.3, P3003.16.1
F 1488—03	Specification for Coextruded Composite Pipe	Table P3002.1(1), Table P3002.1(2), Table P3002.2
F 1667—03	Specification for Driven Fasteners, Nails, Spikes, and Staples	R905.2.5
F 1807—04	Specification for Metal Insert Fittings Utilizing a Copper Crimp Ring for SDR9 Cross-linked Polyethylene (PEX) Tubing	Table M2101.1, Table P2904.6, P2904.9.1.4.2
F 1866—98	Specification for Poly (Vinyl Chloride) (PVC) Plastic Schedule 40 Drainage and DWV Fabricated Fittings	Table P3002.4
F 1960—04	Specification for Cold Expansion Fittings with PEX Reinforcing Rings for Use with Cross-linked Polyethylene (PEX) Tubing	Table M2101.1, Table P2904.6, P2904.9.1.4.2
F 1974—04	Specification for Metal Insert Fittings for Polyethylene/Aluminum/Polyethylene and Crosslinked Polyethylene/Aluminum/Crosslinked Polyethylene Composite Pressure Pipe	Table P2904.6
F 1986—00a	Multilayer Pipe Type 2, Compression Joints for Hot and Cold Drinking Water Systems	Table P2904.4, Table P2904.5, Table P2904.6
F 2006—00	Standard/Safety Specification for Window Fall Prevention Devices for Non-Emergency Escape (Egress) and Rescue (Ingress) Windows	R613.2
F 2080—04	Specification for Cold-expansion Fittings with Metal Compression-Sleeves for Crosslinked Polyethylene (PEX) Pipe	P2904.6, P2904.9.1.4.2

ASTM—continued

F 2090—01A	Specification for Window Fall Prevention Devices —with Emergency Escape (Egress) Release Mechanisms	613.2
F 2098—01	Standard Specification for Stainless Steel Clamps for SDR9 PEX Tubing to Metal Insert Fittings	Table M2101.1
F 2389—04	Standard for Pressure-rated Polypropylene (PP) Piping Systems	Table M2101.1, Table P2904.4, Table P2904.5, Table P2904.6, P2904.10.1

AWS

American Welding Society
550 N. W. LeJeune Road
Miami, FL 33126

Standard reference number	Title	Referenced in code section number
A5.8—04	Specifications for Filler Metals for Brazing and Braze Welding	P3003.5.1, P3003.10.1, P3003.11.1

AWPA

American Wood-Preservers' Association
P.O. Box 5690
Granbury, Texas 76049

Standard reference number	Title	Referenced in code section number
C1—00	All Timber Products—Preservative Treatment by Pressure Processes	R902.2
C33—00	Standard for Preservative Treatment of Structural Composite Lumber by Pressure Processes	R324.1
M4—02	Standard for the Care of Preservative-treated Wood Products	R319.1.1, R320.1.2, R320.3.1
U1—04	USE CATEGORY SYSTEM: User Specification for Treated Wood Except Section 6 Commodity Specification H	R319.1, R324.1.7, R402.1.2, R504.3, Table R905.8.5

AWWA

American Water Works Association
6666 West Quincy Avenue
Denver, CO 80235

Standard reference number	Title	Referenced in code section number
C104—98	Standard for Cement-Mortar Lining for Ductile-iron Pipe and Fittings for Water	P2904.4
C110—98	Standard for Ductile-iron and Gray-iron Fittings, 3 Inches through 48 Inches, for Water	Table P2904.6, Table P3002.3
C115—99	Standard for Flanged Ductile-iron Pipe with Ductile-iron or Gray-iron Threaded Flanges	Table P2904.4
C151/A21.51—02	Standard for Ductile-iron Pipe, Centrifugally Cast, for Water	Table P2904.4
C153—00	Standard for Ductile-iron Compact Fittings for Water Service	Table P2904.6.1
C510—00	Double Check Valve Backflow Prevention Assembly	Table P2902.3
C511—00	Reduced-Pressure Principle Backflow Prevention Assembly	Table P2902.3, P2902.3.5, P2902.5.1

CGSB

Canadian General Standards Board
Place du Portage 111, 6B1
11 Laurier Street
Gatineau, Quebec, Canada KIA 1G6

Standard reference number	Title	Referenced in code section number
37-GP—52M—(1984)	Roofing and Waterproofing Membrane, Sheet Applied, Elastomeric	R905.12.2
37-GP—56M—(1980)	Membrane, Modified Bituminous, Prefabricated and Reinforced for Roofing— with December 1985 Amendment	Table R905.11.2
CAN/CGSB-37.54—95	Polyvinyl Chloride Roofing and Waterproofing Membrane	R905.13.2

CISPI

Cast Iron Soil Pipe Institute
5959 Shallowford Road, Suite 419
Chattanooga, TN 37421

Standard reference number	Title	Referenced in code section number
301—04	Standard Specification for Hubless Cast Iron Soil Pipe and Fittings for Sanitary and Storm Drain, Waste and Vent Piping Applications . . . . Table P3002.1(1), Table P3002.1(2), Table P3002.2, Table P3002.3, Table 3002.4, P3005.2.9	
310—04	Standard Specification for Coupling for Use in Connection with Hubless Cast Iron Soil Pipe and Fittings for Sanitary and Storm Drain, Waste, and Vent Piping Applications . . . . Table P3002.1, Table P3002.2, P3003.6.3	

CPSC

Consumer Product Safety Commission
4330 East West Highway
Bethesda, MD 20814-4408

Standard reference number	Title	Referenced in code section number
16 CFR Part 1201— (1977)	Safety Standard for Architectural Glazing R308.1.1, R308.3	
16 CFR Part 1209— (1979)	Interim Safety Standard for Cellulose Insulation R316.3	
16 CFR Part 1404— (1979)	Cellulose Insulation R316.3	

CSA

Canadian Standards Association
5060 Spectrum Way, Suite 100
Mississauga, Ontario, Canada L4W 5N6

Standard reference number	Title	Referenced in code section number
CSA Requirement 3—88	Manually Operated Gas Valves for Use in House Piping Systems . . . . Table G2420.1.1	
8-93 (Revision 1, 1999)	Requirements for Gas Fired Log Lighters for Wood Burning Fireplaces —with Revisions through January 1999 . . . . G2433.1	
0325.0—92	Construction Sheathing (Reaffirmed 1998) R503.2.1	
0437-Series—93	Standards on OSB and Waferboard (Reaffirmed 2001) R503.2.1, R803.2.1	
A 257.1M—92	Circular Concrete Culvert, Storm Drain, Sewer Pipe and Fittings . . . . Table P3002.2	
A 257.2M—92	Reinforced Circular Concrete Culvert, Storm Drain, Sewer Pipe and Fittings . . . . Table P3002.2	
A 257.3M—92	Joints for Circular Concrete Sewer and Culvert Pipe, Manhole Sections, and Fittings Using Rubber Gaskets P3003.7	
101/I.S.2/A440—05	Specifications for Windows, Doors and Unit Skylights R308.6.9, R613.4, N1102.4.2	
B45.1—02	Ceramic Plumbing Fixtures . . . . Table P2701.1, P2711.1, P2712.1	
B45.2—02	Enameled Cast Iron Plumbing Fixtures . . . . Table 2701.1, P2711.1	
B45.3—02	Porcelain Enameled Steel Plumbing Fixtures . . . . Table P2701.1, Table P2702.2, P2711.1	
B45.4—02	Stainless Steel Plumbing Fixtures . . . . Table P2701.1, P2711.1, P2712.1	
B45.5—02	Plastic Plumbing Fixtures . . . . Table P2701.1, P2711.2, P2712.1	
B45.9—02	Macerating Systems and Related Components P3007.1, P3007.2.1	
B64.1.1—01	Vacuum Breakers, Atmospheric Type (AVB) . . . . Table P2902.3, P2902.3.4	
B64.1.2—01	Vacuum Breakers, Pressure Type (PVB) . . . . Table P2902.3, P2902.3.4	
B64.2—01	Vacuum Breakers, Hose Connection Type (HCVB) . . . . Table P2902.3, P2902.3.2	
B64.2.1—01	Vacuum Breakers, Hose Connection Type (HCVB) with Manual Draining Feature . . . . Table P2902.3, P2902.3.2	
B64.2.1.1—01	Vacuum Breakers, Hose Connection Dual Check Type (HCDVB) . . . . Table P2902.3, P2902.3.2	
B64.2.2—01	Vacuum Breakers, Hose Connection Type (HCVB) with Automatic Draining Feature . . . . Table P2902.3, P2902.3.2	
B64.3—01	Backflow Preventers, Dual Check Valve type with Atmospheric Port (DCAP) . . . . Table P2902.3, P2902.3.3, P2902.5.1	
B64.4—01	Blackflow Preventers, Reduced Pressure Principle Type (RP) . . . . Table P2902.3, P2902.3.5, P2902.5.1	
B64.4.1—01	Backflow Preventers, Reduced Pressure Principle Type for Fire Systems (RPF) . . . . Table P2902.3, P2902.3.5	
B64.5—01	Backflow Preventers, Double Check Valve Type (DCVA) . . . . Table P2902.3, P2902.3.6	
B64.5.1—01	Backflow Preventers, Double Check Valve Type for Fire Systems (DCVAF) . . . . Table P2902.3, P2902.3.6	
B64.6—01	Backflow Preventers, Dual Check Valve Type (DuC) . . . . Table P2902.3	
B64.7—01	Vacuum Breakers, Laboratory Faucet Type (LFVB) . . . . Table P2902.3, P2902.3.2	
B125—01	Plumbing Fittings . . . . Table P2701.1, P2702.2, P2708.3, P2722.1, P2722.2, P2722.3	

B 125.1—05	Plumbing Supply Fittings. P2902.4.1
B137.1—02	Polyethylene Pipe, Tubing and Fittings for Cold Water Pressure Services. . . . . .Table P2904.4, Table P2904.6
B137.2—02	PVC Injection-moulded Gasketed Fittings for Pressure Applications . . . . .Table P2904.6
B137.3—02	Rigid Poly (Vinyl Chloride) (PVC) Pipe for Pressure Applications . . . . .Table P2904.4, P3003.9.2, P3003.14.2
B137.5—02	Cross-linked Polyethylene (PEX) Tubing Systems for Pressure Applications . . . Table P2904.4, Table P2904.5, Table P2904.6
B137.6—02	CPVC Pipe, Tubing and Fittings For Hot and Cold Water Distribution Systems. . . . . .Table P2904.4, Table P2904.5, Table 2904.6
B137.8—02	Polybutylene (PB) Piping for Pressure Applications . . . . Table 2904.4, Table P2904.5, Table 2904.6
B137.9—02	Polyethylene/Aluminum/Polyethylene Composite Pressure-Pipe Systems. . . . . .Table 2904.4.1
B137.10—02	Crosslinked Polyethylene/Aluminum/Crosslinked Polyethylene Composite Pressure-Pipe Systems . . . . .Table P2904.4.1, Table P2904.5, Table M2101.1
B137.11—02	Polypropylene (PP-R) Pipe and Fittings for Pressure Applications . . . . Table P2904.4, Table 2904.5, Table P2904.6
B181.1—02	ABS Drain, Waste and Vent Pipe and Pipe Fittings. . . . . Table P3002.1(1), Table P3002.1(2), Table P3002.2, Table P3002.4, P3003.3.2, P3003.8.2
B181.2—02	PVC Drain, Waste and Vent Pipe and Pipe Fittings. . . . . Table P3002.1(1), Table P3002.1(2), Table P3002.2, Table P3002.3, P3003.9.2, P3003.14.2
B181.3—02	Polyolefin Laboratory Drainage Systems. . . . . .Table P3002.2, P3003.16.1
B182.2—02	PVC Sewer Pipe and Fittings (PSM Type). . . . . Table P3002.1(1), Table P3002.1(2), Table P3002.2, Table P3002.3
B182.4—02	Profile PVC Sewer Pipe & Fittings . . . . .Table P3002.2, Table P3002.3
B602—02	Mechanical Couplings for Drain, Waste and Vent Pipe and Sewer Pipe P3003.3.1, P3003.6.3, P3003.7, P3003.8.1, P3003.14.1, P3003.15, P3003.17
LC3—00	Appliance Stands and Drain Pans. P2801.5
CAN/CSA A257.3M—92	Joints for Circular Concrete Sewer and Culvert Pipe, Manhole Sections, and Fittings Using Rubber Gaskets. P3003.3.5
CAN/CSA B137.9—99	Polyethylene/Aluminum/Polyethylene Composite Pressure Pipe Systems. . . . . .Table P2904.4.1
CAN/CSA B137.10M—02	Crosslinked Polyethylene/Aluminum/Polyethylene Composite Pressure Pipe Systems. . . . . .Table P2904.4.1, Table P2904.5, Table M2101.1

CSSB

Cedar Shake & Shingle Bureau
515 116th Avenue, NE, Suite 275
Bellevue, WA 98004-5294

Standard reference number	Title	Referenced in code section number
CSSB—97	Grading and Packing Rules for Western Red Cedar Shakes and Western Red Shingles of the Cedar Shake and Shingle Bureau . . . . .R702.6, R703.5, Table R905.7.4, Table R905.8.5	

DASMA

Door and Access Systems Manufacturers
Association International
1300 Summer Avenue
Cleveland, OH 44115-2851

Standard reference number	Title	Referenced in code section number
108—2002	Standard Method for Testing Garage Doors: Determination of Structural Performance Under Uniform Static Air Pressure Difference R613.5	

DOC

United States Department of Commerce
100 Bureau Drive Stop 3460
Gaithersburg, MD 20899

Standard reference number	Title	Referenced in code section number
PS 1—95	Construction and Industrial Plywood. . . . . .R404.2.1, Table R404.2.3, R503.2.1, R604.1, R803.2.1	
PS 2—92	Performance Standard for Wood-based Structural-use Panels . . . . R404.2.1, Table R404.2.3, R503.2.1, R604.1, R803.2.1	
PS 20—99	American Softwood Lumber Standard. R404.2.1, R502.1, R602.1, R802.1	

DOTn

Department of Transportation
400 Seventh St. S.W.
Washington, DC 20590

Standard reference number	Title	Referenced in code section number
49 CFR, Parts 192.281(e) & 192.283 (b)	Transportation of Natural and Other Gas by Pipeline: Minimum Federal Safety Standards	G2414.6.1

FEMA

Federal Emergency Management Agency
500 C Street, SW
Washington, DC 20472

Standard reference number	Title	Referenced in code section number
TB-2—93	Flood-resistant Materials Requirements	R324.1.7
FIA-TB-11—01	Crawlspace Construction for Buildings Located in Special Flood Hazard Area	R408.7

FM

Factory Mutual Global Research
Standards Laboratories Department
1151 Boston Providence Turnpike
Norwood, MA 02062

Standard reference number	Title	Referenced in code section number
4450—(1989)	Approval Standard for Class 1 Insulated Steel Deck Roofs—with Supplements through July 1992	R906.1
4880—(2001)	American National Standard for Evaluating Insulated Wall or Wall and Roof/Ceiling Assemblies, Plastic Interior Finish Materials, Plastic Exterior Building Panels, Wall/Ceiling Coating Systems, Interior or Exterior Finish Systems	R314.3

GA

Gypsum Association
810 First Street, Northeast, Suite 510
Washington, DC 20002-4268

Standard reference number	Title	Referenced in code section number
GA-253—99	Recommended Standard Specification for the Application of Gypsum Sheathing	Table R602.3(1)

HPVA

Hardwood Plywood & Veneer Association
1825 Michael Faraday Drive
Reston, Virginia 20190-5350

Standard reference number	Title	Referenced in code section number
HP-1—2000	The American National Standard for Hardwood and Decorative Plywood	R702.5

ICC

International Code Council
5203 Leesburg Pike, Suite 600
Falls Church, VA 22041

Standard reference number	Title	Referenced in code section number
IBC—06	International Building Code®	R110.2, R322.1, R324.1, R324.1.5, R403.1.8, R1001.8.2, G2402.3
ICC EC—06	ICC Electrical Code®—Administrative Provisions	R107.3, G2402.3
IEBC—06	International Existing Building Code®	R101.2, G2401.1
IECC—06	International Energy Conservation Code®	R104.11
IFC—06	International Fire Code®	R102.7, G2402.3, G2412.2, G2423.1
IFGC—06	International Fuel Gas Code®	R104.11, G2401.1

IMC—06	International Mechanical Code®	R104.11, M2106.1, G2402.3
IPC—06	International Plumbing Code®	R104.11, G2402.3
IPSDC—06	International Private Sewage Disposal Code®	R324.1.6
IPMC—06	International Property Maintenance Code®	R102.7
SBCCI SSTD 10—99	Standard for Hurricane Resistant Construction	R301.2.1.1

ISO

International Organization for Standardization
1, rue de Varembé, Case postale 56
CH-1211 Geneva 20, Switzerland

Standard reference number	Title	Referenced in code section number
15874—2002	Polypropylene Plastic Piping Systems for Hot and Cold Water Installations	Table M2101.1

MSS

Manufacturers Standardization Society of the Valve and Fittings Industry
127 Park Street, Northeast
Vienna, VA 22180

Standard reference number	Title	Referenced in code section number
SP-58—93	Pipe Hangers and Supports—Materials, Design and Manufacture	G2418.2

NAIMA

North American Insulation Manufacturers Association
44 Canal Center Plaza, Suite 310
Alexandria, VA 22314

Standard reference number	Title	Referenced in code section number
AH 116 06—02	Fibrous Glass Duct Construction Standards, Fifth Edition	M1601.1.1

NCMA

National Concrete Masonry Association
2302 Horse Pen Road
Herndon, VA 20171-3499

Standard reference number	Title	Referenced in code section number
TR 68-A—75	Design and Construction of Plain and Reinforced Concrete Masonry and Basement and Foundation Walls	R404.1
TR 68B(2001)	Basement Manual Design and Construction Using Concrete Masonry	R404.1

NFPA

National Fire Protection Association
Batterymarch Park
Quincy, MA 02269

Standard reference number	Title	Referenced in code section number
13—02	Installation of Sprinkler Systems	R317.1
31—01	Installation of Oil-burning Equipment	M1801.3.1, M1805.3
58—04	Liquefied Petroleum Gas Code	G2412.2, G2414.6.2
70—05	National Electrical Code	E3301.1, E3301.2, E4201.1, Table E4203.2, E4204.3, E4204.4
72—02	National Fire Alarm Code	R313.1
85—04	Boiler and Construction Systems Hazards Code	G2452.1
211—03	Chimneys, Fireplaces, Vents and Solid Fuel Burning Appliances	R1002.5
259—04	Test Method for Potential Heat of Building Materials	R314.2.5
286—00	Standard Methods of Fire Tests for Evaluating Contribution of Wall and Ceiling Interior Finish to Room Fire Growth	R314.3, R315.4
501—03	Standard on Manufactured Housing	R202
853—03	Standard for the Installation of Stationary Fuel Cell Power Systems	M1903.1

NFRC

National Fenestration Rating Council, Inc.
8484 Georgia Avenue, Suite 320
Silver Spring, MD 20910

Standard reference number	Title	Referenced in code section number
100—2001	Procedure for Determining Fenestration Product U-factors–Second Edition	N1101.5
200—2001	Procedure for Determining Fenestration product Solar Heat Gain Coefficients and Visible Transmittance at Normal Incidence—Second Edition	N1101.5
400—2001	Procedure for Determining Fenestration Product Air Leakage	N1102.4.2

NSF

NSF International
789 N. Dixboro Road
Ann Arbor, MI 48105

Standard reference number	Title	Referenced in code section number
14—2003	Plastic Piping System Components and Related Materials	P2608.3, P2907.3
42—2002e	Drinking Water Treatment Units—Anesthetic Effects	P2907.1, P2907.3
44—2004	Residential Cation Exchange Water Softeners	P2907.1, P2907.3
53—2002e	Drinking Water Treatment Units—Health Effects	P2907.1, P2907.3
58—2004	Reverse Osmosis Drinking Water Treatment Systems	P2907.2, P2907.3
61—2003e	Drinking Water System Components—Health Effects	P2608.5, P2722.1, P2903.9.4, P2904.4, P2904.5, P2904.6, P2907.3

SAE

Society of Automotive Engineers
400 Commonwealth Drive
Warrendale, PA 15096

Standard reference number	Title	Referenced in code section number
J 78—(1998)	Steel Self-drilling Tapping Screws	R505.2.4, R603.2.4, R804.2.4

SMACNA

Sheet Metal & Air Conditioning Contractors National Assoc., Inc.
4021 Lafayette Center Road
Chantilly, VA 22021

Standard reference number	Title	Referenced in code section number
SMACNA—03	Fibrous Glass Duct Construction Standards (2003)	M1601.1.1

TMS

The Masonry Society
3970 Broadway, Suite 201-D
Boulder, CO 80304

Standard reference number	Title	Referenced in code section number
402—05	Building Code Requirements for Masonry Structures	R404.1, R606.1, R606.1.1, R606.12.1, R606.12.2.2.1, R606.11.2.2.2, R606.12.3.1
602—05	Specification for Masonry Structures	R404.1, R606.1, R606.1.1, R606.12.1, R606.12.2.2.1, R606.12.2.2.2, R606.12.3.1

TPI

Truss Plate Institute
583 D'Onofrio Drive, Suite 200
Madison, WI 53719

Standard reference number	Title	Referenced in code section number
TPI 1—2002	National Design Standard for Metal-plate-connected Wood Truss Construction	R502.11.1, R802.10.2,

UL

Underwriters Laboratories, Inc.
333 Pfingsten Road
Northbrook, IL 60062

Standard reference number	Title	Referenced in code section number
17—94	Vent or Chimney Connector Dampers for Oil-fired Appliances—with Revisions through September 1999	M1802.2.2
58—96	Steel Underground Tanks for Flammable and Combustible Liquids—with Revisions through July 1998	M2201.1
80—96	Steel Tanks for Oil-burner Fuel—with Revisions Through June 2003	M2201.1
103—2001	Factory-built Chimneys for Residential Type and Building Heating Appliances —with Revisions through December 2003	R202, R1005.3, G2430.1
127—99	Factory-built Fireplaces—with Revisions through November 1999	R1001.11, R1004.1, R1004.4, R1005.4, G2445.7
174—04	Household Electric Storage Tank Water Heaters—with Revisions through October 1999	M2005.1
181—96	Factory-made Air Ducts and Air Connectors—with Revisions through May 2003	M1601.2, M1601.3.1
181A—98	Closure Systems for Use with Rigid Air Ducts and Air Connectors— with Revisions through December 1998	M1601.2, M1601.3.1
181B—95	Closure Systems for Use with Flexible Air Ducts and Air Connectors— with Revisions through August 2003	M1601.2, M1601.3.1
217—1997	Single and Multiple Station Smoke Alarms—with Revisions Through January 2004	R313.1
325—02	Standard for Door, Drapery, Gate, Louver and Window Operations and Systems —with Revisions through March 2003	R309.6
343—97	Pumps for Oil-Burning Appliances—with Revisions through May 2002	M2204.1
441—96	Gas Vents—with Revisions through December 1999	G2426.1
508—99	Industrial Control Equipment	M1411.3.1
536—97	Flexible Metallic Hose—with Revisions through June 2003	M2202.3
641—95	Type L, Low-temperature Venting Systems—with Revisions through April 1999	R202, R1001.11.5, M1804.2.4, G2426.1
651—05	Schedule 40 and Schedule 80 Rigid PVC Conduit and Fittings	G2414.6.3
726—98	Oil-fired Boiler Assemblies—with Revisions through January 2001	M2001.1.1, M2006.1, G2425.1
727—98	Oil-fired Central Furnaces—with Revisions through January 1999	M1402.1
729—03	Oil-fired Floor Furnaces	M1408.1
730—03	Oil-fired Wall Furnaces	M1409.1
732—95	Oil-fired Storage Tank Water Heaters—with Revisions through January 1999	M2005.1
737—96	Fireplaces Stoves—with Revisions through January 2000	M1414.1
790—04	Standard Test Methods for Fire Tests of Roof Coverings	R902.1
795—99	Commercial-Industrial Gas Heating Equipment	G2442.1, G2452.1
834—04	Heating, Water Supply, and Power Boilers-Electric	M2001.1.1
896—93	Oil-burning Stoves—with Revisions through May 2004	M1410.1
959—01	Medium Heat Appliance Factory-built Chimneys	R1005.6
923—02	Microwave Cooking Appliances—with Revisions through January 2003	M1504.1
1040—96	Fire Test of Insulated Wall Construction—with Revisions through June 2001	R314.3
1256—02	Fire Test of Roof Deck Construction	R906.1
1261—01	Electric Water Heaters for Pools and Tubs—with Revisions through June 2004	M2006.1
1453—04	Electronic Booster and Commercial Storage Tank Water Heaters	M2005.1
1479—03	Fire Tests of Through-Penetration Firestops	R317.3.1.2
1482—98	Solid-fuel Type Room Heaters—with Revisions through January 2000	R1002.2, R1002.5, M1410.1
1715—97	Fire Test of Interior Finish Material—with Revisions through March 2004	R314.4
1738—93	Venting Systems for Gas-burning Appliances, Categories II, III and IV—with Revisions through December 2000	G2426.1
1777—04	Standard for Chimney Liners	R1003.1.8, R1003.11.1, M1801.3.4, G2425.12, G2425.15.4
1995—98	Heating and Cooling Equipment—with Revisions through August 1999	M1402.1, M1403.1, M1407.1
2158A—96	Outline of Investigation for Clothes Dryer Transition Duct	M1502.4

ULC

Underwriters' Laboratories of Canada
7 Crouse Road
Scarborough, Ontario, Canada M1R 3A9

Standard reference number	Title	Referenced in code section number
S 102—1988	Standard Methods for Test for Surface Burning Characteristics of Building Materials and Assemblies—with 2000 Revisions	R316.2

WDMA

Window & Door Manufacturers Association
1400 East Touhy Avenue, Suite 470
Des Plaines, IL 60018

Standard reference number	Title	Referenced in code section number
101/I.S2/A440—05	Specifications for Windows, Doors and Unit Skylights	R308.6.9, R613.4, N1102.4.2

APPENDIX A (IFGS)

SIZING AND CAPACITIES OF GAS PIPING

(This appendix is informative and is not part of the code. This appendix is an excerpt from the 2006 *International Fuel Gas Code,* coordinated with the section numbering of the *International Residential Code.*)

A.1 General piping considerations. The first goal of determining the pipe sizing for a fuel gas piping system is to make sure that there is sufficient gas pressure at the inlet to each appliance. The majority of systems are residential and the appliances will all have the same, or nearly the same, requirement for minimum gas pressure at the appliance inlet. This pressure will be about 5-inch water column (w.c.) (1.25 kPa), which is enough for proper operation of the appliance regulator to deliver about 3.5-inches water column (w.c.) (875 kPa) to the burner itself. The pressure drop in the piping is subtracted from the source delivery pressure to verify that the minimum is available at the appliance.

There are other systems, however, where the required inlet pressure to the different appliances may be quite varied. In such cases, the greatest inlet pressure required must be satisfied, as well as the farthest appliance, which is almost always the critical appliance in small systems.

There is an additional requirement to be observed besides the capacity of the system at 100-percent flow. That requirement is that at minimum flow, the pressure at the inlet to any appliance does not exceed the pressure rating of the appliance regulator. This would seldom be of concern in small systems if the source pressure is $^1/_2$ psi (14-inch w.c.) (3.5 kPa) or less but it should be verified for systems with greater gas pressure at the point of supply.

To determine the size of piping used in a gas piping system, the following factors must be considered:

(1) Allowable loss in pressure from point of delivery to equipment.

(2) Maximum gas demand.

(3) Length of piping and number of fittings.

(4) Specific gravity of the gas.

(5) Diversity factor.

For any gas piping system, or special appliance, or for conditions other than those covered by the tables provided in this code, such as longer runs, greater gas demands or greater pressure drops, the size of each gas piping system should be determined by standard engineering practices acceptable to the code official.

A.2 Description of tables

A.2.1 General. The quantity of gas to be provided at each outlet should be determined, whenever possible, directly from the manufacturer's gas input Btu/h rating of the appliance that will be installed. In case the ratings of the appliances to be installed are not known, Table G2413.2 shows the approximate consumption (in Btu per hour) of certain types of typical household appliances.

To obtain the cubic feet per hour of gas required, divide the total Btu/h input of all appliances by the average Btu heating value per cubic foot of the gas. The average Btu per cubic foot of the gas in the area of the installation can be obtained from the serving gas supplier.

A.2.2 Low pressure natural gas tables. Capacities for gas at low pressure [less than 2.0 psig (13.8 kPa gauge)] in cubic feet per hour of 0.60 specific gravity gas for different sizes and lengths are shown in Table G2413.4(1) for iron pipe or equivalent rigid pipe, in Table G2413.4(3) for smooth wall semi-rigid tubing, in Table G2413.4(5) for corrugated stainless steel tubing and in Table G2413.4(7) for polyethylene plastic pipe. Tables G2413.4(1), G2413.4(3), G2413.4(5) and G2413.4(7) are based upon a pressure drop of 0.5-inch w.c. (125 Pa). In using these tables, an allowance (in equivalent length of pipe) should be considered for any piping run with four or more fittings [see Table A.2.2].

A.2.3 Undiluted liquefied petroleum tables. Capacities in thousands of Btu per hour of undiluted liquefied petroleum gases based on a pressure drop of 0.5-inch w.c. (125 Pa) for different sizes and lengths are shown in the *International Fuel Gas Code.* See Appendix A of that code.

A.2.4 Natural gas specific gravity. Gas piping systems that are to be supplied with gas of a specific gravity of 0.70 or less can be sized directly from the tables provided in this code, unless the code official specifies that a gravity factor be applied. Where the specific gravity of the gas is greater than 0.70, the gravity factor should be applied.

Application of the gravity factor converts the figures given in the tables provided in this code to capacities for another gas of different specific gravity. Such application is accomplished by multiplying the capacities given in the tables by the multipliers shown in Table A.2.4. In case the exact specific gravity does not appear in the table, choose the next higher value specific gravity shown.

TABLE A.2.2
EQUIVALENT LENGTHS OF PIPE FITTINGS AND VALVES

		SCREWED FITTINGS[1]				90° WELDING ELBOWS AND SMOOTH BENDS[2]					
		45°/Ell	90°/Ell	180° close return bends	Tee	$R/d = 1$	$R/d = 1\frac{1}{3}$	$R/d = 2$	$R/d = 4$	$R/d = 6$	$R/d = 8$
k factor =		0.42	0.90	2.00	1.80	0.48	0.36	0.27	0.21	0.27	0.36
L/d' ratio[4] n =		14	30	67	60	16	12	9	7	9	12
Nominal pipe size, inches	Inside diameter d, inches, Schedule 40[6]	L = Equivalent Length In Feet of Schedule 40 (Standard-Weight) Straight Pipe[6]									
$\frac{1}{2}$	0.622	0.73	1.55	3.47	3.10	0.83	0.62	0.47	0.36	0.47	0.62
$\frac{3}{4}$	0.824	0.96	2.06	4.60	4.12	1.10	0.82	0.62	0.48	0.62	0.82
1	1.049	1.22	2.62	5.82	5.24	1.40	1.05	0.79	0.61	0.79	1.05
$1\frac{1}{4}$	1.380	1.61	3.45	7.66	6.90	1.84	1.38	1.03	0.81	1.03	1.38
$1\frac{1}{2}$	1.610	1.88	4.02	8.95	8.04	2.14	1.61	1.21	0.94	1.21	1.61
2	2.067	2.41	5.17	11.5	10.3	2.76	2.07	1.55	1.21	1.55	2.07
$2\frac{1}{2}$	2.469	2.88	6.16	13.7	12.3	3.29	2.47	1.85	1.44	1.85	2.47
3	3.068	3.58	7.67	17.1	15.3	4.09	3.07	2.30	1.79	2.30	3.07
4	4.026	4.70	10.1	22.4	20.2	5.37	4.03	3.02	2.35	3.02	4.03
5	5.047	5.88	12.6	28.0	25.2	6.72	5.05	3.78	2.94	3.78	5.05
6	6.065	7.07	15.2	33.8	30.4	8.09	6.07	4.55	3.54	4.55	6.07
8	7.981	9.31	20.0	44.6	40.0	10.6	7.98	5.98	4.65	5.98	7.98
10	10.02	11.7	25.0	55.7	50.0	13.3	10.0	7.51	5.85	7.51	10.0
12	11.94	13.9	29.8	66.3	59.6	15.9	11.9	8.95	6.96	8.95	11.9
14	13.13	15.3	32.8	73.0	65.6	17.5	13.1	9.85	7.65	9.85	13.1
16	15.00	17.5	37.5	83.5	75.0	20.0	15.0	11.2	8.75	11.2	15.0
18	16.88	19.7	42.1	93.8	84.2	22.5	16.9	12.7	9.85	12.7	16.9
20	18.81	22.0	47.0	105.0	94.0	25.1	18.8	14.1	11.0	14.1	18.8
24	22.63	26.4	56.6	126.0	113.0	30.2	22.6	17.0	13.2	17.0	22.6

continued

TABLE A.2.2—continued
EQUIVALENT LENGTHS OF PIPE FITTINGS AND VALVES

		MITER ELBOWS[3] (No. of miters)					WELDING TEES		VALVES (screwed, flanged, or welded)			
		1-45°	1-60°	1-90°	2-90°[5]	3-90°[5]	Forged	Miter[3]	Gate	Globe	Angle	Swing Check
k factor =		0.45	0.90	1.80	0.60	0.45	1.35	1.80	0.21	10	5.0	2.5
L/d' ratio[4] n =		15	30	60	20	15	45	60	7	333	167	83
Nominal pipe size, inches	Inside diameter d, inches, Schedule 40[6]	L = Equivalent Length In Feet of Schedule 40 (Standard-Weight) Straight Pipe[6]										
$^1/_2$	0.622	0.78	1.55	3.10	1.04	0.78	2.33	3.10	0.36	17.3	8.65	4.32
$^3/_4$	0.824	1.03	2.06	4.12	1.37	1.03	3.09	4.12	0.48	22.9	11.4	5.72
1	1.049	1.31	2.62	5.24	1.75	1.31	3.93	5.24	0.61	29.1	14.6	7.27
$1^1/_4$	1.380	1.72	3.45	6.90	2.30	1.72	5.17	6.90	0.81	38.3	19.1	9.58
$1^1/_2$	1.610	2.01	4.02	8.04	2.68	2.01	6.04	8.04	0.94	44.7	22.4	11.2
2	2.067	2.58	5.17	10.3	3.45	2.58	7.75	10.3	1.21	57.4	28.7	14.4
$2^1/_2$	2.469	3.08	6.16	12.3	4.11	3.08	9.25	12.3	1.44	68.5	34.3	17.1
3	3.068	3.84	7.67	15.3	5.11	3.84	11.5	15.3	1.79	85.2	42.6	21.3
4	4.026	5.04	10.1	20.2	6.71	5.04	15.1	20.2	2.35	112.0	56.0	28.0
5	5.047	6.30	12.6	25.2	8.40	6.30	18.9	25.2	2.94	140.0	70.0	35.0
6	6.065	7.58	15.2	30.4	10.1	7.58	22.8	30.4	3.54	168.0	84.1	42.1
8	7.981	9.97	20.0	40.0	13.3	9.97	29.9	40.0	4.65	222.0	111.0	55.5
10	10.02	12.5	25.0	50.0	16.7	12.5	37.6	50.0	5.85	278.0	139.0	69.5
12	11.94	14.9	29.8	59.6	19.9	14.9	44.8	59.6	6.96	332.0	166.0	83.0
14	13.13	16.4	32.8	65.6	21.9	16.4	49.2	65.6	7.65	364.0	182.0	91.0
16	15.00	18.8	37.5	75.0	25.0	18.8	56.2	75.0	8.75	417.0	208.0	104.0
18	16.88	21.1	42.1	84.2	28.1	21.1	63.2	84.2	9.85	469.0	234.0	117.0
20	18.81	23.5	47.0	94.0	31.4	23.5	70.6	94.0	11.0	522.0	261.0	131.0
24	22.63	28.3	56.6	113.0	37.8	28.3	85.0	113.0	13.2	629.0	314.0	157.0

For SI: 1 foot = 305 mm, 1 degree = 0.01745 rad.

Note: Values for welded fittings are for conditions where bore is not obstructed by weld spatter or backing rings. If appreciably obstructed, use values for "Screwed Fittings."

1. Flanged fittings have three-fourths the resistance of screwed elbows and tees.

2. Tabular figures give the extra resistance due to curvature alone to which should be added the full length of travel.

3. Small size socket-welding fittings are equivalent to miter elbows and miter tees.

4. Equivalent resistance in number of diameters of straight pipe computed for a value of $(f - 0.0075)$ from the relation $(n - k/4f)$.

5. For condition of minimum resistance where the centerline length of each miter is between d and $2^1/_2 d$.

6. For pipe having other inside diameters, the equivalent resistance may be computed from the above n values.

Source: Crocker, S. *Piping Handbook*, 4th ed., Table XIV, pp. 100-101. Copyright 1945 by McGraw-Hill, Inc. Used by permission of McGraw-Hill Book Company.

TABLE A.2.4
MULTIPLIERS TO BE USED WITH TABLES G2413.4(1)
THROUGH G2413.4(8) WHERE THE SPECIFIC GRAVITY
OF THE GAS IS OTHER THAN 0.60

SPECIFIC GRAVITY	MULTIPLIER	SPECIFIC GRAVITY	MULTIPLIER
0.35	1.31	1.00	0.78
0.40	1.23	1.10	0.74
0.45	1.16	1.20	0.71
0.50	1.10	1.30	0.68
0.55	1.04	1.40	0.66
0.60	1.00	1.50	0.63
0.65	0.96	1.60	0.61
0.70	0.93	1.70	0.59
0.75	0.90	1.80	0.58
0.80	0.87	1.90	0.56
0.85	0.84	2.00	0.55
0.90	0.82	2.10	0.54

A.2.5 Higher pressure natural gas tables. Capacities for gas at pressures of 2.0 psig (13.8 kPa) or greater in cubic feet per hour of 0.60 specific gravity gas for different sizes and lengths are shown in Table G2413.4(2) for iron pipe or equivalent rigid pipe, Table G2413.4(4) for semi-rigid tubing, Table G2413.4(6) for corrugated stainless steel tubing and Table G2413.4(8) for polyethylene plastic pipe.

A.3 Use of capacity tables

A.3.1 Longest length method. This sizing method is conservative in its approach by applying the maximum operating conditions in the system as the norm for the system and by setting the length of pipe used to size any given part of the piping system to the maximum value.

To determine the size of each section of gas piping in a system within the range of the capacity tables, proceed as follows. (also see sample calculations included in this Appendix).

(1) Divide the piping system into appropriate segments consistent with the presence of tees, branch lines and main runs. For each segment, determine the gas load (assuming all appliances operate simultaneously) and its overall length. An allowance (in equivalent length of pipe) as determined from Table A.2.2 shall be considered for piping segments that include four or more fittings.

(2) Determine the gas demand of each appliance to be attached to the piping system. Where Tables G2413.4(1) through G2413.4(8) are to be used to select the piping size, calculate the gas demand in terms of cubic feet per hour for each piping system outlet.

(3) Where the piping system is for use with other than undiluted liquefied petroleum gases, determine the design system pressure, the allowable loss in pressure (pressure drop), and specific gravity of the gas to be used in the piping system.

(4) Determine the length of piping from the point of delivery to the most remote outlet in the building/piping system.

(5) In the appropriate capacity table, select the row showing the measured length or the next longer length if the table does not give the exact length. This is the only length used in determining the size of any section of gas piping. If the gravity factor is to be applied, the values in the selected row of the table are multiplied by the appropriate multiplier from Table A.2.4.

(6) Use this horizontal row to locate ALL gas demand figures for this particular system of piping.

(7) Starting at the most remote outlet, find the gas demand for that outlet in the horizontal row just selected. If the exact figure of demand is not shown, choose the next larger figure left in the row.

(8) Opposite this demand figure, in the first row at the top, the correct size of gas piping will be found.

(9) Proceed in a similar manner for each outlet and each section of gas piping. For each section of piping, determine the total gas demand supplied by that section.

When a large number of piping components (such as elbows, tees and valves) are installed in a pipe run, additional pressure loss can be accounted for by the use of equivalent lengths. Pressure loss across any piping component can be equated to the pressure drop through a length of pipe. The equivalent length of a combination of only four elbows/tees can result in a jump to the next larger length row, resulting in a significant reduction in capacity. The equivalent lengths in feet shown in Table A.2.2 have been computed on a basis that the inside diameter corresponds to that of Schedule 40 (standard-weight) steel pipe, which is close enough for most purposes involving other schedules of pipe. Where a more specific solution for equivalent length is desired, this may be made by multiplying the actual inside diameter of the pipe in inches by $n/12$, or the actual inside diameter in feet by n (n can be read from the table heading). The equivalent length values can be used with reasonable accuracy for copper or brass fittings and bends although the resistance per foot of copper or brass pipe is less than that of steel. For copper or brass valves, however, the equivalent length of pipe should be taken as 45 percent longer than the values in the table, which are for steel pipe.

A.3.2 Branch length method. This sizing method reduces the amount of conservatism built into the traditional Longest Length Method. The longest length as measured from the meter to the furthest remote appliance is only used to size the initial parts of the overall piping system. The Branch Length Method is applied in the following manner:

(1) Determine the gas load for each of the connected appliances.

(2) Starting from the meter, divide the piping system into a number of connected segments, and determine the length and amount of gas that each segment would carry assuming that all appliances were operated simultaneously. An allowance (in equivalent length of pipe) as determined from Table A.2.2

should be considered for piping segments that include four or more fittings.

(3) Determine the distance from the outlet of the gas meter to the appliance furthest removed from the meter.

(4) Using the longest distance (found in Step 3), size each piping segment from the meter to the most remote appliance outlet.

(5) For each of these piping segments, use the longest length and the calculated gas load for all of the connected appliances for the segment and begin the sizing process in Steps 6 through 8.

(6) Referring to the appropriate sizing table (based on operating conditions and piping material), find the longest length distance in the first column or the next larger distance if the exact distance is not listed. The use of alternative operating pressures and/or pressure drops will require the use of a different sizing table, but will not alter the sizing methodology. In many cases, the use of alternative operating pressures and/or pressure drops will require the approval of both the code official and the local gas serving utility.

(7) Trace across this row until the gas load is found or the closest larger capacity if the exact capacity is not listed.

(8) Read up the table column and select the appropriate pipe size in the top row. Repeat Steps 6, 7 and 8 for each pipe segment in the longest run.

(9) Size each remaining section of branch piping not previously sized by measuring the distance from the gas meter location to the most remote outlet in that branch, using the gas load of attached appliances and following the procedures of Steps 2 through 8.

A.3.3 Hybrid pressure method. The sizing of a 2 psi (13.8 kPa) gas piping system is performed using the traditional Longest Length Method but with modifications. The 2 psi (13.8 kPa) system consists of two independent pressure zones, and each zone is sized separately. The Hybrid Pressure Method is applied as follows.

The sizing of the 2 psi (13.8 kPa) section (from the meter to the line regulator) is as follows:

(1) Calculate the gas load (by adding up the name plate ratings) from all connected appliances. (In certain circumstances the installed gas load may be increased up to 50 percent to accommodate future addition of appliances.) Ensure that the line regulator capacity is adequate for the calculated gas load and that the required pressure drop (across the regulator) for that capacity does not exceed $^3/_4$ psi (5.2 kPa) for a 2 psi (13.8 kPa) system. If the pressure drop across the regulator is too high (for the connected gas load), select a larger regulator.

(2) Measure the distance from the meter to the line regulator located inside the building.

(3) If there are multiple line regulators, measure the distance from the meter to the regulator furthest removed from the meter.

(4) The maximum allowable pressure drop for the 2 psi (13.8 kPa) section is 1 psi (6.9 kPa).

(5) Referring to the appropriate sizing table (based on piping material) for 2 psi (13.8 kPa) systems with a 1 psi (6.9 kPa) pressure drop, find this distance in the first column, or the closest larger distance if the exact distance is not listed.

(6) Trace across this row until the gas load is found or the closest larger capacity if the exact capacity is not listed.

(7) Read up the table column to the top row and select the appropriate pipe size.

(8) If there are multiple regulators in this portion of the piping system, each line segment must be sized for its actual gas load, but using the longest length previously determined above.

The low pressure section (all piping downstream of the line regulator) is sized as follows:

(1) Determine the gas load for each of the connected appliances.

(2) Starting from the line regulator, divide the piping system into a number of connected segments and/or independent parallel piping segments, and determine the amount of gas that each segment would carry assuming that all appliances were operated simultaneously. An allowance (in equivalent length of pipe) as determined from Table A.2.2 should be considered for piping segments that include four or more fittings.

(3) For each piping segment, use the actual length or longest length (if there are sub-branchlines) and the calculated gas load for that segment and begin the sizing process as follows:

(a) Referring to the appropriate sizing table (based on operating pressure and piping material), find the longest length distance in the first column or the closest larger distance if the exact distance is not listed. The use of alternative operating pressures and/or pressure drops will require the use of a different sizing table, but will not alter the sizing methodology. In many cases, the use of alternative operating pressures and/or pressure drops may require the approval of the code official.

(b) Trace across this row until the appliance gas load is found or the closest larger capacity if the exact capacity is not listed.

(c) Read up the table column to the top row and select the appropriate pipe size.

(d) Repeat this process for each segment of the piping system.

A.3.4 Pressure drop per 100 feet method. This sizing method is less conservative than the others, but it allows the designer to immediately see where the largest pressure drop occurs in the system. With this information, modifications can be made to bring the total drop to the critical appliance within the limitations that are presented to the designer.

Follow the procedures described in the Longest Length Method for Steps (1) through (4) and (9).

For each piping segment, calculate the pressure drop based on pipe size, length as a percentage of 100 feet (30 480 mm), and gas flow. Table A.3.4 shows pressure drop per 100 feet (30 480 mm) for pipe sizes from $^1/_2$ inch (12.7 mm) through 2 inch (51 mm). The sum of pressure drops to the critical appliance is subtracted from the supply pressure to verify that sufficient pressure will be available. If not, the layout can be examined to find the high drop section(s) and sizing selections modified.

Note: Other values can be obtained by using the following equation:

$$\text{Desired Value} = MBH \times \sqrt{\frac{\text{Desired Drop}}{\text{Table Drop}}}$$

For example, if it is desired to get flow through $^3/_4$-inch (19.1 mm) pipe at 2 inches/100 feet, multiple the capacity of $^3/_4$-inch pipe at 1 inch/100 feet by the square root of the pressure ratio:

$$147\,MBH \times \sqrt{\frac{2''w.c.}{1''w.c.}} = 147 \times 1.414 = 208\,MBH$$

$$(MBH = 1000\ \text{Btu/h})$$

A.4 Use of sizing equations. Capacities of smooth wall pipe or tubing can also be determined by using the following formulae:

(1) High Pressure [1.5 psi (10.3 kPa) and above]:

$$Q = 181.6 \sqrt{\frac{D^5 \cdot \left(P_1^2 - P_2^2\right) \cdot Y}{C_r \cdot fba \cdot L}}$$

$$= 2237\, D^{2.623} \left[\frac{\left(P_1^2 - P_2^2\right) \cdot Y}{C_r \cdot L}\right]^{0.541}$$

(2) Low Pressure [Less than 1.5 psi (10.3 kPa)]:

$$Q = 1873 \sqrt{\frac{D^5 \cdot \Delta H}{C_r \cdot fba \cdot L}}$$

$$= 2313\, D^{2.623} \left(\frac{\Delta H}{C_r \cdot L}\right)^{0.541}$$

where:

Q = Rate, cubic feet per hour at 60°F and 30-inch mercury column

D = Inside diameter of pipe, in.

P_1 = Upstream pressure, psia

P_2 = Downstream pressure, psia

Y = Superexpansibility factor = 1/supercompressibility factor

C_r = Factor for viscosity, density and temperature*

$$= 0.00354\, ST \left(\frac{Z}{S}\right)^{0.152}$$

Note: See Table 402.4 for Y and C_r for natural gas and propane.

S = Specific gravity of gas at 60°F and 30-inch mercury column (0.60 for natural gas, 1.50 for propane), or = 1488μ

T = Absolute temperature, °F or = $t + 460$

t = Temperature, °F

Z = Viscosity of gas, centipoise (0.012 for natural gas, 0.008 for propane), or = 1488μ

fba = Base friction factor for air at 60°F (CF=1)

L = Length of pipe, ft

ΔH = Pressure drop, in. w.c. (27.7 in. H_2O = 1 psi)

(For SI, see Section G2413.4)

A.5 Pipe and tube diameters. Where the internal diameter is determined by the formulas in Section G2413.4, Tables A.5.1 and A.5.2 can be used to select the nominal or standard pipe size based on the calculated internal diameter.

TABLE A.3.4
THOUSANDS OF Btu/h (MBH) OF NATURAL GAS PER 100 FEET OF PIPE AT
VARIOUS PRESSURE DROPS AND PIPE DIAMETERS

PRESSURE DROP PER 100 FEET IN INCHES W.C.	PIPE SIZES (inch)					
	$^1/_2$	$^3/_4$	1	$1^1/_4$	$1^1/_2$	2
0.2	31	64	121	248	372	716
0.3	38	79	148	304	455	877
0.5	50	104	195	400	600	1160
1.0	71	147	276	566	848	1640

For SI: 1 inch = 25.4 mm, 1 foot = 304.8 mm.

TABLE A.5.1
SCHEDULE 40 STEEL PIPE STANDARD SIZES

NOMINAL SIZE (in.)	INTERNAL DIAMETER (in.)	NOMINAL SIZE (in.)	INTERNAL DIAMETER (in.)
$^1/_4$	0.364	$1^1/_2$	1.610
$^3/_8$	0.493	2	2.067
$^1/_2$	0.622	$2^1/_2$	2.469
$^3/_4$	0.824	3	3.068
1	1.049	$3^1/_2$	3.548
$1^1/_4$	1.380	4	4.026

TABLE A.5.2
COPPER TUBE STANDARD SIZES

TUBE TYPE	NOMINAL OR STANDARD SIZE inches	INTERNAL DIAMETER inches
K	$^1/_4$	0.305
L	$^1/_4$	0.315
ACR (D)	$^3/_8$	0.315
ACR (A)	$^3/_8$	0.311
K	$^3/_8$	0.402
L	$^3/_8$	0.430
ACR (D)	$^1/_2$	0.430
ACR (A)	$^1/_2$	0.436
K	$^1/_2$	0.527
L	$^1/_2$	0.545
ACR (D)	$^5/_8$	0.545
ACR (A)	$^5/_8$	0.555
K	$^5/_8$	0.652
L	$^5/_8$	0.666
ACR (D)	$^3/_4$	0.666
ACR (A)	$^3/_4$	0.680
K	$^3/_4$	0.745
L	$^3/_4$	0.785
ACR	$^7/_8$	0.785
K	1	0.995
L	1	1.025
ACR	$1^1/_8$	1.025
K	$1^1/_4$	1.245
L	$1^1/_4$	1.265
ACR	$1^3/_8$	1.265
K	$1^1/_2$	1.481
L	$1^1/_2$	1.505
ACR	$1^5/_8$	1.505
K	2	1.959
L	2	1.985
ACR	$2^1/_8$	1.985
K	$2^1/_2$	2.435
L	$2^1/_2$	2.465
ACR	$2^5/_8$	2.465
K	3	2.907
L	3	2.945
ACR	$3^1/_8$	2.945

A.6 Use of sizing charts. A third method of sizing gas piping is detailed below as an option that is useful when large quantities of piping are involved in a job (e.g., an apartment house) and material costs are of concern. If the user is not completely familiar with this method, the resulting pipe sizing should be checked by a knowledgeable gas engineer. The sizing charts are applied as follows.

(1) With the layout developed according to Section R106.1.1 of the code, indicate in each section the design gas flow under maximum operation conditions. For many layouts, the maximum design flow will be the sum of all connected loads. However, in some cases, certain combinations of appliances will not occur simultaneously (e.g., gas heating and air conditioning). For these cases, the design flow is the greatest gas flow that can occur at any one time.

(2) Determine the inlet gas pressure for the system being designed. In most cases, the point of inlet will be the gas meter or service regulator, but in the case of a system addition, it could be the point of connection to the existing system.

(3) Determine the minimum pressure required at the inlet to the critical appliance. Usually, the critical item will be the appliance with the highest required pressure for satisfactory operation. If several items have the same required pressure, it will be the one with the greatest length of piping from the system inlet.

(4) The difference between the inlet pressure and critical item pressure is the allowable system pressure drop. Figures A.6(a) and A.6(b) show the relationship between gas flow, pipe size and pipe length for natural gas with 0.60 specific gravity.

(5) To use Figure A.6(a) (low pressure applications), calculate the piping length from the inlet to the critical utilization equipment. Increase this length by 50 percent to allow for fittings. Divide the allowable pressure drop by the equivalent length (in hundreds of feet) to determine the allowable pressure drop per hundred feet. Select the pipe size from Figure A.6(a) for the required volume of flow.

(6) To use Figure A.6(b) (high pressure applications), calculate the equivalent length as above. Calculate the index number for Figure A.6(b) by dividing the difference between the squares of the absolute values of inlet

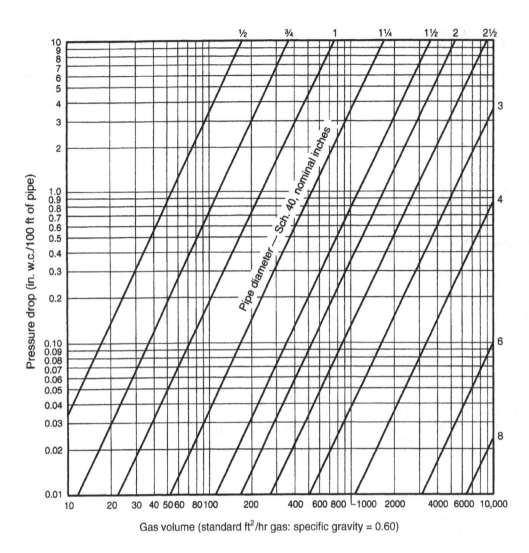

FIGURE A.6(a)
CAPACITY OF NATURAL GAS PIPING, LOW PRESSURE (0.60 WC)

and outlet pressures by the equivalent length (in hundreds of feet). Select the pipe size from Figure A.6(b) for the gas volume required.

A.7 Examples of piping system design and sizing

A.7.1 Example 1: Longest length method. Determine the required pipe size of each section and outlet of the piping system shown in Figure A.7.1, with a designated pressure drop of 0.5-inch w.c. (125 Pa) using the Longest Length Method. The gas to be used has 0.60 specific gravity and a heating value of 1,000 Btu/ft³ (37.5 MJ/m³).

Solution:

(1) Maximum gas demand for Outlet A:

$$\frac{\text{Consumption (rating plate input, or Table G 2413.2 if necessary)}}{\text{Btu of gas}} =$$

$$\frac{35,000 \text{ Btu per hour rating}}{1,000 \text{ Btu per cubic foot}} = 35 \text{ cubic feet per hour} = 35 \text{ cfh}$$

Maximum gas demand for Outlet B:

$$\frac{\text{Consumption}}{\text{Btu of gas}} = \frac{75,000}{1,000} = 75 \text{ cfh}$$

Maximum gas demand for Outlet C:

$$\frac{\text{Consumption}}{\text{Btu of gas}} = \frac{35,000}{1,000} = 35 \text{ cfh}$$

Maximum gas demand for Outlet D:

$$\frac{\text{Consumption}}{\text{Btu of gas}} = \frac{100,000}{1,000} = 100 \text{ cfh}$$

(2) The length of pipe from the point of delivery to the most remote outlet (A) is 60 feet (18 288 mm). This is the only distance used.

(3) Using the row marked 60 feet (18 288 mm) in Table G2413.4(1):

(a) Outlet A, supplying 35 cfh (0.99 m³/hr), requires ³/₈-inch pipe.

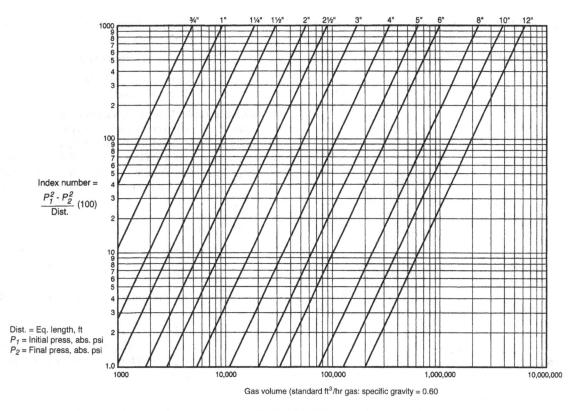

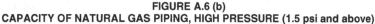

Gas volume (standard ft³/hr gas: specific gravity = 0.60

Index number =

$$\frac{P_1^2 - P_2^2}{\text{Dist.}} (100)$$

Dist. = Eq. length, ft
P_1 = Initial press, abs. psi
P_2 = Final press, abs. psi

FIGURE A.6 (b)
CAPACITY OF NATURAL GAS PIPING, HIGH PRESSURE (1.5 psi and above)

(b) Outlet B, supplying 75 cfh (2.12 m³/hr), requires ³/₄-inch pipe.

(c) Section 1, supplying Outlets A and B, or 110 cfh (3.11 m³/hr), requires ³/₄-inch pipe.

(d) Section 2, supplying Outlets C and D, or 135 cfh (3.82 m³/hr), requires ³/₄-inch pipe.

(e) Section 3, supplying Outlets A, B, C and D, or 245 cfh (6.94 m³/hr), requires 1-inch pipe.

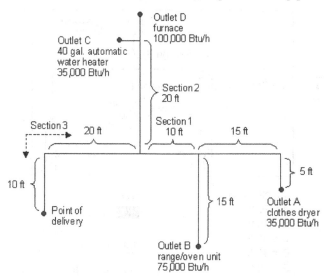

FIGURE A.7.1
PIPING PLAN SHOWING A STEEL PIPING SYSTEM

(4) If a different gravity factor is applied to this example, the values in the row marked 60 feet (18 288 mm) of Table G2413.4(1) would be multiplied by the appropriate multiplier from Table A.2.4 and the resulting cubic feet per hour values would be used to size the piping.

Section A.7.2 through A7.4 note: These examples are based on tables found in the International Fuel Gas Code.

A.7.2 Example 2: Hybrid or dual pressure systems. Determine the required CSST size of each section of the piping system shown in Figure A.7.2, with a designated pressure drop of 1 psi (6.9 kPa) for the 2 psi (13.8 kPa) section and 3-inch w.c. (0.75 kPa) pressure drop for the 13-inch w.c. (2.49 kPa) section. The gas to be used has 0.60 specific gravity and a heating value of 1,000 Btu/ft³ (37.5 MJ/ m³).

Solution

(1) Size 2 psi (13.8 kPa) line using Table 402.4(16).

(2) Size 10-inch w.c. (2.5 kPa) lines using Table 402.4(14).

(3) Using the following, determine if sizing tables can be used.

(a) Total gas load shown in Figure A.7.2 equals 110 cfh (3.11 m³/hr).

(b) Determine pressure drop across regulator [see notes in Table 402.4 (16)].

(c) If pressure drop across regulator exceeds ³/₄ psig (5.2 kPa), Table 402.4 (16) cannot be used. Note: If pressure drop exceeds $\frac{3}{4}$ psi (5.2 kPa), then a larger regulator must be selected or an alternative sizing method must be used.

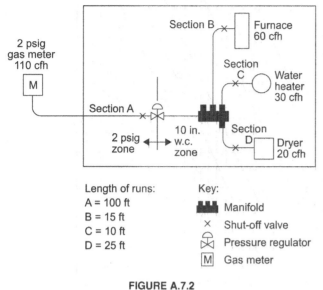

Length of runs:
A = 100 ft
B = 15 ft
C = 10 ft
D = 25 ft

Key:
 Manifold
✕ Shut-off valve
⊕ Pressure regulator
Ⓜ Gas meter

FIGURE A.7.2
PIPING PLAN SHOWING A CSST SYSTEM

(d) Pressure drop across the line regulator [for 110 cfh (3.11 m³/hr)] is 4-inch w.c. (0.99 kPa) based on manufacturer's performance data.

(e) Assume the CSST manufacturer has tubing sizes or EHDs of 13, 18, 23 and 30.

(4) Section A [2 psi (13.8 kPa) zone]

   (a) Distance from meter to regulator = 100 feet (30 480 mm).

   (b) Total load supplied by A = 110 cfh (3.11 m³/hr) (furnace + water heater + dryer).

   (c) Table 402.4 (16) shows that EHD size 18 should be used.

     Note: It is not unusual to oversize the supply line by 25 to 50 percent of the as-installed load. EHD size 18 has a capacity of 189 cfh (5.35 m³/hr).

(5) Section B (low pressure zone)

   (a) Distance from regulator to furnace is 15 feet (4572 mm).

   (b) Load is 60 cfh (1.70 m³/hr).

   (c) Table 402.4 (14) shows that EHD size 13 should be used.

(6) Section C (low pressure zone)

   (a) Distance from regulator to water heater is 10 feet (3048 mm).

   (b) Load is 30 cfh (0.85 m³/hr).

   (c) Table 402.4 (14) shows that EHD size 13 should be used.

(7) Section D (low pressure zone)

   (a) Distance from regulator to dryer is 25 feet (7620 mm).

   (b) Load is 20 cfh (0.57 m³/hr).

   (c) Table 402.4(14) shows that EHD size 13 should be used.

A.7.3 Example 3: Branch length method. Determine the required semi-rigid copper tubing size of each section of the piping system shown in Figure A.7.3, with a designated pressure drop of 1-inch w.c. (250 Pa) (using the Branch Length Method). The gas to be used has 0.60 specific gravity and a heating value of 1,000 Btu/ft³ (37.5 MJ/m³).

Solution

(1) Section A

   (a) The length of tubing from the point of delivery to the most remote appliance is 50 feet (15 240 mm), A + C.

   (b) Use this longest length to size Sections A and C.

   (c) Using the row marked 50 feet (15 240 mm) in Table 402.4(8), Section A, supplying 220 cfh (6.2 m³/hr) for four appliances requires 1-inch tubing.

(2) Section B

   (a) The length of tubing from the point of delivery to the range/oven at the end of Section B is 30 feet (9144 mm), A + B.

   (b) Use this branch length to size Section B only.

   (c) Using the row marked 30 feet (9144 mm) in Table 402.4(8), Section B, supplying 75 cfh (2.12 m³/hr) for the range/oven requires $\frac{1}{2}$-inch tubing.

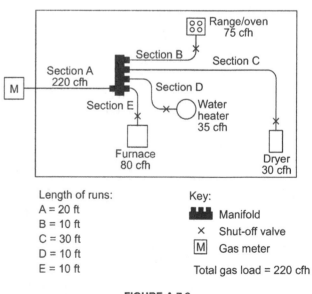

Length of runs:
A = 20 ft
B = 10 ft
C = 30 ft
D = 10 ft
E = 10 ft

Key:
Manifold
✕ Shut-off valve
Ⓜ Gas meter

Total gas load = 220 cfh

FIGURE A.7.3
PIPING PLAN SHOWING A COPPER TUBING SYSTEM

(3) Section C

  (a) The length of tubing from the point of delivery to the dryer at the end of Section C is 50 feet (15 240 mm), A + C.

  (b) Use this branch length (which is also the longest length) to size Section C.

  (c) Using the row marked 50 feet (15 240 mm) in Table 402.4(8), Section C, supplying 30 cfh (0.85 m³/hr) for the dryer requires $^3/_8$-inch tubing.

(4) Section D

  (a) The length of tubing from the point of delivery to the water heater at the end of Section D is 30 feet (9144 mm), A + D.

  (b) Use this branch length to size Section D only.

  (c) Using the row marked 30 feet (9144 mm) in Table 402.4(8), Section D, supplying 35 cfh (0.99 m³/hr) for the water heater requires $^3/_8$-inch tubing.

(5) Section E

  (a) The length of tubing from the point of delivery to the furnace at the end of Section E is 30 feet (9144 mm), A + E.

  (b) Use this branch length to size Section E only.

  (c) Using the row marked 30 feet (9144 mm) in Table 402.4(8), Section E, supplying 80 cfh (2.26 m³/hr) for the furnace requires $^1/_2$-inch tubing.

A.7.4 Example 4: Modification to existing piping system. Determine the required CSST size for Section G (retrofit application) of the piping system shown in Figure A.7.4, with a designated pressure drop of 0.5-inch w.c. (125 Pa) using the branch length method. The gas to be used has 0.60 specific gravity and a heating value of 1,000 Btu/ft³ (37.5 MJ/m³).

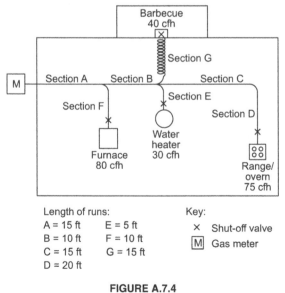

Length of runs:
A = 15 ft E = 5 ft
B = 10 ft F = 10 ft
C = 15 ft G = 15 ft
D = 20 ft

Key:
× Shut-off valve
M Gas meter

FIGURE A.7.4
PIPING PLAN SHOWING A MODIFICATION
TO EXISTING PIPING SYSTEM

Solution

(1) The length of pipe and CSST from the point of delivery to the retrofit appliance (barbecue) at the end of Section G is 40 feet (12 192 mm), A + B + G.

(2) Use this branch length to size Section G.

(3) Assume the CSST manufacturer has tubing sizes or EHDs of 13, 18, 23 and 30.

(4) Using the row marked 40 feet (12 192 mm) in Table 402.4(13), Section G, supplying 40 cfh (1.13 m³/hr) for the barbecue requires EHD 18 CSST.

(5) The sizing of Sections A, B, F and E must be checked to ensure adequate gas carrying capacity since an appliance has been added to the piping system (see A.7.1 for details).

A.7.5 Example 5: Calculating pressure drops due to temperature changes. A test piping system is installed on a warm autumn afternoon when the temperature is 70°F (21°C). In accordance with local custom, the new piping system is subjected to an air pressure test at 20 psig (138 kPa). Overnight, the temperature drops and when the inspector shows up first thing in the morning the temperature is 40°F (4°C).

If the volume of the piping system is unchanged, then the formula based on Boyle's and Charles' law for determining the new pressure at a reduced temperature is as follows:

$$\frac{T_1}{T_2} = \frac{P_1}{P_2}$$

where:

$T_1 =$ Initial temperature, absolute ($T_1 + 459$)

$T_2 =$ Final temperature, absolute ($T_2 + 459$)

$P_1 =$ Initial pressure, psia ($P_1 + 14.7$)

$P_2 =$ Final pressure, psia ($P_2 + 14.7$)

$$\frac{(70+459)}{(40+459)} = \frac{(20+14.7)}{(P_2+14.7)}$$

$$\frac{529}{499} = \frac{34.7}{(P_2+14.7)}$$

$$(P_2+14.7) \times \frac{529}{499} = 34.7$$

$$(P_2+14.7) = \frac{34.7}{1.060}$$

$$P_2 = 32.7 - 14.7$$

$$P_2 = 18 \ psig$$

Therefore, the gauge could be expected to register 18 psig (124 kPa) when the ambient temperature is 40°F (4°C).

A7.6 Example 6: Pressure drop per 100 feet of pipe method. Using the layout shown in Figure A.7.1 and $\Delta H =$ pressure drop, in w.c. (27.7 in. H₂O = 1 psi), proceed as follows:

(1) Length to A = 20 feet, with 35,000 Btu/hr.

For $^1/_2$-inch pipe, $\Delta H = {}^{20\ \text{feet}}/_{100\ \text{feet}} \times 0.3$ inch w.c. $= 0.06$ in. w.c.

(2) Length to B = 15 feet, with 75,000 Btu/hr.

For $^3/_4$-inch pipe, $\Delta H = {}^{15\ \text{feet}}/_{100\ \text{feet}} \times 0.3$ inch w.c. $= 0.045$ in. w.c.

(3) Section 1 = 10 feet, with 110,000 Btu/hr. Here there is a choice:

For 1 inch pipe: $\Delta H = {}^{10\ \text{feet}}/_{100\ \text{feet}} \times 0.2$ inch w.c. $= 0.02$ in w.c.

For $^3/_4$-inch pipe: $\Delta H = {}^{10\ \text{feet}}/_{100\ \text{feet}} \times [0.5$ inch w.c. $+ {}^{(110,000\ \text{Btu/hr-}104,000\ \text{Btu/hr})}/_{(147,000\ \text{Btu/hr-}104,000\ \text{Btu/hr})} \times (1.0$ inches w.c. $- 0.5$ inch w.c.$)] = 0.1 \times 0.57$ inch w.c. ≈ 0.06 inch w.c.

Note that the pressure drop between 104,000 Btu/hr and 147,000 Btu/hr has been interpolated as 110,000 Btu/hr.

(4) Section 2 = 20 feet, with 135,000 Btu/hr. Here there is a choice:

For 1-inch pipe: $\Delta H = {}^{20\ \text{feet}}/_{100\ \text{feet}} \times [0.2$ inch w.c. $+ {}^{(\Delta 14,000\ \text{Btu/hr})}/_{(\Delta 27,000\ \text{Btu/hr})} \times \Delta 0.1$ inch w.c.$)] = 0.05$ inch w.c.$)]$

For $^3/_4$-inch pipe: $\Delta H = {}^{20\ \text{feet}}/_{100\ \text{feet}} \times 1.0$ inch w.c. $= 0.2$ inch w.c.)

Note that the pressure drop between 121,000 Btu/hr and 148,000 Btu/hr has been interpolated as 135,000 Btu/hr, but interpolation for the ¾-inch pipe (trivial for 104,000 Btu/hr to 147,000 Btu/hr) was not used.

(5) Section 3 = 30 feet, with 245,000 Btu/hr. Here there is a choice:

For 1-inch pipe: $\Delta H = {}^{30\ \text{feet}}/_{100\ \text{feet}} \times 1.0$ inches w.c. $= 0.3$ inch w.c.

For $1^1/_4$-inch pipe: $\Delta H = {}^{30\ \text{feet}}/_{100\ \text{feet}} \times 0.2$ inch w.c. $= 0.06$ inch w.c.

Note that interpolation for these options is ignored since the table values are close to the 245,000 Btu/hr carried by that section.

(6) The total pressure drop is the sum of the section approaching A, Sections 1 and 3, or either of the following, depending on whether an absolute minimum is needed or the larger drop can be accommodated.

Minimum pressure drop to farthest appliance:

$\Delta H = 0.06$ inch w.c. $+ 0.02$ inch w.c. $+ 0.06$ inch w.c. $= 0.14$ inch w.c.

Larger pressure drop to the farthest appliance:

$\Delta H = 0.06$ inch w.c. $+ 0.06$ inch w.c. $+ 0.3$ inch w.c. $= 0.42$ inch w.c.

Notice that Section 2 and the run to B do not enter into this calculation, provided that the appliances have similar input pressure requirements.

For SI units: 1 Btu/hr = 0.293 W, 1 cubic foot = 0.028 m^3, 1 foot = 0.305 m, 1 inch w.c. = 249 Pa.

SIZING OF VENTING SYSTEMS SERVING APPLIANCES EQUIPPED WITH DRAFT HOODS, CATEGORY I APPLIANCES, AND APPLIANCES LISTED FOR USE WITH TYPE B VENTS

(This appendix is informative and is not part of the code. This appendix is an excerpt from the 2006 International Fuel Gas Code, coordinated with the section numbering of the International Residential Code.)

EXAMPLES USING SINGLE APPLIANCE VENTING TABLES

Example 1: Single draft-hood-equipped appliance.

An installer has a 120,000 British thermal unit (Btu) per hour input appliance with a 5-inch-diameter draft hood outlet that needs to be vented into a 10-foot-high Type B vent system. What size vent should be used assuming (a) a 5-foot lateral single-wall metal vent connector is used with two 90-degree elbows, or (b) a 5-foot lateral single-wall metal vent connector is used with three 90-degree elbows in the vent system?

Solution:

Table G2428.2(2) should be used to solve this problem, because single-wall metal vent connectors are being used with a Type B vent.

(a) Read down the first column in Table G2428.2(2) until the row associated with a 10-foot height and 5-foot lateral is found. Read across this row until a vent capacity

greater than 120,000 Btu per hour is located in the shaded columns labeled "NAT Max" for draft-hood-equipped appliances. In this case, a 5-inch-diameter vent has a capacity of 122,000 Btu per hour and may be used for this application.

(b) If three 90-degree elbows are used in the vent system, then the maximum vent capacity listed in the tables must be reduced by 10 percent (see Section G2428.2.3 for single appliance vents). This implies that the 5-inch-diameter vent has an adjusted capacity of only 110,000 Btu per hour. In this case, the vent system must be increased to 6 inches in diameter (see calculations below).

122,000 (0.90) = 110,000 for 5-inch vent
From Table G2428.2(2), Select 6-inch vent
186,000 (0.90) = 167,000; This is greater than the required 120,000. Therefore, use a 6-inch vent and connector where three elbows are used.

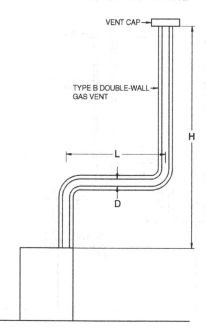

For SI: 1 foot = 304.8 mm, 1 British thermal unit per hour = 0.2931 W.

Table G2428.2(1) is used when sizing Type B double-wall gas vent connected directly to the appliance.

Note: The appliance may be either Category I draft hood equipped or fan-assisted type.

FIGURE B-1
TYPE B DOUBLE-WALL VENT SYSTEM SERVING A SINGLE APPLIANCE WITH A TYPE B DOUBLE-WALL VENT

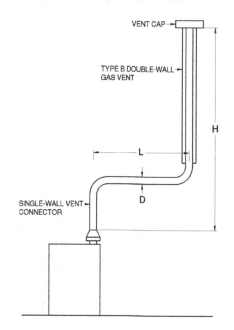

For SI: 1 foot = 304.8 mm, 1 British thermal unit per hour = 0.2931W.

Table G2428.2(2) is used when sizing a single-wall metal vent connector attached to a Type B double-wall gas vent.

Note: The appliance may be either Category I draft hood equipped or fan-assisted type.

FIGURE B-2
TYPE B DOUBLE-WALL VENT SYSTEM SERVING A SINGLE APPLIANCE WITH A SINGLE-WALL METAL VENT CONNECTOR

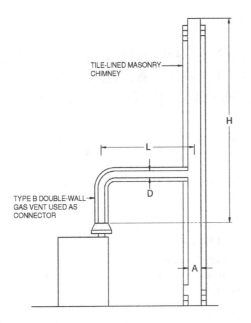

Table 504.2(3) of the *International Fuel Gas Code* is used when sizing a Type B double-wall gas vent connector attached to a tile-lined masonry chimney.

Note: "A" is the equivalent cross-sectional area of the tile liner.

Note: The appliance may be either Category I draft hood equipped or fan-assisted type.

FIGURE B-3
VENT SYSTEM SERVING A SINGLE APPLIANCE
WITH A MASONRY CHIMNEY OF TYPE B
DOUBLE-WALL VENT CONNECTOR

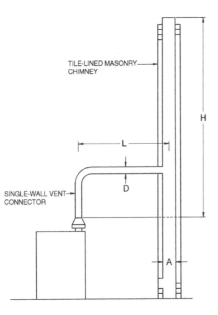

Table 504.2(4) of the *International Fuel Gas Code* is used when sizing a single-wall vent connector attached to a tile-lined masonry chimney.

Note: "A" is the equivalent cross-sectional area of the tile liner.

Note: The appliance may be either Category I draft hood equipped or fan-assisted type.

FIGURE B-4
VENT SYSTEM SERVING A SINGLE APPLIANCE
USING A MASONRY CHIMNEY AND A
SINGLE-WALL METAL VENT CONNECTOR

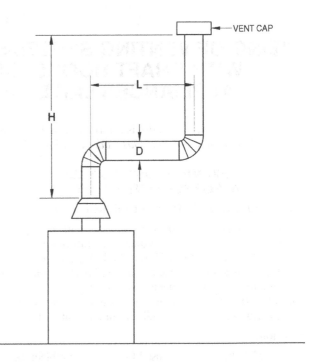

Asbestos cement Type B or single-wall metal vent serving a single draft-hood-equipped appliance [see Table 504.2(5) of the *International Fuel Gas Code*].

FIGURE B-5
ASBESTOS CEMENT TYPE B OR SINGLE-WALL
METAL VENT SYSTEM SERVING A SINGLE
DRAFT-HOOD-EQUIPPED APPLIANCE

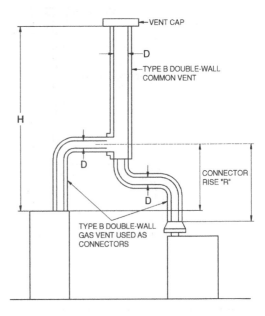

Table G2428.3(1) is used when sizing Type B double-wall vent connectors attached to a Type B double-wall common vent.

Note: Each appliance may be either Category I draft hood equipped or fan-assisted type.

FIGURE B-6
VENT SYSTEM SERVING TWO OR MORE APPLIANCES
WITH TYPE B DOUBLE-WALL VENT AND TYPE B
DOUBLE-WALL VENT CONNECTOR

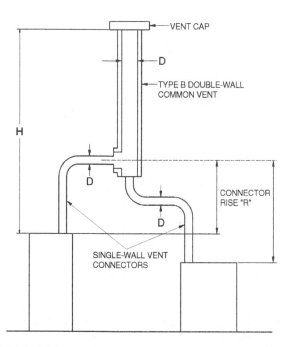

Table G2428.3(2) is used when sizing single-wall vent connectors attached to a Type B double-wall common vent.

Note: Each appliance may be either Category I draft hood equipped or fan-assisted type.

FIGURE B-7
VENT SYSTEM SERVING TWO OR MORE APPLIANCES
WITH TYPE B DOUBLE-WALL VENT AND
SINGLE-WALL METAL VENT CONNECTORS

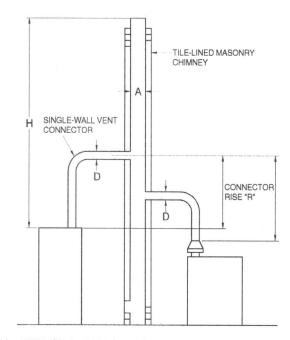

Table G2428.3(4) is used when sizing single-wall metal vent connectors attached to a tile-lined masonry chimney.

Note: "A" is the equivalent cross-sectional area of the tile liner.

Note: Each appliance may be either Category I draft hood equipped or fan-assisted type.

FIGURE B-9
MASONRY CHIMNEY SERVING TWO OR MORE APPLIANCES
WITH SINGLE-WALL METAL VENT CONNECTORS

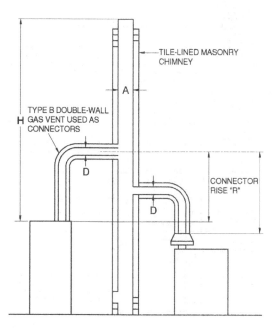

Table G2428.3(3) is used when sizing Type B double-wall vent connectors attached to a tile-lined masonry chimney.

Note: "A" is the equivalent cross-sectional area of the tile liner.

Note: Each appliance may be either Category I draft hood equipped or fan-assisted type.

FIGURE B-8
MASONRY CHIMNEY SERVING TWO OR MORE APPLIANCES
WITH TYPE B DOUBLE-WALL VENT CONNECTOR

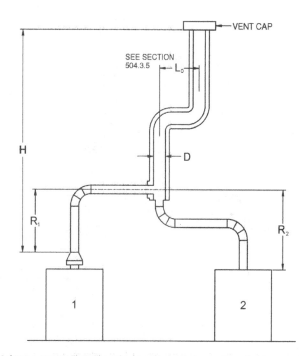

Asbestos cement Type B or single-wall metal pipe vent serving two or more draft-hood-equipped appliances [see Table 504.3(5) of the *International Fuel Gas Code*].

FIGURE B-10
ASBESTOS CEMENT TYPE B OR SINGLE-WALL
METAL VENT SYSTEM SERVING TWO OR MORE
DRAFT-HOOD-EQUIPPED APPLIANCES

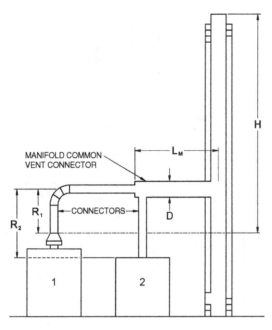

Example: Manifolded Common Vent Connector L_M shall be no greater than 18 times the common vent connector manifold inside diameter; i.e., a 4-inch (102 mm) inside diameter common vent connector manifold shall not exceed 72 inches (1829 mm) in length (see Section G2428.3.4).

Note: This is an illustration of a typical manifolded vent connector. Different appliance, vent connector, or common vent types are possible. Consult Section G2426.3.

FIGURE B-11
USE OF MANIFOLD COMMON VENT CONNECTOR

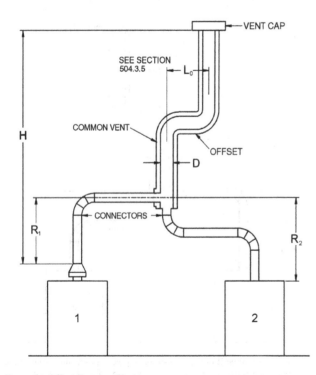

Example: Offset Common Vent

Note: This is an illustration of a typical offset vent. Different appliance, vent connector, or vent types are possible. Consult Sections G2428.2 and G2428.3.

FIGURE B-12
USE OF OFFSET COMMON VENT

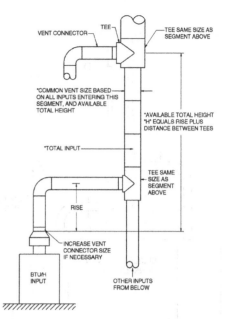

Vent connector size depends on:
- Combined inputs
- Rise
- Available total height "H"
- Table G2428.3(1) connectors

Common vent size depends on:
- Input
- Available total height "H"
- Table G2428.3(1) common vent

FIGURE B-13
MULTISTORY GAS VENT DESIGN PROCEDURE
FOR EACH SEGMENT OF SYSTEM

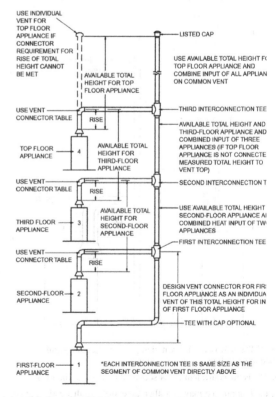

Principles of design of multistory vents using vent connector and common vent design tables (see Sections G2428.3.11 through G2428.3.13).

FIGURE B-14
MULTISTORY VENT SYSTEMS

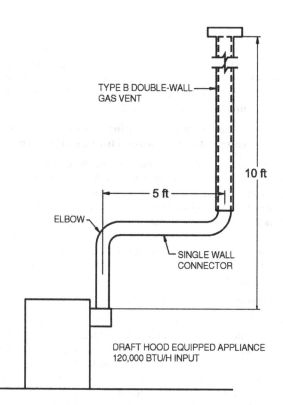

For SI: 1 foot = 304.8 mm, 1 British thermal unit per hour = 0.2931 W.

FIGURE B-15 (EXAMPLE 1)
SINGLE DRAFT-HOOD-EQUIPPED APPLIANCE

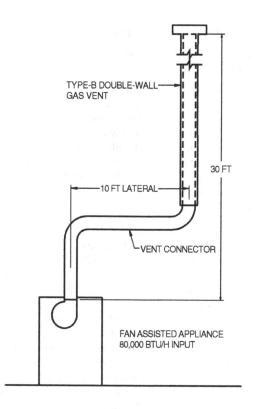

For SI: 1 foot = 304.8 mm, 1 British thermal unit per hour = 0.2931 W.

FIGURE B-16 (EXAMPLE 2)
SINGLE FAN-ASSISTED APPLIANCE

Example 2: Single fan-assisted appliance.

An installer has an 80,000 Btu per hour input fan-assisted appliance that must be installed using 10 feet of lateral connector attached to a 30-foot-high Type B vent. Two 90-degree elbows are needed for the installation. Can a single-wall metal vent connector be used for this application?

Solution:

Table G2428.2(2) refers to the use of single-wall metal vent connectors with Type B vent. In the first column find the row associated with a 30-foot height and a 10-foot lateral. Read across this row, looking at the FAN Min and FAN Max columns, to find that a 3-inch-diameter single-wall metal vent connector is not recommended. Moving to the next larger size single wall connector (4 inches), note that a 4-inch-diameter single-wall metal connector has a recommended minimum vent capacity of 91,000 Btu per hour and a recommended maximum vent capacity of 144,000 Btu per hour. The 80,000 Btu per hour fan-assisted appliance is outside this range, so the conclusion is that a single-wall metal vent connector cannot be used to vent this appliance using 10 feet of lateral for the connector.

However, if the 80,000 Btu per hour input appliance could be moved to within 5 feet of the vertical vent, then a 4-inch single-wall metal connector could be used to vent the appliance. Table G2428.2(2) shows the acceptable range of vent capacities for a 4-inch vent with 5 feet of lateral to be between 72,000 Btu per hour and 157,000 Btu per hour.

If the appliance cannot be moved closer to the vertical vent, then Type B vent could be used as the connector material. In this case, Table G2428.2(1) shows that for a 30-foot-high vent with 10 feet of lateral, the acceptable range of vent capacities for a 4-inch-diameter vent attached to a fan-assisted appliance is between 37,000 Btu per hour and 150,000 Btu per hour.

Example 3: Interpolating between table values.

An installer has an 80,000 Btu per hour input appliance with a 4-inch-diameter draft hood outlet that needs to be vented into a 12-foot-high Type B vent. The vent connector has a 5-foot lateral length and is also Type B. Can this appliance be vented using a 4-inch-diameter vent?

Solution:

Table G2428.2(1) is used in the case of an all Type B vent system. However, since there is no entry in Table G2428.2(1) for a height of 12 feet, interpolation must be used. Read down the 4-inch diameter NAT Max column to the row associated with 10-foot height and 5-foot lateral to find the capacity value of 77,000 Btu per hour. Read further down to the 15-foot height, 5-foot lateral row to find the capacity value of 87,000 Btu per hour. The difference between the 15-foot height capacity value and the 10-foot height capacity value is 10,000 Btu per hour. The capacity for a vent system with a 12-foot height is equal to the capacity for a 10-foot height plus $^2/_5$ of the difference between the 10-foot and 15-foot height values, or 77,000 + $^2/_5$ (10,000) = 81,000 Btu per hour. Therefore, a 4-inch-diameter vent may be used in the installation.

EXAMPLES USING COMMON VENTING TABLES

Example 4: Common venting two draft-hood-equipped appliances.

A 35,000 Btu per hour water heater is to be common vented with a 150,000 Btu per hour furnace using a common vent with a total height of 30 feet. The connector rise is 2 feet for the water heater with a horizontal length of 4 feet. The connector rise for the furnace is 3 feet with a horizontal length of 8 feet. Assume single-wall metal connectors will be used with Type B vent. What size connectors and combined vent should be used in this installation?

Solution:

Table G2428.3(2) should be used to size single-wall metal vent connectors attached to Type B vertical vents. In the vent connector capacity portion of Table G2428.3(2), find the row associated with a 30-foot vent height. For a 2-foot rise on the vent connector for the water heater, read the shaded columns for draft-hood-equipped appliances to find that a 3-inch-diameter vent connector has a capacity of 37,000 Btu per hour. Therefore, a 3-inch single-wall metal vent connector may be used with the water heater. For a draft-hood-equipped furnace with a 3-foot rise, read across the appropriate row to find that a 5-inch-diameter vent connector has a maximum capacity of 120,000 Btu per hour (which is too small for the furnace) and a 6-inch-diameter vent connector has a maximum vent capacity of 172,000 Btu per hour. Therefore, a 6-inch-diameter vent connector should be used with the 150,000 Btu per hour furnace. Since both vent connector horizontal lengths are less than the maximum lengths listed in Section G2428.3.2, the table values may be used without adjustments.

In the common vent capacity portion of Table G2428.3(2), find the row associated with a 30-foot vent height and read over to the NAT + NAT portion of the 6-inch-diameter column to find a maximum combined capacity of 257,000 Btu per hour. Since the two appliances total only 185,000 Btu per hour, a 6-inch common vent may be used.

Example 5a: Common venting a draft-hood-equipped water heater with a fan-assisted furnace into a Type B vent.

In this case, a 35,000 Btu per hour input draft-hood-equipped water heater with a 4-inch-diameter draft hood outlet, 2 feet of connector rise, and 4 feet of horizontal length is to be common vented with a 100,000 Btu per hour fan-assisted furnace with a 4-inch-diameter flue collar, 3 feet of connector rise, and 6 feet of horizontal length. The common vent consists of a 30-foot height of Type B vent. What are the recommended vent diameters for each connector and the common vent? The installer would like to use a single-wall metal vent connector.

Solution: - [Table G2428.3(2)]

Water Heater Vent Connector Diameter. Since the water heater vent connector horizontal length of 4 feet is less than the maximum value listed in Section G2428.3.2, the venting table values may be used without adjustments. Using the Vent Connector Capacity portion of Table G2428.3(2), read down the Total Vent Height (H) column to 30 feet and read across the 2-foot Connector Rise (R) row to the first Btu per hour rating in the NAT Max column that is equal to or greater than the water heater input rating. The table shows that a 3-inch vent connector has a maximum input rating of 37,000 Btu per hour. Although this is greater than the water heater input rating, a 3-inch vent connector is prohibited by Section G2428.3.17. A 4-

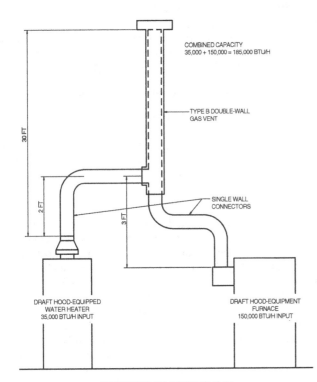

FIGURE B-17 (EXAMPLE 4)
COMMON VENTING TWO DRAFT-HOOD-EQUIPPED APPLIANCES

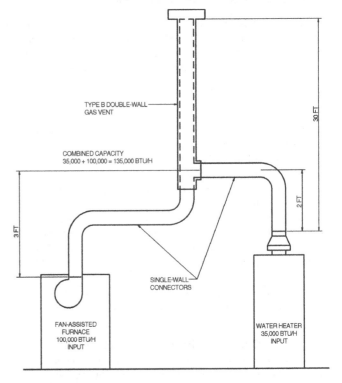

FIGURE B-18 (EXAMPLE 5A)
COMMON VENTING A DRAFT HOOD WITH A FAN-ASSISTED FURNACE INTO A TYPE B DOUBLE-WALL COMMON VENT

2006 INTERNATIONAL RESIDENTIAL CODE®

inch vent connector has a maximum input rating of 67,000 Btu per hour and is equal to the draft hood outlet diameter. A 4-inch vent connector is selected. Since the water heater is equipped with a draft hood, there are no minimum input rating restrictions.

Furnace Vent Connector Diameter. Using the Vent Connector Capacity portion of Table G2428.3(2), read down the Total Vent Height (*H*) column to 30 feet and across the 3-foot Connector Rise (*R*) row. Since the furnace has a fan-assisted combustion system, find the first FAN Max column with a Btu per hour rating greater than the furnace input rating. The 4-inch vent connector has a maximum input rating of 119,000 Btu per hour and a minimum input rating of 85,000 Btu per hour. The 100,000 Btu per hour furnace in this example falls within this range, so a 4-inch connector is adequate. Since the furnace vent connector horizontal length of 6 feet does not exceed the maximum value listed in Section G2428.3.2, the venting table values may be used without adjustment. If the furnace had an input rating of 80,000 Btu per hour, then a Type B vent connector [see Table G2428.3(1)] would be needed in order to meet the minimum capacity limit.

Common Vent Diameter. The total input to the common vent is 135,000 Btu per hour. Using the Common Vent Capacity portion of Table G2428.3(2), read down the Total Vent Height (*H*) column to 30 feet and across this row to find the smallest vent diameter in the FAN + NAT column that has a Btu per hour rating equal to or greater than 135,000 Btu per hour. The 4-inch common vent has a capacity of 132,000 Btu per hour and the 5-inch common vent has a capacity of 202,000 Btu per hour. Therefore, the 5-inch common vent should be used in this example.

Summary. In this example, the installer may use a 4-inch-diameter, single-wall metal vent connector for the water heater and a 4-inch-diameter, single-wall metal vent connector for the furnace. The common vent should be a 5-inch-diameter Type B vent.

Example 5b: Common venting into a masonry chimney.

In this case, the water heater and fan-assisted furnace of Example 5a are to be common vented into a clay tile-lined masonry chimney with a 30-foot height. The chimney is not exposed to the outdoors below the roof line. The internal dimensions of the clay tile liner are nominally 8 inches by 12 inches. Assuming the same vent connector heights, laterals, and materials found in Example 5a, what are the recommended vent connector diameters, and is this an acceptable installation?

Solution:

Table G2428.3(4) is used to size common venting installations involving single-wall connectors into masonry chimneys.

Water Heater Vent Connector Diameter. Using Table G2428.3(4), Vent Connector Capacity, read down the Total Vent Height (*H*) column to 30 feet, and read across the 2-foot Connector Rise (*R*) row to the first Btu per hour rating in the NAT Max column that is equal to or greater than the water heater input rating. The table shows that a 3-inch vent connector has a maximum input of only 31,000 Btu per hour while a 4-inch vent connector has a maximum input of 57,000 Btu per hour. A 4-inch vent connector must therefore be used.

Furnace Vent Connector Diameter. Using the Vent Connector Capacity portion of Table G2428.3(4), read down the Total Vent Height (*H*) column to 30 feet and across the 3-foot Connector Rise (*R*) row. Since the furnace has a fan-assisted combustion system, find the first FAN Max column with a Btu per hour rating greater than the furnace input rating. The 4-inch vent connector has a maximum input rating of 127,000 Btu per hour and a minimum input rating of 95,000 Btu per hour. The 100,000 Btu per hour furnace in this example falls within this range, so a 4-inch connector is adequate.

Masonry Chimney. From Table B-1, the equivalent area for a nominal liner size of 8 inches by 12 inches is 63.6 square inches. Using Table G2428.3(4), Common Vent Capacity, read down the FAN + NAT column under the Minimum Internal Area of Chimney value of 63 to the row for 30-foot height to find a capacity value of 739,000 Btu per hour. The combined input rating of the furnace and water heater, 135,000 Btu per hour, is less than the table value, so this is an acceptable installation.

Section G2428.3.13 requires the common vent area to be no greater than seven times the smallest listed appliance categorized vent area, flue collar area, or draft hood outlet area. Both appliances in this installation have 4-inch-diameter outlets. From Table B-1, the equivalent area for an inside diameter of 4 inches is 12.2 square inches. Seven times 12.2 equals 85.4, which is greater than 63.6, so this configuration is acceptable.

Example 5c: Common venting into an exterior masonry chimney.

In this case, the water heater and fan-assisted furnace of Examples 5a and 5b are to be common vented into an exterior masonry chimney. The chimney height, clay tile liner dimensions, and vent connector heights and laterals are the same as in Example 5b. This system is being installed in Charlotte, North Carolina. Does this exterior masonry chimney need to be relined? If so, what corrugated metallic liner size is recommended? What vent connector diameters are recommended?

Solution:

According to Section 504.3.20 of the *International Fuel Gas Code*, Type B vent connectors are required to be used with exterior masonry chimneys. Use Table 504.3(7) of the *International Fuel Gas Code* to size FAN+NAT common venting installations involving Type-B double wall connectors into exterior masonry chimneys.

The local 99-percent winter design temperature needed to use Table 504.3(7) can be found in the ASHRAE *Handbook of Fundamentals*. For Charlotte, North Carolina, this design temperature is 19°F.

Chimney Liner Requirement. As in Example 5b, use the 63 square inch Internal Area columns for this size clay tile liner. Read down the 63 square inch column of Table 504.3(7a) of the *International Fuel Gas Code* to the 30-foot height row to find that the combined appliance maximum input is 747,000 Btu per hour. The combined input rating of the appliances in this installation, 135,000 Btu per hour, is less than the maximum value, so this criterion is satisfied. Table 504.3(7b), at a 19°F design temperature, and at the same vent height and internal area used above, shows that the minimum allowable input rating of a space-heating appliance is 470,000 Btu per hour. The furnace input rating of 100,000 Btu per hour is less than this minimum value. So this criterion is not satisfied, and an alternative venting design needs to be used, such as a Type B vent shown in Example 5a or a listed chimney liner system shown in the remainder of the example.

According to Section G2428.3.15, Table G2428.3(1) or G2428.3(2) is used for sizing corrugated metallic liners in masonry chimneys, with the maximum common vent capacities reduced by 20 percent. This example will be continued assuming Type B vent connectors.

Water Heater Vent Connector Diameter. Using Table G2428.3(1), Vent Connector Capacity, read down the Total Vent Height (H) column to 30 feet, and read across the 2-foot Connector Rise (R) row to the first Btu/h rating in the NAT Max column that is equal to or greater than the water heater input rating. The table shows that a 3-inch vent connector has a maximum capacity of 39,000 Btu/h. Although this rating is greater than the water heater input rating, a 3-inch vent connector is prohibited by Section G2428.3.17. A 4-inch vent connector has a maximum input rating of 70,000 Btu/h and is equal to the draft hood outlet diameter. A 4-inch vent connector is selected.

Furnace Vent Connector Diameter. Using Table G2428.3(1), Vent Connector Capacity, read down the Vent Height (H) column to 30 feet, and read across the 3-foot Connector Rise (R) row to the first Btu per hour rating in the FAN Max column that is equal to or greater than the furnace input rating. The 100,000 Btu per hour furnace in this example falls within this range, so a 4-inch connector is adequate.

Chimney Liner Diameter. The total input to the common vent is 135,000 Btu per hour. Using the Common Vent Capacity Portion of Table G2428.3(1), read down the Vent Height (H) column to 30 feet and across this row to find the smallest vent diameter in the FAN+NAT column that has a Btu per hour rating greater than 135,000 Btu per hour. The 4-inch common vent has a capacity of 138,000 Btu per hour. Reducing the maximum capacity by 20 percent (Section G2428.3.15) results in a maximum capacity for a 4-inch corrugated liner of 110,000 Btu per hour, less than the total input of 135,000 Btu per hour. So a larger liner is needed. The 5-inch common vent capacity listed in Table G2428.3(1) is 210,000 Btu per hour, and after reducing by 20 percent is 168,000 Btu per hour. Therefore, a 5-inch corrugated metal liner should be used in this example.

Single-Wall Connectors. Once it has been established that relining the chimney is necessary, Type B double-wall vent connectors are not specifically required. This example could be redone using Table G2428.3(2) for single-wall vent connectors. For this case, the vent connector and liner diameters would be the same as found above with Type B double-wall connectors.

TABLE B-1
MASONRY CHIMNEY LINER DIMENSIONS WITH CIRCULAR EQUIVALENTS[a]

NOMINAL LINER SIZE (inches)	INSIDE DIMENSIONS OF LINER (inches)	INSIDE DIAMETER OR EQUIVALENT DIAMETER (inches)	EQUIVALENT AREA (square inches)
4 × 8	$2^1/_2 \times 6^1/_2$	4	12.2
		5	19.6
		6	28.3
		7	38.3
8 × 8	$6^3/_4 \times 6^3/_4$	7.4	42.7
		8	50.3
8 × 12	$6^1/_2 \times 10^1/_2$	9	63.6
		10	78.5
12 × 12	$9^3/_4 \times 9^3/_4$	10.4	83.3
		11	95
12 × 16	$9^1/_2 \times 13^1/_2$	11.8	107.5
		12	113.0
		14	153.9
16 × 16	$13^1/_4 \times 13^1/_4$	14.5	162.9
		15	176.7
16 × 20	13 × 17	16.2	206.1
		18	254.4
20 × 20	$16^3/_4 \times 16^3/_4$	18.2	260.2
		20	314.1
20 × 24	$16^1/_2 \times 20^1/_2$	20.1	314.2
		22	380.1
24 × 24	$20^1/_4 \times 20^1/_4$	22.1	380.1
		24	452.3
24 × 28	$20^1/_4 \times 20^1/_4$	24.1	456.2
28 × 28	$24^1/_4 \times 24^1/_4$	26.4	543.3
		27	572.5
30 × 30	$25^1/_2 \times 25^1/_2$	27.9	607
		30	706.8
30 × 36	$25^1/_2 \times 31^1/_2$	30.9	749.9
		33	855.3
36 × 36	$31^1/_2 \times 31^1/_2$	34.4	929.4
		36	1017.9

For SI: 1 inch = 25.4 mm, 1 square inch = 645.16 mm².

a. Where liner sizes differ dimensionally from those shown in Table B-1, equivalent diameters may be determined from published tables for square and rectangular ducts of equivalent carrying capacity or by other engineering methods.

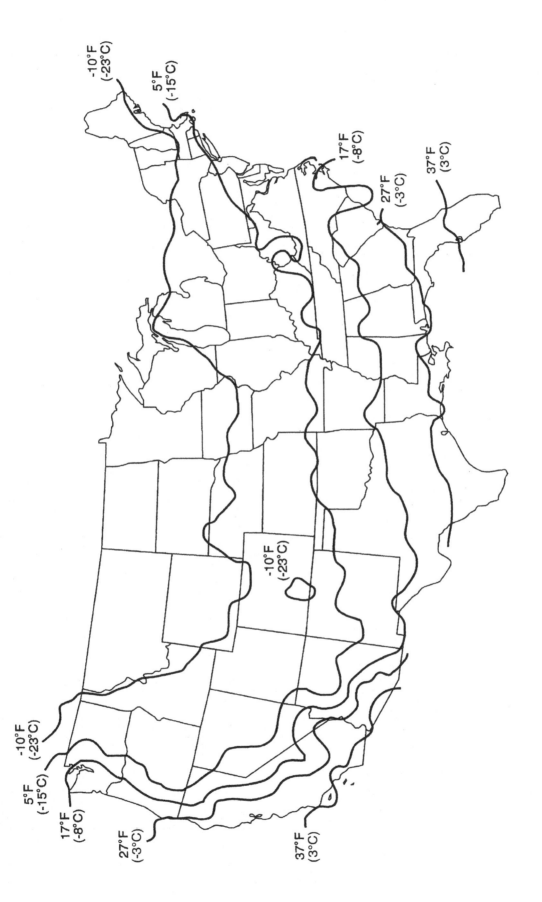

FIGURE B-19

EXIT TERMINALS OF MECHANICAL DRAFT AND DIRECT-VENT VENTING SYSTEMS

(This appendix is informative and is not part of the code. This appendix is an excerpt from the 2006 *International Fuel Gas Code,* coordinated with the section numbering of the *International Residential Code.*)

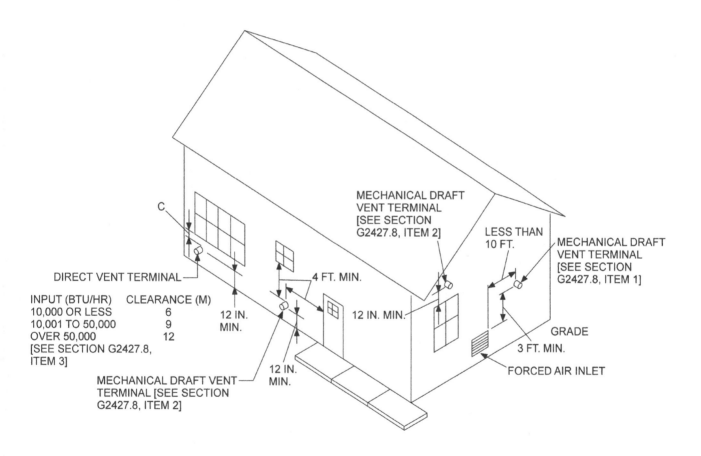

C

MECHANICAL DRAFT
VENT TERMINAL
[SEE SECTION
G2427.8, ITEM 2]

LESS THAN
10 FT.

MECHANICAL DRAFT
VENT TERMINAL
[SEE SECTION
G2427.8, ITEM 1]

DIRECT VENT TERMINAL

INPUT (BTU/HR)	CLEARANCE (M)
10,000 OR LESS	6
10,001 TO 50,000	9
OVER 50,000	12

[SEE SECTION G2427.8,
ITEM 3]

12 IN.
MIN.

4 FT. MIN.

12 IN. MIN.

GRADE

3 FT. MIN.

FORCED AIR INLET

MECHANICAL DRAFT VENT
TERMINAL [SEE SECTION
G2427.8, ITEM 2]

12 IN.
MIN.

For SI: 1 inch = 25.4 mm, 1 foot = 304.8 mm, 1 British thermal unit per hour = 0.2931 W.

APPENDIX C
EXIT TERMINALS OF MECHANICAL DRAFT AND DIRECT-VENT VENTING SYSTEMS

RECOMMENDED PROCEDURE FOR SAFETY INSPECTION OF AN EXISTING APPLIANCE INSTALLATION

(This appendix is informative and is not part of the code. This appendix is an excerpt from the 2006 International Fuel Gas Code, coordinated with the section numbering of the International Residential Code.)

The following procedure is intended as a guide to aid in determining that an appliance is properly installed and is in a safe condition for continuing use.

This procedure is predicated on central furnace and boiler installations, and it should be recognized that generalized procedures cannot anticipate all situations. Accordingly, in some cases, deviation from this procedure is necessary to determine safe operation of the equipment.

(a) This procedure should be performed prior to any attempt at modification of the appliance or of the installation.

(b) If it is determined there is a condition that could result in unsafe operation, the appliance should be shut off and the owner advised of the unsafe condition.

The following steps should be followed in making the safety inspection:

1. Conduct a check for gas leakage. (See Section G2417.6.)

2. Visually inspect the venting system for proper size and horizontal pitch and determine there is no blockage or restriction, leakage, corrosion and other deficiencies that could cause an unsafe condition.

3. Shut off all gas to the appliance and shut off any other fuel-gas-burning appliance within the same room. **Use the shut-off valve in the supply line to each appliance.**

4. Inspect burners and crossovers for blockage and corrosion.

5. **Applicable only to furnaces.** Inspect the heat exchanger for cracks, openings or excessive corrosion.

6. **Applicable only to boilers.** Inspect for evidence of water or combustion product leaks.

7. Insofar as is practical, close all building doors and windows and all doors between the space in which the appliance is located and other spaces of the building. Turn on clothes dryers. Turn on any exhaust fans, such as range hoods and bathroom exhausts, so they will operate at maximum speed. Do not operate a summer exhaust fan. Close fireplace dampers. If, after completing Steps 8 through 13, it is believed sufficient combustion air is not available, refer to Section G2407 of this code for guidance.

8. Place the appliance being inspected in operation. **Follow the lighting instructions.** Adjust the thermostat so appliance will operate continuously.

9. Determine that the pilot(s), where provided, is burning properly and that the main burner ignition is satisfactory by interrupting and reestablishing the electrical supply to the appliance in any convenient manner. If the appliance is equipped with a continuous pilot(s), test the pilot safety device(s) to determine if it is operating properly by extinguishing the pilot(s) when the main burner(s) is off and determining, after 3 minutes, that the main burner gas does not flow upon a call for heat. If the appliance is not provided with a pilot(s), test for proper operation of the ignition system in accordance with the appliance manufacturer's lighting and operating instructions.

10. Visually determine that the main burner gas is burning properly (i.e., no floating, lifting, or flashback). Adjust the primary air shutter(s) as required.

If the appliance is equipped with high and low flame controlling or flame modulation, check for proper main burner operation at low flame.

11. Test for spillage at the draft hood relief opening after 5 minutes of main burner operation. Use a flame of a match or candle or smoke.

12. Turn on all other fuel-gas-burning appliances within the same room so they will operate at their full inputs. **Follow lighting instructions for each appliance.**

13. Repeat Steps 10 and 11 on the appliance being inspected.

14. Return doors, windows, exhaust fans, fireplace dampers and any other fuel-gas-burning appliance to their previous conditions of use.

15. **Applicable only to furnaces.** Check both the limit control and the fan control for proper operation. Limit control operation can be checked by blocking the circulating air inlet or temporarily disconnecting the electrical supply to the blower motor and determining that the limit control acts to shut off the main burner gas.

16. **Applicable only to boilers.** Determine that the water pumps are in operating condition. Test low water cutoffs, automatic feed controls, pressure and temperature limit controls, and relief valves in accordance with the manufacturer's recommendations to determine that they are in operating condition.

APPENDIX E

MANUFACTURED HOUSING USED AS DWELLINGS

SECTION AE101
SCOPE

AE101.1 General. These provisions shall be applicable only to a manufactured home used as a single dwelling unit installed on privately owned (nonrental) lots and shall apply to the following:

1. Construction, alteration and repair of any foundation system which is necessary to provide for the installation of a manufactured home unit.

2. Construction, installation, addition, alteration, repair or maintenance of the building service equipment which is necessary for connecting manufactured homes to water, fuel, or power supplies and sewage systems.

3. Alterations, additions or repairs to existing manufactured homes. The construction, alteration, moving, demolition, repair and use of accessory buildings and structures and their building service equipment shall comply with the requirements of the codes adopted by this jurisdiction.

These provisions shall not be applicable to the design and construction of manufactured homes and shall not be deemed to authorize either modifications or additions to manufactured homes where otherwise prohibited.

Exception: In addition to these provisions, new and replacement manufactured homes to be located in flood hazard areas as established in Table R301.2(1) of the *International Residential Code* shall meet the applicable requirements of Sections R324 of the *International Residential Code*.

SECTION AE102
APPLICATION TO EXISTING MANUFACTURED HOMES AND BUILDING SERVICE EQUIPMENT

AE102.1 General. Manufactured homes and their building service equipment to which additions, alterations or repairs are made shall comply with all the requirements of these provisions for new facilities, except as specifically provided in this section.

AE102.2 Additions, alterations or repairs. Additions made to a manufactured home shall conform to one of the following:

1. Be certified under the National Manufactured Housing Construction and Safety Standards Act of 1974 (42 U.S.C. Section 5401, et seq.).

2. Be designed and constructed to conform with the applicable provisions of the National Manufactured Housing Construction and Safety Standards Act of 1974 (42 U.S.C. Section 5401, et seq.).

3. Be designed and constructed in conformance with the code adopted by this jurisdiction.

Additions shall be structurally separated from the manufactured home.

Exception: A structural separation need not be provided when structural calculations are provided to justify the omission of such separation.

Alterations or repairs may be made to any manufactured home or to its building service equipment without requiring the existing manufactured home or its building service equipment to comply with all the requirements of these provisions, provided the alteration or repair conforms to that required for new construction, and provided further that no hazard to life, health or safety will be created by such additions, alterations or repairs.

Alterations or repairs to an existing manufactured home which are nonstructural and do not adversely affect any structural member or any part of the building or structure having required fire protection may be made with materials equivalent to those of which the manufactured home structure is constructed, subject to approval by the building official.

Exception: The installation or replacement of glass shall be required for new installations.

Minor additions, alterations and repairs to existing building service equipment installations may be made in accordance with the codes in effect at the time the original installation was made subject to approval of the building official, and provided such additions, alterations and repairs will not cause the existing building service equipment to become unsafe, insanitary or overloaded.

AE102.3 Existing installations. Building service equipment lawfully in existence at the time of the adoption of the applicable codes may have their use, maintenance or repair continued if the use, maintenance or repair is in accordance with the original design and no hazard to life, health or property has been created by such building service equipment.

AE102.4 Existing occupancy. Manufactured homes which are in existence at the time of the adoption of these provisions may have their existing use or occupancy continued if such use or occupancy was legal at the time of the adoption of these provisions, provided such continued use is not dangerous to life, health and safety.

The use or occupancy of any existing manufactured home shall not be changed unless evidence satisfactory to the building official is provided to show compliance with all applicable provisions of the codes adopted by this jurisdiction. Upon any change in use or occupancy, the manufactured home shall cease to be classified as such within the intent of these provisions.

AE102.5 Maintenance. All manufactured homes and their building service equipment, existing and new, and all parts thereof shall be maintained in a safe and sanitary condition. All device or safeguards which are required by applicable codes or by the Manufactured Home Standards shall be maintained in conformance with the code or standard under which it was installed. The owner or the owner's designated agent shall be responsible for the maintenance of manufactured homes, accessory buildings, structures and their building service equipment. To determine compliance with this subsection, the building official may cause any manufactured home, accessory building or structure to be reinspected.

AE102.6 Relocation. Manufactured homes which are to be relocated within this jurisdiction shall comply with these provisions.

SECTION AE201
DEFINITIONS

AE201.1 General. For the purpose of these provisions, certain abbreviations, terms, phrases, words and their derivatives shall be construed as defined or specified herein.

ACCESSORY BUILDING. Any building or structure, or portion thereto, located on the same property as a manufactured home which does not qualify as a manufactured home as defined herein.

BUILDING SERVICE EQUIPMENT. Refers to the plumbing, mechanical and electrical equipment including piping, wiring, fixtures and other accessories which provide sanitation, lighting, heating ventilation, cooling, fire protection and facilities essential for the habitable occupancy of a manufactured home or accessory building or structure for its designated use and occupancy.

MANUFACTURED HOME. A structure transportable in one or more sections which, in the traveling mode, is 8 body feet (2438 body mm) or more in width or 40 body feet (12 192 body mm) or more in length or, when erected on site, is 320 or more square feet (30 m²), and which is built on a permanent chassis and designed to be used as a dwelling with or without a permanent foundation when connected to the required utilities, and includes the plumbing, heating, air-conditioning and electrical systems contained therein; except that such term shall include any structure which meets all the requirements of this paragraph except the size requirements and with respect to which the manufacturer voluntarily files a certification required by the secretary (HUD) and complies with the standards established under this title.

For mobile homes built prior to June 15, 1976, a label certifying compliance to the Standard for Mobile Homes, NFPA 501, ANSI 119.1, in effect at the time of manufacture is required. For the purpose of these provisions, a mobile home shall be considered a manufactured home.

MANUFACTURED HOME INSTALLATION. Construction which is required for the installation of a manufactured home, including the construction of the foundation system, required structural connections thereto and the installation of on-site water, gas, electrical and sewer systems and connections thereto which are necessary for the normal operation of the manufactured home.

MANUFACTURED HOME STANDARDS. The Manufactured Home Construction and Safety Standards as promulgated by the United States Department of Housing and Urban Development.

PRIVATELY OWNED (NONRENTAL) LOT. A parcel of real estate outside of a manufactured home rental community (park) where the land and the manufactured home to be installed thereon are held in common ownership.

SECTION AE301
PERMITS

AE301.1 Initial installation. A manufactured home shall not be installed on a foundation system, reinstalled or altered without first obtaining a permit from the building official. A separate permit shall be required for each manufactured home installation. When approved by the building official, such permit may include accessory buildings and structures and their building service equipment when the accessory buildings or structures will be constructed in conjunction with the manufactured home installation.

AE301.2 Additions, alterations and repairs to a manufactured home. A permit shall be obtained to alter, remodel, repair or add accessory buildings or structures to a manufactured home subsequent to its initial installation. Permit issuance and fees therefor shall be in conformance with the codes applicable to the type of work involved.

An addition made to a manufactured home as defined in these provisions shall comply with these provisions.

AE301.3 Accessory buildings. Except as provided in Section AE301.1, permits shall be required for all accessory buildings and structures and their building service equipment. Permit issuance and fees therefor shall be in conformance with the codes applicable to the types of work involved.

AE301.4 Exempted work. A permit shall not be required for the types of work specifically exempted by the applicable codes. Exemption from the permit requirements of any of said codes shall not be deemed to grant authorization for any work to be done in violation of the provisions of said codes or any other laws or ordinances of this jurisdiction.

SECTION AE302
APPLICATION FOR PERMIT

AE302.1 Application. To obtain a manufactured home installation permit, the applicant shall first file an application in writing on a form furnished by the building official for that purpose. At the option of the building official, every such application shall:

1. Identify and describe the work to be covered by the permit for which application is made.

2. Describe the land on which the proposed work is to be done by legal description, street address or similar description that will readily identify and definitely locate the proposed building or work.

3. Indicate the use or occupancy for which the proposed work is intended.

4. Be accompanied by plans, diagrams, computations and specifications and other data as required in Section AE302.2.

5. Be accompanied by a soil investigation when required by Section AE502.2.

6. State the valuation of any new building or structure or any addition, remodeling or alteration to an existing building.

7. Be signed by permittee, or permittee's authorized agent, who may be required to submit evidence to indicate such authority.

8. Give such other data and information as may be required by the building official.

AE302.2 Plans and specifications. Plans, engineering calculations, diagrams and other data as required by the building official shall be submitted in not less than two sets with each application for a permit. The building official may require plans, computations and specifications to be prepared and designed by an engineer or architect licensed by the state to practice as such.

Where no unusual site conditions exist, the building official may accept approved standard foundation plans and details in conjunction with the manufacturer's approved installation instructions without requiring the submittal of engineering calculations.

AE302.3 Information on plans and specifications. Plans and specifications shall be drawn to scale on substantial paper or cloth and shall be of sufficient clarity to indicate the location, nature and extent of the work proposed and shown in detail that it will conform to the provisions of these provisions and all relevant laws, ordinances, rules and regulations. The building official shall determine what information is required on plans and specifications to ensure compliance.

SECTION AE303
PERMITS ISSUANCE

AE303.1 Issuance. The application, plans and specifications and other data filed by an applicant for permit shall be reviewed by the building official. Such plans may be reviewed by other departments of this jurisdiction to verify compliance with any applicable laws under their jurisdiction. If the building official finds that the work described in an application for a permit and the plans, specifications and other data filed therewith conform to the requirements of these provisions and other data filed therewith conform to the requirements of these provisions and other pertinent codes, laws and ordinances, and that the fees specified in Section AE304 have been paid, the building official shall issue a permit therefor to the applicant.

When the building official issues the permit where plans are required, the building official shall endorse in writing or stamp the plans and specifications APPROVED. Such approved plans and specifications shall not be changed, modified or altered without authorization from the building official, and all work shall be done in accordance with the approved plans.

AE303.2 Retention of plans. One set of approved plans and specifications shall be returned to the applicant and shall be kept on the site of the building or work at all times during which the work authorized thereby is in progress. One set of approved plans, specification and computations shall be retained by the building official until final approval of the work.

AE303.3 Validity of permit. The issuance of a permit or approval of plans and specifications shall not be construed to be a permit for, or an approval of, any violation of any of the provisions of these provisions or other pertinent codes of any other ordinance of the jurisdiction. No permit presuming to give authority to violate or cancel these provisions shall be valid.

The issuance of a permit based on plans, specifications and other data shall not prevent the building official from thereafter requiring the correction of errors in said plans, specifications and other data, or from preventing building operations being carried on thereunder when in violation of these provisions or of any other ordinances of this jurisdiction.

AE303.4 Expiration. Every permit issued by the building official under these provisions shall expire by limitation and become null and void if the work authorized by such permit is not commenced within 180 days from the date of such permit, or if the work authorized by such permit is suspended or abandoned at any time after the work is commenced for a period of 180 days. Before such work can be recommenced, a new permit shall be first obtained, and the fee therefor shall be one-half the amount required for a new permit for such work, provided no changes have been made or will be made in the original plans and specifications for such work, and provided further that such suspension or abandonment has not exceeded one year. In order to renew action on a permit after expiration, the permittee shall pay a new full permit fee.

Any permittee holding an unexpired permit may apply for an extension of the time within which work may commence under that permit when the permittee is unable to commence work within the time required by this section for good and satisfactory reasons. The building official may extend the time for action by the permittee for a period not exceeding 180 days upon written request by the permittee showing that circumstances beyond the control of the permittee have prevented action from being taken. No permit shall be extended more than once.

AE303.5 Suspension or revocation. The building official may, in writing, suspend or revoke a permit issued under these provisions whenever the permit is issued in error or on the basis of incorrect information supplied, or in violation of any ordinance or regulation or any of these provisions.

SECTION AE304
FEES

AE304.1 Permit fees. The fee for each manufactured home installation permit shall be established by the building official.

When permit fees are to be based on the value or valuation of the work to be performed, the determination of value or valuation under these provisions shall be made by the building official. The value to be used shall be the total value of all work required for the manufactured home installation plus the total value of all work required for the construction of accessory buildings and structures for which the permit is issued as well as all finish work, painting, roofing, electrical, plumbing, heating, air conditioning, elevators, fire-extinguishing systems and any other permanent equipment which is a part of the accessory building or structure. The value of the manufactured home itself shall not be included.

AE304.2 Plan review fees. When a plan or other data are required to be submitted by Section AE302.2, a plan review fee shall be paid at the time of submitting plans and specifications for review. Said plan review fee shall be as established by the building official. Where plans are incomplete or changed so as to require additional plan review, an additional plan review fee shall be charged at a rate as established by the building official.

AE304.3 Other provisions.

AE304.3.1 Expiration of plan review. Applications for which no permit is issued within 180 days following the date of application shall expire by limitation, and plans and other data submitted for review may thereafter be returned to the applicant or destroyed by the building official. The building official may extend the time for action by the applicant for a period not exceeding 180 days upon request by the applicant showing that circumstances beyond the control of the applicant have prevented action from being taken. No application shall be extended more than once. In order to renew action on an application after expiration, the applicant shall resubmit plans and pay a new plan review fee.

AE304.3.2 Investigation fees: work without a permit.

AE304.3.2.1 Investigation. Whenever any work for which a permit is required by these provisions has been commenced without first obtaining said permit, a special investigation shall be made before a permit may be issued for such work.

AE304.3.2.2 Fee. An investigation fee, in addition to the permit fee, shall be collected whether or not a permit is then or subsequently issued. The investigation fee shall be equal to the amount of the permit fee required. The minimum investigation fee shall be the same as the minimum fee established by the building official. The payment of such investigation fee shall not exempt any person from compliance with all other provisions of either these provisions or other pertinent codes or from any penalty prescribed by law.

E304.3.3 Fee refunds.

AE304.3.3.1 Permit fee erroneously paid or collected. The building official may authorize the refunding of any fee paid hereunder which was erroneously paid or collected.

AE304.3.3.2 Permit fee paid when no work done. The building official may authorize the refunding of not more than 80 percent of the permit fee paid when no work has been done under a permit issued in accordance with these provisions.

AE304.3.3.3 Plan review fee. The building official may authorize the refunding of not more than 80 percent of the plan review fee paid when an application for a permit for which a plan review fee has been paid is withdrawn or canceled before any plan reviewing is done.

The building official shall not authorize the refunding of any fee paid except upon written application by the original permittee not later than 180 days after the date of the fee payment.

SECTION AE305
INSPECTIONS

AE305.1 General. All construction or work for which a manufactured home installation permit is required shall be subject to inspection by the building official, and certain types of construction shall have continuous inspection by special inspectors as specified in Section AE306. A survey of the lot may be required by the building official to verify that the structure is located in accordance with the approved plans.

It shall be the duty of the permit applicant to cause the work to be accessible and exposed for inspection purposes. Neither the building official nor this jurisdiction shall be liable for expense entailed in the removal or replacement of any material required to allow inspection.

AE305.2 Inspection requests. It shall be the duty of the person doing the work authorized by a manufactured home installation permit to notify the building official that such work is ready for inspection. The building official may require that every request for inspection be filed at least one working day before such inspection is desired. Such request may be in writing or by telephone at the option of the building official.

It shall be the duty of the person requesting any inspections required either by these provisions or other applicable codes to provide access to and means for proper inspection of such work.

AE305.3 Inspection record card. Work requiring a manufactured home installation permit shall not be commenced until the permit holder or the permit holder's agent shall have posted an inspection record card in a conspicuous place on the premises and in such position as to allow the building official conveniently to make the required entries thereon regarding inspection of the work. This card shall be maintained in such position by the permit holder until final approval has been issued by the building official.

AE305.4 Approval required. Work shall not be done on any part of the manufactured home installation beyond the point indicated in each successive inspection without first obtaining the approval of the building official. Such approval shall be given only after an inspection has been made of each successive step in the construction as indicated by each of the inspections required in Section AE305.5. There shall be a final inspection and approval of the manufactured home installation, including connections to its building service equipment, when completed and ready for occupancy or use.

AE305.5 Required inspections.

AE305.5.1 Structural inspections for the manufactured home installation. Reinforcing steel or structural framework of any part of any manufactured home foundation system shall not be covered or concealed without first obtaining the approval of the building official. The building official, upon notification from the permit holder or the permit holder's agent, shall make the following inspections and shall either approve that portion of the construction as completed or shall notify the permit holder or the permit holder's agent wherein the same fails to comply with these provisions or other applicable codes:

1. Foundation inspection: To be made after excavations for footings are completed and any required reinforcing steel is in place. For concrete foundations, any required forms shall be in place prior to inspection. All materials for the foundation shall be on the job, except where concrete from a central mixing plant (commonly termed "transit mixed") is to be used, the concrete materials need not be on the job. Where the foundation is to be constructed of approved treated wood, additional framing inspections as required by the building official may be required.

2. Concrete slab or under-floor inspection: To be made after all in-slab or underfloor building service equipment, conduit, piping accessories and other ancillary equipment items are in place but before any concrete is poured or the manufactured home is installed.

3. Anchorage inspection: To be made after the manufactured home has been installed and permanently anchored.

AE305.5.2 Structural inspections for accessory building and structures. Inspections for accessory buildings and structures shall be made as set forth in this code.

AE305.5.3 Building service equipment inspections. All building service equipment which is required as a part of a manufactured home installation, including accessory buildings and structures authorized by the same permit, shall be inspected by the building official. Building service equipment shall be inspected and tested as required by the applicable codes. Such inspections and testing shall be limited to site construction and shall not include building service equipment which is a part of the manufactured home itself. No portion of any building service equipment intended to be concealed by any permanent portion of the construction shall be concealed until inspected and approved. Building service equipment shall not be connected to the water, fuel or power supply or sewer system until authorized by the building official.

AE305.5.4 Final inspection. When finish grading and the manufactured home installation, including the installation of all required building service equipment, is completed and the manufactured home is ready for occupancy, a final inspection shall be made.

AE305.6 Other inspections. In addition to the called inspections specified above, the building official may make or require other inspections of any construction work to as certain compliance with these provisions or other codes and laws which are enforced by the code enforcement agency.

SECTION AE306
SPECIAL INSPECTIONS

AE306.1 General. In addition to the inspections required by Section AE305, the building official may require the owner to employ a special inspector during construction of specific types of work as described in this code.

SECTION AE307
UTILITY SERVICE

AE307.1 General. Utility service shall not be provided to any building service equipment which is regulated by these provisions or other applicable codes and for which a manufactured home installation permit is required by these provisions until approved by the building official.

SECTION AE401
OCCUPANCY CLASSIFICATION

AE401.1 Manufactured homes. A manufactured home shall be limited in use to use as a single dwelling unit.

AE401.2 Accessory buildings. Accessory buildings shall be classified as to occupancy by the building official as set forth in this code.

SECTION AE402
LOCATION ON PROPERTY

AE402.1 General. Manufactured homes and accessory buildings shall be located on the property in accordance with applicable codes and ordinances of this jurisdiction.

SECTION AE501
DESIGN

AE501.1 General. A manufactured home shall be installed on a foundation system which is designed and constructed to sustain within the stress limitations specified in this code and all loads specified in this code.

> **Exception:** When specifically authorized by the building official, foundation and anchorage systems which are constructed in accordance with the methods specified in Section AE600 of these provisions, or in the United States Department of Housing and Urban Development Handbook, *Permanent Foundations for Manufactured Housing,* 1984 Edition, Draft, shall be deemed to meet the requirements of this Appendix E.

AE501.2 Manufacturer's installation instructions. The installation instructions as provided by the manufacturer of the manufactured home shall be used to determine permissible points of support for vertical loads and points of attachment for anchorage systems used to resist horizontal and uplift forces.

AE501.3 Rationality. Any system or method of construction to be used shall admit to a rational analysis in accordance with well-established principles of mechanics.

SECTION AE502
FOUNDATION SYSTEMS

AE502.1 General. Foundation systems designed and constructed in accordance with this section may be considered as a permanent installation.

AE502.2 Soil classification. The classification of the soil at each manufactured home site shall be determined when required by the building official. The building official may require that the determination be made by an engineer or architect licensed by the state to conduct soil investigations.

The classification shall be based on observation and any necessary tests of the materials disclosed by borings or excavations made in appropriate locations. Additional studies may be necessary to evaluate soil strength, the effect of moisture variation on soil-bearing capacity, compressibility and expansiveness.

When required by the building official, the soil classification design bearing capacity and lateral pressure shall be shown on the plans.

AE502.3 Footings and foundations. Footings and foundations, unless otherwise specifically provided, shall be constructed of materials specified by this code for the intended use and in all cases shall extend below the frost line. Footings of concrete and masonry shall be of solid material. Foundations supporting untreated wood shall extend at least 8 inches (203 mm) above the adjacent finish grade. Footings shall have a minimum depth below finished grade of 12 inches (305 mm) unless a greater depth is recommended by a foundation investigation.

Piers and bearing walls shall be supported on masonry or concrete foundations or piles, or other approved foundation systems which shall be of sufficient capacity to support all loads.

AE502.4 Foundation design. When a design is provided, the foundation system shall be designed in accordance with the applicable structural provisions of this code and shall be designed to minimize differential settlement. Where a design is not provided, the minimum foundation requirements shall be as set forth in this code.

AE502.5 Drainage. Provisions shall be made for the control and drainage of surface water away from the manufactured home.

AE502.6 Under-floor clearances—ventilation and access. A minimum clearance of 12 inches (305 mm) shall be maintained beneath the lowest member of the floor support framing system. Clearances from the bottom of wood floor joists or perimeter joists shall be as specified in this code.

Under-floor spaces shall be ventilated with openings as specified in this code. If combustion air for one or more heat-producing appliances is taken from within the under-floor spaces, ventilation shall be adequate for proper appliance operation.

Under-floor access openings shall be provided. Such openings shall be not less than 18 inches (457 mm) in any dimension and not less than 3 square feet (0.279 m²) in area and shall be located so that any water supply and sewer drain connections located under the manufactured home are accessible.

SECTION AE503
SKIRTING AND PERIMETER ENCLOSURES

AE503.1 Skirting and permanent perimeter enclosures. Skirting and permanent perimeter enclosures shall be installed only where specifically required by other laws or ordinances. Skirting, when installed, shall be of material suitable for exterior exposure and contact with the ground. Permanent perimeter enclosures shall be constructed of materials as required by this code for regular foundation construction.

Skirting shall be installed in accordance with the skirting manufacturer's installation instructions. Skirting shall be adequately secured to assure stability, to minimize vibration and susceptibility to wind damage, and to compensate for possible frost heave.

AE503.2 Retaining walls. Where retaining walls are used as a permanent perimeter enclosure, they shall resist the lateral displacements of soil or other materials and shall conform to this code as specified for foundation walls. Retaining walls and foundation walls shall be constructed of approved treated wood, concrete, masonry or other approved materials or combination of materials as for foundations as specified in this code. Siding materials shall extend below the top of the exterior of the retaining or foundation wall or the joint between siding and enclosure wall shall be flashed in accordance with this code.

SECTION AE504
STRUCTURAL ADDITIONS

AE504.1 General. Accessory buildings shall not be structurally supported by or attached to a manufactured home unless engineering calculations are submitted to substantiate any proposed structural connection.

> **Exception:** The building official may waive the submission of engineering calculations if it is found that the nature of the work applied for is such that engineering calculations are not necessary to show conformance to these provisions.

SECTION AE505
BUILDING SERVICE EQUIPMENT

AE505.1 General. The installation, alteration, repair, replacement, addition to or maintenance of the building service equipment within the manufactured home shall conform to regulations set forth in the Manufactured Home Standards. Such work which is located outside the manufactured home shall comply with the applicable codes adopted by this jurisdiction.

SECTION AE506
EXITS

AE506.1 Site development. Exterior stairways and ramps which provide egress to the public way shall comply with applicable provisions of this code.

AE506.2 Accessory buildings. Every accessory building or portion thereof shall be provided with exits as required by this code.

SECTION AE507
OCCUPANCY, FIRE SAFETY AND ENERGY CONSERVATION STANDARDS

AE507.1 General. Alterations made to a manufactured home subsequent to its initial installation shall conform to the occupancy, fire-safety and energy conservation requirements set forth in the Manufactured Home Standards.

SECTION AE600
SPECIAL REQUIREMENTS FOR FOUNDATION SYSTEMS

AE600.1 General. Section AE600 is applicable only when specifically authorized by the building official.

SECTION AE601
FOOTINGS AND FOUNDATIONS

AE601.1 General. The capacity of individual load-bearing piers and their footings shall be sufficient to sustain all loads specified in this code within the stress limitations specified in this code. Footings, unless otherwise approved by the building official, shall be placed level on firm, undisturbed soil or an engineered fill which is free of organic material, such as weeds and grasses. Where used, an engineered fill shall provide a minimum load-bearing capacity of not less than 1,000 psf (48 kN/m²). Continuous footings shall conform to the requirements of this code. Section AE502 of these provisions shall apply to footings and foundations constructed under the provisions of this section.

SECTION AE602
PIER CONSTRUCTION

AE602.1 General. Piers shall be designed and constructed to distribute loads evenly. Multiple section homes may have concentrated roof loads which will require special consideration. Load-bearing piers may be constructed utilizing one of the methods listed below. Such piers shall be considered to resist only vertical forces acting in a downward direction. They shall not be considered as providing any resistance to horizontal loads induced by wind or earthquake forces.

1. A prefabricated load-bearing device that is listed and labeled for the intended use.

2. Mortar shall comply with ASTM C 270 Type M, S or N; this may consist of one part portland cement, one-half part hydrated lime and four parts sand by volume. Lime shall not be used with plastic or waterproof cement.

3. A cast-in-place concrete pier with concrete having specified compressive strength at 28 days of 2,500 psi (17 225 kPa).

Alternate materials and methods of construction may be used for piers which have been designed by an engineer or architect licensed by the state to practice as such.

Caps and leveling spacers may be used for leveling of the manufactured home. Spacing of piers shall be as specified in the manufacturer's installation instructions, if available, or by an approved designer.

SECTION AE603
HEIGHT OF PIERS

AE603.1 General. Piers constructed as indicated in Section AE602 may have heights as follows:

1. Except for corner piers, piers 36 inches (914 mm) or less in height may be constructed of masonry units, placed with cores or cells vertically. Piers shall be installed with their long dimension at right angles to the main frame member they support and shall have a minimum cross-sectional area of 128 square inches (82 560 mm²). Piers shall be capped with minimum 4-inch (102 mm) solid masonry units or equivalent.

2. Piers between 36 and 80 inches (914 mm and 2032 mm) in height and all corner piers over 24 inches (610 mm) in height shall be at least 16 inches by 16 inches (406 mm by 406 mm) consisting of interlocking masonry units and shall be fully capped with minimum 4-inch (102 mm) solid masonry units or equivalent.

3. Piers over 80 inches (2032 mm) in height may be constructed in accordance with the provisions of Item 2 above, provided the piers shall be filled solid with grout and reinforced with four continuous No. 5 bars. One bar shall be placed in each corner cell of hollow masonry unit piers or in each corner of the grouted space of piers constructed of solid masonry units.

4. Cast-in-place concrete piers meeting the same size and height limitations of Items 1, 2 and 3 above may be substituted for piers constructed of masonry units.

SECTION AE604
ANCHORAGE INSTALLATIONS

AE604.1 Ground anchors. Ground anchors shall be designed and installed to transfer the anchoring loads to the ground. The load-carrying portion of the ground anchors shall be installed to the full depth called for by the manufacturer's installation directions and shall extend below the established frost line into undisturbed soil.

Manufactured ground anchors shall be listed and installed in accordance with the terms of their listing and the anchor manufacturer's instructions and shall include means of attachment of ties meeting the requirements of Section AE605. Ground anchor manufacturer's installation instructions shall include the amount of preload required and load capacity in various types of soil. These instructions shall include tensioning adjustments which may be needed to prevent damage to the manufactured home, particularly damage that can be caused by frost heave. Each ground anchor shall be marked with the manufacturer's identification and listed model identification number which shall be visible after installation. Instructions shall accompany each listed ground anchor specifying the types of soil for which the anchor is suitable under the requirements of this section.

Each approved ground anchor, when installed, shall be capable of resisting an allowable working load at least equal to 3,150 pounds (14 kN) in the direction of the tie plus a 50 percent overload [4,725 pounds (21 kN) total] without failure. Failure shall be considered to have occurred when the anchor moves more than 2 inches (51 mm) at a load of 4,725 pounds (21 kN) in the direction of the tie installation. Those ground anchors which are designed to be installed so that loads on the anchor are other than direct withdrawal shall be designed and installed to resist an applied design load of 3,150 pounds (14 kN) at 40 to 50 degrees from vertical or within the angle limitations specified by the home manufacturer without displacing the tie end of the anchor more than 4 inches (102 mm) horizontally. Anchors designed for connection of multiple ties shall be capable of resisting the combined working load and overload consistent with the intent expressed herein.

When it is proposed to use ground anchors and the building official has reason to believe that the soil characteristics at a given site are such as to render the use of ground anchors advisable, or when there is doubt regarding the ability of the ground anchors to obtain their listed capacity, the building official may require that a representative field installation be made at the site in question and tested to demonstrate ground anchor capacity. The building official shall approve the test procedures.

AE604.2 Anchoring equipment. Anchoring equipment, when installed as a permanent installation, shall be capable of resisting all loads as specified within these provisions. When the stabilizing system is designed by an engineer or architect licensed by the state to practice as such, alternative designs may be used, providing the anchoring equipment to be used is capable of withstanding a load equal to 1.5 times the calculated load. All anchoring equipment shall be listed and labeled as being capable of meeting the requirements of these provisions. Anchors as specified in this code may be attached to the main frame of the manufactured home by an approved $^3/_{16}$-inch-thick (4.76 mm) slotted steel plate anchoring device. Other anchoring devices or methods meeting the requirements of these provisions may be permitted when approved by the building official.

Anchoring systems shall be so installed as to be permanent. Anchoring equipment shall be so designed to prevent self-disconnection with no hook ends used.

AE604.3 Resistance to weather deterioration. All anchoring equipment, tension devices and ties shall have a resistance to deterioration as required by this code.

AE604.4 Tensioning devices. Tensioning devices, such as turnbuckles or yoke-type fasteners, shall be ended with clevis or welded eyes.

SECTION AE605
TIES, MATERIALS AND INSTALLATION

AE605.1 General. Steel strapping, cable, chain or other approved materials shall be used for ties. All ties shall be fastened to ground anchors and drawn tight with turnbuckles or other adjustable tensioning devices or devices supplied with the ground anchor. Tie materials shall be capable of resisting an allowable working load of 3,150 pounds (14 kN) with no more than 2 percent elongation and shall withstand a 50 percent overload [4,750 pounds (21 kN)]. Ties shall comply with the weathering requirements of Section AE604.3. Ties shall connect the ground anchor and the main structural frame. Ties shall not connect to steel outrigger beams which fasten to and intersect the main structural frame unless specifically stated in the manufacturer's installation instructions. Connection of cable ties to main frame members shall be $^5/_8$-inch (15.9 mm) closed-eye bolts affixed to the frame member in an approved manner. Cable ends shall be secured with at least two U-bolt cable clamps with the "U" portion of the clamp installed on the short (dead) end of the cable to assure strength equal to that required by this section.

Wood floor support systems shall be fixed to perimeter foundation walls in accordance with provisions of this code. The minimum number of ties required per side shall be sufficient to resist the wind load stated in this code. Ties shall be evenly spaced as practicable along the length of the manufactured home with the distance from each end of the home and the tie nearest that end not exceeding 8 feet (2438 mm). When continuous straps are provided as vertical ties, such ties shall be positioned at rafters and studs. Where a vertical tie and diagonal tie are located at the same place, both ties may be connected to a single anchor, provided the anchor used is capable of carrying both loadings. Multisection manufactured homes require diagonal ties only. Diagonal ties shall be installed on the exterior main frame and slope to the exterior at an angle of 40 to 50 degrees from the vertical or within the angle limitations specified by the home manufacturer. Vertical ties which are not continuous over the top of the manufactured home shall be attached to the main frame.

SECTION AE606
REFERENCED STANDARDS

ASTMC 270-04 Specification for Mortar
for Unit Masonry AE602

NFPA 501-03 Standard on Manufactured
Housing AE201

APPENDIX F

RADON CONTROL METHODS

SECTION AF101
SCOPE

AF101.1 General. This appendix contains requirements for new construction in jurisdictions where radon-resistant construction is required.

Inclusion of this appendix by jurisdictions shall be determined through the use of locally available data or determination of Zone 1 designation in Figure AF101.

SECTION AF102
DEFINITIONS

AF102.1 General. For the purpose of these requirements, the terms used shall be defined as follows:

SUBSLAB DEPRESSURIZATION SYSTEM (Passive). A system designed to achieve lower sub-slab air pressure relative to indoor air pressure by use of a vent pipe routed through the conditioned space of a building and connecting the sub-slab area with outdoor air, thereby relying on the convective flow of air upward in the vent to draw air from beneath the slab.

SUBSLAB DEPRESSURIZATION SYSTEM (Active). A system designed to achieve lower sub-slab air pressure relative to indoor air pressure by use of a fan-powered vent drawing air from beneath the slab.

DRAIN TILE LOOP. A continuous length of drain tile or perforated pipe extending around all or part of the internal or external perimeter of a basement or crawl space footing.

RADON GAS. A naturally-occurring, chemically inert, radioactive gas that is not detectable by human senses. As a gas, it can move readily through particles of soil and rock and can accumulate under the slabs and foundations of homes where it can easily enter into the living space through construction cracks and openings.

SOIL-GAS-RETARDER. A continuous membrane of 6-mil (0.15 mm) polyethylene or other equivalent material used to retard the flow of soil gases into a building.

SUBMEMBRANE DEPRESSURIZATION SYSTEM. A system designed to achieve lower-sub-membrane air pressure relative to crawl space air pressure by use of a vent drawing air from beneath the soil-gas-retarder membrane.

SECTION AF103
REQUIREMENTS

AF103.1 General. The following construction techniques are intended to resist radon entry and prepare the building for post-construction radon mitigation, if necessary (see Figure AF102). These techniques are required in areas where designated by the jurisdiction.

AF103.2 Subfloor preparation. A layer of gas-permeable material shall be placed under all concrete slabs and other floor systems that directly contact the ground and are within the walls of the living spaces of the building, to facilitate future installation of a sub-slab depressurization system, if needed. The gas-permeable layer shall consist of one of the following:

1. A uniform layer of clean aggregate, a minimum of 4 inches (102 mm) thick. The aggregate shall consist of material that will pass through a 2-inch (51 mm) sieve and be retained by a $^1/_4$-inch (6.4 mm) sieve.

2. A uniform layer of sand (native or fill), a minimum of 4 inches (102 mm) thick, overlain by a layer or strips of geotextile drainage matting designed to allow the lateral flow of soil gases.

3. Other materials, systems or floor designs with demonstrated capability to permit depressurization across the entire sub-floor area.

AF103.3 Soil-gas-retarder. A minimum 6-mil (0.15 mm) [or 3-mil (0.075 mm) cross-laminated] polyethylene or equivalent flexible sheeting material shall be placed on top of the gas-permeable layer prior to casting the slab or placing the floor assembly to serve as a soil-gas-retarder by bridging any cracks that develop in the slab or floor assembly and to prevent concrete from entering the void spaces in the aggregate base material. The sheeting shall cover the entire floor area with separate sections of sheeting lapped at least 12 inches (305 mm). The sheeting shall fit closely around any pipe, wire or other penetrations of the material. All punctures or tears in the material shall be sealed or covered with additional sheeting.

AF103.4 Entry routes. Potential radon entry routes shall be closed in accordance with Sections AF103.4.1 through AF103.4.10.

AF103.4.1 Floor openings. Openings around bathtubs, showers, water closets, pipes, wires or other objects that penetrate concrete slabs or other floor assemblies shall be filled with a polyurethane caulk or equivalent sealant applied in accordance with the manufacturer's recommendations.

AF103.4.2 Concrete joints. All control joints, isolation joints, construction joints and any other joints in concrete slabs or between slabs and foundation walls shall be sealed with a caulk or sealant. Gaps and joints shall be cleared of loose material and filled with polyurethane caulk or other elastomeric sealant applied in accordance with the manufacturer's recommendations.

AF103.4.3 Condensate drains. Condensate drains shall be trapped or routed through nonperforated pipe to daylight.

AF103.4.4 Sumps. Sump pits open to soil or serving as the termination point for sub-slab or exterior drain tile loops shall be covered with a gasketed or otherwise sealed lid. Sumps used as the suction point in a sub-slab depressurization system shall have a lid designed to accommodate the vent pipe. Sumps used as a floor drain shall have a lid equipped with a trapped inlet.

AF103.4.5 Foundation walls. Hollow block masonry foundation walls shall be constructed with either a continuous course of solid masonry, one course of masonry grouted solid, or a solid concrete beam at or above finished ground surface to prevent passage of air from the interior of the wall into the living space. Where a brick veneer or other masonry ledge is installed, the course immediately below that ledge shall be sealed. Joints, cracks or other openings around all penetrations of both exterior and interior surfaces of masonry block or wood foundation walls below the ground surface shall be filled with polyurethane caulk or equivalent sealant. Penetrations of concrete walls shall be filled.

AF103.4.6 Dampproofing. The exterior surfaces of portions of concrete and masonry block walls below the ground surface shall be dampproofed in accordance with Section R406 of this code.

AF103.4.7 Air-handling units. Air-handling units in crawl spaces shall be sealed to prevent air from being drawn into the unit.

> **Exception:** Units with gasketed seams or units that are otherwise sealed by the manufacturer to prevent leakage.

AF103.4.8 Ducts. Ductwork passing through or beneath a slab shall be of seamless material unless the air-handling system is designed to maintain continuous positive pressure within such ducting. Joints in such ductwork shall be sealed to prevent air leakage.

Ductwork located in crawl spaces shall have all seams and joints sealed by closure systems in accordance with Section M1601.3.1.

AF103.4.9 Crawl space floors. Openings around all penetrations through floors above crawl spaces shall be caulked or otherwise filled to prevent air leakage.

AF103.4.10 Crawl space access. Access doors and other openings or penetrations between basements and adjoining crawl spaces shall be closed, gasketed or otherwise filled to prevent air leakage.

AF103.5 Passive submembrane depressurization system. In buildings with crawl space foundations, the following components of a passive sub-membrane depressurization system shall be installed during construction.

> **Exception:** Buildings in which an approved mechanical crawl space ventilation system or other equivalent system is installed.

AF103.5.1 Ventilation. Crawl spaces shall be provided with vents to the exterior of the building. The minimum net area of ventilation openings shall comply with Section R408.1 of this code.

AF103.5.2 Soil-gas-retarder. The soil in crawl spaces shall be covered with a continuous layer of minimum 6-mil (0.15 mm) polyethylene soil-gas-retarder. The ground cover shall be lapped a minimum of 12 inches (305 mm) at joints and shall extend to all foundation walls enclosing the crawl space area.

AF103.5.3 Vent pipe. A plumbing tee or other approved connection shall be inserted horizontally beneath the sheeting and connected to a 3- or 4-inch-diameter (76 mm or 102 mm) fitting with a vertical vent pipe installed through the sheeting. The vent pipe shall be extended up through the building floors, terminate at least 12 inches (305 mm) above the roof in a location at least 10 feet (3048 mm) away from any window or other opening into the conditioned spaces of the building that is less than 2 feet (610 mm) below the exhaust point, and 10 feet (3048 mm) from any window or other opening in adjoining or adjacent buildings.

AF103.6 Passive subslab depressurization system. In basement or slab-on-grade buildings, the following components of a passive sub-slab depressurization system shall be installed during construction.

AF103.6.1 Vent pipe. A minimum 3-inch-diameter (76 mm) ABS, PVC or equivalent gas-tight pipe shall be embedded vertically into the sub-slab aggregate or other permeable material before the slab is cast. A "T" fitting or equivalent method shall be used to ensure that the pipe opening remains within the sub-slab permeable material. Alternatively, the 3-inch (76 mm) pipe shall be inserted directly into an interior perimeter drain tile loop or through a sealed sump cover where the sump is exposed to the sub-slab aggregate or connected to it through a drainage system.

The pipe shall be extended up through the building floors, terminate at least 12 inches (305 mm) above the surface of the roof in a location at least 10 feet (3048 mm) away from any window or other opening into the conditioned spaces of the building that is less than 2 feet (610 mm) below the exhaust point, and 10 feet (3048 mm) from any window or other opening in adjoining or adjacent buildings.

AF103.6.2 Multiple vent pipes. In buildings where interior footings or other barriers separate the sub-slab aggregate or other gas-permeable material, each area shall be fitted with an individual vent pipe. Vent pipes shall connect to a single vent that terminates above the roof or each individual vent pipe shall terminate separately above the roof.

AF103.7 Vent pipe drainage. All components of the radon vent pipe system shall be installed to provide positive drainage to the ground beneath the slab or soil-gas-retarder.

AF103.8 Vent pipe accessibility. Radon vent pipes shall be accessible for future fan installation through an attic or other area outside the habitable space.

> **Exception:** The radon vent pipe need not be accessible in an attic space where an approved roof-top electrical supply is provided for future use.

AF103.9 Vent pipe identification. All exposed and visible interior radon vent pipes shall be identified with at least one label on each floor and in accessible attics. The label shall read:"Radon Reduction System."

AF103.10 Combination foundations. Combination basement/crawl space or slab-on-grade/crawl space foundations shall have separate radon vent pipes installed in each type of foundation area. Each radon vent pipe shall terminate above the roof or shall be connected to a single vent that terminates above the roof.

AF103.11 Building depressurization. Joints in air ducts and plenums in unconditioned spaces shall meet the requirements of Section M1601. Thermal envelope air infiltration requirements shall comply with the energy conservation provisions in Chapter 11. Fireblocking shall meet the requirements contained in Section R602.8.

AF103.12 Power source. To provide for future installation of an active sub-membrane or sub-slab depressurization system, an electrical circuit terminated in an approved box shall be installed during construction in the attic or other anticipated location of vent pipe fans. An electrical supply shall also be accessible in anticipated locations of system failure alarms.

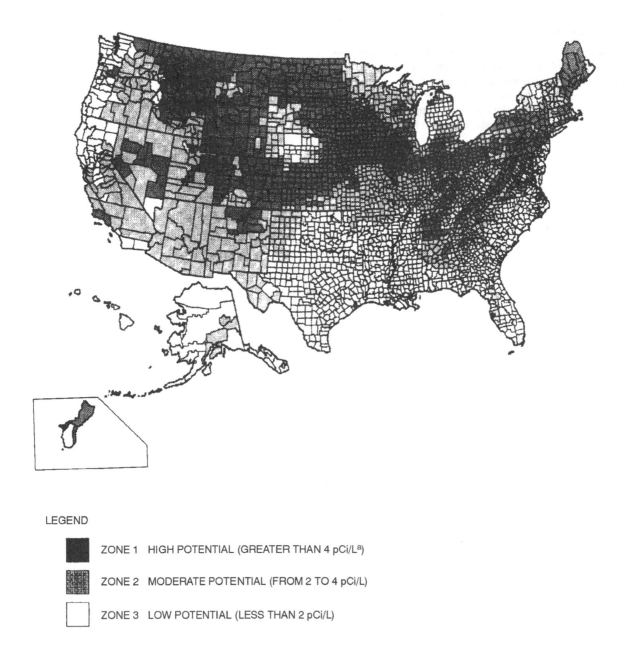

LEGEND

ZONE 1   HIGH POTENTIAL (GREATER THAN 4 pCi/L[a])

ZONE 2   MODERATE POTENTIAL (FROM 2 TO 4 pCi/L)

ZONE 3   LOW POTENTIAL (LESS THAN 2 pCi/L)

a. pCi/L standard for picocuries per liter of radon gas. EPA recommends that all homes that measure 4 pCi/L and greater be mitigated.

The United States Environmental Protection Agency and the United States Geological Survey have evaluated the radon potential in the United States and have developed a map of radon zones designed to assist building officials in deciding whether radon-resistant features are applicable in new construction.

The map assigns each of the 3,141 counties in the United States to one of three zones based on radon potential. Each zone designation reflects the average short-term radon measurement that can be expected to be measured in a building without the implementation of radon control methods. The radon zone designation of highest priority is Zone 1. Table 1 of this appendix lists the Zone 1 counties illustrated on the map. More detailed information can be obtained from state-specific booklets (EPA-402-R-93-021 through 070) available through State Radon Offices or from U.S. EPA Regional Offices.

**FIGURE AF101
EPA MAP OF RADON ZONES**

TABLE AF101(1)
HIGH RADON POTENTIAL (ZONE 1) COUNTIES[a]

ALABAMA	**CONNECTICUT**	Morgan	Wabash	Thomas	Cass	Washington
Calhoun	Fairfield	Moultrie	Warren	Trego	Hillsdale	Watonwan
Clay	Middlesex	Ogle	Washington	Wallace	Jackson	Wilkin
Cleburne	New Haven	Peoria	Wayne	Washington	Kalamazoo	Winona
Colbert	New London	Piatt	Wells	Wichita	Lenawee	Wright
Coosa		Pike	White	Wyandotte	St. Joseph	Yellow Medicine
Franklin	**GEORGIA**	Putnam	Whitley		Washtenaw	
Jackson	Cobb	Rock Island		**KENTUCKY**		**MISSOURI**
Lauderdale	De Kalb	Sangamon	**IOWA**	Adair	**MINNESOTA**	Andrew
Lawrence	Fulton	Schuyler	All Counties	Allen	Becker	Atchison
Limestone	Gwinnett	Scott		Barren	Big Stone	Buchanan
Madison		Stark	**KANSAS**	Bourbon	Blue Earth	Cass
Morgan	**IDAHO**	Stephenson	Atchison	Boyle	Brown	Clay
Talladega	Benewah	Tazewell	Barton	Bullitt	Carver	Clinton
	Blaine	Vermilion	Brown	Casey	Chippewa	Holt
CALIFORNIA	Boise	Warren	Cheyenne	Clark	Clay	Iron
Santa Barbara	Bonner	Whiteside	Clay	Cumberland	Cottonwood	Jackson
Ventura	Boundary	Winnebago	Cloud	Fayette	Dakota	Nodaway
	Butte	Woodford	Decatur	Franklin	Dodge	Platte
COLORADO	Camas		Dickinson	Green	Douglas	
Adams	Clark	**INDIANA**	Douglas	Harrison	Faribault	**MONTANA**
Arapahoe	Clearwater	Adams	Ellis	Hart	Fillmore	Beaverhead
Baca	Custer	Allen	Ellsworth	Jefferson	Freeborn	Big Horn
Bent	Elmore	Bartholomew	Finney	Jessamine	Goodhue	Blaine
Boulder	Fremont	Benton	Ford	Lincoln	Grant	Broadwater
Chaffee	Gooding	Blackford	Geary	Marion	Hennepin	Carbon
Cheyenne	Idaho	Boone	Gove	Mercer	Houston	Carter
Clear Creek	Kootenai	Carroll	Graham	Metcalfe	Hubbard	Cascade
Crowley	Latah	Cass	Grant	Monroe	Jackson	Chouteau
Custer	Lemhi	Clark	Gray	Nelson	Kanabec	Custer
Delta	Shoshone	Clinton	Greeley	Pendleton	Kandiyohi	Daniels
Denver	Valley	De Kalb	Hamilton	Pulaski	Kittson	Dawson
Dolores		Decatur	Haskell	Robertson	Lac Qui Parle	Deer Lodge
Douglas	**ILLINOIS**	Delaware	Hodgeman	Russell	Le Sueur	Fallon
El Paso	Adams	Elkhart	Jackson	Scott	Lincoln	Fergus
Elbert	Boone	Fayette	Jewell	Taylor	Lyon	Flathead
Fremont	Brown	Fountain	Johnson	Warren	Mahnomen	Gallatin
Garfield	Bureau	Fulton	Kearny	Woodford	Marshall	Garfield
Gilpin	Calhoun	Grant	Kingman		Martin	Glacier
Grand	Carroll	Hamilton	Kiowa	**MAINE**	McLeod	Granite
Gunnison	Cass	Hancock	Lane	Androscoggin	Meeker	Hill
Huerfano	Champaign	Harrison	Leavenworth	Aroostook	Mower	Jefferson
Jackson	Coles	Hendricks	Lincoln	Cumberland	Murray	Judith Basin
Jefferson	De Kalb	Henry	Logan	Franklin	Nicollet	Lake
Kiowa	De Witt	Howard	Marion	Hancock	Nobles	Lewis and Clark
Kit Carson	Douglas	Huntington	Marshall	Kennebec	Norman	Liberty
Lake	Edgar	Jay	McPherson	Lincoln	Olmsted	Lincoln
Larimer	Ford	Jennings	Meade	Oxford	Otter Tail	Madison
Las Animas	Fulton	Johnson	Mitchell	Penobscot	Pennington	McCone
Lincoln	Greene	Kosciusko	Nemaha	Piscataquis	Pipestone	Meagher
Logan	Grundy	Lagrange	Ness	Somerset	Polk	Mineral
Mesa	Hancock	Lawrence	Norton	York	Pope	Missoula
Moffat	Henderson	Madison	Osborne		Ramsey	Park
Montezuma	Henry	Marion	Ottawa	**MARYLAND**	Red Lake	Phillips
Montrose	Iroquois	Marshall	Pawnee	Baltimore	Redwood	Pondera
Morgan	Jersey	Miami	Phillips	Calvert	Renville	Powder River
Otero	Jo Daviess	Monroe	Pottawatomie	Carroll	Rice	Powell
Ouray	Kane	Montgomery	Pratt	Frederick	Rock	Prairie
Park	Kendall	Noble	Rawlins	Harford	Roseau	Ravalli
Phillips	Knox	Orange	Republic	Howard	Scott	Richland
Pitkin	La Salle	Putnam	Rice	Montgomery	Sherburne	Roosevelt
Prowers	Lee	Randolph	Riley	Washington	Sibley	Rosebud
Pueblo	Livingston	Rush	Rooks		Stearns	Sanders
Rio Blanco	Logan	Scott	Rush	**MASS.**	Steele	Sheridan
San Miguel	Macon	Shelby	Russell	Essex	Stevens	Silver Bow
Summit	Marshall	Steuben	Saline	Middlesex	Swift	Stillwater
Teller	Mason	St. Joseph	Scott	Worcester	Todd	Teton
Washington	McDonough	Tippecanoe	Sheridan		Traverse	Toole
Weld	McLean	Tipton	Sherman	**MICHIGAN**	Wabasha	Valley
Yuma	Menard	Union	Smith	Branch	Wadena	Wibaux
	Mercer	Vermillion	Stanton	Calhoun	Waseca	

a. EPA recommends that this county listing be supplemented with other available State and local data to further understand the radon potential of Zone 1 area.

(continued)

TABLE AF101(1)—continued
HIGH RADON POTENTIAL (ZONE 1) COUNTIES[a]

Yellowstone National Park

NEBRASKA
Adams
Boone
Boyd
Burt
Butler
Cass
Cedar
Clay
Colfax
Cuming
Dakota
Dixon
Dodge
Douglas
Fillmore
Franklin
Frontier
Furnas
Gage
Gosper
Greeley
Hamilton
Harlan
Hayes
Hitchcock
Hurston
Jefferson
Johnson
Kearney
Knox
Lancaster
Madison
Nance
Nemaha
Nuckolls
Otoe
Pawnee
Phelps
Pierce
Platte
Polk
Red Willow
Richardson
Saline
Sarpy
Saunders
Seward
Stanton
Thayer
Washington
Wayne
Webster
York

NEVADA
Carson City
Douglas
Eureka
Lander
Lincoln
Lyon
Mineral
Pershing
White Pine

NEW HAMPSHIRE
Carroll

NEW JERSEY
Hunterdon
Mercer
Monmouth
Morris
Somerset
Sussex
Warren

NEW MEXICO
Bernalillo
Colfax
Mora
Rio Arriba
San Miguel
Santa Fe
Taos

NEW YORK
Albany
Allegany
Broome
Cattaraugus
Cayuga
Chautauqua
Chemung
Chenango
Columbia
Cortland
Delaware
Dutchess
Erie
Genesee
Greene
Livingston
Madison
Onondaga
Ontario
Orange
Otsego
Putnam
Rensselaer
Schoharie
Schuyler
Seneca
Steuben
Sullivan
Tioga
Tompkins
Ulster
Washington
Wyoming
Yates

N. CAROLINA
Alleghany
Buncombe
Cherokee
Henderson
Mitchell
Rockingham
Transylvania
Watauga

N. DAKOTA
All Counties

OHIO
Adams
Allen
Ashland
Auglaize
Belmont
Butler
Carroll
Champaign
Clark
Clinton
Columbiana
Coshocton
Crawford
Darke
Delaware
Fairfield
Fayette
Franklin
Greene
Guernsey
Hamilton
Hancock
Hardin
Harrison
Holmes
Huron
Jefferson
Knox
Licking
Logan
Madison
Marion
Mercer
Miami
Montgomery
Morrow
Muskingum
Perry
Pickaway
Pike
Preble
Richland
Ross
Seneca
Shelby
Stark
Summit
Tuscarawas
Union
Van Wert
Warren
Wayne
Wyandot

PENNSYLVANIA
Adams
Allegheny
Armstrong
Beaver
Bedford
Berks
Blair
Bradford
Bucks
Butler
Cameron
Carbon
Centre
Chester
Clarion
Clearfield
Clinton
Columbia
Cumberland
Dauphin
Delaware
Franklin
Fulton
Huntingdon
Indiana
Juniata
Lackawanna
Lancaster
Lebanon
Lehigh
Luzerne
Lycoming
Mifflin
Monroe
Montgomery
Montour
Northampton
Northumberland
Perry
Schuylkill
Snyder
Sullivan
Susquehanna
Tioga
Union
Venango
Westmoreland
Wyoming
York

RHODE ISLAND
Kent
Washington

S. CAROLINA
Greenville

S. DAKOTA
Aurora
Beadle
Bon Homme
Brookings
Brown
Brule
Buffalo
Campbell
Charles Mix
Clark
Clay
Codington
Corson
Davison
Day
Deuel
Douglas
Edmunds
Faulk
Grant
Hamlin
Hand
Hanson
Hughes
Hutchinson
Hyde
Jerauld
Kingsbury
Lake
Lincoln
Lyman
Marshall
McCook
McPherson
Miner
Minnehaha
Moody
Perkins
Potter
Roberts
Sanborn
Spink
Stanley
Sully
Turner
Union
Walworth
Yankton

TENNESEE
Anderson
Bedford
Blount
Bradley
Claiborne
Davidson
Giles
Grainger
Greene
Hamblen
Hancock
Hawkins
Hickman
Humphreys
Jackson
Jefferson
Knox
Lawrence
Lewis
Lincoln
Loudon
Marshall
Maury
McMinn
Meigs
Monroe
Moore
Perry
Roane
Rutherford
Smith
Sullivan
Trousdale
Union
Washington
Wayne
Williamson
Wilson

UTAH
Carbon
Duchesne
Grand
Piute
Sanpete
Sevier
Uintah

VIRGINIA
Alleghany
Amelia
Appomattox
Augusta
Bath
Bland
Botetourt
Bristol
Brunswick
Buckingham
Buena Vista
Campbell
Chesterfield
Clarke
Clifton Forge
Covington
Craig
Cumberland
Danville
Dinwiddie
Fairfax
Falls Church
Fluvanna
Frederick
Fredericksburg
Giles
Goochland
Harrisonburg
Henry
Highland
Lee
Lexington
Louisa
Martinsville
Montgomery
Nottoway
Orange
Page
Patrick
Pittsylvania
Powhatan
Pulaski
Radford
Roanoke
Rockbridge
Rockingham
Russell
Salem
Scott
Shenandoah
Smyth
Spotsylvania
Stafford
Staunton
Tazewell
Warren
Washington
Waynesboro
Winchester
Wythe

WASHINGTON
Clark
Ferry
Okanogan
Pend Oreille
Skamania
Spokane
Stevens

W. VIRGINIA
Berkeley
Brooke
Grant
Greenbrier
Hampshire
Hancock
Hardy
Jefferson
Marshall
Mercer
Mineral
Monongalia
Monroe
Morgan
Ohio
Pendleton
Pocahontas
Preston
Summers
Wetzel

WISCONSIN
Buffalo
Crawford
Dane
Dodge
Door
Fond du Lac
Grant
Green
Green Lake
Iowa
Jefferson
Lafayette
Langlade
Marathon
Menominee
Pepin
Pierce
Portage
Richland
Rock
Shawano
St. Croix
Vernon
Walworth
Washington
Waukesha
Waupaca
Wood

WYOMING
Albany
Big Horn
Campbell
Carbon
Converse
Crook
Fremont
Goshen
Hot Springs
Johnson
Laramie
Lincoln
Natrona
Niobrara
Park
Sheridan
Sublette
Sweetwater
Teton
Uinta
Washakie

a. EPA recommends that this county listing be supplemented with other available State and local data to further understand the radon potential of Zone 1 area.

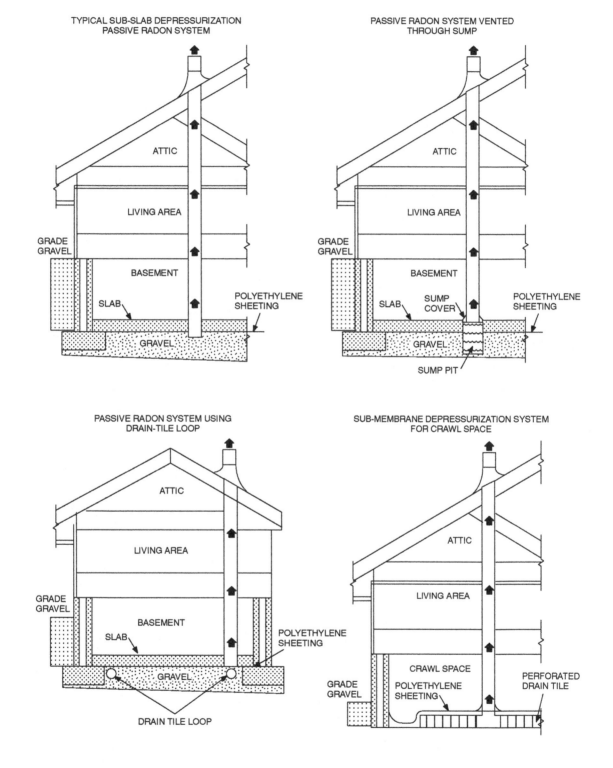

TYPICAL SUB-SLAB DEPRESSURIZATION
PASSIVE RADON SYSTEM

PASSIVE RADON SYSTEM VENTED
THROUGH SUMP

ATTIC

LIVING AREA

GRADE
GRAVEL

BASEMENT

SLAB

POLYETHYLENE
SHEETING

GRAVEL

ATTIC

LIVING AREA

GRADE
GRAVEL

BASEMENT

SLAB

SUMP
COVER

POLYETHYLENE
SHEETING

GRAVEL

SUMP PIT

PASSIVE RADON SYSTEM USING
DRAIN-TILE LOOP

SUB-MEMBRANE DEPRESSURIZATION SYSTEM
FOR CRAWL SPACE

ATTIC

LIVING AREA

GRADE
GRAVEL

BASEMENT

SLAB

POLYETHYLENE
SHEETING

GRAVEL

DRAIN TILE LOOP

ATTIC

LIVING AREA

CRAWL SPACE

GRADE
GRAVEL

POLYETHYLENE
SHEETING

PERFORATED
DRAIN TILE

FIGURE AF102
RADON-RESISTANT CONSTRUCTION DETAILS FOR FOUR FOUNDATION TYPES

APPENDIX G

SWIMMING POOLS, SPAS AND HOT TUBS

SECTION AG101
GENERAL

AG101.1 General. The provisions of this appendix shall control the design and construction of swimming pools, spas and hot tubs installed in or on the lot of a one- or two-family dwelling.

SECTION AG102
DEFINITIONS

AG102.1 General. For the purposes of these requirements, the terms used shall be defined as follows and as set forth in Chapter 2.

ABOVE-GROUND/ON-GROUND POOL. See "Swimming pool."

BARRIER. A fence, wall, building wall or combination thereof which completely surrounds the swimming pool and obstructs access to the swimming pool.

HOT TUB. See "Swimming pool."

IN-GROUND POOL. See "Swimming pool."

RESIDENTIAL. That which is situated on the premises of a detached one- or two-family dwelling or a one-family townhouse not more than three stories in height.

SPA, NONPORTABLE. See "Swimming pool."

SPA, PORTABLE. A nonpermanent structure intended for recreational bathing, in which all controls, water-heating and water-circulating equipment are an integral part of the product.

SWIMMING POOL. Any structure intended for swimming or recreational bathing that contains water over 24 inches (610 mm) deep. This includes in-ground, above-ground and on-ground swimming pools, hot tubs and spas.

SWIMMING POOL, INDOOR. A swimming pool which is totally contained within a structure and surrounded on all four sides by the walls of the enclosing structure.

SWIMMING POOL, OUTDOOR. Any swimming pool which is not an indoor pool.

SECTION AG103
SWIMMING POOLS

AG103.1 In-ground pools. In-ground pools shall be designed and constructed in conformance with ANSI/NSPI-5 as listed in Section AG108.

AG103.2 Above-ground and on-ground pools. Above-ground and on-ground pools shall be designed and constructed in conformance with ANSI/NSPI-4 as listed in Section AG108.

SECTION AG104
SPAS AND HOT TUBS

AG104.1 Permanently installed spas and hot tubs. Permanently installed spas and hot tubs shall be designed and constructed in conformance with ANSI/NSPI-3 as listed in Section AG108.

AG104.2 Portable spas and hot tubs. Portable spas and hot tubs shall be designed and constructed in conformance with ANSI/NSPI-6 as listed in Section AG108.

SECTION AG105
BARRIER REQUIREMENTS

AG105.1 Application. The provisions of this chapter shall control the design of barriers for residential swimming pools, spas and hot tubs. These design controls are intended to provide protection against potential drownings and near-drownings by restricting access to swimming pools, spas and hot tubs.

AG105.2 Outdoor swimming pool. An outdoor swimming pool, including an in-ground, above-ground or on-ground pool, hot tub or spa shall be surrounded by a barrier which shall comply with the following:

1. The top of the barrier shall be at least 48 inches (1219 mm) above grade measured on the side of the barrier which faces away from the swimming pool. The maximum vertical clearance between grade and the bottom of the barrier shall be 2 inches (51 mm) measured on the side of the barrier which faces away from the swimming pool. Where the top of the pool structure is above grade, such as an above-ground pool, the barrier may be at ground level, such as the pool structure, or mounted on top of the pool structure. Where the barrier is mounted on top of the pool structure, the maximum vertical clearance between the top of the pool structure and the bottom of the barrier shall be 4 inches (102 mm).

2. Openings in the barrier shall not allow passage of a 4-inch-diameter (102 mm) sphere.

3. Solid barriers which do not have openings, such as a masonry or stone wall, shall not contain indentations or protrusions except for normal construction tolerances and tooled masonry joints.

4. Where the barrier is composed of horizontal and vertical members and the distance between the tops of the horizontal members is less than 45 inches (1143 mm), the horizontal members shall be located on the swimming pool side of the fence. Spacing between vertical members shall not exceed $1^3/_4$ inches (44 mm) in width. Where there are decorative cutouts within vertical members, spacing within the cutouts shall not exceed $1^3/_4$ inches (44 mm) in width.

5. Where the barrier is composed of horizontal and vertical members and the distance between the tops of the horizontal members is 45 inches (1143 mm) or more, spacing between vertical members shall not exceed 4 inches (102 mm). Where there are decorative cutouts within vertical members, spacing within the cutouts shall not exceed $1^3/_4$ inches (44 mm) in width.

6. Maximum mesh size for chain link fences shall be a $2^1/_4$-inch (57 mm) square unless the fence has slats fastened at the top or the bottom which reduce the openings to not more than $1^3/_4$ inches (44 mm).

7. Where the barrier is composed of diagonal members, such as a lattice fence, the maximum opening formed by the diagonal members shall not be more than $1^3/_4$ inches (44 mm).

8. Access gates shall comply with the requirements of Section AG105.2, Items 1 through 7, and shall be equipped to accommodate a locking device. Pedestrian access gates shall open outward away from the pool and shall be self-closing and have a self-latching device. Gates other than pedestrian access gates shall have a self-latching device. Where the release mechanism of the self-latching device is located less than 54 inches (1372 mm) from the bottom of the gate, the release mechanism and openings shall comply with the following:

   8.1. The release mechanism shall be located on the pool side of the gate at least 3 inches (76 mm) below the top of the gate; and

   8.2. The gate and barrier shall have no opening larger than $1/_2$ inch (13 mm) within 18 inches (457 mm) of the release mechanism.

9. Where a wall of a dwelling serves as part of the barrier, one of the following conditions shall be met:

   9.1. The pool shall be equipped with a powered safety cover in compliance with ASTM F 1346; or

   9.2. Doors with direct access to the pool through that wall shall be equipped with an alarm which produces an audible warning when the door and/or its screen, if present, are opened. The alarm shall be listed in accordance with UL 2017. The audible alarm shall activate within 7 seconds and sound continuously for a minimum of 30 seconds after the door and/or its screen, if present, are opened and be capable of being heard throughout the house during normal household activities. The alarm shall automatically reset under all conditions. The alarm system shall be equipped with a manual means, such as touch pad or switch, to temporarily deactivate the alarm for a single opening. Deactivation shall last for not more than 15 seconds. The deactivation switch(es) shall be located at least 54 inches (1372 mm) above the threshold of the door; or

   9.3. Other means of protection, such as self-closing doors with self-latching devices, which are approved by the governing body, shall be acceptable so long as the degree of protection afforded

is not less than the protection afforded by Item 9.1 or 9.2 described above.

10. Where an above-ground pool structure is used as a barrier or where the barrier is mounted on top of the pool structure, and the means of access is a ladder or steps:

   10.1. The ladder or steps shall be capable of being secured, locked or removed to prevent access; or

   10.2. The ladder or steps shall be surrounded by a barrier which meets the requirements of Section AG105.2, Items 1 through 9. When the ladder or steps are secured, locked or removed, any opening created shall not allow the passage of a 4-inch-diameter (102 mm) sphere.

AG105.3 Indoor swimming pool. Walls surrounding an indoor swimming pool shall comply with Section AG105.2, Item 9.

AG105.4 Prohibited locations. Barriers shall be located to prohibit permanent structures, equipment or similar objects from being used to climb them.

AG105.5 Barrier exceptions. Spas or hot tubs with a safety cover which complies with ASTM F 1346, as listed in Section AG107, shall be exempt from the provisions of this appendix.

SECTION AG106
ENTRAPMENT PROTECTION FOR SWIMMING POOL AND SPA SUCTION OUTLETS

AG106.1 General. Suction outlets shall be designed to produce circulation throughout the pool or spa. Single-outlet systems, such as automatic vacuum cleaner systems, or multiple suction outlets, whether isolated by valves or otherwise, shall be protected against user entrapment.

AG106.2 Suction fittings. Pool and spa suction outlets shall have a cover that conforms to ANSI/ASME A112.19.8M, or an 18 inch × 23 inch (457 mm by 584 mm) drain grate or larger, or an approved channel drain system.

Exception: Surface skimmers

AG106.3 Atmospheric vacuum relief system required. Pool and spa single- or multiple-outlet circulation systems shall be equipped with atmospheric vacuum relief should grate covers located therein become missing or broken. This vacuum relief system shall include at least one approved or engineered method of the type specified herein, as follows:

1. Safety vacuum release system conforming to ASME A112.19.17; or

2. An approved gravity drainage system.

AG106.4 Dual drain separation. Single or multiple pump circulation systems shall be provided with a minimum of two suction outlets of the approved type. A minimum horizontal or vertical distance of 3 feet (914 mm) shall separate the outlets. These suction outlets shall be piped so that water is drawn through them simultaneously through a vacuum-relief-protected line to the pump or pumps.

AG106.5 Pool cleaner fittings. Where provided, vacuum or pressure cleaner fitting(s) shall be located in an accessible position(s) at least 6 inches (152 mm) and not more than 12 inches (305 mm) below the minimum operational water level or as an attachment to the skimmer(s).

SECTION AG107
ABBREVIATIONS

AG107.1 General.

ANSI—American National Standards Institute
11 West 42nd Street, New York, NY 10036

ASME—American Society of Mechanical Engineers
Three Park Avenue
New York, NY 10016-5990

ASTM—ASTM International
100 Barr Harbor Drive, West Conshohocken, PA 19428

NSPI—National Spa and Pool Institute
2111 Eisenhower Avenue, Alexandria, VA 22314

UL—Underwriters Laboratories, Inc.
333 Pfingsten Road
Northbrook, Illinois 60062-2096

SECTION AG108
STANDARDS

AG108.1 General.

ANSI/NSPI

ASTM

ASME

UL

APPENDIX H
PATIO COVERS

SECTION AH101
GENERAL

AH101.1 Scope. Patio covers shall conform to the requirements of this appendix chapter.

SECTION AH102
DEFINITION

Patio covers. One-story structures not exceeding 12 feet (3657 mm) in height. Enclosure walls shall be permitted to be of any configuration, provided the open or glazed area of the longer wall and one additional wall is equal to at least 65 percent of the area below a minimum of 6 feet 8 inches (2032 mm) of each wall, measured from the floor. Openings shall be permitted to be enclosed with (1) insect screening, (2) approved translucent or transparent plastic not more than 0.125 inch (3.2 mm) in thickness, (3) glass conforming to the provisions of Section R308, or (4) any combination of the foregoing.

SECTION AH103
PERMITTED USES

AH103.1 General. Patio covers shall be permitted to be detached from or attached to dwelling units. Patio covers shall be used only for recreational, outdoor living purposes and not as carports, garages, storage rooms or habitable rooms.

SECTION AH104
DESIGN LOADS

AH104.1 General. Patio covers shall be designed and constructed to sustain, within the stress limits of this code, all dead loads plus a minimum vertical live load of 10 pounds per square foot (0.48 kN/m^2) except that snow loads shall be used where such snow loads exceed this minimum. Such covers shall be designed to resist the minimum wind loads set forth in Section R301.2.1.

SECTION AH105
LIGHT AND VENTILATION/EMERGENCY EGRESS

AH105.1 General. Exterior openings required for light and ventilation shall be permitted to open into a patio structure conforming to Section AH101, provided that the patio structure shall be unenclosed if such openings are serving as emergency egress or rescue openings from sleeping rooms. Where such exterior openings serve as an exit from the dwelling unit, the patio structure, unless unenclosed, shall be provided with exits conforming to the provisions of Section R311 of this code.

SECTION AH106
FOOTINGS

AH106.1 General. In areas with a frostline depth of zero as specified in Table R301.2(1), a patio cover shall be permitted to be supported on a slab on grade without footings, provided the slab conforms to the provisions of Section R506 of this code, is not less than 3.5 inches (89 mm) thick and the columns do not support live and dead loads in excess of 750 pounds (3.34 kN) per column.

SECTION AH107
SPECIAL PROVISIONS FOR ALUMINUM SCREEN ENCLOSURES IN HURRICANE-PRONE REGIONS

AH107.1 General. Screen enclosures in hurricane-prone regions shall be in accordance with the provisions of this Section.

AH107.1.1 Habitable spaces. Screen enclosures shall not be considered habitable spaces.

AH107.1.2 Minimum ceiling height. Screen enclosures shall have a ceiling height of not less than 7 feet (2134 mm).

AH107.2 Definitions.

SCREEN ENCLOSURE. A building or part thereof, in whole or in part self-supporting, and having walls of insect screening and a roof of insect screening, plastic, aluminum, or similar lightweight material.

AH107.3 Screen enclosures.

AH107.3.1 Thickness. Actual wall thickness of extruded aluminum members shall be not less than 0.040 inches (1.02 mm).

AH107.3.2 Density. Screen density shall be a maximum of 20 threads per inch by 20 threads per inch mesh.

AH107.4 Design.

AH107.4.1 Wind load. Structural members supporting screen enclosures shall be designed to support minimum wind loads given in Table AH107.4(1) and AH107.4(2). Where any value is less than 10 psf (0.479 kN/m^2) use 10 psf (0.479 kN/m^2).

AH107.4.2 Deflection limit. For members supporting screen surfaces only, the total load deflection shall not exceed $l/60$. Screen surfaces shall be permitted to include a maximum of 25 percent solid flexible finishes.

AH107.4.3 Importance factor. The wind factor for screen enclosures shall be 0.77 in accordance with Table 1604.5 of the *International Building Code*.

AH107.4.4 Roof live load. The minimum roof live load shall be 10 psf (0.479 kN/m^2).

AH107.5 Footings. In areas with a frost line is zero, a screen enclosure shall be permitted to be supported on a concrete slab on grade without footings, provided the slab conforms to the provisions of Section R506, is not less than $3^{1}/_{2}$ inches (89 mm) thick, and the columns do not support loads in excess of 750 pounds (3.36 kN) per column.

TABLE AH107.4(1)
DESIGN WIND PRESSURES FOR ALUMINUM SCREEN ENCLOSURE FRAMING
WITH AN IMPORTANCE FACTOR OF 0.77[a, b, c]

LOAD CASE	WALL	Basic Wind Speed (mph)											
		100		110		120		130		140		150	
		Exposure Category Design Pressure (psf)											
		C	B	C	B	C	B	C	B	C	B	C	B
A[d]	Windward and leeward walls (flow thru) and windward wall (non-flow thru) L/W = 0-1	12	8	14	10	17	12	19	14	23	16	26	18
A[d]	Windward and leeward walls (flow thru) and windward wall (non-flow thru) L/W = 2	13	9	16	11	19	14	22	16	26	18	30	21
B[e]	Windward: Non-gable roof	16	12	20	14	24	17	28	20	32	23	37	26
B[e]	Windward: Gable roof	22	16	27	19	32	23	38	27	44	31	50	36
	ROOF												
All[f]	Roof-screen	4	3	5	4	6	4	7	5	8	6	9	7
All[f]	Roof-solid	12	9	15	11	18	13	21	15	24	17	28	20

For SI: 1 mile per hour = 0.44 m/s, 1 pound per square foot = 0.0479 kPa, 1 foot = 304.8 mm.

a. Values have been reduced for 0.77 Importance Factor in accordance with Table 1604.5 of the *International Building Code*.

b. Minimum design pressure shall be 10 psf in accordance with Section 1609.1.2 of the *International Building Code*.

c. Loads are applicable to screen enclosures with a mean roof height of 30 feet or less. For screen enclosures of different heights the pressures given shall be adjusted by multiplying the table pressure by the adjustment factor given in Table AH107.4(2).

d. For Load Case A flow thru condition the pressure given shall be applied simultaneously to both the upwind and downwind screen walls acting in the same direction as the wind. The structure shall also be analyzed for wind coming from the opposite direction. For the non-flow thru condition the screen enclosure wall shall be analyzed for the load applied acting toward the interior of the enclosure.

e. For Load Case B the table pressure multiplied by the projected frontal area of the screen enclosure is the total drag force, including drag on screen surfaces parallel to the wind, which must be transmitted to the ground. Use Load Case A for members directly supporting the screen surface perpendicular to the wind. Load Case B loads shall be applied only to structural members which carry wind loads from more than one surface.

f. The roof structure shall be analyzed for the pressure given occurring both upward and downward.

TABLE AH107.4(2)
HEIGHT ADJUSTMENT FACTORS

MEAN	EXPOSURE	
Roof Height (ft)	B	C
15	1	0.86
20	1	0.92
25	1	0.96
30	1	1.00
35	1.05	1.03
40	1.09	1.06
45	1.12	1.09
50	1.16	1.11
55	1.19	1.14
60	1.22	1.16

For SI: 1 foot = 304.8 mm.

APPENDIX I
PRIVATE SEWAGE DISPOSAL

SECTION AI101
GENERAL

AI101.1 Scope. Private sewage disposal systems shall conform to the *International Private Sewage Disposal Code.*

EXISTING BUILDINGS AND STRUCTURES

SECTION AJ101
PURPOSE AND INTENT

AJ101.1 General. The purpose of these provisions is to encourage the continued use or reuse of legally existing buildings and structures. These provisions are intended to permit work in existing buildings that is consistent with the purpose of the *International Residential Code*. Compliance with these provisions shall be deemed to meet the requirements of the *International Residential Code*.

AJ101.2 Classification of work. For purposes of this appendix, all work in existing buildings shall be classified into the categories of repair, renovation, alteration and reconstruction. Specific requirements are established for each category of work in these provisions.

AJ101.3 Multiple categories of work. Work of more than one category may be part of a single work project. All related work permitted within a 12-month period shall be considered a single work project. Where a project includes one category of work in one building area and another category of work in a separate and unrelated area of the building, each project area shall comply with the requirements of the respective category of work. Where a project with more than one category of work is performed in the same area or in related areas of the building, the project shall comply with the requirements of the more stringent category of work.

SECTION AJ102
COMPLIANCE

AJ102.1 General. Regardless of the category of work being performed, the work shall not cause the structure to become unsafe or adversely affect the performance of the building; shall not cause an existing mechanical or plumbing system to become unsafe, hazardous, insanitary or overloaded; and unless expressly permitted by these provisions, shall not make the building any less conforming to this code or to any previously approved alternative arrangements than it was before the work was undertaken.

AJ102.2 Requirements by category of work. Repairs shall conform to the requirements of Section AJ301. Renovations shall conform to the requirements of Section AJ401. Alterations shall conform to the requirements of Section AJ501 and the requirements for renovations. Reconstructions shall conform to the requirements of Section AJ601 and the requirements for alterations and renovations.

AJ102.3 Smoke detectors. Regardless of the category of work, smoke detectors shall be provided where required by Section R313.2.1.

AJ102.4 Replacement windows. Regardless of the category of work, when an existing window, including sash and glazed portion is replaced, the replacement window shall comply with the requirements of Chapter 11.

AJ102.5 Flood hazard areas. Work performed in existing buildings located in a flood hazard area as established by Table R301.2(1) shall be subject to the provisions of Section R105.3.1.1.

AJ102.6 Equivalent alternatives. These provisions are not intended to prevent the use of any alternate material, alternate design or alternate method of construction not specifically prescribed herein, provided any alternate has been deemed to be equivalent and its use authorized by the building official.

AJ102.7 Other alternatives. Where compliance with these provisions or with this code as required by these provisions is technically infeasible or would impose disproportionate costs because of structural, construction or dimensional difficulties, other alternatives may be accepted by the building official. These alternatives may include materials, design features and/or operational features.

AJ102.8 More restrictive requirements. Buildings or systems in compliance with the requirements of this code for new construction shall not be required to comply with any more restrictive requirement of these provisions.

AJ102.9 Features exceeding *International Residential Code* requirements. Elements, components and systems of existing buildings with features that exceed the requirements of this code for new construction, and are not otherwise required as part of approved alternative arrangements or deemed by the building official to be required to balance other building elements not complying with this code for new construction, shall not be prevented by these provisions from being modified as long as they remain in compliance with the applicable requirements for new construction.

SECTION AJ103
PRELIMINARY MEETING

AJ103.1 General. If a building permit is required at the request of the prospective permit applicant, the building official or his designee shall meet with the prospective applicant to discuss plans for any proposed work under these provisions prior to the application for the permit. The purpose of this preliminary meeting is for the building official to gain an understanding of the prospective applicant's intentions for the proposed work, and to determine, together with the prospective applicant, the specific applicability of these provisions.

SECTION AJ104
EVALUATION OF AN EXISTING BUILDING

AJ104.1 General. The building official may require an existing building to be investigated and evaluated by a registered design professional in the case of proposed reconstruction of any portion of a building. The evaluation shall determine the existence of any potential nonconformities with these provisions, and shall provide a basis for determining the impact of the proposed changes on the performance of the building. The evaluation shall use the following sources of information, as applicable:

1. Available documentation of the existing building.

   1.1. Field surveys.

   1.2. Tests (nondestructive and destructive).

   1.3. Laboratory analysis.

Exception: Detached one- or two-family dwellings that are not irregular buildings under Section R301.2.2.2.2 and are not undergoing an extensive reconstruction shall not be required to be evaluated.

SECTION AJ105
PERMIT

AJ105.1 Identification of work area. The work area shall be clearly identified on all permits issued under these provisions.

SECTION AJ201
DEFINITIONS

AJ201.1 General. For purposes of this appendix, the terms used are defined as follows.

ALTERATION. The reconfiguration of any space, the addition or elimination of any door or window, the reconfiguration or extension of any system, or the installation of any additional equipment.

CATEGORIES OF WORK. The nature and extent of construction work undertaken in an existing building. The categories of work covered in this Appendix, listed in increasing order of stringency of requirements, are repair, renovation, alteration and reconstruction.

DANGEROUS. Where the stresses in any member; the condition of the building, or any of its components or elements or attachments; or other condition that results in an overload exceeding 150 percent of the stress allowed for the member or material in this code.

EQUIPMENT OR FIXTURE. Any plumbing, heating, electrical, ventilating, air conditioning, refrigerating and fire protection equipment, and elevators, dumb waiters, boilers, pressure vessels, and other mechanical facilities or installations that are related to building services.

LOAD-BEARING ELEMENT. Any column, girder, beam, joist, truss, rafter, wall, floor or roof sheathing that supports any vertical load in addition to its own weight, and/or any lateral load.

MATERIALS AND METHODS REQUIREMENTS. Those requirements in this code that specify material standards; details of installation and connection; joints; penetrations; and continuity of any element, component or system in the building. The required quantity, fire resistance, flame spread, acoustic or thermal performance, or other performance attribute is specifically excluded from materials and methods requirements.

RECONSTRUCTION. The reconfiguration of a space that affects an exit, a renovation and/or alteration when the work

area is not permitted to be occupied because existing means of egress and fire protection systems, or their equivalent, are not in place or continuously maintained; and/or there are extensive alterations as defined in Section AJ501.3.

REHABILITATION. Any repair, renovation, alteration or reconstruction work undertaken in an existing building.

RENOVATION. The change, strengthening or addition of load-bearing elements; and/or the refinishing, replacement, bracing, strengthening, upgrading or extensive repair of existing materials, elements, components, equipment and/or fixtures. Renovation involves no reconfiguration of spaces. Interior and exterior painting are not considered refinishing for purposes of this definition, and are not renovation.

REPAIR. The patching, restoration and/or minor replacement of materials, elements, components, equipment and/or fixtures for the purposes of maintaining those materials, elements, components, equipment and/or fixtures in good or sound condition.

WORK AREA. That portion of a building affected by any renovation, alteration or reconstruction work as initially intended by the owner and indicated as such in the permit. Work area excludes other portions of the building where incidental work entailed by the intended work must be performed, and portions of the building where work not initially intended by the owner is specifically required by these provisions for a renovation, alteration or reconstruction.

SECTION AJ301
REPAIRS

AJ301.1 Materials. Except as otherwise required herein, work shall be done using like materials or materials permitted by this code for new construction.

   AJ301.1.1 Hazardous materials. Hazardous materials no longer permitted, such as asbestos and lead-based paint, shall not be used.

   AJ301.1.2 Plumbing materials and supplies. The following plumbing materials and supplies shall not be used:

     1. All-purpose solvent cement, unless listed for the specific application;

     2. Flexible traps and tailpieces, unless listed for the specific application; and

     3. Solder having more than 0.2 percent lead in the repair of potable water systems.

AJ301.2 Water closets. When any water closet is replaced with a newly manufactured water closet, the replacement water closet shall comply with the requirements of Section P2903.2.

AJ301.3 Safety glazing. Replacement glazing in hazardous locations shall comply with the safety glazing requirements of Section R308.1.

AJ301.4 Electrical. Repair or replacement of existing electrical wiring and equipment undergoing repair with like material shall be permitted.

   Exceptions:

     1. Replacement of electrical receptacles shall comply with the requirements of Chapters 33 through 42.

2. Plug fuses of the Edison-base type shall be used for replacements only where there is no evidence of overfusing or tampering per the applicable requirements of Chapters 33 through 42.

3. For replacement of nongrounding-type receptacles with grounding-type receptacles and for branch circuits that do not have an equipment grounding conductor in the branch circuitry, the grounding conductor of a grounding type receptacle outlet shall be permitted to be grounded to any accessible point on the grounding electrode system, or to any accessible point on the grounding electrode conductor, as allowed and described in Chapters 33 through 42.

SECTION AJ401
RENOVATIONS

AJ401.1 Materials and methods. The work shall comply with the materials and methods requirements of this code.

AJ401.2 Door and window dimensions. Minor reductions in the clear opening dimensions of replacement doors and windows that result from the use of different materials shall be allowed, whether or not they are permitted by this code.

AJ401.3 Interior finish. Wood paneling and textile wall coverings used as an interior finish shall comply with the flame spread requirements of Section R315.

AJ401.4 Structural. Unreinforced masonry buildings located in Seismic Design Category D_2 or E shall have parapet bracing and wall anchors installed at the roofline whenever a reroofing permit is issued. Such parapet bracing and wall anchors shall be of an approved design.

SECTION AJ501
ALTERATIONS

AJ501.1 Newly constructed elements. Newly constructed elements, components and systems shall comply with the requirements of this code.

Exceptions:

1. Openable windows may be added without requiring compliance with the light and ventilation requirements of Section R303.

2. Newly installed electrical equipment shall comply with the requirements of Section AJ501.5.

AJ501.2 Nonconformities. The work shall not increase the extent of noncompliance with the requirements of Section AJ601, or create nonconformity with those requirements which did not previously exist.

AJ501.3 Extensive alterations. When the total area of all the work areas included in an alteration exceeds 50 percent of the area of the dwelling unit, the work shall be considered as a reconstruction and shall comply with the requirements of these provisions for reconstruction work.

Exception: Work areas in which the alteration work is exclusively plumbing, mechanical or electrical shall not be included in the computation of total area of all work areas.

AJ501.4 Structural. The minimum design loads for the structure shall be the loads applicable at the time the building was constructed, provided that no dangerous condition is created. Structural elements that are uncovered during the course of the alteration and that are found to be unsound or dangerous shall be made to comply with the applicable requirements of this code.

AJ501.5 Electrical equipment and wiring.

AJ501.5.1 Materials and methods. Newly installed electrical equipment and wiring relating to work done in any work area shall comply with the materials and methods requirements of Chapters 33 through 42.

Exception: Electrical equipment and wiring in newly installed partitions and ceilings shall comply with all applicable requirements of Chapters 33 through 42.

AJ501.5.2 Electrical service. Service to the dwelling unit shall be a minimum of 100 ampere, three-wire capacity and service equipment shall be dead front having no live parts exposed that could allow accidental contact. Type "S" fuses shall be installed when fused equipment is used.

Exception: Existing service of 60 ampere, three-wire capacity, and feeders of 30 ampere or larger two- or three-wire capacity shall be accepted if adequate for the electrical load being served.

AJ501.5.3 Additional electrical requirements. When the work area includes any of the following areas within a dwelling unit, the requirements of Sections AJ501.5.3.1 through AJ501.5.3.5 shall apply.

AJ501.5.3.1 Enclosed areas. Enclosed areas other than closets, kitchens, basements, garages, hallways, laundry areas and bathrooms shall have a minimum of two duplex receptacle outlets, or one duplex receptacle outlet and one ceiling or wall type lighting outlet.

AJ501.5.3.2 Kitchen and laundry areas. Kitchen areas shall have a minimum of two duplex receptacle outlets. Laundry areas shall have a minimum of one duplex receptacle outlet located near the laundry equipment and installed on an independent circuit.

AJ501.5.3.3 Ground-fault circuit-interruption. Ground fault circuit interruption shall be provided on newly installed receptacle outlets if required by Chapters 33 through 42.

AJ501.5.3.4 Lighting outlets. At least one lighting outlet shall be provided in every bathroom, hallway, stairway, attached garage and detached garage with electric power to illuminate outdoor entrances and exits, and in utility rooms and basements where these spaces are used for storage or contain equipment requiring service.

AJ501.5.3.5 Clearance. Clearance for electrical service equipment shall be provided in accordance with Chapters 33 through 42.

AJ501.6 Ventilation. All reconfigured spaces intended for occupancy and all spaces converted to habitable or occupiable space in any work area shall be provided with ventilation in accordance with Section R303.

AJ501.7 Ceiling height. Habitable spaces created in existing basements shall have ceiling heights of not less than 6 feet 8 inches (2032 mm). Obstructions may project to within 6 feet 4 inches (1930 mm) of the basement floor. Existing finished ceiling heights in nonhabitable spaces in basements shall not be reduced.

AJ501.8 Stairs.

AJ501.8.1 Stair width. Existing basement stairs and handrails not otherwise being altered or modified shall be permitted to maintain their current clear width at, above, and below existing handrails.

AJ501.8.2 Stair headroom. Headroom height on existing basement stairs being altered or modified shall not be reduced below the existing stairway finished headroom. Existing basement stairs not otherwise being altered shall be permitted to maintain the current finished headroom.

AJ501.8.3 Stair landing. Landings serving existing basement stairs being altered or modified shall not be reduced below the existing stairway landing depth and width. Existing basement stairs not otherwise being altered shall be permitted to maintain the current landing depth and width.

SECTION AJ601
RECONSTRUCTION

AJ601.1 Stairways, handrails and guards.

AJ601.1.1 Stairways. Stairways within the work area shall be provided with illumination in accordance with Section R303.6.

AJ601.1.2 Handrails. Every required exit stairway that has four or more risers, is part of the means of egress for any work area, and is not provided with at least one handrail, or in which the existing handrails are judged to be in danger of collapsing, shall be provided with handrails designed and installed in accordance with Section R311 for the full length of the run of steps on at least one side.

AJ601.1.3 Guards. Every open portion of a stair, landing or balcony that is more than 30 inches (762 mm) above the floor or grade below, is part of the egress path for any work area, and does not have guards or in which the existing guards are judged to be in danger of collapsing, shall be provided with guards designed and installed in accordance with Section R312.

AJ601.2 Wall and ceiling finish. The interior finish of walls and ceilings in any work area shall comply with the requirements of Section R315. Existing interior finish materials that do not comply with those requirements shall be removed or shall be treated with an approved fire-retardant coating in accordance with the manufacturer's instructions to secure compliance with the requirements of this section.

AJ601.3 Separation walls. Where the work area is in an attached dwelling unit, walls separating dwelling units that are not continuous from the foundation to the underside of the roof sheathing shall be constructed to provide a continuous fire separation using construction materials consistent with the existing wall or complying with the requirements for new

structures. Performance of work shall be required only on the side of the wall of the dwelling unit that is part of the work area.

AJ601.4 Ceiling height. Habitable spaces created in existing basements shall be permitted to have ceiling heights of not less than 6 feet 8 inches (2032 mm). Obstructions may project to within 6 feet 4 inches (1930 mm) of the basement floor. Existing finished ceiling heights in nonhabitable spaces in basements shall not be reduced.

APPENDIX K

SOUND TRANSMISSION

SECTION AK101
GENERAL

AK101.1 General. Wall and floor-ceiling assemblies separating dwelling units including those separating adjacent townhouse units shall provide air-borne sound insulation for walls, and both air-borne and impact sound insulation for floor-ceiling assemblies.

SECTION AK102
AIR-BORNE SOUND

AK102.1 General. Air-borne sound insulation for wall and floor-ceiling assemblies shall meet a Sound Transmission Class (STC) rating of 45 when tested in accordance with ASTM E 90. Penetrations or openings in construction assemblies for piping; electrical devices; recessed cabinets; bathtubs; soffits; or heating, ventilating or exhaust ducts shall be sealed, lined, insulated or otherwise treated to maintain the required ratings. Dwelling unit entrance doors, which share a common space, shall be tight fitting to the frame and sill.

SECTION AK103
STRUCTURAL-BORNE SOUND

AK103.1 General. Floor/ceiling assemblies between dwelling units or between a dwelling unit and a public or service area within a structure shall have an Impact Insulation Class (IIC) rating of not less than 45 when tested in accordance with ASTM E 492.

SECTION AK104
REFERENCED STANDARDS

APPENDIX L
PERMIT FEES

TOTAL VALUATION	FEE
$1 to $ 500	$24
$501 to $2,000	$24 for the first $500; plus $3 for each additional $ 100 or fraction thereof, to and including $2,000
$2,001 to $40,000	$69 for the first $2,000; plus $11 for each additional $1,000 or fraction thereof, to and including $40,000
$40,001 to $100,000	$487 for the first $40,000; plus $9 for each additional $1,000 or fraction thereof, to and including $100,000
$100,001 to $500,000	$1,027 for the first $100,000; plus $7 for each additional $1,000 or fraction thereof, to and including $500,000
$500,001 to $1,000,000	$3,827 for the first $500,000; plus $5 for each additional $1,000 or fraction thereof, to and including $1,000,000
$1,000,001 to $5,000,000	$6,327 for the first $1,000,000; plus $3 for each additional $1,000 or fraction thereof, to and including $5,000,000
$5,000,001 and over	$18,327 for the first $ 5,000,000; plus $1 for each additional $1,000 or fraction thereof

APPENDIX M

HOME DAY CARE—R-3 OCCUPANCY

SECTION AM101
GENERAL

AM101.1 General. This appendix shall apply to a home day care operated within a dwelling. It is to include buildings and structures occupied by persons of any age who receive custodial care for less than 24 hours by individuals other than parents or guardians or relatives by blood, marriage, or adoption, and in a place other than the home of the person cared for.

SECTION AM102
DEFINITIONS

EXIT ACCESS. That portion of a means of egress system that leads from any occupied point in a building or structure to an exit.

SECTION AM103
MEANS OF EGRESS

AM103.1 Exits required. If the occupant load of the residence is more than nine, including those who are residents, during the time of operation of the day care, two exits are required from the ground-level story. Two exits are required from a home day care operated in a manufactured home regardless of the occupant load. Exits shall comply with Section R311.

AM103.1.1 Exit access prohibited. An exit access from the area of day-care operation shall not pass through bathrooms, bedrooms, closets, garages, fenced rear yards or similar areas.

Exception: An exit may discharge into a fenced yard if the gate or gates remain unlocked during day-care hours. The gates may be locked if there is an area of refuge located within the fenced yard and more than 50 feet (15 240 mm) from the dwelling. The area of refuge shall be large enough to allow 5 square feet (0.5 m²) per occupant.

AM103.1.2 Basements. If the basement of a dwelling is to be used in the day-care operation, two exits are required from the basement regardless of the occupant load. One of the exits may pass through the dwelling and the other must lead directly to the exterior of the dwelling.

Exception: An emergency and escape window complying with Section R310 and which does not conflict with Section AM103.1.1 may be used as the second means of egress from a basement.

AM103.1.3 Yards. If the yard is to be used as part of the day-care operation it shall be fenced.

AM103.1.3.1 Type of fence and hardware. The fence shall be of durable materials and be at least 6 feet (1529 mm) tall completely enclosing the area used for the day-care operations. Each opening shall be a gate or door equipped with a self-closing and self-latching device to be installed at a minimum of 5 feet (1528 mm) above the ground.

Exception: The door of any dwelling which forms part of the enclosure need not be equipped with self-closing and self-latching devices.

AM103.1.3.2 Construction of fence. Openings in the fence, wall or enclosure required by this section shall have intermediate rails or an ornamental pattern that do not allow a sphere 4 inches (102 mm) in diameter to pass through. In addition, the following criteria must be met:

1. The maximum vertical clearance between grade and the bottom of the fence, wall or enclosure shall be 2 inches (51 mm).

2. Solid walls or enclosures that do not have openings, such as masonry or stone walls, shall not contain indentations or protrusions except for tooled masonry joints.

3. Maximum mesh size for chain link fences shall be $1^1/_4$-inches (32 mm) square unless the fence has slats at the top or bottom which reduce the opening to no more than $1^3/_4$ inches (44 mm). The wire shall not be less than 9 gage [(0.148 in.) (3.8 mm)].

AM103.1.3.3 Decks. Decks that are more than 12 inches (305 mm) above grade shall have a guard in compliance with Section R312.

AM103.2 Width and height of an exit. The minimum width of a required exit is 36 inches (914 mm) with a net clear width of 32 inches (813 mm). The minimum height of a required exit is 6 feet 8 inches (2032 mm).

AM103.3 Type of lock and latches for exits. Regardless of the occupant load served, exit doors shall be openable from the inside without the use of a key or any special knowledge or effort. When the occupant load is 10 or less, a night latch, dead bolt or security chain may be used, provided such devices are openable from the inside without the use of a key or tool and mounted at a height not to exceed 48 inches (1219 mm) above the finished floor.

AM103.4 Landings. Landings for stairways and doors shall comply with Section R311 except that landings shall be required for the exterior side of a sliding door when a home day-care is being operated in a Group R-3 Occupancy.

SECTION AM104
SMOKE DETECTION

AM104.1 General. Smoke detectors shall be installed in dwelling units used for home day-care operations. Detectors shall be installed in accordance with the approved manufacturer's instructions. If the current smoke detection system in the dwelling is not in compliance with the currently adopted code for smoke detection, it shall be upgraded to meet the currently adopted code requirements and Section AM103 before daycare operations commence.

AM104.2 Power source. Required smoke detectors shall receive their primary power from the building wiring when that wiring is served from a commercial source and shall be equipped with a battery backup. The detector shall emit a signal when the batteries are low. Wiring shall be permanent and without a disconnecting switch other than those required for over-current protection. Required smoke detectors shall be interconnected so if one detector is activated, all detectors are activated.

AM104.3 Location. A detector shall be located in each bedroom and any room that is to be used as a sleeping room and centrally located in the corridor, hallway or area giving access to each separate sleeping area. When the dwelling unit has more than one story, and in dwellings with basements, a detector shall be installed on each story and in the basement. In dwelling units where a story or basement is split into two or more levels, the smoke detector shall be installed on the upper level, except that when the lower level contains a sleeping area, a detector shall be installed on each level. When sleeping rooms are on the upper level, the detector shall be placed at the ceiling of the upper level in close proximity to the stairway. In dwelling units where the ceiling height of a room open to the hallway serving the bedrooms or sleeping areas exceeds that of the hallway by 24 inches (610 mm) or more, smoke detectors shall be installed in the hallway and in the adjacent room. Detectors shall sound an alarm audible in all sleeping areas of the dwelling unit in which they are located.

VENTING METHODS

(This appendix is informative and is not part of the code.
This appendix provides examples of various of venting methods.)

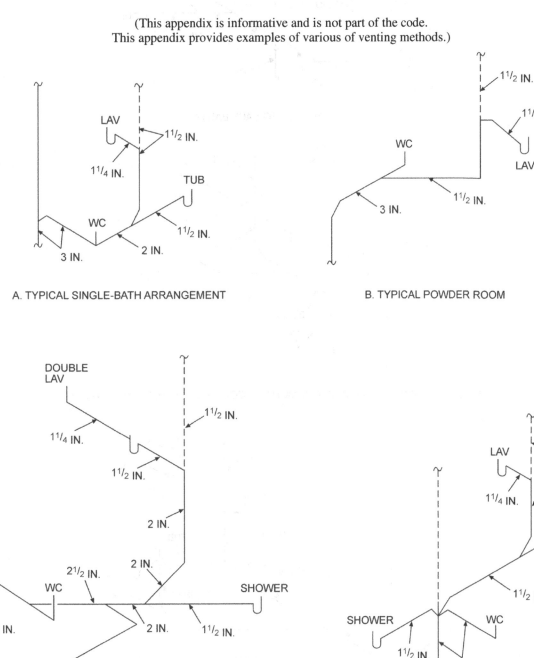

A. TYPICAL SINGLE-BATH ARRANGEMENT

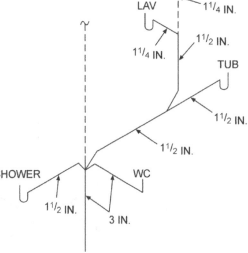

B. TYPICAL POWDER ROOM

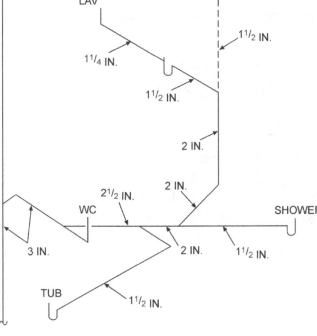

C. MORE ELABORATE SINGLE-BATH
ARRANGEMENT

D. COMBINATION WET- AND STACK-VENTING
WITH STACK FITTING

For SI: 1 inch = 25.4.

**FIGURE N1
TYPICAL SINGLE-BATH WET-VENT ARRANGEMENTS**

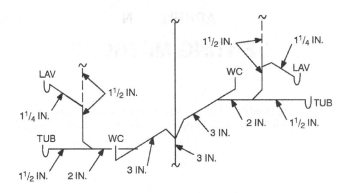

A. TYPICAL BACK-TO-BACK BATHS

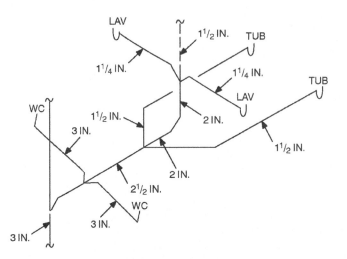

B. DOUBLE BATHS WITH FIXTURES ON COMMON HORIZONTAL BRANCH, COMMON WET VENT

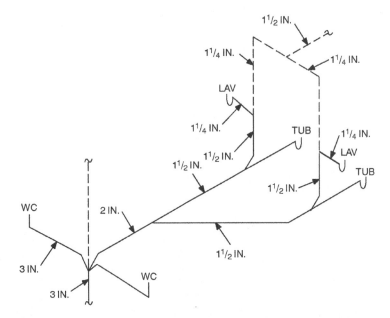

C. DOUBLE BATHS WITH WASTE FIXTURES ON COMMON HORIZONTAL BRANCH, INDIVIDUAL WET VENTS

For SI: 1 inch = 25.4.

FIGURE N2
TYPICAL DOUBLE-BATH WET-VENT ARRANGEMENTS

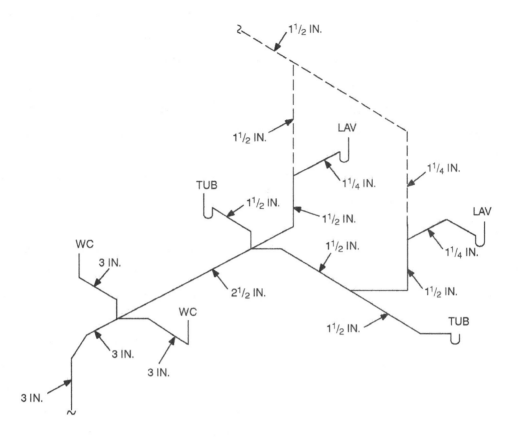

For SI: 1 inch = 25.4 mm.

FIGURE N3
TYPICAL HORIZONTAL WET VENTING

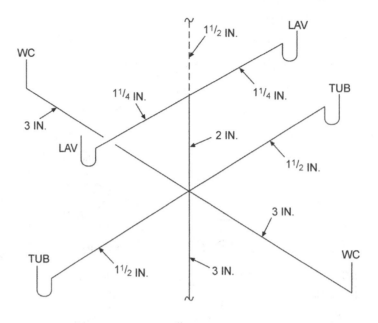

A. VERTICAL WET VENTING

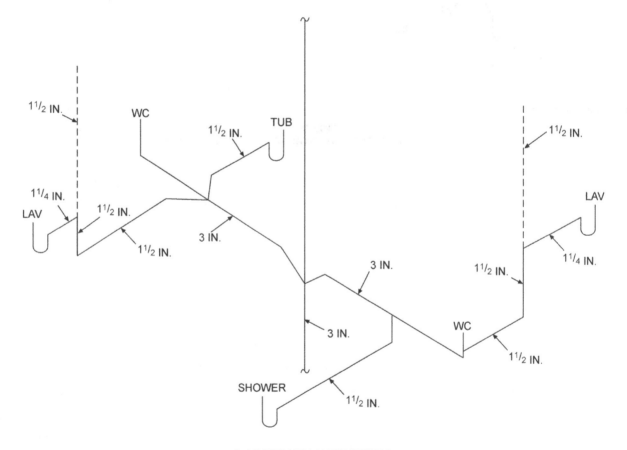

B. HORIZONTAL WET VENTING

For SI: 1 inch = 25.4 mm.

**FIGURE N4
TYPICAL METHODS OF WET VENTING**

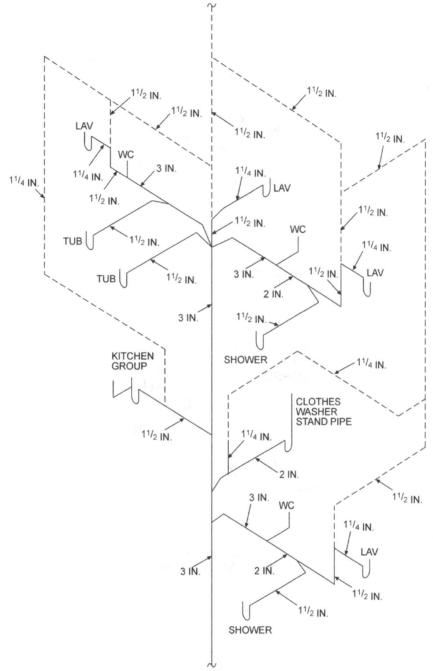

For SI: 1 inch = 25.4 mm.

FIGURE N5
SINGLE STACK SYSTEM FOR A TWO-STORY DWELLING

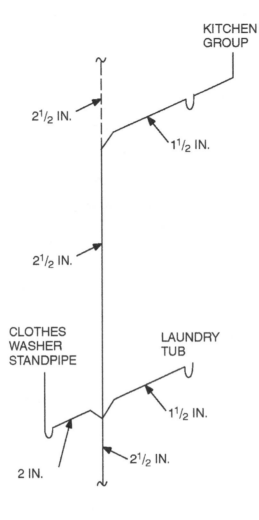

For SI: 1 inch = 25.4 mm.

FIGURE N6
WASTE STACK VENTING

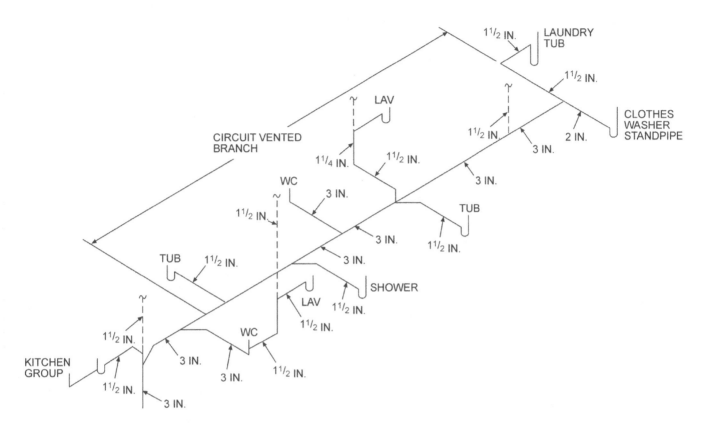

For SI: 1 inch = 25.4 mm.

FIGURE N7
CIRCUIT VENT WITH ADDITIONAL NONCIRCUIT VENTED BRANCH

GRAY WATER RECYCLING SYSTEMS

Note: Section P2601.2 of the International Residential Code *requires all plumbing fixtures that receive water or waste to discharge to the sanitary drainage system of the structure. To allow for the use of a gray water recycling system, Section P2601.2 of the* International Residential Code *should be revised to read as follows:*

P2601.2 Connections. Plumbing fixtures, drains and appliances used to receive or discharge liquid wastes or sewage shall be directly connected to the sanitary drainage system of the building or premises, in accordance with the requirements of this code. This section shall not be construed to prevent indirect waste systems.

> **Exception:** Bathtubs, showers, lavatories, clothes washers and laundry trays are not required to discharge to the sanitary drainage system where those fixtures discharge to an approved gray water recycling system.

SECTION AO101
GENERAL

AO101.1 Scope. The provisions of this appendix shall govern the materials, design, construction and installation of gray water systems for flushing of water closets and urinals and for subsurface landscape irrigation [see Figures AO101.1(1) and AO101.1(2)].

AO101.2 Definition. The following term shall have the meaning shown herein.

GRAY WATER. Waste discharged from lavatories, bathtubs, showers, clothes washers and laundry trays.

AO101.3 Permits. Permits shall be required in accordance with Section R105 of the *International Residential Code.*

AO101.4 Installation. In addition to the provisions of Section AO101, systems for flushing of water closets and urinals shall comply with Section AO102 and systems for subsurface landscape irrigation shall comply with Section AO103. Except as provided for in Appendix O, all systems shall comply with the provisions of the *International Residential Code.*

AO101.5 Materials. Above-ground drain, waste and vent piping for gray water systems shall conform to one of the standards listed in Table P3002.1(1) of the *International Residential Code.* Gray water underground building drainage and vent pipe shall conform to one of the standards listed in Table P3002.1(2) of the *International Residential Code.*

AO101.6 Tests. Drain, waste and vent piping for gray water systems shall be tested in accordance with Section P2503 of the *International Residential Code.*

AO101.7 Inspections. Gray water systems shall be inspected in accordance with Section P2503 of the *International Residential Code.*

AO101.8 Potable water connections. Only connections in accordance with Section AO102.3 shall be made between a gray water recycling system and a potable water system.

AO101.9 Waste water connections. Gray water recycling systems shall receive the waste discharge only of bathtubs, showers, lavatories, clothes washers and laundry trays.

AO101.10 Filtration. Gray water entering the reservoir shall pass through an approved filter such as a media, sand or diatomaceous earth filter.

> **AO101.10.1 Required valve.** A full-open valve shall be installed downstream of the last fixture connection to the gray water discharge pipe before entering the required filter.

AO101.11 Collection reservoir. Gray water shall be collected in an approved reservoir constructed of durable, nonabsorbent and corrosion-resistant materials. The reservoir shall be a closed and gas-tight vessel. Access openings shall be provided to allow inspection and cleaning of the reservoir interior.

AO101.12 Overflow. The collection reservoir shall be equipped with an overflow pipe of the same diameter as, or larger than, the influent pipe for the gray water. The overflow shall be indirectly connected to the sanitary drainage system.

AO101.13 Drain. A drain shall be located at the lowest point of the collection reservoir and shall be indirectly connected to the sanitary drainage system. The drain shall be the same diameter as the overflow pipe required in Section AO101.12.

AO101.14 Vent required. The reservoir shall be provided with a vent sized in accordance with Chapter 31 of the *International Residential Code* and based on the diameter of the reservoir influent pipe.

SECTION AO102
SYSTEMS FOR FLUSHING WATER
CLOSETS AND URINALS

AO102.1 Collection reservoir. The holding capacity of the reservoir shall be a minimum of twice the volume of water required to meet the daily flushing requirements of the fixtures supplied with gray water, but not less than 50 gallons (189 L). The reservoir shall be sized to limit the retention time of gray water to a maximum of 72 hours.

AO102.2 Disinfection. Gray water shall be disinfected by an approved method that uses one or more disinfectants such as chlorine, iodine or ozone.

AO102.3 Makeup water. Potable water shall be supplied as a source of makeup water for the gray water system. The potable water supply shall be protected against backflow in accordance with Section P2902 of the *International Residential Code.* A full-open valve shall be located on the makeup water supply line to the collection reservoir.

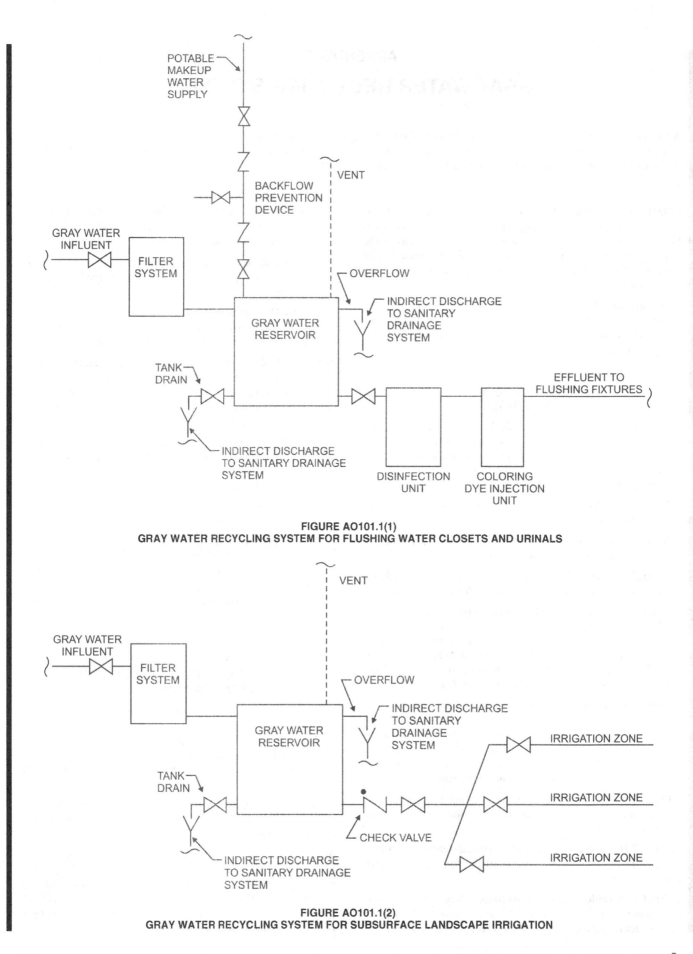

FIGURE AO101.1(1)
GRAY WATER RECYCLING SYSTEM FOR FLUSHING WATER CLOSETS AND URINALS

FIGURE AO101.1(2)
GRAY WATER RECYCLING SYSTEM FOR SUBSURFACE LANDSCAPE IRRIGATION

AO102.4 Coloring. The gray water shall be dyed blue or green with a food grade vegetable dye before such water is supplied to the fixtures.

AO102.5 Materials. Distribution piping shall conform to one of the standards listed in Table P2904.5 of the *International Residential code*.

AO102.6 Identification. Distribution piping and reservoirs shall be identified as containing nonpotable water. Piping identification shall be in accordance with Section 608.8 of the *International Plumbing Code®*.

SECTION AO103
SUBSURFACE LANDSCAPE
IRRIGATION SYSTEMS

AO103.1 Collection reservoir. Reservoirs shall be sized to limit the retention time of gray water to a maximum of 24 hours.

AO103.1.1 Identification. The reservoir shall be identified as containing nonpotable water.

AO103.2 Valves required. A check valve, and a full-open valve located on the discharge side of the check valve, shall be installed on the effluent pipe of the collection reservoir.

AO103.3 Makeup water. Makeup water shall not be required for subsurface landscape irrigation systems. Where makeup water is supplied, the installation shall be in accordance with Section AO102.3.

AO103.4 Disinfection. Disinfection shall not be required for gray water used for subsurface landscape irrigation systems.

AO103.5 Coloring. Gray water used for subsurface landscape irrigation systems shall not be required to be dyed.

AO103.6 Estimating gray water discharge. The system shall be sized in accordance with the demands per day per occupant based on the type of fixtures connected to the gray water system. The discharge shall be calculated by the following equation:

$$C = A \times B \qquad \text{(Equation AO-1)}$$

A = Number of occupants:
Number of occupants shall be determined by the actual number of occupants but not less than two occupants for 1 bedroom and one occupant for each additional bedroom.

B = Estimated flow demands for each occupant:
25 gallons per day (95 Lpd) per occupant for showers, bathtubs and lavatories and 15 gallons per day (57 Lpd) per occupant for clothes washers or laundry trays.

C = Estimated gray water discharge based on the total number of occupants.

AO103.7 Percolation tests. The permeability of the soil in the proposed absorption system shall be determined by percolation tests or permeability evaluation.

AO103.7.1 Percolation tests and procedures. At least three percolation tests shall be conducted in each system area. The holes shall be spaced uniformly in relation to the bottom depth of the proposed absorption system. More percolation tests shall be made where necessary, depending on system design.

AO103.7.1.1 Percolation test hole. The test hole shall be dug or bored. The test hole shall have vertical sides and a horizontal dimension of 4 inches to 8 inches (102 mm to 203 mm). The bottom and sides of the hole shall be scratched with a sharp-pointed instrument to expose the natural soil. All loose material shall be removed from the hole and the bottom shall be covered with 2 inches (51 mm) of gravel or coarse sand.

AO103.7.1.2 Test procedure, sandy soils. The hole shall be filled with clear water to a minimum of 12 inches (305 mm) above the bottom of the hole for tests in sandy soils. The time for this amount of water to seep away shall be determined and this procedure shall be repeated if the water from the second filling of the hole seeps away in 10 minutes or less. The test shall proceed as follows: Water shall be added to a point not more than 6 inches (152 mm) above the gravel or coarse sand. Thereupon, from a fixed reference point, water levels shall be measured at 10-minute intervals for a period of 1 hour. Where 6 inches (152 mm) of water seeps away in less than 10 minutes, a shorter interval between measurements shall be used, but in no case shall the water depth exceed 6 inches (152 mm). Where 6 inches (152 mm) of water seeps away in less than 2 minutes, the test shall be stopped and a rate of less than 3 minutes per inch (7 s/mm) shall be reported. The final water level drop shall be used to calculate the percolation rate. Soils not meeting the requirements of this section shall be tested in accordance with Section AO103.7.1.3.

AO103.7.1.3 Test procedure, other soils. The hole shall be filled with clear water, and a minimum water depth of 12 inches (305 mm) shall be maintained above the bottom of the hole for a 4-hour period by refilling whenever necessary or by use of an automatic siphon. Water remaining in the hole after 4 hours shall not be removed. Thereafter, the soil shall be allowed to swell not less than 16 hours or more than 30 hours. Immediately after the soil swelling period, the measurements for determining the percolation rate shall be made as follows: Any soil sloughed into the hole shall be removed, and the water level shall be adjusted to 6 inches (152 mm) above the gravel or coarse sand. Thereupon, from a fixed reference point, the water level shall be measured at 30-minute intervals for a period of 4 hours, unless two successive water level drops do not vary by more than 0.62 inch (16 mm). At least three water level drops shall be observed and recorded. The hole shall be filled with clear water to a point not more than 6 inches (152 mm) above the gravel or coarse sand whenever it becomes nearly empty. The water level shall not be adjusted during the three measurement periods except to the limits of the last measured water level drop. When the first 6 inches (152 mm) of water seeps away in less than 30 minutes, the time interval between measurements shall be 10 minutes and the test run for 1 hour. The water depth shall not exceed 5 inches (127 mm) at any time during the measurement period. The drop that occurs during the final measurement period shall be used in calculating the percolation rate.

AO103.7.1.4 Mechanical test equipment. Mechanical percolation test equipment shall be of an approved type.

AO103.7.2 Permeability evaluation. Soil shall be evaluated for estimated percolation based on structure and texture in accordance with accepted soil evaluation practices. Borings shall be made in accordance with Section AO103.7.1 for evaluating the soil.

AO103.8 Subsurface landscape irrigation site location. The surface grade of all soil absorption systems shall be located at a point lower than the surface grade of any water well or reservoir on the same or adjoining property. Where this is not possible, the site shall be located so surface water drainage from the site is not directed toward a well or reservoir. The soil absorption system shall be located with a minimum horizontal distance between various elements as indicated in Table AO103.8. Private sewage disposal systems in compacted areas, such as parking lots and driveways, are prohibited. Surface water shall be diverted away from any soil absorption site on the same or neighboring lots.

TABLE AO103.8
LOCATION OF GRAY WATER SYSTEM

ELEMENT	MINIMUM HORIZONTAL DISTANCE	
	HOLDING TANK (feet)	IRRIGATION DISPOSAL FIELD (feet)
Buildings	5	2
Property line adjoining private property	5	5
Public water main	10	10
Seepage pits	5	5
Septic tanks	0	5
Streams and lakes	50	50
Water service	5	5
Water wells	50	100

For SI: 1 foot = 304.8 mm.

AO103.9 Installation. Absorption systems shall be installed in accordance with Sections AO103.9.1 through AO103.9.5 to provide landscape irrigation without surfacing of gray water.

AO103.9.1 Absorption area. The total absorption area required shall be computed from the estimated daily gray water discharge and the design-loading rate based on the percolation rate for the site. The required absorption area equals the estimated gray water discharge divided by the design-loading rate from Table AO103.9.1.

AO103.9.2 Seepage trench excavations. Seepage trench excavations shall be a minimum of 1 foot (305 mm) to a maximum of 5 feet (1524 mm) wide. Trench excavations shall be spaced a minimum of 2 feet (610 mm) apart. The

soil absorption area of a seepage trench shall be computed by using the bottom of the trench area (width) multiplied by the length of pipe. Individual seepage trenches shall be a maximum of 100 feet (30 480 mm) in developed length.

TABLE AO103.9.1
DESIGN LOADING RATE

PERCOLATION RATE (minutes per inch)	DESIGN LOAD FACTOR (gallons per square foot per day)
0 to less than 10	1.2
10 to less than 30	0.8
30 to less than 45	0.72
45 to 60	0.4

For SI: 1 minute per inch = 2.362 s/mm; 1 gallon per square foot = 40.743 L/m^2.

AO103.9.3 Seepage bed excavations. Seepage bed excavations shall be a minimum of 5 feet (1524 mm) wide and have more than one distribution pipe. The absorption area of a seepage bed shall be computed by using the bottom of the trench area. Distribution piping in a seepage bed shall be uniformly spaced a maximum of 5 feet (1524 mm) and a minimum of 3 feet (914 mm) apart, and a maximum of 3 feet (914 mm) and a minimum of 1 foot (305 mm) from the sidewall or headwall.

AO103.9.4 Excavation and construction. The bottom of a trench or bed excavation shall be level. Seepage trenches or beds shall not be excavated where the soil is so wet that such material rolled between the hands forms a soil wire. All smeared or compacted soil surfaces in the sidewalls or bottom of seepage trench or bed excavations shall be scarified to the depth of smearing or compaction and the loose material removed. Where rain falls on an open excavation, the soil shall be left until sufficiently dry so a soil wire will not form when soil from the excavation bottom is rolled between the hands. The bottom area shall then be scarified and loose material removed.

AO103.9.5 Aggregate and backfill. A minimum of 6 inches (152 mm) of aggregate ranging in size from $^1/_2$ to $2^1/_2$ inches (13 mm to 64 mm) shall be laid into the trench below the distribution piping elevation. The aggregate shall be evenly distributed a minimum of 2 inches (51 mm) over the top of the distribution pipe. The aggregate shall be covered with approved synthetic materials or 9 inches (229 mm) of uncompacted marsh hay or straw. Building paper shall not be used to cover the aggregate. A minimum of 9 inches (229 mm) of soil backfill shall be laid above the covering.

AO103.10 Distribution piping. Distribution piping shall be not less than 3 inches (76 mm) in diameter. Materials shall comply with Table AO103.10. The top of the distribution pipe shall be not less than 8 inches (203 mm) below the original surface. The slope of the distribution pipes shall be a minimum of 2 inches (51 mm) and a maximum of 4 inches (102 mm) per 100 feet (30 480 mm).

AO103.11 Joints. Distribution pipe shall be joined in accordance with Section P3003 of the *International Residential Code.*

TABLE AO103.10
DISTRIBUTION PIPE

MATERIAL	STANDARD
Polyethylene (PE) plastic pipe	ASTM F 405
Polyvinyl chloride (PVC) plastic pipe	ASTM D 2729
Polyvinyl chloride (PVC) plastic pipe with pipe stiffness of PS 35 and PS 50	ASTM F 1488

APPENDIX P
FIRE SPRINKLER SYSTEM

The provisions contained in this appendix are not mandatory unless specifically referenced in the adopting ordinance.

AP101 Fire sprinklers. An approved automatic fire sprinkler system shall be installed in new one- and two-family dwellings and townhouses in accordance with Section 903.3.1 of the *International Building Code*.

ICC *INTERNATIONAL RESIDENTIAL CODE* ELECTRICAL PROVISIONS/NATIONAL ELECTRICAL CODE CROSS-REFERENCE

This table is a cross-reference of the *International Residential Code*, Chapters 33 through 42, and the 2005 *National Electrical Code* (NFPA 70-2005).

International Residential Code		National Electrical Code
CHAPTER 33	**GENERAL REQUIREMENTS**	
SECTION E3301	**GENERAL**	
E3301.1	Applicability	None
E3301.2	Scope	90.2
E3301.3	Not covered	90.2
E3301.4	Additions and alterations	None
SECTION E3302	**BUILDING STRUCTURE PROTECTION**	
E3302.1	Drilling and notching	None
E3302.2	Penetrations of fire-resistance-rated assemblies	300.21
E3302.3	Penetrations of firestops and draftstops	300.21
SECTION E3303	**INSPECTION AND APPROVAL**	
E3303.1	Approval	110.2
E3303.2	Inspection required	None
E3303.3	Listing and labeling	110.3
SECTION E3304	**GENERAL EQUIPMENT REQUIREMENTS**	
E3304.1	Voltages	110.4
E3304.2	Interrupting rating	110.9
E3304.3	Circuit characteristics	110.10
E3304.4	Protection of equipment	110.11
E3304.5	Unused openings	110.12(A)
E3304.6	Integrity of electrical equipment	110.12(C)
E3304.7	Mounting	110.13(A)
E3304.8	Energized parts guarded against accidental contact	110.27(A)
E3304.9	Prevent physical damage	110.27(B)
E3304.10	Equipment identification	110.21
E3304.11	Identification of disconnecting means	110.22
SECTION E3305	**EQUIPMENT LOCATION AND CLEARANCES**	
E3305.1	Working space and clearances	110.26
Figure E3305.1	Working space and clearances	110.26(A)
	Footnote 1	110.26(F)(1)(a)
	Footnote 2	110.26(A)(3) and 110.26(E)
	Footnote 3	110.26(B)
	Footnote 4	230.70(A), 240.24(D) and 240.24(E)
	Footnote 5	110.26(D)
E3305.2	Working clearances for energized equipment and panelboards	110.26(A)(1), (2), & (3)
E3305.3	Clearances over panelboards	110.26(F)(1)(a)
E3305.4	Location of clear spaces	110.26(B), 230.70(A) and 240.24(D) & (E)
E3305.5	Access and entrance to working space	110.26(C)(1)
E3305.6	Illumination	110.26(D)
E3305.7	Headroom	110.26(E)
SECTION E3306	**ELECTRICAL CONDUCTORS AND CONNECTIONS**	
E3306.1	General	Articles 110, 300 and 310

INDEX

EDITORIAL CHANGES – SECOND PRINTING

Page 6, Section 109.1.5.2: entire section is deleted.

Page 47, Table R302.1: column 4, row 4 now reads . . . 2 feet

Page 56, Section R314.3: line 6 to the end now reads . . . not more than 450 when tested in the maximum thickness intended for use in accordance with ASTM E 84. Loose-fill-type foam plastic insulation shall be tested as board stock for the flame spread index and smoke-developed index.

Page 107, Table R503.2.1.1(1): column 4, row 10 now reads . . . 130

Page 137, Table R602.10.1: column 4, row 5 now reads . . . Located in accordance with Section R602.10 and at least every 25 feet on center but not less than 16% of braced wall line for Method 3 or 25% of braced wall line for Methods 2, 4, 5, 6, 7 or 8.

Page 137, Table R602.10.1: column 4, row 6 now reads . . . Located in accordance with Section R602.10 and at least every 25 feet on center but not less than 30% of braced wall line for Method 3 or 45% of braced wall line for Methods 2, 4, 5, 6, 7 or 8.

Page 267, Figure R802.5.1: figure has been replaced.

Page 293, Section R905.4.3: line 2 now reads . . . ASTM D 226, Type I or ASTM D 4869, Type I or II.

Page 294, Section R905.5.3: line 2 now reads . . . ASTM D 226, Type I or ASTM D 4869, Type I or II.

Page 294, Section R905.6.3: line 2 now reads . . . ASTM D 226, Type I or ASTM D 4869, Type I or II.

Page 294, Section R905.7.3: line 2 now reads . . . ASTM D 226, Type I or ASTM D 4869, Type I or II.

Page 295, Section R905.8.3: line 2 now reads . . . ASTM D 226, Type I or ASTM D 4869, Type I or II.

Page 298, Section R905.13.2: line 3 now reads . . . 6754, ASTM D 6878, or CAN/CGSB 37.54.

Page 325, Section M1305.1.3: line 2 now reads . . . ances requiring access shall be provided with an opening

Page 331, Section M1401.5: line 4 now reads . . . tion R324.1.5.

Page 334, Section M1415.1: line 2 now reads . . . accordance with Section R1002.

Page 336, Section M1506.2: line 2 now reads . . . **M1507.2 Recirculation of air.** Exhaust air from bathrooms

Page 338, Section M1601.3.8: line 3 now reads . . . located or installed in accordance with Section R324.1.5.

Page 349, Section M1805.1: now reads . . . **M1805.1 General.** Masonry and factory-built chimneys shall be built and installed in accordance with Sections R1003 and R1005, respectively. Flue lining for masonry chimneys shall comply with Section R1003.11.

Page 353, Section M2001.4: line 4 now reads . . . accordance with Section R324.1.5.

Page 365, Section G2403 (202): Connector definition is deleted.

Page 399, Section G2427.6 (503.6): line 2 now reads . . . tions G2427.6.1 through G2427.6.10. (See Section G2403,

Page 399, Section G2427.6.3 (503.6.3): entire section is deleted.

Page 422, Section G2451.2 (630.2): now reads . . . **G2451.2 (630.2) Support.** Infrared radiant heaters shall be fixed in a position independent of gas and electric supply lines. Hangers and brackets shall be of noncombustible material.

Page 481, Table E3502.2: row 4 now reads . . . 1,500 volt-amperes × total number of 20-ampere-rated small appliance and laundry circuits.

Page 506, Section E3801.4.5: Exception, line 3 now reads . . . countertop in construction designed for the physically

Page 526, Section E3809.2: now reads . . . **E3809.2 Loading and protection.** The ampere load of flexible cords serving fixed appliances shall be in accordance with Table E3809.2. This table shall be used in conjunction with applicable end use product standards to ensure selection of the proper size and type. Where flexible cord is approved for and used with a specific listed appliance, it shall be considered to be protected where applied within the appliance listing requirements.

Page 558, AWPA: new standard added now reads . . . C1—100 All Timber Products—Preservative Treatment by Pressure Processes R902.2

Page 560: new standard added DASMA

Page 565, UL: new standard added now reads . . . 651—05 Schedule 40 and Schedule 80 Rigid PVC Conduit and Fittings G2414.6.3

EDITORIAL CHANGES – THIRD PRINTING

Page 242, Table R703.7.3: column 5 heading now reads . . . footnote b

Page 250, Table R802.5.1(1): For Rafter Spacing of 12 inches on center, under a Dead Load of 10 psf, under 2 × 4 the following entries now read . . . Douglas Fir Larch #1, 11-1; Douglas Fir Larch #2, 10-1

Page 298, Section 905.13.2: line 3 now reads . . . 6754, ASTM D 6878, or CGSB CAN/CGSB 37.54.

Page 325, Section M1301.1.1.1: line 4 now reads . . . installed in accordance with Section R324.1.5.

Page 341, Section M1701.6: line 4 now reads . . . Section R324.1.5.

Page 402, Section G2427.7.12: line 3 now reads . . . G2427.6.10.

Page 403, Section G2427.10.2: now reads . . . with Sections G2427.10.2.1 through G2427.10.2.4.

Page 411, Section G2428.2.9: Exception deleted

Page 417, Section G2428.3.16: Exception deleted

EDITORIAL CHANGES – FOURTH PRINTING

Page 41, Section R301.2.1.2: last sentence now reads . . . Glazed opening protection for windborne debris shall meet the requirements of the Large Missile Test of ASTM E 1996 and ASTM E 1886 referenced therein.

Page 41, Section R301.2.1.2: line 11 of the exception now reads . . . R301.2(2) or ASCE 7.

Page 43, Section R301.2.2.1.1: line 6 now reads . . . Section 1613.5.2 of the *International Building Code.*

Page 43, Section R301.2.2.1.1: line 10 now reads . . . 1613.5 of the *International Building Code.* The value

Page 43, Section R301.2.2.1.1: line 11 now reads . . . of S_{DS} determined according to Section 1613.5 of the

Page 46, Table R301.5, footnote e now reads . . . See Section R502.2.2 for decks attached to exterior walls.

Page 57, Section R314.5.4: line 3 now reads . . . required by Section R408.4 and where entry is made only

Page 57, Section R314.7: line 3 now reads . . . dance with Section R320.5.

Page 63, Section R324.1.7: Item 2, line 3 now reads . . . shall conform to the provisions of FEMA/FIA-TB-2.

Page 63, Section R324.2: line 6 now reads . . . accordance with Sections R324.2.1 through R324.2.3.

Page 64, Section R324.3: last sentence now reads . . . Buildings and structures constructed in whole or in part in coastal high-hazard areas shall be designed and constructed in accordance with Sections R324.3.1 through R324.3.6.

Page 81, Table R404.1.1(1): footnote e now reads . . . Wall construction shall be in accordance with either Table R404.1.1(2), Table R404.1.1(3), Table R404.1.1(4), or a design shall be provided.

Page 83, Table R404.1.1(3): column 5, row 1 now reads . . . SC, ML-CL and inorganic CL soils

Page 185, Figure R606.11(3): Foundation for wood floor illustration, deleted HEIGHT 8 FT MAX arrow on left side of illustration.

Page 213, Table R611.7(11): column 1, row 3 now reads . . . D_0^d or D_1^d

Page 250, Table R802.5.1(1): For Rafter Spacing of 12 inches on center, under a Dead Load of 10 psf, under 2 × 4 the following entries now read . . . Douglas Fir Larch #2, 10-10

Page 256, Table R802.5.1(4): For Rafter Spacing of 12 inches on center, under a Dead Load of 10 psf, under 2 × 10 the following entries now read . . . Spruce-pine-fir #2, 20-2; Spruce-pine-fir #3, 15-3

Page 319, Table N1102.1: column 11, row 1 header now reads . . . CRAWL SPACE^c WALL *R*-VALUE

Page 336, SECTION M1506: Section title now reads M1506.1 Ducts

Page 347, Section M1801.3.3: line 2 now reads . . . with a cleanout opening complying with Section R1103.17.

Page 355, Section M2101.6: line 3 now reads . . . the provisions of Sections R508.8, R602.6, R602.6.1 and

Page 355, Section M2102.6: line 4 now reads . . . R802.7. Holes in cold-formed, steel-framed, load-bearing

Page 355, Section M2102.6: line 6 now reads . . . R507.2, R603.2 and R804.2. In accordance with the provisions

Page 359, Section M2201.6: line 4 now reads . . . Section R324 or shall be anchored to prevent flotation, collapse

Page 369, Section G2404.7: line 5 now reads . . . tant construction requirements of Section R324.

Page 369, 9, Section G2404.7: line 9 of the Exception now reads . . . requirements of Section R324.

Page 393, Section G2420.5: last sentence added to the Exception, Piping from the shutoff valve to within 3 feet (914 mm) of the appliance connection shall be sized in accordance with Section 402.

Page 419, Section G2439.5.1: last sentence added, The maximum length of the exhaust duct does not include the transition duct.

Page 431, Section P2705.1: Item 7, last line now reads . . . in accordance with Section R324.1.5.

Page 496, Table E3605.5.3: row 1, headings, Copper should be centered over 1st and 2nd columns, Aluminum or Copper-Clad Aluminum should be centered over 3rd and 4th columns.

EDITORIAL CHANGES – FIFTH PRINTING

Page 41, Section R301.2.1.2: line 2 of the exception now reads . . . thickness of $^7/_{16}$ inch (11 mm) and a maximum span of

Page 72, Section R403.1.6: last sentence now reads . . . R505.3.1 or R603.3.1.

Page 118, Table R505.3.2(1): column 1, row 6 now reads . . . 550S162-97

Page 119, Table R505.3.2(2): column 1, row 6 now reads . . . 550S162-97

Page 347, Section M1801.3.3: line 2 now reads . . . with a cleanout opening complying with Section R1003.17.

EDITORIAL CHANGES – SIXTH PRINTING

Page 2, Section R104.10: line 8 now reads . . . not lessen health, life and fire safety or structural requirements.

Page 54, Section R311.5.4: now reads . . . **R311.5.4 Landings for stairways.** There shall be a floor or landing at the top and bottom of each stairway. A flight of stairs shall not have a vertical rise larger than 12 feet (3658 mm) between floor levels or landings. The width of each landing shall not be less than the width of the stairway served. Every landing shall have a minimum dimension of 36 inches (914 mm) measured in the direction of travel.

Exception: A floor or landing is not required at the top of an interior flight of stairs, including stairs in an enclosed garage, provided that a door does not swing over the stairs.

Page 82, Table R404.1.1(2): Table head now reads . . . TABLE R404.1.1(2) 8-INCH MASONRY FOUNDATION WALLS WITH REINFORCING WHERE d > 5 INCHES[a, c]

Page 83, Table R404.1.1(3): Table head now reads . . . TABLE R404.1.1(3) 10-INCH MASONRY FOUNDATION WALLS WITH REINFORCING WHERE d > 6.75 INCHES[a, c]

Page 83, Table R404.1.1(3): footnote c now reads . . . c. Vertical reinforcement shall be Grade 60 minimum. The distance, d, from the face of the soil side of the wall to the center of vertical reinforcement shall be at least 6.75 inches.

Page 84, Table R404.1.1(4): Table head now reads . . . TABLE R404.1.1(4) 12-INCH MASONRY FOUNDATION WALLS WITH REINFORCING WHERE d > 8.75 INCHES[a, c]

Page 84, Table R404.1.1(4): footnote c now reads . . .c. Vertical reinforcement shall be Grade 60 minimum. The distance, d, from the face of the soil side of the wall to the center of vertical reinforcement shall be at least 8.75 inches.

Page 98, Figure R502.2 Floor Construction: section number reference on left side of illustration now reads . . . 2 IN. CLEAR-ANCE—SEE SECTION R1001.11

Page 107, Table R503.2.1.1(1): column 2, row 10 now reads . . . $^{23}/_{32}$, $^3/_4$

Page 129, Figure R602.3(2) Framing Details: replaced as shown

Page 190, Section R608.2.2: line 3 now reads . . . R608.2.1, the minimum area of horizontal reinforcement

Page 244, Section R802.3.2: line 6 now reads . . . accordance with Table R802.5.1(9) and butted joists shall

Page 307, Section R1003.15.1: line 10 now reads . . . areas of clay flue linings are shown in Tables R1003.14(1)

Page 307, Section R1003.15.1: line 11 now reads . . . and R1003.14(2) or as provided by the manufacturer or as

Page 361, Section M2301.5: line 3 now reads . . . P2902.5.5.

Page 491, Section E3602.1: line 3 now reads . . . cord-and-plug-connected loads of up to 1,440 volt-amperes or

Page 613, Section AH104.1: line 6 and 7 now reads . . . designed to resist the minimum wind loads set forth in Section R301.2.1.

EDITORIAL CHANGES – SEVENTH PRINTING

Page 272, Table R804.3.1(1): column 1, last row now reads . . .1200S162-97

Page 273, Table R804.3.1(2): column 1, last row now reads . . .1200S162-97

Page 274, Table R804.3.1(3): column 1, last row now reads . . .1200S162-97

Page 275, Table R804.3.1(4): column 1, last row now reads . . .1200S162-97

Page 276, Table R804.3.1(5): column 1, last row now reads . . .1200S162-97

Page 277, Table R804.3.1(6): column 1, last row now reads . . .1200S162-97

Page 278, Table R804.3.1(7): column 1, last row now reads . . .1200S162-97

Page 279, Table R804.3.1(8): column 1, last row now reads . . .1200S162-97

Page 613, Section AH105.1: line 8 now reads . . .exits conforming to the provisions of Section R311 of this

Page 241, Figure R703.7.2.2: added "NOT" and deleted "WITH FASTENERS" from left side call-out, which now reads: "STEEL ANGLE NOT ATTACHED TO STUD."

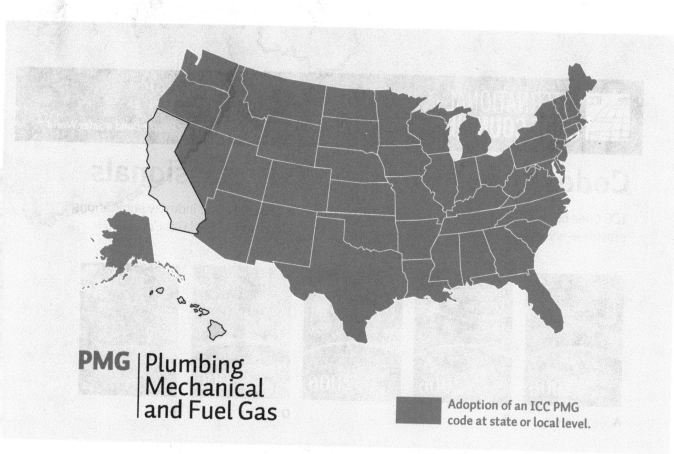

FULL-COLOR FIGURES AND PHOTOS!
SIGNIFICANT CHANGES TO THE IPC®/IMC®/ IFGC®, 2006

Authored by code experts **Bob Guenther, Robert Konyndyk and Stephen Van Note,** this new release helps you easily identify key changes from the 2003 to 2006 IPC®, IMC®, and IFGC® and read analysis of the effect each change has on application. This title focuses squarely on changes that are utilized frequently, have had a change in application, or have special significance including new technologies. A straightforward analysis of the impact of these changes will help familiarize building officials, plans examiners, inspectors, contractors, design professionals, and others apply new code provisions effectively. The full-color text includes hundreds of photos and illustrations. (156 pages)
#7202S06

FLASH CARDS
Helpful study tools that provide code users, students and exam candidates with an effective, time-tested method for study and information retention. Each set includes easy-to-use Flash Cards that are prepared and reviewed by code experts to ensure accuracy and quality.

2006 IFGC® (140 cards)
#1621S06

2006 IMC® (150 cards)
#1321S06

2006 IPC® (150 cards)
#1221S06

Question

When the building official issues a permit, the construction documents shall be approved, in writing or by stamp, as _____.

Answer

Reviewed for Code Compliance

STUDY COMPANIONS

ICC's Study Companions are the ideal way to master the code for everyday application or to prepare for exams. Each comprehensive study guide provides practical learning assignments for independent study. Study Companions can be used for independent study or as part of instructor-led programs in the workplace, college courses, or vocational training programs.

The Study Companions are organized into study sessions and include learning objectives, key points for review, code text and commentary applicable to the specific topic. Each study session includes a helpful quiz that allows the learner to measure their level of knowledge. Each Study Companion's answer key lists the code section referenced in each question for further information.

2006 INTERNATIONAL PLUMBING CODE® STUDY COMPANION

Addresses numerous topics of the 2006 IPC, including general regulations, fixtures, water supply and distribution, sanitary drainage, and vents.
- 12 study sessions
- 25-question quiz at end of each study session
- 300 total study questions
- 371 pages
#4217S06

2006 INTERNATIONAL MECHANICAL CODE® STUDY COMPANION

Covers all of the general topics found in the 2006 IMC®, including equipment and appliance installation, exhaust and duct systems, combustion air, chimneys and vents, specific appliances, and refrigeration.
- 16 study sessions
- 25-question quiz at end of each study session
- 400 total study questions
- 336 pages
#4317S06

ORDER YOURS TODAY! 1-800-786-4452 | www.iccsafe.org

People Helping People Build a Safer World™

PROFESSIONAL DEVELOPMENT OPPORTUNITIES
Plumbing/Mechanical/Fuel Gas (PMG)

TRAINING AND EDUCATION

Bring ICC's powerful, high-impact training to your group and show them that you're serious about their professional growth and mastery of interpreting and applying applicable I-Codes®. The Code Council offers Seminars, Institutes, and Webinars designed to keep you up to date on existing and emerging issues. Continuing Education Units (CEUs) are awarded and recognized by many states.

Our curriculum includes the following seminars:

Plumbing
- 2006 IPC Fundamentals
- 2006 IPC Performing Commercial Plumbing Inspections
- 2006 IRC Performing Residential Plumbing Inspections
- 2006 IPC Update
- 2006 IPC Transition from the 1997 UPC

Mechanical
- 2006 IMC Fundamentals
- 2006 IMC Performing Commercial Mechanical Inspections
- 2006 IRC Performing Residential Mechanical Inspections
- 2006 IMC Update
- 2006 IMC Trans fro the 1997 UMC

Fuel Gas
- 2006 IFGC Fundamentals
- 2006 IFGC Update

Comprehensive
- 2006 IPC/IMC/IFGC Significant Changes
- 2006 IRC Fundamentals Mechanical, Fuel Gas, Plumbing and Electrical Provisions

CERTIFICATION AND TESTING

Meet your career goals with ICC Certification for government officials and private sector inspectors, and testing for contractors. Our offerings are available online and in-person, including:

- Residential Plumbing Inspector
- Commercial Plumbing Inspector
- Plumbing Plans Examiner

- Residential Mechanical Inspector
- Commercial Mechanical Inspector
- Mechanical Plans Examiner

FOR A COMPLETE EXAM LISTING: www.iccsafe.org/certification

FOR MORE INFORMATION: www.iccsafe.org/training

8-62301-31

Approve plumbing products with a name you have come to trust.

ICC-ES PMG Listing Program

When it comes to approving plumbing, mechanical, or fuel gas (PMG) products, ask manufacturers for their ICC-ES PMG Listing. Our listing ensures code officials that a thorough evaluation has been done and that a product meets the requirements in both the codes and the standards. ICC-ES is the name code officials prefer when it comes to approving products.

FOR DETAILS! 1-800-423-6587, x5478 | www.icc-es.org/pmg

ICC EVALUATION
SERVICE

8-61804-35

Don't Miss Out On Valuable ICC Membership Benefits. Join ICC Today!

Join the largest and most respected building code and safety organization. As an official member of the International Code Council®, these great ICC® benefits are at your fingertips.

EXCLUSIVE MEMBER DISCOUNTS

ICC members enjoy exclusive discounts on codes, technical publications, seminars, plan reviews, educational materials, videos, and other products and services.

TECHNICAL SUPPORT

ICC members get expert code support services, opinions, and technical assistance from experienced engineers and architects, backed by the world's leading repository of code publications.

FREE CODE–LATEST EDITION

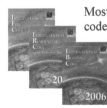

Most new individual members receive a free code from the latest edition of the International Codes. New corporate and governmental members receive one set of major International Codes (Building, Residential, Fire, Fuel Gas, Mechanical, Plumbing, Private Sewage Disposal).

FREE CODE MONOGRAPHS

Code monographs and other materials on proposed International Code revisions are provided free to ICC members upon request.

ICC BUILDING SAFETY JOURNAL®

A subscription to our official magazine is included with each membership. The magazine offers insightful articles authored by world-renowned code experts, plus code interpretations, job listings, event calendars, and other useful information. ICC members may also enjoy a subscription to an electronic newsletter.

PROFESSIONAL DEVELOPMENT

Receive "Member Only Discounts" for on-site training, institutes, symposiums, audio virtual seminars, and on-line training! ICC delivers educational programs that enable members to transition to the I-Codes, interpret and enforce codes, perform plan reviews, design and build safe structures, and perform administrative

functions more effectively and with greater efficiency. Members also enjoy special educational offerings that provide a forum to learn about and discuss current and emerging issues that affect the building industry.

ENHANCE YOUR CAREER

ICC keeps you current on the latest building codes, methods, and materials. Our conferences, job postings, and educational programs can also help you advance your career.

CODE NEWS

ICC members have the inside track for code news and industry updates via e-mails, newsletters, conferences, chapter meetings, networking, and the ICC website (www.iccsafe.org). Obtain code opinions, reports, adoption updates, and more. Without exception, ICC is your number one source for the very latest code and safety standards information.

MEMBER RECOGNITION

Improve your standing and prestige among your peers. ICC member cards, wall certificates, and logo decals identify your commitment to the community and to the safety of people worldwide.

ICC NETWORKING

Take advantage of exciting new opportunities to network with colleagues, future employers, potential business partners, industry experts, and more than 40,000 ICC members. ICC also has over 300 chapters across North America and around the globe to help you stay informed on local events, to consult with other professionals, and to enhance your reputation in the local community.

For more information about membership
or to join ICC, visit www.iccsafe.org/members
or call toll-free 1-888-ICC-SAFE (422-7233), x33804

INTERNATIONAL CODE COUNCIL®

People Helping People Build a Safer World™

8-61804-34

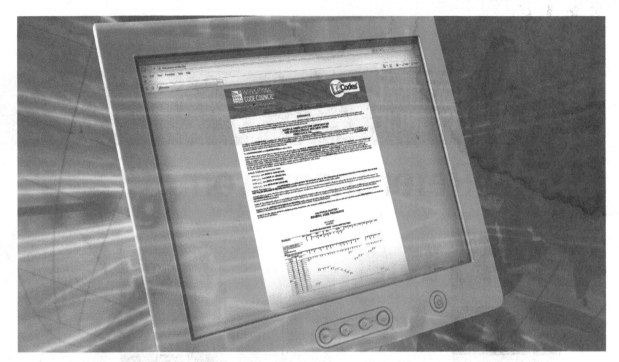